ENCYCLOPEDIA OF

BIOPROCESS TECHNOLOGY: FERMENTATION, BIOCATALYSIS, AND BIOSEPARATION

VOLUME 3

WILEY BIOTECHNOLOGY ENCYCLOPEDIAS

Encyclopedia of Bioprocess Technology: Fermentation, Biocatalysis, and Bioseparation
Edited by Michael C. Flickinger and Stephen W. Drew

Encyclopedia of Molecular Biology
Edited by Thomas E. Creighton

Encyclopedia of Cell Technology
Edited by Raymond E. Spier

Encyclopedia of Ethical, Legal, and Policy Issues in Biotechnology
Edited by Thomas J. Murray and Maxwell J. Mehlman

ENCYCLOPEDIA OF BIOPROCESS TECHNOLOGY: FERMENTATION, BIOCATALYSIS, AND BIOSEPARATION

EDITORIAL BOARD

ENCYCLOPEDIA OF

BIOPROCESS TECHNOLOGY: FERMENTATION, BIOCATALYSIS, AND BIOSEPARATION

VOLUME 3

Michael C. Flickinger
University of Minnesota
St. Paul, Minnesota

Stephen W. Drew
Merck and Co., Inc.
Rahway, New Jersey

A Wiley-Interscience Publication
John Wiley & Sons, Inc.
New York / Chichester / Weinheim / Brisbane / Singapore / Toronto

2143u751

This book is printed on acid-free paper. ♾

Published simultaneously in Canada.

For ordering and customer service, call 1-800-CALL-WILEY.

Library of Congress Cataloging-in-Publication Data:

Flickinger, Michael C.
The encyclopedia of bioprocess technology : fermentation, biocatalysis, and bioseparation / Michael C. Flickinger, Stephen W. Drew.
p. cm.
Includes index.
ISBN 0-471-13822-3 (alk. paper)
1. Biochemical engineering--Encyclopedias. I. Drew, Stephen W., 1945- . II. Title.
TP248.3.F57 1999
660.6'03--dc21 99-11576
CIP

Printed in the United States of America.

10 9 8 7 6 5 4 3 2 1

ENCYCLOPEDIA OF

BIOPROCESS TECHNOLOGY: FERMENTATION, BIOCATALYSIS, AND BIOSEPARATION

VOLUME 3

G

β-GALACTOSIDASE, ENZYMOLOGY

TORU NAKAYAMA
Tohoku University
Sendai, Japan
TERUO AMACHI
Kyoto University
Kyoto, Japan

KEY WORDS

Functional modification of bioactive compounds
Galactooligosaccharides
β-Galactosidase
Galactosyllactose
Glycosylhydrolase family
Lactase
Lactose hydrolysis
Microbial enzyme
Transgalactosylation
Reaction mechanism

OUTLINE

INTRODUCTION

β-Galactosidase (lactase, EC 3.2.1.23) catalyzes the hydrolysis and transgalactosylation reactions of β-D-galactopyranosides, such as lactose. The enzyme occurs in a wide variety of organisms including microorganisms, plants, and animals. The application of β-galactosidase in bioprocess technology has been achieved exclusively with microbial enzymes, which have long been used for the hydrolysis of lactose for increasing the digestibility of milk or for improving the functional properties of dairy products. During the past decade, another potential application of the enzyme has also been developed; the β-galactosidase-catalyzed transgalactosylation has proved to be useful for structural and functional modifications of food materials, medicines, and other biologically active compounds. In this article, we survey the enzymology and application in bioprocess technology of the microbial β-galactosidases with special emphasis on application of their transgalactosylation.

BRIEF SURVEY OF β-GALACTOSIDASE ENZYMOLOGY

Microbial Distribution and Purification

β-Galactosidase occurs in a variety of microorganisms including yeasts, molds, bacteria, actinomycetes, and archaebacteria (Table 1). β-Galactosidase is known as a marker enzyme for coliform bacteria, which are indicators of the fecal pollution of water.

Microbial β-galactosidases have been purified by a combination of several conventional techniques, such as salting-out fractionation, ion exchange, gel filtration, hydroxyapatite, and hydrophobic interaction chromatographies. The affinity chromatography of β-galactosidase uses *p*-aminophenyl- (or *p*-aminobenzyl-)β-D-thiogalactopyranoside agarose (1), which is commercially available and has been used for the purification of the *Escherichia coli lacZ* β-galactosidase, its recombinant derivatives, and the other β-galactosidases (2).

Molecular Properties, Substrates, and Inhibitors

The specificity and molecular properties of the microbial β-galactosidases differ significantly with the source of the enzyme. For example, the *Saccharopolyspora rectivirgula* β-galactosidase is highly specific for β-D-galactosides (3), whereas the *Sulfolobus solfataricus* enzyme shows broad substrate specificity and can efficiently hydrolyze β-glycosides other than β-galactosides as well (4). The catalytic competence and specificity of transgalactosylation also depend on the enzyme (see following). The reported native molecular weights and subunit structures of β-

Table 1. Examples of β-Galactosidase-Producing Microorganisms

Yeasts
Candida pseudotropicalis, Cryptococcus laurentii, Kluyveromyces (Saccharomyces) lactis, K. fragilis, K. marxianus, Lipomyces sp., *Torulopsis sphaerica, T. versalitis*
Molds
Altenaria altenata, Aspergillus awamori*, A. cellulosae*, A. foetidus*, A. nidulans, A. niger, A. oryzae, A. phoenicis*, A. terreus,* A. wentii*, Aspergillus* sp., *Chaetomium globosum*, C. cochlioides*, C. funicola*, C. thermophile* var. *coprophile*, Fusarium maniliforme, F. oxysporum* var. *lini, Geotricum candida*, Humicola grisea* var. *thermoidea*, H. lanuginosa*, Macrophomina phaseoli, Malbranchea pulchella* var. *sulfurea*, Mucor miehei*, M. pusillus*, Mucor* sp.*, *Mucor muccedo*, M. javanicus*, Neurospora crassa, Paecilomyces varioti, Penicillium* sp., *P. multicolor, P. canescens, P. citrinum, P. luteum*, P. chrysogenum*, P. frequentans*, P. cycropium*, P. toxidarium*, P. glaucum*, P. notatum*, P. roqueforti*, Phycomyces blakeleeanus, Rhizopus acidus*, R. niveus*, R. nigricans*, R. delemar*, R. javanicus*, R. formosciensis*, R. chinensis*, Scopulariopsis* sp., *Sclerotium tuliparum, Spicaria* sp., *Sporotrichum* sp.*, *S. thermophile*, Sterigmatomyces elviae, Thermomyces lanuginosus, Torula thermophila*, Trichoderma viride**
Basidiomycetes
*Corticium rolfsii, Culvularia inaequalis, Pycnoporus cinnabarinus, Sporobolomyces singulalis**
Bacteria
Gram-negative
Aeromonas cavie, A. formicans, Agrobacterium rediobacter, Bacteroides polypragmatus, Buttiauxella agrestis, Enterobacter (Aerobacter) cloacae, Escherichia coli, Fibrobacter succinogenes, Klebsiella pneumoniae, Rhizobium meliloti, R. trifolii, Shigella dysenteriae*, Thermotoga maritima, Thermus* sp., *Treponema phagedenis, Xanthomonas campestris, X. manihotis*
Gram-positive
Arthrobacter sp., *Bacillus acidocaldarius, B. circulans, B. coagulans, B. macerans, B. megaterium, B. subtilis, B. stearothermophilus,* alkalophilic *Bacillus, Bifidobacterium* sp., *B. bifidum*, B. longum, Clostridium acetobutylicum, Corynebacterium murisepticum, Lactobacillus delbruckii* subsp. *bulgaricus, L. bulgaricus, L. casei, L. helveticus, L. murinus, L. plantarum, L. sake, Lactococcus lactis, Leuconostoc citrovorum, L. lactis, Streptococcus salivarius* subsp. *thermophilus, S. thermophilus, S. (Diplococcus) pneumoniae, Thermoanaerobacter* sp., *Thermoanaerobacterium (Clostridium) thermosulfurigenes*
Actinomycetes
Actinomyces viscosus, Nocardia sp., *Saccharopolyspora rectivirgula, Streptomyces lividans*, S. venezuelae*, S. violaceus**
Archaebacteria
Caldariella acidophila, Sulfolobus solfataricus, Pyrococcus woesei, Haloferax alicantei

Note: Microorganisms listed include not only those whose β-galactosidase is characterized or sequenced but also those that have been only shown to produce β-galactosidase(s) based on the observed hydrolytic and transgalactosylation activities (indicated by an asterisk; May 1998).

galactosidases range, respectively, from 19,000 to 630,000 (the *Kluyveromyces lactis* enzyme [5]) and from monomer (e.g., the *S. rectivirgula* enzyme [3]) to heterooctamer (the *E. coli ebg* enzyme [6]). The bacterial β-galactosidases generally show optimum pHs at neutral regions whereas most mold enzymes show them at acidic regions, which in some cases reach pH 2 (7). Some β-galactosidases of the *lacZ* family require essential monovalent and divalent ions for their activities; the *S. rectivirgula* β-galactosidase is a multimetal enzyme that requires multiple divalent ions for its maximum thermostability and activity (8). However, some other β-galactosidases (e.g., the *Rhizobium meliloti* enzyme [9]) do not show such metal ion requirements. These observations strongly suggest the structural diversity of the microbial β-galactosidases.

Many chromogenic, fluorogenic, and luminogenic substrates that are specific for β-galactosidase have been designed (Table 2). D-Galactose, D-galactosylamines, β-D-thiogalactosides, and their analogs and derivatives serve as specific inhibitors of some microbial β-galactosidases (Table 3).

Structural Classification

Henrissat has compared all the available sequences of glycosyl hydrolases using hydrophobic cluster analysis and classified these enzymes into families (48). Many microbial β-galactosidases have thus far been cloned and sequenced and have been classified, on the basis of sequence similarities, into at least four categories according to Henrissat's classification (Table 4).

Glycosyl Hydrolase Family 2. β-Galactosidases belonging to this glycosyl hydrolase family, typified by the *E. coli lacZ* β-galactosidase, consist of a large subunit protein of approximately 1,000 amino acids and show very high sequence similarities to the animal β-glucuronidases, suggesting that the β-galactosidases of family 2 have a close evolutionary relationship with the β-glucuronidases (48). Many β-galactosidases of this family require monovalent and divalent metal ions for maximum activity (e.g., the *E. coli lacZ, Kluyveromyces lactis, Streptococcus thermophilus*, and *Saccharopolyspora rectivirgula* enzymes).

Table 2. Chromogenic, Fluorogenic, and Luminogenic Substrates for β-Galactosidase Gal, β-D-Galactopyranoside

A. Chromogenic Substrates

Substrate	Reported molar absorption coefficient of chromophore[a], M^{-1} cm^{-1}	Reference
2-Nitrophenyl-Gal (ONPG)	$\Delta\epsilon_{420}$ = 3000 (pH 7.0)	10
	$\Delta\epsilon_{410}$ = 2100 (pH 7.0)	11
	$\Delta\epsilon_{410}$ – 4500 (pH 8.6)	11
	$\Delta\epsilon_{410}$ = 4600 (pH 9.2)	11
4-Nitrophenyl-Gal (PNPG)	$\Delta\epsilon_{347,iso}$ = 290.4[b]	12
	$\Delta\epsilon_{405}$ = 13400 (pH 7.2)	3
	$\Delta\epsilon_{405}$ = 8900 (pH 7.0)	11
	$\Delta\epsilon_{405}$ = 18300 (pH 8.6)	11
3,4-Dinitrophenyl-Gal	$\Delta\epsilon_{400}$ = 15924 (pH 7.0[c])	12
2-Chloro-4-nitrophenyl-Gal		13
X-Gal[d]	Blue insoluble product	Suitable for histochemical identification
VBzTM-Gal[e]	Red product, $\Delta\epsilon_{520}$ = 55000	Suitable for histochemical identification; Refs. 14 and 15
VLM-Gal[f] and derivative	$\Delta\epsilon_{545}$ = 32500 (VLM-Gal)	15
VQM-Gal[g] and derivative	$\Delta\epsilon_{520}$ = 42500 (VQM-Gal)	15
6-Bromo-2-naphthyl-Gal	Product visualized by subsequent reactions	Suitable for activity staining in the gel; Refs. 16 and 17

B. Fluorogenic Substrates

Substrate	Excitation (nm)	Emission (nm)	Reference
FMG[h], FDG[i]	485	530	18
4-Hydroxy-4-methylcoumarin Gal	365	450	19
7-Hydroxycoumarine-4-acetate Gal	370	455	19

C. Substrates for Luminescent Determination

Substrate	Comments	Reference
D-Luciferin-*O*-Gal	Bioluminogenic	20
AMPGD[j]	Chemiluminogenic	21
Lumi-Gal 530™	Chemiluminogenic	22

[a]The value depends on assay conditions used. See reference for further details.
[b]Isosbestic point.
[c]Calculated from the reported equation: $\Delta\epsilon_{400} = (1.637 \pm 0.007) \times 10^4/(1 + [H^+]/10^{-5.45\pm0.02})$.
[d]5-Bromo-4-chloro-3-indolyl β-D-galactopyranoside.
[e](2-(2-(4-(β-D-Galactopyranosyloxy)-3-methoxyphenyl)-vinyl)-3-methylbenzothiazoliumtoluene-4-sulfonate).
[f]4-[2-(4-β-D-galactopuyranosyloxy-3-methoxyphenyl)-vinyl]-1-methylquinoliniumiodide.
[g]2-[2-(4-β-D-galactopuyranosyloxy-3-methoxyphenyl)-vinyl]-1-methylquinoliniumiodide.
[h]Fluorescein mono-β-D-galactoside.
[i]Fluorescein di-β-D-galactoside.
[j](3-(4-Methoxyspiro[1,2-dioxetane-3,2′-tricyclo[3.2.1.1$^{3.7}$][decane]-4-yl)phenyl-β-D-galactopyranoside.

Glycosyl Hydrolase Family 1. The family 1 β-galactosidases include the enzyme of an archaeon, *Sulfolobus solfataricus* (48). This enzyme is essentially a β-glycosidase that can efficiently hydrolyze β-galactosides as well as the other β-glycosides (4). The *S. solfataricus* enzyme, consisting of 489 amino acid residues (49), differs in subunit size from the family 2 of enzymes and exhibits no sequence similarity to that family. Interestingly, the lactase/phlorizin hydrolase, a mammalian β-galactosidase playing a central role in lactose digestion in the small intestine, belongs to family 1 (48). Thus, it might be possible that family 2 β-galactosidases evolved to the animal β-glucuronidases whereas family 1 enzymes evolved to the digestive lactase enzymes functioning in the mammalian intestines. A comparison of stereo- and primary structures suggested that, along with β-glucosidases, cellulases, xylanases, and glycanases, the β-galactosidases belonging to families 1 and 2 form a superfamily, the 4/7 superfamily, which is characteristic of a common eightfold β/α architecture with two conserved, catalytically important glutamates near the C-terminal ends of β-strands four and seven (50).

The *lacG* β-Galactosidases. Strains of *Arthrobacter* sp. and *Bacillus stearothermophilus* produce β-galactosidases belonging to a recently proposed β-galactosidase family, the *lacG* family, which shows no sequence similarity to enzyme families 1 and 2 (51).

Glycosyl Hydrolase Family 35. *Xanthomonas manihotis*, which is a gram-negative phytopathogenic bacterium, *Arthrobacter* sp., *Bacillus circulans*, and *Aspergillus niger* produce β-galactosidases with strong sequence similarities to animal β-galactosidases belonging to family 35 of glycosyl hydrolases (52–54). Domain 1 of the *Xanthomonas*

Table 3. Specific β-Galactosidase Inhibitors

Inhibitor	Enzyme[a]	Type[b]	K_i and other properties[c]	Reference
D-Galactose	1	C	34 mM	23
	2	C	2.32 mM	3
	3	C	350 mM	24
	4	N	40 mM	25
D-Galactal	1	C	14 μM	26
D-Galactosamine	2	C	2.7 mM	3
2-Amino-D-galactopyranose[d]	1	C	1 mM	27
Galactobiose [β-D-Gal*p*-(1,4)-D-Gal]	1	C	0.1 mM	28
β-D-Thiogalactosides	1	C	Isopropyl-, 0.085 mM	29
	2	C	Methyl-, 12 mM Isopropyl-, 15 mM	3
β-D-Galactopyranosyl trimethylammonium bromide[d]	1	C	1.4 mM	30
C-(β-D-Galactopyranosylmethyl) amine[d] and derivatives	1	C	7.8 μM	31
Furanoses[d]	1		L-Ribose, 0.03 mM D-Lyxose, 0.09 mM	23
D-Galactonolactones[d]	1		D-Galactonolactone, 0.25 mM	23
Amino sugars and amino alcohols[d]	1			27
Galactostatin and derivatives[d]	1	C	4 nM	32
Galactose-type imino sugar[d]	4		4 nM	33
1,5-Dideoxy-1,5-imino-D-galactitol[d]	1		13 μM	34
5-Amino-5-deoxy-D-galactopyranoside	1		45 nM	34
2′-Amino-2′-deoxymethyl β-lactoside	1			35
N-Bromoacetyl-β-D-galactopyranosylamine	1	IRR		36
β-D-Galactopyranosylmethyl-*p*-nitrophenyltriazene	4	A		37
(1/2,5,6)-2-(3-Azibutylthio)-5,6-epoxy-3-cyclohexen-1-*ol*	1	A	29 mM	38
Conduritol C *cis*-epoxide	1	A		39
2-Deoxy-2-fluoro-β-D-galactopyranosyl fluoride	1	A	K_i, 1.3 mM; k_i, 3.2 min^{-1}	40
	4	A	K_i, 5.4 mM; k_i, 2.5 min^{-1}	
	5	A	K_i, 1.3 mM; k_i, 0.8 min^{-1}	
2′,4′-Dinitrophenyl 2-deoxy-2-fluoro-β-D-galactopyranoside	1	A	K_i, 0.71 mM; k_i, 2.4 min^{-1}	41
Diazomethyl β-D-galactopyranosyl ketone	4	A	K_i, 30 mM; k_i, 0.56 min^{-1}	42
2-Nitro-1-(4,5-dimethoxy-2-nitrophenyl) ethyl	1	IRR	Photoreversible thiol label	43
Phytic acid	1	U	3.46 mM	44
L-Histidine				45
Calystegin B1	6	C	1.6 μM	46
Isoflavones and their glycosides	6	C/N	2′,3′,4′,7-Tetrahydroxy-isoflavone, 26 μM	47
	6	C/N	Genistein 4′,7-di-*O*-α-L-rhamnoside, 4 μM	

[a]Sources of enzymes: 1, *Escherichia coli (lacZ)*; 2, *Saccharopolyspora rectivirgula*; 3, *Streptococcus thermophilus*; 4, *Aspergillus oryzae*; 5, *A. niger*; and 6, bovine liver. For the assay conditions, see References.
[b]Type of inhibition: C, competitive; N, noncompetitive; U, uncompetitive; IRR, irreversible; A, affinity labeling reagent.
[c]K_i, Inhibition constant; k_i, inactivation rate constant.
[d]Observed strong inhibitions by these compounds argue for the occurrence of a half-chair cationic transition state during catalysis.

manihotis β-galactosidase shows similarity to regions within catalytic domains of cellulases and xylanases belonging to family 10 of glycosyl hydrolases.

It should be mentioned that the strain of *Arthrobacter* sp. can produce β-galactosidases of different enzyme families, that is, families 1 and 35 as well as the *lacG* family (51,52,55). Cumulative knowledge of amino acid sequences and stereostructures of the microbial β-galactosidases will provide important information for further insights into the classification and evolutionary and structure–function relationships of the microbial β-galactosidases.

MOLECULAR PROPERTIES AND CURRENT STATUS IN BIOPROCESS TECHNOLOGY OF SOME MICROBIAL β-GALACTOSIDASES

Although many β-galactosidase-producing microorganisms are known (see Table 1), only a limited number of microorganisms have been selected as sources of the enzymes used in industry. Most of these are of fungal origin, such as *Kluyveromyces lactis, K. fragilis, Aspergillus niger*, and *A. oryzae*. These fungi were chosen mainly because they can inexpensively produce the β-galactosidase and are generally recognized as safe (GRAS) as food additives. Recent extensive studies led to discovering many other fungal and bacterial enzymes that are applicable in bioprocess technology. In this section, we describe the molecular properties and current status in bioprocess technology of some microbial β-galactosidases (Table 5).

Fungal Enzymes

The *Kluyveromyces lactis, K. fragilis, Aspergillus niger*, and *A. oryzae* β-galactosidases, the commercially available GRAS enzymes, are currently used in industry for manufacturing milk and dairy products (see following). The *Pae-*

Table 4. List of GenBank Accession Numbers of the Microbial β-Galactosidase Sequences (May 1998) with Possible Assignment to the Glycosyl Hydrolase Families

Microorganism (enzyme)	GenBank accession number	Possible assignment[a]
Aspergillus niger (*lacA*)	A00968, L06037 S37150	35
Actinobacillus pleuropneumoniae (*lacZ*)	U62625	2
Arthrobacter sp. (*lacZ*)	U12334	2
	U78028	35
	U17417	*lacG*
Bacillus circulans (*bgaA*)	L03424	*lacG*
(*bgaB*)	L03425	*lacG* or 35
(*bgaC*)	D88750	35
B. stearothermophilus	M13466	*lacG*
Clostridium acetobutylicum (*cbgA*)	M35107	2
C. perfringens (*pbg*)	D49537	
C. (*Thermoanaerobacter*) *thermosulfurigenes* (*lacZ*)	M57579	2
Enterobacter cloacae	D42077	2
Escherichia coli (*lacZ*)	V00296	2
(*ebg*)	X52031	2
Kluyveromyces lactis (*LAC4*)	M84410	2
Lactobacillus acidophilus (*lacL*)	AB004867	2
(*lacM*)	AB004868	2
L. delbrueckii subsp. *bulgaricus*	M23530	2
Lactococcus lactis (*lacZ*)	U60828	2
Leuconostoc lactis	M92281	2
Pyrococcus woesei	AF043283	1
Rhizobium meliloti (*lacZ*)	L20757	
Saccharopolyspora rectivirgula	D86429	2
Streptococcus thermophilus (*lacZ*)	M63636	2
Sulfolobus solfataricus	M34696, X15950, X15372	1
Thermotoga maritima	U08186	2
Thermus sp.	Z93773	*lacG*
Xanthomonas manihotis (*Bga*)	L35444	35

[a]For details, see text.

cilomyces varioti β-galactosidase is a thermostable extracellular enzyme with a pH optimum at the acidic region and is potentially applicable to the processing of acid whey. The *A. oryzae*, *Cryptococcus laurentii*, and *Sterigmatomyces elviae* enzymes exhibit high transgalactosylation activities and are used for the production of the "galacto-oligosaccharides," which are derived from lactose (see following).

Kluyveromyces lactis. Among the fungal enzymes listed in Table 5, biochemical and genetic studies have been the most intensive with the *K. lactis* enzyme. The *K. lactis* β-galactosidase is encoded by the *LAC4* gene, whose regulation involves the galactose/lactose induction and catabolite repression systems (64). The *K. lactis* enzyme consists of a subunit protein with an M_r of 117,618 (64) and is probably a dimer judging from its sedimentation coefficient under nondenaturing conditions (56), although higher aggregates are also known. The deduced amino acid sequence of the β-galactosidase is very similar to that of the *E. coli lacZ* enzyme; thus, the enzyme belongs to the family 2. However, Maxilact, a commercial preparation of the *K. lactis* enzyme, was shown to contain four enzyme forms of β-galactosidase, whose native molecular weights were estimated to be 19,000, 41,000, 550,000, and 630,000, respectively (5). The biochemical relevance of these observations to the reported subunit size (M_r, 117,618) and structure remains to be clarified.

Aspergillus niger. *A. niger* produces three forms of β-galactosidases with M_r of 124,000, 150,000, and 173,000, respectively, which can be separated by ion exchange, gel filtration, and hydrophobic interaction chromatographies (57). These isoforms share very similar isoelectric points, amino acid compositions, and kinetic parameters. The observed multiplicity seems to vary with the culture conditions and is probably related to differences in the carbohydrate content. The amino acid sequence of the *A. niger* β-galactosidase exhibits similarity to animal β-galactosidases belonging to the glycosyl hydrolase family 35 (65) (see Table 4).

Crypotococcus laurentii and Strerigmatomyces elviae. *C. laurentii* and *S. elviae* have been found to produce β-galactosidases that have high galactosyl transfer activities (2,62). The *C. laurentii* cells entrapped in the calcium alginate gels are currently used for industrial-scale production of 4′-galactosyllactose-containing "galactooligosaccharides" (see following).

The *C. laurentii* and *S. elviae* β-galactosidases share the following similar characteristics: *(1)* occurrence in the cell wall fractions of the yeast cells; *(2)* homodimer of a subunit protein with an M_r of 86,000–100,000; *(3)* optimum pH for hydrolysis reaction at pH 4.3–5.0; *(4)* high transgalactosylation activity that yields 4′-galactosyllactose as a main product from lactose; and *(5)* strong inhibition by Hg^{2+} but

Table 5. Molecular Properties of Some Microbial β-Galactosidases of Biotechnological Interest

Enzyme	Native M_I[a] (subunit structure)	Optimum pH[b]	Transfer activity[c]	Metal ion requirements	Localization[d]	Comments
			Fungal enzymes			
Kluyveromyces lactis	nr (probably dimeric)	7.1[lac,onpg]	+	Mn^{2+}/Na^{+}	I	Subunit M_r, 117, 618; Refs. 56 and 57
K. fragilis	203,000 (nr)	6.8[onpg]	+	Co^{2+}, Mg^{2+}, Mn^{2+}/K^{+}	I	Ref. 58
Aspergillus niger	124,000–173,000 (nr)	3.0[lac]	+	nr	I	Three multiple forms; Ref. 57
A. oryzae	105,000 (homodimer)	4.5[lac]	+ +	no	E	Refs. 59 and 60
Paecilomyces varioti	94,000 (nr)	3.5	+	no	E	Thermostable enzyme; Ref. 61
Sterigmatomyces elviae	170,000 (homodimer)	4.5–5.0[onpg]	+ +	no	I (cell wall)	Thermostable enzyme; Ref. 2
Cryptococcus laurentii	200,000 (homodimer)	4.3[onpg]	+ +	no	I (cell wall)	Ref. 62
			Bacterial enzymes			
Escherichia coli	465,000 (homotetramer)	7.0[onpg]	+	Mg^{2+}, Mn^{2+}/Na^{+}	I	The *lacZ* enzyme
Bacillus circulans	160,000 (monomer)	6.1[lac]	+ +	no	I	Isozyme 2; Ref. 63
Saccharopolyspora rectivirgula	145,000 (monomer)	7.1[lac]	+ +	Mn^{2+}/Na^{+}	E	Thermostable enzyme; Ref. 3
Streptococcus thermophilus	282,000 (probably dimeric)	7.0	+ +	Mg^{2+}/K^{+}, Na^{+}	I	Ref. 24

Note: nr, not reported; no, no requirement.
[a]Molecular weight under nondenaturing conditions.
[b]Optimum pH for hydrolysis of lactose[lac] or 2-nitrophenyl-β-D-galactopyranoside[onpg].
[c]+, Transgalactosylation activity has been reported; + +, very high transgalactosylation activity has been reported.
[d]I, intracellular, E, extracellular.

not by *p*-chloromercuribenzoic acid. However, the *S. elviae* enzyme is distinct from the *C. laurentii* enzyme in terms of thermostability; the *S. elviae* enzyme is stable up to 80 °C for 1 h although *S. elviae* is a mesophile; in contrast, the *C. laurentii* enzyme is inactivated after incubation at 58 °C for 10 min. The *S. elviae* enzyme exhibits sevenfold higher β-D-fucosidase activity than β-D-galactosidase activity (2).

Bacterial and Actinomycetous Enzymes

Escherichia coli. The *E. coli lacZ* β-galactosidase is the best characterized β-galactosidase; its three-dimensional structure was already known (66). The regulation mechanism of the transcription of the *lacZ* β-galactosidase gene was established as the operon model. The extensive structural and mechanistic studies of the *E. coli lacZ* enzyme have provided the basis for the detailed insights into β-galactosidase catalysis. However, the usefulness of this enzyme in bioprocess technology has been examined only on a laboratory scale; the *E.coli lacZ* enzyme stops short of making a contribution in the food industry, mainly because of the unacceptability of the bacterium in uses related to food. Rather, the *E. coli lacZ* β-galactosidase and its gene are essential tools in genetic engineering, immunochemical, and molecular biology studies.

The *E. coli lacZ* β-galactosidase is a tetrameric protein consisting of identical subunits of a 1,023-amino acid polypeptide chain, which folds into five sequential domains with an extended segment at the amino terminus (66). The enzyme has four active sites per tetramer, each of which is made up of elements from two different subunits. All the catalytically important amino acid residues are located in domain 3 of the subunit. The enzyme requires the essential divalent and monovalent cations for its maximum activity, but the role of these metal activators in the enzyme mechanism is not yet known in detail. The proposed mechanism of the hydrolysis and transgalactosylation reactions catalyzed by the *E. coli lacZ* β-galactosidase are reviewed in another section. The "α-complementation" phenomenon would be worth noting as a unique characteristic of this enzyme (67,68). A fragment of amino-terminal peptide, which consists of approximately 50 amino acids encoded by a plasmid, is capable of intraallelic (α-) complementation with a defective form of the β-galactosidase encoded by the host. The α-complementation has been used for designing plasmid vectors that permit histochemical identification of recombinant clones in genetic engineering studies.

The *E. coli* cells produce a second type of β-galactosidase, the *ebg* β-galactosidase ("*ebg*" denotes "evolved β-galactosidase"), which was found as a gene product that confers a lactose-positive phenotype to $lacZ^{-}$ mutants. The enzyme is encoded by two distinct structural genes of the *ebg* operon, *ebgA* and *ebgC* genes that encode polypeptide chains of M_r 110,000 and 20,000, respectively (6). The sequence of the *ebgA* gene shows 50% nucleotide identity with that of the *lacZ* β-galactosidase gene, indicating that the *ebg* β-galactosidase has a strong evolutionary relationship with the *lacZ* β-galactosidase (69). The active form of the *ebg* β-galactosidase is an $\alpha_4\beta_4$ heterooctamer, which is a 1:1 complex of the *ebgA* and *ebgC* gene products (6). Because the wild-type *ebg* β-galactosidase is too catalytically feeble to allow the *E. coli* cells to grow on lactose, spontaneous mutations must occur in the *ebg* β-galactosidase genes before growth can be sustained; when the $lacZ^-$ mutants are placed under selection pressure with lactose as a sole carbon source, strains producing mutant *ebg* enzyme with enhanced catalytic competence are selected. Thus, although the *ebg* β-galactosidase has made no significant contribution to bioprocess technology to date, the *ebg* system has been extensively studied as an interesting model for the biology of acquisitive evolution and the chemistry of the evolution of catalytic function.

Bacillus circulans. The commercial preparation of *B. circulans* β-galactosidase contains two isozymes, isozymes 1 and 2, which are reported to be monomeric with M_r of 240,000 and 160,000, respectively (63). The enzyme preparation could be applicable to enzymatic hydrolysis of lactose in skim milk (70). The isozymes are different from each other in terms of substrate specificity and the ability to catalyze transgalactosylation. Isozyme 2 has an extremely high transgalactosylation activity; the reaction of the enzyme with 4.56% (w/v) lactose yields the transgalactosylation products (galactooligosaccharides) consisting of di-, tri-, tetra-, and pentasaccharides, and the amount of the oligosaccharides reaches the maximum, which is 41% (w/w) of the total sugar in the reaction mixture. In contrast, isozyme 1 produces oligosaccharides whose amount is only 6% (w/w) of the total sugar at maximum under the same conditions. Thus, isozyme 2 of *B. circulans* β-galactosidase is used for the production of "galactooligosaccharides" as well as for the functional modification of food materials, medicines, and other bioactive compounds by transgalactosylation. Two β-galactosidase-encoding genes, *bgaA* and *bgaB*, have been identified in *B. circulans* (see Table 4) and are potential candidates that code for these isozymes. Relevance of these genes to the isozymes, however, still remains to be established. Recently, *B. circulans* has also been found to have a third β-galactosidase, *bgaC*, which is an *exo*-β-1,3-galactosidase belonging to glycosyl hydrolase family 35 (53). The commercial preparation of *B. circulans* β-galactosidase does not contain the activity of this third β-galactosidase (53). The recombinant *bgaC* β-galactosidase also showed transgalactosylation activity, which is useful for the synthesis of β-1,3-linked galactosyl oligosaccharides (71).

Streptococcus thermophilus. *S. thermophilus* is one of the dairy lactic streptococci that are used as starter cultures for a variety of industrial dairy fermentations such as those during the yogurt-making processes. The primary function of these dairy lactic streptococci is the rapid conversion of lactose into lactic acid, an essential process in dairy fermentation (72). Two distinct biochemical pathways involved in lactose utilization by the lactic streptococci have been established. One of these pathways uses the lactose permease system for the transport of unmodified lactose, which is then hydrolyzed by an intracellular β-galactosidase. The other pathway is mediated by the phosphoenolpyruvate-dependent phosphotransferase system, in which lactose is phosphorylated during translocation and the resultant lactose-6-phosphate is hydrolyzed by phospho-β-galactosidase (EC 3.2.1.85). In *S. thermophilus* cells, the lactose permease system plays a central role in lactose transport across the bacterial cell membrane, and the β-galactosidase functions as a key enzyme in the metabolism of the internalized lactose in dairy lactic fermentation (72). The *S. thermophilus* β-galactosidase is promising for industrial hydrolysis of milk lactose, judging from its thermostability and established safety as a food additive. Also, the enzyme is used as a catalyst for the industrial production of "galactooligosaccharides" from lactose because of its high transgalactosylation activity.

The *S. thermophilus* β-galactosidase consists of a subunit protein of M_r 116,860 (73). However, its quaternary structure is not yet firmly established. The enzyme exists in three distinct forms, whose apparent native molecular weights are 204,000, 186,000, and 282,000, and one of which is converted to another in a time-, temperature-, and concentration-dependent manner (24). The native molecular weight of 500,000–600,000 has also been reported (72). Structural analysis of the *S. thermophilus* β-galactosidase genes showed that the β-galactosidase is a family 2 enzyme with 48%, 35%, and 32.5% amino acid sequence identity to the *Lactobacillus bulgaricus, Escherichia coli*, and *Klebsiella pneumoniae* enzymes, respectively. The enzyme requires monovalent and divalent metal ions for its maximum activity, and the specificity of the metal activation varies with the substrate (24). The enzyme shows high transgalactosylation activity to produce a variety of disaccharides as main products from lactose (74), in contrast to the *Aaspergillus oryzae* β-galactosidase, for example, producing tri-, tetra-, and pentasaccharides as main transfer products from lactose.

Saccharopolyspora rectivirgula. A strain of *S. rectivirgula*, a thermophilic actinomycete, produces an extracellular thermostable β-galactosidase with high transgalactosylation activity. The *S. rectivirgula* β-galactosidase is a monomeric enzyme with a molecular weight of 145,000 and $S^{\circ}_{20,w}$ of 7.1*S* (3). When the enzyme reacts with 1.75 M lactose at 70 °C and pH 7.0 for 22 h, it yields oligosaccharides in the maximum yield (other than lactose) of 41% (w/w). The enzyme is stable at pH 7.2 up to 60 °C (for 4 h in the presence of 10 μM $MnCl_2$) or 70 °C (for 22 h in the presence of 1.75 M lactose and 10 μM $MnCl_2$). Thus the enzyme is applicable to an immobilized enzyme system at high temperatures (>60 °C) for efficient production of the oligosaccharides from lactose. This monomeric enzyme has eight specific binding sites for divalent metal ions. These sites are classified as follows: a very tight (class I) site for Ca^{2+}, three tight (class II) sites consisting of two Ca^{2+}-specific sites (class II_{Ca}) and one Mn^{2+}-specific site (class II_{Mn}; K_d for Mn^{2+}, 2.0 nM), and four loose (class III) sites for Mn^{2+} (K_d, 1.2 μM) and Mg^{2+} (K_d, 2 μM). The class II Mn^{2+} is catalytically important in maintaining the native structure essential for activity. Occupation of class III sites by Mg^{2+} or Mn^{2+} is of physiological importance to attain sufficient thermostability by which this extracellular β-galactosidase remains active for a prolonged time at elevated temperatures, as was observed during the growth of *S. rectivirgula* (8).

MECHANISM OF THE *ESCHERICHIA COLI lacZ* β-GALACTOSIDASE-CATALYZED REACTIONS

The mechanism of the hydrolysis and transgalactosylation reactions has been the most intensively studied with the *Escherichia coli lacZ* β-galactosidase. Figure 1 shows the proposed kinetic mechanism of the *E. coli lacZ* β-galactosidase (75). In this mechanism, hydrolysis and transgalactosylation occur at the same catalytic site and share a common galactosyl enzyme intermediate. Extensive kinetic studies have also suggested the presence of two distinct subsites near the active site of the *lacZ* enzyme, the "galactose site" and "glucose site," that respectively bind galactose and glucose portions of the lactose molecule (76,77). The glucose site shows affinities to a variety of nucleophilic compounds, such as sugars, alcohols, and thiols. The presence of such a glucose site is important in the kinetic behaviors of hydrolysis and transgalactosylation of lactose.

Kinetic Mechanism of Reactions with ONPG

The kinetic mechanism of the *Escerichia coli lacZ* β-galactosidase-catalyzed reactions can be most simply shown by the reaction with 2-nitrophenyl β-D-galactopyranoside (ONPG), which is a synthetic substrate whose aglycone serves as a good leaving group (75). The ONPG (termed S in Fig. 1) binds to the enzyme to form a Michaelis complex ES (step a). The release of the aglycone, 2-nitrophenolate, occurs very rapidly (k_2 = 2,100 s^{-1} at 25 °C and pH 7.0) to yield a galactosyl enzyme intermediate (step b), which partitions into subsequent hydrolysis (c) and transgalactosylation (g, h). For the hydrolysis reaction, the water molecule reacts with the galactosyl enzyme intermediate with k_3 of 1,200 s^{-1}, yielding galactose and free enzyme (degalactosylation; step c). For the transgalactosylation, a galactosyl acceptor (termed Nu) present in the reaction mixture binds to a glucose site of the enzyme with an affinity of K_i'' to form a ternary complex (step g), followed by the formation of the galactosyl adduct (Gal-Nu; step h). In the transgalactosylation reaction, the apparent rate of the 2-nitrophenol release is increased, decreased, or unchanged with increasing acceptor concentrations, depending on the nature of the acceptor and the relative magnitude of k_2, k_3, and k_6. These observations have been kinetically analyzed to estimate the K_i'' and k_6 values of a variety of acceptors, which are the measures of how well they bind at the glucose site and how readily they react to form a galactosyl adduct, respectively. Mathematical considerations of such analyses are described by Deschavanne et al. (76) and Huber et al. (77).

Kinetic Mechanism of Reactions with Lactose

The reaction with lactose, a natural substrate, proceeds with a mechanism that is somewhat different from that of the ONPG reaction. Steps d, e, and f, which are shown with shadowing in Figure 1, must be considered instead of step b, because glucose, a product of lactose hydrolysis, has an affinity (K_i') to the glucose site, in contrast to 2-nitrophenolate, which has no affinity to the enzyme (75); released glucose can reversibly bind to form a ternary complex (step e), and then reacts with galactose to yield allolactose. During the lactose hydrolysis, allolactose is only transiently formed because it is also a substrate of the enzyme. The reported rate constants (at 25 °C, pH 7.0) are k_4 = 60 s^{-1}, k_5 = 380 s^{-1}, and k_3 = 1,200 s^{-1} (75). When high concentrations of lactose are reacted, lactose also serves as a galactosyl acceptor (Nu), and the transfer product is galactosyllactose. The galactosyllactose can, in turn, serve as a galactosyl acceptor to produce tetrasaccharides. In this manner, a mixture of oligosaccharides consisting of di-, tri-, and higher saccharides is produced from lactose.

Role of Active Site Residues in Catalysis

Extensive chemical modification, affinity labeling, and site-directed mutagenesis studies have been carried out to find functional amino acid residues at the active site of the *Escherichia coli lacZ* β-galactosidase; these have shown that Glu461, Glu537, and Tyr503 are important for en-

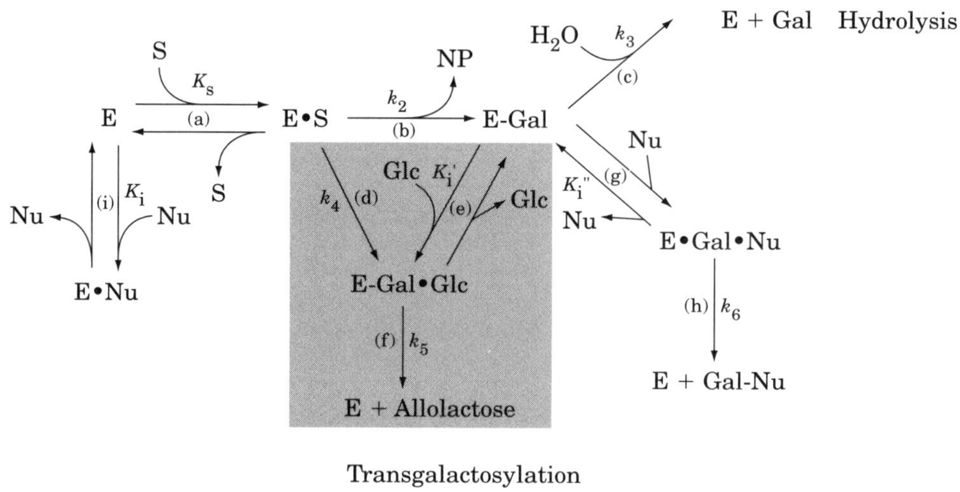

Figure 1. Proposed kinetic mechanism of hydrolysis and transgalactosylation reactions catalyzed by the *E. coli lacZ* β-galactosidase. *E*, enzyme; *S*, substrate; *NP*, 2-nitrophenol; *Gal*, galactose; *Glc*, glucose; *Nu*, galactosyl acceptor. Shadowed portion (steps *d*, *e*, *f*) indicates the pathway that must be considered, instead of step *b*, for the mechanism of the reaction with lactose. This mechanism also shows that the Nu also binds to free enzyme to form a deadend complex whose dissociation constant is *Ki* (step *i*).

zyme activity. All these catalytically important residues are located in the third domain of the enzyme subunit and, indeed, are shown to be located at the active site by X-ray crystallographic studies (66). However, some argument still exists on the details of the role(s) of each active site residue. Here we briefly describe the currently proposed roles of these amino acid residues in β-galactosidase catalysis (Fig. 2).

The glycosidic oxygen atom of the enzyme-bound substrate undergoes protonation, which facilitates aglycon departure. Intensive kinetic studies with site-directed mutants (17,78) suggest that the Glu461 serves as a Brønsted general acid–base catalyst at this step, although some evidence also argues for the role of Tyr503 being the general acid–base catalyst (79,80). The elimination of the aglycon probably produces an oxocarbonium ion intermediate that is planar at its anomeric end. This reaction is assumed because the planar and positively charged analogs of galactose serve as strong inhibitors of the *E. coli lacZ* β-galactosidase (75) (see Table 3). Affinity-labeling studies provided evidence suggesting that Glu537 is a nucleophile that attacks the anomeric carbon of the cationic intermediate (41), yielding a galactosyl enzyme in which the galactose moiety is covalently bound to the enzyme through an ester linkage. The water (or acceptor) then reacts with the covalent intermediate to produce galactose (or galactosyl adduct) and free enzyme. Although the enzyme requires monovalent and divalent ions for its activity, the role of the metal activators is not yet known. The bound Mg^{2+} (1 atom/active site; K_d, 0.65 μM [25]) has been suggested to act as an electrophilic catalyst that accelerates aglycone departure (12), whereas there is alternative evidence suggesting that it is only important in maintaining the proper active site structure (30,75). The Glu416, His418, and Glu461 are ligands to the bound Mg^{2+} ion (66).

APPLICATIONS OF β-GALACTOSIDASE-CATALYZED REACTIONS IN BIOPROCESS TECHNOLOGY

Hydrolysis

Cow milk generally contains 4.5–5.0% (w/v) lactose, whose low digestibility, low solubility, and low sweetness pose the following problems, the so-called lactose problems, in the milk and dairy industries (81).

1. Lactose intolerance is widespread in the world's adult population, with 70% being unable to metabolize large quantities of lactose in milk because of lack of intestinal lactase.
2. The high lactose content in nonfermented milk products, such as ice cream and condensed milk, often causes lactose crystallization during preservation. The crystals must not be greater than 10 μm in length, or they would make the products sandy.
3. The whey and whey permeate, which are the lactose-containing by-products from cheese manufacturing, have posed a serious problem as environmental pollutants, although these are potentially good carbohydrate sources for foods and animal feeds (82).

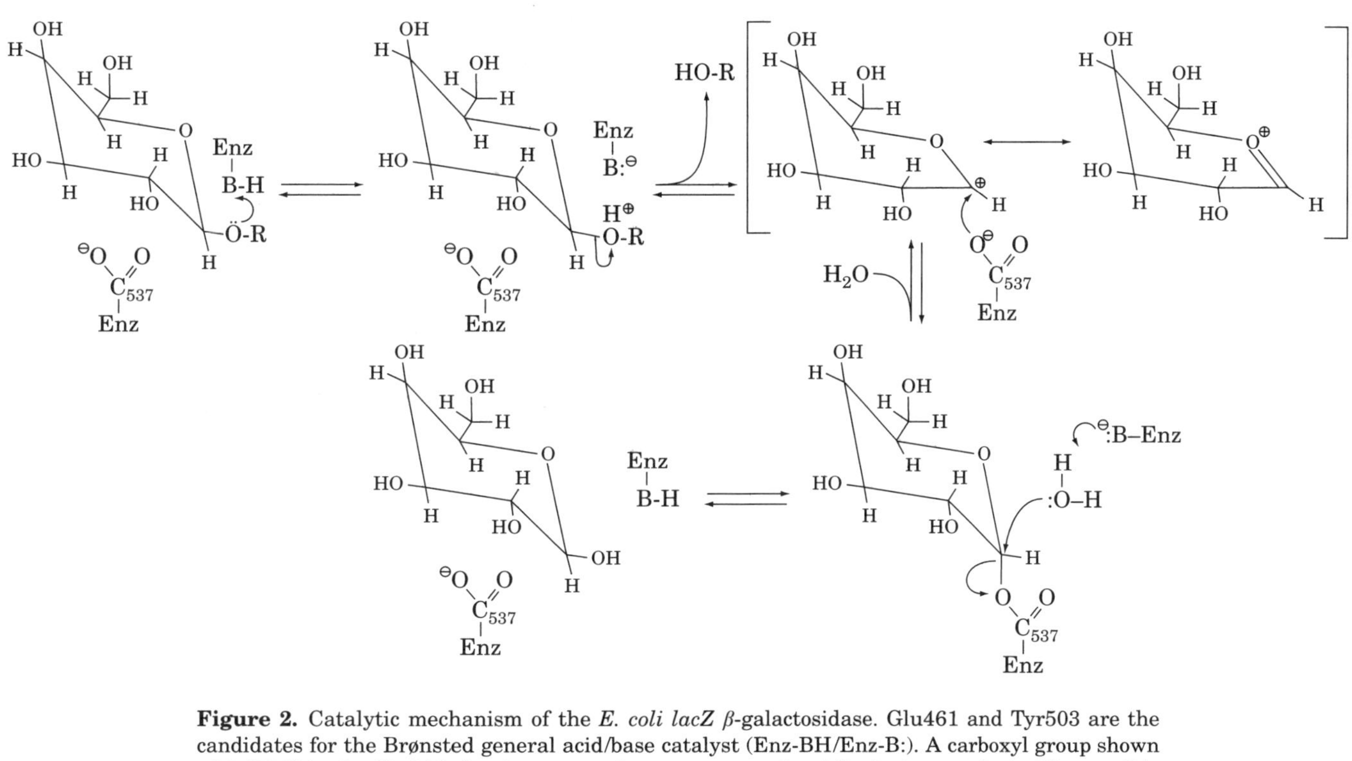

Figure 2. Catalytic mechanism of the *E. coli lacZ* β-galactosidase. Glu461 and Tyr503 are the candidates for the Brønsted general acid/base catalyst (Enz-BH/Enz-B:). A carboxyl group shown with "537" is the Glu537 that is suggested to act as a nucleophile during catalysis. The possible resonance of the planar cationic intermediate is shown with *square brackets*.

All these problems can be circumvented by decomposing lactose in the products by the action of β-galactosidase (83,84). For the purpose of alleviating lactose maldigestion, low-lactose milk is currently produced in industry by the continuous or batchwise treatment of milk with fungal β-galactosidases. Fungal β-galactosidase tablets have also been devised for this purpose and are commercially available (85,86). Enzymatic lactose hydrolysis can be considered to be a unit operation that can be integrated into the transformation of a wide range of whey and whey permeates (87). Hydrolysis of lactose produces a sugar mixture, glucose plus galactose, that is four to five times sweeter and three to four times more soluble than the parent disaccharide. The sugar mixture can serve as a condensed sugar syrup that is used as a sweetener or intermediate food product for manufacturing bakery and confectionary products, ice cream, and other specialities, permitting the efficient utilization of wheys and whey permeates (87).

Application of Immobilized Enzyme Technology in Industrial Lactose Hydrolysis

It is important to achieve industrial lactose hydrolysis with minimum cost. Immobilization of β-galactosidases is a means to reduce the cost of industrial lactose hydrolysis. Methods of immobilization and the properties of the immobilized β-galactosidases have been summarized (84). Baret (87) pointed out the following practical aspects in attaining the best performance of the immobilized enzyme system for the processing of whey and its permeates: (*1*) the nature of the substrate wheys and permeates, (*2*) the design of the reactor system, and (*3*) the operating strategy. The key objective of these considerations is to maximize the amount of hydrolyzed lactose per unit weight of catalyst, in other words, to maximize the half-life of the system. Semiindustrial-scale investigations were carried out to examine these aspects using immobilized *Aspergillus niger* and *A. oryzae* β-galactosidases that are covalently bound on a porous silica carrier:

1. Omission of demineralization of permeates had no adverse effect on the performance and stability of the systems. The need for demineralization is related to the application of the end product. However, controlling the levels of suspended solids and colloids in wheys was specifically important during operations in a fixed-bed reactor.
2. Among the fixed-bed, fluidized-bed, and stirred-tank reactors, the fixed-bed reactor gave the best kinetic results, in which the minimum reactor volume was necessary.
3. When no particular care is taken, the efficiency of the immobilized enzyme system immediately decreases because of deposition of material within the bed, the formation of a coating around the particles, microbial contamination, channeling, plugging, and other reasons. Operating strategy must therefore be considered to minimize these problems. The strategy required includes the removal of solids or colloidal substances in substrates, sanitation of substrates, introduction of the cycling of cleaning/sanitation procedures of the reactor during the operation, and programming the operating temperatures. These strategies greatly enhanced the activity and stability of the reactors.

It should be emphasized that the prevention against microbial contamination during operation is of specific importance for the processing of wheys and their permeates, which are nutritionally rich and excellent growth media for microorganisms. Methods for sanitizing the substrates, reactor, and other possible sources of microbial contamination were evaluated. Dilute acetic acid, which has been commonly used in the laboratory for disinfection of the immobilized enzyme matrices, served as only a poor sanitizer under industrial conditions. The substituted diethylenetriamines were effective for the disinfection of the immobilized enzyme matrices without causing loss of enzyme activity and have been used in large-scale operations.

The technology obtained from these large-scale examinations was transferred to fully industrial-scale operations; a plant processing 20,000 L/day of sweet whey has thus been operated, in which the hydrolysis reactor can process 1,000 L/h of nondemineralized acidified sweet whey using 40 kg of the immobilized β-galactosidase matrices (87). The processed products have been used as ingredients in the food industries.

Production of Lactose-Related Oligosaccharides by Transgalactosylation

It was shown in the 1950s that oligosaccharides are formed during the hydrolysis of lactose in milk with the fungal β-galactosidases (88,89). These oligosaccharides, the so-called galactooligosaccharides (GO), which are the products of the intermolecular transgalactosylation of lactose, have long been regarded as unwanted by-products in lactose hydrolysis (90). The *Escherichia coli lacZ* and *ebg* β-galactosidases were also shown to have transgalactosylation activity; their physiological significance has attracted the attention of biochemists because one of the transgalactosylated products, allolactose, serves as an inducer for the lactose operon of the bacterium (91).

During the past decade, GO has been shown to have some beneficial effects in maintaining human health (see later) and thus has been developed as a food material with health-giving functions. For the efficient production of GO from lactose, many attempts have been made to find microorganisms producing the GO or the β-galactosidase with high transgalactosylation activity. To date, a large number of such microorganisms have been found, among which *Bacillus circulans* (63), *Streptococcus thermophilus* (74), *Saccharopolyspora rectivirgula* (3), *Aspergillus oryzae* (92), *Cryptococcus laurentii* (62), and *Sterigmatomyces elviae* (2) produce enzymes with particularly high transgalactosylation activities. Also, the cells of yeasts, such as *Lipomyces* sp. (93), and those of the lactic acid bacteria, such as *Bifidobacterium bifidum* (94), can produce GO abundantly from lactose. Table 6 shows the identified structures of GO obtained with several enzymes and microbial cells. The structures of GO depend on the enzymes or microorganisms used as catalysts. The *B. circulans, S. rectivirgula, C.*

Table 6. Structures of Oligosaccharides Produced from Lactose by the Transgalactosylation Catalyzed by Microbial β-Galactosidases and Cells

Structure	Enzyme									Microorganism		
	1	2	3	4	5	6	7	8	9	10	11	12
β-D-Gal*p*-(1→6)-D-Glc (allolactose))	●	●			●	●	●		●	●		●
β-D-Gal*p*-(1→3)-D-Glc						●	●		●	●		●
β-D-Gal*p*-(1→2)-D-Glc						●	●		●			
β-D-Gal*p*-(1→6)-D-Gal		●			●	●	●			●		●
β-D-Gal*p*-(1→3)-D-Gal						●	●					
β-D-Gal*p*-(1→6)-β-D-Gal*p*-(1→4)-D-Glc (6′-galactosyllactose)								●		●		
β-D-Gal*p*-(1→3)-β-D-Gal*p*-(1→4)-D-Glc (3′-galactosyllactose)								●			●	●
β-D-Gal*p*-(1→4)-β-D-Gal*p*-(1→4)-D-Glc (4′-galactosyllactose)			●	●					●		●	
β-D-Gal*p*-(1→6)[β-D-Gal*p*-(1→4)]-D-Glc					●			●	●			●
β-D-Gal*p*-(1→2)[β-D-Gal*p*-(1→4)]-D-Glc									●			
β-D-Gal*p*-(1→2)[β-D-Gal*p*-(1→6)]-D-Glc									●			●
β-D-Gal*p*-(1→3)[β-D-Gal*p*-(1→6)]-D-Glc									●			
β-D-Gal*p*-(1→4)-β-D-Gal*p*-(1→6)-D-Glc									●			
β-D-Gal*p*-(1→6)-β-D-Gal*p*-(1→6)-D-Glc					●							
β-D-Gal*p*-(1→4)-β-D-Gal*p*-(1→3)-D-Glc									●			
β-D-Gal*p*-(1→4)-β-D-Gal*p*-(1→2)-D-Glc									●			
β-D-Gal*p*-(1→6)-β-D-Gal*p*-(1→6)-D-Gal					●							
β-D-Gal*p*(1→6)-β-D-Gal*p*-(1→6)-β-D-Gal*p*-(1→4)-D-Glc					●			●				
β-D-Gal*p*-(1→3)-β-D-Gal*p*-(1→6)-β-D-Gal*p*-(1→4)-D-Glc								●				
β-D-Gal*p*-(1→6)-β-D-Gal*p*-(1→3)-β-D-Gal*p*-(1→4)-D-Glc								●				
β-D-Gal*p*-(1→3)-β-D-Gal*p*-(1→3)-β-D-Gal*p*-(1→4)-D-Glc												●
β-D-Gal*p*-(1→6)-β-D-Gal*p*-(1→6)-β-D-Gal*p*-(1→6)-β-D-Gal*p*-(1→4)-D-Glc								●				
β-D-Gal*p*-(1→3)-β-D-Gal*p*-(1→3)-β-D-Gal*p*-(1→3)-β-D-Gal*p*-(1→4)-D-Glc												●
β-D-Gal*p*-(1→3)-β-D-Gal*p*-(1→3)-β-D-Gal*p*-(1→3)-β-D-Gal*p*-(1→3)-β-D-Gal*p*-(1→4)-D-Glc												●
β-D-Gal*p*-(1→3)-β-D-Gal*p*-(1→3)-β-D-Gal*p*-(1→3)-β-D-Gal*p*-(1→3)-β-D-Gal*p*-(1→3)-β-D-Gal*p*-(1→4)-D-Glc												●

Note: Closed circles indicate that the corresponding sugar has been identified in the transfer products. Enzymes are 1, *Escherichia coli* (Ref. 95); 2, *Streptococcus thermophilus* (Ref. 74); 3, *Sterigmatomyces elviae* (Ref. 2); 4, *Cryptococcus laurentii* (Ref. 62); 5, *Kluyveromyces lactis* (Ref. 96); 6, *K. fragilis* (Refs. 97,98); 7, *Aspergillus niger* (Ref. 98); 8, *A. oryzae* (Ref. 92); 9, *Bacillus circulans* (Ref. 99). Microorganisms are 10, *Penicillium chrysogenum* (Ref. 100); 11, *Chaetomium globosum* (Ref. 101); 12, *Bifidobacterium bifidum* (Ref. 94).

laurentii, S. elviae, and *A. oryzae* enzymes as well as *B. bifidum* cells produce from lactose the oligosaccharides with a general structure of Gal-$(Gal)_n$-Glc* as main products, where Gal and Glc* indicate the galactose and reducing terminal glucose residues and n the integer ($n \geq 1$), whereas the *S. thermophilus* and *E. coli lacZ* enzymes mainly produce disaccharides (Gal-Glc*) (Table 6).

Currently, GO is industrially produced by means of the *A. oryzae* and *S. thermophilus* β-galactosidases (102). In the reported system, lactose (80%, w/v) is first reacted with the *A. oryzae* enzyme at 65 °C for 4 h (103). The content of GO in the reaction mixture reaches 31% of the total sugar. The remaining lactose is further converted to the other disaccharidal transfer products by means of the *S. thermophilus* β-galactosidase (104). Industrial production of 4′-galactosyllactose is also achieved with *C. laurentii* cells entrapped in calcium alginate gels, which is used batchwise, yielding 4′-galactosyllactose whose content reaches 35% of total sugar (105). The immobilized cells can be used repeatedly, more than 30 times.

The lactose-related oligosaccharides that can be produced by the action of β-galactosidase also include the lactitol oligosaccharides, which are produced by the *A. oryzae* β-galactosidase-catalyzed transgalactosylation to lactitol [β-galactopyranosyl-(1→4)-D-glucitol], a reduced form of lactose (106).

Functional Properties of Lactose-Related Oligosaccharides

GO is of low sweetness with excellent taste quality and is stable under acidic conditions during food processing. Thus, the oligosaccharides serve as food materials potentially applicable to a wide variety of food manufacturing and cooking uses. GO also has beneficial physiological effects on human health, as described in the following subsections.

GO Is a Functional Sweetener. Ingested GO passes through the human small intestine without being digested; GO is therefore a low-calorie sweetener. *Streptococcus mutans*, a causative agent for dental caries, cannot metabolize GO or produce from GO the insoluble glucan that is responsible for dental plaque formation. Thus, GO is also a sweetener beneficial for prevention of dental caries.

GO Can Improve Intestinal and Fecal Conditions. The observation that GO can improve intestinal and fecal conditions has been explained in terms of the selective promotion by GO of bifidobacterial growth in human colons. The most dominant indigenous bacteria in the neonatal gut are the bifidobacteria, whose predominance in the gut has been suggested to cause beneficial effects in maintaining human health; these bacteria, as well as their metabolites, provide protection from infection and facilitate the normal functions of the gut. However, the bifidobacterial population in the human intestine decreases with aging processes, and the etiological relevance of this fact to the increased incidence of intestinal diseases in the middle- and advanced-aged generations has been implicated. Many in vitro experiments showed that GO specifically promotes the growth of the intestinal bifidobacteria and lactobacilli (e.g., see Refs. 2, 99 and 107). Ito et al. (108) further examined the in vivo effects of the ingestion of transgalactosylated disaccharides (TD), which were derived from lactose by the action of *Streptococcus thermophilus* β-galactosidase, on fecal microflora and their metabolism with 12 middle-aged male volunteers. The results showed that TD ingestion resulted in a specific increase in the fecal populations of bifidobacteria and lactobacilli, as well as a decrease in those of the genera *Bacteroides* and *Candida*. It also lowered fecal pH and caused a decrease in fecal concentrations of the products of amino acid catabolism, such as isobutyric and isovaleric acids, ammonia, *p*-cresol, and indole. These results suggest that the ingested TD reached the colon and selectively promoted the growth of bifidobacteria and lactobacilli to depress the microbial decomposition of amino acids. The GO containing di-, tri-, tetra-, or higher oligosaccharides showed a similar growth promotion of the fecal bifidobacteria and affected the metabolism of fecal microflora (109). It should be noted that the lactitol oligosaccharides also showed a selective promotion of bifidobacterial growth and improved the metabolism of fecal microflora (110).

Effect on Intestinal Absorption of Ca^{2+} and Lipid Metabolism. Osteoporosis, a loss of bone mass causing an increased risk of bone fracture, is associated with the aging processes and in particular with menopause in women; estrogen deficiency resulting from menopause results in more rapid bone resorption than formation, resulting in a net loss of bone mineral density. Estrogen deficiency also causes hyperlipemia, which produces a higher risk of coronary heart disease. The effects of GO ingestion on calcium absorption, prevention of bone loss, and total serum cholesterol content were examined (111,112). Ovariectomized Wistar rats fed on a diet containing GO (5% in place of sucrose in the control diet) absorbed calcium more efficiently than those on the control diet, and the bone ash weight and tibia calcium content of the GO-ingesting rats were significantly higher than those of the controls. GO also lowered the total concentration of serum cholesterol. It is suggested that the ingested GO is metabolized in the large intestine by intestinal bacteria to enhance the production of volatile acids, such as propionate, which facilitates intestinal calcium absorption and lowers the total serum cholesterol. These observations argue for possible prevention by GO against osteoporosis and hyperlipemia caused by postmenopausal estrogen deficiency.

Functional Modification of Various Compounds by Transgalactosylation

Glycosylation is a powerful means for structural and functional modification of medicines and many other bioactive compounds; it enhances their solubilities, physicochemical stabilities, biological half-lives, membrane permeabilities, and intestinal absorption and also improves their taste qualities. In the biosynthesis of glycosides and oligosaccharides, sugar nucleotides, such as UDP-glucose, serve as substrates for enzymes that catalyze glycosidic linkage formation. However, the use of these systems in biotransformations has been limited by the restricted availability of the corresponding enzymes and substrates. An alternative is to use transglycosylation catalyzed by glycosidases; there exists a variety of glycosidases with different specificities, and their substrates are inexpensively available. Because the transfer product is always a substrate for the glycosidase, the success of the procedure as a preparative method in part depends on the rate of hydrolysis of the product being slower than that of the glycosyl donor. In practice, these conditions can be attained rather readily. Thus, there has been a steadily growing interest in the glycosidases for glycoside and oligosaccharide synthesis. In this context, β-galactosidases with high transgalactosylation activities have also been used for obtaining the galactosyl adducts of sucrose, cyclodextrins, oligosaccharides, glycosides, amino acids, peptides, alcohols, vitamins, and other bioactive compounds (Table 7). We here describe two interesting examples of synthesis of galactosides using β-galactosidases in further detail.

Synthesis of β-Galactosylthiamin. Thiamine (vitamin B_1) occurs in its phosphorylated forms in nature and plays important roles in sugar metabolism as a cofactor. 5′-*O*-(β-Galactopyranosyl)-thiamine has been produced by the *Aspergillus oryzae* β-galactosidase-catalyzed transgalactosylation using ONPG as a galactosyl donor (121). In contrast to thiamine having a specific odor and strong tongue-pricking taste, the β-galactosylthiamine obtained was odorless and mildly sweet with no such stimulative taste, suggesting its usefulness as a food additive and medicine.

One-Pot Synthesis of Sialyl Oligosaccharides Using a Multienzyme System. Sialyl oligosaccharides are important recognition elements on the surface of cells and serve as ligands of some cell adhesion molecules. Herrmann et al. described a new efficient synthesis of α(2,6)sialyl *N*-acetyllactosamine and analogs by combined use of the *Bacillus circulans* β-galactosidase and pig liver α(2,6)sialyltransferase (116). In this system, the enzymes work sequentially in one pot to form a sialyl *N*-acetyllactosamine. CMP-sialic acid, a substrate for the sialyltransferase, is generated in situ with a coupled recycling enzyme system. The β-galactosidease-catalyzed transgalactosylation proceeds irreversibly because the galactosyl adduct is irreversibly sialylated with the sialyltransferase to form a sialyl *N*-acetyllactosamine, which is no longer a substrate for the β-galactosidase.

CONCLUDING REMARKS

To explore more efficient applications of β-galactosidase in bioprocess technology, efforts are still being made to dis-

Table 7. Synthesis of Galactosyl Adducts by Means of β-Galactosidase-Catalyzed Transgalactosylation

Galactosyl acceptor	Product	Galactosyl donor[a]	Catalyst[b]	Reference
	Sugars, glycosides, and derivatives			
Sucrose	Isoraffinose	1	1	113
Branched cyclodextrins	Mono- and di-β-Gal products at their side chains	1	2, 3, 4	114
Maltopentaose	4^5-*O*-β-Gal-maltopentaose	1	2	115
GlcNAc	Sialyl oligosaccharides	1	2	116
GlcNAc and derivatives	β-Gal-(1→4 or 1→6)-β-GlcNAc-pNP	1	2	117
GlcNAc and derivatives	β-Gal-(1→3)-β-GlcNAc	1	1, 5	118
GalNAc	β-Gal(1→3)-β-GalNAc	1	1, 5	119
α-Gal-OMe	β-Gal-(1→3 or 1→6)-α-Gal-OMe	2	1	120
1-β-Ara-cytosine	3′-*O*-β-Gal- and 3′-*O*-β-Gal-(1→)-β-Gal-1-β-Ara-cytosines	1	6	121
Rubsoside	13-*O*-β-Glc-19-*O*-[β-Gal-(1→4 or 1→6)β-Glc]- and 13-*O*-[β-Gal-(1→4)-β-Glc]-19-*O*-β-Glc-steviols	1	1, 2	122
β-Glc-pNP	β-Gal-(1→3, 1→4, or 1→6)-β-Glc-pNP	1	2	122
Moranoline	β-Gal-(1→2, 1→4, or 1→6)-moranoline	1	2	123
β-GlcNAc (1-2 or 1-6)-Man	β-Gal-(1→4)-β-GlcNAc-(1→2 or 1→6)-Man	1	2, 7	124
	Amino acids and peptides			
N-Protected-L-serine methyl esters	3-*O*-β-Gal-L-Ser derivatives	1	1	125
N-Protected dipeptide esters	β-(1→3)-digalactosyl peptides	1	1	126
	Alcohols			
Alcohols	Alkyl β-Gals	1	1	127,128
2-Fluoroethanol	2-Fluoroethyl-β-Gal and derivatives	1	8	129
Allyl alcohol	Allyl β-Gal	1	9	130
	Vitamins and other bioactive compounds			
Thiamine	5′-*O*-β-Gal-thiamin	3	3	121
Kojic acid	Galactosylkojic acid	1	2	132
Oximes	β-Gal-oxime derivatives	1, 3	3	133

Note: Gal, D-Galactopyranosyl or D-galactopyranose; Glc, D-glucopyranosyl; Ara, D-arabinofuranosyl; GlcNac, *N*-acetyl-D-glucosamine; Man, D-mannopyranose; Me, methyl; Et, ethyl; pNP, *p*-nitrophenyl. (See also Table 6 for the galactooligosaccharide synthesis, in which lactose serves as both galactosyl donor and acceptor.)

[a]1, lactose; 2, PNPG; 3, ONPG.

[b]1, *Escherichia coli* (*lacZ*); 2, *Bacillus circulans*; 3, *Aspergillus oryzae*; 4, *Penicillium multicolor*; 5, bovine; 6, *Sporobolomyces singularis*; 7, *Diplococcus pneumoniae*; 8, *Kluyveromyces lactis*; 9, *Streptococcus thermophilus*.

cover a novel microbial β-galactosidase in nature. In addition, attempts are being made through genetic and protein engineering approaches to improve the specificity and enhance the stability of known β-galactosidases. These attempts, along with basic mechanistic and structural studies, will further enhance our understanding of the molecular mechanism and the structure–function relationship of β-galactosidase.

BIBLIOGRAPHY

1. E. Steers, Jr., P. Cuatrecasas, and H.B. Pollard, *J. Biol. Chem.* **246**, 196–200 (1971).
2. N. Onishi and T. Tanaka, *Appl. Environ. Microbiol.* **61**, 4026–4030 (1995).
3. M. Nakao, M. Harada, Y. Kodama, T. Nakayama, Y. Shibano, and T. Amachi, *Appl. Microbiol. Biotechnol.* **40**, 657–663 (1994).
4. D. Grogan, *Appl. Environ. Microbiol.* **57**, 1644–1649 (1991).
5. A. Mbuyi Kalala, A.G. Schnek, and J. Leonis, *Eur. J. Biochem.* **178**, 437–443 (1988).
6. A.C. Elliott, S. K, M.L. Sinnott, P.J. Smith, J. Bommuswamy, Z. Guo, B.G. Hall, and Y. Zhang, *Biochem. J.* **282**, 155–164 (1992).
7. K. Sakaguchi and T. Yamaguchi, *J. Ferment. Technol.* **51**, 750–752 (1973).
8. M. Harada, M. Inohara, M. Nakao, T. Nakayama, A. Kakudo, Y. Shibano, and T. Amachi, *J. Biol. Chem.* **269**, 22021–22026 (1994).
9. M. Leahy, S. Fanning, F. O'Gara, and D. Sheehan, *Biochem. Soc. Trans.* **19**, 19S (1991).
10. I.S. Dunn and P.A. Jennings, *Prot. Eng.* **5**, 441–446 (1992).
11. J.P. Richard, R.E. Huber, S. Lin, C. Heo, and T.L. Amyes, *Biochemistry* **35**, 12377–12386 (1996).
12. T. Selwood and M.L. Sinnott, *Biochem. J.* **268**, 317–323 (1990).
13. D.R. Hwang and M.E. Scott, *Bioorg. Chem.* **21**, 284–293 (1993).

14. B.W. Bainbridge, N. Mathias, R.G. Price, A.C. Richardson, J. Sandhu, and B.V. Smith, *FEMS Microbiol. Lett.* **80**, 319–324 (1991).
15. I. Pocsi, S.A. Taylor, A.C. Richardson, B.V. Smith, and R.G. Price, *Biochim. Biophys. Acta* **1163**, 54–60 (1993).
16. R.P. Erickson and E. Steers, Jr., *J. Bacteriol.* **102**, 79–84 (1970).
17. I.S. Dunn and P.A. Jennings, *Prot. Eng.* **4**, 205–213 (1990).
18. Z. Huang, *Biochemistry* **30**, 8530–8540 (1991).
19. N. Bagget, M.A. Case, P.R. Darby, and C.J. Gray, *Carbohydr. Res.* **197**, 295–301 (1990).
20. R. Geiger, E. Schneider, K. Wallenfels, and W. Miska, *Biol. Chem. Hoppe-Seyler* **373**, 1187–1191 (1992).
21. V.K. Jain and I.T. Magrath, *Anal. Biochem.* **199**, 119–124 (1991).
22. E.G. Beale, E.A. Deeb, R.S. Handley, H. Akhavan-Tafti, and A.P. Schaap, *BioTechniques* **12**, 320–323 (1992).
23. R.E. Huber and R.L. Brockbank, *Biochemistry* **26**, 1526–1531 (1987).
24. J. Smart and B. Richardson, *Appl. Microbiol. Biotechnol.* **26**, 177–185 (1987).
25. H. Shukla and M. Chaplin, *Enzyme Microb. Technol.* **15**, 297–299 (1993).
26. D.F. Wentworth and R. Wolfenden, *Biochemistry* **13**, 4715–4720 (1974).
27. R.E. Huber and M.T. Gaunt, *Can. J. Biochem.* **60**, 608–612 (1982).
28. H. Nakano, S. Takenishi, and Y. Watanabe, *Agric. Biol. Chem.* **51**, 2267–2269 (1987).
29. C.G. Cupples, J.H. Miller, and R.E. Huber, *J. Biol. Chem.* **265**, 5512–5518 (1990).
30. G.S. Case, M.L. Sinnott, and J.-P. Tenu, *Biochem. J.* **133**, 99–104 (1973).
31. J.N. BeMiller, R.J. Gilson, R.W. Myers, M.M. Santoro, and M.P. Yadav, *Carbohydr. Res.* **250**, 93–100 (1993).
32. Y. Miyake and M. Ebata, *Agric. Biol. Chem.* **52**, 1649–1654 (1988).
33. Y. Ichikawa and Y. Igarashi, *Tetrahedron Lett.* **36**, 4585–4586 (1995).
34. G. Legler and S. Pohl, *Carbohydr. Res.* **155**, 119–129 (1986).
35. K. Bock and K. Adelhorst, *Carbohydr. Res.* **202**, 131–149 (1990).
36. J. Yariv, K.J. Wolson, J. Hildesheim, and S. Blumberg, *FEBS Lett.* **15**, 24–26 (1971).
37. T. Mega, T. Nishijima, and T. Ikenaka, *J. Biochem.* **107**, 641–644 (1990).
38. R.E. Huber, *Carbohydr. Res.* **214**, 35–41 (1991).
39. M. Herrchen and G. Legler, *Eur. J. Biochem.* **138**, 527–531 (1984).
40. S.G. Withers, K. Rupitz, and I.P. Street, *J. Biol. Chem.* **263**, 7929–7932 (1988).
41. J.C. Gebler, R. Aebersold, and S.G. Withers, *J. Biol. Chem.* **267**, 11126–11130 (1992).
42. J.N. BeMiller, R.J. Gilson, R.W. Myers, and M.M. Santoro, *Carbohydr. Res.* **250**, 101–112 (1993).
43. R. Golan, U. Zehavi, M. Naim, A. Patchornik, and P. Smirnoff, *Biochim. Biophys. Acta* **1293**, 238–242 (1996).
44. J. Inagawa, I. Kiyosawa, and T. Nagasawa, *Agric. Biol. Chem.* **51**, 3027–3032 (1987).
45. R.A. Field, A.H. Haines, E.J. Chrystal, and M.C. Luszniak, *Biochem. J.* **274**, 885–889 (1991).
46. N. Asano, A. Kato, K. Oseki, H. Kizu, and K. Matsui, *Eur. J. Biochem.* **229**, 369–376 (1995).
47. K. Nishiyama, S. Esaki, I. Deguchi, N. Sugiyama, and S. Kamiya, *Biosci. Biotechnol. Biochem.* **57**, 107–114 (1993).
48. B. Henrissat, *Biochem. J.* **280**, 309–316 (1991).
49. M.V. Cubellis, C.C. Rozzo, P. Motecucchi, and M. Rossi, *Gene* **94**, 89–94 (1990).
50. J. Jenkins, L. Lo-Leggio, G. Harris, and R. Pickersgill, *FEBS Lett.* **362**, 281–285 (1995).
51. K.R. Gutshall, D.E. Trimbur, J.J. Kasmir, and J.E. Brenchley, *J. Bacteriol.* **177**, 1981–1988 (1995).
52. K. Gutshall, K. Wang, and J.E. Brenchley, *J. Bacteriol.* **179**, 3064–3067 (1997).
53. Y. Ito and T. Sasaki, *Biosci. Biotechnol. Biochem.* **61**, 1270–1276 (1997).
54. C.H. Taron, J.S. Benner, L.J. Hornstra, and E.P. Guthrie, *Glycobiology* **5**, 603–610 (1995).
55. D.E. Trimber, K.R. Gutshall, P. Prema, and J.E. Brenchley, *Appl. Environ. Microbiol.* **60**, 4544–4552 (1994).
56. R.C. Dickson, L.R. Dickson, and J.S. Markin, *J. Bacteriol.* **137**, 51–61 (1979).
57. F. Widmer and J.-L. Leuba, *Eur. J. Biochem.* **100**, 559–567 (1979).
58. T. Uwajima, H. Yagi, and O. Terada, *Agric. Biol. Chem.* **36**, 570–577 (1972).
59. Y. Tanaka, A. Kagamiishi, A. Kiuchi, and T. Horiuchi, *J. Biochem.* **77**, 241–247 (1975).
60. T. Mega and Y. Matsushima, *J. Biochem.* **85**, 335–341 (1979).
61. C.T. Kelly, M.R. O'Mahoney, and W.M. Fogarty, *Appl. Microbiol. Biotechnol.* **27**, 383–388 (1988).
62. K. Ohtsuka, A. Tanoh, O. Ozawa, T. Kanematsu, T. Uchida, and R. Shinke, *J. Ferment. Bioeng.* **70**, 301–307 (1990).
63. Z. Mozaffar, K. Nakanishi, R. Matsuno, and T. Kamikubo, *Agric. Biol. Chem.* **48**, 3053–3061 (1984).
64. O. Poch, H. L'Hote, V. Dallery, F. Debeaux, R. Fleer, and R. Sodoyer, *Gene* **118**, 55–63 (1992).
65. V. Kumar, S. Ramakrishnan, T. Teeri, J. Knowles, and B. Hartley, *Biotechnology* **10**, 82–85 (1992).
66. R.H. Jacobson, X.-J. Zhang, R.F. Dubose, and B.W. Matthews, *Nature* **369**, 761 (1994).
67. I. Zabin, *Mol. Cell. Biochem.* **49**, 87–96 (1982).
68. A. Ullman, *Bioessays* **14**, 201–205 (1992).
69. H.W. Stokes, P.W. Betts, and B.G. Hall, *Mol. Biol. Evol.* **2**, 469–477 (1985).
70. K. Nakanishi, R. Matsuno, K. Torii, K. Yamamoto, and T. Kamikubo, *Enzyme Microb. Technol.* **5**, 115–120 (1983).
71. H. Fujimoto, M. Miyasato, Y. Ito, T. Sasaki, and K. Ajisaka, *Glycoconj. J.* **15**, 155–160 (1998).
72. W.M. de Vos and G. Simons, *Biochimie (Paris)* **70**, 461–473 (1988).
73. C.J. Schroeder, C. Robert, G. Lenzen, L.L. McKay, and A. Mercenier, *J. Gen. Microbiol.* **137**, 369–380 (1991).
74. N.A. Greenberg and R.R. Mahoney, *Food Chem.* **10**, 195–204 (1983).
75. R.E. Huber, M.N. Gupta, and S.K. Khare, *Int. J. Biochem.* **26**, 309–318 (1994).

76. P.J. Deschavanne, O.M. Viratelle, and J.M. Yon, *J. Biol. Chem.* **253**, 833–837 (1978).
77. R.E. Huber, M.T. Gaunt, and K.L. Hurlburt, *Arch. Biochem. Biophys.* **234**, 151–160 (1984).
78. J.P. Richard, R.E. Huber, C. Heo, T.L. Amyes, and S. Lin, *Biochemistry* **35**, 12387–12401 (1996).
79. M. Ring, I.M. Armitage, and R.E. Huber, *Biochem. Biophys. Res. Commun.* **131**, 675–680 (1985).
80. M. Ring and R.E. Huber, *Arch. Biochem. Biophys.* **283**, 342–350 (1990).
81. T.P. Shukla, *CRC Crit. Rev. Food Technol.*, 325–356 (1975).
82. G. Moulin and P. Galzy, *Biotechnol. Gen. Eng. Rev.* **2**, 347–374 (1984).
83. V. Gekas and M. Lopez-Leiva, *Process Biochem.* **2**, 2–12 (1985).
84. N.A. Greenberg and R.R. Mahoney, *Process Biochem.* 2–8 (February/March 1981).
85. M.S. Medow, K.D. Thek, L.J. Newman, S. Berezin, M.S. Glassman, and S.M. Schwarz, *Am. J. Dis. Child.* **144**, 1261–1264 (1990).
86. A.E. Klingerman, *Am. J. Clin. Nutr.* **51**, 890–892 (1990).
87. J.L. Baret, *Methods Enzymol.* **136**, 411–423 (1987).
88. M. Aronson, *Arch. Biochem. Biophys.* **39**, 370–378 (1952).
89. J.H. Pazur, *Science* **117**, 355–356 (1953).
90. J.E. Prenosil, E. Stuker, and J.R. Bourne, *Biotechnol. Bioeng.* **30**, 1019–1025 (1987).
91. B.G. Hall, *J. Bacteriol.* **150**, 132–140 (1982).
92. T. Toba, A. Yokota, and S. Adachi, *Food Chem.* **16**, 147–162 (1985).
93. M. Dombou, H. Nakajima, and T. Komaki, *Tokkaihei*, 1–112978 (1989).
94. V. Dumortier, J. Montreuil, and S. Bouquelet, *Carbohydr. Res.* **201**, 115–123 (1990).
95. R.E. Huber, G. Kurz, and K. Wallenfels, *Biochemistry* **15**, 1994–2001 (1976).
96. N. Asp, A. Burvall, A. Dahlqvist, P. Hallgren, and A. Lundblad, *Food Chem.* **5**, 147–153 (1980).
97. J.H. Pazur, C.L. Tipton, T. Budovich, and J.M. Marsh, *J. Am. Chem. Soc.* **80**, 119–121 (1958).
98. T. Toba and S. Adachi, *J. Dairy Sci.* **61**, 33–38 (1978).
99. S. Yanahira, T. Kobayashi, T. Suguri, M. Nakakoshi, S. Miura, H. Ishikawa, and I. Nakajima, *Biosci. Biotechnol. Biochem.* **59**, 1021–1026 (1995).
100. A. Ballio and S. Russi, *J. Chromatogr.* **4**, 117–129 (1960).
101. P.A.J. Gorin, R.H. Haskin, and D.W.S. Westlake, *Can. J. Chem.* **44**, 2083–2086 (1966).
102. K. Matsumoto, Y. Kobayashi, S. Ueyama, T. Watanabe, R. Tanaka, T. Kan, A. Kuroda, in I. Karube ed., *Japanese Technology Reviews*, Gordon and Breach, New York, 1992, pp. 90–106.
103. K. Matsumoto, Tokkaishou, 63–109789 (1988).
104. K. Matsumoto, *Denpun Kagaku* **36**, 123–129 (1989).
105. O. Ozawa, *Food Chemicals* **6**, 75–81 (1988).
106. S. Yanahira, T. Suguri, T. Yakabe, Y. Ikeuchi, G. Hanagata, and E. Deya, *Carbohydr. Res.* **232**, 151–159 (1992).
107. M. Dombou, H. Yamamoto, H. Nakajima, K. Tomita, and T. Komaki, *Denpun Kagaku* **38**, 365–367 (1991).
108. M. Ito, M. Kimura, Y. Deguchi, A. Miyamori-Watanabe, T. Yajima, and T. Kan, *J. Nutr. Sci. Vitaminol. (Tokyo)* **39**, 279–288 (1993).
109. M. Ito, Y. Deguchi, K. Matsumoto, M. Kimura, N. Onodera, and T. Yajima, *J. Nutr. Sci. Vitaminol. (Tokyo)* **39**, 635–640 (1993).
110. S. Yanahira, M. Morita, S. Aoe, T. Suguri, I. Nakajima, and E. Deya, *J. Nutr. Sci. Vitaminol. (Tokyo)* **41**, 83–94 (1995).
111. O. Chonan and M. Watanuki, *J. Nutr. Sci. Vitaminol. (Tokyo)* **41**, 95–104 (1995).
112. O. Chonan, K. Matsumoto, and M. Watanuki, *Biosci. Biotechnol. Biochem.* **59**, 236–239 (1995).
113. K. Suyama, S. Adachi, T. Toba, T. Sohma, C.-J. Hwang, and T. Itoh, *Agric. Biol. Chem.* **50**, 2069–2075 (1986).
114. S. Kitahata, K. Fujita, Y. Takagi, K. Hara, H. Hashimoto, T. Tanimoto, and K. Koizumi, *Biosci. Biotechnol. Biochem.* **56**, 242–245 (1992).
115. S. Kitahata, K. Fujita, K. Hara, and H. Hashimoto, *Agric. Biol. Chem.* **55**, 2433–2434 (1991).
116. G.F. Herrmann, Y. Ichikawa, C. Wandrey, F.C.A. Gaeta, J.C. Paulson, and C.-H. Wong, *Tetrahedron Lett.* **34**, 3091–3094 (1993).
117. T. Usui, S. Kubota, and H. Ohi, *Carbohydr. Res.* **244**, 315–323 (1993).
118. L. Hedbys, E. Johansson, K. Mosbach, P.O. Larsson, A. Gunnarsson, S. Svensson, and H. Lonn, *Glycoconjugates* **6**, 161–168 (1989).
119. L. Hedbys, E. Johansson, K. Mosbach, and P.O. Larsson, *Carbohydr. Res.* **186**, 217–223 (1989).
120. V. Kery, S. Kucar, M. Matulova, and J. Haplova, *Carbohydr. Res.* **209**, 83–87 (1991).
121. Y. Suzuki and K. Uchida, *Biosci. Biotechnol. Biochem.* **58**, 639–643 (1994).
122. S. Kitahata, H. Ishikawa, T. Miyata, and O. Tanaka, *Agric. Biol. Chem.* **53**, 2923–2928 (1989).
123. T. Murata, S. Akimoto, M. Horimoto, and T. Usui, *Biosci. Biotechnol. Biochem.* **61**, 1118–1120 (1997).
124. M. Kojima, T. Seto, Y. Kyotani, H. Ogawa, S. Kitagawa, K. Mori, S. Maruo, T. Ohgi, and Y. Ezure, *Biosci. Biotechnol. Biochem.* **60**, 694–696 (1996).
125. H. Fujimoto, M. Isomura, T. Miyazaki, I. Matsuo, R. Walton, T. Sakakibara, and K. Ajisaka *Glycoconj. J.* **14**, 75–80 (1997).
126. D. Cantacuzene and S. Attal, *Carbohydr. Res.* **211**, 327–331 (1991).
127. S. Attal, S. Bay, and D. Cantacuzene, *Tetrahedron* **48**, 9251–9260 (1992).
128. D.H. Crout, D.A. MacManus, and P. Critchley, *J. Chem. Soc. Perkin Trans.* **1**, 1865–1868 (1990).
129. D.E. Stevenson, R.A. Stanley, and R.H. Furneaux, *Biotechnol. Bioeng.* **42**, 657–666 (1993).
130. D.E. Stevenson, A.D. Woolhouse, R.H. Furneaux, D. Batcheler, and C.T. Eason, *Carbohydr. Res.* **256**, 185–188 (1994).
131. D.E. Stevenson and R.H. Furneaux, *Carbohydr. Res.* **284**, 279–283 (1996).
132. M.A. Hassan, F. Ismail, S. Yamamoto, H. Yamada, and K. Nakanishi, *Biosci. Biotechnol. Biochem.* **59**, 543–545 (1995).
133. M. Pozo and V. Gotor, *J. Chem. Soc. Perkin Trans.* **9**, 1001–1002 (1993).

See also PROTEIN EXPRESSION, SOLUBLE.

GAS HOLD-UP

ANDREAS LÜBBERT
Martin Luther University
Wittenberg, Germany

KEY WORDS

Aeration
Bubble size
Fermenter design
Foam problems
Gas bubbles
Multiphase flow
Oxygen transfer
Specific interfacial area
Void fraction

OUTLINE

Basic Definitions
Biotechnological Relevance of the Gas Holdup
Fluid–Dynamic Details
Measurement Techniques
Conclusion

BASIC DEFINITIONS

Cultivation media in biotechnology are basically multiphase fluid systems with a continuous liquid phase in which cells and gas bubbles are dispersed. In such systems, the gas holdup ϵ is defined as the fraction of the gas phase of the total volume V_D of the dispersion considered.

$$\epsilon = \frac{V_G}{V_D} \tag{1}$$

V_G is the total gas volume within V_D (i.e., the gas holdup is a volume-related quantity). When considering the liquid phase, the gas bubbles are void areas. Thus, the gas holdup is sometimes called the void fraction.

When the entire culture within a bioreactor is considered, we then more precisely refer to the corresponding holdup as the total or global gas holdup. However, we may also speak about spatial distributions of the gas phase, particularly in systems where the gas phase is not quasi-homogeneously distributed. We then prefer to consider local gas holdups ϵ_l, which are defined by the differential quotient

$$\epsilon_l = \frac{dV_G}{dV_D} \tag{2}$$

Usually we assume that we can consider the dispersion to be a quasi-homogeneous fluid. This is a reasonable assumption when discussing technical reactor sizes, where the bubbles are usually very small as compared to the reactor dimensions. However, when we discuss effects on small scales, we should keep in mind that in gas–liquid dispersions the gas phase is composed of individual bubbles of volume V_j. It does not make sense to consider gas holdups when looking for the effects on scales in the order of the mean bubble size. On such small scales, the gas–liquid two-phase flow system must be treated in a discontinuous way. Hence, discussions about the gas holdup in bioreactors are more or less restricted to large scales as compared to the bubble size. In most practical cases we think of global gas holdups when discussing dispersions within bioreactors.

When we consider the individual bubbles of volume V_j, we may rewrite the total gas holdup as

$$\epsilon = \frac{V_G}{V_D} = \frac{\Sigma V_j}{V_D} \tag{3}$$

where the sum is drawn over all bubbles in the entire volume V_D of the dispersion. The volume V_j characterizes the bubble size. However, it is often more convenient to use diameter instead, since the diameter can most easily be estimated by visual inspection of the dispersion. Unfortunately, the bubble diameter is not well defined, since different bubbles of the same volume might have different forms. This problem has been circumvented by defining the so-called equivalent diameter d_j, which is the diameter of a sphere with the bubble's volume V_j. We can then write the gas volume V_G as

$$V_G = \Sigma V_j = \frac{\pi}{6} \Sigma d_j^3 \tag{4}$$

Bubbles have several functions in bioreactors, the most important being the transport of a gaseous reactant from the dispersed gas phase across the bubble surfaces into the liquid phase. In this respect, the transport cross section A available for this mass transfer is of primary importance. A is the total interfacial area between the gas and the liquid phase. This is, apart from the top surface of the dispersion, the sum of all bubble surfaces

$$A = \Sigma A_j \approx \pi \Sigma d_j^2 \tag{5}$$

where A_j is the surface area of the jth bubble. In biochemical reaction engineering, one prefers to discuss the transport cross section as a specific, and in this particular case a volume-related, quantity. This specific interfacial area a is defined as

$$a = \frac{A}{V_D} = \frac{\Sigma A_j}{V_D} \approx \frac{\pi \Sigma d_j^2}{V_D} \tag{6}$$

With these definitions we are able to relate the gas holdup to the quantities of primary interest in biochemical reaction engineering. Starting from equation 1, we obtain

$$\epsilon = \frac{V_G}{V_D} = \frac{AV_G}{V_D A} = \frac{A}{V_D} \frac{\frac{\pi}{6} \Sigma d_j^3}{\pi \Sigma d_j^2} = a \frac{\frac{\pi}{6} \Sigma d_j^3}{\pi \Sigma d_j^2} \tag{7}$$

Because the bubbles in biotechnologically relevant dispersions are not of uniform size, such expressions are of little practical use. One is more interested in characterizing the bubbles by some mean bubble diameter. In light of the preceding discussion it is convenient, as proposed by Sauter, to take the quantity

$$d_s = \frac{\Sigma d_j^3}{\Sigma d_j^2} \tag{8}$$

as such a mean bubble diameter. The sums are drawn over all bubbles j. In the literature, d_s is referred to as the Sauter diameter.

With the Sauter diameter d_s, we obtain a simple very important practical relation between the relevant gas-phase characteristics of a bubble dispersion:

$$\epsilon = a\frac{d_s}{6} \quad \text{or} \quad a = \frac{6\epsilon}{d_s} \tag{9}$$

which essentially states that the important specific interfacial area a is proportional to the gas holdup ϵ, the proportionality constant being dependent in a most simple way from the Sauter mean bubble diameter d_s. In other words, the specific interfacial area can increase by increasing the gas holdup, and is more effective, when the mean bubble size is kept smaller.

Assuming that the volume V_L of the liquid phase of a culture within a bioreactor is given, the higher the gas holdup ϵ, the larger the total volume V_D of the dispersion. Usually, in discussions about gas-phase behavior within bioreactors, the solid phase, in particular the cells, can be attributed to the liquid phase. In reactors where the reactor horizontal cross-sectional area A_{RCS} is uniform with respect to the height H_D of the dispersion, we can write the gas holdup as a function of the height H_D of the fully aerated liquid dispersion and the height H_L of the liquid without aeration, i.e., when there are no gas bubbles present.

$$\epsilon = \frac{V_G}{V_D} = \frac{V_D - V_L}{V_D} = \frac{(H_D - H_L)A_{RCS}}{H_D A_{RCS}} = \frac{(H_D - H_L)}{H_D} \tag{10}$$

This relative enhancement of the culture level in the reactor by aeration, or simply by the presence of gas bubbles, is of practical importance, since, as discussed later, one must monitor this level during the operation of the bioreactor. In cases where it is possible to interrupt the gas throughput through the reactor and to remove the bubbles, this equation may be used to measure the total gas holdup. If such a degassing is possible, this is the most simple measurement of the total gas holdup in dispersions.

From a global view, a higher gas holdup of the dispersion leads to a lower mean density ρ_D of the dispersion. When the gas phases can be assumed to be quasi-homogeneously distributed, the density ρ_D can be represented as a linear function of the gas holdup ϵ:

$$\rho_D = (1 - \epsilon)\rho_L \tag{11}$$

where ρ_L is the density of the liquid phase. This has an immediate consequence for the static pressure at a particular height H within the reactor. Since the static pressure drop Δp over a height interval ΔH is dependent on the density of the dispersion, it is possible to directly relate the static pressure drop and the gas holdup

$$\Delta p = \rho_D g \Delta H \tag{12}$$

$$\epsilon = 1 - \frac{\Delta p}{\rho_L g \Delta H} \tag{13}$$

which is a relationship that is often used to measure the gas holdup in multiphase reactors. Here, g is the gravitational acceleration.

The gas holdup can also be considered in terms of the aeration rate. The aeration rate is often represented by the gas throughput Q_G relative to the cross-sectional area A_{RCS} of the reactor. This quantity is referred to as the superficial gas velocity w_{sg}, since it formally has the dimension of a velocity

$$w_{sg} = \frac{Q_G}{A_{RCS}} \tag{14}$$

When the mean bubble velocity in the laboratory coordinate system is w_{bl}, then with

$$w_{bl} = \frac{H_D}{\tau_B} \tag{15}$$

where τ_B is the mean residence time of the bubbles in the reactor, and

$$Q_G = \frac{V_G}{\tau_B} \tag{16}$$

we obtain

$$\epsilon = \frac{V_G}{V_D} = \frac{V_G}{A_{RCS}H_D} = \frac{V_G\tau_B}{\tau_B A_{RCS} H_D} = \frac{w_{sg}\tau_B}{V_D} = \frac{w_{sg}}{w_{bl}} \tag{17}$$

In other words, the gas holdup is proportional to the superficial gas velocity, the proportionality constant being the reciprocal value of the mean bubble rise velocity in the laboratory coordinate system. The superficial gas velocity is the variable that can be manipulated. The effective bubble velocity w_{bl}, however, depends on the fluid-dynamic properties within the reactor, as well of the broth rheology. Equation 17 has some practical consequences, as discussed later.

BIOTECHNOLOGICAL RELEVANCE OF THE GAS HOLDUP

Of all the properties that determine the performance of a bioreactor with respect to a biochemical conversion process, it is the gas–liquid mass transfer capacity or capability that is the most important. This mass transfer capacity is primarily determined by the specific gas–liquid interfacial area a that the bioreactor is able to provide. As discussed, this quantity is primarily related to the gas holdup. When there is no gas holdup, then there is no mass transfer.

The mass transfer rate is usually described by simple transport equation analogues to Fick's first or Ohm's law, where it is assumed that the mass current through a region that represents a resistance for the mass transfer is drawn by some driving force. In the gas–liquid mass transfer between gas bubbles and the bulk liquid, the essential mass transfer resistance is assumed to be the liquid-side boundary layer at the physical bubble surface, and the driving force is the difference between the dissolved gas concentrations at both sides of this boundary layer. In this simple picture, the mass transfer resistance is characterized by the product k_La, where a is the specific interfacial area, which by equation 9 is proportional to the gas holdup. The mass transfer coefficient k_L is defined by this simple transport law. In the particular case of the oxygen transfer from a bubble into its surrounding liquid, the transport resistance is essentially the liquid-side boundary layer around the bubble. At the physical interfacial area the liquid is assumed to be saturated with oxygen (oxygen concentration O^*), while on the bulk liquid side we assume the mean dissolved oxygen concentration O present. The oxygen transfer rate OUR can then be written as

$$OUR = k_La\Delta O \propto \epsilon \quad (18)$$

where $\Delta Q = O^* - O$ is the mean driving force.

Although practically most important, the mass transfer effects that are of interest in biotechnology are not restricted to oxygen gas–liquid mass transfer. The transfer of CO_2 is also of importance, particularly in anaerobic systems. Beer production may serve as an important example of this transfer. Here, however, the mass transfer goes in reverse (i.e., CO_2 is mainly transferred from the liquid to the gas bubbles).

The straightforward way to increase the gas holdup ε, and thus the mass transfer rate, is to increase the volume of gas sparged per time unit ($Q_G = dV_G/dt$) into the dispersion. Considering the specific interfacial area a, it would be preferable to use a sparger that produces smaller bubbles or to apply strong mechanical agitation of the dispersion to redisperse larger bubbles. By equation 9, a smaller mean bubble diameter d_s leads to a higher specific mass transfer area a. These simple arguments, however, have their clear practical limits, as the following discussion will show.

The quality of the gas holdup may be defined by the size of the bubbles, characterized by the mean (Sauter) diameter d_s and by some surface properties of the bubbles. The size of a bubble determines its rise velocity with respect to the surrounding bulk liquid. When this relative velocity decreases, the bubble more closely follows the liquid motion. Since the liquid phase is operated in a batch mode within most of the practically used bioreactors, the residence time of the bubbles within the bioreactor becomes higher with smaller bubble size. Very small bubbles may even stay within the bioreactor for a very long time before leaving the dispersion at its top surface. Additionally, the escape of the bubbles from the dispersion may be delayed or even suppressed by surface active molecules dissolved in the liquid phase, which stabilize the boundary layer. Both effects increase the gas holdup, often in an undesirable way.

Long residence times of very small bubbles are undesirable when they exceed the mean time spans during which the bubbles are able to positively contribute to the oxygen transfer. To get an idea of the time constant of the mass transfer, consider that single bubbles with an equivalent diameter of 5 mm can be assumed to have a mass transfer time constant of roughly 5 min. Consequently, when such bubbles stay within the reactor for periods much longer than 5 min, they no longer actively contribute to mass transfer. They must then be regarded as dead volume, which essentially reduces the effective reactor volume (i.e., it reduces the productivity).

When we are dealing with very long gas residence times, the dispersion accumulates a great amount of gas, and we consider foamy behavior of the dispersion. Such situations are extremely undesirable, since with continuing gas supply, the gas holdup becomes larger and larger until the headspace of the reactor becomes full and the dispersion enters the gas vent line. There it is usually stopped at the off-gas filter. This filter quickly becomes clogged by the material dissolved or dispersed in the liquid (e.g., the cells or dissolved proteins). Since the gas throughput is usually controlled, the reactor pressure will be enhanced to hold the throughput. The pressure may quickly rise, and eventually some part of the reactor, most often the tubing by which substrate or base is supplied, bursts. To prevent such accidents, in general, the gas holdup in a bioreactor must be kept under control. This is most often done by measuring the height of the dispersion with some level sensor and by reducing the gas holdup with chemical additives (antifoam agents) when the maximal level is reached.

Antifoam agents increase the coalescence behavior of the gas bubbles, thus leading to a larger bubble size. Larger bubbles are much more easily disengaging from the dispersion, so that the gas holdup drops. Soya oil or silicone-based solutions most often serve as antifoam agents. However, their addition to biological cultures must be well dosed. One argument for keeping the amount of antifoam agents within the medium low is that these molecules, as they are surface active, most often negatively influence many downstream processing procedures. Also, the addition of too much antifoam agent may drastically reduce the gas holdup in a bioreactor. This discussion shows that, with respect to mass transfer, there are competing requirements. Thus, the addition of antifoam must be optimized to obtain a maximum performance of the mass transfer in the reactor while avoiding the problems with foam.

There is another problem associated with high gas holdups or foam formation. It is well known that microbial cells are in some sense surface active as well; that is, they easily attach themselves to bubble surfaces and are then transported upward toward the surface of the dispersion. This flotation effect is a negative influence on reactor performance when a stable foam layer develops on top of the dispersion. Then, a significant amount of the biomass might become captured in the foam layer. In particular, in the beginning of a cultivation (after inoculation) when the biomass concentration is low, the biomass can become nearly quantitatively trapped in the foam layer. Then, consequently, one does not find a significant biomass concentration in the bulk. After addition of an antifoam agent, the biomass appears back in the bulk. Hence, it is necessary to avoid these kinds of gas holdup effects by initially adding a sufficient, but not too large, amount of antifoam agent to the medium before inoculation.

While the small-scale shear flow that appears in the surroundings of the bubbles when they rise in the liquid does not have any negative effect on microbes, there are some problems with aerated cultures of shear-sensitive mammalian cells. It was observed that some cell lines die when their cultures are aerated with air bubbles, while they grow without problems when they are indirectly supplied with oxygen (e.g., by means of oxygen diffusion through silicone tubes submerged in the culture medium). The current explanation of this phenomenon is that it is not the shear flow around the rising bubbles that damages the cells but the sudden disruption of the liquid film that appears immediately before the gas bubble is disengaged into the reactor's headspace. When a cell happens to be within this bursting film, then it is subject to extreme shear rates that will most probably destroy the cell membrane. Similar effects are also to be expected during bubble coalescence or redispersion effects, since in such situations the new bubble surfaces are formed very quickly. Thus, very high local velocity gradients appear on the scale of the diameters of these cells.

Another important function of the dispersed gas phase in a bioreactor is mixing. In bubble column bioreactors, the bubbles are the only agitators. The individual bubble agitates the liquid medium locally; however, when the total gas holdup is high enough, eddies with diameters up to the reactor diameter are induced within the dispersion. With higher gas holdups, excellent mixing can be obtained, at least in dispersions with a lower liquid-phase viscosity. Mixing by means of the rising gas bubbles is not only important in bubble columns; even in industrial-scale stirred tank bioreactors, a considerable part of mixing is performed by the gas phase, and this effect is proportional to the gas holdup. This point will be discussed in more detail in the next section.

FLUID–DYNAMIC DETAILS

Practically, the gas holdup within a biotechnical cultivation broth is first of all a function of the aeration rate Q_G. Bioreactors are usually aerated with compressed air, which is dispersed into the culture medium by means of a sparger. The mass transfer via the top surface of the culture is most often comparably small and can thus be neglected. In stirred tank reactors, gas can be drawn into the dispersion by an impeller. This is the case when the impeller is operated in an unbaffled vessel where whirling occurs. However, such operational modes are generally avoided in technical bioreactors. Hence, when there is no aeration, the gas holdup is negligible.

In mechanically stirred bioreactors, which are used in most practical cases in biochemical engineering, a second influence parameter gains importance. This is the mechanical power P_D transferred into the dispersion. For this case of practical importance, we obtain

$$\epsilon \propto \left(\frac{P_D}{V_D}\right)^{0.47} w_{sg}^{0.53} \qquad (19)$$

as an experimental or engineering correlation, where both (P_D/V_D) and w_{sg} are power-related quantities. The quantity w_{sg} is related to the power P_c released by the rising bubbles into the dispersion. This power was originally supplied by the air compressor against the static pressure of the liquid at the sparger location.

$$P_c = Q_G \Delta p = w_{sg} W \qquad (20)$$

where Δp is the pressure difference between the surface of the dispersion and the position of the sparger and W is the weight of the dispersion above the sparger (i.e., essentially the weight of the liquid phase). The point to note is that the mechanical power pneumatically introduced into the dispersion by the gas is proportional to the superficial gas velocity w_{sg}. In bubble column bioreactors this is the only form of mechanical power input.

What is the reason for this power dependency of the gas holdup? With the mechanical power, two fluid–mechanical effects relevant to the gas holdup appear:

- The first effect is that with higher energy inputs, the velocities of the circulatory fluid flows within the reactors become more intense. Hence, a larger number of the bubbles within the flow are recirculated. Thus, the mean residence time increases and, consequently, by equation 17, so does the gas holdup. In particular, intensive eddy motions with high local flow velocities may trap some smaller bubbles within the dispersion for some time.
- The second effect supports the first one. At higher power input levels, the chaotic eddy flow motions lead to higher shear rates in the liquid phase on the scale of the bubble size. This reduces the mean bubble diameter, as characterized by the Sauter diameter, due to redispersion. Consequently, even more bubbles closely follow the liquid motions and thus the gas holdup increases.

In stirred tank reactors, one also induces a global circulation of the dispersion by means of the agitator to keep the mean gas residence time high and thus to increase the gas holdup and the specific interfacial area a or mass transfer coefficient $k_L a$.

To obtain a high enough gas holdup or specific interfacial area a in dispersions that are not foaming, one might come to the conclusion that the increase of the superficial gas velocity will help to solve the problem. It is true that with higher gas throughputs we usually increase the gas holdup; however, this has its limits, at least with respect to its quality characterized by the mean bubble diameter d_s or the specific interfacial area a.

Two competing processes come into play:

1. With higher gas throughputs we increase the energy flow into the liquid-phase motion, thus the dispersion of gas bubbles or, better, the redispersion by shear rates increases.
2. On the other hand, with higher holdups the number density of the bubbles increases and consequently the number of collisions between bubbles. This results in an increased coalescence probability and hence an increase in the mean bubble size. It is simple to understand that the coalescence probability is directly proportional to the probability that the bubbles meet each other, and this is proportional to the gas holdup. Larger bubbles most probably disengage from the dispersion. In this way the gas holdup is affected.

Consequently, the mean bubble diameter depends on the relative rates of coalescence and redispersion. Both are critically dependent on the composition of the medium, particularly, the concentrations of surface active agents. It is unlikely that this problem can be treated quantitatively in the near future, at least not for a real biochemical cultivation broth, where the composition with respect to the surface active components varies with time in a complex, virtually unpredictable way. The current knowledge is essentially of qualitative nature.

According to equation 17 we can increase the gas holdup with the superficial gas velocity, provided the residence time of the gas phase or the corresponding mean bubble velocity w_{bl} relative to the reactor coordinates remains practically constant. At higher gas throughput, the circulation flows in the reactor are intensified and, hence, so is the mean residence time of the bubbles. This leads to a lower mean bubble-rise velocity and an increased holdup. When the holdup increases too much, the bubble coalescence probability exceeds the redispersion probability. Consequently, the bubble size increases. At higher viscosity the mean bubble-rise velocity is lower, and, according to equation 17, large gas holdups and larger bubbles appear at even lower superficial velocities.

On the other hand, in dispersions with liquid phases of higher viscosity, one observes drift of the gas holdup toward higher values with a long time constant. This effect can be understood from the fact that due to the smaller slip velocity between bubble and bulk liquid, the probability of small bubbles escaping from the dispersion at its top surface becomes lower. Thus, very small bubbles accumulate in the dispersion, leading to an increased holdup. The bubble size distributions tail toward small equivalent bubble diameter values.

The effective bubble rise velocity w_{bl} does not sensitively depend on the superficial gas velocity w_{sg} for bubbles in the relevant Sauter diameter region of some millimeters. Hence, with larger superficial gas velocity w_{sg} the gas holdup will rise according to equation 17. Then, coalescence appears to a considerable extent. Big bubbles appear, and the dispersion becomes quite inhomogeneous with respect to bubble size. At very high superficial gas velocities the reactor will become flooded with gas.

At small gas holdups, the interfacial area that can be produced in a stirred tank reactor is essentially independent of the agitator type, and practically only dependent on the effectively introduced mechanical power. Higher mechanical power inputs by means of the agitator lead to a more intense redispersion of coalesced bubbles; the mean bubble size becomes smaller and the gas holdup higher. However, it is not possible to increase the gas holdup to arbitrarily high values, since agitators are restricted in the amount of gas they are able to disperse. When a critical gas throughput is exceeded, they become flooded. An essential criterion for an agitation system is the maximal gas throughput it can disperse. In other words, with respect to gas holdup and mass transfer considerations, it is straightforward to primarily distinguish different agitators by their ability to disperse more or less gas.

MEASUREMENT TECHNIQUES

To control the gas holdup, at least in the sense of keeping it below a critical value, it must be measured. Many techniques have been proposed in the literature; however, only a few of them can routinely be applied in biotechnical production practice.

The most convenient method to estimate the gas holdup in large reactors is to measure static pressure differences Δp between sites in the reactor at different heights and to relate Δp to the gas holdup ϵ with equation 12. The liquid-phase density ρ_L can be estimated reliably in simple density measurements. Some pressure sensors on the market are sensitive and also robust enough for measurements in production reactors.

The most straightforward way to proceed is to use equation 10 and register the height H_D of the surface of the dispersion. In any case, it is necessary to realize when it exceeds a limit H_{Dmax}. This can be done by various level meters or sensors. Since in most real bioprocesses the oxygen supply cannot be switched off and it is not possible to remove the bubbles, the volume of liquid phase must be determined indirectly. The most simple way is to make use of the culture weight, which is often measured even at production reactors and some estimate on the density of the nongas phase of the dispersion.

The most simple level monitor is a conductivity probe, which makes use of the fact that biotechnical culture media are electrically conductive. Hence, a single electrode placed at the critical height in the headspace of the bioreactor, connected with some counterelectrode in the dispersion (e.g., the wall of the reactor) via a voltage source, would suffice. Only when the dispersion reaches the electrode does the current loop close. The flowing electric current then indicates that the gas holdup reached its maximal value.

The same measurement principle can be used to measure local gas holdups. Then, however, the probe must be pointlike (i.e., small as compared to the bubble size). Whenever the point is within a bubble, the current flow between the electrodes is interrupted. The ratio of the sum of the time intervals Δt_j when the current is interrupted to the total measurement time t_{tm} is exactly the local gas holdup when the total measurement time t_{tm} was not too short. All other pointlike sensors that can distinguish between gas and liquid phases can be used in the same way. One example is the total light reflection probe.

There are other probes for local gas hold measurements with larger measurement volumes, such as capacitance sensors, which measure the local gas hold-up by introducing a small capacitor into the dispersion, so that the dispersion forms the dielectric. The dielectric constant is dependent on the gas holdup within this capacitor. After some calibration, this sensor can be used to monitor the local gas holdup, provided it can be made certain that the holdup between the plates of the capacitor is representative of the holdup of its surroundings.

CONCLUSION

In bioreactors, air is dispersed primarily to provide a sufficiently large specific interfacial area a for oxygen mass transfer into the liquid phase of the culture medium. Although a, or its product with the mass transfer coefficient k_L, the k_La value, is the key quantity from the point of view of allowing a high oxygen transfer rate, the gas holdup is a very important quantity that must be kept under control in practically all bioreactors. Too little gas holdup does not allow for a sufficient mass transfer; too much gas holdup leads to problems with the reactor operation, in particular when due to foam formation. The gas holdup can be controlled by the gas throughput and, in stirred tanks, also by the stirrer speed. However, when foam is formed, it might become necessary to destroy it. Mechanical foam destroyers are applied in a few cases; however, in most bioreactors, foam is combated by chemical antifoam agents. Since the gas holdup and hence the total volume of the dispersion may become too high when foam is being formed, most reactors are equipped with a level control. This impressively underlines the importance attributed to the gas holdup in biotechnology.

See also FERMENTATION MONITORING, DESIGN AND OPTIMIZATION; FERMENTER DESIGN; MASS TRANSFER.

GENE TRANSFER, GRAM-POSITIVE BACTERIA

IRINA BAGYAN
University of Connecticut Health Center
Farmington, Connecticut

SIMON CUTTING
Royal Holloway University of London
Egham, Surrey, U.K.

KEY WORDS

Bacillus subtilis
Conjugation
DNA-mediated transformation
Plasmids
Protoplast transfection
Transduction

OUTLINE

INTRODUCTION

A variety of methods are available for gene transfer in gram-positive bacteria, and the choice of technique depends on the nature of the experiment, recipient strain, and donor DNA. For naturally transformable species such as *Bacillus*, transformation of competent cells is usually used for the introduction of chromosomal or plasmid DNA. For strains lacking satisfactory levels of natural competence methods such as polyethylene glycol–mediated protoplast transformation have been successfully used, although these are currently being replaced by the more straightforward technique of electroporation. In addition, transduction and conjugation can be used for DNA transfer in gram-positive bacteria. We outline here the major principles of each mode of gene transfer and list particular applications with different gram-positive examples.

TRANSFORMATION

Transformation is defined as the transfer of free exogenous DNA into a cell to permanently alter its heredity. Four modes of transformation exist in gram-positive bacteria: (*1*) transformation of naturally competent bacteria; (*2*) transformation of cells with artificially induced competence, (*3*) polyethylene glycol–mediated protoplast transformation, and (*4*) electrotransformation (electroporation).

Transformation of Naturally Competent Bacteria

Transformation of naturally competent bacteria involves the active uptake by a cell of free extracellular DNA and the heritable incorporation of its genetic information. Natural transformation was first detected in *Streptococcus pneumoniae*, where the ability of this bacterium to be transformed by exogenous DNA was used to prove that DNA is the genetic material (1).

Natural competence is a physiologically and genetically determined property of a particular strain. This physiological state is dependent on the expression of specific genes whose products provide the necessary state for competence and subsequent DNA uptake. *Bacillus subtilis* is the most intensively studied naturally transformable species. Almost 40 genes (*com* genes) required for competence have now been identified in *B. subtilis*. About half of them regulate the development of the competent state, and the remainder are involved in DNA uptake and the transformation process (late *com* genes).

In *B. subtilis*, the competent state develops postexponentially during the transition from exponential growth to stationary phase. Culture density is one of the physiological conditions controlling the development of competence. As cells grow to high density, two extracellular peptide pheromones accumulate in the culture medium; they cause increased transcription of the *comS* gene. Synthesis of ComS, in turn, leads to activation of the ComK transcription factor, which directs expression of the late *com* genes (2–4). In addition to these cell density signals, there are also nutritional signals that control activation of ComK (5). Similar mechanisms exist in *Streptococcus*: in *S. pneumoniae*, competence develops in exponentially growing cells at a critical cell density (10^7–10^8 cells/mL) (6). Here, cellular secretion of a small 17-residue peptide is responsible for inducing cells to develop competence (7,8). In *B. subtilis*, only 1–10% of the cell population becomes competent whereas in *Streptococcus* spp. (e.g., *S. pneumoniae* and *S. sanguis*) up to 100% of the cells in a culture become competent (9).

DNA uptake mechanisms in *B. subtilis* and *S. pneumoniae* are also similar (for review, see Refs. 10 and 11). In brief, double-stranded DNA noncovalently binds to the surface of a competent cell (approximately 50 [*B. subtilis*] and 30–80 [*S. pneumoniae*] DNA-binding sites are used per competent cell). The transfer of DNA into the cell initiates with the generation of double-strand breaks. One strand is degraded while the other enters the cell (12); thus, DNA enters in a single-stranded form. In the case of transformation by homologous DNA, single-stranded DNA can integrate into the recipient chromosome (or resident plasmid) by homologous recombination. In *B. subtilis*, if plasmid DNA contains regions homologous to the recipient genome, the efficiency of transformation will be greater than 10^6 transformants per microgram of DNA. In the case of transformation by a nonhomologous plasmid, following entry of single-stranded plasmid DNA the plasmid must first resume a double-stranded form to replicate. If the transforming plasmid is monomeric, two complementary strands should independently enter the cell to produce a double-stranded plasmid. For this reason transformation by plasmid monomers is very inefficient in *B. subtilis*. Most plasmid preparations from gram-positive (or gram-negative) bacteria contain plasmid multimers and can yield transformants in *B. subtilis* at a level of 10^3 to 10^5 transformants per microgram of DNA. Plasmid DNA isolated from Rec^- bacteria contains a much lower amount of multimers and will lead to 100 to 1000-fold lower transformation frequencies.

To bypass the requirement for plasmid multimers in *B. subtilis*, a so-called plasmid marker rescue transformation system has been developed. This system involves the uptake of donor DNA by competent cells and the subsequent recombination with a partially homologous resident plasmid (13–16). A simple procedure can be used for the preparation of competent cells of *S. pneumoniae* (17). Methods for transformation of other naturally competent streptococci have also been developed, for example, *S. sanguis* Challis (18), *S. milleri* (19), and *S. mutans* (20).

For *B. subtilis*, the most commonly used method of preparation of competent cells remains the classical two-step procedure of Dubnau and Davidoff-Abelson (21). Laboratory techniques for *Bacillus* transformation (as well as other molecular biological methods for *Bacillus*) can be found in Harwood and Cutting's work (22). Transformation of naturally competent cells is the method of choice for introduction of DNA into *B. subtilis*. Applications include allele replacement and the introduction of chromosomal markers and plasmids. Transformation has been used for fine structure genetic mapping in *B. subtilis*. This method, however, is limited by the length of DNA taken up by a competent cell. The average size of transforming DNA (about 30 kb) limits the recombinational distance between two markers and hence the genetic distance that can be established. Nonsaturating concentrations of transforming DNA (less than 50 ng/mL in *B. subtilis*) should be used in mapping experiments to avoid congression (i.e., the transformation of the same competent cell by two genetically unlinked markers as a result of independent DNA uptake events). However, congression can be used as a strain construction tool, in that a genetic marker can be introduced into a recipient chromosome by selection for an unlinked marker (22). For this purpose, however, much higher DNA concentrations (e.g., 1–5 μg/mL in *B. subtilis*) must be used (to get 2–3% congression).

Transformation Using Artificially Induced Competence

Some bacterial species can be artificially manipulated to allow the uptake of exogenous DNA. This so-called chemically induced competence can be achieved typically by treatment of cells with calcium or some other cation and is used in the standard *Escherichia coli* transformation protocol. This is the method of choice for transformation of *E. coli*. Among gram-positive bacteria, *Staphylococcus aureus* has been transformed by this technique (23,24). In this procedure, cells collected at early exponential growth phase are washed with a NaCl solution, suspended in a $CaCl_2$-containing buffer, and incubated with DNA. The transformation efficiency has been shown to be dependent on the pH of the buffer, the presence of divalent cations, and the growth phase of the culture. Transformation efficiency is generally higher in nuclease-minus mutants (25). Also, efficient development of artificial competence requires that the cells are lysogenic for phage φ11 (24). Certain phage-expressed products are able to enhance competence; for example, both phage tails (26) and the product of early gene 31 (24) have been reported to enhance competence. Efficiencies of 10^4 transformants per microgram of DNA can be achieved with this system. The main advantage of the artificial competence procedure for *S. aureus* is that it is effective for transformation of chromosomal DNA whereas protoplast transformation is not. This technique has been used extensively for genetic mapping where linkages representing large distances (100–150 kb) of the chromosome can be defined by the uptake of very large chromosomal fragments by the cell (27).

Protoplast Transformation

Protoplasts are generated by treatment of bacteria with muralytic enzymes (lysozyme or other, depending on the species) to degrade the peptidoglycan cell wall in the presence of an osmotic stabilizer. When exposed to high concentrations of polyethylene glycol (PEG) in the presence of DNA, protoplasts take up the DNA and if allowed to recover on suitably rich media can regenerate the cell wall. Addition of PEG can also cause protoplast fusion.

Use of protoplast techniques has advantages and disadvantages. Protoplasts allow the uptake of double-stranded plasmid DNA, enabling plasmid monomers to be used. However, protoplasts are very fragile, burst easily, and are extremely sensitive to traces of detergent on glassware. In addition, the method is generally considered laborious; protoplast regeneration can be extremely slow (2–6 days) and occurs at a low frequency. Finally, complete media are required, precluding the direct selection for prototrophic markers. Before the introduction of electroporation methods, protoplast transformation had been widely used for transformation of gram-positive species that are not naturally competent.

Protoplast transformation techniques for gram-positive bacteria were first developed for *Streptomyces* and *Bacillus*. PEG-induced transformation of *Streptomyces* protoplasts by plasmid DNA was first reported in 1978 (28) and for *B. subtilis* in 1979 (29). The *B. subtilis* protoplast transformation technique is still used for many *Bacillus* species, such as *B. anthracis* (30), *B. amyloliquefaciens* (31), *B. stearothermophilus* (32,33), and *B. licheniformis* (34).

In *Streptomyces* spp., transformation and transfection of protoplasts are important tools for gene transfer, being very efficient, even with large constructs up to at least 40 kb, and more than 10^7 transformants per microgram of DNA can be obtained, as has been described for *S. lividans* and *S. coelicolor* (35). Numerous protoplast transformation techniques have been developed for other *Streptomyces* species: *S. peucetius* (36), 10^6 transformants/μg DNA; *S. tendae* (37), 10^6 to 10^7 transformants/μg DNA; *S. clavuligerus* (38), 10^5 transformants/μg DNA; *S. venezuelae* (39), 10^7 transformants/μg DNA; *S. niveus* (40), 10^6 transformants/μg DNA; and *S. lincolnensis* (41), 10^3 transformants/μg DNA. *Streptomyces* protoplasts are more easily manipulated than those of other bacterial genera and exhibit reproducibly high levels of regeneration. Indeed, this ease of use has enabled protoplast fusion to be used not only for strain construction but also for genetic mapping in *Streptomyces* (35).

Protocols have been developed for protoplast transformation in a wide range of other gram-positive species. PEG-mediated protoplast transformation methods (giving up to 10^6 transformants/μg DNA) are available for different Actinomycetes such as *Micromonospora olivasterospora* (42), *Amycolatopsis (Nocardia) orientalis* (43), and *Rhodococcus* spp. (44). A method for PEG-mediated transformation of the mycelium has been developed for *Amycolatopsis (Nocardia) mediterranei* that yields 10^6 transformants/μg DNA without the need for protoplast isolation (45).

Staphylococcus protoplasts prepared with lysostaphin and stabilized with sucrose and Mg^{2+} are currently being used for PEG-mediated plasmid transformation (46) and protoplast fusion (47). Protoplast fusion has also been used for chromosomal mapping in *Staphylococcus aureus* (48) and also for plasmid transfer. There has also been reported a method for PEG-mediated plasmid transformation of a temperature-sensitive peptidoglycan-deficient mutant of *S. aureus*, giving transformation frequencies comparable with those obtained with protoplasts prepared with lysostaphin treatment (49). PEG-mediated protoplast transformation has been developed for different streptococci, including *S. lactis* (50), *S. faecalis* (51,52), *S. faecium* (53), *S. thermophilus* (54), and *S. pyogenes* (55).

Early techniques used protoplast transformation for introduction of plasmid DNA into mycobacteria (56). Electroporation techniques have subsequently proved to be more effective, although an efficient method for transformation of *Mycobacterium smegmatis* protoplasts has been described (57). As with mycobacteria, electroporation has now replaced protoplast transformation techniques in the lactic acid bacteria and *Listeria* spp. (58–60). PEG-mediated protoplast transformation was the first method of transformation reported for *Clostridium perfringens* (61,62). Before the development of electroporation techniques, protoplast transformation was used extensively for the introduction of DNA into *Clostridium acetobutylicum* (63,64), the most recent method reported giving up to 10^6 transformants/μg DNA (65).

Electroporation

Application of a high-voltage electric field for a very short period of time (microseconds) causes transient structural changes in lipid bilayers and biological membranes, leading to reversible pore formation (electroporation) and allowing the passive influx of DNA molecules. Electroporation as a tool for cell transformation was first used in the early 1980s for the transformation of animal cells (66). Subsequently, the technique was utilized for prokaryotes (e.g., for transformation of *Bacillus cereus* protoplasts [67]).

Transformation of intact bacterial cells by electroporation was first reported in 1987 for *Lactococcus lactis* (68) and *Lactobacillus casei* (69). A number of reports on electroporation of different gram-positive bacteria appeared shortly thereafter: *Enterococcus faecalis* (70), *Streptococcus thermophilus* (71), *Mycobacterium smegmatis*, and BCG (72), *Clostridium perfringens* (73), and a number of other gram-positive species—*Lactobacillus acidophilus, Bacillus cereus, Leuconostoc* spp., and *Listeria* spp. (74).

The first attempts to transform gram-positive bacteria using electroporation were not very efficient, and for a large number of gram-positive species the technique has been extensively refined. The most important parameters affecting electroporation efficiency appear to be field strength and duration and shape of the pulse (75). Other parameters involve the cell culture growth conditions (e.g., media composition and phase of cell growth at the time of harvesting for electroporation) and transformation conditions (e.g., final concentration of cells and DNA in the transformation mixture, ionic strength, and the pH of electroporation buffer). Optimal conditions of course, will differ from strain to strain.

Some strains can give frequencies of 10^7 to 10^8 transformants/μg DNA, and in these strains library construction is a distinct possibility. Most strains, however, give frequencies suitable for the exchange of shuttle vectors between strains. However, frequency of electrotransformation of gram-positive bacteria is generally lower than that observed with *E. coli*. Possibly, the dense composition of the gram-positive cell wall presents a more complex barrier to entry of DNA. Several protocols were developed to overcome this problem, such as incorporation of low concentrations of the cell wall synthesis inhibitor penicillin-G in the growth medium substantially increased the efficiency of transformation of *Listeria monocytogenes* (to 10^6 transformants/μg DNA) (76). Gentle treatment of lactococcal cells with muralytic enzymes before electroporation (77) and pretreatment of *Clostridium perfringens* cells with a peptidase lysostaphin (78) also raised the efficiency of electrotransformation. In the last two cases, muralytic enzymes partially digested the cell wall, providing access of plasmid DNA to the cell membrane pores subsequently created by electroporation.

Electroporation has now become the method of choice for the transformation of a wide range of gram-positive bacteria. For example, in *Clostridium* electroporation has superseded the standard method of protoplast transformation used previously. Electroporation methods have been reported for *Clostridium perfringens* (10^7 transformants/μg DNA) (79–81), *C. beijerinckii* (82), *C. acetobutylicum* (83), and *Clostridium thermosaccharolyticum* (155). Electroporation is also a method of choice for lactic acid bacteria (*Lactobacillus, Lactococcus*) and has replaced the more time-consuming and frequently unreliable protoplast transformation techniques (84–87).

In *Mycobacterium smegmatis*, electroporation is relatively efficient, yielding little more than 10^5 transformants/μg DNA (88). In other mycobacterial species similar yields have been reported: *Mycobacterium intracellulare* (10^6 transformants/μg DNA) (89), *M. vaccae* (10^3–10^5 transformants/μg DNA) (90), and *M. bovis* BCG (91,92). It was shown that elevation of electroporation temperature to 37 °C improves the efficiency of transformation of slow-growing *Mycobacterium* spp. (93).

For *Listeria*, electroporation yields as high as 10^6 transformants/μg DNA have been reported for *L. monocytogenes, L. innocua*, and *L. ivanovii*, and somewhat lower yields for *L. seeligeri* (94).

Efficient electroporation techniques have also been developed for some actinomycetes, for example, *Actinomyces viscosus* and *Actinomyces naeslundii* (10^2–10^7 transformants/μg DNA) (95), *Brevibacterium flavum* and *Corynebacterium glutamicum* (up to 5×10^7 transformants/μg DNA) (96), *Brevibacterium methylicum* (10^5 transformants/μg DNA) (97), and *Rhodococcus fascians* (10^5 transformants/μg DNA) (98). A method for electroporation of *Streptomyces parvulus* and *S. vinaceus* has been reported (10^5 transformants/μg DNA) (99).

Electroporation has also been used for transformation in the *Streptococcus* species: *Streptococcus pyogenes* (10^7 transformants/μg DNA) (100), *S. thermophilus* (7×10^5 transformants/μg DNA) (101), *Enterococcus faecalis* (10^5 transformants/μg DNA) (102), *S. parasanguis* (103), *S. agalactiae* (10^4 transformants/μg DNA) (104), and *S. suis* type 2 (10^7 transformants/μg DNA) (105). Employed later than for other gram-positive bacteria, electroporation techniques have been developed for the staphylococci, such as *Staphylococcus aureus* (4×10^8 transformants/μg DNA) (106), *S. staphylolyticus*, and *S. carnosus* (10^5 transformants/μg DNA) (107).

Electroporation has been applied to many *Bacillus* species, such as *Bacillus thuringiensis* (108), *B. amyloliquefaciens* (109), *B. cereus* (110), *B. subtilis* (111), and *B. anthraecis* (112). However, electroporation is relatively inefficient for the *Bacillus* species and transformation frequencies are low (no more than 10^5 transformants/μg DNA) for most of the species including *B. subtilis* (111,113,114). For *B. subtilis* the maximum electrotransformation frequency reported was 9×10^4 transformants/μg DNA (115). *B. thuringiensis*, however, is an exception, with transformation frequencies up to 10^7 transformants/μg DNA being reported (116). Recently, an efficient method for electroporation of *B. brevis* cells has been developed. It gives 10^7 up to 10^8 transformants/μg DNA (117). Electroporation techniques have also been reported for transformation of *Leuconostoc paramesenteroides* (118) and species of Bifidobacterium (119). Electroporation is rapidly becoming the method of choice for the majority of gram-positive bacteria.

TRANSDUCTION

Gene transfer mediated by a bacteriophage is termed transduction; two types of transduction can be defined, generalized and specialized.

In the process of generalized transduction, any gene has a more or less equal probability of being transferred. During phage infection of the donor bacterium some phage heads incorporate a "headful" of bacterial DNA rather than phage DNA, and upon lysis the bacteria release phage particles carrying bacterial DNA. The phage particle adsorbs to the recipient bacterium and injects the bacterial DNA fragment, which can recombine into the recipient genome by homologous recombination.

During specialized transduction ("restricted transduction"), only genes adjacent to the lysogenic prophage can be transduced as a result of erroneous excision of the prophage. During excision, adjacent bacterial DNA can be excised together with the prophage (it is also possible for some prophage DNA to remain unexcised). After packaging the newly formed phage particle contains excised bacterial DNA, which can be transferred to other bacteria during the life cycle of the virus.

Generalized Transduction

Several transducing phages have been described for *B. subtilis*. The generalized transducing phage PBS1 is the most widely used for genetic analysis. The phage is pseudolysogenic, however, and although it generates turbid plaques on a plate seeded with *B. subtilis* it does not enter a true lysogenic state. PBS1 transduction is used for long-range genetic mapping because a single transducing particle encompasses more than 200 kb (about 10% of the host genome) and can be used to obtain an approximate position of a gene on the chromosome. Following this, DNA-mediated transformation of naturally competent cells is used to obtain a more accurate position of the gene of interest. For rapid mapping of genes, a kit of nine auxotrophic strains has been created (120). Together, these strains contain 22 auxotrophic markers spaced at relatively even intervals around the chromosome. A PBS1-transducing lysate is prepared from the strain containing the mutation to be mapped (which should have the wild-type allele of the auxotrophic markers carried by the kit strains) and used to infect each kit strain with selection for Aux^+ and screening for the mutant phenotype. An alternative and perhaps easier approach is to use the large collection of "silent" chromosomal insertions of Tn*917* (121). Here, a transducing lysate is made from the Tn*917* insertion strain and used to infect the strain with the unknown mutation, followed by selection for MLS resistance (encoded by the transposon) and screening for correction of the mutant phenotype. For a review of genetic mapping in *Bacillus*, see Ref. 122.

A PBS1 lysate prepared from the donor strain (donor lysate) is used to infect a motile recipient strain. PBS1 infects cells by adsorption to the flagella, so it can only infect motile cells. The procedures for preparation of phage lysates and generalized transduction can be found in Refs. 122 and 123.

Generalized transduction is a useful tool for genetic analysis of closely linked markers in *Staphylococcus aureus* (124,125). Transduction is limited by the size of the phage genome (45 kb for φ11); linkages representing large (100–150 kb) distances are being mapped by transformation of competent cells. Phage-based gene transfer systems serve as efficient tools of gene transfer in mycobacteria. Generalized transduction by the phage 13 is used for marker transfer to *Mycobacterium smegmatis* (126). For protocols, see Ref. 127.

Generalized transducing streptomycete bacteriophage FP43 has been used to transduce plasmid pRHB101 into species of *Streptomyces, Streptoverticillium*, and *Saccharapolyspora* (128–131). pRHB101 contains a 7.7-kb segment of FP43 DNA (*hft* segment) cloned into plasmid PlJ702 (130). The transducing particles contain linear concatemers of pRHB101, which form circular monomers in the transductants.

Specialized Transduction

Two specialized transducing phages have been extensively used as cloning vectors in *B. subtilis*: φ105 and SPβ (132). Both phages were designed for cloning *B. subtilis* genes by *trans* complementation (133). Cloning systems based on the phage φ105 have been extensively reviewed (133).

SPβ is a large temperate phage (126 kb) that can tolerate up to 10 kb of extra DNA. It has been used for complementation studies (123) and for transferring *lacZ* fusions between strains (134). Methods have been developed for integration of foreign DNA into the SPβ prophage by homologous recombination, after which DNA can be transferred to other strains by transduction. Procedures for preparation of SPβ lysates and SPβ transductions can be found in Ref. 123.

Phage vectors derived from the temperate phage φC31 have been developed for *Streptomyces*. Integration may involve *att* sites of the phage and chromosome, or, for the vectors in which the *att* site has been deleted, the phage can integrate by homologous recombination, either using a preexisting prophage to form a double lysogen or at the chromosomal site homologous to a DNA fragment cloned into the vector (35).

CONJUGATION

Conjugation requires cell–cell contact and provides a useful method for the interspecies and intergenera plasmid transfer within gram-positive bacteria, especially for strains lacking efficient transformation systems. Conjugative plasmids are very common among gram-positive bacteria. Usually, the conjugative plasmid itself is transferred, and in addition small plasmids can also be transferred. Some gram-positive conjugational plasmids are known to promote transfer of mobilizable plasmids across genus barriers. The conjugative streptococcal plasmid pAMβ1 is efficient in the mobilization of small plasmids in streptococci and lactococci (135). Large conjugative *Bacillus* plasmids, such as pLS20, can mobilize pUB110 for interspecies transfer (136). The majority of *Streptomyces* plasmids are self-transmissible. Conjugational crosses are useful means of strain construction in *Streptomyces* (35). Conjugative *Staphylococcus aureus* plasmids can efficiently mobilize those small plasmids that can form relaxation complexes (137). The most efficient conjugation system exists in *Enterococcus faecalis*, termed *pheromone-mediated plasmid exchange* (138,139). Here, plasmid-free enterococci excrete short oligopeptide pheromones (clump-inducing agents) that induce plasmid-containing cells to express cell surface proteins (aggregation substance) that allow the donor to bind to a receptor on the surface of the recipient and promote cell–cell contact. The plasmid is then transferred to the recipient cell and represses further pheromone synthesis. This transfer system is extremely efficient, even in matings conducted in liquid. Other conjugation systems are not so efficient and can promote DNA transfer only when the cells are forced into close contact by being placed on a filter or an agar surface.

Rare examples of conjugative mobilization of chromosomal markers exist in gram-positive bacteria. Conjugation in *Streptomyces* has been used for mapping of chromosomal genes (35,140). A conjugative transposon Tn*925* from *E. faecalis* not only can mediate its own transfer in *B. subtilis* but also can mobilize the transfer of chromosomal genes (141). Until recently, it was believed that none of the gram-positive conjugation systems are capable of Hfr-like transfer of chromosomal genes. However, such a mechanism was recently proposed to occur in mycobacterium smegmatis (142).

Conjugative transposons (the elements transposing from a DNA molecule in one cell to a DNA molecule in another cell by a conjugational mechanism but that are incapable of autonomous replication) were first isolated from streptococcal strains associated with human disease. The most carefully studied are Tn*916* from *E. faecalis* (143), Tn*3701* from *Streptococcus pyogenes* (144), and Tn*1545* (145) and Tn*5253* from *S. pneumoniae* (146). Conjugative transposons are useful shuttle vectors, especially when the gene to be transferred has to be maintained at the same copy number as the chromosome. The cloned DNA can be inserted into Tn916 at a *Bst*X1 site without disrupting the function of the transposon (147). The chimeric transposon is then introduced into the gram-positive cell by electroporation or protoplast transformation (the method of transfer should not lead to the rearrangement of large incoming DNA molecules). The chimeric transposon will then insert into the host chromosome. If it subsequently proves problematic to transform DNA directly into the strain of interest, an intermediate host (e.g., *B. subtilis*) can be used. This host is transformed with DNA of the *E. coli* plasmid containing Tn*916* with the gene of interest cloned into it, and the strain will then act as a conjugative donor to any other gram-positive strain. Thus, the chimeric transposon will be transferred from the intermediate host to the strain of interest by conjugational mating on a solid surface (agar or filter). As an example of its recent application, Tn*916* was reported to be transferred to *Leuconostoc lactis* LM2301 (148), which opens the possibility of genetic studies in this *Leuconostoc* species for which no efficient conjugal or transformation system has been described previously. A series of Tn*916*-based shuttle vectors for analysis of transcriptional regulation in gram-positive bacteria has also been developed (149).

Conjugative transfer can occur even between gram-negative (*E. coli*) and gram-positive bacteria and can be exploited for transfer to gram-positive bacteria of the DNA cloned in *E. coli*. The first shuttle vector of this type (pAT187) was constructed by Trieu-Cuot et al. (150). It contains replication functions for both *E. coli* (from pBR322) and gram-positive bacteria (from pAMβ1) and an origin of transfer (*oriT*) from the gram-negative broad host range plasmid pRK2. Donor strains of *E. coli* must also harbor pRK2 or some other IncP-type plasmid or chromosomally inserted DNA fragments carrying IncP *tra* functions. Subsequently, other versatile mobilizable vectors have been developed (151–154). Such shuttle vectors can be transferred in filter matings to a wide variety of gram-positive bacteria, including *Bacillus, Clostridium, Streptomyces, Enterococcus, Streptococcus, Staphylococcus, Listeria, Nocardia*, and *Rhodococcus* (150,151,155,156). Improved procedures for conjugative transfer from *E. coli* to *Staphylococcus aureus* and *Listeria monocytogenes* have also been developed involving subinhibitory concentrations of cell wall synthesis inhibitors (penicillins) in the mating medium to increase transfer frequency (157). Conjugative transfer from *E. coli* to *Mycobacterium smegmatis* has also been described (158,159).

Conjugation has proved to be extremely valuable to gram-positive strains that lack a natural competence system and have proved refractory to protoplast transformation and electroporation.

BIBLIOGRAPHY

1. O.T. Avery, C.M. MacLeod, and M. McCarty, *J. Exp. Med.* **79**, 137–159 (1944).
2. R. Magnuson, J. Solomon, and A.D. Grossman, *Cell* **77**, 207–216 (1994).
3. J.M. Solomon, R. Magnuson, A. Srivastava, and A.D. Grossman, *Genes Dev.* **9**, 547–558 (1995).
4. D. van Sinderen, A. Luttinger, L. Kong, D. Dubnau, G. Venema, and L. Hamoen, *Mol. Microbiol.* **15**, 455–462 (1995).
5. D. Dubnau, *Microbiol. Rev.* **55**, 395–424 (1991).
6. A. Tomasz and R.D. Hotchkiss, *Proc. Natl. Acad. Sci. U.S.A.* **51**, 480–487 (1964).
7. A. Tomasz and J.L. Mosser, *Proc. Natl. Acad. Sci. U.S.A.* **55**, 58–66 (1966).
8. L.S. Havarstein, G. Coomaraswamy, and D.A. Morrison, *Proc. Natl. Acad. Sci. U.S.A.* **92**, 11144–11149 (1995).
9. H.O. Smith and D.B. Danner, *Annu. Rev. Biochem.* **50**, 41–68 (1981).
10. M.G. Lorenz, and W. Wackernagel, *Microbiol. Rev.* **58**, 563–602 (1994).
11. D. Dubnau, in A.L. Sonenshein, J.A. Hoch, and R. Losick eds., Bacillus subtilis *and Other Gram-Positive Bacteria: Biochemistry, Physiology, and Molecular Genetics*, American Society for Microbiology, Washington, D.C., 1993, pp. 555–584.
12. V. Mejean and J.-P. Ceaverys, *Biol. Chem.* **268**, 5594–5599 (1993).
13. S. Contente and D. Dubnau, *Plasmid* **2**, 555–571 (1979).
14. Y. Weinrauch and D. Dubnau, *J. Bacteriol.* **154**, 1077–1087 (1983).
15. Y. Weinrauch and D. Dubnau, *J. Bacteriol.* **169**, 1205–1211 (1987).
16. P. Haima, S. Bron, and G. Venema, *Mol. Gen. Genet.* **223**, 185–191 (1990).
17. R.D. Porter and W.R. Guild, *J. Bacteriol.* **97**, 1033–1035 (1969).
18. G. Pozzi, R.A. Musmanno, P.M. Lievens, M.R. Oggioni, P. Plevani, and R. Manganelli, *Res. Microbiol.* **141**, 659–670 (1990).
19. A.E. Jacob, W.A. Horton, and D.B. Drucker, *Microbios* **60**, 167–175 (1989).
20. G.R. Shah and P.W. Caufield, *Anal. Biochem.* **214**, 343–346 (1993).
21. D. Dubnau and R. Davidoff-Abelson, *J. Mol. Biol.* **56**, 209–221 (1972).
22. C.R. Harwood and S.M. Cutting eds., *Molecular Biological Methods for Bacillus*, Wiley, New York, 1990.
23. M. Lindberg, J.-E. Sjostrom, and T. Johanson, *J. Bacteriol.* **109**, 844–847 (1972).
24. L. Rudin, J.-E. Sjostrom, M. Lindberg, and L. Philipson, *J. Bacteriol.* **118**, 155–164 (1974).
25. G.S. Omenn and J. Friedman, *J. Bacteriol.* **101**, 921–924 (1970).

26. M.P. Jackson, J. DeSena, J. Lednicky, B. Mepherson, R. Haile, R.G. Garrison, and M. Pogolsky, *Infect. Immun.* **39**, 939–947 (1983).
27. P.A. Pattee and D.S. Neveln, *J. Bacteriol.* **124**, 201–211 (1975).
28. M.J. Bibb, J.M. Ward, and D.A. Hopwood, *Nature (Lond.)* **274**, 398–400 (1978).
29. S. Chang and S.N. Cohen, *Mol. Gen. Genet.* **168**, 111–115 (1979).
30. S.-I. Mahino, C. Sasakawa, I. Uchida, N. Terakado, and M. Yoshikawa, *FEMS Microbiol. Lett.* **44**, 45–48 (1987).
31. J. Vehmaanpera, *FEMS Microbiol. Lett.* **49**, 101–105 (1988).
32. T. Imanaka, M. Fujii, I. Aramori, and S. Aiba, *J. Bacteriol.* **149**, 824–830 (1982).
33. M. Zhang, H. Nakai, and T. Imanaka, *Appl. Environ. Microbiol.* **54**, 3162–3164 (1988).
34. K.K. Jensen and M.F. Hulett, *J. Gen. Microbiol.* **135**, 2283–2287 (1989).
35. D.A. Hopwood, M.J. Bibb, K.K. Chater, T. Kieser, C.J. Bruton, H.M. Kieser, D.J. Lydiate, C.P. Smith, J.M. Ward, and H. Schrempf, *Genetic Manipulation of* Streptomyces: *A Laboratory Manual*, John Innes, Norwich, U.K. 1985.
36. J.S. Lampel and W.R. Strohl, *Appl. Environ. Microbiol.* **51**, 126–131 (1986).
37. P. Engel, *Appl. Environ. Microbiol.* **53**, 1–3 (1987).
38. M. Garcia-Dominguez, J.F. Martin, B. Mahro, A.L. Demain, and P. Liras, *Appl. Environ. Microbiol.* **53**, 1376–1381 (1987).
39. J. Anne, L. Van Mellaert, and H. Eyssen, *Appl. Microbiol. Biotechnol.* **32**, 431–435 (1990).
40. H.A. Hussain and D.A. Ritchie, *J. Appl. Bacteriol.* **71**, 422–427 (1991).
41. Z. Jandova and P. Tichy, *Folia Microbiol. (Praha)* **37**, 181–187 (1992).
42. M. Hasegawa, *Gene* **115**, 85–91 (1992).
43. P. Matsushima, M.A. McHenney, and R.H. Baltz, *J. Bacteriol.* **169**, 2298–2300 (1987).
44. M.E. Singer and W.R. Finnerty, *J. Bacteriol.* **170**, 638–645 (1988).
45. J. Madon and R. Hutter, *J. Bacteriol.* **173**, 6325–6331 (1991).
46. J. Polak and R.P. Novick, *Plasmid* **7**, 152–162 (1982).
47. P. Schaeffer, B. Cami, and R.D. Hotchkiss, *Proc. Natl. Acad. Sci. U.S.A.* **73**, 2151–2155 (1976).
48. M.L. Stahl and P.A. Pattee, *J. Bacteriol.* **154**, 395–405 (1983).
49. D.T. Nieuwlandt and P.A. Pattee, *J. Bacteriol.* **171**, 4906–4913 (1989).
50. M.A. Sanders and M.A. Nicholson, *Appl. Environ. Microbiol.* **53**, 1730–1736 (1987).
51. M.D. Smith, *J. Bacteriol.* **162**, 92–97 (1985).
52. R. Wirth, F.Y. An, and D.B. Clewell, *J. Bacteriol.* **165**, 831–836 (1986).
53. W. Chen, K. Ohmiga, and S. Shimizu, *Appl. Environ. Microbiol.* **52**, 612–616 (1986).
54. A. Mercenier, C. Robert, D.A. Romero, I. Castellino, P. Slos, and Y. Lemoine, *Biochimie (Paris)* **70**, 567–577 (1988).
55. D.M. Catt and J.D. Jollick, *Microbios* **68**, 189–207 (1991).
56. W.R. Jacobs, Jr., M. Tuckman, and B.R. Bloom, *Nature (Lond.)* **327**, 532–535 (1987).
57. S.A. Naser, C.M. McCarthy, G.B. Smith, and A.K. Tupponce, *Curr. Microbiol.* **27**, 153–156 (1993).
58. B.M. Chassy, *FEMS Microbiol. Rev.* **46**, 297–312 (1987).
59. K. Watanaba, Y. Kakita, Y. Nakashima, and F. Miake, *Biosci. Biotechnol. Biochem.* **56**, 1859–1862 (1992).
60. M.F. Vicente, F. Baquero, and J.C. Perez-Diaz, *Plasmid* **18**, 89–92 (1987).
61. D.L. Heefner, C.H. Squires, R.J. Evans, B.J. Kopp, and M.J. Yarus, *J. Bacteriol.* **159**, 460–464 (1984).
62. C.H. Squires, D.L. Heefner, R.J. Evans, B.J. Kopp, and M.J. Yarus, *J. Bacteriol.* **159**, 465–471 (1984).
63. L. Podvin, G.R. Reysset, J. Hubert, and M. Sebald, in J.M. Hardie and S.P. Boriello eds., *Anaerobes Today*, Wiley, Chichester, U.K., 1988, pp. 135–140.
64. G. Reysset, J. Hubert, L. Podvin, and M. Sebald, *Biotechnol. Technol.* **2**, 199–204 (1988).
65. H. Azeddoug, J. Hubert, and G. Reysset, *J. Gen. Microbiol.* **138**, 1371–1378 (1992).
66. E. Neumann, M. Schaefer-Ridder, Y. Wang, and P.H. Hofschneider, *EMBO J.* **1**, 841–845 (1982).
67. N. Shivarova, W. Forster, H.-E. Jacob, R. Grigorova, *Z. Allg. Mikrobiol.* **23**, 595–599 (1983).
68. S.K. Harlander, in J.J. Ferretti and R. Curtiss, eds., *Streptococcal Genetics*, American Society for Microbiology, Washington, D.C., 1987, pp. 229–233.
69. B.M. Chassy and J.L. Flickinger, *FEMS Microbiol. Lett.* **44**, 173–177 (1987).
70. S. Fiedler and R. Wirth, *Anal. Biochem.* **170**, 38–44 (1988).
71. G.A. Somkuti and D.H. Steinberg, *Biochimie (Paris)* **70**, 579–585 (1988).
72. S.B. Snapper, L. Lugosi, A. Jekkel, R.E. Melton, T. Kieser, B.R. Bloom, and W.R. Jacobs, Jr., *Proc. Natl. Acad. Sci. U.S.A.* **85**, 6987–6991 (1988).
73. S.P. Allen and H.P. Blaschek, *Appl. Environ. Microbiol.* **54**, 2322–2324 (1988).
74. J.B. Luchansky, P.M. Muriana, and T.R. Klaenhammer, *Mol. Microbiol.* **2**, 637–646 (1988).
75. A. Mercenier and B. Chassy, *Biochimie (Paris)* **70**, 503–517 (1988).
76. S.F. Park and G.S.A.B. Stewart, *Gene* **94**, 129–132 (1990).
77. I.B. Powell, M.G. Achen, A.J. Hillier, and B.E. Davidson, *Appl. Environ. Microbiol.* **54**, 655–660 (1988).
78. P.T. Scott and J.I. Rood, *Gene* **82**, 327–333 (1989).
79. S.P. Allen and H.P. Blaschek, *FEMS Microbiol. Lett.* **58**, 217–220 (1990).
80. J.T. Rood and S.T. Cole, *Microbiol. Rev.* **55**, 621–648 (1991).
81. M.K. Phillips-Jones, *FEMS Microbiol. Lett.* **54**, 221–226 (1990).
82. G.A. Birrer, W.R. Chesbro, and R.M. Zsigray, *Appl. Microbiol. Biotechnol.* **41**, 32–38 (1994).
83. S.Y. Lee, L.D. Mermelstein, G.N. Bennett, and E.T. Papoutsakis, *Ann. N.Y. Acad. Sci.* **665**, 39–51 (1992).
84. T.W. Aukrust, M.B. Brurberg, and I.F. Nes, *Methods Mol. Biol.* **47**, 201–208 (1995).
85. D.A. McIntyre and S.K. Harlander, *Appl. Environ. Microbiol.* **55**, 2621–2626 (1989).
86. D.A. McIntyre and S.K. Harlander, *Appl. Environ. Microbiol.* **55**, 604–610 (1989).
87. C.M. Rush, L.M. Hafner, and P. Timms, *J. Med. Microbiol.* **41**, 272–278 (1994).
88. S.B. Snapper, R.E. Melton, S. Mustafa, T. Kieser, and W.R. Jacobs, Jr., *Mol. Microbiol.* **4**, 1911–1919 (1990).
89. B.I. Marklund, D.P. Speert, and R.W. Stokes, *J. Bacteriol.* **177**, 6100–6105 (1995).

90. T.R. Garbe, J. Barathi, S. Barnini, Y. Zhang, C. Abou-Zeid, D. Tang, R. Mukherjee, and D.B. Joung, *Microbiology* **140**, 133–138 (1994).

91. A. Aldovini, R.H. Husson, and R.A. Young, *J. Bacteriol.* **175**, 7282–7289 (1993).

92. S.N. Cho, J.H. Hwang, S. Park, Y. Chong, S.K. Kim, C.Y. Song, and J.D. Kim, *Yonsei Med. J.* **39**, 141–147 (1998).

93. B.J. Wards and D.M. Collins, *FEMS Microbiol. Lett.* **145**, 101–105 (1996).

94. J.E. Alexander, P.W. Andrew, D. Jones, and I.S. Roberts, *Lett. Appl. Microbiol.* **10**, 179–181 (1990).

95. M.K. Yeung and C.S. Kozelsky, *J. Bacteriol.* **176**, 4173–4176 (1994).

96. A.A. Vertes, M. Inui, M. Kobayashi, Y. Kurusu, and H. Yukawa, *Res. Microbiol.* **144**, 181–185 (1993).

97. J. Nesvera, J. Hochmannova, M. Patek, A. Stroglova, and V. Becvarova, *Appl. Microbiol. Biotechnol.* **40**, 864–866 (1994).

98. J. Desomer, P. Dhaese, and M. Van Montagu, *Appl. Environ. Microbiol.* **56**, 2818–2825 (1990).

99. C. Mazy-Servais, D. Baczkowski, and J. Dusart, *FEMS Microbiol. Lett.* **151**, 135–138 (1997).

100. D. Simon and J.J. Ferretti, *FEMS Microbiol. Lett.* **66**, 219–224 (1991).

101. G.A. Somkuti, D.K. Solaiman, T.L. Johnson, and D.H. Steinberg, *Biotechnol. Appl. Biochem.* **13**, 238–245 (1991).

102. G.M. Dunny, L.N. Lee, and D.J. LeBlanc, *Appl. Environ. Microbiol.* **57**, 1194–1201 (1991).

103. J.C. Fenno, A. Shaikh, and P. Fives-Taylor, *Gene* **130**, 81–90 (1993).

104. M.L. Ricci, R. Manganelli, C. Berneri, G. Orefici, and G. Pozzi, *FEMS Microbiol. Lett.* **119**, 47–52 (1994).

105. H.E. Smith, H.J. Wisselink, U. Vecht, A.L. Gielkans, and M.A. Smits, *Microbiology* **141**, 181–188 (1995).

106. S. Schenk and R.A. Laddaga, *FEMS Microbiol. Lett.* **73**, 133–138 (1992).

107. J. Augustin and F. Gotz, *FEMS Microbiol. Lett.* **54**, 203–207 (1990).

108. L. Masson, G. Prefontaine, and R. Brousseau, *FEMS Microbiol. Lett.* **51**, 273–277 (1989).

109. J. Vehmaanpera, *FEMS Microbiol. Lett.* **52**, 165–169 (1989).

110. B.H. Belliveau and J.T. Trevors, *Appl. Environ. Microbiol.* **55**, 1649–1652 (1989).

111. P. Bridigi, E. De Rossi, M.L. Bertarini, G. Riccardi, and D. Matteuzzsi, *FEMS Microbiol. Lett.* **57**, 135–138 (1990).

112. C.P. Quinn and B.N. Dancer, *J. Gen. Microbiol.* **136**, 1211–1215 (1990).

113. H. Kusaoke, Y. Hayashi, Y. Kadowaki, and H. Kimoto, *Agric. Biol. Chem.* **53**, 2441–2446 (1989).

114. I.R. McDonald, P.W. Riley, R.J. Sharp, and A.J. McCarthy, *J. Appl. Bacteriol.* **79**, 213–218 (1995).

115. M. Ohse, K. Takahashi, Y. Kadowaki, and H. Kusaoke, *Biosci. Biotechnol. Biochem.* **59**, 1433–1437 (1995).

116. W. Schurter, M. Geiser, and D. Mathe, *Mol. Gen. Genet.* **218**, 177–181 (1989).

117. A. Okamoto, A. Kosugi, Y. Koizumi, F. Yanagida, and S. Udaka, *Biosci. Biotechnol. Biochem.* **6**, 202–203 (1997).

118. S. David, G. Simons, and W.M. De Vos, *Appl. Environ. Microbiol.* **55**, 1483–1489 (1989).

119. A. Argnani, R.J. Leer, N. van Luijk, and P.H. Pouwels, *Microbiology* **142**, 109–114 (1996).

120. R.A. Dedonder, J.A. Lepesant, J. Lepesant-Kejzlarova, A. Billault, M. Steinmetz, and F. Kunst, *Appl. Environ. Microbiol.* **33**, 989 (1977).

121. M.A. Vandeyar and S.A. Zahler, *J. Bacteriol.* **167**, 530–534 (1986).

122. S.M. Cutting and V. Azevedo, in K.W. Adolph ed., *Methods in Molecular Genetics*, Academic Press, Orlando, Fla., 1995, pp. 323–338.

123. S.M. Cutting and Vander Horn, in C.R. Harwood and S.M. Cutting eds., *Molecular Biological Methods for* Bacillus, Wiley, Chichester, U.K., 1990, pp. 27–74.

124. H.L. Ritz and J.N. Baldvin, *Proc. Soc. Exp. Biol. Med.* **110**, 667–671 (1962).

125. C.J. Schroeder and P.A. Pattee, *J. Bacteriol.* **157**, 533–537 (1984).

126. C.V. Sunder Raj and T. Ramakrishnan, *Nature (Lond.)* **228**, 280–281 (1970).

127. N.D. Connell, *Methods Cell Biol.* **45**, 107–125 (1994).

128. R.H. Baltz and M.A. McHenney, in C.L. Hershberger, S.W. Queener, and G. Hegeman eds., *Genetics and Molecular Biology of Industrial Microorganisms*, American Society for Microbiology, Washington, D.C., 1989, pp. 163–167.

129. P. Matsushima, M.A. McHenney, and R.H. Baltz, *J. Bacteriol.* **171**, 3080–3084 (1989).

130. M.A. McHenney and R.H. Baltz, *J. Bacteriol.* **170**, 2276–2282 (1988).

131. M.A. McHenney and R.H. Baltz, *J. Antibiot. (Tokyo)* **42**, 1725–1727 (1989).

132. S.A. Zahler, in D. Dubnau ed., *The Molecular Biology of the Bacilli*, Academic Press, New York, 1982, pp. 269–305.

133. J. Errington, in C.R. Harwood and S.M. Cutting eds., *Molecular Biological Methods for* Bacillus, Wiley, Chichester, U.K., 1990, pp. 175–220.

134. P. Youngman, in C.R. Harwood and S.M. Cutting eds., *Molecular Biological Methods for* Bacillus, Wiley, Chichester, U.K., 1990, pp. 221–266.

135. D. van der Lelie, H.A.B. Wosten, S. Bron, L. Oskam, and G. Venema, *J. Bacteriol.* **172**, 47–52 (1990).

136. T.M. Koehler and C.B. Thorne, *J. Bacteriol.* **169**, 5271–5278 (1987).

137. S.J. Projan, and G.L. Archer, *J. Bacteriol.* **171**, 1841–1845 (1989).

138. G.M. Dunny, B. Brown, and D.B. Clewell, *Proc. Natl. Acad. Sci. U.S.A.* **75**, 3479–3483 (1978).

139. D. Clewell, *Microbiol. Rev.* **45**, 409–436 (1981).

140. D.A. Hopwood, in J.R. Norris and D.W. Ribbons eds., *Methods in Microbiology*, vol. 7B, Academic Press, London, 1972, pp. 29–158.

141. O.R. Torres, R.Z. Korman, S.A. Zahler, and G.M. Dunny, *Mol. Gen. Genet.* **225**, 395–400 (1991).

142. L.M. Parsons, C.S. Jankowski, and K.M. Derbyshire, *Mol. Microbiol.* **28**, 571–582 (1998).

143. A.E. Franke and D.B. Clewell, *J. Bacteriol.* **145**, 494–502 (1981).

144. C. Le Bougwenec, G. De Cespedes, and T. Horaud, *J. Bacteriol.* **170**, 3930–3936 (1988).

145. P. Courvalin and C. Carlier, *Mol. Gen. Genet.* **205**, 291–297 (1986).

146. P. Ayoubi, A.O. Kilic, and M.N. Vijayakumar, *J. Bacteriol.* **173**, 1617–1622 (1991).

147. M. Norgren, M.G. Caparon, and J.R. Scott, *Infect. Immun.* **57**, 3846–3850 (1989).

148. M. Zuniga, I. Pardo, and S. Ferrer, *FEMS Microbiol. Lett.* **135**, 179–185 (1996).
149. R.I. Geist, N. Okada, and M.G. Caparon, *J. Bacteriol.* **175**, 7561–7570 (1993).
150. P. Trieu-Cuot, C. Carlier, P. Martin, and P. Courvalin, *FEMS Microbiol. Lett.* **48**, 289–294 (1987).
151. D.R. Williams, D.I. Young, and M. Young, *J. Gen. Microbiol.* **136**, 819–826 (1990).
152. M. Bierman, R. Logan, K. O'Brien, E.T. Seno, R.N. Rao, and B.E. Schoner, *Gene* **116**, 43–49 (1992).
153. T. Smokvina, P. Mazodier, F. Boccard, C.J. Thompson, and M. Guerineau, *Gene* **94**, 53–59 (1990).
154. A. Schafer, A. Tauch, W. Jager, J. Kalinowski, G. Thierbach, and A. Puhler, *Gene* **145**, 69–73 (1994).
155. P. Mazodier, R. Petter, and C. Thompson, *J. Bacteriol.* **171**, 3583–3585 (1989).
156. T. Voeykova, L. Emelyanova, V. Tabakov, and N. Mkrtumyan, *FEMS Microbiol. Lett.* **162**, 47–52 (1998).
157. P. Trieu-Cuot, E. Derlot, and P. Courvalin, *FEMS Microbiol. Lett.* **109**, 19–23 (1993).
158. R. Lazrag, S. Clavel-Seres, H.L. David, and D. Roulland-Dussoix, *FEMS Microbiol. Lett.* **69**, 135–138 (1990).
159. E.P. Gormley and J. Davies, *J. Bacteriol.* **173**, 6705–6708 (1991).

See also ENERGY METABOLISM, MICROBIAL AND ANIMAL CELLS; ENZYMES, DIRECTED EVOLUTION; METABOLITES, PRIMARY AND SECONDARY; MUTAGENESIS; PROTEIN EXPRESSION, SOLUBLE.

GENETIC INSTABILITY

JOHN CULLUM
MATTHIAS REDENBACH
LB Genetik, University of Kaiserslautern
Kaiserslautern, Germany
DASLAV HRANUELI
PLIVA d.d., Research Institute
Zagreb, Croatia

KEY WORDS

Mutation
Scale-up
Selection
Strain degeneration

OUTLINE

INTRODUCTION

The description *genetic instability* is used in slightly different ways by different authors. Some would consider it synonymous with a high mutation rate. This can affect all genes in the organism or only particular genes and may occur only in particular growth conditions. Ideally the mutation rate should be measured as the number of new mutants arising per genome per generation. In practice it is often difficult or impossible to measure the mutation rate per generation. For example the presence of multiple genomes in fragments of a mycelial organism often prevents quantitation. In many cases the proportion of mutants in a population is measured, and this quantity depends on both the mutation rate and the selective forces acting on the population. For most organisms the frequency of spontaneous mutants for a particular gene in the population is under 10^{-4}. An operational definition of genetic instability is that the frequency of mutants is considerably higher than this. In *Streptomyces*, genes are considered unstable if spontaneous mutants are present at 0.1% per spore or higher (1). Defining genetic instability in terms of the number of mutants in the population is useful for industrial microorganisms because the mutations that can cause problems during fermentation are those that accumulate to significant levels in the population. Genetic instability occurs in brewer's yeast (*Saccharomyces cerevisiae*). In this case, *petite* mutants, which are defective in mitochondrial function, occur at high frequency (about 1% of colonies). However, the mutants grow much slower than wild type and do not cause any problems in fermentations. Such genetic instability is not very important for industrial applications. In contrast, many industrial strains undergo a rather ill-defined process of "strain degeneration" in which undesirable properties such as reduced production level accumulate. This is normally countered by careful strain maintenance and testing. Strain degeneration likely reflects the fact that desirable properties for production (e.g., high levels of secondary metabolites) probably have a negative influence on the ability of the strain to compete, so mutants outgrow it. This will happen even at low mutation rates because of the presence of strong selective forces.

Little literature is available about genetic instability in industrial strains. This reflects that such data can be valuable proprietary knowledge, and the fact that genetic instability is often ignored as a phenomenon during fermentation development. One important industrial organism in which some studies have been carried out is *Streptomyces rimosus*, the oxytetracycline producer (2). Some high producers are genetically unstable, and the instability can be recognized by the frequent appearance of morphological mutants. Two classes of mutants with undesirable properties have been recognized: those in class I retain resistance to the antibiotic but have much lower production, whereas those in class II have also lost resistance. Because high levels of oxytetracycline are produced during industrial fermentation, class II mutants are irrelevant to the production process; they are killed by the antibiotic. However, class I mutants survive, and competition experiments show that they outgrow the parent production strain. If care is taken with inoculum preparation it is possible to reduce problems from genetic instability to a manageable level. Experience shows that the way in which the strain is grown can have large effects on genetic instability. Thus, if the strain is propagated via spores rather than as mycelial fragments, then a much higher frequency of mutants occurs. This is shown in Figure 1 in which plates are shown from cultures propagated as mycelial fragments and compared with those from spores. The dark brown colonies are the high producers; it can be seen that a variety of lighter variants occur—most of these are class I variants. Another important factor is the medium. More instability is seen on malt agar than on soya-mannitol agar. A third factor is growth temperature: higher temperatures (e.g., 37 °C) lead to much higher levels of genetic instability, compared with lower temperatures (e.g., 28 °C). Thus, careful choice of growth conditions can be used to control instability.

MUTATION AND SELECTION

The proportion of mutants in the population is governed both by the frequency of generation of mutants and by selection effects acting on the mutants. In general in batch processes both of these parameters will vary throughout the process, making simple models such as have been used for chemostat populations invalid. If the population is small compared with the mutation rate, then the frequency of mutations will be governed by the time needed to generate the first mutant and there will be a lot of variation between parallel cultures. In most industrial fermentations the cultures should be large enough to make stochastic effects unimportant. However, large fluctuations in yield in small cultures (e.g., at the shake flask stage) might point to genetic instability being present. Experimental studies with chemostats show that mutants usually displace the initial population within 100 generations. The biological basis for this behavior is that strains are not well adapted to growth in the chemostat, so very strong selection pressures result. This tends to limit the usefulness of continuous culture for many industrial applications despite the many desirable features for plant utilization. Unless the selected properties in the continuous culture conditions are the same as those desired, genetic instability will cause problems. One area in which this may be true is in biomass production. In some cases continuous culture can be used with recycling of biomass. In this case there is little growth, so the number of generations remains relatively small.

Usually during scale-up of batch processes one hopes to alter the physiological conditions between the large and small fermenters as little as possible. Thus, scale-up can be considered, at a first approximation, as just growing the culture for more generations. For example, scaling up from 1 L to 200,000 L corresponds to about 18 more generations. This increase can lead to genetic instability becoming a problem. An example of genetic instability occurring during scale-up was provided by the bicyclomycin fermentation using *Streptomyces sapporonensis* (3). There was a drastic reduction in yield compared with shake flasks on scaling up to 4,000 L in a pilot plant, and this was shown to be due to the accumulation of low-producing mutants. The mutants could be recognized on the basis of colony morphology because they were deficient in aerial mycelium production. There is very little information in the literature on the frequency of genetic instability causing problems during scale-up. This is partly because it is not gen-

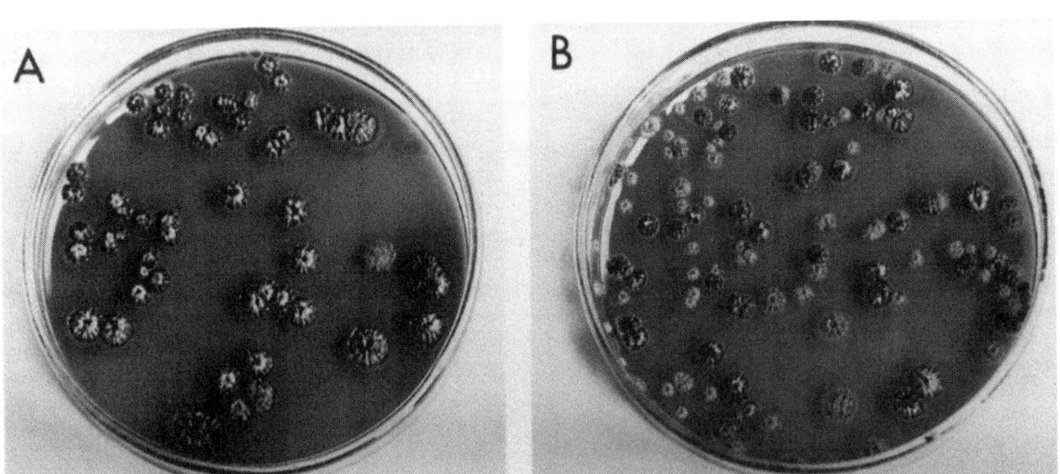

Figure 1. Colonies of an *S. rimosus* strain selected for high oxytetracycline production. (**A**) was propagated via spores on agar whereas (**B**) was propagated as mycelial fragments in liquid culture. The pigmentation of the colonies correlates with their oxytetracycline production. *Source*: Figure modified from Ref. 2 by kind permission of the American Society for Microbiology.

erally recognized as a problem and it is unusual to do genetic studies on the population of cells in a fermenter. We suspect that genetic instability is not uncommon in industrial fermentations, but that the empirical optimization of yield will result in growth conditions that reduce the instability to acceptable levels. However, this would have the consequence that the conditions will probably be suboptimal for product yield by the nonmutated cells. This suggests that it would be valuable to monitor the proportion of mutants so as to provide an extra parameter for the fermentation development. As most large scale fermentations are carried out in a multistage process with the inoculum being introduced to a series of three or more fermenters, on increasing scale it would be possible, for instance, to try to optimize the genetic stability in the smaller fermenters and then optimize the yield in the final production fermenter. This sort of strategy depends on methods for detecting mutants, which will be discussed later.

A genetic approach to increased stability can also be taken. This will to some extent occur automatically in a strain selection program because obviously unstable strains will be rejected. A second approach that has been used in some cases is to carry out genetic crosses between unstable high producers and stable low producers to obtain high producers with improved stability. In principle it should be possible to approach the problem in a directed rational way. If the mechanism for generation of the mutants is known, it ought to be possible to obtain derivatives with a lower mutation rate. However, this will usually involve some disruption of the DNA processing in the cell and is likely to lead to lower viability and production. A second approach is to introduce a system that selects against mutants. Thus, if mutation is caused by deletions it ought to be possible to move a selectable gene into the deletogenic region and thus ensure that the mutants will have a growth disadvantage (4). In many cases the strategies to combat instability involve manipulation of the selection effects rather than the mutation rate.

The use of immobilized cells or biomass recycling should greatly reduce the potential for genetic instability because little growth will occur. However, even in these cases there may be high mutagenesis rates in a subpopulation of stationary-phase cells (see later) and differential growth of a subpopulation. This has been demonstrated in colonies on agar, but the role of genetic instability in industrial applications using nongrowing cells is not yet known.

BACTERIA

Many different types of mutations can occur, and the different classes often involve different molecular mechanisms. Mutations can be classified into point mutations or DNA rearrangements. Point mutations were originally defined as mutations that map as a single point in genetic crosses. In molecular terms they are usually taken as mutations that affect a single base or a small number of bases. In contrast, DNA rearrangements affect a larger number of bases. One type of point mutation is a base change in which a single base in the DNA is changed to another base. In a *transition*, a purine is substituted for another purine (i.e., A → G or G → A), whereas a *transversion* substitutes a purine by a pyrimidine (i.e., A → C, A → T, G → C, G → G). The other important type of point mutation is the addition or deletion of a small number of bases in the DNA. These are often referred to as frameshift mutations because the addition or deletion of one or two bases in a protein coding area causes a shift of the reading frame such that all distal codons will be read wrongly. Important DNA rearrangements in bacteria are deletions (loss of a DNA segment), insertions (insertion of a segment of DNA into another sequence), and inversions (a segment of DNA is inverted with respect to the neighboring sequences). The effect of any mutation depends on its location. In bacteria the genes are usually closely packed with little space in between. This means that most point mutations will occur in genes or their control regions and therefore have a genetic effect. If deletions or insertions affect genes, they will usually result in inactivation. In some cases insertions can carry promoters and activate neighboring genes after insertion. Inversions will have a clear cut phenotypic effect only if the inversion end points lie in genes or operons. However, as the transcription direction of genes with respect to the replication origin is highly nonrandom it is to be expected that many inversions will have subtle effects on the growth of strains. A major problem in understanding genetic changes that might occur in genetic instability is that the functions of about two-thirds of the genes in a microorganism are unknown. Thus even if a particular mutation in a specific gene can be associated with genetic instability, it may be difficult to understand why.

The mutation frequency in a strain may be altered by genetical or environmental factors. Mutator mutations can lead to a greatly enhanced rate of mutagenesis. Many mutator mutations lie in genes involved in DNA repair and may increase mutation rates up to 1,000-fold. Most mutator strains grow more slowly than their nonmutator parent strains. However, despite this, they win against nonmutator strains in chemostat competition experiments. This demonstrates the importance of mutation in conditions with high selection and indicates that inappropriate selection strategies could lead to enrichment of mutator strains. In *E. coli* and *Salmonella* a subpopulation of cells in stationary phase can enter a hypermutable state. This leads to the creation of new mutants in the absence of growth and can produce mutants that can grow and be enriched in the absence of growth of other cells in the population. It is likely that such mutagenic states exist in most microorganisms. It is thus important to culture and store industrial microorganisms in conditions that minimize the chance to enter such a state.

In some cases there are programmed mutation events that switch between expression of different sets of genes. One well known example is that of phase variation in *Salmonella* strains, in which a site-specific recombination system causes inversion of a segment of DNA. The invertible segment carries a promoter that reads out of it and can activate a neighboring operon; inversion leads to switching off this operon. In *Salmonella* this leads to a switch in the expression between two flagellar antigens that probably helps to defeat the host's immune system, but similar systems might control other functions in other bacteria. In some cases it is known that certain regions of the chromosome are particularly subject to instability and may not be suitable sites for introducing foreign DNA during strain construction. The replication terminator region of the *E. coli* chromosome is subject to a high frequency of DNA rearrangements. The end regions of the linear chromosomes of *Streptomyces* are also subject to genetic instability (5). This can lead to large deletions. (Deletions of over 1 Mb of the 8-Mb chromosome can occur without very dramatic phenotypic effects!) There can also be tandem DNA amplification of chromosomal sequences: the sequences can vary from 2 to 250 kb in size, and the copy number can be over 100. The amplifications may be accompanied by deletion of neighboring chromosomal sequences up to the chromosome end. These unstable regions may affect secondary metabolite production (Table 1). The mutations are usually deleterious for growth, compared with wild-type strains, but may give an advantage in comparison with high-production parents. The mutants can often be recognized by changes in colony morphology.

It may be difficult to detect mutations in bacteria because they may not show very obvious morphological or auxotrophy properties, and it is likely that the phenomenon of genetic instability is often not recognized. If the changes involve large-scale genome rearrangements, then pulsed-field gel electrophoresis (PFGE) may prove useful. PFGE can separate linear DNA fragments in size ranges up to about 10 Mb (bacterial genomes range in size from about 0.6 to 10 Mb). It is often possible to find rarely cutting enzymes that restrict the whole genome into 5–30 fragments that can be conveniently separated by PFGE. The gels can also be used for Southern blotting to detect specific DNA sequences. PFGE can be used to screen strains for DNA rearrangements (deletions, duplications, and sometimes inversions).

Phage infection can be a major problem in some industrial fermentations. In addition to lytic phages that kill all the cells, there are many lysogenic phages that maintain themselves within the host cells either by integrating into the host chromosome or by establishing a plasmid state. Lysogenic phages exist within some industrial strains (e.g., the phages RP2 and RP3 of *S. rimosus*, [6]) and may cause changes in the behavior of a strain. Some phages (e.g., phage Mu in *E. coli*) can integrate into random sites in the chromosome and inactivate genes, in other words, they are transposable elements).

It is possible that some changes observed in bacteria are of an epigenetic nature, that is, they are stable changes that do not depend on changes in the DNA sequence. One candidate for such an effect is the mycelial fragment size in *Streptomyces* fermentations. In many species fast growth is associated with larger fragment sizes. If a continuous culture in a chemostat is started from fast-growing mycelium, then large fragments are stably maintained at a variety of dilution rates. However, if the chemostat is started from a slow-growing culture entering stationary phase, then a small fragment size is maintained over a range of dilution rates. It is possible to switch from small fragments to large fragments by a short period of growth at high dilution rate, and from small fragments to large fragments by a short period at a low dilution rate. At dilution rates between these critical rates the form depends on the history of the culture and remains stable for many generations (7).

Table 1. Genetic Instability Affecting Secondary Metabolite Production in *Streptomyces*

Species	Product	Effect
S. rimosus	Oxytetracycline	Deletion or amplification of genes
S. fradiae	Tylosin	Amplifications and deletions in biosynthesis cluster
S. griseus	Streptomycin	Deletion of A-factor gene, loss of streptomycin synthesis
S. ambofaciens	Spiromycin	DNA amplification leads to loss of production

PLASMIDS

Plasmids are common in most groups of bacteria. Most plasmids are circular DNA molecules, but linear plasmids are present in some species. Natural plasmids may be of significance in some industrial strains, but the most important role of plasmids is for constructing recombinant strains. Natural plasmids come in a large range of sizes (2–1,000 kb) and copy numbers (1–500/chromosome) and have been used to construct vectors with different copy numbers for different applications. Two important classes of genetic instability that can affect plasmids are segregational instability (plasmid loss), and structural instability (DNA rearrangements in the plasmid). Low copy number plasmids have systems to ensure that plasmid copies are distributed to both daughter cells on division. Many natural plasmids carry conjugation systems that would reintroduce the plasmid into any member of the population that had lost it. However, biological containment considerations and the desire to obtain small vectors make this undesirable for recombinant plasmids. It is likely that many high copy number plasmids do not need an active system to segregate them to daughter cells but instead rely on the low statistical chance of producing plasmid-free progeny. (If there are N copies of a plasmid at cell division then the chance of producing a plasmid-free cell is 2^{-N}; for $N = 20$ this is 10^{-6}.) However, the situation is complicated if plasmid multimers are produced (i.e., circles containing two or more copies of the plasmid as a direct repeat). The copy number control determines the number of replication origins, so a plasmid dimer would have only half the number of molecules per cell as a monomer. Because multimers are generated by homologous recombination systems, plasmids often have site-specific recombination systems that break down multimers into monomers and result in a higher segregational stability (e.g., the *xer* system of ColE1 [8]). It is thus important to ensure that the stability-enhancing systems of natural plasmids are retained during vector construction; many common laboratory cloning vectors have lost such systems because they were constructed with minimal-size replicons. Choice of fermentation conditions can make a large difference in the segregational stability of plasmids. The choice of limiting nutrient in chemostat cultures can radically affect plasmid loss. Choice of host may also have a significant effect; at present this can be tested only empirically because the basis for differences is unknown. The stability of plasmids is often affected by the growth rate of cultures. There are cases in which plasmids are more stable at high growth rates. Inocula for fermenters are typically prepared in several stages involving increasing volumes. An improvement in stability can be obtained by avoiding slow growth during inoculum preparation (i.e., transferring to a larger volume before the culture enters stationary state).

The replication systems of plasmids may differ significantly from that of the chromosome (9), and this may lead to a significantly higher frequency of mutation including DNA rearrangements. Some plasmids have a rolling-circle replication mode with single-stranded DNA intermediates. This system is more prone to DNA rearrangements than other replication systems. DNA rearrangements can include deletions, insertions, duplications, and inversions. The chromosomal replication control system is a positive-control system in which every replication origin in the cell is activated once in the cell cycle, and all origins are activated almost simultaneously. In contrast most plasmids have negative-control systems, which initiate replication at an origin at times spread through the cell cycle. A random pool mechanism functions, which means that replication may be initiated on a particular copy of the origin several times in a cell cycle, while another copy is not replicated. This means that any mutated copy of a plasmid with a small advantage for selection for replication can rapidly replace a parent plasmid. DNA rearrangements in plasmids are usually easily detected using restriction digests and agarose gel electrophoresis.

The selection of segregational or structural mutants can lead to rapid overgrowth of the original population. A plasmid-carrying strain may grow more slowly than one that has lost the plasmid, but this is by no means universally true for natural plasmids. However, it is likely that most constructs for industrial use will negatively affect the growth of the host strain. Although larger derivatives of a plasmid are often less stable than small derivatives, it is not clear that this is mainly due to the size per se. It seems much more the case that the presence of promoters on the plasmid can lead to selection against the insert. In some plasmids, transcription from insert promoters into the vector can lead to stability problems, possibly by interfering with vector replication. Therefore it has become common practice to insert transcriptional terminators into vectors to prevent read-through from the insert into the vector; this was one key step in constructing successful vectors for *Bacillus*. It is also desirable only to transcribe the cloned gene when product is required because this often reduces selection against the insert. Various inducible promoter systems have been developed, such as the recombinant *tac* promoter, which combines the desirable strength of the *trp* operon promoter with the control of the *lac* promoter (induction with IPTG). Other promoters are the lambda P_R and P_L promoters, which can be controlled by temperature by using a temperature-sensitive repressor. Another possibility is to use a promoter from phage T7 that can be transcribed only when the special T7 RNA polymerase is expressed. By using a combination of plasmid stability systems and control of the transcription of cloned genes it is possible to achieve good results in *E. coli*. However, even in *E. coli* there is a need for considerable empirical work, and many surprises will occur. In other bacterial systems the state of knowledge about plasmid stability is considerably worse, and a good portion of luck is necessary to achieve success.

YEAST AND FUNGI

Many of the same considerations apply to yeast and fungi as to bacteria. However, the more complex eukaryotic nature of these organisms leads to some further mechanisms for generating mutants. Eukaryotes have a number of linear chromosomes. Many yeast and fungi are haploid, but there are also diploid and dikaryotic (i.e., two separate haploid nuclei in the same cytoplasm) strains. There can be changes in the number of chromosomes (either in the number of copies of the genome present, "ploidy"; or in the loss or gain of copies of single chromosomes, "aneuploidy"). There can also be transfer of parts of chromosomes to other chromosomes (translocation). Most yeast and fungi also have sexual systems in which meiosis takes place. Such sexual cycles are usually associated with stress conditions (e.g., nutrient stress) and may be avoidable by suitable strain maintenance conditions. Meiosis can lead to changes in the genetic make up of cultures when diploid or dikaryotic organisms are used. Although many of the present important industrial strains are haploid, this may become more of a problem in the future as a wider range of organisms is used in biotechnology. Sexual systems usually involve the interaction of two different mating types; some organisms have only two sexes, but many fungi have several mating types. The mating types may be able to change: the best-characterized example of this is the mating-type switch in *S. cerevisiae* in which it is possible for a strain to switch backward and forward between the two mating types, a and α. There are strains that switch at high frequencies (homothallic strains) due to the presence of a gene, *HOM*, encoding an endonuclease, whereas other strains switch only at low frequency (heterothallic). The ability to switch between mating types can thus be seen as a sort of genetic instability and may have consequences for fermentation if one mating type is selected under fermenter conditions.

Transposable elements are also present in yeast and fungi. One class that has not been found in bacteria the retrotransposons, which resemble retroviruses. They transpose via an RNA intermediate and transcribe into neighboring genes. The example of *Ty1* in yeast shows that this can activate genes by providing a new promoter. This ability to activate as well as inactivate genes could be relevant to genetic stability. Such elements are present in many copies in the cell (>30 copies), which makes it unlikely that they can be removed by genetic manipulation. Apart from the ability to change gene expression patterns by transposition, it is possible that mutations in control proteins will result in new patterns. The *areA* gene in *Aspergillus nidulans* controls ammonium repression for a large number of genes involved in the catabolism of various nonfavored nitrogen sources (10). Many mutations in *areA* are known that differentially change the pattern of control (i.e., some genes are silenced and some are overexpressed). In one case a duplication within the *areA* gene resulted in another gene becoming subject to *areA* control. This mutational flexibility would allow easy evolution of new expression patterns and could contribute to genetic instability.

Another feature of eukaryotes that makes them genetically more complex than bacteria is the possession of mitochondria, which also have their own genome (mtDNA). In *S. cerevisiae* there are very frequent deletions in the mtDNA that cause a loss of mitochondrial function. The organisms can still grow, but the slower growth results in small colonies (which is why they are called *petite* mutants). These mutants account for 1% of colonies or more, depending on the strain. Another genetic phenomenon caused by mtDNA is senescence (11) in the fungus *Podospora anserina*. This species cannot grow indefinitely vegetatively and, after a certain amount of mycelial growth, death occurs. The senescence is caused by the accumulation of a high number of copies of a plasmid formed by excision from the mtDNA. Such senescence phenomena may cause problems in the maintenance of industrial strains.

Problems of genetic instability can also arise in culture collections. For example, it is by no means uncommon for the secondary metabolite spectrum of a filamentous fungus to change radically if it is examined with the same strain from a culture collection after a gap of several years. Although this problem can arise with any organism, many bacteria can be kept in nongrowing states (e.g., lyophilization, storage at -70 °C in glycerol), which limits the scope for selection to act. However, most filamentous fungi cannot be maintained in this way. In view of the number of different strains that must be maintained, there is probably no real answer to the problem. It is likely that adaptive mutation occurs in most microorganisms as a response to certain stresses and might be one cause for such strain degeneration. Experience can guide the choice of media and storage conditions to try to minimize these effects.

Epigenetic effects can also be important in yeast and fungi. The DNA of most eukaryotes contains some 5-methyl cytosine, which is formed by methylation after DNA replication; this usually occurs at CG dinucleotides, and so the complementary strand can also be methylated. If a methylated sequence is replicated, two hemimethylated daughter molecules (i.e., each methylated in only one strand) are produced. Because DNA methylase enzymes usually prefer such hemimethylated substrates this results in an inheritance of the methylation patterns. Methylation of genes is associated with their nonexpression, but it is not clear how the de novo methylation arises. In some cases the introduction of foreign genes into fungi can result in their inactivation by methylation. The reason why this occurs is not known. In some cases methylation seems to be targeted at duplicate copies of a gene in an organism, that is, introduction of an extra copy of a gene will result in inactivation of either the introduced copy or the original copy. The methylation may also lead to mutation. 5-Methyl cytosine can be spontaneously deaminated to produce thymine and, in the absence of repair, this gives a C $\rightarrow$ T transition. This can result in a "RIP effect" in some fungi in which a very high mutation rate is observed after introduction of an extra copy of a gene such that one copy is inactivated by multiple C $\rightarrow$ T transitions (12). A further epigenetic effect that has been observed in *S. cerevisiae* is the existence of prions (13). These are proteins of altered structure that can catalyze the change of copies of normal structure to the altered structure. This can function as a sort of "infective" process and is a potential source of instability.

ANIMAL AND PLANT CELLS

Animal cells have been used for the growth of viruses for the production of vaccines and in more recent years for the production of recombinant proteins. Plant cells are not yet used on an industrial scale, but there has been interest in this possibility. Smaller-scale cultivation of animal cells has been used extensively in laboratories. Most primary cell cultures have a limited life span in the sense that they can divide only a limited number of times. However, immortalized cell lines, which do not have this limitation, are of most interest for large-scale culture. These are tumour cell lines that have often changed considerably in their properties as compared with the original cells from which they derive. Cultured animal cells are often genetically unstable and undergo frequent karyotype changes. These include polyploidy (changes in the number of chromosome sets from diploidy to triploidy and tetraploidy), aneuploidy (loss or gain of copies of particular chromosomes), and frequent chromosomal rearrangements (deletions, inversions, and translocations).

Chinese hamster ovary (CHO) cells are an immortalized cell line that has been used to produce recombinant proteins at a 10,000-L scale (14). They possess good growth properties for large-scale culture and pose less safety problems with respect to viruses than cell lines more closely related to humans. An important strategy to obtain increased production of recombinant products is to work with high copy numbers of the introduced genes. This can be done by using a cell line that lacks dihydrofolate reductase (DHFR) and transforming with the required gene coupled to the DHFR gene. The antibiotic methotrexate inhibits DHFR, and selection for growth on increasing levels of DHFR results in the amplification of the DHFR gene and coupled sequences. The amplifications can have a complex structure and be either extrachromosomal or integrated in the chromosome. Continued selection usually results in integrated copies, but the positions of integration vary more or less randomly between independent clones.

Control of stability requires the careful construction of master cell banks and working cell banks and the avoidance of too many generations of growth. Various methods can be used to check genetic stability. One method is to use Southern blots. However, this does not show heterogeneity in the population. A better method is to use karyotype-based methods. In particular, fluorescent in situ hybridization (FISH) can locate the amplified sequences and show heterogeneity. If chromosome-specific probes are used it is possible to monitor the copy number of a chromosome in interphase nuclei. For instance, Figure 2 shows a normal diploid cell and a cell carrying five copies of a chromosome. If metaphase cells are used, details of chromosome rearrangements can be monitored with appropriate probes. It would normally be necessary to carry out FISH and measurements of productivity at several times during growing cultures. Flow cytometry can be used to show variations in DNA content in a population of cells that can indicate the extent of polyploidy and, to some extent, aneuploidy.

The fact that the amplifications are selected with methotrexate and that lower producers might be expected to overgrow the best producers would suggest that stability would be enhanced if methotrexate were present in the medium. However, there are data showing an increased instability with methotrexate (15). One explanation is that methotrexate leads to a reduction of thymidine synthesis, which could allow more dUTP incorporation into the DNA. The resulting repair and strand cutting might stimulate instability. In fact an experiment with a heterogeneous population of clones in the absence of methotrexate suggested that such cultures may be surprisingly stable in the absence of selection.

VIRUSES

Viruses are propagated for vaccine manufacture. This includes the use of live attenuated viruses and killed viruses. Various culture systems are used, such as cell cultures and chicken eggs. Viruses show a bewildering range of different genome structures (single- and double-stranded DNA or RNA) with specialized replication methods. Virus replication may be much less accurate than host replication (especially for RNA viruses) and thus show a high mutation rate. Genetic instability has vital biological safety considerations because a reversion of attenuated viruses to higher pathogenicity or changes that reduce the efficiency of killing in killed-virus vaccines could be disastrous. Any release of viruses from the cells used to culture the virus could also have serious implications. The main weapon against these problems is to reduce the number of growth cycles as much as possible. It is also important to monitor the cultures and product.

CONSEQUENCES FOR GOOD MANUFACTURING PRACTICE

Good manufacturing practice aims to guarantee reproducibility of production runs and avoid product contamination. The details are dependent on the organism and product, but there are some general considerations:

1. A master cell bank (MCB) is derived from a single colony or single cell. It consists of many replicates kept in the most stable form possible (e.g., lyophilized or frozen at −70 °C or in liquid nitrogen). Each ampoule is opened only once. It is important to make sufficient replicates because generation of a new MCB will require an application to the regulatory authority. It is also usual to store the material at more than one location in case of destruction (e.g., freezer problems, fire).

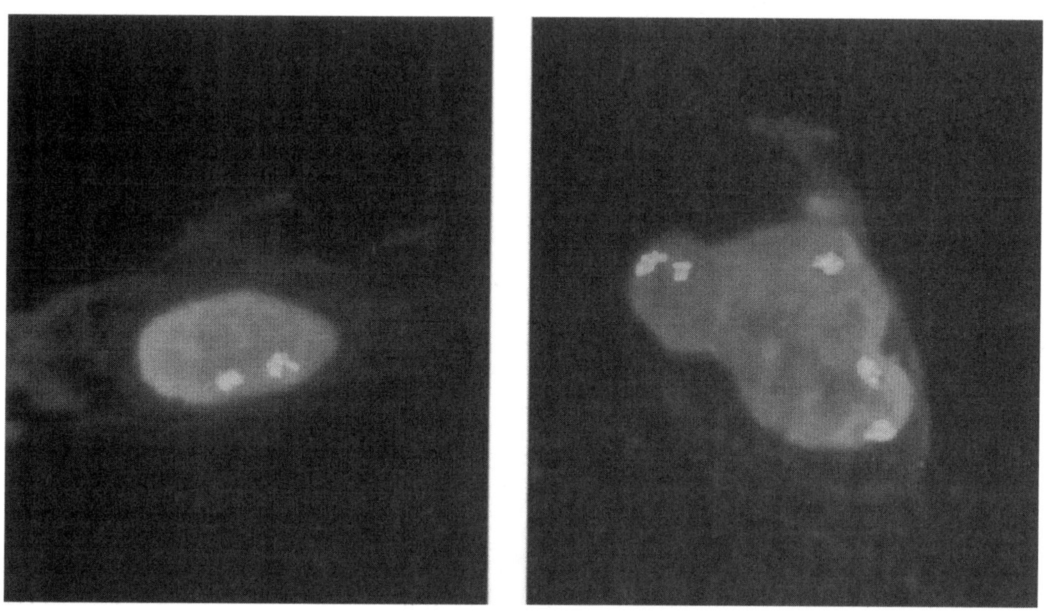

Figure 2. Fluorescent in situ hybridization (FISH) of human fibroblast cells with a probe specific for repetitive sequences of chromosome 1. One interphase nucleus shows the expected two signals for a diploid cell whereas the other nucleus is aberrant with five copies of the chromosome. *Source*: Photograph kindly provided by Dr. Harry Scherthan, Department of Human Biology, University of Kaiserslautern.

2. The working cell bank (WCB) is derived from the MCB and is used to inoculate a production batch.
3. Detailed records of both MCBs and WCBs must be kept, and all possible efforts must be made to limit the number of growth cycles (number of passages) to minimize the effects of genetic instability.
4. Both MCBs and WCBs must be rigorously tested. The tests for MCBs will be particularly exhaustive and depend on the process. A genotypic characterization using various DNA tests is usual (e.g., DNA fingerprinting to confirm the strain, and testing the structure of recombinant DNA constructs). There will also be phenotypic tests and tests that normal production levels occur. There may be tests for contamination (e.g., viruses in cell culture).

These procedures aim to confirm the reproducibility of the starter cultures and the production process. As long as the build up of any variants is fairly reproducible, genetic instability during the fermentation will not conflict with these aims but will lead to a reduction in production compared with the ideal case. However, any evidence for genetic instability and the occurrence of heterogeneity during production fermentations is likely to lead to problems with the regulatory authorities. It is likely that this also contributes to the reluctance to consider genetic instability in many fermentation sections of companies.

BIBLIOGRAPHY

1. J. Cullum, J. Altenbuchner, F. Flett, and W. Piendl, *Gen. Eng. Rev.* **4**, 59–78 (1986).
2. B. Gravius, T. Bezmalinović, D. Hranueli, and Cullum J, *Appl. Environ. Microbiol.* **59**, 2220–2228 (1993).
3. T. Miyoshi, M. Iseki, T. Konomi, and H. Imanaka, *J. Antibiot.* **33**, 480–487 (1980).
4. P. Kaiser, F. Flett, and J. Cullum, *Bio/Technology* **10**, 570–573 (1992).
5. M. Redenbach, F. Flett, W. Piendl, I. Glocker, U. Rauland, O. Wafzig, R. Kliem, P. Leblond, and J. Cullum, *Mol. Gen. Genet.* **241**, 255–262 (1993).
6. H. Rausch, M. Vešligaj, D. Počta, G. Biuković, J. Pigac, J. Cullum, H. Schmieger, and D. Hranueli, *J. Gen. Microbiol.* **139**, 2517–2524 (1993).
7. D. Noack, *J. Basic Microbiol.* **28**, 101–106 (1988).
8. S.D. Colloms, P. Sykora, G. Szatmari, and D.J. Sherratt, *J. Bacteriol.* **172**, 6973–6980 (1990).
9. G. del Solar, R. Giralso, M.J. Ruiz-Echevarría, M. Espinosa, and R. Díaz-Orejas, *Microbiol. Mol. Biol. Rev.* **62**, 434–464 (1998).
10. N.M. Crawford and H.N. Arst, Jr., *Ann. Rev. Genet.* **27**, 115–146 (1993).
11. A.J.F. Griffiths, *Ann. Rev. Genet.* **26**, 351–372 (1992).
12. E.U. Selker, *Ann. Rev. Genet.* **24**, 579–613 (1990).
13. R.B. Wickner, *Ann. Rev. Genet.* **30**, 109–139 (1996).
14. W.R. Arathoon and J.R. Birch, *Science* **232**, 1390–1395 (1986).
15. F.M. Wurm. in H. Hauser and R. Wagner eds., *Mammalian Cell Biotechnology in Protein Production*, Walter de Gruyter, Berlin, 1997, pp. 87–118.

GLUCOSIDASES

RYUICHI MATSUMO
Kyoto University
Kyoto, Japan

KEY WORDS

Alkyl glucoside
Condensation
Reverse hydrolysis
Surfactant
Transfer reaction
Two-phase system

OUTLINE

INTRODUCTION

β-Glucosidase has the potential for practical use in the saccharification of cellulosic materials and the synthesis of aryl or alkyl glucosides in a nonaqueous reaction medium. The latter topic, especially synthesis in an aqueous–organic two-phase system, has been described in detail.

β-Glucosidase (β-D-glucoside glucohydrolase, EC 3.2.1.21) catalyzes the cleavage of the β-glucosidic ether bond of aryl and alkyl glucosides. Aryl glucosides are, in general, better substrates than alkyl glucoside. β-Glucosidase from some origins can also hydrolyze β-galactosides and β-fucosides. Its hydrolytic potential in practical use would be as a component of cellulases for saccharification of cellulose.

The enzyme also has a synthetic potential for aryl or alkyl glucosides through reverse hydrolysis (condensation) or transfer reactions. Research concerning β-glucosidase-catalyzed synthesis of glucosides in a nonaqueous phase has recently been active, although its industrial application has not been reported yet.

HYDROLYTIC REACTIONS

Saccharification of Cellulose

β-Glucosidase is one of the key enzymes for saccharification of cellulose. The enzymatic saccharification proceeds through synergistic actions of *endo*-glycanase, *exo*-glycanase, and β-glucosides, and β-glucosidase catalyzes the last step of the saccarification, hydrolysis of cellobiose to glucose. β-Glucosidase favors shorter cellooligosaccharides, especially cellobiose.

β-Glucosidase activity of cellulase from *Trichoderma reesei*, which is one of the most promising microorganisms for cellulose saccharification on an industrial scale, is still weak, so addition of β-glucosidase from other origins for the cellulose saccharification process has been attempted, resulting in a significant increase in cellulose conversion (1). Reinforcement of β-glucosidase activity of *T. reesei* cellulase by gene technology has also been investigated (2).

Hydrolysis of Alkyl Glucoside

Alkyl glucosides, which are substrates for β-glucosidase, have an amphiphilic structure with hydrophilic and hydrophobic moieties and are a group of nonionic surfactants. The phase diagram for a ternary mixture of octyl β-D-glucoside–buffer(water)–octanol was prepared, and it included the regions of micellar solution, microemulsion, liquid crystalline phase, and an undefined structure (3). In the aqueous monophasic system, octyl β-D-glucoside exists as monomers at concentrations below its critical micellar concentration (CMC), and it forms aggregates at concentrations above the CMC, while the monomer concentration remains at the CMC. Figure 1 shows the initial reaction rates for β-glucosidase-catalyzed hydrolysis of octyl β-D-glucoside at various substrate concentrations (4). For the lower substrate concentrations (< ca. 20 mmol/L), the relation between the reaction rate and substrate concentration followed a classical Michaelis–Menten scheme, but the reaction rate remained constant at the higher substrate concentrations. As shown in the insert of Fig. 1, the relation could be divided into two regions, and two lines intersected at 19 mmol/L, which corresponded to the CMC of the system. This indicates that octyl β-D-glucoside monomer is the actual substrate for the enzyme.

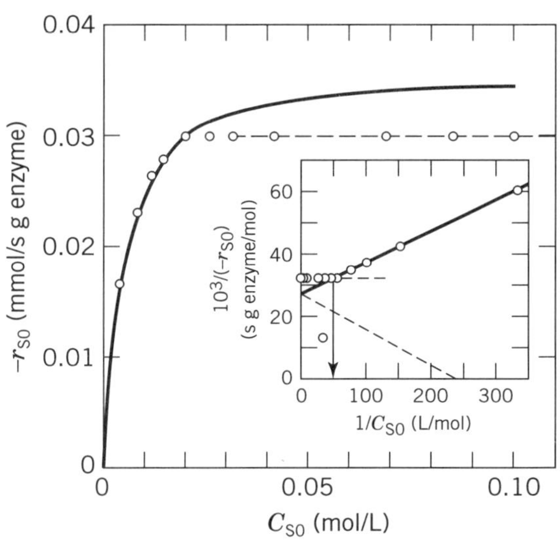

Figure 1. β-Glucosidase-catalyzed hydrolysis of octyl β-D-glucoside. *Source*: Adapted from Ref. 3.

SYNTHESIS OF ALKYL GLUCOSIDES

β-Glucosidase-catalyzed synthesis of aryl or alkyl glucosides in a nonaqueous system has been an active area of research, as have lipase-catalyzed acylation and proteinase-catalyzed peptide synthesis.

Chemical synthesis of alkyl glucosides is well developed and used at the commercial level. However, the processes are complicated by the presence of identically reactive hydroxyl groups and by the toxicity of some of the catalysts used. β-Glucosidase is effective for synthesis through transglycosylation (5,6) or condensation (7–14). Although the reaction rate of condensation is slower and hence a longer time is required to reach equilibrium compared with transglycosylation, condensation has some advantages, such as cheaper substrates and easier purification of the product. Alkyl glucoside synthesis through transglycosylation is kinetically controlled, but the yield of alkyl glucoside through transglycosylation was almost equal to that through condensation (5). On the basis of substrate costs, condensation would be a choice for alkyl glucoside synthesis.

Equation 1 shows the alkyl glucoside synthesis through β-glucosidase-catalyzed condensation between glucose and alcohol.

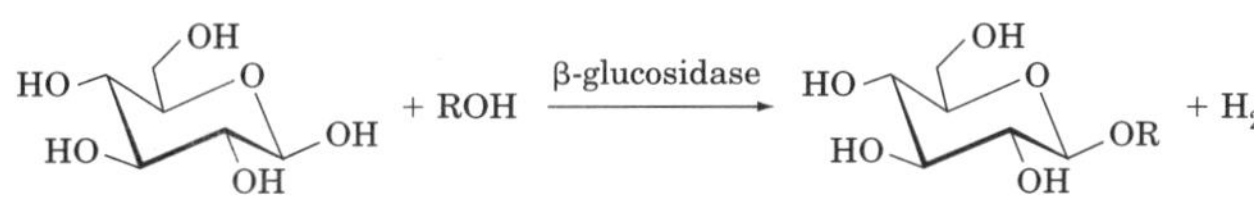

(1)

where ROH indicates aliphatic alcohol. Thermodynamically, hydrolysis predominates over synthesis in an aqueous reaction medium. One approach to shift the equilibrium is to add a water-miscible solvent, which decreases the thermodynamic water activity and thereby favors synthesis. An alternative approach is to add a water-immiscible solvent. In this case, a hydrophobic product is extracted to the organic phase so that more product is synthesized.

One-Phase System

Aryl and alkyl glucosides were synthesized in a mixture of acetonitrile and water through condensation using β-glucosidase from almonds (7). The synthetic reaction was entirely regio- and stereoselective. The enzyme had a pronounced affinity for the alcohols containing a phenyl group. The ratio of water to acetonitrile was a crucial factor for both the product yield and the enzyme stability, a 9:1 (v/v) acetonitrile–water mixture being found to be the best. It was also reported that the enzyme could synthesize alkyl β-D-galactosides.

Two-Phase System

In the synthesis of alkyl glucosides in an aqueous–organic two-phase system, one of the substrates, alcohol, itself is used as an organic phase. The aqueous phase includes glucose and free (9,10,12,13) or immobilized (5,8,11) β-glucosidase. This synthesis has mostly been used for alkyl glucoside synthesis. The water activity a_w or the water content in the system plays an important role in both the reaction rate and the equilibrium yield in alkyl glucoside synthesis (5,8–11). An a_w of at least 0.67 was needed for synthesis of octyl β-D-glucoside. The reaction rate increased, but the equilibrium yield decreased with increasing a_w. A high yield of octyl glucoside in a short time was realized by controlling the a_w during the condensation reaction (11). The optimum water content was higher for the synthesis of glucoside with a shorter alkyl residue (8,10). Physicochemical understanding for these results have not been established yet.

Figure 2 shows the transient changes in the yield of alkyl β-D-glucosides, Y_P, through β-glucosidase-catalyzed condensation between glucose and alcohols at 60 °C in the two-phase system (13). The volume ratio of organic to aqueous phase was fixed at 16.7. The length of the alkyl chain significantly affected both the equilibrium yield and the reaction rate. The longer the alkyl chain, the lower were both the yield and the rate. The figure also shows that the enzyme could catalyze the condensation between glucose and secondary alcohols, although the yield of 2-alkyl glucoside was lower than that of 1-alkyl glucoside.

In the two-phase system, the product partitioned to the organic phase. Because it is known (15) that β-glucosidase catalyzes the condensation of glucose molecules to form oligosaccharides, mainly β-glucobioses, at higher glucose concentrations, the undesirable condensation would also occur in the system as well as the intended condensation between glucose and alcohol in the system. Figure 3 summarizes phenomena assumed to occur in the two-phase system for alkyl glucoside synthesis. Under the following assumptions, *(1)* the condensations occur only in the aqueous phase containing the enzyme; *(2)* the product P, alkyl glucoside, can partition to the organic phase with a partition coefficient P_{app}; *(3)* the glucose, G, and produced β-glucobioses, D, locate only in the aqueous phase due to their high hydrophilicity; *(4)* the aqueous phase is always saturated with an alcohol, A, that is, the concentration of the alcohol in the aqueous phase, $C_{A,aq}$, is given by its solubility in water, and *(5)* the water concentration, C_W, is kept constant during the reaction, equations for estimation of the equilibrium yields of alkyl glucoside, Y_P, and β-glucobioses, Y_D, through β-glucosidase-catalyzed condensation in the two-phase system have been derived (13):

$$Y_P = \frac{X}{4K_DC_{G0}/C_W}\left\{\left[(1 + X)^2 + \frac{8K_DC_{G0}}{C_W}\right]^{1/2} - (1 + X)\right\} \quad (2)$$

$$Y_D = 1 - \frac{C_W(1 + X)}{4K_DC_{G0}} \cdot \left\{\left[(1 + X)^2 + \frac{8K_DC_{G0}}{C_W}\right]^{1/2} - (1 + X)\right\} \quad (3)$$

where

$$X = (1 + \alpha P_{app})K_PC_{P,aq}/C_W \quad (4)$$

C_{G0} is the initial glucose concentration in the aqueous phase, $C_{P,aq}$ is the alkyl glucoside concentration in the aqueous phase, α is the volume ratio of aqueous to organic phase, and K_P and K_D are the equilibrium constants for the alkyl glucoside and β-glucobiose formation, respectively. The water concentration can be calculated by $C_W = -6.43C_{G0} + 54.58$, where both C_W and C_{G0} are in units of moles per liter.

The equilibrium constant for formation of β-glucobioses, K_D, was determined to be 1.17. The equilibrium constant for formation of 1-alkyl β-D-glucosides with various alkyl chains, K_P, was evaluated for the experimentally observed

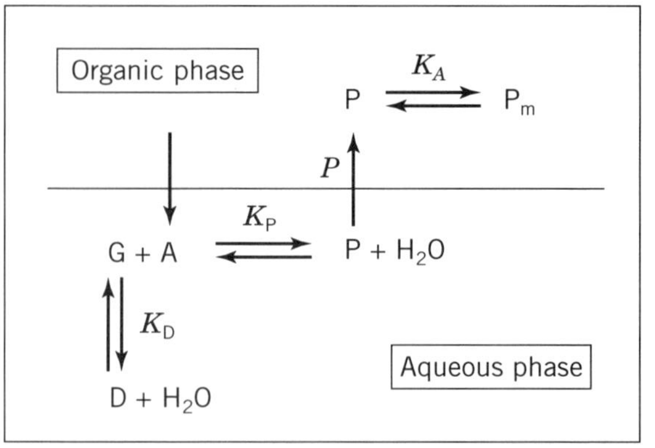

Figure 3. Reaction scheme supposed to occur in the synthesis of alkyl β-D-glucoside through β-glucosidase-catalyzed condensation in an aqueous–organic two-phase system. A, D, G, P, and Pm denote alcohol, β-glucobioses, glucose, alkyl glucoside, and the m-meric aggregate of alkyl glucoside, respectively. K_A is the association constant for the m-meric aggregate, K_D is the equilibrium constant for the formation of β-glucobioses, K_P is the equilibrium constant for the alkyl glucoside formation, and P is the partition coefficient of the monomeric alkyl glucoside.

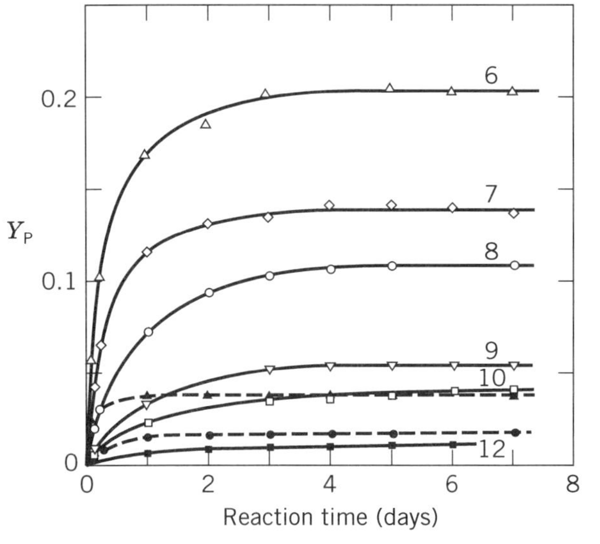

Figure 2. Kinetics for the synthesis of alkyl β-D-glucosides through β-glucosidase-catalyzed condensation of glucose and various alcohols in a two-phase system. The numbers represent the alkyl chain lengths of glucosides produced. The open symbols represent 1-alkyl β-D-glucosides, and the closed ones 2-alkyl β-D-glucosides.

Y_P values according to equations 2 and 4 (Fig. 4). The K_P value was independent of alkyl chain length and was 1.90. It was also shown that alkyl glucoside with any alkyl chain formed a dimeric aggregate in the alcohol phase with an association constant K_A of 4.2 L/mol, which was independent of alkyl chain length, and that the partition coefficient of the monomeric alkyl glucoside, P, increased exponentially with the alkyl chain length, n ($\ln P = 1.08n - 5.17$). The apparent partition coefficient, P_{app}, can be calculated by

$$P_{app} = P(1 + 2K_A P C_{P,aq}) \qquad (5)$$

Using these equations and parameters, we can calculate the equilibrium yields, Y_P and Y_D, under any conditions. The volume ratio of aqueous to organic phase, α, is an important factor affecting the yields. At higher α values, the equilibrium yields of alkyl glucosides are higher, while the equilibrium yields of β-glucobioses become lower. These results indicate that higher α values are preferable for alkyl glucoside synthesis. However, the concentrations of alkyl glucosides in their corresponding alcohol phase, $C_{P,org}$, decrease with an increase in α. The initial glucose concentration, C_{G0}, is another important factor. The yield, Y_D, of β-glucobioses, undesirable products in alkyl glucoside synthesis, largely increases at higher C_{G0}. Although the yield of alkyl glucoside, Y_P, also increases with increasing C_{G0}, the extent is not as great as that for the yield of β-glucobioses.

A Heterogeneous System

Alkyl β-D-glucoside could be synthesized in a heterogeneous system where solid glucose and β-glucosidase are added in a buffer-saturated alcohol (14). In such a system, a very viscous phase appeared, and it could be easily separated from the alcohol phase after the condensation reaction. Because the enzyme in the viscous phase was expected to be still active, the phase was repeatedly used for the synthesis of alkyl glucoside. Figure 5 shows 1-octyl β-D-glucoside synthesis by repeated use of the viscous phase. The enzyme was added only in the first run, and no further addition was made in subsequent batches. In each batch, glucose was added in an amount equal to that of the product in the previous batch. In this system, the overall yield of alkyl glucoside could be significantly improved without immobilization of the enzyme. The reason for the increase in the overall yield by the consecutive synthesis can be explained by

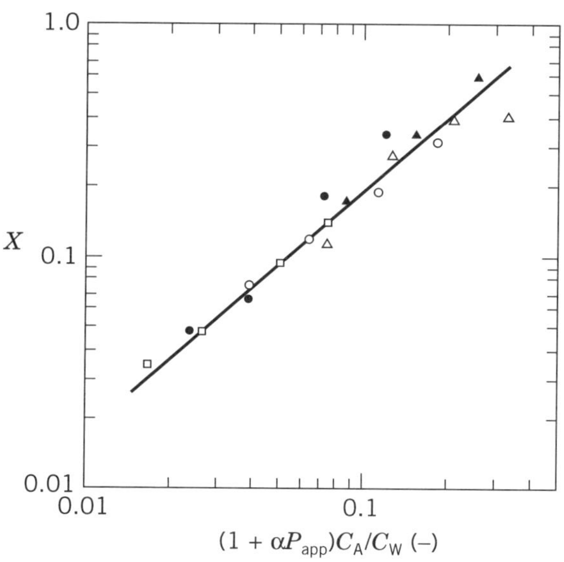

Figure 4. Estimation of the equilibrium constant, K_P, for the formation of 1-alkyl β-D-glucosides at 60 °C. X in the ordinate is defined by equation 4. The X values that satisfied the experimental equilibrium yields of alkyl glucosides in equation 2 were determined numerically. The K_P value was determined according to equation 4. The keys are the same as those in Figure 2.

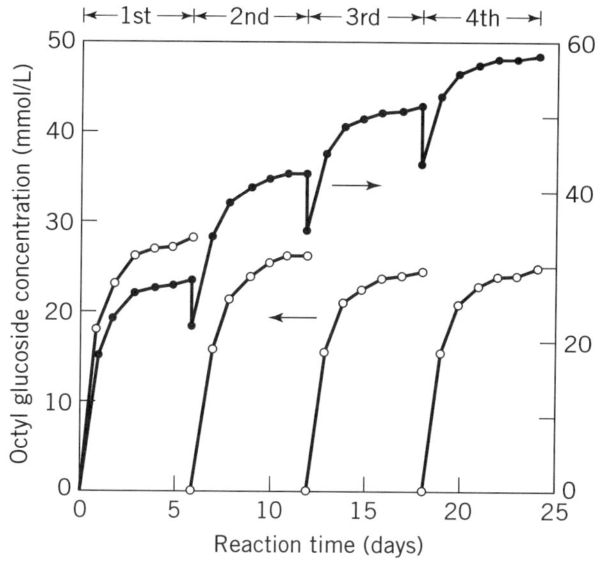

Figure 5. Repeated use of β-glucosidase for the synthesis of 1-octyl β-D-glucoside in the heterogeneous system. The enzyme was added only in the first batch, and no further addition was made in subsequent batches.

$$Y_{P,n} = \frac{\sum_{i=1}^{n} (VC_{P,i})}{VC_{G0} + \sum_{i=1}^{n} (VC_{P,i})} = \frac{\sum_{i=1}^{n} C_{P,i}}{C_{G0} + \sum_{i=1}^{n} C_{P,i}} \qquad (6)$$

where $Y_{P,n}$ is the overall yield of alkyl glucoside at the end of nth batch, $C_{P,i}$ is the product concentration at the end of ith batch, C_{G0} is the initial glucose concentration in the first batch, based on the total volume of reaction medium, V, which is nearly equal to the volume of the alcohol phase, and n is the number of cycles. Equation 6 indicates that the overall yield, $Y_{P,n}$, asymptotically approaches unity when n increases.

BIBLIOGRAPHY

1. B.C. Stockton, D.J. Mitchell, K. Grohmann, and M.E. Himmel, *Biotechnol. Lett.* **13**, 57–62 (1991).

2. C. Barnett, R.M. Berka, and T. Fowler, *Bio/Technology* **9**, 562–567 (1991).
3. J. Chopineau, D. Thomas, and M.-D. Legoy, *Eur. J. Biochem.* **183**, 459–463 (1989).
4. J. Chopineau, M. Ollivon, and M.-D. Legoy, *Ann. N.Y. Acad. Sci.* **672**, 566–571 (1995).
5. E.M. Vulfson, R. Patel, J.E. Beecher, A.T. Andrews, and B.A. Law, *Enzyme Microb. Technol.* **12**, 950–954 (1990).
6. N. Mitsuo, H. Takeichi, and T. Satoh, *Chem. Pharm. Bull.* **32**, 1183–1187 (1984).
7. G. Vic, J. Biton, D.L. Beller, J.-M. Michel, and D. Thomas. *Biotechnol. Bioeng.* **46**, 109–116 (1995).
8. E.N. Vulfson, R. Patel, and B.A. Law, *Biotechnol. Lett.* **12**, 397–402 (1990).
9. V. Laroute and R.M. Willemot, *Biotechnol. Lett.* **14**, 167–174 (1992).
10. Z. Chahid, D. Montet, M. Pina, and J. Graille, *Biotechnol. Lett.* **14**, 281–284 (1992).
11. G. Ljunger, P. Adlercreutz, and B. Mattiasson: *Enzyme Microb. Technol.* **16**, 751–755 (1994).
12. Z. Chahid, D. Montet, M. Pina, F. Bonnot, and J. Graille, *Biotechnol. Lett.* **16**, 795–800 (1994).
13. C. Panintrarux, S. Adachi, Y. Araki, Y. Kimura, and R. Matsuno, *Enzyme Microb. Technol.* **17**, 32–40 (1995).
14. C. Panintrarux, S. Adachi, and R. Matsuno, *J. Mol. Cat. B: Enzymatic* **1**, 165–172 (1996).
15. K. Ajisaka, H. Nishida, and H. Fujimoto, *Biotechnol. Lett.* **9**, 243–248 (1987).

GLUTAMIC ACID.

See AMINO ACIDS, GLUTAMATE; AMINO ACIDS, PRODUCTION PROCESSES; GLUTAMIC ACID PRODUCING MICROORGANISMS; PRODUCTION OF L-GLUTAMIC ACID.

GLUTAMIC ACID PRODUCING MICROORGANISMS

SHUKUO KINOSHITA
Kyowa Hakko Kogyo Co., Ltd.
Tokyo, Japan

KEY WORDS

Amino acids
Cell membrane
Corynebacterium
Corynebacterium glutamicum
Glutamic acid
Morphology
Taxonomy

OUTLINE

Introduction
Changes in Ideas Regarding Coryneform Bacteria and the Genus *Corynebacterium* in *Bergey's Manual*
Taxonomic Position of Glutamic Acid Producing Bacteria
Morphological and Physiological Changes of Glutamic Acid Producing Bacteria
Bibliography

INTRODUCTION

Amino acid fermentation is now the key industry supplying natural amino acids on an industrial scale. The discovery of a glutamic acid producing bacterium by Kinoshita et al. in 1956 was the start of the subsequent development of amino acid production by regulation of biosynthetic metabolism. At that time, a two-step L-glutamic acid production process through reductive amination of α-ketoglutaric acid had been studied. Instead, screening of glutamic acid producing microorganisms was carried out using a glutamic acid requiring strain of *Leuconostoc mesenteroides* to directly produce glutamic acid by fermentation from carbohydrate, and glutamic acid producing bacteria were isolated from nature and some habitats with high frequency. The glutamic acid producing bacterium first discovered was reported as *Micrococcus glutamicus* in 1958 (1), and other glutamic acid producing coryneform species, all patent strains, were subsequently isolated by many scientists. They were subjected to experiments, and industrial production began with various new species of the genera *Brevibacterium, Corynebacterium, Microbacterium*, and *Arthrobacter* (Table 1).

Table 1. Glutamic Acid Producing Bacteria

Strain	Isolator
Corynebacterium glutamicum ATCC12032	Kyowa Hakko
Brevibacterium divaricatum ATCC14020	Wei-chuan Foods
Corynebacterium lilium ATCC15990	International Mineral & Chemistry
Corynebacterium callunae ATCC15991	International Mineral & Chemistry
Brevibacterium saccharolyticum ATCC14066	Ajinomoto
Brevibacterium fravum ATCC14067	Ajinomoto
Brevibacterium lactofermentum ATCC13869	Ajinomoto
Brevibacterium roseum ATCC13825	Ajinomoto
Corynebacterium herculis ATCC13868	Herculis Powder
Brevibacterium thiogenitalis ATCC19240	Takeda
Microbacterium ammoniaphilum ATCC15354	Asahi Kasei

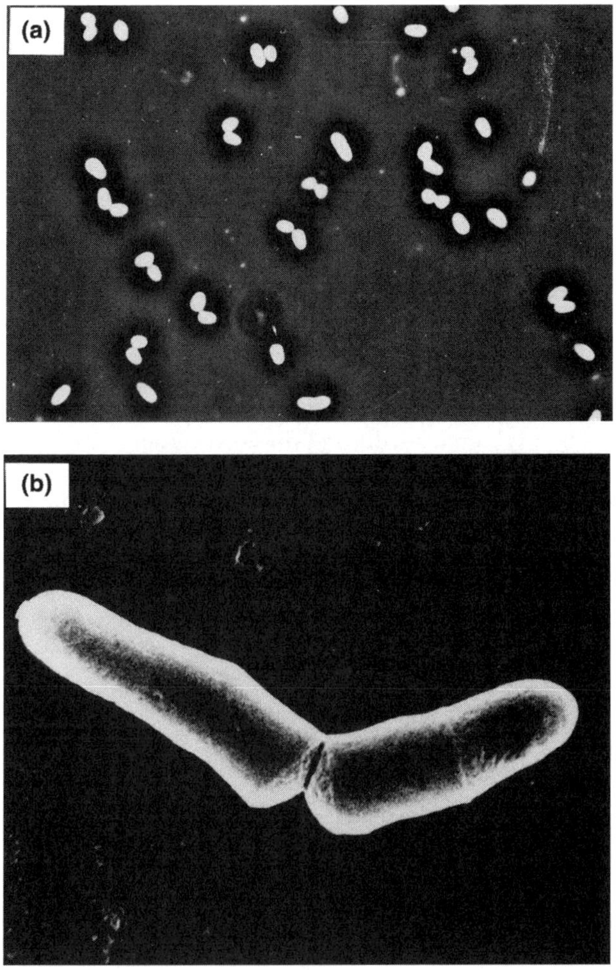

Figure 1. Morphologies of *Corynebacterium glutamicum* ATCC 13032^T: (**a**) Phase-contrast micrograph; (**b**) scanning electron micrograph.

CHANGES IN IDEAS REGARDING CORYNEFORM BACTERIA AND THE GENUS *CORYNEBACTERIUM* IN *BERGEY'S MANUAL*

The genus *Corynebacterium* was proposed by Lehmann and Neumann in 1896 (2) to accommodate the diphtheria bacillus. Thereafter, various isolated strains not only of pathogenic bacteria for animal and plants but also of saprophytic bacteria were identified as members of the genus *Corynebacterium*. In this way, many bacterial species with diverse properties were simultaneously classified into this genus. Under the circumstances, Jehsen (3) introduced a generic term of *coryneform* to clarify this genus definition. The word *coryneform* was coined by Oerskov in 1923 (4) to explain the morphology of ray fungi. Jensen used the word *coryneform* to describe one of the more characteristic morphological features, V formation (Fig. 1), particularly on cell division of aerobic and anaerobic gram-positive, non-acid-fast, pleomorphic, nonbranching, rod-shaped, nonsporogenous bacilli. However, it was difficult to distinguish between coryneform bacteria and strains with more or less

Table 2. Taxonomic Position of the Genus *Corynebacterium* and Related Genera in *Bergey's Manual*

Bergey's Manual Edition (year)	Position of the genus *Corynebacterium* and related taxa
Determinative Bacteriology Fifth (1939)	Order II. Actinomycetales
	Family I. Mycobacteriaceae
	Genus I. *Corynebacterium*
	Genus II. *Mycobacterium*
	Family II. Actinomycetaceae
	Genus I. *Leptotrichia*
	Genus II. *Erysipelothrix*
	Genus III. *Proactinomyces*
	Genus IV. *Actinomyces*
Determinative Bacteriology Sixth (1948)	Order I. Eubacteriales
	Family VIII. Corynebacteriaceae
	Genus I. *Corynebacterium*
	Genus II. *Listeria*
	Genus III. *Erysipelothrix*
Determinative Bacteriology Seventh (1957)	Order IV. Eubacteriales
	Family XII. Corynebacteriaceae
	Genus I. *Corynebacterium*
	Genus II. *Listeria*
	Genus III. *Erysipelothrix*
	Genus IV. *Microbacterium*
	Genus V. *Cellulomonas*
	Genus VI. *Arthrobacter*
Determinative Bacteriology Eighth (1974)	Part 17. Actinomycetes and related organisms
	Coryneform Group of Bacteria
	Genus I. *Corynebacterium*
	a. Human and animal parasites and pathogens
	b. Plant pathogenic corynebacteria
	c. Nonpathogenic corynebacteria
	Genus II. *Arthrobacter*
	Genera incertae sedis
	Brevibacterium
	Microbacterium
	Genus III. *Cellulomonas*
	Genus IV. *Kurthia*
Systematic Bacteriology Vol. 2 (1986)	Section 15. Irregular, nonsporing, gram-positive rods
	Genus *Corynebacterium*
	Plant pathogenic species of Corynebacterium
	Genus *Gardnerella*
	Genus *Arcanobacterium*
	Genus *Arthrobacter*
	Genus *Brevibacterium*
	Genus *Curtobacterium*
	Genus *Caseobacter*
	Genus *Microbacterium*
	Genus *Aureobacterium*
	Genus *Cellulomonas*
	Genus *Agromyces*
	Genus *Arachnia*
	Genus *Rothia*
	Genus *Propionibacterium*
	Genus *Eubacterium*
	Genus *Acetobacterium*
	Genus *Lachnospira*
	Genus *Butyrivibrio*
	Genus *Thermoanaerobacter*
	Genus *Actinomyces*
	Genus *Bifidobacterium*

Table 2. Taxonomic Position of the Genus *Corynebacterium* and Related Genera in *Bergey's Manual* *(continued)*

Determinative Bacteriology Ninth (1994)	Group 20. Irregular, nonsporing, gram-positive rods
	Genus *Acetobacterium*
	Genus *Acetogenium*
	Genus *Actinomyces*
	Genus *Acromicrobium*
	Genus *Agromyces*
	Genus *Arachnia*
	Genus *Arcanobacterium*
	Genus *Arthrobacter*
	Genus *Aureobacterium*
	Genus *Bifidobacterium*
	Genus *Brachybacterium*
	Genus *Brevibacterium*
	Genus *Butyrivibrio*
	Genus *Caseobacter*
	Genus *Cellulomonas*
	Genus *Clavibacter*
	Genus *Coriobacterium*
	Genus *Corynebacterium*
	Genus *Curtobacterium*
	Genus *Eubacterium*
	Genus *Exiguobacterium*
	Genus *Falcivibrio*
	Genus *Gardnerella*
	Genus *Jonesia*
	Genus *Lachnospira*
	Genus *Microbacterium*
	Genus *Mobiluncus*
	Genus *Pimerobacter*
	Genus *Propionibacterium*
	Genus *Rarobacter*
	Genus *Rothia*
	Genus *Rubrobacter*
	Genus *Sphaerobacter*
	Genus *Terrabacter*
	Genus *Thermoanaerobacter*

similar morphological characters, and it was not possible to clearly classify them. In the fifth edition of *Bergey's Manual of Determinative Bacteriology* (Table 2), the genus *Corynebacterium* belonged to the family Mycobacteriaceae of the order Actinomycetales (1939). However, in the sixth edition (1948), the family Corynebacteriaceae was created under the order Eubacterial, and this family included the genera *Corynebacterium, Listeria*, and *Erysipelothrix*. Furthermore, in the seventh edition (1957), the genera *Microbacterium, Cellulomonas*, and *Arthrobacter* were included in the family Corynebacteriaceae, in addition to the previously mentioned three genera.

The generic term *coryneform group of bacteria* first appeared in Part 17, "Actinomycetes and Related Organisms," in the eighth edition (1974), and the genera *Corynebacterium, Arthrobacter, Cellulomonas*, and tentatively, the genus *Kurthia* were included in this group. In addition, the genera *Brevibacterium* and *Microbacterium* were included, though as *incertae sedis*. Most of these genera were heterogeneous in composition. The genera *Listeria* and *Erysipelothrix*, which were not coryneform bacteria, were moved to the gram-positive, asporogenous, rod-shaped bacteria group in the eighth edition and treated as genera of uncertain affiliation. The classification and identification of coryneform bacteria have long been difficult and confusing. As a result, new isolated strains were identified in any genera as new species. However, with the application of modern taxonomic methods, such as cell wall analysis, nucleic acid base composition, and DNA relatedness, the relationships of coryneform bacteria and traditionally associated taxa had been clarified (Table 3). The definition of the genus *Corynebacterium* should now be restricted to those microorganisms that contain *meso*-A_2pm, acetylated muramic acid, mycolic acid with ca. 22 to 38 carbon atoms, and arabinose and galactose in the cell wall. In addition, these species have dihydrogenated menaquinones with either eight or nine isoprene units. In this respect, the genus

Table 3. Chemotaxonomic Characteristics of Coryneform Bacteria

Genus	Peptidoglycan		Major menaquinone	Mycolic acid	Acyl type	GC content (%)
	Diamino acid	Type				
Corynebacterium	meso-A_2pm	A1γ				
	L-Lys	A3α, A4α, B1α	8, 9,	+, −	Glycolyl	52–76
	D-Orn	B2β	10, 11, 12		Acetyl	
	L-DAB	B2γ				
Brevibacterium	meso-A_2pm	A1γ	7			
	LL-A_2pm	A3γ	8, 9, 10	+, −	Glycolyl	54–73
	L-Lys	A3α, A4α, B1β	11, 12		Acetyl	
	D-Orn	B2β				
Microbacterium	meso-A_2pm	A1γ				
	L-Lys	B1α, B1β	8, 9,	+, −	Glycolyl	58–73
	D-Orn	B2β	11, 12		Acetyl	
Cellulomonas	L-Orn	A4β	9	−	Acetyl	66–73
Arthrobacter	meso-A_2pm	A1γ				
	LL-A_2pm	A3γ	8, 9,	+, −	Glycolyl	54–76
	L-Lys	A3α, A4α	12, 13, 14		Acetyl	
	D-Orn	B2β				

Table 4. Reclassification of Coryneform Bacteria and Related Taxa

Peptidoglycan						
Diamino acid	Type	Acyl type	Major menaquinone	Mycolic acid	GC content (mol%)	Genera
meso-A_2pm	A1γ	Acetyl	9(H2), 8(H2)	C22–C38	51–65	*Corynebacterium*
meso-A_2pm	A1γ	Glycolyl	8(H2)	C34–C54	63–72	*Rhodococcus*
meso-A_2pm	A1γ	Glycolyl	9(H2)	C44–C60	60–65	*Gordona*
meso-A_2pm	A1γ	Glycolyl	8(H4)	C49–C69	64–72	*Nocardia*
meso-A_2pm	A1γ	Glycolyl	8(H2)	C60–C90	66–73	*Mycobacterium*
meso-A_2pm	A2γ		9	C64–C78	67–72	*Tsukamurella*
meso-A_2pm	A1γ	Acetyl	8(H2), 7(H2)	—	63–66	*Brevibacterium*
LL-A_2pm	A3γ	Acetyl	8(H4)	—	68–73	*Pimelobacter*
LL-A_2pm	A3γ	Acetyl	8(H4)	—	66–68	*Nocardioides*
LL-A_2pm	A3γ		8(H4)	—	72	*Terrabacter*
L-Lys	A3α, A4α	Acetyl	9(H2), 9, 8	—	59–69	*Arthrobacter*
L-Lys	B1α, B1β	Glycolyl	11, 12	—	69–75	*Microbacterium*
L-Lys	A3α	Acetyl	7	—	53–56	*Exigunobacterium*
L-Lys	A3α	Acetyl	8	—	67	*Rubrobacter*
L-Lys	A3α	Acetyl	9(H4)	—	52	*Arcanobacterium*
L-Orn	A4β	Acetyl	9(H4)	—	71–76	*Cellulomonas*
D-Orn	B2β	Acetyl	9	—	68–73	*Curtobacterium*
D-Orn	B2β	Glycolyl	11, 12, 13	—	66–70	*Aureobacterium*
L-DAB	B2γ	Acetyl	9, 10	—	63–78	*Clavibacter*
L-DAB	B2γ		10	—	60–63	*Rathayibacter*

Table 5. G+C contents of DNA and DNA–DNA Relatedness between *Corynebacterium glutamicum* ATCC13032^T and Other Glutamate Producing Bacteria

Strain	G+C content (mol%)	DNA relatedness to KY9002^T (%)		
		Ref. 10	Ref. 11	Ref. 12
Corynebacterium glutamicum ATCC13032^T	53.4	100	100	100
Corynebacterium lilium ATCC15990^T	52.9		93.3	
Brevibacterium divaricatum ATCC14020^T	54.4	93.4	77.9	
"Brevibacterium flavum" ATCC14067	54.6		87.2	
"Brevibacterium lactofermentum" ATCC13665	54.4	102.1	82.7	
"Brevibacterium immariophilum" ATCC14068	54.6	76.2		105
"Brevibacterium roseum" ATCC13825	54.6	73.5		108
"Brevibacterium saccharolyticum" ATCC14066	54.3	77.3		109
"Brevibacterium ammoniagenes" ATCC13745	nd	75.6		
"Corynebacterium acetoacidophilum" ATCC13870	52.9	76.2		87
"Brevibacterium thiogenitalis" ATCC19240	54.3			106
"Corynebacterium herculis" ATCC13868	54.2			96
"Microbacterium ammoniaphilum" ATCC15354	54.4	78.4		
Arthrobacter sp. NCIB942	nd			95
Brevibacterium sp. ATCC14902	nd			93
Microbacterium sp. ATCC 15283	nd			104
Corynebacterium callunae ATCC15991^T	51.2		36.7	47
Corynebacterium ammoniagenes ATCC6871^T	54.6			17

Brevibacterium was distinguished from the genus *Corynebacterium* in that it does not contain mycolic acid and polysaccharide components such as arabinose and galactose. Furthermore, the genera *Arthrobacter* and *Microbacterium* were defined as a group of microorganisms that are additionally different in the diamino acid and major menaquinone (Table 4). As a result, the genus *Corynebacterium*, defined in chemotaxonomic characteristics, includes most animal pathogenic and parasitic species belonging to section a, and some saprophytic species (e.g., *Corynebacterium glutamicum*) belonging to section c in the eighth edition. On the other hand, all plant pathogens and many saprophytic species belonging to section b were excluded from this genus. *Bergey's Manual* was newly started in 1986 with a new title, *Bergey's Manual of Systematic Bacteriology*, in which the genus *Corynebacterium* and the related genera are included in section 15, and, consequently, the term *coryneform bacteria* disappeared from the index.

TAXONOMIC POSITION OF GLUTAMIC ACID PRODUCING BACTERIA

The strain isolated as accumulating a remarkable amount of glutamic acid was initially named *Micrococcus glutam-*

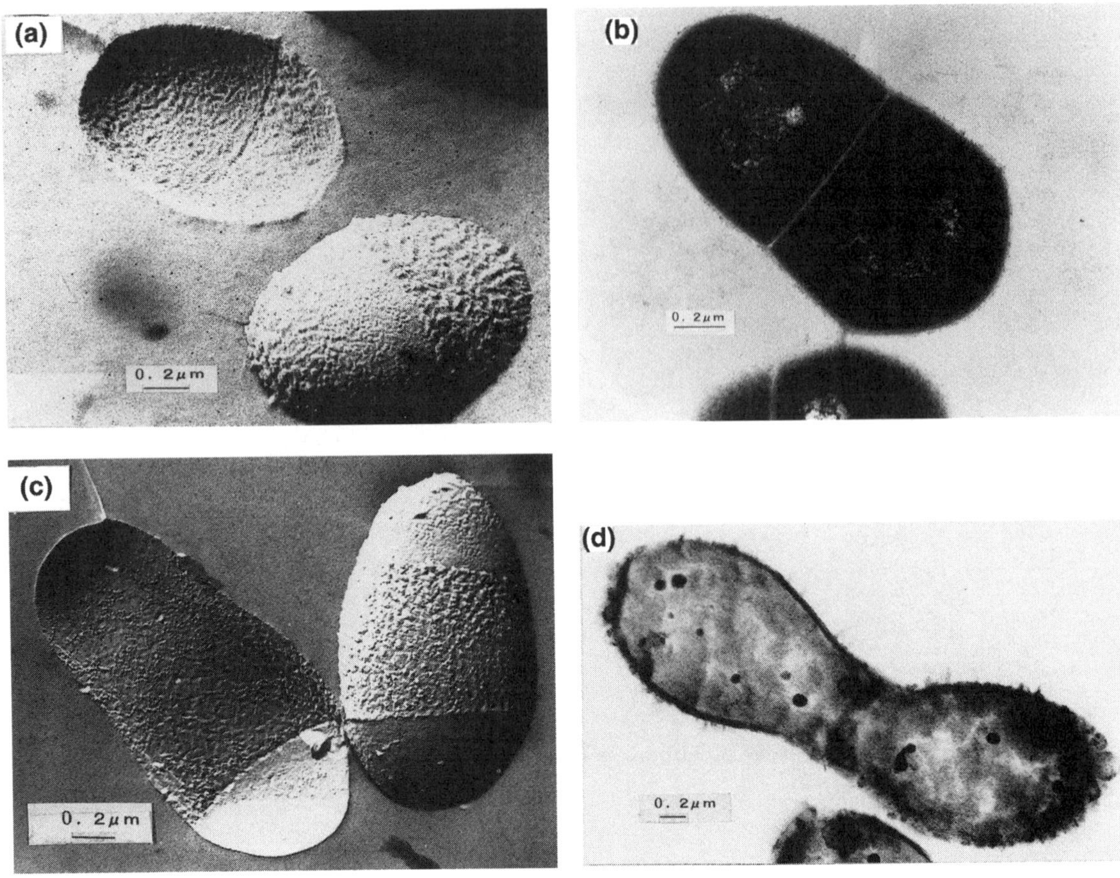

Figure 2. Electron micrographs of *Corynebacterium glutamicum* ATCC 13032^T in biotin-sufficient and biotin-deficient cultivation: (**a**) freeze-fracturing electron micrograph of the biotin-sufficient cells, (**b**) cross-sectioning electron micrograph of the biotin-sufficient cells, (**c**) freeze-fracturing electron micrograph of the biotin-deficient cells, (**d**) cross-sectioning electron micrograph of the biotin-deficient cells.

icus nov. sp. by Kinoshita et al. (1), with ATCC 13032 as a type strain. This strain was later revealed to belong to the genus *Corynebacterium* based on chemotaxonomic characteristics and reclassified to *C. glutamicum* by Abe et al. (6). On the other hand, many other glutamic acid producing species, patent strains, reported as different species than *C. glutamicum*, were found to be so-called coryneform bacteria. They were also found to have very similar chemotaxonomic properties based on studies by Keddie (7), Bousfield (8), and Yamada and Komagata (9), and were revealed to be bacteria belonging to *Corynebacterium* in the strict sense. Furthermore, as shown in Table 5, Komatsu and Kaneko (10) showed in 1980 that eight strains, *C. acetoacidophilum* ATCC 13870, *Brevibacterium lactofermentum* ATCC 13655, *B. divaricatum* 14020^T, *B. sacchalolyticum* ATCC 14066, *B. roseum* ATCC 13825, *B. immariophilum* ATCC 14068, *B. ammoniagenes* ATCC 13745, and *Microbacterium ammoniaphilum* ATCC 15354, exhibited 73 to 102% DNA binding to the type strain of *C. glutamicum*. However, as no measures had been taken to make taxonomical changes for these strains, they were subjected to studies using these names. In 1991, Liebl et al. (11) showed that five strains, *B. divaricatum* 14020^T, *B. flavum* ATCC 14067, *B. lactofermentum* ATCC 13665 and ATCC 13869, and *C. lilium* ATCC 15991^T were closely related to the type strain of *C. glutamicum* based on the high levels of DNA relatedness, and were thus transferred to the same species as *C. glutamicum*. *C. acetoacidophilum*, *C. acetoglutamicum*, and *C. acetophilum* have been reported to produce remarkable amounts of glutamic acid from acetic acid as a carbon source other than carbohydrate. While *C. acetoacidophilum* ATCC 13870 was confirmed to be genomically similar to *C. glutamicum* (Table 5), the taxonomic positions of the other two strains are unknown. Furthermore, hydrocarbon-assimilating bacteria *Corynebacterium hydrocarboclastus* ATCC 15592, which produce a remarkable amount of glutamic acid from hydrocarbons, exhibited more than 70% DNA binding to the type strain of *Rhodococcus erythropolis* (not shown).

MORPHOLOGICAL AND PHYSIOLOGICAL CHANGES OF GLUTAMIC ACID PRODUCING BACTERIA

C. glutamicum has some characteristic features, including that it undergoes V-formation (Fig. 1) by snapping division

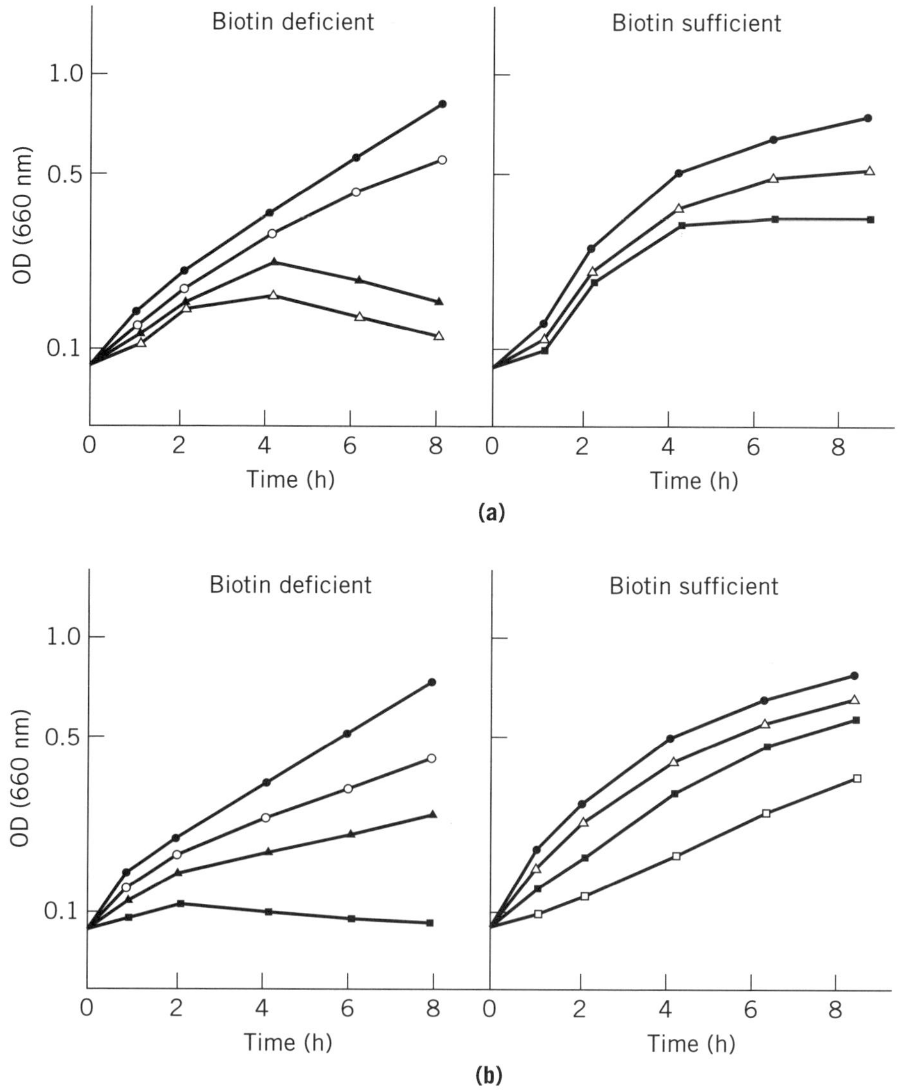

Figure 3. Effect of lysozyme and SDS on growth of *Corynebacterium glutamicum* ATCC 13032^{T} in biotin-sufficient and biotin-deficient cultivation: (**a**) Effect of lysozyme on the growth of *C. glutamicum* ATCC13032. Cells were inoculated into media supplemented with 0 μg/mL (●), 12.5 μg/mL (○), 25 μg/mL (▲), 50 μg/mL (△), and 100 μg/mL (□) of lysozyme; (**b**) effect of SDS on the growth of *C. glutamicum* ATCC13032. Cells were inoculated into media supplemented with 0 μg/mL (●), 10 μg/mL (○), 20 μg/mL (▲), 30 μg/mL (△), 40 μg/mL (■), and 70 μg/mL (□) of SDS.

upon cell division, and requires biotin for growth. However, glutamic acid production was observed only in a very narrow range of biotin content, that is, at suboptimal concentration (2.5–5.0 μg/mL) for cell growth. In 1962, Somerson et al. (13) found that the addition of penicillin was very effective, even in the presence of excessive amounts of biotin. Therefore, it is widely known that in order to accumulate a remarkable amount of glutamic acid extracellularly, low levels of biotin concentration, or penicillin treatment at high levels of biotin concentration, are required. However, the detail of their mechanisms is not known, although there have been reports attributing the accumulation of glutamic acid due to the change in the membrane permeability. At a low concentration of biotin, cells swelled and elongated. According to electron microscopic observation, outer layers changed grossly and lost ordered structures, and many blebs were observed at low concentrations of biotin (Fig. 2). It was observed that glutamic acid producing bacteria have an increased sensitivity to lyzozyme or SDS treatment when cultured at low concentrations of biotin (Fig. 3).

BIBLIOGRAPHY

1. S. Kinoshita, K. Nakayama, and S. Akita, *Bull. Agric. Chem. S. Jpn.* **22**, 176 (1958).
2. K.B. Lehmann, and R. Neumann, *Atlas und Grundriss der Bakteriologie unt Lehrbuch der speciellen bakteriologischen Diagnostik*, 1st ed., J. F. Lehmann, Munchen, 1896.
3. H.L. Jensen, *Annu. Rev. Microbiol.* **6**, 77–90 (1952).
4. J. Oerskov, *Investigations into the Morphology of the Ray Fungi*, Levin and Munksgaard, Copenhagen, 1923.
5. *Bergey's Manual of Determinative Bacteriology*, Williams & Wilkins, Baltimore, Md., 1974.
6. S. Abe, K. Takayama, and S. Kinoshita, *J. Gen. Appl. Microbiol.* **13**, 279–301 (1967).
7. R.M. Keddie and G.L. Cure, *J. Appl. Bacteriol.* **42**, 229–252 (1977).
8. I.J. Bousfield, *J. Gen. Microbiol.* **71**, 441–455 (1972).
9. K. Yamada and K. Komagata, *J. Gen. Appl. Microbiol.* **18**, 399–416 (1972).
10. Y. Komatsu and T. Kaneko, *Rep. Ferment. Res. Inst.* **55**, 1–5 (1980).

11. W. Lieble, M. Ehmann, W. Ludwig, and K.H. Schleifer, *Int. J. Syst. Bacteriol.* **41**, 255–260 (1991).
12. K. Ochiai, M. Suzuki, and K. Ando, *Nippon Nogeikagaku Kaishi* **72**, suppl. 2A2, p. 6 (1998).
13. U.S. Pat. 3,080,297 (March 5, 1963), N.L. Somerson and T. Phillips (to Merck & Co., Inc.).

See also AMINO ACIDS, PRODUCTION PROCESSES; ENERGY METABOLISM, MICROBIAL AND ANIMAL CELLS; FERMENTATION MONITORING, DESIGN AND OPTIMIZATION; METABOLITES, PRIMARY AND SECONDARY.

GLYCOSYLATION OF RECOMBINANT PROTEINS

DAVID C. JAMES
KYM N. BAKER
University of Kent
Canterbury, U.K.

KEY WORDS

Glycan
Glycosylation
Mass spectrometry
Oligosaccharide
Posttranslational modification
Recombinant protein

OUTLINE

INTRODUCTION

Recombinant proteins produced by animal cell expression systems can be subject to numerous posttranslational modifications that can result in a structurally heterogeneous product. Glycosylation is the most extensive of all posttranslational modifications, and for proteins destined for therapeutic use, this source of molecular diversity may be of particular significance. For example, distinct protein glycoforms may possess different properties such as biological activity, receptor binding, stability, immunogenicity, and clearance rate in vivo (1). The carbohydrate components of glycoproteins can also play crucial roles in protein folding, oligomer assembly, and secretion processes. Recombinant protein heterogeneity therefore represents an important issue for the biopharmaceutical industry and regulatory authorities, with regard to product characterization and definition, consistency, potency, and purity (2–4). Authorities such as the U.S. Food and Drug Administration are now demanding increasingly sophisticated carbohydrate analyses as part of the "well-characterized" product or process validation. Therefore, within the biotechnology/biopharmaceutical industries there is now considerable interest in technologies enabling recombinant glycoprotein analysis and control of glycoprotein heterogeneity.

Although there has been much progress in the field of glycobiology in recent years, with a plethora of published information on biological properties of carbohydrates (5,6), methods of glycan analysis (7–10), cellular biosynthetic pathways, and the molecular cell biology of the key enzymes involved (11–18), this article highlights key bioprocess factors that may influence the glycosylation of a recombinant protein produced by an animal expression system.

Oligosaccharide Structures Present on Recombinant Glycoproteins

Although there is considerable diversity in the types of protein–carbohydrate conjugates that occur naturally, recombinant glycoproteins produced by eukaryotic expression systems are mainly associated with N-linked glycans (N-glycosidic bond to the amide group of an asparagine residue within the consensus sequence Asn-X-Ser/Thr) or O-linked glycans (O-glycosidic bond to the hydroxyl side group of either a serine or threonine residue). Carbohydrates are also components of the glycophosphatidylinositol (GPI) anchor used to secure some proteins to cell membranes.

The molecular diversity of recombinant glycoproteins arises from two major sources: variable glycosylation site occupancy, also called macroheterogeneity (i.e., the presence or absence of glycans), and microheterogeneity (variation in carbohydrate structures at individual glycosylation sites). In the case of the former, the presence of sequence motifs or amino acids necessary for glycosylation within the polypeptide sequence does not guarantee their glycosylation. For example, at the cellular level N-glycosylation site occupancy at Asn-X-Ser/Thr sequons may be related to several factors including oligosaccharyldolichol substrate availability, oligosaccharyl transferase activity, rate of protein synthesis, competition with protein folding, and other cotranslational events (19,20). Protein structural or polypeptide sequence factors may also be significant; for example, the target glycosylation sequon Asn-X-Thr has been found to be 40 times more susceptible to glycosylation by oligosaccharyltransferase than the Asn-X-Ser sequence (21).

Recombinant glycoproteins should most appropriately be considered as populations of individual "glycoforms," as illustrated for recombinant IFN-γ produced by Chinese hamster ovary (CHO) cells in Figure 1. In general terms, the relative abundance and distribution of protein glycoforms within a glycoprotein population are a function of the following:

1. The host cell and expression system employed
2. The interaction of the host cell and recombinant glycoprotein product with the bioprocess environment

N-Linked Glycans

All mature N-linked glycans contain a common trimannosyl core sugar structure, a pentasaccharide consisting of three mannose (Man) residues and two *N*-acetylglucosamine (GlcNAc) residues (Manα1-6(Manα1-3)Manβ1-4GlcNAcβ1-4GlcNAc-(Asn)). This can form part of simple oligomannose structures or be extensively modified by the addition of other GlcNAc, galactose (Gal), fucose (Fuc), and sialic acid (*N*-acetyl- or *N*-glycolylneuraminic acids; NeuAc or NeuGc) residues. N-linked chains can be classed into three main subgroups: high-mannose type, complex type, and hybrid type (Fig. 2). High-mannose-type glycans contain α-linked mannose chains bound to both arms of the trimannosyl core. Complex-type chains contain no extra mannose residues and have α-linked GlcNAc residues bound to the trimannosyl core, with Gal and NeuAc residues added subsequently. Commonly, there may be α1-6- or α1-3-linked Fuc residues added to the core (Asn-linked) GlcNAc. Hybrid-type chains contain both α-linked Man residues bound to the α1-6-linked Man arm of the trimannosyl core, while complex-type chains are bound to the α1-3-linked Man arm. However, numerous possible variations are commonly found. Following are the basic sources of *N*-glycan heterogeneity:

1. The degree of branching of complex type chains (number of antennae). For example, the complex chain illustrated in Figure 2 contains two branching chains from the trimannosyl core and is thus termed a biantennary-complex-type chain. Typically, complex N-linked glycans may be binantennary, triantennary, or tetraantennary.

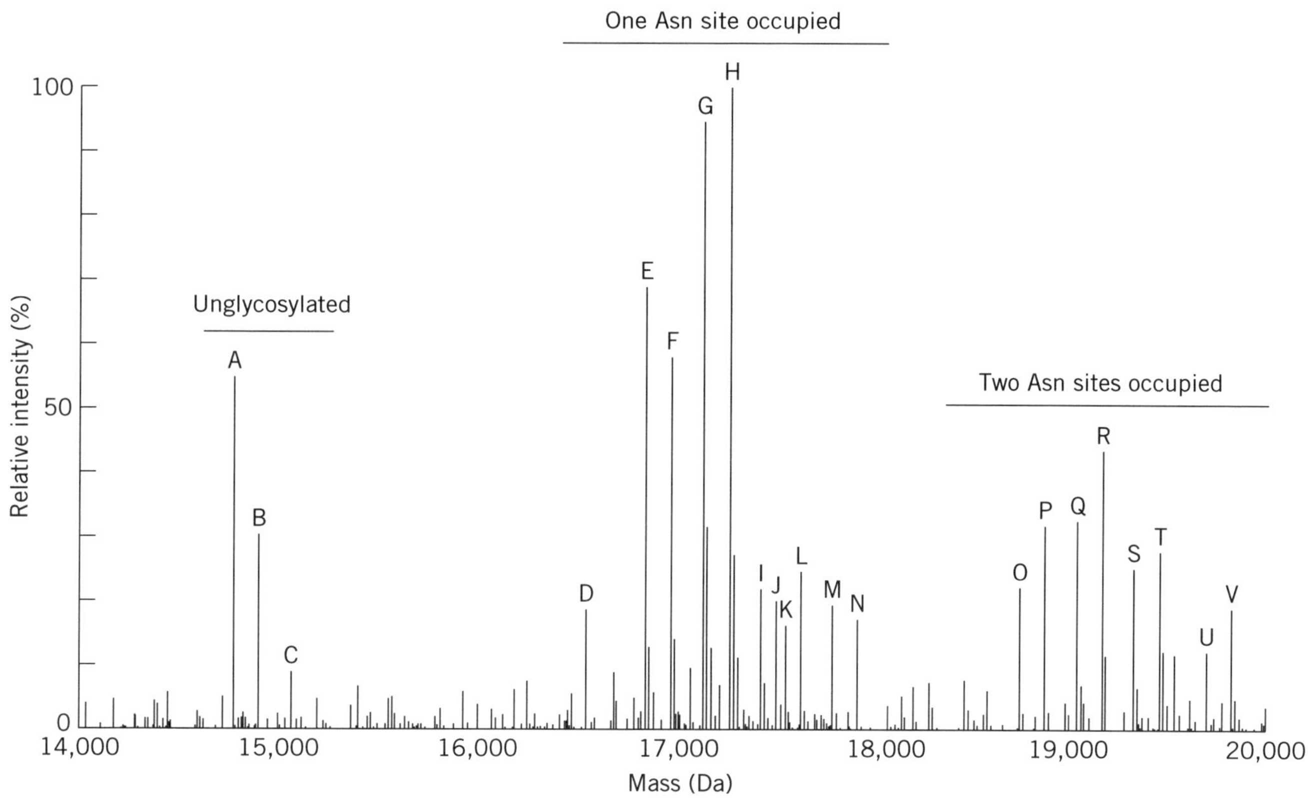

Figure 1. The heterogeneity of recombinant human IFN-γ produced by Chinese hamster ovary cells. Each component of the deconvoluted electrospray mass spectrum represents a different IFN-γ protein varying in number of glycosylation sites occupied, structure of *N*-glycans at individual sites, and C-terminal proteolytic clipping. *Source*: From Ref. 22.

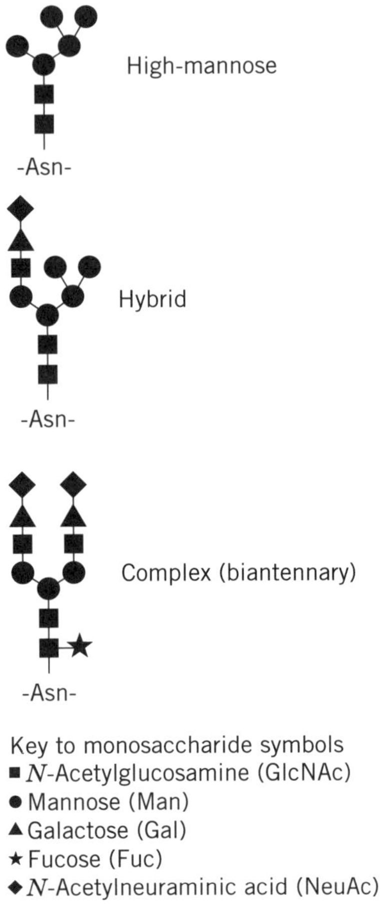

Figure 2. Common *N*-glycan structures found on glycoproteins.

2. The presence or absence of terminal sialic acid residues linked to antennae.
3. Variation in the type of sialic acid; N-acetyl and N-glycolyl variants are the most common.
4. The presence or absence of a 'bisecting' GlcNAc attached to the trimannosyl core β-linked mannose (which prevents branching)
5. The presence or absence of terminal poly-*N*-acetyllactosamine repeats (Galβ1-4GlcNAcβ1-3-).
6. Unusually, sulfation at GlcNAc residues or phosphorylation at Man residues.

O-Linked Glycans

O-linked glycosylation is extensive in structural proteins such as proteoglycans. Although the majority of O-linked glycans are bound to either a serine or threonine residue core, small glycan structures can also be O-linked to the side chain of hydroxylysine or hydroxyproline. *O*-Glycans may occur as a single *N*-acetylgalactosamine (GalNAc) residue, disaccharides NeuAcα2-6GalNAc/Galβ1-3GalNAc, or larger glycans of up to 10 residues attached to the core GalNAc containing Gal, Fuc, and GlcNAc residues (Fig. 3).

Glycan Biosynthesis

The synthesis and elongation of glycan structures is an ordered, rather than random, process that begins in the rough endoplasmic reticulum and is completed in the *trans*-Golgi network. In the case of *N*-glycans, efficient transport of recombinant glycoprotein through the constitutive secretory pathway of eukaryotic cell expression systems is a prerequisite for high-level production of recombinant protein and modification of N-linked oligosaccharides. In the case of the latter, the ordered cellular localization and activity of glycosidase and glycosyltransferase enzymes define the sequence and structure of *N*-glycans associated with the secreted protein (11,14–16,23–25). This dynamic process is mediated by the trafficking of secretory vesicles between distinct subcellular compartments and by molecular interactions within those compartments (26).

N-glycosylation may itself be necessary for efficient secretion of some proteins (27,28). Recent studies have shown that secreted or membrane-targeted proteins become concentrated in an intermediate compartment (IC) between the endoplasmic reticulum (ER) and the *cis*-Golgi (29,31). Although the nature of this compartment remains obscure at present and may not be solely a vesicular transport system but a microtubule-directed system (32), distinct molecular markers of the IC, such as the membrane-bound mannose-binding lectin ERGIC53 (28,33,34) or homologs such as p58 (35), which cycles between the ER, IC, and *cis*-Golgi (36), may function to specifically regulate the sorting of glycoproteins.

As shown in Figure 4, biosynthesis of N-linked glycans begins with the cotranslational en bloc transfer of a large, high-mannose oligosaccharide terminating in glucose (Glc) residues ($Glc_3Man_9GlcNAc_2$) from a dolichol pyrophosphate–linked precursor to an appropriate Asn residue in the nascent polypeptide chain. This process, catalyzed by the oligosaccharyltransferase (OST) complex, occurs in the lumen of the ER. This is followed by the subsequent removal of the three terminal glucose residues via an α-glucosidase (steps 1–3, Fig. 4), a quality-control process that has been shown to play an important part in the correct folding and assembly of some secreted proteins through interactions with the molecular chaperones calnexin and calreticulin (37,38).

Subsequent trimming of mannose residues by ER and Golgi-specific α-mannosidases is followed by addition of terminal GlcNAc, Gal, and sialic acid and core fuc residues by glycosyltransferases utilizing nucleotide-sugar substrates imported into the lumen of Golgi compartments by specific transporters (antiporters) (39). Control of glycosylation reactions is complex and relies upon a multiplicity of molecular, cellular, and physiological interactions such as relative distribution of enzyme activities, substrate availability, and rate of protein transport (25). In some circumstances, local protein structure may serve to modify glycan processing. For example, it is well documented that terminal galactosylation and sialylation of IgG *N*-glycans in the Fc region at Asn_{297} is restricted by steric hindrance (40,41). However, in general, key control points that contribute to *N*-glycan heterogeneity are those catalyzed by the following:

1. ER and Golgi mannosidase I, GlcNAc transferase I (conversion of high-mannose to complex-type glycans)

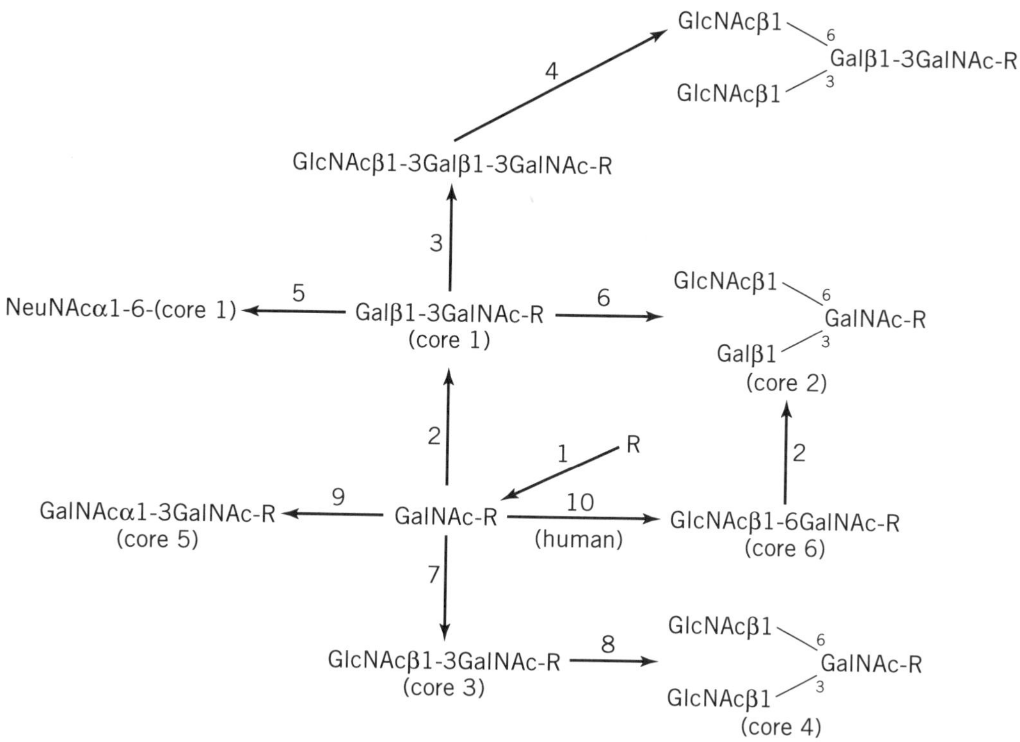

Figure 3. Biosynthesis of *O*-glycans. R represents Ser or Thr.

2. GlcNAc transferases II, IV, V (degree of branching of *N*-glycans)
3. GlcNAc transferase III (bisecting GlcNAc)
4. α-2,3 and α-2,6 sialyltransferases (degree of sialylation and linkage)
5. CMP-NeuAc-4-hydroxylase (conversion of NeuAc to NeuGc)

Mathematical modeling of this complex process has recently been attempted to permit comparison of strategies to obtain specific protein glycoform distributions (20,42).

The biosynthesis of O-linked glycans is similar to that of *N*-glycans and undergoes similar competition and regulation of its glycosyltransferases. It has been well reviewed by Schachter and Brockhausen (43), and four of the basic core structures and known biosynthetic pathways are outlined in Figure 3. Although many of the glycosyltransferases involved in *N*-glycan processing have been identified and cloned, most of the *O*-glycan transferases, with the exception of the core 2 GlcNAc-transferase, have not yet been cloned. This is in part due to the low percentage of glycans present on recombinant proteins that are O-linked.

GLYCOPROTEIN ANALYSIS

Underpinning our ability to control protein glycosylation is the necessity to monitor and in some cases completely characterize the recombinant glycoprotein product. More routine profiling of glycoform populations may also be useful to ensure reproducible product quality, batch-to-batch consistency, and product stability. Recent advances in analytical technology now permit relatively rapid and detailed assessment of glycoprotein and glycan structure by a variety of different methodologies, yielding information of varying complexity (9,44).

The choice of analytical strategy employed for characterization of a recombinant glycoprotein will depend on factors such as level of structural information required (e.g., site-specific glycosylation, oligosaccharide structure or monosaccharide composition, etc.), sample composition, and critically, the amount of protein available for analysis. Although some techniques, such as mass spectrometry, are now proving indispensable for many applications, it is a combination of complimentary techniques that will often provide the total analytical solution (45,46). The principal techniques currently employed are highlighted in the following sections.

Mass Spectrometry

Increasingly prevalent among current analytical strategies are mass spectrometric methods. Since the early 1980s, the emergence of "soft" ionization methods for biopolymer characterization by mass spectrometry (MS), such as fast atom bombardment (FAB), electrospray ionization (ESI), and matrix-assisted laser desorption/ionization (MALDI), has provided a formidable weapon in the armory of the analytical biochemist. There are several excellent reviews describing the application of this technology for protein and peptide analysis in detail (7,47–49).

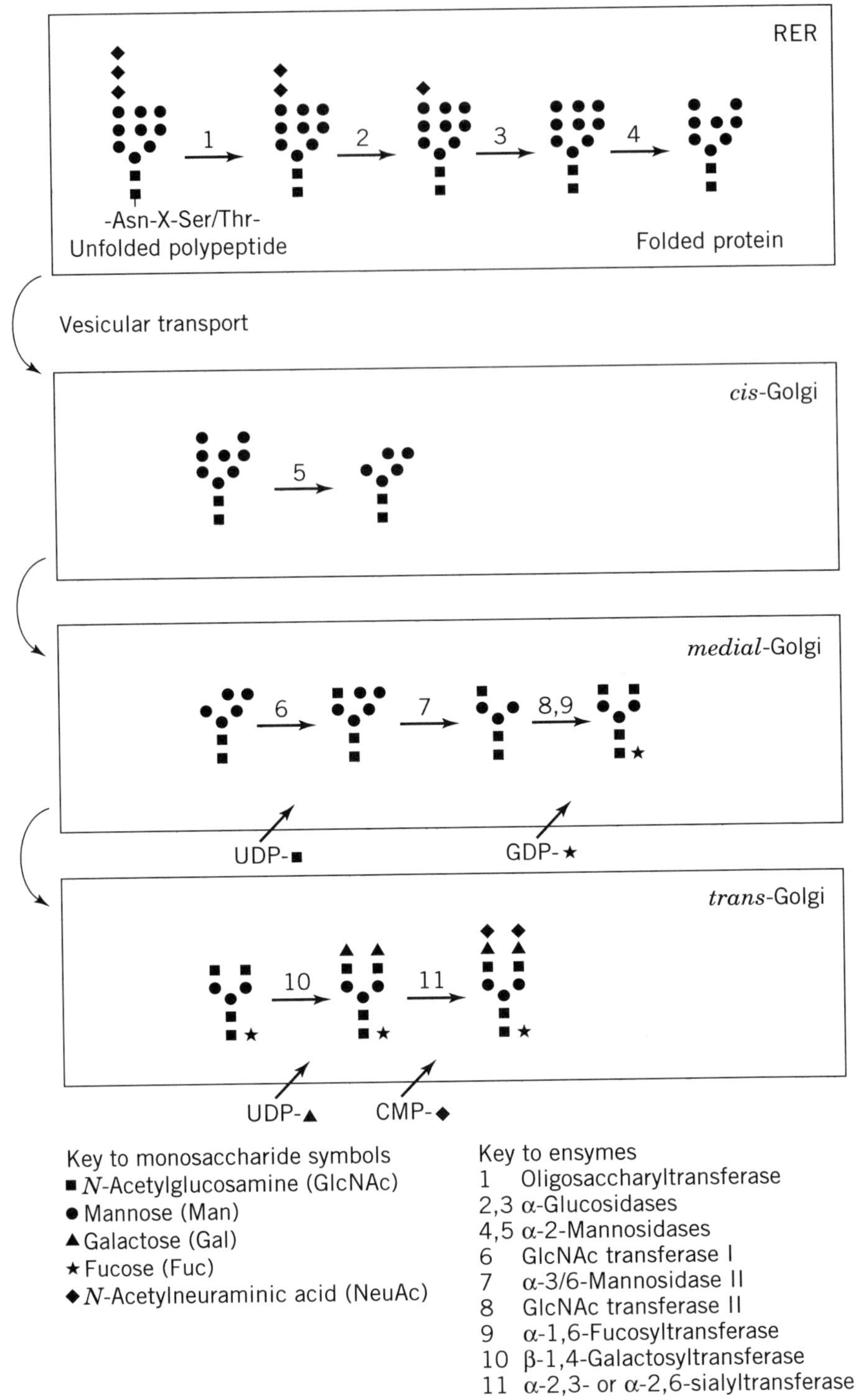

Figure 4. Biosynthesis of *N*-glycans. Schematic diagram illustrating the major enzymatic steps leading to the synthesis of a typical complex biantennary glycan.

Fast Atom Bombardment Mass Spectroscopy. FAB-MS and its variant, liquid secondary ion MS (LSI-MS) have an upper mass limit of ~15 and ~25 kDa, respectively, and thus can be used only to directly characterize smaller glycoconjugates such as glycopeptides (50,51). Because molecules tend to fragment in the ion source, this technique has been employed successfully to provide detailed sequence, linkage, and branching pattern information for N-linked glycans (52); however, derivatization of the carbohydrates by permethylation, peracetylation, or periodate oxidation is frequently necessary to obtain strong signals and provide linkage information. This technique is not generally amenable to rapid analysis of mixtures of glycopeptides or oligosaccharides, and successful operation of instrumentation and interpretation of spectra requires substantial specialist training. It is reviewed in detail by Peter-Katalinic (53).

Matrix-Assisted Laser Desorption/Ionization Mass Spectroscopy. MALDI-MS is a relatively simple technique. Biopolymers (theoretically up to 300 kDa) are cocrystallized on a metal surface in the presence of a molar excess (typically >5000:1) of a low molecular weight, strongly UV absorbing matrix. Biopolymer and matrix ions are desorbed by a short laser pulses (typically at 337 nm, 5 ns), and the gas phase ions are electrostatically directed into a time-of-flight (TOF) mass analyzer. Analyte resolution (width of a peak at half height divided by the mass/charge (m/z) ratio) using MALDI TOF-MS (typically <500) is $\pm 0.1\%$ for continuous ion extraction MALDI or >0.01% for MALDI with delayed ion extraction (54). This technique can be used to identify and characterize heterogeneous samples of glycopeptides (55–57), and released oligosaccharides (enzymatically or chemically) (58–61), and for the simultaneous sequencing of oligosaccharides in conjunction with exoglycosidase digestion (62,63). Recent reports highlight the use of MALDI-MS to rapidly monitor the site-specific glycosylation of a recombinant glycoprotein by interfacing it with HPLC-based peptide mapping (64). There are few examples of MALDI-MS glycoform profiling using whole glycoproteins, a result of the extensive heterogeneity of natural and recombinant glycoproteins. However, glycoproteins of less than 20 kDa, with perhaps only a single occupied glycosylation site, may be amenable to more detailed analysis. For example, variably sialylated forms of CHO-cell-derived recombinant interleukin 4 (~17 kDa) have been observed by Tsarbopoulos et al. (65).

Electrospray Ionization Mass Spectroscopy. ESI, typically in combination with a single- or triple-quadrupole mass analyzer, is capable of determining the molecular weight of whole recombinant glycoproteins or glycopeptides up to 100 kDa, with a mass accuracy of $\pm 0.01\%$ at a resolution routinely in excess of 1,000. Ionization of biopolymers in volatile solvents results in the production of multiply charged molecular ions (averaging 1 proton/kDa), and conditions are sufficiently mild that noncovalent interactions can often be observed because covalent bonds are not disrupted. The resulting mass spectrum of multiply charged molecular ions is then "deconvoluted" to yield a single zero-charge peak. The advantage of ESI-MS is that it may be coupled on-line with liquid chromatography (LC) or capillary electrophoresis (CE) to analyze complex samples such as proteolytic digests at low picomole to femtomole sensitivity.

The increased resolution and softer ionization afforded by ESI-MS makes this technique the most attractive method for direct analysis of recombinant glycoprotein populations because individual glycoforms can often be resolved. For example, heterogeneous glycoform populations of recombinant human interferon γ have been analyzed by James et al. (22) (Fig. 1). ESI-MS interfaced with LC has proved to be the method of choice for direct, sensitive analysis of protein digests. Selective, on-line identification of glycopeptides within these complex mixtures can be achieved by the use of operating conditions that enhance the formation of oxonium fragment ions (e.g., $HexNAc^+ = 204$ Da), such as collision-induced dissociation (66). Isolated glycopeptides may be analyzed in detail by ESI-MS directly (67,68) or in combination with exoglycosidase digestions to provide a more extensive site-specific structural analysis of N- and O-linked oligosaccharides (69). Although ESI-MS may be interfaced with CE for analysis of glycoproteins and glycopeptides, some CE separation protocols rely on acidic buffers, which are compatible with ESI-MS but unsuitable for separation of sialylated glycoconjugates (70).

Methods for glycoconjugate analysis by MALDI-MS and ESI-MS have been reviewed in detail by Burlingame (8).

HPLC

Liquid chromatographic techniques have been extensively employed to separate underivatized and derivatized oligosaccharides released enzymatically or chemically (e.g., by hydrazinolysis [71]) from glycoproteins. These approaches can most conveniently be described as being based on oligosaccharide size or charge.

Low-resolution separations based on hydrodynamic volume, such as those employing gel permmeation chromatography (e.g., Bio-Gel P4), were utilized in combination with limited exoglycosidase digestion of oligosaccharides to provide structural information (72). However, this approach has largely been superceded by the development of high-resolution normal-phase separations using amide-based media, in combination with simple procedures to fluorescently label oligosaccharides at the reducing terminus (73,74). This technology may also be used in combination with exoglycosidase sequencing and for analysis of glycoproteins in polyacrylamide gels (63,75).

Charge-based separations, distinguishing between differently sialylated glycans, have typically involved resolution of borohydride-reduced oligosaccharides by high-pH anion exchange chromatography with pulsed amperometric detection (HPAE-PAD) (76), although effective separation of fluorophore-derivatized glycans has also been demonstrated with amine-based columns now available commercially (73).

Capillary Electrophoresis

CE is proving itself to be a versatile analytical tool for (glyco)protein analysis both by itself (77,78) and in combination with other analytical techniques such as mass spectrometry (70,79). There are now several examples of glycoprotein analysis by CE, from analysis of intact glycoproteins (80–83), glycopeptides (84) and released oligosaccharides (85–87), to individual monosaccharides (88). These examples employ a variety of separation mechanisms (e.g., capillary zone electrophoresis, micellar electrokinetic capillary chromatography, capillary isoelectric-focusing) designed to enhance the resolution of the specific analyte(s) in question.

Gel Electrophoresis of Carbohydrates

Separation of released oligosaccharides in polyacrylamide gels after derivatization with fluorescent labels at the reducing terminus provides a simple, inexpensive method for profiling oligosaccharide populations. Limited structural information is obtained in combination with exoglycosidase digestions (89,90).

CONTROLLING GLYCOSYLATION

Given the ability to detect and analyze glycoproteins in detail, how can recombinant protein glycosylation be controlled? There are five main options currently available:

1. Choice of host cell
2. Genetic engineering of the host cell
3. Genetic engineering of the recombinant glycoprotein
4. Control of the host cell environment
5. Remodeling of protein glycans in vitro

Choice of Host Cell

Different host cell types are known to vary in their ability to glycosylate recombinant proteins. As there is now a profusion of expression systems available (mammalian, insect, or yeast cells in culture; transgenic animals and plants), initial choice of expression system is crucial and will, to a large extent, govern the type of glycosylation pattern obtained. Common bacterial expression systems such as *Escherichia coli* have no capacity to glycosylate proteins in either N- or O-linked conformations.

Yeast. Most yeast strains tend to hypermannosylate (the addition of a large number of mannose residues to the core trimannosyl oligosaccharide) recombinant proteins. We may infer that this modification would speed clearance of the glycoprotein from the bloodstream by hepatic and macrophage mannose receptors (91,92), and it has been shown to reduce the efficacy of recombinant hepatitis B vaccine (93). It is possible to use mutant yeast strains (e.g. mnn-9 or the temperature-sensitive ngd-29) that do not hypermannosylate recombinant proteins. In these strains, N-glycosylation is confined to core oligosaccharide residues containing up to eight mannose residues (93,94).

Insect Cell Lines. The first steps in *N*-glycan biosynthesis appear to be highly conserved throughout all eukaryotic cells, including insect cells, such as those commonly derived from the lepidopteran insect *Spodoptera frugiperda* (*Sf* 9). For example, Kawar et al. (95) recently demonstrated that *Sf* 9 cells encode and express an α1,2-mannosidase, similar to other eukaryotic enzymes, that releases mannose from Man_9, $GlcNAc_2$. In addition, an α1,2-mannosidase has been isolated from virus-infected *Sf* 21 cells; it acts preferentially on Man-6 oligosaccharides (96,97). The authors suggest that this is a pivotal step in glycan processing that determines whether complex or hybrid structures are formed.

However, numerous studies have shown that baculovirus-infected insect cells differ markedly from mammalian cells in subsequent glycan processing capability. These perceived limitations in glycosylation have hindered the wide acceptance of insect cell systems for large-scale production of human therapeutics. Whereas mammalian cells are generally capable of complex-type *N*-glycan processing, insect cells produce proteins with truncated glycans, notably trimannosyl core structures ($Man_3GlcNac_2$) plus or minus fucose (55,98–100). The molecular basis for the accumulation of trimannosyl *N*-glycans on proteins secreted by some insect cells (notably *Sf*-derived cell lines) has recently been elucidated. After initial transfer of *N*-acetylglucosamine onto Man-5 by a GlcNAc transferase I identified in insect cells (101,102), subsequent removal of mannose and *N*-acetylglucosamine by a Golgi-specific α-mannosidase II (103,104) and β-*N*-acetylglucosaminidase (105–107) occurs, yielding the trimannosyl core structure. In some insect cell lines, specifically those derived from *Estigmena acrea*, β-*N*-acetylglucosaminidase activity is absent, and consequently, *N*-glycans bearing terminal *N*-acetylglucosamine residues are present (107,108). Further *N*-glycan processing is cell line specific. For example, several studies report the inability of *Spodoptera*- (*Sf* 9), *Estigmena*-, and *Trichoplusia*-derived cell lines to sialylate *N*-glycans (100,108–110). However, *Mamestra brassicae* and *Manduca sexta* cells have been reported as being capable of sialylating the *N*-glycans of recombinant human plasminogen (111). This group also reported that *Sf*21 cells were capable of synthesizing complex *N*-glycans (112), but that this was dependent upon time postinfection (113)—complex *N*-glycans appearing at later stages of the infection cycle.

Mammalian Cell Lines. Although phylogenetically more similar to humans and generally capable of complex-type oligosaccharide synthesis, commonly used cell lines derived from rodent tissues maintain some key differences in glycan processing.

Firstly, murine cell lines (e.g., hybridomas, mouse–human heterohybridomas, C127 cells, and NS0 cells) express the enzyme α-1,3-galactosyltransferase (α-1,3-GalT), which leads to the synthesis of glycans terminating in Galα1,3-Galβ1,4-GlcNAc motifs (114,115). As more than 1% of human serum IgG is directed against this epitope (116,117), its association with a therapeutic protein would be undesirable, resulting in rapid clearance. In humans, apes, old world monkeys, and CHO cell lines the α-1,3-GalT gene has become inactivated through frameshift mutations (118,119).

Secondly, in contrast to human cells, murine cell lines have a tendency to confer *N*-glycolylneuraminic acid (NeuGc), an oncofetal antigen and immunogenic derivative of *N*-acetylneuraminic acid (NeuAc). The hydroxylase enzyme that converts CMP-NeuAc to CMP-NeuGc is cytosolic (120) and has recently been cloned (121). NeuGc has been found to be associated with the oligosaccharides of an IgM produced by a human–mouse hybridoma, where it was twice as prevalent as NeuAc (122). Similarly, comparison of glycans associated with a chimeric IgG produced by CHO and NS0 cells revealed modification of NS0-derived glycans with NeuGc, and not NeuAc, which was linked to enhanced rate of plasma clearance (123). Proteins produced by CHO cells have a low NeuGc content, typically 1–2% of total sialic acids (124), and are generally not immunogenic (125), although higher levels of substitution with NeuGc may elicit an immune response (126). CHO and baby hamster kidney (BHK) cell lines lack the functional α-2,6-sialyltransferase (ST) enzyme found in human and murine cells and synthesize exclusively α-2,3-linked terminal sialic acids. This alteration has been addressed by transfection of CHO and BHK cells with an α-2,6-ST gene, as discussed later.

Lastly, although not of immunological significance, another variable *N*-glycan modification that may be associated with recombinant proteins derived from CHO cells is the addition of poly-*N*-acetyllactosamine repeats (Galβ1-4GlcNAcβ1-3-) to terminal Gal residues on the antennae of complex oligosaccharides (126–129).

Transgenic Animals. There are very few studies describing the glycosylation of recombinant proteins produced in the mammary gland of transgenic animals. The few reports that exist indicate that glycan processing is altered. James et al. (54) compared the site-specific glycosylation of recombinant human interferon-γ (IFN-γ) produced by CHO cells, *Sf* 9 insect cells, and the mammary gland of transgenic mice. In the latter case, *N*-glycans at one site (Asn_{97}) were predominantly truncated or of the high-mannose type, whereas *N*-glycans at the other site (Asn_{25}) were of the complex, sialylated variety similar to those associated with IFN-γ from CHO cells.

Similar data has recently been reported for recombinant antithrombin produced in transgenic goat milk, where one of three N-linked sites (Asn_{155}) in the transgenic-derived protein was unaccountably associated with high-mannose-type glycans, not complex glycans as in the natural human protein (130). Furthermore, the recombinant antithrombin contained a GalNAc for Gal substitution on some *N*-glycans, was highly fucosylated and less sialylated than natural human antithrombin, and contained both NeuAc and NeuGc.

The total extent of glycosylation of a recombinant protein produced by transgenic animals may also differ from that of its natural counterpart. A recombinant erythropoietin fusion protein produced in transgenic mouse and rabbits was less glycosylated overall than the molecule produced in CHO cells (131). In addition, recombinant human bile salt–stimulated lipase (BSSL) derived from transgenic mouse milk exhibited a significantly lower degree of O-glycosylation than was found for the native enzyme or for recombinant CHO or C127 cell–derived BSSL (132). In both cases, however, biological activity was not compromised.

Transgenic Plants. Transgenic plants are now being evaluated as a viable option for therapeutic protein production (133–136). However, not surprisingly, *N*-glycan processing in plant cells is substantially different from that obtained in mammalian cell systems. *N*-Glycans universally lack terminal sialic acid residues, and potentially allergenic residues such as core α1-3-linked fucose (137), xylose or β1-2-mannose residues have been detected. Recombinant erythropoietin (EPO) produced in transgenic tobacco cells was found to have no biological activity in vivo (138), possibly because of its lack of sialylation and consequently high clearance rate.

Genetic Engineering of the Host Cell

As already outlined, the glycosylation of recombinant glycoproteins is generally host cell specific, being largely dependent on the complement of specific glycosyltransferases available for glycan processing. Thus, the majority of efforts to augment glycan processing in various host cells by genetic means has involved introduction of a cloned heterologous glycosyltransferase with a desirable processing activity that is absent in the parental host cell. It should be stated, however, that successful introduction of a new biosynthetic activity into a host cell by this route depends on the preexisting capability of the cell to synthesize the requisite nucleotide sugar substrate and the ability to transport it into the Golgi lumen using a specific nucleotide sugar transporter. For this reason, our ability to engineer the protein glycosylation capabilities of yeast and insect cells is restricted at present.

Potential targets for genetic alteration of *N*-glycans of recombinant proteins produced by mammalian cells are summarized in Table 1. In the case of CHO and BHK cells, which both lack a functional α-2,6-sialyltransferase (α-2,6-ST), terminal sialylation of *N*-glycans has received the most attention. This is primarily for two reasons, *(1)* the potential for modulation of the pharmacokinetic behavior of a therapeutic protein product: the relationship between terminal sialylation of *N*-glycans and rate of protein clearance from the bloodstream (mediated by hepatic asialoglycoprotein receptors) is now well documented (149–151), and *(2)* the production of recombinant proteins with "authentic" humanlike *N*-glycans consisting of a mixture of α2-3- and α2-6-linked sialic acids. However, experiments involving cotransfection of cells with cloned α-2,6-ST cDNA have met with variable success: although α2-6-linked sialic acids have been shown to occur in all cases, the absolute level (mole per mole) of recombinant protein sialylation was either not determined (152), unaffected (140), or increased in some transfectants but not others (153). A soluble form of the α-2,6-ST expressed in BHK cells did not modify a model secreted recombinant glycoprotein, indicating that functional activity is dependent upon correct localization of the heterologous transferase (154).

Recent reports highlight the potential use of CHO cells expressing heterologous fucosyltransferases for the production of bioactive oligosaccharides, specifically Golgi membrane and secreted forms of fucosyltransferase III for Lewis (a) synthesis (146), and fucosyltransferase VI for sialyl Lewis X synthesis (145). In the latter case, expression of the introduced glycosyltransferase was controlled using antisense technology.

There have also been recent attempts to engineer baculovirus-infected insect cell expression systems for improved recombinant protein glycosylation. This has been carried out directly, for example, via the stable expression of β1-4-galactosyltransferase in *Sf* 9 insect cells (147), which resulted in the presence of corresponding terminal galactose residues on glycans associated with human tissue plasminogen activator secreted after recombinant baculovirus infection. Alternatively, the glycosylation of proteins produced in insect cells may be improved by the use of baculovirus expression vectors employing the immediate-early promoter element, permitting the production of recombinant glycoproteins with more extensively processed N-linked oligosaccharides (155).

Table 1. Targets for Genetic Engineering of Recombinant Protein Glycosylation in Animal Cells

Protein[a]	Purpose	Function	Reference
	Mammalian cells		
α2,6-Sialyltransferase ↑	Increase α2,6-sialylation	Increase half-life, authenticity	140
Cytosolic sialidase ↓	Eliminate extracellular digestion of *N*-glycans	Increase half-life	141
CMP-Neu5Ac hydroxylase ↓	Eliminate NeuGc	Reduce protein antigenicity	121,142
GlcNAc transferase III ↑	Add bisecting GlcNAc	Enhance bioactivity	143
α1,3-Gal transferase ↓	Eliminate Gal α1,3-Gal	Reduce protein antigenicity	144
Fuc transferase VI ↑	Synthesize sialyl Lewis X	Synthesise biologically active oligosaccharides	145
Fuc transferase III ↑	Synthesize Lewis (a)	Synthesise biologically active oligosaccharides	146
	Insect cells		
β1,4 Gal transferase ↑	Increase glycan complexity	Mimic mammalian proteins	147
β1,2 GlcNAc transferase I ↑	Increase glycan complexity	Mimic mammalian proteins	107
	Transgenic animals		
α1,2-Fuc transferase ↑	Increase α1,2-fucosylation	Synthesise biologically active oligosaccharides	148

[a]A desired up- or downregulation of the gene is indicated by an arrow.

There is only one example where the transgenic animal mammary gland has been engineered for altered glycan processing. Prieto et al. (148) expressed human α-1,2-fucosyltransferase (α-1,2-Fuc T), which generates the H blood group antigen, in transgenic mice. Milk from these animals contained soluble α-1,2-Fuc T, as well as 2′-fucosyllactose (Fucα1-2Galβ1-4Glc) and modified glycoproteins containing the H antigen.

Genetic Engineering of the Recombinant Glycoprotein

There is now a well-documented link between local protein structure and glycosylation at a particular site (5), in other words, glycosylation of a protein produced in a single cell type can be different at individual glycosylation sites. This may arise due to steric hindrance of glycosyltransferases, as in the case of the Fc *N*-glycans of IgG (40,41), or other peptide–carbohydrate interactions. Thus, by utilizing protein engineering technology such as site-directed mutagenesis, the possibility arises that glycosylation sites associated with undesirable glycans (e.g., high-mannose) may be deleted, new sites may be created, or local protein structure may be redesigned to alter glycan processing at a glycosylation site. Two current examples are highlighted: engineered forms of tissue plasminogen activator (tPA) and IgG.

Recombinant human tPA produced by CHO cells has three potential N-glycosylation sites at Asn_{117}, Asn_{184} (variably occupied), and Asn_{448} (156), and one O-linked site at Thr_{61}, which is associated with a single fucose residue. *N*-Glycans at Asn_{117} are invariably of the high-mannose type, in contrast to the complex-type *N*-glycans at the other Asn sites, and are responsible for promoting the clearance of tPA from circulation via interactions with macrophage and hepatic mannose receptors (157). Thus, second-generation tPA molecules, TNK-tPA, have been mutated to remove Asn_{117} (N117Q), with the introduction of a new glycosylation site, Asn_{103} (T103N). The latter mutation restores fibrin binding activity to the wild-type tPA molecule. Significantly, the TNK-tPA variant exhibits complex-type *N*-glycans at Asn_{103} and has a 10-fold increase in its serum half-life (158–160).

Human IgG is glycosylated at Asn_{297} in the Fc region, predominantly with a mixture of variably galactosylated, minimally sialylated biantennary oligosaccharides. Numerous studies have shown that variations in the structure of these glycans can serve to modulate antibody effector functions (40,161) via noncovalent interactions between carbohydrate and local amino acid residues and Fc receptors (41,162). Lund et al. (161) mutated a number of amino acid residues proximal to the Asn_{297} of a recombinant IgG3 produced in CHO cells; specifically, residues 241, 243, 264, 265, and 301 were replaced with alanine in each case. These mutations resulted in increased galactosylation and sialylation relative to the wild-type oligosaccharide chains, although there was a reduction in antibody effector functions in some cases.

Control of the Host Cell Environment

For most recombinant proteins produced by mammalian cells in culture, the cell environment itself can contribute to product molecular heterogeneity. The following factors may all influence the structural diversity of glycans associated with recombinant therapeutic proteins or antibodies:

1. Serum
2. Ammonium
3. Extracellular glycosidases
4. Method of cell culture

Serum. In light of recent bovine spongiform encephalitis epidemics and the desire for control of a fully defined cell culture environment, there is a trend in industry to remove serum from culture media. However, the presence or absence of serum has been shown to affect protein glycosylation. For an IgG1 monoclonal antibody produced by hybridomas, Patel et al. (163) demonstrated that ascites-derived protein differed from serum-free-culture-derived material only with respect to the content of sialic acid. IgG1 derived from culture in serum-containing media had an intermediate sialic acid content and a lower incidence of outer-arm galactosylation than the other two preparations. However, for antibodies produced by CHO cells, *N*-glycan galactosylation may be improved in serum-containing medium (164). For other types of recombinant protein, serum content of the medium may also be significant. Adaptation of BHK-21 cells producing a recombinant mutant IL-2 from a serum-containing to a serum-free environment resulted in changes to *N*-glycan antennarity, sialylation, sialylation, and fucosylation, as well as an increase in the overall degree of protein glycosylation (165).

Ammonium. Increases in the concentration of the toxic catabolite ammonia during cell culture have been linked to various changes in protein glycosylation. Borys et al. (166) demonstrated that exogenous additions of ammonium (as ammonium chloride) up to 9 mM decreased the overall glycosylation of recombinant mouse placental lactogen 1 produced by CHO cells in a pH-dependent manner, this effect being most pronounced at elevated pH (8.0). Subsequently, it was shown that ammonium ion concentrations above 2 mM specifically inhibited the sialylation of recombinant granulocyte-macrophage colony-stimulating factor produced by CHO cells (167). This effect was thought to be mediated by an ammonium-induced elevation of the acidic *trans*-Golgi compartment pH, resulting in reduced sialyltransferase activity in vivo, a concept originally explored by Thorens and Vassalli (168).

In parallel with this work, it was demonstrated that an artificial increase in ammonium ion concentration induced an accumulation of UDP-*N*-acetylhexosamines in mammalian cells, with a concomitant decline in growth rate (169); in fact the “U ratio” (UTP:UDP-*N*-acetylhexosamine) was shown to be a sensitive indicator of cell growth rate. Subsequently, increases in the antennarity of *N*-glycans associated with a human mutant IL-2 produced by BHK-21 cells were also correlated with an increased cellular concentration of UDP-*N*-acetylhexosamines, as artificially induced by exogenous glutamine (7.5 mM) or ammonium (15 mM) (170). In another report, similar increases in the cellular UDP-*N*-acetylhexosamine pool were induced by incubation of BHK cells with glutamine and uridine (171), which increased *N*-glycan antennarity but not sialylation. This suggests that the effect of ammonium on sialylation alone is due to another mechanism—probably changes in *trans*-Golgi pH, as discussed above. In BHK cells, it is now likely that ammonium ions elevate cellular UDP-*N*-acetylhexosamine levels by direct incorporation of ammonium into fructose-6-phosphate to yield a synthetic precursor of UDP-GlcNAc, glucosamine-6-phosphate; the reaction being catalyzed by glucosamine-6-phosphate isomerase (172). Presumably UDP-GlcNAc transport into the Golgi is stimulated, increasing the availability of substrate for resident GlcNAc transferases III and IV involved in *N*-glycan branching.

In practice, therefore, excessive ammonium concentrations in cell culture should be avoided. Alternatively, use of the glutamine synthetase (GS) expression system (173) avoids this issue. Ammonia is utilized enzymatically by the transfected selectable marker, GS, and does not therefore accumulate during cell culture.

Extracellular Glycosidases. Sialidase, β-galactosidase, β-hexosaminidase, and fucosidase have been identified in cell culture supernatant from CHO and other mammalian cell lines (174–177). At neutral pH the most active is sialidase, which has been purified from CHO cell culture supernatant (178). Gu et al. (179) have demonstrated that sialylation of recombinant human IFN-γ produced by CHO cells decreased in parallel with loss in cell viability in batch culture. Therefore elimination of the sialidase from CHO cells constitutes a viable target for cell engineering (141). However, the fucosidase purified from CHO cell culture supernatant could not release fucose from the recombinant CHO cell–produced glycoproteins gp120 or soluble CD4 with the Fucα1-6GlcNAc linkage, or from human serum α(1)-acid glycoprotein with the Fucα1-3GlcNAc linkage (180).

Method of Cell Culture. The type of culture used to produce a recombinant protein affects its glycosylation. For example, during batch and fed-batch cultures of antibodies produced by NS0 cells, and during recombinant human IFN-γ production by CHO cells, there was a progressive increase in the proportion of high-mannose (predominantly $Man_5GlcNAc_2$) and truncated *N*-glycans associated with the recombinant protein (181,182). The increased appearance of high-mannose glycans suggests a change in the glycan processing capability of cells as a function of time in culture (in this case presumably affecting GlcNAc transferase I activity, by as yet unknown mechanisms). This may also be true for truncated *N*-glycans (lacking terminal monosaccharides), although at present it is more likely that these derive from the action of extracellular glycosidases, as previously mentioned.

These considerations emphasize the necessity to harvest product from batch cultures prior to significant cell lysis. Furthermore, As a means of ensuring reproducible product quality, batch-to-batch consistency, and stability, deleterious changes to recombinant glycoprotein product heterogeneity may be averted by implementation of appropriate monitoring technologies capable of rapid quantitative analyses of intact glycoprotein populations, enabling control of bioreactor operations (e.g., to avoid late harvest of undersialylated product).

Few reports describe the effect of long-term continuous culture on the processing and modification of complex recombinant proteins. Goldman et al. (183) have shown quantitatively that both long-term fluidized-bed perfusion and batch cultures of CHO cells produce recombinant human IFN-γ with a consistent and highly comparable degree of glycosylation, monitored as glycosylation site occupancy and level of *N*-glycan sialylation. However, as previously mentioned, the sialylation of recombinant IFN-γ produced in stirred-tank batch culture declined significantly during the stationary and death phase. Differences in the sialylation of recombinant human tissue kallikrein produced by CHO cells from microcarrier culture and from a serum-free suspension cell recycle process have been reported (130), in which a reduced degree of *N*-glycan sialylation was observed with the microcarrier process.

Other Factors. Other factors may affect glycosylation of recombinant glycoproteins produced by animal cells in culture to a limited extent, although such effects are likely to be host cell and recombinant protein specific. For example, mild hypoxia and elevated pCO_2 have minimal effects on the glycosylation of CHO cell–derived tPA (184,185). In perfusion culture of BHK-21 cells producing a recombinant mutant IL-2, nutrient limitations (glucose, amino acids, dO_2) led to short-term changes in macroheterogeneity, but microheterogenity was largely unchanged (165).

Oligosaccharide Synthesis In Vitro

Advances in the chemical synthesis of oligosaccharides, notably using cloned glycosyltransferases, now permit the synthesis or modification of *N*-glycan-type structures (186,187). Thus, postharvest remodeling of *N*-glycans on purified proteins is now possible (e.g., to fully sialylate *N*-glycans for improved protein half-life in vivo (188), although cost constraints may limit the commercial application of this approach.

SUMMARY

The glycosylation of a potential recombinant therapeutic protein will contribute to a set of product characteristics relating to product efficacy, including bioactivity and/or pharmacokinetic behavior, although these will be protein specific. Initial choice of expression system is of crucial significance and will define the structure of oligosaccharides associated with individual protein glycoforms, although other factors implicit in large-scale bioprocess development may also be significant. At present, the most realistic goal may be a consistent, reproducible, humanlike (albeit heterogeneous) glycosylation pattern. However, an ideal production process would perhaps result in a single, defined protein glycoform, rather than a mixture of glycoform variants, each with varying characteristics. This may be considered as analogous to chiral separation of small-molecule drugs, where there is a necessity for optically pure molecules for medical use.

Such control of recombinant protein glycosylation may possibly be achieved at a number of levels: *(1)* genetic control of glycan processing in the host cell, *(2)* engineering recombinant proteins for a defined glycosylation status, *(3)* control of cell physiology in a bioreactor environment for enhanced production of specific glycoforms, and *(4)* downstream processing to selectively recover specific glycoforms. Incorporation of any of these developments into a bioprocess must involve a consideration of clinical efficacy and potency, technical feasibility, cost of goods, and regulatory requirements.

Central to these issues is our ability to quantitively monitor and completely characterize heterogeneous biological products. Glycoproteins can now be analyzed by a diverse array of techniques, yielding structural information of varying complexity. Recent advances in analytical methodology permit relatively rapid, detailed assessment of glycan structure, glycosylation site-occupancy, and particularly sialylation (extent, type of sialic acid, linkage). More routine fingerprinting of glycoform populations may also be useful to ensure reproducible product quality, batch-to-batch consistency, and stability.

BIBLIOGRAPHY

1. N. Jenkins and E.M.A. Curling, *Enzyme Microb. Technol.* **16**, 354–364 (1994).
2. M.J. Geisow, *Trends Biotechnol.* **10**, 333–335 (1992).
3. N. Jenkins, R.B. Parekh, and D.C. James, *Nat. Biotechnol.* **14**, 975–981 (1996).
4. D.T.Y. Liu, *Trends Biotechnol.* **10**, 114–120 (1992).
5. P.M. Rudd, and R.A. Dwek, *Crit. Rev. Biochem. Mol. Biol.* **32**, 1–100 (1997).
6. A. Varki, *Glycobiology* **3**, 97–130 (1993).
7. J.S. Andersen, B. Svensson, and P. Roepstorff, *Nat. Biotechnol.* **14**, 449–457 (1996).
8. A.L. Burlingame, *Curr. Opin. Biotechnol.* **7**, 4–10 (1996).
9. R.A. Dwek, C.J. Edge, D.J. Harvey, M.R. Wormald, and R.B. Parekh, *Ann. Rev. Biochem.* **62**, 65–100 (1993).
10. D.C. James, *Cytotechnology* **22**, 17–24 (1996).
11. C. Abeijon and C.B. Hirschberg, *Trends Biochem. Sci.* **17**, 32–36 (1992).
12. M.C. Field and L.J. Wainwright, *Glycobiology* **5**, 463–472 (1995).
13. C.B. Hirschberg, C. Abeijon, K. Yanagisawa, E. Mandon, A. Orellana, P.W. Robbins, M. Ishihara, Z. Wei, and S. Swiedler, *Glycoconjugate J.* **10**, 310–310 (1993).
14. D.H. Joziasse, *Glycobiology* **2**, 271–277 (1992).
15. R. Kleene and E.G. Berger, *Biochim. Biophys. Acta* **1154**, 283–325 (1993).
16. C. Rabouille, N. Hui, F. Hunte, R. Kieckbusch, E.G. Berger, G. Warren, and T. Nilsson, *J. Cell Sci.* **108**, 1617–1627 (1995).
17. P. Stanley, and E. Ioffe, *FASEB J.* **9**, 1436–1444 (1995).
18. D.H. Vandeneijnden and D.H. Joziasse, *Curr. Opin. Struct. Biol.* **3**, 711–721 (1993).
19. M. Shelikoff, A.J. Sinskey, and G. Stephanopoulos, *Cytotechnology* **15**, 195–208 (1994).
20. M. Shelikoff, A.J. Sinskey, and G. Stephanopoulos, *Biotechnol. Bioeng.* **50**, 73–90 (1996).
21. Y. Gavel and G. Vonheijne, *Prot. Eng.* **3**, 433–442 (1990).
22. D.C. James, M.H. Goldman, M. Hoare, N. Jenkins, R.W.A. Oliver, B.N. Green, and R.B. Freedman, *Protein. Sci.* **5**, 331–340 (1996).
23. K.J. Colley, *Glycobiology* **7**, 1–13 (1997).
24. R. Kornfeld and S. Kornfeld, *Ann. Rev. Biochem.* **54**, 631–664 (1985).
25. A. Varki, *Trends Cell Biol.* **8**, 34–40 (1998).
26. H.R.B. Pelham, *Cell Struct. Funct.* **21**, 413–419 (1996).

27. S. Cottet, and B. Corthesy, *Eur. J. Biochem.* **246**, 23–31 (1997).
28. K. Fiedler, and K. Simons, *Cell* **81**, 309–312 (1995).
29. W.E. Balch, J.M. McCaffery, H. Plutner, and M.G. Farquhar, *Cell* **76**, 841–852 (1994).
30. J. Krijnselocker, M. Ericsson, P.J.M. Rottier, and G. Griffiths, *J. Cell Biol.* **124**, 55–70 (1994).
31. M. Mizuno and S.J. Singer, *Proc. Natl. Acad. Sci.* U.S.A. **90**, 5732–5736 (1993).
32. J.F. Presley, N.B. Cole, T.A. Schroer, K. Hirschberg, K.J.M. Zaal, and J. LippincottSchwartz, *Nature* **389**, 81–85 (1997).
33. C. Arar, V. Carpentier, J.P. Lecaer, M. Monsigny, A. Legrand, and A.C. Roche, *J. Biol. Chem.* **270**, 3551–3553 (1995).
34. C. Itin, A.C. Roche, M. Monsigny, and H.P. Hauri, *Mol. Biol. Cell* **7**, 483–493 (1996).
35. U. Lahtinen, U. Hellman, C. Wernstedt, J. Saraste, and R.F. Pettersson, *J. Biol. Chem.* **271**, 4031–4037 (1996).
36. C. Itin, M. Foguet, F. Kappeler, J. Klumperman, and H.P. Hauri, *Biochem. Soc. Trans.* **23**, 541–544 (1995).
37. C. Hammond and A. Helenius, *Curr. Opin. Cell Biol.* **7**, 523–529 (1995).
38. A. Helenius, E.S. Trombetta, D.N. Hebert, and J.F. Simons, *Trends Cell Biol.* **7**, 193–200 (1997).
39. C.B. Hirschberg, *Glycobiology* **7**, 169–171 (1997).
40. R. Jefferis and J. Lund, *Chem. Immunol.* **65**, 111–128 (1997).
41. M.R. Wormald, P.M. Rudd, D.J. Harvey, S.C. Chang, I.G. Scragg, and R.A. Dwek, *Biochemistry* **36**, 1370–1380 (1997).
42. P. Umana and J.E. Bailey, *Biotechnol. Bioeng.* **55**, 890–908 (1997).
43. H. Schachter and I. Brockhausen, in H.J. Allen and E.C. Kisailus eds., *Glycoconjugates, Composition, Structure and Function*, Marcel Dekker, New York, pp. 263–332.
44. P. Jackson and J.T. Gallagher, *A Laboratory Guide to Glycoconjugate Analysis* Berkhauser Verlag, Basel, 1997.
45. M.E. Hemling, M.A. Mentzer, C. Capiau, and S.A. Carr, in A.L. Burlingame and S.A. Carr eds., *Mass Spectrometry in the Biological Sciences*, Humana Press, Totowa, N.J., 1996, pp. 307–331.
46. G.D. Roberts, W.P. Johnson, S. Burman, K.R. Anumula, and S.A. Carr, *Anal. Chem.* **67**, 3613–3625 (1995).
47. P. Roepstorff, K.H. Schram, J.S. Andersen, K. Rafn, T. Baldursson, J. Kroll, K. Poulsen, J. Knudsen, and K. Kristiansen, *Mol. Biotechnol.* **4**, 1–12 (1995).
48. G. Siuzdak, *Proc. Nat. Acad. Sci. U.S.A.* **92**, 646–646 (1995).
49. J.T. Stults, *Curr. Opin. Struct. Biol.* **5**, 691–698 (1995).
50. A. Dell, K.H. Khoo, M. Panico, R.A. McDowell, A.T. Etienne, A.J. Reason, and H.R. Morris, in M. Fukuda and A. Kobata eds., *Glycobiology: A 222Practical Approach*, IRL Press, Oxford, 1994, pp. 187.
51. A. Dell, A.J. Reason, K.H. Khoo, M. Panico, R.A. McDowell, and H.R. Morris, *Methods Enzymol.* **230**, 108–132 (1994).
52. A. Dell, H.R. Morris, R.L. Easton, M. Panico, M. Patankar, S. Oehninger, R. Koistinen, H. Koistinen, M. Seppala, and G.F. Clark, *J. Biol. Chem.* **270**, 24116–24126 (1995).
53. J. Peter-Katalinic, *Mass Spectrom. Rev.* **13**, 77–98 (1994).
54. M.L. Vestal, P. Juhasz, and S.A. Martin, *Rapid Commun. Mass Spectrom.* **9**, 1044–1050 (1995).
55. D.C. James, R.B. Freedman, M. Hoare, O.W. Ogonah, B.C. Rooney, O.A. Larionov, V.N. Dobrovolsky, O.V. Lagutin, and N. Jenkins, *Bio / Technology* **13**, 592–596 (1995).
56. D.J. Kroon, J. Freedy, D.J. Burinsky, and B. Sharma, *J. Pharm. Biomed. Anal.* **13**, 1049–1054 (1995).
57. C.W. Sutton, J.A. Oneill, and J.S. Cottrell, *Anal. Biochem.* **218**, 34–46 (1994).
58. D.J. Harvey, *J. Chromatogr. A* **720**, 429–446 (1996).
59. D.I. Papac, J.B. Briggs, E.T. Chin, and A.J.S. Jones, *Glycobiology* **8**, 445–454 (1998).
60. D.I. Papac, E.T. Chin, J.B. Briggs, and A.J.S. Jones, *Glycobiology* **6**, 101–101 (1996).
61. D.I. Papac, A. Wong, and A.J.S. Jones, *Anal. Chem.* **68**, 3215–3223 (1996).
62. B. Kuster, T.J.P. Naven, and D.J. Harvey, *J. Mass Spectrom.* **31**, 1131–1140 (1996).
63. P.M. Rudd, G.R. Guile, B. Kuster, D.J. Harvey, G. Opdenakker, and R.A. Dwek, *Nature* **388**, 205–207 (1997).
64. B.J. Harmon, X.J. Gu, and D.I.C. Wang, *Anal. Chem.* **68**, 1465–1473 (1996).
65. A. Tsarbopoulos, M. Karas, K. Strupat, B.N. Pramanlk, T.L. Nagabhushan, and F. Hillenkamp, *Anal. Chem.* **66**, 2062–2070 (1994).
66. S.A. Carr, M.J. Huddleston, and M.F. Bean, *Prot. Sci.* **2**, 183–196 (1993).
67. E.A. Kragten, A.A. Bergwerff, J. Vanoostrum, D.R. Muller, and W.J. Richter, *J. Mass Spectrom.* **30**, 1679–1686 (1995).
68. R.S. Rush, P.L. Derby, D.M. Smith, C. Merry, G. Rogers, M.F. Rohde, and V. Katta, *Anal. Chem.* **67**, 1442–1452 (1995).
69. P.A. Schindler, C.A. Settineri, X. Collet, C.J. Fielding, and A.L. Burlingame, *Protein Sci.* **4**, 791–803 (1995).
70. J.F. Kelly, S.J. Locke, L. Ramaley, and P. Thibault, *J. Chromatogr. A* **720**, 409–427 (1996).
71. T.P. Patel, and R.B. Parekh, *Methods Enzymol.* **230**, 57–66 (1994).
72. C.J. Edge, T.W. Rademacher, M.R. Wormald, R.B. Parekh, T.D. Butters, D.R. Wing, and R.A. Dwek, *Proc. Natl. Acad. Sci. U.S.A.* **89**, 6338–6342 (1992).
73. J.C. Bigge, T.P. Patel, J.A. Bruce, P.N. Goulding, S.M. Charles, and R.B. Parekh, *Anal. Biochem.* **230**, 229–238 (1995).
74. G.R. Guile, P.M. Rudd, D.R. Wing, S.B. Prime, and R.A. Dwek, *Anal. Biochem.* **240**, 210–226 (1996).
75. P.M. Rudd, and R.A. Dwek, *Curr. Opin. Biotechnol.* **8**, 488–497 (1997).
76. M.R. Hardy and R.R. Townsend, *Methods Enzymol.* **230**, 208–225 (1994).
77. M.V. Novotny, *Methods Enzymol.* **271**, 319–347 (1996).
78. G. Teshima and S.L. Wu, *Methods Enzymol.* **271**, 264–293 (1996).
79. D. Figeys, A. Ducret, and R. Aebersold, *J. Chromatogr. A* **763**, 295–306 (1997).
80. S. Hoffstetterkuhn, G. Alt, and R. Kuhn, *Electrophoresis* **17**, 418–422 (1996).
81. D.C. James, R.B. Freedman, M. Hoare, and N. Jenkins, *Anal. Biochem.* **222**, 315–322 (1994).
82. K.G. Moorhouse, C.A. Rickel, and A.B. Chen, *Electrophoresis* **17**, 423–430 (1996).
83. D.E. Morbeck, B.J. Madden, and D.J. McCormick, *J. Chromatogr. A* **680**, 217–224 (1994).

84. R.S. Rush, P.L. Derby, T.W. Strickland, and M.F. Rohde, *Anal. Chem.* **65**, 1834–1842 (1993).
85. K. Kakehi and S. Honda, *J. Chromatogr. A* **720**, 377–393 (1996).
86. G. Okafo, L. Burrow, S.A. Carr, G.D. Roberts, W. Johnson, and P. Camilleri, *Anal. Chem.* **68**, 4424–4430 (1996).
87. P.L. Weber and S.M. Lunte, *Electrophoresis* **17**, 302–309 (1996).
88. A. Guttman, *J. Chromatogr. A* **763**, 271–277 (1997).
89. P. Jackson, *Mol. Biotechnol.* **5**, 101–123 (1996).
90. P. Jackson, *Anal. Biochem.* **216**, 243–252 (1994).
91. K.K.S. Ng, K. Drickamer, and W.I. Weis, *J. Biol. Chem.* **271**, 663–674 (1996).
92. M.E. Taylor and K. Drickamer, *J. Biol. Chem.* **268**, 399–404 (1993).
93. P.J. Kniskern, A. Hagopian, P. Burke, L.D. Schulz, D.L. Montgomery, W.M. Hurni, C.Y. Ip, C.A. Schulman, R.Z. Maigetter, D.E. Wampler, D. Kubek, R.D. Sitrin, D.J. West, R.W. Ellis, and W.J. Miller, *Vaccine* **12**, 1021–1025 (1994).
94. L. Lehle, A. Eiden, K. Lehnert, A. Haselbeck, and E. Kopetzki, *FEBS Lett.* **370**, 41–45 (1995).
95. Z. Kawar, A. Herscovics, and D.L. Jarvis, *Glycobiology* **7**, 433–443 (1997).
96. D.J. Davidson, R.K. Bretthauer, and F.J. Castellino, *Biochemistry* **30**, 9811–9815 (1991).
97. J.X. Ren, R.K. Bretthauer, and F.J. Castellino, *Biochemistry* **34**, 2489–2495 (1995).
98. H.D. Klenk, *Cytotechnology* **20**, 139–144 (1996).
99. E. Kretzschmar, R. Geyer, and H.D. Klenk, *Biol. Chem.* Hoppe-Seyler **375**, 323–327 (1994).
100. M. Manneberg, A. Friedlein, H. Kurth, H.W. Lahm, and M. Fountoulakis, *Protein Sci.* **3**, 30–38 (1994).
101. F. Altmann, G. Kornfeld, T. Dalik, E. Staudacher, and J. Glossl, *Glycobiology* **3**, 619–625 (1993).
102. V. Kubelka, F. Altmann, G. Kornfeld, and L. Marz, *Arch. Biochem. Biophys.* **308**, 148–157 (1994).
103. F. Altmann and L. Marz, *Glycoconjugate Journal* **12**, 150–155 (1995).
104. D.L. Jarvis, D.A. Bohlmeyer, Y.F. Liao, K.K. Lomax, R.K. Merkle, C. Weinkauf, and K.W. Moremen, *Glycobiology* **7**, 113–127 (1997).
105. F. Altmann, H. Schwihla, E. Staudacher, J. Glossl, and L. Marz, *J. Biol. Chem.* **270**, 17344–17349 (1995).
106. I. Vandie, A. Vantetering, H. Bakker, D.H. Vandeneijnden, and D.H. Joziasse, *Glycobiology* **6**, 157–164 (1996).
107. R. Wagner, H. Geyer, R. Geyer, and H.D. Klenk, *J. Virol.* **70**, 4103–4109 (1996).
108. O.W. Ogonah, R.B. Freedman, N. Jenkins, K. Patel, and B.C. Rooney, *Bio / Technology* **14**, 197–202 (1996).
109. T.A. Hsu, N. Takahashi, Y. Tsukamoto, K. Kato, I. Shimada, K. Masuda, E.M. Whiteley, J.Q. Fan, Y.C. Lee, and M.J. Betenbaugh, *J. Biol. Chem.* **272**, 9062–9070 (1997).
110. P.C. Kulakosky, M.L. Shuler, and H.A. Wood, *In Vitro Cell. Dev. Biol. Animal* **34**, 101–108 (1998).
111. D.J. Davidson and F.J. Castellino, *Biochemistry* **30**, 6689–6696 (1991).
112. D.J. Davidson, M.J. Fraser, and F.J. Castellino, *Biochemistry* **29**, 5584–5590 (1990).
113. D.J. Davidson and F.J. Castellino, *Biochemistry* **30**, 6167–6174 (1991).
114. C.A.K. Borrebaeck, A.C. Malmborg, and M. Ohlin, *Immunol. Today* **14**, 477–479 (1993).
115. D.M. Sheeley, B.M. Merrill, and L.C.E. Taylor, *Anal. Biochem.* **247**, 102–110 (1997).
116. R.M. Hamadeh, G.A. Jarvis, U. Galili, R.E. Mandrell, P. Zhou, and J.M. Griffiss, *J. Clin. Invest.* **89**, 1223–1235 (1992).
117. J.A. Lavecchio, A.D. Dunne, and A.S.B. Edge, *Transplantation* **60**, 841–847 (1995).
118. R.D. Larsen, C.A. Riveramarrero, L.K. Ernst, R.D. Cummings, and J.B. Lowe, *J. Biol. Chem.* **265**, 7055–7061 (1990).
119. D.F. Smith, R.D. Larsen, S. Mattox, J.B. Lowe, and R.D. Cummings, *J. Biol. Chem.* **265**, 6225–6234 (1990).
120. E.A. Muchmore, M. Milewski, A. Varki, and S. Diaz, *J. Biol. Chem.* **264**, 20216–20223 (1989).
121. T. Kawano, S. Koyama, H. Takematsu, Y. Kozutsumi, H. Kawasaki, S. Kawashima, T. Kawasaki, and A. Suzuki, *J. Biol. Chem.* **270**, 16458–16463 (1995).
122. T.J. Monica, S.B. Williams, C.F. Goochee, and B.L. Maiorella, *Glycobiology* **5**, 175–185 (1995).
123. J. Marzowski, W.C. Wang, A.R. Flesher, and H.V. Raff, *Biotechnol. Bioeng.* **46**, 399–407 (1995).
124. C.H. Hokke, A.A. Bergwerff, G.W.K. Vandedem, J. Vanoostrum, J.P. Kamerling, and J.F.G. Vliegenthart, *FEBS Lett.* **275**, 9–14 (1990).
125. A. Noguchi, C.J. Mukuria, E. Suzuki, and M. Naiki, *Nephron* **72**, 599–603 (1996).
126. A. Noguchi, C.J. Mukuria, E. Suzuki, and M. Naiki, *J. Biochem.* **117**, 59–62 (1995).
127. K.Y. Do and R.D. Cummings, *J. Biol. Chem.* **268**, 22028–22035 (1993).
128. K.Y. Do, D.F. Smith, and R.D. Cummings, *Biochem. Biophys. Res. Commun.* **173**, 1123–1128 (1990).
129. M. Hummel, H.C. Hedrich, and A. Hasilik, *Eur. J. Biochem.* **245**, 428–433 (1997).
130. E. Watson, B. Shah, L. Leiderman, Y.R. Hsu, S. Karkare, H.S. Lu, and F.K. Lin, *Biotechnol. Prog.* **10**, 39–44 (1994).
131. T. Edmunds, S.M. VanPatten, J. Pollock, E. Hanson, R. Bernasconi, E. Higgins, P. Manavalan, C. Ziomek, H. Meade, J.M. McPherson, and E.S. Cole, *Blood* **91**, 4561–4571 (1998).
132. V.P. Korhonen, M. Tolvanen, J.M. Hyttinen, M. UusiOukari, R. Sinervirta, L. Alhonen, M. Jauhiainen, O.A. Janne, and J. Janne, *Eur. J. Biochem.* **245**, 482–489 (1997).
133. M. Stromqvist, J. Tornell, M. Edlund, A. Edlund, T. Johansson, K. Lindgren, L. Lundberg, and L. Hansson, *Transgenic Res.* **5**, 475–485 (1996).
134. J.K.C. Ma, A. Hiatt, M. Hein, N.D. Vine, F. Wang, P. Stabila, C. Vandolleweerd, K. Mostov, and T. Lehner, *Science* **268**, 716–719 (1995).
135. L. Miele, *Trends Biotechnol.* **15**, 45–50 (1997).
136. M.D. Smith and B.R. Glick, *Food Technol. Biotechnol.* **35**, 183–191 (1997).
137. G.C. Whitelam, *J. Sci. Food Agric.* **68**, 1–9 (1995).
138. F. Altmann, V. Tretter, V. Kubelka, E. Staudacher, L. Marz, and W.M. Becker, *Glycoconjugate Journal* **10**, 301–301 (1993).
139. S. Matsumoto, K. Ikura, M. Ueda, and R. Sasaki, *Plant Mol. Biol.* **27**, 1163–1172 (1995).

140. E. Grabenhorst, A. Hoffmann, M. Nimtz, G. Zettlmeissl, and H.S. Conradt, *Eur. J. Biochem.* **232**, 718–725 (1995).

141. J. Ferrari, R. Harris, and T.G. Warner, *Glycobiology* **4**, 367–373 (1994).

142. A. Gregoire, A. Visvikis, A. Marc, and J.-L. Goergen, in O.-W. Merten, P. Perrin, and B. Griffiths eds., *New Developments and New Applications in Animal Cell Technology: Proceedings of the 15th ESACT Meeting*, 1998, pp. 191–193.

143. A.R. Sburlati, P. Umana, E.G.P. Prati, and J.E. Bailey, *Biotech. Prog.* **14**, 189–192 (1998).

144. J.P. Soulillou, *Biodrugs* **8**, 29–32 (1997).

145. E.G.P. Prati, P. Scheidegger, A.R. Sburlati, and J.E. Bailey, *Biotechnol. Bioeng.* **59**, 445–450 (1998).

146. J. Costa, E. Grabenhorst, M. Nimtz, and H.S. Conradt, *J. Biol. Chem.* **272**, 11613–11621 (1997).

147. J.R. Hollister, J.H. Shaper, and D.L. Jarvis, *Glycobiology* **8**, 473–480 (1998).

148. P.A. Prieto, P. Mukerji, B. Kelder, R. Erney, D. Gonzalez, J.S. Yun, D.F. Smith, K.W. Moremen, C. Nardelli, M. Pierce, Y.S. Li, X. Chen, T.E. Wagner, R.D. Cummings, and J.J. Kopchick, *J. Biol. Chem.* **270**, 29515–29519 (1995).

149. E.S. Cole, E.H. Nichols, L. Poisson, M.L. Harnois, and D.J. Livingston, *Fibrinolysis* **7**, 15–22 (1993).

150. A.R. Flesher, J. Marzowski, W.C. Wang, and H.V. Raff, *Biotechnol. Bioeng.* **47**, 405–405 (1995).

151. B. Smedsrod, and M. Einarsson, *Thromb. Haemostasis* **63**, 60–66 (1990).

152. S.L. Minch, P.T. Kallio, and J.E. Bailey, *Biotechnol. Prog.* **11**, 348–351 (1995).

153. L. Monaco, A. Marc, A. EonDuval, G. Acerbis, G. Distefano, D. Lamotte, J.M. Engasser, M. Soria, and N. Jenkins, *Cytotechnology* **22**, 197–203 (1996).

154. E. Grabenhorst, J. Costa, and H.S. Conradt, *Mol. Biol. Cell* **7**, 2583–2583 (1996).

155. D.L. Jarvis and E.E. Finn, *Nat. Biotechnol.* **14**, 1288–1292 (1996).

156. M.W. Spellman, L.J. Basa, C.K. Leonard, J.A. Chakel, J.V. Oconnor, S. Wilson, and H. Vanhalbeek, *J. Biol. Chem.* **264**, 14100–14111 (1989).

157. M. Otter, M.M. Barrettbergshoeff, and D.C. Rijken, *J. Biol. Chem.* **266**, 13931–13935 (1991).

158. C.R. Benedict, C.J. Refino, B.A. Keyt, R. Pakala, N.F. Paoni, G.R. Thomas, and W.F. Bennett, *Circulation* **92**, 3032–3040 (1995).

159. B.A. Keyt, N.F. Paoni, C.J. Refino, L. Berleau, H. Nguyen, A. Chow, J. Lai, L. Pena, C. Pater, J. Ogez, T. Etcheverry, D. Botstein, and W.F. Bennett, *Proc. Nat. Acad. Sci. U.S.A.* **91**, 3670–3674 (1994).

160. N.F. Paoni, B.A. Keyt, C.J. Refino, A.M. Chow, H.V. Nguyen, L.T. Berleau, J. Badillo, L.C. Pena, K. Brady, F.M. Wurm, J. Ogez, and W.F. Bennett, *Thromb. Haemostasis* **70**, 307–312 (1993).

161. J. Lund, N. Takahashi, J.D. Pound, M. Goodall, and R. Jefferis, *J. Immunol.* **157**, 4963–4969 (1996).

162. R. Jefferis, J. Lund, and M. Goodall, *Immunol. Lett.* **44**, 111–117 (1995).

163. T.P. Patel, R.B. Parekh, B.J. Moellering, and C.P. Prior, *Biochem. J.* **285**, 839–845 (1992).

164. M.R. Lifely, C. Hale, S. Boyce, M.J. Keen, and J. Phillips, *Glycobiology* **5**, 813–822 (1995).

165. M. Gawlitzek, H.S. Conradt, and R. Wagner, *Biotechnol. Bioeng.* **46**, 536–544 (1995).

166. M.C. Borys, D.I.H. Linzer, and E.T. Papoutsakis, *Biotechnol. Bioeng.* **43**, 505–514 (1994).

167. D.C. Andersen and C.F. Goochee, *Biotechnol. Bioeng.* **47**, 96–105 (1995).

168. B. Thorens and P. Vassalli, *Nature* **321**, 618–620 (1986).

169. T. Ryll, U. Valley, and R. Wagner, *Biotechnol. Bioeng.* **44**, 184–193 (1994).

170. M. Gawlitzek, U. Valley, and R. Wagner, *Biotechnol. Bioeng.* **57**, 518–528 (1998).

171. S.I. Grammatikos, U. Valley, M. Nimtz, H.S. Conradt, and R. Wagner, *Biotechnol. Prog.* **14**, 410–419 (1998).

172. A. Cayli, M. Wirth, and R. Wagner, in W. Merten, P. Perrin, and B. Griffiths eds., *New Developments and New Applications in Animal Cell Technology: Proceedings of the 15th ESACT Meeting*, Kluwer Academic, Dordrecht, 1998, pp. 157–160.

173. C.R. Bebbington, G. Renner, S. Thompson, D. King, D. Abrams, and G.T. Yarranton, *Bio / Technology* **10**, 169–175 (1992).

174. M.J. Gramer and C.F. Goochee, *Biotechnol. Bioeng.* **43**, 423–428 (1994).

175. M.J. Gramer, C.F. Goochee, V.Y. Chock, D.T. Brousseau, and M.B. Sliwkowski, *Bio / Technology* **13**, 692–698 (1995).

176. E. Munzert, R. Heidemann, H. Buntemeyer, J. Lehmann, and J. Muthing, *Biotechnol. Bioeng.* **56**, 441–448 (1997).

177. E. Munzert, J. Muthing, H. Buntemeyer, and J. Lehmann, *Biotechnol. Prog.* **12**, 559–563 (1996).

178. T.G. Warner, J. Chang, J. Ferrari, R. Harris, T. McNerney, G. Bennett, J. Burnier, and M.B. Sliwkowski, *Glycobiology* **3**, 455–463 (1993).

179. X.J. Gu, B.J. Harmon, and D.I.C. Wang, *Biotechnol. Bioeng.* **55**, 390–398 (1997).

180. M.J. Gramer, D.V. Schaffer, M.B. Sliwkowski, and C.F. Goochee, *Glycobiology* **4**, 611–616 (1994).

181. A.D. Hooker, M.H. Goldman, N.H. Markham, D.C. James, A.P. Ison, A.T. Bull, P.G. Strange, I. Salmon, A.J. Baines, and N. Jenkins, *Biotechnol. Bioeng.* **48**, 639–648 (1995).

182. D.K. Robinson, C.P. Chan, C.Y. Ip, P.K. Tsai, J. Tung, T.C. Seamans, A.B. Lenny, D.K. Lee, J. Irwin, and M. Silberklang, *Biotechnol. Bioeng.* **44**, 727–735 (1994).

183. M.H. Goldman, D.C. James, M.H. Rendall, A.P. Ison, M. Hoare, and A.T. Bull, *Biotechnol. Bioeng.*, **60**, 596–607 (1998).

184. R. Kimura and W.M. Miller, *Biotechnol. Prog.* **13**, 311–317 (1997).

185. A.A. Lin, R. Kimura, and W.M. Miller, *Biotechnol. Bioeng.* **42**, 339–350 (1993).

186. G.F. Herrmann, Y. Ichikawa, C. Wandrey, F.C.A. Gaeta, J.C. Paulson, and C.H. Wong, *Tetrahedron Lett.* **34**, 3091–3094 (1993).

187. Y. Nakahara, S. Shibayama, Y. Nakahara, and T. Ogawa, *Carbohydr. Res.* **280**, 67–84 (1996).

188. M. Gilbert, R. Bayer, A.M. Cunningham, S. Defrees, Y.H. Gao, D.C. Watson, N.M. Young, and W.W. Wakarchuk, *Nat. Biotechnol.* **16**, 769–772 (1998).

See also CHINESE HAMSTER OVARY CELLS, RECOMBINANT PROTEIN PRODUCTION; PROTEIN GLYCOLYSATION.

GLYOXYLATE BYPASS, REGULATION

DAVID C. LAPORTE
STEPHEN P. MILLER
SATINDER K. SINGH
University of Minnesota
Minneapolis, Minnesota

KEY WORDS

Bypass
Glyoxylate
Isocitrate dehydrogenase
Isocitrate lyase
Kinase
Krebs' cycle
Phosphatase
Phosphorylation

OUTLINE

INTRODUCTION

Overview

The glyoxylate bypass of the bacterium *Escherichia coli* has been used as a model system for the study of metabolic regulation. *E. coli* is an ideal organism for these studies because of the powerful genetic tools that can be used to achieve experimental manipulations that are not possible in many other systems. The long-term goal of these projects is to extract fundamental principles of metabolic control from a detailed analysis of how each regulatory mechanism functions and how these mechanisms interact to control the entire process.

The Glyoxylate Bypass

Adaptation of *E. coli* to growth on acetate or fatty acids requires the induction of the enzymes of the glyoxylate bypass: isocitrate lyase and malate synthase (1–4). This pathway is essential for growth under these conditions because it prevents the net loss of the entering carbon as CO_2 in the Krebs' cycle (Fig. 1) (5,6).

The glyoxylate bypass is regulated at two levels. Coarse control of the glyoxylate bypass is achieved by regulating the transcription of *aceBAK*. The *aceBAK* operon encodes malate synthase, isocitrate lyase, and isocitrate dehydrogenase (IDH) kinase/phosphatase, respectively. Once induced, the flow of isocitrate through the bypass is regulated by the phosphorylation of IDH, the Krebs' cycle enzyme that competes with isocitrate lyase. This phosphorylation cycle is catalyzed by a bifunctional kinase/ phosphatase. Both of these regulatory mechanisms are discussed later.

REGULATION OF TRANSCRIPTION

Regulation of the Glyoxylate Bypass Operon

The regulatory circuits that control *aceBAK* are illustrated in Figure 2.

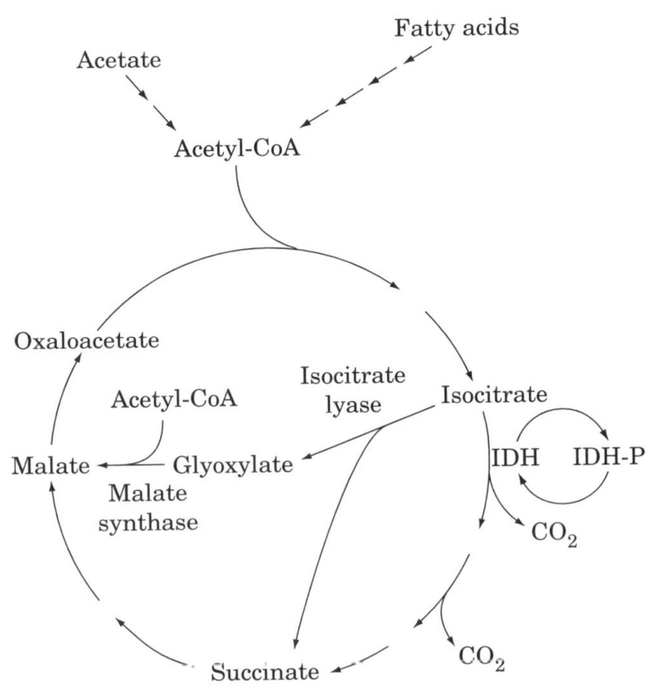

Figure 1. The Krebs' cycle and the glyoxylate bypass. The glyoxylate bypass is catalyzed by isocitrate lyase and malate synthase. The flux through this pathway is regulated by phosphorylation of isocitrate dehydrogenase (IDH).

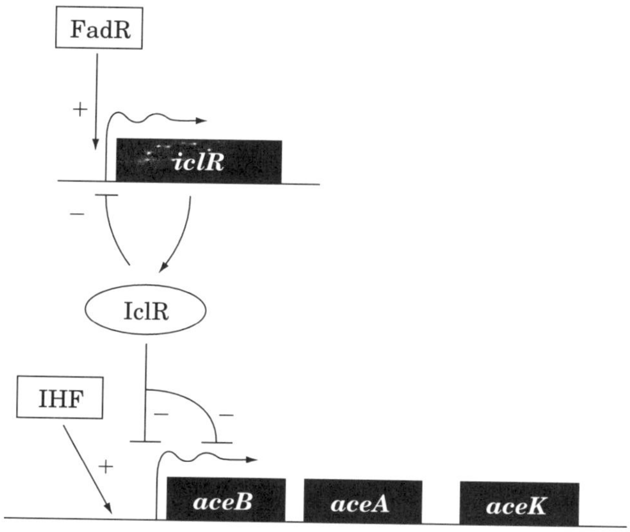

Figure 2. Regulation of the glyoxylate bypass operon. *aceBAK* encodes malate synthase, isocitrate lyase, and IDH kinase/phosphatase, respectively. This operon is repressed by IclR and activated by IHF. The gene encoding IclR is repressed by its own product and activated by FadR.

The *aceBAK* operon appears to respond to metabolic stress. Full expression is observed only during growth on acetate or fatty acids. However, the presence of these substrates is not sufficient for induction because the operon is repressed by the simultaneous presence of preferred carbon sources such as glucose or pyruvate (1–4). Poor carbon sources produce a partial induction of *aceBAK* even though the glyoxylate bypass is not required for growth on these substrates. This partial induction appears to be a response to general metabolic stress because there is a roughly linear relationship between the culture doubling time and the extent of induction when the different growth conditions are compared. Finally, partial induction of the operon occurs upon growth into the stationary phase on rich media (6).

The *aceBAK* operon is transcribed from a single, $\sigma 70$-type promoter (3). The activity of this promoter is controlled by a repressor protein, IclR, which binds to a site that overlaps the "−35" box of the promoter (3,7,8). Preliminary evidence indicates that IclR represses transcription by competing with RNA polymerase (unpublished). The mechanism that regulates IclR activity has not been determined.

A Remote Operator for IclR

We have recently identified a remote site for IclR. This operator lies within the coding region of *aceB*. Preliminary evidence indicates that binding of IclR to this site does not directly repress *aceBAK* expression. Rather, this site appears to act cooperatively with the promoter-proximal operator, a site that does repress this operon (unpublished). Our working model proposes that this cooperativity results from simultaneous binding of IclR at both sites, with looping out of the intervening DNA. This type of cooperativity between remote sites has been most extensively characterized for the *lac* operon but appears to occur in other operons as well (9–11).

The Role of Integration Host Factor

Integration host factor (IHF) is a sequence-specific DNA binding protein that binds to many sites in the *E. coli* chromosome. This histonelike protein has been implicated in a wide variety of processes including plasmid replication, lambda phage recombination, and transcriptional activation and repression (12–14).

We have identified a binding site for IHF upstream of the *aceBAK* promoter. IHF activates *aceBAK* expression under inducing conditions by preventing IclR repression. In contrast, IHF does not appear to affect IclR activity under repressing conditions (15). The carbon-source specificity of the IHF effect amplifies regulation of *aceBAK*.

IHF can affect transcription by a variety of mechanisms. The most clearly established mechanism is one in which IHF binds to a site well removed from the promoter, bending the DNA at an angle of about 140° and so bringing a remote activator site near the promoter. Similar mechanisms have been proposed for IHF in processes as diverse as plasmid replication and recombination (12–14,16–18).

IHF may also act by direct interactions with other proteins. For example, IHF has been suggested to activate transcription at three promoters by direct contact with RNA polymerase, although this remains to be established (13,14). IHF can also repress transcription by binding to sites that overlap a promoter. In these systems, IHF probably acts by competing with RNA polymerase (19).

IHF plays a novel role in the regulation of *aceBAK* expression. In their recent review, Goosen and Van de Putte have suggested that IHF may not participate in transcriptional regulation (13). This is clearly not true for *aceBAK*, which requires IHF for full carbon-source sensitivity. However, the mechanism by which IHF plays this role must still be determined.

Our working model proposes that IHF and IclR bind independently near the *aceBAK* promoter under repressing conditions (Fig. 3). Under inducing conditions, we suspect that IclR undergoes a structural change that causes the binding of IHF and IclR to become competitive. IclR repression is relieved under these conditions because the IHF site is occupied.

Regulation of *iclR* Expression

The gene encoding IclR is subject to repression by its own product. IclR achieves this autorepression by binding to an operator that overlaps the promoter of *iclR*. This provides a feedback control mechanism that helps maintain a constant level of IclR (20).

The expression of *iclR* is also regulated by FadR. FadR was initially identified because it represses the genes encoding the enzymes of fatty acid degradation (21). It was subsequently shown to activate the transcription of *fabA*, a gene whose product participates in unsaturated fatty acid biosynthesis (22). FadR activates *iclR* expression and may provide a link between *aceBAK* and fatty acids, a carbon source that also requires the glyoxylate bypass (L. Gui, A. Sunnarborg, and D.C. LaPorte, unpublished, 1996).

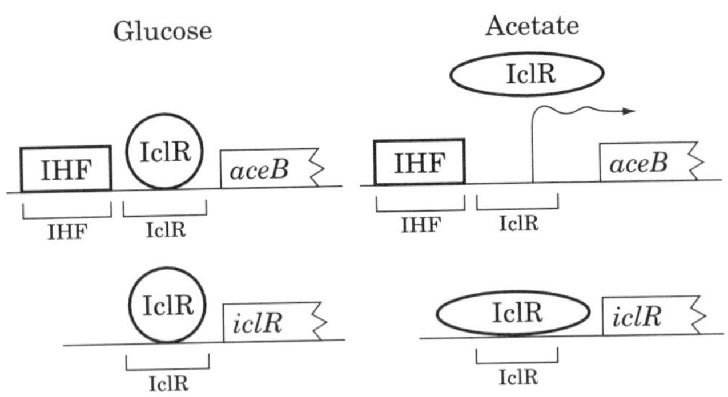

Figure 3. Model for the role of IHF in regulation of IclR. IclR binds to sites that overlap the promoters of *aceBAK* and *iclR*. During growth on glucose (repressing conditions), IHF and IclR can bind simultaneously near the *aceBAK* promoter. Adaptation to acetate (inducing) produces a conformational change in IclR. This conformational change prevents IclR from binding when IHF is bound but has no effect on IclR binding in the absence of IHF. Repression of *aceBAK* is released on acetate because of the presence of an IHF site. Repression of *iclR* is constitutive because this promoter does not include an IHF site.

Differential Control by IclR

Perhaps the most unusual feature of this system is the ability of IclR to differentially regulate *aceBAK* and *iclR*. Whereas IclR-mediated repression of *aceBAK* undergoes a striking response to the carbon source, repression of *iclR* is nearly constant. This difference is probably an advantage to the cell because induction of *iclR* expression on acetate would oppose induction of *aceBAK*. IHF, which is required for the carbon-source sensitivity of IclR, appears to be responsible for differential control by this repressor. *aceBAK* responds to the carbon source because it has an associated site for IHF (see earlier). In contrast, *iclR* does not respond to the carbon source because it does not have a site for IHF (15).

Another Role for IclR?

In addition to the regulation of *aceBAK* and *iclR*, IclR also appears to regulate *acs*, the gene encoding acetyl coenzyme A synthetase in *E. coli*. Mutations in *iclR* reduced consumption of acetate mediated by Acs (23). However, the precise role of IclR in regulating *acs* remains uncertain.

IclR Is a Member of a Unique Protein Family

IclR has lent its name to a family of transcriptional regulatory proteins. This family is characterized by about 25% identity throughout their sequences, with particularly high homology in a region near the N-terminus. This is the region where we identified sequences that appear to match the consensus for helix-turn-helix DNA-binding motifs (24). An "IclR signature" in the C-terminal regions of these proteins has also been defined in the PROSITE database (25).

The IclR family is distinct from other families of transcriptional regulators in bacteria. It is broadly distributed, currently including 19 members found in 12 different Gram-positive and Gram-negative bacterial species; in some cases, the function of the protein is known. Others were identified during genomic sequencing projects, and their functions remain to be identified. The proteins of known function regulate a wide variety of catabolic genes. IclR also fits this pattern because the glyoxylate bypass is essential for the catabolism of acetate (24,26–31 and the NCBI database at *www.ncbi.nlm.nih.gov*).

REGULATION OF PROTEIN PHOSPHORYLATION

Fine Control: The Role of IDH Phosphorylation

Fine control of the glyoxylate bypass is achieved by the phosphorylation of IDH, the Krebs' cycle enzyme that competes with isocitrate lyase. This phosphorylation cycle is catalyzed by a bifunctional protein, IDH kinase/phosphatase (32–34).

Structural Basis of the Regulation of IDH

IDH is a member of a large family of dehydrogenases. This family, comprising at least 17 proteins, includes members from prokaryotes, yeast, and humans. Although some of these enzymes act on isocitrate, others act on isopropyl malate or tartarate. Members of this family exhibit extensive sequence homology, particularly among residues that have been implicated in substrate binding and catalysis for *E. coli* IDH. These enzymes lack the Rossman fold, which is an important structural feature of many other dehydrogenases (35–40).

The phosphorylated form of IDH is completely inactive (33). Inactivation results from the phosphorylation of Ser-113, an active site residue that forms a hydrogen bond with isocitrate in dephospho-IDH (41,42). Phosphorylation has little effect on the three-dimensional structure of IDH. Rather, it blocks isocitrate binding primarily by electrostatic and steric repulsion (43–46).

The mechanism by which phosphorylation controls IDH is a striking contrast with regulatory phosphorylation of glycogen phosphorylase, the only other phosphoprotein whose structure has been extensively characterized. The phosphorylation site of glycogen phosphorylase is remote from the active site. Phosphorylation of this enzyme activates catalysis by inducing a conformational change (47,48).

Structure and Mechanism of IDH Kinase/Phosphatase

IDH kinase/phosphatase is expressed from a single gene and appears to have a single active site that catalyzes the kinase and phosphatase reactions (4,33,34,49–51). This active site also appears to catalyze an ATPase activity that is more efficient then either the kinase or phosphatase reaction (52). Our working model proposes that the IDH phosphatase reaction results from the kinase back reaction tightly coupled to ATP hydrolysis (reviewed in Ref. 34) (Fig. 4).

Regulation of the IDH Kinase/Phosphatase

The IDH phosphorylation cycle is controlled by a wide variety of metabolites. Examples include isocitrate, NADP, 3-

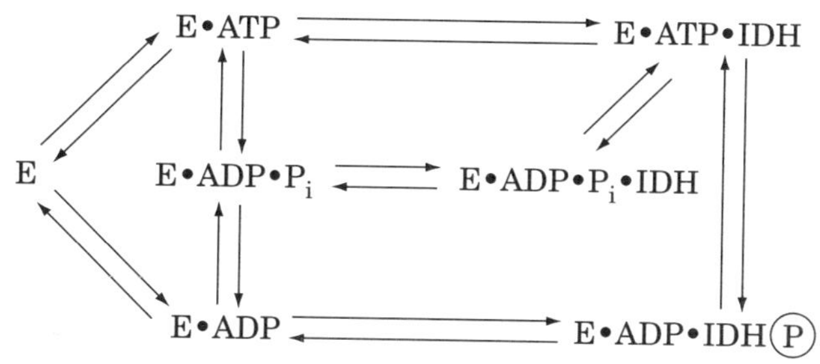

Figure 4. Model for the catalytic mechanism of IDH kinase/phosphatase. This model proposes that the kinase and phosphatase reactions occur at the same active site. The intrinsic ATPase activity of IDH kinase/phosphatase, which occurs even in the absence of IDH, also occurs at this site. The phosphatase reaction results from the back reaction of IDH kinase coupled to the hydrolysis of ATP. E, IDH kinase/phosphatase; IDH, isocitrate dehydrogenase; IDH-P, the phosphorylated form of IDH.

phosphoglycerate, pyruvate, and AMP. There appear to be two distinct mechanisms by which these ligands regulate IDH kinase/phosphatase (Fig. 5) (4,34,50,53–55).

Access to the phosphorylation site of IDH is controlled by the substrates of this enzyme. When these substrates bind to IDH, they induce a conformational change, narrowing the active site cleft and burying the site of phosphorylation. Because isocitrate binds only to dephospho-IDH, this ligand inhibits only IDH kinase. Isocitrate activates IDH phosphatase by preventing product inhibition by isocitrate (56; S.P. Miller, E.J. Karschnia, A.M. Dean, and D.C. LaPorte, unpublished, 1998).

A second mechanism for regulating this cycle results from a conformational equilibrium in IDH kinase/phosphatase. The equilibrium between the kinase and phosphatase conformations is controlled by metabolites that bind preferentially to the phosphatase conformation. Preferential binding activates this activity and inhibits IDH kinase (S.P. Miller, E.J. Karschnia, C. Romfo, and D.C. LaPorte, unpublished, 1998).

CONTROL OF METABOLISM

Growth on Glucose

The glyoxylate bypass serves no useful purpose during growth on carbon sources such as glucose or pyruvate. Under these conditions, *aceBAK* is repressed by IclR. Although a basal level of IDH kinase/phosphatase is present, the conditions favor the dephosphorylation of IDH. Pyruvate, which activates IDH phosphatase and inhibits IDH kinase, appears to be the primary regulator of this phosphorylation cycle under these conditions (6,34).

Growth on Acetate

Adaptation to growth on acetate requires the induction of the glyoxylate bypass operon. The mechanisms responsible for transcriptional regulation have already been discussed.

During steady-state growth on acetate, about 75% of the IDH is converted to the phosphorylated form. Because phospho-IDH is completely inactive, partial phosphorylation of IDH inhibits the Krebs' cycle and so forces isocitrate through the glyoxylate bypass (54,57). Under these conditions, isocitrate must be partitioned between the Krebs' cycle for the production of energy and the glyoxylate bypass for the production of metabolic intermediates. The primary regulators of IDH kinase/phosphatase during growth on acetate appear to be isocitrate, 3-phosphoglycerate, and AMP. Isocitrate and 3-phosphoglycerate act as indicators of the levels of metabolites, whereas AMP reflects the energy needs of the cell (34).

Metabolic Transitions

The flux through the glyoxylate bypass responds dramatically to the culture conditions. For example, addition of glucose to a culture growing on acetate virtually eliminates flux through this pathway (57,58). Inhibition of the bypass is probably advantageous because, when glucose is available, the glyoxylate bypass is no longer required to replenish metabolites used for biosynthesis. Inhibition of the bypass results, in part, from the rapid dephosphorylation of IDH, increasing this enzyme's activity and thus diverting isocitrate through the Krebs' cycle (54,55). IDH kinase/phosphatase catalyzes the rate-limiting step in this transition (55). Long-term adaptation to the carbon source results from control of *aceBAK* transcription.

The response of the IDH phosphorylation cycle during these metabolic transitions is amplified by two mechanisms. Zero-order ultrasensitivity results when one of the modification enzymes (e.g., kinase or phosphatase) is saturated by its protein substrate. Multistep effects, which can either enhance or reduce sensitivity, result when an effector controls more than one step in a protein modification cascade (53). The glyoxylate bypass provided the first experimental demonstrations of these effects. Both phenomena have since been found in other systems.

Study of the glyoxylate bypass led to the discovery of another mechanism for sensitivity enhancement: the branch point effect. This effect results from the competition between enzymes for a limiting amount of a common substrate. During growth on acetate, isocitrate levels are nearly saturating for IDH but are within the pseudo-first-order range of isocitrate lyase. A consequence of this difference in saturation is that the flux carried by isocitrate lyase is ultrasensitive to the phosphorylation state of IDH. Both the theoretical and experimental studies responsible for identifying this phenomena were the products of studies on the glyoxylate bypass (54).

FUTURE DIRECTIONS

The glyoxylate bypass has proven to be a very useful model for metabolic control. The basic control mechanisms have all been identified, with the important exception of the mechanism that controls the IclR repressor. We also have considerable understanding of metabolite levels, flux rates, and how these factors are controlled. These studies have led to the identification of several fundamental principles of metabolic control.

Although the basic control processes have been identified, much remains to be learned concerning their mechanisms of action. How, for example, does IHF control the carbon-source sensitivity of IclR? Are the models that we

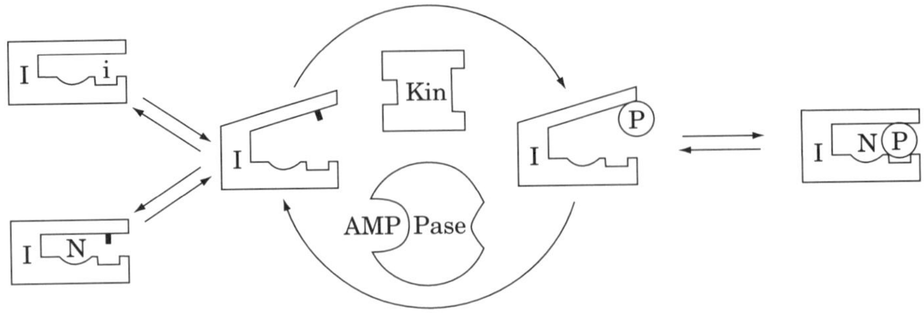

Figure 5. Model for the regulation of the IDH phosphorylation cycle. This model proposes that IDH kinase/phosphatase can exist in two conformations, one favoring the kinase reaction (Kin) and the other favoring the phosphatase (Pase). This conformational equilibrium is proposed to be controlled by metabolites such as AMP, which bind preferentially to the phosphatase form. IDH (I) can exist in either the open or closed conformation, depending on whether the substrates isocitrate (i) or $NADP^+$/NADPH (N) are bound. The open conformation is a substrate for IDH kinase/phosphatase, whereas the closed conformation is not. The site of phosphorylation, Ser-113, is indicated with a tick mark for dephospho-IDH. Phosphoserine-113 is indicated by Ⓟ.

proposed for the catalytic mechanism and regulation of IDH kinase/phosphatase correct and, if so, what are their structural bases? As so often happens in biology, the more we know, the more we realize how much we do not know.

BIBLIOGRAPHY

1. C.B. Brice and H.L. Kornberg, *J. Bacteriol.* **96**, 2185–2186 (1968).
2. S.R. Maloy and W.D. Nunn, *J. Bacteriol.* **149**, 173–180 (1982).
3. T. Chung, D.J. Klumpp, and D.C. LaPorte, *J. Bacteriol.* **170**, 386–392 (1988).
4. D.C. LaPorte, P.E. Thorsness, and D.E. Koshland, Jr. *J. Biol. Chem.* **260**, 10563–10568 (1985).
5. H.L. Kornberg and N.B. Madsen, *Biochim. Biophys. Acta* **24**, 651–653 (1951).
6. H.L. Kornberg, *Biochem. J.* **99**, 1–11 (1966).
7. B. Pan, I. Unnikrishnan, and D.C. LaPorte, *J. Bacteriol.* **178**, 4704 (1996).
8. D. Negre, J.-C. Cortay, A. Galinier, P. Sauve, and A.J. Cozzone, *J. Mol. Biol.* **228**, 23–29 (1992).
9. R. Schleif, *Ann. Rev. Biochem.* **61**, 199–223 (1992).
10. A.M. Friedman, T.O. Fischmann, and T.A. Steitz, *Science* **268**, 1721–7 (1995).
11. A.E. Chakerian and K.S. Matthews, *Mol. Microbiol.* **6**, 963–8 (1992).
12. D.I. Friedman, *Cell* **55**, 545–554 (1988).
13. N. Goosen and P. van de Putte, *Mol. Microbiol.* **16**, 1–7 (1995).
14. M. Freundlich, N. Ramani, E. Mathew, A. Sirko, and P. Tsui, *Mol. Microbiol.* **6**, 2557–2563 (1992).
15. E. Resnik, B. Pan, N. Ramani, M. Freundlich, and D.C. LaPorte, *J. Bacteriol.* **178**, 2715– (1996).
16. S.D. Goodman, S.C. Nicholson, and H.A. Nash, *Proc. Natl. Acad. Sci. U.S.A.* **89**, 11910–11914 (1992).
17. L. Moitoso de Vargas, S. Kim, and A. Landy, *Science* **244**, 1457–1461 (1989).
18. T.T. Stenzel, P. Patel, and D. Bastia, *Cell* **49**, 709–717 (1987).
19. P. Tsui, L. Huang, and M. Freundlich, *J. Bacteriol.* **173**, 5800–5807 (1991).
20. L. Gui, A.R. Sunnarborg, B. Pan, and D.C. Laporte, *J. Bacteriol.* **178**, 321– (1996).
21. W.D. Nunn, *Microbiol. Rev.* **50**, 179–192 (1986).
22. M.F. Henry and J.E.J. Cronan, *J. Biol. Chem.* **222**, 843–849 (1991).
23. S. Shin, S.G. Song, D.S. Lee, J.G. Pan, and C. Park, *FEMS Microbiol. Lett.* **146**, 103–108 (1997).
24. A. Sunnarborg, D. Klumpp, T. Chung, and D.C. LaPorte, *J. Bacteriol.* **172**, 2642–2649 (1990).
25. A. Bairoch, P. Bucher, and K. Hofmann, *Nucl. Acids Res.* **25**, 217–221 (1997).
26. A.A. DiMarco, B. Averhoff, and L.N. Ornston, *J. Bacteriol.* **175**, 4499–4506 (1993).
27. A. Ferrandez, J.L. Garcia, and E. Diaz, *J. Bacteriol.* **179**, 2573–2581 (1997).
28. M.A. Prieto, E. Diaz, and J.L. Garcia, *J. Bacteriol.* **178**, 111–120 (1996).
29. C.P. Smith and K.F. Chater, *J. Mol. Biol.* **204**, 569–580 (1988).
30. M.R. Barnes, W.A. Duetz, and P.A. Williams, *J. Bacteriol.* **179**, 6145–6153 (1997).
31. C.S. Harwood, N.N. Nichols, M.K. Kim, J.L. Ditty, and R. Parales, *J. Bacteriol.* **176**, 6479–6488 (1994).
32. M. Garnak and H.C. Reeves, *J. Biol. Chem.* **254**, 7915–7920 (1979).
33. D.C. LaPorte and D.E. Koshland, Jr., *Nature* **300**, 458–460 (1982).
34. D.C. LaPorte, *J. Cell. Biochem.* **51**, 14–18 (1993).
35. J.R. Cupp and L. McAlister-Henn, *J. Biol. Chem.* **267**, 16417–16423 (1992).
36. Y.O. Kim, I.U. Oh, H.S. Park, J. Jeng, B.J. Song, and T.L. Huh, *Biochem. J.* **308**, 63–68 (1995).
37. A. Ishii, M. Suzuki, T. Sahara, Y. Takada, S. Sasaki, and N. Fukunaga, *J. Bacteriol.* **175**, 6873–6880 (1993).
38. P.A. Tipton and B.S. Beecher, *Arch. Biochem. Biophys.* **313**, 15–21 (1994).
39. K. Miyazaki and T. Oshima, *Protein Eng.* **7**, 401–403 (1994).
40. R. Chen, A. Greer, and A.M. Dean, *Proc. Natl. Acad. Sci. U.S.A.* **92**, 11666–11670 (1995).

41. J.H. Hurley, A.M. Dean, D.E.J. Koshland, and R.M. Stroud, *Biochemistry* **30**, 8671–8678 (1991).
42. B.L. Stoddard, A. Dean, and D.E.J. Koshland, *Biochemistry* **32**, 9310–9316 (1993).
43. A.M. Dean, M.H. Lee, and D.E.J. Koshland, *J. Biol. Chem.* **264**, 20482–20486 (1989).
44. P.E. Thorsness and D.E.J. Koshland, *J. Biol. Chem.* **262**, 10422–10425 (1987).
45. A.M. Dean and D.E.J. Koshland, *Science* **249**, 1044–1046 (1990).
46. J.H. Hurley, A.M. Dean, P.E. Thorsness, D.E.J. Koshland, and R.M. Stroud *J. Biol. Chem.* **265**, 3599–3602 (1990).
47. M.F. Browner and R.J. Fletterick, *Trends Biochem. Sci.* **17**, 66–71 (1992).
48. S.R. Sprang, K.R. Acharya, E.J. Goldsmith, D.I. Stuart, K. Varvill, R.J. Fletterick, N.B. Madsen, and L.N. Johnson, *Nature* **336**, 215–221 (1988).
49. D.C. LaPorte and T. Chung, *J. Biol. Chem.* **260**, 15291–15297 (1985).
50. D.C. LaPorte, C.S. Stueland, and T. Ikeda, *Biochimie* **71**, 1051–1057 (1989).
51. C.S. Stueland, T.P. Ikeda, and D.C. LaPorte, *J. Biol. Chem.* **264**, 13775–13779 (1989).
52. C.S. Stueland, K.R. Eck, K.T. Stieglbauer, and D.C. LaPorte, *J. Biol. Chem.* **262**, 16095–16099 (1987).
53. D.C. LaPorte and D.E. Koshland, Jr., *Nature* **305**, 286–290 (1983).
54. D.C. LaPorte, K. Walsh, and D.E. Koshland, Jr., *J. Biol. Chem.* **259**, 14068–14075 (1984).
55. C.S. Stueland, K. Gorden, and D.C. LaPorte, *J. Biol. Chem.* **263**, 19475–19479 (1988).
56. J. Finer-Moore, S.E. Tsutakawa, D.B. Cherbavaz, D.C. LaPorte, D.E. Koshland, Jr., and R.M. Stroud, *Biochemistry* (1997).
57. K. Walsh and D.E.J. Koshland, *J. Biol. Chem.* **260**, 8430–8437 (1985).
58. K. Walsh and D.E.J. Koshland, *J. Biol. Chem.* **259**, 9646–9654 (1984).

GOOD MANUFACTURING PRACTICE (GMP) AND GOOD INDUSTRIAL LARGE SCALE PRACTICE (GLSP)

BETH H. JUNKER
Merck Research Laboratories
Rahway, New Jersey

KEY WORDS

Biologics
Drugs
GLSP
GMP
Nonconformances
Performance testing
Qualification
Regulatory guidelines
Validation

OUTLINE

INTRODUCTION

Validation is the process of "establishing documented evidence that a system will do what it purports to do," a simple definition of validation given by the U.S. Food and Drug Administration (FDA) in 1977 (1). The progression of an organization with respect to its validation status has been described by a Validation Capability/Maturity Model (2). It starts with the "validation unaware," then shifts to the "validation aware" once knowledge of the theory is gained. The organization is then "validation active" when it is forced to participate by regulations and customer demand. It moves to the highest level of "validation enthusiast" when it possesses practical experience, recognizes benefits and limitations, and encourages others.

An overview of validation has been presented as a complement to distinct limits in the range of product testing. Validation was developed because of product failures that were not detected as a result of product testing (3). One specific event was a sepsis outbreak caused by improperly autoclaved vials not detected despite sterility testing of a statistically significant number of vials (4). Other authors have compiled an early history of good management practice (GMP) for bulk pharmaceutical drugs (5), early writings on interpretation of validation guidelines (6), and a history of validation in the United States (1). The approaches to issues in biological and pharmaceutical manufacturing change as better analytical techniques evolve, improvements in equipment design and operation are made, and pilot approaches are tested and proven in time.

Bulk pharmaceutical chemicals [BPCs] are made by chemical synthesis, fermentation (recombinant or secondary metabolite), and/or enzymatic reactions. For BPCs, FDA guidelines clearly state that the level of GMP rigor can be modulated based on the position of the step in the overall process (early steps requiring less depth) and the specific product (for some products, early steps might require more depth). A specific evaluation of the FDA's approach to BPCs has been documented with the emphasis placed on determining at what step focused GMP attention should be given (7). Although the discussion involves a specific processing step, the accompanying issue is the validation of the equipment, portion of the facility, and operating/testing personnel being used for that processing step.

Biologics are products made from living organisms that are complex proteins, viruses, antigens, or antibodies. Production may use recombinant cells, animal or insect cells, plants, or live animals. The guidelines for biologics GMPs include a history of GMPs for biologics (8), a series of 1989 papers by former FDA inspectors (9–11), and a description of differences in validation between conventional pharmaceuticals and biologics (12). Commentary concerning how the guidelines for drug inspections in effect at the time (13) applied to inspections of biopharmaceutical manufacturing facilities by the FDA also was published with clear emphasis on purification, formulation, filling, and lyophilization (14). These papers provided insight as to what the FDA was looking for regarding biologics facility design, validation, and operation. Although these guidelines now are almost a decade old, the areas cited still should be addressed, but in a manner consistent with current GMP guidelines where changes have evolved. As a result of these writings, specific guidelines for biotech facilities were developed (15–18).

There are clear differences in approaches to GMPs between bulk finishing (tableting, granulation, sterile filling) and bulk synthesis (fermentation, chromatography) for both biologics and drugs (19). The focus areas are (*1*) raw material variability, specifically natural sources from cornsteep liquor to fetal calf serum, and the potential for uncontrolled contaminants; (*2*) product tracking and yield, specifically use of activity assays rather than mass assays, and the resulting inability to conduct material balances; (*3*) step yields for which it is difficult to account for step activity losses; and (*4*) process complexity encompassing several processing steps (19). Recent progress in defining a well-characterized biopharmaceutical utilizing advanced analytical techniques (20) has assisted in providing regulatory consistency based on product characteristics rather than whether the product was approved by Center for Drug Evaluation and Research (CDER) or Center for Biologics Evaluation and Research (CBER) (21). It has also permitted processing equipment, procedure, and manufacturing site changes to occur late in the clinical development of a product (22).

Direct and indirect information is available to companies about the scope and nature of validation for both GMP and good industrial large-scale practice (GLSP). Direct statements are present in the Code of Federal Regulations (CFR), specifically Sections 21 CFR 210 and 21 CFR 211, which focus on drugs and pharmaceuticals, and Section 21 CFR 600, which adds additional regulations relevant for biologics. These chapters discuss outcomes rather than methods to achieve them and consequently are subject to interpretation by each company for each type of process. This thematic approach might be analogous to varied interpretations of Bible verses by individuals based on their individual lifetime experiences and goals with no single interpretation universally applicable. Interpretations of the regulations can also be found in various published "Points to Consider" and "Guidelines" documents authored by FDA officials, often with substantial input from industry (23–27). CDER and CBER executive secretariat staff maintain a list of guidelines and provide them readily on request. Public conference presentations, panel discussions, and publications by FDA inspectors also offer interpretations, but, as each situation can be different, it can be difficult for inspectors to speak specifically (or answer a quick question) and for audience members to directly apply the approaches presented. The FDA publication entitled *Human Drug cGMP Notes* provides some guidance regarding specific policy questions. Finally, the major indirect method for companies to obtain information is by reviewing public information about 483 violations and other reports from inspections of other companies. These summaries (obtainable from the "The Pink Sheet" and "The Gold Sheet" published by F-D-C Reports) usually contain two- to three-sentence quotes out of situational context and thus are subject to possible misinterpretation and misapplication. Perhaps the most valuable information concerning "reduction to practice" arises from individual companies who present and publish their specific approaches to GMP and GLSP. Implementation has been a challenge for companies as they evolve from research to a development focus; specifically of note is an early paper published by Biogen outlining these issues for interferon (28).

FACILITY REGULATORY GUIDELINES

Good Manufacturing Practices

The following of current good manufacturing practice (cGMP) regulations, as well as validation execution, is not a complete set of rules but rather a philosophy (29) that is largely dependent on the process and the company. The goal of validation is to satisfy Section 211.42 of 21CFR, which states that the facility and equipment "must be appropriate and suitable for the intended function" (30). Although some may insist that all equipment used in a process must be validated, many others have debated about just how much validation is sufficient. It is most efficient to judge the need for additional validation through scientific justification for the knowledge demonstrated (31). Nevertheless, there can be a "keeping up with the Joneses" philosophy both within and among companies that can lead to wasteful testing.

Validation activities can form a significant percentage of the time and money expenditure in a new pharmaceutical plant. Initially validation was essentially a repeat of installation, construction, and start-up testing and was considered a redundant task whose goal was merely to ensure contractors performed their jobs correctly (30). Often key senior management may not fully appreciate validation scope, benefits, limitations, and intensity. This lack of appreciation is most detrimental if the plant cannot attain full start-up until commissioning is complete; delayed construction and start-up schedules then compress validation, timelines (29). Appropriate validation testing will expose nonconformances and deficiencies in design, construction, and operation (29). Such revelations can be troublesome, particularly if they erupt in waves, but they can be helpful overall if the end user is permitted to evaluate the problem and to choose to accept operation with the limitation present. Thus, it may not be necessary to resolve every variation or discrepancy with a modification.

The validation master plan is essentially the validation specification and can be as important as the plant specification itself. It is a structured, detailed work plan that, although approved initially by the various subgroups involved, should be a working document linking all aspects of validation from start to finish (29). It might contain a description of the facility, especially its purpose, function, and limitations. A preliminary equipment and utility list, possibly stating what type of testing is required for each item as well as listing preliminary acceptance criteria, is useful for identifying needs for validation test equipment as well as constructing preliminary manpower estimates. Applicable policies and standard operating procedures (SOPs), both approved and those yet to be written, can be referenced, along with floor plans marked with clean/used equipment, raw material/product, and personnel flow patterns. Finally, a sample protocol might be drafted at this point to permit agreement on a common template.

There are several segments and types of validation testing, each of which encompasses various activities depending on the practices of the facility. Grouping of individual activities within these sections is less important than completing the required testing to ensure proper and consistent equipment operation for the processes (31). Duplication of testing among the various validation phases also should be avoided (31). The major categories of validation testing are briefly described next. Most of the initial validation phases lead to subsequent proof of product or process specific reliability.

The design qualification (DQ) focuses on a GMP review to identify potential design weaknesses that may impact qualification/validation before ordering equipment and constructing the plant. One approach is to conduct this GMP review using guidewords, such as *contamination, containment, validation, instrumentation, cleaning*, and *sterilization*, to examine each aspect of the design (32). This exercise is best done by representatives of the unit (either internal or outside the company) that is going to be conducting the validation. Some project leaders recommend a different unit conduct the validation than executed the design or construction to maintain appropriate checks and balances. Others believe that "design/build" strategies that encompass validation give a single-source responsibility to a single contractor and avoid fingerpointing when errors are uncovered (33). Potential GMP omissions that are uncovered during this exercise then can be evaluated by the validation/design team. Appropriate solutions may involve additional procedures, extended validation testing, or design modification.

Installation qualification (IQ) provides documented evidence that the equipment was both built and installed correctly according to design criteria and manufacturers' recommendations. Its primary focus is on documentation collection and review including manuals, purchase orders, pressure test data, passivation logs, spare parts lists, SOPs, hazard operability (HAZOP) results, calibrations and loop checks, and as-built piping and instrument diagrams (P&ID)/isometric drawings. IQ ensures that the ordered equipment was received and is safe to operate. Recommendations for the scope and collection of these validation commissioning documents (VCDs) have been documented (34). There is a balance between requesting sufficient documentation and cataloging of excess, redundant, and unnecessary documents (29). Results of factory acceptance testing (FAT) as well as site acceptance testing (SAT) also can be included in the IQ. Typically, the IQ is essentially a documentation review (30) with some not willing even to power up the equipment until the IQ has been executed, reviewed, or even approved. Such constraints might hamper timely identification and resolution of equipment problems, however.

Operational qualification (OQ) provides documented evidence that the equipment and plant operate correctly and have the sufficient range of capabilities over the expected span of operational parameters (29). OQ testing centers on functional tests of equipment features including alarm and interlock verification. It includes draft SOP suitability testing, some aspects of start-up testing, and overall system capacity checks for utilities (30). Often IQ, and in some cases OQ, can be performed by mechanical contractors, and portions of these studies might be performed at the vendor site before shipment of the equipment. The OQ testing needs to be completed satisfactorily to test the limits of the system at a minimum on the aspects on which later performance qualification (PQ) testing will be done.

PQ provides documented evidence that the plant repeatedly can produce product within established specifications and quality attributes (29). It involves extensive testing of the process as well as the process equipment. This testing encompasses many types of qualification and validation, including steam-in-place (SIP), clean-in-place (CIP), sterilizing and decontamination autoclave load patterns, glasswasher load patterns, environmental testing under processing conditions, and analysis of utility quality. It also includes process validation with example protocol strategies available (35). Triplicate runs generally are utilized to show reproducibility but can become extensive when combined with upper and lower processing boundary condition values (1). The concept of a worst-case scenario is used to minimize testing, but the worst case should be within the boundaries of normal operating conditions previously validated for critical parameters and not necessarily an "edge-of-failure" value (1). Before or at least concurrent with performance qualification, key analytical methods utilized should be validated (36).

During facility operation, substantial additional validation effort is required. This effort includes ongoing monitoring of utility and environmental quality, periodic (typically annual) ongoing validation challenges, and revalidation or equivalency testing prompted by equipment or process modification. Appropriate pages and previous acceptance criteria from original protocols might be used in lieu of authoring new protocols in some instances. In other cases, ongoing validation protocols might be reused provided they are written in a generic fashion referencing umbrella policy documents or SOPs for specific details subject to alteration. Once the scope of the ongoing validation effort is documented and agreed on, the impact of the planned workload on both equipment downtime and manpower can be evaluated.

Commissioning, the proper documentation of facility construction and installation, is an attractive alternative to validation for noncritical utilities that are not in product contact, such as cooling fluids or instrument air. Key aspects and strategies for commissioning have been documented (37). Every activity performed by construction, installation, or start-up personnel is documented although a formal qualification protocol may not be followed. For full effectiveness, commissioning should include challenge and capacity testing (30). Interestingly, one author notes that when noncritical "commissioned" utilities fail or are problematic, their critical importance becomes obvious (30).

The extent of validation is notably dependent on the degree to which in-process controls and end-product testing can ensure quality and safety. Validation thus is most extensive for biologics products (complex therapeutic proteins and vaccines) and less extensive for well-characterized proteins and secondary metabolites (21,38). References are available concerning general validation issues (38), potential validation problems and pitfalls (29,39), process validation (6), process validation life cycle approaches (3), and validation streamlining (40). The details of validation stages, including example regulatory internal and external references, sample systems to be examined, and typical schedules also are available for a model case (41). In addition, a benchmarking survey on current practices in the validation of aseptic processing as well as investigation and additional testing when results are unsatisfactory is available; although it is oriented toward aseptic filling, its approaches might be common to prior aseptic processing steps (42).

Validation protocols should cross-reference but not include items subject to change, such as equipment data sheets and equipment specifications, so that as data sheets are updated by engineering, the protocol is automatically updated (29,43). Protocols may be written and even approved before installation of the equipment, especially when the facility operations group already is familiar with the item. For computer equipment, a computer system validation and system life cycle approach has been adopted (44).

Good Industrial Large-Scale Practices

GLSP (or GILSP) started as the title to a section added to the Guidelines for Research Involving Recombinant DNA Molecules in 1991 (45–47). The GLSP level of containment was recommended for large-scale research or production of biological materials that have an extended history of large-scale safe use (48). The major difference between GLSP and BL1-LS was the reduction of the requirement for a closed system with inactivation of organisms, elimination of the need for exhaust gas treatment, and control of aerosols by procedure not physical devices (47). It referred to a level of physical containment, less rigorous than BL1-LS, which was recommended for recombinant organisms, composed of a nonpathogenic host cell free of adventitious agents and possessing a history of safe use and a well-characterized vector/insert free of harmful sequences. The resulting recombinant organism should be safe to humans and the environment; specifically, the inserted rDNA should not make disease-causing substances more resistant to drugs. The new GLSP section in these guidelines required (47): (*1*) institutional procedures for control of production of recombinant organisms (i.e., an evaluation that health and safety issues are addressed); (*2*) SOPs and training to ensure proper culture handling and maintenance of clean workplaces; (*3*) suitable protective clothing and facility/equipment containment; (*4*) proper handling of environmental discharges; (*5*) aerosol control when necessary; and (*6*) spill containment procedures for large losses. Additional guidelines focus on good lab practices, such as omitting smoking, mouth pipetting, and eating/drinking, as well as the establishment of internal accident reporting procedures.

An outline of BL1 through BL3 issues is available (47) in which the LS designation is added for large scale, typically greater than 10 L, to the corresponding laboratory designation. For BL1-LS, a closed system to reduce the escape of organisms is required, with a validated inactivation procedure before removing material from the closed system. Aerosol production is minimized through design (e.g., sealed sampling devices with filters). Systems are sterilized before maintenance, and an emergency spill plan is in effect. BL2-LS requires all BL1-LS requirements as well as prevention of aerosol generation, enclosure of the closed system within a secondary containment device (e.g., biosafety cabinet), and use of rotating mechanical seals to prevent leakage. Devices should be installed to monitor containment integrity during operation. The closed system should be tested for organism release with a nonrecombinant organism before use and the system should be permanently identified with a biohazard symbol. For BL3-LS, all BL1-LS and BL2-LS requirements are applicable as well as operating the production equipment at the lowest possible internal pressure to ensure containment integrity. The area surrounding the closed system must contain airlock access, readily decontaminated surfaces, sealed penetrations to permit decontamination by gas and reduce cross-contamination into other areas, and hand-washing facilities with valves that are not hand controlled. Finally, there should be controlled airflow into the facility and negative pressure with respect to surrounding areas.

Generally speaking, GLSP guidelines cover National Institutes of Health (NIH) requirements for working with recombinant DNA cultures; however cGMP guidelines may require product protection at the higher BL2 or BL3 levels for some aspects of the process. Specifically negative pressure can be required for BL2 to protect personnel while positive pressure can be required for GMP to protect product. This conflict has been resolved in one instance using inlet HEPA filtration and higher-pressure surrounding areas to provide a barrier as well as physical segregation of live virus areas (49).

A general outline of safety and containment issues and practices at various levels is available that includes examples of monitoring of organism release and suggestions for what equipment and where on the equipment to monitor (50). A goal distinction can be made spanning the range from product to personnel protection (50): "low" avoids contamination of product cultures, "minimal" targets typical laboratory levels attainable when good laboratory practices are utilized, and "prevent" aims for no release in the air, on surfaces, or on personnel.

MAJOR FOCUSES OF FACILITY PERFORMANCE VALIDATION

Steam-in-Place (SIP)

Strategies. Typically, SIP refers to vessel sterilization, most commonly that of fermenters and associated tankage. It can also refer to sterilization of other items such as dead-end filters in their housings, transfer lines, and cross-flow microfiltration skids. The basic sequence for a SIP procedure involves air removal, steam injection, condensate removal, and the breaking of vacuum with sterile air (51). Typically, the vessel contents are heated with venting until 95 °C, at which time backpressure is applied with continued jacket heating until the vessel temperature is slightly above 100 °C. Steam then can be introduced through various internal ports, such as the sparger line, by turning on associated steam supplies to these ports, creating a forward flow of steam into the vessel. Alternatively, internal ports, such as addition ports without subsurface dip tubes, can be opened to external traps or cracked valves creating a backward flow of steam out from the vessel. Vessel backpressure may be applied by passing the exhaust through a trap or through a backpressure control valve that permits fine adjustments of the sterilization hold temperature. To assure proper equipment operation, backpressure transmitter values should be compared against vessel pressure gauge readings and should be in agreement with expected values for the sterilization hold temperature. An outline of example sterilization procedures for bioreactor, media sterilization filters, and inoculum transfer lines is available (52). In addition, a useful reference describing sterilization theory and practice has been published (53).

The extent of sterilization is calculated using the parameter F_0, the magnitude of which is directly influenced by the area under the sterilization hold temperature versus time curve. The definition and examples of the use of F_0 in the design of sterilization cycles is available (54,55). The F_0 value is a measure of the equivalent kill of a test microorganism, typically *Bacillus stearothermophilus* spores for saturated steam applications, at the vessel sterilization temperature relative to that at 121 °C. Although commonly calculated including only those temperatures at or above 121 °C, calculation beginning at lower temperatures below 121 °C can be appropriate for larger vessels, heat-sensitive media, or difficult-to-heat loads. Accounting for this heat-up time can substantially augment the value of F_0, although the precise amount of this augmentation is complicated by the need to measure D-values for *B. stearothermophilus* spores in the medium under study at various temperatures. A target F_0 thus may be calculated on the basis of temperature mapping results from the system under study and the D-value of spore suspension in the medium to be sterilized.

For most SIP systems, the expected temperature and time required for sterilization are known or can be estimated based on prior experience of the operations group. Thus, production cycle safety factors need not be excessive and might be reduced to as little as a 5-min extension of the hold time or a 1 °C rise in sterilization hold temperature. Conversely, for transfer line sterilizations where F_0 accumulation rises to above 100 min within 5 min post-steam introduction, it can be beneficial to extend the production cycle sterilization time to 20–30 min.

Effectiveness Studies. Effectiveness of SIP cycles is evaluated overall by the rate of foreign contamination occurring in the specific vessel for the specific process over time. Because inoculated cultures might overgrow low-level contaminants, a media challenge, preferably with the same medium to be used during processing, is often conducted. Often a sterility medium is substituted for the process medium, but care should be taken that the medium ingredients are of similar quality as typical raw materials, that current CIP procedures can remove them adequately, and that the vapor pressure and solids content of the medium reflect the problematic aspects of the simulated process. A media hold test typically lasts 7–10 days and uses a growth temperature of 37 °C for favoring bacterial contaminants but a lower temperature of 25–30 °C to favor fungal contaminants. A compromise temperature of 33–35 °C might be appropriate if only one set of tests is desired.

For definitive proof of sterilization performance, a spore-spiked media challenge can be undertaken. Spore challenges, containing 10^{-6} spores, are used in the form of strips for nonsubmerged locations and ampoules for submerged locations. When large particles are involved, spores can be mixed into gels typical of particle sizes, sterilized, then minced to expose spores during sterility testing of the challenges. Successful kill of these challenges provides a sterility assurance of at least a 10^{-6} log reduction of spores. Challenges are placed, along with thermocouples, at the hardest-to-heat locations (i.e., those with the lowest accumulated F_0) based on the temperature distribution study results. Examples of challenged locations include (*1*) condensate return lines for agitator seal, sample valve, addition assemblies, and bottom valve, (*2*) sparger and vent air filters, and (*3*) liquid and headspace of vessel. Typically, the side of the equipment not exposed to direct steam is the side challenged. A summary of spore challenge methodologies is available (56,57); many of these methodologies have been established by the food and pharmaceutical industries.

During processing, sterile samples are obtained from individual vessels on average once per day depending on the facility practices. Sampling might be more frequent if the vessel contents are to be transferred to other sterilized fermenters in the near future. Sampling might be less frequent if containment (either product, personnel, or environment) is an issue. Samples are evaluated for culture purity for an inoculated cultivation and for sterility for an uninoculated cultivation. Aliquots of samples can be subcultured in nutrient-rich broth, incubated at 37 °C to promote growth of bacterial contaminants, then streaked on a nutrient-rich agar and reincubated at 37 °C. Subcultures can be examined microscopically using Gram stain techniques. In some cases, aliquots can be placed into phenol red–containing medium and any color change noted (typically yellow if contamination is present). Loopfuls of samples can be directly plated onto nutrient agar conductive to bacterial or fungal growth and incubated at 37 °C or a lower temperature of about 25 °C, respectively. False positives on sterility testing have been noted at 0.25% (58). Automated sterility analysis may reduce the number of false positives as well as speed up testing by incorporating fluorescence or other color change endpoints based on metabolic end products (58).

SIP sterilization of hydrophobic sparger or vent filters can be complicated by insufficient air removal. Specifically, dual filters positioned in series on an SIP system may not attain sufficient air removal or condensate drainage due to suboptimal placement of steam condensate piping. Filter integrity testing after the production SIP cycle (which may have the hold time extended by adding a safety factor relative to the fractional test cycle) is required to ensure SIP did not damage the filter. Although in some cases the filters might be heat damaged if the sterilization cycle is unusually long or hot, typically damage is caused by a large pressure differential across the filter, often caused by suboptimal cooldown procedures poststerilization that might not be reviewed during SIP cycle development (59).

Sterilizing filters also may need to be integrity tested after batch use depending on the GMP issues associated with the type of product. Recording of filter lot numbers as well as integrity testing of filters after use can be an important aid in investigating contamination. Post-use integrity testing can be complicated by the difficulty of removing certain materials from the filter that can alter the bubble point or diffusion rate. These issues are best reviewed on a case-by-case basis with the filter manufacturer, who may have measured bubble points or diffusion points for different solutions in addition to those for water. Pre-use integrity testing often is conducted if the vendor testing certification is not sufficient for process use, although passing results do not ensure subsequent filter integrity throughout the batch. Dual series filters can be used to minimize the need for pre-use integrity testing but sterilization strategies need to be adapted to accommodate this arrangement. Post-use integrity testing can be done by removing the filter from its housing, removing the entire housing from the equipment with the filter intact, or testing the filter while located on the equipment with appropriate tie-ins. The latter two methods provide proof of proper installation of the filter if that is required.

Investigations of Failed Results

Most SIP cycle deficiencies can be corrected via SOP (most desired) or equipment modifications (less desired). Equipment modifications might be difficult to accomplish for an automatically sterilized vessel. It should be confirmed that the act of testing is not inducing failures before implementing corrections, however. Investigation strategies for SIP failures might also apply to autoclave load pattern testing failures.

Failures during SIP validation testing might be caused by test procedure faults rather than proposed SIP cycle deficiencies. The biological indicator or thermocouple may be blocking steam/condensate flow, particularly if large spore strips are used in small process lines. Because spore suspensions and strips are incubated in nutrient medium and inspected visually for cloudiness daily, the contaminant cultured should be identified to genus and species if possible. Contamination from handling can result in growth of *Staphylococcus* instead of *B. stearothermophilus* (60). Contamination after testing by contact with unkilled spores from positive controls or from bags used to enclose spore vials might also occur.

The most frequent causes of contamination during fermenter operation (61) include construction materials, mechanical seals, valves, poor or overly complex design, operator error, instrumentation (e.g., calibration, valve failure), process air, transfer/feed lines, contaminated inoculum, incorrect batch or continuous media sterilization, and inadequate procedures. A mechanical checklist for returning a fermenter to service after an initial or repeated contamination can be useful. Factors important in fermenter contamination (62,63) as well as design issues and sterility risks (64) have been documented.

Incorrect specification of procedures often can be responsible for contaminations. Specifically, direct charges of powders through the vessel manway or handhole need to be carefully rinsed into the tank to avoid accumulation of dry powders on the tank upper side wall. Homogenization of clumpy hydrophobic media powders should be considered and sieving of large particles from protein meals (e.g., cornmeal) should be conducted. Media that foam on heat up during sterilization (e.g., medium that contains tomato paste and soy flour) should be avoided as these excessively rich laboratory media, developed for screening and initial growth of a wide variety of organisms, may not be necessary for growth of the desired culture. Use of modified production tank medium for the seed tank stage(s) should be explored instead of using laboratory shake flask seed medium (65).

A monthly contamination tracking system can be used to present trends in the most probable cause of contamination. These might be classified as (*1*) a mechanical problem involving an equipment failure such as leaky valve, leaky mechanical seal, gasket, probe breakage, or incorrect temperature calibration; (*2*) an operational problem in which SOP procedures were not followed; (*3*) a cross-contamination in which contamination detection was not timely and material transferred into a sterilized tank subsequently was found to be contaminated; and (*4*) a process problem in which the procedures specified were not optimal such as incorrect batch sheet instructions concerning sterilization hold time or temperature. These monthly reports should be circulated to the operations group at all management levels and mechanical staff, as well as reviewed by quality personnel.

Contaminating organisms should be identified both in a preliminary fashion by microscopic observation and Gram stain and more definitively using commercial biochemical testing kits or fatty acid analyses. A record of the types of contaminants as well as ages and processes should be maintained and evaluated at least yearly for trends. Yearly contaminants can be classified into categories by process, age of contamination (e.g., after sterilization [AS], after inoculation [AI], <48 h, 48–120 h or >120 h), and type of organism (cocci, gram-positive, gram-negative rod, fungal, or miscellaneous). The sterility investigation experience of the facility then can be utilized to determine the likely areas where the contamination may have entered. Example guidelines suggest that (*1*) gram-positive rods might result from incomplete media sterilization, improper inlet air filter installation, or air filter integrity failure; (*2*) gram-negative rods might result from incomplete water sterilization (perhaps of filter-sterilized nutrient additions), jacket leakage into the vessel, or water present in the inlet air to the vessel; (*3*) cocci might result from human contact during either laboratory seed preparation or seed fermentor inoculation; and (*4*) fungal contaminants, although rare, might result from improper inlet air sterilization as well as inadequate cleaning of cracks/crevices from fermenter ports.

Clean-in-Place (CIP)

Strategies. The goal of cleaning validation is to satisfy 21 CFR 211.67 in which it is stated that written equipment cleaning procedures should be established and validated. It should be demonstrated that cleaning procedures consistently reduce product and cleaning agent residuals to predetermined levels that are acceptable for the processing steps. CIP was first developed by the dairy industry (66), and this industry has conducted substantial research into effective strategies and appropriate CIP skid design. One major decision important for CIP skid design (67) is the issue of reuse of acid–base cleaning solutions, which must be weighed against concerns of cross-contamination and resoiling of equipment. A second major decision is the degree of dismantling and manual cleaning operationally acceptable for each piece of equipment. The amount of time expected to elapse before initiation of cleaning and the length of time equipment can sit idle before recleaning need to be established on the basis of reasonable expectations of operating practices as well as process requirements.

Key factors in cleaning effectiveness are the discharge of cleaning solutions at their operating pH (if acceptable to site environmental officials) to avoid precipitation of soil, adequate rinsing of cleaning solutions through all internals, adequate contact time of cleaning agent/rinses, adequate velocity for contact of cleaning agent/rinses, maintenance of optimal temperature during cleaning solution hold time or recirculation time, and proper concentration of cleaning solution. The drainability of equipment is another key factor in ensuring cleaning effectiveness. Common problem areas include minor dead legs of a few milliliters that are not rinsed through during the cleaning process, circulation pumps that are not completely drainable, shadowed areas in which impingement of a spray is prevented by vessel internals, and crevices from bolts and fittings. In cases where failed batches are pasteurized or sterilized in the vessel, dilution of the batch with water can assist subsequent cleaning efforts.

Often there are multiple points of entry of CIP liquid into a vessel (e.g., spray balls, air inlet, and vent and transfer lines) and liquid is returned by the reactor drain (67). Spray balls can be fixed or rotating, removable during processing or remaining in batch contact. High pressures and flow rates can be required for proper impingement for best cleaning effectiveness leading to heavy utility loads, particularly for the water rinses. Although in many applications two spray balls are located at the vessel top, additional spray balls can be aimed at hard-to-reach locations such as impeller undersides or behind baffles. The acceptable level of flooding of the vessel needs to be determined as the filling of the vessel to at least the lower impeller may enhance cleaning of lower baffles and bottom mechanical portion of the agitator (67).

A typical cleaning scenario (68) may include several steps depending on the product requirements. These steps are an initial thorough rinse with lower-quality warm water to remove loose soil (particularly at the batch level), a caustic-based detergent solution to hydrolyze biological materials, a rinse with deionized water, an acid-based detergent solution to remove traces of mineral salts, a second rinse with deionized water, and final rinses with higher-purity water. In some cases, a water boilout (heating of water to 80–90 °C) is conducted before the final water rinses. The acid-based detergent typically is not necessary for fermentation applications, especially if deionized water used for cleaning, and can be done on an as-needed basis with 0.5 wt% nitric acid (67). Dilute phosphoric acid and sodium hydroxide can be used in lieu of acid/caustic-based detergents for some applications, although the presence of a surfactant can greatly enhance soil removal. Finally, an empty vessel sterilization, with condensate flowing out from the bottom valve, can aid in soil removal.

For cleaning of nonvessels, the ability to document effectiveness becomes increasingly more difficult. Membrane filters often are restricted to single product use primarily because it is difficult to ensure total removal of the prior product without destructive testing. Filters can be reused for the same product, but sanitization/sterilization procedures are of concern to minimize unwanted bioburden. For homogenizers and cell disrupters, cleaning and visual inspection of the head are likely to require substantial manual disassembly and rebuilding.

In the case of chromatography columns, significant resin expense suggests reuse is especially desirable for multiple products, but product-specific testing for removal is exceedingly difficult. More commonly, chromatographic media are dedicated to one product but equipment (e.g., columns, piping, pumps) is used for multiple products (69). Chromatographic media have been reused for multiple cycles (more than 100) of the same product, but validation of number of multiple uses is required (70). Specifically, the media and equipment must be compatible with validated cleaning processes and consideration must be given to which substances bind reversibly and irreversibly to the resin (69). Cleaning and sanitization can occupy about 20% of each cycle period, which encourages the use of larger versus smaller columns (69). Blank runs are conducted followed by a reduced-cycle cleaning run and evaluation of the contaminant breakthrough potential (69). Finally, quantification of the functional stability of the resin for specific cleaning agents can be accomplished by measuring retention times of various test proteins (69).

Effectiveness Studies. Hardest-to-clean locations should be identified based on prior experience with the existing vessel, initial cleaning studies of a new vessel, or the best judgment of the processing group based on similar vessel designs. The criteria for sample site location should be documented. Some examples of common locations include the air sparger, underside of impellers, batch level, sample valve, baffles, lifting holes for shaft, fermenter dome, Ingold ports/plugs, threaded fittings, and bolted internals and piping dead legs on addition ports. The use of worst-case scenarios, while a popular approach, may lead to overly rigorous cleaning that is operationally burdensome for many nonworse-case processes although the amount of testing can be reduced for a multiuse facility.

Visual inspection, while subjective, is the most powerful asset available for the prompt analysis of cleaning effectiveness. Training of inspectors properly concerning where and how to look can greatly improve the likelihood of identification of problem areas. Chrome mirrors with extended arms, powerful flashlights, and clean white cloths can be utilized to aid inspection. Tank imperfections should be documented ahead of time to avoid mistaking tank welds, external coiling coil welds, or tank intrinsic stains as cleaning failures. Unfortunately, visual inspection may not be possible for some pieces of equipment without excessive mechanical dismantling (e.g., disk-stack centrifuges, packed chromatography columns). In some cases, the limit of visual detection has been determined using spiked coupons of typical process soils. For example, the visual detection limit was 100 μg/ 25-cm^2 swab area for an active pharmaceutical ingredient isolated powder (71).

The sampling of rinse water indirectly evaluates equipment cleanliness, but it assumes that the residual soil is soluble and thus totally removed in the rinse water. Sometimes rinse water samples are taken throughout the cleaning process so the progression of cleaning can be evaluated; this testing can be detrimental to sensitive assay equipment such as total organic carbon (TOC) analyzers that are not designed for excessively soiled samples. Solution pH can be especially useful to indirectly detect cleaning agent residuals because most cleaning agents are concentrated acids or bases in addition to carbon-based surfactants (Table 1). In addition, a 100 ppm solution of CIP100 corresponds to a TOC of 4.8 ppm (66); other results are available for other concentrations (Table 1).

Direct sampling of the vessel interior is accomplished by a consistent surface swabbing procedure, typically a 25-cm$_2$ area (5-cm × 5-cm square area or equivalent). Recoveries depend on the type of contaminant and type of surface being swabbed (72). Specifically, recovery data for a purified recombinant DNA protein (containing phosphate and lactic acid at pH 4.5) varied from 89% for stainless steel to 62% for polypropylene to 55% for glass (72). Recovery data for stainless steel varied from 67% for whole-cell homogenate to 82% for formulated protein (containing mannitol, phosphate, and lactic acid at pH 4.5) and 105% for purified protein (72). Other recovery data for three types of media using different types of materials (e.g., stainless steel, Nalgene, Teflon, polypropylene, Pyrex) are available (Table 2).

Table 1. Ability of Various Dilutions of Cleaning Agents CIP100 and CIP200 to Pass USP Chemical Tests

Solution	pH	Conductivity (μmho/cm)	TOC (ppm)
CIP100			
1% (10,000 ppm)	12.6	1,927	25.7
0.05% (5,000 ppm)	11.4	581	13.4
0.01% (1,000 ppm)	10.7	123	3.5
0.001% (100 ppm)	9.4	8.6	0.57
0.0001% (10 ppm)	6.5	2.9	0.16
CIP200			
1% (10,000 ppm)	2.0	7,530	—
0.05% (5,000 ppm)	2.7	872	9.7
0.01% (1,000 ppm)	3.2	179	2.6
0.001% (100 ppm)	4.1	21.4	0.41
0.0001% (10 ppm)	5.1	2.1	0.15

Rinse water and swab samples can be tested for TOC, protein content, specific product, residual cleaning agent, and overall rinse water quality roughly equivalent to that of the water source used for cleaning (e.g., pH, conductivity, TOC, bioburden, limulus amoebocyte lysate [LAL]). Product-specific testing can include ELISA, SDS-PAGE, and other protein-specific assays. TOC analysis, which detects all carbon-based residuals, is becoming a low-cost, rapid-turnaround, low-level detection method for applications when a product-specific test is not required. A description of the method and analyzer and its limitations is available (66). The sample TOC content must be within the range of instrument calibration for accurate results. For equipment that is to be used wet and without subsequent sterilization or sanitization, a bioburden study might be beneficial using both rinse water and swabbing tests. Such a study should be done after the equipment has been sitting idle for a period of time greater than or equal to permitted idle periods during processing. Procedures have been documented for bioburden (73) and LAL (74) testing.

A survey of sampling methods (specifically visual, rinse water, and swab) and assay methods used for early and later phase processing was conducted for multiple companies as well as details of methods in place at Chiron presented (75). Case studies of cleaning validation have been published by Regeneron (76) and Immunex (77). As processing progresses into downstream isolation steps, methods for analysis must have greater product specificity (66), but for upstream steps nonspecific assays such as TOC are likely to be sufficient. In some cases this requirement adds substantial cost and sampling burdens. Acceptance criteria are affected by the sensitivity of analytical equipment in part as well as the amount of residue tolerable in the equipment for the specific processing step. Guidelines for carryover levels of product have been established for specific products by some companies such as Eli Lilly (71).

Cleanability studies using different types of materials can be conducted. The general procedure is to submerge a

Table 2. Recovery Study Results (Fractional Recovery) for Media from Three Processes Tested in Absence of cells

Material of coupon	Recombinant yeast/chemically defined medium 1	Recombinant *E. coli*/chemically defined medium 2	Mammalian cell culture/Williams E medium
Stainless steel	0.86	0.80	0.92
Nalgene	0.28	0.58	1.07
Teflon	0.35	0.65	1.11
Polypropylene	0.36	0.86	1.05
Pyrex	0.26	0.61	0.76

coupon in the soil matrix, evaporate to dryness, and then process it through cleaning solutions/rinses either in stagnant trays or using impingement (Table 3). Each surface is then visually evaluated and swab samples submitted for TOC testing. Dye removal studies can be conducted, but these may not be representative because of different rinsing characteristics (67). Challenge studies also can be developed to document log reduction of TOC, bioburden (based on indicator organisms), or endotoxin by spreading a known amount of material on the equipment. Usually more material is spiked than typically is present as a contaminant so as to have a quantifiable amount remaining at the conclusion of the simulated cleaning process. It is not clear that soil removal is linear between these two rather disparate concentration ranges.

Validation of a backup cleaning agent is recommended in case material from the primary supplier becomes unavailable. This assignment might be accomplished by demonstration of equivalency during the initial validation or as part of a later changeover program. Testing might include cleanability studies on test coupons and test equipment using worst-case scenarios or conducting equivalency runs with a new agent in actual processing equipment.

Investigations of Failed Results. Investigation of failed cleanout procedures generally center on procedural issues rather than testing interferences as in the case of SIP testing. Common deficiencies include insufficient hold time or hold temperature for cleaning agent, inadequate contact of cleaning agent or rinse water with all vessel inner surfaces and internals, improper amount of cleaning agent used, and inappropriate cleaning pretreatment steps. Sufficient time must be devoted to training and clarification of procedures with operating staff. For example, the equipment must be properly vented to allow complete rinsing. Because the rinse water quality depends on both the cleaning effectiveness and the quality of the source water, proper water system operation also may impact the investigation.

Laboratory Autoclave Load Pattern Testing

Load pattern testing should be conducted for typical loads used for laboratory-sterilized materials that are to be transferred to sterilized process equipment. Items to be sterilized should be placed in the autoclave such that they are not touching and there is adequate room for steam to surround each object. Temperature distribution testing of the load should be conducted. Even if the accumulated F_0 appears satisfactory, spore challenges should still be pursued because spores will be killed efficiently only if moist heat is present (78). Assuming both the temperature distributions and spore challenge test results are adequate, a production run should be performed with the actual material to be sterilized. As reasonable, each element in the load should be tested for sterility either through a test usage for equipment or by incubating the liquid load and then testing sterility either through a test usage for equipment or by incubating the liquid load and then testing sterility samples. This production load sterility testing is particularly important to test equipment wrapping closures, ability of personnel to execute validated load patterns, and the successful sterilization of a solids-containing medium.

For dry materials, use of the prevac cycle is most desirable to remove air from the equipment promptly. If a prevac cycle is not available or not considered reliable due to equipment failure or fear of disrupting product closures, then a small amount of water (10 mL in a 4-L bottle) may have to be added to increase the amount of steam generated within the bottle. Sterilization of filter manifolds for nutrient addition is complicated by the fact that hydrophobic filters may not be wet enough to generate sufficient steam for spore kill. This is particularly an issue with autoclaving of capsule filters in series because of poor heat transfer through the plastic filter body and poor steam penetration through the tubing ends. Dual filters in series need to be tested carefully, especially when more than one set is used in a manifolded arrangement that limits steam

Table 3. Cleanability Study Results: Residual TOC after Cleaning of Dried Medium to Point of Visual Cleanliness up to a 10-min rise via Impingement

Material of coupon	Recombinant yeast/chemically defined medium 1	Recombinant *E. coli*/chemically defined medium 2	Mammalian cell culture/Williams E medium
Stainless steel	0.10	0.13	0.08
Nalgene	0.16	0.08	0.07
Teflon	0.09	0.09	0.09
Polypropylene	0.10	0.10	0.08
Pyrex	0.01	0.14	0.10

penetration. After each production cycle run of a validated autoclave load pattern, any filters in the load need to be integrity tested to ensure that no damage has occurred during autoclaving. A complete reference on integrity testing of membrane filters is available (79).

For liquid cycles, liquid heat-up times are greatly affected by the container size, fill volume, type of liquid, and material of construction of the container. There is a conflict between safety directives to remove glass from processing areas by replacing glass with plastic containers and the substantially longer heat-up times observed with plastic (e.g., Nalgene) carboys because of their low heat transfer coefficient. Decontamination of liquids or equipment should be conducted in a different autoclave unit than the sterilizing autoclave. Decontamination load patterns also may undergo load pattern validation, although the heat sensitivity of the individual organism should be considered during validation. Although worst-case studies using *Bacillus stearothermophilus* may be appropriate, typical BL1 or BL2 organisms or viruses may exhibit a substantially lower heat sensitivity, thus reducing the necessary decontamination load cycle time. This potential for reduction may become particularly important for larger liquid volume loads.

Generally, the vacuum break filter is steam sterilized separately before the autoclave load. Periodically, after a specific number of autoclave loads have been sterilized, the filter sterilization cycle is run again. Because most vacuum break filters are not maintained under positive air pressure between the conclusion of the filter sterilization cycle and the beginning of the autoclave chamber cycle, then this filter does not necessarily remain sterile once the autoclave door is opened.

Utilities

Periodic monitoring of utility quality can be a critical factor to ensure product quality. Monitoring focuses on water, steam, and air systems in product contact. Frequencies can vary from daily for water systems to monthly for clean steam systems. In a few applications, sampling just before, during, and/or after system use is warranted. Action limits should be established on the basis of product quality requirements, while alert limits should be established based on historical data (if available) or a comfortable safety margin. Because the alert limit level does not impact product quality but rather serves to warn of the possibility for a quality problem to occur, it should be set considering input from the operations group regarding prior or expected system behavior.

By far the major testing focus is on water systems because impurity levels can become concentrated by the large volumes used in processing. Testing may include USP chemical analysis, bioburden, and endotoxin depending on the application. Although sampling of water systems is complicated by multiple use points, in many cases sampling may be rotated among use points so long as each use point is sampled on a regular basis. Typically, at least one bioburden and one endotoxin sample are taken daily from at least one point of use. Recent changes to the United States Pharmacopeia (USP) affect purified water and water for injection (WFI) testing regimens (80). The goal of these modifications was to reduce required USP chemical testing to conductivity, pH, and TOC (66) analysis. Before this change, numerous chemical tests were required, some of which were qualitative in nature with end points that were difficult to interpret. The impact of these new USP chemical testing procedures needs to be understood so it can be evaluated for existing water systems (81). Newly available on-line TOC, pH, and resistivity meters can ensure proper system operation according to these new USP chemical testing guidelines.

Bioburden evaluation is conducted to determine both the number and type of organism present. Water systems are subject to the formation of biofilms in which bacteria attach themselves to piping crevices or cracks. They create a polysaccharide extracellular matrix that promotes long-term attachment of both the original and subsequently adhering bacteria. References on biofilm formation mechanisms and control are available (82–84). Like weeds on a lawn, once present biofilms are difficult to remove so it is best not to permit them to form initially.

LAL analysis for endotoxin, while primarily a test conducted on clean steam and WFI systems, can also be helpful for monitoring operation of USP purified water systems. In fact, LAL and TOC and bioburden results may be examined together to investigate excursions. All material present in water that is quantified as TOC is not necessarily endotoxin because the endotoxin assay is specific for lipopolysaccharide. All endotoxin should be detected in the less-specific TOC assay, however. Using the *E. coli* endotoxin standard, TOC levels were measured for 0.25, 0.5, and 1.0 EU/mL solutions and a linear relationship was obtained:

$$\text{TOC (mg/L)} = 72.2\ \text{LAL (EU/ml)} - 5.6 (r^2 = 0.99) \quad (1)$$

Consequently, for an endotoxin limit of 0.25 EU/ml (a typical action limit for WFI systems), the TOC measured in the same sample might be as high as 14 mg/L (14 ppm). Thus, TOC analysis can be a good indicator of rising endotoxin levels and may serve as confirmation of endotoxin assay results. Similarly, a rise in bioburden from 1 to 100 cfu/mL may not coincide with a noticeable rise in TOC because 1 to 100 cfu/mL corresponds to a range of 10^{-4} to 10^{-6} ppm TOC substantially lower than usually detected (assuming 10^{-12} g dcw/cell and 50% carbon content).

Controlled Environments

Controlled environments ensure that there is documented control of air quality that is appropriate for each processing step. Establishment of classified areas should be examined carefully to ensure that classification at the proposed level is reasonable while the area is in operation. Specifically, certain areas that produce powders, mist, or steam may not be able to achieve class 100,000 conditions if samples were taken during truly worst-case dynamic operation. Such areas might include rooms containing glasswashers, autoclaves, fermenters, or raw material subdivision operations.

A comprehensive monitoring program should be developed to include items such as viable airborne particles, nonviable airborne particles, pressure differentials, airflow direction, temperature, relative humidity, and surface microbial contamination present on staff, equipment, work surfaces, walls, doors, floors, and curtains (85,86). Monitoring should be conducted using viable and nonviable air samplers, contact plates, surface swabs, and settling plates. A regular monitoring schedule should be instituted, with some tests taken during simulated or actual dynamic conditions (while work is occurring in the area). When bringing an area in and out of GMP service, initially there might be an intensive amount of monitoring, perhaps 3–5 consecutive days, to show consistency of air quality before initiation of GMP processing. A summary of environmental monitoring program characteristics, including additional references, is available (87).

Evaluation criteria for a laminar flow hood or biosafety cabinet are dependent on the critical nature of the hood, specifically whether there is a subsequent sterilizing step downstream for a sterile product. Monitoring during actual operation sometimes is considered using either exposure plates or an air sampler. If monitoring is conducted directly within the work area, then it may interfere with the technician's ability to properly manipulate materials. Conversely, if monitoring is moved to occur in a corner, then testing is not actually sampling air exposed to product. Motorized air samplers interfere with desired airflow patterns in a biosafety cabinet or laminar flow hood by creating air currents and noise that can be a distraction to the technician. Settling plates give a qualitative outcome because a defined amount of air is not sampled (85). These plates may selectively attract larger particles and may need to be exposed for a far longer time than the length of the actual manipulation (85).

Action and alert limits should be set based on the nature of the step, the ability of subsequent steps to remove or propagate contaminants, and the product exposure risk (open versus closed system) (85). Although a typical alert limit is 80% of the action limit, the alert limit may be set lower to ensure a comfortable quality safety margin. Data should be trended over a year to detect seasonal variations. Often data do not monotonically increase toward the action or alert limit but rather spike to higher levels because of a system disturbance.

For action and/or alert limit excursions, identification of the microorganism class can be useful (e.g., gram-positive or gram-negative rod, cocci, fungal) but identification down to genus and species can be expensive and burdensome. Such detailed identification may be useful if documented evidence is needed to support negligible cross-contamination of cultures during open transfers of multiple processes in the same hood. For any action limit excursion, there is a need to evaluate what processes may have been affected and what was the potential quality impact. Typically, such an event is considered a processing abnormality, and the resulting impact is evaluated by operational personnel in conjunction with relevant quality groups.

Corrective actions implemented in response to an excursion might include improved cleaning regimens, rotation of disinfectants (88), altered sanitization frequencies, monitoring of ultraviolet light intensities, and revised gowning procedures for the area. Testing of disinfectant effectiveness (e.g., disinfectant type and contact time) against a challenge organism also might be useful in troubleshooting reasons for elevated bioburden levels (Table 4). Documented retraining or simple observation of operating personnel using the area, sampling personnel, or sample analysis personnel also may be useful. Additional testing to determine if the action limit value is a one-time-only or reproducible occurrence is key to the investigation. Such testing may consist of repeated sampling of the same locations or increased sampling locations vertically and horizontally within the area to localize the problem. Often quality groups may not support additional sampling because of retesting concerns, but such sampling may be the only way to effectively and scientifically troubleshoot a problem. If the additional sampling plan is defined before execution, then randomness can be minimized. The initial action or alert limit value should not be discarded; rather, all the data should be considered together in evaluating the most probable reason for the excursion.

Constant Temperature Units

Validation testing of constant temperature units centers on documentation of uniform temperature distribution and required humidity level throughout the chamber. Testing is relatively straightforward and inexpensive. Although the requirements for uniformity depend on the type of unit and its particular application, typical examples of set-point deviations might include ±1.0 °C for an incubator, ±2.0 °C for a refrigerator, and ±10 °C for a low-temperature freezer. Testing with the unit loaded can identify problem areas such as reduced air circulation in a low-temperature freezer and excessive heating from a rotary shaker motor in an incubator cabinet or room. Portions of the unit not satisfying the testing requirements can be excluded from use of the process unit if repairs are not made.

Product Changeover

The purpose of a documented product changeover procedure is to ensure that equipment, personnel, and facility

Table 4. Disinfectant Effectiveness Against Microbial Challenge

Disinfectant (dilution ratio, v/v)	1 min	5 min	10 min	45 min
	Staphylococcus epidermis			
Klenzyme (1:64)	520	480	425	455
CIP100 (1:1280)	420	490	505	375
CIP100 (1:11)	345	345	295	245
Vesphene (1:128)	<10	<10	0	0
LpH (1:256)	25	<10	0	0
3M Compublend (1:2)	<10	<10	0	0
	Aspergillus niger			
Klenzyme (1:64)	48	43	38	46
CIP100 (1:1280)	57	55	46	47
CIP100 (1:11)	34	26	30	27
Vesphene (1:128)	0	0	0	0
LpH (1:256)	0	0	0	0
3M Compublend (1:2)	0	0	0	0

are ready for the next product (89). The major expense of time, money, and effort is expended in preparing the equipment. Equipment cleaning is evaluated using equivalency testing to ensure that existing procedures can clean the next product, with revalidation recommended if major changes in procedures are required. Disposables that may be difficult to clean, such as tubing, filters, or gaskets, may be replaced, although the mechanical support necessary to replace every diaphragm in product contact may be excessive. An evaluation of existing equipment SIP/CIP cycles, autoclave, and glasswasher load patterns, as well as manual equipment-washing procedures, to determine sufficiency for the new process should be conducted. This review should be followed by a cataloging of any changes in equipment, particularly those necessitating revalidation, that might be needed for the next process. The facility preparation spans from the removal of raw materials and equipment used in the previous process from the area to updating or revision of batch records and SOPs. Personnel need to be trained for the new process using batch sheets, newly developed load patterns, SIP or CIP cycles, and altered SOPs.

Surface testing of equipment and the facility might be conducted to document the removal of prior product, particularly for areas in which open transfers are to be conducted. Surface testing after the first run of the new process (typically a practice or engineering run) might be conducted to document the ability of existing procedures to acceptably clean a new product from the area. This minimized testing may only be appropriate if (*1*) similarity of processing steps and their components (e.g., media, host cell organism, contaminating organisms) can be scientifically justified, with the ability to do this successfully varying according to the processing step; (*2*) similarity with respect to TOC recoveries either by comparison testing or scientific justification is documented; and (*3*) similarity with respect to cleanability by comparison testing or scientific justification is documented.

INVESTIGATIONS OF NONCONFORMANCES

There is a tension between facility productivity and possibility of nonconformance. The results of nonconformance can be ranked according to their effect on productivity, schedule, manpower, maintenance, and capital or retrofit costs. As written in the BPC guide for inspections (90), the recommended starting point for inspectors is to review product failures (and by extension failed batches), excursions in monitoring results, and failed validation testing. It is thus of utmost importance to determine the most probable cause of failure, complete any necessary investigation reports, and document any resulting corrective actions. Guidelines for investigation of testing and validation failures are available (60,91–93). Pressure to obtain a most probable or assignable cause should not be imposed, and the cause should be permitted to be listed as "unknown" if truly no one cause emerges as most likely during the investigation. A higher-than-normal percentage of unknown assignments then might prompt reevaluation of investigative or testing procedures and eventually aid in obtaining the real trends in assignable causes for the facility. Failures believed to be caused by poor equipment design, inadequate procedures, or suboptimal process development should precipitate the appropriate alterations, which should be evaluated according to established change control procedures. Investigations are always ongoing, and, although hypotheses can be formed quickly within the short window permitted for an investigation, the accuracy of any hypothesis can be tested and revised in the weeks and months that follow. Recognition of the scientific nature of any investigation of a nonconformance should be featured prominently in any investigation SOPs.

BIBLIOGRAPHY

1. K. Chapman, *Pharm. Technol.* **15**, 82–96 (1991).
2. C. Tayler, *Pharm. Eng.* **16**, 50–58 (1996).
3. D. Gold, *PDA J Pharm. Sci. Technol.* **50**, 55–60 (1996).
4. C. Bruch, *Bull. Parenter. Drug Assoc.* **28**, 105–121 (1974).
5. D. Larkin, *Pharm. Eng.* **9**, 27–30 (1989).
6. Y. Chiu, *Pharm. Technol.* **12**, 132–138 (1988).
7. C. Martinez, *Pharm. Eng.* **14**, 8–14 (1994).
8. M. Beatrice, *Pharm. Eng.* **10**, 29–35 (1991).
9. D. Hill and M. Beatrice, *Pharm. Eng.* **9**, 35–41 (1989).
10. D. Hill and M. Beatrice, *Biopharm. (Jilove)* **2**, 20–26 (1989).
11. D. Hill and M. Beatrice, *Biopharm. (Jilove)* **2**, 28–32 (1989).
12. T. Schoemaker, *Pharm. Eng.* **11**, 23–27 (1991).
13. *Guide to Inspection of Bulk Pharmaceutical Chemical Manufacturing*, CDER, FDA, Washington, D.C., November 1987.
14. H. Avallone, *Pharm. Eng.* **9**, 40–48 (1989).
15. *Interferon Test Procedures: Points to Consider in the Production and Testing of Interferon Intended for Investigational Use in Humans*, Office of Biologics, FDA, Washington, D.C., 1983.
16. *Supplement to the Points to Consider in the Production and Testing of New Drugs and Biologicals Produced by Recombinant DNA Technology: Nucleic Acid Characterization and Genetic Stability*, CBER, FDA, Washington, D.C., April 1992.
17. *Points to Consider in the Characterization of Cell Lines Used to Produce Biologicals*, CBER, FDA, Washington, D.C., May 1993.
18. *Points to Consider in the Manufacture and Testing of Monoclonal Antibody Products for Human Use*, CBER, FDA, Washington, D.C., 1994.
19. S. Fitzpatrick, A. Ma'ayan, and J. Wagget, *Chem. Eng. Prog.* **86**, 26–31 (1990).
20. M. Stern, *Gen. Eng. News* **18**, 10, 32 (1998).
21. L. Little, *BioPharm*, **10**, 8–9 (1997).
22. T. Orr, *Gen. Eng. News* **18**, 12–14 (1998).
23. *Guideline on Sterile Drug Products Produced by Aseptic Processing*, Center for Drugs and Biologics and Office of Regulatory Affairs, FDA, Washington, D.C., June 1987.
24. *Guideline on the Preparation of Investigational New Drug Products*, CDER, FDA, Washington, D.C., March 1991.
25. *Guideline on the General Principles of Process Validation*, Center for Drugs and Biologics, FDA, Washington, D.C., May 1987.
26. *Guideline for Inspections of Biotechnology Manufacturing Facilities*, FDA, Washington, D.C., 1991.
27. *Guide to Inspections of Validation of Cleaning Processes*, FDA, Washington, D.C., 1993.

28. W. Kelley, *Pharm. Technol.* **10**, 48–54 (1986).
29. P. James, *Pharm. Eng.* **18**, 72–82 (1998).
30. L. Angelucci, *Pharm. Eng.* **18**, 40–44 (1998).
31. G. Sofer, J. McEntire, and J. Akers, *Biopharm (Jilove)* **8**, 60–63 (1995).
32. N. Fletcher, *Pharm. Eng.* **9**, 23–27 (1989).
33. A. Signore, *Pharm. Eng.* **18**, 72–82 (1998).
34. D. Tisak and R. Koster, *J. Valida. Technol.* **3**, 394–397 (1997).
35. C. DeSain, *Biopharm (Jilove)* **5**, 22–24 (1992).
36. C. DeSain, *Biopharm (Jilove)* **5**, 30 (1992).
37. W. Wheeler, *Pharm. Eng.* **4**, 48–56 (1994).
38. R.G. Werner, H. Langlouis-Gau, F. Walz, H. Allgaier, and H. Hoffman, *Arzneim.-Forsch.* **38**, 855–862 (1988).
39. J. Akers, J. McEntire, and G. Sofer, *Biopharm (Jilova)* **7**, 8–11 (1994).
40. ISPE San Francisco/Bay Area Chapter, *Pharm. Eng.* **18**, 8–24 (1998).
41. J. Adamson, *Pharm. Eng.* **12**, 16–22 (1992).
42. J. Agalloco and J. Akers, *J. Pharm. Sci. Technol.* **51**, S1–S23 (1997).
43. J. Bauers and J. Hargroves, *Pharm. Eng.* **16**, 36–42 (1996).
44. J. Alford and F. Cline, *Biopharm (Jilova)* **3**, 32–36 (1990).
45. Guidelines for research involving recombinant DNA molecules: notice, *Fed. Regist.* **51**, 16958–16985 (May 7, 1986).
46. Recombinant DNA research: actions under the guidelines; Notice, *Fed. Regist.* **56**, 33174–33183 (July 18, 1991).
47. J. Van Houten and D.O. Fleming, *J. Ind. Microbiol.* **11**, 209–215 (1993).
48. J. Van Houten, in M. Ladisch and A. Bose eds., *Harnessing Biotechnology for the 21st Century: Proceedings of the Ninth International Biotechnology Symposium*, American Chemical Society, Washington, D.C., 1992, pp. 415–418.
49. M. Hamers, *Biotechnology* **11**, 561–570 (1993).
50. G. Muijs, in C. Collins and A. Beale eds., *Safety in Industrial Microbiology and Biotechnology*, Butterworth-Heinemann, Oxford, 1992, pp. 214–238.
51. A. Shahidi, R. Torregrossa, and Y. Zelmanovich, *Pharm. Eng.* **15**, 72–83 (1995).
52. Y. Chisti, *Chem. Eng. Prog.* **88**, 80–85 (1992).
53. F. Bader, in A. Demain and N. Solomon eds., *Manual of Industrial Microbiology*, American Society for Microbiology, Washington, D.C., 1986, pp. 345–362.
54. M. Akers, I. Attia, and K. Avis, *Pharm. Technol. Int.* **1**, 45–49 (1978).
55. M. Enzinger, A. DeVisser, and M. Townsend, *Pharm. Technol. Int.* **1**, 41–43 (1978).
56. M. Enzinger, M. Goodsir, M. Korozynski, F. Parham, and M. Schneier, Technical Monograph No. 1, Parental Drug Association, Inc., Philadelphia, Pa., (1978).
57. *Guidance for Industry for Submission Documentation for Sterilization Process Validation*, CDER, FDA, Washington, D.C., November 1994.
58. M. Reisman, *Pharm. Cosm. Qual.* 22 (1998).
59. D. Berman, T. Myers, and C. Suggy, *J. Parenter. Sci. Technol.* **40**, 119–121 (1986).
60. R. Borghese, *Biopharm (Jilova)* **6**, 28–32 (1993).
61. R. Manfredini, L. Golzi Saporiti, and V. Cavallera, *Proc. Conf. Chem. Biotechnol. Biol. Active Nat. Prod.* 1981, pp. 320–391.
62. C. Perkowski, *Biopharm (Jilova)*, 62–65 (1987).
63. C. Perkowski, *J. Parenter. Sci. Technol.* **44**, 113–117 (1990).
64. M.C. Sharma and A.K. Gurtu, in S. Neidleman and A. Laskin eds., *Advances in Applied Microbiology*, vol. 39, Academic Press, New York, 1993, pp. 1–27.
65. B. Junker, Z. Mann, P. Gailliot, K. Byrne, and J. Wilson, *Biotechnol. Bioeng.* **60**, 580–588 (1998).
66. K. Jenkins, A. Vanderwielen, J. Armstrong, L. Leonard, G. Murphy, and N. Piros, *PDA J. Pharm. Sci. Technol.* **50**, 6–15 (1996).
67. Y. Christi and M. Moo-Young, *J. Ind. Microbiol.* **13**, 201–207 (1994).
68. R. Wisiniewski and C. Burman, in A. Lubiniecki and S. Vargo eds., *Regulatory Practice for Biopharmaceutical Production*, Wiley-Liss, New York, 1994, pp. 407–445.
69. N. Adner and G. Sofer, *Biopharm (Jilova)* **7(3)**, 16–21 (1994).
70. R. Seely, H. Wight, H. Fry, S. Rudge, and G. Slaff, *Biopharm (Jilova)* **7**, 41–48 (1994).
71. G.L. Fourman and M.V. Mullen, *Pharm. Technol.* **17**, 54–60 (1993).
72. S. Lombardo, P. Inampudi, A. Scotton, G. Ruezinsky, R. Rupp, and S. Nigam, *Biotechnol. Bioeng.* **48**, 513–519 (1995).
73. C. Amador, P. Deschenes, M. Dorminy, B. Dudley, B. Gordon, C. Lampe, R. McGarrah, J. Prulello, and R. Wild, *J. Parenter. Sci. Technol.* **44**, 324–331 (1990).
74. D. Held, R. Mehigh, C. Wooge, S. Crump, and W. Kappel, *Biopharm (Jilova)* **10**, 32–37 (1997).
75. P. McArthur and M. Vasilevsky, *Pharm. Eng.* **15**, 24–31 (1995).
76. P. Inampudi, S. Lombardo, G. Ruezinsky, T. Baltrus, J. Dugger, P. Remsen, R. Rupp, and S. Nigam, *Ann. N.Y. Acad. Sci.* **782**, 363–374 (1996).
77. J. Geigert, R. Klinke, K. Carter, and A. Vahratian, *PDA J. Pharm. Sci. Technol.* **48**, 236–240 (1994).
78. T. Macek, in G. Phillips and W. Miller eds., *Industrial Sterilization*, Duke Univ. Press, Durham, 1972, pp. 19–34.
79. J. Brantley and J. Martin, *Gen. Eng. News* **17**, 24, 40 (1997).
80. USP-NF Fifth Supplement, 3442–3444, 3464–3467, 3482–3483, 3547–3555 (1996).
81. W. Collentro, *Ultrapure Water* **11**, 22–29 (1994).
82. F. Riedewald, *Pharm. Eng.* **17**, 8–20 (1997).
83. M. Mittelman, *Microcontamination* **3**, 51–55, 70 (1985).
84. J. Lukanich, *Chem. Process.* **61**, 78 (1998).
85. N. Roscioli, C. Renshaw, A. Gilbert, C. Kerry, and P. Probst, *Biopharm (Jilova)* **9**, 32–40 (1996).
86. J. Barta, *Biopharm (Jilova)* **11**, 84 (1998).
87. B. Immel, *Biopharm (Jilova)* **11**, 46–49 (1998).
88. D. Conner and M. Eckman, *Pharm. Technol.* **16**, 148–160 (1992).
89. J. Odum, *Pharm. Eng.* **15**, 8–20 (1995).
90. *Guide to Inspections of Bulk Pharmaceutical Chemicals*, CDER, May 1994.
91. S. Kuwahara, *Biopharm (Jilova)* **9**, 24–29 (1996).
92. S. Kuwahara, *Biopharm (Jilova)* **9**, 40–45 (1996).
93. J. Sandberg, *Med. Device Diagn. Ind.* **17**, 102–106 (1995).

See also ANTIBODY PURIFICATION; ASTM STANDARDS FOR BIOTECHNOLOGY; CLEANING, CLEANING VALIDATION.

H

HEATING, VENTILATION, AND AIR-CONDITIONING

DENNIS DOBIE
Fluor Daniel
Marlton, New Jersey

KEY WORDS

Air changes
cGMP
Classification
Containment
Control system
Ductwork
HEPA
Room seals
Space pressurization
Terminal control units

OUTLINE

INTRODUCTION

Biological products are derived from living sources, which presents special challenges for a heating, ventilation, and air-conditioning (HVAC) system serving a facility that processes and handles these products. The most important challenge is to prevent microbial contamination, which to a large measure is controlled by the HVAC system by utilizing classified spaces, providing containment, and maintaining pressurization. Biological facilities are subject to federal regulation. The Food and Drug Administration's (FDA's) Center for Biologic Evaluation and Research (CEBR) is the primary regulating agency in the United States and its principal concern is the prevention of cross-contamination between different products produced in a multiproduct facility (1). Bioprocess procedures are generally conducted in environments that are cleaner than standard buildings. The HVAC system assumes a large part of the responsibility in maintaining these clean environments. Air used to condition a space must be highly filtered, properly directed, and able to maintain pressure relationships within and between adjacent spaces.

HVAC DESIGN PROCESS

Initial Effort

The HVAC design process begins with meetings with process engineers, architects, and representatives from the owner or facility user. The process and instrument diagrams (P&IDs) are reviewed, and a general understanding of the process is conveyed to all interested parties. Operation of the facility is reviewed, and any plans for future additions or modifications are discussed.

Basis of Design

After the initial meeting, a written basis of design is produced that describes the regulations and codes that will govern the design. Spaces are defined by function, and temperature and humidity requirements are determined. Room classifications are listed and adjacency of spaces and pressure relationships are documented. Any unusual or unique facility requirements must also be designed into the HVAC system at this time, such as emergency backup or redundancy for HVAC systems. This is also the stage of the design process during which alternate studies are conducted to compare options for the HVAC system. The cost of a backup or redundant HVAC supply system may be compared with the cost of product loss or experiment interruption should temperatures or airflow go out of control or specification. Heat recovery from exhaust systems and thermal storage are examples of other potential areas for study (2).

Airflow diagrams are produced that show areas served by a particular air handling system including supply, return, exhaust, and transfer air between spaces. The basis of design also describes major equipment to be used and the level of quality of components and construction material.

Cost Estimate

From the basis of design and a plan of the area, a factored estimate can be developed for budgeting purposes. Preparation of the estimate usually includes a dollar per cubic feet per minute (CFM) quote or a budget quote from vendors, as well as previous history to determine air handling unit costs. Estimates are also made for ductwork, as derived from pounds per CFM history and historic cost data. Costs for fans, high-efficiency particulate air (HEPA) filters, terminal boxes, and building management systems are also included. The level of effort expended to produce the basis of design will affect the accuracy of the estimate. Estimates from a basis of design can range from 15 to 25% accuracy. The estimate at this stage of the project is usually developed for capital appropriation.

Document Preparation

After owner or user approval of the basis of design, final plans and specifications are prepared using manual or computer-aided design tools for preparation of plans and sections. The utilization of three-dimensional design systems for document preparation is becoming increasingly popular. In a three-dimensional system, an electronic model is built that includes all disciplines, including HVAC, architectural, structural, electrical, piping, equipment, and so on. From the model interference, checks can be made that indicate physical conflict. These conflicts are resolved in the design process, avoiding potentially expensive field modifications and possible project delays.

Bidding and Award

When plans and specifications are completed they are sent to HVAC contractors for bid and eventual award. Design and build is also becoming a very popular option, in which design and construction are performed by a single entity with potential cost and schedule savings (3). When awarding a contract the engineer usually meets with the contractor to review the bid for completeness and technical accuracy, and to review the contractor's history on previous projects performing similar work. Since this is specialized HVAC work, it is necessary to have a contractor with sufficient resources and similar experience to avoid delays and potential rework that could increase costs.

REGULATION AND CODE CONSIDERATION

Code Compliance

Codes affect separation of air systems, gowning of people, airflow rates, exhaust requirements, construction materials, and many other aspects of an HVAC system for bioprocess facilities. If codes are not met, the result can be a danger to people working in the facility, products that do not meet specifications, and failure to license the facility.

Classified Spaces

Clean rooms provide clean air in aseptic spaces and are defined or classified in the United States by Federal Standard 209E of September 1992. The European Economic Community (EEC) published guidelines called "Guide to Good Manufacturing Practice for Medical Products in the European Community," which are more stringent than U.S. FDA regulations (4). Room classifications are established by measurement of the number of particles 0.5 μm and larger that are contained in 1 ft^3 of sampled air. Generally class 100 to 100,000 rooms are used in the bioprocess industry. Rooms may be classified as clean at class 1 or 10 for other applications, particularly in the microchip industry. Table 1 (derived from Federal Standard 209E) shows the air cleanliness classes.

EEC guidelines classify spaces by alpha character, as indicated in Table 2, and Table 3 shows the U.S. and EEC comparison.

Laboratory Spaces

Laboratory spaces are associated with bioprocess facilities. These classifications range from BL-1 through BL-4, with BL-1 being the least hazardous. BL-1 compares to a school science laboratory, while BL-4 labs must provide ultimate containment and require glove boxes and double HEPA filtration. These laboratory classifications were formulated in the United States by the National Institutes of Health (NIH) and are published as guidelines in the *Federal Register*. The guidelines contain requirements for filtration of air, recirculation of air, containment, and biosafety cabinet requirements. There are various classifications of biosafety cabinets, depending on the level of containment required. The American Society of Heating, Refrigerating, and Air-Conditioning Engineers, Inc. (ASHRAE) guide should be consulted and owner requirements considered when selecting the required class and type of cabinet (5).

Other Regulations

Other regulations and guidelines may include those published by insuring agencies and corporate or plant guidelines formulated from specific industry experience and current good manufacturing practices (cGMPs) as published in the *U.S. Code of Federal Regulations* parts 210 and 211. The cGMPs apply to manufacture, holding, or processing of drugs to ensure safety, purity, and quality of the product. cGMPs regulate space temperature, and humidity and require a pressure differential between adjacent rooms of different cleanliness levels and a minimum air change rate of 20 air changes per hour. The regulation also addresses class 100 requirements at filling lines and microbiological testing sites, in addition to laminar flow control equipment parameters and testing at fill lines. Good practice usually places class 100 areas in class 10,000 backgrounds, which have between 35 and 60 air changes per hour in the space, depending on the activity and particulate generation within the space. Walls, floors, and ceilings for cGMP areas are to be constructed of smooth, cleanable surfaces, impervious to sanitizing solutions and resistant to chipping, flaking, and oxidizing. Horizontal ducts, pipes, or crevices that foster dust accumulation and cannot be easily cleaned are not permitted.

Figure 1 is a sample graphic depiction of an air classification drawing that is prepared to communicate to the design team and reviewing agencies the classification of each area. This is usually a full-size drawing (scale: 1/8 ft = 1 ft) and uses various shading symbols to identify each classified area.

TEMPERATURE, HUMIDITY, AND AIRFLOW

Space Temperature

Generally areas are designed to provide room temperatures from 67 and 77 °F (19 and 25 °C) with a control point of 72 °F (22 °C). Control of humidity from 40 to 55% is necessary for personal comfort, to prevent corrosion, to control microbial growth, and to reduce the possibility of static electricity. Calculations are made to determine internal heat gains from lights, people, and equipment in addition to transmission gains from adjacent spaces to determine if supply airflow quantities are adequate to provide design space temperature conditions. These calculations are compared with airflow quantities required by classification to establish the minimum air required to satisfy both the space cooling load requirements and air cleanliness classification. Fan heat from recirculating fans can also be a large heat contributor in clean spaces.

Heat-loss calculations must also be made to determine heat loss through walls, roof, and floor. It may be necessary to add additional heat to a reheat coil to overcome these losses. No credit should be taken for process heat gain in this calculation, since the process could be dormant and the space would still need to be maintained at proper temperature. Lower space temperatures may be required where people are very heavily gowned and would be uncomfortable at "normal" room conditions. Humans are generally the largest contributors to airborne particulate in clean spaces; personal discomfort may cause perspiration, which significantly adds to the generation of airborne particulate. To achieve colder room conditions, the temperature of the supply air to the rooms may have to be colder than the normal 50–55 °F (10–13 °C), resulting in a greater dehumidification demand on the air-conditioning system. The amount of dehumidification required can be deter-

Table 1. Airbone Particulate Cleanliness Classes

Class		Measured particle size (μm)									
		0.1		0.2		0.3		0.5		5	
SI	English	m^3	ft^3	m^3	ft^3	m^3	ft^3	m^3	ft^3	m^3	ft^3
M 1.5	1	1,240	35	265	7.5	106	3	35.5	1	—	—
M 2.5	10	12,400	350	2,650	75	1,060	30	353	10	—	—
M 3.5	100	—	—	26,500	750	10,600	300	3,530	100	—	—
M 4.5	1,000	—	—	—	—	—	—	35,300	1,000	247	7
M 5.5	10,000	—	—	—	—	—	—	353,000	10,000	2,470	70
M 6.5	100,000	—	—	—	—	—	—	3,530,000	100,000	24,700	700

Note: Limits listed are maximum concentrations of particle sizes equal to and larger than the measured particle sizes shown.

Table 2. Classification of Clean Areas Defined by the European Economic Community

Grade	Maximum permitted number of particles per m³ equal to or above 0.5 μm	Maximum permitted number of viable microorganisms per m³
A	3,500	<1
	Laminar airflow work station	
B	35,000	5
C	350,000	100
D	3,500,000	500

- Laminar airflow speed: 0.3 m/s for vertical flow 0.45 m/s for horizontal flow.
- Air changes per hour for grades B, C, and D should be higher than 20.
- U.S. Federal Standard 209E corresponding particle counts are
  Class 100 (Grades A and B)
  Class 10,000 (Grade C)
  Class 100,000 (Grade D)
- It is accepted that sometimes conformity with particle counts may not be met at point of fill due to generation of particles or droplets from the product itself.

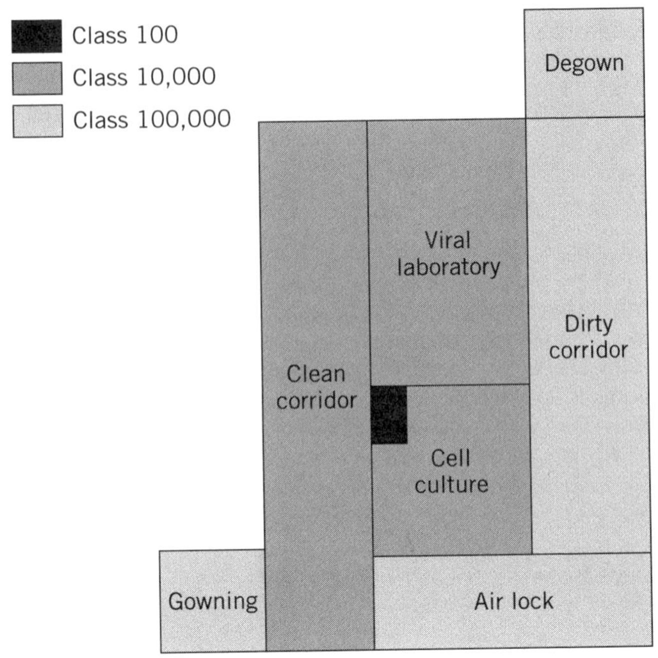

Figure 1. Air Classification Drawing.

mined by use of a standard psychrometric chart by plotting room, mixed air, outdoor air, and coil conditions. Formulas to determine cooling coil loads are available from most coil manufacturers and air handling unit vendors (6). Cooling coil duty in British thermal units per hour (Btuh; *W*) is determined from the enthalpy difference between entering and exiting cooling coil air dry and wet bulb conditions.

Space Humidity

In some cases where products sensitive to moisture are handled, the room relative humidity requirement may be as low as 15 to 20% and may require the use of chemical dehumidifiers. It is necessary to establish this criteria early in the design process to allow selection of dehumid-

Table 3. Air Cleanliness Classification for Aseptically Produced Products

Operation	Parameters	United States	European Economic Community
Nonsterilized product or container (U.S.: controlled area)	Clean room type	Class 100,000	Grade C (Class 10,000)
	Maximum particle size 0.5 μm or above	100,000/ft³ (3,531,000/m³)	350,000/m³ (9,900/ft³)
	Maximum viable organisms	2.5/ft³ (88.3/m³)	100/m³ (2.8/ft³)
	Airflow rate	Minimum 20 air changes/h	Higher than 20 air changes/h
	Space pressurization	0.05 in water	Positive
Sterilized product or container (U.S.: critical area)	Clean room type	Class 100 environment in class 10,000 background	Grade A (class 100) environment in grade B (class 100) background
	Maximum particle size 0.5 μm or above	100/ft³ (3,531/m³)	3,500/m³ (99.1/ft³)
	Maximum viable organisms	0.1/ft³ (3.53/m³)	Class 100 room with grade A area: 1/m³ (0.03/ft³) Grade B background 5/m³ (0.14/ft³)
	Airflow rate	90 ft/min ± 20% (45.7 m/s)	Grade A: laminar flow work station Vertical: 0.3 m/s (59 ft/min) Horizontal: 0.45 m/s (88.6 ft/min) Grade B: higher than 20 air changes/h
	Space pressurization	0.05 in water	Positive

ification equipment and add allowances to the construction-cost budget estimate. The lower the required temperatures and humidity, the more sophisticated the HVAC equipment becomes, resulting in higher start-up and operating costs. Range of temperature control also affects the cost of the control and HVAC delivery system. The closer the space temperature tolerances are, the more sensitive the controls must be to react. To achieve close temperature tolerances, the air distribution system must be more extensive and elaborate to prevent hot or cold spots. An HVAC room balance, shown in Table 4, is a sample of a preliminary design document that indicates room temperature and humidity design conditions, in addition to dimensions. Table 4 also lists room classification air changes per hour, supply, exhaust, return, and transfer air requirements. This information is useful in communicating conditions to the design team and reviewing agencies. The table should be kept up to date, because it is useful to installing contractors and the construction supervision team as a check on ultimate system performance. When the project is near completion this information is usually transferred to the construction documents for ease of validation and agency approval.

Calculation of Air Changes

Table 4 indicates a typical range of air change rates generally used to achieve the desired room cleanliness classifications and to meet federal and local regulations. These air change rates vary widely in actual practice due to the level of activity, number and type of particulate generators in a room (such as people and equipment), and room size and quality of air distribution. It is generally best to use historic data to establish airflows, which is usually done with significant input from the owner based on past experience or preference. There is nothing sacred about an air change rate as long as minimum airflow rates required by code are maintained. The goal is to achieve desired particulate cleanliness levels and stay at or above a 20 air changes/h minimum.

To calculate airflow requirements from room air change rates, the following formula should be used:

$$\text{air flow (CFM)} = \frac{\text{room volume (ft}^3\text{)} \cdot \text{Number of air changes/h}}{60}$$

or (metrically)

$$\text{air flow (m}^3\text{/h)} = \frac{\text{room volume (m}^3\text{)} \cdot \text{Number of air changes/h}}{60}$$

Adjacency

In the early stages of design, the HVAC engineer must work with the architect, process engineer, user, and layout planner, if one is involved, to establish adjacency of spaces. It is suggested that a computerized drafting system with block layouts be used to allow the frequent manipulation that is required to arrive at an optimum arrangement. During this exercise the layout is primarily determined by process function and spaces are classified as clean, dirty, or contained. A contained space is used to contain a product or material that could be spilled into an occupied space or find its way into the building air system. From an HVAC standpoint it is desirable to keep similarly classified areas as physically close to each other as possible so they can be connected to the same air handling system, thereby minimizing duct runs, cost, and air system complexity. It is also imperative that spaces be arranged to allow people to move around without disrupting the cleanliness or containment of the spaces.

Ultimately, as the layout develops, the suites or groups of spaces will be established and groupings of rooms can be made for assignment to separate air handling systems. It is not desirable to mix dirty and clean systems or suites that may allow the possibility of cross-contamination from one suite to another. Leaks can develop in a filter, or some source of contamination could find its way through the air

Table 4. HVAC Room Balance Sheet

Room name and number	Room temperature	Relative humidity (%)	Area (ft²)	Room height (ft)	Air changes per hour	Hood, general exhaust	Required return CFM	Total supply CFM	Transfer air Out	Transfer air In
Viral lab #1—100 (BL2, class 10K)	68	50	544	9	40	1,230	1,984	3,264	170	120
C/C lab #1—101 (BL2, class 10K)	68	50	453	9	40	345	2,323	2,718	170	120
Gowning 102 (BL2, class 10K)	68	50	125	9	40	0	385	750	365	0
Viral lab #2—103 (BL2, class 10K)	68	50	544	9	40	1,230	1,984	3,264	170	120
Clean corridor 104 (BL2, Class 10K)	68	50	1,300	9	40	0	6,905	7,800	1,320	425
Dirty corridor 105 (BL2, class 100K)	68	50	700	9	20	0	3,300	2,100	0	1,250
Degown airlock 106 (BL2, class 100K)	68	50	80	9	20	0	480	240	0	240

supply or return systems, providing a source for cross-contamination.

PRESSURIZATION

Requirement and Theory

Pressurization of spaces is required to keep products from being contaminated by particulate or to protect people from contact with harmful substances by physical means or inhalation. Where very severe potential exposure exists, air packs and breathing apparatus are employed.

Pressurization can best be explained by thinking of air leaking from a balloon. Pressure is built up inside the balloon and air escapes from a small leak in the balloon. If the hole stays the same size, a greater pressure differential between the inside and outside of the balloon will allow the air to escape at a higher velocity, which can be termed a higher level of pressurization. As the air in the balloon is depleted, the pressure differential drops, the rate of air escaping diminishes, and the velocity decreases. The same principle applies in bioprocess facilities where rooms are made as tight as possible to contain air, allowing the air handling system to build up or reduce pressure in rooms. These conditions are maintained by airflow, or leakage, between adjacent spaces caused by pressure differentials created by the air supply, return, and exhaust systems. Where major demarcations of pressure are required, air locks are used. These are small rooms with controlled airflows acting as barriers between spaces. Standard 209E recommends maintaining pressures of 0.05 in water (12 Pa) between adjacent spaces with doors closed. During facility operation as doors are opened the design differential is greatly reduced, but air must continue to flow from the higher to lower pressure space, even though at a reduced flow rate. To maintain a differential of 0.05 in water (12 Pa), a velocity of approximately 900 ft/min (4.7 m/s) should be maintained through all openings or leaks in the room, such as cracks around door openings. In theory the actual required velocity is less, but in actual practice it is prudent to use 900 ft/min. Table 5 gives the velocity pressures airflows from 0.01 to 0.20 in water (2.4–48 Pa), which can be used in approximating airflow between rooms. Limits on pressure drops must be set during the design (usually 0.05 in water) and observed. The pressure differential exerts a force on the door that can be calculated. If the force is too great (0.15 in water/36 Pa), the door may not close fully or may be difficult to open. This is particularly important in large complex facilities where many levels of pressurization may be required. Many facilities now use sliding doors, and it is essential that the seals be carefully designed to allow minimum leakage and proper containment or pressurization.

Table 5. Velocity Pressure to Velocity Conversion: Standard Air

VP	V	VP	V	VP	V	VP	V
0.01	401	0.06	981	0.11	1328	0.16	1602
0.02	566	0.07	1060	0.12	1387	0.17	1651
0.03	694	0.08	1133	0.13	1444	0.18	1699
0.04	801	0.09	1201	0.14	1499	0.19	1746
0.05	896	0.10	1266	0.15	1551	0.20	1791

Note: VP, velocity pressure; V, velocity.

Pressurization Diagram

Figure 2 indicates levels of higher to lower pressurization with arrows showing the direction of airflow, (leakage) between the spaces. In actual practice the calculated airflow quantities are approximate since in constructing a facility it is very difficult to build everything absolutely airtight. A very small unexpected leak or a seal that is not as tight as expected in design can affect the pressurization and airflow required to maintain the design value of differential. Some flexibility must be built into the air handling system capacity and controls to compensate for these variations.

One technique to establish pressurization in separate rooms is to have a common reference for the control sensors. This common reference is usually a mechanical room or an interstitial space where the pressure is not affected by changes in the controlled spaces. It may be necessary to build an elaborate zero pressurization chamber to house the common sensor to negate the effects of wind or cross-currents. Referencing spaces to each other can lead to problems if doors are frequently opened and upsets in pressure occur. The controls may never settle down and tend to constantly modulate, a problem termed *hunting*.

Room Seals and Doors

In most facilities the openings around the doors between rooms are where leaks occur due to pressure differentials between rooms. In making rooms tight any room openings

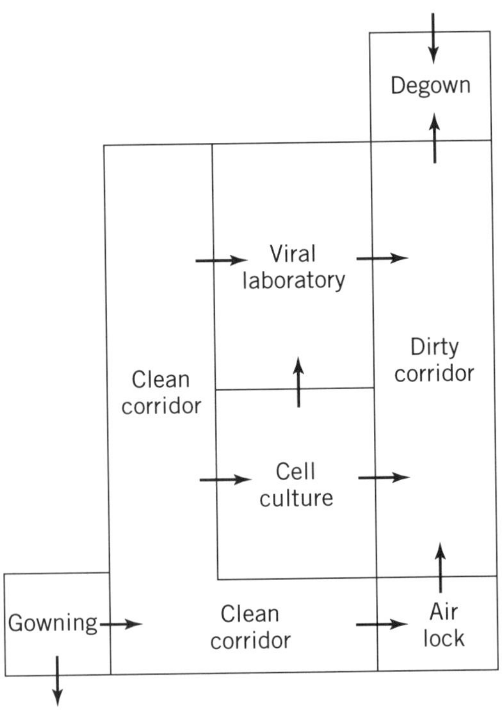

Figure 2. Pressurization Diagram.

must be sealed with a proper sealant that will not promote growth of organisms and can be easily cleaned. Areas to be sealed include ceiling tiles, lighting fixtures, pipe penetrations, telephone outlet penetrations, and any cracks or openings that appear in the structure. A typical door would have the following dimensions and crack area at the perimeter: door size, 3 ft (0.915 m) wide by 7 ft (2.41 m) high; cracks at top and sides, 1/8 in (3.2 mm) with an undercut of 1/4 in (6 mm). The calculated area around the door is equal to 0.24 ft^2 (0.024 m^2). To achieve 0.05 in water (12 Pa) pressure differential across the door, approximately 215 CFM (375 m^2/h) of airflow through the cracks is required. Door seals around the top and sides are usually made of closed cell neoprene and should generally be used to reduce the crack area. To reduce the undercut, a drop-type seal, which is commercially available, should be used. The drop type is preferred to a wipe type, since it will not mar or leave residue on the floor. Air used for pressurization must be accounted for in system calculations. Air through cracks or openings is accounted for as transfer air and shown in the HVAC room balance table.

Of particular importance is exhaust air from equipment and hoods that may be on or off at different times during occupied periods. These variations must be dynamically compensated for to maintain room pressurization. To maintain the required balance, numerous systems are employed using manual and automatic dampers, constant and variable volume air control boxes, and elaborate airflow sensing devices. These components are combined with control systems and sensing devices to ensure that the room pressurization is maintained.

AIR HANDLING SYSTEMS

Constant Volume Systems

After block layouts, adjacency layouts, room design criteria, and pressurization levels have been established, the number and type of air handling systems can be selected. Air handling systems most commonly used are constant and variable volume with and without reheat coils for control. For controlled spaces the most reliable system is a constant volume system with terminal reheat (CVRH). This system is not used in less critical or commercial applications. In a terminal reheat system the air leaving the cooling coil is set at a fixed temperature, and the terminal reheat responds to a space thermostat, turning on heat to satisfy the load. This can waste energy, since air is cooled and then reheated. Reheat systems require energy to both simultaneously heat and cool the air, as can easily be demonstrated on a psychrometric chart. Many energy codes prohibit this practice for comfort applications, however, where close control of temperature and humidity is required for process areas the energy conservation requirement is waived. The advantages of reheat systems are that humidity is always controlled (since dehumidification always takes place at the cooling coil) and each space or zone that needs temperature control can easily be accommodated by adding a reheat coil and thermostat. Another advantage of the CVRH system is that airflow is constant, which makes balancing and pressurization easier to maintain. A reheat system is probably the simplest and easiest of all systems to understand and maintain. Figure 3 is a typical depiction of an airflow diagram for a CVRH system.

Variable Air Volume Systems

A variable air volume (VAV) system is generally used in administrative areas and some storage spaces where pressure control is not critical, humidity control is not essential, and some variations in space temperature can be tolerated. The VAV system works by delivering a constant-temperature air supply to spaces with reductions in airflow as cooling loads diminish. This reduction in air results in a reduced cooling effect, allowing the airflow or cooling to match the actual room load, which provides the design space temperature condition. This eliminates the energy used for reheat and saves fan energy, because the total amount of air moved is reduced. Some form of perimeter heating must be supplied for spaces with exterior walls or large roof heat losses. The perimeter heating can be baseboard radiation or some form of air heating using heating coils. Finned radiation or convection heating devices should not be used in clean spaces, since they are not easily cleaned and allow places for unwanted particulate buildup. Combinations of systems can be used, especially if variable quantities of supply and exhaust air are required for fume hoods or intermittent exhausts.

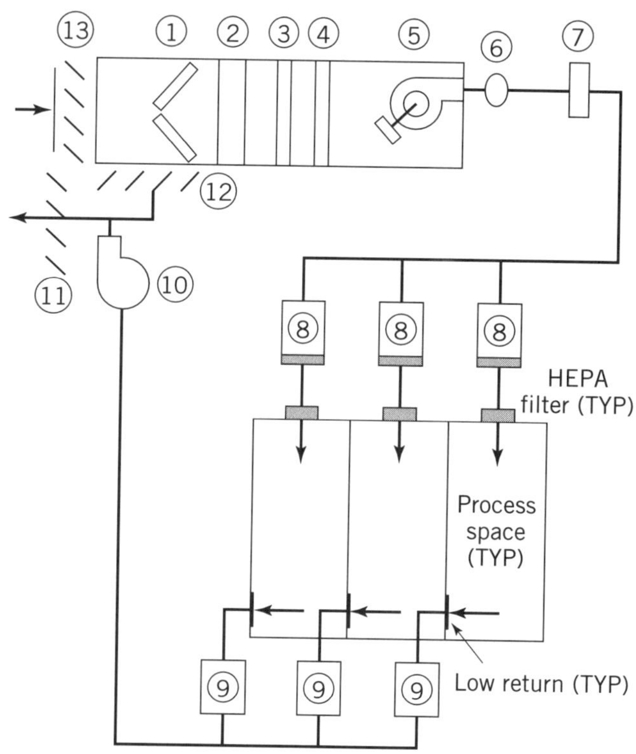

Figure 3. Airflow diagram: 1, 35% filters; 2, 85% filters; 3, preheat coil; 4, cooling coil; 5, fan section; 6, humidifier; 7, air monitoring device; 8, constant volume reheat box; 9, constant volume box; 10, return fan; 11, relief air damper; 12, return air damper; 13, outdoor air damper.

Recirculation Systems

In bioprocess facilities large quantities of air may be required to promote unidirectional flow and air cleanliness. This is particularly true in a class 100 space. In many cases the large quantities of air exceed the requirements for cooling, so it is desirable and possible to recirculate air within the space and only pass enough air through the air handling unit to perform the heating or cooling. In many cases the recirculating systems are supplied as a separate package installation. Numerous manufacturers provide these packages, which include supports, fans, and filters. Some vendors provide fans that are self-contained with each 2-ft by 4-ft filter module.

Airflow Sheets

The airflow sheets should be developed on full-size drawings using computer drafting. They should list each space and show air quantities supplied, returned, and exhausted from each space. They also must show air transferred into and out of spaces, and, while quantities should be shown, they will probably require field modification to attain pressurization. The airflow sheet becomes a useful tool for transfer of information to the owner or user, for agency reviews, for transmission of information to HVAC designers, and for other engineering disciplines. These documents are also invaluable to construction contractors and for system checking by construction managers and balancing contractors. Airflow sheets provide a pictorial description of each air system and show how the elements comprising the system are related.

DEHUMIDIFIERS

Need for Dehumidification

In many cases low relative humidity is required, such as in drying or powder areas where hygroscopic materials are handled and would absorb unwanted moisture from the air. A chemical dehumidifier may be required to obtain the desired low humidity conditions. The need for a dehumidifier can be determined by utilizing a psychrometric chart, which will establish a leaving-coil dew point. Normal chilled water is supplied at 42–44 °F (5.6–6.7 °C) for use in cooling coils. Depending on the number of coil rows and fin spacing, a minimum dew point of about 50–52 °F (10–11.1 °C) can be obtained. This results in a minimum room relative humidity of approximately 50% at 70 °F (21.1 °C). To obtain lower relative humidity conditions chemical dehumidification is employed.

Chemical Dehumidification

Chemical dehumidifiers are commercially available air handling units that contain a sorbent material (desiccant) that can be a solid or liquid. Wet dehumidifiers use absorbents that change physically during the process. Lithium salt solutions are generally used to remove moisture from conditioned air and are then regenerated by heat, usually using a steam heat exchanger. Dry dehumidifiers use adsorbents that do not experience phase changes during the process. Silica gel and activated alumna are generally used. A rotating wheel is commonly used to remove moisture from the conditioned air. The wheel is regenerated by passing heated outdoor air over the wheel to dry it out. Steam or electric coils are usually employed for regeneration. Depending on the amount of dehumidification required and the amount of outdoor air (usually with a high moisture content), it may be best to combine the dehumidifier with a conventional air handling unit and only dehumidify a small portion of the air or just the outdoor air. The dehumidifier has a high initial cost compared with a conventional air handling unit. The size should be optimized to do only the required duty with an appropriate safety factor. Knowledgeable vendors in this specialized area should be consulted to find the best combination of dehumidification equipment, system arrangement, and control for the application. These systems also require considerable physical space, energy consumption, and service—important criteria to be considered in system selection.

HUMIDIFIERS

Air handling systems that bring in quantities of outdoor air for makeup require the addition of moisture to the airstream to provide desired space humidity levels. There are many commercially available devices for providing humidity to air, but the most commonly used are steam grid humidifiers. It is important to use clean steam (7), not plant steam, which may contain boiler chemicals and impurities from deteriorating piping and equipment. Humidifiers are controlled by modulation of a steam valve at the humidifier, and include a chamber to prevent condensation and water droplets in the duct. The valve is controlled by a signal located in the return or exhaust airstream or in a room humidstat. A high-limit stat is placed in the duct downstream of the humidifier to override the controlling stat and prevent condensation in the duct. Placement of the humidifier in the duct is critical and must follow the manufacturer's recommendations to prevent condensation and provide proper dispersion space. Duct downstream of the humidifier is usually constructed of stainless steel to prevent rust or corrosion should water be present.

AIR HANDLING UNIT APPLICATION

Air Handling Unit Construction

Conventional air handling units consist of filters, coils, and fans in a metal casing, with an insulation liner applied to the inside of the casing. For bioprocess applications the unit casing must be a sandwich of metal with insulation between the metal sheets to provide a smooth, cleanable interior surface that does not foster the growth of organisms. Chilled water or propylene glycol solutions are generally used for cooling and dehumidification. Direct expansion refrigerant, in which the refrigerant is in the air unit coil, may be used, but these systems are less reliable than chilled water or glycol and are more difficult to control in the narrow air temperature ranges required. Units should contain good access doors, view ports, electrical convenience outlets, and interior lighting for maintenance. The casings should be tightly sealed and designed for pressures that are higher than commercial applications due to the generally high system air pressures required for bioprocess applications. All sealants and lubricants exposed to the airstream should be food grade to minimize the chance of air contamination. Units designated as draw-through have the coils on the suction side of the fan. Blow-through units have the coils on the discharge side of the fan and have the advantage of a filter downstream of the coils, reducing potential contamination of the supply duct system. On blow-through units an air distribution plate must be installed to properly distribute air evenly over the filter and coils. It should be noted that if the unit contains a HEPA filter and the air is ducted directly to the clean space without a terminal HEPA filter, the duct should be constructed of a clean, smooth, nonflaking material, usually stainless steel.

Air Handling Unit Location

It is best to locate units indoors for ease of service and to extend the life of the units. They should also be located as close as possible to the main rooms they are serving to minimize larger and longer duct runs. Airflow measuring devices and humidifiers must also be installed, and unit location must allow enough straight run of duct for installation and proper operation of these devices. Location of outdoor air-inlet louvers must be carefully considered. Intakes are generally located on the building sidewall high off the ground to minimize dust intake. Intakes should also be away from truck docks or parking lots, where undesirable fumes and particulate are generated. In locating inlets the prevailing winds should also be considered, and any nearby exhausts or fume concentrations should be avoided to prevent recirculation of exhaust air back into the supply system.

RETURN AND EXHAUST FAN SELECTION AND LOCATIONS

Return Fans

Return and exhaust fans are integral parts of the air handling systems and also must be shown on airflow diagrams. Return fans are used on systems with long duct returns or that have return system pressure drops greater than 0.5 in water (120 Pa). This allows proper total system balance and minimizes suction pressure required from the supply fan. If a return fan is not used, the capacity of the supply fan can be overextended and it may be difficult to limit and properly control the amount of outside air being admitted to the unit. Outside air fluctuations are also more susceptible to exterior wind conditions. Return fans are also needed when required to provide a negative pressure in rooms that require containment. Return fans can be of standard centrifugal type or an in-line type, which works nicely for installation directly into return ducts in crowded equipment rooms. Return fans may also be required to handle varying quantities of air or provide a constant flow of air at varying pressure conditions. To achieve these conditions some form of damper control, inlet vane, or variable frequency drive motor control is generally used.

Exhaust Fans

Building exhausts are generally collected and ducted to exhaust fans in groups or clusters. Exhaust fans should be located as near to the building discharge as possible since this keeps the duct under a negative pressure and any leaks will be into the duct, and not contaminated air from the duct into an occupied space or mechanical room. For this reason roof locations of fans are preferred, even though this may make service difficult in severe weather conditions. When fans are located in mechanical rooms or interstitial spaces, it is essential to tightly seal the discharge duct before it exits the building in a roof vent or wall louver. Roof penetrations should be kept to a minimum to prevent leaks. Fumes and toxic exhausts should be extended through the roof and terminated well above the roof line in a suitable stackhead. Designs for stackheads are found in most duct manuals and are used to direct exhaust vertically at high velocities to enhance dispersion of fumes and prevent water from finding its way into the exhaust system (8). Extremely toxic or dangerous active biological agents may require HEPA filtration or other treatment, such as incineration, before exhaust to the atmosphere.

Redundancy

Another important consideration occurs when return or exhaust fans are used as part of the containment system. Should the fans be essential to maintaining containment, it may be desirable to have a backup fan or redundant system. This is essential if loss of containment can be harmful to humans or would result in an expensive loss of product. Airflow switches, which give a warning in case of fan system failures, are also desirable options for critical systems. The airflow sensing method to prove flow is preferred to an electrical motor indication since the motor could be running with a broken fan belt and the operator would not know that the fan is not moving air.

HEPA FILTERS

HEPA Filter Description

HEPA filters are used in the final cleaning stages of air to remove very fine particles. Rough and intermediate filters that are not as efficient as HEPAs are used to remove larger particulate matter. This prolongs the life of the more expensive HEPAs, significantly reducing the cost of filter element replacement. A HEPA filter, by definition, removes 99.97% of particles of 0.3-μm size. HEPA-type filters called ultraparticulate arrestences (ULPAs) can be even more efficient and are commercially available at a greater cost. Pressure drop across filters is rated for nominal flow at new clean conditions and is usually about 1 in (25.4 mm), and the systems are designed to change out filters at approximately 2 in (50.8 mm) or double the clean ratings, although filters can be operated at higher pressure drops, which some owners prefer. The higher operating pressures result in a greater use of electrical energy to power the fan to overcome the increased resistance that it must operate against.

Location

HEPA filters are usually located downstream of heating and cooling coils in air handling systems since coils are potential sources of contamination. The most popular HEPA filter location is in the room ceiling using standard laminar flow outlets nominally 24 in by 48 in (0.6 m by 1.2 m). These outlets contain manual control dampers, testing sample ports, a diffusion panel, and a HEPA filter element. They are commercially available in permanent or throw-away units where the entire assembly is disconnected from the flexible duct supply terminal and replaced when the filter is dirty. Sealing of filters to frames is an installation problem and is best solved by using a filter frame with a gel-like seal into which the filter fits. This material does not support biological growth, and these filters are offered by several manufacturers.

Bag-In Bag-Out Housings

Where danger of exposure to filter changing maintenance personnel exists, filters are placed in special enclosures called bag-in bag-out housings. These units are quite expensive and employ elaborate sealing methods and a double bag arrangement that allows changing of the filter element(s) without ever exposing the changer to the dirty filter or its collected contaminant.

Laminar Flow Workstations

HEPAs are also used in class 100 workstations in single or double filtration of air to provide sufficient cleaning before unidirectional (laminar) air distribution to the work space. These workstations are commercially available in horizontal or vertical flow patterns generally recirculating within the clean space. Filters are commercially available in many standard sizes and come with many frame arrangements with different methods used to form a seal between the filter element and the frame.

TERMINAL AIR CONTROL DEVICES

In most bioprocess applications the volume of airflow must be controlled to ensure proper air changes and provide air for pressurization. The choices to control airflow are (*1*) a variable flow responding to some pressure or temperature actuated signal or (*2*) constant volume to provide a uniform quantity of air under varying upstream pressure conditions. Many commercial devices, usually called boxes, or terminal control units, are available that utilize a variety of air dampening designs—from a simple blade type to pneumatic bladder type and conical type. Each type has its own application and associated cost implications. The published damper design and test data for flow and repeatability should be studied against the suspected operation and matched to the proper application. Dampers are controlled from pneumatically controlled operators or electric motors responding to electric or electronic signals. Pneumatic dampers generally react more quickly to changing airflow requirements. Dampers and controls should be selected for linearity of response, repeatability, and percentage of error when compared with desired flow. The damper is generally housed in a device called a volume control box, which contains deflecting baffles and flow-measuring or pressure-sensing devices that are used for flow control. Accuracy of flow control boxes, combined with the controlling system, must be aggregated to ensure that the actual flow will be within acceptable limits to control pressurization. The accuracy of control devices can vary between 5 and 10% of maximum flow. This built-in inaccuracy must be calculated when establishing flow differential rates. For commercial applications, control boxes are lined with insulating acoustical material, which should be avoided on boxes used for critical or controlled areas to avoid dirt catchers or host locations for unwanted growth.

Some vendors offer complete control box systems, including controls, supply and return boxes, and a system to allow tracking. *Tracking* is a term used to describe when the control system monitors the supply flow and automatically adjusts the return or exhaust flow to maintain a fixed CFM differential between supply and return or exhaust. This maintains the space pressure differential.

AIR TERMINAL OUTLETS

Supply Terminals

In clean spaces the desired distribution of air is unidirectional or acts as a piston of air that reduces turbulence and eddy currents in the airflow. This piston of air carries particulate from the ceiling to the floor return and helps to prevent airborne particulate matter from recirculating and contaminating the work space. In most cases it is desirable to recirculate air within a space through a filter since the return air has less particulate than typical outdoor air and does not require extensive heating and cooling. Air terminals should be selected of materials that are nonflaking, nonoxidizing, and are easily wiped clean.

Location

Placement of outlets should be above work areas since the cleanest air should be introduced at these spaces. If the product at the workstation is hazardous to humans the air should be supplied at the operator's back and pulled into the work area to ensure that flow is away from the operator and any particulate will be captured in the airstream. Containment velocity at the inlet of a biosafe hood or enclosure should be about 100 ft/min (0.508 m/s). Airflow near hoods or enclosures should avoid any air disturbance that could cause eddy currents or air turbulence. These currents will upset capture patterns of the hood and make verification of hood face velocity difficult. Many of the cabinet manufacturers have extensively tested capture velocities and provide excellent data on the actual operating conditions. When locating terminals and enclosures it is important to consider movement of personnel within the space and identify frequently opened doors, which also tend to upset airflow patterns. Should the operator pose a threat to the product, the airflow pattern should be reversed, with the air supplied behind the product and blowing past the product to avoid contamination from the operator.

Return Terminals

Return terminals are also an important consideration and are generally located low in the walls for clean rooms. In class 10,000 to 100,000 rooms low cleanable wall registers are generally used. In cleaner areas low return wall systems, termed *air walls*, are used. The air wall is an almost continuous opening at the base of the wall with the air ducted up in the wall system and collected for return to the air handling system. Air wall inlets are generally located not more than 15 ft (4.5 m) in plain view from a supply terminal to reduce the likelihood of turbulence. Figure 4 shows a typical return air wall detail.

DUCT MATERIALS, PRESSURE, AND CLEANABILITY

Materials of Construction

Unlined galvanized steel ductwork is used in rectangular, round, and elliptical (or flat oval) configurations for the majority of the systems. Because galvanized duct can flake off or rust, it should not be used downstream of the HEPA filters to avoid contamination from the duct system itself. When the HEPA filter is located upstream of the room terminal and a long run of duct is present, the material of choice for the duct is stainless steel, but this is expensive and its use should be minimized. Many systems may also be fumigated or cleaned in place, and the duct material chosen should not be affected by the cleaning agent.

Duct Pressures

Ductwork systems in bioprocess facilities tend to have higher system pressure requirements than commercial systems due to extensive use of filters, volume control devices, and physically complex arrangements. The duct system pressures must be calculated and clearly stated on the contract documents to allow the fabricator to provide the proper metal thickness and construction methods for the required system pressures (9). System pressures will also change as the system is operated with filters that get dirty or space pressure conditions that vary. Duct systems must allow for these pressure fluctuations, and the fans may require speed controls, inlet vanes, or variable pitch blades to match the varying flow and pressure conditions.

Cleanability

Cleanability of duct systems is important to ensure that if an installed system gets dirty or contaminated it can be cleaned. In the design stage care must be taken to locate access doors in the duct, where they can be easily reached without compromising a process or violating a controlled space. It is best to minimize entry of facility system maintenance personnel into the clean spaces. All sealed duct shipped to the site should have only end seals broken, and then quickly resealed, during final installation. In very critical applications the duct is factory cleaned and sealed before shipment to the site. This step removes the oil and other contaminants present during duct construction but is expensive. It may be difficult to find sheet-metal fabricators willing to do this work, since they are not always set up for such procedures and it makes shipping the completed product difficult.

SYSTEM OPERATING PROCEDURES

Operating Procedure

When the airflow sheets are completed it is appropriate to finalize the system operating procedure. This is usually a written document stating how systems are turned on and off, and it gives a description of their frequency of operation. For clean spaces general practice is to operate the air system 15–30 min for purging before beginning production or use of the space. In some cases it is necessary to operate a system continuously to ensure that pressurization or containment is always maintained. When analyzing system operation it is important to assess how an unexpected system shutdown or flow reduction will affect an adjacent sys-

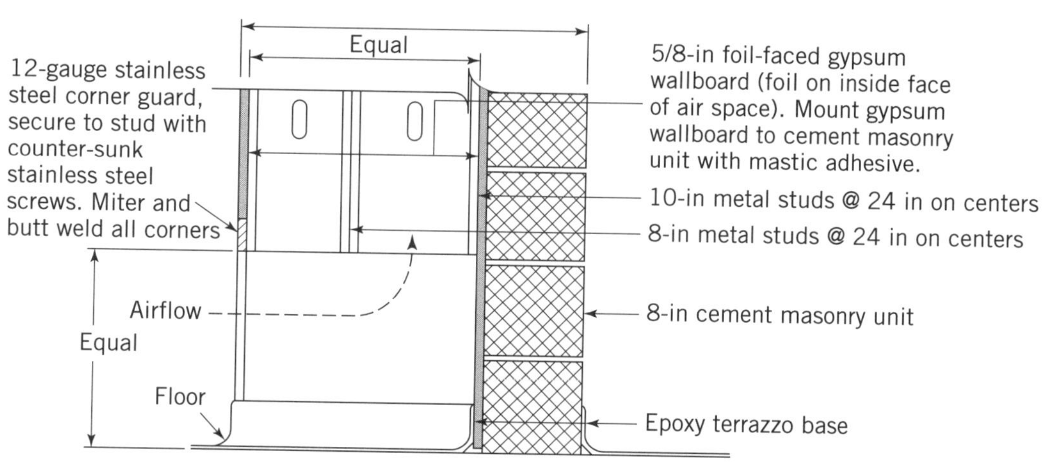

Figure 4. Return air wall detail.

tem or areas. Exhaust fans and fume hoods that may have intermittent or varying flows should be clearly noted in the operating procedure. It is essential that laboratory and production personnel clearly understand and follow the established procedures; otherwise, the air system may go out of balance if an exhaust or return system is inadvertently shut down.

System Alarms and Emergency Measures

Alarms that sound to indicate loss of pressurization are valuable features and essential in the HVAC design of critical areas. At this stage of design, procedures should be developed for the air system to accommodate a product spill or accident in a contained space. The ramifications of a spill on the air system, controlled space, and adjacent operations must be evaluated. Cleanup procedures could include fumigation of the air system, which would require operation of a relief connection to the ductwork for venting the fumigant.

Operations Review

The entire step of establishing operating procedures should ask as many what if questions as possible. System operation should ensure that the HVAC systems can maintain cleanliness and containment requirements. These conditions must be maintained while operating as close to normal as possible, as well as during periods of unexpected problems. Generally the reviewing agencies ask many of these questions; therefore, it is essential that the design team be fully prepared with thoughtful, responsive answers.

EMERGENCY ELECTRICAL POWER

An essential step in the HVAC design process is coordination with the electrical design team. Motor lists for HVAC equipment must be prepared and reviewed with the electrical design team. The need for motors designated for emergency power, variable speed, reduced voltage starting, or other special characteristics must be communicated to the electrical designers early in the design process. The sizing of the emergency generator can be greatly affected by motors required on emergency power from the HVAC system. Fans, equipment, or sensing devices that require interlocks must also be picked up by the electrical designers. The motor list must be kept up to date from project inception through commissioning. The motor list is useful for a reviewing agency, a valuable tool in training plant operators, and a great aid in understanding the HVAC system.

BUILDING CONTROL AND AUTOMATION SYSTEMS

Sequence of Operation

The automatic control system that controls and monitors the HVAC system is called by many names: the automatic temperature control system (ATC), the energy management and control system (EMCS), the building automation system (BAS), or the building management system (BMS). The first element in the design of the system is the development of a sequence of operation, which is a written description of the HVAC and related systems operation. A separate sequence is usually written for each air handling system, describing the complete operation of the system from control of coils and humidifiers to control of the room temperature and humidity. Starting and stopping of the air handling unit fans is outlined, along with interlocking of exhaust or return fans in relation to the main air system fan operation. Generally all fans operate at the same time, which is necessary to maintain pressurization. The sequence also addresses abnormal occurrences such as a smoke detection alarm or failure of an exhaust fan. The sequence describes what happens to system components during an abnormal occurrence. It may be necessary to shut a supply fan down if a major exhaust fan should fail to prevent or minimize the loss of pressurization. The sequence also describes any energy management strategies to be included in the system, such as a night temperature setback or reduced ventilation and exhaust rates during unoccupied periods.

System Architecture

The control system of choice for major facilities, and even for some small systems, is a direct digital control (DDC) system. Most major control system vendors and many of the smaller vendors offer DDC systems that are similar but contain internal differences. The systems are computer based and have the ability to communicate within and outside the system by coded digital signals. System architecture refers to the major components of the DDC system and their interrelationship. The architecture is developed by determining what components are initially required, what may be required in the future, and how the system may expand as additional requirements are added. Figure 5 shows a typical DDC system architecture.

Points List

After the sequence of operation is completed and the airflow diagrams are defined, the next step is to develop the alarm, control, and monitoring points list. This is an all-inclusive list of points that are to be connected to the DDC system. There are two major types of points: digital and analog. A digital point is simpler, generally less expensive, and works on a simple on–off or contact principle. Digital points are used to start and stop fans, indicate an on–off condition, or anything that requires only a single contact. An analog point is used to measure variables such as temperature, pressure, and flow rate. These points generally use 4- to 20-mA signals that provide varying signals in response to the parameter measured. The electronic signals used by the BAS may be transduced from variable pneumatic or pressure signals. The points list should include analog control points such as cooling coil valves and room temperatures. Monitoring points can be digital or an-

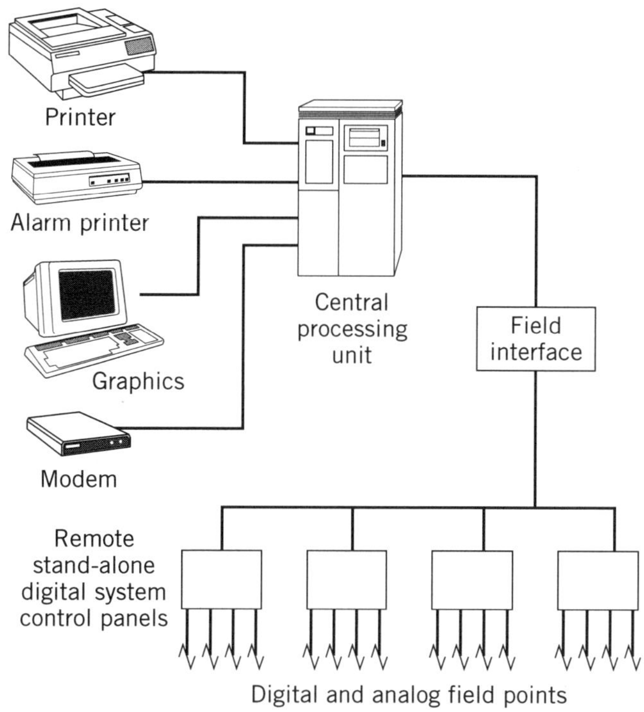

Figure 5. Direct Digital Control System Architecture.

alog, and can include fan run, room temperature indication, damper position, and room pressure indication. Alarm points can be either digital or analog and can include smoke detection in an air handling unit system, high or low environmental chamber temperature, high room humidity, or loss of room pressurization. Sensors located in clean spaces should be unobtrusive and installed in a cGMP acceptable manner.

Estimate of System Cost

The automatic control and monitoring system is a major cost element in the overall HVAC system for a bioprocess facility. After the points list is developed a good estimate can be made for the system cost. Several estimating numbers can be used in providing an educated guess of the cost, with a general range from a low of $500/point to as high as $1,200/point. Larger systems have lower dollar-per-point costs because the cost of the computers, monitors, and printers are diluted over many points. Smaller systems have higher costs because this dilution effect does not come into play. Systems with a higher percentage of analog points over digital points also tend to raise dollar-per-point costs. Control wiring is also a significant component of the BAS cost, and long runs of wiring tend to raise the dollar-per-point cost. When a floor plan is completed, the system architecture is established, and the points list is completed, this information should be given to one or more DDC vendors for a budget estimate. Vendors will make estimates based on the information provided, take off wiring runs, and suggest panel locations in arriving at the budget cost. At the completion of this exercise the cost of this major element can be locked in. Should the cost be greater than desired some points can be eliminated or the scope of the system reduced to bring the cost within the budget. When a system is bid it is recommended that component unit pricing be obtained from the vendor to lock in future costs should additions or modifications be desired. It is essential that vendors give classes and instructions on their systems before final turnover to the owner's operators.

Future System Flexibility

The trend is that after a DDC system is installed the number of points grows as the user becomes familiar with the system. Additional points may be added for maintenance, such as filter change out pressure devices. Run time logs set up in the system software are useful for lubrication schedules, belt checks, and so on. This requires software to log in and integrate data and to print out notices at preselected intervals of run time. Another useful feature is the phone modem, which would allow an operator to call in from a remote location and check on the system operation. This feature can save a trip to the facility if an alarm situation develops. Someone at the plant, such as the security guard who received the alarm, could place a call to the operator at home. The operator could dial into the BMS by modem and possibly analyze the problem by reading some key parameters. He or she could then determine the severity of the problem and either ignore it until the next shift or instruct someone on-site to take corrective action. Even though the original budget may not allow all features that are ultimately desired, it is good to look toward the future in anticipation of system expansion and ensure that the purchased system is capable of desired expansion.

TESTING, BALANCING, AND CLEANING

Testing and balancing for bioprocess systems is more critical than for commercial systems due to the very complex systems and higher pressures encountered. Associations of testing and balancing agencies publish excellent standard forms and procedures that are used in balancing air and water systems. Air balancing consists of setting terminal devices and air-moving equipment to deliver the proper quantities of air. For bioprocess facilities the additional requirement of establishing pressure differentials between adjacent spaces must also be set. These differentials are obtained by adjusting airflows, smoke tests, taking pressure readings, and setting controls. This effort can take some time as each facility is different and each room has different leakage characteristics that affect pressurization. As part of the balancing it may be discovered that duct systems or rooms are not as tight as designed and may require additional sealing to obtain the required pressure differentials. Recall that airflows shown on drawings are design values and generally require minor adjustment to achieve the required pressure differentials. A simple solution to many pressurization problems is to keep increasing outdoor air to the system. This can lead to problems if design values are exceeded, with heating or cooling coils not meeting this need, resulting in off-design room temperatures and humidity levels. The air handling unit coils will use available cooling capacity to condition excessive quantities of outdoor air, resulting in room supply temperature higher than as designed.

In these cases the rooms or systems should be tightened to get closer to design values. The optimum time to balance is when few construction workers or facility personnel are in the spaces. The balance should be done with all doors closed, since opening and closing result in system pressure upsets and make balancing difficult. Before the system is turned over to the owner, the rooms and ductwork system must be thoroughly cleaned to allow certification of room cleanliness conditions. During construction the work areas should be broom-cleaned after each day's work and open duct systems kept closed. If followed during construction, these procedures will result in an easier final cleanup before turnover.

VALIDATION

When a bioprocess facility is to be validated, the validating agency will peruse the HVAC documentation and should communicate with the design engineers to establish the validation protocol as it relates to the HVAC system. If the design is proper, the contractor has properly installed the system, and the components perform as specified, the systems should be easily validatable. The validator will follow a master plan and protocols to verify the actual system installation and operation against design values and intent (10). The physical parameters reported by the BAS system will be verified by measurements using calibrated instruments to verify accuracy. Airflow measurements will be made at supply and exhaust outlets, as well as traverses across the face of hoods, to verify proper flows and capture patterns. Filters will be physically tested or challenged to uncover leaks in the filter media or at seals between the holding frame and elements. Any leaks will be field corrected. Samples of air will also be taken by the validator to ensure that the cleanliness requirement for the controlled space is actually maintained by the installed operating air system. If problems occur they are usually in the form of pressurization deficiencies or filter leaks. Even after the mechanical contractor has completed his or her work, a hole for a pipe could be cut in a room and not properly sealed, resulting in a change in room pressurization. Filter elements may not be properly sealed with the frame, resulting in leaks, or small holes may be discovered in the filter element as a result of handling or damage during installation. Should any of these problems occur, and are not easily resolvable by the validator, it may be necessary for the HVAC contractor and engineer to lend assistance in making the system valid. Another important criterion at this stage is the calibration of airflow monitors, which should be installed in supply and return systems to critical areas to ensure that airflow rates are as designed. Temperature and humidity sensors at critical areas should also be checked for accuracy at this time by actually reading space conditions and checking against values reported by the BMS.

SUMMARY

The HVAC system is an essential component in the successful operation of a bioprocess facility and must be given a proper amount of attention during the planning and design stage. The space requirements for air walls, control boxes, extensive duct runs, and large air handling units must be established early in the design process to prevent unsatisfactory compromises later on. The owner and operators must work closely with the architect and engineer design team to agree on the HVAC system philosophy and the proposed operation of the system. The amount of energy management and control, as well as system redundancy, are among important economic decisions that should be agreed on during this design process. Compromises made during the design development should make economic sense, be mutually beneficial, and be understood by affected members of the project team. It is best from a cost perspective to resolve HVAC issues as early in the design process as possible to avoid design delays and more costly changes later in the design process. This avoids delays in the design process and minimizes the effect of changes on other engineering disciplines. All parties involved must believe in the HVAC system philosophy to avoid costly and time-consuming retrofits during construction or at commissioning, validation, or after the facility begins operating.

BIBLIOGRAPHY

1. A.J. Shahidi, R. Torregrossa, and Y. Zelmanovich, *Pharm. Eng.* **5**, 72–82 (1995).
2. B.K. Lyderson, N.A. D'Elia, and K.L. Nelson, *Bioprocess Engineering Systems: Equipment and Facilities*, Wiley, New York, 1994, pp. 643–668.
3. J. Olsztynski, *PM Eng.* **2**, 20–21 (1996).
4. M. delValle, *Pharm. Eng.* **15**, 14–22 (1995).
5. *Technical Committee 9.10, HVAC Applications, ASHRAE Handbook*, American Society of Heating, Refrigerating, and Air-Conditioning Engineers, Inc., Atlanta, Ga. 1995, pp. 13.6–13.8.
6. Carrier Corporation, *System Design Manual*, vol. 1, *Load Estimating*, Carrier Air Conditioning Company, Syracuse, N.Y., 1972, pp. I-115–I-151.
7. B.K. Lyderson, N.A. D'Elia, and K.L. Nelson, *Bioprocess Engineering Systems: Equipment and Facilities*, Wiley, New York, 1994, pp. 588–593.
8. *Committee on Industrial Ventilation, Industrial Ventilation*, American Conference of Governmental Hygienists, Inc., Lansing, Mich., 1992, pp. 5–53.
9. *Technical Committee 5.2, Fundamentals, ASHRAE Handbook*, American Society of Heating, Refrigerating, and Air-Conditioning Engineers, Inc., Atlanta, Ga. 1993, pp. 32.1–32.40.
10. B.K. Lyderson, N.A. D'Elia and K.L. Nelson, *Bioprocess Engineering Systems: Equipment and Facilities*, Wiley, New York, 1994, pp. 2, 747–781.

HEMICELLULASES

Liisa Viikari
Maija Tenkanen
Kaisa Poutanen
VTT Biotechnology and Food Research
Espoo, Finland

KEY WORDS

Baking
Bleaching
Feed
Food
Hemicellulases
Hemicellulose
Mannan
Pulping
Xylan
Xylanases

OUTLINE

INTRODUCTION

During the past decade, the industrial use of hemicellulases, especially xylanases, has grown remarkably. Hemicellulases form a group of enzymes that degrade different hemicelluloses in annual plants and woody materials. Originally, hemicellulases were studied for use in the total hydrolysis of waste liquors containing hemicelluloses or for fractions derived from lignocellulosics. Today, hemicellulases are used for partial or restricted hydrolysis of fibrous materials containing hemicelluloses. Thus, hemicellulases have found uses in processing of hemicellulose-containing plant materials in the food, feed, and pulp and paper industries. The market for hemicellulases has been growing since the beginning of the 1990s due to a better understanding of the role of hemicelluloses in processing of the raw materials and to increased knowledge of these enzymes. Several commercial hemicellulases with different properties are now available from enzyme producers.

Hemicelluloses are natural components in cereals, and their properties and performance in food and feed raw materials can be improved by the action of enzymes. Exogenous enzymes offer alternatives to chemical additives in baking, where the application of microbial enzymes is expected to increase. Most of the added enzyme types are already present at low levels in the grain raw material. Commercial enzyme preparations principally consist of the same enzyme types but are usually of microbial origin. Enzymes in food and feed applications may also be used for upgrading the quality of the raw material. In the pulp and paper industry, the partial hydrolysis of fiber hemicelluloses by hemicellulases leads to improved processes and decreased chemical consumption (i.e., cleaner technology). The role of hemicellulases has been clearly demonstrated in improving the bleachability of pulps, whereas in other applications their impact is less significant.

Most of the enzymes studied in applications have been hemicellulase mixtures. For a successful application of hemicellulases a good understanding of both the substrate and the enzyme complex is needed. A reason for the sometimes fluctuating results on the effects of enzymes in various applications is that the preparations tested may contain various activities, including enzymes other than hemicellulases. Commercial enzyme preparations may vary considerably in composition and ratio of various activities, depending on the source.

This chapter highlights progress in the use of hemicellulases and their properties and exploitation in different areas (e.g., food, feed, pulp and paper) where the applications are commercially used.

STRUCTURE AND COMPOSITION OF HEMICELLULOSES

Hemicelluloses are heteropolysaccharides consisting of several polymers and varying in their monosaccharide composition, glycosidic linkage composition, substitution pattern, and degree of polymerization (1). They are often reported to be chemically associated (e.g., cross-linked) to other polysaccharides, proteins, or lignin. The structural characteristics affect physical properties, such as solubility, viscosity, crystallinity, and porosity and thus the enzymatic degradability.

The definition of the term *hemicellulose* is not very clear. It mainly refers to plant cell wall polysaccharides that can be extracted by alkaline solutions. Some hemicelluloses are also soluble in water. Hemicelluloses are usually classified according to the main sugar residue in the backbone, for example, xylans (D-xylose), mannans (D-mannose), galactans (D-galactose), and glucans (D-glucose), the first two being the main groups of hemicelluloses. Depending on the origins, hemicelluloses are substituted, to varying degrees, by side groups (1). In cereals in particular, the term *pentosan* is often used as a synonym for *hemicelluloses*, especially for xylans. Hemicelluloses present in cereals and woody materials are discussed in the following. Pectic substances consisting of polygalacturonic acid, arabinans, galactans, and arabinogalactans, and noncellulosic β-glucans, which sometimes are included in hemicelluloses, are not considered in this review.

Hemicelluloses in Cereals

In cereals, hemicelluloses consist of arabinoxylans, arabinogalactans, and xyloglucans, of which arabinoxylans are most abundant. The content of arabinoxylans in cereals is reported to be 1.4–8.5%, with water-soluble fractions contributing 1.5–4.3%. Wheat and rye brans have the highest content of arabinoxylans. The main component of aleurone cell walls of cereals is arabinoxylan, and the cell walls of grain endosperms also contain significant amounts of arabinoxylan.

The arabinoxylans of annual plants have a linear backbone of (1,4)-β-D-xylopyranosyl units, which is mainly substituted with terminal α-L-arabinofuranosyl substituents attached to carbon 2 and/or carbon 3 of the xylosyl residues (1). Thus, xylans may be mono- or double-substituted by arabinose. The xylan backbone often contains other substituents as well, such as 4-*O*-methyl-α-D-glucuronic, acetic, or ferulic acids. The ratio of arabinose to xylose may be used as a simple indicator of the degree of substitution in arabinoxylans and varies between 0.54 and 0.70. The structure of arabinoxylans in annual plants varies among different species and parts. Thus, even arabinoxylans in aleurone and endosperm have different structures.

Hemicelluloses in Woody Materials

Wood is composed mainly of cellulose, hemicellulose, and lignin. Depending on the wood species, about 20–30% of dry weight of wood is hemicellulose. Hemicelluloses are associated with cellulose and lignin in woods. Not only the relative amounts but also the chemical composition of these polysaccharides in softwood and hardwood vary. Other polysaccharides, such as pectic compounds, are also present in wood in minor quantities. The distribution of carbohydrates in the wood fibers varies depending on the wood species and growing conditions (2). In general, in softwoods the xylan content of the innermost cell wall layer (the warty layer), is very high, but otherwise the xylan is relatively uniformly distributed throughout the fiber walls. In hardwoods, the outermost layers of fiber walls are rich in xylan. In softwoods the glucomannan content increases steadily from the outer to inner cell walls.

In wood, the two most common hemicelluloses are xylans and glucomannans. The hardwoods contain mainly xylan, whereas the amount of glucomannan in softwoods is approximately double compared with that of xylan. Hardwood xylan is composed of β-D-xylopyranosyl units, which may contain 4-*O*-methyl-α-D-glucuronic acid and acetyl side groups. 4-*O*-methylglucuronic acid is linked to the xylan backbone by *O*-(1,2) glycosidic bonds, and the acetic acid is esterified at the carbon 2 and/or 3 hydroxyl group. The softwood xylan is arabino-4-*O*-methylglucuronoxylan, in which the xylan backbone is substituted at carbon 2 and 3 with 4-*O*-methyl-α-D-glucuronic acid and α-L-arabinofuranosyl residues, respectively. Softwood galactoglucomannan has a backbone of β-(1,4)-linked β-D-glucopyranosyl and β-D-mannopyranosyl units, which are partially substituted by α-D-galactopyranosyl and acetyl groups (2).

HEMICELLULASES

Due to the complex structure of hemicelluloses, several different enzymes are needed for their enzymatic degradation and modification (Fig. 1). The two main glycanases that depolymerize the hemicellulose backbone are *endo*-1,4-β-D-xylanase and *endo*-1,4-β-D-mannanase. Small oligosaccharides are further hydrolyzed by 1,4-β-D-xylosidase, 1,4-β-D-mannosidase, and 1,4-β-D-glucosidase. The side groups are removed by α-L-arabinosidase, α-D-glucuronidase, and α-D-galactosidase. Esterified side groups are liberated by acetyl xylan esterase and acetyl glucomannan esterase (3).

Hemicellulases are produced by many species of bacteria and fungi, as well as by several plants. Many microorganisms produce a multiple pattern of these enzymes to degrade the plant materials efficiently. After identification of the most desirable enzymes for a potential application, the production of the enzyme can be improved by genetic engineering. Today, most commercial hemicellulase preparations are produced by genetically modified *Trichoderma* or *Aspergillus* strains.

Most microbial hemicellulases studied are active in slightly acidic conditions between pH 4 and 6 and at temperatures below 70 °C. More thermophilic, acidic, and alkalophilic hemicellulases are of great interest in many applications. Xylanases that are stable and function efficiently at high temperatures can be produced by thermophilic bacteria especially. Several xylanase genes, encoding proteins active at temperatures from 75 °C up to 95 °C (pH 6–8), have been isolated. The most thermophilic xylanases so far described are produced by species of the extremely thermophilic bacterium *Thermotoga* (4). Only a few ther-

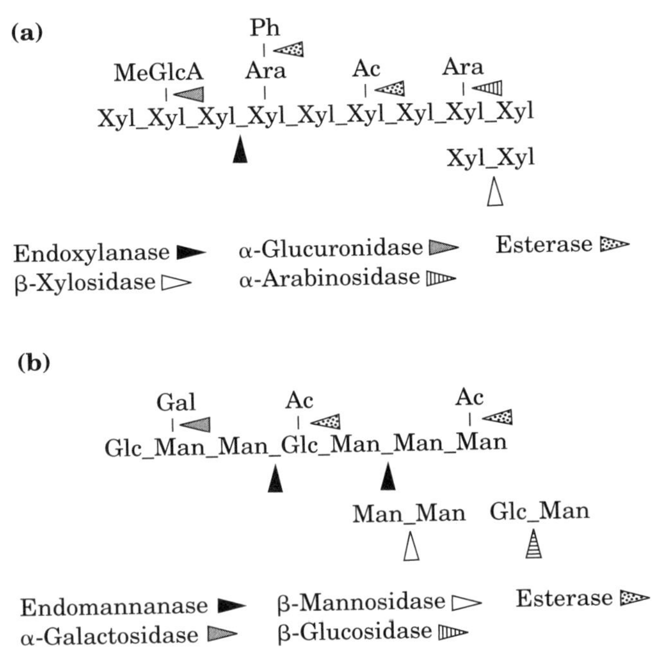

Figure 1. Enzymes participating in the hydrolysis of xylans (**a**) and glucomannans (**b**): Ph, phenolic groups; Ac, acetyl; Ara, arabinose; MeGlcA, methyl-glucuronic acid; Xyl, xylose, Gal, galactose; Glc, glucose; Man, mannose.

mophilic mannanases have so far been characterized (5,6). Xylanases and mannanases with alkaline pH optima have been detected especially in alkalophilic *Bacillus* species.

Xylanases

Since xylan is the most abundant hemicellulose in most raw materials, a major part of the published work on hemicellulases deals with the properties and mode of action, as well as applications of xylanases (7–9). Endoxylanases (1,4-β-D-xylan xylanohydrolases, EC 3.2.1.8) catalyze the random hydrolysis of 1,4-β-D-xylosidic linkages in xylans (Fig. 1). Most xylanases are rather small proteins (molecular mass around 20 kDa) with a basic isoelectric point (pl 8–10). Another group of xylanases has also been identified. These xylanases, are somewhat larger (molecular mass over 40 kDa) with acidic isoelectric point (pl 3–5). This grouping of xylanases is in consonance with the classification of glycanases based of hydrophobic cluster and sequence analysis (10). The xylanases belonging to the two identified xylanase groups (glycosyl hydrolase families 10 and 11, formerly F and G) differ from each other also with respect to their catalytic properties (11). Xylanases with high Mr/low pl (family 10) seem to exhibit greater catalytic versatility than the low Mr/high pl xylanases (family 11) and thus they are, for example, able to more efficiently hydrolyze highly substituted xylans.

The three-dimensional structures of different low molecular mass xylanases belonging to family 11 have been determined (12–14). These enzymes have very similar structures; they are ellipsoidal, well-packed molecules having diameters between 30 and 45 Å, and do not contain any separate substrate-binding domain. Some bacterial xylanases, however, have been found to contain either a cellulose-binding domain (15) or a xylan-binding domain (16). Only one crystal structure of family 10 xylanases has been solved (17). Figure 2 shows the three-dimensional structures of a catalytic domain of the xylanase A (family 10) of *Streptomyces lividans* and the xylanase II (family 11) of *Trichoderma reesei*.

Most of the xylanases characterized act randomly and are able to hydrolyze different types of xylans, showing only differences in the spectrum of end products. The main products formed from the hydrolysis of xylans are xylobiose, xylotriose, and substituted oligomers of two to four xylosyl residues. The chain length and the structure of the substituted products depend on the mode of action of the individual xylanase (11). Some xylanases, however, have a rather strict substrate specificity. A unique xylanase that requires a glucuronic acid substituent in the xylan backbone is produced by *Bacillus subtilis* (18). On the other hand, xylanases that require at least 24 unsubstituted xylose residues for action are produced by *Talaromyces emersonii* (19).

Mannanases

Endomannanases (1,4-β-D-mannan mannanohydrolase, EC 3.2.1.78) catalyze the random hydrolysis of β-D-1,4 mannopyranosyl linkages within the main chain of mannans and various polysaccharides consisting mainly of mannose, such as glucomannans, galactomannans, and galactoglucomannans (Fig. 1). Mannanases are generally larger proteins than xylanases (Mr 30–90 kDa) and have acidic isoelectric points. They also seem to be a more heterogeneous group of enzymes than xylanases (i.e., no clear groups based on biochemical properties have been identified). According to protein sequences, the known mannanases belong to three different glycosyl hydrolase families (families 5, 26, and 44) (10).

The mannanase of *T. reesei*, has been found to have a multidomain structure similar to several cellulolytic enzymes, that is, the protein contains a catalytic core domain that is separated by a linker from a cellulose-binding domain (20). The first three-dimensional structure of mannanase was recently published (21).

The main hydrolysis products from galactomannans and glucomannans are mannobiose, mannotriose, and various mixed oligosaccharides. The hydrolysis yield is dependent on the degree of substitution, as well as on the distribution of the substituents (22). The hydrolysis of glucomannans is also affected by the glucose to mannose ratio. Some mannanases are able to hydrolyze not only the β-1,4-bond between two mannose units but also the bond between the adjacent glucose and mannose units (23,24). Several endoglucanases are also able to hydrolyze galactoglucomannans.

Other Hemicellulases

Enzymes needed for further hydrolysis of the short oligomeric compounds produced by endoenzymes from hemicelluloses are β-xylosidase (1,4-β-D-xyloside xylohydrolase, EC 3.2.1.37), β-mannosidase (1,4-β-D-mannoside mannohydrolase, EC 3.1.1.25), and β-glucosidase (Fig. 1). β-Xylosidases and β-mannosidase catalyze the hydrolysis of xylo- and manno-oligosaccharides, respectively, by removing successive xylose or mannose residues from the nonreducing termini. Exoglycanases are generally larger proteins than endoglycanases, with molecular weights above 100 kDa, and they are often built up by two or more subunits.

Side groups that are still attached to oligosaccharides after the hydrolysis of xylans and mannans by xylanase or mannanase, respectively, restrict the action of β-xylosidase and β-mannosidase. The hydrolysis stops at a substituted sugar unit. The β-xylosidase of *T. reesei* is unable to hydrolyze the linkage adjacent to a substituted xylose unit carrying a 1,3-linked α-L-arabinofuranosyl substituent, whereas the linkage is easily hydrolyzed if the substituent is 1,2-linked α-L-arabinofuranosyl or 4-*O*-methyl-α-D-glucuronic acid (25). Similarly, the 1,6-linked α-D-galactopyranosyl side group restricts the action of β-mannosidase of *A. niger*, which is able to only slowly hydrolyze the linkage adjacent to the substituted mannose unit.

The side groups connected to xylan and glucomannan main chains are removed by α-glucuronidase, α-arabinosidase (α-L-arabinofuranoside arabinofuranohydrolase, EC 3.2.1.55), and α-D-galactosidase (α-D-galactoside galactohydrolase, EC 3.2.1.22) (Fig. 1). Acetyl substituents bound to hemicellulose are removed by esterases. There are clearly different types of side group cleaving enzymes. Some of them are able to hydrolyze only substituted short-chain oligomers, which first must be produced by the backbone depolymerizing endoenzymes (xylanases and man-

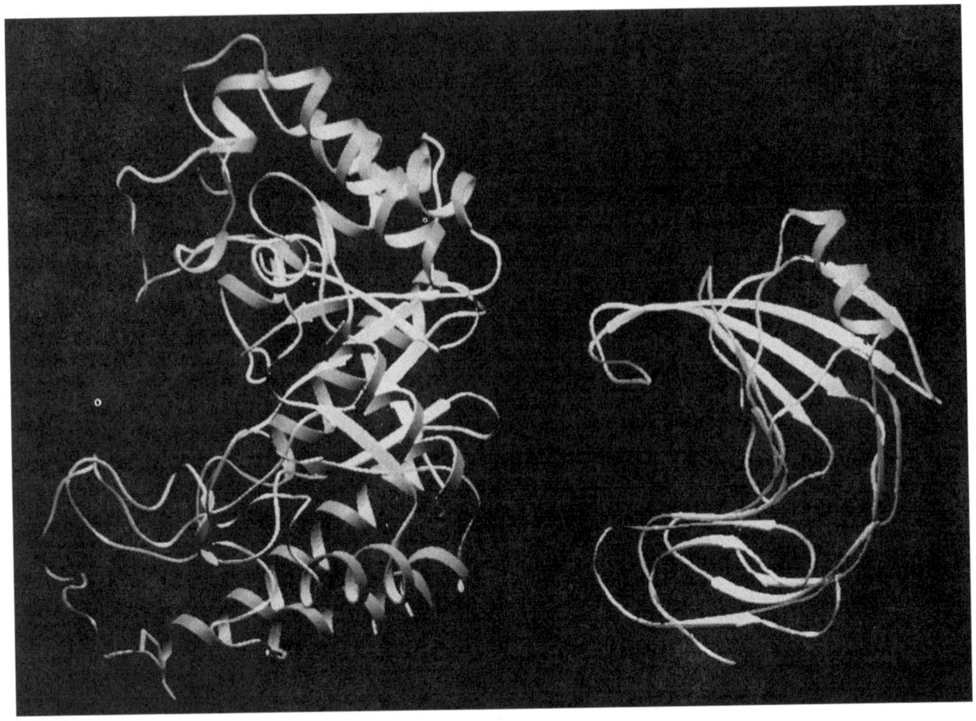

Figure 2. The three-dimensional structures of the catalytic domain of the xylanase A of *Streptomyces lividans* and the xylanase II from *Trichoderma reesei*. *Source*: Courtesy of Professor Juha Rouvinen.

nanases). Others are capable of also attacking intact polymeric substrates. Most accessory enzymes of the latter type, however, prefer oligomeric substrates. The synergism between different hemicellulolytic enzymes is also observed by the accelerated action of endoglycanases in the presence of accessory enzymes. Some accessory enzymes, such as an arabinosidase and an esterase of *Pseudomonas fluorescens* subs. *cellulosa* and an acetyl xylan esterase of *Trichoderma reesei* have been found to contain a multidomain structure with a separate cellulose binding domain (26).

Evaluation of Hemicellulases

Enzyme activity is usually determined using isolated natural substrates. Isolation of a substrate from complex raw materials, however, often leads to chemical and structural modification of the substrate. Thus, in hemicelluloses, some of the side groups may be removed during chemical extraction, or the average degree of polymerization may change. Also, the physical status of the substrate in the activity assay may be different from that in the practical application. Thus, the solubility of the substrate in the activity assay is usually high, whereas in fiber-bound matrix the substrate is usually insoluble and/or may be also chemically linked to other components. Although the substrates in activity assays are not identical to the practical raw materials, they are chosen to correspond to the conditions under which the enzyme will be used. The main advantage and justification for the use of isolated substrates is that they give reproducible assay figures. Furthermore, they are more readily available and consistent and give the opportunity to compare different types of enzymes at various places.

It is well recognized that comparison of enzyme activities is complicated by the utilization of different analysis methods (27). Hence, in practical applications, the action of enzymes must be compared on the actual substrate. In these tests, the liberation of sugars and/or oligosaccharides into solution is the direct measurement of the action of the enzyme. This is usually carried out either by reducing sugar analysis or by HPLC. Several indirect methods to correlate the activity with the actual effect in the practical use, such as viscosity determination, have been developed. However, the most reliable method for understanding the effects of different enzymes is to compare the enzymes and dosages in the final application.

APPLICATION OF HEMICELLULASES

The major applications of hemicellulases in the food, feed, and pulp and paper industries are described in Table 1.

Food

Enzymes have long been recognized as useful tools for improving the processing behavior or properties of cereal foods, and their use is continuously increasing. The main enzymes used are amylases acting on starch, but hemicel-

Table 1. Benefits of Using Hemicellulases in Different Industrial Areas

Application area	Effect	Benefit
Food	Dough rheology	Improved handling
	Loaf volume	Quality and yield
	Staling rate	Improved storage
Feed	Nutritive value	Cost savings
	Feed conversion rate	Cost savings
Pulp and paper	Bleachability	Chemical savings, quality

lulases have also been used in cereal processing. The main area in which hemicellulases are used is baking, where dough handling and bread quality can be improved by the action of enzymes.

Hemicellulases in Cereal Processing. Endogenous and added enzymes both have an important effect on the quality of cereal foods, especially in baking. Although enzymes act on molecular level, they can induce remarkable changes in both the microstructure and the functional properties of cereal foods. Progress has been made recently in the analysis of enzymes present in flours. The presence of endogenous enzymes has been suggested to be the cause of the variation in liquid loss among wheat doughs made from different cultivars (28). Differences in α-arabinofuranosidase, β-xylosidase, and endoxylanase levels have been detected in various European wheat varieties. Although present only at low levels, these enzymes may contribute to the bread-making performance of flours (29).

Enzymes have been shown to differ in their abilities to both solubilize insoluble substrates and further hydrolyze the released fragments. The chemical composition, architecture, and physical properties of the surrounding tissue of the cereal cell walls are all of importance if selective solubilization of the polysaccharides is attempted. Physical inaccessibility of the substrate may retard or hinder the reactions, and several enzymes may be needed. As compared with xylanases, less is known about the other enzymes acting on the hemicelluloses. The role of ferulic acid in cereals is interesting. Ferulic acid is esterified to the arabinose side groups of arabinoxylans, and it has been suggested that it plays an important role in the architecture of the cell walls and in the cross-linking of arabinoxylan molecules. Very recently, it was reported that ferulic acid esterase influenced the solubilization of β-glucans and arabinoxylans from barley cell walls (30).

Baking. Flour, water, yeast, and salt are the basic components of bread and other baked foods. It is, however, common practice to improve the processing of dough and/or the quality of the baked products by adding other ingredients. Process criteria are especially related to dough handling (machinability) and process yield (water retention, proofing times). The product quality is mostly evaluated on the basis of appearance (volume, color) and eating quality (crumb elasticy, flavor) (31). Various bread improving components, such as emulsifiers, oxidizing agents, and enzymes, have been shown to influence bread quality, especially when lower-quality flours with a low gluten content are used.

A small amount of arabinoxylans (pentosans) is present in wheat and rye flour. Insoluble arabinoxylan especially hinder the development of wheat gluten, which is vital for the formation of the loaf structure. White flour contains 2.5–3%, whole-meal wheat flour about 5%, and rye as much as 8% arabinoxylan, which has an important role in the water binding of dough. Arabinoxylans can bind nearly 10 times their weight of water and may account for approximately one-third of the water-binding capacity of dough. By controlled enzymatic hydrolysis of xylans, the dough becomes easier to handle and the resulting bread has a larger loaf volume and an improved crumb structure (32).

To understand the reactions of hemicellulases, several reports have concentrated on studying the mechanisms of action of xylanases and other hemicellulases. The release of high molecular weight water-soluble arabinoxylans from water-inextractable cell wall material has been shown. The specific viscosities of extracts from xylanase-treated doughs were 40–65% higher than those from untreated dough, because the average apparent viscosity of the extracted arabinoxylans remained high (33).

It has been suggested that the effect of xylanase on bread volume improvement can be ascribed to redistribution of water from the xylan phase to the gluten (continuous phase). Water distribution is of crucial importance in the dough and bread. The transfer of water from one component to another, due to the decreased water-binding capacity of a partially hydrolyzed polymer, is suggested to be a major reason behind changed functional properties (34). The increase in gluten volume fraction gives the gluten more extensibility, which eventually results in better oven spring (31).

Xylanases have also been found to improve wheat dough fermentation stability (35), counterbalance the variable flour quality in wheat bread baking (33), retard staling, and improve the texture of high-fiber bread (36). They are also useful in the manufacture of low-fat biscuits (37).

Animal Nutrition and Feed

A part of the energy content in many cereals is locked up in the form of nonstarch polysaccharides, undigestible for several animals. Therefore, selected enzymes, such as hemicellulases, can be added to break down these polymers, leading to increased metabolizable energy. In addition, degradation of hemicelluloses often liberates starch and protein, masked by the cell walls in the cereals, increasing their utilization. In some cereals, a large part of arabinoxylans is soluble and causes high viscosity in the small intestine of some animal species. Addition of xylanases decreases viscosity and improves feed utilization.

Hemicellulases in Animal Nutrition. Animal feed is predominantly composed of plant material, and cereals form the major part of diets for both poultry and pigs. Some of these raw materials, such as rye and barley, are known to be less suitable than maize and wheat. This is emphasized by their gross energy and metabolizable energy in the target species. Thus barley, with a gross energy similar to that of wheat, has, in practice, an approximately 10% lower energy availability than wheat (38). This has previously limited its value as a feed component as compared with the more digestible materials. The decreased nutritive value has been associated with the presence of arabinoxylans in rye and wheat. Feed utilization and digestion can often be increased by the addition of external enzymes to the feed.

Endogenous enzymes play a crucial part in an animal's own digestion of feed components. External enzymes added directly to feeds act as supplements to the normal digestive enzymes already found in animals. The external enzymes carry out their work on the feed materials present in the animal's digestive tract. They can therefore be considered as an extension of the animal's own digestive enzyme system, especially in the case of young animals whose digestive systems are not fully mature. Enzymes can be added to animal feeds either to degrade soluble fiber, which has an antinutritional effect, or to supplement the animal's own digestive enzymes. In the latter case, supplementation of the feed with amylase or protease is the current practice. β-Glucanases are also used, especially for the hydrolysis of barley β-glucans.

The content of arabinoxylan, β-glucan, and other nonstarch polysaccharides in feed raw materials does not remain constant, and varies with variety and also with the climatic conditions under which the raw material is grown. Prediction of the enzyme response has prove to be complicated. Different methods, such as viscosity measurements or determination of arabinoxylan level, have been studied as possible predictors of the nutritive quality with varying success (39). At present, however, no rapid predictive test exists that can be applied to raw material evaluation, and in vivo methods are still the best.

The desired pH optimum of the enzyme preparations are around 6–6.5, prevailing in the small intestine, the area in which the majority of the enzyme action of xylanases is believed to occur. The stability of the enzyme both in processing and in feed is important. Enzymes must be resistant to the heat processing of the feeds and also to the transition in the digestive system to the point of action. In practical feed-mill conditions, the addition of steam in pelleting is responsible for the loss of enzyme activity. Commercial preparations may be protected by being absorbed to a carrier or coated with a steam-resistant material that hinders access of the steam to the enzyme product. The intrinsic thermal resistance of an enzyme may thus be improved by a protective system. As the majority of xylanases used in feed applications have very similar intrinsic thermal stabilities, the thermal stability of different products is mainly a function of the efficiency of the carrier and/or coating technology used to protect the enzyme product. The loss of activity can be to some extent avoided by using dried enzymes, which are applied after pelleting (38).

Poultry Feed. Wheat has been found to have a variable energy content, depending on local growth conditions. While there has been found to be no correlation between the level of total nonstarch polysaccharides in wheat, a significant negative correlation between the level of water-soluble nonstarch polysaccharides and the average metabolizable energy (AME) has been demonstrated. It has been observed that supplementation of a wheat-based broiler diet with xylanase preparations improved the AME value of the wheat used by 7–10% (40). An increased release of proteins and nonstarch carbohydrates from the cells has been verified by chemical and microscopic analysis (41). The addition of xylanase enzymes can therefore be used to improve the nutritive value of wheat-based diets. This is now common practice in Western Europe and other areas of the world where wheat is available as a feed cereal.

Enzyme supplementation of wheat-containing diets is, however, cost sensitive, and the dosages used may not always be those that give the maximum effect. In practice, addition levels vary from 300 to 500 g of enzyme preparation per ton of feed for broilers (42). Too-high enzyme dosages have, in some cases, led to an imbalance between protein and energy.

Due to the inherent variations in biological systems, small fluctuations in results from place to place have been observed (38). This should be contrasted with the more traditional enzyme applications, where the response is much more predictable and can even be mathematically modeled. It has been observed that wheat with a lower AME than average responds more to the enzyme supplementation, thus introducing another source of variation in the response. Despite this, however, xylanases are now routinely used in wheat-containing diets for broilers, and considerable efforts are being made to understand their mode of action and predict their effects. Similar results have also been obtained with rye-based diets (43). The effects obtained with commercial enzymes may sometimes be due to the action of several enzymes, as the commercial preparations often contain several polysaccharide-degrading activities.

As compared with broilers, laying hens and turkeys are more mature animals with a more developed digestive tract, better able to cope with the nonstarch polysaccharides in cereals and vegetable sources. It has been observed that the advantage of the extra energy liberated by the enzyme supplementation of the feed can be gained by changing the feeding strategy. As the enzyme effect has been found to decrease during the growth of the birds, aging, a feeding strategy where enzyme dose is reduced as the feeding program continues, may provide a more cost-effective route for improvement of feed efficiency and performance.

Pig Feed. Pig feeds containing very high levels of wheat are also less common, as the formulation of pig feeds tends to be multicereal based. According to some experiments, however, increasing levels of xylanase supplementation, up to 600 g of the commercial enzyme per ton of feed (containing 20–35% barley and 30–35% wheat), resulted in improvement of performance, as measured by feed conversion ratio and live weight gain (38).

Pig feeds normally contain more by-products than poultry feeds because of the overall higher digestive efficiency of the pig compared with poultry. These by-products can also be upgraded by enzyme supplementation and, in some instances, offer higher potentials than the materials from which they are derived. The presence of noncereal feed raw materials in the complex diet (e.g., rapeseed meal, sugar beet pulp, sunflower meal) has been found to diminish the enzyme effect, as the substrates present were not degradable by the xylanase used. This implies that attention must be paid to the raw material composition of pig diets if the maximum enzyme effect is to be achieved (38).

Pulp and Paper

In the pulp and paper industry, hemicellulases are mainly used to improve the bleachability of pulps. In this application, xylanases are the most important group of enzymes, although mannanases may also facilitate the bleaching of certain types of pulps (44). In addition, xylanases have been present in enzyme preparations used to improve the drainage (water removal) or deinking of recycled fibers. The mechanisms in these applications are, however, not clear, and they are obviously not based solely on the action of hemicellulases.

Pulping and Bleaching Processes. The main aim in chemical pulping is to remove lignin and to separate the wood fibers from each other to render them suitable for the paper-making process. In the pulping process, the lignified middle lamella located between the wood fibers is solubilized by various chemicals. Today, the predominant pulping method is the kraft process, where the cooking liquors are incinerated and the cooking chemicals recycled.

In bleaching, the primary goal is to remove the low amount of residual lignin present in the pulp after cooking without decreasing the molecular weight of cellulose. Lignin in unbleached pulps typically represents only about 1% of the dry weight. During pulping, however, lignin is chemically modified and condensed, resulting in poorly degradable structures. Cooking and bleaching are separate process phases, differing from each other with respect to the selectivity of the chemicals used. In the bleaching processes, lignin is sequentially degraded and extracted in several phases. Bleaching sequences are generally composed of at least five phases. Previously, the bleaching of chemical pulps was carried out with elemental chlorine and chlorine dioxide. In Europe today pulp is bleached by elemental chlorine free (ECF) or totally chlorine free (TCF) bleaching processes, where oxygen, ozone, or peroxide is used.

Extensive modifications of hemicelluloses take place during pulping processes. During conventional kraft cooking, part of the hemicelluloses is first solubilized in the cooking liquor. At the start of the kraft process xylan in wood is partly solubilized by the alkaline cooking liquid and many of the side groups are cleaved off. It has recently been observed that the majority of the 4-*O*-methylglucuronic acid side groups are converted in the early phases of the kraft cook to hexenuronic acid (45). In the later phases of the process, when the alkalinity of the cooking liquor decreases, part of the solubilized xylan is relocated onto the cellulose fibers (46). Although glucomannan is the main hemicellulose in softwood, the bulk of glucomannans are dissolved and degraded during kraft pulping. Reprecipitation of glucomannan has not been reported to take place to the same extent as xylans. Thus, the relative amount of xylan is increased in pine kraft pulp as compared to pine wood (2).

In addition to xylan, lignin is also partially readsorbed on the fibers. Lignin has been reported to be linked to hemicelluloses, forming lignin–carbohydrate complexes (LCC). Furthermore, hemicelluloses seem to physically restrict the passage of high molecular mass lignin out of the pulp fiber cell wall, and thus the removal of hemicelluloses, especially xylan, can be expected to enhance the extractability of residual lignin from pulps.

Modification of Fibers by Hemicellulases. The effect of hemicellulases in bleaching is based on the modification of pulp hemicelluloses, enhancing the removal of lignin in chemical bleaching. It has been proposed that the action of xylanases is due to the partial hydrolysis of reprecipitated xylan or to removal of xylan from the lignin-carbohydrate (LC) complexes. Both mechanisms would lead to enhanced diffusion of entrapped lignin from the fiber wall. In softwood kraft fibers, removal of xylan by xylanases was found to uncover lignin (47). Thus, it can be expected that removal of xylan improves the extractability of lignin by exposing lignin surfaces.

The partial hydrolysis of xylan or glucomannan may also degrade and improve the extractability of LCC. Both softwood and hardwood kraft pulps have been reported to contain LCC in which carbohydrates and lignin may be connected to each other by ether or glycosidic linkages (48). However, no direct evidence for the type of linkage(s) existing between hemicelluloses and lignin has yet been presented. The action of xylanases on both reprecipitated and LC xylan in enhancing bleachability suggests that it is probably not only the type but also the location of the xylan that is important in the mechanism of xylanase-aided bleaching.

Hemicellulases in Bleaching Processes. The use of xylanases in different bleaching sequences uniformly leads to a reduction in chemical consumption (8,44). The benefits obtained by enzymes are dependent on the type of bleaching sequence used, as well as on the residual lignin content (measured as the kappa number) of the pulp. Originally, xylanases were studied to reduce the consumption of elemental chlorine. Later, xylanases were combined with various ECF and TCF bleaching sequences to improve the otherwise lower brightness of pulp or to decrease the bleaching costs. In chlorine bleaching, an average reduction of 25% in active chlorine consumption in prebleaching, or a reduction of about 15% in total chlorine consumption has been reported both in laboratory scale and in mill trials. As a result, the concentration of chlorinated compounds, measured as adsorbable organic halogen (AOX), in the bleaching effluent during mill trials was reduced by 15–20% (8). Today, xylanases are industrially used both in ECF and TCF sequences. In ECF sequences the enzymatic step is often implemented due to the limiting chlorine dioxide production capacity. The use of enzymes allows bleaching to higher brightness values when chlorine gas is not used. In TCF sequences, the advantage of the enzymatic step is due to improved brightness, maintenance of fiber strength, and savings in bleaching costs.

The amount of enzyme needed for the bleaching is also a key parameter with respect to both enzyme cost and yield loss. Generally, it seems that although the hydrolysis (the solubilization of carbohydrates) increases as a function of the enzyme dosage used, no further benefits can be obtained to the bleachability after a certain limit (49). Thus, to maximize the positive effect of the enzyme on the pulp kappa number and brightness, and simultaneously to minimize the yield loss, laboratory-scale experiments are needed to optimize the enzyme dosage.

Enzymatic pretreatment has been shown to be fully compatible with existing industrial equipment, which is a considerable advantage of this method, especially when

Table 2. Cost Savings with Xylanases in Bleaching

Chemical savings	
Chlorine dioxide	$4 /ton
Costs	
Enzyme	$1.5 /ton
Acid	$0–0.2 /ton
Net savings	$2.3–2.5 /ton pulp

Source: From Roehm Enzymes, Finland.

compared with some other competing technologies. Xylanases are typically mixed with water before being added to the unbleached pulp by a shower bar and allowed to react in the high-density storage tank for 1 or 2 h before the subsequent chemical bleaching steps. Presently a number of mills in North America and Scandinavia use enzymes continuously.

The approximate price of xylanase in 1997 was less than $2/ton pulp. Estimations for the capital cost of enzyme delivery and pH adjustment varied from $10,000 to $1,000,000 in 1995. The potential economic benefits of enzyme bleaching are significant to the pulp and paper industry. Calculation of the relative economic benefits in an ECF sequence reveals that the reduction of the chlorine dioxide consumption leads to savings of at least $2/ton pulp (Table 2). The costs of oxygen-based chemicals (ozone, peroxide) are even higher and the respective savings even more pronounced.

FUTURE OUTLOOK

Today, hemicellulases are used in three major industrial sectors: in food as baking aids, in feed for increasing the nutritional value, and in the pulp and paper industry for saving chemical costs. Most applications are based on the action of xylanases, and several commercial xylanase preparations with different properties are available. The future developments of hemicellulases aim at more efficient and stable enzymes. Obviously, enzymes with different pH and higher-temperature optima and stability are desired for most applications. Thermal stability has been identified as one of the key properties in all industrial sectors. The use of monocomponent enzyme products, alone or in optimized combinations for targeted use, will also become more common. Tailored side group cleaving enzymes will be available, in addition to the present major endoglycanases. Already most of the industrial enzymes are produced by strains developed by genetic engineering. These techniques are expected to lead to more cost-efficient enzyme products.

BIBLIOGRAPHY

1. G.O. Aspinall, *Adv. Carbohydr. Chem.* **14**, 429–468 (1959).
2. E. Sjöström, *Wood Chemistry: Fundamentals and Application*, 2nd ed., Academic Press, San Diego, Calif., 1993.
3. R.F.H. Dekker, in T. Higuchi ed., *Biosynthesis and Biodegradation of Wood Components*, Academic Press, Orlando, Fla., 1985, pp. 505–533.
4. H.D. Simpson, U.R. Haufler, and R.M. Daniel, *Biochem. J.* **277**, 413–417 (1991).
5. P.A. Bicho, T.A. Clark, K. Mackie, H.W. Morgan, and R.M. Daniel, *Appl. Microbiol. Biotechnol.* **36**, 337–343 (1991).
6. G.D. Duffaud, C.M. McCutchen, P. Leduc, K.N. Parker, and R.M. Kelly, *Appl. Environ. Microbiol.* **63**, 169–177 (1997).
7. M.P. Coughlan and G.P. Hazlewood, *Biotechnol. Appl. Biochem.* **17**, 259–289 (1993).
8. L. Viikari, A. Kantelinen, J. Sundquist, and M. Linko, *FEMS Microbiol. Rev.* **13**, 335–350 (1994).
9. A. Sunna and G. Antranikian, *Crit. Rev. Biotechnol.* **17**, 39–67 (1997).
10. B. Henrissat and A. Bairoch, *Biochem. J.* **293**, 781–788 (1993).
11. P. Biely, M. VrŠanská, M. Tenkanen, and D. Kluepfel, *J. Biotechnol.* **57**, 151–166 (1997).
12. H. Okada, *Adv. Prot. Res.* **12**, 81 (1989).
13. A. Törrönen, A. Harkki, and J. Rouvinen, *EMBO J.* **13**, 2493 (1994).
14. A. Törrönen and J. Rouvinen, *Biochemistry* **34**, 847 (1995).
15. K. Sakka, Y. Kojima, T. Kondo, S. Karita, K. Ohmiya, and K. Shimada, *Biosci. Biotech. Biochem.* **57**, 273–277 (1993).
16. F. Shareck, C. Roy, M. Yaguchi, R. Morosoli, and D. Kluepfel, *Gene* **107**, 75 (1991).
17. U. Derewenda, L. Swenson, R. Green, Y. Wei, R. Morosoli, F. Shareck, D. Kluepfel, and Z.S. Derewenda, *J. Biol. Chem.* **269**, 20811–20814 (1994).
18. K. Nishitani and D.J. Nevins, *J. Biol. Chem.* **266**, 6539 (1991).
19. M.G. Tuohy, J. Puls, M. Clayssens, M. VrŠanská, and M.P. Coughlan, *Biochem. J.* **290**, 515 (1993).
20. H. Stålbrand, A. Saloheimo, J. Vehmaanperä, B. Henrissat, and M. Penttilä, *Appl. Environ. Microbiol.* **61**, 1090–1097 (1995).
21. M. Hilge, S.M. Gloor, W. Rypniewski, O. Sauer, T.D. Heightman, W. Zimmerman, K. Winterhalter, and K. Piontek, *Structure* **6**, 1433–1444 (1998).
22. B.V. McCleary, in G.F. Leatham and M.E. Himmel eds., *Enzymes in Biomass Conversion*, ACS Symp. Ser. 460, 1991, American Chemical Society, Washington, D.C., pp. 437–449.
23. I. Kusakabe, G.G. Park, N. Kumita, T. Yasui, and K. Murakami, *Agric. Biol. Chem.* **52**, 519 (1988).
24. M. Tenkanen, M. Makkonen, M. Perttula, L. Viikari, and A. Teleman, *J. Biotechnol.* **57**, 191–204 (1997).
25. M. Tenkanen, E. Luonteri, and A. Teleman, *FEBS Lett.* **399**, 303–306 (1996).
26. G.P. Hazlewood and H.J. Gilbert, in J. Visser, G. Beldman, M.A. Kusters-vanSomeren, and A.G.J. Voragen eds., *Xylans and Xylanases*, Elsevier, Amsterdam, 1992, pp. 259–273.
27. M.J. Bailey, P. Biely, and K. Poutanen, *J. Biotech.* **23**, 257–270 (1992).
28. G. Beldman, D. Osuga, and J.R. Whitaker, *J. Cereal Sci.* **23**, 169–180 (1996).
29. G. Cleemput, W. Bleukx, M. van Oort, M. Hessing, and J.A. Delcourt, *J. Cereal Sci.* **22**, 139–145 (1995).
30. J. Moore, C.W. Bamforth, P.A. Kroon, B. Bartolomé, and G. Williamson, *Biotechnol. Lett.* **18**, 1423–1426 (1996).
31. J. Maat, M. Roza, J. Verbakel, H. Stam, M.J. Santos da Silva, M. Bosse, M.R. Egmond, M.L.D. Hagemans, R.F.M. v. Gorcom, J.G.M. Hessing, C.A.M.J.J. v.d. Hondel, and C. v. Rotterdam, in J. Visser, G. Beldman, M.A. Kusters-van Someren, and A.G.J. Voragen eds., *Xylans and Xylanases*, Elsevier, Amsterdam, 1992, pp. 349–360.

32. H. Gruppen, F.J.M. Kormelink, and A.G.J. Voragen, *J. Cereal Sci.* **18**, 129–143 (1993).
33. X. Rouau, M.L. El-Hayek, and D. Moreau, *J. Cereal Sci.* **19**, 259–272 (1994).
34. K. Poutanen, *Trends Food Sci. Technol.* **8**, 300–306 (1997).
35. B. Sprössler, *Getreide. Mehl Brot* **50**, 281–283 (1996).
36. T. Laurikainen, H. Härkönen, K. Autio, and K. Poutanen, *J. Sci. Food. Agric.* **76**, 239–249 (1998).
37. U.S. Pat. 5,108,764 (April 28, 1992) S.A.S. Craig, L.R. Beehler, R.T. Deihl, H. Levine, A.M. Magliacano, P.R. Mathewson, M.S. Otterburn, L. Slade, and P. Verduin (to Nabisco Brands, Inc.).
38. W.D. Cowan, in T. Godfrey and S. West eds., *Industrial Enzymology*, 2nd ed., Macmillan, London, 1996, pp. 69–101.
39. M. Choct, G. Annison, and R.P. Trimble, *Proc. Austr. Poultry Sci. Symp.* **5**, 78 (1993).
40. G. Annison, *J. Agric. Food Chem.* **39**, 1252–1256 (1991).
41. A. Tervilä-Wilo, T. Parkkonen, A. Morgan, M. Hopeakoski-Nurminen, K. Poutanen, P. Heikkinen, and K. Autio, *J. Cereal Sci.* **24**, 215–225 (1996).
42. B. Carr, M. Lessire, T.H. Nguyen and M. Larbier, *Proceedings of the 19th World's Poultry Congress*, Amsterdam, 1992, pp. 411–415.
43. M.R. Bedford and H.L. Classen, in J. Visser, G. Beldman, M.A. Kusters-van Someren, and A.G.J. Voragen eds., *Xylans and Xylanases*, Elsevier, Amsterdam, 1992, pp. 361–370.
44. A. Suurnäkki, M. Tenkanen, J. Buchert, and L. Viikari, *Adv. Biochem. Eng.* **57**, 261–287 (1997).
45. A. Teleman, V. Harjunpää, M. Tenkanen, J. Buchert, T. Hausalo, T. Drakenberg, and T. Vuorinen, *Carbohydr. Res.* **272**, 55–71 (1995).
46. S. Yllner and B. Enström, *Svensk Papperstidn.* **59**, 229–232 (1956).
47. J. Buchert, G. Carlsson, L. Viikari and G. Ström, *Holzforschung* **50**, 69–70 (1996).
48. G. Gellerstedt and E.-L. Lindfors, *Proc. Int. Pulp Bleaching Conf., vol. 1*, SPCI, Stockholm, 1991, p. 73.
49. J. Buchert, M. Ranua, A. Kantelinen, and L. Viikari, *Appl. Microbiol. Biotechnol.* **37**, 825–829 (1992).

See also ENZYMES, PULP AND PAPER PROCESSING; HEMICELLULOSE CONVERSION.

HEMICELLULOSE CONVERSION

G.T. TSAO
NINGJUN CAO
C.S. GONG
Purdue University
West Lafayette, Indiana

KEY WORDS

2,3-Butanediol
Ethanol
Hemicellulose
Pretreatment
Xylitol
Xylose

OUTLINE

INTRODUCTION

Hemicellulose is a plant cell wall polysaccharide that is localized mostly in cell wall middle lamella. It is in close association with cellulose, lignin, and pectic materials. This close association contributes to the rigidity and flexibility of the plant cell wall. Hemicellulose is the third-most abundant polymer in nature. In some plants, it comprises up to 40% of the total dry material. Unlike cellulose, hemicellulose exhibits a wide diversity in both structure and constitution. The degree of polymerization of hemicellulose is usually less than 200. The type and amount of hemicellulose vary widely, depending upon plant material, type of tissue, stage of growth, and so on. For those reasons, it is difficult to obtain a typical sugar composition of a typical hemicellulose.

Hemicelluloses are heteropolysaccharides that are composed of various hexoses (e.g., glucose, mannose, and galactose), pentoses (D-xylose and L-arabinose), uronic acids, acetic acid, and other minor sugars. Thus, by definition, hemicelluloses are short branched-chain heteropolysaccharides of mixed hexosans and pentosans that are easily hydrolyzed (1). The most common form of hemicellulose is composed of xylose polymer (xylan) that is found in large quantity in annual plants and deciduous trees and in smaller amounts in conifers. Xylan of grasses and cereals is generally characterized by the presence of L-arabinose, which is linked as a single unit side chain to a D-xylose backbone. Hydrolysis of hemicellulose in annual plants, agricultural wastes, and hardwood yields glucose, D-xylose, L-arabinose, and other minor sugars. Table 1 summarizes the amounts of hemicellulose in different plant materials; Table 2 summarizes the sugar composition of selected biomass materials after the hydrolysis.

HEMICELLULOSE HYDROLYSIS

Hemicellulose can be hydrolyzed to its sugar constituents either enzymatically or chemically. Chemical hydrolysis of hemicellulose is much easier to accomplish than the hydrolysis of cellulose due to the heterogeneous nature of hemicellulose, its chemical composition, and its low degree of polymerization. Most acids are good hydrolytic agents. The most common method of acid hydrolysis uses dilute mineral acids. Frequently during acid hydrolysis, xylose is

Table 1. Biomass Constiuents

Type of material	Hemicellulose (%)	Cellulose (%)	Lignins (%)
	Monocotyledons		
Stems	25 ~ 50	25 ~ 40	10 ~ 30
Leaves	80 ~ 85	15 ~ 20	–
Fibers	5 ~ 20	80 ~ 95	–
	Woods		
Hardwood (angiosperms)	24 ~ 40	40 ~ 55	18 ~ 25
Softwood (gymnosperms)	25 ~ 35	45 ~ 50	25 ~ 35
	Papers		
Newspaper	25 ~ 40	40 ~ 55	18 ~ 30
Wastepaper	10 ~ 20	60 ~ 70	5 ~ 10
Waste fibers	20 ~ 30	60 ~ 80	2 ~ 10

Source: From Ref. 2.

Table 2. Hemicellulose Neutral Carbohydrate Content of Agriculture Residues

	% of total sugars			
Plant Residues	Xylose	Arabinose	Glucose	Others
Corn residues				
Cobs	65.1	9.6	25.3	–
Leaves	59	9.4	29.1	2.5
Stalks	70.5	9.0	14.5	5.9
Husks	53.5	12.3	32.6	1.6
Pith	71.5	9.8	15.7	3
Fibers	63.8	6.6	26.8	2.8
Wheat straw	57.9	9.1	28.1	5
Soybean				
Stalks and leaves	59.9	6.6	6.1	27.4
Hulls	26.6	12.7	21	39.7
Sunflower				
Stalks	60.6	2.2	32.6	4.6
Pith	10.7	11.8	63.5	14
Flax Straw	64.6	12.8	1.2	21.4
Sweet clover hays	49.3	21.9	8.9	9.9
Peanut hulls	46.3	5	46.6	2.1
Sugarcane bagasse	59.5	14.5	26	–

Source: From Ref. 3.

degraded rapidly to furfural and condensation byproducts. Some degradation products are inhibitory to microorganisms. The toxic effect of those chemicals on cell growth and ethanol formation of *Pichia stipitis* and *Saccharomyces cerevisiae* are shown in Tables 3 and 4 (4). The use of a very dilute acid (less than 0.2% acid by weight), shorter reaction time, a lower reaction temperature, and rapid removal of hydrolytic agents are preferred to prevent the decomposition of sugars, particularly xylose. In recent years, a great deal of research has focused on the dilute acid hydrolysis of agricultural residues and wood products to obtain hemicellulose sugars (5,6).

BIOCONVERSION OF HEMICELLULOSE SUGARS

When cellulosic materials are subjected to hydrolysis, a mixture of monosaccharides is produced. The predominant sugars released are glucose, xylose, and to a lesser extent, arabinose, galactose, and mannose. When microorganisms are exposed to this sugar mixture, the phenomena of diauxic growth and differential rates of sugar utilization are often observed. The differential utilization of sugars by a given microorganism is dictated by sugar uptake, utilization rates, and by the degree of catabolite repression, which affects specific enzyme activity and biosynthesis. As a rule, microorganisms prefer glucose over galactose, followed by xylose and arabinose.

Because hemicellulose is abundant in nature and renewable, extensive research has been undertaken to convert hemicellulose-derived carbohydrates, particularly xylose, into useful products. In this article, the emphasis is on the utilization of hemicellulose-derived xylose into ethanol, xylitol, and 2,3-butanediol. Other products such as solvents (acetone and butanol) (7) and single-cell proteins (8) have been reviewed.

Table 3. The Effect of Potential Inhibitors on *Pichia stipitis* Growth and Ethanol Production from Xylose

Compound[a]	Concentration (g/L)	% Inhibition (Growth)	% Inhibition (EtOH)
Furfural	2	99	95
HMF	5	98.6	91.4
Acetate	10	36	69
HBA	0.75	70	84
SGA	0.75	62	80
Vanillin	0.5	88	89

Source: From Ref. 9.
[a]HMF, hydroxymethylfurfural; HBA, hydroxybenzaldehyde; SGA, siringaldehyde.

Table 4. The Effects of Potential Inhibitors on *S. cerevisiae* Growth and Ethanol Production from Glucose

Compound[a]	Concentration (g/L)	% Inhibition (Growth)	% Inhibition (EtOH)
Furfural	2	90	89
HMF	5	89	95
Acetate	10	48	27
HBA	0.75	53	37
SGA	0.75	61	54
Vanillin	0.5	51	30

Source: From Ref. 9.
[a]HMF, hydroxymethylfurfural; HBA, hydroxybenzaldehyde; SGA, siringaldhyde.

ETHANOL

Ethanol is the most studied metabolic product from xylose by fermentation. Ethanol can be used as a "neat" fuel substitute for gasoline. It can also be used as an oxygenated fuel extender by converting ethanol to ethyl tertiary-butyl ether (ETBE). Using ethanol as fuel and fuel extender provides energy security and diversity, improves global competitiveness, reduces trade deficits, revitalizes the agricultural industry, and promotes energy independence. It also improves the environment by improving urban air quality and by reducing the threat of global warming. Currently the production of ethanol from biomass is not competitive. Many research efforts were made to achieve the economical production of ethanol from biomass (for review see Ref. 9). In order to increase ethanol production efficiency from biomass, full and efficient utilization of hemicellulose carbohydrates is essential.

Conversion of Xylose to Ethanol by Yeasts

Yeasts are facultative organisms that have the ability to produce energy for their own use from suitable organic compounds under either aerobic or anaerobic conditions. Under aerobic conditions, sugars are metabolized to carbon dioxide and water for the production of energy and cell constituents. Under anaerobic conditions, most sugars are metabolized to ethanol by a process known as alcohol fermentation. Although many facultatively fermentative yeasts utilize xylose as carbon source for growth, the ability of those yeasts to produce ethanol from xylose is limited (10).

Yeast strains that utilize xylose often produce xylitol from xylose extracellularly as a normal metabolic activity, but only a few can produce significant quantities of ethanol. The prominent strains that produce ethanol from xylose include *Kluveromyces maxianus* (11), *Pachysolen tannophilus, Candida shihatae*, and *Pichia stipitis*. However, the efficient production of ethanol from xylose is limited by the regulation of dissolved oxygen as well as by the imbalance of cofactors in the metabolic pathway during xylose utilization. In recent years, there have been many efforts carried out to improve yeast strains to produce ethanol from xylose more efficiently.

Metabolic Pathway. Ethanol fermentation by yeasts from xylose can be divided into four distinctive steps. The first step is the reduction of xylose to xylitol, mediated by NADPH/NADH-linked xylose reductase (XR). This is followed by the oxidation of xylitol to xylulose, mediated by NAD-linked xylitol dehydrogenase (XDH). Xylulose-5-phosphate, the key intermediate, is generated from the phosphorylation of xylulose by xylulose kinase. Xylulose is then channeled into the pentose phosphate pathway for further metabolism (Fig. 1). In contrast, a great majority of bacteria convert xylose into its ketoisomer, xylulose, mediated by xylose isomerase (commonly known as glucose isomerase) as the initial step in xylose metabolism.

Ethanol Yield and Carbon Balance. Starting with 6 molecules of xylose, 10 molecules of ethanol can be produced using the combination of two pathways (Fig. 1). The net equation for the reactions leading to the production of ethanol from xylose is

$$6 \text{ xylose to } 10 \text{ ethanol} + 10\ CO_2$$

These values are the theoretical maximum yields (same as for glucose fermentation) that represent the maximum value, which can be achieved only if no sugar is assimilated to cell mass or oxidized through the TCA cycle. Of the total carbon contained in xylose, two-thirds goes to ethanol, and

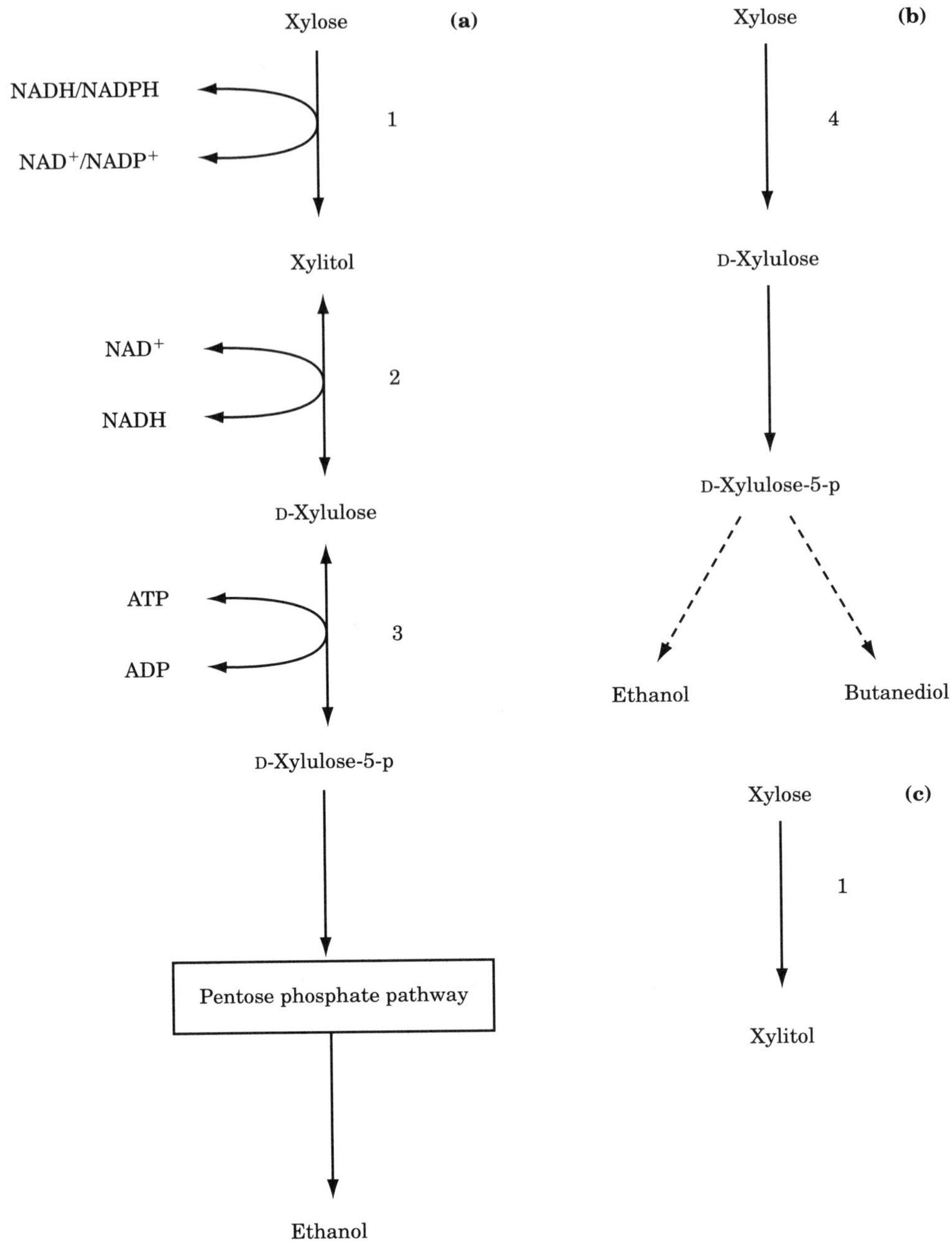

Figure 1. Abbreviated pathways for xylose utilization by microorganisms. (**a**) Yeast pathway; (**b**) bacterial pathway; (**c**) yeast xylitol pathway. (1) Xylose reductase; (2) xylitol dehydrogenase; (3) xylulokinase; (4) xylose isomerase.

the other one-third is lost as CO_2. On a weight basis, the yield of ethanol from both xylose and glucose is 51%.

Yeasts That Ferment Xylose to Ethanol. Following are the general characteristics of yeasts that ferment xylose to ethanol.

1. Oxygen requirement (12).
   a. Oxygen is required for the efficient uptake of xylose.
   b. Under low dissolved oxygen (DO) conditions, the electron transport system is not able to oxidize NADH efficiently, this causes the imbalance of $NADH/NAD^+$ that leads to the accumulation of xylitol.
   c. An increase in DO will enhance cell growth and xylose fermentation.
   d. When oxygen is in excess, TCA cycle activity is enhanced, which results in excessive cell growth and causes the reassimilation of ethanol produced.
   e. The reassimilation of ethanol leads to the accumulation of acetaldehyde and acetic acid.
   f. The optimal DO gives the lowest xylitol accumulation with the highest ethanol yield.
2. Enzyme cofactor imbalance
   a. Xylose reductase and xylitol dehydrogenase have different cofactor requirements.

   b. Different nitrogen sources (organic vs. inorganic) affects the level of xylitol dehydrogenase activity (13).
   c. Different DOs affect the balance of cofactors.
3. Low ethanol tolerance compared with that in *S. cerevisiae*
   a. Fermentation is inhibited at an ethanol concentration of 42 to 45 g/L.
4. Xylitol, glycerol, arabitol, and acetate are the common byproducts.
5. Higher sensitivity to potential fermentation inhibitors than glucose-fermenting yeast strains during glucose fermentation (Tables 3 and 4).
6. Specific ethanol productivity from xylose is at least one order of magnitude lower than that in glucose fermentation by *S. cerevisiae*.

Natural Yeast Strains. *P. tannophilus, C. shehatae*, and strains of *P. stipitis* are three of the most intensively studied naturally occurring yeast species that produce a significant quantity of ethanol from xylose.

Pachysolen tannophilus. *P. tannophilus* is an unusual yeast that was originally isolated by Wickerham (14) from wood sulfite liquor. This yeast was the first naturally occurring yeast species that was found to produce a significant quantity of ethanol from xylose (15). More detailed studies of this yeast were reported by Slininger et al. (16). As described in the literature, ethanol production from xylose by *P. tannophilus* exhibited the following characteristics:

1. A small amount of oxygen was required for optimal ethanol production.
2. A lag phase of about 20 h was observed before the accumulation of ethanol.
3. Ethanol yield was reported to be 0.34 g ethanol per g xylose consumed (68% of the theoretical value).
4. Optimal temperature for cell growth and ethanol production was 32 °C.
5. Optimal pH for cell growth was 2.5–4.5, and optimal pH for ethanol production was 2.5.
6. Ethanol is reassimilated under high DO conditions.
7. Ethanol production was subjected to substrate (55 g/L xylose) and product (30 g/L ethanol) inhibition.
8. Xylitol was accumulated during fermentation.

Candida shehatae. The ability of *C. shehatae* to produce ethanol from xylose was first reported by du Preez and van der Walt (17) in 1983. Since then, many studies have been conducted relating to the properties of this yeast. As described in the literature (18–20), ethanol production from xylose by *C. shehatae* exhibited the following characteristics:

1. A small amount of DO is required for maximal ethanol production.
2. Cell growth was inhibited at an ethanol concentration of 37.5 g/L, and cell viability was significantly lower in the presence of up to 50 g/L ethanol.
3. Glucose and mannose are the preferred substrate in a sugar mixture.
4. Ethanol yield can be as high as 84% of the theoretical value (0.43 g ethanol per g xylose consumed).
5. The maximum ethanol production rate was 0.48 g/g/h at 50 g/L xylose.
6. Xylitol was accumulated under very limited DO conditions due to the accumulation of NADH.
7. Under high DO conditions, the ethanol produced was reassimilated.

Pichia stipitis. *P. stipitis* is the most effective natural yeast for the conversion of xylose to ethanol. This yeast species shares many characteristics with its close relative *C. shehatae*. Toivola et al. (21) performed a systemic screening program with type strains of some 200 yeast species and identified *P. stipitis* as one of the yeast species that produces ethanol from xylose. There are many studies that explored the properties of this yeast in relation to its oxygen requirement, ethanol tolerance, enzyme cofactors balance, and so on. As described in the literature (22,23), ethanol production from xylose by *P. stipitis* exhibited the following characteristics:

1. Trace oxygen was required to sustain cell growth and maintenance.
2. *P. stipitis* has a higher ethanol (64 g/L) tolerance than other xylose-fermenting yeasts.
3. Compared with *C. shehatae, P. stipitis* has a lower specific ethanol productivity but a higher ethanol yield.
4. Up to 57 g/L of ethanol can be accumulated by *P. stipitis*.
5. Xylitol production by *P. stipitis* was lower than other xylose-fermenting yeasts.
6. The optimum temperature for ethanol production was 25 °C.
7. The optimum pH range for growth and fermentation was 4–7.
8. Ethanol was reassimilated at high DO levels, and acetate was formed when ethanol was metabolized.

Genetically Modified Yeasts. Attempts to modify the xylose-fermenting pathway in *S. cerevisiae* using the xylose isomerase gene from various bacterial sources have not been successful (24) in spite of previous reports showing the ability of glucose-fermenting yeast strains, *S. cerevisiae* (25) and *Schizosaccharomyces pombe* (26), to produce ethanol from xylulose or from xylose in the presence of xylose isomerase (glucose isomerase) (27) (Fig. 2). Other approaches of improving yeast strains through genetic recombination have met with some encouraging results.

Saccharomyces cerevisiae. Kotter and Ciriacy (28) studied xylose utilization of a *S. cerevisiae* transformant that expressed two key enzymes (xylose reductase and xylitol dehydrogenase) derived from *P. stipitis* in xylose metabo-

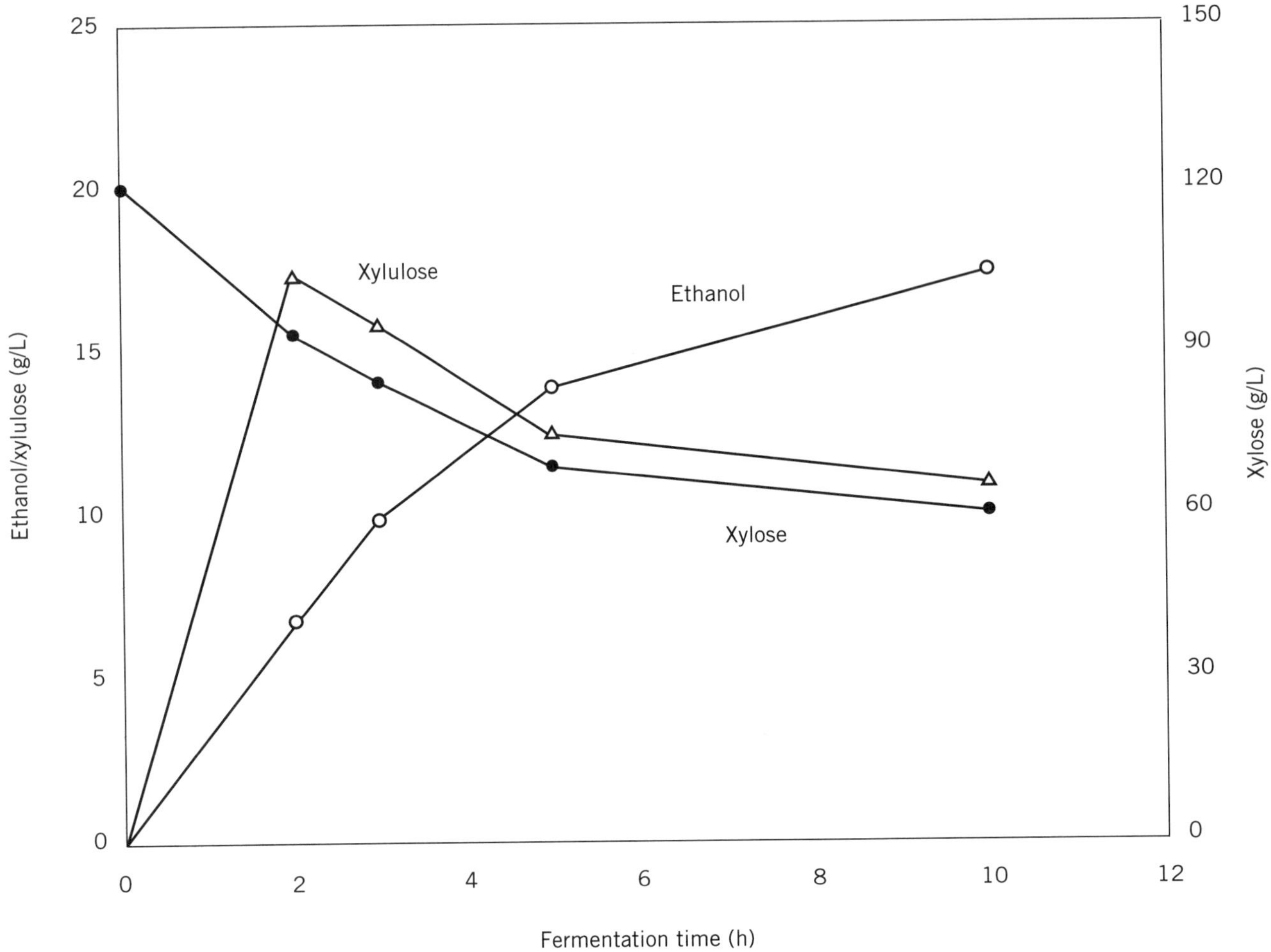

Figure 2. Production of ethanol from xylose by Baker's yeast in the presence of xylose isomerase. *Source*: From Ref. 27.

lism. Under fermentative conditions, the transformant converts only half of available xylose to xylitol and ethanol. The acquired ability to ferment xylose was interpreted to be a result of the duel cofactor dependence of XR and the generation of NADPH by the pentose phosphate pathway. The limitation of xylose utilization in the transformant was likely caused by an insufficient capacity of the non-oxidative pentose pathway, as indicated by the accumulation of sedoheptulose-7-phosphate and the absence of fructose-1,6-diphosphate and pyruvate accumulation.

Recombinant Yeast Saccharomyces 1400 (pLNH33). The recombinant yeast strain *Saccharomyces* 1400 (pLNH33) was developed using the high-ethanol-tolerance *Saccharomyces* yeast 1400 (29) as the host and cloned with the xylose reductase and xylitol dehydrogenase genes from *P. stipitis*, and the xylulokinase gene from *S. cerevisiae*. *Saccharomyces* 1400 (pLNH33) was shown to ferment both sugars in a 1:1 mixture of glucose (52.8 g/L) and xylose (56.3 g/L) (Fig. 3) to ethanol under microaerobic condition in a relatively high yield (84% of theoretical value) (30). A final ethanol concentration obtained was 50 g/L after 48 h of incubation. This recombinant was also shown to produce ethanol from glucose and xylose that were derived from corn fiber during simultaneous saccharification and fermentation process in the presence of a fungal cellulase (31).

Conversion of Xylose to Ethanol by Bacteria

A wide range of bacterial species utilizes D-xylose and L-arabinose as carbon and energy sources. In most case, the direct isomerization of aldopentoses to their corresponding ketoses is the first step in pentose metabolism. For example, D-xylose is converted into D-xylulose, and L-arabinose is converted into L-ribulose. L-Ribulose can be converted into D-xylulose by epimerase. D-xylulose is the key intermediate for further metabolism. For this reason, most bacterial species can readily utilize L-arabinose as well.

Bacterial species, particularly those belonging to *Klebsiellae and Erwinia*, and *Escherichia coli*, are well known for their ability to metabolize hexoses and pentoses to produce either neutral compounds (butanediol, acetoin, and ethanol) or mixed acids and ethanol: Research on the production of ethanol from pentoses by bacteria has revolved around the improvement of such bacteria through genetic recombination.

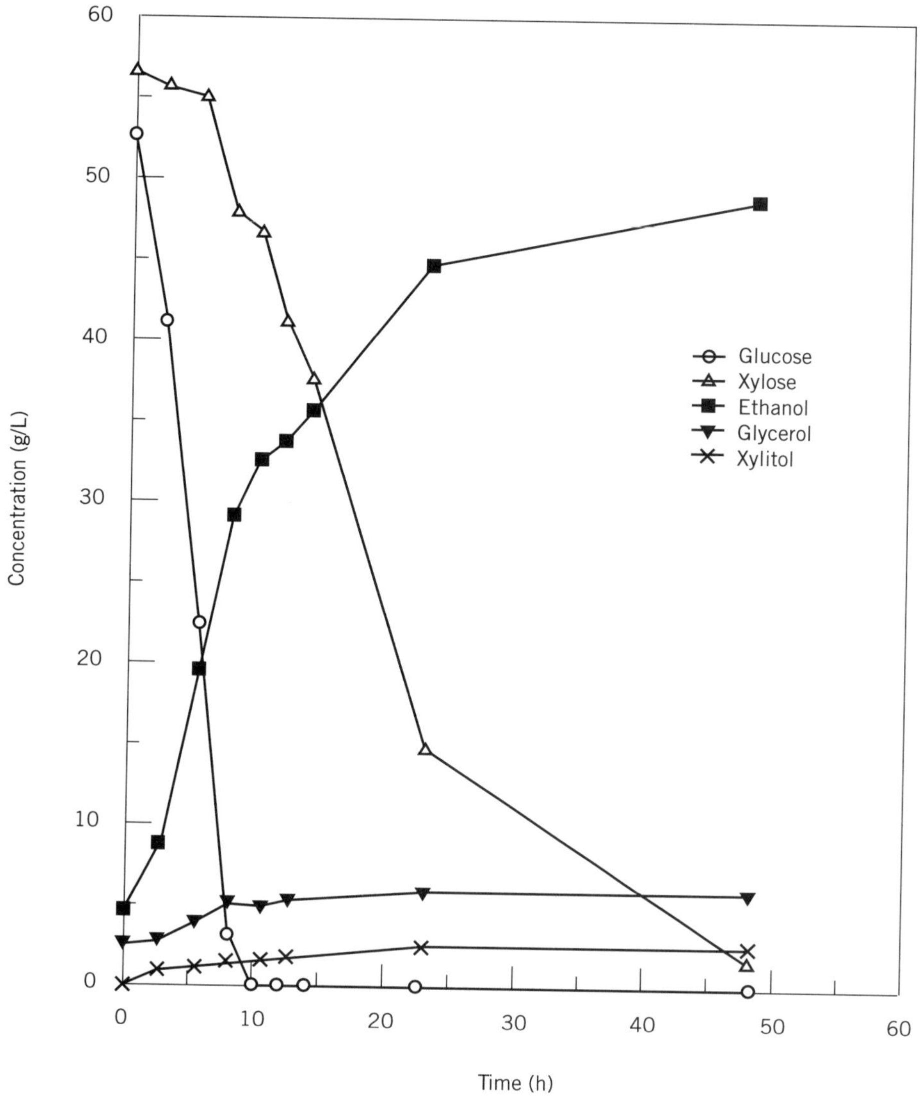

Figure 3. Fermentation of a glucose–xylose mixture by recombinant yeast, *Saccharomyces* 1400 (pLNH33). *Source*: From Ref. 30.

***Klebsiella oxytoca* M5A1.** Pyruvate decarboxylase and alcohol dehydrogenase genes encoding the ethanol pathway from *Zymomonas mobilis* were transferred into *K. oxytoca* (32) The transformant, *K. oxytoca* M5A1, was able to divert pyruvate from the normal fermentative pathway to ethanol production. *K. oxytoca* M5A1 was able to produce ethanol from both glucose and xylose in excess of 40 g/L with an efficiency of 0.48 g of ethanol per g xylose and 0.5 g of ethanol per g glucose. The maximal volumetric productivity for both sugars was similar (ca. 2 g/L).

***Klebsiella planticola*.** Multicopy plasmids containing the *Z. mobilis* pyruvate decarboxylase gene were inserted into *K. planticola*. The transformant was shown to produce 31.6 g/L ethanol from a mixture of sugars (79.6 g/L) that included xylose (33). The yield of acids (formate, acetate, lactate) and butanediol was reduced significantly.

***Escherichia coli*.** Pyruvate decarboxylase of *Z. mobilis* was transferred into *E. coli* (34). The transformant was able to produce 39.2 g/L ethanol from 80 g/L xylose with an indicated yield of 96% of the theoretical value. The maximal volumetric productivity was 0.7 g ethanol per liter per hour in batch fermentation (35).

***Zymomonas mobilis*.** Two operons encoding xylose assimilation enzymes (xylose isomerase and xylulokinase) and pentose phosphate pathway enzymes (transaldolase, and transketolase) were constructed and introduced into *Z. mobilis* (36). The transformant, CP4 (pZB5), was able to not only grow but also ferment xylose anaerobically into ethanol. This transformant also acquired the ability to ferment glucose and xylose simultaneously with a slower rate of xylose than glucose utilization. Anaerobic fermentation using equal amounts of glucose (25 g/L) and xylose (25 g/L) in mixture gave an ethanol yield of 25 g/L ethanol with 95% of the theoretical yield within 30 h.

Conversion of Xylose to Ethanol by Mycelial Fungi

Several filamentous fungi belonging to the genera *Fusarium, Mucor, Rhizopus*, and *Monilia* are known to ferment both glucose and xylose to ethanol (37). The early work on xylose metabolism in *Fusarium lini* indicated the potential for ethanol production from xylose by this fungus (38). Overall, the slow rate of ethanol production from xylose by fungi is undesirable because it would add to capital costs in industrial application. However, because additional ethanol can be obtained from xylose, the use of mold for this purpose is an option.

XYLITOL

Xylitol is a five-carbon sugar alcohol that occurs naturally in small quantities (less than 0.9 g/100g) in many fruits and vegetables that constitute part of human diet. It is the only so-called second-generation polyol sweetener that is allowed to have specific health claims in some world markets. Unfortunately, it is one of the most expensive polyol sweeteners. Availability and cost of production are the obstacles impeding the increased use of xylitol. On an industrial scale, xylitol is produced through chemical reduction of xylose derived from hemicellulosic hydrolyzate. Birch hydrolyzate is the most common substrate. The chemical process is expensive because of the high temperature and high pressure required for hydrogenation of xylose. Furthermore, extensive steps for separation and purification add to the cost.

Xylitol can be formed as a metabolic intermediary of D-xylose utilization by a wide range of microorganisms. It is converted from xylose either by NADPH-dependent aldehyde reductase, or from xylulose by NADH-dependent xylitol dehydrogenase (Fig. 1). Many natural yeast strains have the ability to produce a significant quantity of xylitol from xylose extracellularly as a normal metabolic activity. The prominent strains that produce xylitol include *Candida* sp. (39), *Candida boidinii* (40), *Candida guillermondii* (41), *Candida parapsilosis* (42), *Candida tropicalis* (43), and *Debaryomyces hansenii* (44). However, xylose is an expensive substrate for xylitol production. Recent advances in obtaining xylose-rich hemicellulose hydrolyzates from lignocellulosic materials can provide ample substrate for the biological production of xylitol inexpensively. As a result, xylitol can be potentially produced from such materials as an option for effective utilization of lignocellulosic biomass.

Usually the weight yield of xylitol produced from xylose by naturally occurring yeast species is around 75 to 80%. Two of the reasons for the relatively low yield are the growth of cells and the need for energy maintenance. Ideally, xylitol overaccumulation would occur if the pathway from xylitol to xylulose was cut off, but this is unrealistic because of the redox balance. The next option is to reduce the activity of xylitol dehydrogenase. Therefore, the mutation–selection technique, based on differential xylose–xylitol utilization, can be applied to obtain high-xylitol-producing organisms. Based on this principle, a mutant strain from a proficient xylose-utilizing *Candida* yeast was selected; it had xylitol weight yields as high as 90% from xylose (45).

The production of xylitol from hemicellulosic materials has been considered as an alternative approach in the utilization of the xylose fraction of biomass substrates and the production liquid of fuels and chemical feedstocks. Over the years, many studies have been conducted utilizing the hemicellulose portion of agricultural residues for xylitol production. The materials studied include sugarcane bagasse (39,46), *Eucalyptus* wood (47), and rice straw (48).

2,3-BUTANEDIOL

Butanediol (butylene glycol) is a colorless and odorless liquid with a high boiling point (180–184 °C) and a low freezing point (60 °C). The heating value of butanediol (27,198 J/g) is similar to that of ethanol (29,055 J/g) and methanol (22,081 J/g). It can be dehydrated to methylethyl ketone (MEK) and used as an octane booster for gasoline or as high-grade aviation fuel. MEK can be dehydrated to 1,3-butadiene and dimerized to styrene. Therefore, butanediol has a diverse industrial usage particularly as a polymeric feedstock in addition to its use for manufacturing butadiene or antifreeze (49).

Among many isomers, 2,3-butanediol is the only one that can be produced by microorganisms. Bacterial species of *Klebsielleae* are known to metabolize hexoses and pentoses derived from hemicellulose to produce 2,3-butanediol, acetoin, and ethanol as metabolic products. Other groups of enterobacteria such as *Erwinia* also utilize hemicellulose-derived carbohydrates to produce mixed acids and 2,3-butanediol. Research on the production of butanediol and its immediate precursor, acetoin, from renewable biomass has revolved around the microorganisms *Klebsiella oxytoca (K. pneumonia)* and *Enterobacter aerogenes (Aerobacter aerogenes)*.

Butanediol-producing bacteria utilize all the major sugars (hexoses and pentoses) to produce butanediol in high yield and high concentration (up to 10%, w/v) under optimum conditions (e.g., DO, pH, etc.) (49,50). Three enzymes, α-acetolactate synthetase (ALS), α-acetolactate decarboxylase (ALDC), and acetoin reductase (AR), are directly responsible for the accumulation of 1 mole of butanediol from 2 moles of pyruvate (51). In *K. terrigena* and *E. aerogenes*, these enzymes are encoded by the gene *budB* (ALS), *budA* (ALDC), and *budC* (AR), respectively. The three genes are organized in an operon and activated by transcriptional products of the *budR* gene (52). The expression of the operon is optimal under anaerobic conditions at low pH environments and in the presence of a low concentration of acetate. The metabolic pathway leading to butanediol formation is shown in Figure 4.

DO has a profound effect on butanediol production and end-products distribution. Oxygen supply is important because it determines the flow of carbon source via the respiratory pathway versus the butanediol-producing fermentative pathway. The role of oxygen supply on biomass and butanediol accumulations has been studied extensively. Butanediol yield may be optimized by minimizing the available oxygen. However, with a small oxygen transfer rate, the cell yield will be low, causing the total reaction rate to slow down. In the presence of excessive oxygen, metabolism leads to the oxidation of NADH to NAD^+. A high NAD^+/NADH ratio leads to excessive biomass accu-

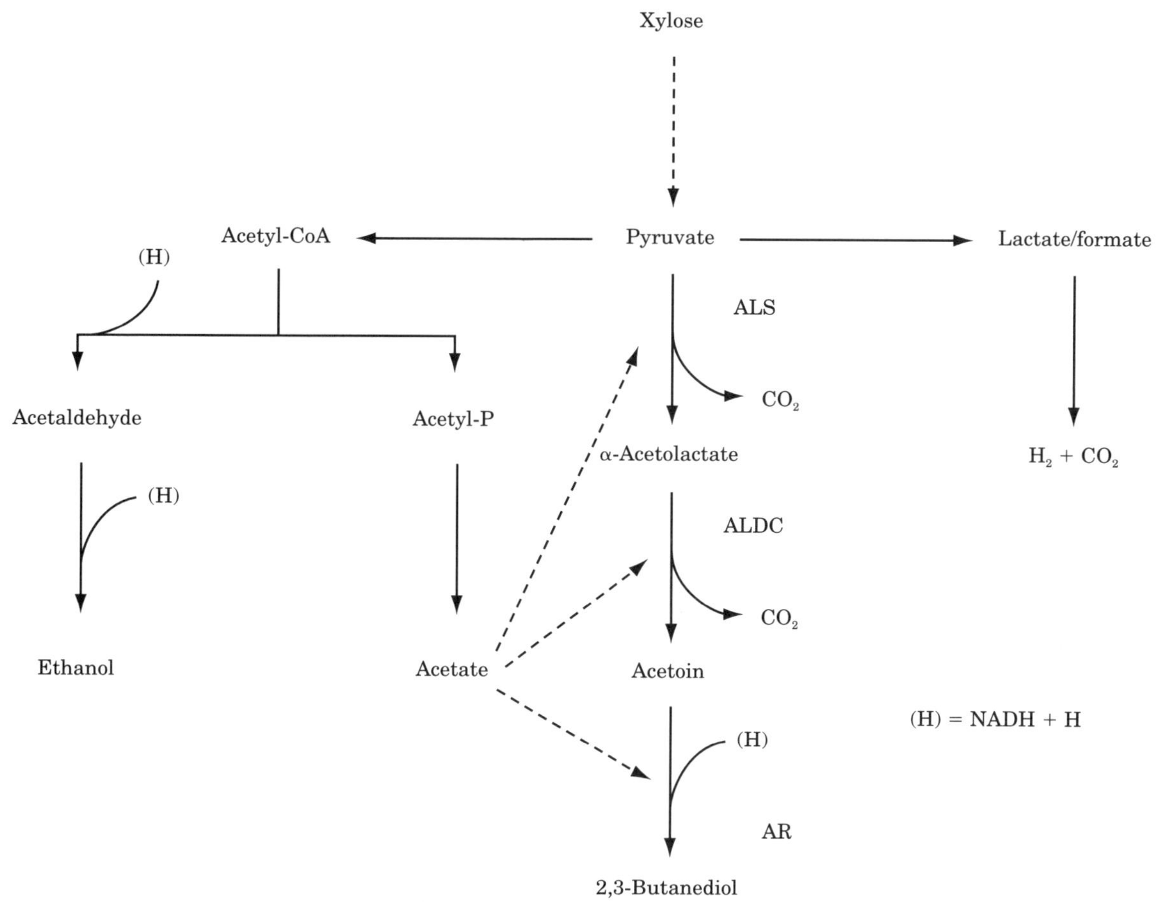

Figure 4. The metabolic pathway leading to the production of 2,3-butanediol and the role of acetic acid. ALS, α-acetolactate synthetase; ALDC, α-acetolactate decarboxylase; AR, acetoin reductase.

mulation. There is a general agreement among the investigators that the oxygen supply rate is perhaps the most critical criterion for the performance of butanediol fermentation (see Ref. 53 for detail).

In addition to DO, the pH environment also exerts great influence on the product distribution and yield. In enterobacteria, the type and ratio of fermentation products formed depend on two biochemical constraints: the need to maintain redox balance, and the requirement for maintaining the pH of the medium in a physiological range. Therefore, in a different pH environment, different relative amounts of end products are formed, resulting in an increased proportion of neutral compounds and decreased pH of the medium (53). Under a neutral or high-pH environment, bacteria tend to utilize a carbon source for cell growth and organic acid production. It has been shown that when the pH is lower than 6, over 50% of the pyruvate was channeled into butanediol production. Under this condition, enzymes of the butanediol pathway can constitute as much as 2.5% of the total protein in *E. aerogenes* (54).

Zeng et al. (55), investigated the effect of pH on growth and product formation from glucose by *E. aerogenes* in a continuous culture operation and found the pH range of 5.5–6.5 is optimal for butanediol and acetoin production. A similar pH optimum was also observed in *K. oxytoca*. In general, the biomass concentration increases steadily with increased pH. At high pH, butanediol concentration decreases and there is a concomitant increase of acetic acid production. Acetic acid has a dual role in the regulation of butanediol formation. It serves as activator for butanediol accumulation at low concentration. At a concentration of 10 g/L or higher, it inhibits butanediol production (56). The strength of acetic acid inhibition depends on the concentration of its undissociated form, HAc. and the concentration of HAc is in turn determined by the pH.

The production of butanediol from lignocellulosic materials has been considered as an alternate approach in the conversion of biomass substrates to liquid fuels and chemical feedstocks. Over the years, there have been many studies utilizing agricultural residues for butanediol production. The materials studied include hardwood hemicellulose fraction (57,58), wheat and barley straws (59), corn stover (59), and corn cobs (60).

BIBLIOGRAPHY

1. T.E. Timell, *Adv. Carbohyd. Chem.* **19**, 247–253 (1964).
2. E.B. Cowling and T.K. Kirk, *Biotechnol. Bioeng. Symp.* **6**, 95–123 (1976).

3. C.S. Gong, L.F. Chen, M.C. Flickinger, and G.T. Tsao, *Adv. Biochem. Eng. Biotechnol.* **20**, 93–118 (1981).
4. J.P. Delgenes, R. Moletta, and J.M. Navarro, *Enzyme Microb. Technol.* **19**, 220–225 (1996).
5. R. Torget, P. Werdene, M. Himmel, and K. Grohman, *Appl. Biochem. Biotechnol.* **24/25**, 115–126 (1990).
6. D.J. Schnell, R. Torget, A. Power, P.J. Walter, K. Grohman, and N.D. Hinman, *Appl. Biochem. Biotechnol.* **28/29**, 87–97 (1991).
7. E.T. Papoutsakis, in O. Kitani and C.W. Hall eds., *Biomass Handbook*, Gordon and Breach Science Publishers, New York, 1989, pp. 271–286.
8. J.H. Litchfield, in M. Moo-Young ed., *Biomass Conversion Technology: Principles and Practice*, Pergamon Press, Oxford, 1987, pp. 113–122.
9. L. Olsson and B. Hahn-Hagerdal, *Enzyme Microb. Technol.* **18**, 312–331 (1996).
10. J.A. Barnett, *Adv. Charbohydr. Chem. Biochem.* **32**, 125–234 (1976).
11. A. Margaritis and P. Bajpai, *Appl. Environ. Microbiol.* **44**, 1039–1041 (1982).
12. J.M. Laplace, J.P. Delgenes, R. Moletta, and J.M. Navarro, *Appl. Microbiol. Biotechnol.* **36**, 158–162 (1991).
13. S. Palnitkar and A. Lachke, *Can. J. Microbiol.* **38**, 258–260 (1992).
14. L.J. Wickerham, in J. Lodder ed., *The Yeasts*, North Holland, Amsterdam, 1971, p. 448–454.
15. H. Schneider, P.Y. Wang, Y.K. Chan, and R. Maleszka, *Biotechnol. Lett.* **3**, 89–92 (1981).
16. P.J. Slininger, R.J. Bothast, J.E. van Cauwenberge, and C.P. Kurtzman, *Biotechnol. Bioeng.* **24**, 371–384 (1982).
17. J.C. du Preez and J.P. van der Walt, *Biotechnol. Lett.* **5**, 357–362 (1983).
18. J.C. du Preez, M. Bosch, and B.A. Prior, *Appl. Microbiol. Biotechnol.* **25**, 521–525 (1987).
19. J.C. du Preez, B. van Driessel, and B.A. Prior, *Appl. Microbiol. Biotechnol.* **30**, 53–58 (1989).
20. J.R. Kastner, M. Ahmad, W.J. Jones, and R.S. Roberts. *Biotechnol. Bioeng.* **40**, 1282–1285 (1992).
21. A. Toivola, D. Yarrow, E. van den Bosch, J.P. van Dijken, and W.A. Scheffers, *Appl. Environ. Microbiol.* **47**, 1221–1223 (1984).
22. J.C. DuPreez, B. van Driessel, and B.A. Prior, *Appl. Microbiol. Biotechnol.* **30**, 53–58 (1989).
23. P.J. Slininger, L.E. Branstrator, R.J. Bothast, W.R. Okos, and M.R. Ladisch, *Biotechnol. Bioeng.* **37**, 973–960 (1991).
24. R. Amore, M. Wilhelm, and C.P. Hollenberg, *Appl. Microbiol. Biotechnol.* **30**, 351–357 (1989).
25. C.S. Gong, L.F. Chen, M.C. Flickinger, L.C. Chiang, and G.T. Tsao, *Appl. Environ. Microbiol.* **41**, 430–436 (1981).
26. P.Y. Wang, C. Shopsis, and H. Schneider, *Biochem. Biophys. Res. Commun.* **94**, 248–254 (1980).
27. L.C. Chiang, C.S. Gong, L.F. Chen, and G.T. Tsao, *Appl. Environ. Microbiol.* **42**, 284–289 (1981).
28. P. Kotter and M. Ciriacy, *Appl. Microbiol. Biotechnol.* **38**, 776–783 (1993).
29. M.T. D'Amore, C.J. Panchal, J. Russell, and G.G. Stewart, *Crit. Rev. Biotechnol.* **9**, 287–305 (1990).
30. M.S. Krishnan, N.W.Y. Ho, and G.T. Tsao, in B.C. Saha, and J. Woodward eds., *Fuels and Chemicals from Biomass*, ACS Symposium Series 666, American Chemical Society, Washington, D.C., 1997, pp. 74–92.
31. M. Moniruzzaman, B.S. Dien, C.D. Skory, Z.D. Chen, R.B. Hespell, N.W.Y. Ho, B.E. Dale, and R.J. Bothast, *World J. Microbiol. Biotechnol.* **13**, 341–346 (1997).
32. K. Ohta, D.S. Beall, J.P. Mejia, K.T. Shanmugam, and L.O. Ingram, *Appl. Environ. Microbiol.* **57**, 2810–2815 (1991).
33. J.S. Tolan and R.K. Finn, *Appl. Environ. Microbiol.* **53**, 2039–2044 (1987).
34. F. Alterthum and L.O. Ingram, *Appl. Environ. Microbiol.* **55**, 1943–1948 (1989).
35. K. Ohta, F. Alterthum, and L.O. Ingram, *Appl. Environ. Microbiol.* **56**, 463–465 (1990).
36. M. Zhang, C. Eddy, K. Deanda, M. Finkelstein, and S. Picataggio, *Science* **267**, 240–243 (1995).
37. P. Ueng and C.S. Gong, *Enzyme Microb. Technol.* **4**, 169–171 (1982).
38. M.G. White and J.J. Williams, *Biochem. J.* **22**, 583–591 (1928).
39. L.F. Chen and C.S. Gong, *J. Food Sci.* **50**, 226–228 (1985).
40. E. Winkelhausen, P. Pittman, S. Kuzmanova, and T.W. Jeffries, *Biotechnol. Lett.* **18**, 753–758 (1996).
41. V. Meyrial, J.P. Delgenes, R. Moletta, and J.M. Navarro, *Biotechnol. Lett.* **13**, 281–286 (1991).
42. V. Nolleau, L. Preziosi-Belloy, J.P. Delgenes, and J.M. Navarro, *Curr. Microbiol.* **27**, 191–197 (1993).
43. H. Horitsu, Y. Yahashi, K. Takamizawa, K. Kawai, T. Suzuki, and N. Watanable, *Biotechnol. Bioeng.* **40**, 1085–1091 (1992).
44. J.C. Roseiro, M.A. Peito, F.M. Girio, and T. Amaral-Collaço, *Arch. Microbiol.* **156**, 484–490 (1991).
45. J. Lu, L.B. Tsai, C.S. Gong, G.T. Tsao, *Biotechnol. Lett.* **17**, 167–170 (1995).
46. M.J. Pfeifer, S.S. Silva, M.G.A. Felipe, I.C. Roberto, and I.M. Mancilha, *Appl. Biochem. Biotechnol.* **57/58**, 423–430 (1996).
47. J.C. Parajo, H. Dominguez, and J.M. Dominguez, *Bioprocess Eng.* **13**, 125–131 (1995).
48. I.C. Roberto, M.G.A. Felipe, I.M. Mancilha, M. Vitola, S. Sato, and S.S. Silva, *Bioresource Technol.* **51**, 255–257 (1995).
49. S.K. Garg and A. Jain, *Biosource Technol.* **51**, 103–109 (1995).
50. N.B. Jansen, and G.T. Tsao, *Adv. Biochem. Eng. Biotechnol.* **27**, 85–99 (1983).
51. R.J. Magee and N. Kosaric, *Adv. Appl. Microbiol.* **32**, 89–161 (1987).
52. D. Mayer, V. Schlensog, and A. Bock, *J. Bacteriol.* **177**, 5261–5269 (1995).
53. I.R. Booth, *Microbiol. Rev.* **49**, 356–378 (1985).
54. L. Johansen, K. Bryn, and F.C.J. Stormer, *J. Bacteriol.* **123**, 1124–1130 (1975).
55. A.P. Zeng, H. Biebl, and W.D. Deckwer, *Appl. Microbiol. Biotechnol.* **33**, 485–489 (1990).
56. E.K.C. Yu and J.N. Saddler, *Appl. Environ. Microbiol.* **44**, 777–784 (1982).
57. F.R. Frazer, and T.A. McCaskey, *Enzyme Microb. Technol.* **13**, 110–115 (1991).
58. B.P. Grover, S.K. Garg, and J. Verma, *World J. Microbiol. Biotechnol.* **6**, 328–332 (1990).
59. E.K.C. Yu, L. Deschatelets, N. Levitin, and J.N. Saddler, *Biotechnol. Lett*, **6**, 611–614 (1984).
60. N. Cao, Y. Xia, C.S. Gong, and G.T. Tsao, *Appl. Biochem. Biotechnol.* **63/65**, 129–139 (1997).

HUMAN AND PRIMATE CELL LINES

SALLY WARBURTON
ECACC
Wiltshire, United Kingdom

KEY WORDS

Adenovirus
Bioreactor
Carcinoma
Cytokine
Diploid
Glycosylation
Immortalization
Oncogene
Transformation
Vaccine

OUTLINE

INTRODUCTION

Before the advent of monoclonal antibodies and genetically manipulated cell lines, the major use of primate and human cells was as a substrate for viral vaccine manufacture (e.g., polio, rubella, and measles in human diploid cells MRC 5 or WI 38 [1]).

It was not until the mid 1970s that it was considered that continuous cell lines may also prove useful for vaccine production. At the same time, there was a great deal of interest in the use of interferon in clinical studies. The major source was from primary human leukocyte cultures, but the amount produced was very limited.

Exploitation of animal cell lines in the field of biotechnology as opposed to bacteria, filamentous fungi, and yeast has occurred relatively recently because of complex culture media, slow population doubling times, and low productivity levels of cells. The ability of animal cells to glycosylate proteins in an active form, unlike bacteria, has led to new developments in manufacturing, purification, and recombinant DNA technology in the engineering of cells for production of therapeutic proteins. Strict safety standards must be set for any biological manufacturing process involving eukaryotic protein production, research into the adaptation of suitable cell lines, the latest bioreactor development, immortalization methods, and cell products are discussed in this article.

HUMAN CELL LINES

Human Diploid Cells

In the 1950s there was an urgent need to produce a vaccine to combat polio, which led to the use of primary monkey kidney cells for growing the virus. This was successful until it was discovered after a large amount of vaccine had been released that monkeys were infected with wild simian viruses (e.g., SV40) (2).

The unreliability of primary cell culture led to the development of human diploid cells, such as MRC 5 and WI 38 (1), and human fetal lung fibroblasts, which were not infected by SV 40 and are now the preferred substrate.

Human diploid cells are particularly susceptible to viruses pathogenic to humans (e.g., polio, echo, arbo, rhino, measles, rubella, and coxsackie A9) and are very valuable for their isolation and development of vaccines (1–3).

These two cell lines are nontumorigenic, essential when considering using human diploid cells for live virus vaccines in humans (4); have a finite life span (because they have no malignant characteristics) (5); and begin to senesce after passage (5). These cells must be cryopreserved to prevent cross-contamination, genetic drift, microbial contamination, and maintenance of low passage stocks (6–9).

MRC 5 and WI 38 are free of latent and oncogenic viruses and are genetically stable; therefore, they are the permitted cell lines used for producing products such as vaccines for human use (5), although MRC 5 is preferred because of its higher replication rate and better tolerance to environmental changes (1).

Human Carcinomas

Useful in the study of cancer and drug therapy, human carcinomas are also useful models for the study of normal and pathophysiological events in vivo (10). Two human colon cell lines particularly useful are CaCo 2 and HT 29 because they undergo enterocyte differentiation in vitro (11). They have been widely used for studies on the barrier properties of the human intestine for drug transport and the cytoprotective activity of potential drugs and their pharmaceutical formulations (12).

A rotating wall vessel (RVW) designed to simulate cell cultivation conditions on earth in the microgravity of space has been set up using these cell lines (13). Reduced shear forces allow the cells to grow in three-dimensional form, to differentiate and to mimic the in vivo tissues. This type of bioreactor could provide an environment for cell cultures, enabling novel cell aggregate configurations that may be usefully applied in the field of pathogenesis of infection and tissue engineering.

PRIMATE CELL LINES

Vero

The Vero cell line was established in 1962 to overcome the problem of using primary monkey kidney cells that were infected with the SV 40 virus (14–16). The use of primary tissue for poliovirus vaccine became more difficult because of international agreements on the conservation of wild animals. The Institute Merieux in France extensively tested the Vero cell line, demonstrating that Vero cells show no tumorigenicity, contain no aberrant viruses, and sustain efficient proliferation of various vaccine viruses (16). As a result, killed polio vaccine using Vero cells instead of primary monkey kidney cells has been available in France since 1984. Production is carried out on microcarriers in large-scale bioreactors.

Vero is a classical cell line that is still used as a capable substrate for virus reproduction of a wide variety of completely unrelated viruses, such as arboviruses, hemorrhagic fever viruses, rubella, measles, herpes, and growth of rickettsiae. (Unfortunately, Vero cells do not culture wild-type influenza viruses, which still need to be cultured on primary monkey kidney cells for identification.) Vero cells have been one of the most powerful basic resources for the entire field of animal virology in the past quarter of a century and will be a major cell substrate for virology and biotechnology in the years to come.

Other primate cell lines, such as COS 1 and COS 7 African green monkey kidney cells transformed with the SV 40 virus, have been found along with Vero cells to express HIV-1-like particles once transfected (17). This important development offers an alternative to the use of live virus vectors for the production and evaluation of AIDS vaccines based on noninfectious particles.

Table 1 summarizes some of the versatile human and primate cell lines that are currently being used in the biotechnology industry and the products they have to offer.

IMMORTALIZATION

Primary cell lines can be maintained in vitro for a limited period. As cells approach senescence (34), they are unable to synthesize DNA. They reach a crisis point where cell division ceases and the cells die, this is known as apoptosis, or programmed cell death.

The ability to develop immortalized cells in vitro is a powerful tool for the biological investigator. The immortalization of cells establishes continuous cell lines that play an important role in the study of the biology of cell growth, differentiation, and senescence. Large-scale culture of immortal cell lines enables the production of large quantities of DNA or any cell product for research or for the pharmaceutical industry.

The development of the biotechnology industry and the requirement for eukaryotic cells in the post-translational processing of protein products of cells has placed a great deal of pressure on the scientific and regulatory communities to provide continuous cell lines from specific cell lineages that are safe to use and with phenotypic properties relevant to their tissue of origin.

Table 2 provides the most up-to-date examples of the in vitro methods used in the immortalization of primary cultures of various cell types.

BIOREACTORS AND PRODUCTS

Bioreactors are used to obtain the maximum quantity of protein required. Fermenters are used for cell lines that grow in suspension or for high-density production of cells (e.g., HIV-1 and HIV-2 in human T lymphoblastoid cells). Fixed-bed bioreactors are advantageous because they provide close cell-to-cell contact, which is important for virus transfer (18). Porous glass spheres in fixed-bed systems retain nonadherent cells and provide large quantities of HIV.

For attached cell lines (e.g., 293, human embryo kidney) (29), solid-support bioreactors have been developed either by a stirred-tank reactor using microcarriers (55) or hollow-fiber culture systems (56). For example, the polio vaccine is prepared in Vero cells grown on microcarriers in 1,000-L tanks (16).

Medium replacement is important when growing cells on a solid support or in suspension to sustain growth and protein production. A constant supply of growth factors must be supplied while reducing buildup of inhibitory metabolites such as lactic acid and ammonia (29). Serum-free formulations must be considered when dealing with extracellular protein purification from mammalian cells. Other environmental aspects to be considered for efficient management of bioreactors are low shear forces, high transfer rates of oxygen and carbon dioxide, and a controlled microenvironment.

Examples of new bioreactors are the Technomouse, which combines the principle of nutrient and oxygen perfusion via silicone membranes and organlike cultivation of human cells (long-term primary tissue culture is possible) (56). The VERAX process (57) uses a fluidized bed of macroporous collagen carriers, which is particularly suitable for the production of secreted recombinant products, such as Aujeszky herpesvirus in Vero cells.

Products

Interferon was the first cytokine to be made from cultured animal cells, and it is now prepared routinely at more than 95% purity, showing that scale-up to 10,000 L is possible and economically viable (26). Originally produced from human white blood cells, only small amounts could be obtained. NAMALWA, a transformed B-lymphoblastoid cell line, produces plentiful amounts of interferon and is safe to use on humans. The final product (Wellferon) is of high purity, correctly glycosylated, and contains at least six distinct α-interferons, unlike that produced from *Escherichia coli* by recombinant DNA technology (27).

Fibronectin, an effective factor for the growth of anchorage-dependant cells, is produced by HUH 6 (Table 1), a human hepatoblastoma cell line. At more than 90%

Table 1. Human and Primate Cell Lines

Cell line name	Cell line type	Product	Use	Reference
U 937 THP 1 JM, C 8166 JHAN	Human T-cell lymphoblast	Supports growth of HIV-1 and HIV-2	Vaccine production and research	18–22
RPMI 2650	Human carcinoma	Growth factors autocrine-stimulator, TGF-α, and TFG-β	Regulation of cell growth and productivity in vitro of human epithelial cells	23
HUH clone 5	Human hepatoblastoma	Fibronectin	Attachment factor in serum-free media	24
NAWALWA	Human lymphoblastoid	IPSF, IIα	Improves immuno-globulin productivity from human hybridomas	25
NAMALWA	Human lymphoblast	Interferon α when infected with sendai virus and sodium butyrate	Treatment of hairy cell leukemia potentiates cytotoxic T cell and natural killer cell response	26,27
293	Human embryo kidney	Supports growth of adenovirus vectors that carry reporter genes into neuronal cells Acetylcholin, esterase	Gene therapy in cystic fibrosis Structure and function analysis	28
293 transformed with human parathyroid calcium receptor	Human embryo kidney	Recombinant protein (similar to human plasma factor X)	Structural studies and clinical evaluation and growth of viruses	29–31
HepG 2 PLC/PRF/5 Hep3B	Human hepatocellular carcinomas	Human plasma proteins, e.g., transferrin, plasminogen, α-fetoprotein, albumin (PLC/PRF/5-albumin not produced)	Experimental models for investigation of plasma protein biosynthesis and synthesis of hepatitis B surface antigen HbsAg (not HepG2)	32,33
Vero	African green monkey kidney	Aujeszky virus and oral polio virus	Vaccine production	16

TFG = transforming growth factor.

purity, its biological activity is the same as commercial fibronectin from hamster kidney cells, BHK 21 (24), but only 2 L of HUH-6 will produce up to 27 mg of protein.

SAFETY CONSIDERATIONS

Any cell line used for the propagation of a specific product for the pharmaceutical industry must be vigorously controlled. Fully authentic cell banks provide that safety measure and are stored in liquid nitrogen for indefinite periods. We only need to recall the first cell line to be isolated, HeLa in 1950, deemed to be the single most important tool of biomedical research in the past 50 years. It was sent all over the world, resulting in mass cross-contamination and millions of dollars of medical research being lost.

Guidelines have been set for the acceptability of using human diploid cells for the production of live virus vaccines in humans (58) to ensure that there are no extraneous microbial contaminants or transforming factors (59,60). The increasing use of animal cells for the production of recombinant proteins and vaccines means that standardization in this field is ever more important to ensure the provision of safe and reliable therapeutic diagnostic reagents (61,62).

A further major concern is the long-term risk of malignancy represented by heterogeneous contaminating DNA, especially if it were to contain potentially oncogenic sequences (e.g., hepatitis B vaccine produced by recombinant DNA techniques). The World Health Organization (WHO) has published strict guidelines on the use of cell lines for vaccine production (63).

FUTURE TRENDS

Gene Therapy

Recently, treatment of human genetic disorders has been mediated by gene transfer technology, especially for monogenic disorders such as cystic fibrosis. Research has been carried out into the feasibility of using a defective nonreplicating recombinant adenovirus vector to transfer a reporter gene (e.g., β-galactosidase) into neuronal cells (28). Human embryonal kidney cells 293 were used as a packaging cell and cotransfected with the vector, which could then be inserted into cystic fibrosis patients. Gene therapy replaces the specific gene sequence into a target cell (28).

If this is successful, this may lead to an improvement in controlling monogenic diseases in which only a minority of cells are required for genetic alteration in order to replace a deficient enzyme.

Bioartificial Liver

Liver failure is the major cause of mortality in the United States, with 30,000 deaths per year. There are no artificial devices available, transplants are effective but donors are limited.

Table 2. In Vitro Methods

Cell line type	Method of transformation	Uses	References
Human umbilical vein endothelial cells	Low-level radiation, oncogenes, SV40, spontaneous somatic cell hybridization with a human osteosarcoma	Von Willebrand factor, prostacyclin (e.g., ECV304 and EA.hy926)	35–39
HPEC A1, human placental cells	Plasmic pRNS1 lipofection	Endothelial characteristics after 80 cell divisions	40–43
Human B lymphocytes	Epstein-Barr virus	Express normal B-cell markers, e.g., HLA class I and II antigens, tissue typing, Genetic research	44,45
Human dermal fibroblasts	SV40 T antigen	Oncogenic transformation, aging, senescence	48
Monkey kidney epithelium 47/17/3, 42/17/8, ori-11/7	Genetic engineering, oncogenes, polyoma large T, H-ras	Alternative to primary cell culture	49–51
Human embryo kidney 293	Sheared human adenovirus 5 DNA	Sensitive to human adenoviruses; isolation of transformation defective host range mutants of adenovirus type 5 and titration of adeno type 21	52,53
293N3S	Produced by passaging 293 in nude mice	Large-scale production of viruses in bioreactors	54

A temporary support to support biotransformation and detoxification functions is required to sustain patients waiting for a donor or until the liver has repaired itself.

Development of a hollow fiber reactor using hepatocyte entrapment with rabbit and rat models has been successful, and scale-up is now being developed for human clinical trials (64).

CONCLUSION

The products and cell lines of human and primate origin discussed in this article have contributed to the general acceptance of using animal cells as the substrate for production of therapeutic proteins.

Many proteins are now obtained by expressing the gene concerned in Chinese hamster ovary cells by recombinant DNA technology, but human and primate cell lines are used as well. Many proteins will always be made more simply in bacteria, yeast, or insect cells and in the future by chemical synthesis. But until then, where glycosylation of proteins is necessary and chemical refolding of bacterial products is not always possible for full pharmaceutical activity, mammalian cells will still be used, especially with the constant improvement in protein-free and serum-free formulations and improvement in higher cell densities through bioreactor research.

BIBLIOGRAPHY

1. J.P. Jacobs, C.M. Jones, and J.P. Baille, *Nature* **227**, 168–170 (1970).
2. L. Hayflick, S.A. Plotkin, T.W. Norton, and H. Koprowski, *Am. J. Hyg.* **75**, 240 (1962).
3. D. Ikie, *Prog. Immunobiol.* **2**, 305 (1965).
4. J.P. Jacobs, D.I. Magrath, A.J. Garrett, and G.C. Schild, *J. Biol. Standardisation* **9**, 331–342 (1981).
5. J.B. Griffiths, in R.E. Spier and J.B. Griffiths eds., *Animal Cell Biotechnology*, Academic Press, London, 1985, pp. 3–11.
6. A. Doyle, C.B. Morris, and W.J. Armitage, *Cryopreservation of Animal Cells: Upstream Processes, Equipment and Techniques*, Liss, New York, 1988, pp. 1–17.
7. G.N. Stacey, *Folia Microbiol* **42**, 113–116 (1997).
8. A. Doyle, B.J. Bolton, *Animal Cell Technology*, New York, Wiley, pp. 244–271.
9. R.J. Hay, Y.A. Reld, P.R. McClintock, T.R. Chen, and M.L. Macey, *J. Cell. Biochem.* **Suppl. 24**, 107–130 (1996).
10. P. Arturson, *Crit. Rev. Therapeutic Drug Carrier Systems* **8**, 305–330 (1991).
11. M. Pinto, M.D. Appay, and P. Simon-Assman, *Biol. Cell.* **44**, 193–196 (1982).
12. J. Fogh, W.C. Wright, and J.D. Loveless, *J. Natl. Cancer. Inst.* **58**, 209–214 (1977).
13. J.R. Latigo, T.P. Battle, R. Cook, P. Freemantle, and G.N. Stacey, in M.J. Corrondo, B. Griffiths, and L.P. Moreira eds., *From Vaccines to Genetic Medicine*, ESACT, Kluwer, London, 1996, pp. 385–389.
14. P.D. Minor, in R.E. Spiers ed., *Animal Cell Technology: Products of Today, Prospects for Tomorrow*, ESACT, Butterworth-Heinemann, Oxford, U.K., 1994, pp. 741–748.
15. K. Shaw and N. Nathanson, *Am. J. Epidemiol.* **103**, 1–12 (1976).
16. B.J. Montagnon, B. Fanget and J.C. Vincent-Falquet, in B. Simizu and T. Terasima eds., *Vero Cells: Origin, Properties and Biochemical Applications*, Chiba Univ., Tokyo, 1988, pp. 181–210.
17. Haynes, *AIDS Res. Hum. Retroviruses* **7**, 17–27 (1991).
18. J.B. Clark and J.B. Griffiths, in R.E. Spiers ed., *Production of Biologicals from Animal Cells in Culture*, ESACT, Butterworth-Heinemann, Oxford, U.K., 1990, pp. 519–522.
19. C. Sundstrom and K. Nilsson, *Int. J. Cancer* **17**, 565–577 (1976).
20. P. Ralph, M. Moore, and K. Nilsson, *J. Exp. Med.* **143**, 1528–1533 (1976).
21. H.S. Koren, S.J. Anderson, and J.W. Larrick, *Nature* **279**, 328–331 (1979).
22. D.G. Fischer, M.C. Pike, H.S. Koren, and R. Snyderman, *J. Immunol.* **125**, 463–465 (1980).

23. M. Clynes and M. Dooley, in R.E. Spiers ed., *Production of Biologicals from Animal Cells in culture*, ESACT, Butterworth-Heinemann, Oxford, U.K., 1990, pp. 519–522.
24. K. Nagamine, M. Shiraishi, Z. Kong, K. Shinohara, and H. Murakami, in R.E. Spiers ed., *Production of Biologicals from Animal Cells in culture*, ESACT, Butterworth-Heinemann, Oxford, U.K., 1990, pp. 167–169.
25. H. Murakami, in R.E. Spiers ed., *Production of Biologicals from Animal Cells in culture*, ESACT, Butterworth-Heinemann, Oxford, U.K., 1990, pp. 735–740.
26. N.B. Finter, in R.E. Spiers ed., *Production of Biologicals from Animal Cells in culture*, ESACT, Butterworth-Heinemann, Oxford, U.K., 1990, pp. 3–12.
27. R.G. Werner, in R.E. Spiers ed., *Animal Cell Technology: Products of Today, Prospects for Tomorrow*, ESACT, Butterworth-Heinemann, Oxford, U.K., 1994, pp. 369–375.
28. S.D. Sivasubramaniam and G.N. Stacey, in M.J.T. Corrondo, B. Griffiths, and L.P. Moreira eds., *From Vaccines to Genetic Medicine*, ESACT, Kluwer, London, 1996, pp. 385–389.
29. J. Shiloach, J. Kaufman, L. Trinh, and C. Kemp, in M.J.T. Corrondo, B. Griffiths, and L.P. Moreira eds., *From Vaccines to Genetic Medicine*, ESACT, Kluwer, London, 1996.
30. F.L. Graham, *J. Gen. Virol.* **68**, 937–940 (1987).
31. A.E. Rudolf, M.P. Mullane, R. Porche-Sorbet, and J.P. Miletich, *Protein Express. Purif.* **10**, 373–378 (1997).
32. B.B. Knowles, C.C. Howe, and D.P. Aden, *Science* **209**, 497–499 (1980).
33. D.P. Aden, A. Fogel, I. Damjanor, S. Plotkin, and B.B. Knowles, *Nature (London)* **282**, 615 (1979).
34. L. Hayflick and P.S. Moorhead, *Exp. Cell. Res.* **25**, 585 (1961).
35. N.A. Punchard, D. Watson, R. Thompson, and M. Shaw, in R.I. Freshney and M.G. Freshney eds., *Culture of Immortalized Cells*, Wiley, New York, 1996, pp. 203–238.
36. K. Takahasi and Y. Sawasaki, *In Vitro Cell Dev. Biol.* **26**, 265 (1990).
37. K. Takahasi and Y. Sawasaki, *In Vitro Cell Dev. Biol.* **27A**, 766 (1991).
38. K. Takahasi and Y. Sawasaki, *In Vitro Cell Dev. Biol.* **28A**, 380 (1992).
39. O. Hohenarter, C. Schmatz, and H. Katinger, in M.J. Carrondo, J.B. Griffiths, and J.L.P. Moreira eds., *Animal Cell Technology: From Vaccines to Genetic Medicine*, ESACT, Kluwer, The Netherlands, 1997, pp. 9–11.
40. P. Litzkas, K.K. Jha, and H.L. Ozer, *Mol. Cell. Biol.* **4**, 2549–2552 (1984).
41. M. Teifel, L.T. Heine, S. Milbredt, and P. Friedl, *Eur. J. Biochem.* 1996, submitted for publication.
42. M. Schutz and P. Friedl, *Eur. J. Cell. Biol.* 1996, submitted for publication.
43. M. Schultz and P. Friedl, in M.J. Corrando, J.B. Griffiths, and J.L.P. Moreira eds., *Animal Cell Technology: From Vaccines to Genetic Medicine*, ESACT, Kluwer, The Netherlands, 1997, pp. 749–753.
44. A.H. Davies, F.J. Evans, A.B. Rickinson, J. Huddleston, and A.N. Emery, in R.E. Spiers ed., *Production of Biologicals from Animal Cells in Culture*, ESACT, Butterworth-Heinemann, Oxford, U.K., 1990.
45. B.J. Bolton and N.K. Spur, in R.I. Freshney and M.G. Freshney eds., *Culture of Immortalised Cells*, Wiley, New York, 1996, pp. 283–298.
46. C. Macdonald and B. Bolton, in A. Doyle and J.B. Griffiths eds., *Mammalian Cell Culture*, Wiley, W. Sussex, U.K., 1997, pp. 154–167.
47. M.A. Epstein and B.G. Achong, *The Epstein Barr Virus*, William Heinemann, London, 1986, pp. 2–11.
48. L.V. Mayne, T.N.C. Price, K. Moorwood, and J.F. Burke, in R.I. Freshney and M.G. Freshney eds., *Culture of Immortalized Cells*, Wiley, New York, 1996, pp. 77–94.
49. H. Land, L.F. Parada, and R.A. Weinberge, *Nature* **304**, 596–598 (1983).
50. M. Rassoulzadegan, A. Cowrie, A. Carr, F. Cuzin, R. Kamen, and N. Glaichenhaus, *Nature* **300**, 713–716 (1982).
51. J. Clarke, C. Macdonald, U. Kreuzburg-Duffy, and J.B. Griffiths, in R. Spier, J.B. Griffiths, and W. Berthold eds., *Animal Cell Technology: Products of Today, Prospects for Tomorrow*, ESACT, Butterworth-Heinemann, Oxford, U.K., 1994, pp. 50–51.
52. F.L. Graham, J. Smiley, W.C. Russell, and R. Naim, *J. Gen. Virol.* **36**, 59–72 (1977).
53. T. Harrison, F. Graham, and J. Williams, *Virology* **77**, 319–329 (1977).
54. F.L. Graham, *J. Gen. Virol.* **68**, 937–940 (1987).
55. S. Reuveny, in A.S. Lubiniecki ed., *Large Scale Mammalian Cell Culture Technology*, Dekker, New York, 1990, pp. 271–341.
56. T. Omasa, *Biotechnol. Bioeng.* **48**, 673–680 (1995).
57. T. Marique, D. Malarme, P. Stragier, and J. Werenne, in R.E. Spiers ed., *Animal Cell Technology: Products of Today, Prospects for Tomorrow*, ESCAT, Butterworth-Heinemann, Oxford, U.K., 1994, pp. 273–277.
58. J.P. Jacobs, D.I. Magrath, A.J. Garrett, and G.C. Schild, *J. Biol. Standard.* **9**, 331–342 (1981).
59. R.J. Hay, Y.A. Reid, P.R. McClintock, T.R. Chen, and M.L. Macey, *J. Cell. Biochem.* **Suppl. 24**, 107–130 (1996).
60. R.J. Hay, *Anal. Biochem.* **171**, 225–237 (1988).
61. G.N. Stacey, *Folia. Microbiol.* **42**, 113–116 (1997).
62. A. Doyle and B.J. Bolton, in Wiley, London, 1992, pp. 243–271.
63. J. Petricciani, in R.E. Spier and J.B. Griffiths eds., *Animal Cell Biotechnology*, vol. 3, 1988, pp. 14–24.
64. M.V. Peshwa, S.L. Nyberg, F.J. Wu, B. Amiot, F.B. Cerra, and W.S. Hu, in R.E. Spiers ed., *Animal Cell Technology: Products of Today, Prospects for Tomorrow*, ESACT, Butterworth-Heinmann, Oxford, U.K., 1994, pp. 273–277.
65. A. Nagel, E. Effenberger, S. Koch, L. Lubbe, and U. Marx, in R.E. Spiers ed., *Animal Cell Technology: Products of Today, Prospects for Tomorrow*, ESACT, Butterworth-Heinmann, Oxford, U.K., 1994, pp. 296–304.

HYBRIDOMA, ANTIBODY PRODUCTION

FRANCISCO J. CASTILLO
Berlex Biosciences
Richmond, California

KEY WORDS

Antibody
Ascites
Bioreactor
Hybridoma
MAb

Monoclonal
Myeloma

OUTLINE

INTRODUCTION

The creation of hybridomas capable of producing monoclonal antibodies (MAbs) was accomplished by Koller and Milstein in 1975 (1). Two decades later, the number of MAbs used in research, purification, and immunoassay applications is large and has sustained continued growth (2–5). The medical applications for MAbs have not lived up to expectations, lagging behind other biologicals, because only a relatively small number have received approval by the FDA (6–8) (Fig. 1), four for therapeutic and five for in vivo diagnostic use (Table 1). However, this could change rapidly in the near future. In 1996, of the 12 biologics approved by the FDA, four were MAbs (9). Two MAbs were approved in 1997 (10), one of which (Zenapax) was the first approved humanized antibody. Three Biologic License Applications (BLAs) for MAbs were also filed in 1997 (11–13), and one was filed in May 1998 (14), for a total of six under review by FDA at that time. In addition, of the 350 biopharmaceutical products in development, more than 70 are MAbs, and several are in pivotal clinical trials near completion (15–18) (Fig. 2). Thus, the number of fillings of BLAs with the FDA and international regulatory agencies for MAb-based products could see significant growth in the future, to be followed by corresponding approvals.

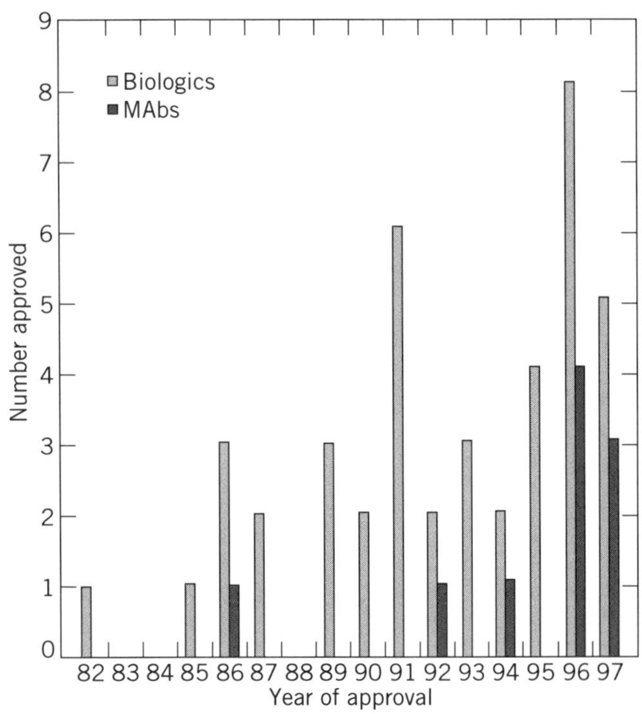

Figure 1. MAbs and other biologics approved by the FDA.

MAbs are routinely produced in vivo and in vitro. For in vivo production, mice (mostly BALB/c) are the preferred species. The abdominal cavity of pristane-primed mice is used to incubate injected hybridoma cells, which grow and secrete MAb that concentrates in the ascitic fluid. The MAb-rich ascites is collected by repeated needle aspiration (tapping) of the animals (19,20).

With the advent of mammalian cell culture scale-up for the production of recombinant proteins during the 1980s (21–26), concomitant with the high expectations regarding the potential uses and demand for MAbs (27–31), the cultivation of hybridoma cells was intensively studied. This lead to the development of a variety of bioreactors, culture media, and culture processes (31–44), many of which, when properly combined and used, easily generate required quantities of high-quality material at relatively low cost.

Genetically engineered MAbs, such as chimeric, humanized, MAb fragments and fusion proteins, are produced in mammalian and insect cells, microorganisms, and transgenic animals and plants (45–57).

Production of human MAbs using human cells still presents important challenges, although progress has been made (58–60). More significant advances have been made with the use of bacteriophage display technique and the development of transgenic mice, which express the human genes for antibody production (61,62). These animals when inoculated with antigens produce only human antibodies. Their esplenocytes can be isolated and fused with murine myelomas to produce stable hybridomas that produce human MAbs, although containing murine glycosylations instead of human. The time and procedures needed to generate these hybridomas are the same as for conventional murine hybridomas.

As clearly seen in Table 1, very different processes for MAb production have been made to work in cost-effective, validated fashion. The choice will depend on the intended application, scale required, personnel experience, and available infrastructure. With the available technology, even with the present limited level of understanding of murine hybridoma physiology, setting up a culture process for MAb production should be a relatively straightforward exercise.

HYBRIDOMA GROWTH AND MAb PRODUCTION

The performance of a process involving cultivation of hybridomas or any other cells will be the result of the com-

Table 1. Monoclonal Antibodies Approved by the FDA (by May 1998)

Product	Company	Indication	Approval date	Production method
Orthoclone OKT®3	Ortho Biotech, (Raritan, NJ)	Reversal of acute kidney transplant rejection	June 1986	Ascites (63)
		Reversal of heart and liver transplant rejection	June 1993	Ascites
OncoScint CR/OV	Cytogen (Princeton, NJ)	Detection, staging, and follow-up of colorectal and ovarian cancers	December 1992	Airlift fermentation
ReoPro	Centocor (Malvern, PA)	Antiplatelet prevention of blood clots in the setting of high-risk PTCA	December 1994	Extended perfusion (spin-filter) (44)
	Eli Lilly (Indianapolis, IN)	Prevention of cardiac ischemic complications in patients undergoing PTCA	November 1997	
CEA-Scan	Immunomedics (Morris Plains, NJ)	Metastatic colorectal cancer imaging	June 1996	Ascites
MyoScint	Centocor (Malvern, PA)	Myocardial necrosis imaging agent	July 1996	Extended perfusion (spin filter)
Verluma	Boehringer lingelheim (Ridgefield, CT) NeoRx (Seattle, WA)	Small cell lung cancer imaging agent	August 1996	
ProstaScint	Cytogen (Princeton, NJ)	Recurrent prostate cancer imaging agent	October 1996	Extended perfusion (hollow fiber) (64)
Rituxan	Idec Pharmaceuticals (San Diego, CA) Genentech, Inc. (S. San Francisco, CA)	Low-grade non-Hodgkin's lymphoma	November 1997	Suspension culture
Zenapax	Hoffman-La Roche, Inc. (Nutley, NJ)	Prevention of acute graft rejection in kidney transplant	December 1997	Suspension culture

Source: Based on Ref. 9.
Note: PTCA = percutaneous transluminal coronary angioplasty

bination of three main factors: the cell line, the culture medium, and the bioreactor conditions.

The Hybridoma Cell

Preparation of Hybridomas. Hybridomas are prepared by fusing splenocytes (antibody-producing spleen B lymphocytes) from antigen-treated animals and myelomas (1). Splenocytes do not propagate indefinitely in vitro but myelomas do, so when both types of cells are fused, the hybridomas formed are able to propagate and produce the antibodies. Many detailed protocols for preparation of hybridomas have been described in the literature (2,65–69).

There are two preferred and widely used murine myelomas that descend from a transplantable murine (BALB/c) plasmacytoma, MOPC 21, that was adapted to in vitro culture and renamed P3K (70). These two myelomas are P3-X63-Ag8.653 (71) and NSO/1 (65), and both have lost the ability to produce the heavy and light chains of the original MOPC 21 antibody (IgG1). A third nonproducing cell, Sp2/0-Ag14 (72), was obtained through several steps (73) and is itself a hybridoma derived from the fusion of myeloma P3-X63-Ag8X and splenocytes from a BALB/c mouse (74). The use of any of these cells guarantees that the antibodies produced by hybridomas derived from them will correspond only to those specified by the genes from the splenocytes used in the fusion. Monoclonality, however, will only be ensured when the proper procedures are used to clone and subclone the hybrids produced (68,75–79). Cloning and subcloning are also required to select stable clones that retain the antibody class and the specific productivity (see "Hybridoma Stability").

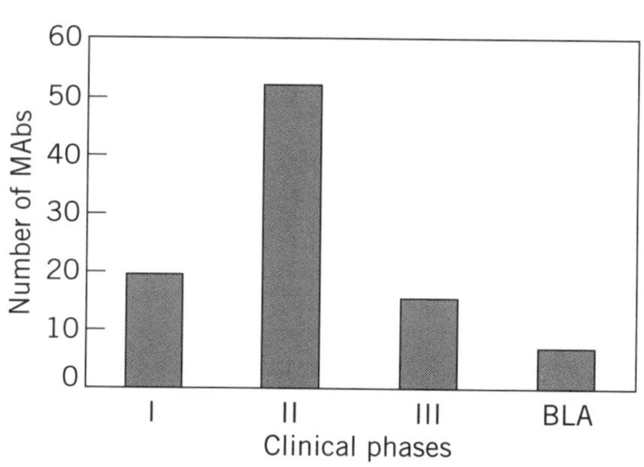

Figure 2. Number of MAbs on test at different stages of clinical trials.

Preparation and Characterization of Hybridoma Banks. After clone selection, it is important to ensure the availability of homogeneous and healthy cell stocks, and this is accomplished by preparing cell banks. For industrial applications, cell banks are prepared at several stages: research, master cell bank (MCB), manufacturer's working cell bank (MWCB), and end-of-production cells (EPC).

MCBs are intended for the preparation of MWCBs, which in turn, are prepared to ensure a supply of homogeneous cell seed for manufacturing purposes and for the life of the product, and EPC banks are for quality control and validation purposes (80,81). The successful preparation and maintenance of cell banks requires very healthy, exponentially growing cells, the use of fresh medium and appropriate quality and quantity of cryopreservant (usually, dimethylsulfoxide [DMSO]) at time of freeze, and storage in monitored liquid nitrogen freezers. Detailed procedures and protocols for preparation, freezing, and storage of hybridoma stock banks are well described in the literature (82–84). Characterization of the banks is required to ensure identity, consistent performance, and lack of contaminating adventitious agents. The characterization includes recoverability of the cells upon thawing, identification of cell species, growth and production kinetics of the cells, and testing for the presence of endogenous and adventitious infectious agents (Table 2).

Hybridoma Stability. Somatic variation (antibody class shift) resulting in structural MAb variants produced by hybridomas have been reported (85–87). This phenomenon occurs as a normal evolutionary event of the humoral immune response, but once stable hybridoma clones are selected its frequency is low. The most frequent instability problem in hybridomas seems to be a decrease in MAb yield that correlates with the emergence of nonproducing cells (88–94). Nonproducing hybridomas arise with the loss of chromosomes 12 for the heavy and 16 and 6 for the light, λ, and κ chains, respectively (95–98). Loss of chromosomes occurs with higher frequency shortly after fusion, so it is very important to clone and subclone to ensure the selection of stable clones. Interspecies hybrids such as murine/human or murine/rabbit have increased chromosome segregation compared to murine/murine, which results in more frequent instability problems (35,99).

For extended cultivation processes, it is important to evaluate the performance of the cells beyond the intended incubation time and to compare the quantity and quality of the MAb produced during the runs and to prepare EPC banks as mentioned above (100).

The Culture Medium

The culture medium has to provide, at a minimum, an energy source and all nutritional requirements of the cells.

Table 2. Characterization of Hybridoma Banks

Test	MCB	MWCB	EPC
Sterility	+	+	+
Mycoplasma	+	+	+
Growth and production kinetics	+	+	+
Authenticity	+	+	+
Species-specific viruses	|		
Retroviruses	+[a]	−	+[a]
Adventitious viruses in vitro assay	+	−	+
Adventitious viruses in vivo assay	+	−	+

Source: Ref. 80 and 81.

[a]Retrovirus testing is not required for murine hybridomas (81).

Energy Sources. Hybridoma cells use glucose and glutamine as the preferred sources of energy. Evidence suggests that glucose and glutamine use by mammalian cells in culture is under reciprocal regulation (101–103). Hybridomas are derived from tumor cells (myelomas) and as such metabolize glucose mainly through glycolysis (104). Many hybridomas generate lactate in quasi-stoichiometric levels compared to the glucose used; the lactate accumulates in the medium, and in later stages of the culture it can be used by the cells (105,106). Glucose utilization rates range from less than 1 to more than 10 pmol/cell/day (43); the sugar is required by hybridomas and growth does not occur in its absence.

Glutamine is also essential for growth and maintenance of cultures, and its utilization rate, which ranges from less than 1 to more than 5 pmol/cell/day (43), depends on the concentration. Glutamine addition has been shown to increase the longevity of cultures, which translates into higher MAb yields (107–109). Recent studies have shown that glucose and glutamine consumption can be redirected to minimize lactate and ammonia production by using a rational medium design and a feeding strategy. Application of such approaches in fed-batch culture has led to very high levels of cells and MAb (110).

Pyruvate is both produced and consumed by hybridomas at different rates (line dependent), but its addition to the medium is not required for growth or MAb production.

Nutritional Requirements

Amino Acids. Mouse cells, like other animal cells, are unable to synthesize certain (essential) amino acids, and these have to be provided in the medium. Nonessential amino acids can be synthesized by the hybridomas, but are frequent components of some media formulations (i.e., Ham's F-12, and others [111]). Addition of these amino acids results in lower energy requirements by the cells because anabolic amino acid metabolism is eliminated. Depending on the cell line, this supplementation may also result in higher specific growth rates.

Lipids. Lipids are essential structural elements of all cell membranes, and one type, the phospholipids, are the most abundant components of membranes (112). Starting from ethanolamine and fatty acids, hybridomas synthesize phosphatidyl ethanolamine (113), from which they can then derive other phospholipids (114,115). Hybridomas respond differently to lipid supplements, and specific effects on cell and MAb yields have been reported (116).

Vitamins and Inorganic Ions. These components are required by hybridomas and mammalian cells in general. Sufficient amounts are included in most media formulations (111), and limitations are not frequently observed under normal non-high-density growth conditions.

Oxygen. Hybridoma specific oxygen consumption rates have been reported to range from 1 to 40 pg/cell/h (117–119). Oxygen is one of the main limiting factors in the maintenance of viable populations in high-density perfusion cultures. In most instances, hybridoma cells will be unaffected by a wide range of oxygen concentrations. However, high oxygen concentrations (i.e., 100% or more of air saturation) can inhibit growth rates and induce cell lysis in later culture stages, presumably through the production of damaging oxygen radicals. In some hybridomas, the optimal concentrations of dissolved oxygen for growth and MAb production are different (118,120).

Carbon Dioxide. Most hybridomas will initiate growth well in an air atmosphere without additional CO_2. However, CO_2 produced by the cells has a significant effect on the culture pH. Because of accumulation of CO_2 and lactic acid, the pH of batch cultures without base addition may drop to values below 6.5.

Basal Media. Most media used in mammalian cell culture are based on the balanced salt solution. For hybridoma work, Dulbecco's Modified Eagles Medium (DMEM) and a 1:1 mixture of DMEM and Ham's F-12 (111) are commonly used and contain the nutritional requirements discussed. These formulations have been modified often to increase cell and MAb yields.

Supplements. A universal supplement used in cell culture is animal serum. Serum contains nutrients and many factors that directly or indirectly stimulate hybridoma growth.

Basal media supplemented with serum (usually 10% or less fetal bovine serum [FBS]) will allow exuberant growth of murine hybridomas in either stationary (i.e., T flasks) or suspension (shake flasks or spinners), and cultures in simple batch mode without optimization will yield, in most cases, from 1 to 3 $\times$ 10^6 cells/mL and from 10 to over 100 mg/L of MAb. If needed, serum depleted of immunoglobulins can be used to minimize their contamination of the product (121).

Serum can be replaced by combinations of supplements such as insulin, transferrin, ethanolamine, and selenium, which are employed in serum-free formulations (122,123). Proteins and other contaminants are reduced when utilizing serum-free formulations and eliminated with protein-free formulations. Studies on serum-free and protein-free media are abundant in the literature (122–129), and several convenient commercial formulations are available (Table 3). Addition of basic proteins and spermine, a low molecular weight polyamine, to serum-free media has been found to stimulate IgM production by a human hybridoma line (130).

Most murine hybridomas adapt readily to one or more of these media, with growth kinetics, cell, and product yields often as good or better than those in serum-containing media. In addition, when done properly, the cells respond well to freezing in serum-free and protein-free media (123,128,130), as shown in Table 4. Additional advantages in using serum-free and protein-free media include the reduced risk for introducing adventitious agents of animal origin, such as retroviruses and prions, which is a significant concern if the MAb product is intended for therapeutic applications (131–133).

Table 3. Commercially Available Serum-Free and Protein-Free Media for Hybridoma Growth and Production of MAbs

Media	Protein content (mg/L)	Availability as liquid (L) or powder (P)	Suppliers
UltraDOMA	30	L	Bio-Whittaker[a]
UltraDOMA-PF	0	L, P	Bio-Whittaker
Hybridoma-SFM	20	L	Gibco BRL[b]
PFHM-II	0	L, P	Gibco BRL
CCM1	2100	L, P	HyClone[c]
SFX mab	low	L	HyClone
PFmab	0	L	HyClone
HB 101	780	L, P	Irvine Scientific[d]
HB 102	780	L, P	Irvine Scientific
HB 104	780	L, P	Irvine Scientific
HB PRO	1	L, P	Irvine Scientific
HB GRO	50	L, P	Irvine Scientific
Ex-Cell 610	11	L, P	JRH Biosciences[e]
Ex-Cell 620	11	L, P	JRH Biosciences
QBSF-51	45	L	Sigma[f]
QBSF-52	45	L	Sigma
QBSF-53	408	L	Sigma
QBSF-55	65	L	Sigma
QBSF-56	428	L	Sigma
Hybri-Max	0	L	Sigma

[a]Bio-Whittaker, Inc., 8830 Biggs Ford Rd., Walkersville, MD, 21793; U.S. tel. 800-638-8147/301-898-7025; fax 301-845-8338; international fax 301-845-8291.
[b]Gibco BRL, Life Technologies, Inc., P.O. Box 9418, Gaithersburgh, MD, 20898; U.S. tel. 800-828-6686; fax 800-331-2286; international fax 301-258-8238.
[c]HyClone Laboratories, Inc., 1725 South HyClone Rd., Logan, UT 84321-6212, U.S. tel. 800-492-5663; fax 800-533-9450; international tel. 801-753-4584; International Fax 801-753-4589.
[d]Irvine Scientific, 2511 Daimler Street, Santa Clara, CA 92705-5588; U.S. tel. 800-577-6097/800-437-5706; fax 714-261-6522.
[e]JRH Biosciences, 13804 W. 107th Street, Lenexa, KS 66215-0848, U.S. tel. 800-255-6032; international tel. 913-469-5580; fax 913-469-5584.
[f]Sigma Chemical Company, P.O. Box 14508, St Louis, MO 63178-9916; U.S. tel. 800-325-3010/800-848-7791; fax. 800-325-5052; international tel. 314-771-5750; international fax. 314-771-5757.

Bioreactors

A large variety of in vitro systems are used to grow hybridomas, including multiwell plates, Petri dishes, T flasks, multiplate containers, hollow fiber cartridges, roller bottles, shake flasks, spinner flasks, airlift fermentors, and stirred-tank fermentors (30–44). The bioreactors provide the vessel and the physicochemical environment, which includes, among the most important factors, pH, temperature, agitation, gas exchange, redox potential, and osmolarity.

Effect of pH. Hybridomas can tolerate pH extremes (i.e., 6.4 to 8) if adapted gradually. Uptake and utilization of nutrients are affected by pH; for instance, glucose and glutamine utilization rates increase when the pH is alkaline (Fig. 3). Cultures evolve toward acidity because of the production of lactic acid and CO_2. To control the pH, sodium bicarbonate is frequently used as a buffering agent and is initially added to media in concentrations of grams per liter. Organic buffers such as 2-[4-(2-hydroxyethyl)-1-piperazinyl] ethanesulfonic acid (HEPES), 2-morpholinoethanesulfonic acid (MES), 3-morpholino-propanesulfonic acid (MOPS) or others (134) are also used at concentrations of 10 mM or higher.

Table 4. Growth and MAb Production by Three Hybridomas Cultivated in 2-L Spinner Flasks Using Serum-Free and Protein-Free Media

	Line A		Line B		Line C	
Media	SF[a]	PF	SF	PF	SF	PF
Maximum viable cell concentration ($\times 10^6$/mL)	1.2	1.5	1.9	1.9	1.6	1.5
Specific growth rate (h^{-1})	0.038	0.022	0.04	0.043	0.041	0.033
Doubling time (h)	18.3	31.5	17.3	16.1	16.9	21
Maximum MAb concentration (mg/L)	74	108	62	76	90	106

Source: F.J. Castillo and J. Thrift, unpublished results.

Note: All hybridomas were derived from BALB/c splenocytes. Line A was derived from fusion with P3-X63Ag8.653 myeloma (71) and produces IgM. Lines B and C were derived from fusion with P3NS1/1-Ag4-1 myeloma (65) and produce IgG1 and IgG2a, respectively.

[a]Data correspond to averages obtained in 8 and 6 consecutive passages, respectively in serum-free (SF) medium containing bovine transferrin and insulin and protein-free (PF) medium containing chelated iron to replace transferrin. The adapted cells had been frozen in corresponding media formulations supplemented only with 10% DMSO as cryoprotectant, and were grown as described upon thawing.

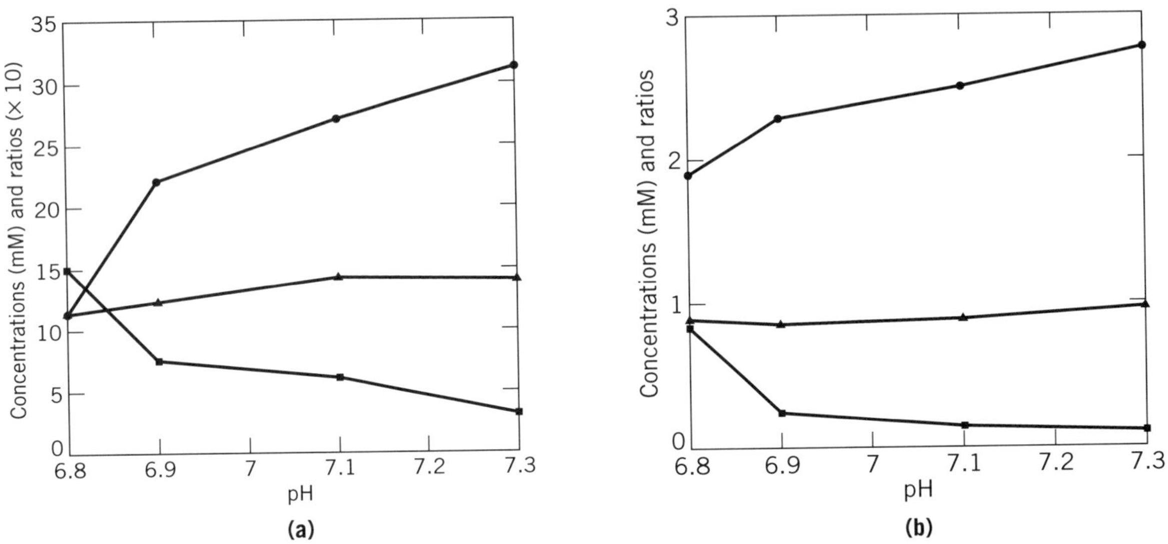

Figure 3. Effect of pH on the utilization of glucose and glutamine and the production of lactate and ammonia by line XMMLY-B. The hybridoma was grown in chemostat culture at a dilution rate of 0.017 h^{-1}. Data were collected at each pH after reaching steady states. (**a**) Glucose concentration (■), lactate concentration (●), ratio of lactate produced to glucose utilized (▲). (**b**) Glutamine concentration (■), ammonia concentration (●), ratio of ammonia produced to glutamine utilized (▲).

Maximum specific growth rates are usually observed at pH values between 6.8 and 7.3, but growth and MAb production are frequently optimal at different pH values. A production process may consist of a growth phase in which the pH is controlled at a certain value, followed by a production phase where the pH is adjusted to the value that maximizes product secretion.

The quality of the secreted MAb can be affected by the pH in the medium either by facilitating chemical changes or, when low, by providing a favorable environment for the action of released acidic proteases (106,135–137). Thus, the effect of pH on growth, productivity, MAb yields, and MAb quality should be determined to maximize yields and product consistency.

Effect of Temperature. Optimal temperature for myeloma and hybridoma growth is 37 °C, and the cells recover from excursions to slightly higher temperatures. Lower temperatures cause decreases in growth and MAb production (138), but the cells are not affected and recover promptly when placed again at 37 °C. Myelomas and hybridomas can be maintained at room temperature for several days, which is convenient for shipments of cells in culture. For permanent storage a liquid nitrogen freezer is essential. A properly frozen hybridoma stock, upon thawing and placed in culture with fresh medium, will usually resume growth within 48 hours.

Effect of Agitation and Gas Bubbles. Concerns about shear sensitivity of mammalian cells prompted the development of airlift fermenters for use with mammalian cells (139) and hybridoma growth (140,141) that are used for commercial production of MAbs (Table 1). However, experience has shown that each line is different (142), shear sensitivity depends on culture conditions (143–145), most hybridomas are not especially sensitive to shear stress in culture, and some can stand very harsh agitation. Bubble-cell interactions cause damage, but shear-protective agents such as the polyol Pluronic F-68, which is frequently added to cell cultures, can minimize the damaging effects of bubble disengagement (144–145).

Redox Potential. Although the oxidation–reduction potential (redox potential) is routinely measured on-line in microbial fermentations, the same is not frequently done in cell cultures. Little understanding of correlations between the redox potential on growth or MAb production exist (146), but optimal values may be defined and applied in the future.

Effect of Osmolarity. Hybridomas, like other mammalian cells, grow uninhibited in osmolalities ranging from 260 to 300 mOsmol/kg H_2O (approximate osmolarities of 260 to 300 mmol/L [147]) and commercial basal media are designed to have these values by adjusting the NaCl content. High osmolarity has been reported to inhibit growth rate and cell yields (148), but also to induce higher MAb secretion at later stages of cultivation (149–153). Some amino acids and analogs act as osmoprotective agents for hybridomas and can even increase specific antibody productivity (154).

Growth Kinetics

Requirements for hybridoma growth are a viable inoculum and proper extracellular environment, including essential nutrients and suitable physicochemical conditions.

When all these conditions are met, hybridomas, like all other eukaryotic cells in culture, have kinetics of growth similar to microbial cells and also obey the exponential growth law (155,156). Hybridomas are not anchorage dependent and grow as well in nonagitated cultures, such as multiwell plates, Petri dishes, T flasks and hollow fiber bioreactors as they do in agitated containers, including roller bottles, shake flasks, spinner flasks, or fermenters. Hybridoma cells grow monodispersed, and in batch cultures, in most media, they reach concentrations of several million per milliliter, with viabilities of well over 90% throughout the growth phase.

When growth data are plotted as viable cell counts against time, a typical batch growth curve of hybridoma may contain the six phases observed in microbial culture: lag, accelerating growth, exponential growth, decelerating growth, stationary, and decline (156). A lag period is generally observed when starting cultures using cells thawed from frozen stocks. Upon thawing, the cells go through an initial period of adaptation, but once exponential growth is attained no lag period will occur upon subculturing provided that the cells are passaged to fresh medium while still growing exponentially. If a culture is maintained beyond the exponential phase, a lag often is observed upon passage, and such cells are said to have been overgrown. The longer cells are maintained in the spent medium beyond the exponential phase, the more likely that a lag period will occur in the next culture (157). Ultimately, there will be a point of overgrowth past which all cells will die either by apoptosis or by necrosis (158), induced by starvation or toxic metabolite buildup (159–164), and the culture can no longer be recovered.

During exponential growth, doubling times for most hybridomas are between 14 and 23 h, corresponding to maximum specific growth rates ranging from 0.05 to 0.03 h^{-1}.

Hybridomas have been grown in chemostat cultures at dilution rates ranging from critical (Dc), which is equivalent to maximum specific growth rate (156), to less than 10% of Dc. Viability decreases with dilution rate (Fig. 4**a**), and a common observation with all hybridomas is that an intrinsic minimum specific growth rate of approximately 0.02 h^{-1} exists for all of them (164,165) (Fig. 4**b**). A minimum specific growth rate was also reported for other cells and attributed to nutritional limitations (166). Death of hybridomas at low dilution rates may be mostly by apoptosis induced by nutritional limitation(s) and reduced protein synthesis. A practical implication of this phenomenon in a continuous culture process is that the fraction of dead cells will increase with decreases in the dilution rate (purge rate in high-density perfusions [156]). The ratio of viable/dead cells will have effects on MAb concentrations and homogeneity and on the longevity of production runs, all of which impact product yields.

MAb Production Kinetics

MAbs are produced by viable cells (167). Patterns observed for MAb production by hybridomas are the same as those seen with microbial cells for product formation (168), and most can be defined either as growth-associated, partly growth-associated, non-growth-associated, or negatively growth-associated (169–178).

Figure 5 shows the growth and MAb production of two hybridomas in batch cultures. One clearly has growth-associated production (**a**), and the second (**b**) has a non-growth-associated pattern with more that 90% of the MAb produced after growth has ceased.

Specific productivities of several hybridomas at different dilution rates in continuous chemostat cultures are shown in Figure 6. The pattern of line XMMEN-A was non-growth-associated, whereas that of all other hybridomas tested in the group was growth-associated. The specific pattern is very important in defining the productivity profile of continuous cultures, and depending on the system and its mode of operation, the bleed or harvest rate will have to be adjusted for maximum productivity (156). In high-density cultures, productivity can also be affected by other factors, external or autocrine (179), which may not be obvious or predictable in low-density cultures and that need to be evaluated under operation conditions.

The validity and usefulness of applying the microbial growth and non-growth-associated concepts to the production of proteins by animal cells, including hybridomas, has been questioned (180). However, it is clear that for two cell lines with different productivity profiles (as shown in Fig. 5), the optimal harvest times will be very different, and this will have a significant impact on the corresponding operations.

MAb Production Processes

In Vitro. MAb yields by most hybridomas in simple batch cultures without optimization are in the 10 to 100

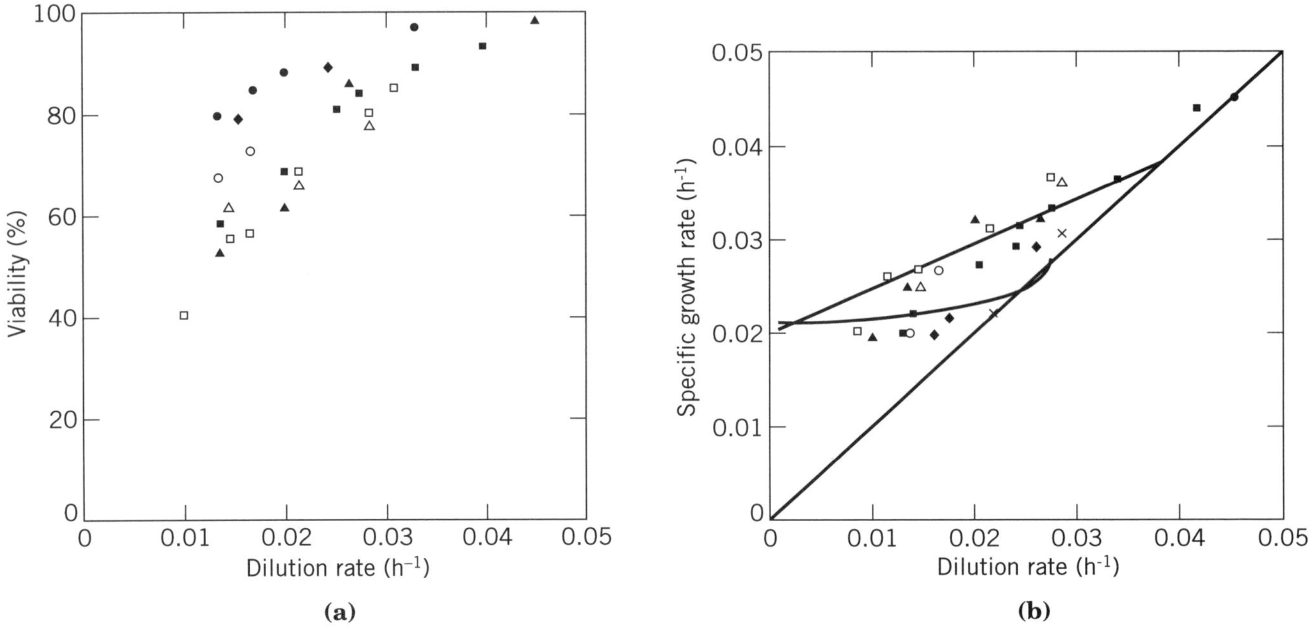

Figure 4. Changes in viability (**a**) and specific growth rates (**b**) with dilution rate in chemostat culture of several hybridomas. Cell lines (Table 5): XMMEN-A (○), XMMME-C (●), XMMLY-B (△), B-17 (▲), XMMCO-E (□), XMMLY-F (■), XMMME-D (x), XMMID-G (♦).

mg/L range. Production of milligram quantities can be easily accomplished using T flasks, spinner flasks, shaker flasks, or roller bottles. Gram quantities require either more of these containers and more labor or the use of larger vessels such as fermenters. Productive batch cultures need a minimum of conditions, which include an exponentially growing inoculum, a buffered medium with an initial pH adjusted between 7.2 and 7.5, and the temperature of incubation regulated at 37 ± 0.5 °C. Growth kinetics, specific MAb productivities, and other characteristics of several hybridomas grown with these conditions in batch suspension cultures and in different vessels are complied in Table 5.

Fed-batch culture of hybridoma growth continues to evolve, and very high MAb yields, in the order of grams per liter, have been reported recently (110,181–184). This may become the process of choice in future commercial operations.

High-density continuous processes can be done at many scales with available preassembled equipment. The systems vary in their complexity, instrumentation, and requirements for infrastructural support. Some are easy to set up and operate, but others require special training and experience. All demand attention in order to be used routinely in successful ways, but they are reliable and productive when understood and operated properly (39–44).

In Vivo. Production in ascites has the advantage of high concentration, and it is widely used. It is commonly used in the production of small quantities of MAb for research and diagnostic purposes; two of the approved MAbs for medical use are made using ascites technology (Table 1). Contract suppliers can provide ascites at different scales from a few milligrams to kilograms per year. Facilities designed for production of up to kilogram quantities of MAbs under cGMP are operated for contract manufacturing by companies such as Charles River in Massachusetts and North Carolina and Genzyme Transgenics. Although clearly not the preferred method of production for therapeutic MAbs, as explicitly stated by the Committee for Proprietary Medicinal Products (185), validated and cost-effective ascites processes exist, and several MAbs in advanced clinical trials are made in ascites.

Emerging in vivo processes involve the use of transgenic animals. Reports of high yields in the milk of goats and other animals coupled with effective separation and purification techniques may make this a competitive technology for certain applications (186). However, for regulatory and strategic reasons, it is advisable to adopt in vitro processes for production of pharmaceutical-grade MAbs. The use of animals for production of MAbs for therapeutic applications will have to deal with, among other complications, the growing concerns of potential bovine spongiform encephalopathy or other prion contaminations (133, 187,188).

Production of antibodies in plants has also been done (50) and may represent a viable alternative for some applications.

MAb Recovery and Purification

MAbs are secreted by hybridomas, and when using protein-free media, the cultures contain high MAb/total protein ratios. In high-density, hollow-fiber perfusion cultures, MAb levels corresponding to more than 40% of the total protein have been reported (43). For some applications, a cell-free harvest of this relative composition may

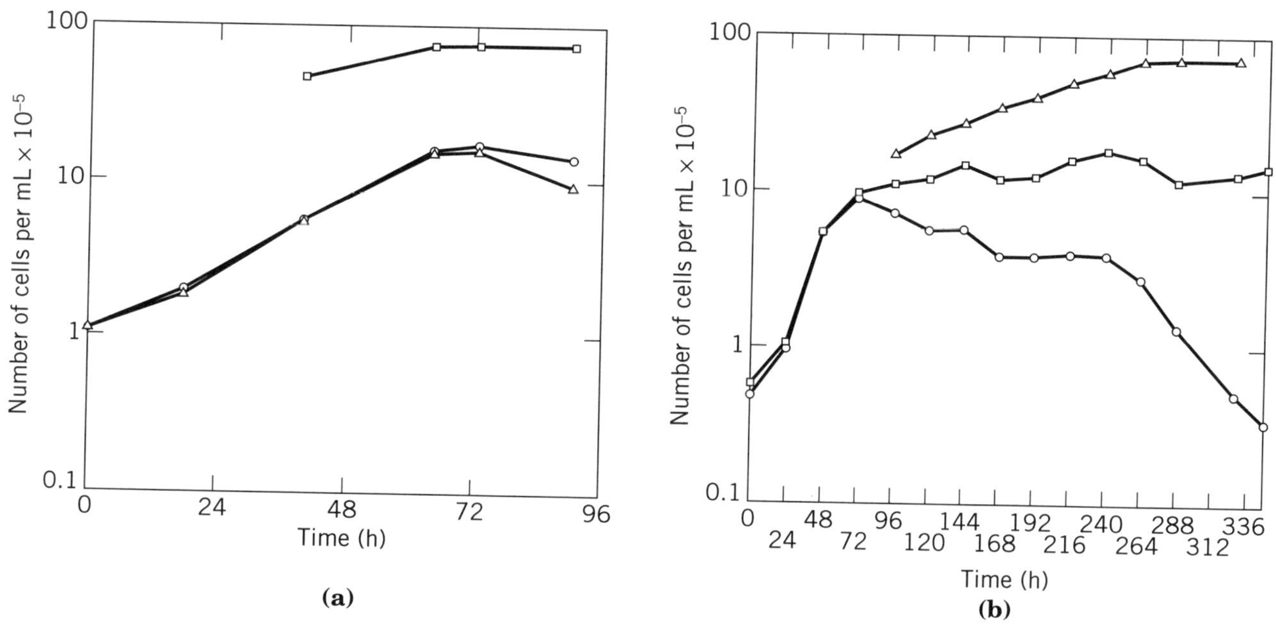

Figure 5. Growth curves and MAb production by two hybridoma lines (**a**, XMMME-C, and **b**, XMMEN-A) with growth-associated and non-growth-associated productivity patterns. Total (○), viable (△), and MAb concentration per mg/L (□).

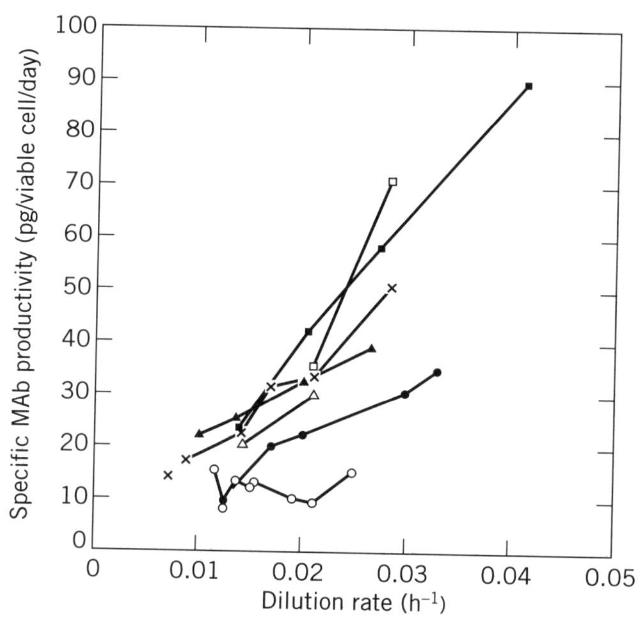

Figure 6. Growth-associated and non-growth-associated productivity patterns of several hybridoma cells grown in continuous chemostat cultures. Symbols same as in Figure 4.

be used as such without further processing to remove other proteins present (although contaminating cellular proteases may degrade the MAb upon prolonged incubation).

MAb recovery traditionally starts with cell removal by centrifugation or filtration. New processes aim at obviating separation steps by applying the entire culture to a fluidized bed of chromatography beads (189). Purification of MAbs involves usually two to three steps, which may include combinations of precipitation or column processes such as ion-exchange separation; affinity separations (protein A or dye ligand); or hydroxylapatite, hydrophobic, mix-mode, and gel filtration (190–192)

PROCESS CHANGES

Process development is a multidisciplinary endeavor aimed at maximizing yields and minimizing costs while maintaining product consistency. Traditionally, there has been reluctance in industry to publish results on the development of processes for biopharmaceuticals, including those for MAbs (34). This reluctance reflects the need to maintain advantage over competitors.

Significant advances have taken place in the understanding of hybridoma physiology and in the development of highly productive MAb processes. Through the years, a wealth of information from academia and industry became available in publications and through other sources. The development and implementation of an MAb production process benefits today from this information, and systematic approaches to follow have been described (34,193).

Many approved biologicals are manufactured by obsolete or suboptimal processes. For instance, OKT3, the first therapeutic MAb approved, is made with ascites technology. It was introduced into the market 12 years ago and continues to be manufactured by the same process. There are several reasons for this apparent conservatism in the biopharmaceutical industry.

First, there is the pressure to expedite manufacturing for introduction of a biological product into the market. A recent example of this pressure was that confronted by Chiron Corporation for the production of Betaseron (recombinant interferon-β). After approval of the product in

Table 5. Batch Culture of Hybridoma Lines in Suspension

Line	Parental myeloma[a]	MAb type	Bioreactor	Maximum specific growth rate (h^{-1})	Doubling time (h)	Specific MAb productivity[b] (pg/viable cell/day)		
						A	B	C
XMMEN-A	P3-X63Ag8.653	IgM	2-L spinner	0.05	14	15	13	N/D
XMMEN-A	P3-X63Ag8.653	IgM	40-L stirred fermenter	0.05	14	20	15	19
XMMEN-A	P3-X63Ag8.653	IgM	40-L airlift fermenter	0.018	37	24	15	18
A-72	P3-X63Ag8.653	IgM	2-L spinner	0.05	14	15	14	15
A-78	P3-X63Ag8.653	IgM	2-L spinner	0.05	14	13	14	13
XMMLY-B	P3NSI/1-Ag4-1	IgG1	2-L spinner	0.043	16	52	12	4
B-17	P3NSI/1-Ag4-1	IgG1	2-L spinner	0.033	21	76	10	N/D
B-17	P3NSI/1-Ag4-1	IgG1	40-L stirred fermenter	0.023	29	>80	20	22
XMMME-C	P3-X63/Ag8	IgG2a	2-L spinner	0.043	16	40	2	N/D
XMMME-D	P3NSI/1-Ag4-1	IgG1	2-L spinner	0.038	18	90	5	N/D
XMMCO-E	P3NSI/1-Ag4-1	IgG2b	2-L spinner	0.035	20	90	5	N/D
XMMLY-F	P3NSI/1-Ag4-1	IgG2a	2-L spinner	0.046	15	70–90	50–75	40
XMMID-G	Sp2/0	IgG2a	2-L spinner	0.035	20	100-12	10	15
XMMID-G	Sp2/0	IgG2a	40-L stirred fermenter	0.036	19	100-15	10	10

Note: N/D = no data.
[a]All splenocytes used in fusions were of BALB/c origin.
[b]A = during exponential growth; B = at peak viable cell density; C = during stationary phase.

July 1993 for the treatment of relapsing-remitting multiple sclerosis (194), the company had an initial shortage of manufacturing capacity. It was forced to establish a computerized lottery system to supply between 12,000 and 20,000 patients the first year until capacity was brought up in line with the demand (195). Obviously, with such pressing needs, there is little incentive to introduce process changes that may produce further delays.

A second and very important cause has been the difficulty of introducing process changes after pivotal phase III clinical testing or after commercialization of a product. Regulatory approval of process changes could only be obtained after the results of extensive and costly testing in toxicology and efficacy provided guarantees that the changes were not altering significant properties of the product (196). Thus, in several instances, introduction of process changes has either been abandoned or postponed as long as the difficulties and costs implied could not be offset by the potential gains.

More recently, the established dogma in the production of biopharmaceuticals that "the process defines the product" has lost its absolutism. This has been an evolutionary process guided by caution. New powerful techniques are being used for characterization of biologics, and experience gained with the clinical use of these products has helped redefine the relevance of microheterogeneity and the potential problems with contaminants such as residual DNA, proteins, and endogenous retroviruses (197–198). As a result, regulatory agencies have gradually adopted a less-stringent view on the inalterability of the process and the dedication of facilities and have shifted toward a more flexible position based on the ability to demonstrate equivalence of the products made by different processes and at different sites. New regulations and guidelines encourage process improvements when capable of producing equivalent product (199–201). However, it should be kept in mind that culture conditions may significantly affect the properties of MAbs (202,203), and efforts should be made to identify these effects as early as possible in order to guide the selection of the most advantageous technology for a process (204).

CONCLUSIONS

With the present accessibility to relevant, practical information, availability of hybridoma-specific culture media, alternative bioreactor technology, and multiple purification methods, in most cases a simple MAb production process can be easily implemented.

BIBLIOGRAPHY

1. G. Kohler and C. Milstein, *Nature* **256**, 495–497 (1975).
2. E. Harlow and D. Lane, *Antibodies: A Laboratory Manual*, Cold Spring Harbor Laboratory, New York, 1988.
3. O.W. Merten, in R.E. Spier and J.B. Griffiths eds., *Animal Cell Biotechnology*, vol. 4, Academic Press, London, 1990, pp. 257–315.
4. M.J. Perry, in J.R. Birch and E.S. Lennox eds., *Monoclonal Antibodies: Principles and Applications*, Wiley-Liss, New York, 1995, pp. 107–120.
5. C.R. Hill, in J.R. Birch and E.S. Lennox eds., *Monoclonal Antibodies: Principles and Applications*, Wiley-Liss, New York, 1995, pp. 121–136.
6. T.A. Waldmann, *Science* **252**, 1657–1662 (1991).
7. K.J. Morrow, *Gen. Eng. News.* **15**, 1–24 (1995).
8. D.A. Scheinberg and P.B. Chapman, in J.R. Birch and E.S. Lennox, *Monoclonal Antibodies: Principles and Applications*, Wiley-Liss, New York, 1995, pp. 45–105.
9. *Approved Biotechnology Drugs and Vaccines*, Pharmaceutical Research and Manufacturers of America, 1998.
10. A.F. Holmer, *New Drug Approvals in 1997*, doc. #3020, Pharmaceutical Research and Manufacturers of America, 1998.

11. *Neoprobe plans response to FDA requests for additional information under RIGscan ® CR49 BLA marketing application*, 1997, *http://www.neoprobe.com*
12. D. Strickland, *BioWorld Today* **8**, 1–2 (1997).
13. R. Osborne, *BioWorld Today* **8**, 1–2 (1997).
14. R. Osborne, *BioWorld Today* **9**, 2 (1998).
15. A.F. Holmer, *350 Biotechnology Medicines in Development*, doc. #3004, Pharmaceutical Research and Manufacturers of America, 1998.
16. K.J. Morrow, *Gen. Eng. News.* **17**, 1–21 (1997).
17. L. Seachrist, *BioWorld Today*, **8**, 1–3 (1997).
18. A. DePalma, *Gen. Eng. News.* **18**, 18–30 (1998).
19. J.P. Chandler, in S.S. Seaver ed., *Commercial Production of Monoclonal Antibodies: A Guide for Scale-Up*, Dekker, New York, 1987, pp. 75–92.
20. N.J. Hoogenraad and C.J. Wraight, *Methods Enzymol.* **121**, 375–381 (1986).
21. B.K. Lydersen ed. *Large Scale Cell Culture Technology*, Hanser Publishers, Munich, 1987.
22. A. Prokop and M.Z. Rosenberg, in A. Fiechter ed., *Advances in Biochemical Engineering*, vol. 39, Springer-Verlag, Berlin, 1989, pp. 29–71.
23. B. Griffiths, in R.I. Freshney ed., *Animal Cell Culture: A Practical Approach*, IRL Press, Oxford, U.K., 1989, pp. 33–69.
24. A.S. Lubiniecki ed. *Large Scale Mammalian Cell Culture Technology*, Dekker, New York, 1990.
25. R. Spier, in R.E. Spier and J.B. Griffiths eds., *Animal Cell Biotechnology*, vol. 5, Academic Press, London, 1992, pp. 1–46.
26. J.R. Birch and S.J. Froud, *Biologicals* **22**, 127–133 (1994).
27. D.E. Yelton, D.H. Margulies, B. Diamond, and M.D. Scharff, in R.H. Kennett, T.J. McKearn, and K.B. Bechtol eds., *Monoclonal Antibodies Hybridomas: A New Dimension in Biological Analyses*, Plenum Press, New York, 1980, pp. 3–17.
28. J.A. Roth ed. *Monoclonal Antibodies in Cancer: Advances in Diagnosis and Treatment*, Futura, Mount Kisco, N.Y., 1986.
29. A. Mizrahi ed. *Monoclonal Antibodies: Production and Application. Advances in Biotechnological Processes*, vol 11, Liss, New York, 1989.
30. *Proc. Int. Symp. on Monoclonal Antibodies for Therapy, Prevention and In Vivo Diagnosis of Human Disease, Utrech. Dev. Biol. Standard.*, vol. 71, 1989.
31. K. James, in R.E. Spier and J.B. Griffiths eds., *Animal Cell Biotechnology*, vol. 4, Academic Press, London, 1990, pp. 205–253.
32. S.S. Seaver ed., *Commercial Production of Monoclonal Antibodies: A Guide for Scale-Up*, Dekker, New York, 1987.
33. A. Mizrrahi ed. *Upstream Processes: Equipments and Techniques, Advances in Biotechnological Processes*, vol 7, Liss, New York, 1988.
34. C. Harbour, J.P. Barford, and K.S. Low, in A. Flechter ed., *Advances in Biochemical Engineering/Biotechnology*, vol. 37, Springer-Verlag, Berlin, 1988, pp. 1–40.
35. O.W. Merten, in A.O.A. Miller ed., *Advanced Research on Animal Cell Technology*, Kluwer, Dordrecht, The Netherlands, 1989, pp. 367–398.
36. C.P. Marquis, C. Harbour, J.P. Barford, and K.S. Low, *Cytotechnology* **4**, 69–76 (1990).
37. J.R. Birch, K. Lambert, R. Boraston, P.W. Thompson, and M.A. Boss, *Biotech'85* 1–3 (1985).
38. D.H. Randerson, *J. Biotechnol.* **2**, 241–255 (1985).
39. K. Nilsson, W. Scheirer, H.W.D. Katinger, and K. Mosbach, *Methods Enzymol.* **121**, 352–360 (1986).
40. S.F. de St. Groth, *Methods Enzymol.* **121**, 360–375 (1986).
41. J. Shevitz, S. Reuveny, T.L. LaPorte, and G.H. Cho, in A. Mizrahi ed., *Monoclonal Antibodies: Production and Applications*, Liss, New York, 1989, pp. 81–105.
42. S. Reuveny and A. Lazar, in A. Mizrahi ed., *Monoclonal Antibodies: Production and Applications*, Liss, New York, 1989, pp. 45–80.
43. F.J. Castillo, L.J. Mullen, J.C. Thrift, and B.C. Grant, *Ann. N.Y. Acad. Sci.* **665**, 72–80 (1992).
44. Y.M. Deo, M.D. Mahadevan, and R. Fuchs. *Biotechnol. Prog.* **12**, 57–64 (1996).
45. V.T. Oi and S.L. Morrison, *Biotechniques* **4**, 214–221 (1986).
46. M. Better, P.C. Chang, R.R. Robinson, and A.H. Horwitz, *Science* **240**, 1041–1043 (1988).
47. M. Better and A.H. Horwitz, *Methods Enzymol.* **178**, 476–496 (1989).
48. M. Sollazzo, R. Billetta, and M. Zanetti, in J.M. Cruse and R.E. Lewis eds., *The Year in Immunology 1989–90. Molecules and Cells of Immunity*, Karger, Basel, 1990, pp. 162–184.
49. A. Hiatt, R. Cafferkey, and K. Bowdish, *Nature* **342**, 76–78 (1989).
50. J. Hodgson, *Biotechnology* **9**, 421–425 (1991).
51. A. Wright, S.U. Shin, and S.L. Morrison, *Crit. Rev. Immunol.* **12**, 125–168 (1992).
52. J.J. Trill, A.R. Shatzmann, and S. Ganguly, *Curr. Opinion Biotechnol.* **6**, 553–560 (1995).
53. J.R. Birch and E.S. Lennox, *Monoclonal Antibodies, Principles and Applications*, Wiley-Liss, New York, 1995.
54. R. Irving and P. Hudson, in H. Zola ed., *Monoclonal Antibodies: The Second Generation*, Bios Scientific Publishers, Oxford, U.K., 1995, pp. 119–139.
55. D.M. Fishwild, S.L. O'Donnell, T. Bengoechea, D.V. Hudson, F. Harding, S.L. Bernhard, D. Jones, R.M. Kay, K.M. Higgins, S.R. Schramm, and N. Lonberg, *Nat. Biotechnol.* **14**, 845–851 (1996).
56. S. Paul, *Antibody Engineering Protocols. Methods in Molecular Biology*, vol. 51, Humana Press, Totowa, N.J., 1995.
57. M.W. Young, W.B. Okita, E.M. Brown, and J.M. Curling, *BioPharm* **10**, 34–38 (1997).
58. G. Winter and C. Milstein, *Nature* **349**, 293–299 (1991).
59. D.M. Knight, M. McDonough, M.A. Moore, D. Abercrombie, R. Siegel, and J. Ghrayeb, *Hum. Antibod. Hybridomas* **3**, 129–136 (1992).
60. T. Lindl, *Cytotechnology* **21**, 183–193 (1996).
61. M. Neuberger and M. Bruggemann, *Nature* **386**, 25–26 (1997).
62. R. Sherman-Gold, *Genet. Eng. News* **17**, 4–35 (1997).
63. W.W. Lime, in A.S. Lubiniecki and S.A. Vargo eds., *Regulatory Practice for Biopharmaceutical Production*, Wiley-Liss, New York, 1994, pp. 235–253.
64. S.L. Danheiser, *Gen. Eng. News* **16**, 1–36 (1996).
65. G. Galfre and C. Milstein, *Methods in Enzymol.* **73**, 3–46 (1981).
66. J.J. Langone and H.V. Vunakis, *Methods Enzymol.* **92**, (1983).
67. A. Osterhaus and F. Uytdehaag, in R.E. Spier and J.B. Griffiths eds., *Animal Cell Biotechnology*, vol. 2, Academic Press, London, 1985, pp. 49–69.

68. J.W. Goding, *Monoclonal Antibodies: Principles and Practices*, Academic Press, London, 1986.
69. W.C. Davis, *Monoclonal Antibody Protocols. Methods in Molecular Biology*, vol. 45, Humana Press, Totowa, N.J., 1995.
70. K. Horibata and A.W. Harris, *Exp. Cell Res.* **60**, 61–77 (1970).
71. J.F. Kearney, A. Radbruch, B. Liesegang, and K. Rajewsky. *J. Immunol.* **123**, 1548–1550 (1979).
72. M. Shulman, C.D. Wilde, and G. Kohler, *Nature* **276**, 269–270 (1978).
73. G. Kholer and C. Milstein, *Eur. J. Immunol.* **6**, 511–519 (1976).
74. G. Kholer and C. Milstein, *Nature* **256**, 495–497 (1975).
75. J.M. Davis, *Methods Enzymol.* **121**, 307–322 (1986).
76. C.I. Civin and M.L. Banquerigo, *Methods Enzymol.* **121**, 322–327 (1986).
77. M.B. Rittenberg, A. Buenafe, and M. Brown, *Methods Enzymol.* **121**, 327–331 (1986).
78. A. Boeye, *Methods Enzymol.* **121**, 332–340 (1986).
79. X. Li, K. Abdi, and S.J. Mentzer, *Hybridoma* **11**, 645–652 (1992).
80. Points to Consider in the Characterization of Cell Lines Used to Produce Biologicals, FDA, Washington, D.C., 1993.
81. Points to Consider in the Manufacture and Testing of Monoclonal Antibody Products for Human Use, FDA, Washington, D.C., 1997.
82. F.J. Castillo, in J.A. Asenjo and J.C. Merchuk, *Bioreactor System Design*, Dekker, New York, 1995, pp. 13–45.
83. C.B. Morris, in J.G. Day and M.R. McLellan eds., *Methods in Molecular Biology*, vol. 38, Humana Press, Totowa, N.J., 1995, pp. 179–187.
84. R.J. Hay, in J.C. Hunter-Cevera and A. Belt eds., *Maintaining Cultures for Biotechnology and Industry*, Academic Press, San Diego, Calif., 1996, pp. 161–178.
85. D.E. Yelton, W.D. Cook, and M.D. Scharff, *Transplant. Proc.*, **12**, 439–443 (1980).
86. P. Thammana and M.D. Scharff, *Eur. J. Immunol.* **13**, 614–619 (1983).
87. G. Spira, H.L. Aguila, and M.D. Scharff, *J. Immunol.* **140**, 2675–2680 (1988).
88. F. Melchers, M. Potter, and N.L. Warner, in F. Melchers, M. Potter, and N. Warner eds., *Lymphocyte Hybridomas: Current Topics in Microbiology and Immunology*, vol. 81, Springer-Verlag, Berlin, 1978, pp. IX–XXIII.
89. G.L. Altshuler, R. Dilwith, J. Sowek, and G. Belfort, *Biotechnol. Bioeng. Symp. No. 17*, 1986, pp. 725–736.
90. K.K. Frame and W-S. Hu, *Biotechnol. Bioeng.* **35**, 496–476 (1990).
91. S.S. Ozturk and B.O. Palsson, *Hybridoma* **9**, 167–175 (1990).
92. G.M. Lee, A. Varma, B.O. Palson, *Biotechnol. Prog.* **7**, 72–75 (1991).
93. A-P. Zeng, *Biotechnol. Bioeng.* **50**, 238–247 (1996).
94. R. Hubbard, in A. Wiseman ed., *Topics in Enzyme and Fermentation Biotechnology 7*, Ellis Horwood, New York, 1983, pp. 196–263.
95. P. D'Eustachio, D. Pravtcheva, K. Marcu, and F.H. Ruddle, *J. Exp. Med.* **151**, 1545–1560 (1980).
96. D. Swan, P. D'Eustachio, L. Leinwand, J. Seidman, D. Keithley, and F.H. Ruddle, *Proc. Natl. Acad. Sci. USA* **76**, 2735–2739 (1979).
97. T. Meo, H.J. Johnson, V. Beechey, J. Andrews, J. Peters, and A.G. Searle, *Proc. Nat. Acad. Sci. USA* **77**, 550–553 (1980).
98. P. D'Eustachio, A.L. Bothwell, T.K. Takaro, D. Baltimore, and F.H. Ruddle, *J. Exp. Med.* **153**, 793–800 (1981).
99. R. Westerwoudt, in J.J. *Methods Enzymol.* **121**, 3–18 (1986).
100. F.J. Castillo, L.J. Mullen, B.C. Grant, J. DeLeon, J.C. Thrift, L.W. Chang, J.M. Irving, and D.J. Burke, *Dev. Biol. Stand.* **83**, 55–64 (1994).
101. L. Reitzer, B. Wice, and D. Kennell, *J. Biol. Chem.* **254**, 2669–2676 (1979).
102. H. Zielke, C. Zielke, and P. Ozand, *Fed. Proc.* **43**, 121–125 (1984).
103. D. Barngrover, J. Thomas, and M. Thilly, *J. Cell. Sci.* **78**, 173–189 (1985).
104. A. Fiechter and F.K. Gmunder, in A. Fiechter ed., *Advances in Biochemical Engineering / Biotechnology*, vol. 39, Springer-Verlag, Berlin, 1989, pp. 1–28.
105. K. Low and C. Harbour, *Dev. Biol. Stand.* **60**, 73–79 (1985).
106. D. Velez, S. Reuveny, L. Miller, and J. Macmillan, *J. Immunol. Methods* **86**, 45–52 (1986).
107. O.T. Ramirez and R. Mutharasan, *Biotechnol. Bioeng.* **36**, 839–848 (1990).
108. Y.J. Schenider and A. Lavoix, *J. Immunol. Methods* **129**, 251–268 (1990).
109. M.C. Flickinger, N.K. Goebel, T. Bibile, and S. Boyce-Jacino, *J. Biotechnol.* **22**, 201–226 (1992).
110. L. Xie and D.I. Wang, *Biotechnol. Bioeng.* **51**, 725–729 (1996).
111. R.G. Ham and W.L. McKeehan, *Methods Enzymol.* **58**, 44–93 (1979).
112. G. van Meer, *Ann. Rev. Cell. Biol.* **5**, 247–275 (1989).
113. H. Murakami, H. Masui, G.H. Sato, N. Sueoka, T.P. Chow and T. Kanosueoka, *Proc. Nat. Acad. Sci. USA* **79**, 1158–1162 (1982).
114. J. Darnell, H. Lodish, and D. Baltimore, *Molecular Cell Biology*, Scientific American Books, New York, 1986.
115. B. Alberts, D. Bray, J. Lewis, M. Raff, K. Roberts, and J.D. Watson, *Molecular Biology of the Cell*, Garland Publishing, New York, 1994.
116. S. Savonniere, N. Zeghari, L. Miccoli, S. Muller, M. Maugras, and M. Donner, *J. Biotechnol.* **48**, 161–173 (1996).
117. R. Boraston, P.W. Thompson, S. Graland, and J.R. Birch, *Dev. Biol. Stand.* **55**, 103–111 (1984).
118. W.M. Miller, C.R. Wilke, and H.W. Blanch, *J. Cell. Physiol.* **132**, 524–530 (1987).
119. D. Wolpart, D. Kirwan, and J. Gainer, *Biotechnol. Bioeng.* **36**, 630–635 (1990).
120. H.A. Phillips, J.M. Scharer, N.C. Bols, and M. Moo-Young, *Biotechnol. Lett.* **9**, 745–750 (1987).
121. P.A. Underwood, *Methods Enzymol.* **121**, 301–306 (1986).
122. H. Murakami, in M. Umeda, H. Koyama, J. Minowada, and M. Oishi, eds., *Biotechnology of Mammalian Cells*, Japan Scientific Societies Press/Springer-Verlag, Berlin, 1987, pp. 119–136.
123. H. Murakami, in A. Mizrahi ed., *Monoclonal Antibodies: Production and Applications*, Liss, New York, 1989, pp. 107–141.
124. F.J. Darfler, *In Vitro Cell. Dev. Biol.* **26**, 769–778 (1990).
125. E. Mariani, A.R. Mariani, M.C.C. Monaco, E. Lalli, M. Vitale, and A. Facchini, *J. Immunol. Methods* **145**, 175–183 (1991).
126. H.R. Maurer, *Cytotechnology* **5**, 1 (1991).
127. R.J. Ertola, A.M. Giulietti, and F.J. Castillo, in J.A. Asenjo and J.C. Merchuck eds., *Bioreactor System Design*, Dekker, New York, 1995, pp. 89–137.

128. M.J. Keen and T.W. Steward, *Cytotechnology* **17**, 203–211 (1995).
129. Y.M. Qi, P.F. Greenfield, and S. Reid, *Cytotechnology* **21**, 95–109 (1996).
130. Y. Miyazaki, S. Nishimoto, T. Sasaki, and T. Sugahara, *Cytotechnology* **26**, 111–118 (1998).
131. *Proc. Int. Meeting on Transmissible Spongiform Encephalopathies-Impact on Animal and Human Health, Heidelberg, Dev. Biol. Standard* vol. 80, 1993.
132. D.M. Ascher, *Transmissible Spongiform Encephalopathies: Towards Risk Assessment and Management*, FDA Update, CBER, FDA, Washington, D.C., 1997.
133. German Federal Ministry of Health, *Guidelines on Safety Measures in Connection with Medicinal Products Containing Body Materials Obtained from Cattle, Sheep, or Goats for Minimizing the Risk of Transmission of BSE and Scrapie*, German Federal Bulletin no. 40, Feb. 26, 1994.
134. K. Nagira, M. Hayashida, M. Shiga, K. Sasamoto, K. Kina, K. Osada, T. Sugahara, and H. Murakami, *Cytotechnology* **17**, 117–125 (1995).
135. D.W. Karl, M. Donovan, and M.C. Flickinger, *Cytotechnology* **3**, 157–169 (1990).
136. R.B. Kratje, W. Lind, and R. Wagner, *J. Biotechnol.* **32**, 107–125 (1994).
137. G.K. Sureshkumar and R. Mutharasan, *Biotechnol. Bioeng.* **37**, 292–295 (1991).
138. H.W.D. Katinger, W. Sheirer, and E. Kromer, *Ger. Chem. Eng.* **2**, 31–38 (1979).
139. J.R. Birch, P.W. Thompson, K. Lambert, and R. Boraston, ACS Meeting, Philadelphia, 1984.
140. J.R. Birch, R. Boraston, and L. Wood, *Trends Biotechnol.* **3**, 162–166 (1985).
141. J.F. Petersen, L.V. McIntire, and E.T. Papoutsakis, *Biotechnol. Prog.* **6**, 114–120 (1990).
142. A. Handa-Corrigan, in R.E. Spier and J.B. Griffiths eds. *Animal Cell Biotechnology*, vol. 4, Academic Press, London, 1990, pp. 123–132.
143. R.E. Spier and J.P. Whiteside, in R.E. Spier and J.B. Griffiths eds., *Animal Cell Biotechnology*, vol. 4, Academic Press, London, 1990, pp. 134–148.
144. I. Abu-Reesh and F. Kargi, *Enzyme Microb. Technol.* **13**, 913–919 (1991).
145. J.D. Michaels, A.K. Mallik, and E.T. Papoutsakis, *Biotechnol. Bioeng.* **51**, 399–409 (1996).
146. S.B. Pluschkell and M.C. Flickinger, *Cytotechnology* **19**, 11–26 (1996).
147. I. Reich, R. Schnaare, and E.T. Sugita, in J.E. Hoover ed., *Remington: The Science and Practice of Pharmacy*, Philadelphia College of Pharmacy and Science, Philadelphia, 1995, pp. 613–627.
148. K. Oyaas, T.M. Berg, O. Bakke, and D.W. Levine, in R.E. Spier, J.B. Griffiths, J. Stephanne, and P.J. Crooy eds., *Advances in Animal Cell Biology and Technology for Bioprocesses*, Butterworth, London, 1989, pp. 212–220.
149. S.K.W. Oh, P. Vig, F. Chua, W.K. Teo, M.G.S. Yap, *Biotechnol. Bioeng.* **42**, 601–610 (1993).
150. S.Y. Park and G.M. Lee, *Biotechnol. Bioeng.* **48**, 699–705 (1995).
151. X. Yang, G.W. Oehlert, and M.C. Flickinger, *Biotechnol. Bioeng.* **50**, 184–196 (1996).
152. S.K.W. Oh, F.K.F. Chua, and A.B.H. Choo, *Biotechnol. Bioeng.* **46**, 525–535 (1995).
153. J.S. Ryu and G.M. Lee, *Biotechnol. Bioeng.* **55**, 565–570 (1997).
154. K. Oyaas, T.E. Ellingsen, N. Dyrset, and D.W. Levine, *Cytotechnology* **17**, 143–152 (1995).
155. J.R. Birch and S.J. Pirt, *J. Cell Sci.* **7**, 661–670 (1970).
156. S.J. Pirt, *Principles of Microbe and Cell Cultivation*, Blackwell Scientific Publications, Oxford, U.K., 1975.
157. A. Martial, M. Dardenne, J.M. Engasser, and A. Marc, *Cytotechnology* **5**, 165–171 (1991).
158. G. Majno and I. Joris, *Am. J. Pathol.* **146**, 3–15 (1995).
159. R.P. Sing, M. Al-Rubeal, C.D. Gregory, and A.N. Emery, *Biotechnol. Bioeng.* **44**, 720–726 (1994).
160. F. Franek, *Biotechnol. Bioeng.* **45**, 86–90 (1995).
161. S. Mercille and B. Masiie, *Biotechnol. Bioeng.* **44**, 1140–1154 (1994).
162. F. Franek and K. Chladkova-Sramkova, *Cytotechnology* **18**, 113–117 (1995).
163. F. Franek and K. Sramkova, *Cytotechnology* **21**, 81–89 (1996).
164. D.E. Martens, E.M. Sipkeme, C.D. de Goojier, E.C. Beuvery, and J. Tramper, *Biotechnol. Bioeng.* **48**, 49–65 (1995).
165. R. Boraston, P. Thompson, K.S. Garland, and J. Birch, *Dev. Biol. Stand.* **55**, 103–111 (1984).
166. M.G. Tovey and D. Brouty-Boye, *Exp. Cell. Res.* **118**, 383–388 (1979).
167. J.M. Savinell, G.M. Lee, and B.O. Palsson, *Bioprocess Eng.* **4**, 231–234 (1989).
168. D.I.C. Wang, C.L. Cooney, A.L. Demain, P. Dunnill, A.E. Humphrey, and M.D. Lilly, *Fermentation and Enzyme Technology*, Wiley, New York, 1979.
169. O.W. Merten, S. Reiter, G. Himmler, W. Scheirer, and K. Katinger, *Dev. Biol. Stand.* **60**, 219–227 (1985).
170. O.W. Merten, G.E. Palfi, G. Klement, and F. Steindl, in R.E. Spier and J.B. Griffiths eds., *Modern Approaches to Animal Cell Technology*, Butterworth, Somerset, U.K., 1987, pp. 381–396.
171. O.W. Merten, *Cytotechnology* **1**, 113–121 (1988).
172. E. Suzuki and D.F. Ollis, *Biotechnol. Prog.* **6**, 231–236 (1990).
173. M.C. Flickinger, N.K. Goebel, and M.A. Bohn, *Bioprocess Eng.* **5**, 155–164 (1990).
174. O.W. Merten, H. Keller, L. Cabanie, M. Leno, and M. Hardefelt, *Cytotechnology* **4**, 77–89 (1990).
175. P.M. Hayter, N.F. Kirkby, and R.E. Spier, *Enzyme Microb. Technol.* **14**, 454–461 (1992).
176. O.W. Merten, D. Moeurs, H. Keller, M. Leno, G.E. Palfi, L. Cabanie, and E. Couve, *Biotechnol. Bioeng.* **44**, 753–764 (1994).
177. M. Harigae, M. Matsumura, and H. Kataoka, *J. Biotechnol.* **34**, 227–235 (1994).
178. G.G. Banik and C.A. Heath, *Biotechnol. Bioeng.* **48**, 289–300 (1995).
179. W.R. Kidwell, *Bio / Technology*, **7**, 462–463 (1989).
180. J.P. Barford, P.J. Phillips, C.P. Marquis, and C. Harbour, *Cytotechnology* **21**, 133–148 (1996).
181. T.A. Bibila and D.K. Robinson, *Biotechnol. Prog.* **11**, 1–13 (1995).
182. H.P.J. Bonarius, V. Hatzimanikatis, K.P.H. Meesters, C.D. de Gooijer, G. Schmid, and J. Tramper, *Biotechnol. Bioeng.* **50**, 299–318 (1996).
183. G.S. Oh, T. Izuishi, T. Inoue, W.S. Hu, and T. Yoshida, *J. Ferment. Bioeng.* **81**, 329–336 (1996).

184. R.K. Biener, W. Waldraff, W. Noe, J. Haas, M. Horwaldt, and E.D. Gilles, *Ann. N.Y. Acad. Sci.* **782**, 272–285 (1996).
185. Commission of the European Communities, *Tibtech* **6**, G5–G8 (1988).
186. *Production of Recombinant Antibodies in the Milk of Transgenic Animals*, Genzyme Transgenics Corp., Framingham, Mass., 1996.
187. J. Hodgson, *Bio/Technology* **8**, 990 (1990).
188. R.H. Kimberlin, *Dev. Biol. Stand.* **75**, 75–82 (1991).
189. J. Thommes, M. Halfar, S. Lenz, and A.R. Kula, *Biotechnol. Bioeng.* **45**, 205–211 (1995).
190. T.C. Ransohoff and H.L. Levine, in R. Seetharam and S.K. Sharma eds., *Purification and Analysis of Recombinant Proteins*, Dekker, New York, 1991, pp. 213–235.
191. P. Gagnon, *Purification Tools for Monoclonal Antibodies*, Validated Biosystems, Tucson, Ariz., 1996.
192. P. Gagnon, *Gen. Eng. News.* **17**, 6–33 (1997).
193. H. Murakami, *Cytotechnology* **3**, 3–8 (1990).
194. *Gen. Eng. News.* **13**, 1–25 (1993).
195. *Wall Street Journal*, **11**, 1, 8 (Sept. 2, 1993).
196. C.N. Durfor and C.L. Scribner, *Ann. N.Y. Acad. Sci.* **665**, 356–363 (1992).
197. D.E. Wierenga, J. Cogan, and J.C. Petricciani, *Biologicals* **23**, 221–224 (1995).
198. J.C. Petricciani and F.N. Horaud, *Biologicals* **23**, 233–238 (1995).
199. G. Schaffner, M. Haase, and S. Giess, *Biologicals*, **23**, 253–259 (1995).
200. *FDA Guidance Concerning Demonstration of Comparability of Human Biological Products, Including Therapeutic Biotechnology-derived Products*, FDA, Washington, D.C., April 1996.
201. M.A. Schenerman and K. Phillips, *BioPharmaceuticals* **10**, 20–72 (1997).
202. B.J. Moellering, J.L. Tedesco, R.R. Townsend, M.R. Hardy, R.W. Scott, and C.P. Prior, *BioPharmaceuticals* **3**, 30–38 (1990).
203. B.L. Maiorella, J. Winkelhake, J. Young, B. Moyer, R. Bauer, M. Hora, J. Andys, J. Thomson, T. Patel, and R. Parekh, *Bio/Technology* **11**, 387–391 (1993).
204. L.C. Chen, *Gen. Eng. News* **17**, 1–10 (1997).

See also AFFINITY FUSIONS, GENE EXPRESSION ; AMMONIA TOXICITY, ANIMAL CELLS; ANTIBODY PURIFICATION; CULTURE MEDIA, ANIMAL CELL, LARGE SCALE PRODUCTION.

HYPOXIA, EFFECTS ON ANIMAL CELLS

DIANE L. HEVEHAN
WILLIAM M. MILLER
Northwestern University
Evanston, Illinois

KEY WORDS

Antioxidants
Gene expression
Metabolism
O_2-sensing
pO_2
Reactive oxygen species
Stress proteins

OUTLINE

INTRODUCTION

Oxygen (O_2) plays a primary role in cell metabolism and viability. By serving as the terminal electron acceptor in oxidative phosphorylation, O_2 is important for energy production. Oxygenation is a major consideration in the design of bioreactors for cultivating cells. Although mammalian cells do not consume O_2 rapidly (the specific O_2 uptake rate, q_{O_2}, is typically 0.05–0.5 $\times$ 10^{-9} mmol cell^{-1} h^{-1}) (1,2), O_2 can be a limiting factor in part because of its poor solubility in aqueous media. Oxygen demand by cells is regulated by environmental conditions such as O_2 tension (pO_2), pH, cell type, cell density, and nutrient depletion, all of which may vary throughout the culture period. To accommodate increasing O_2 demands as cultures increase in cell density and proliferative state, a number of bioreactor configurations have been developed (3): agitation, use of microcarriers for adherent cells, and sparging can effectively enhance the driving force for O_2 transfer.

Although much attention has been focused on overcoming extracellular resistances to O_2 transfer, intracellular events further complicate the development of optimal oxygenation strategies. During aerobic metabolism, the reduction of molecular O_2 to water may yield three potentially damaging reactive oxygen species (ROS): the superoxide anion (O_2^-), the hydroxyl radical ($OH\cdot$), and hydrogen peroxide (H_2O_2). ROS are also produced independently of respiration by O_2-consuming reactions in the cytosol (4) as well as in the culture medium. ROS can damage lipids, proteins, sugars, and DNA. Such damage, collectively known as oxidative stress, can lead to damage of the cell membrane and induction of cell death if it is not counteracted by radical-scavenging defense systems. This dual nature of O_2 should be considered when developing mammalian cell culture strategies, and provides an impetus to use controlled hypoxic conditions to optimize reactor productivity (5). The extent of hypoxia can be characterized by the extent to which q_{O_2} is decreased relative to that for cells without O_2 limitation. Moderate hypoxia not only will decrease q_{O_2} but may also lower ROS concentrations and increase metabolic efficiency (6). Insights into the physiological responses of cells to hypoxia will facilitate the use of pO_2 as a manipulatable parameter during cell culture, as well as provide increased understanding of the consequences of an interruption in the O_2 supply.

It would not be possible for a cell to respond to O_2 deprivation and O_2 toxicity without an inherent ability to sense alterations in pO_2. Indeed, cells have evolved complex regulatory mechanisms to ensure metabolic efficiency and protection against oxidative damage. As the O_2 supply changes, cells in vivo and in vitro adapt through multiple physiological responses, ranging from altered energy metabolism and production of antioxidants to changes in growth and differentiation. O_2 deprivation elicits homeostatic responses on the systemic, local, and intracellular levels. Adaptation at the systemic level involves a systemwide increase in O_2 delivery to the tissues through regulation of breathing, red blood cell mass (erythropoiesis), and vasodilation. Locally, the tissue response to hypoxia entails proliferation of capillaries (angiogenesis). Adaptation to O_2 deprivation at the cellular level involves a shift from oxidative phosphorylation and gluconeogenesis to anaerobic glycolysis, as well as the expression of stress-specific proteins. Whether at the organismal or cellular level, many of the effects noted here are mediated through a cascade of events that are initiated by the sensing of pO_2 and propagated by a signal which ultimately leads to hypoxic regulation of specific gene expression. The proteins coded for by these genes facilitate cellular survival at low pO_2 and act to restore normal pO_2.

This article focuses on the current understanding of O_2-dependent responses of animal cells, particularly those exposed to hypoxic conditions. The effects of pO_2 on cells in a wide range of culture systems are addressed first. Subsequently, the molecular mechanisms by which hypoxia mediates its effects—O_2-sensing, signal transduction, and gene expression—are examined. Mechanisms used by cells in vivo have provided the groundwork for uncovering many of the mechanistic details that underlie the hypoxic responses of cultured cells. Although different cells exhibit similar coping mechanisms during periods of O_2 deprivation, their specific adaptation to hypoxia is individually tailored according to cell type and function, as well as the extent of hypoxia.

CELL RESPONSES TO pO_2 AND HYPOXIA

Upon first encountering O_2 deprivation, cells respond rapidly by altering several properties, such as reducing proliferation, shifting to anaerobic glycolysis, and inhibiting protein synthesis. Survival under sustained hypoxia requires a second phase of adaptive responses, which are characterized by extensive cellular and metabolic reprogramming. The selective expression of stress proteins and other specific proteins underlies these secondary processes. Upregulated expression of genes involved in glycolysis and downregulated expression of those involved in gluconeogenesis or the TCA cycle is a prime example of this adaptive mechanism. Immediate and adaptive responses to changes in pO_2 are discussed in the following sections.

Growth, Viability, and Differentiation

The effects of dissolved oxygen concentration (DO) on cell growth in batch culture have been evaluated by many investigators (7–12). The DO value for maximal cell concentration in batch culture is typically about 50% of air saturation (50% DO) but may vary considerably with cell type. Lower values of DO, 5–35% (13) and 0.5% (6), have been reported for maximal steady-state viable cell concentrations in continuous hybridoma culture. Besides differences between cell lines, poor batch growth at low DO could result from added stress from low initial cell concentrations, glucose depletion, or insufficient time for the cells to adapt to low DO.

Steady-state cell viability has been shown to increase as the DO is decreased to 0.1–0.25% DO in continuous culture (6,13). Enhanced viability at lower DO may reflect less oxidative stress imposed on the cells. The greater production of ROS associated with progressively greater DO could damage cellular components and induce cell death. Mammalian cells are able to tolerate the presence of ROS because of the action of antioxidants. However, damage ensues when antioxidant defenses are overwhelmed by ROS. Indeed, in a murine hybridoma culture, increased DNA fragmentation correlated with increasingly hyperoxic culture conditions from 200% to 476% DO (14). Although maximum cell growth in batch culture usually occurs below 100% DO, some cell lines have been adapted to grow at near-maximal rates at DO values as high as 200% DO (15). CHO cells have been adapted to 380% DO (16) and, more recently, a murine hybridoma was adapted for growth at 150% DO by gradually exposing the cells to higher DO levels in continuous culture (17). In both cases, resistance to O_2 toxicity has been attributed to increased activity of antioxidant enzymes during the adaptation process.

An obvious way to avoid oxidative stress is to reduce the amount of O_2 available to the cells, as was demonstrated for hybridomas (6). Low pO_2 stimulates the proliferation of other cell types, particularly fibroblasts seeded at low density (18–20). The proliferation rate and lifetime (i.e., number of population doublings before senescence) of WI-38 and IMR-90 lung embryo fibroblast cells were increased by reducing the gas-phase O_2 concentration from 20% to 2% (19). Based on these findings, Falanga and Kirsner (21) set out to establish fibroblast cultures from single cells, which normally fail to proliferate under 20% O_2. Low pO_2 stimulated and maintained single-cell proliferation, which is consistent with fibroblast proliferation during hypoxic situations in vivo (e.g., during wound healing).

Increased clonal growth of cells under O_2 concentrations less than that of the ambient atmosphere suggest that pO_2 may play a role in directing developmental processes. In other words, pO_2 in tissues may regulate not only cell growth and death but also cell differentiation. This hypothesis and the putative pO_2 gradients that exist within the bone marrow microenvironment provide the rationale for investigating the effects of pO_2 in hematopoietic cell culture. Earlier work showed that a decrease in the gas-phase O_2 tension from 20% to 5% O_2, which better approximates the in vivo environment, enhances the size and number of hematopoietic colonies originating from precursor cells in semisolid culture (22–25). The documented effects of pO_2 in liquid hematopoietic cultures have been controversial, with optimal production reported under 20% O_2 for murine bone marrow cultures (26) and 5% O_2 for cord blood mononuclear cell cultures (27,28). For cord blood mononuclear cell cultures, Smith and Broxmeyer (29) reported comparable cell numbers for mature cells of various lineages under 5% and 20% O_2. However, for cultures of $CD34^+$ peripheral blood cells, Laluppa (30) obtained greater production of megakaryocytes and erythroid cells under 20% O_2, while granulocyte production was greater under 5% O_2. In contrast, most reports are consistent in that low pO_2 is superior for the production and maintenance of progenitor cells of all lineages (27–31). As a result, progenitor cell and mature cell production may be regulated by pO_2 in an inverse manner (30,31). Thus, while low pO_2 may be favorable during the early part of cell culture, higher pO_2 later in culture may enhance the differentiation of certain lineages.

Oxygen also plays a role in the development of other tissues. For example, placental epithelial stem cells tend to proliferate when cultured under 5% O_2 and to differentiate under 20% (32). The use of three-dimensional tissue culture models has facilitated the characterization of embryoid bodies (EBs) obtained from embryonic stem cells. EBs were found to grow equally well and remain highly viable when cultured under 20% or 1% O_2 for the first 5 days of EB differentiation (33). In contrast, embryonic stem cell plating efficiency and differentiation into the hematopoietic lineage are enhanced under 5–7% O_2 (34,35).

Antioxidants and ROS

Mammalian cells have evolved a line of defense to limit the damage caused by ROS generated during aerobic metabolism. This antioxidant defense system includes glutathione, vitamin E, and enzymes such as superoxide dismutase (SOD), catalase, glutathione-*S*-transferase (GST), and glutathione peroxidase (GPX). In addition, extracellular antioxidants such as albumin, urate, and billirubin also protect cells against oxidative damage.

High pO_2 in culture may increase the production of ROS to levels that overwhelm inherent antioxidant defenses. The associated detrimental effects can be overcome by stepwise adaptation to increases in pO_2. Adaptation of hybridoma cells to 100% DO from 10% was associated with an increase in the specific activities of the antioxidant enzymes GPX, GST, and SOD (17). Similarly, adaptation of CHO cells to growth under 99% O_2 was associated with increased specific activities of SOD, catalase, and GPX (16). Thus, enhanced tolerance likely is caused in part by upregulation of the synthesis of these enzymes. For example, a *cis*-acting regulatory sequence that is responsive to ROS, the antioxidant responsive element (ARE), has been found to activate the genes encoding the enzymes GST and NAD(P)H:quinone reductase (36). However, the activity of some antioxidant enzymes may decrease again above 100% DO (17), possibly through inactivation by ROS.

Decreased colony formation at higher pO_2 in semisolid culture suggests that hematopoietic cells are susceptible to oxidative stress. Indeed, addition of antioxidants such as monothioglycerol and β-mercaptoethanol to high-pO_2 cultures yields colonies similar in size and number to those in low-pO_2 cultures (22,23,37,38). H_2O_2 and superoxide concentrations have been found to increase throughout the duration of bone marrow cell cultures (39). Progenitor cell production was increased by adding catalase, an H_2O_2 scavenger, or mannitol, an hydroxyl radical scavenger. In contrast, even though superoxide levels were decreased, the effects of SOD addition were negligible.

Rather than adding antioxidants or upregulating their synthesis through adaptation, an alternate strategy to reduce ROS damage is to decrease ROS generation by operating at low pO_2. Although low pO_2 can be beneficial in promoting proliferation, as already stated, and suppressing apoptosis (see following), it may actually sensitize cells to injury during reoxygenation. Involvement of ROS in reoxygenation injury has been demonstrated by decreased damage in tissues treated with antioxidants during reoxygenation (40). ROS involvement is also shown by increased glutathione levels during reoxygenation of CHO cells from severe hypoxia or anoxia (41). Glutathione levels increase more extensively after more severe hypoxic exposure and increase to a much greater extent after an abrupt increase in pO_2 than during gradual reoxygenation. Sensitization to oxidative damage could not be attributed to a decrease in glutathione levels during hypoxia. In fact, there is evidence to contradict a contribution of decreased antioxidant levels to sensitization to oxidative stress. Some antioxidant enzymes, including SOD, GST, and catalase, are part of a family of stress-related proteins that may be induced during hypoxia (42–44). Upregulated expression of these enzymes during O_2 deprivation may anticipate their role of detoxifying ROS produced during reoxygenation (45).

It has been suggested that intracellular ROS are mediators of apoptosis and may even trigger cell death if present at elevated levels (46). In support of this hypothesis, the features of cells undergoing apoptosis closely resemble those of cells stressed by oxidative damage in terms of their morphology, physiology, biochemistry, and gene expression (47). Intracellular ROS may induce cell death in several ways, from direct damage of DNA to protein oxidation. If oxidants induce apoptosis, a natural extension would be that antioxidants inhibit apoptosis. Indeed, antioxidants such as *N*-acetylcysteine (48), *N*-acetylcarnitine (49), idebenone (50), and trolox (analog of vitamin E) (51) have been shown to inhibit apoptosis. Antioxidant protection against cell death is not confined only to direct oxidant-induced apoptosis but also to receptor- and growth factor-mediated forms of apoptosis. This suggests that oxidation may be a common mechanism in the progression of many different forms of apoptosis. Studies correlating oxidative damage with the induction of apoptosis have led to the identification of another potential antioxidant, the Bcl-2 protein. The antioxidant role of Bcl-2 is suggested by its ability to protect cells from H_2O_2-induced and menadione (a quinone compound that generates O_2^-)-induced oxidative death (52,53). Overexpression of the Bcl-2 protein has also been shown to suppress apoptosis triggered by glucocorticoids, irradiation, chemotherapeutic agents, growth factor deprivation, and tumor necrosis factor alpha (TNF-α) (47,54). For these reasons, Bcl-2 is thought to inhibit cell death by decreasing the generation of ROS, which would in turn inhibit intracellular oxidation reactions critical to progression of the apoptotic program. However, Bcl-2 may have other functions that are unrelated to its putative antioxidant properties. Under hypoxic conditions with low levels of ROS, Bcl-2 is still able to protect cells from apoptotic death (55–57).

Stress-Related Proteins

Redirection of protein synthesis in response to hypoxia is part of a more general mammalian cell response to stresses such as temperature elevation and substrate deprivation. Such adverse conditions elicit expression of specific sets of stress proteins, namely, heat shock proteins (HSPs), glucose-regulated proteins (GRPs), oxygen-regulated proteins (ORPs), and hypoxia-associated proteins (HAPs). Stress proteins play an important role in adaptation to a new environment, as well as increased resistance to the same or other stresses. Which sets of proteins are upregulated varies with the cell type and the type of stress involved. However, recruitment of a specific family of proteins is not restricted to a single stimulus, as would be implied by its name.

The best-characterized of the stress proteins are HSPs, in particular the HSP70 family, the members of which are produced by many cell types as a protective response to high temperature, glucose deprivation, hypoxia, ischemia, the presence of heavy metal salts, and reoxygenation (58). These proteins function as chaperones that decrease protein denaturation and facilitate protein folding, processing, and trafficking. They are believed to help regulate biological processes including embryogenesis, differentiation, growth, and metabolism (59). GRPs have been less extensively characterized, but like HSPs, their response is not stress specific in regards either to their stimuli for upregulation or to their protection against other stresses (60). Investigations into stress protein induction in response to hypoxia and ischemia have implicated the upregulation of HSPs, GRPs, and ORPs. It should be noted that many of the ORPs have in fact been identified as a subset of the GRPs (61).

The effects of O_2 deprivation and subsequent reoxygenation on the potential coordinated regulation of HSPs and GRPs have been examined in CHO cells (61). The data indicate that while GRP induction is a sustained response to prolonged O_2 deprivation, HSP induction is only transient. Furthermore, when cells are reintroduced to an O_2-containing environment, the GRPs are repressed whereas the HSPs are again temporarily induced. Recently, an antioxidant role for both the constitutive and inducible forms of HSP70 has been confirmed (62). Overexpression of the HSP70s in rat cardiomyocytes confers a marked increase in tolerance to oxidative challenges including exposure to H_2O_2, hydroxyl radical, menadione, and hypoxia and reoxygenation. This finding is consistent with reports that oxidative stress induces expression of a number of HSPs, and that preexposure to mild hyperthermia or moderate oxidative stress significantly increases resistance of lymphocytes challenged with high doses of ROS (63). Interestingly, HSP32 (heme oxygenase-1) is also expressed in response to prolonged (~12 h) hypoxia (64).

Redirected protein synthesis is further demonstrated by the selective expression of ORP150 in cultured rat astrocytes exposed to hypoxia, but not in cells exposed to heat shock, H_2O_2, Co^{2+}, 2-deoxyglucose, or tunicamycin (65). Interestingly, the human analog of ORP150 is expressed in response to a wider range of conditions—exposure to hypoxia, 2-deoxyglucose, or tunicamycin, but not to heat shock—in the HeLa and human astrocytoma U373 cell lines (66). The wider expression range may be a characteristic of established cell lines because ORP150 induction is also restricted to hypoxia in human mononuclear phagocytes. As discussed earlier, many stress proteins upregulated by exposure to hypoxia are also involved in the response to other stresses. The term HAPs has been coined to describe a unique set of endothelial cell stress proteins that are distinct from HSPs, GRPs, and most ORPs. Like ORP150 in primary cells, these HAPs are preferentially upregulated by exposure to hypoxia, while other stresses such as heat and glucose deprivation do not induce them (67). Moreover, HAPs do not offer any cross-protection against other stresses. The glycolytic enzyme, GAPDH, has been identified as a HAP. GAPDH expression in the nuclear fraction of endothelial cells is upregulated mostly at the transcriptional level, and suggests a nonglycolytic role for this enzyme (68,69). In contrast, the GAPDH mRNA content in rat cardiomyocytes remains unchanged during hypoxia (64).

Although the molecular mechanisms underlying the regulation of stress proteins by hypoxia and reoxygenation are still unclear, it is known that their expression is, at least in part, regulated at the transcriptional level. Thus far, most insight into these regulatory mechanisms has been gained from studies of the HSP genes. Transient exposure of cultured myogenic cells to hypoxia has been shown to transcriptionally activate the HSP70 gene through binding of a nuclear transcription factor, heat shock transcription factor (HSTF), to distinct *cis*-acting regions, heat shock elements (HSEs), within the 5′ promoter (70). Furthermore, it was found that heat shock and hypoxia induce the HSP70 gene through the same regulatory HSTF–HSE complex. However, only one of two functional HSEs is required for heat shock inducibility, whereas both are needed for hypoxic induction (71). The activation of HSTF is highly dependent on cellular redox potential. Thiol-reducing agents inhibit HSTF activity and the expression of HSP genes in HeLa cells challenged by heat or other stresses (72), which is consistent with a report that the antioxidant *N*-acetyl-L-cysteine decreases the induction of HSP32 mRNA and protein in rat cardiomyocytes exposed to hypoxia (64).

Cell Metabolism

O_2 is a substrate for ATP generation and a modulator of the oxidation–reduction potential of cell culture media. Thus, pO_2 is a major determinant of the metabolic profile of a cell, affecting both nutrient consumption (oxygen, glucose, and glutamine) and waste production (lactate and ammonia) (73). Steady-state metabolic rates, including q_{O_2}, are not greatly affected by pO_2 between 10% and 100% DO (15–150 mm Hg). pO_2 dependence above 10 mm Hg is reflected primarily by changes in the oxidative phosphorylation ratios: cytoplasmic [ATP]/[ADP][P_i] and intramitochondrial [NAD^+]/[NADH] (74). As pO_2 and the respiration rate decrease, the rate of ATP synthesis declines. The resulting excess in ATP consumption over synthesis decreases the [ATP]/[ADP][P_i] ratio, which stimulates respiration (74). If pO_2 is held steady at the lower value, the [ATP]/[ADP][P_i] ratio will decline until a new steady state is attained. The rapid turnover of ATP allows these changes to occur within a few seconds and, hence, respiration appears unaffected by changes in pO_2. Moreover, a decrease in pO_2 may also lower the [NAD^+]/[NADH] ratio, which also enhances respiration (74). However, below a pO_2 of about 10 mm Hg, changes in the [ATP]/[ADP][P_i] and [NAD^+]/[NADH] ratios are not sufficient to maintain baseline respiration rates (74), and steady-state q_{O_2} decreases with pO_2 in a Michaelis–Menten-type manner with $K_{O_2} < 1$ mm Hg for hybridomas (6,13). Prolonged exposure to severe hypoxia (80% decrease in q_{O_2}) or anoxia has been shown to suppress CHO-K1 cell O_2 uptake, as evidenced by a transiently (~24 h) lower value of q_{O_2} at the same pO_2 value during reoxygenation (41).

To compensate for reduced aerobic metabolism during O_2 deprivation, the flux through glycolysis increases, with most of the glucose being converted to lactate. This phenomenon, referred to as the Pasteur effect, has been observed by many investigators (5,9,10,13,73,75). Glutamine metabolism is also O_2 dependent. Glutamine is an essential nitrogen and carbon source for the biosynthesis of metabolic intermediates and for energy production via the TCA cycle; ammonia and alanine are its major nitrogen-containing by-products. When excess O_2 is available, cultweed cells typically obtain most of their metabolic energy from glutamine oxidation (13). In contrast, when O_2 is limiting, cells favor glucose over glutamine for energy production. Because glucose metabolism is diverted from providing metabolic intermediates to providing energy at low pO_2, glutamine is then recruited to a greater extent for cell mass production. This is illustrated by an increase in glutamine utilization and a diminishing yield of NADH from glutamine at very low DO (6,13). The greater glutamine consumption rate also makes more nitrogen available for biosynthesis, which is consistent with higher specific production rates for ammonia and alanine (13,73). Similarly, Ozturk and Palsson (13) reported a large increase in the specific consumption rate of most amino acids under anoxic conditions in continuous hybridoma culture.

Interestingly, the metabolic alterations observed under hyperoxic conditions may resemble those occurring under hypoxia. A shift to anaerobic metabolism has been reported in hyperoxic cultures of CHO cells (76), hybridomas (17), WI-38 cells (9), and human lymphocytes (77). Hyperoxia has also been reported to increase the glutamine consumption rate (17). At first thought, it is counterintuitive that there would be an increasing reliance on glycolysis at high DO. However, the increased glucose consumption rate may result from O_2 toxicity. Enhanced glycolysis may compensate for impaired mitochondrial function under hyperoxia, as well as provide additional energy to prevent and repair oxidative damage.

Although an increase in anaerobic metabolism to make up for energy deficits is useful for surviving O_2 limitation, the greatest defense against hypoxia would be a suppression of energy demand. This strategy has been perfected by cells from the highly anoxia-tolerant aquatic turtle. In response to anoxia, ATP demand is decreased by 90% in aquatic turtle hepatocytes and by 50% in brain cortical cells (78).

Protein Synthesis

Protein turnover accounts for a large fraction of the ATP demand during normoxia (78). Therefore, inhibition of protein synthesis and degradation under O_2-limiting conditions would conserve energy and contribute to maintaining the ATP supply–demand balance. Indeed, an important feature and one of the first signs of hypoxia in many cell types is a rapid and dramatic decline in overall protein biosynthesis (78–81), which has been attributed to a block in translation (78,80). Protein synthesis rapidly recovers to control levels on reoxygenation. Protein degradation is increased by ~30% under hypoxia in NHIK 3025 cervical carcinoma cells (79), but is decreased by 94% in aquatic turtle hepatocytes (78). The rapid reduction of protein synthesis that occurs when O_2 is limited does not appear to be in response to blockage of the electron transport chain or to the associated change in redox state, but rather has been attributed to the presence of an O_2 sensor that senses the lack of molecular O_2 or O_2 by-products under hypoxic conditions.

Most reports on the effects of DO on protein production by cultured cells have been limited to antibody production by hybridomas and lymphocytes. The effects of pO_2 on the specific antibody production rate (q_{Ab}) are cell line specific, and the optimal DO level for cell growth is often different than that for antibody production. Over the range from 1% to 100% DO, q_{Ab} may be greater at low DO (7,75), greater at intermediate DO (6), or largely unaffected by DO (13,17,82,83). q_{Ab} may be decreased below 1% DO (6,84). However, other investigators have reported no change in q_{Ab} from anoxic exposure in continuous or perfusion culture (13,85). It must be emphasized that these conclusions are limited to the conditions tested. For example, although q_{Ab} for the CC9C10 hybridoma cell line was similar for 10–100% DO, q_{Ab} was 50% greater at 125% DO (17). Also, cells grown in serum-free media (SFM) appear to be more sensitive to low DO. Three hybridoma cell lines cultured in SFM exhibited a 50–85% lower q_{Ab} for DO $\leq$ 7.5%, while the same cell lines grown with serum showed no decrease in q_{Ab} even at 3% DO (82).

Much less has been reported regarding the effects of pO_2 or hypoxia on recombinant protein production. The limited data available suggest that q_P is generally greater at higher DO, but the DO range over which the transition occurs is cell line dependent. Chotigeat et al. reported an increase in steady-state follicle-stimulating hormone (FSH) production with DO by CHO cells in perfused microcarrier culture; q_{FSH} was threefold greater at 60–90% DO than at 10% DO (86). Similarly, Wang et al. reported 50–240% greater q_{EPO} by L-929 cells at 75% DO in batch microcarrier culture as compared to that at 7.5–30% DO, with a larger increase in q_{EPO} at 75% DO for cells cultured in SFM (87). In contrast, q_{tPA} by BHK cells was unchanged at 7.5–75% DO in batch microcarrier culture and then increased with DO up to 225–300% DO for cells in serum-containing and serum-free media, although q_{tPA} did fall off at 420% DO (12). Cell growth was inhibited for DO > 75% of air saturation, so that the maximum tPA titer was obtained at 225% and 150% DO for cells in serum-containing and serum-free media, respectively. At the other end of the DO spectrum, Lin et al. examined the effects of O_2 deprivation on tPA production by CHO cells in perfused microcarrier culture (5). q_{tPA} did not decrease from that at 33% DO during exposure to mild hypoxia (40–60% decrease in q_{O_2}), but was affected by more severe O_2 deprivation. q_{tPA} decreased by 80% during the first day of severe hypoxia (80% decrease in q_{O_2}), and then recovered to control levels during the next 2 days of severe hypoxia. Upon reoxygenation, q_{tPA} transiently (3 days) increased to twice the control value. During anoxic exposure, q_{tPA} gradually decreased by 80% over 2 days and then remained at that value even for 2 days after reoxygenation (5).

The attraction in using mammalian cells for recombinant protein expression lies with their ability to carry out complex posttranslational processing, which is often an important determinant of a protein's activity in vivo. Few studies have addressed the effect of pO_2 or hypoxia on protein glycosylation. Regoeczi et al. (88) examined the in vivo glycosylation of serum proteins obtained from rats placed in a hypobaric chamber at 380 mm Hg (80 mm Hg pO_2). The only significant difference in transferrin glycosylation was a greater proportion of molecules fucosylated (fucosylation index) in hypoxic vs. control animals (28% vs. 24%; $p = .02$). In contrast, the IgG fucosylation index was lower in hypoxic animals (10% vs. 20%; $p = .012$). Chotigeat et al. (86) examined the effects of DO on the isoelectric focusing (IEF) gel profile of FSH produced by CHO cells in perfused microcarrier culture. FSH acidity increased with DO, as evidenced by an increase in the fraction of FSH isoforms with pI $\leq$ 4.5 from 65% at 10% DO to 88% at 90% DO. The shift of FSH isoforms to lower pI was attributed to more extensive sialylation at higher DO values, which is consistent with a fivefold greater sialyltransferase activity in cells at 90% DO vs. that at 10% DO. Lin et al. (5) examined the specific activity and gel electrophoresis profiles of tPA produced by slowly growing CHO cells in perfused microcarrier culture. Exposure to hypoxia or anoxia and subsequent reoxygenation did not significantly affect tPA specific activity or site occupancy (fraction of tPA molecules glycosylated at three sites [68 kDa] vs. two sites [65 kDa]). However, during reoxygenation from anoxia, an extra tPA band appeared at 70 kDa. This band, which may have resulted from disrupted signal peptide cleavage or altered glycosylation, disappeared after 3 days at 33% DO.

Growth Factors and Their Receptors

The effects of O_2 on hematopoietic cell growth and differentiation may be caused by differences in the production of growth factors or their receptors with pO_2. Oxygen-regulated production of the cytokine erythropoietin (EPO) and of several angiogenic growth factors has been well established. There is evidence for specific effects of pO_2 on individual cell types. For example, macrophages produce cytokines in an O_2-dependent manner (89), while lymphocytes and macrophages that are cocultured under 5% O_2 cooperate to produce a factor that stimulates proliferation of erythroid precursor cells (90). There is also evidence that cord blood mononuclear cells produce more interleukin 6 (IL-6) under 5% compared to 20% O_2 (91).

Changes in growth factor receptor expression in response to low pO_2 may exhibit different patterns than those for the respective ligands. For example, while low pO_2 upregulates the transcription and synthesis of transforming growth factor β_1 (TGF-β) in dermal fibroblasts, culture under 2% O_2 for 24 h decreases receptor binding of [^{125}I]TGF-β by 65% and TGF-β type II receptor mRNA levels by 90% (92). Fibroblasts cultured under 2% O_2 also showed decreased receptor binding of [^{125}I]EGF (epidermal growth factor) that was reversible after 12 h in culture under 20% O_2 (92). For both receptors, decreased ligand binding was in large part the result of a lower number of binding sites. Exposure to hypoxia has also been shown to reversibly decrease [^{125}I]EGF binding to and the number of available EGF receptors (EGFR) on hepatocytes, without affecting total EGFR protein content (93). Interestingly, there was no effect of hypoxia on [^{125}I]insulin binding.

MOLECULAR MECHANISMS MEDIATING OXYGEN EFFECTS

The metabolic and cellular changes that accompany hypoxia are primarily the result of differential gene expression. The mechanisms underlying these gene-based responses involve hypoxia sensing, signal transduction, and ultimately expression of the appropriate gene.

Oxygen Sensing

Despite the advances being made toward identifying O_2-regulated cellular functions, the nature of the O_2 sensor itself and the path of the signal from the sensor to the machinery that controls protein expression and function remain unclear. Among the molecules that can bind to and react with molecular O_2 are heme proteins. By analogy to hemoglobin, heme proteins are thought to change conformational state between oxygenated and deoxygenated forms, thus regulating their function depending on pO_2. Involvement of a ferroheme protein in the O_2-sensing process, originally suggested by Goldberg et al. (94), is supported by several lines of evidence. Although much of the insight into mechanisms of O_2-sensing stems from studies on EPO-producing hepatocytes and on carotid body type I cells, it seems that a heme-based O_2 sensor system is present in most cell types (95).

The response to hypoxia can be mimicked, both in magnitude and direction, by exposure to cobaltous ions or iron-chelating agents (96) or blocked by carbon monoxide, hallmarks of the involvement of a heme-containing protein. Expression of hypoxia-inducible genes (Table 1) is generally enhanced in cells treated with desferrioxamine (DF), an iron chelator. Iron chelators prevent synthesis of functional heme proteins by inhibiting the incorporation of ferrous atoms into the heme molecule. Cells with a nonfunctioning O_2 sensor are unable to detect O_2 present in the environment and are likely to interpret their state as hypoxic. Furthermore, because iron catalyzes the conversion of H_2O_2 to OH· and OH^- via the Fenton reaction, iron chelators may significantly alter levels of chemical messengers in the O_2-sensing pathway and lead to an hypoxic-like response (97). The heme protein hypothesis is further substantiated by experiments in which cobalt and nickel substitute for iron in the heme moiety. These transition metals exhibit a characteristic low binding affinity for O_2, thereby locking the heme protein in the deoxy conformation and sending an hypoxic signal. All the hypoxia-induced genes examined have been shown to be stimulated under such conditions (Table 1), while hypoxia-inhibited genes are also inhibited by Co^{2+} and DF. Conversely, the effects of hypoxia can be abrogated by using carbon monoxide (CO), which substitutes for O_2, to lock heme in the oxy conformation (94). Studies on expression of the phosphoenolpyruvate carboxykinase (PCK) gene in primary rat hepatocytes (139) further confirm the CO–O_2 sensor relationship, although in this case it acts conversely: PCK gene expression is inhibited under hypoxia, but is stimulated under elevated pO_2 or in the presence of CO (Table 1). Parallel effects of CO on O_2-sensing are evident in the carotid body, where ventilation is regulated (142). The current through K^+ channels in type I cells is rapidly downregulated by lowering pO_2. Addition of CO to hypoxic type I cells reverses the downregulation of K^+ currents. Collectively, these data implicate a heme protein that changes conformation on binding O_2 or CO as the primary O_2 sensor.

It is likely that the O_2 sensor is a membrane-bound, multisubunit cytochrome *b* protein that binds O_2 and reduces it to ROS such as superoxide and H_2O_2, which then in turn leads to downstream signaling of gene transcription. Numerous findings have contributed to this hypothesis. Absorption spectra of hepatoma cells and carotid body cells coincide with those of type *b*-like cytochromes that undergo a shift in conformation in response to changing pO_2 (143,144). Among this group of *b*-cytochromes are those that are part of the NADPH–oxidase enzyme complex, which generates H_2O_2 in an O_2-dependent manner. During hypoxia, the H_2O_2-generating activity of the oxidase would be inhibited, leading to lower H_2O_2 levels (4). In confirmation, the rate of H_2O_2 production by hepatoma cells decreased as pO_2 decreased (145). A regulatory role for H_2O_2 in the coupling of environmental pO_2 to intracellular events has been demonstrated: adding exogenous H_2O_2 to hypoxic cultures negated the expected increase in EPO production, while adding antioxidants or catalase to high-pO_2 cultures (to reduce H_2O_2) increased EPO mRNA levels (146). H_2O_2 may have several modes of action in transducing the pO_2 signal to the transcriptional machinery in O_2-responsive genes. Thus, it seems likely that changes in environmental pO_2 that regulate cellular functions are sensed by an H_2O_2-generating heme protein. Recent findings, however, indicate that hypoxic induction of genes for vascular endothelial growth factor (VEGF), aldolase, and tyrosine hydroxylase (TH) does not involve NADPH oxidase (103,147). In particular, in PC-12 cells DF and Co^{2+} do not interfere with H_2O_2 generation by the O_2 sensor, but rather affect TH gene expression by a mechanism further downstream from H_2O_2 formation (103).

O_2-sensing also appears to involve the regulation of nucleotide cyclases, heme-containing proteins that catalyze the conversion of intracellular ATP/GTP to cyclic AMP/GMP (148). Hypoxia decreases cAMP and cGMP levels, which in turn may alter stimulation of protein phosphorylation pathways. It is not known whether these nucleotide cyclases actually sense extracellular pO_2 levels, similar to the mechanism proposed for oxidases, or if they are regulated by a signal generated further downstream from the O_2 sensor. For example, H_2O_2 can affect activation of guanylate cyclase and alter cGMP levels (4). In this way, during hypoxia, lower concentrations of H_2O_2 could contribute to modifying the phosphorylation status of certain proteins.

Contrary to these observations, it is also possible that heme proteins may not be involved in the hypoxic regulation of certain genes, such as endothelial GAPDH (69). If a heme protein is involved, it possesses different structural or functional properties from those involved in other hypoxia-responsive systems as regulation of endothelial GAPDH (e.g., by CO) differs from that of EPO or VEGF (see Table 1).

Signal Transduction

Protein Phosphorylation. A decrease in H_2O_2 concentration could transduce the hypoxic signal through several different pathways, most likely involving protein phosphorylation and redox potential. Protein phosphorylation is an important intracellular transduction mechanism because of its regulation of several transcription factors. A heme protein in the bacterium *Rhizobium melliloti* senses hypoxia and activates its kinase domain, leading to phosphorylation of transcription factors and subsequent gene expression (149). In eukaryotes, NADPH oxidase, the putative O_2 sensor that operates in some cell types, becomes phosphorylated on activation (150), suggesting that the dependence of O_2-sensing on a phosphorylation cascade is conserved between bacteria and mammals. Tyrosine phosphorylation has been implicated in the hypoxic activation of NF-κB DNA-binding activity (151) and in the transcriptional and posttranscriptional induction of the VEGF gene (152,153). The DNA-binding activity of hypoxia-inducible factor-1 (HIF-1; see next section) is also dependent on its phosphorylation status. Treatment with phosphatase or with kinase inhibitors prevented hypoxia-inducible binding and EPO gene expression, demonstrating the requirement for HIF-1 to be in a phosphorylated form (154).

Table 1. Oxygen-Regulated Gene Expression

Gene	Stimulus: Hypoxia	Stimulus: DF	Stimulus: Co^{2+}	CO reversal[a]	Regulation: Hypoxia-inducible DNA element	Regulation: mRNA stability element	References
Miscellaneous							
Erythropoietin (EPO)	+	+	+	Y	HIF-1	P	94,97–102
Tyrosine hydroxylase (TH)	+	+	+		HIF-1	Y	103–105
Transferrin (TF)	+				HIF-1		106
Inducible nitic oxide synthase	+				HIF-1		107
Heme oxygenase 1	+				HIF-1		108
Xanthine oxidase I	+						109
Retrotransposon VL30	+				HIF-1		110
Hypoxia-inducible factor-1α (HIF-1α)	+	+	+				111
Hypoxia-inducible factor-1β (HIF-1β)	+	+	+				111
Interleukin 2 (IL-2)	+						112
Interleukin 4 (IL-4)	+						112
Interferon-γ (IFN-γ)	+						112
Ornithine decarboxylase	+						113
Interleukin 10 (IL-10)	−						112
Angiogenic growth factors							
Vascular endothelial growth factor (VEGF)	+	+	+	Y	HIF-1	Y	96,114–118
Platelet-derived growth factor-β (PDGF-β)	+	+	+				96,119,120
Transforming growth factor-β (TGF-β)	+						121
Acidic/basic fibroblast growth factor (a/bFGF)	+		+				120
Tumor necrosis factor-α (TNF-α)	+				NF-IL-6 (P)		122
Endothelin	+						123
Interleukin 6 (IL-6)	+				NF-IL-6		112,124,125
Interleukin 1 (IL-1)	+						126
Interleukin 8 (IL-8)	+				NF-IL-6 (P)		124,127
Placental growth factor (PLGF)	−	−	−				96
Glucose transporters							
GLUT-1	+	+	+	Y	HIF-1	Y	128–131
GLUT-2	−	−	−				129
GLUT-3	+	+	+				129
Metabolic Enzymes							
Lactate dehydrogenase-A (LDH-A)	+	+	+		HIF-1		129,132–135
Glyceraldehyde-3-phosphate dehydrogenase (GAPDH)	+	+	+	N			69
Aldolase A	+	+	+		HIF-1		129,135,136
Enolase 1	+				HIF-1		135,136
Phosphoglycerate kinase 1 (PGK-1)	+	+	+		HIF-1		132,136
Phosphofructokinase L (PFK-L)	+	+	+		HIF-1		132,136
Pyruvate kinase	+						136,137
Hexokinase	+						137
Triose phosphate isomerase	+						138
Phosphoenolpyruvate carboxykinase (PCK)	−		−	Y			139,140
Tyrosine aminotransferase	−						141

Note: DF, desferrioxamine; +, induced gene expression; −, suppressed gene expression; Y, yes; N, no; P, probable, but not proven; HIF-1, hypoxia-inducible factor-1-binding motif; NF-IL-6, nuclear factor–interleukin-6-binding motif.
[a]Reversal of hypoxic effects by carbon monoxide (CO).

Redox Potential. Although hypoxia-mediated enhancement of DNA-binding activity of several transcription factors could reflect changes in the phosphorylation state of these proteins, it could also reflect changes in the cellular redox state. Cellular redox potential is governed in part by the ratio of reduced (SH) to oxidized (S–S) thiols that results from the surrounding pO_2. Under normoxic conditions, the equilibrium is maintained by H_2O_2 either through direct oxidation of reduced glutathione (GSH) to oxidized glutathione (GSSG) or through formation of the hydroxyl radical, which can oxidize glutathione and protein thiol groups (4). Under hypoxic conditions, the decreasing H_2O_2 concentration is accompanied by a shift in the equilibrium toward the reduced forms of glutathione and protein thiols. This shift in redox state can regulate the activity of K^+ channels and the DNA-binding activity of transcription factors (4).

Several studies have coupled changes in transcription factor activity with perturbations in the redox state of the cell. Transcription factors AP-1, NF-κB, and p53, as well as protein factors such as heat shock proteins (HSPs), are redox sensitive. An ubiquitous nuclear protein called Ref-1 serves as a redox factor that stimulates AP-1 DNA-binding activity through single conserved cysteine residues in the c-Fos and c-Jun proteins that comprise the AP-1 transcription complex (155). AP-1 activity is enhanced by antioxidants and reducing agents (155,156), whereas it is inhibited by H_2O_2 (157) and sulfhydryl blockers (155). Furthermore, data suggest that the c-*fos* and c-*jun* family of genes are regulated at the transcriptional level as well, with exposure to hypoxia inducing mRNA expression (114). However, in these latter studies AP-1 activity was not assessed. In contrast to AP-1, NF-κB activity is induced by oxidative stress: H_2O_2 can activate NF-κB, while treatment with thiol reagents or other antioxidants abolishes its activity (156,158). A conflicting report has shown enhanced NF-κB activity under hypoxic conditions (151). Like AP-1, it seems that NF-κB also requires reduction of specific cysteine residues in its DNA-binding domain, since its association with DNA in vitro is inhibited by alkylating or oxidizing agents (159). However, there is currently no evidence linking modification of NF-κB thiol groups to regulation of its biological activity. This fact demonstrates the complex nature of regulation of cellular signaling through redox chemistry.

The functions of HSPs are multifold: they protect cells from stress from heat, ROS, and sulfhydryl reagents because they aid protein folding, processing, and transport. Expression of HSPs is dependent on binding of heat shock transcription factors (HSFs) to their promoter region. The relevance of the cellular redox state in the heat shock response is demonstrated by the inhibition of HSF activity and HSP gene expression by thiol reducing agents (72). The induction of HSP expression with reoxygenation is consistent with these findings (61).

The role of redox status on the basic helix-loop-helix (bHLH) group of proteins, a broad group of transcription factors that share common helix-loop-helix regions required for dimerization, is also of considerable interest. Binding and activation of USF, a bHLH transcription factor, proceeds only when two cysteine residues in the HLH dimer interface are reduced (160). This mechanism also implicates redox regulation in the assembly of other transcription factors, in particular of the activation of HIF-1, a heterodimer consisting of two bHLH transcription factors, the HIF-1α and HIF-1β subunits (111).

Of the factors mentioned previously, HIF-1 is probably the most relevant to low-pO_2 studies because it is an ubiquitous transcription factor (see Table 1) whose function is restricted to the hypoxic regulation of a broad range of physiologically relevant genes (161–163). HIF-1 DNA-binding activity and the protein levels of HIF-1α and HIF-1β each increase exponentially as cells are exposed to decreasing pO_2 in vitro within the physiological range. A half-maximal response was seen between 1.5% and 2% O_2 in the gas phase, with a maximal response at 0.5% O_2 (124). Maintaining the proper redox ratio is essential for the hypoxic activation of HIF-1. Changing the redox state of cells through treatment with oxidizing agents impairs hypoxia signaling mechanisms by disrupting expression of HIF-1α and HIF-1 DNA-binding activity (164). HIF-1 must remain in the reduced state to enable binding to DNA. No cysteine residues were found in the basic DNA-binding domain of either HIF-1α or HIF-1β. However, one cysteine was found in the HLH region of HIF-1β and several in the PAS (Per-ARNT-Sim; named for the domain found in the *Drosophila* transcription factors Period and Single-minded and the mammalian aryl hydrocarbon receptor nuclear translocator protein) domain of each subunit (164). Interaction between the HLH and PAS domains is necessary for heterodimerization of HIF-1α and HIF-1β and for subsequent DNA binding. The redox-sensitive activation of HIF-1 has since been attributed primarily to the HIF-1α subunit (165).

Another factor that has been shown to activate transcription of specific target genes in hypoxic cells is C/EBPβ. C/EBPβ shows a structural feature akin to the AP-1 complex with a cysteine residue similarly positioned as in c-Fos and c-Jun that is critical for DNA-binding activity (124,155). C/EBPβ protein binds to an hypoxia-inducible element called the NF-IL-6 site in a tissue-specific manner (124,125). Although this site is distinct from the HIF-1-binding motif, the cellular processes leading to gene transcription may still be related. Given the analogy between the actions of HIF-1 and C/EBPβ and the implication that redox potential plays a role in both HIF-1- and C/EBPβ-mediated hypoxia-inducing mechanisms, the NF-IL-6 site becomes another important locus for induced gene expression under O_2 deprivation. The presence of tissue-specific factors regulates whether transcription can be driven by HIF-1 or C/EBPβ.

In addition to increased transcription rates, gene expression during hypoxia is also regulated by the stability of mRNA. Redox may also contribute to this posttranscriptional mode of gene regulation. Czyzyk-Krzeska (4) reported that binding of hypoxia-inducible protein factors to tyrosine hydroxylase mRNA stability elements requires that they be in the reduced state.

Thus, posttranslational signaling mechanisms that involve both phosphorylation and sulfhydryl group modifications contribute to the transcriptional and posttranscriptional response to hypoxia. HIF-1 activation best exemplifies these modes of regulation. In addition, activation of HIF-1 may depend on ligand binding. Association of HSP90 with HIF-1α allows this subunit to change conformation, thereby enabling dimerization with HIF-1β and subsequent DNA binding of the HIF-1 complex (166). Considering this information, it now becomes logical to extend the previous hypothesis suggesting that the signaling cascade begins with a heme protein O_2 sensor and is propagated by an intermediate, such as H_2O_2. Intermediate molecules are likely to target certain transcription factors, such as HIF-1, and regulate their binding to the respective genes.

Gene Expression

Transcriptional Regulation. The identification of the aforementioned transcription factors has facilitated our understanding of O_2 regulation of genes and consequently advanced our understanding of the physiological adaptation to hypoxia. Despite a significant reduction in total mRNA synthesis when cells are deprived of O_2, transcription of several mammalian genes that encode for various hormones, cytokines, and enzymes is markedly increased. In particular, low pO_2 induces expression of genes controlling erythropoiesis, ventilation, angiogenesis, and energy metabolism. Furthermore, the commonality among the regulatory elements present in these genes suggest that they are part of a widespread O_2-sensing system. Although beyond the scope of this article, cellular adaptation to the other extreme, hyperoxia, also involves the upregulation of certain genes which, in this case, lead to detoxification of ROS (36).

EPO. Progress in understanding O_2-dependent gene expression has mostly stemmed from characterizing hypoxia-induced production of erythropoietin (EPO) because the effects are more dramatic for this cytokine than for the other O_2-sensitive genes studied thus far. EPO is the key hormone in erythropoiesis: synthesized primarily in the adult kidney and fetal liver, it travels to hematopoietic tissues where it binds to its receptor on erythroid progenitor cells, protecting them from apoptosis and allowing them to proliferate and differentiate into mature red blood cells. Under the hypoxic stress of anemia or reduced blood oxygenation, EPO production is as much as 1000 fold greater than normoxic levels (167). By increasing the number of red blood cells, EPO enhances the O_2-carrying capacity of the blood. Studies on the transcriptional control of the EPO gene have been greatly facilitated by the use of the cultured human fetal hepatoma cells Hep3B and HepG2 (168). Increased EPO production under hypoxia in these model systems mirrors the in vivo response. There is a close correlation between the level of hypoxia and the production rates of EPO protein and EPO mRNA in these cells.

Regulation of transcription rate involves interactions between *trans*-acting elements (transcription factors) and *cis*-acting elements within the promoters and enhancers of regulated genes. The best-studied mechanism of hypoxic regulation is the binding of HIF-1 to DNA, and is best exemplified by regulation of the EPO gene. The *cis* element most critical to hypoxic induction of EPO transcription is located in the 3′-flanking region of the gene, 120 bp downstream from the polyadenylation site. The 3′ enhancer includes three sequences that contribute to the hypoxic response: (*1*) a 5′ portion containing the highly conserved HIF-1-binding site, (*2*) the middle portion, which does not bind specific proteins but is necessary for the hypoxic induction of transcription, and (*3*) a 3′ portion containing highly conserved nuclear receptor-binding half-sites that resemble known steroid and thyroid hormone response elements. A variety of hormones act by binding to nuclear receptors. In particular, the binding of a specific nuclear receptor, orphan nuclear factor 4, to the EPO enhancer amplifies the transcriptional activation by HIF-1, demonstrates a marked impact on the hypoxic induction of the EPO gene and contributes to tissue specificity of this response (169).

The importance of the 3′ enhancer had previously misled some to believe that it was solely responsible for hypoxic induction of the EPO gene. However, transfection studies with the enhancer revealed that mRNA synthesis of a reporter gene was only induced about 10 fold, which is well below the induced level of endogenous EPO mRNA (170). This discrepancy was resolved through identification of O_2-sensitive elements within the promoter region, 5′ to the EPO gene (171). The 3′ enhancer and 5′ promoter act synergistically to achieve at least a 40-fold stimulation of transcription of the EPO gene under hypoxia, commensurate with endogenous EPO mRNA levels. Similar to the enhancer region, the EPO promoter contains nuclear hormone response half-sites. Alone, they do not contribute to hypoxic induction, but they are needed to amplify the induction signal. It has also been proposed that the EPO gene may be negatively regulated by higher pO_2. Other elements in the EPO promoter including GATA sites and ribonucleoprotein-binding sites may contribute to the inhibition of a constitutive activator protein and lead to repressed transcription under normoxia (167). During hypoxia, when the negative regulators are reduced, they could be displaced from their corresponding DNA-binding sites and allow transcription to proceed. This finding is consistent with studies that suggest that under high pO_2 the O_2 sensor generates greater amounts of ROS that lead to the inhibition of EPO gene expression (146).

More recently, other transcription factors have been implicated in transducing the signal from the HIF-1 site to the transcriptional initiator. The non-DNA-binding transcriptional activator P300/CBP binds to HIF-1α and enhances hypoxic induction by increasing the response of the EPO enhancer (172). P300/CBP is a multifunctional protein that, in addition to regulating several tissue-specific enhancers, plays a role in controlling the cell cycle and differentiation pathways. The HIF-1-binding site within the EPO enhancer also binds constitutively to a smaller protein complex consisting of the activating transcription factor-1 (ATF-1) and the cAMP-responsive element-1 (CREB-1) (173). During hypoxia, these two transcription factors are displaced by HIF-1. Thus, HIF-1-dependent induction of gene expression involves interactions between different transcription factors and sequences. It is likely that these interactions direct the specificity of the hypoxic response, that is, which genes or cells are affected.

Several findings have led to the conclusion that HIF-1-mediated induction of gene expression is not restricted to the EPO gene or to cells originating from the kidney or liver. Transfection experiments in many different cell lines using the EPO 3′ enhancer coupled to reporter genes have demonstrated the universality of this O_2-sensing and signal transduction mechanism (174). Factors with similar affinity for specific DNA sequences within the enhancer were activated regardless of whether or not the cells were EPO producing (98,175). Furthermore, mutated EPO enhancers, which are known to prevent HIF-1 binding, abolished hypoxic induction of mRNA synthesis of the reporter genes. O_2 regulation of other genes in an HIF-1-dependent fashion is confirmed by the observation that the key regulatory sites for many other O_2-regulated genes contain the HIF-1-binding consensus sequence (see Table 1) with the core motif CGTG (161). Among those to be discussed next are the genes encoding TH, VEGF, GLUT-1, and several glycolytic enzymes.

TH. The most immediate response to hypoxia at the whole-body level is the regulation of ventilation. The sensitivity of type I cells in the carotid body to a decrease in pO_2 is greater than in other O_2-sensitive cells: inhibition of conductance through K^+ channels, membrane depolarization, increase in intracellular calcium, and release of neurotransmitters occur rapidly. Neurotransmitters convey the hypoxic signal to the sinus nerve. Hypoxia stimulates the protein activity, as well as the fast induction of gene expression, of tyrosine hydroxylase (TH), the rate-limiting enzyme in the synthesis of the neurotransmitter dopamine. The molecular mechanisms by which hypoxia induces increased TH gene expression have mostly been explored in PC-12 (pheochromocytoma cell line) cells, which exhibit similar O_2-dependent characteristics as the scarce type I primary cells (104). A decrease of O_2 levels (to 15% O_2 in the gas phase) produces an hypoxic effect. When PC-12 cells are exposed to 10% O_2, TH mRNA levels reach a maximum within 6 h and are sustained at a high level for up to 53 h thereafter (4). The initial increase in TH mRNA is primarily attributed to a rapid induction of transcription, whereas maintenance of TH mRNA during long-term exposure to hypoxia involves mechanisms that enhance mRNA stability (see later section).

The *cis* elements required for hypoxic induction of TH are located in a promoter region containing HIF-1, AP-1, and AP-2 sites (105). As for the EPO gene, binding of the HIF-1 complex to its site is required for hypoxic induction, but other interactions are also involved. Binding of c-Fos/c-Jun heterodimers to the AP-1 site within the TH promoter was shown to increase in hypoxic cells (105). Mutation of the AP-1 site, which prevented binding of these transcription factors, inhibited hypoxic induction. Another similarity to regulation of the EPO gene is the possibility that nuclear receptors may also play a role in cooperation with HIF-1 for TH (176). Moreover, as with EPO mRNA, decreases in H_2O_2 levels caused by the addition of antioxidants or catalase induced TH mRNA expression. This suggests that H_2O_2 serves as an intermediate during the hypoxic regulation of both the TH and EPO genes.

Angiogenic Growth Factors. Angiogenesis is a fundamental process not only in development but also in wound healing, recovery from ischemic injury, and the pathogenesis of tumor growth. The common denominator among these processes is the hypoxic microenvironment in which they usually occur. The formation of new capillaries increases blood flow to the tissues, allowing delivery of an adequate supply of O_2. Control of vascular functions and blood vessel proliferation under hypoxic conditions has been associated with a group of O_2-regulated genes that encode several growth factors important for stimulating growth and migration to sites of ischemic injury (see Table 1). Some of these genes that have been reported to be inducible by hypoxia share homologous sequences and similar regulatory mechanisms to those operating in the EPO and TH genes.

Of the angiogenic cytokines, VEGF, which is believed to be the central mediator of blood vessel formation, has garnered the most attention. VEGF is produced by many different cell types and tissues, and its expression is significantly amplified by hypoxia. Hypoxic induction of VEGF mRNA has been demonstrated both in vivo and in cultured cells (115). The preponderance of data suggest that VEGF regulation employs similar O_2-sensing, signal transduction, and gene expression mechanisms as both EPO and TH (Table 1). Transfection experiments using reporter genes localized the *cis*-acting element containing the well-conserved HIF-1-binding site to a 28-bp sequence, 900 bp upstream of the transcriptional start site (116,177). Minchenko et al. (178) previously reported that two O_2-responsive elements existed in the 5′ and 3′ regions flanking the VEGF gene. AP-1- and AP-2-binding sites were also identified within the VEGF promoter (114,177).

The C/EBPβ DNA-binding factor has been found to mediate IL-6 induction by hypoxia in endothelial cells and vascular smooth muscle cells (124,125). Transfection of IL-6 promoter–reporter gene constructs revealed the NF-IL-6-binding site as the hypoxia-inducible element, suggesting that a similar mechanism may account for hypoxic induction of other genes that contain NF-IL-6 sites, such as TNF-α and IL-8.

Genes Involved in Energy Metabolism. To maintain viability and function, cells have carefully choreographed strategies of nutrient utilization. During periods of O_2 deprivation, when the O_2-dependent tricarboxylic acid cycle and oxidative phosphorylation are impaired, cells adapt by generating ATP primarily through anaerobic glycolysis. Because the glycolytic pathway produces much less ATP than oxidative metabolism, the glycolytic capacity of the cells must be enhanced under hypoxic conditions. The rate of glycolysis can be increased by transporting more glucose into the cell, by upregulating the activity and synthesis of glycolytic enzymes, and by downregulating expression of gluconeogenic enzymes that consume large amounts of ATP. As discussed next, hypoxia governs expression of the genes involved in all three modes of adaptation.

In the first mode, glucose uptake increases to compensate for lower ATP yields. The expression of glucose transporters has been found to increase in response to several environmental stimuli. Most notably, hypoxia induces insulin-independent glucose transporter 1 (GLUT-1) protein and mRNA both in vivo and in various cell lines (179). The induction of GLUT-1 contributes to sustained energy levels under hypoxic stress in muscle cells, endothelial cells, adipocytes, HeLa cells, and HT 1080 (fibrosarcoma) cells. Increases in glucose uptake and lactate production are consistent with increases in glucose transport (128,180). It is likely that hypoxic regulation of the GLUT-1 gene, whose enhancer contains two phorbol ester response elements and a cAMP response element, depends on synergy between multiple *cis*-acting elements and transcription factors (95). In addition to GLUT-1, its isoform GLUT-3 is also induced by hypoxia in the HepG2 hepatoma cell line, whereas GLUT-2 expression is inhibited (129). The isoform-specific regulation of glucose transport coincides with the decrease in the apparent K_m for glucose observed under hypoxia: GLUT-3, which demonstrates the most dramatic induction by hypoxia, also has the lowest apparent K_m for glucose (128,180). Increased steady-state levels of GLUT-1 mRNA in response to hypoxia are not only the result of increased transcription; GLUT-1 is also subject to regulation through mechanisms that stabilize mRNA.

The remaining two modes by which pO_2 alters glucose metabolism involve the regulation of a number of glycolytic and gluconeogenic enzymes. As expected, the response to hypoxia is opposite in direction depending on the pathway with which the enzyme is associated (see Table 1). In hypoxic surroundings, increased expression of rate-limiting glycolytic enzymes would shift metabolism toward glycolysis. However, in a well-oxygenated environment in selected tissues, enhanced expression of PCK would favor gluconeogenesis.

In general, the three groups of genes—those for glucose transporters, glycolytic enzymes, and gluconeogenic enzymes—exhibit patterns of expression in response to hypoxia that are consistent with their response to transition metal ions (e.g., cobalt) or iron chelators (e.g., DF), as has been shown for other O_2-regulated genes (Table 1). Some of these genes possess *cis*-acting elements, which convey this response and cross-compete with the EPO enhancer for binding to HIF-1 (132). This suggests that hypoxic regulation of EPO and certain metabolic genes are linked by a similar O_2-sensing and signal transduction mechanism. Indeed, the HIF-1-binding consensus sequence has been found in several of these genes (Table 1).

Because induction of LDH-A activity by hypoxia is one of the greatest among the glycolytic enzymes, its hypoxia-inducible elements have been studied in detail. Expression of the mouse *LDH-A* gene in response to hypoxia is driven by cooperation among three domains, which are arranged similarly to those observed in the EPO 3′ enhancer. There is (*1*) an HIF-1-binding site, which is required but not sufficient for hypoxic induction of gene expression, (*2*) an adjacent sequence, which is located in an analogous position as a critical region in the EPO enhancer, and (*3*) a cAMP response element (CRE) (133). At a minimum, in transfection experiments the HIF-1 site must be accompanied by either of the other two domains to confer hypoxic inducibility. Furthermore, the CRE-binding protein is highly homologous to P300, which binds to HIF-1α (see earlier discussion). Despite the substantial induction of LDH-A mRNA by hypoxia, no effect on its isozyme LDH-B was observed (129). Therefore, the distinctive pattern of response to hypoxia by the genes involved in energy metabolism is both isoform- and isozyme-specific, as in the cases of GLUT-1 (and GLUT-3) and LDH-A, respectively. Finally, hypoxic induction of LDH and GAPDH, which are used as a measure of viability and as an internal control for electrophoretic purposes, respectively, may lead to the misinterpretation of results collected in the laboratory if the differential effects of pO_2 on their expression are not considered.

Posttranscriptional Regulation. In addition to the mRNA transcription rate, the decay rate should not be overlooked; it, too, is an important contributing factor in increasing gene expression during hypoxia. A potential advantage associated with this bimodal regulation of gene expression is conservation of energy. Initially, hypoxia increases specific mRNAs through transcriptional induction, but, during longer exposure times, mechanisms conferring mRNA stability may take over. This phenomenon is illustrated by the increase in TH mRNA stability during long-term hypoxia, which allows maintenance of TH mRNA levels without the energy demand that accompanies synthesis of new mRNA molecules (104). Studying posttranscriptional regulation by hypoxia is complicated by the fact that many inhibitors of transcription nonspecifically stabilize mRNA. For this reason, although hypoxic regulation of EPO mRNA stability is expected, it has yet to be confirmed. In contrast, using nonstabilizing transcription blockers and other techniques, hypoxic regulation at the posttranscriptional level has been demonstrated for VEGF and GLUT-1 mRNAs (152), in addition to TH mRNA (104).

Both stability and instability elements may be involved in the regulation of steady-state EPO, TH, VEGF, and GLUT-1 mRNA levels. These *cis*-acting sequences within mRNA could direct secondary structure, act as binding sites for *trans*-acting regulatory proteins, or serve as target sites of nucleases (4). The hypoxia-regulated TH mRNA stability element has been identified as a pyrimidine-rich tract in the 3′ untranslated region (UTR) termed the hypoxia-inducible protein-binding sequence (HIPBS). Oxygen regulation by protein binding to this site was confirmed first by mutations that prevent protein binding and consequently decrease TH mRNA half-life and second by the ability of HIPBS to confer stability on reporter mRNA. The HIPBS closely resembles sequences in the 3′ UTR of both EPO and VEGF mRNAs (153,181). Moreover, similar or the same protein factors as those that bind to HIPBS in TH mRNA also bind to the homologous sequences in EPO and VEGF mRNAs (4). The most common determinants of mRNA stability in mammalian cells are AU-rich elements (182). In addition to three HIPBS motifs, the 3′ UTR region of VEGF mRNA contains multiple AU-rich stabilizing and destabilizing elements that also appear to mediate stability by forming hypoxia-regulated complexes (111,155). Thus, in combination with transcriptional upregulation, the presence of O_2-regulated mRNA stability and instability elements could account for enhanced gene expression under hypoxic conditions.

ACKNOWLEDGMENTS

Supported by NSF Grant BES-9402030. DLH was supported in part by Carcinogenesis Training Grant (NIH CA09560).

BIBLIOGRAPHY

1. W.F. McLimans, L.E. Blumenson, and K.V. Tunnah, *Biotechnol. Bioeng.* **10**, 741 (1968).
2. R.J. Fleischaker and A.J. Sinskey, *Eur. J. Appl. Microbiol. Biotechnol.* **12**, 193 (1981).
3. M.C. Borys and E.T. Papoutsakis, in J.B. Griffiths, A. Doyle, and D.G. Newell, eds., *Protocols in Cell and Tissue Culture*, Wiley, Chichester, 1993, p. 1.

4. M.F. Czyzyk-Krzeska, *Respir. Physiol.* **110**, 99 (1997).
5. A.A. Lin, R. Kimura, and W.M. Miller, *Biotechnol. Bioeng.* **42**, 339 (1993).
6. W.M. Miller, C.R. Wilke, and H.W. Blanch, *J. Cell. Physiol.* **132**, 524 (1987).
7. A. Mizrahi, G.V. Vosseller, Y. Yagi, and G.E. Moore, *Proc. Soc. Exp. Biol. Med.* **139**, 118 (1972).
8. S. Reuveney, D. Velez, J.D. Macmillan, and L. Miller, *J. Immunol. Methods* **86**, 53 (1986).
9. A.K. Balin, D.B.P. Goodman, H. Rasmussen, and V.J. Cristofalo, *J. Cell. Physiol.* **89**, 235 (1976).
10. D.G. Kilburn, M.D. Lilly, D.A. Self, and F.C. Webb, *J. Cell. Sci.* **4**, 25 (1969).
11. G.W. Taylor, J.P. Kondig, S.C. Nagle, Jr., and K. Higuchi, *Appl. Microbiol.* **21**, 928 (1971).
12. J. Wang, H. Honda, H. Watanabe, and T. Kobayashi, *J. Ferment. Bioeng.* **79**, 579 (1995).
13. S.S. Ozturk and B.Ø. Palsson, *Biotechnol. Prog.* **6**, 437 (1990).
14. M.A. Cacciuttolo, L. Trinh, J.A. Lumpkin, and G. Rao, *Free Radical Biol. Med.* **14**, 267 (1993).
15. A.R. Oller, C.W. Buser, M.A. Tyo, and W.G. Thilly, *J. Cell. Sci.* **14**, 43 (1989).
16. P. van der Valk, J.J.P. Gille, A.B. Oostra, E.W. Roubos, T. Sminia, and H. Joene, *Cell Tissue Res.* **239**, 61 (1985).
17. C.C.H. Jan, A. Petch, N. Huzel, and M. Butler, *Biotechnol. Bioeng.* **54**, 153 (1997).
18. W.G. Taylor, A. Richter, J. Evans, and K.K. Sanford, *Exp. Cell. Res.* **86**, 152 (1974).
19. L. Packer and C. Fuehr, *Nature (Lond.)* **267**, 423 (1977).
20. A.K. Balin, A.J. Fisher, and D.M. Carter, *J. Exp. Med.* **160**, 152 (1984).
21. V. Falanga and R.S. Kirsner, *J. Cell. Physiol.* **154**, 506 (1993).
22. J. Katahira and H. Mizoguchi, *Int. J. Cell Cloning* **5**, 420 (1987).
23. I.N. Rich and B. Kubanek, *Br. J. Haematol.* **52**, 588 (1982).
24. H.E. Broxmeyer, S. Cooper, and T. Gabig, *Ann. N.Y. Acad. Sci.* **554**, 177 (1989).
25. H.E. Broxmeyer, S. Cooper, L. Lu, M.E. Miller, C.D. Langenfeld, and P. Ralph, *Blood* **76**, 323 (1990).
26. M.G. Cipolleschi, P.D. Sbarba, and M. Olivotto, *Blood* **82**, 2031 (1993).
27. M.R. Koller, J.G. Bender, W.M. Miller, and E.T. Papoutsakis, *Exp. Hematol.* **20**, 264 (1992).
28. M.R. Koller, J.G. Bender, E.T. Papoutsakis, and W.M. Miller, *Ann. N.Y. Acad. Sci.* **665**, 105 (1992).
29. S. Smith and H.E. Broxmeyer, *Br. J. Haematol.* **63**, 29 (1986).
30. J.A. LaIuppa, E.T. Papoutsakis, and W.M. Miller, *Exp. Hematol.* **26**, 835 (1998).
31. M.G. Cipolleschi, G. D'Ippolito, P.A. Bernabei, R. Caporale, R. Nannini, M. Mariani, M. Fabbiani, P. Rossi-Ferrini, M. Olivotto, and P.D. Sbarba, *Exp. Hematol.* **25**, 1187 (1997).
32. O. Genbacev, Y. Zhou, J.W. Ludlow, and S.J. Fisher, *Science* **277**, 1669 (1997).
33. M. Gassmann, J. Fandrey, S. Bichet, M. Wartenberg, H.H. Marti, C. Bauer, R.H. Wenger, and H. Acker, *Proc. Natl. Acad. Sci. U.S.A.* **93**, 2867 (1996).
34. M.V. Wiles, *Methods Enzymol.* **225**, 900 (1993).
35. A.J. Potocnik, P.J. Nielsen, and K. Eichmann, *EMBO J.* **13**, 5274 (1994).
36. T.H. Rushmore, M.R. Morton, and C.B. Pickett, *J. Biol. Chem.* **266**, 11632 (1991).
37. T.R. Bradley, G.S. Hodgson, and M. Rosendaal, *J. Cell Physiol.* **97**, 522 (1978).
38. Y. Ishikawa and T. Ito, *Eur. J. Haematol.* **40**, 126 (1988).
39. R.C. Meagher, A.J. Salvado, and D.G. Wright, *Blood* **72**, 273 (1988).
40. H. de Groot and M. Brecht, *Biol. Chem.* **372**, 35 (1991).
41. A.A. Lin and W.M. Miller, *Biotechnol. Bioeng.* **40**, 505 (1992).
42. S. Hannah, K. Mecklenburgh, I. Rahman, G.J. Bellingan, A. Greening, C. Haslett, and E.R. Chilvers, *FEBS Lett.* **372**, 233 (1995).
43. M. Hermes-Lima and K.B. Storey, *Am. J. Physiol.* **265**, R646 (1993).
44. G.J. Moffat, A.W. McLaren, and C.R. Wolff, *J. Biol. Chem.* **269**, 16397 (1994).
45. D.K. Das, R.M. Engelman, and Y. Kimura, *Cardiovasc. Res.* **27**, 578 (1993).
46. C. Richter, V. Gogvadze, R. Laffranchi, R. Schlapbach, M. Schweizer, M. Suter, P. Walter, and M. Yaffee, *Biochim. Biophys. Acta* **1271**, 67 (1995).
47. C.M. Payne, C. Bernstein, and H. Bernstein, *Leuk. Lymphoma* **19**, 43 (1995).
48. W. Malorni, R. Rivabene, M.T. Santini, and G. Donelli, *FEBS Lett.* **327**, 75 (1993).
49. N. Ramakrishnan, D.E. McClain, and G.N. Catravas, *Int. J. Radiat. Biol.* **63**, 693 (1993).
50. R.R. Ratan, T.H. Murphy, and J.M. Baraban, *J. Neurochem.* **62**, 376 (1994).
51. V.J. Forrest, Y. Kang, D.E. McClain, D.H. Robinson, and N. Ramakrishnan, *Free Radical Biol. Med.* **16**, 675 (1994).
52. D.M. Hockenberry, Z.N. Oltvai, X. Yin, C.L. Milliman, and S.J. Korsmeyer, *Cell* **75**, 241 (1993).
53. S.J. Korsmeyer, X. Yin, Z.N. Oltvai, D.J. Veis-Novack, and G.P. Linette, *Biochim. Biophys. Acta* **1271**, 63 (1995).
54. A.F.G. Slater, C.S.I. Nobel, and S. Orrenius, *Biochim. Biophys. Acta* **1271**, 59 (1995).
55. S. Shimizu, Y. Eguchi, H. Kosaka, W. Kamiike, H. Matsuda, and Y. Tsujimoto, *Nature (Lond.)* **374**, 811 (1995).
56. M.D. Jacobson and M.C. Raff, *Nature (Lond.)* **374**, 814 (1995).
57. R.J. Muschel, E.J. Bernahard, L. Garza, W.G. McKenna, and C.J. Koch, *Cancer Res.* **55**, 995 (1995).
58. L.W. Browder, M. Pollock, R.W. Nickells, J.J. Heikkila, and R.S. Winning, in M. DiBerardino and L.D. Etkin, eds., *Developmental Biology: A Comprehensive Synthesis. Genomic Adaptability in Cell Specialization*, Plenum Press, New York, 1989, p. 97.
59. J.R. Subjeck and T. Shyy, *Am. J. Physiol.* **250**, C1 (1986).
60. A.S. Lee, *Curr. Opin. Cell Biol.* **4**, 267 (1992).
61. J.J. Sciandra, J.R. Subjeck, and C.S. Hughes, *Proc. Natl. Acad. Sci. U.S.A.* **81**, 4843 (1984).
62. K.Y. Chong, C.C. Lai, S. Lille, C. Chang, and C.Y. Su, *J. Mol. Cell Cardiol.* **30**, 599 (1998).
63. M. Marini, F. Frabetti, D. Musiani, and C. Franceschi, *Int. J. Radiat. Biol.* **70**, 337 (1996).
64. D.R. Borger and D.A. Essig, *Am. J. Physiol.* **274**, H965 (1998).
65. K. Kuwabara, M. Matsumoto, J. Ikeda, O. Hori, S. Ogawa, Y. Maeda, K. Kitagawa, N. Imuta, T. Kinotshita, D.M. Stern, H. Yanagi, and T. Kamada, *J. Biol. Chem.* **271**, 5025 (1996).

66. J. Ikeda, S. Kaneda, K. Kuwabara, S. Ogawa, T. Kobayashi, M. Matsumoto, T. Yura, and H. Yanagi, *Biochem. Biophys. Res. Commun.* **230**, 94 (1997).
67. M. Tucci, R.J. McDonald, R. Aaronson, K.K. Graven, and H.W. Farber, *Am. J. Physiol.* **271**, L341 (1996).
68. K.K. Graven and H.W. Farber, *New Horiz.* **3**, 208 (1995).
69. K.K. Graven, R.J. McDonald, and H.W. Farber, *Am. J. Phys.* **274**, C347 (1998).
70. I.J. Benjamin, B. Kroger, and R.S. Williams, *Proc. Natl. Acad. Sci. U.S.A.* **87**, 6263 (1990).
71. R. Mestril, S. Chi, M.R. Sayen, and W.H. Dillmann, *Biochem. J.* **298**, 561 (1994).
72. L.E. Huang, H. Zhang, S.W. Bae, and A.Y.C. Liu, *J. Biol. Chem.* **269**, 30718 (1994).
73. W.M. Miller, C.R. Wilke, and H.W. Blanch, *Bioprocess Eng.* **3**, 103 (1988).
74. D.F. Wilson, *Experientia (Basil)* **46**, 1160 (1990).
75. C. Zupke, A.J. Sinskey, and G. Stephanopoulos, *Appl. Microbiol. Biotechnol.* **44**, 27 (1995).
76. P. van der Valk, J.J.P. Gille, L.H.W. van der Plas, J.F. Jongkind, A. Verkerk, A.W.T. Konings, and H. Joenje, *Free Radical Biol. Med.* **4**, 345 (1988).
77. A.L. Sagone, Jr., *Am. J. Hematol.* **18**, 269 (1985).
78. P.W. Hochachka, L.T. Buck, C.J. Doll, and S.C. Land, *Proc. Natl. Acad. Sci. U.S.A.* **93**, 9493 (1996).
79. E.O. Pettersen, N.O. Juul, and O.W. Ronning, *Cancer Res.* **46**, 4346 (1986).
80. S. Tinton, Q. Tran-Nguyen, and P. Buc-Calderon, *J. Biochem.* **249**, 121 (1997).
81. S.M. Kraggerud, J.A. Sandvik, and E.O. Pettersen, *Anticancer Res.* **15**, 683 (1995).
82. T. Ogawa, M. Kamihira, H. Yoshida, S. Iijima, and T. Kobayashi, *J. Ferment. Bioeng.* **74**, 372 (1992).
83. S.S. Ozturk and B.Ø. Palsson, *Biotechnol. Prog.* **7**, 481 (1991).
84. S. Mercille and B. Massie, *Cytotechnology* **15**, 117 (1994).
85. Y. Shi, D.D.Y. Ryu, and S.H. Park, *Biotechnol. Bioeng.* **42**, 430 (1993).
86. W. Chotigeat, Y. Watanapokasin, S. Mahler, and P.P. Gray, *Cytotechnology* **15**, 217 (1994).
87. J. Wang, M. Hata, Y.S. Park, S. Iijima, and T. Kobayashi, *J. Ferment. Bioeng.* **78**, 321 (1994).
88. E. Regoeczi, M. Kay, P.A. Chindemi, O. Zaimi, and K.L. Suyama, *Biochem. Cell Biol.* **69**, 239 (1991).
89. I.N. Rich, *Exp. Hematol.* **14**, 746 (1986).
90. R. Pennathur-Das and L. Levitt, *Blood* **69**, 899 (1987).
91. M.R. Koller, J.G. Bender, E.T. Papoutsakis, and W.M. Miller, *Blood* **80**, 403 (1992).
92. V. Falanga, H. Takagi, P.I. Ceballos, and J.B. Pardes, *Exp. Cell Res.* **213**, 80 (1994).
93. T. Hirose, H. Terajima, A. Yamauchi, K. Kinoshita, K. Furuke, T. Gomi, Y. Kawai, S. Tsuyuki, Y. Nakamura, I. Ikai, T. Taniguchi, T. Inamoto, and Y. Yamaoka, *J. Hepatol.* **27**, 1081 (1997).
94. M.A. Goldberg, S.P. Dunning, and H.F. Bunn, *Science* **242**, 1412 (1988).
95. H.F. Bunn and R.O. Poyton, *Physiol. Rev.* **76**, 839 (1996).
96. J.M. Gleadle, B.L. Ebert, J.D. Firth, and P.J. Ratcliffe, *Am. J. Physiol.* **268**, C1362 (1995).
97. V.T. HO and H.F. Bunn, *Biochem. Biophys. Res. Commun.* **223**, 175 (1996).
98. G.L. Wang and G.L. Semenza, *Proc. Natl. Acad. Sci. U.S.A.* **90**, 4304 (1993).
99. I.J. Rondon, S.B. Scandurro, R.B. Wilson, and B.S. Beckman, *FEBS Lett.* **359**, 267 (1995).
100. M. Goldberg, C.C. Gaut, and H.F. Bunn, *Blood* **77**, 271 (1991).
101. G.L. Wang and G.L. Semenza, *Blood* **82**, 3610 (1993).
102. B. Jiang, G.L. Semenza, C. Bauer, and H.H. Marti, *Am. J. Physiol.* **271**, C1172 (1996).
103. S.L. Kroll and M.F. Czyzyk-Krzeska, *Am. J. Physiol.* **274**, C167 (1998).
104. M.F. Czyzyk-Krzeska, B.A. Furnari, E.E. Lawson, and D.E. Millhorn, *J. Biol. Chem.* **269**, 760 (1994).
105. M.L. Norris and D.E. Millhorn, *J. Biol. Chem.* **270**, 23774 (1995).
106. A. Rolfs, I. Kvietikova, M. Gassmann, and R.H. Wenger, *J. Biol. Chem.* **272**, 20055 (1997).
107. G. Melillo, T. Musso, A. Sica, L.S. Taylor, G.W. Cox, and L. Varesio, *J. Exp. Med.* **182**, 1683 (1995).
108. P.J. Lee, B.-H. Jiang, B.Y. Chin, N.V. Iyer, J. Alam, G.L. Semenza, and A.M.K. Choi, *J. Biol. Chem.* **272**, 5275 (1997).
109. L.S. Terada, D.M. Guidot, J.A. Leff, I.R. Willingham, M.E. Hanley, D. Piermattei, and J.E. Repine, *Proc. Natl. Acad. Sci. U.S.A.* **89**, 3362 (1992).
110. S.D. Estes, D.L. Stoler, and G.R. Anderson, *J. Virol.* **69**, 6335 (1995).
111. G.L. Wang, B. Jiang, E.A. Rue, and G.L. Semenza, *Proc. Natl. Acad. Sci. U.S.A.* **92**, 5510 (1995).
112. A. Naldini, F. Carraro, S. Silvestri, and V. Bocci, *J. Cell. Physiol.* **173**, 335 (1997).
113. L.D. Longo, S. Packianathan, J.A. McQueary, R.B. Stagg, C.V. Byus, and C.D. Chain, *Proc. Natl. Acad. Sci. U.S.A.* **90**, 692 (1993).
114. M.A. Goldberg and T.J. Schneider, *J. Biol. Chem.* **269**, 4359 (1994).
115. A. Minchenko, T. Bauer, S. Salceda, and J. Caro, *Lab. Invest.* **71**, 374 (1994).
116. Y. Liu, S.R. Cox, T. Morita, and S. Kourembanas, *Circ. Res.* **77**, 638 (1995).
117. K.P. Claffey, S. Shih, A. Mullen, S. Dziennis, J.L. Cusik, K.R. Abrams, S.W. Lee, and M. Detmar, *Mol. Biol. Cell* **9**, 469 (1998).
118. J.A. Forsythe, B.-H. Jiang, N.V. Iyer, F. Agani, S.W. Leung, R.D. Koos, and G.L. Semenza, *Mol. Cell. Biol.* **16**, 4604 (1996).
119. S. Kourembanas, R.L. Hannan, and D.V. Faller, *J. Clin. Invest.* **86**, 670 (1990).
120. K. Kuwabara, S. Ogawa, M. Matsumoto, S. Koga, M. Clauss, D.J. Pinsky, P. Lyn, J. Leavy, L. Witte, J. Joseph-Silverstein, M.B. Furie, G. Torcia, F. Cozzolino, T. Kamada, and D.M. Stern, *Proc. Natl. Acad. Sci. U.S.A.* **92**, 4606 (1995).
121. V. Falanga, S.W. Qian, D. Danielpour, M.H. Katz, A.B. Roberts, and M.B. Sporn, *J. Invest. Dermatol.* **97**, 634 (1991).
122. P. Ghezzi, C.A. Dinarello, M. Bianchi, M.E. Rosandich, J.E. Repine, and C.W. White, *Cytokine* **3**, 189 (1991).
123. S. Kourembanas, P.A. Marsden, L.P. McQuillan, and D.V. Faller, *J. Clin. Invest.* **88**, 1054 (1991).

124. S.-F. Yan, I. Tritto, D. Pinsky, H. Liao, J. Huang, G. Fuller, J. Brett, L. May, and D. Stern, *J. Biol. Chem.* **270**, 11463 (1995).

125. S.-F. Yan, Y.S. Zou, M. Mendelsohn, Y. Gao, Y. Naka, S.D. Yan, D. Pinsky, and D. Stern, *J. Biol. Chem.* **272**, 4287 (1997).

126. R.S. Shreeniwas, S. Koga, M. Karakurum, D. Pinsky, E. Kaiser, J. Brett, B.A. Wolitzky, C. Norton, J. Plocinski, W. Benjamin, D.K. Burns, A. Goldstein, and D. Stern, *J. Clin. Invest.* **90**, 2333 (1992).

127. M. Karakurum, R. Shreeniwas, J. Chen, D. Pinsky, S.-D. Yan, M. Anderson, K. Sunouchi, J. Major, T. Hamilton, K. Kuwabara, A. Rot, R. Nowygrod, and D. Stern, *J. Clin. Invest.* **93**, 1564–1570 (1994).

128. J.D. Loike, L. Cao, J. Brett, S. Ogawa, S.C. Silverstein, and D. Stern, *Am. J. Physiol.* **263**, C326 (1992).

129. B.L. Ebert, J.M. Gleadle, J.F. O'Rourke, S.M. Bartlett, J. Poulton, and P.J. Ratcliffe, *Biochem. J.* **313**, 809 (1996).

130. I. Stein, M. Neeman, D. Shweiki, A. Itin, and E. Keshet, *Mol. Cell. Biol.* **15**, 5363 (1995).

131. B.L. Ebert, J.D. Firth, and P.J. Ratcliffe, *J. Biol. Chem.* **270**, 29083 (1995).

132. J.D. Firth, B.L. Ebert, C.W. Pugh, and P.J. Ratcliffe, *Proc. Natl. Acad. Sci. U.S.A.* **91**, 6496 (1994).

133. J.D. Firth, B.L. Ebert, and P.J. Ratcliffe, *J. Biol. Chem.* **270**, 21021 (1995).

134. H.H. Marti, H.H. Jung, J. Pfeilschifter, and C. Bauer, *Eur. J. Physiol.* **429**, 216 (1994).

135. G.L. Semenza, B.-H. Jiang, S.W. Leung, R. Passantino, J.-P. Concorde, P. Maire, and A. Giallongo, *J. Biol. Chem.* **271**, 32525 (1996).

136. G.L. Semenza, P.H. Roth, H.-M. Fang, and G.L. Wang, *J. Biol. Chem.* **269**, 23757 (1994).

137. D. Wolfle and K. Jungermann, *Eur. J. Biochem.* **151**, 299 (1985).

138. K.A. Webster, *Mol. Cell. Biochem.* **77**, 19 (1987).

139. T. Kietzmann, H. Schmidt, K. Unthan-Fechner, I. Probst, and K. Jungemann, *Biochem. Biophys. Res. Commun.* **2**, 792 (1993).

140. T. Kietzmann, H. Schmidt, I. Probst, and K. Jungermann, *FEBS Lett.* **311**, 251 (1992).

141. M. Nauck, D. Wolfle, N. Katz, and K. Jungermann, *Eur. J. Biochem.* **119**, 657 (1981).

142. J.R. López-López and C. González, *FEBS Lett.* **299**, 251 (1992).

143. A. Gorlach, G. Holtermann, W. Jelkmann, J.T. Hancock, S.A. Jones, O.T.G. Jones, and H. Acker, *Biochem. J.* **290**, 771 (1993).

144. A.R. Cross, L. Henderson, O.T.G. Jones, M.A. Delpiano, J. Hentschel, and H. Acker, *Biochem. J.* **272**, 743 (1990).

145. J. Fandrey, S. Frede, and W. Jelkmann, in M.C. Hogan, ed., *Oxygen Transport to Tissue*, vol. XVI, Plenum Press, New York, 1994, p. 591.

146. J. Fandrey, S. Frede, and W. Jelkmann, *Biochem. J.* **303**, 507 (1994).

147. R.H. Wenger, H.H. Marti, S. Schuerer-Maly, I. Kvietikova, C. Bauer, M. Gassmann, and F.E. Maly, *Blood* **87**, 756 (1996).

148. C.T. Taylor, S.J. Lisco, C.S. Awtrey, and S.P. Colgan, *J. Pharmacol Exp. Ther.* **284**, 568 (1998).

149. M.A. Gilles-Gonzalez, G. Gonzalez, M.F. Perutz, L. Kiger, M.C. Marden, and C. Poyart, *Biochemistry* **33**, 8067 (1994).

150. S.J. Chanock, J.E. Benna, R.M. Smith, and B.M. Babior, *J. Biol. Chem.* **269**, 24519 (1994).

151. A.C. Koong, E.Y. Chen, and A.J. Giaccia, *Cancer Res.* **54**, 1425 (1994).

152. D. Mukhopadhyay, L. Tsiokas, X.-M. Zhou, D. Foster, J.S. Brugge, and V.P. Sukhatme, *Nature (Lond.)* **375**, 577 (1995).

153. A.P. Levy, N.S. Levy, and M.A. Goldberg, *J. Biol. Chem.* **271**, 2746 (1996).

154. G.L. Wang, B. Jiang, and G.L. Semenza, *Biochem. Biophys. Res. Commun.* **216**, 669 (1995).

155. C. Abate, L. Patel, F.J. Rauscher, III, and T. Curran, *Science* **249**, 1157 (1990).

156. H. Schenk, M. Klein, W. Erdbrugger, W. Droge, and O.K. Schulze, *Proc. Natl. Acad. Sci. U.S.A.* **91**, 1672 (1994).

157. M. Meyer, R. Schreck, and P.A. Baeuerle, *EMBO J.* **12**, 2005 (1993).

158. R. Schreck, P. Rieber, and P.A. Baeuerle, *EMBO J.* **10**, 2247 (1991).

159. M.B. Toledano and W.J. Leonard, *Proc. Natl. Acad. Sci. U.S.A.* **88**, 4328 (1991).

160. P. Pognonec, H. Kato, and R.G. Roeder, *J. Biol. Chem.* **267**, 24563 (1992).

161. R.H. Wenger and M. Gassmann, *Biol. Chem.* **378**, 609 (1997).

162. K. Guillemin and M.A. Krasnow, *Cell* **89**, 9 (1997).

163. N.V. Iyer, L.E. Kotch, F. Agani, S.W. Leung, E. Laughner, R.H. Wenger, M. Gassmann, J.D. Gearhart, A.M. Lawler, A.Y. Yu, and G.L. Semenza, *Genes Dev.* **12**, 149 (1998).

164. G.L. Wang, B.-H. Jiang, and G.L. Semenza, *Biochem. Biophys. Res. Commun.* **212**, 550 (1995).

165. L.E. Huang, Z. Arany, D.M. Livingston, and H.F. Bunn, *J. Biol. Chem.* **271**, 32253 (1996).

166. P.J. Kallio, I. Pongratz, K. Gradin, J. McGuire, and L. Poellinger, *Proc. Natl. Acad. Sci. U.S.A.* **94**, 5667 (1997).

167. J. Fandrey, *Respir. Physiol.* **101**, 1 (1995).

168. M.A. Goldberg, G.A. Glass, J.M. Cunningham, and H.F. Bunn, *Proc. Natl. Acad. Sci. U.S.A.* **84**, 7972 (1987).

169. D.L. Galson, T. Tsuchiya, D.S. Tendler, L.E. Huang, Y. Ren, T. Ogura, and H.F. Bunn, *Mol. Cell. Biol.* **15**, 2135 (1995).

170. V.A. Ho, A. Acquaviva, E. Duh, and H.F. Bunn, *J. Biol. Chem.* **270**, 10084 (1995).

171. K.L. Blanchard, A.M. Acquaviva, D.L. Galson, and H.F. Bunn, *Mol. Cell. Biol.* **12**, 5373 (1992).

172. Z. Arany, L.E. Huang, R. Eckner, S. Bhattacharya, C. Jiang, M.A. Goldberg, and H.F. Bunn, *Proc. Natl. Acad. Sci. U.S.A.* **93**, 12969 (1996).

173. I. Kvietikova, R.H. Wenger, H.H. Marti, and M. Gassman, *Nucleic Acids Res.* **23**, 4550 (1995).

174. P.H. Maxwell, M.K. Osmond, C.W. Pugh, and P.J. Ratcliffe, *Proc. Natl. Acad. Sci. U.S.A.* **90**, 2423 (1993).

175. C.W. Pugh, B.L. Ebert, O. Ebrahim, and P.J. Ratcliffe, *Biochim. Biophys. Acta* **1217**, 297 (1994).

176. L.H. Fossom, C.R. Sterling, and A.W. Tank, *Mol. Pharmacol.* **42**, 898 (1992).

177. A.P. Levy, N.S. Levy, S. Wegner, and M.A. Goldberg, *J. Biol. Chem.* **270**, 13333 (1995).
178. A.S. Minchenko, S. Salceda, T. Bauer, and J. Caro, *Cell. Mol. Biol. Res.* **40**, 35 (1994).
179. N. Bashan, E. Burdett, H.S. Hundal, and A. Klip, *Am. J. Physiol.* **262**, C682 (1992).
180. M. Mueckler, *Eur. J. Biochem.* **219**, 713 (1994).
181. I.J. Rondon, L.A. Macmillan, B.S. Beckman, M.A. Goldberg, T. Schneider, H.F. Bunn, and J. Malter, *J. Biol. Chem.* **266**, 16594 (1991).
182. C. Chen and A. Shyu, *Trends Biol. Sci.* **20**, 465 (1995).

See also MASS TRANSFER.

I

INOCULUM PREPARATION

CRAIG J.L. GERSHATER
S. B. Pharmaceuticals
Harlow, Essex
England

KEY WORDS

Culture assessment
Fermentation development
Inoculum development
Microbial inoculum
Seed stage fermentation
Stock culture

OUTLINE

INTRODUCTION

Inoculum preparation is the part of the fermentation development process that secures the microbial genotype for experimental and/or production purposes by providing a viable biomass capable of high productivity. Microbial processes generally consist of two distinct elements: a biotic factor the biomass and an abiotic factor (the growth environment). Hence, in applied microbiology terms

$$\text{pheotype} = \text{genotype} + \text{environment}$$

where the phenotype represents the desired attribute (accumulation of biomass or a valuable product of that biomass), the genotype represents the collection of metabolically useful genes and associated gene products, and the environment provides a system (principally a fermenter) of control over gene expression, that is, growth and product formation. At some point in the fermentation development program, the biotic and abiotic elements must be linked; this will be defined as the supply of consistent biomass to the fermenter plus growth medium. The growth and productive phases of fermentation are, of course, the subject of intense investigatory work; much has been written about the strict control requirements of the fermenter system, including the response of the microorganism to the nutritional and environmental factors. However, the requirements for the inoculum and the influence the inoculum can have on the outcome of the fermentation has received only scant attention (in terms of literature citations). A detailed assessment of the effect of the inoculum on microbial productivity can be justified, particularly in the case of nonrecombinant microorganisms, on the basis that commercially useful microbes will be isolated directly or indirectly from an ecosystem very remote from the fermentation laboratory, where the influences on growth are unknown.

Microbial Ecology

The microbial metabolic attribute to be exploited has evolved as a result of environmental influences unknown to the fermentation scientist. There is a case to be made for the basic ecology of the microorganism to be understood prior to culture preservation and inoculum preparation. A simple classification of organisms isolated from soil provides a loose context against which culture maintenance and development may be defined; soil organisms may be divided into two basic categories (1,2), the attributes of which are summarized in the following:

1. Indigenous, or autochthonous species, which have lower reproductive rates and are specialized for a particular habitat or more complex/intractable substrate. These are known as K strategists and include ascomycetes, basidiomycetes, and some actinomycetes.
2. Invaders, or allochthonous species, which have high reproductive rates and tend to function as pioneer species utilizing simple soluble substrates. The zymogenous fungi within this group include the sugar-fungi *Mucor, Rhizopus, Trichoderma, Penicillium*, and *Aspergillus*; organisms such as these may be activated by the presence of fresh substrate (breaking spore dormancy). These types of organisms, sometimes termed r strategists, possess good competitive abilities, which (along with some K strategists) include the ability to produce toxins (antibiotics), which is of obvious commercial interest.

Cognizance of the background ecology of the microorganism may permit a more appropriate culture storage and recovery method to be devised to maximize the chances of the organism successfully colonizing the next stage in the fermentation process.

Inoculum Development Conundrum

Process optimization must endeavor to replicate those factors responsible for the expression of the desired attribute in the totally artificial environment of monoculture; nutrient sufficiency (feast); and relatively stable physical parameters, such as temperature, pH, and oxygen availability. Although relatively few studies on the influence of inoculum on final stage fermentations have been made, it is certain that the inoculum preparation stage of most fermentations will affect the expression of the desired phenotype, whether qualitatively, quantitatively, or temporally. Hence, inoculum development is faced with a fundamental problem; the factors tested in seed stage or inoculum preparation investigations can only be assessed retrospectively in various high-productivity, final stage fermentations, remote from the seed stage influence. The conclusions drawn from inoculum development studies must deconvolute effects due to seed from those due to final stage fermentations. This article therefore seeks to explore how inoculum preparation can be incorporated into standard fermentation development programs.

Factors Critical to Inoculum Preparation

To adequately define the essential requirements for inoculum preparation, it is necessary to identify the elements that make up the process:

- Culture storage
- Stock culture
- Culture assessment
- Inoculum development
- Inoculation criteria

These elements will now be discussed and placed in the context of an overall inoculum strategy for fermentation development.

CULTURE STORAGE

Culture storage is only successful if the subsequent growth and productivity of the microorganism is consistent with expectations. Culture storage methodologies have been described in several authoritative texts (3–7) and are reviewed only briefly here.

Subculture

Depending on how cultures are received for fermentation development, some growth stage on agar will generally be required. The method of subculture relies on transferring colonies from one solid (depleted) medium to a fresh medium. This method is only suitable for short-term storage. Each transfer to fresh medium provides a comparatively nutrient replete environment, adaptation to which may select for an altered phenotypic expression. To minimize this possibility, minimal media are selected and the numbers of subcultures carried out are kept to a minimum. Minimal media agars are prepared in test tubes or screw-top Universal bottles, with the agars slanted to maximize surface area. Colonies from a master plate can be streaked onto the agar slants, allowed to grow for a minimum period, and then stored for a few weeks.

Desiccation

Removal or immobilization of water from the culture prevents metabolic activity, which is the key requirement of preservation. Drying methods are suitable for spore-forming fungi and streptomycetes; the principle is based on providing a solid, inert surface onto which spore suspensions can be dried. The methods differ in terms of the solid support that is used, which can include sterile soil, silica gel, dry sand, filter-paper disks, and various types of inert beads (glass, plastic, or porcelain). A suspension of spores (or possibly vegetative culture) is added to the sterile inert support, and after aseptic stoppering of the tube or vessel the mixture is allowed to dry at 25 °C. This method provides a mechanism to store, typically, spore suspensions for perhaps months, but it is generally not recommended for long-term storage of process critical stock cultures.

Freeze-Drying (Lyophilization)

Freeze-drying removes water from a frozen culture by sublimation under a vacuum. It is suitable for the long-term preservation of many types of culture and is particularly suitable for the preservation of vegetatively grown cells. Bacteria, while frequently stored by lyophilization, may be susceptible to damage by random mutation (7,8). The benefits of freeze-drying cultures is that the lyophials are relatively easy to store, transport, and identify. The cultures may be stable for up to 10 years. The disadvantages include the need to purchase a relatively expensive freeze-drier and the risk of mutation occurring in stored bacteria, which may be minimized by storing the lyophials in the dark at 5 °C.

Freezing

Freezing the culture effectively deprives the cell of water and slows all metabolic and chemical activity. Intracellular water deprivation arises as a result of osmotic pressure induced at temperatures between −5 °C and −15 °C, the temperature at which ice forms in extracellular fluid but the intracellular space remains unfrozen (9). Under these circumstances there is solute enrichment of the extracellular fluid, and water leaves the cell by osmosis, freezes outside, and the volume fraction of the cell is thus reduced. The rate of freezing will affect the size distribution of the ice crystals and with slow cooling will tend to result in considerable water loss from the cell and large extracellular crystals; rapid freezing (−200 °C/min, obtained by plunging the sample in liquid nitrogen) will result in smaller extracellular ice crystals and some intracellular ice crystal formation. It is necessary to determine the correct freezing rate for the particular cell under investigation. The properties and composition of the growth medium qualitatively affect the success of cryopreservation. Recovery from cryopreservation may also be critical due to an effect known as recrystallization, which refers to small crystals rearranging themselves into potentially more damaging larger crystals, leading to cell rupture. This may typically happen at temperatures between −100 °C and −20 °C. The potentially damaging effects are minimized by rapid thawing.

To minimize the general effect of ice crystal damage, high-viscosity cryopreservants are often added to the cell suspension (in the case of vegetative culture storage); these may be of two basic types, intracellular (membrane penetrating) or extracellular (nonmembrane penetrating). The intracellular agents used include dimethylsulfoxide (DMSO) and glycerol. Glycerol (added at 5–10% v/v to the final culture dispensing fluid) has the effect of reducing lesions due to osmotic shock by reducing the size of ice crystals. Extracellular cryopreservants include various carbohydrates, such as sucrose, mannitol, dextrin, and starch; 20% powdered skim milk and 12% sucrose are widely used.

The principal advantage of freezing as a method of culture storage is the relatively long duration of culture stability. The lower the temperature, the longer the duration. At −20 °C it is reasonable to consider that microbial cultures will remain stable for 1 to 2 years; at −80 °C, cultures should remain stable for 5 years or more, indefinite storage is obtained at temperatures of less than −80 °C using liquid nitrogen. The vapor phase of liquid nitrogen is −156 °C, and the liquid phase is −196 °C; extreme care must be exercised when recovering culture vials from the liquid phase, because leakage of liquid nitrogen into vials may result in explosion of the sealed ampoules due to gas expansion at elevated temperatures. The main disadvantages of freezing for culture storage are the relatively high capital costs for compressor freezers (particularly those for less than −80 °C) and secure purchasing and management procedures for maintaining liquid nitrogen stocks and reliability of electrical supplies. In addition, frozen ampoules are not easily transported.

Fungal culture storage may require a combination of all the methods described earlier for long-term storage, dependent on the morphological characteristics of the strain in question; for example, aconidial strains may be impossible to preserve by lyophilization (10), but may be stored on silica gel under liquid nitrogen. Methods appropriate to bacterial culture storage using both drying and liquid nitrogen have been reviewed (11).

STOCK CULTURE

A fermentation development or production process involving the culturing of commercially significant microorganisms demands that a robust system be in place for stock culture generation and maintenance. Variability of stock culture performance cannot be tolerated because culture variation masks the influence of nutritional or environmental factors within a development program.

What Is a Stock Culture?

The stock culture represents the current microbial investment in a fermentation program, which is manifested in both qualitative and quantitative terms. Qualitatively, the stock culture defines the current genotypic status of a commercially significant microorganism (i.e., its product formation capacity in a defined environment). Quantitatively, the stock culture is represented by a carefully controlled holding of cultivar, which will be drawn on for development and production purposes. At some critical point this holding must be replenished with equivalent stock culture or be replaced by an improved stock. Awareness of the microbial ecology of the culture may be important given that the cultivar will be required to propagate within a new environment (the ecological niche represented by fresh fermentation medium) by rapid colonization. If, for example, the culture of interest is considered to be an allochthonous species (r strategist) capable of colonizing fresh medium from spore inoculum, then the stock culture should be evaluated as a spore-based stock culture.

Management of the microbial culture resources is the most important facet of a microbiological bioprocess enterprise. Each bioprocess organization defines its own protocols to secure the microbial investment; however, the principal objective must be to ensure that the bioprocess enterprise is never placed at risk due to stock culture mismanagement.

Stock Culture Generation

At various stages in a microbial development program it will be necessary to generate stock culture. The method chosen is dictated in part by the preservation protocol appropriate for the fermentation program. In my laboratory, with a relatively high turnover in new cultures for fermentation development, vegetative stock cultures are often generated and stored at −80 °C. The stock generation model for cryopreservation is shown in Figure 1.

The model in Figure 1 assumes that the culture to be laid down to stock has been evaluated or in some other way nominated for long-term storage. Hence, prior to long-term culture storage, working stocks of the same culture will have been available. Working stock cultures may similarly be prepared using the model described earlier, but perhaps with far fewer vials, or they may be preserved using one of the other methods described previously.

The generation of fresh stocks may be dictated by several factors that may or may not be interrelated. As the stock culture holding is depleted, it must be replenished with fresh stocks with performance criteria as close to the existing stocks as possible. Stock culture management protocols determine at what point this should be; for critical bioprocesses (main line production) this may be initiated when 60% to 70% of the existing stock has been utilized. With less critical bioprocesses, stock culture replenishment may commence when 70% to 80% of the existing stock is depleted. Fresh stocks are also required when there is qualitative change in the culture component of the bioprocess. This may result from a deliberate attempt to improve the culture by strain improvement or culture reselection; alternatively, it may result from an inexplicable,

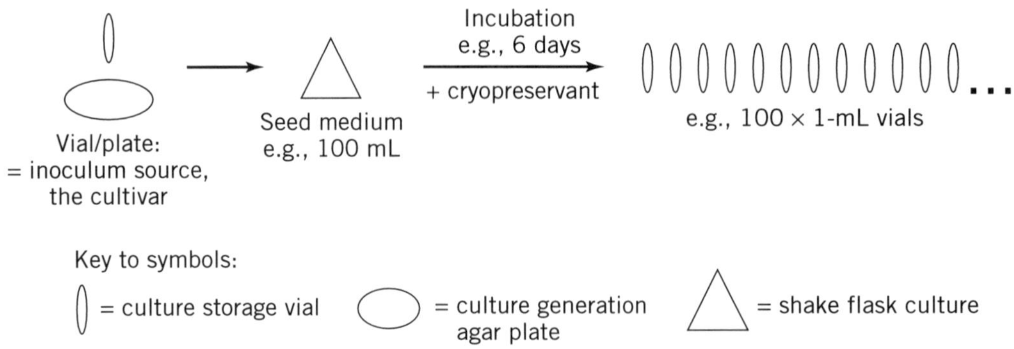

Figure 1. Stock culture generation models. Inoculum source for the first-stock generation is usually culture grown on agar plates; this is typically a plug (agar plugs are aseptically cut through the colony center) inoculated to liquid seed medium. The seed medium provenance, selected for vegetative growth, resides within relevant literature or, in the case of screening derived cultures, from original cultivation protocols. The incubation conditions for the culture are determined from the literature or from preexisting protocols. Dispensing of adequately grown vegetative culture to long-term culture vials is carried out under strict aseptic techniques; the number of vials dispensed to must be restricted to prevent the heterogeneous distribution of culture between first and last vials in the series.

irreversible deterioration in culture performance. If fresh stocks are to be generated from an improved strain, then the working stocks evaluated as part of the strain improvement process serve as the progenitor to the stock culture generation model previously described. If, however, the need to generate new stocks arises from severe culture deterioration, then a previous generation stock must be used for current stock culture generation. This deterioration may be accounted for by a reduction in culture viability, culture productivity performance, or microbial or viral contamination.

As new stock cultures are prepared, a family tree of related strains will be generated. To replenish stocks from an existing stock culture, a vial of the parental working stock can be used to generate a new daughter stock. Within the stock culture model each time a new generation stock culture is prepared after testing, a portion of that stock holding must be set aside for permanent retention. In my laboratory, at each stage in the stock maintenance program a portion of the existing stock is nominated as the master or "crown–jewel" stock and is held inviolable for process operations but represents a temporal marker in the history of a culture development program. To keep track of culture usage and family history, a suitable nomenclature must be devised. We use a simple decimal notation system in which the integer before the decimal point represents the current generation number and the figure after the decimal point represents the sibling stock number for the current generation.

Eventually even master or crown-jewel stocks may become depleted or obsolete, but within the active lifetime of a bioprocess, the master stock can only be drawn on in the event of a catastrophic failure of the main working stocks due to either an inexplicable loss of viability or performance or mechanical failure of the entire culture preservation system. To provide additional protection to the culture storage facility, a decision may be made to preserve split stocks not just in separate freezers but also on different sites, if possible. This type of stock culture management can be determined within an organization's disaster planning protocols.

CULTURE ASSESSMENT

The generation of new stocks imposes a profound influence on the bioprocess, and performance criteria must be reestablished at each stage in the culture maintenance program. As with any biological system, the inherent variability of the culture must be assessed prior to any development or production processes being committed to the new stock culture. The elements of culture maintenance that must be in place before an assessment of culture variability can be carried out include the following:

- Adequate numbers of culture vials (previously found to be monoculture)
- A suitable seed culture fermentation protocol
- A final or production phase fermentation protocol
- Suitable response variables to the final stage fermentation
- Validated assays for quantitatively determining the levels of the response variables
- Full supporting documentation to all of the elements mentioned

Culture variability must be addressed directly by the use of suitable statistical evaluation procedures. These procedures are largely based on a system of replication at each stage of the evaluation so as to obtain sufficient data describing the spread of variability and the contribution to total variation by each critical stage in the evaluation protocol. If the culture is being generated and evaluated as part of a good management practice (GMP) process, the stringency of these requirements may be greater. Figure 2

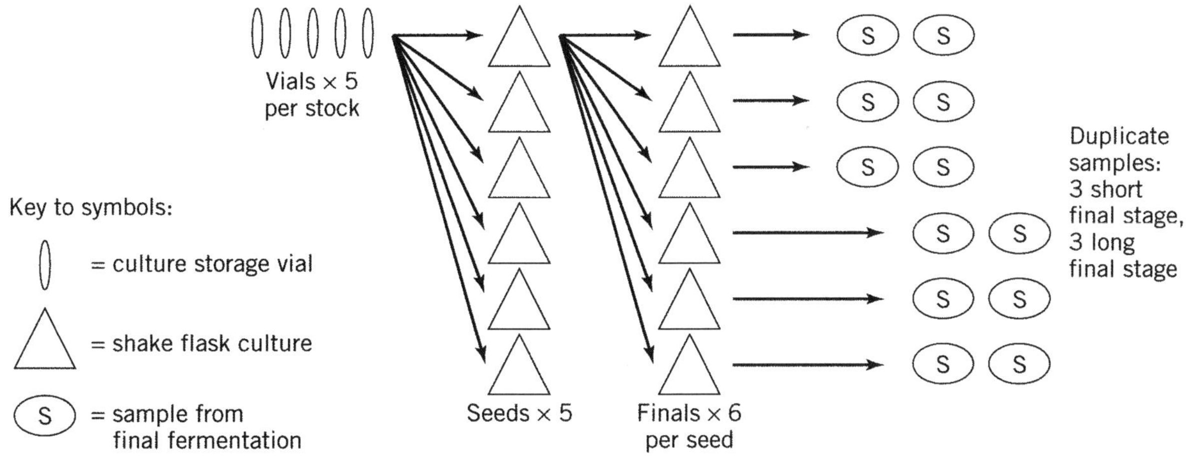

Figure 2. Culture variability testing model. To ensure adequate data is generated for a statistical analysis of variance to be undertaken, appropriate numbers of elements at each stage in the procedure must be selected. In this (model) case, five vials of stock culture are each used to inoculate each of five seed-media shake flasks. Each seed culture, after incubation to generate adequate biomass, is used to inoculate each of six final stage fermentations and will be assessed for reproducibility of growth. Half of the final stage shake flasks (three fermentation flasks) are incubated for a relatively short time and then duplicate sampled; the remaining half are incubated for a relatively longer time and then duplicate sampled. Each sample can be the source of duplicate or triplicate assays to enhance the analysis of variance (not shown in the model).

broadly defines the type of testing that may be carried out using a newly generated or existing stock.

This model permits replication sufficient to explore the contribution to the total variation of each of the stages capable of being a source of variation. This is sometimes known as a nested design, and data arising are analyzed using standard analysis of variance (ANOVA) procedures. Not only are data generated that capture the inherent variation associated with vial-to-vial variation, but the relative robustness of the remaining system can also be evaluated. Data from this kind of trial may be summarized in the form of statistical tables, but perhaps of more use would be data summarized using graphical plots. Figure 3 represents the actual data obtained for a reproducibility trial exploring the vial-to-vial variability of a fungal culture producing an antibiotic.

INOCULUM DEVELOPMENT

As stated earlier, the key to successful inoculum preparation lies in the outcome in the final stage of the fermentation, hence, evaluation of seed protocols is always retrospective. The factors that can influence the inoculum and therefore the final-stage productivity are multifarious. This chapter has already identified certain key functions that may or may not influence the development of a successful seed culture, such as, microbial ecology, culture storage, and variability. In addition, some of the factors that may need to be considered in developing a inoculum preparation strategy include:

- Seed medium
- Seed incubation conditions
- Microbial growth dynamics
- Optimum transfer criteria (qualitative inoculum variable)
- Volume of inoculum on transfer (biomass concentration, quantitative inoculum variable)
- Number of seed-stage passages
- Assessing inoculum development

Several of these influences have been investigated in the past, but often in isolation from one another. The recognition that inoculum development can play a key role in fermentation development programs was recognized over 20 years ago; in a remark attributed to Hockenhull but reported by Calam (12), he stated that "once a fermentation had started it can be made worse but not better." Despite this recognition, the bioprocess industry (high-added-value products), appears to have been very relaxed about setting resources aside to investigate the influence of inoculum on final stage fermentation productivities. This is perhaps somewhat in contrast to the fermentation food industries, where starter culture methods have been developed to provide high quality and consistent product.

In the dairy industry, factors such as correct carbon and nitrogen sources, vitamin and mineral supplements, and the addition of phage inhibitory agents, antioxidants, and neutralizers have all been considered important in defining the correct formulation for bulk starter media for lactic acid bacteria (13) in cheese production.

In the case of alcohol fermentation, relatively little work appears to have been done to investigate the influence of inoculum on alcohol yields. When reported, the studies are often limited to just one aspect of the inoculum process, for example, that inoculum levels may vary greatly with *Sac-*

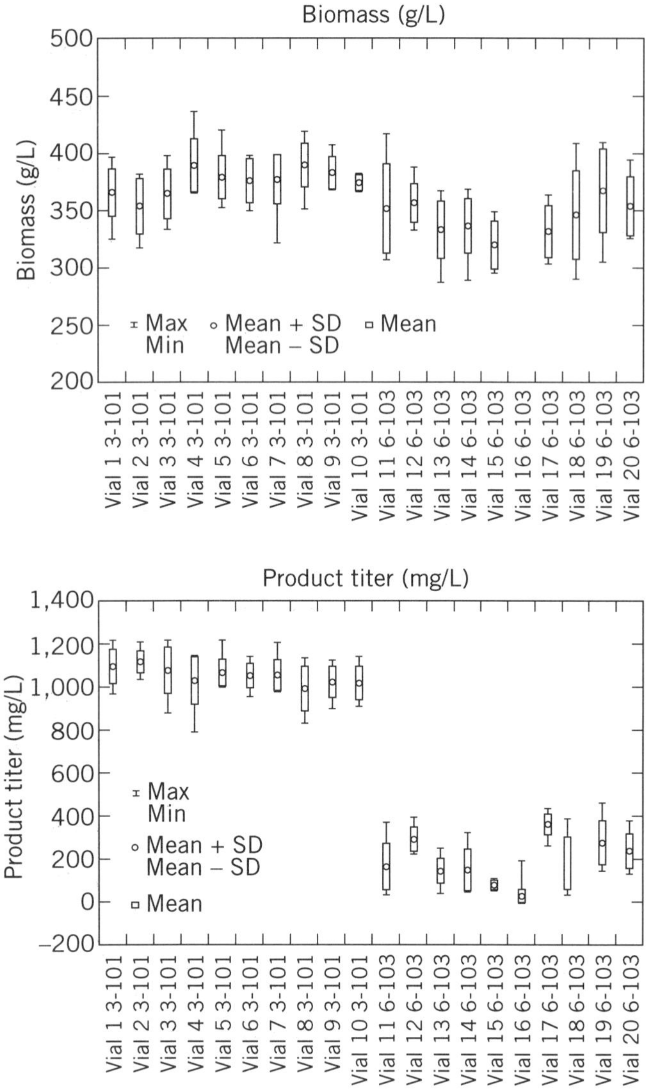

Figure 3. Reproducibility data from an antibiotic-producing culture. A fungal culture subject to a fermentation development program generated a number of reisolates that were tested in a reproducibility trial similar to one described in the model (Fig. 2). The two reisolates were called 3-101 and 6-103. Preliminary testing in single shake-flask experiments had indicated that although 3-101 was superior to 6-103, the difference was not great, therefore, working stocks of both cultures were prepared. In terms of biomass the two cultures appear very similar, although the variability of culture 6-103 was greater than that of 3-101. Biomass data was unavailable for vial 16, but this did not prevent an ANOVA of the data. In terms of product titer, 3-101 was clearly far superior to 6-103, both in terms of overall titre as well as vial-to-vial variation. The ANOVA for these experiments revealed that vial-to-vial variation was the greatest contributor to variation of 6-103 performance, and that factor tended to mask the true culture productivity.

charomyces cerevisiae, at between 0.05 and 60% (14). More recent studies with yeast fermentations have attempted to define the effect of inoculum on growth and ethanol production (15), including investigating at which point in the microbial growth curve inoculation should take place—in this case exponentially growing cells were favored. In addition, these studies were carried out while keeping factors such as medium formulation, medium sterilization, and culture growth conditions all constant, while varying inoculum size, under these circumstances inoculum levels of 10% were found to generate maximum biomass and, hence, ethanol.

Fermentations devoted to solvent formation have also been investigated to determine the influence of inoculum. The production of acetone, butanol, and ethanol by anaerobic fermentation using *Clostridium butylicum* was found to be adversely affected by repeated subculturing. This can be partly ameliorated by incorporating heat shocking and subculturing into an inoculum development program wherein the effect of heat shocking the culture to 75 °C for 1.5 min may reduce the degenerative effects of subculturing (16). Again, more recent studies have confirmed the deleterious effect of repeated subculture with saccharolytic clostridia (17). Using *Clostridium acetobutylicum* it was recognized that there would be an abrupt decrease in solventogenesis after 10–12 successive transfers incorporating heat shock; further, after 25 transfers, the culture failed to form spores, and after heat shock failed to grow at all. The dramatic decrease in solventogenesis on subculture was found to be associated with deficiencies in acetate/butyrate CoA transferase and acetoacetate decarboxylase activity. The addition of exogenous acetic and butyric acids maintained solventogenesis to at least the 40th transfer.

The influence of inoculum development for fungal cultures has been investigated. The production of γ-linolenic acid (GLA) by shake flask culture of the sporulation-deficient strain of *Cunninghamella echinulta* was found to be profoundly affected by the use of fragmented mycelia, increasing biomass and GLA titre (18).

Most investigations into the influence of inoculum involving fungal and filamentous bacterial cultures have been carried out for the production of various bioactive metabolites at the high-value-added end of the bioprocess industry, some of the issues of inoculum preparation here will be now considered in more detail.

Multiple Growth Cycles

The generation of inoculum for large-scale bioactive metabolite production by fermentation generally necessitates a buildup of active biomass through multiple growth cycles. This is illustrated in Figure 4 for a typical 3,000-L pilot-plant fermentation.

In developing the multiple-growth cycle seed train, several experimental objectives must be met. The first of these is the seed medium for culture growth. There are virtually no references to seed stage medium development in the literature (19,20), however, some general principles do apply. The seed medium formulation should seek to minimize nutrient shift effects while supporting good vegetative growth over a short time scale. The overriding proviso for seed medium formulation is that high productivity is supported in the final stage medium. In the case of bioactive metabolite fermentations, the seed stage fermentation should not result in production of the metabolite of inter-

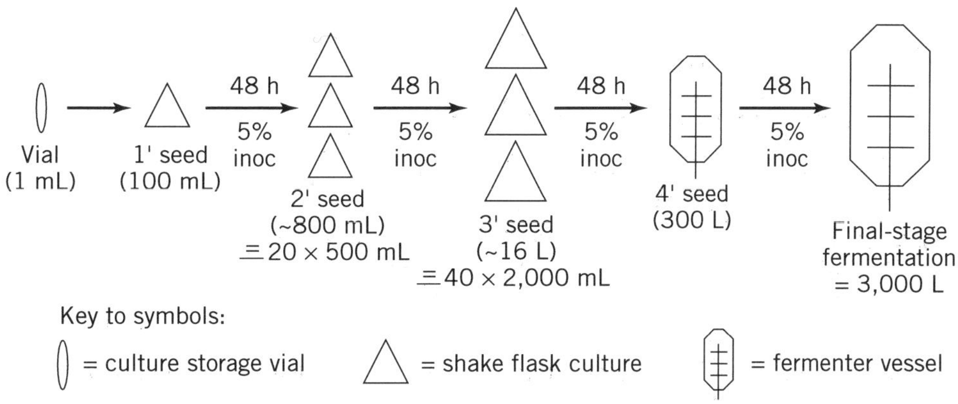

Figure 4. Typical multiple growth cycle fermentation seed train. The illustration indicates the logistics in preparing the inoculum for a 3,000-L final stage fermentation. In the example given, the seed train is defined by having a quaternary seed-stage inoculate the final-stage fermentation. The incubation times defined for each stage of the fermentation are designed to yield optimal inoculum to the next stage in the process. Throughout the seed train, the shock of culture transfer to nutrient feast is minimized by maintaining a relatively short residence in each seed fermentation, thus minimizing the risk of transferring from famine to feast at each stage. Inoculum transfer volumes impose logistic constraints on the seed train such that at the tertiary seed stage 40 2-L shake flasks must be inoculated. This represents a risk to the system in terms of contamination and culture heterogeneity, but small-scale fermenters can be used to overcome these problems.

est. Bioactive metabolites may be growth inhibitory to the producing strain (and hence are secondary metabolites), and their presence in the seed stage broth may complicate or inhibit the final stage product accretion. Sterilization control profiles for the seed stage medium, just as for the final stage medium, should ensure that Maillard reactions (browning) adversely affecting protein and glucose availability are minimized. Sterilization cycles should also be designed such that vitamin destruction is minimized and so the formulation of the seed medium will seek to prevent trace element precipitation (e.g., phosphate and magnesium combinations and iron complexes formed on heating or pH shifts).

The overall inoculation protocol seeks to define a quality standard for the inoculum. Earlier studies had identified that larger volumes of inocula of *Trichoderma reesei* (in the range 0.5 to 5.0% v/v) ended to reduce subsequent lag phases (21). These sorts of studies were extended (22) using *Penicillium chrysogenum* for penicillin production; it had been established that the spore concentration in shake-flask seed culture was critical to growth and antibiotic production. Above 10^4–10^5 spores/mL, growth of the production cultures was good, critical enzyme levels were high, and penicillin production was good. Below 10^3 spores/mL, the fermentations were suboptimal in every respect. In further studies (23), *Penicillium* spores were seeded at different concentrations into seed medium, and samples were removed at regular intervals and examined for growth on agar for numbers of viable centers. At some critical point in the fermentation, the number of viable growth centers fell by up to 85%, presumably due to aggregation and increased branching frequency in the hyphae. These observations have been confirmed by image-analysis work (24), which demonstrates a morphology shift from dense pellet formations to more dispersed forms with inocula above 10^5 spores/mL. These data tend to confirm the notion of a quality curve for inoculum development, as illustrated in Figure 5.

INOCULATION/SEED TRANSFER CRITERIA

Multistage inoculum transfers and the success criteria associated with those transfers must be determined experimentally. Each stage in the process can be examined individually, but at some point the seed train protocol must

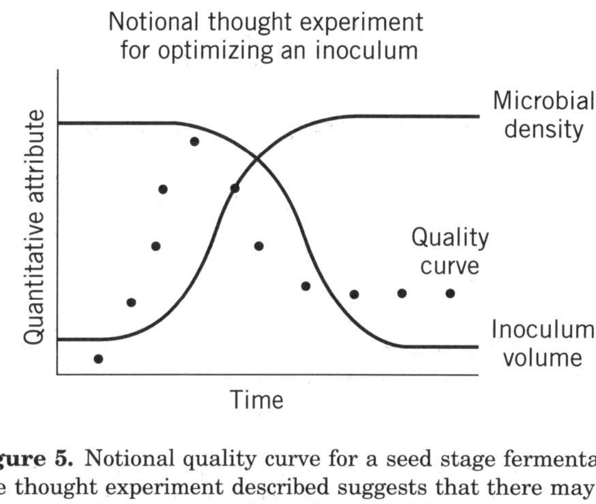

Figure 5. Notional quality curve for a seed stage fermentation. The thought experiment described suggests that there may be a compromise between inoculum volume and the qualitative attributes associated with the inoculum. Determining the notional quality curve requires an investigation into the optimization of the desired response variable in the subsequent fermentation stage. This may be achieved by carrying out timed transfers and measuring the physiological status of the organism at each time point.

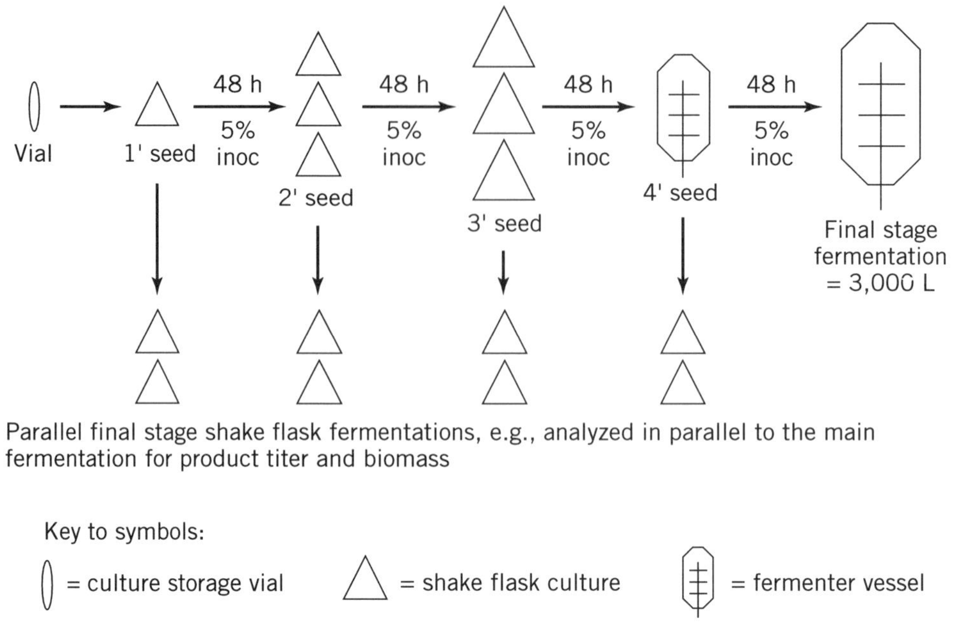

Figure 6. Inoculation–seed transfer experimental design. Final stage shake flask fermentations running in parallel to the main seed train, inoculated from seed stage broth generated in the inoculum train. Principal objective with this type of design is to compare productivity between each final stage fermentation to help determine if there are effects due to passage or some other seed incubation condition.

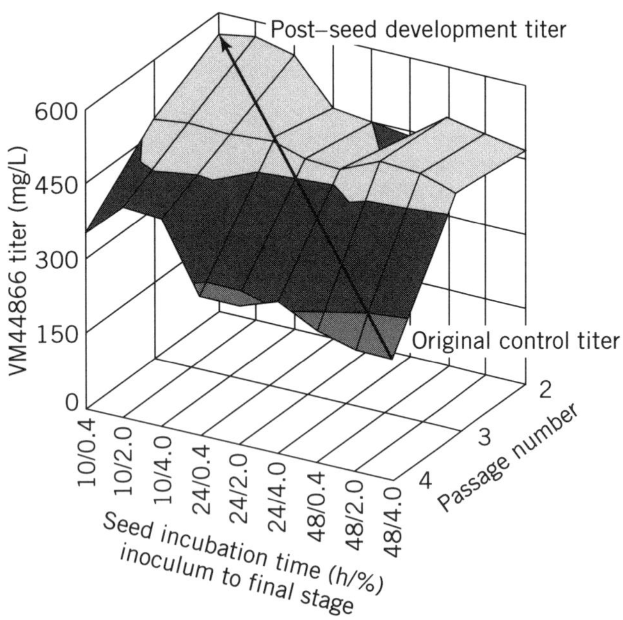

Figure 7. Response surface plot for milbemycin production under the influence of seed stage development factors. The response surface plot shows the final stage titers of milbemycin obtained under the seed inoculum conditions indicated by the three-dimensional plots. These data show that percentage inoculum levels into the final stages were not a key factor. However, passage number was significant, with quaternary seed stages giving rise to the lowest final stage titers. Similarly, juvenile seed was optimal for all conditions tested.

be established. As already indicated, seed stage assays may be expensive in terms of numbers and time required for full analysis, and care must exercised to ensure that seed stage data correlate with final stage fermentation productivity. The use of complex media may impose additional variation on a multistage process, and wherever additional variation is suspected, greater replication in the data-gathering phase will be required.

The principal physiological markers that can be used for seed transfer criteria include:

- Biomass density or an analyte reflecting that variable (ATP)
- Gross morphology attributes (particularly for filamentous bacteria and fungi)
- Respiratory gas analysis from aerated fermenters (carbon dioxide evolution rate)
- Macromolecular titre (DNA, RNA)
- Final stage productivity

Final stage productivity can only be determined by inoculating final-stage fermentations from each seed stage and analyzing for productivity at each stage. The format for this type of experiment is shown in Figure 6.

INOCULUM PREPARATION STUDIES

There are very few studies reported in the literature examining the complete inoculum preparation process from initial inoculation to final stage productivity. *Streptosporangium* cultures that are capable of producing useful secondary metabolites, including chloramphenicol (25), dihydrosinefungin (26), and endothelin-converting enzyme inhibitors (27), have been investigated for various inoculum stage conditions (28). The factors examined with respect to the inoculum stage of *Streptosporangium* cultures include inoculum storage, media sources, and inoculum size on two antibiotic fermentations. Three culture preservation methods were examined: spores, slant tubes, and frozen agar plugs. When each of these sources were tested in different media and analyzed for growth, slant transfer tubes were found to be the most effective for maintaining high growth concentrations in the final stage. Examining

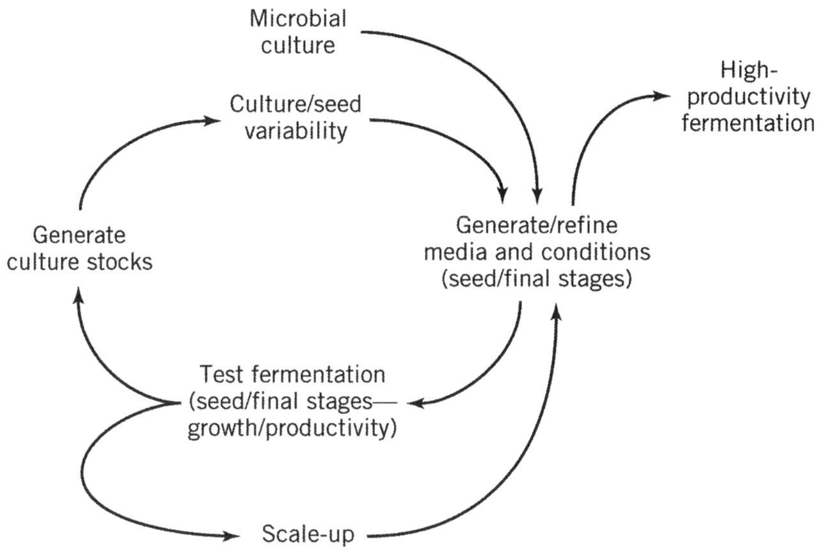

Figure 8. Fermentation development model incorporating seed stage investigations. The model shown assumes the process starts with receipt of the microbial culture. Seed and final stage media are identified initially. Fermentations are carried out to assess growth, and if adequate to the purpose, then culture stocks may be generated. These stocks must be tested for variability before further studies can commence. Seed and final stage fermentation conditions can then be refined with the newly generated and tested stock culture. The fermentation development process continues until performance criteria for the system are met, the process is scaled up, and if necessary further refined. Finally a high-productivity fermentation can be defined using a well-characterized and genotypically secure process.

incubation conditions with various *Streptosporangium* cultures indicated that aeration and growth phase of the inoculum may affect the time course of the final stage fermentation, particularly if that fermentation stage is relatively short. These influences may be less pronounced if longer final stage fermentations are required.

The improved production of milbemycin (a macrolide with anthelminthic properties) from *Streptomyces hygroscopicus* as a result of intensive seed stage development has been reported (29). In this work an attempt was made to examine several interacting factors affecting seed stage development and subsequent final stage production of milbemycin. Time-course fermentation profiles of the seed stage revealed that a major component of the seed medium, dextrin, was not utilized, and the other major carbohydrate component, glucose, could be reduced fourfold without affecting the productivity of the final stage fermentation. Aside from medium composition, the other seed development factors investigated included:

- Age of culture at transfer
- Temperature of seed-stage incubation
- Inoculum level
- Passage number of seed stages

Inoculum age was found to have a profound influence on final stage titers, with more juvenile seeds (10 h as compared to 48 h for the control) proving beneficial to milbemycin production. Inoculum levels could be halved from 4% v/v without loss of productivity. Passage number was found to be critical in seed media containing dextrin, as these seed fermentations showed a marked decrease in final stage titers in the tertiary and quaternary seed stages. Studies investigating the combined effects of seed incubation times, percentage inoculum into each seed stage or the final stage fermentations, and passage number were undertaken. Figure 7 shows the response surface plot summary of that work, where by examining the various seed stage factors, overall the titer of the final stage fermentation was increased fourfold.

INOCULUM DEVELOPMENT SUMMARY

This chapter has attempted to describe the various influences that may be examined when trying to define the optimal inoculum-development protocols. As indicated from the literature, this part of the fermentation process has been largely neglected, and yet the evidence indicates (from the relatively few studies undertaken) that inoculum preparation may have a profound effect on the outcome of a fermentation development process. Figure 8 describes how inoculum development may be incorporated into the overall fermentation development strategy for a bioprocess enterprise.

BIBLIOGRAPHY

1. R. Campbell, *Microbial Ecology*, 2nd ed., Blackwell Scientific Publications 1983, pp. 82, 90.
2. M. Alexander, *Introduction to Soil Microbiology*, 2nd ed., Wiley, New York, 1977, pp. 16–17.
3. L.T. Chang and R.P. Elander, in A.L. Demain and N.A. Solomon eds., *Manual of Industrial Microbiology and Biotechnology*, American Society of Microbiology, Washington, D.C., 1986, pp. 49–55.
4. J.J.S. Snell, in B.E. Kirsop and J.J.S. Snell eds., *Maintenance of Microorganisms: A Manual of Laboratory Methods*, Academic Press, Inc, London, 1984, pp. 11–21.
5. D. Smith and A.H.S. Onions, *The Preservation and Maintenance of Living Fungi*, Commonwealth Mycological Institute, Kew, U.K., 1983, pp. 5–30.
6. A. Dietz and B.W. Churchill, in M. Moo-Young ed., *Comprehensive Biotechnology: The Principles, Applications, and Regulations of Biotechnology in Industry, Agriculture, and Medicine*, vol. 1, Pergamon Press, Oxford, 1985, pp. 37–49.
7. A. Dietz, in H.-J. Rehm and G. Reed eds., *Biotechnology: A Comprehensive Treatise in 8 Volumes*, vol. 1, Verlag Chemie, Weinheim, 1981, pp. 411–434.
8. M.J. Ashwood-Smith and E. Grant, *Cryobiology* **13**, 206–213 (1976).
9. C. Herbert, P. Roux-Salembien, and J.-M. Egly, *Biofutur* **128**, 3–11 (1993).

10. C. Wilson, *Fungal Genet. Newsletter* **33**, 47–48 (1986).
11. I.J. Bousfield, in B.E. Kirsop, and J.J.S. Snell eds., *Maintenance of Microorganisms: A Manual of Laboratory Methods*, Academic Press, Inc, London, 1984, pp. 63–68.
12. C.T. Calam, *Process Biochem.* 7–12 (April 1976).
13. W.E. Whitehead, J.W. Ayers, and W.E. Sandine, *J. Dairy Sci.* **76**, 2344–2353 (1993).
14. P. Strehaino, M. Mota, and G. Goma, *Biotechnol. Lett.*, **5**, 135–140 (1983).
15. G. Birol and K. Ozergin-Ulgen, *Turk. J. Chem.* **19**, 250–257 (1995).
16. J.R. Gapes, V.F. Larsen, and I.S. Maddox, *J. Appl. Bact.* **55**, 363–265 (1983).
17. O. Assobhei, A. El Kanouni, M. Ismaili, M. Loutfi, and H. Petitdemange, *J. Ferment. Bioeng.* **85**, 209–212 (1998).
18. H.-C. Chen and T.-M. Liu, *Enz. Microbiol. Technol.* **21**, 137–142 (1997).
19. G.R. Hunt and R.W. Stieber, in A.L. Demain and N.A. Solomon eds., *Manual of Industrial Microbiology and Biotechnology*, American Society of Microbiology, Washington, D.C., 1986, pp. 32–40.
20. S. John Pirt, *Principles of Microbial and Cell Cultivation*, Blackwell Scientific, Oxford, 1975, pp. 118, 196–198.
21. D.E. Brown and M.A. ZainudeEn, *Biotechnol. Bioeng.* **20**, 1045–1061 (1978).
22. G.M. Smith and C.T. Calam, *Biotechnol. Lett.* **2**, 261–266 (1980).
23. C.T. Calam and G.M. Smith, *FEMS Microbiol. Lett.*, **10**, 231–234 (1981).
24. K.T. Tucker and C.R. Thomas, *Biotechnol. Lett.*, **14**, 1071–1074 (1992).
25. A. Tamura, I. Takeda, S. Naruto, and Y. Yoshimura, *J. Antibiot.* **24**, 270 (1971).
26. R. Cooper, P. Das, C. Federbush, R. Mierzwa, M. Patel, P. Birendra, and I. Truumees, *J. Indust. Microbiol.* **5**, 1–8 (1990).
27. Y. Tsurumi, N. Ohata, T. Iwamoto, S. Kyoto, and S. Okuhara, *J. Antibiot.* **47**, 619–630 (1994).
28. G. Platas, J. Collado, H. Martinez, M. Arrese, P. Pelaez, and M.T. Diez, *Microbiologia SEM* **13**, 193–200 (1997).
29. S.R.C. Warr, C.J.L. Gershater, and S.J. Box, *J. Indust. Microbiol.* **16**, 295–300 (1996).

INSECT CELL CULTURE, PROTEIN EXPRESSION

JAMES VAUGHN
ARS, USDA
Beltsville, Maryland

KEY WORDS

Baculoviruses
Expression vectors
Glycosylation
Insect cell lines
Insect glycoproteins
Nutrient requirements
Oxygen requirements
Protein processing
rDNA
Recombinant proteins

OUTLINE

INTRODUCTION

Biotechnology has made feasible the production of many complex proteins of importance in human or animal medicine in cultures of single cells, either microbial or animal. Animal cells (eukaryotic cells) can accomplish the post-translation modifications of these protein products that are essential for their biological function. For this reason, they are often preferred for production to the more easily grown microbial cells (prokaryotic cells), which cannot perform these modifications. Among the several types of eukaryotic cells available as hosts for gene expression, insect cells have some advantages that make them the host of choice. They can be subcultured indefinitely without being "transformed" eliminating the possible risk of the presence of cancer-promoting substances. Insect cells do not require a suitable substrate for attachment, but grow readily in suspension, making them easier to grow in commercial-scale reactors. They are not likely to contain viruses that would present a hazard to humans or other warm-blooded animals. The vector used to transfer the foreign gene to the cell, a baculovirus, is derived from a group of viruses that infect only arthropods and therefore have no risk for higher animals. This virus contains several very active promoters and so the yields of expressed protein can be very high, as much as 20–250 times that of mammalian cells. The baculovirus–insect cell system has been used to express approximately 500 gene products in insect cells, and 95% of the expressed recombinant proteins have been biologically active (1). Methods for scaling up production to 100-L bioreactors have been developed and successfully tested. In this article, the baculovirus system and the expression of heterologous genes in insect cells are reviewed with emphasis on optimizing the expression and scale-up of production systems. For detailed reviews of baculovirus genetics and the construction of vectors, readers are referred to articles by Frieson and Miller, (2) Luckow and Summers (3), and Kool and Vlak (4).

BACULOVIRUSES

Classification

The viruses used for the construction of expression vectors are pathogens of insects, and the early interest in them was their potential for the biocontrol of insect pests. They are classified in the family *Baculoviridae*. Baculoviruses have rod-shaped nucleocapsids containing a single molecule of circular, supercoiled, double-stranded DNA and are enveloped. The family contains two genera, the genus *Nucleopolyhedrovirus* (NPV) and the genus *Granulovirus* (GV) (5). Figure 1 illustrates the structure of typical members of each of these genera. At this time the expression vectors have been constructed using two species of the genus *Nucleopolyhedrovirus, Autographa californica* NPV, abbreviated AcMNPV, and *Bombyx mori* NPV (BmNPV). The NPV are characterized by the formation in the nucleus of infected cells of paracrystalline, proteinaceous occlusion bodies (OB) in which the virions are occluded. The occluded virion, known as the occlusion-derived virus (ODV), may contain a single nucleocapsid (SNPV) or multiple numbers (MNPV). This virion, protected from the environment by the OB, is the means by which the virus is spread horizontally in nature. A second virion phenotype is also produced during replication. This virion is primarily a single nuclear capsid in a loose-fitting envelope obtained as the nucleocapsid buds from the plasma membrane and is referred to as budded virus (BV) or extracellular virus (ECV) (Fig. 2). This envelope contains terminal peplomers consisting of a single viral glycoprotein inserted into the plasma membrane during viral replication. Dispersion of the virus among internal body tissues is accomplished by the virions of this second phenotype, which is the phenotype used to infect cells in culture.

Members of this family of viruses have never been found in animals outside the Arthropoda. Within the arthropods, these viruses predominantly infect insects and within the insects the majority have been found in Lepidoptera, the moths and butterflies. Other than insects, the virus has been found in diseased shrimp, prawns, and crabs. The host range of a virus is usually limited to closely related species within a genus or a family. However, AcMNPV is the most notable exception in that it has been reported to infect species in 10 families (6).

Replication

Infection begins when a permissive host consumes food contaminated with the viral occlusion bodies. The proteinaceous bodies are dissolved by the alkaline contents of the insect midgut, and the released virions invade the cells of the midgut epithelium by fusion of the viral envelops to the plasma membrane of the cells. Following fusion, the released nucleocapsids move through the cytoplasm to the cell nucleus and gain entry through the nuclear pores (7,8). Inside the nucleus, the nucleocapsids are uncoated and replication begins.

The synthesis of new viral proteins occurs in a cascade-like manner. In the baculoviruses the proteins are divided temporally into immediate early (α), early (β), late (γ), and hyperexpressed late (δ). No viral proteins associated with the nucleocapsid are required for the initiation of mRNA production (transcription) or early protein synthesis (translation). Immediate early genes are transcribed by the host RNA polymerase II, and transcription begins very soon after infection. The presence of α-gene products initiates the transcription of β genes. The synthesis of the α and β proteins usually occurs within 3–6 h after infection and has been defined as the time up to the beginning of viral DNA synthesis.

At around 6 h post infection, the infected cells exhibit hypertrophy of the nucleus and the accumulation of a granular chromatin known as the virogenic stroma. This stage coincides with the beginning of viral DNA synthesis, believed to be directed by an α-amanitin-resistant RNA polymerase, and the transcription of the late or γ genes. The γ polypeptides include most of the structural proteins of the virions. Assembly of virions begins in the virogenic stroma about 12–18 h post infection. The nucleocapsids assembled during this period bud from the cell nucleus and are transported to the plasma membrane at sites where the viral protein, gp 64, has been inserted. Here the nucleocapsid buds from the cell and acquires its envelope.

Beginning about 18 h post infection, the δ genes, such as those for polyhedrin and p10, are transcribed and were originally referred to as very late genes. These two genes are transcribed until cell death and now are referred to as "hyperexpressed" genes. Also at this time, the nucleocapsids no longer bud from the cell nucleus, but rather become enveloped within the nucleus and are enclosed in the polyhedrin, forming the occlusion bodies or polyhedra that give these viruses their name, "nuclear polyhedrosis virus" (Fig. 3). Because the ODV are not needed to infect cell cultures and the polyhedrin is synthesized in large amounts, this gene site was the first to be used in constructing expression vectors. Additional information on the Baculoviridae may be obtained in *The Biology of Baculoviruses*, Vols. I and II (9), and in the review of Adams et al. (6).

THE CULTURE SYSTEM

Choosing a Cell Line

Two cell lines, the Sf-21AE (10) and Sf-9, a clone of the Sf-21AE (11), from the ovaries of the fall armyworm *Spodoptera frugiperda* have been widely used in studies with vectors constructed in the AcMNPV. Vectors constructed in the BmNPV are replicated in the Bm5 cell line (12). The BmNPV infects only the silkworm and thus far has been limited to cell lines from that insect. The AcMNPV infects 32 insect species (13) with cell lines available from many of them (Table 1). In addition to lines from several different species, there are different lines from a variety of tissue sources, such as embryo, ovary, blood cells, and fat body. These represent a wealth of potentially valuable host material in which foreign genes could be expressed. The Sf-9 cell line is used for the majority of studies, principally because it is available from commercial sources or from the American Type Culture Collection; it grows well in suspension, and many of the commercial cell culture media were developed using that line or its uncloned parent, Sf-21AE, to evaluate cell growth.

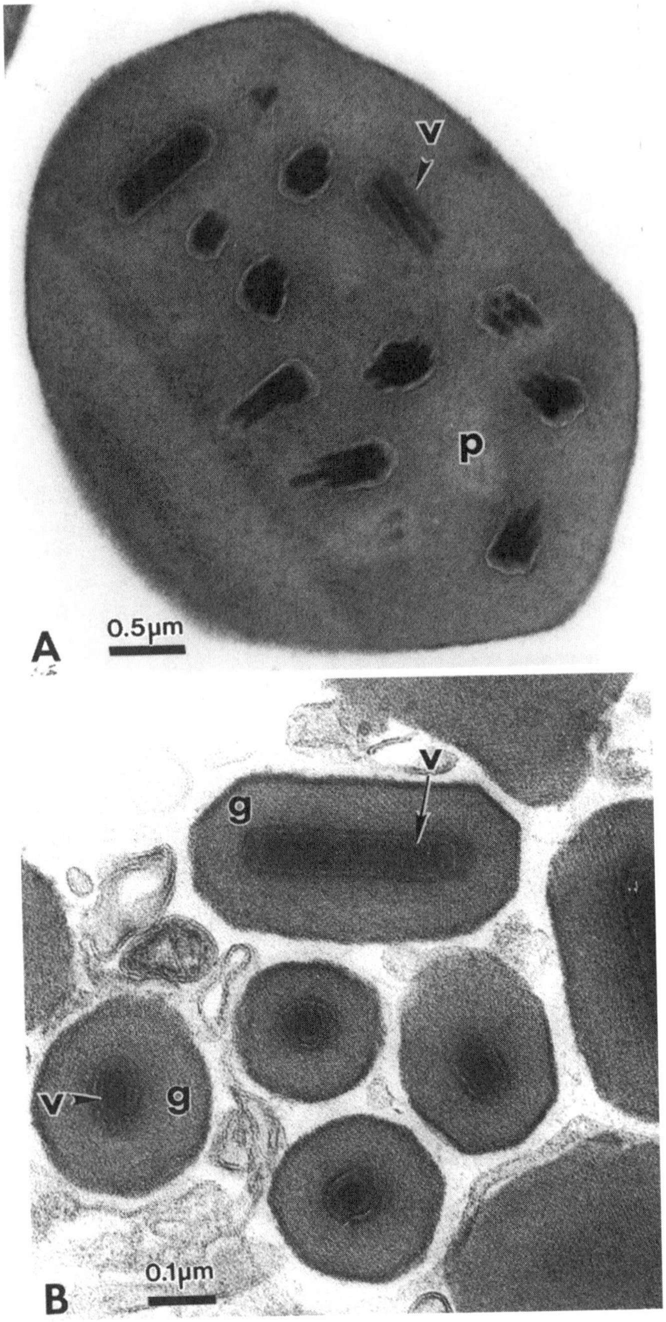

Figure 1. (**a**) Electron micrograph of a section of a polyhedra of a virus from the genus *Nucleopolyhedrovirus*. (**b**) Electron photomicrograph of granules of a virus from the genus *Granulovirus*. *p*, polyhedrin matrix; *v*, occluded virus particles; *g*, granulin matrix.

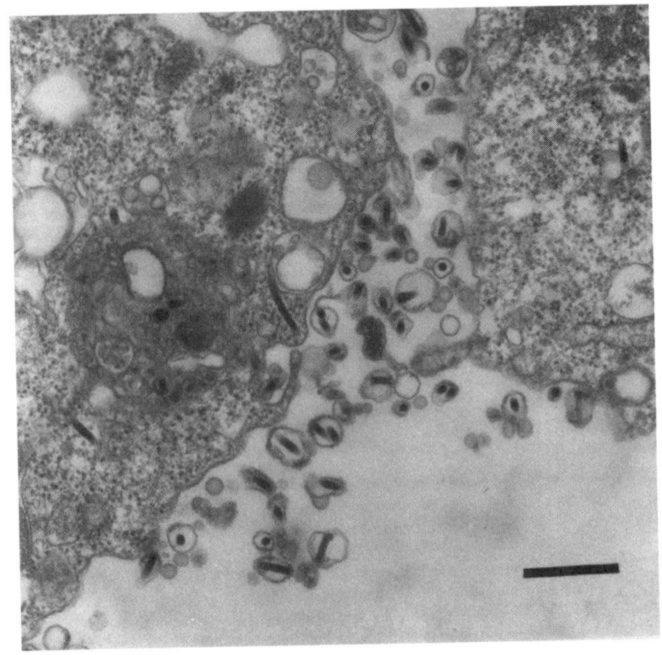

Figure 2. Release of budded virus of the gypsy moth NPV from infected gypsy moth ovarian cell lines. *Bar* 0.5 μm.

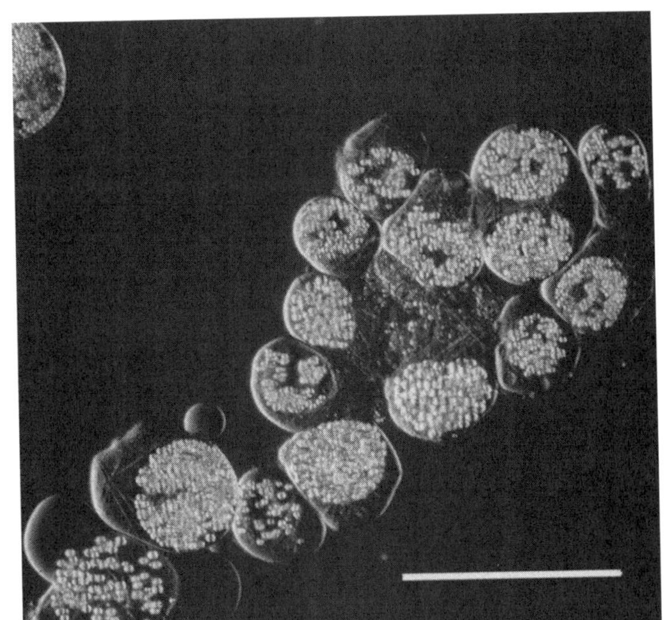

Figure 3. Formation of polyhedra in the nuclei of gypsy moth fat body cells, IPLB-LdFB, infected with LdMNPV. Viewed with Nomarski differential interference contrast optics. *Bar* 100 μm.

For ease in scaling up from laboratory studies to commercial production, cells that grow well in suspension without clumping, have relatively short population doubling times, and achieve high saturation densities are the most desirable. In media containing low levels of the surfactant Pluronic F-68, cells of the two spodopteran lines grow well in suspension in vessels with a capacity of at least 30 L. Population doubling times have been in the range of 18–24 h, and saturation densities of 5–8 $\times$ 10^6 cells/mL were achieved in batch systems. Saturation density was reached in 6–8 days depending on the level of inoculum used. The Tn-368 cell line from the cabbage looper, *Trichoplusia ni* (16), and the line Mb-0503 from the cabbage moth, *Mamestra brassicae* (17), also grow well in sus-

Table 1. Insect Cell Lines Susceptible to ACMNPV

Cell line	Tissue of origin	Species of origin	Reference
IPLB-Sf-21	Pupal ovary	*Spodoptera frugiperda*	10
IPLB-Sf-21AE, clone SF-9	Pupal ovary	*Spodoptera frugiperda*	11
IZD-MB0503	Blood cells	*Mamestra brassicae*	14
NIAS-MaBr-85	Larval fat body	*Mamestra brassicae*	14
Tn-368	Adult ovary	*Trichoplusia ni*	14
IPLB-TN-RR	Embryo	*Trichoplusia ni*	15
BTI-TN5B1-4	Embryo	*Trichoplusia ni*	15
IPLB-LD64BA	Pupal ovary	*Lymantria dispar*	14
IPLB-LdElt	Embryo	*Lymantria dispar*	15
UFL-AG-286	Embryo	*Anticarsia gemmatalis*	15
BCIRL-HZ-AM3	Pupal ovary	*Helicoverpa zea*	15
BCIRL-HV-AM1	Pupal ovary	*Heliothis virescens*	14
UCR-SE-1	Neonate larvae	*Spodoptera exigua*	15
NIAS-LeSe-11	Larval fat body	*Leucania separata*	15
IPRI-MD-108	Blood cells	Malocosoma disstria	14
FPMI-MS-5	Neonate larvae	*Manduca sexta*	15
IPRI-CF-1	Neonate larvae	*Choristoneura fumiferana*	15
IAFEs-1	Ovary	*Euxoa scandens*	15

pension and have been used by several investigators for the expression of genes in AcMNPV.

A recently developed cell line, TN-5B1-4 from *T. ni* (18), grows attached to a substrate and may be a line of choice in the production of expressed gene products secreted from the cells or where profusion is required for optimum yields. However, the cells are not obligatory substrate dependent, and substrains have been adapted to culture in suspension systems. These cells have been patented and are also available from commercial sources.

Product Yield. The levels of recombinant proteins produced in these few cell lines range from extraordinarily high, 1,000 mg/L, to less than 5 mg/L. The choice of the cell line used in production appears to be an important reason for the wide range in yields. Hink and his colleagues were among the first to examine the effect of the host cell line on product yield (19). Recombinant AcMNPVs expressing either the gene for pseudorabies virus glycoprotein (gp50T), human plasminogen (HPg), or β-galactosidase (β-gal) were tested in 23 different cell lines. The lines were selected to enable comparisons among lines from insects with different levels of susceptibility to AcMNPV and cloned lines with their parent line as well as the level of expression of each of the three genes in a cell line.

As might have been expected, no single line produced the highest yield from all three genes. Susceptibility of the insect for the wild-type virus was not a guide for selecting cell lines. Generally high yields corresponded to a high percentage of wild-type infectivity, but a high percentage of infectivity did not always result in high yields of expressed protein. *M. brassicae* is reported to be very susceptible to AcMNPV; however, four of the six lines from this insect had low sensitivity to the wild-type virus. One of these four cell lines did show expression of a recombinant gene. One line of the six had 78% of the cells infected by the wild-type virus and produced the second highest levels of β-gal but only moderate levels of HPg. The yields of expressed protein from two *T. ni* cell lines, with an almost fourfold difference in the percentage of cells infected by wild-type virus, corresponded to percent of infected cells except for gp50T, which was produced equally by both cell lines. Two lines from embryos of the gypsy moth, an insect not susceptible to the AcMNPV, were 70–78% infected with wild-type virus. Both produced moderate yields of HPg and gp50T. One, LdElta, produced more β-gal than Sf-21AE but not Sf-9. Betenbaugh et al. (20) also compared the yields of β-gal from Sf-9 and LdElta and found that LdElta produced slightly more β-gal activity on a per milliliter basis and on a per milligram of protein basis. LdElta also produced 38% more of the rotavirus protein VP4 than did Sf-9. Results of Hink et al. (19) were also mixed when the infectivity of cloned lines was compared with that of the parent uncloned line. The two clones, Sf-9 and Cl-15, of the *S. frugiperda* line Sf-21AE were several times more sensitive to the wild-type virus. However, the parent Sf-21 produced higher yields of HPg and β-gal than either clone. The clone of the *Spodoptera exigua* line, although equally susceptible to the wild-type virus, did not produce as high yields of any of the recombinant proteins as did the parent.

The foregoing studies were done in suspension culture, but similar comparisons have been made in static cultures in which differences in the effect of cell contact inhibition among cell lines must be taken into account. Cell growth in static cultures is often inhibited when the cell density becomes such that the cells are in contact with each other. Wickham et al. (21) screened eight cell lines in static cultures for the production of β-gal. In five of the eight cell lines, the effect of contact inhibition was moderate to severe, causing a three-to sixfold decrease in yield when the cells were infected at confluency. The production of BV also declined with increasing cell density. The cell lines Sf-21AE and Sf-9 were the least affected. These cells tend to attach lightly to a substrate and continue to grow even after confluency is reached. The cell line BTI-TN-5B1-4 was most affected.

When the density effects were eliminated by inoculating each cell line at an appropriate cell density, the

BTI-Tn-5B1-4 and the BTI-TnM cell lines produced 216 and 236 international units (IU) of β-gal/mL culture, respectively, compared to 97.2 and 104 for Sf-9 and Sf-21AE, respectively. Yields from the other cell lines tested were between these values. On a per cell basis, 5B1-4 and the *M. brassicae* line Mb0507 produced more than 1,000 IU per million cells and Sf-9 produced only 204 IU per million cells. The amount of β-gal produced per microgram of cellular protein by 5B1-4 and Tn M cells was 4.5 to 5.0 times greater than for Sf-21AE and Sf-9 cells.

The β-gal produced by the vector cell systems tested was retained within the cells. To determine if expression in a vector cell system in which the product was secreted into the surrounding medium was different, Wickham et al. (22) studied the expression of a gene for secreted alkaline phosphatase (SEAP) using the same eight cell lines as hosts. The cells were infected at a multiplicity of infection (MOI) of 10 and the cultures were harvested after 3 days. The 5B1-4 cells produced 5 to 10 times more SEAP per milliliter than any of the other lines, and 77% of the SEAP was secreted. The Mb0503 cell line produced the least SEAP per milliliter, but 85% was secreted. In Sf-9 cells only 59% of the SEAP was secreted. If secreted products are important factors in production or recovery of expressed proteins, then the choice of cell line should take this into consideration.

Posttranslational Processing. The capability of extensive protein processing by insect cells is particularly important in the preparation of human and animal vaccines. However, knowledge of the biosynthesis and catabolism of glycoprotein glycans from invertebrates, in particular from insects, has been described as "at best fragmentary" (23). The use of the baculovirus expression system to produce foreign proteins resulted in expanded interest in the posttranslational processing of proteins by insects. Cultured insect cells can perform O-linked and N-linked glycosylation, phosphorylation, mryistoylation, and palmitylation, folding, or cleavage of certain proteins to their active state. Contrary to early reports, some insect glycoproteins do contain sialic acid, and cultured cells can process nascent glycoproteins to sialylated, complex structures (23). Native proteins from the normal host animal differ slightly from those proteins produced in mammalian cells from other animals. Thus, proteins processed in insect cells would be expected to differ from the native protein. Glycoproteins expressed in Sf-9 cells are usually smaller than their native counterpart, which results from differences in glycosylation because the unglycosylated molecules are the same size as the native unglycosylated molecules. However, most of the more than 500 gene products expressed in insect cells that undergo co- and posttranslation processing are very similar in biological activity or antigenicity to the native protein. Glycosylation and other posttranslational modifications of foreign proteins by cultured insect cells are reviewed by Marz et al. (23) and Luckow (24).

Results from baculovirus–cell studies are inconsistent if not contradictory. Kuroda et al. (25) showed that a fowl plague virus hemagglutinin expressed in Sf-9 cells lacked galactose and sialic acid. However, a human plasminogen expressed in Sf-21AE cells contained complex type sialic acid-containing side chains normally found in this protein (26). This was the first documentation that these complex type sialic acid side chains could be produced in proteins from insect cells. As Sf-9 is a cloned line derived from Sf-21AE it is not clear what caused the differences. In studies where O-glycosylation is required, results are similarly unclear. The native human insulin receptor expressed in Sf-9 was not detectably glycosylated (27), however, the pseudorabies viral protein, gp50, was glycosylated (28).

Studies on the phosphorylation of proteins expressed in Sf-9 cells reveal considerable variation in the levels and sites involved. In his review of the subject, Jarvis (29) noted that in his work the simian virus 40 large tumor antigen (SVT ag) expressed in Sf-9 cells was phosphorylated at different sites than in mammalian cells, but that others had reported no difference in the phosphorylation patterns of the p40^{x} protein in Sf-9 and in human lymphoid cells. Jarvis also noted that in a study by Hoss et al. (30) those phosphorylation sites in SVT ag protein recognized by cytoplasmic kinases were better utilized than sites recognized by nuclear kinases. The SVT ag protein and several other proteins were also acetylated in Sf-9 cells. The only processing pathway that seems to be missing in the spodopteran cells, Sf-21AE and Sf-9, is one for α-amidation (29).

Jarvis summarized his review by saying that the available results indicate that lepidopteran cells can adequately process foreign proteins with the exception of the α-amidation and some cleavages at sites containing basic amino acids. This conclusion is based upon studies with only one cell line and a single clone of the line. The differences already demonstrated in the capabilities of the parent line and its clone in glycosylation should provide an inducement to test additional cell lines. For example, cell lines from the insect fat body are more difficult to culture, but this organ is very important for the processing of metabolites in the insect. Baculoviruses replicate very efficiently in fat body cells, and lines from this tissue might very well contain processing pathways that are even better than those in the two lines used thus far.

Choosing a Medium

The first insect cell lines were developed by Grace (31) using a medium he developed. This medium became the standard for the culture of cells from lepidopterous insects, and most media developed later were based on it. One of the first modifications of Grace's medium was the TNM-FH medium of Hink (16), who added lactalbumin hydrolysate and yeast extract to the Grace formulation. Hink used this medium to establish the Tn-368 line from the cabbage looper. This was followed by a medium developed by Gardiner and Stockdate (32) that contained tryptose broth. This medium, originally called BML-TC/10, is sold today by several suppliers under the name "TC-100." A more defined modification of the Grace medium, IPL-41, was developed in 1981 by Weiss and colleagues (33). Several trace metals and vitamin B_{12} were added to the Grace formulation. Maltose replaced some of the sucrose, and the levels of several amino acids were changed. All these media are sold by several media suppliers either as dry powders or as liquids. Generally all the media require supplementation with 5–10% fetal bovine serum (FBS). When the media are used in suspension cultures, methyl cellulose or Pluronic F-68 is added. Most of the commonly used lepidopteran cell lines will grow well in any of these media, and a cell line seems to adapt easily from one to another.

The requirement for FBS presents some problems when insect cells are to be used for the production of gene expression products. It is expensive, the quality varies from lot to lot, and the supply is sometimes unreliable. There is also the possibility of introducing bovine viruses or mycoplasmas into the medium with the serum. Finally, the serum proteins complicate the downstream recovery and purification of the expressed protein. For these reasons, there has been a great interest in the development of protein-free media. The IPL-41 medium was the starting point for the development of most of the protein-free medium. Maiorella et al. (34) developed the first such formulation. The FBS was omitted and the IPL-41 was supplemented with 0.4% ultrafiltered yeast extract and a lipid emulsion containing cod liver oil, fatty acid methyl esters, cholesterol, and α-tocopherol (added as an antioxidant and free radical scavenger).

The population doubling times and the maximum cell densities of Sf-9 cells grown in this medium were equal to those of the cells in serum-supplemented IPL-41. Yields of macrophage colony-stimulating factor from cells infected with a baculovirus containing the gene for this protein were equivalent in both the IPL-41-FBS-supplemented medium and the serum-free medium. There are a number of protein-free media available from commercial supplies that are, presumably, similar in composition to the medium of Maiorella et al., and the yields of foreign gene products from cells grown in these media generally have been comparable to yields in FBS-supplemented media.

Reporting on the development of the serum-free medium SF900, Weiss et al. (35) showed that Sf-9 cells reached higher maximum cell densities and had shorter population doubling times on the serum-free medium than on Grace's medium plus 10% FBS. Similar results were obtained with the TN-368 cell line, where the final cell density was almost twofold higher in the serum-free medium compared to growth in Grace's + FBS. The production of β-gal in serum-free medium was 152% of that in Grace's medium + serum. A maximum of 500,000 units/mL was produced in 6 days in shaker flask cultures.

The growth of Sf-21AE cells in stirred tank bioreactors was compared using serum-supplemented medium (TNM-FH, TC-100, TC-100 + 2.5 g/L tryptose phosphate broth, IPL-41, and IPL-41 + 2.5 g/L tryptose phosphate broth) and serum-free medium (Ex-CELL 400, EX-CELL 401, and the Sf900) (36). Similar final cell densities and population doubling times were obtained with the TC-100, the IPL-41, and the serum-free media. The additional supplementation of either TC-100 or IPL-41 with tryptose phosphate broth gave superior growth compared to all other media.

Hink and colleagues (19) compared yields of gp50T, HPg, and β-gal from the cell lines Sf-21AE, Sf-9, TN-368, and EAA in the serum-free medium of Maiorella et al. or FBS-supplemented TNM-FH, IPL-41, and TC-100. The results did not consistently favor either serum-supplemented or serum-free media. For example, Sf-9 cells produced more HPg in serum-free medium but more β-gal in serum-supplemented medium. Results with Sf-21AE were reversed—more HPg with serum-supplemented and more β-gal with serum-free medium. These studies were done in static culture systems with the cells attached to the growth surface.

When EX-CELL 400 was used as the serum-free medium, the β-gal produced by either Sf-21AE, Sf-9, or TN-368 was one-eighth to one-third lower than the levels produced by these cells in TC-100 + FBS (37). These studies were also done in static cultures. Comparison of the replication of wild-type virus in the two media indicted very similar levels of BV produced in either serum-supplemented or serum-free medium. Similar numbers of polyhedra were produced in each test system and the specific activity, as determined by bioassay, was the same as or higher in the serum-free medium than in the serum-supplemented medium.

Although it is clear that the choice of medium is an important factor in optimizing either the replication of wild-type baculoviruses or the expression of recombinant gene products, the available data give little guidance in choosing a medium. The currently available protein-free media appear to provide adequate nutritional support for the production of recombinant gene products. With the advantages such media provide in the downstream harvesting of the gene product, they are the most desirable. As the physiology and metabolism of insect cells in culture become better understood, improved media formulations will certainly be developed. At this time, testing various combinations of cell lines and media is the best approach to optimizing yields.

Preparing a Recombinant Baculovirus

Recombinant baculoviruses, because of the large size of their genome, are prepared using a multistep process to insert the foreign gene into the viral genome. Beginning with a bacterial plasmid, the first step is to select or construct a suitable transfer vector and insert the desired foreign gene. Then, susceptible insect cells are simultaneously infected with the transfer vector and wild-type AcMNPV DNA (transfection). In the infected cell, the two DNAs recombine and produce the new recombinant virus. The progeny from this virus must then be isolated and separated from the wild-type virus progeny produced in cells infected only by the parental DNA. This step is typically accomplished by infecting a monolayer of cells and reisolating the virus from well-isolated foci or plaques. Usually, three cycles of plaque purification are required to ensure the complete separation of the recombinant and wild-type viruses. Recombinant viruses have been prepared most often from the AcMNPV and less frequently from BmNPV, but the same or similar methods should be useful in preparing recombinants from other NPVs. For excellent detailed discussions of preparing recombinant baculoviruses, see Refs. 24, 38.

Selecting a Transfer Vector or Plasmid. The transfer vectors have three features in common: (*1*) they are bacterial plasmids that have an origin of replication and an antibiotic-resistant marker to aid in selection when the plasmid is amplified in *Escherichia coli*; (*2*) the plasmid also has a suitable viral promoter and a suitable restriction enzyme site for cloning downstream from the promoter site; and (*3*) the cloning site is flanked by native AcMNPV DNA to promote homologous recombination. When the polyhedrin promoter is used in constructing the vector, the nature and the amount of the 5′ upstream sequence of the gene in the native DNA has an important effect on the expression of the foreign gene (39,40). Transfer vectors used with some baculovirus are of three types: (*1*) vectors to express a single foreign gene, (*2*) coexpression vectors that contain the gene for a marker protein, such as the *Escherichia coli LacZ* gene, and a gene for the desired protein, and (*3*) dual expression vectors containing genes for two foreign proteins. Many of these vectors are described in the literature and can be purchased from commercial sources (e.g., Clontech, Palo Alto, Calif.; PharMingen, San Diego, Calif.; Invitogen, San Diego, Calif.; Life Sciences Inc, Gaithersburg, Md.; Stratagene, San Diego, Calif.). Descriptions of several baculovirus vectors, some commercially available, are given by López-Ferber et al. (38).

The first, and perhaps most commonly used, cloning site was the polyhedrin gene site. This gene is not essential for maintenance of the virus in cell culture and is abundantly expressed in the infected cell, often to as much as 50% of the total protein produced in an infected cell. Another nonessential gene site often used is the p10 gene, which codes for a fibrillar protein with a relative molecular weight of 10,000. It also is expressed in the late phase of virus replication in copious quantities. These two gene promoters have been used together to construct vectors for multiple gene expression by a number of researchers (e.g., 41,42).

Transfection. Transfection is the next critical step in the development of a recombinant baculovirus. Procedures used should be designed to optimize the recombination between the vector and the viral DNA. The commonly used coprecipitation of DNA with calcium phosphate requires inexpensive reagents that are simple to prepare. The calcium ions and DNA are mixed together and allowed to form a fine precipitate, which when combined with cells will bind to the cell surface and facilitate the transfer of the DNA into the cell. DEAE-dextran, poly-L-lysine polyomithine, and polybrene have also been used to transfect insect cells. The efficiency may sometimes be improved by treating the cells with dimethylsulfoxide or glycerol to osmotically shock them. Typically, only 0.1–1.0% of the progeny viruses are recombinants, requiring that large numbers of plaques must be screened to locate the few containing recombinant viruses. Cationic lipids can also be used in transfection. One such preparation, Lipofectin, is commercially available. A procedure for using this material was described by Trotter and Wood (43), who typically obtained virus populations containing 15% recombinants with the Lipofectin procedure using the supercoiled form of DNA.

Kitts et al. (44) reported that the proportion of recombinant progeny viruses can be increased to about 25% by using linearized viral DNA instead of the circular DNA. DNAs that can be linearized at a specific point are prepared by introducing a unique enzyme recognition site. The vector has sufficient sequences from both sides of the restriction site on the viral DNA to facilitate homologous recombination. Recombination closes the cut and restores circularity to the viral DNA, thereby permitting replication, as it is believed that linear DNA cannot be replicated. In this system, the number of progeny viruses is greatly reduced, making the location of recombinant virus much easier.

PRODUCTION PROTOCOLS

Choosing a Bioreactor

The insect cell lines Tn-368, Sf-21AE, and Sf-9 do not require attachment to a substrate to grow, and thus suspension culture systems are favored for the production of recombinant gene products. In all suspension systems the cells must be kept uniformly distributed throughout the medium to provide equal access to nutrients and prevent the localized buildup of toxic metabolites. The dissipation of the energy required to maintain this needed homogeneity can cause significant cell damage. The larger the system, the more of a problem will be damage resulting from mechanical agitation. Animal cells and insect cells in particular, because of their relatively large size and the lack of protective cell wall, are more susceptible to mechanical damage than microorganisms. A number of systems have been developed to overcome this problem and provide a means for the economic production of insect viruses and gene expression products.

Laboratory Scale. For research purposes, shake flasks or spinner flasks are usually the systems of choice. Shake flask culture systems are simple, require no expensive equipment, and provide an excellent environment for growth of insect cells. Neutra et al. (45) reported that shake flasks with working volumes up to 120–150 mL supported growth of the Sf-9 cells to a final density of 5×10^6 cells/mL with a doubling time of 20 h. The cultures were incubated on an orbital shaker at 100 rpm. Working volumes of 25–45% of the flask volume were found to be most satisfactory. Cultures infected with an engineered baculovirus containing the *E. coli LacZ* marker gene produced about 600 mg/L of β-gal.

Spinner flasks also are easy to use and require little expensive equipment. They have the added advantage that the culture vessels used can be easily modified to add aeration equipment or the incorporation of electrodes for monitoring pH or other parameters. This permits the use of larger working volumes and allows the quantitative monitoring of some culture conditions. Although there are several different designs for culture vessels, the basic design is one in which a magnetic bar or paddle is suspended in the growth chamber and driven by an external magnet. Standard spinner flasks are available with fluid volumes from 25 mL to 36 L. Many designs have side arms that provide access without removing the cap. Larger vessels have ports in the lids for pH and O_2 probes, medium addition and removal, and gas inlets.

Working volumes in spinner flasks range from 32% to 80% of capacity and rotation speeds of the stirring bar are from 50 to 100 rpm. In working volumes less than 500 mL, oxygen transfer from the headspace of the flask into the agitated culture medium is sufficient to meet the requirements of the insect cells (46). Cameron et al. (47) found that with agitation of 100 rpm oxygen levels became limiting at working volumes of 1000 mL. Therefore, in these small cultures sparging is not needed and the only shear forces affecting the cells result from the agitation of the medium. Eberhard and Schügerl (48) demonstrated the importance of agitation speed in designing culture systems. In cultures with a working volume of 100 mL in 1000-mL flasks at an agitation speed of 35 rpm, cells settled to the bottom and growth was reduced as a result of localized depletion of oxygen and nutrients. Final cell densities increased with increased agitation speed, up to 100 rpm, until oxygen levels were no longer growth limiting. Kioukia et al. (49) compared the growth of Sf-9 cells in 100-mL spin flasks, 50-mL working volume, stirred at speeds of 80, 120, and 150 rpm. Optimum cell growth before infection was at 120 rpm. Viral infection was enhanced at 150 rpm but, as in the noninfected cultures, cell damage occurred. It appears that 100 to 120 rpm is adequate for nonaerated spinner cultures if there is sufficient headspace in the vessel. Culture volumes greater than 1 L require aeration and thus present additional problems in bioreactor design and operation.

Scale-Up. Once the need for culture volumes exceeds 1 L, providing sufficient oxygen for cell growth and virus replication becomes a serious problem. Hink and Strauss (50) reported that the dissolved oxygen (DO) level in a 2-L bioreactor decreased from 100% to 2% in 24 h during the growth of the Tn-368 cells. The amount of sparging required to maintain the DO level at 100% caused excessive medium foaming and vacuolation of the cells. Infected cells consume oxygen at almost twice the rate of uninfected cells, thus increasing the supply problem. Tramper et al. (51) attempted to identify the sources of shear stress and to establish the level at which Sf-9 cells were damaged. They found that the critical shear stress at which cell viability began to decline was 1–4 $N\,m^{-2}$. This result agreed with the data of Hink and Strauss for the TN-368 cells. They found, as did Hink and Strauss, that cell death occurred at stirrer speeds greater than 220 rpm. They also observed that cell survival in the foam was very low and that the number of viable cells decreased with every bubble burst. Their conclusion was that the dispersion of gas in the medium by means of stirring and sparging sufficient to provide the needed oxygen would result in a larger death rate than the growth rate of the cells and that membrane bioreactors or oxygen supplied through semipermeable tubing might be required to solve this problem.

The effects of hydrodynamic forces on insect cells in culture were reviewed by Chalmers (52), and his conclusions are discussed here briefly. Cells in sparged, agitated culture systems are subject to shear forces associated with gas bubbles under three conditions: in the region where the bubbles are injected into the medium, during the free rise of the bubbles to the surface, and at the air—medium interface at the top of the vessel. The preponderance of data reviewed by Chalmers indicated that most of the cell death occurred at the air–medium interface. During the rise of the bubbles through the medium, cells became attached to a bubble and death occurred when the bubble ruptured at the surface. According to studies by Trinh et al. (53), on the average 1,200 Sf-9 cells were killed per bubble burst when suspended in TNM-FH medium without protective additives. Only 292 cells killed were present on the bubble film; the remainder was believed to be in the thin layer of medium surrounding the bubble. No cells were killed when the medium contained 0.1% Pluronic F-68.

Protective additives are important components of media for insect cells grown in suspension culture systems. Vaughn (54) used 0.15% methyl cellulose for the first culture of lepidopteran cells in spinners. Hink and Strauss (50) found that 0.3% methyl cellulose allowed them to increase the impeller speed from 100 to 220 rpm in a 4-L stirred tank fermentor. However, the most successful protectant has been Pluronic F-68. Maiorella et al. (34) and Murhammer and Goochee (55) used it for the culture of Sf-9 cells in an agitated, sparged bioreactor and in an airlift bioreactor. Murhammer and Goochee reported that population doubling times and final cell densities were not affected by concentrations up to 0.5% Pluronic F-68. Maximum viability of 95–99% was achieved with 0.1%. This concentration also was used by Maiorella et al. and is the concentration most frequently used for insect cell cultures.

Murhammer and Goochee suggested that Pluronic F-68 protected cells by interacting with the cell membrane and that this interaction was the basis for the protective mechanism. Photographs presented by Chalmers (52) show that in the presence of 0.2% Pluronic F-68, no cells attach to gas bubbles in the culture medium. Therefore, much of the killing of cells when the bubbles rupture at the air–medium interface was avoided. The mechanism that prevents the cell–bubble interaction is not known, but effects have been quite striking. The development of scale-up protocols for the growth of insect cells on a commercially feasible scale now is concentrated on the use of the standard stirred tank reactors or airlift fermentors rather than the development of specialized vessels.

Sparged Stirred Vessels. The sparged, stirred tank fermentors used extensively for many types of microbial fermentations can be operated inexpensively and their scale-up is straightforward. Glass fermentors with working volumes of a few liters to stainless steel tanks with capacity up to 8,000 L have been used for a number of years for the production of biologicals in mammalian cells. The most notable examples are the production of foot-and-mouth virus vaccine, interferon, and monoclonal antibodies. These vessels, used to culture animal cells, are round bottomed and fitted with agitators having marine propellers or one or more Rushton-type turbine stirrers. The stirrers may be driven directly or by external magnetic drives. To protect the cells from contact with air bubbles and the damage that occurs when those bubbles break, some specialized aeration systems have been designed in which the air is introduced into the medium stream inside rotating stainless steel mesh cages that exclude the cells.

With the discovery of the protective effect of Pluronic F-68, most insect cell culture is now done in the more typical sparged, stirred tank. Overton and Kost (56) reported the use of 36-L stirred vessels to produce human synovial phospholipase A_2 in Sf-9 cells. The vessels were fitted with a ring sparger with 2-mm holes, and DO level was maintained at 50% by oxygen sparging. The expression level of purified product from this system ranged between 1.0 and 1.5 mg/L and was comparable to that from shake flasks. Silberklang et al. (57) have described the scale-up of a stirred tank system of 70 L to produce antistasin, an anticoagulant protein from a leech, in Sf-9 cells. They used an open-pipe (5-mm-diameter), sparged Chemap fermentor equipped with hydrofoil impellers. The key to their success in scale-up, they considered, was the use of a strain of the Sf-9 cells that had been well adapted to suspension growth by serial passage in small spinner cultures for 6–8 weeks before freezing the cells in the master cell bank.

Airlift Bioreactors. Airlift bioreactors have been used for some time to culture fragile cells such as hybridoma cells. These vessels consist of two tubes, one inside the other. Air is sparged in the bottom of one tube, the riser. As the air bubbles rise to the top of the riser the difference in density between the medium in that tube and the second tube, the downcomer, results in circulation of the medium and cells without the need for mechanical stirring. Possible damage to fragile cells from the turbulence and resulting shear forces in the eddies at the ends of the stirrer blades is avoided. Insect cells were considered to be more sensitive to these forces than most other animal cells and so many investigators studied their growth in the airlift bioreactor.

The airlift bioreactors have been modified for use with attachment-dependent cells such as the BTI-TN-5B1-4 line by adding glass beads or other materials to the downcomer tube, which is modified to retain them. The beads form a packed bed to which the cells attach and are then bathed in oxygenated medium. This method avoids the shear forces the cells encounter when the beads are maintained in suspension in stirred tanks. A model system describing the effects of such parameters as the ratio of riser area to downcomer area, airflow rate, and DO to the growth of the 5B1-4 cells has been developed by Kumar and Shuler (58).

Maiorella et al. (34) used an airlift bioreactor in their early studies on the scale-up of Sf-9 cells for recombinant protein production. Pluronic F-68 was added to the medium to protect the cells from damage by the bursting bubbles. It also limited the buildup of foam so that no additional antifoam was needed. Bubble size was controlled at 0.5–1.0 cm in diameter by selection of a sparger with the appropriate orifice diameter. A sparging rate of 420–1260 mL/min was used as this provided an adequate supply of oxygen but did not cause excessive turbulence or foaming in a 21-L vessel. Cell growth was equal to that in 100-mL shake flasks, and the production of recombinant human macrophage-stimulating factor reached 40 mg/L in 4 days.

Rice et al al. (59) compared the growth and production of recombinant protein by Sf-9 cells in an 8-L stirred tank vessel and in a 6-L airlift vessel and found that the cell growth was almost identical in the two systems. Population doubling time in the airlift was 22.5 h and in the stirred tank 21.9 h. Maximum density of the Sf-9 cells in the airlift was 4.94×10^6 cells/mL compared to 4.91×10^6 in the stirred tank. Yields, specific activity of the recombinant human protein kinase, were also virtually identical. They also considered the relevant costs of each system. In the laboratory the cost of providing temperature control, DO control, and other operating expenses were about equal. However, Rice and his colleagues estimated that the cost of scaling up the airlift system to 36 or 45 L would be 98% more than that of scaling up the stirred tank system to the same level. Other examples of the use of these systems for the production of recombinant proteins from insect cells can be found in the proceedings of a workshop on the subject (60).

Establishing Culture Conditions

Sterility. All animal cells must be grown under sterile conditions. The very rapid growth rate of most microorganisms compared to that of animal cells means that even the presence of small numbers of microorganisms at the beginning of the fermentation will result in the overgrowth of the animal cells by the microorganisms. This will cause the loss of the fermentation or at least a significant reduction in the yield and additional problems in product recovery.

Insect cell culture media generally contain heat-labile components and, therefore, must be filter sterilized. Liquid media obtained from commercial media suppliers are sterile and the sterility need only be maintained by use of proper microbiological technique. If the media are prepared on site, pressure filtration, using compressed air or nitrogen, through a filter stack consisting of a depth prefilter, a 0.45-μm membrane filter, and finally a 0.22- or 0.01-μm membrane filter held in a stainless steel filter holder, is recommended. In the preparation of small batches of medium the filtered medium is usually collected aseptically into sterile vessels and held refrigerated until needed. In this situation it is feasible to run sterility checks before the batch of medium is used. When large volumes of medium are required for a single production run, it may be most practical to prepare and sterilize the medium just before use. Clarifying filters need not be sterile and may be separate from the sterilizing filters. Sterilizing filters should be located as near to the fermentors as practical and should be tested for integrity before use. Venders provide methods and equipment for this test.

Small fermentors can be assembled and sterilized by autoclaving, assuming an autoclave large enough to hold them is available. All parts of the system must be thoroughly cleaned and sanitized using one of several detergents sold for this purpose. To prepare the vessel for autoclaving, all access ports should either be fitted with the correct tube or probe or plugged. Oxygen and pH probes should be prepared according to instructions provided by the manufacturers. Glass connectors should be secured, the tips plugged with a nonabsorbent material such as glass wool and covered with aluminum foil. The vessel should be filled with water or phosphate-buffered saline to the working level. The sterilizing time will depend on the size of the vessel. The vessel must be vented at all times and should not be removed from the autoclave until the vessel has cooled somewhat to avoid drawing nonsterile air into it. The vessel can then be closed, placed in a sterile hood, if possible, and allowed to cool completely. The water is then aseptically removed and the culture medium added.

Fermentation vessels too large or too heavy to be placed in an autoclave are steam sterilized in place. Proper design of the system is very important to achieve sterilization of such equipment. Steam vapor must contact all the parts of the system to be sterilized. Steam must enter the vessel in such a manner as to replace all the cold air, and the resulting condensate must be adequately drained from the system. Pockets and crevices in the system must be avoided or minimized. Perhaps most important, standard operating procedures must be developed and the operating personnel well trained in those procedures. Sterilization is discussed in more detail by Bergman (61).

Oxygen Supply. Insect cells require oxygen to grow and will rapidly deplete the dissolved oxygen in the culture medium. Oxygen availability is critical to optimum yields of both wild-type virus and recombinant protein production. Determining a suitable level of dissolved oxygen and developing ways to maintain that level have been the subjects of considerable research. The oxygen utilization rate (OUR) for insect cells, 3.5–9.9 $\times$ 10^{-17} mol $cell^{-1}$ s^{-1}, is slightly higher than that of many other animal cells. Moreover, the rate increases to nearly double by 12–18 h following virus infection and then declines to near that of uninfected cells. This period of maximum OUR corresponds to the time of γ-gene expression and the production of virions.

The infected insect cells are also more fragile than uninfected cells. Thus, simply increasing the sparging rate or the agitation may be counterproductive. One possible alternative may be lowering the incubation temperature from 27 °C to 25 °C, which increased the oxygen solubility and lowered the OUR of Sf-9 cells by 26% without a reduction in yield of recombinant β-gal (62). However, DO levels between 20% and 60% will support optimum cell growth and virus replication and have been achieved in most bioreactors with sparging raters that do not cause significant cell damage. See the review by Taticek et al. (63).

Feeding Strategies. Bioreactors may be operated in several ways related to the way nutrients are supplied to the cells. Batch, fed-batch, semicontinuous, and perfusion methods have all been tested for growing insect cells. Stirred tank bioreactors operated in a batch mode are the most common microbiological fermentations. They are simple, inexpensive, and easy to scale up. In this system medium is added to the bioreactor and inoculated with the desired cells and then run until cell growth or product production stops. During this time the cell nutrients are gradually depleted and metabolic waste products accumulate in the tank. pH may change as a result of the accumulation of such waste products as ammonia or lactic acid. One of these changes can eventually become growth limiting. However, a baculovirus infection ultimately results in death and lysis of the cell, and this may terminate the fermentation before any of the metabolic factors become limiting. Therefore, the batch method, using a carefully designed starting medium, may provide optimum yields with the least cost.

Additional cell growth or product production may be obtained with the fed-batch system, if either nutrient depletion or waste accumulation has terminated cell function before the lysis of infected cells. Limited studies on the accumulation of ammonia in cultures of insect cells indicates that ammonia does not accumulate to toxic levels unless the glucose is depleted (64,65). The addition of ammonia in the form of NH_4Cl or $(NH_4)_2SO_4$ up to a concentration of 10 mM was not inhibitory to cells of either the Sf-9 or BTI-EAA lines. Growth did slow somewhat at a concentration of 30 mM, indicating that these cell lines are considerably less sensitive to ammonia than many mammalian cells (64). If adequate concentrations of glucose were present, the concentration of ammonia found in both these studies did not approach these levels (64,65). Similarly, insect cells produce only small amounts of lactic acid during growth in an adequately oxygenated culture. Bédard et al. found that lactate was consumed by Sf-9 cells at the end of the growth period (64). They attribute the difference between lactate production in mammalian and insect cells to the presence of a complete TCA cycle in insect cells but not in many mammalian cells.

The advantage of the fed-batch system appears to be from the replacement of depleted nutrients. Feeding can be done simply by adding additional complete medium to the culture. Lindsay and Betenbaugh (66) found that the resuspension of high cell density cultures of Sf-9 cells in fresh medium at the time of infection resulted in a fourfold increase in yield of the recombinant β-gal. However, this is somewhat inefficient because many of the components in the medium are not severely depleted during a normal growth cycle; for example, it is unlikely that any vitamins or inorganic salts become growth limiting. If the growth-limiting nutrients can be identified, only they need to be supplemented.

Few of the amino acids in a medium are depleted more than 50%. The need for amino acids varies among insect cell lines. Mitsuhashi (67) determined the amino acid requirements of cell lines from six different insect species. Only five, cystine, histidine, isoleucine, methionine, and threonine, were essential for all six cell lines. In the study of Bédard et al. (64) only methionine, of these five, was consumed by Sf-9 and BTI-EAA cells. However, the nonessential amino acids glutamine, glutamate, and aspartate were extensively depleted. No amino acid was depleted to growth-limiting levels.

Wang et al. (68) found that the specific yield of a recombinant protein from Sf-9 cells was increased 100% by feeding glucose and glutamine to high cell density cultures immediately after infection. The amount added returned the levels to approximately those in fresh medium. Lee and Park (69) demonstrated that the addition of Yeastolate to a culture of Sf-9 cells in stationary growth phase would stimulate further growth and increase the final cell density about 3.5 fold, but no information on the critical components of the Yeastolate was given.

Semicontinuous culture is another way to maintain suitable levels of oxygen and nutrients to support optimum virus replication or recombinant protein production in high cell density cultures. In these systems, cells are grown in one vessel and when high cell densities are reached, a portion of the cell suspension is transferred to one or more new vessels, fed with fresh medium, and inoculated with virus for production. Fresh medium is also added to the cell culture vessel to continue the production of cells. At the time the virus or recombinant protein is harvested some medium may be left in the production vessel to inoculate the next batch of cells. Such systems have been used with TN-368 cells (50) and with Sf-9 cells (70) for the production of wild-type AcMNPV. A similar two-stage system was used with *B. mori* BM-5 cells for the production of recombinant proteins (71).

Perfusion culture systems have supported the highest final cell densities, $>10^7$ cells/mL, achieved with insect cells. In perfusion cultures the fresh medium flows through the culture vessel at a constant rate and all the cells are retained in the vessel during the entire process. This method requires that the cells be removed from the effluent either by filtration or entrapment, as in hollow fibers. This method reduces the amount of headspace needed in a fermentor because the medium can be oxygenated before entering the fermentor. Cells are exposed to a steady-state environment, and higher cell densities per unit volume are possible. The perfusion is either "open," in which the medium passes through the bioreactor and is collected, or "closed" in which the medium is reused after oxygen has been added and, if necessary, the pH adjusted.

Deutschmann and Jäger (36) were able to maintain Sf-21AE in exponential growth for 16 days in a perfused 1.4-L stirred tank. The maximum cell density achieved was 5.5×10^7 viable cells/mL. Perfusion was started after 72 h of cultivation and maintained at a constant rate for the duration of the fermentation. Oxygen was provided through a polypropylene membrane, 5 m/L working volume. As the cell density increased, available oxygen was increased by increasing pressure in the vessel and finally by aerating with pure oxygen. When the cell density reached the maximum, DO could not be maintained, in part because of membrane fouling, and limited oxygen was given as the reason growth stopped.

Weiss and Vaughn (72) produced wild-type AcMNPV in Sf-21AE cells in a perfusion system using a 3-L Moduculture vessel equipped with Biospin filters. Cultures with a cell density of 1×10^5 cells/mL were inoculated at a MOI of 0.005 plaque-forming units (PFU)/mL. Medium was perfused in the open mode at 600 mL/day for 6 days. A 98% infection resulted with a yield of 1×10^8 polyhedra/mL. A total of 6.6 L of spent medium containing 2.4×10^9 BV/mL was produced, a 10-fold increase over the yield from batch spinner cultures.

Caron et al. (73) established Sf-9 cells in a 4-L stirred tank perfusion system using a custom-designed tangential flow filter to separate cells and medium and reoxygenated the used medium, eliminating the need for sparging the culture. The culture was incubated in the batch mode until the cell density reached 3×10^6 cells/mL. Perfusion then began at a rate of 1.4–2.8 mL/min. When the culture reached a cell density of about 1×10^7 cells/mL, the volume was reduced to 1 L, the cells infected with the recombinant virus, and serum-free medium added to return the volume to 4 L. During the production phase the culture was perfused with medium containing only 0.5% or 1.0% serum instead of 10% FBS. Cell density increased from 4×10^6 to 15×10^6 cells/mL in 3 to 4 days. However, the average protein production in this system was reported to be only 76% of that in shaker flask cultures. The lower level was believed to result from the serum level during the infection stage and not from the conditions of culture during the production phase.

Infection Strategies

Passage Effects. As described earlier, baculoviruses produce two different virions during their replication: the budded virus (BV), released from the cell by budding through the plasma membrane, and the virion enclosed in the polyhedra, the occlusion-derived virus (ODV). The ODV released from the occlusion body under alkaline conditions, pH above 9.5, is highly infectious through the gut of a target insect, but is infectious to cells in culture only at extremely high concentrations and therefore is of little use in cell culture studies. The BV are very infectious to cell cultures and are the standard means of infecting cells with either wild-type or recombinant viruses. Continued passage in insect cell culture of a baculovirus using the BV results in the gradual decline in polyhedra yield and infectivity of the polyhedra produced (74).

To avoid these passage effects, it is standard practice to use only low-passage virus in vitro studies. A master bank of virus inoculum is prepared, usually from the blood of insects fed polyhedra, and stored frozen. To prepare this material, susceptible insects in the midlarval stage of development are fed a dose of polyhedra sufficient to kill 90% of them. When the infected larvae begin to show signs of disease, they are bled by cutting a rear leg and collecting the blood in ice-cold cell culture medium. The collection in cold medium is required to prevent enzymatic oxidation of phenols to toxic quinones in the hemolymph upon exposure to air. In addition, antioxidants may be added to the medium. The diluted blood is centrifuged at low speed to remove the blood cells and the supernatant filtered through a sterile 0.45-μm membrane filter. The filtered blood is divided into small volumes and frozen at -80 °C. This blood is used to prepare the subsequent passages in cell culture. Generally a large amount of the third-passage in vitro virus is prepared and becomes the working stock. With the AcMNPV, this allows the buildup of seed inoculum and a production run of several viral replications before significant passage effects develop. Not all viruses will remain stable for this many in vitro passages. Slavicek et al. (75) found that the wild-type gypsy moth NPV was not stable for more than one in vitro passage. However, they were able to select a strain with greater stability for cell culture studies.

Wickham et al. (76) demonstrated the generation and accumulation of defective interfering (DI) particles within high in vitro passage baculovirus. DI particles build up in in vitro culture systems as a result of continued passage of the virus at high MOIs. These particles, identified microscopically by their shorter length, were lacking large sections of their DNA. Typical of DI particles, they caused large reductions in the amount of recombinant protein produced in cells infected with MOIs of more than 0.01 plaque-forming unit/cell. To avoid the accumulation of defective particles the cells must be inoculated with dilute virus suspensions, low MOI. In a system where the residual culture is left in the production vessels to inoculate incoming cells, the level of infectious virus gradually increases with each run, and with each passage the proportion of defected particles also increases. To delay the buildup of DI particles, De Gooijer et al. (77) proposed a mathematical model for the infection of insect cells with NPV in a semicontinuously or continuously operated two-stage reactor. Such a reactor was assembled and the model tested as a semicontinuous system (78). The system consisted of two 1- or 2-L reactors connected in sequence. Cells were grown in the first reactor and used to inoculate the second, the production reactor. The culture in the production vessel was established with 300 mL of cell suspension and additional cell suspension was added for the next 3 days until the working volume of 800 mL was reached. The production reactor was then operated as a batch reactor for 1 day. Production of β-gal was 35 U/mL for 7 runs compared to an average of 22 U/mL when operated in a simple batch mode for 12 consecutive runs of 5 days each. As predicted in the model, the amount of infectious virus produced began to decline at this point in each system, as did the production of β-gal, as the result of the buildup of DI particles.

MOI and Cell Density Effects. The MOI used may also affect the total yield of virus or recombinant protein. Dougherty et al. (79) found that with a MOI of 5, 98% of the Tn-368 cells were infected in 24 h and at a MOI of 20, 99% were infected. Brown and Faulkner (80) found that in monolayer cultures of Tn-368 cells the yield of wild-type virus was constant for increasing MOIs from 0.001 to 1.0, increased with increasing MOI until 10, and then began to decline. They suggested that at higher MOIs at which cells were infected by several particles that damage to the cell membrane caused the decrease in virus yield. Licari and Bailey (81) reported that in monolayer cultures of Sf-9 cells infected in early log phase of growth that the yield of β-gal was constant for MOIs between 0.1 and 10, but lower at 0.01 and 100. However, when monolayers were inoculated in the late log phase the yield increased with increased MOI from 0.1 to 100. It is not clear why MOIs greater than that required to infect 100% of the cells should increase yield. In suspension culture, studies to optimize the production of recombinant protein from Sf-9 cells demonstrated that MOI did not affect yield and could be chosen on the basis of technical needs or economic objectives without sacrificing yield (80). A very high MOI of 580 plaque-forming units/mL to infect high density cell cultures, 4–5 $\times 10^6$ cells/mL, combined with medium replacement was used for optimum yields in an airlift fermentor (82). To obtain sufficient virus to inoculate a large bioreactor at these high MOIs requires concentration of culture supernatants, which may not be economical unless the increase in yield is substantial.

The time of infection also has significant effect on the yield of recombinant protein obtained. Vaughn (83) demonstrated that monolayer cultures that had reached confluency at the time of inoculation produced little or no progeny virus. This was later confirmed by Wood et al. (84), whose data suggested that contact inhibition was the cause. The optimal time for infection is related to the MOI used (81,85). It is a very narrow period and is dependent on the cell density and the growth stage of the cells at the time of inoculation. Maiorella et al. (34) found that the yield of recombinant protein was reduced by 60% if cultures with high cell densities were infected in the stationary phase rather than the log phase. King et al. (86) obtained similar results but also noted that in their studies only 50% of the cells were infected at the MOI used. They implied that this was because the uninfected cells continued to grow and depleted the nutrients. Once infected with a baculovirus, insect cells do not divide. In cultures with low cell densities inoculated with a low MOI, the uninfected cells will continue to divide and the progeny cells become infected by viruses produced from viral replication in the cells infected with the original inoculum. The cycle will continue until all cells are infected or until the nutrients are depleted. This is the common practice in batch culture systems. An advantage is that less seed virus must be produced outside the bioreactor. However, there is less control over the passage of virus, and the time required to complete the fermentation may be extended. If the protein produced is degraded by enzymes released from lysing cells, yields may be lower under these conditions even if the amount produced is high.

ACKNOWLEDGMENTS

My thanks to Dr. Jean R. Adams (retired) of this laboratory for kindly providing the electron micrographs in Figures 1 and 2 and to Dr. Dwight E. Lynn of this laboratory for the photograph used in Figure 3.

BIBLIOGRAPHY

1. R.M. Patterson, J.K. Selkirk, and B.A. Merrick, *Environ. Health Perspect.* **103**, 756–759 (1995).
2. P.D. Frieson and L.K. Miller, in W. Doerfler and P. Boehm, eds., *The Molecular Biology of Baculoviruses*, Springer-Verlag, Berlin, 1986, pp. 31–49.
3. V.A. Luckow and M.D. Summers, *Biotechnology* **6**, 47–55 (1988).
4. M. Kool and J.M. Viak, *Arch. Virol.* **130**, 1–16 (1993).
5. L.E. Volkman, G.W. Blissard, B.A. Keddie, R. Possee, and D.A. Theilmann, in F.A. Murphy, C.M. Fauquet, D.H.L. Bishop, S.A. Ghabrial, A.W. Jarvis, G.P. Martelli, M.A. Mayo, and M.D. Summers, eds., *Virus Taxonomy: Sixth Report of the International Committee on Taxonomy of Viruses*, Springer-Verlag, New York, 1995, pp. 104–113.
6. J.R. Adams, J.T. McClintock, and J.A. Couch, in J.R. Adams and J.R. Bonami, eds., *Atlas of Invertebrate Viruses*, CRC Press, Boca Raton, 1991, pp. 87–226.
7. R.R. Granados and K.A. Lawler, *Virology* **108**, 297–308 (1981).
8. D.A. Hammer, T.J. Wickman, M.L. Shuler, H.A. Wood, and R.R. Granados, in M.L. Shuler, H.A. Wood, R.R. Granados, and D.A. Hammer, eds., *Baculovirus Expression Systems and Biopesticides*, Wiley-Liss, New York, 1995, pp. 103–119.
9. R.R. Granados and B.A. Federici, eds., *The Biology of the Baculoviruses*, vols. I, II, CRC Press, Boca Raton, 1986.

10. J.L. Vaughn, R.H. Goodwin, G.J. Tompkins, and P. McCawley, *In Vitro* **13**, 213–217 (1977).
11. M.D. Summers and G.G. Smith, Texas Agriculture Experiment Station Bulletin No. 1555, 1987.
12. T.D.C. Grace, *Nature (Lond.)* **216**, 613 (1967).
13. A. Gröner, in R.R. Granados and B.A. Federici, eds, *The Biology of Baculoviruses*, vol. I, *Biological Properties and Molecular Biology*, CRC Press, Boca Raton, 1986, pp. 177–202.
14. W.F. Hink and D.R. Bezanson, in E. Kurstak, ed., *Techniques in Setting Up and Maintenance of Tissue and Cell Cultures*, Elsevier Biomedical, County Clare, Ireland, 1985, pp. C111/1–C111/30.
15. W.F. Hink and R.L. Hall, in J. Mitsuhashi, ed., *Invertebrate Cell System Applications*, vol. II, CRC Press, Boca Raton, 1989, pp. 269–293.
16. W.F. Hink, *Nature (Lond.)* **226**, 466–467 (1970).
17. H.G. Miltenburger, P. David, U. Mahr, and W. Zipp, *Z. Angew. Entomol.* **82**, 306–323 (1977).
18. R.R. Granados, L. Guoxun, A.C.G. Derksen, and K.A. McKenna, *J. Invertebr. Pathol.* **64**, 260–266 (1994).
19. W.F. Hink, D.R. Thomsen, A.L. Meyer, D.J. Davidson, and F.J. Castellino, *Biotechnol. Prog.* **7**, 9–14 (1991).
20. M.J. Betenbaugh, L. Balog, and P.-S. Lee, *Biotechnol. Prog.* **7**, 462–467 (1991).
21. T.J. Wickham, T. Davis, R.R. Granados, M.L. Shuler, and H.A. Wood, *Biotechnol. Prog.* **8**, 391–396 (1992).
22. T.R. Davis, T.J. Wickham, K.A. McKenna, R.R. Granados, M.L. Shuler, and H.A. Wood, *In Vitro Cell. Dev. Biol.* **29A**, 388–390 (1993).
23. L. Marz, F. Altmann, E. Staudacher, and V. Kubelka, in J. Montreuil, J.F.G. Vliegenthart, and H. Schachter eds., *Glycoproteins*, Elsevier, Amsterdam, 1995, pp. 543–563.
24. V.A. Luckow, in M.L. Shuler, H.A. Wood, R.R. Granados, and D.A. Hammer eds., *Baculovirus Expression Systems and Biopesticides*, Wiley-Liss, New York, 1995, pp. 51–90.
25. K. Kuroda, H. Geyer, R. Geyer, W. Doerfler, and H.-D. Klenk, *Virology* **174**, 418–429.
26. D.J. Davidson, M.J. Fraser, and F.J. Castellino, *Biochemistry* **29**, 5548–5590 (1990).
27. J. Sissom and L. Ellis, *Biochem. J.* **261**, 119–126 (1989).
28. D.R. Thomsen, L.E. Post, and A.P. Elhammer, *J. Cell. Biochem.* **43**, 67–79 (1990).
29. D.L. Jarvis, in M.F.A. Goosen, A.J. Daugulis, and P. Faulkner, eds., *Insect Cell Culture Engineering*, Dekker, New York, 1993, pp. 195–219.
30. A. Hoss, I. Moarefi, K.-H. Scheidtmann, L. Cisek, J.L. Corden, I. Dornreiter, A.K. Arthur, and E. Fanning, *J. Virol.* **64**, 4799–4807 (1990).
31. T.D.C. Grace, *Nature (Lond.)* **195**, 788–789 (1962).
32. G.R. Gardiner and H. Stockdale, *J. Invertebr. Pathol.* **25**, 363–370 (1975).
33. S.A. Weiss, G.C. Smith, S.S. Kalter, and J.L. Vaughn, *In Vitro* **17**, 495–502 (1981).
34. B. Maiorella, D. Inlow, A. Shauger, and D. Harano, *Biotechnology* **6**, 1406–1410 (1988).
35. S.A. Weiss, W.G. Whitford, and G. Godwin, in M.J. Fraser, ed., *Proceedings, Eighth International Conference on Invertebrate and Fish Tissue Culture*, Tissue Culture Association, Columbia, Md., 1991, pp. 153–159.
36. S.M. Deutschmann and V. Jäger, *Enzyme Microb. Technol.* **16**, 506–512 (1994).
37. P. Wang, R.R. Granados, and M.L. Shuler, *J. Invertebr. Pathol.* **59**, 46–53 (1992).
38. M. López-Ferber, W.P. Sisk, and R.D. Possee, in C.D. Richardson, ed., *Baculovirus Expression Protocols*, Humana Press, Totowa, N.J., 1995, pp. 25–63.
39. V.A. Luckhow and M.D. Summers, *Virology* **167**, 56–71 (1988).
40. Y. Matsuura, R.D. Possee, H.A. Overton, and D.H.L. Bishop, *J. Gen. Virol.* **68**, 1233–1250 (1987).
41. T.J. French and P. Roy, *J. Virol.* **64**, 1530–1536 (1990).
42. U. Weyer and R.D. Possee, *J. Gen. Virol.* **72**, 2967–2974 (1991).
43. K.M. Trotter and H.A. Wood, in C.D. Richardson, ed., *Baculovirus Expression Vectors*, Humana Press, Totowa, N.J., 1995, pp. 97–105.
44. P.A. Kitts, M.D. Ayers, and R.D. Possee, *Nucleic Acids Res.* **18**, 5667–5672 (1990).
45. R. Neutra, B.-Z. Levi, and Y. Shoham, *Appl. Microbiol. Biotechnol.* **37**, 74–78 (1992).
46. J.L. Vaughn and R.H. Goodwin, in J.A. Romerger, ed., *Beltsville Symposium on Agricultural Research, vol. I, Virology in Agriculture*, Alleheld Osmun, Monclair, N.J., 1977, pp. 109–116.
47. I.R. Cameron, R.D. Possee, and D.H.L. Bishop, *Trends Biotechnol.* **7**, 66–70 (1989).
48. U. Eberhard and K. Schügerl, *Dev. Biol. Stand.* **66**, 325–330 (1985).
49. N. Kioukia, M. Al-Rubeal, Z. Zang, A.N. Emery, A.W. Nienow, and C.R. Thomas, *Biotechnol. Lett.* **17**, 7–12 (1995).
50. W.F. Hink and E.M. Strauss, in E. Kurstak, K. Maramorosch, and A. Dübendorfer, eds., *Invertebrate Systems In Vitro*, Elsevier/North Holland, Amsterdam, 1980, pp. 27–33.
51. J. Tramper, J.B. Williams, and D. Joustra, *Enzyme Microb. Technol.* **8**, 33–36 (1986).
52. J.J. Chalmers, in M.L. Shuler, H.A. Wood, R.R. Granados, and D.A. Hammer, eds., *Baculovirus Expression Systems and Biopesticides*, Wiley-Liss, New York, 1995, pp. 175–204.
53. K. Trinh, M. Garcia-Briones, F. Hink, and J.J. Chalmers, *Biotechnol. Bioeng.* **43**, 37–45 (1994).
54. J.L. Vaughn, in *Proc. Second Int. Colloq. Invertebr. Tissue Cult., Tremezzo, Como*, Instituto Lombardo, Milan, 1967, pp. 119–125.
55. D.W. Murhammer and C.F. Goochee, *Biotechnology* **6**, 1411–1418 (1988).
56. L.K. Overton and T.A. Kost, in M.L. Shuler, H.A. Wood, R.R. Granados, and D.A. Hammer, eds., *Baculovirus Expression Systems and Biopesticides*, Wiley-Liss, New York, 1995, pp. 233–242.
57. M. Silberklang, K. Ramasubramanyan, S.L. Gould, A.B. Lenny, T.C. Seamen, S. Wang, G.R. Hunt, B. Junker, K.E. Mazina, M.R. Tota, O. Palyha, and D. Jain, in M.L. Shuler, H.A. Wood, R.R. Granados, and D.A. Hammer, eds., *Baculovirus Expression Systems and Biopesticides*, Wiley-Liss, New York, 1995, pp. 205–231.
58. A. Kumar and M.L. Shuler, *Biotechnol. Prog.* **11**, 412–419 (1995).
59. J.W. Rice, N.B. Ranki, T.M. Gurganus, C.M. Marr, J.B. Barna, M.M. Walters, and D.J. Burns, *Biotechniques* **15**, 1052–1059 (1993).
60. J.M. Vlak, E.J. Schaeger, and A.R. Bernard, *Baculovirus and Recombinant Protein Production Processes*, Editiones Roche, Basel, Switzerland, 1992.
61. D.G. Bergmann, in A.S. Lubinlechi, ed., *Large-Scale Mammalian Cell Culture Technology*, Dekker, New York, 1990, pp. 417–449.

62. S. Reuveny, Y.J. Kim, C.W. Kemp, and J. Shiloach, *Appl. Microbiol. Biotechnol.* **38**, 619–623 (1993).
63. R.A. Taticek, D.A. Hammer, and M.L. Shuler, in M.L. Shuler, H.A. Wood, R.R. Granados, and D.A. Hammer, eds., *Baculovirus Expression Systems and Biopesticides*, Wiley-Liss, New York, 1995, pp. 131–174.
64. C. Bédard, R. Tom, and A. Kamin, *Biotechnol. Prog.* **9**, 615–624 (1993).
65. L. Öhman, J. Ljunggren, and L. Häggström, *Appl. Microbiol. Biotechnol.* **43**, 1006–1013 (1995).
66. D.A. Lindsay and M.J. Betenbaugh, *Biotechnol. Bioeng.* **39**, 614–618 (1992).
67. J. Mitsuhashi, in J. Mitsuhashi, ed., *Invertebrate Cell System Applications*, vol. I, CRC Press, Boca Raton, 1989, pp. 3–20.
68. M.-Y. Wang, S. Kwong, and W.E. Bentley, *Biotechnol. Prog.* **9**, 355–361 (1993).
69. S.H. Lee and T.H. Park, *Biotechnol. Lett.* **16**, 327–332 (1994).
70. M. Klöppinger, G. Fertig, E. Fraune, and H.G. Miltenburger, *Cytotechnology* **4**, 271–278 (1990).
71. J.L. Zhang, N. Kalogerakis, L.A. Behie, and K. Iatrou, *Biotechnol. Bioeng.* **42**, 357–366 (1993).
72. S.A. Weiss and J.L. Vaughn, in R.R. Granados and B.A. Federici, eds., *The Biology of Baculoviruses*, vol. II, CRC Press, Boca Raton, 1986, pp. 63–87.
73. A.W. Caron, R.L. Tom, A.A. Kamen and B. Massie, *Biotechnol. Bioeng.* **43**, 881–891 (1994).
74. E.A. MacKinnon, J.F. Henderson, D.B. Stoltz, and P. Faulkner, *J. Ultrastruct. Res.* **49**, 419–435, (1974).
75. J.M. Slavicek, M.J. Mercer, M.E. Kelly, and N. Hayes-Plazolles, *J. Invertebr. Pathol.* **67**, 153–160 (1996).
76. T.J. Wickham, T. Davis, R.R. Granados, D.A. Hammer, M.L. Shuler, and H.A. Wood, *Biotechnol. Lett.* **13**, 483–488 (1991).
77. C.D. de Gooijer, R.H.M. Koken, F.L.J. van Lier, M. Kool, J.M. Vlak, and J. Tramper, *Biotechnol. Bioeng.* **40**, 537–548 (1992).
78. F.L.J. van Lier, J.P.T.W. van den Hombergh, C.D. de Gooijer, M.M. den Boer, J.M. Vlak, and J. Tramper, *Enzyme Microb. Technol.* **18**, 460–466 (1996).
79. E.M. Dougherty, R.M. Weiner, J.L. Vaughn, and C.F. Reichelderfer, *Appl. Environment. Microbiol.* **41**, 1166–1172 (1981).
80. M. Brown and P. Faulkner, *J. Invertebr. Pathol.* **26**, 251–257 (1975).
81. P. Licari and J.E. Bailey, *Biotechnol. Bioeng.* **37**, 238–246 (1991).
82. J.E. Lazarte, P.-F. Tosi, and C. Nicolau, *Biotechnol. Bioeng.* **40**, 214–217 (1992).
83. J.L. Vaughn, *J. Invertebr. Pathol.* **28**, 233–237 (1976).
84. H.A. Wood, L.B. Johnston, and J.P. Burand, *Virology* **119**, 245–254 (1982).
85. J.F. Power, S. Reid, K.M. Radford, P.F. Greenfield, and L.K. Nielsen, *Biotechnol. Bioeng.* **44**, 710–719 (1994).
86. G.A. King, A.J. Daugulis, P. Faulkner, and M.F.A. Goosen, *Biotechnol. Prog.* **8**, 567–571 (1992).

See also Ammonia toxicity, animal cells; Chinese hamster ovary cells, recombinant protein production; Culture media, animal cell, large-scale production ; Expression systems, *E. coli*; Expression systems, mammalian cells; Glycosylation of recombinant proteins; Insect cells and larvae, gene expression systems; Protein expression, soluble.

INSECT CELLS AND LARVAE, GENE EXPRESSION SYSTEMS

Michael J. Betenbaugh
Eric Ailor
Erik M. Whiteley
Johns Hopkins University
Baltimore, Maryland
Tsu-An Hsu
Vaccine Pharmaceutical R&D, Merck & Co., Inc.
West Point, Pennsylvania

KEY WORDS

Baculovirus
Bioreactor
Cell culture
Expression system
Glycosylation
Insect
Larva
Recombinant DNA

OUTLINE

INTRODUCTION

Insects and other invertebrates include more than half of the animal species that have been identified on the planet. A number of these invertebrates are valuable to the clothing industry where they are used in the production of silk. From an agricultural perspective, however, insects are a major pest of crops and other plant life. Developing methods to control the proliferation of insects has been a continuing aim of the agricultural and forest service industries. As with other animal species, these hosts are capable of maintaining and transmitting viral infections. One approach that has been considered for many years to control insect proliferation is the application of insect-specific viruses. As a result, a substantial body of knowledge has evolved over the past 20 years concerning the structure and function of insect viruses. Through these biological advances, it is now possible to alter the genetic content of these insect-specific viruses and incorporate heterologous genes from other organisms. The development of genetically altered insect viruses has had a tremendous impact on the development of enhanced viral biopesticides and on the production of recombinant DNA products of biotechnological interest. Baculoviruses such as the *Autographica californica* nuclear polyhedrosis virus (AcMNPV or AcNPV) and the *Bombyx mori* nuclear polyhedrosis virus (BmNPV) can be modified to incorporate genes from mammalian, insect, plant, and viral sources, among others. Insect cells or insect larvae are then infected with these altered viruses to generate recombinant proteins for applications such as structure and function studies, diagnostics, and vaccines. To generate the large quantities of recombinant protein and virus needed for biotechnological and biopesticide applications, efficient and scaleable production processes have been developed. These processes must be particularly sensitive to the characteristics of the baculovirus expression vector system (BES or BEVS), which includes both a cell growth and infection stage. Various insect cell hosts, viral vectors, and expression methodologies are being explored along with alternative media formulations and different bioreactor operating strategies. These efforts are examined in the following sections.

INSECT CELL LINES

Insect cell lines have been established from nearly 100 different species within the following seven orders of Arthropoda: Lepidoptera, Diptera, Homoptera, Hymenoptera, Orthoptera, Hemiptera, and Coleoptera. These lines are isolated from eggs or embryonic, larval, or adult tissues such as hemocytes, ovaries, imaginal discs, testes, midgut, and fat body using methods that are now well known (1). The procedures typically involve the lysis of tissue followed by suspension in a rich serum-containing medium in tissue culture flasks. The nutrients are replaced periodically, and the cultures are monitored until cell division is observed. The cells are then subcultured after one to several weeks to obtain a new cell line. Eventually, the cell lines are screened for desired properties such as short doubling times, virus susceptibility, recombinant gene expression, or posttranslational processing. Clonal isolates can be obtained that have better growth or gene expression properties, as was the case in the development of the Sf-9 clone from the established cell line (IPLB-Sf-21) of the fall armyworm, *Spodoptera frugiperda*. Both of these cell lines are now widely used for the replication of AcMNPV and production of recombinant proteins.

The screening of established cell lines led to the identification of a *Trichoplusia ni* (cabbage looper) cell line (BTI-TN5B1-4), with the trade name High Five™ (Invitrogen, Inc.), which often provides recombinant protein yields superior to those of Sf-9 and Sf-21 (2). Originally an attachment-dependent cell line, High Five cells have been adapted to suspension growth in many laboratories. Adaptation to suspension is aided by the addition of 50–2,000 U/cm^3 heparin to prevent clumping, but the heparin should be removed prior to viral infection and expression of products (3,4). Shown in Figure 1 is a micrograph of healthy and AcMNPV-infected Sf-9 and High Five cells. Commercial serum-free media have been developed for High Five, Sf-21, and Sf-9 cell lines. Because the Sf-9 cell line produces higher levels of the extracellular virus, this cell line is often used for virus production, whereas the High Five are used for heterologous gene expression. Although High Five and Sf are the cell lines of choice for many applications, recombinant protein production from AcMNPV has been achieved in a number of other cell lines such as those isolated from *Heliothis, Lymantria, Plutella*, and *Trichoplusia* (5,6). A smaller variety of cell lines from *Bombyx mori* (silkworm) and *Lymantria dispar* (gypsy moth) are available for infection with BmNPV and *Lymantria dispar* nuclear polyhedrosis virus (LdNPV), respectively.

Cell lines are usually cultured at the laboratory scale in polystyrene culture flasks or roller bottles. Those cell lines that are not anchorage dependent can be grown in shaker flasks or small suspension-culture flasks. Cells are monitored for number density and passaged one to two times per week by subculturing into fresh medium. Attachment-dependent cells must be dislodged by agitation, scraping, or enzymatic treatment. Following a lag phase, the cells grow logarithmically with doubling times between 0.5 and 2 days. Maximum cell densities are typically in the range of 10^6 cells/mL but may rise as high 10^7 cells/mL in some cases. Cell size is in the range of 10 to 20 μm, and culture temperatures are typically between 25 and 30 °C. Stocks of cells can be cryopreserved in liquid nitrogen using medium and dimethylsulfoxide (DMSO). Recovery is achieved by a rapid thaw and can be enhanced by using a serum-bearing medium prior to freezing the cells.

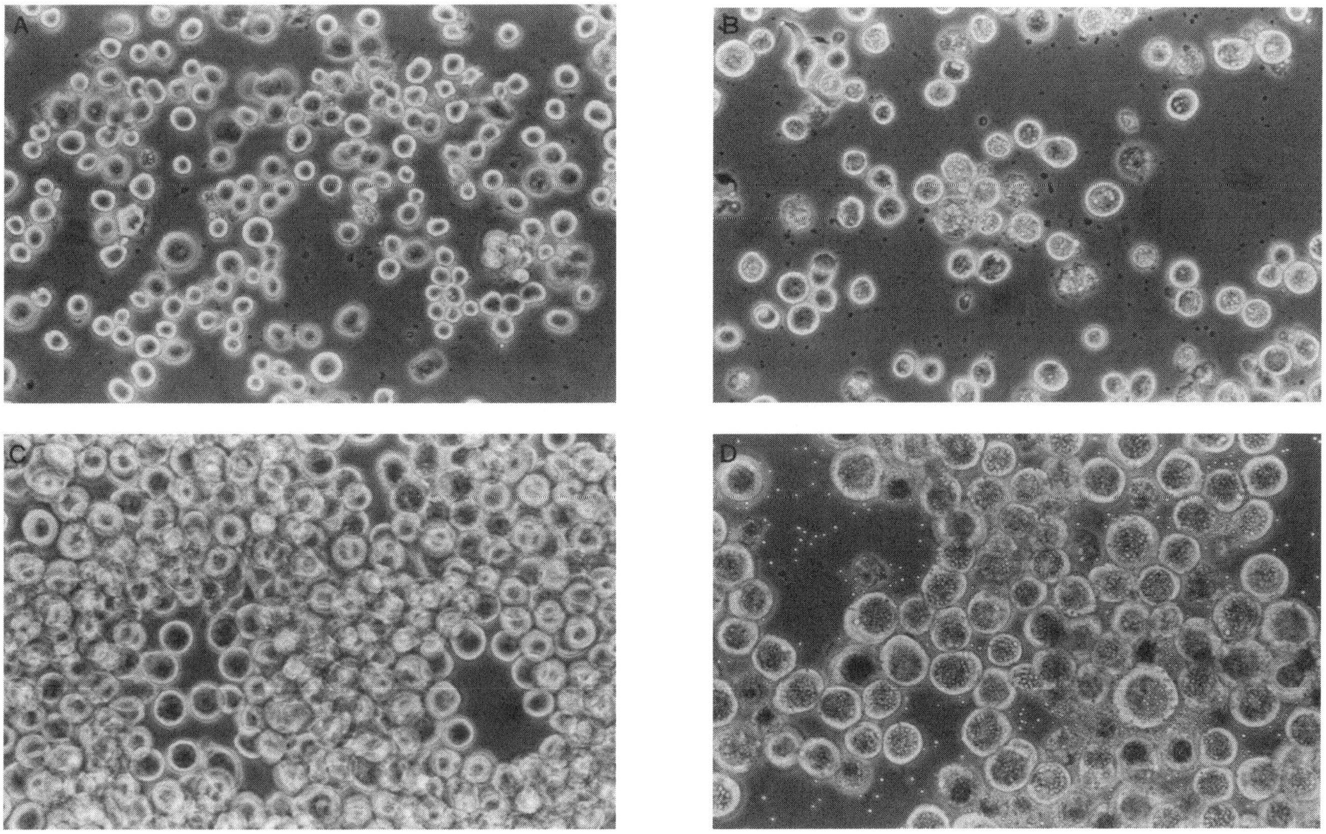

Figure 1. Insect cell morphology. (**a**) Healthy Sf-9 cells grown in insect culture medium. (**b**) Sf-9 infected with wild-type AcMNPV baculovirus. (**c**) Healthy BTI-TN5B1-4. (**d**) BTI-TN5B1-4 infected with wild-type AcMNPV baculovirus. Photographs taken at 32×.

VIRUSES

The most widely studied and utilized insect viruses are the baculoviruses, a helper-independent eukaryotic DNA vector that can infect a wide range of lepidopteran (butterflies and moths) insects and insect cells. "Baculo" refers to the rod-shaped capsid that is characteristic of these viruses, which are 40–50 nm in diameter and 200–400 nm in length. The two most widely studied baculoviruses, AcMNPV and BmNPV, contain DNAs that are approximately 130 kb in length. The relatively large size of the virus allows for the encapsulation of large pieces of foreign DNA in addition to the essential viral DNA.

The infection of a single cell by the baculovirus is illustrated in Figure 2. Baculoviruses are deposited in two forms, budded and occluded, in the infected cells (7,8). At the initial stages of infection, membrane-enveloped viruses bud from the plasma membrane of an infected host 10–14 h after infection in a form commonly referred to as extracellular virus (EV), nonoccluded virus (NOV), or budded virus (BV). A second form of the virus, the occluded virus (OV), forms in the nucleus of an infected cell beginning at 20–24 h after viral infection. This intracellular virus is enveloped and encased within a crystalline viral protein matrix called polyhedrin. These protein crystals, called occlusion bodies, polyhedra, or polyhedral inclusion bodies (PIBs), may be as large as 5 μm in diameter and contain many virions (Fig. 3). The OV encased within the polyhedrin protein matrix may contain multiple nucleocapsids within a single envelope for multiply embedded nuclear polyhedrosis viruses (MNPVs), or a single nucleocapsid per envelope for single nuclear polyhedrosis viruses (SNPVs). In addition to nuclear polyhedrosis virus (NPV), a second type of occluded baculovirus, the granulosis virus (GV), includes one or, very rarely, two virions embedded in a much smaller protein matrix called a capsule or granule. Most baculoviruses used as biopesticides and all baculoviruses used in biotechnology are NPVs, although a few GVs have shown potential for future applications.

It is the OV that is responsible for transmission of the virus from insect to insect, as shown in the baculovirus life cycle depicted in Figure 4 (8). After death and decomposition of an infected insect, the polyhedra are deposited on vegetation where they are ingested by other feeding insect larvae. Transported with food to the midgut, the polyhedra are dissolved in the alkaline environment, subsequently releasing virions into the lumen of the midgut. These viruses then fuse to the epithelial cells where they cause an infection in the midgut tissue. The infected cells then release BV, which are responsible for secondary infection of other tissues by spreading through the hemolymph and trachea. The infected larva continues to feed for 5–7 days until it becomes listless. The larval cuticle melanizes, and the muscular structure dissolves in a process called melt-

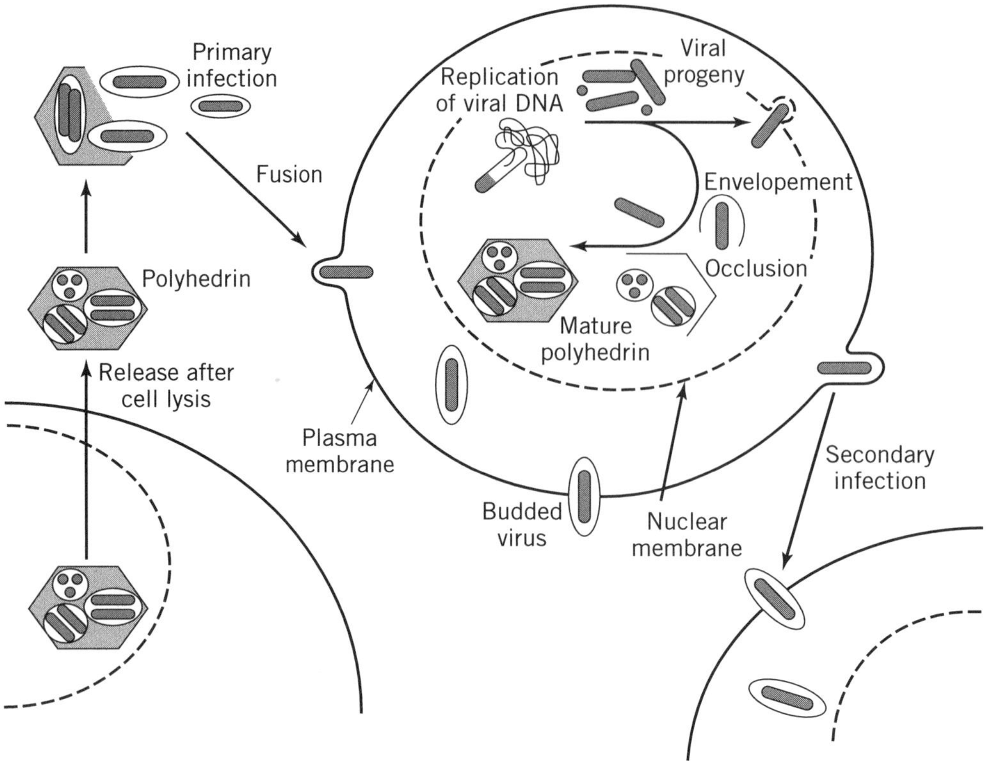

Figure 2. Single-cell replication life cycle of the natural baculovirus, AcMNPV. Primary infection by occluded virus occurs upon disintegration of the polyhedrin matrix and subsequent fusion to the plasma membrane. Budded virus produced within the cell proceed to initiate the secondary infection process. *Source*: Adapted from figure appearing in *Baculovirus Expression Systems and Biopesticides*, M.L. Shuler, H.A. Wood, R.R. Granados, and D.A. Hammer, Wiley-Liss, 1995. Reprinted by permission of John Wiley & Sons, Inc.

ing or wilting. The larva eventually ruptures, releasing about 25% of its dry weight in the form of PIBs, which are spread over vegetation for consumption by other insects and reinitiation of the infection cycle.

BACULOVIRUS GENE EXPRESSION

Baculovirus gene expression of viruses such as AcMNPV occurs over four phases: immediate early (α), early (β), late (γ), and very late (δ). During the immediate early phase, viral genes are expressed using host cell factors. The immediate early gene products then control the expression of the early genes, whose products are responsible for cytoskeletal rearrangement and nucleus enlargement. Extensive viral DNA replication occurs about 5–7 h after infection, followed by expression of the late genes. The late phase extends from 6 until 20–24 h postinfection and involves the assembly of nucleocapsids, the transport and release of budded virus (BV), and the degradation of the host genome. The formation of occluded viruses occurs during the very late phase, beginning about 20–24 h postinfection. The occluded viral particles can accumulate for 3–5 days until the infected cells lyse.

Two genes abundantly expressed during the very late phase are the polyhedrin and p10 genes. The product of the polyhedrin gene (*polh*) is the 29-kDa protein that forms the crystalline matrix of the occlusion particles protecting encased virus nucleocapsids. The product of the p10 gene is a 10-kDa protein that forms large fibrous arrays in the nucleus and occasionally in the cytoplasm. In 1983, it was observed that viruses containing deletions in the polyhedrin gene were stable and could be replicated in cells in culture (9). Though the cells infected with these viruses lack the characteristic occlusion particles, production of BV was unaffected. Furthermore, these BV could be collected from the cell culture media and used to infect other cells in order to passage the altered virus. Because the polyhedrin gene is nonessential for BV infection and propagation in cell culture, the construction of a recombinant baculovirus can be accomplished by replacing the polyhedrin gene with a foreign gene such as interferon-β (10). In this way, the foreign gene, now under the control of the very strong polyhedrin promoter, is expressed, instead of the polyhedrin gene at the very late phase of AcMNPV infection. Similarly, it was observed that disruption of the p10 gene resulted in disruption of the fibrous matrix, but the formation of BV was not affected. Subsequently, recombinant baculovirus constructs containing foreign genes inserted under the control of the p10 promoter were used to produce heterologous proteins in place of the p10 protein (11). Because the polyhedrin and p10 are very strong late promoters, with as much as 50% of the total stained protein consisting of polyhedrin (12), these promoters have

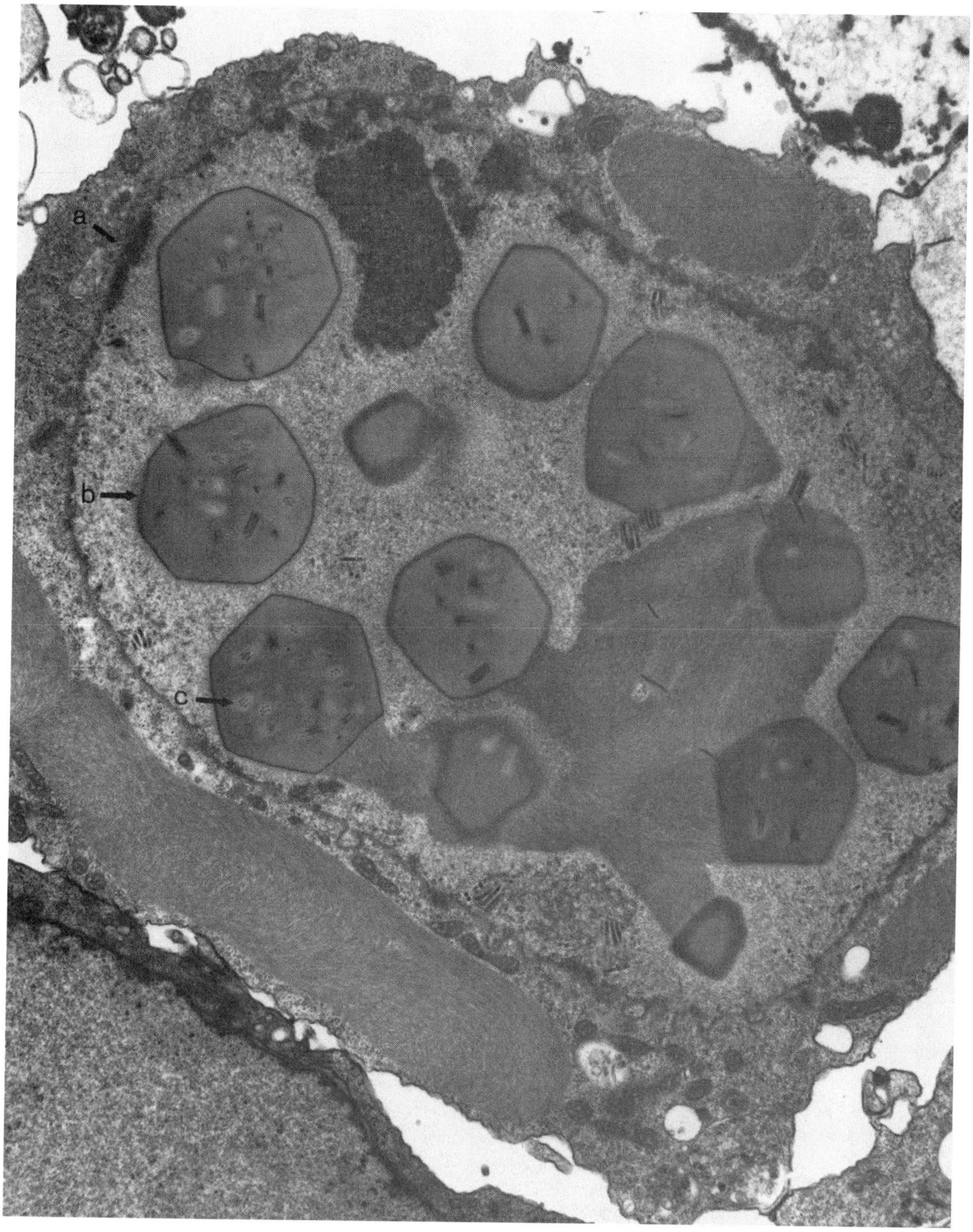

Figure 3. Electron micrograph of AcMNPV infection in *S. frugiperda*. (a) Nuclear membrane, (b) polyhedrin inclusion bodies, and (c) virions at 48 h postinfection (magnification 15,000×). *Source*: Micrograph courtesy of G. Williams.

the potential for very high rates of foreign gene expression. One disadvantage of this system is that production of heterologous protein is accomplished late in the viral infection, at which time host cell function is deteriorating. Furthermore, the insect host eventually succumbs to the infection, thus requiring a new batch of insect cells to be grown and infected each time recombinant protein is to be produced.

In an effort to avoid the problems associated with insect cell death due to virus infection, a number of vectors have been developed that allow the stable integration of foreign DNA into insect cell hosts. Stable expression of heterologous proteins using these integrated vectors has been achieved in several insect cell lines including Sf-9, *Drosophila* S2, and lines from *Aedes albopictus* (13). However, expression levels are in most cases many times lower than

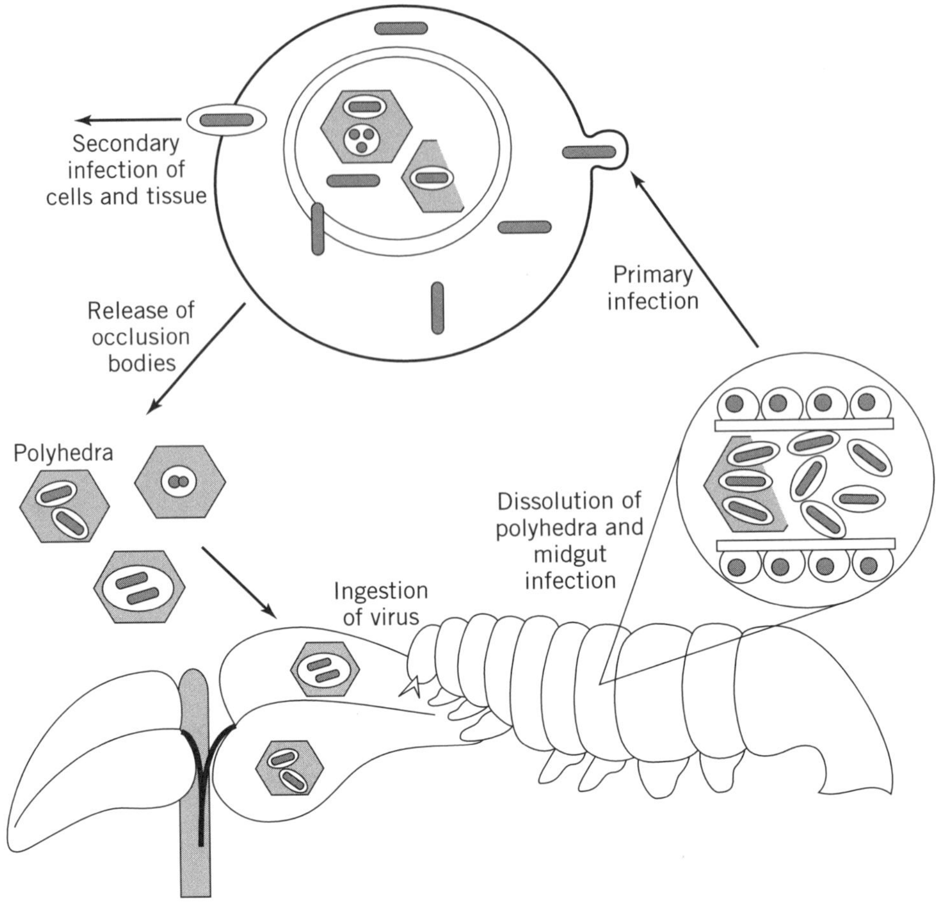

Figure 4. Life cycle of the natural baculovirus, AcMNPV. *Source*: Adapted from figure appearing in *Baculovirus Expression Systems and Biopesticides*, M.L. Shuler, H.A. Wood, R.R. Granados, and D.A. Hammer, Wiley-Liss, 1995. Reprinted by permission of John Wiley & Sons, Inc.

with baculovirus, so these systems have yet to approach the popularity of BES. Stable expression systems are likely to find niche applications when lower expression is acceptable in return for desirable host processing functions.

CONSTRUCTION OF RECOMBINANT BACULOVIRUSES

More than 500 different baculoviruses of the Baculoviridae family are currently known. For use as a vector to deliver foreign genes into insect cells, the most extensively used is AcMNPV (14). BmNPV (15) has also been used but to a much lesser extent, and work is currently in progress to develop expression systems based on other baculoviruses including LdNPV and granulosis viruses. Techniques for manipulating AcMNPV to generate recombinant baculoviruses is discussed here because this subtype is the most widely investigated and the procedures are similar for many other baculoviruses. AcMNPV is a large, double-stranded, circular DNA virus for which the complete nucleotide sequence has been determined (16). The large size of the baculovirus DNA prohibits the typical recombinant DNA manipulations used for cloning with bacterial plasmids. Consequently, an alternative approach involving plasmid transfer vectors and allelic replacement (homologous recombination) was developed and remains the most popular method for foreign gene insertion into the baculovirus. Many improvements have been made for generating recombinant baculoviruses, but almost all are based on homologous recombination.

The first step in allelic replacement involves the manipulation of a transfer plasmid in *Escherichia coli* by cloning and subcloning techniques to insert a foreign gene downstream from a baculovirus promoter, typically those for polyhedrin or p10. The promoter and gene are also flanked by baculovirus DNA derived from a nonessential locus such as polyhedrin or p10 genes. The recombinant transfer plasmid is then mixed with native baculovirus DNA, and the mixture is transfected into insect cells using calcium phosphate or cationic lipids. A small fraction of the baculoviruses will undergo homologous recombination with the transfer plasmid to create a recombinant baculovirus containing a heterologous gene under the control of the appropriate promoter (Fig. 5). The recombinant baculovirus constructs may be identified by examining for the absence or presence of a particular baculovirus or plasmid function due to the insertion of the foreign gene.

The initial screening technique involved identification of plaques of insect cells containing occlusion-minus (occ^-) infections as a result of the deletion of the polyhedrin gene from the original baculovirus DNA. Occlusion-positive plaques include an opaque light refractile characteristic that can be visually distinguished from the characteristics of occ^- plaques under the microscope. However, picking an occ^- clone is a formidable task for inexperienced researchers because young nonrecombinant plaques often show no

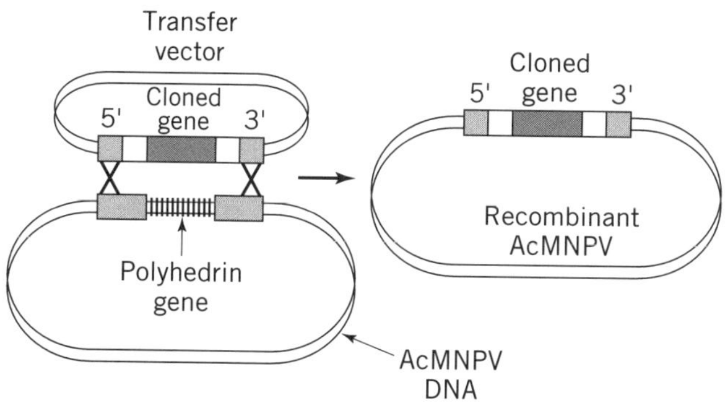

Figure 5. Homologous recombination. A baculovirus gene locus (usually p10 or polyhedrin) is targeted for homologous recombination using a transfer vector that includes flanking homologous sequences and, in some cases, encoding a selectable marker. The recombination event (represented here by X's) results in the replacement of sequences flanked by the homologous areas with the sequence of the foreign gene.

sign of "occlusion," and recombination frequencies can be as low as 0.1%. These difficulties resulted in the time consuming and tedious isolation of recombinant baculovirus through multiple rounds of plaque purification. The first modification of the system was the use of the bacterial gene *lacZ* to produce chromogenic recombinant plaques that enhance plaque purification (17,18). Significant progress in isolating recombinant baculovirus was achieved by derivatizing the circular, double-stranded wild-type baculovirus into a linear form through the introduction of a unique restriction site (Bsu36I) at the polyhedrin locus (19). A higher proportion of recombinant viruses was obtained when the linearized baculovirus genome was cotransfected with a transfer vector, because the infectivity of the linearized virus alone was substantially reduced below wild-type levels. More recently, a faster approach that can generate pure recombinant baculovirus in 7–10 days was reported (20). This technique uses a novel baculovirus shuttle vector (bacmid) that can replicate in *E. coli* and infect susceptible lepidopteran insects. The homologous recombination event and screening for the recombined bacmid occur entirely in *E. coli*. A recombinant baculovirus is then generated by introducing the composite bacmid into insect cells in culture and harvesting the virus from the medium. This approach eliminates the need for plaque purification because only the recombinant baculovirus is generated during the initial infection.

Other expression vectors are available that provide for the expression of up to four genes in the same baculovirus construct. These constructs may include one or more polyhedrin, p10, or heat shock promoters to control expression of the foreign genes. Infection of insect cells with several viruses, each containing multiple genes, can provide simultaneous expression of numerous proteins in one insect cell. Constructs containing baculovirus promoters such as IE1 and core that are activated earlier in the infection are also available (21). Specialized vectors that include histidine tags, which help facilitate protein purification, as well as gp64 or melittin fusions, for enhancing protein secretion, are also available. For a laboratory not experienced with BEVS, it is advisable to employ the linear baculovirus genome with a lethal deletion and selectable marker or the *E. coli* bacmid shuttle vector as the initial method for generating recombinants. Many kits are now commercially available that include the transfer vectors and baculovirus DNA for generating recombinant baculovirus constructs.

POSTTRANSLATIONAL PROCESSING

Because insects are eukaryotic organisms, the derived cell lines are capable of performing many of the posttranslational processes including secretion, folding, and glycosylation observed in higher organisms. In fact, insect cells are often the expression system of choice for the rapid generation of functional eukaryotic proteins. However, the posttranslational processing may not be identical to that performed in the native environment. It may be worthwhile to compare the protein expressed in the insect cells with the native protein to determine if significant differences in posttranslational processing exist that will detract from subsequent structural or functional analysis. Some of the posttranslational processing events that are observed in insect cells are described in the following sections.

Glycosylation

A majority of secreted and membrane proteins are glycosylated, and the attached oligosaccharides can play critical roles in cellular targeting, structural stability, resistance to proteolysis, immunogenicity, and circulatory half-life. The oligosaccharide residues are added to the protein at specific amino acids during processing in the endoplasmic reticulum (ER). Subsequently, these oligosaccharides are trimmed and modified during processing in the ER and Golgi compartments. Both N-linked and O-linked glycosylation have been observed in insect cells.

During N-glycosylation in both insect and mammalian cells an initial $Glc_3Man_9GlcNAc_2$ group is added to the Asn residue of the Asn-X-Ser/Thr sequence. Glucose and mannose residues are then trimmed from the $Glc_3Man_9GlcNAc_2$ group. However, the level of trimming and subsequent modifications are likely to be different in insect cells than those observed in mammalian cells. Recent evidence indicates that these differences may depend significantly on the particular insect cell line used. Many initial studies of *N*-glycans from heterologous glycoproteins obtained from Sf-9 or Sf-21 cells indicated the presence of only high-mannose type ($Man_{9\text{-}5}GlcNAc_2$) or short, truncated, paucimannosidic ($Man_{3\text{-}2}GlcNAc_2$) oligosaccharides, sometimes including an $\alpha(1,6)$-linked fucose (22,23). In contrast, studies with recombinant human plasminogen indicated that insect cells could synthesize complex N-linked oligosaccharides (24). Several more recent studies on heterologous glycoproteins have indicated that certain insect cell lines can synthesize hybrid and complex oligosaccharides. Analysis of *N*-glycans of heterologous interferon-γ secreted from *Estigmena acrea* (Ea-4) confirmed the presence of complex, hybrid, paucimannosidic, and high-mannose N-linked oligosaccharides (25). In contrast, *Spodoptera frugiperda* (Sf-9) cells were observed to produce only hybrid, paucimannosidic, and high-mannose *N*-glycans. A complete description of the *N*-glycans for intracellular and secreted IgG_{2a} from High Five cells indicated the secretion of complex, hybrid, and paucimannosidic glycoforms (26), along with the intracellular retention of glycoforms that were principally high mannose. A comparison of the *N*-glycans of IgG from High Five cells with those attached to an IgG_{3a} from the mammalian NS/0 cells reveals significant similarities and some differences in the glycosylation properties of the two cell lines (Table 1). Based on the intracellular and extracellular *N*-glycans of IgG, a putative pathway for oligosaccharide processing in *T. ni* cells has been proposed (Fig. 6). The initial mammalian ER pathway, including trimming of the terminal glucose and mannose residues, seems operative. Insect cell processing diverges from mammalian cell processing in the pathways leading to the paucimannosidic structures, with the critical step likely to be the cleavage of the GlcNAc from the α-(1,3) branch by a possible Golgi β-*N*-acetylglucosaminidase (28). Other significant differences in insect cell processing are the α-(1-3)-linked addition of fucose, also observed in the processing pathway of plants, and the absence of any attached sialic acids (25,26,29). Recently, glycoprotein processing has been manipulated in insect cells by the expression of glycosylation pathway enzymes in order to generate glycoforms even more similar to those produced by mammalian cells (21,30).

O-glycosylation, which involves the addition of carbohydrates to hydroxylated amino acids such as serine or threonine residues, has been observed for several recombinant proteins produced by baculovirus-infected insect cells. A pseudorabies viral glycoprotein, gp50, and extracellular domains of the respiratory syncytial virus protein were found to contain the attached monosaccharide GalNAc and lesser amounts of the disaccharide Galβ1-3GalNAc (31). In mammalian cells, only the disaccharide form was observed in the presence or absence of attached sialic acids. Activity levels of the relevant glycosyltransferases in the insect cells and mammalian cells supported the particular distribution patterns observed.

Other Posttranslational Processes

Many proteins expressed in insect cells are directed to the appropriate cellular compartment including the nucleus, cytoplasm, organelles, plasma membrane, or extracellular space. For secreted proteins, a short (15–40 amino acids) signal peptide is present for translocating polypeptide from the cytoplasm into the endoplasmic reticulum (ER). This signal peptide, whether from mammalian, plant, or yeast sources, is usually cleaved correctly by the signal peptidases in insect cells. However, signal peptide processing is not always efficient, especially for peptides from plants or other distant eukaryotes. In fact, a complete absence of processing has been observed in at least one case (32). Replacement of the original signal peptide with another signal peptide from a protein such as melittin of honeybee venom (33) or the gp67 and EGT proteins (7) from AcMNPV can help to alleviate this limitation in some cases. Sometimes, a substantial fraction of proteins destined for secretion will be retained intracellularly in incorrectly folded forms or as aggregates. Coexpression of chaperones or folding enzymes can reduce aggregation and enhance secretion of recombinant proteins that rely on such folding assistance factors (34). Other posttranslational processing steps including phosphorylation, fatty acid acylation, α-amidation, and N-terminal amidation have also been observed in insect cells or insect larvae.

APPLICATIONS

Baculovirus-infected insect cells often provide high levels of expression of heterologous proteins, leading to the use of this system across a broad spectrum of applications and fields. BES has been used for the production of proteins for structure and function studies, vaccines, multisubunit assemblies, and diagnostic reagents (35). Furthermore, insect cells have been exploited as tools for examining the role of proteins in physiological processes. The application of BES products as therapeutics is also under consideration, though it is not currently a major application of the technology. Finally, the use of baculoviruses as biopesticides, the original motivation for studying these viruses, represents a major thrust of the technology (4,36). Wild-type baculoviruses have been used as efficacious pest control agents for many years but have a limited host range and can take between 5 and 10 days to debilitate insect larvae. A number of genetic modifications, mostly centered on replacement of polyhedrin or p10 genes with heterologous insect toxins, have been implemented to lower viral killing time and reduce the duration of pest feeding. Some of the heterologous genes that have been considered are listed in Table 2. A major concern in the development of baculovirus pesticides is the potential safety of these agents. However, extensive studies have concluded that baculoviruses have a very low range of infectivity and will not infect beneficial insects or mammals. Inclusion of a foreign gene under the control of a baculovirus promoter appears to present little danger as well.

LARGE-SCALE PROCESSING: CELL CULTURE OR LARVAL PRODUCTION

Structure and functional characterization studies and diagnostic reagents often require product levels beyond the scale of small shaker flasks. Biopesticides will require huge amounts of the final product for use in field trials. The production of these large quantities of heterologous proteins or baculoviruses will require the use of large-scale cell culture or insect larvae. Insect cell culture provides an environment in which the process parameters can be carefully controlled and monitored (4,7). It also allows for the utilization of a single cell line with more uniform posttranslational processing than can be achieved in larvae. Purification of proteins from cell culture medium, especially serum-free medium, is usually easier than from larvae of infected insects. Larvae will include significant levels of other insect proteins, and the presence of larval proteases can lead to a degraded final product. In contrast, the use of larvae may provide substantial cost advantages

Table 1. Comparison of Glycosylation Patterns of IgG Produced in Insect Cells (*Trichoplusia ni*) and Mammalian Cells (NS/0)

	Insect IgG lysate (mol %)	Insect IgG supernatant (mol %)	Humanized murine IgG NS/0 (mol %)
High-mannose structures			
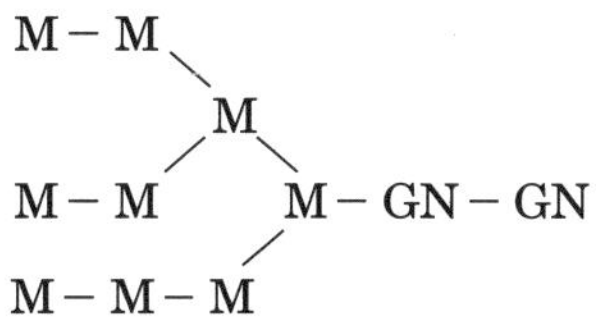	3.0	–	–
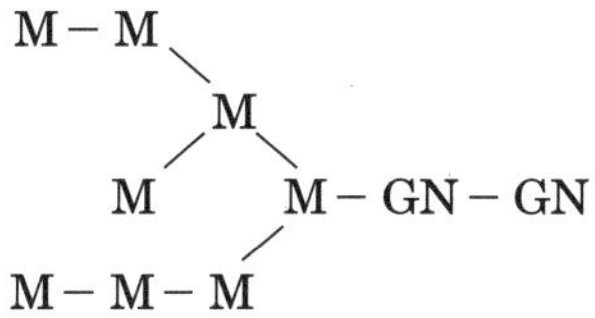	12.6	–	1
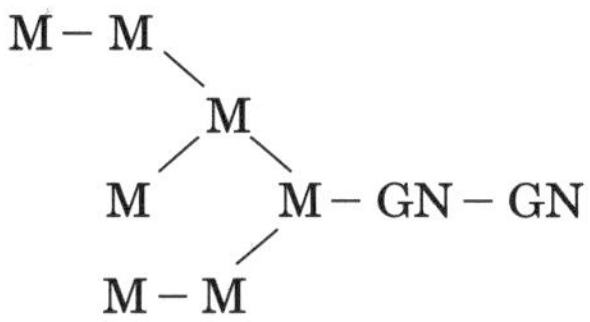	9.9	–	2
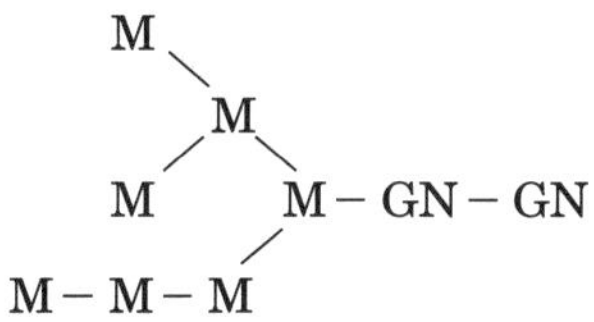	19.2	–	2
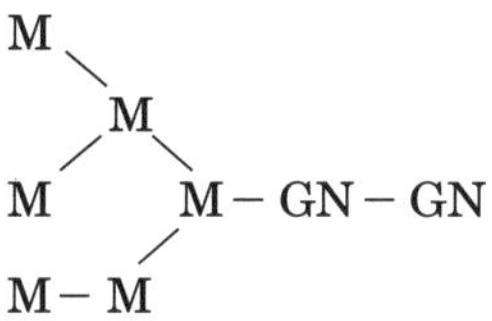	2.6	–	7
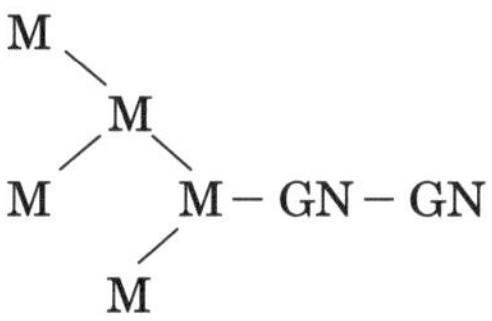	2.0	–	23
Hybrid structures			
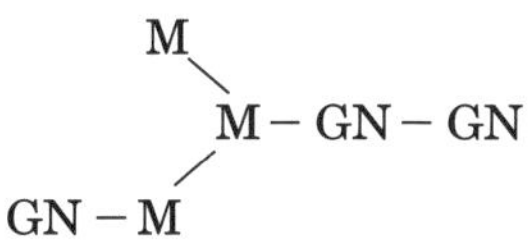	1.2	1.6	2
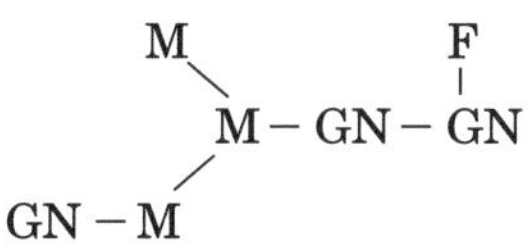	14.2		7
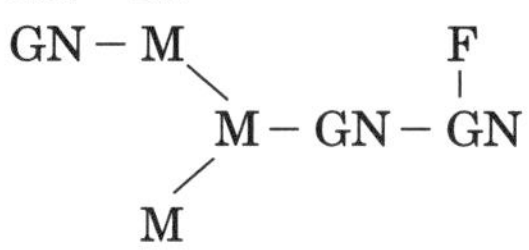	0.3	2.7	–
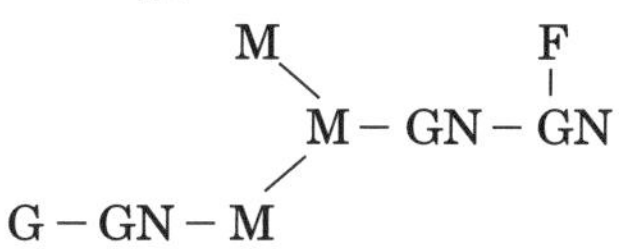	–	–	4

Table 1. Comparison of Glycosylation Patterns of IgG Produced in Insect Cells (*Trichoplusia ni*) and Mammalian Cells (NS/0) *(continued)*

	Insect IgG lysate (mol %)	Insect IgG supernatant (mol %)	Humanized murine IgG NS/0 (mol %)
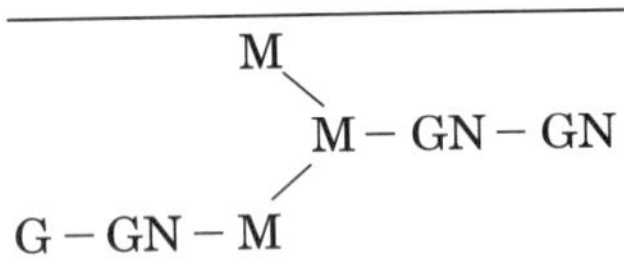	–	–	2
	Low-mannose structures		
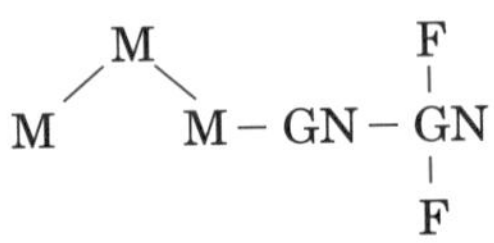	11.8	18.0	–
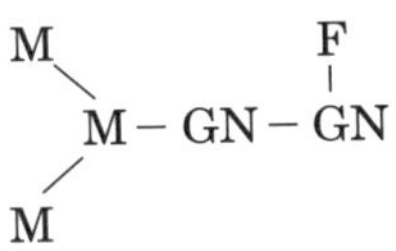	9.5	16.7	–
	Complex structures		
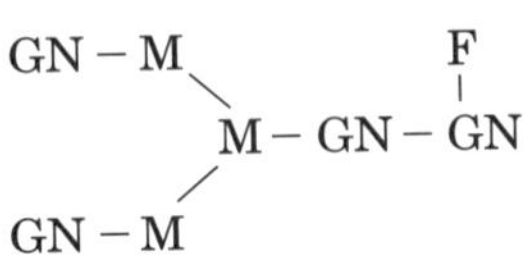	5.6	16.0	22
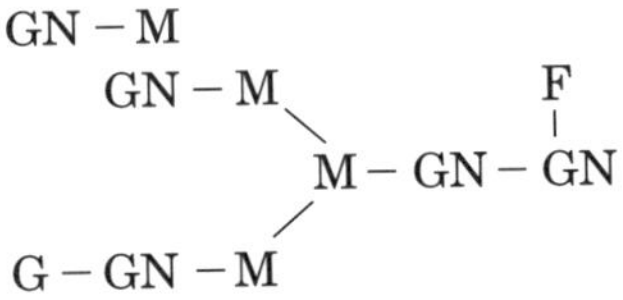	8.1	14.2	9
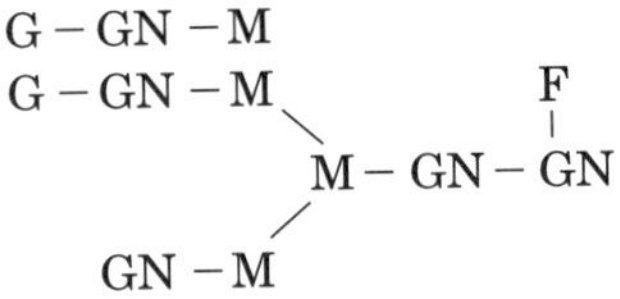	–	5.3	10
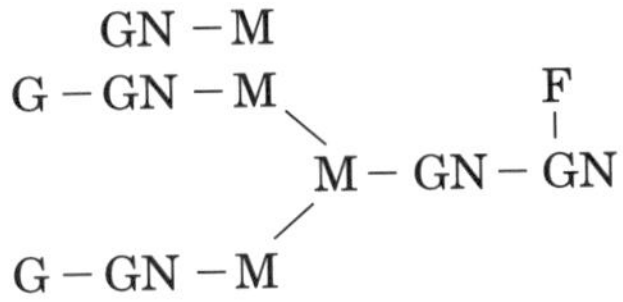	–	–	5

Source: Data for *T. ni* from Ref. 26; data for mammalian cells from Ref. 27.

for the production of recombinant proteins. The cost of commercial media used in cell culture can become prohibitive for large volumes, and a significant capital investment is required for purchasing large bioreactors. Some posttranslational processing events such as glycosylation, secretion, and carboxy-terminal amidation also may be more efficient in larvae (37). These considerations may warrant the consideration of insect larvae as an alternative method for the production of recombinant proteins.

Cell Culture Processes

Increasing the volume and mass of production from a small 100-mL spinner flask to a large 100-L bioreactor is not a simple linear process (13). Many of the process parameters, especially the delivery of O_2 and mixing, change when production is moved to a large bioreactor. Even so, large-scale production of proteins in BES has been achieved in 100-L bioreactors while maintaining cell growth kinetics and productivity. Issues that are relevant to cell culture scale-up are addressed next.

Bioreactors: Stirred-Tank versus Airlift

Both airlift and agitated–sparged (stirred-tank) bioreactors have been successfully used for the production of recombinant proteins and virus in larger scales. Stirred-tank bioreactors typically consist of glass, carbonate, or stainless steel vessels containing a marine or paddle impeller driven by an overhead motor (Fig. 7). Temperature, oxygen, and pH can be controlled. Temperature control can be maintained by immersion in a water jacket, using circulating cooling and heating coils, or by enclosure in a controlled incubator. Stirred-tank bioreactors typically require the injection of air or oxygen into the bioreactor at a position below the impeller. Air can also be supplied to the head space in order to limit the amount of oxygen sparged directly into the bioreactor. An alternative to the stirred-

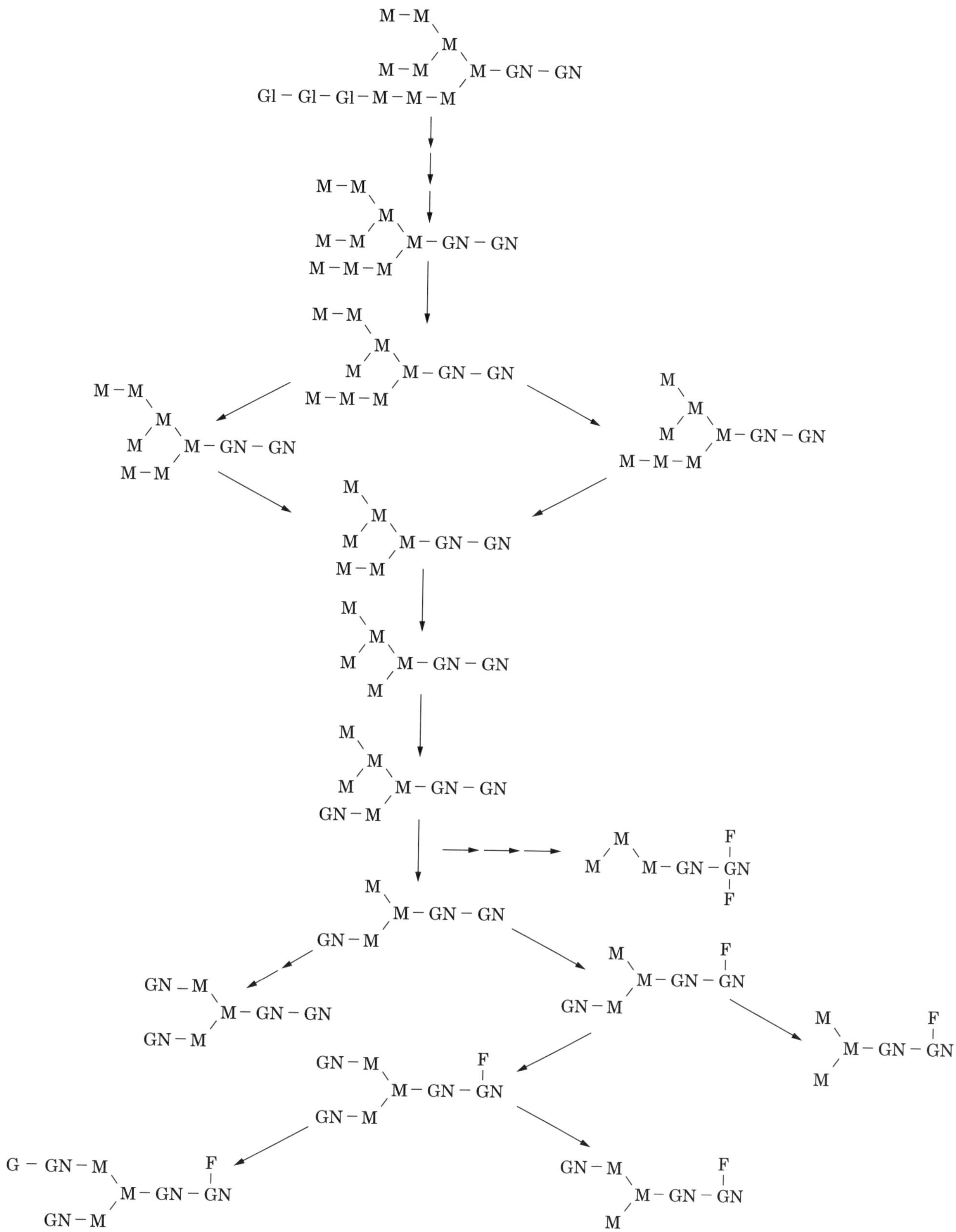

Figure 6. *N*-glycan processing pathway in *T. ni* insect cells. The oligosaccharide chain consists of the following saccharide units: F, fucose; Gl, glucose; GN, *N*-acetylglucosylamine; M, mannose. (Figure shown as published in *Differential N-Glycan Patterns of Secreted and Intracellular IgG Produced in Trichopulsia ni Cells*, T.A. Hsu et al., *The Journal of Biological Chemistry*, 1997. Reprinted by permission of The American Society for Biochemistry & Molecular Biology.

Table 2. Genes Utilized to Increase the Insecticidal Effects of Recombinant Baculoviruses

Name of gene	Effect(s) of expressed gene on host insect	% increase in insect killing
Diurectic hormone	Reduction of hemolymph volume	20
Juvenile hormone esterase	Feeding cessation and blackening	—
Bacillus thuringiensis toxin	Feeding cessation	$(2 \times LD_{50})^{a}$
Scorpion toxin (AaIT)	Paralysis, body tremors, and dorsal arching	30–40
Mite toxin (Txp-1)	Paralysis	30–40
Wasp toxin (antigen 5)	Premature melanization and low weight gain	—
AcNPV deletion mutant lacking ecdysteroid UDP-glucosyltransferase gene (egt)	Reduces larval feeding times	20% reduction in ST_{50}[b]

[a]LD_{50}, the dose necessary to kill 50% of the larval population.
[b]ST_{50}, the time necessary to kill 50% of the larval population.

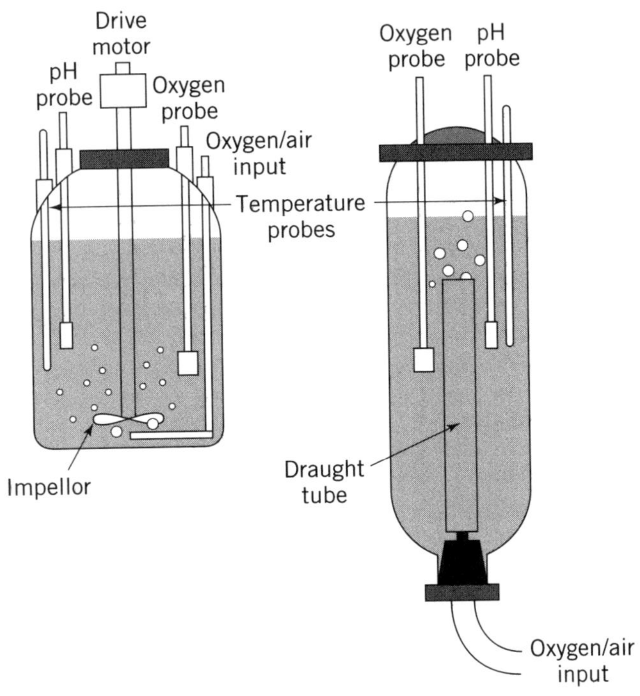

Figure 7. Configuration of stirred-tank and airlift reactors.

tank bioreactor is the airlift bioreactor in which liquid circulation through the bioreactor is driven by the lifting force created by injection of air or oxygen through the bottom of the bioreactor (Figure 7). This bioreactor includes an inner draft tube that carries the air/oxygen and liquid medium up through the bioreactor to the surface of the medium in order to set up circular flow of medium within the bioreactor.

A side-by-side comparison of a 6-L airlift vessel and an 8-L stirred-tank vessel (5.5-L working volume) containing Sf-9 cells in serum-free medium showed similar growth rates, cell viabilities, and cell densities throughout the cell growth period (38). Doubling times were about 22 h in both cases, and the growth curves were similar to those achieved in 100-mL spinner flasks, demonstrating the scaleability of each approach. Most importantly, production of a recombinant protein (kinase C-η) was virtually identical for both vessels in terms of protein yield and activity. Consequently, the principal consideration for choosing a production vessel is likely to depend on availability and cost. Stirred-tank vessels of many different sizes are readily available, whereas airlift scale-up requires that a constant height-to-circumference ratio be maintained in order to allow airlift function. This can lead to very tall vessels for large bioreactor volumes. The stirred-tank fermentation system can be implemented without a commercial vendor, if necessary, by purchasing a large motor and vessel, whereas an airlift system is usually purchased as a single unit from a commercial source.

Though stirred-tank and airlift are the principal bioreactor configurations used, some variations in these two types are available. Clumping of attachment-dependent cell lines in suspension, such as with High Five, is common, so bioreactor configurations that allow attachment-dependent cell growth and infection have been developed (3,39). These attached cell lines are typically grown in roller bottles or on supporting particles such as glass beads or microcarriers. The supporting particles are then incorporated into immobilized perfusion or airlift bioreactor configurations. In addition, bioreactors in which oxygen is supplied through membrane tubing have been utilized, though not extensively for large-scale cell culture (40). As the size of the bioreactor increases, the amount of tubing required for efficient mass transport increases significantly.

Process Monitoring Techniques

Cell growth and infection in insect cell bioreactors are usually monitored by following total and viable cell numbers by means of trypan blue staining and counting with a hemacytometer. Cell viability typically drops after several days after baculovirus infection of an insect cell culture. Given the labor-intensive and invasive nature of this assay, other techniques have been considered for monitoring cell growth and infection, both off-line and on-line (40). Coulter™ particle counters and optical density (OD) measurements taken using a fiber optic probe can speed the cell counting process considerably during the cell growth period. Alternatively, growth can be monitored on-line through indirect measurements of cellular respiration such as oxygen uptake rate (OUR) or CO_2 production rates (CPR). Monitoring the infection process is much more challenging, though several techniques are under development to supplement trypan blue staining. OUR can be used to follow the infection process because of the transient increase in oxygen consumption that occurs during the initial stages of the baculovirus infection. Other tools that have been used to monitor the infection process include epifluorescence microscopy of live and dead cells using fluorescence stains, as well as flow cytometric monitoring of DNA to track virus production (41). Fluorometric assays can be used to examine protein production directly in stained cells and will supplement conventional bioassays and SDS-PAGE analysis.

Oxygen Supply

One of the critical nutrients that must be supplied to an insect cell culture during the growth and infection stages is oxygen. Oxygen consumption rates per cell range from 10^{-17} to 10^{-18} mol/cell s and can increase by 30% or more during the initial stages of infection. Absence of sufficient aeration and oxygen supply during the infection stage can inhibit the viral infection and lower recombinant protein yields several fold (42). Consequently, it is important to monitor the dissolved oxygen level in large-scale insect cell bioreactors using commercially available probes. Some researchers have suggested that dissolved oxygen tension (DOT) be maintained at about 50% saturation to reduce the required oxygen sparging that is potentially lethal to insect cells (43). However, it is more important to ensure that the DOT always remain above 20% air saturation to prevent any limitations during the infection period.

Cell Damage During Sparging

Oxygen supply to the bioreactors is typically provided through sparging of oxygen or air into the bottom of a stirred-tank or airlift bioreactor. The oxygen is transferred from the gas to the liquid as the bubbles rise up the bioreactor. In order to provide sufficient mass transfer area, the air is bubbled through small openings. Impeller rotation can enhance bubble breakup further to increase the surface area available for mass transfer. The resulting bubble size is typically 1–5 mm in diameter, and the Reynolds number of a rising bubble can vary from 100 to 1,000 in conventional bioreactors (44). It is now known that the sparging of oxygen or air can cause significant damage to insect cells as well as mammalian cells in suspension (45–47). This cell damage is principally due to the adhesion of cells to bursting gas bubbles at the gas–liquid interface. In contrast, agitation is likely to be less of a factor in causing cell damage in many sparged bioreactor environments. Suspended cells can often tolerate high agitation rates as long as the gas–liquid interface is removed (48,49). Cell damage from agitation may occur in some cases due to the formation of Kolmogorov eddies on a length scale comparable to the size of a cell (48,49).

High-speed video microscopy studies have been used to observe directly cell–bubble interactions during the bubble rise through an insect cell culture medium (46,47). Cells attach independently or in clumps to a rising bubble and are carried upward through the fluid (Fig. 8). These cells then accumulate in high concentrations in the foam layer at the surface of the medium. The rupture of bubbles at the gas–liquid interface leads to the formation of upward and downward liquid jets that generate large shear stresses sufficient to damage or kill insect cells attached to the bubbles. The cell concentration in the jets is higher than in the medium, and more than 75% of these cells are killed by bubble rupture and jet formation. Approximately 1,200 Sf-9 cells can be killed per bubble rupture (44).

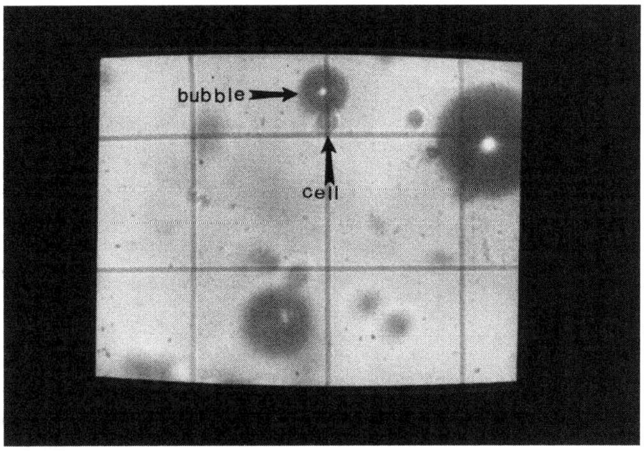

Figure 8. Insect cell attachment to bubbles rising through culture medium. Grid spacing is 100 μm. *Source*: Photograph courtesy of J. Chalmers.

Much of the cell damage due to bubble rupture can be prevented by the addition of surfactant polymers such as Pluronic F-68 at concentrations in the range of 0.1 to 0.2% (46,50–52). The addition of Pluronic F-68 prevents the adhesion of cells to bubbles and eliminates the damage to cells caused by bubble breakup at the liquid surface. The polymeric surfactants may prevent cell adherence to bubbles by lowering the interfacial tension at the gas–liquid interface. Most commercial serum-free cell culture media for insect and mammalian cells include Pluronic F-68, although other surfactants such as methylcellulose, serum, carboxymethylcellulose, bovine serum albumin, and dextrans have been considered as well. Unfortunately, only a limited selection of surfactants can be used because the interaction of the surfactants with the cells may affect membrane permeability and cause cell lysis (44).

Media Formulation

Media formulations are derived from elements originally identified in the insect hemolymph and include salts, vitamins, amino acids, sugars, organic acids, and trace elements (53,54). A number of proprietary commercial serum-free media are available for insect cell growth, including Express 5 and Sf900II from Gibco-BRL, and Excell 401 and 405 from JRH Biosciences. These media are derivations of noncommercial media such as Grace's, TNM-FH, TC-100, and IPL-41 that were supplemented with 5–20% fetal calf serum and used for many years in the cultivation of insect cells. The serum-free media are supplemented with lipids (cholesterol, α-tocopherol acetate, cod liver oil fatty acid methyl esters, β-sitosterol, and stigmasterol), yeast extract (yeastolate), and Pluronic F-68 in addition to other proprietary alterations (55,56). Antibiotics such as penicillin G and streptomycin are often added to the medium during cell culture.

Media sugars, including glucose, maltose, and fructose, can be partially or completely consumed during insect cell growth and recombinant protein production. Also, a significant fraction of the amino acids, including glutamate, glutamine, serine, asparagine, threonine, aspartate, and arginine, can be consumed during insect cell growth. However, much of this consumption may be devoted to energy use by the cells rather than the production of biomass (54,55). In contrast, a significant increase in the level of alanine is realized during Sf-9 cell growth experiments in IPL-41 medium; the amino acid may represent a sink for ammonia released during amino acid catabolism. Other amino acids remain relatively balanced throughout the insect cell growth experiment. Because the level of amino acids present in IPL-41 medium is far in excess of that required for biomass production, media alterations that reduce these amino acid levels may lead to more efficient nutrient utilization (56). Also, a more detailed characterization of the yeast extract components critical to cell growth may eventually result in the development of a truly defined medium for insect cell culture (53,56). The consumption rates of these nutrient components may change slightly but not significantly following baculovirus infection (43).

Cell Culture By-products

A significant amount of carbon dioxide is exhausted due to nutrient oxidation. Metabolites including lactate and other organic acids (pyruvate, α-ketoglutarate, fumarate) can also accumulate during growth and infection. The production of lactate from Sf-9 cells is generally lower than that observed with mammalian cells and poses no inhibitory effect on growth, although, lactate levels can increase significantly following DO limitations (57). The metabolites secreted in cell culture may change significantly depending on the cell line. Ammonia may be produced or consumed, depending on the particular cell type and cell culture conditions. In fact, much higher levels of lactate and/or ammonia are generated by *T. ni* and *B. mori* cells when compared to Sf-9 cells (43,58). In addition to toxic metabolites, the accumulation of proteases has been noted during the late viral infection stage and may be correlated with recombinant protein degradation (43). The most common approach to prevent protein degradation is to include a cocktail of protease inhibitors with the viral infection medium and during the protein purification process.

Most insect culture media are not buffered with carbon dioxide, so there is no need for a CO_2 environment for growth. The optimal pH for growth of most insect cells is between 6.2 and 6.3, though the pH of Sf-9 cultures in an uncontrolled medium can reach a minimum value between 5.9 and 6.1 (40). Because pH variations in the medium will affect both the growth and infection stages (59), pH control is suggested for optimal productivity in large-scale cell cultures. Osmolarity is another variable that should be considered, although, varying the osmolarity from 350 to 385 mosm/kg did not alter the insect cell growth rate more than 10% (60). Most media osmolarities range from 320 to 375 mosm/kg.

Batch Production Techniques

Recombinant baculovirus and heterologous proteins have been produced using batch, fed-batch, and continuous production modes. However, batch production of recombinant proteins remains the method of choice for most users. Typical processing of insect cells involves the batch cultivation of cells in a small suspension vessel, followed by a gradual scale-up of the cultures into progressively larger bioreactors. The cells are grown to a maximum cell density in one vessel and then pumped through steam-sterilized lines to increasingly larger vessels containing fresh medium. The optimal growth temperature is usually between 27 and 29 °C, with cell doubling times between 17 and 24 h, depending on the cell line and medium utilized. Some reports have recorded that High Five cells have a shorter doubling time (13–17 h) than Sf-9 cells (20–24 h) (58). In the final production-scale vessel, the cells are infected with virus during the exponential growth phase, and the virus or recombinant protein is collected at the conclusion of the infection stage (typically between 48 and 100 h postinfection).

Virus Stocks, Multiplicity of Infection, and Time of Infection

To initiate the infection, an aliquot of virus is added to the cell culture broth. Virus stocks are made from a single viral isolate obtained by plaque assay or end-point dilution. The plaque assay involves the infection of a monolayer of cells with a low concentration of virus, followed by an overlay of agarose-containing medium. Eventually, an isolated region of infected cells, called a plaque, is identified, and an agarose plug is collected at this plaque location for generating viral stocks. With the end-point dilution method, the tissue culture medium is diluted to the point at which any infection must result from a single infectious virus. Large stocks of the virus are obtained by infecting a monolayer or suspension cell culture with the agarose plug (from a plaque) or tissue culture medium (from the end-point dilution). The cell culture medium is then collected after 4–5 days, however, it may be necessary to repeat the infection several times in order to generate a viral stock of sufficient titer. The viral titer is obtained by determining the number of plaque-forming units (PFU) per milliliter, using the plaque assay, or by finding the 50% tissue culture–infectious dose ($TCID_{50}$) per milliliter using end-point dilution. The $TCID_{50}$ dose per milliliter is then multiplied by 0.69 to determine the number of plaque-forming units per milliliter. Several master viral stocks are frozen at −70 °C for long-term storage, with one master stock kept at 4 °C in order to generate working viral stocks.

Cells are then infected at a desired multiplicity of infection (MOI), which represents the number of plaque-forming units per viable cell, from the working viral stock. In small shaker-flask cultures, cells are typically infected at an MOI greater than 1 during the mid to late exponential phase of growth in order to generate a synchronous infection. Infecting the cells at the highest possible cell density and cell viability while still in the exponential growth phase provides high volumetric productivities. However, an efficient infection is also dependent on the presence of adequate nutrient levels in the medium. If the cells are infected at a point near the end of the exponential phase or during the stationary phase, insufficient substrates will be available for the viral infection, and low recombinant product yields will result (42).

In larger bioreactors, infection at a high MOI (>1) may not be feasible or desirable due to the requirements for large amounts of virus stock. If cells are infected at a lower MOI, then the infection of the cells will not be synchronous, and some insect cell growth will occur prior to secondary infection by BV generated during the initial infection (60). If the MOI is too low or cell density too high, the cells will grow to the late exponential or stationary phase prior to complete infection, and product yields will be lower. Consequently, it is important to optimize both the time of infection (TOI) and MOI to ensure that all cells are infected by virus and that sufficient substrates remain for virus and/or protein production (61). This optimization is facilitated by the use of mathematical models that consider substrate consumption during both cell growth and viral infection (62,63). With these models, protein yields can be achieved using MOIs as low as 0.0001 that are equivalent to those obtained with infections at a MOI of 5 (61). Infections at low MOI and low cell density may be very effective for cell lines such as Sf-9, but the approach may not be as successful for cell lines such as TN-5-B1-4, which produce much lower levels of extracellular virus.

Fed-Batch, Semibatch, and Sequential-Batch Production

It is desirable to grow the cells to a high cell density to maximize the productive capacity of the bioreactor. Studies have shown that Sf-9 cell concentrations can reach 3–5 $\times$ 10^6 cells/mL in batch culture for a typical medium. Supplementation of glucose, glutamine, and/or yeastolate can be used to increase the maximum cell density by at least 50% and enhance recombinant productivity. These nutrients can be added continuously in a fed-batch mode or in a single dose near the end of the initial growth period (Fig. 9). As an alternative, fresh medium can be fed continuously in a perfusion mode to replace the spent medium exiting through a hollow fiber device (0.4–0.6 vol/day) to provide cell densities up to 10 $\times$ 10^6 cells/mL of bioreactor volume (64). However, the perfusion of fresh medium results in a slightly lower cell mass yield on a per milliliter of total medium basis (4.1 $\times$ 10^6 cells/mL total medium) as compared with nutrient supplementation (6.7 $\times$ 10^6 cells/mL total medium). Another approach that has been used to increase cell density is the replacement of 80% of the medium for several consecutive days near the end of the growth stage. Using the same hollow fiber technology (0.2-μm pore size), most of the spent media is removed, and then fresh medium is added to the bioreactor. This approach can be used to double cell densities up to 11.6 $\times$ 10^6 cells/mL reactor volume (64). Even greater enhancements in cell density can be obtained when nutrient supplementation of glucose, glutamine, and yeastolate is combined with medium replacement (15.8 $\times$ 10^6 cells/mL reactor volume) (64). A similar nutrient supplementation strategy is used to enhance the yields of recombinant proteins during the infection stage because the recombinant product yield is highly dependent on the nutrient levels. One approach that is used to ensure high product yields is to resuspend the cells in fresh medium prior to the infection, or else to add fresh medium to the bioreactor at the beginning and/or during the infection stage (42). The supplementation of limiting nutrients such as glucose, glutamine, and yeastolate can also be used to enhance the infection process (4,65–67), eliminating the costly replacement of the entire medium.

Another semibatch operating strategy used for production of recombinant baculoviruses and heterologous proteins is the utilization of separate growth and infection reactors (Fig. 10) (68). An initial cell growth reactor is followed by single or multiple sequential infection vessels in series. After each cycle, the contents from one infection vessel is pumped to a following infection vessel while the contents of the last infection vessel are harvested for product. The first infection vessel is then filled with medium and cells from the growth reactor and inoculated with virus. The cell growth reactor is provided with fresh medium, and the growth cycle repeated. Some of the cells from the growth stage may remain in the reactor to serve as an inoculum for the next growth cycle (69,70), and a portion of the contents of the first infection reactor can be left to infect a new batch of cells. However, modeling studies have indicated that the productivity of this sequential batch system is expected to decline after a number of cycles due to the buildup of mutant viruses known as defective interfering particles (DIPs) (see later section) (71). The formation of these mutants is due to the "passage" effect (see later section). The recombinant protein productivity of the second and subsequent infection stages must be monitored closely for the build-up of these particles, and the second infection stage must be completely emptied and inoculated with a fresh sample of virus in order to eliminate the DIPs. Semicontinuous operation has also been achieved by maintaining a first-growth reactor in a continuous operating mode and then gradually filling a second reactor (72). Following the filling of the second reactor, virus is added to initiate a batch infection in the second stage. At the end of the infection process, the second stage is emptied in part or completely, and the filling and infection cycle repeated. The medium in the infection reactors may also be supplemented or replaced by separating the cells from spent medium by settling, membrane filtration, or ultracentrifugation. Since the nutrients are often not consumed completely in the infection stage(s), it may not be necessary to replace all the spent media. These sequential-batch processes can be achieved with processing times dictated by the exhaustion of nutrients, decline of cell viability, and accumulation of recombinant proteins.

Continuous Production Systems

The release of BV from infected insect cells into the cell culture medium has led to the design and evaluation of continuous-flow stirred-tank reactors for the production of polyhedra and recombinant proteins. The utilization of a continuous system could reduce production costs by 50% when compared with a batch system (73). Batch systems have the disadvantage of long down times as a result of cleaning and sterilization after each run. Continuous sys-

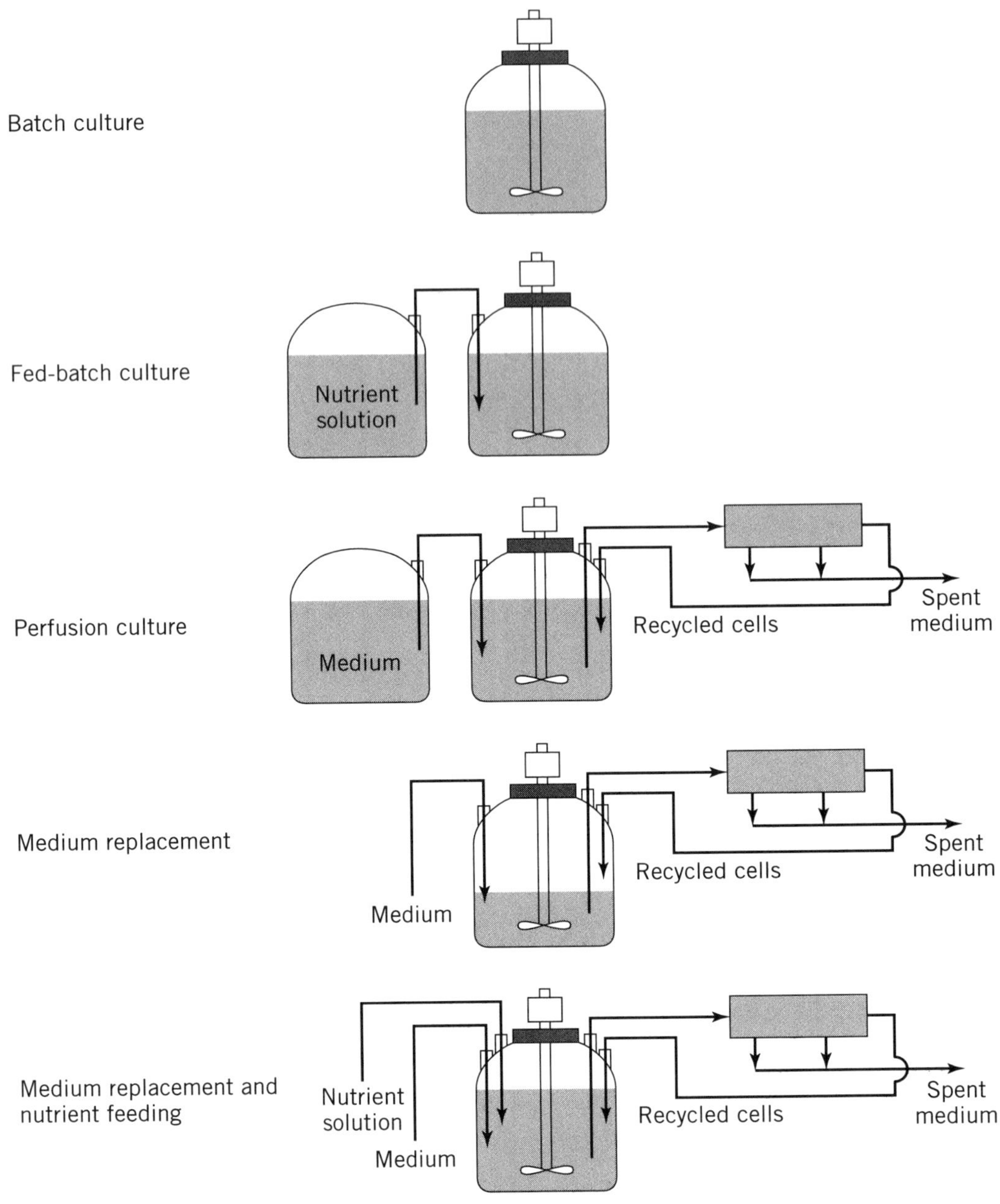

Figure 9. Various reactor configurations for batch and fed-batch production strategies.

tems typically consist of sequential two-stage continuous-flow bioreactors in which the first stage is used for cell growth and the second stage is used for virus production and heterologous gene expression (Fig. 10). The effluent of cells from the first stage feeds the second infection stage, and the effluent of the second stage includes the infected cells and associated baculoviruses and/or recombinant product. The BV generated in the second stage is used to infect the feed from the first reactor. In one study (74), a 60-h mean residence time was maintained in each stage, and production of polyhedra and recombinant protein was sustained for approximately 4 weeks. Because about 200 BVs are produced per cell, there is a gradual accumulation of virus in the second reactor, followed by a severe decline in productivity after 4 weeks due to the formation of DIPs. A model that considers the presence of infectious virus (I-BV), defective virus (DIP), and abortive virus (A-BV) effectively predicted the accumulation of DIPs and the decline of recombinant protein production for a two-stage continuous-infection process (71).

Thus, a continuous reactor system is not likely to be feasible for extended production of baculovirus and recombinant protein. However, such a set-up may be used for production over shorter periods such as a 30-day cycle prior to the accumulation of DIPs (71). This reactor set-up would be analogous to a continuous catalytic reactor that experiences gradual catalyst deactivation during its operation and has to be periodically recharged with new catalyst. This could be accomplished by a complete shut down and restart with fresh virus or else by providing a secondary

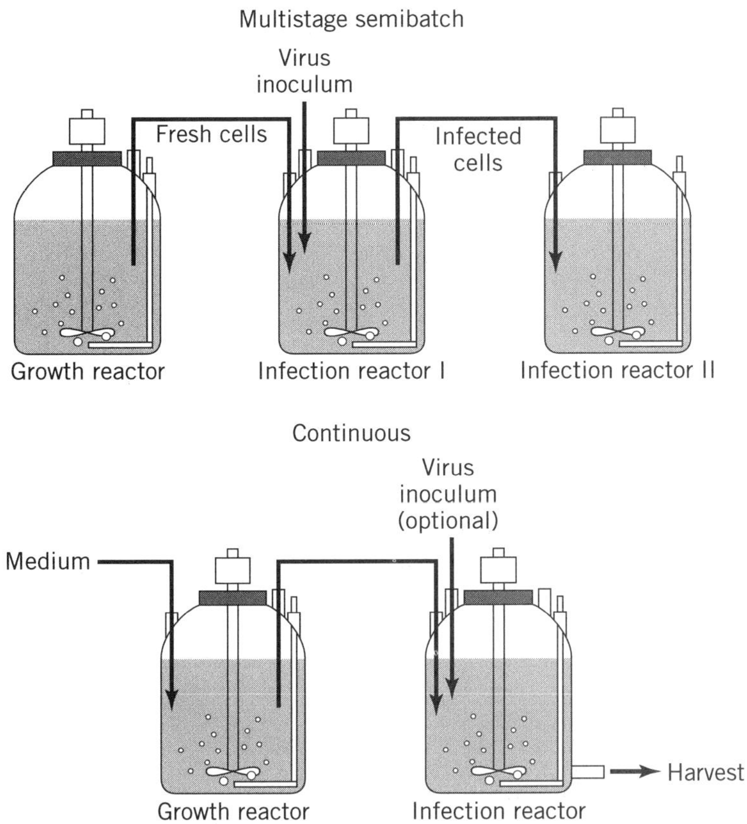

Figure 10. Multistage reactor systems. Multistage reactor systems used for the semibatch and continuous production of recombinant proteins in insect cell culture using AcMNPV. The initial reactor is primarily used for cell growth, whereas subsequent reactors contain the insect cells infected with the virus.

feed of fresh virus to the second infection reactor. A continuous set-up provides for larger processing volumes of insect cell culture and recombinant product than does batch operation, although the yield of recombinant protein in the effluent may be low compared with batch or fed-batch operation. A recent innovation in the continuous culture process has been the sequential combination of a continuous stirred-tank reactor (CSTR) for growth, followed by a tubular reactor for infection. The tubular reactor provides for the continuous introduction of baculovirus at higher MOIs without the negative passage effects present in a two-stage CSTR system (75).

Passage Effects and Mutant Virus Formation

The continuous passage of viruses in cell culture often leads to the formation of mutant viruses such as DIPs (76,77). The DIPs will replicate more rapidly than the normal-size viruses, and the concentration of DIPs can rise to as high as 60% of the total virus stock (76,77). These DIPs average about 200–230 nm in size, as compared with the standard 330-nm-sized virions (77), and can lack up to 40% of the baculovirus genome due to spontaneous deletions during passaging (76). The deleted segment often includes the polyhedrin gene and at least one other gene necessary for virus replication. Consequently, DIP mutants cannot be plaque purified because they replicate only in cells that are also infected by a normal virus. To avoid the passage effect, viral passage number should be monitored, and a low passage master viral titer stock used to generate working viral titer stocks. If contamination with DIPs is suspected, then a plaque assay can be performed to reisolate the infectious virus. Also, low MOI infections can be used to minimize DIP formation, but this approach results in asynchronous infections.

Other mutant viruses, such as ones that result in the generation of a few polyhedra (FP), have also been observed. The FP mutants substantially lower the number of polyhedra produced per infected cell and reduce the number of occluded viruses generated (78). These FP mutants may have a selective advantage in cell culture by producing more budded virus at the expense of polyhedra. Unlike DIPs, FP mutants can be plaque purified. Once again, formation of FP mutants can be minimized by working with low-passage viral titers and utilizing a master viral titer stock. Because FP mutants do not include significant virus levels in the polyhedra, insect-to-insect transmission will not include FP mutants, which are not capable of spreading through polyhedra infections.

Larval Production Systems

Based on productivity comparisons with polyhedra, approximately 100–200 larvae will provide a final protein yield approximately equal to 1 L of cells for an intracellular recombinant protein (7). The use of larvae eliminates the need for high capital investments in large-scale bioreactors but requires the development of systems for insect rearing and infection. Furthermore, purification strategies must be implemented to isolate the proteins from insect parts or hemolymph. Typical species considered include *Spodoptera frugiperda, Trichoplusia ni, Heliothis virescens, Manduca sexta*, and *Hyalaphora cecropia* for AcMNPV infections, and *Bombyx mori* for BmNPV infections (7,37). Lepidopteran eggs are available from commercial suppliers and hatch 5 days to 1 week after incubation at 25–27 °C in a humidified environment under a controlled light cycle. Artificial diets are also available from commercial sources or may be prepared from components in the laboratory. Constituents typically include pinto beans, Brewer's yeast, ascorbic acid, wheat germ, sorbic acid, 14% auromycin, and a vitamin mix. The ingredients are combined with agar and then provided to larvae with periodic replacement to ensure a moist supply of nutrients. It is important to infect the insects at the appropriate developmental stage to provide the maximum yield of baculovirus or heterologous protein. The insects typically molt five or six times as larvae with the last stage being the larval–pupal molt. The intermolt period is the stadium or instar. Optimal infections are typically achieved one or two days after molting into the last larval instar. Infections are accomplished by feeding the insects with polyhedra inclusion bodies (PIBs) or by injecting budded virus (BV) directly into the hemolymph. For recombinant baculoviruses that are occlusion negative, injection of the budded virus is the required method of infection. A small number of mock-infected insects should also be maintained to check for possible contamination problems and to provide a comparison of infected and uninfected larvae. Shown in Figure 11a is a photograph of an infected *T. ni* larva expressing green fluorescent protein as compared with an uninfected larva (79). Figure 11b shows massive expression of β-galactosidase in a *T. ni* larva when infected with AcMNPV containing the heterologous *lacZ* gene under the control of the p10 promoter. At 60 h postinfection the infection has cleared the midgut and entered the fat body. The best time to collect product from infected insects is generally prior to death. For secreted proteins, the product is collected by bleeding the insects to collect hemolymph, whereas nonsecreted proteins are isolated by homogenization of whole insects. Unfortunately, homogenization can result in the release of proteases; these can be minimized by removal of the insect gut before processing (7,37). Alternatively, a particular metabolically active tissue such as the fat body can be removed from the insect and used as a source of the desired protein (7,37).

CONCLUSION

The use of the baculovirus–insect cell system for heterologous gene expression and recombinant biopesticides has grown dramatically since its first application in 1983. Commercially available vectors that provide for the rapid selection and screening of recombinant viruses has made this technology accessible to most biotechnology laboratories. The development of serum-free media, the isolation of alternative insect cell lines, and the implementation of efficient bioreactor operating strategies have increased productivities and lowered costs. As a result, insect cells are often the method of choice for rapid and high-level production of functional eukaryotic proteins. The use of BES is likely to increase even more with the growth of genomics and its associated need for functional characterization of unknown proteins. However, the application of insect cell-derived proteins as pharmaceuticals will continue to be slow due to the different posttranslational modifications in insect cells as compared with mammalian cells. Future efforts will be directed at enhancing posttranslational processing functions and improving protein secretion from

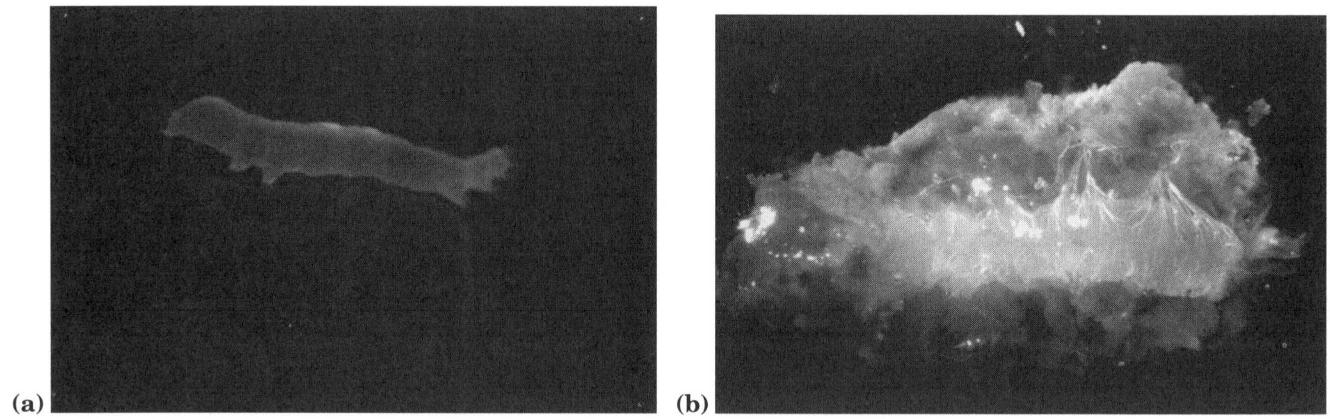

Figure 11. (**a**) Photo of *T. ni* larvae sitting atop a laboratory UV transilluminator table. The upper larva has been infected (72 h) with a recombinant AcMNPV baculovirus that contains the gene for green fluorescent protein (GFPuv) placed under control of the polyhedrin promoter. The brightness is due to the massive overexpression of GFPuv, which has been shown to reach 26% of the total larvae protein. The lower larva (nonglowing) is the uninfected control. *Source*: Photograph courtesy of W. Bentley. (**b**) Micrograph showing AcMNPV-β-galactosidase infection of *T. ni* insect larva at 60 h postinfection. Staining represents β-galactosidase expressed in insect cell tissues. Infection has cleared the midgut and spread to the fat body. *Source*: Micrograph courtesy of G. Williams.

current and novel insect cell hosts. Other promising areas include the use of insect larvae and the development of stable expression systems such as *Drosophila* S2 cells (80). With these advances, insects and insect cells are likely to remain a popular choice for the manufacture of recombinant DNA products.

ACKNOWLEDGMENTS

The authors are grateful to Greg Williams, Jeffrey Chalmers, and William Bentley for graciously providing unpublished photographs for this review. Special thanks to Mike Owens for technical input and assistance.

BIBLIOGRAPHY

1. D.E. Lynn, *Tissue Culture Methods* **12**, 23–29 (1989).
2. T.R. Davis, T.J. Wickham, K.A. McKenna, R.R. Granados, M.L. Shuler, and H.A. Wood, *In Vitro Cell. Dev. Biol.* **29A**, 388–390 (1993).
3. R.A. Taticek, D.A. Hammer, and M.L. Shuler, in M.L. Shuler, H.A. Wood, R.R. Granados, and D.A. Hammer eds., *Baculovirus Expression Systems and Biopesticides*, Wiley-Liss, New York, 1995, pp. 131–174.
4. D.W. Murhammer, *Appl. Biochem. Biotechnol.* **59**, 199–220 (1996).
5. M.J. Betenbaugh, L. Balog, and P-S. Lee, *Biotechnol. Prog.* **7**, 462–467 (1991).
6. W.F. Hink, D.R. Thomsen, D.J. Davidson, A.L. Meyer, and F.J. Castellino, *Biotechnol. Prog.* **7**, 9–14 (1991).
7. D.R. O'Reilly, L.K. Miller, and V.A. Luckow, *Baculovirus Expression Vectors: A Laboratory Manual*, W.H. Freeman and Company, New York, 1992.
8. M.L. Shuler, D.A. Hammer, R.R. Granados, and H.A. Wood, in M.L. Shuler, H.A. Wood, R.R. Granados, and D.A. Hammer eds., *Baculovirus Expression Systems and Biopesticides*, Wiley-Liss, New York, 1995, pp. 1–11.
9. G.E. Smith, M.J. Fraser, and M.D. Summers, *J. Virol.* **46**, 584–593 (1983).
10. G.E. Smith, M.D. Summers, and M.J. Fraser, *Mol. Cell. Biol.* **3**, 2156–2165 (1983).
11. J.M. Vlak, A. Schouten, M. Usmany, G.J. Belsham, E.C. Klinge-Roode, A.J. Maule, J.W.M. Van Lent, and D. Zuidema, *Virology* **179**, 312–320 (1990).
12. V.A. Luckow and M.D. Summers, *Biotechnology* **6**, 47–55 (1988).
13. R.M. Patterson, J.K. Selkirk, A.B. Merreck, *Environ. Health Perspect.* **103**, 756–759 (1995).
14. M.D. Summers and G.E. Smith, *A Manual of Methods for Baculovirus Vectors and Insect Cell Culture Procedures*, Texas Agricultural Experiment Station Bulletin no. 1555, 1987.
15. S. Maeda, T. Kawai, M. Obinata, H. Fujiwara, T. Horiuchi, Y. Saeki, Y. Sato, and M. Furusawa, *Nature* **315**, 592–594 (1985).
16. M.D. Ayres, S.C. Howard, J. Kuzio, M. Lopez-Feriber, and R.D. Possee, *Virology* **202**, 586–605 (1994).
17. G.D. Pennock, C. Shoemaker, and L.K. Miller, *Mol. Cell. Biol.* **4**, 399–406 (1984).
18. J. Vialard, M. Lalumiere, T. Vernet, D. Briedis, G. Alkhatib, D. Henning, D. Levin, and C. Richardson, *J. Virol.* **64**, 37–50 (1990).
19. P.A. Kitts and R.D. Possee, *Biotechniques* **14**, 810–817 (1993).
20. V.A. Luckow, S.C. Lee, G.F. Barry, and P.O. Olins, *J. Virol.* **67**, 4566–4579 (1993).
21. D.L. Jarvis and E.E. Finn, *Nat. Biotechnol.* **14**, 1288–1292 (1996).
22. L. Marz, F. Altmann, E. Staudacher, and V. Kubelka in H.S.J. Montreuil and J.F.G. Vliegenthart eds., *Glycoproteins*, Elsevier Science B.V., 1995, pp. 543–563.
23. F. Altmann, *Trends Glycosci. Glycotechnol.* **40**, 101–114 (1996).
24. D.J. Davidson, M.J. Fraser, and F.J. Castellino, *Biochemistry* **29**, 5584–5590 (1990).
25. O.W. Ogonah, R.B. Freedman, N. Jenkins, K. Patel, and B.C. Rooney, *Biotechnology* **14**, 197–202 (1996).
26. T.-A. Hsu, N. Takahashi, Y. Tsukamoto, K. Kato, I. Shimada, K. Masuda, E.M. Whiteley, J-Q. Fan, Y.C. Lee, and M.J. Betenbaugh, *J. Biol. Chem.* **272**, 9062–9070 (1997).
27. C.C. Yu Ip, W.J. Miller, M. Silberklang, G.E., Mark, R.W. Ellis, L. Huang, J. Glushka, H. Van Halbeek, J. Zhu, and J.A. Alhadeff, *Arch. Biochem. Biophys.* **308**, 387–399 (1994).
28. F. Altmann, H. Schwihla, E. Staudacher, J. Gloosl, and L. Marz, *J. Biol. Chem.* **270**, 17344–17349 (1995).
29. D.L. Jarvis and E.E. Finn, *Virology* **212**, 500–511 (1995).
30. R. Wagner, S. Liedtke, E. Kreztschmar, H. Geyer, R. Geyer, and H-D. Klenk, *Glycobiology* **6**, 165–175 (1996).
31. D.R. Thomson, L.E. Post, and A.P. Elhammer, *J. Cell. Biochem.* **43**, 67–79 (1990).
32. L.C. Iacono-Connors, C.S. Schmaljohn, and J.M. Dalrymple, *Infect. Immun.* **58**, 366–372 (1990).
33. D.C. Tessier, D.Y. Thomas, H.E. Khouri, F. Laliberte, and T. Vernet, *Gene* **98**, 177–183 (1991).
34. M.J. Betenbaugh, E. Ailor, E. Whiteley, P. Hinderliter, and T-A. Hsu, *Cytotechnology* **20**, 149–159 (1996).
35. M. Kidd, V.C. Emery, *Appl. Biochem. Biotechnol.* **42**, 137–159 (1993).
36. S. Maeda, *Curr. Opin. Biotechnol.* **6**, 313–319 (1995).
37. S. Maeda, *Ann. Rev. Entomol.* **34**, 351–372 (1989).
38. J.W. Rice, N.B. Rankl, T.M. Guranuss, C.M. Marr, J.B. Barna, M.M. Walters, and D.J. Burns, *BioTechniques* **15**, 1052–1059 (1993).
39. G.A. King, J. Kuzio, A.J. Daugulis, P. Faulkner, D. Bayly, and M.F.A Goosen, *Biotechnol. Lett.* **10**, 683–688 (1988).
40. S.N. Agathos, *Cytotechnology* **20**, 173–189 (1996).
41. M. Al Rubeai, N. Kioukia, and A.N. Emery, *Monitoring of Insect-Baculovirus Culture Processes by Flow Cytometry*, Cell Culture Engineering IV, San Diego, 1994.
42. D.A. Lindsay and M.J. Betenbaugh, *Biotechnol. Bioeng.* **39**, 614–618 (1992).
43. M.Y. Wang, T.R. Pulliam, M. Valle, V.N. Vakharia, and W.E. Bentley, *J. Biotechnol.* **46**, 243–254 (1996).
44. J.J. Chalmers, in M.L. Shuler, H.A. Wood, R.R. Granados, and D.A. Hammer eds., *Baculovirus Expression Systems and Biopesticides*, Wiley-Liss, New York, 1995, pp. 175–204.
45. J. Tramper, J.B. Williams, and D. Joustra, *Enzyme Microb. Technol.* **8**, 33–36 (1986).
46. A. Handa, A.N. Emery, and R.E. Spier, *Dev. Biol. Standard.* **66**, 241–253 (1987).
47. F. Bavarian, L.S. Fan, and J.J. Chalmers, *Biotechnol. Prog.* **7**, 140–150 (1991).
48. S.K.W. Oh, A.W. Nienow, M. Al-Rubeai, and A.N. Emery, *J. Biotechnol.* **12**, 45–62 (1989).

49. K.T. Kunas and E.T. Papoutsakis, *Biotechnol. Bioeng.* **36**, 476–483 (1990).
50. A. Handa-Corrigan, A.N. Emery, and R.E. Spier, *Enzyme Microb. Technol.* **11**, 230–235 (1989).
51. D.W. Murhammer and C.F. Goochee, *Biotechnology* **6**, 1411–1418 (1988).
52. D.W. Murhammer, *Appl. Biochem. Biotechnol.* **31**, 283–310 (1991).
53. M.J. Fraser, *Curr. Topics Microbiol. Immunol.* **158**, 131–172 (1992).
54. M. Drews, T. Paalme, and R. Vilu, *J. Biotechnol.* **40**, 187–198 (1995).
55. J.P. Ferrance, A. Goel, and M.M. Ataai, *Biotechnol. Bioeng.* **42**, 697–707 (1993).
56. R. Bhatia, G. Jesionowski, J. Ferrance and M.M. Ataai, *Cytotechnology* **20** (1996).
57. W.E. Bentley, M.-Y. Wang, and V. Vakharia, *Ann. N. Y. Acad. Sci.* **745**, 336–359 (1994).
58. T. Sugiura and E. Amman, *Biotechnol. Bioeng.* **51**, 494–499 (1996).
59. P. Licari, and J.E. Bailey, *Biotechnol. Bioeng.* **37**, 238–246 (1990).
60. J. Zhang, N. Kalogerakis, L.A. Behie, and K. Iatrou, *J. Biotechnol.* **33**, 249–258 (1994a).
61. K.T.K. Wong, C.H. Peter, P.F. Greenfield, S. Reid, and L.K. Nielsen, *Biotechnol. Bioeng.* **49**, 659–666 (1995).
62. P. Licari and J.E. Bailey, *Biotechnol. Bioeng.* **39**, 432–441 (1992).
63. J.F. Power, S. Reid, K.M. Radford, P.F. Greenfield, and L.K. Nielson, *Biotechnol. Bioeng.* **44**, 710–719 (1994).
64. R.E. Spier, J.B. Griffiths, W. Berthold, *Animal Cell Technol.* 493–503.
65. B. Nguyen, K. Jarnagin, S. Williams, H. Chan, and J. Barnett, *J. Biotechnol.* **31**, 205–217 (1993).
66. S. Reuveny, Y.J. Kim, C.W. Kemp, and J. Shiloach, *Biotechnol. Bioeng.* **42**, 235–239 (1993).
67. M.-Y. Wang, S. Kwong, and W.E. Bentley, *Biotechnol. Prog.* **9**, 355–361 (1993).
68. W.F. Hink and E.M. Strauss, in E. Kurstak, K. Maramorosch, and A. Dubendorfer eds., *Invertebrate Systems In Vitro*, Elsevier/North-Holland Biomedical Press, Amsterdam, 1980, pp. 27–33.
69. J. Zhang, N. Kalogerakis, L.A. Behie, and K. Iatrou, *Biotechnol. Bioeng.* **42**, 357–366 (1993).
70. J. Zhang, N. Kalogerakis, L.A. Behie, and K. Iatrou, *Biotechnol. Prog.* **10**, 636–643 (1994).
71. C.D. de Gooijer, R.H.M. Koken, F.L.J. van Lier, M. Kool, J.M. Vlak, and J.A. Tramper, *Biotechnol. Bioeng.* **40**, 537–548 (1992).
72. F.L.J. van Lier, J.P.T.W. van den Hombergh, C.D. de Gooijer, M.M. den Boer, J.M. Vlak, and J. Tramper, *Enzyme Microb. Technol.* **18**, 460–466 (1996).
73. J. Tramper and J.M. Vlak, *Ann. N. Y. Acad. Sci.* **469**, 279–288 (1986).
74. F.L.J. van Lier, van G.C.F. Duijnhoven, M.M.J.C.M. de Vaan, J.M. Vlak, and J. Tramper, *Biotechnol. Prog.* **10**, 60–64 (1994).
75. Y.C. Hu, M.Y. Wang, and W.E. Bentley, *Cytotechnology*, **24**, 143–152.
76. M. Kool, J.W. Voncken, F.L.J. van Lier, J. Tramper, and J.M. Vlak, *Virology* **183**, 739–746 (1991).
77. T.J. Wickham, T. Davis, R.R. Granados, D.A. Hammer, M.L. Shuler, and H.A. Wood, *Biotechnol. Lett.* **13**, 483–488 (1991).
78. M.J. Fraser, G.E. Smith, and M.D. Summers, *J. Virol.* **47**, 287–300 (1983).
79. H.J. Cha, M.Q. Pham, G. Rao, and W.E. Bentley, *Biotechnol. Bioeng.* **56**, 238–243 (1997).
80. J.S. Culp, H. Johansen, B. Hellmig, J. Beck, T.J. Matthews, A. Delers, and M. Rosenburg, *Biotechnology* **9**, 173–177 (1991).

See also AMMONIA TOXICITY, ANIMAL CELLS; CHINESE HAMSTER OVARY CELLS, RECOMBINANT PROTEIN PRODUCTION; CULTURE MEDIA, ANIMAL CELL, LARGE SCALE PRODUCTION; EXPRESSION SYSTEMS, *E. COLI*; EXPRESSION SYSTEMS, MAMMALIAN CELLS; GENE TRANSFER, GRAM POSITIVE BACTERIA; GLYCOSYLATION OF RECOMBINANT PROTEINS; INSECT CELL CULTURE, PROTEIN EXPRESSION; PHENYLALANINE.

INSECTICIDES, MICROBIAL PRODUCTION

SHUJI SENDA
TETSO OMATA
YASUO NINOMIYA
Nitto Denko Company
OsakaJapan

KEY WORDS

Asymmetric synthesis
Bond formation
Chiral
Enantiomers
Esterifcation
Hydrolysis
Lipase
Oxidation
Pheromone
Transesterification

OUTLINE

INTRODUCTION

Recently, the use of chiral chemicals in insect pest control has increased. Typically, these chiral chemicals are synthetic pyrethroid insecticides and pheromones, originally derived from natural compounds. The stereochemistry of chiral and bioactive products influences their bioactivity (1). In the case of Japanese beetle and gypsy moth pheromones, only one enantiomer is bioactive, but its antipode inhibits pheromone activity. In the case of olive fruit fly pheromone, one enantiomer is active in male insects, while the other is active in female insects. Thus, it is very important that chiral and bioactive chemicals are synthesized with high optical purity.

Methods of synthesizing chiral chemicals are as follows:

1. Use of natural chiral materials as starting substrates
2. Optical resolution
3. Asymmetric synthesis

Biochemical optical resolution and asymmetric synthesis using enzymes, microorganisms, or tissue cells have been actively researched, because these methods can produce chiral compounds with high optical purity. They have therefore contributed to many useful reviews and reports(2–6).

In this article we discuss recent chiral insecticide synthesis, especially of insect pheromones using microbial processes. We have classified these processes by their reaction types.

HYDROLYSIS REACTIONS

Lipase-Catalyzed Hydrolysis

In this field, enzymatic optical resolution using lipase is the most useful method studied by many researchers. Lipase has the following characteristics:

1. It is available at industrial production scale for a relatively low price.
2. It does not usually require coenzymes for catalytic reactions.
3. It has a remarkably wide substrate specificity.
4. It is stable in organic solvents.
5. It can catalyze both hydrolysis and esterification.

It is for these reasons that the use of lipase has been widely reported.

In the case of hydrolysis using lipase, optically pure (+)-*endo* and (+)-*exo*-brevicomin, the attractant pheromones of the bark beetle, were synthesized in short steps, starting from the α-hydroxy ketone derivative, (*R*)-4-benzyloxy-8-nonen-3-one, which was prepared via enzyme-mediated kinetic resolution of racemic 4-benzyloxy-2,8-nonadien-3-yl propanoate using Lipase OF (from *Candida cylindracea*, Meito Sangyo Co.) in phosphate buffer. The reaction was stopped at 22% yield to obtain high enantiomeric excess. The unused alcohol, (3*RS*,4*S*)-4-benzyloxy-8-nonen-3-ol, was then reused by conversion to ketone (7) (Scheme 1).

The enantioselective hydrolysis of tertiary α-benzyloxy esters [$R\text{-}C(CH_3)(OCH_2Ph)CO_2CH_3$] using the same lipase (Lipase OF) gave various enantiometrically enriched esters and acids on a preparative scale. The positive aspects of this reaction are (*1*) the racemic substrates are readily obtainable, (*2*) the method of the hydrolytic reaction is quite simple, and (*3*) the functional group of the product is adequately protected. Because the optical purity of carboxylic acid was low after the first reaction, a second reaction was needed to obtain an optically pure carboxylic acid. We call these reactions “two-step resolutions.” Using the enantiometrically enriched carboxylic acid obtained above, (−)-frontalin was efficiently synthesized in nine steps to demonstrate the utility of this method (8) (Scheme 2).

Although discrimination of two enantiomers of a secondary alcohol whose hydroxy group is located in the middle of a long hydrocarbon chain is rather difficult, a two-step resolution of 6-methyl-2-heptyn-4-yl acetate using lipase (Lipase A, Amano Pharmaceutical Co.) was accomplished. The first step was enzymatic hydrolysis of racemic acetate. The second step of the reactionary alcohol was acetylation and enzymatic hydrolysis of this ester, with enzymatic hydrolysis and chemical hydrolysis of the anreactionary ester. These enantiomerically enriched alcohols were converted in both enantiomers of (*E*)-6-methyl-2-hepten-4-ol, the aggregation pheromone of the American palm weevil (9) (Scheme 3).

Naoshima et al. showed lipase (Lipase PS from *Pseudomonas cepacia*, Amano Pharmaceutical Co.) catalyzed hydrolysis of the acetate of racemic alkan-2-ol and alkan-3-ol carried out in an acetone–water solvent system. This system led to (*R*)-alcohol with 80–95% ee at 26–35% yield and (*S*)-acetate with 32–69% ee at 40–64% yield. They discussed enantioselectivity of enzymatic reactions using the E-value (enantiomeric ratio) developed by Chen et al. (10). The reaction with a large E-value is more enantioselective. In these hydrolysis reactions in acetone–water solvents, E-values were larger than in other solvent systems (e.g., water and methanol, water and ethylene glycol, water and diethyl ether, and water and hexane). In the case of hydrolysis of dodecan-2-yl acetate, the E-value in the acetone–water solvent was 65 compared to 6–25 in other solvent systems. Similarly, in the case of hydrolysis of dodecan-3-yl acetate, the E-value in the acetone–water solvent was 11, whereas in other solvent systems it was 5. The two-step resolution gave enantiomerically enriched alcohols as well. The increased enantioselectivity observed in the acetone–water reaction system was utilized in the synthesis of optically active natural products possessing alkan-2-ol, alken-2-ol, or alken-3-ol skeletons, such as (2*R*,6*R*,10*R*)-6,10,14-trimethylpentadecan-2-ol (the sex

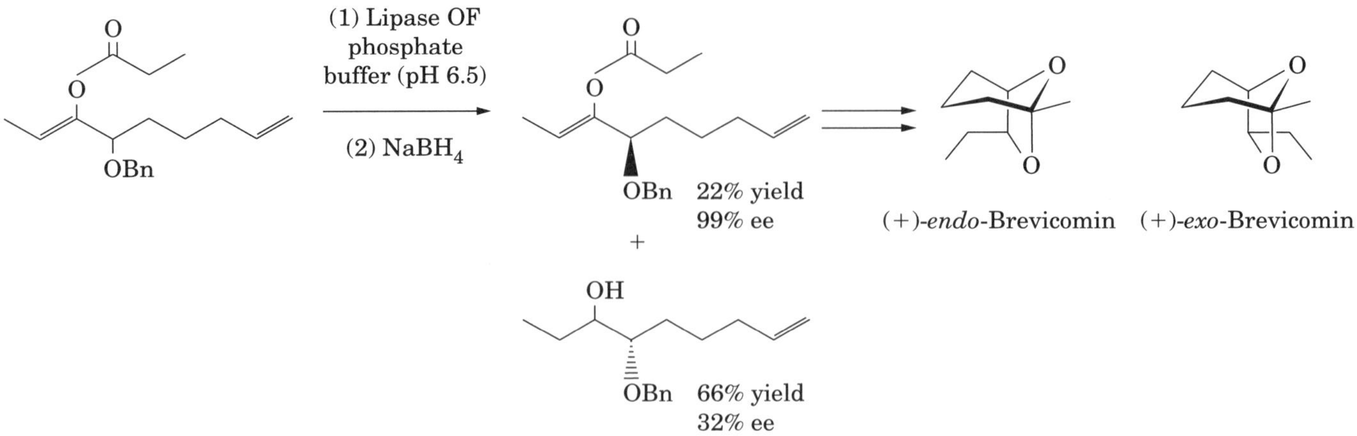

Scheme 1. Hydrolysis of an enol ester (7).

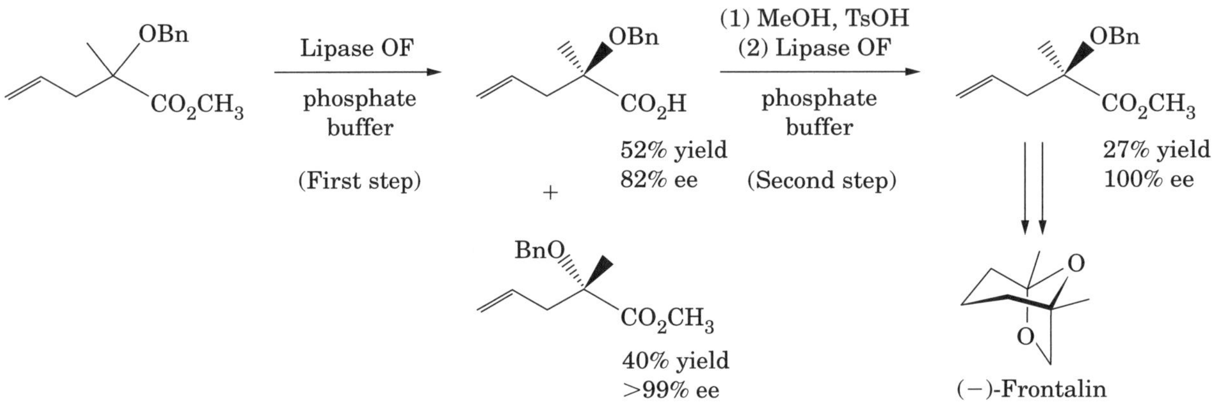

Scheme 2. Hydrolysis of α-benzyloxy acid ester (8).

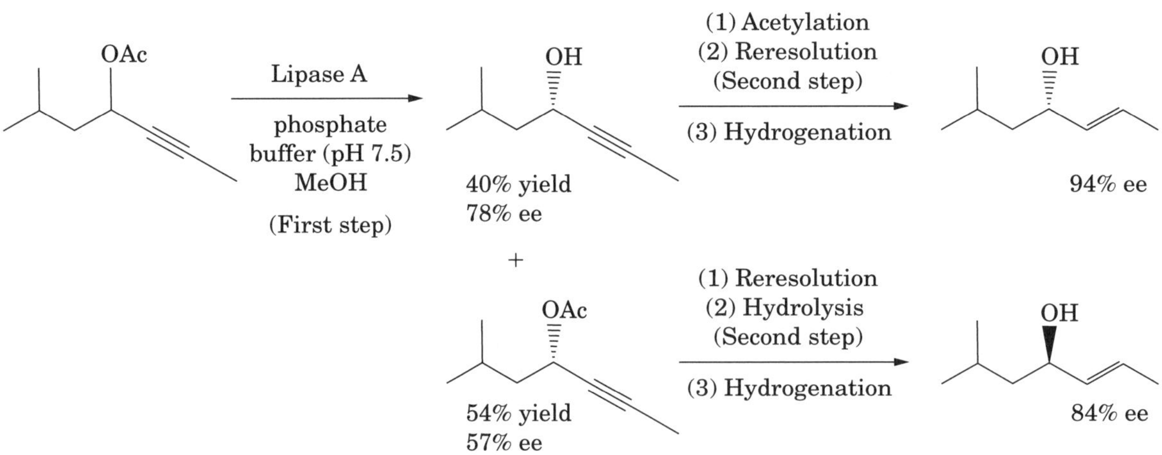

Scheme 3. Hydrolysis of acetyloxy acetylene (9).

pheromone of *Corcyra cephalonica*), (*S*)-sulcatol (the aggregation pheromone of ambrosia beetles), (*S,E*)-10-tridecen-2-yl acetate (the sex pheromone of the hessian fly), and 3-octanol (pheromone of ants) (11–13) (Scheme 4).

Though the optical resolution of racemic starting material is a convenient method for obtaining both enantiomers, the yield of the desired enantiomer cannot exceed 50%. To overcome the low yield, an elaborate procedure, the asymmetric hydrolysis of *meso* substrate, has been developed, an example of which follows. Grandjean et al (14) synthesized optically pure seaweed pheromones dictyopterene A and C′. A key intermediate in this synthesis is

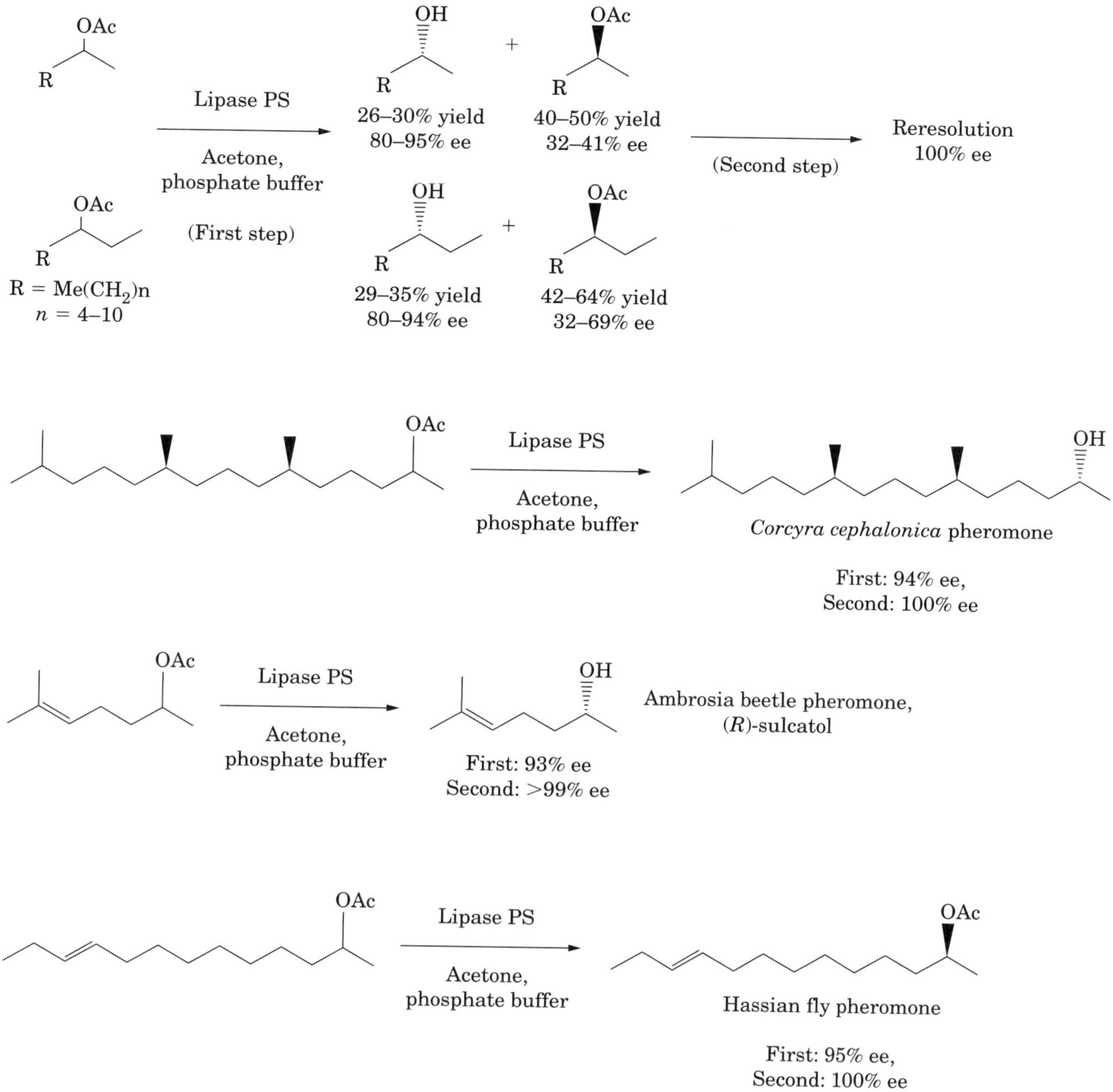

Scheme 4. Hydrolysis of 2- and 3-acetoxyhydrocarbons (11–13).

cis-(1*S*,2*R*)-1-hydroxymethyl-2-butyryloxymethylcyclopropane (>99% ee), which was easily obtained from the enantioselective hydrolysis of the *meso* diester, *cis*-1,2-bis(butyryloxymethyl)cyclopropane, with the pig pancreatic lipase (PPL, Sigma) in 95% yield (14) (Scheme 5).

PPL-catalyzed enantioselective hydrolysis of the *meso* epoxydiol diester *cis*-2,3-epoxybutan-1,4-yl diacetate yielded (2*S*,3*R*)-4-acetoxy-2,3-epoxybutan-1-ol. The yield of this enzymatic hydrolysis was 71% (90% ee). This enantiomerically enriched epoxy alcohol was converted into two naturally occurring epoxides, disparlure (gypsy moth pheromone) and ruby tiger moth pheromone (15) (Scheme 6).

Immobilized lipase was examined for easy separation of product from catalyst. Naoshima et al. synthesized (2*S*,6*S*,10*S*)-6,10,14-trimethylpentadecan-2-ol, the sex pheromone of *Corcyra cephalonica*, in high enantiomerically pure form by the two-step resolution of the 2,2,2-trichloroethyl carbonate of (2*RS*,6*S*,10*S*)-6,10,14-trimethylpentadecan-2-ol with immobilized *Pseudomonas fluorescens* lipase (Amano Pharmaceutical Co.) on glass beads. Catalyst and products were separated by filtration (16).

Other Enzymes as Catalysts for Enantioselective Hydrolysis

Baker's yeast was used as a catalyst for enantioselective hydrolysis. Glanzer et al. prepared (*R*)- and (*S*)-1-alkyn-3-

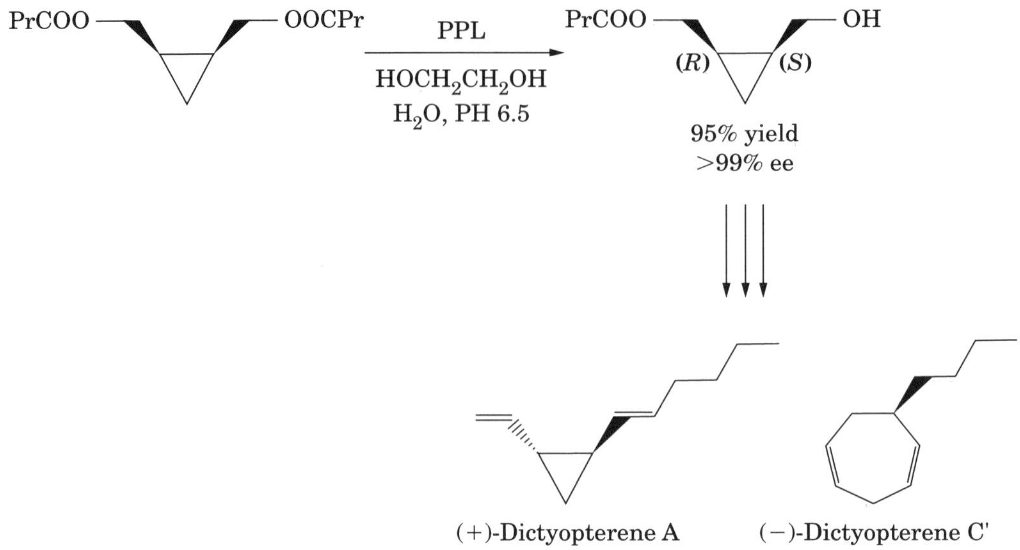

Scheme 5. Hydrolysis of *meso*-cyclopropanediol diester (14).

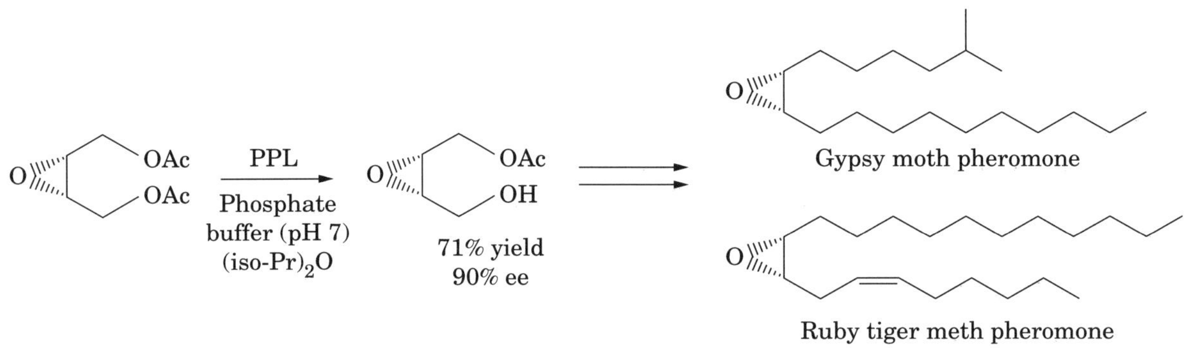

Scheme 6. Hydrolysis of *meso*-epoxydiol diester (15).

ols via enantioselective hydrolysis of their racemic esters. While lipase (AY30, SAM II, AK, P), fermenting microorganisms (*Rhizopus*), and freeze-dried microorganisms (*Brevibacterium*) failed in discriminating between enantiomers, lyophilized cells of baker's yeast gave (S)-alkyn-3-ols and their corresponding (R)-esters with >90% ee (17) (Scheme 7).

1-Benzyloxy-5-oxo-2-hepten-4-yl acetate was transformed to (4R,5S)-1-benzyloxy-2-hepten-4,5-diol in fermenting yeast. Prior to the reduction, the acetate was hydrolyzed. Although the high enantiomeric excess was observed for the diol, the hydroxy acetate arising from the direct reduction of acetoxy ketone was not in a diastereomerically or enantiomerically pure state. Thus, the diol arose from hydrolysis reduction. (4R,5S)-1-Benzyloxy-2-hepten-4,5-diol was transformed into (+)-*endo*-brevicomin with high optical purity (96.5% ee) (18) (Scheme 8).

In one particular case using microbial processes, hydrolysis and the opening of an epoxide occurred at the same time. Treatment of a diastereomeric mixture of 5,6-epoxy-6-methyl-2-heptyl pentanoate with crude immobilized enzyme preparation derived from *Rhodococcus* sp. (Novo SP 409) or whole lyophilized cells of *Rhodococcus erythropolis* (NCIB 11540) in buffer solution led to (2R,5R)-2-(1-hydroxy-1-methylethyl)-5-methyltetrahydrofuran (pityol, pheromone of the elm bark beetle) and its dia-

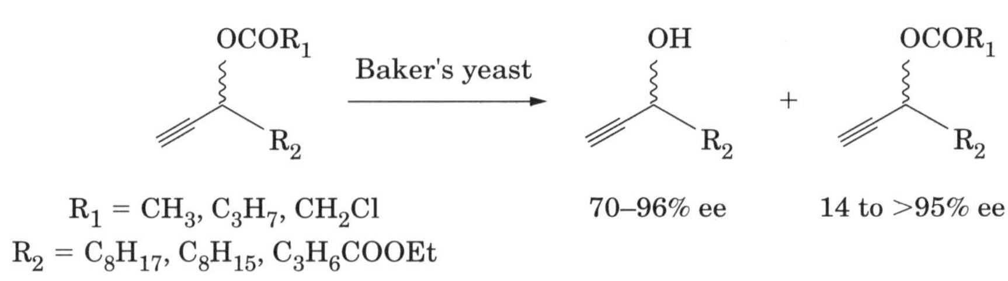

Scheme 7. Hydrolysis of 1-alkyn-3-yl ester using baker's yeast (17).

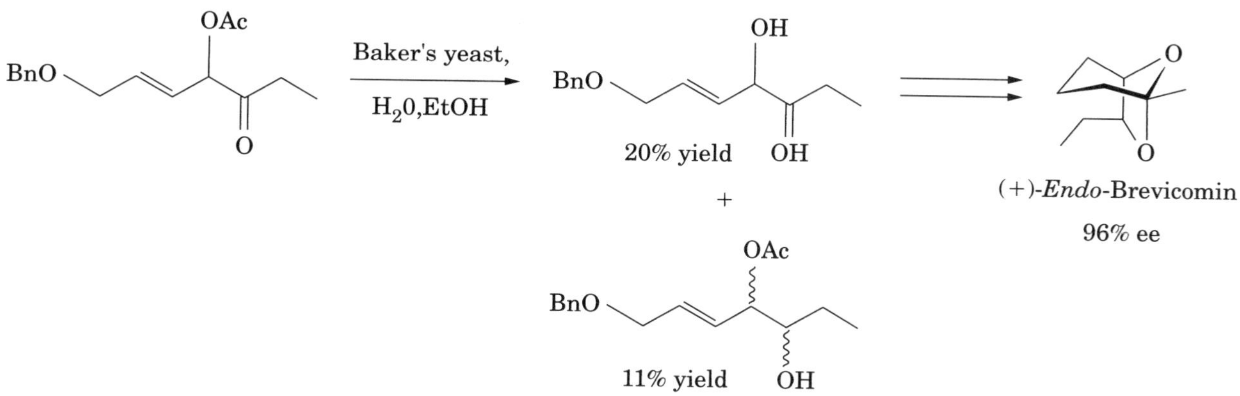

Scheme 8. Hydrolysis and reduction of ketoacetate using baker's yeast (18).

stereomer. Because pityol was not formed from 5,6-epoxy-6-methyl-2-heptanol in absence of a biocatalyst, it was concluded that this epoxide-opening reaction was triggered by an enzyme (19) (Scheme 9).

ESTERIFICATION AND TRANSESTERIFICATION REACTIONS

Lipase as Esterification and Transesterification Catalyst

In many cases, esterification and transesterification using lipase are performed in organic solvent (20). This has important implications for their application, as water is not the ideal reaction medium. In preparative organic synthesis, many compounds dissolve better in organic solvents than in water, and indeed some compounds are unstable in water. Lipase is uniquely stable in organic solvent. 12-Tetrahydropyranyloxy-3-methyldodecanol was resolved via a *Candida rugosa* lipase-catalyzed acetylation with vinyl acetate in diisopropyl ether. (*S*)-enantiomer was acetylated at >96% ee. Some chiral insect pheromones were synthesized from this material (21) (Scheme 10).

An intramolecular version of esterification and transesterification produces lactones. For example, Sugar et al. (22) synthesized the Japanese beetle pheromone (*R*,*Z*)-5-(1-decenyl) oxacyclopentan-2-one by means of the enzymatic enantioselective lactonization of racemic methyl 4-hydroxy-5-tetradecynoate, using porcine pancreatic lipase (PPL) in polar solvent (diethyl ether) at a somewhat low temperature (10–15 °C). The desired (*R*)-lactone was obtained (43% yield, 82% ee). After repetition of the enzymatic lactonization, as shown in Scheme 11, lactone with higher enantiomeric excess (97%) was obtained. The natural enantiomer of Japanese beetle pheromone was prepared from this compound by semihydrogenation.

Fukusaki et al (23) also synthesized the Japanese beetle pheromone by lipase-catalyzed enantioselective acylation and lactonization. An acylation of methyl 4-hydroxy-5-tetradecynoate with succinic anhydride was accomplished by Lipase PS (from *Pseudomonas* sp., Amano Pharmaceutical Co.) in diisopropyl ether. This partially enantiomerically enriched diester was subsequently converted to (*R*)-5-(1-decynyl)oxacyclopentan-2-one (over 99% ee) by lipase-catalyzed enantioselective lactonization with Lipase OF (from *Candida cylindracea*, Meito Sangyo Co.) in phosphate buffer (23). It should be emphasized that even in aqueous solution, a lactonization (not hydrolysis) took place. The sex pheromone of the cupreous chafer beetle, (*R*,*Z*)-5-(1-octeny)oxacyclopentan-2-one, was synthesized in a similar manner. Using lipase-catalyzed enantioselective acylation by Lipase PS (Amano Pharmaceutical Co.) in diisopropyl ether, (*R*)-enantiomer of methyl 4-hydroxy-

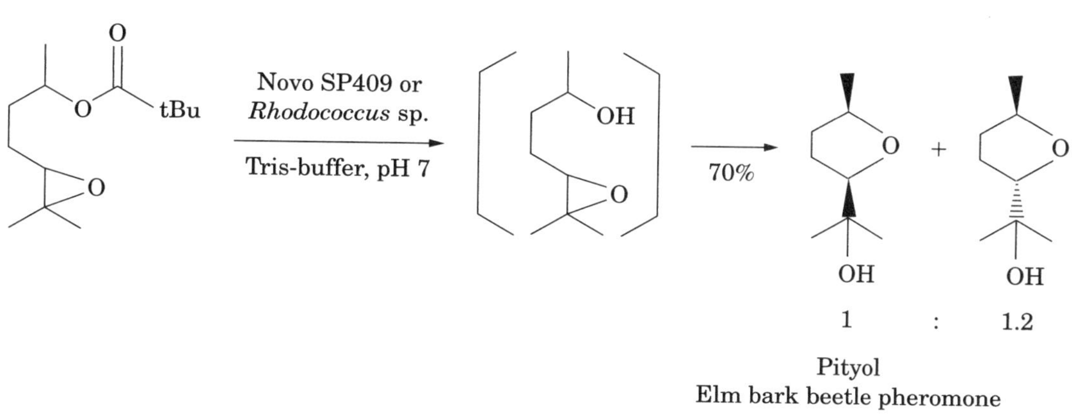

Scheme 9. Enzyme-triggered opening of epoxide (19).

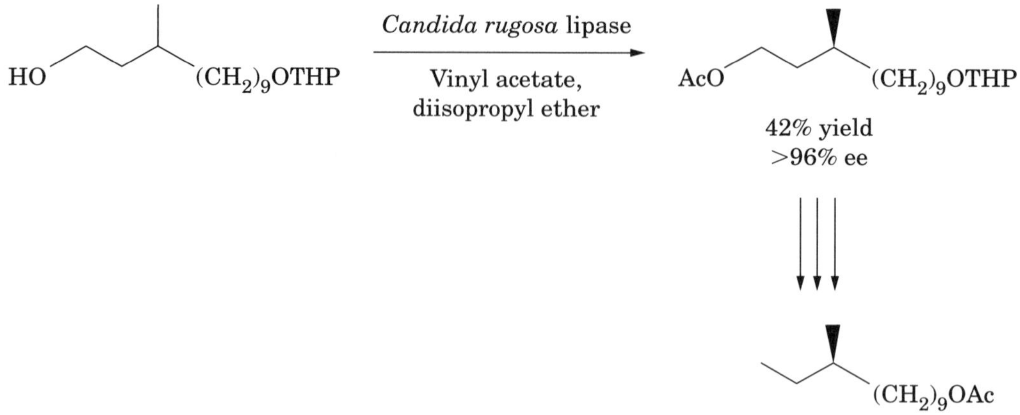

Scheme 10. Transesterification using lipase (21).

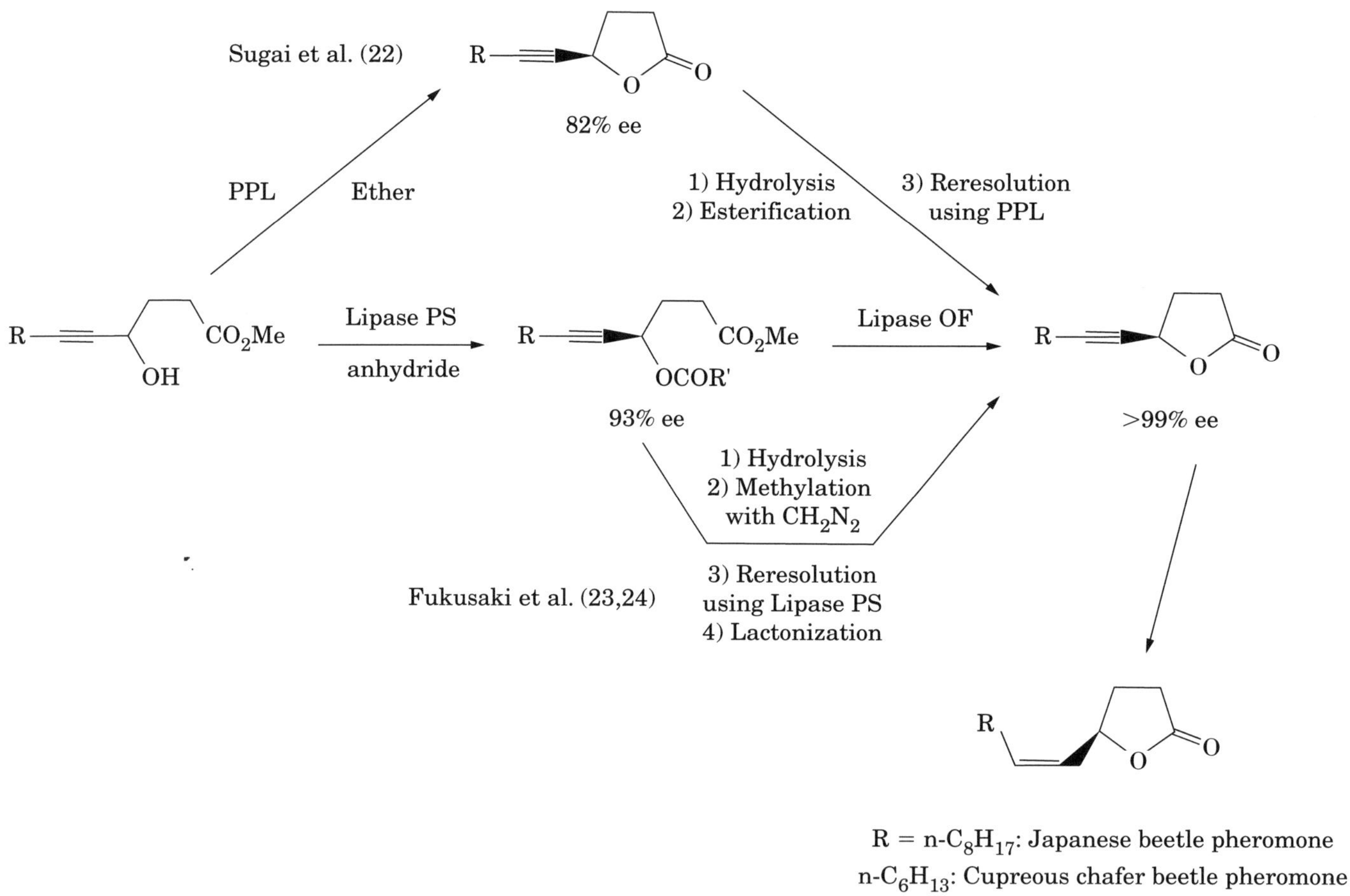

Scheme 11. Lactonization using lipase.

5-dodecynoate was preferentially converted to give (*R*)-acyloxy ester (93% ee). Further enantiomeric purification was carried out by repetition of the enzymatic reaction (24). Both Sugai et al. and Fukusaki et al. employed E-values to evaluate the enantioselectivity of the enzymatic reaction. The enantioselectivity of lactonization by lipase was not so high. The E-value of the reaction obtained by Sugai was 20, and that by Fukusaki was 39. However, the enantioselectivity of acylation by lipase was high. The E-value of an acylation of methyl 4-hydroxy-5-tetradecynoate with succinic anhydride by Lipase PS was 115 (Scheme 11).

Henkel et al. (25) synthesized (5*R*,6*S*) and (5*S*,6*R*)-6-acetoxyhexadecan-5-olide (the major component of mosquito oviposision attractant pheromone) using an enantioselective lipase-catalyzed lactonization and the subsequent acetylation with vinyl acetate (Lipase SP 382 and SP 526, Novo Nordisk A/S). Acetylation was achieved at high temperatures (55–70 °C). In this case, the enantioselectivity of lactonization was also low and the selectivity of acetylation was comparatively high (E-value of 53) (Scheme 12).

Enzymatic resolution using γ- and δ-hydroxysulfone for syntheses of γ- and δ-lactone was attempted, rather than using hydroxycarbonic acid or ester. Jacobs et al. (26) demonstrated the resolution of racemic hydroxysulfone using a lipase (Lipase PS-30). *R*-(+)-4-hexanolide (the sex pheromone of the female dermestid beetle *Trogoderma glabrum*) and (2*R*,5*S*)-2-methyl-5-hexanolide (the sex pheromone of the carpenter bee) were synthesized via an intramolecular acylation of the carbonates derived from these sulfone intermediates, which provided a new and simple route for syntheses of chiral γ- and δ-lactones. However, the enantioselectivity of this enzymatic reaction was not very high (E-values of 32 for 3-hydroxysulfone and 27 for 4-hydroxysulfone) (Scheme 13).

Enzymatic reactions were also efficiently combined with chemically mediated diastereomeric control to set up the enantiomerically enriched building blocks with defined multiple chiral centers. Kim et al. (27) efficiently resolved *anti*-1,2-diol by using lipase PS-catalyzed transesterification and achieving the total synthesis of the bark beetle pheromone (+)- and (−)-*endo*-brevicomin, starting from resolved diols.

Enzymatic differentiation of enantiotopic groups in *meso* compounds (asymmetrization or desymmetrization) provided enantiomerically pure compounds with multiple stereogenic centers in high yield. Mori et al. (28) synthesized (2*R*,3*S*)-2,3-(isopropylidendioxy)-4-hydroxypentan-1-yl acetate, the key intermediate for the synthesis of (+)-*endo*-brevicomin. *Meso*-2,3-isopropylidendioxy-1,4-butanediol was converted to the optically active monoacetate with vinyl acetate and lipase AK (from *Pseudomonas*, Amano Pharmaceutical Co.) in high chemical and optical yield (96% yield, 98.5% ee). The enzymatic esterification of *meso*-2,6-dimethyl-1,7-heptanediol with isopropenyl acetate in the presence of *Pseudomonas cepacia* lipase (Lipase PS30, Amano Pharmaceutical Co.) in organic medium provided the enantiomerically enriched monoester with high enantiomeric excess. In this case, the chemical yield was not very high, but it was more than 50%. This compound was used as the synthetic intermediate of the insect pheromone tribolure (29) (Scheme 14).

To solve the problem of unsatisfactory enantiomeric excess of the product in the course of a kinetic resolution, a sequential use of hydrolysis and acylation with lipase has been demonstrated in order to secure further enantiomerically enriched products. This is one pattern of a two-step resolution. Naoshima et al. (30) synthesized (*E*)-1-ethyl-5-methyl-4-heptenyl acetate, the aggregation pheromone of *Cathartus quadricollis* (square-necked grain beetle) using a two-step lipase reaction. The first reaction was hydrolysis of (*E*)-1-ethyl-5-methyl-4-heptenyl acetate with Lipase PS in a phosphate buffer and acetone. The second reaction was transesterification of the product, (1*R*,4*E*)-alcohol (98% ee) with porcine pancreas lipase

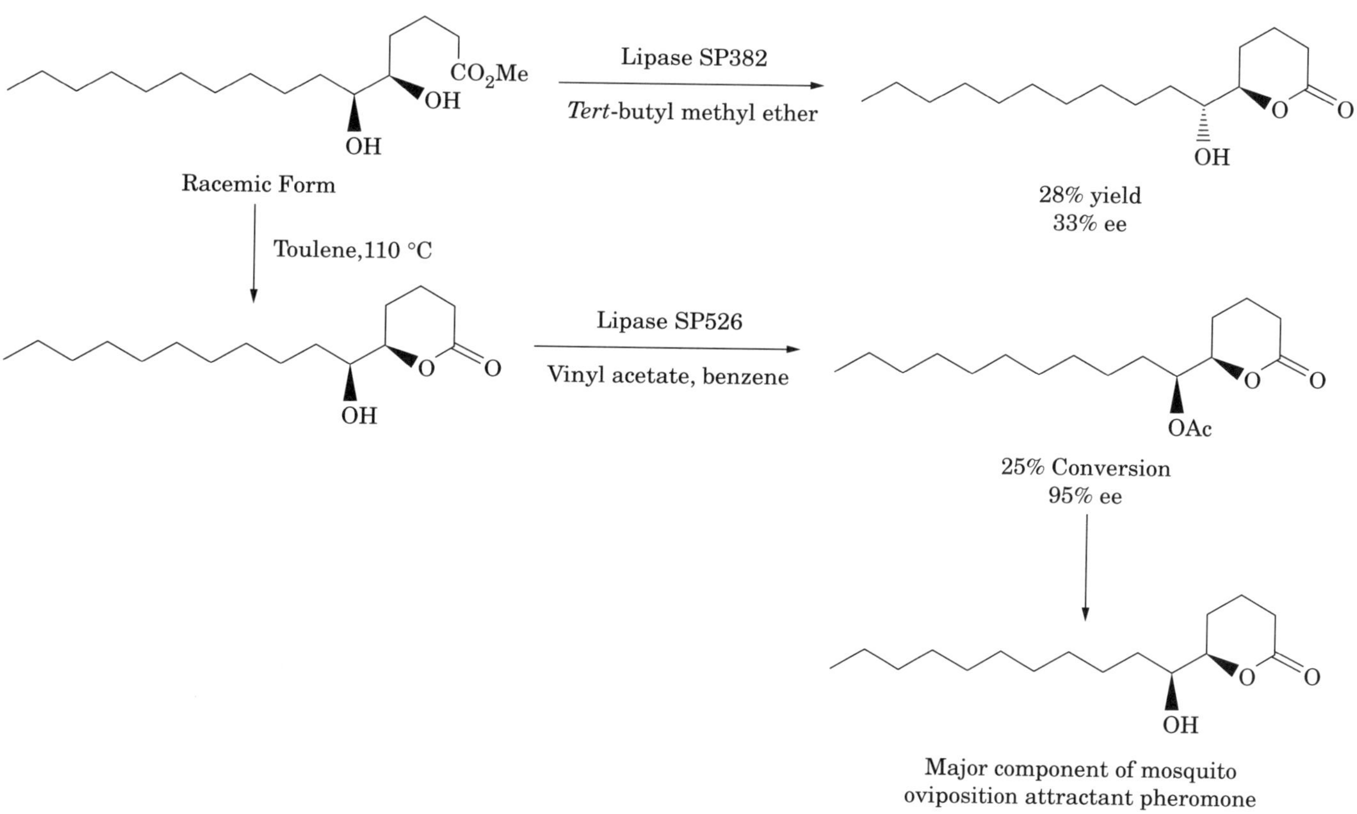

Scheme 12. δ-Lactonization using lipase (25).

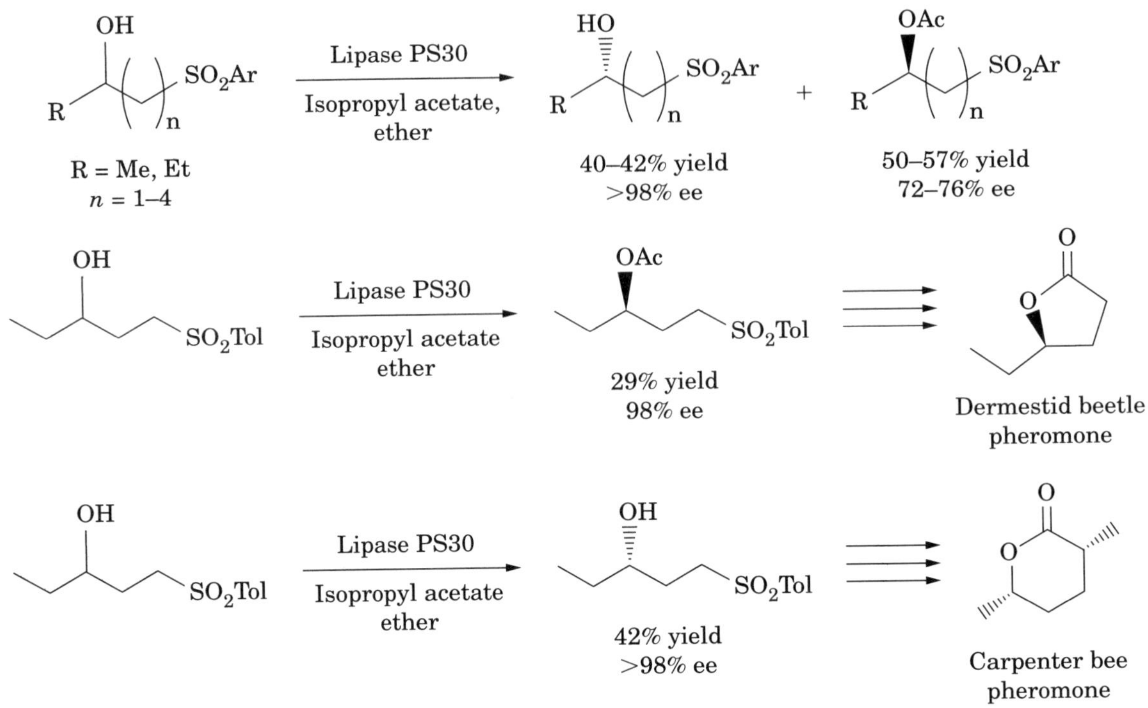

Scheme 13. Enzymatic resolution of hydroxysulfones (26).

(PPL, Sigma) in isopropenyl acetate and ether (30). The second example is also shown in Scheme 15. Fukusaki et al. (31) achieved a yield of enantiomerically pure epoxy alcohol using lipase-catalyzed acylation and hydrolysis (Pancreatin F, Amano Pharmaceutical Co.). In this case, the same lipase immobilized on celite was used in both lipase-catalyzed reactions. These reactions were performed at 5 °C. However, the enantioselectivity of these reactions was not very high: The E-value was 23 for acylation and 20 for hydrolysis. Hence, the recrystallization of corresponding 3,5-dinitrobenzoate of this epoxy alcohol was performed to get an enantiomerically enriched epoxy alcohol. This is the intermediate of disparlure (the gypsy moth pheromone).

(+)-*Exo*, (+)-*endo*-brevicomin, and (−)-frontalin were synthesized via lipase-catalyzed acylation of oxygen heterocyclic structure, 2-(1-hydroxypropyl)-6-methyl-3,4-dihydro-2H-pyran and 2-hydroxymethyl-2,6-dimethyl-3,4-dihydro-2H-pyran. Three enzymatic kinetic resolutions were combined in frontalin synthesis to acquire a high enantiomeric excess of the product. The first reaction was hydrolysis of 2-methoxycarbonyl-2,6-dimethyl-3,4-dihydro-2H-pyran with protease; the second reaction was transesterification with Lipase MY (Meitou Sangyo Co.), and the third reaction was hydrolysis with Lipase P. As a result, enantiomeric excess of the pheromone reached 91% (32) (Scheme 15).

REDUCTIONS

Baker's Yeast as a Catalyst for the Reduction

The reduction of ketones is one of the most established microbial processes and has been extensively investigated since the 1960s, when pioneering studies revealed the stereochemistry of the reduction in a general sense (33,34). Alcohol with a high enantiomeric excess can be obtained using this method.

Acetoacetylated Meldrum's acid (1,3-dioxan-4,6-dione) was found to be reduced enantioselectively (>99% ee) by baker's yeast to the corresponding (*S*)-alcohol. Sato et al. (35) converted this (*S*)-alcohol to *cis*-3,6-dimethyl tetrahydropyran-2-one, a major component of pheromonal secretion of the male carpenter bee (*Xylocopa*).

3-Oxo-6-heptenoic acid salt was reduced to the enantiomerically enriched hydroxy acid (>99% ee) by baker's yeast as a key step. This enantiomerically enriched hydroxy acid was further converted to (*S*)-4-hexanolide, the pheromone antipode of the store grain beetle (36) (Scheme 16).

(*S*)-4-(Phenylthio)-2-pentanol was sufficiently produced with high enantiomeric excess by enzymatic reduction of the corresponding ketone using baker's yeast. Using this chiral compound, highly enantiomerically enriched insect pheromonal compounds having spiroketal structures were synthesized (37) (Scheme 17).

(2*R*,5*S*)-Pityol (the pheromone of the bark beetle, *Pityophthorus pityographus*) was synthesized using two highly stereoselective biocatalytic reactions. 6-Methyl-2-oxo-5-heptene was reduced to (*S*)-alcohol (sulcatol, the pheromone of *Gnathotrichus sulcatus*) (98.5% ee) by baker's yeast. (*S*)-Sulcatol was converted to urethane by way of phenylisocyanate, and urethane then underwent epoxydation by the fungus *Aspergillus niger*. Treatment of this epoxide major product with sodium hydroxide led to formation of (2*R*,5*S*)-pityol (100% ee) (38) (Scheme 18).

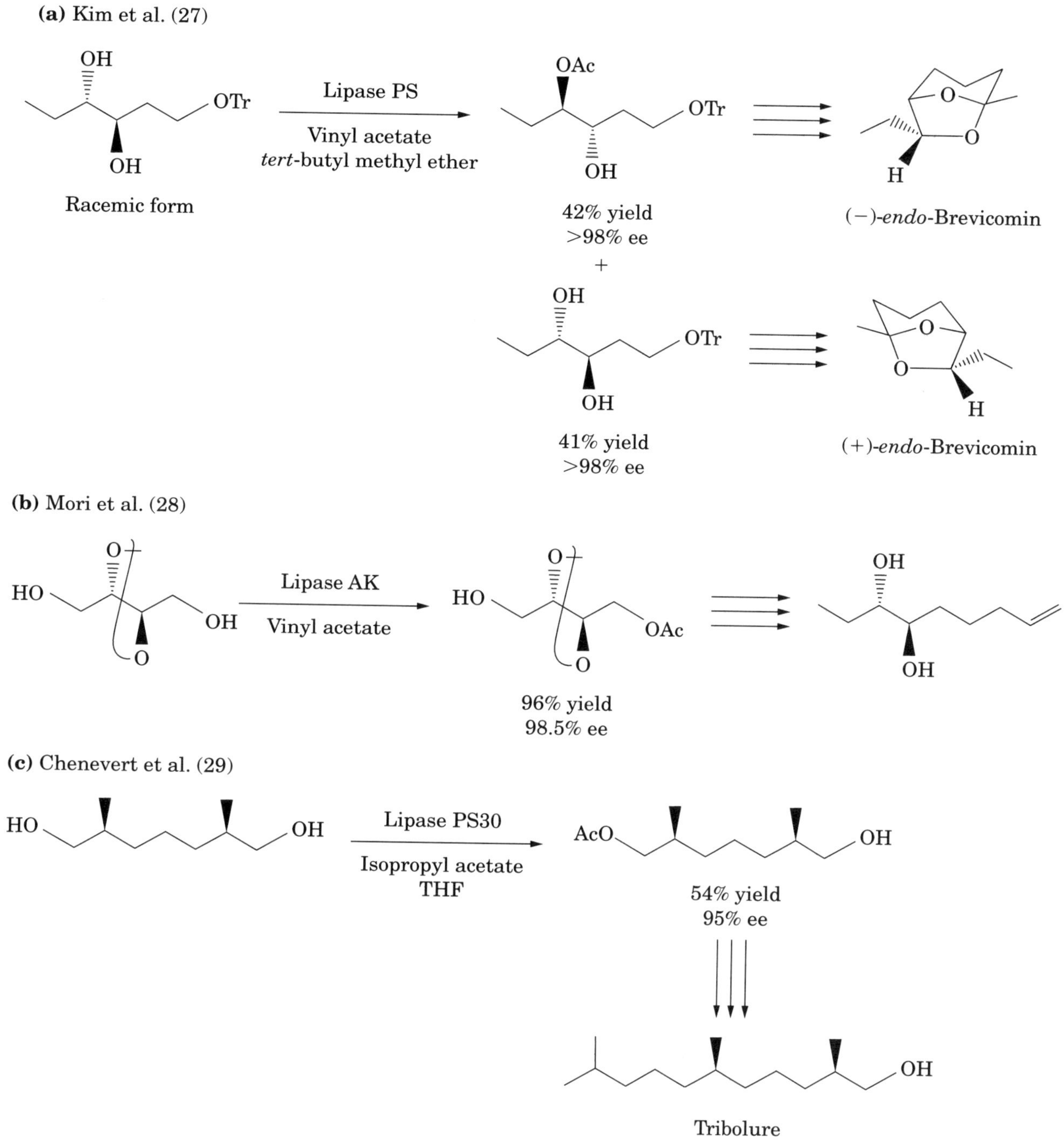

Scheme 14. Enzymatic resolution of the compounds with multistereogenic center.

Other Enzymes as Catalysts for the Reduction

A series of studies using *Thermoanaerobium brockii* alcohol dehydrogenase (TBADH) have been reported. Highly enantioselective reduction of various methyl- and ethylketones having a wide range of functional groups, such as methyl ester, cyano, ethyl ether, phenyl, and chloride as well as double and triple bonds, is carried out using TBADH as a catalyst. Reduction of most of the substrates yields alcohol with an (*S*)-configuration, arising from highly selective hydride attack at the *re* face of the carbonyl groups. (*S*)-(+)-(*Z*)-Tetradec-5-en-13-olide, one of several synergistic aggregation pheromones produced by the male flat grain beetle was prepared from (*S*)-(+)-methyl-8-hydroxynonanoate with a high enantiomeric excess (>99%) in a six-step synthesis (39).

(*S*)-(+)-Methyl 8-hydroxynonanoate, which was obtained via enantioselective reduction of methyl 8-oxononanoate with TBADH, was converted to (*S*)-(+)-ferrulactone II, the aggregation pheromone of the flat grain beetle (40).

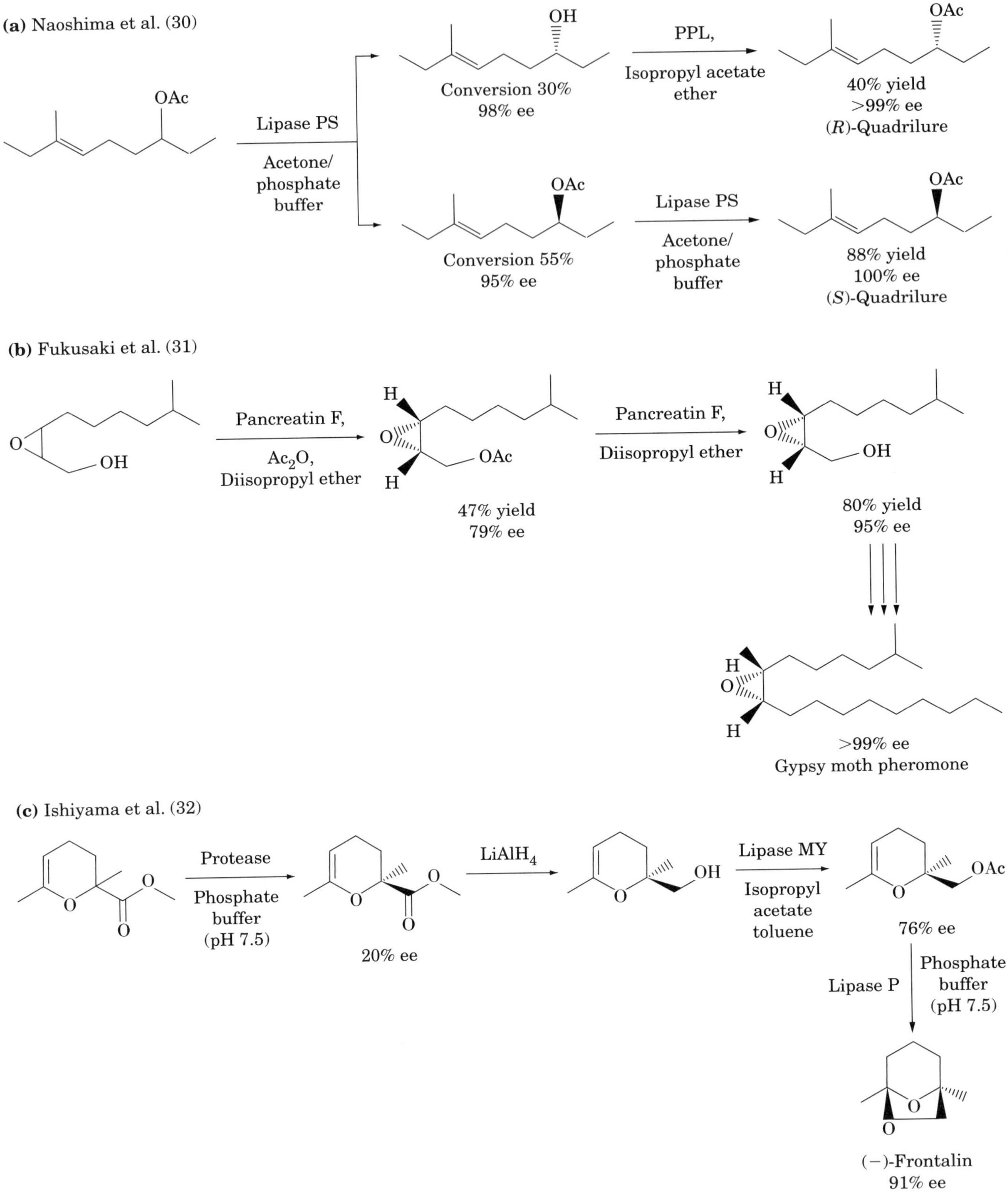

Scheme 15. Hydrolysis and esterification in one synthesis.

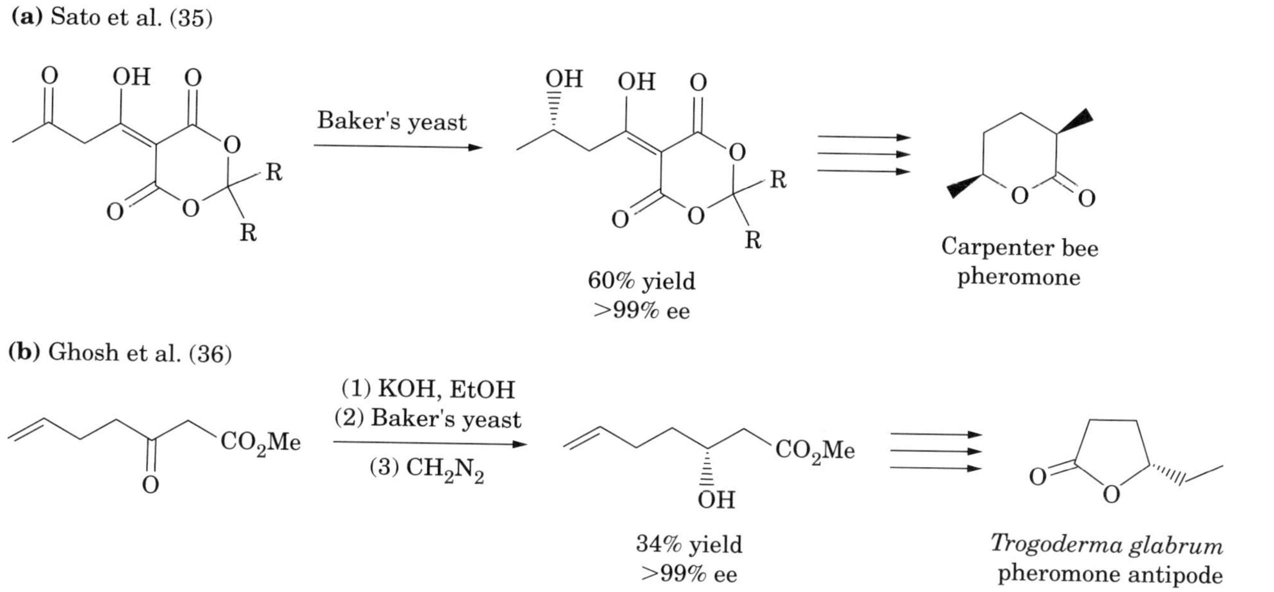

Scheme 16. Reduction using baker's yeast.

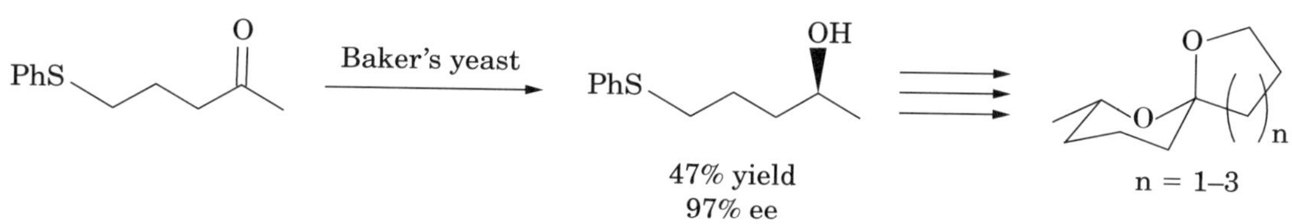

Scheme 17. Reduction of ketosulfide using baker's yeast (37).

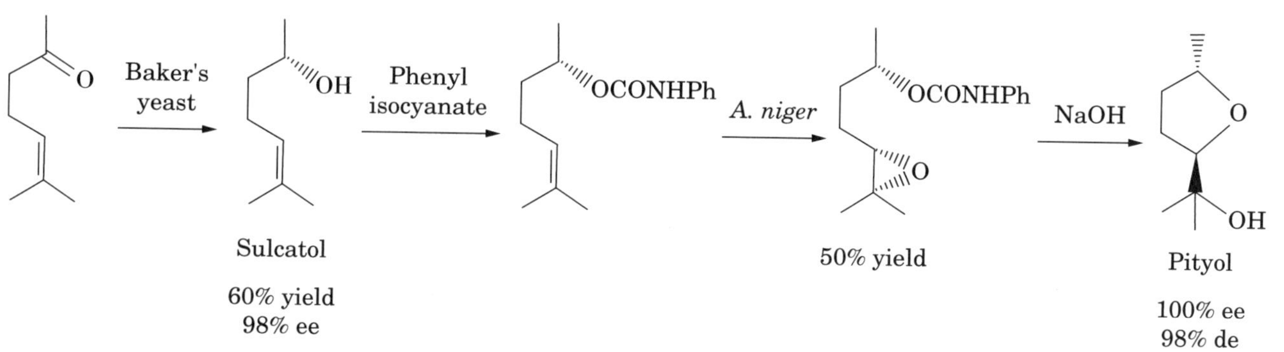

Scheme 18. Reduction and epoxydation using microbiological methods (38).

(2*S*,8*S*)-2,8-Dihydroxynonane, which was characterized by synthetically useful C2 symmetry, was also obtained by the TBADH-catalyzed reaction of nonan-2,8-diol. Keinan et al. (41) synthesized four isomers of 8-methyldec-2-yl propanoate (the western corn rootworm sex pheromone) using this single chiral building block (Scheme 19).

OXIDATION REACTIONS

Microbial Baeyer–Villiger oxidations are important reactions in providing enantiomerically enriched lactones. Asymmetric synthesis of 5-hexadecanolide (oriental hornet pheromone) using a biological Baeyer–Villiger reaction was achieved by two strains of *Acinetobacter*. This allows a three-step synthesis of (*S*)-(−)-5-hexadecanolide, as well as direct access to (*R*)-(+)-2-undecylcyclopentanone, a precursor for chemical synthesis of (*R*)-(+)-5-hexadecanolide. These products were obtained with optical purities of 74 and 95% (42).

Lebreton et al. (43) performed novel microbiological Baeyer–Villiger oxidation using fungi. Racemic cyclobutanone derivatives were oxidized to give an enantiomerically enriched γ-lactone with high enantiomeric excess us-

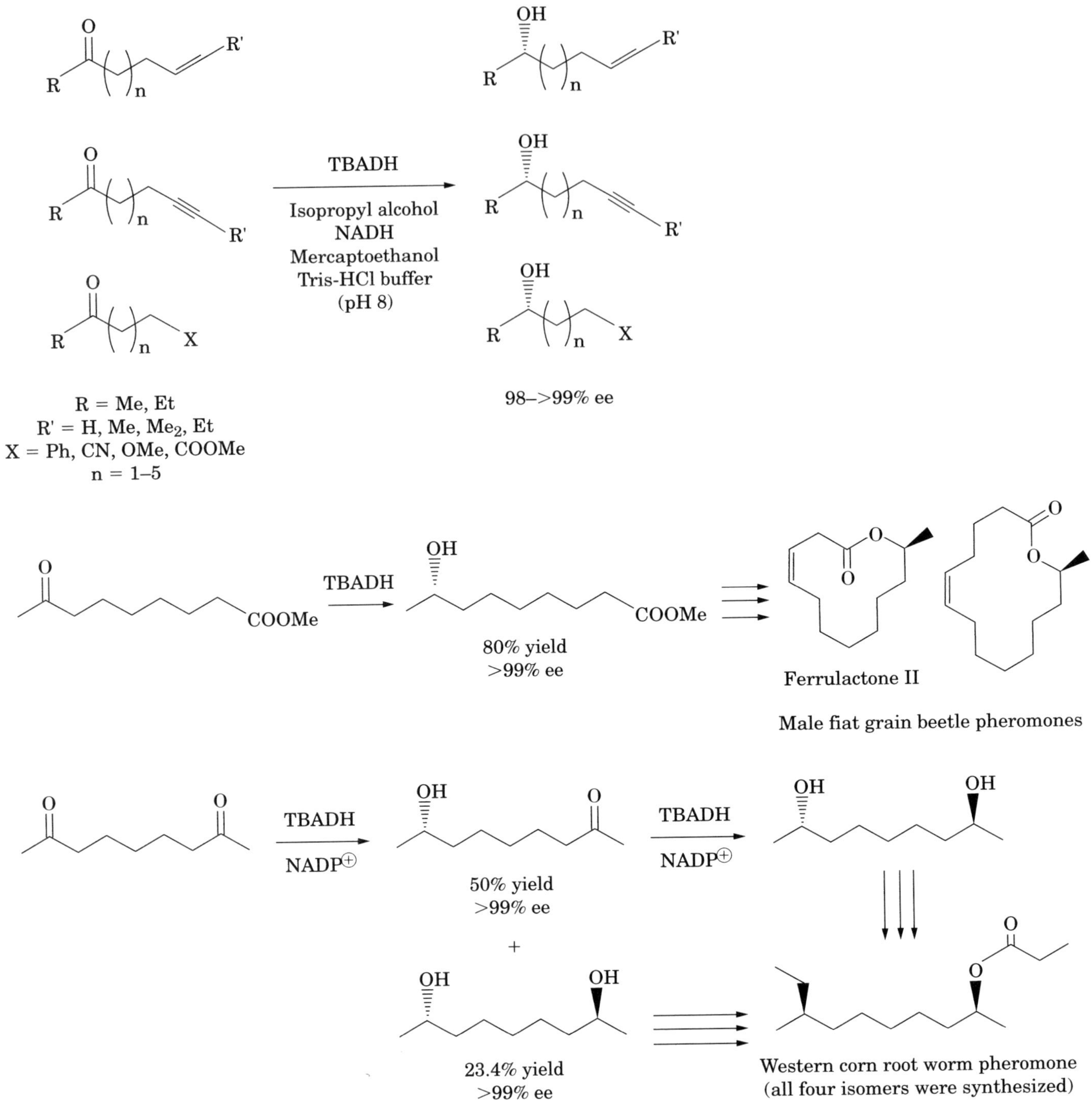

Scheme 19. Reduction using TBADH (39–41).

ing *Cunninghamella echinurata*. They synthesized (+)-multifidine and (+)-viridiene, the major components of a brown algae pheromone (Scheme 20).

C—C BOND FORMATION REACTIONS

Two enzymatic C—C bond formation reactions of carbohydrate-related compounds were demonstrated in pheromone synthesis. The first example is fructose 1,6-diphosphate (FDP) aldolase from rabbit mussle (RAMA). This aldolase was used in the synthesis of (+)-(1*S*,5*R*,7*S*)-*exo*-brevicomin. This enzyme accepts 5-oxohexanal, as well as the corresponding 5-dithioacetal, as its substrate, and catalyzes the stereospecific aldol reaction of dihydroxyacetone phosphate (DHAP) and aldehyde (80–90% yield). After removal of the phosphate group from the aldol adduct, the product was converted into the desired bicyclic system. (+)-*exo*-Brevicomin was obtained by a deoxygenation of the side chain (44).

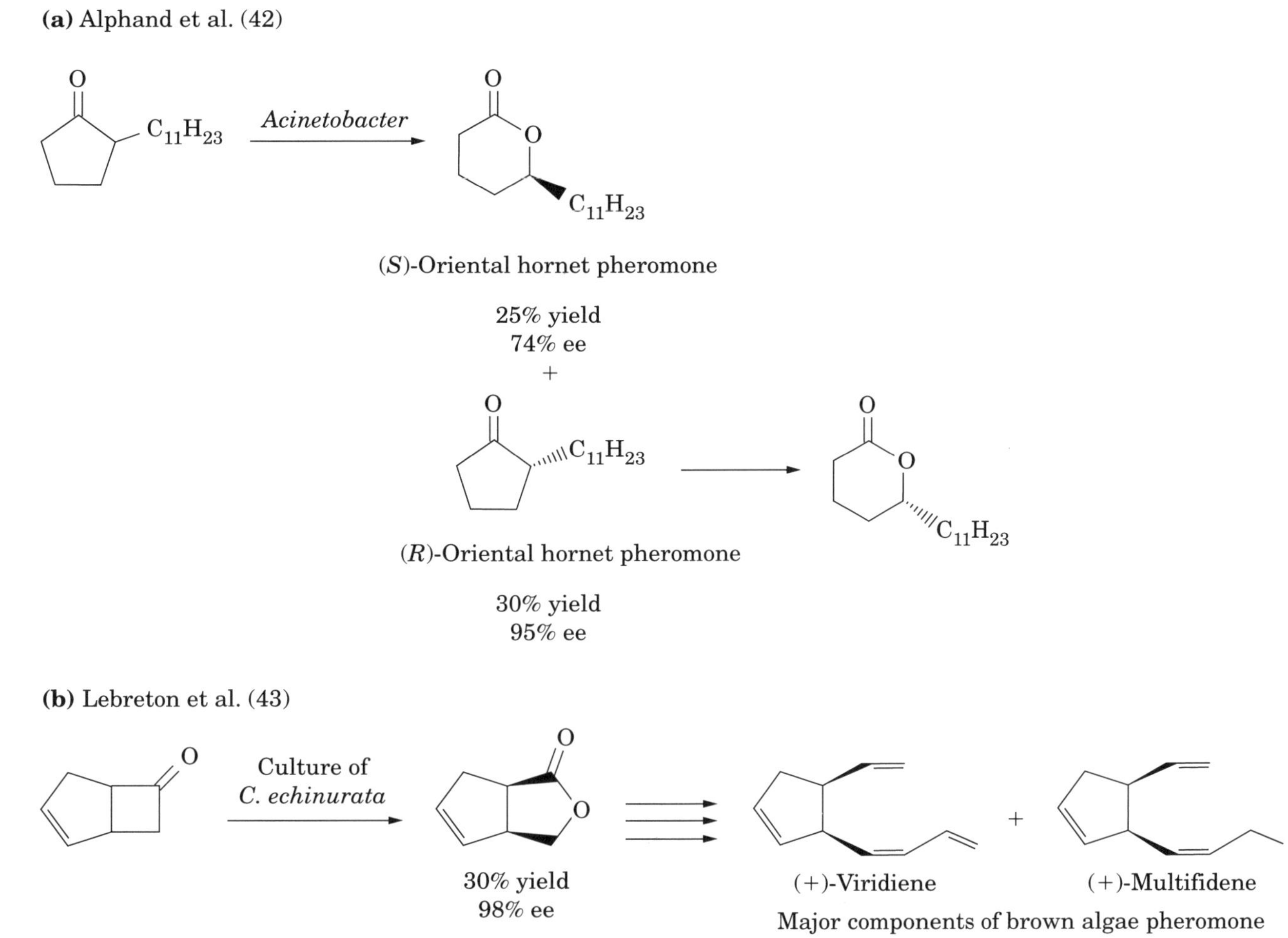

Scheme 20. Microbiological Baeyer–Villiger oxidation.

Myles et al (45) reported that the commercially available enzyme preparation of transketolase catalyzed the condensation of β-hydroxypyruvic acid and 2-hydroxybutylaldehyde to furnish the enantiomerically enriched dihydroxyketone with vicinal diol of *D-threo* configuration. The remarkable advantage in this is the stereoselective formation of vicinal diols of a desirable configuration, even when starting from a racemic 2-hydroxyaldehyde. Indeed, transketolase accepts only the *D*-enantiomer of 2-hydroxyaldehyde and produces the *threo* isomer of the product with very high diastereoselectivity. This dihydroxyketone with *D-threo* configuration was transformed into (+)-*exo*-brevicomin in six steps (Scheme 21).

CONCLUSION

In this article we have introduced about 35 chiral pheromone syntheses using microbial processes. Many microbial processes have been developed in this field, and some of these are already in use. Combination of organic syntheses with microbial processes will provide efficient routes for the enantioselective synthesis of bioactive natural products.

Materials of natural product origin are believed to be more benign to the environment than artificially synthesized compounds. As such, the social demand for reducing artificial insecticide use to protect the environment has promoted development of the use of bioactive natural products, as well as products with structures similar to those of natural products for use as insecticides. It has also promoted development of synthesis technologies of these compounds. Improved technology using microbial reactions allows these chiral compounds to be produced with high chemical and optical purity at a lower price, and clears the way to utilize naturally derived materials as biological insecticides. These advanced insecticides, like synthetic chiral pesticides and pheromones (typical pheromones in this article are listed in the next section), reduce impact on the environment and improve pest control systems.

We did not discuss chiral pyrethroid synthesis in this article, although practical preparations of chiral pyrethroid intermediates using microbial processes have been studied and used in industrial processes (46,47).

PHEROMONE LIST

Ambrosia beetle (*Gnathotrichus sulcatus*) pheromone[sulcatol] (11,38)

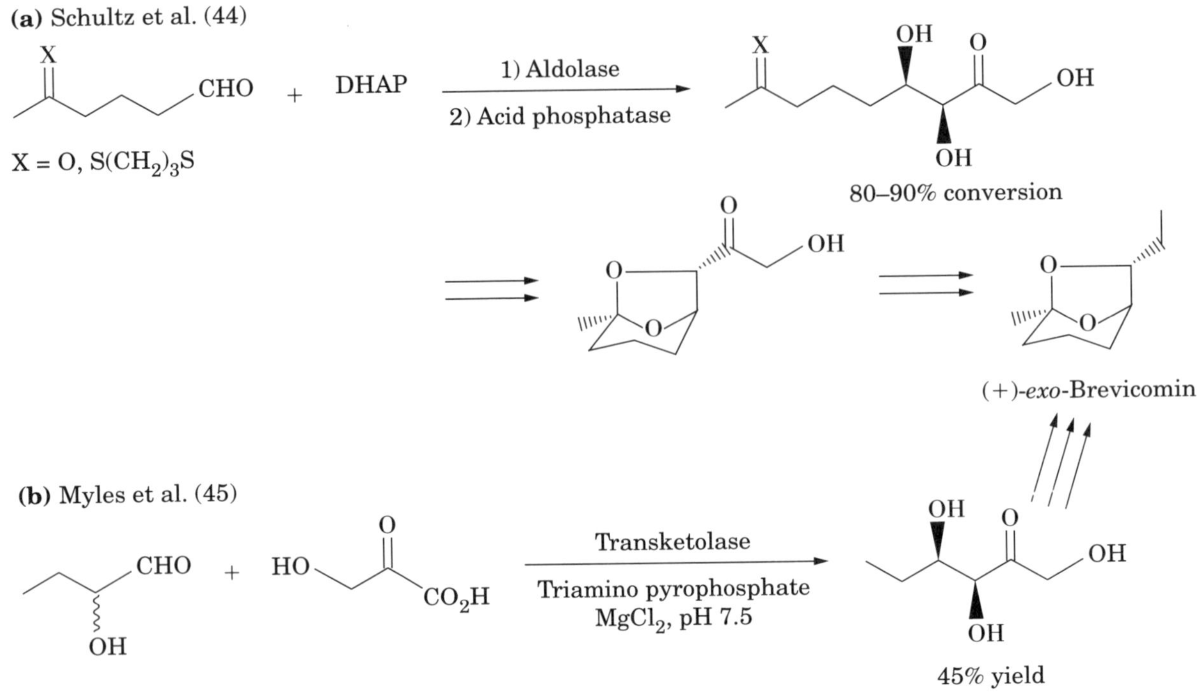

Scheme 21. Enzyme-mediated C—C bond formation.

American palm weevil (*Rhynchophorus palmarum*) pheromone (9)

Bark beetle pheromone (*Dryocoetes autographus, Dendroctonus frontalis*) [endo-brevicomin] (7,18,27,28,32)

Bark beetle pheromone (*Dendroctonus bremicomis*) [exo-brevicomin] (7,32,44,45)

Brown algae pheromone (*Cutleria mulutifida, Chorda tomentosa*) [multifidene] (43)

Brown algae pheromone (*Desmarestia aculeata* and *viridis*) [viridiene] (43)

Carpenter bee (*Xylocopa hirutissima*) pheromone (26,35)

Cupreous chafer beetle (*Anomala cuprea*) pheromone (24)

Dermestid beetle (*Trogoderma glabrum*) pheromone (26)

Elm bark beetle (*Pityophthorus pityographus*) pheromone [pityol] (19,38)

Flat grain beetle (*Cryptolestes ferrugineus*) pheromone [ferrulactone] (39,40)

Flour beetle (*Tribolium confusum* and *castaneum*) pheromone [tribolure] (29)

Gypsy moth pheromone (*Lymantria dispar*) [disparlure] (15,31)

Hessian fly (*Mayetiola destructor*) pheromone (12)

Japanese beetle (*Popillia japonica*) pheromone [japonilure] (22,23)

Oriental hornet (*Vespa orientalis*) pheromone (42)

Ruby tiger moth (*Phragmatobia fuliginosa*) pheromone (15)

Mosquito (*Culex pipients fatigans*) pheromone (25)

Seaweed pheromone [dictyopterene A, C′] (14)

Southern pine beetle (*Dendroctonus frontalis*) pheromone [frontalin] (8,32)

Square-necked

Square-necked grain beetle (*Cathartus quadricollis*) pheromone [quadrilure] (30)

Store grain beetle (*Trogoderma glabrum*) pheromone (36)

Western corn rootworm (*Diabrotica virgifera*) pheromone (41)

BIBLIOGRAPHY

1. K. Mori, in J. ApSimon ed., *The Total Synthesis of Natural Products*, vol. 9, Wiley, New York, 1992, pp. 1–534.
2. J.B. Jones, *Tetrehedron*, **42**, 3351–3403 (1986).
3. K. Faber, *Biotransformations in Organic Chemistry*, 3rd ed., Springer-Verlag, Berlin, 1997.
4. C.-H. Wong and G.M. Whitesides, *Enzymes in Synthetic Organic Chemistry*, Pergamon, Oxford, 1994.
5. K. Drauz and H. Waldmann eds., *Enzyme Catalysis in Organic Synthesis: A Comprehensive Handbook*, VCH, Weinheim, Germany, 1995.
6. K. Mori, *Synlett*, 1097–1109 (1995).
7. K. Matsumoto, N. Suzuki, and H. Ohta, *Tetrahedron Lett.* **31**, 7163–7166 (1990).
8. T. Sugai, H. Kakeya, and H. Ohta, *J. Org. Chem.* **55**, 4643–4647 (1990).
9. K. Mori and K. Ishigami, *Liebigs Ann. Chem.*, 1195–1198 (1992).

10. C.S. Chen, Y. Fujimoto, G. Girdankas, and C.J. Sih, *J. Am. Chem. Soc.* **104**, 7294–7299 (1982).
11. Y. Naoshima, M. Kamezawa, H. Tachibana, Y. Munakata, T. Fujita, K. Kihara, and T. Raku, *J. Chem. Soc. Perkin Trans. 1*, 557–561 (1993).
12. M. Kamezawa, H. Tachibana, T. Ohtani, and Y. Naoshima, *Biosci. Biotechnol. Biochem.* **57**, 1962–1963 (1993).
13. M. Kamezawa, T. Raku, H. Tachibana, T. Ohtani, and Y. Naoshima, *Biosci. Biotechnol. Biochem.* **58**, 598–599 (1994).
14. D. Grandjean, P. Pale, and J. Chuche, *Tetrahedron* **47**, 1215–1230 (1991).
15. J.-L. Brevet and K. Mori, *Synthesis*, 1007–1012 (1992).
16. Y. Naoshima, Y. Munakata, S. Yoshida, and A. Funai, *J. Chem. Soc. Perkin Trans. 1*, 549–553 (1991).
17. B.I. Glanzer, K. Konigsberger, B. Berger, K. Faber, and H. Griengi, *Chem. Phys. Lipids* **54**, 43–48 (1990).
18. G. Pedrocchi-Fantoni and S. Servi, *J. Chem. Soc. Perkin Trans. 1*, 1764–1765 (1991).
19. M. Mishitz, A. Hackinger, I. Francesconi, and K. Faber, *Tetrahedron* **50**, 8661–8664 (1994).
20. A. Klibanov, *Acc. Chem. Res.* **23**, 114–120 (1990).
21. S. Sankaranarayanan, A. Sharma, B.A. Kulkarni, and S. Chattopadhyay, *J. Org. Chem.* **60**, 4251–4254 (1995).
22. T. Sugai, S. Ohsawa, H. Yamada, and H. Ohta, *Synthesis*, 1112–1114 (1990).
23. E. Fukusaki, S. Senda, Y. Nakazono, and T. Omata, *Tetrahedron* **47**, 6223–6230 (1991).
24. E. Fukusaki, S. Senda, Y. Nakazono, and T. Omata, *Biosci. Biotechnol. Biochem.* **56**, 1160–1161 (1992).
25. B. Henkel, A. Kunath, and H. Schick, *Liebigs Ann. Chem.*, 921–923 (1995).
26. H.K. Jacobs, B.H. Mueller, and A.S. Gopalan, *Tetrahedron* **41**, 8891–8898 (1992).
27. M.-J. Kim, G.-B. Choi, J.-Y. Kim, and H.-J. Kim, *Tetrahedron Lett.* **36**, 6253–6256 (1995).
28. K. Mori and H. Kiyota, *Liebigs. Ann. Chem.*, 989–992 (1992).
29. R. Chenevert and M. Desjardins, *J. Org. Chem.* **61**, 1219–1222 (1996).
30. Y. Naoshima, Y. Munakata, and A. Funai, *Biosci. Biotechnol. Biochem.* **56**, 999–1000 (1992).
31. E. Fukusaki, S. Senda, Y. Nakozono, H. Yuasa, and T. Omata, *J. Ferment. Bioeng.* **73**, 280–283 (1992).
32. J. Ishiyama, M. Toi, C. Ichikawa, S. Oi, and Y. Inoue, *J. Chem. Soc. Jap.*, 1162–1166 (1993).
33. V. Prelog, *Pure Appl. Chem.* **9**, 119 (1964).
34. R. MacLeod, H. Prosser, L. Fikentscher, J. Lanyi, and H.S. Mosher, *Biochemistry* **3**, 838 (1964).
35. M. Sato, J. Sakaki, Y. Sugita, T. Nakano, and C. Kaneko, *Tetrahedron Lett.* **31**, 7463–7466 (1990).
36. S.K. Ghosh, S. Chattopadhyay, and V.R. Mamdapur, *Tetrahedron* **47**, 3089–3094 (1991).
37. H. Liu and T. Cohen, *J. Org. Chem.* **60**, 2022–2025 (1995).
38. A. Archelas and R. Furstoss, *Tetrahedron Lett.* **33**, 5241–5242 (1992).
39. E. Keinan, K.K. Seth, R. Lamed, R. Ghirlando, and S.P. Singh, *Biocatalysis* **3**, 57–71 (1990).
40. E. Keinan, S.C. Sinha, and S.P. Singh, *Tetrahedron* **47**, 4631–4638 (1991).
41. E. Keinan, S.C. Sinha, and A. Sinha-Bagchi, *J. Org. Chem.* **57**, 3631–3636 (1992).
42. V. Alphand, A. Archelas, and R. Furstoss, *J. Org. Chem.* **55**, 347–350 (1990).
43. J. Lebreton, V. Alphand, and R. Furstoss, *Tetrahedron Lett.* **37**, 1011–1014 (1996).
44. M. Schultz, H. Waldmann, W. Vogt, and H. Kunz, *Tetrahedron Lett.* **31**, 867–868 (1990).
45. D.C. Myles, P.J. Andrulis, III, and G.M. Whitesides, *Tetrahedron Lett.* **32**, 4835–4838 (1991).
46. H. Danda, T. Nagatomi, A. Maehara, and T. Uemura, *Tetrahedron* **47**, 8701 (1991).
47. S. Masutomo, A. Inoue, K. Kumagai, R. Murai, and S. Mitsuda, *Biosci. Biotechnol. Biochem.* **59**, 720 (1995).

INSULIN, PURIFICATION

Inger Mollerup
Steen Weber Jensen
Per Larsen
Ole Schou
Leo Snel
Novo Nordisk A/S
Bagsvaerd, Denmark

KEY WORDS

Analysis
Characterization
Insulin
Process
Purification

OUTLINE

INTRODUCTION

Until 1982 all insulin preparations available were porcine or bovine insulins prepared by extraction of pancreas. In 1982, Novo launched semisynthetic human insulin, where porcine insulin was enzymatically converted into human insulin; later the same year Eli Lilly and Company launched human insulin of recombinant DNA origin, and

Novo launched human insulin of recombinant origin in 1988. From Eli Lilly and Company two different production methods have been reported. The first method was the expression in *Escherichia coli* of insulin A- and B-chains in two separate systems (1,2), the two chains being isolated and purified separately and later combined chemically, followed by final purification procedures. The second method reported another expression system, also in *E. coli*, where proinsulin was expressed using a tryptophan promotor with a methionine linkage to proinsulin. This linkage was cleaved by cyanogen bromide, followed by folding and formation of the correct disulphide bonds (3,2), and finalized by enzymatic removal of the C-peptide.

This article describes the manufacturing process and analysis of human insulin produced in *Saccharomyces cerevisiae*. A single-chain insulin precursor is expressed and secreted into the medium. After initial recovery, the connecting peptide is cleaved enzymatically in a transpeptidation reaction by the same technology as employed for enzymatic conversion of porcine into human insulin (4). The methods for production of human insulin by recombinant DNA technology have been reviewed by Ladisch (5). The clinical use of human insulin has been reviewed by Heinemann (6), giving an overview of the clinical development resulting from the transition from animal to human insulin and the characteristics of different types of formulation and their impact on blood glucose control.

MANUFACTURING PROCESS

Introduction

In the development of a purification process for insulin, a number of targets must be met. For a product such as insulin to be injected daily, the purity, measured by reverse-phase high-performance liquid chromatography (RP-HPLC), must be very high, and the content of host cell proteins and DNA must be very low. The production takes place on a large scale because the world consumption of insulin is on the order of tons per year. Therefore, the purification processes need to be both robust and performed at a very large scale; the implications for process design were reviewed by Prouty (7). Also the purification process must be adapted to the actual fermentation process.

The insulin production strategies, based on either *E. coli* or yeast technology, include both chemical and enzymatic reactions at different stages of the downstream processing scheme. This leads to the formation of a number of insulin derivatives closely related to insulin and consequently difficult to remove by traditional purification methods. Process-scale RP-HPLC, being a chromatographic separation technique with great separation potential, has been employed for this purpose by Novo Nordisk A/S and other companies (8).

Insulin forms crystals readily (9). This aptitude is utilized throughout the process, both for the purpose of isolating insulin from process streams and for removal of impurities that do not cocrystallize with insulin.

Throughout the process, insulin purity and concentration are monitored by RP-HPLC (see the section "Purity") because this technique has proven to be applicable even in very impure process streams such as the fermentation broth (10).

Expression and Fermentation

Expression and secretion of insulin are achieved by cloning transformants of *S. cerevisiae*. This organism was chosen as host for a number of reasons (11): yeast secretes the insulin precursor into the fermentation broth, making cell disruption unnecessary, the disulfide bonds are established correctly, yeast is an organism easily grown in large fermentors, and yeast is an organism producing no toxic substances.

Investigations of the expression of proinsulin along with a series of insulin precursors where the C-peptide was exchanged by a short peptide (12,13) showed expression of proinsulin in yeast to be too low to be of industrial interest, whereas in *E. coli* the expression is sufficient (1,2). The influence of the composition of the connecting peptide (either amino acid B29 or B30 to A1) on different single-chain insulin precursors containing 1–5 amino acids in the peptide was evaluated. A single-chain precursor with a 3–amino acid connecting peptide between B29 and A1 was chosen for commercial production by Novo Nordisk A/S.

Large-scale fermentations are performed in suitable fermentors (11). The main fermentor is inoculated at first as a fed-batch process; and after 3 days the culture is grown as a continuous fermentation in a simple medium containing yeast extract as its main nitrogen source. The insulin precursor is recovered from the fermentation broth on a continuous basis.

Recovery and Purification

Recovery. The objective of the recovery procedure is to isolate the insulin precursor from the fermentation liquid and obtain a stable intermediate product for further processing. The recovery procedure is outlined in Figure 1, and a view of the recovery plant is shown in Figure 2.

Initially, cell removal is performed by centrifugation of the fermentation broth in disc stack centrifuges. The supernatant is passed through a cation exchange column at low pH, under which conditions insulin is retained on the column. Elution of the insulin precursor takes place by increasing the pH. The eluate is filtered to remove the last traces of yeast particles, whereby the need for containment further downstream is avoided. The product is isolated from the eluate by crystallization, the crystals being removed by centrifugation or filtration. After redissolving the crystals, impurities (mainly host cell proteins) are precipitated at a pH slightly above neutral by addition of ethanol to approximately 60% v/v; at this ethanol concentration insulin is very soluble. The precipitate is removed by filtration, and a second crystallization of insulin results in a product of more than 90% purity, as measured by RP-HPLC, and suited for intermediate storage at −20 °C. This intermediate product still contains impurities from the fermentation broth (e.g., yeast proteins and DNA).

Enzymatic Conversion. The use of enzymatic processes in insulin manufacturing was initiated in 1979, when conversion of porcine insulin into human insulin catalyzed by

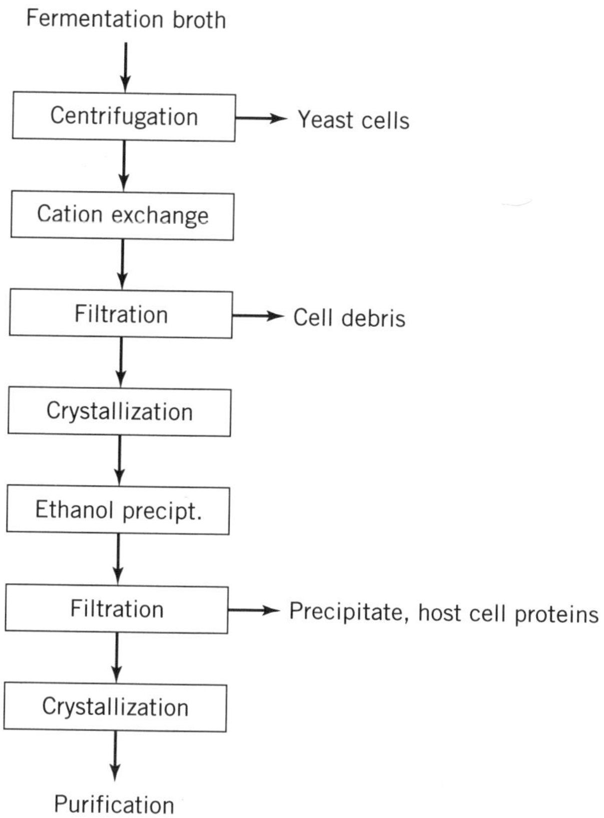

Figure 1. Flow chart of recovery process.

Figure 2. Insulin production plant.

trypsin was demonstrated, resulting in the launch of a semisynthetic human insulin in 1982 (4). Porcine and human insulin differ by only a single amino acid: the C-terminal amino acid (B30) of the B-chain is alanine in porcine insulin and threonine in human insulin. The conversion of porcine into human insulin is achieved by an enzymatic process in which trypsin is used in an organic/aqueous environment to cleave the B-chain after lysine at B29 and to insert threonine instead of alanine (14).

The enzymatic process may be performed as a one-step process, a transpeptidation, or as a two-step process in which the formation of the des-(B30) insulin takes place, and then threonine is coupled to the des-(B30) insulin in the form of a threonine ester. The two-step process is used if a chromatographic purification of the intermediate product des-(B30) insulin is required, but the one-step process is the most simple to perform. In the first industrial human insulin product, launched by Novo in 1982, the conversion was carried out as a one-step process. This enzymatic process utilizing trypsin also proved useful for conversion of single-chain insulin precursors produced by recombinant DNA technology (12–14).

Another enzyme that can be used for the conversion processes is *Achromobacter lyticus* protease (4,13), which compared to trypsin has the benefit of being lysine specific, thus avoiding the risk of cleavage after arginine in position B22; such cleavage may occur when using trypsin. Cleavage sites are shown in Figure 3.

Purification. The purification process of human insulin by recombinant DNA technology from yeast is shown in Figure 4.

After transpeptidation, the crude insulin ester is purified by RP-HPLC at low pH. The purification step is performed on octadecyl-subsituted silica, and the eluent is an ethanol-containing ammonium sulfate buffer at pH 3.0. After chromatography, the human insulin ester is isolated by crystallization. This step removes host cell proteins and trypsin as well as some by-products formed during the transpeptidation reaction (e.g., cyclic des-(B30) insulin). To keep trypsin inactive during the purification step, it is essential that the processing be performed under acid conditions. The purity of the resulting product is typically 90–92%, as measured by RP-HPLC.

After the RP-HPLC process, the human insulin ester is subjected to anion exchange chromatography in a neutral acetate buffer at room temperature. This purification serves the primary purpose of removing any remaining des-(B30) insulin, which is formed as a by-product during the transpeptidation reaction. Des-(B30) insulin is a component with physical properties very similar to those of human insulin. The separation is, therefore, based on the difference in charge between insulin ester and des-(B30) insulin. In addition, this step also reduces a number of other differently charged insulin-related derivatives to a very low level, and host cell proteins are no longer detectable after this step. After collection of the eluate from the column, the product is isolated by crystallization, typically resulting in more than 94% purity, as measured by RP-HPLC.

Hydrolysis. Subsequently, in order to remove the ester group, the human insulin ester is subjected to hydrolysis, resulting in the formation of human insulin. In the choice

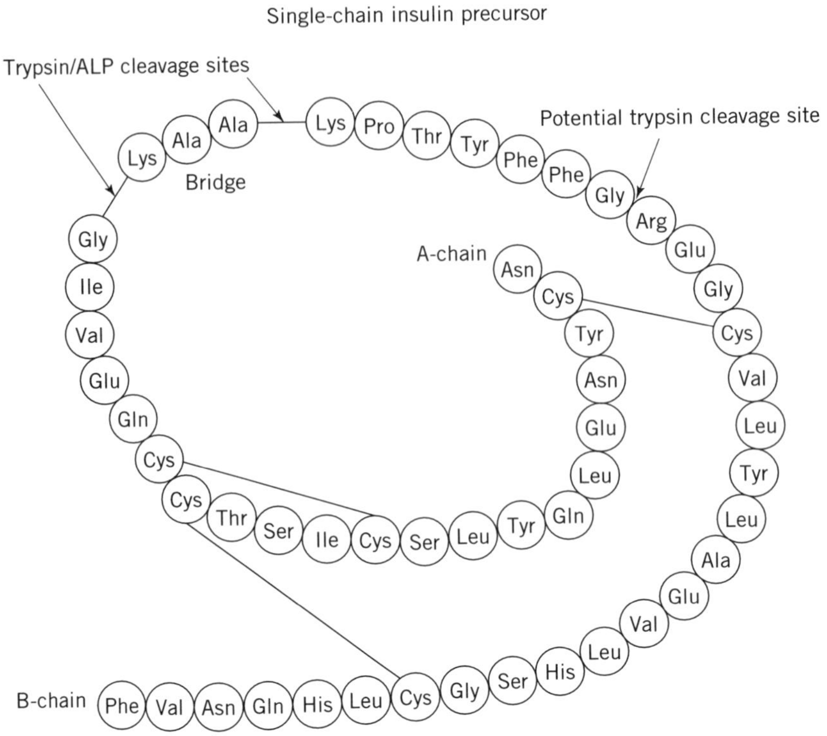

Figure 3. Human insulin precursor. Cleavage sites for trypsin and *Achromobacter lyticus* protease are indicated.

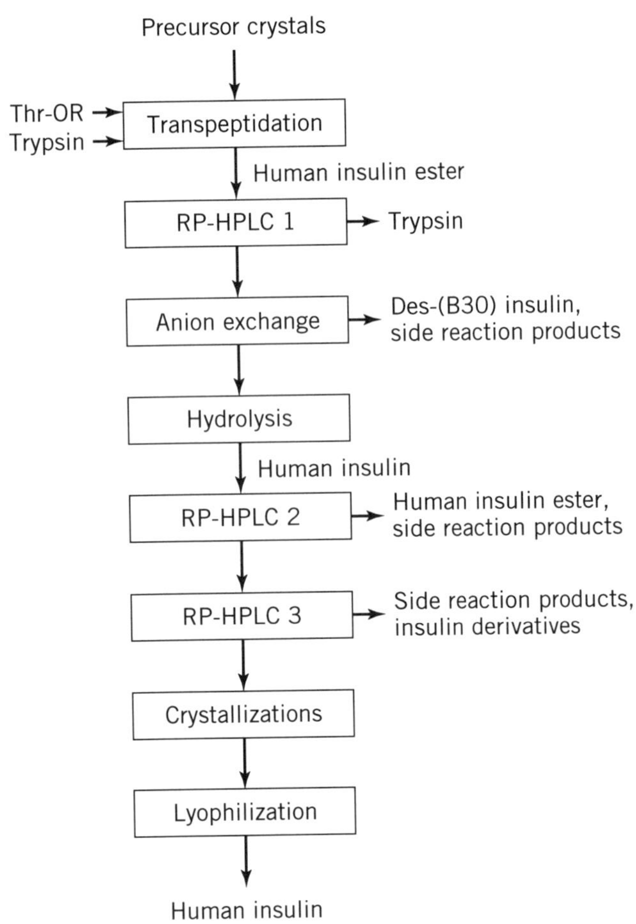

Figure 4. The purification process.

of parameters for the hydrolysis, the relative rates of the desired reaction and the by-product formation must be taken into consideration.

Among the products from the side reactions are cyclic by-products (as single-chain B30-A1 human insulin) and by-products resulting from disulfide-bridge interchange. The parameters finally chosen to balance these different effects resulted in a hydrolysis process performed at high pH and low temperature. It is important to control both pH and temperature very closely because too high pH or temperature may lead to destruction of the disulfide bonds.

At the end of the hydrolysis, the reaction mixture is acidified, and the product is isolated by crystallisation. The purity of this product is typically more than 94%, as measured by RP-HPLC.

Final Purification. After hydrolysis the human insulin must be purified to comply with the specifications for the bulk product. The product contains several impurities such as small amounts of human insulin ester, partially cleaved precursor components, single-chain human insulin, and deamidated insulin. Reverse-phase chromatography at low pH, traditionally used for the purification of peptides, does not suffice to ensure the required purity. This can, however, be achieved by addition of an RP-HPLC step at neutral pH. This purification step (RP-HPLC 2 in Fig. 4) is performed on octadecyl-subsituted silica, and the neutral eluent is based on ethanol and potassium chloride buffered with Tris. After chromatography the human insulin is isolated by crystallization. Primarily this step removes the human insulin ester and reduces the partially cleaved precursor components. The purity of the product is typically more than 99.0%, as measured by RP-HPLC.

The final purification step is also based on reverse-phase chromatography and is performed on octadecyl-subsituted silica at low pH, the eluent being an ethanol-containing ammonium sulfate buffer at pH 3.0 (RP-HPLC 3, Fig. 4). After chromatography the human insulin is isolated by crystallization. Primarily this step removes the single-chain human insulin and reduces the remaining insulin by-products to insignificant levels. The purity of the product is typically more than 99.5%, as measured by RP-HPLC.

Bulk Formulation. After the final chromatographic purification, the product must be conditioned for the subsequent drug product formulation. The target is a product that is stable when stored and with no components unsuitable for the following formulation. Thus residual salts and buffers from the chromatographic steps must be reduced.

During bulk formulation the product must be treated carefully to eliminate decomposition. The main risk factors when handling the product after the final chromatographic purification are formation of aggregates (fibrillation) and deamidation of the insulin molecule (15). Aggregation can be the result of stressing the molecule (i.e., agitation), whereas deamidation primarily results from prolonged exposure to acid conditions (formation of A21 desamido insulin).

Crystallization is used for the final treatment of both animal and human insulins, followed by freeze-drying. Crystallizations are carried out in the presence of zinc; after the last crystallization the crystals are washed to reduce any excess content of zinc. To a wide extent the crystallization processes of insulin bulk production are based upon the development work done in the 1950s (9).

For special drug types and for other purposes where a zinc-free product is needed, alternatives to the crystallization processes can be used. These may include isoelectric precipitation, gel filtration chromatography, or ultra- and diafiltration.

The freeze-drying procedure results in a bulk product with a water content below 10.0% (loss on drying); it is very stable when stored frozen. Storage as a bulk solution is possible, but in the case of insulin it is not feasible due to the demand for very large bulk quantities.

ANALYSIS OF RECOMBINANT HUMAN INSULIN

Introduction

Human insulin is a well defined and well characterized protein, and recombinant human insulin is chemically, physically, and biologically equivalent to both pancreatic human insulin and semisynthetic human insulin (4,16). As purification methods continue to improve, so does the analytical methodology to demonstrate purity and integrity throughout the manufacturing process. The requirement to show process and batch consistency demands an array of analytical techniques to show any presence of adventitious agents, host cell contaminants, process agents, and insulin-related by-products. Moreover, changes of the protein structure during the manufacturing process may influence the biological activity of the insulin product.

Identification

The primary structure of human insulin can be verified by amino acid composition analysis, amino acid sequence analysis, determination of electrophoretic mobility, measurement of retention on RP-HPLC, measurement of molecular mass, and peptide fingerprinting. The first three methods are all well established but of limited specificity.

The need to distinguish between the various insulin products during the manufacturing process was the driving force in the development of RP-HPLC, the separation principle being based on hydrophobic differences (17,18). The RP-HPLC method of choice is based on C18 stationary phases and mobile phases containing acetonitrile or methanol buffers at pH between 2 and 8. The high selectivity and resolving power allow the separation of various insulin species, precursors, and insulin derivatives differing by only one amino acid (Fig. 5).

Determination of molecular mass by mass spectrometry is being established in quality control laboratories as a precise method of identity. Equipped with an electrospray ionization source, the single- or triple-quadropole instruments can determine the mass of insulin within 0.5 Da and clearly distinguish between deamidated insulin and human insulin, which have a difference of only one mass unit.

Fingerprinting or peptide mapping is an excellent tool to confirm the correct amino acid sequence as well as the proper formation of disulfide bonds. The peptide map is generated by fragmentation by an enzyme (preferably *Staphylococcus aureus* V8 protease) cleaving at a specific residue (e.g., glutamic acid), thus giving four fragments easily separated by RP-HPLC. The peptide map of a sample is then compared to the map of an insulin reference standard. The patterns should be identical (Fig. 6). Each fragment peak on the reference map must be characterized at least once by amino acid sequence analysis and mass spectrometry.

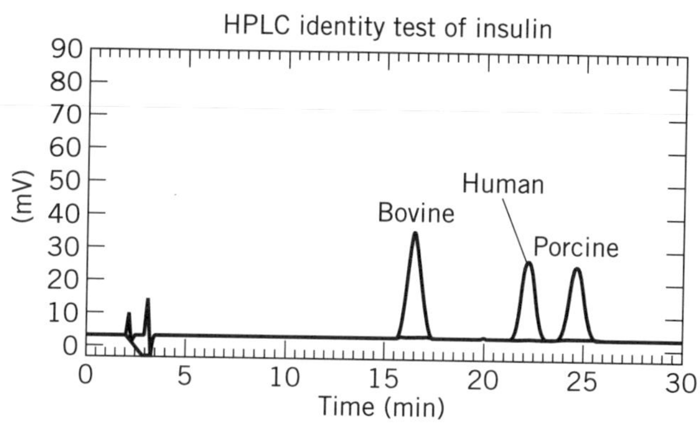

Figure 5. RP-HPLC chromatogram showing the separation of different insulin species differing by only one amino acid. Column: Lichrosorb RP-18, 5 μm; solvent A: 0.2 M sodium sulfate, 0.04 M phosphoric acid in 10% acetonitrile, pH adjusted to 2.5 with ethanolamine; solvent B: 50% acetonitrile; flow rate: 1.0 mL/min; temperature: 40 °C; 50 μg of sample was injected and eluted isocratically at 49% (solvent A) 51% (solvent B).

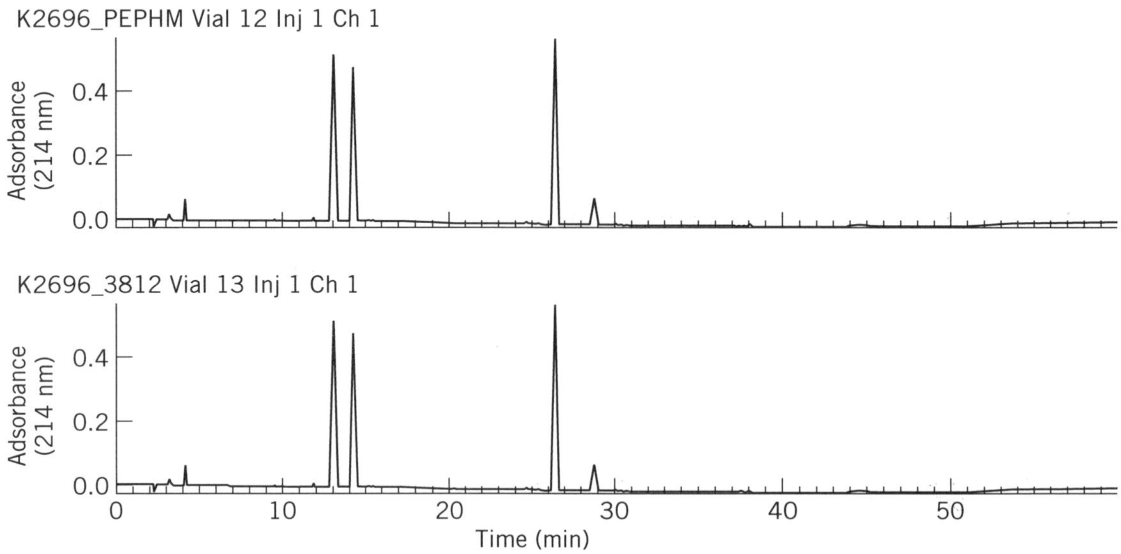

Figure 6. Peptide map of insulin sample (bottom panel) compared to a human insulin reference standard (top panel). Column: Lichorsorb RP-18, 5 μm; solvent A: 0.2 M sodium sulfate, 0.04 M phosphoric acid in 10% acetonitrile, pH adjusted to 2.5 with ethanolamine; solvent B: 50% acetonitrile; flow rate: 1.0 mL/min; temperature: 40 °C; 25 μg of sample was cleared prior to injection and eluted with a gradient.

The secondary and tertiary structures of human insulin can be assessed by nuclear magnetic resonance, circular dichroism, X-ray crystallography, and measurement of biological activity. Among these methods, only bioactivity is measured on a routine basis; the other tests are performed once to fully characterize human insulin.

Purity

Biological products in general require a special assessment to demonstrate purity, safety, and efficacy. The problems associated with this are the following:

- The need to establish batch-to-batch reproducibility because of the biological nature of early stages of the manufacturing process
- The need for chemical and biological tests to assess purity and potency where these can be affected by variation in starting materials, manufacturing process, and product stability
- The need to avoid adventitious agents arising from the starting materials and/or the downstream processes; these can be microbiological or associated with nonmicrobiological contaminants

Routine quality control testing should be designed to take into account the specific nature of the product. Therefore, specifications and methods of analysis are developed for each product. It is generally acknowledged that the final product cannot be completely characterized, and that in-process controls are necessary to prove its consistent quality.

Microbiological Impurities. Each manufacturing step is performed under conditions designed to keep the bioburden low and avoid endotoxins in the recombinant human insulin. Whereas there is an absolute requirement for sterility in the formulated product, a result of below 100 viable counts/g is acceptable for the drug substance. The absence of endotoxins is confirmed by either the Limulus amoebocyt lysate (LAL) test or by a biological test in rabbits. Process validation documents complete removal of potential adventitious virus or other agent.

Chemical Impurities

Contaminants. Host cell proteins, DNA, and process-related chemicals are all impurities that need to be controlled, preferably by validation studies or by in-process/specification testing. Sensitive assays should be developed to detect any residual levels of yeast proteins, trypsin used for the transpeptidation, or DNA to the parts per million limit. For DNA detection the Threshold® DNA assay is much preferred to former hybridization assays, mainly due to its higher sensitivity but also to its robustness. The host cell contaminant assay is based on antibodies generated against the yeast proteins from the production strain lacking the insulin gene. Thus antibodies are raised against proteins ranging in molecular weights from 10 kDa to 100 kDa and thus also covering proteins the size of insulin. Table 1 shows the reduction of DNA and host cell proteins during the purification process.

Absence of trypsin can be shown by a trypsin assay (ELISA or activity), and extensive step reduction studies have proven a more than 20 log reduction of the trypsin level.

Related Impurities. Insulin-related impurities are defined as derivatives generated by deamidation, dimerization, polymerization, peptide cleavage, and disulfide exchange (15), or as intermediate products from the

Table 1. Purity of Human Insulin or Human Insulin Precursor During Downstream Processing as Indicated by RP-HPLC Analysis (%) and Test for Residual Levels of Host Cell Proteins and DNA

	Purity tests		
Process step	HPLC (%)	Yeast proteins (ppm)	DNA (ppm)
1. Yeast fermentation broth			
2. Isolation of precursor	~20	>10,000	1,000
3. Purification of precursor	~90	1,000	<3
4. Conversion to Insulin ester	~85		
5. Purification	~94	<1	<0.5
6. Hydrolysis to insulin	~94		
7. Preparative HPLC	>99.0	<1	<0.5
8. Monocomponent Insulin	>99.0	<1	<0.5

manufacturing process (e.g., insulin precursor, insulin ester).

During human insulin manufacture the purity at different process steps is assessed by RP-HPLC. RP-HPLC on alkylsilane supports has become the method of choice due to the hydrophobic differences between insulin and the insulin-related impurities (10,18,19). Separation on ion exchange HPLC resin has also been reported (18) and compared with RP-HPLC, but the latter is preferred for in-process as well as specification analyses. Figure 7 shows the purity analyses on four different purification process steps.

Deamidation. Human insulin contains six potential deamidation sites: the A-chain residues A5(Gln), A15(Gln), A18(Gln), and A21(Asn), and the B-chain residues B3(Asn) and B4(Gln). The deamidation in acid and neutral environment has been described (15), and it has been shown that the desamido insulins retain almost full biological activity. In order to separate all deamidated forms of insulin, complementary methods might be used on uncoated capillaries (e.g., RP-HPLC and HPCE [20]). The elution order of the various desamido forms can hardly be predicted because it depends on the selectivity of the stationary phase.

Aggregation. Human insulin forms covalent as well as noncovalent dimers and polymers (15). The association to dimers and hexamers (noncovalent) is reversible, and at certain conditions the insulin monomer linearizes and forms fibrils. These fibrils are unwanted reaction products of low solubility and can be avoided by proper manufacturing precautions (15). The covalent insulin dimers and polymers form during handling of insulin solutions and during storage at elevated temperatures.

The method used to quantitate the extent of dimerization/polymerization is gel permeation chromatography. The column support is a silica-based diol, and the mobile phase should contain a high concentration of acetic acid in order to maintain the dissociation state of the analyte.

Assay

The potency of human insulin has been determined by biological assays such as the mouse convulsion assay and the rabbit blood glucose test. RP-HPLC assay data compared

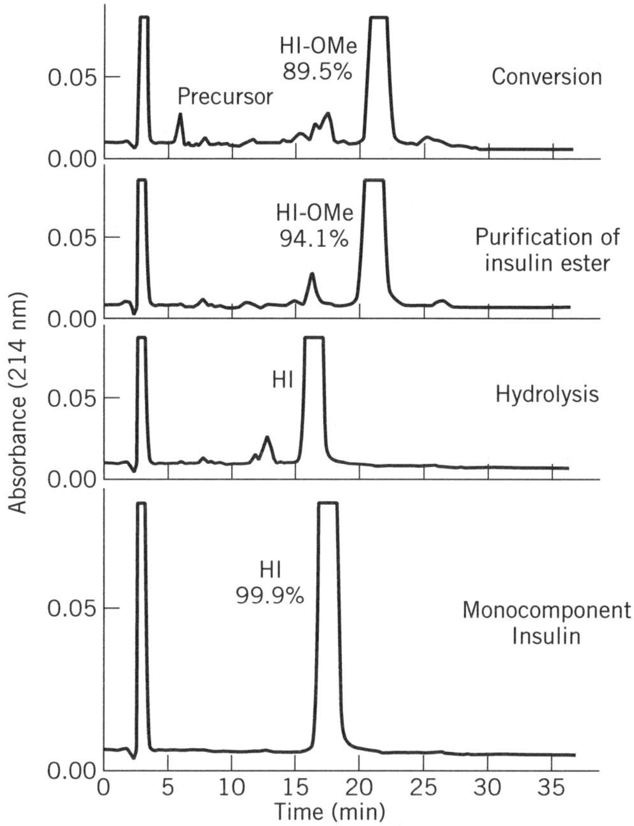

Figure 7. Purification steps after conversion of insulin precursor: RP-HPLC chromatogram of in-process samples from the purification process. Column: Lichrosorb RP-18, 5 μm, 4 × 250 mm; solvent A: 0.2 M sodium sulfate, 0.04 M phosphoric acid in 10% acetonitrile, pH adjusted to 2.5 with ethanolamine; solvent B: 50% acetonitrile; flow rate: 1.0 mL/min; temperature: 40 °C; 50 μg of sample was injected and eluted isocratically at 49% (solution A) 51% (solution B).

with bioassays seemed to justify a chemical assay to express the potency, and such methods are now described in the pharmacopoeias. The resolution of the RP-HPLC assay allows a quantitation of the human insulin content in the presence of even substancial amounts of related substances. Furthermore, the RP-HPLC assay offers superior precision with a standard deviation of only about 1%. This is particularly valuable for the determination of the human insulin content in a sample and for monitoring stability studies on bulk drug substance and drug product.

STABILITY

Human insulin manufactured by the method just described is very stable when stored frozen, protected from light. Hardly any change of quality can be detected even after several years of storage. However, when stored at between 2 and 8 °C, which is the prescribed storage temperature, insulin preparations will gradually degrade while forming neutral deamidation products and dimers (15). Therefore, the shelf life of insulin preparations is limited to 2–3 years.

BIBLIOGRAPHY

1. R.E. Chance, J.A. Hoffmann, E.P. Kroeff, M.G. Johnson, E.W. Schirmer, W.W. Bromer, M.J. Ross, and R. Wetzel, in D.H. Rich and E. Gross eds., *Peptides: Synthesis, Structure, Function*, Pierce Chemical Company, 1981, pp. 721–728.
2. B.H. Frank and R.E. Chance, *Münch. med. Wschr.* **125**, 14–20 (1983).
3. B.H. Frank, J.M. Pettee, R.E. Zimmerman, and P.J. Burck, in D.H. Rich and E. Gross eds., *Peptides: Synthesis, Structure, Function*, Pierce Chemical Company, 1981, pp. 729–738.
4. J. Markussen, U. Damgaard, K.H. Jørgensen, E. Rasmussen, L. Snel, L. Thim, and H.O. Voigt, in *Hormone Drugs. Proceedings of the FDA-USP Workshop on Drug and Reference Standards for Insulins, Somatropins, and Thyroid-Axis Hormones*, 1982, pp. 116–126.
5. M.R. Ladisch and K.L. Kohlmann, *Biotechnol. Prog.* **8**, 469–478 (1992).
6. L. Heinemann and B. Richter, *Diabetes Care* **16**, 90–100 (1993).
7. W.F. Prouty, in Y.H. Chiu and J.L. Gueriguian eds., *Drug Biotechnology Regulation*, Marcel Dekker, New York, 1991, pp. 221–262.
8. E.P. Kroeff, R.A. Owens, E.L. Campbell, R.D. Johnson, and H.I. Marks, *J. Chromatogr.* **461**, 45–61 (1989).
9. J. Schlichtkrull, *Insulin Crystals: Chemical and Biological Studies on Insulin Crystals and Insulin Zinc Suspensions*, Munksgaard, Copenhagen, 1958.
10. L. Snel, U. Damgaard, and I. Mollerup, *Chromatographia* **24**, 329–332 (1987).
11. I.V. Diers, E. Rasmussen, P.H. Larsen, and I.-L. Kjærsig, in Y.H. Chiu and J.L. Gueriguian eds., *Drug Biotechnology Regulation*, Marcel Dekker, New York, 1991, pp. 166–176.
12. L. Thim, L. Snel, K. Norris, and M.T. Hansen, in C.L. Hershberger, S.W. Queener, and G. Hegeman eds., *Genetics and Molecular Biology of Industrial Microorganisms*, American Society for Microbiology, Washington D.C., 1989, pp. 322–328.
13. T. Christensen, H. Dalbøge, and L. Snel, in Y.H. Chiu and J.L. Gueriguian eds., *Drug Biotechnology Regulation*, Marcel Dekker, New York, 1991, pp. 206–221.
14. J. Markussen, *Human Insulin by Trypsic Transpeptidation of Porcine Insulin and Biosynthetic Precursors*, MTP Press, Lancaster, U.K., 1987.
15. J. Brange, *Stability of Insulin: Studies on the Physical and Chemical Stability of Insulin in Pharmaceutical Formulation*, Kluwer Academic, Boston, 1994.
16. R.E. Chance, E.P. Kroeff, and J.A. Hoffmann, in *Insulins, Growth Hormone and Recombinant DNA Technology*, Raven, New York, 1981, pp. 71–85.
17. U. Damgaard and J. Markussen, *Horm. Metab. Res.* **11**, 580–581 (1979).
18. B.S. Welinder and S. Linde, in W.S. Hancock ed., *CRC Handbook of HPLC for the Separation of Amino Acids, Peptides and Proteins*, CRC Press, 1984, pp. 357–362.
19. A. McLead and S.P. Wood, *J. Chromatogr.* **285**, 319–331 (1984).
20. G. Mandrup, *J. Chromatogr.* **604**, 267–281 (1992).

L-ISOLEUCINE

MASAYUKI INUI
MASATO TERASAWA
HIDEAKI YUKAWA
Mitsubishi Chemical Corporation
Ibaraki, Japan

KEY WORDS

ilv Genes
L-Isoleucine
LCR process
L-Leucine
L-Valine

OUTLINE

INTRODUCTION

L-Isoleucine, L-valine, and L-leucine are termed branched-chain amino acids, because the chemical structures and physical characteristics of these amino acids are similar. L-Isoleucine belongs to the aspartic acid group of amino acids, which also contains L-threonine, L-lysine, and L-methionine, because four of six carbon atoms in these amino acids are derived from aspartic acid. On the other hand, L-valine and L-leucine belong to pyruvate group of amino acids, which also include L-alanine, because all carbon atoms in these two amino acids are derived from pyruvate. However, these three amino acids are synthesized by parallel biosynthetic pathways.

L-Isoleucine is one of the essential amino acids and therefore of importance for the pharmaceutical and feed industries. Particularly, the primary use of this amino acid in pharmaceuticals is as part of injectable solutions of amino acids. Because L-isoleucine possesses two asymmetric carbon atoms, there exist four optical isomers, L-, D-, L-allo-, and D-alloisoleucine. Therefore, it is difficult to produce and separate pure L-isoleucine at low cost by chemical synthetic methods, and research has been initiated to produce L-isoleucine by fermentative production methods. Although isolation of microorganisms that excrete amino acids in nature has been attempted and a number of strains that can excrete L-valine and L-leucine have been found, none have been detected that can excrete L-isoleucine.

In the early 1950s, Umbarger et al. studied the isoleucine biosynthetic pathway and showed that the precursor of L-isoleucine could be not only L-threonine, but also D-threonine and α-ketobutyric acid. Based on these results, the production of L-isoleucine using D-threonine and α-

ketobutyric acid as a precursor by fermentation was reported in 1962 (1).

In the late 1960s, the genetic determinants of feedback inhibition of amino acid biosynthesis were elucidated, and it was subsequently proved that feedback inhibition of amino acid biosynthesis was released in mutants resistant to the corresponding amino acid analogue (2). The cumulative result of these studies is that L-isoleucine is now produced on an industrial scale by fermentation.

This article will review the biosynthetic pathway and related enzymes and genes for L-isoleucine synthesis and will also describe microbial production of L-isoleucine.

BIOSYNTHETIC PATHWAY OF ISOLEUCINE AND ITS REGULATION

Elucidation of the pathways to L-isoleucine and L-valine began with the analysis of auxotrophic mutants of *Escherichia coli* and *Salmonella typhimurium*. These mutants usually require both L-isoleucine and L-valine for growth, less frequently only L-isoleucine, and very rarely only L-valine. The phenomenon was finally understood when it was found that the two amino acids are produced by a series of reactions catalyzed by the same enzymes (Fig. 1). At the first step of the pathway to L-isoleucine, α-ketobutyrate reacts with active acetoaldehyde derived from pyruvate by the action of acetohydroxy acid synthase (AHAS), whereas in case of L-valine, two molecules of pyruvate are condensed. The next three steps are catalyzed in parallel by acetohydroxy acid isomeroreductase (AHAIR), dihydroxy acid dehydrase (DH), and transaminase B (TrB). Threonine deaminase (TD) catalyses the deamination of threonine to yield α-ketobutyrate. L-Leucine branches from the L-valine route at the α-ketoisovalerate step.

As found in several biosynthetic pathways, feedback inhibition and repression function in the regulation of the isoleucine-valine biosynthesis pathway (Table 1).

The activity of TD is strongly inhibited by L-isoleucine in most microorganisms. Because it is a competitive inhibitor with L-threonine, the degree of inhibition depends on the concentration of L-threonine. However, it was found that in *Rhodospirillum rubrum*, the inhibition by L-isoleucine of TD is very weak. Although it was reported that in *Pseudomonas aeruginosa* the TD gene is constitutively expressed, the TD gene is generally repressed. In *Neurospora crassa*, TD is repressed by only L-isoleucine, whereas in other organisms, TD is repressed by the presence of L-isoleucine, L-valine, or L-leucine as a multivalent repression (Table 1).

In *E. coli* and *S. typhimurium*, three different isozymes have been demonstrated for AHAS. The activities of two, AHAS I and AHAS III, are inhibited by L-valine. Inhibition by L-isoleucine of AHAS I and AHAS III are not strong compared with L-valine inhibition, and L-leucine does not inhibit them. AHAS I genes (*ilvBN*) are repressed by the presence of both L-valine and L-leucine as a multivalent repression, and AHAS III genes (*ilvIH*) are repressed by L-leucine. Although another isozyme, AHAS II, does not show sensitivity to L-valine, AHAS II genes (*ilvGM*) are repressed by the presence of L-isoleucine, L-valine, and L-leucine. In the strain *E. coli* K-12, lack of AHAS II activity is caused by a natural frameshift mutation in the *ilvG* gene for AHAS II. On the other hand, an active AHAS II (*ilvGM* product) is present in *E. coli* B and W and *S. typhimurium*, and these strains are therefore able to grow in a minimal medium containing L-valine.

In *Bacillus subtilis, P. aeruginosa, Saccharomyces cerevisiae*, and *N. crassa*, some mutants that require L-isoleucine and L-valine show no detectable AHAS activity, suggesting that only one AHAS enzyme may exist in these organisms. However, no genetic evidence concerning the presence or absence of AHAS isozymes has been presented to date. In *Brevibacterium flavum*, gene disruption studies show that only one AHAS gene exists in this organism (3). AHAS enzymes from these organisms are also inhibited by L-valine.

In addition, the biosynthetic pathway leading to the branched-chain amino acids in plants is of interest for herbicide development. AHAS is the target of a variety of herbicides (4,5), and inhibition of AHAIR also was demonstrated to result in some herbicidal effects (6). Furthermore, it is noteworthy that in *Rhizobium meliloti*, the AHAIR gene is required for nodulation (7).

Thus, because (*1*) the common enzymes, which show end-product inhibition and repression, function in the parallel branched-chain amino acids biosynthetic pathway; (*2*) the pathway is located downstream of biosynthetic flows of essential amino acids such as L-threonine, L-lysine, and L-methionine; and (*3*) the central metabolite for production of energy and glucogenesis is linked to this pathway, it presumed that the regulation of this pathway is complicated. Although there are a number of reports analyzing the isoleucine-valine biosynthetic pathway, it is still unclear in *E. coli* and even less so in other organisms.

BIOSYNTHETIC GENES OF ISOLEUCINE

The nucleotide sequences of the *ilv* genes from several organisms have been reported. AHAS enzyme, which is the first enzyme in the parallel isoleucine-valine biosynthetic pathway, is composed of a large and a small subunit in an $\alpha 2\beta 2$ structure in several organisms. It has been suggested that for all isozymes, the small subunit is important for maximal catalytic activity of the large subunit and contributes importantly to the regulation of the activity of the two valine-resistant isozymes in *E. coli*. Three sets of genes for AHAS isozymes have been isolated and sequenced from *E. coli*. These are the *ilvBN* genes for AHAS I, the *ilvGM* genes for AHAS II, and the *ilvIH* genes for AHAS III (8–10). AHAS (*ilvBN*) genes from *Bacillus subtilis* (11), *Brevibacterium flavum* (*ilvLS* genes) (3), *Caulobacter crescentus* (12), *Corynebacterium glutamicum* (13), *Lactococcus lactis* (14), *Mycobacterium avium* (15), and *Streptomyces avermitilis* (16) have been isolated, and all contain genes for a large and a small subunit. However, the presence of AHAS isozymes in these organisms has not been reported. In contrast, the large subunit genes of the AHAS isozymes of the cyanobacterium *Spirulina plantensis* (17) are not adjacent to their corresponding small subunit genes. This is similar to the gene organization observed in *Arabidopsis*

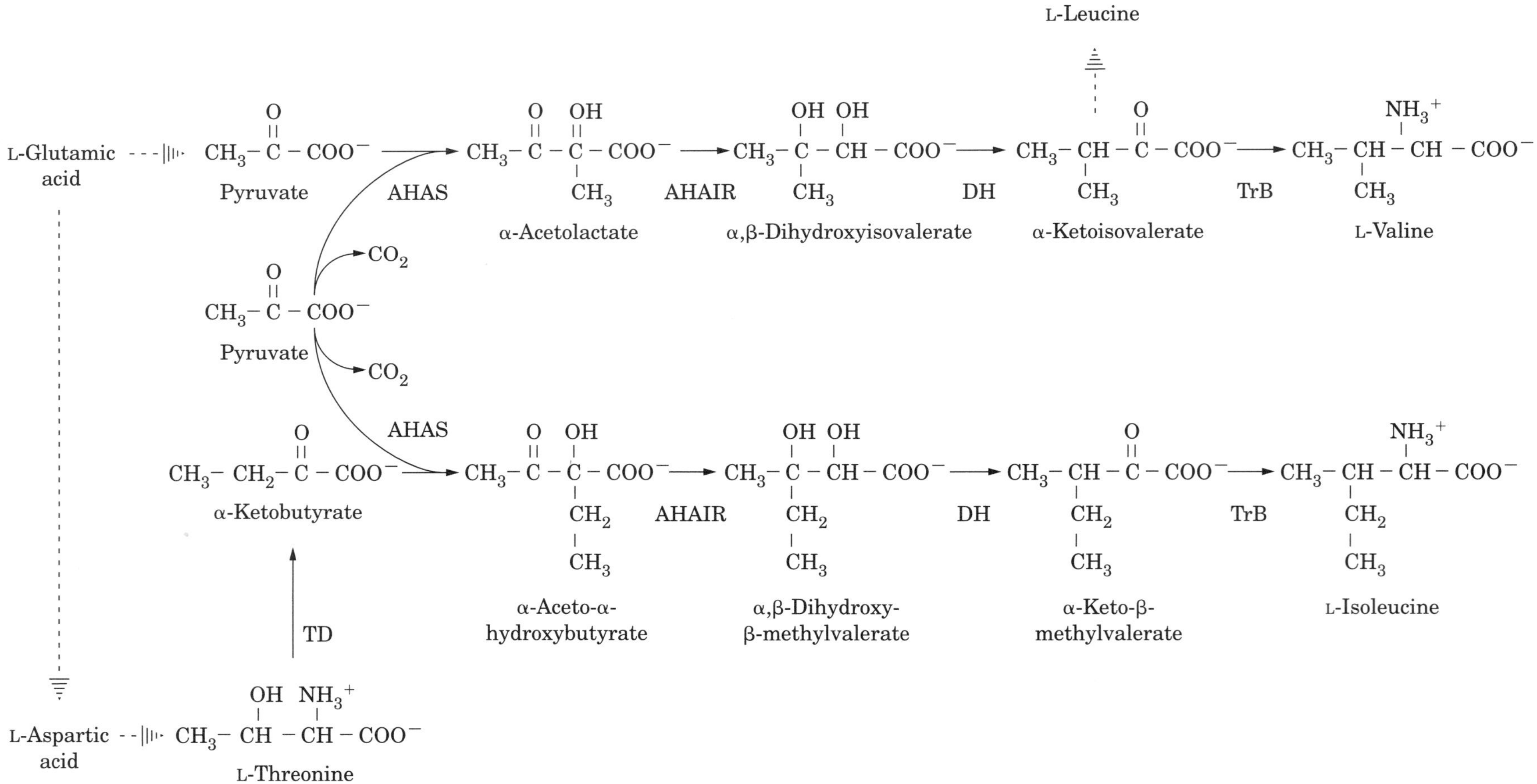

Figure 1. Biosynthesis of isoleucine and valine. TD = threonine deaminase; AHAS = acetohydroxy acid synthases; AHAIR = acetohydroxy acid isomeroreductase; DH = dihydroxy acid dehydrase; TrB = transaminase B.

Table 1. Feedback Inhibition and Repression of Key Enzymes for Branched-Chain Amino Acids Biosynthesis Pathway

Strains	Threonine deaminase		Acetohydroxy acid synthase	
	Inhibition	Repression	Inhibition	Repression
E. coli *S. typhimurium*	ile	ile + val + leu	I; val (ile) II; insensitive III; val (ile)	I; val + leu II; ile + val + leu III; leu
B. subtilis	ile	ile + leu + (val)	val (ile)	val + leu
P. aeruginosa	ile	constitutive	val (ile, leu)	ile + val + (leu)
S. cerevisiae	ile	ile + val + (leu)	val	ile + val + leu
N. crassa	ile	ile	val (ile)	val

Note: ile = L-isoleucine; val = L-valine; leu = L-leucine.
[a]I, II, and III show isozymes of AHAS.

thaliana (18), *Nicotina tabacum* (5), *Brassica napus* (19), yeasts (20), and *Synechococcus* PCC 7942 (21).

AHAIR (*ilvC*) genes, whose enzyme catalyzes the second reaction in the parallel isoleucine-valine biosynthetic pathway, have been isolated from *Br. flavum* (22), *Synechocystis* sp. strain PCC 6803 (23), *B. subtilis* (11), *L. lactis* (14), *Rhizobium meliloti* (7), *E. coli* (24), *S. cerevisiae* (*ILV5* gene) (25), *C. glutamicum* (13), *St. avermitilis* (15), *A. thaliana* (26), *M. avium* (15), and spinach *Spinacia oleracea* L. (27), respectively. In *E. coli* and *S. typhimurium*, it was demonstrated that expression of the *ilvC* genes is induced by the substrates of the AHAIR enzyme, 2-aceto-2-hydroxy-butyrate and 2-acetolactate (28). This substrate induction of *ilvC* expression is mediated by a positive activator encoded by the *ilvY* gene (29). The *ilvC* gene is linked to *ilvY*, but its transcription is divergent. In *R. meliloti*, nucleotide sequence analysis of the *ilvC* region revealed the presence of a short open reading frame, ORF2, adjacent to the *ilvC* coding region but transcribed in the opposite direction. It was proposed that this ORF2 coded for a putative regulatory protein, but this protein has no homology with the *E. coli ilvY* gene product (7). However, in other bacteria such as *Br. flavum, Synechocystis* sp., *B. subtilis, L. lactis, S. cerevisiae*, and spinach, no open reading frames homologous to the *E. coli ilvY* gene or to the *R. meliloti* ORF2 were detected downstream of *ilvC*.

DH (*ilvD*) genes were cloned from *E. coli* (30), *L. lactis* (14), *B. subtilis* (31), and *Ca. crescentus* (32), respectively. In *Ca. crescentus*, the *ilvR* gene, which is transcribed divergently from the *ilvD* gene and which positively regulates the expression of the *ilvD* gene, was found upstream of the *ilvD* region (32). Amino acid sequence comparison revealed that the IlvR protein is a member of the LysR family of transcriptional regulators. So far, in other organisms, such a regulator gene has not been reported.

Although the TrB (*ilvE*) gene was isolated from *E. coli* (30), there has been no report of the corresponding gene from other organisms.

Interestingly, the *ilv* genes in several bacteria are organized in clusters scattered throughout the chromosome. In *E. coli*, the *ilv* genes are organized into three clusters, namely *ilvGMEDAYC, ilvIH*, and *ilvBN* (8,10,24,30). In *L. lactis*, the *ilvDBNCA* genes have been shown to cluster in one region (14). In *B. subtilis* (23), *Br. flavum* (3,22), *C. gultamicum* (13), *M. avium* (15), and *St. avermitilis* (16), the *ilvBNC* genes are also found in one cluster.

PRODUCTION OF L-ISOLEUCINE

There have been a number of reports regarding the production of L-isoleucine by fermentation. As described earlier, it was found that feedback inhibition in the isoleucine-valine biosynthesis could be released by isolation of a mutant resistant to the corresponding amino acid analogue. Attempts to use the analogue-resistant mutant for production of L-isoleucine were then started.

Kisumi et al. reported in *Serratia marcescens* (strain GIHVLAr2795) that a mutant resistant to both isoleucine hydroxamate and α-aminobutyric acid produced 12 g/L of L-isoleucine in 72 h in a medium containing glucose as a carbon source by fermentation (33). Isoleucine hydroxamate was described as an isoleucine antagonist in *Serratia marcescens* (34), and α-aminobutyric acid was known as a valine analogue in *E. coli* (35). This strain had increased activities of TD, AHAS, aspartokinase, and homoserine-dehydrogenase compared with the parental strain and lacked both feedback inhibition and repression of TD.

Komatsubara et al. reported construction of a threonine-producing strain by the transduction of an isoleucine-producing strain of *Serratia marcescens* (36). First, strain GIHVLr6426 was isolated as an isoleucine-hydroxamate-resistant mutant from wild-type *Serratia marcescens*. This mutant had high activities of TD and AHAS enzymes, both of which were insensitive to feedback inhibition. Genetic analysis indicated that strain GIHVLr6426 carried two regulatory mutations located in the *ilv* region. One, designated *ilvA2*(Fr), leads to desensitization of the TD enzyme, and another, designated *ihr-1*, leads to constitutive synthesis of TD and AHAS enzymes (36). Previously, a threonine-producing strain of *Serratia marcescens*, T-693, which carried four regulatory mutations for threonine biosynthesis and had an increased activity of AHAS, was isolated (37). To construct an isoleucine-producing strain, the strain T-693 was used as a recipient in transduction. Strain T-803 was constructed by transferring the two mutations, *ilvA2*(Fr) and *ihr-1*, carried by strain GIHVLr6426 into strain T-693. The resulting strain had the six regulatory mutations, which are *lysCl*(Fr), *thrA*$_1$*l*(Fr), *thrA*$_2$*1*(Fr), *hnr-1, thr-1*, and *ilvA2*(Fr), for threonine and isoleucine biosynthesis and produced about 25 g/L of L-isoleucine in 96 h in a medium containing sucrose and urea by fermentation (36).

Shiio et al. reported L-isoleucine production from glucose by α-amino-β-hydroxyvaleric acid (AHV) resistant mutants of *Br. flavum* (38). Previously, it had been demonstrated that in *Br. flavum*, growth inhibition by AHV, an analogue of threonine, was reversed not only by L-threonine but also partially by L-isoleucine, and that threonine-producing mutants, which were isolated as the analogue-resistant mutants, accumulated a small amount of L-isoleucine (39). It was also reported that the AHV resistant strain ARI-129 of *Br. flavum* produced 11 g/L of L-isoleucine in 2 days by fermentation (38). Although homoserine dehydrogenase, which is one of threonine biosynthetic enzymes, from the AHV resistant strain ARI-129 was insensitive to feedback inhibition by L-isoleucine and L-threonine, no difference was observed in the TD enzyme between the parent strain and the AHV-resistant strain ARI-129. In general, because feedback inhibition of TD enzyme by L-isoleucine is competitive with respect to L-threonine, the degree of the inhibition by L-isoleucine decreases as the concentration of L-threonine increases (40). Therefore, the authors presumed that the AHV-resistant mutant ARI-129 also contained a mutation affecting permeability to L-threonine, and L-isoleucine was overproduced because of an increase in the intracellular level of L-threonine, even though the TD enzyme was not released from feedback inhibition by L-isoleucine. But no experimental evidence relative to the mechanism described above has yet been presented. Furthermore, *O*-methyl-L-threonine (OMT), an analogue of isoleucine, inhibits the growth of the ARI-129 strain and this inhibition can be reversed by addition of L-isoleucine. Hence, OMT-resistant mutants derived from the strain ARI-129 could be isolated (38). The OMT-resistant mutant, strain AORI-126, produced 14.5 g/L of L-isoleucine by fermentation (38). Specific activity of TD enzyme from the AORI-126 strain increased about twofold higher than that of the parent strain and the ARI-129 strain, whereas the degree of inhibition of the enzyme by L-isoleucine was the same among three strains.

Ikeda et al. reported L-isoleucine production from glucose by ethionine-resistant mutants of *Br. flavum* (41). Two types of L-isoleucine-producing mutants, strains 168 and 14083, were isolated from L-threonine producers by their enhanced resistance to ethionine. The L-isoleucine productivity of strains 168 and 14083 were 9.9 g/L and 11.3 g/L in 2 days by fermentation, respectively (41). The 168 mutant was shown to increase the specific activity of the TD enzyme, whereas the effect on 14083 was based on the insensitivity of the TD enzyme to feedback inhibition by L-isoleucine. However, because ethionine is an analogue of methionine, the reason that some ethionine-resistant mutants have the insensitivity of the TD enzyme is mysterious. Moreover, Ikeda et al. showed that, when the optimum fermentation conditions were controlled, including the concentration of acetate and ammonium, the L-isoleucine productivity of the 14083 strain was 33.5 g/L after 77 h of cultivation (42).

Eggeling et al. reported that *Corynebaterium glutamicum* ATCC 13032 produces about 13 g/L of L-isoleucine from a synthetic precursor of 200-mM α-ketobutyric acid in 119 h by fermentation and also produces up to about 19 g/L of L-isoleucine in 100 h in fed-batch cultures (43).

Kase et al. reported that RAM-83, a mutant derived from a threonine-producing strain of *Corynebacterium glutamicum*, when successively endowed with resistivity to such substances as thiaisoleucine, ethionine, 4-azaleucine, and α-aminobutyric acid, produces 9.7 g/L of L-isoleucine with a medium containing 10% molasses as sugar by fermentation (44). Although the thioisoleucine and α-aminobutyric-acid resistant mutations resulted in a significant increase in L-isoleucine productivity, it was not shown which enzyme was released from feedback inhibition by L-isoleucine (44).

Thus, several fermentation processes for the microbial production of L-isoleucine have been developed and, in fact, this amino acid is now produced on an industrial scale. However, the fermentation processes still have some disadvantages because of high carbon consumption during cell growth, a low conversion rate from the carbon source to the final product, and high concentrations of by-products. Because chemical characteristics of branched-chain amino acids such as L-isoleucine, L-leucine, and L-valine are similar, it is very difficult to separate them from each other. Therefore, it is preferable that content of L-leucine and L-valine in the broth should be low to facilitate the selective removal and purification of L-isoleucine.

L-ISOLEUCINE PRODUCTION BY LIVING CELL REACTION PROCESS

We have devised a new process: the living cell reaction (LCR) process using native immobilization cells of *Br. flavum* MJ-233 for L-isoleucine and L-valine production (45,46). This process differs from the usual fermentation method in that nongrowing viable cells are used, and the production of L-isoleucine and L-valine is carried out under conditions of repressed cell division and growth. A minimal medium lacking the essential growth factor, biotin, was used as the reaction mixture in the LCR process. For isoleucine production, when ethanol was the energy source and α-ketobutyric acid (α-KB) was the precursor, the productivity of L-isoleucine was 26 g/L/day. The content of L-isoleucine in total amino acids produced in the mixture was 97%, and the molecular yield from α-KB was 95% (46). This is expected to result in a simplified purification process. Moreover, L-isoleucine could be formed at high yield for several runs by recycling the same cells. However, two unnatural amino acids, which were economically difficult to remove from L-isoleucine by ordinary chromatographic separation, were found as by-products in the LCR process for L-isoleucine production. These were identified as norvaline (Nva) and *O*-ethylhomoserine (*O*-EH) (47). Kisumi et al. reported previously that Nva was formed from threonine or α-aminobutyric acid by *Serratia marcescens* in the fermentation broth of L-isoleucine, and they suggested that Nva formation was closely related to leucine biosynthesis (48,49). We reported that Nva formation was depressed by using a leucine auxotrophic *B. flavum* mutant in the LCR process (47). It was also shown that, in *Br. flavum*, Nva formation was closely related to leucine biosynthesis. On the other hand, Murooka and Harada reported that *O*-EH was specifically formed from ethanol by ethanol-assimilating bacteria (50). Also, Murooka et al. reported that *O*-EH was synthesized from ethanol and *O*-acetyl-homoserine, which is an intermediate of methionine biosynthesis in *Corynebacterium* and *Saccharomyces* (51). We reported that *O*-EH formation was repressed by addition of L-methionine to the reaction mixture in the LCR process using *Br. flavum* (47). However, the homoserine-*O*-acetyltransferase of *Br. flavum* was not subject to either inhibition or repression by addition of L-methionine, and the *O*-EH-forming enzyme, which converts *O*-acetylhomoserine to *O*-EH, was speculated to be repressed by L-methionine (47).

In addition, the LCR process could also be applied to L-valine production from glucose at high yield. In this case, the rate of production of L-valine was 35 g/L/day and the molecular yield from glucose was 80% (46).

Although cloning and expression of *ilv* genes from various organisms has been reported, there is so far no report regarding L-isoleucine production using a recombinant strain that increased or improved isoleucine biosynthetic enzymes by such genetic engineering techniques.

BIBLIOGRAPHY

1. M. Kisumi, *J. Biochem.* **52**, 390–399 (1962).
2. E.A. Adelberg, *J. Bacteriol.* **76**, 326 (1958).
3. M. Inui, A. A. Vertès, M. Kobayashi, Y. Kurusu, and H. Yukawa, *DNA Sequence* **3**, 303–310 (1993).
4. J.V. Schloss and A. Aulabaugh, *Z. Naturforsh., C* **45**, 544–551 (1990).
5. B.J. Mazur, C. F. Chui, and J. K. Smith, *Plant Physiol.* **85**, 1110–1117 (1987).
6. A. Aulabaugh and J. V. Schloss, *Biochemistry* **29**, 2824–2830 (1990).
7. O.M. Aguilar and D.H. Grasso, *J. Bacteriol.* **173**, 7756–7764 (1991).
8. R.C. Wek, C.A. Hauser, and G.W. Hatfield, *Nucleic Acids Res.* **13**, 3995–4010 (1985).
9. R.P. Lawther, D.H. Calhoun, C.W. Adams, C.A. Hauser, J. Gray, and G.W. Hatfield, *Proc. Natl. Sci. USA* **78**, 922–925 (1981).
10. C.H. Squires, M. DeFelice, S.R. Wessler, and J.M. Calvo, *J. Bacteriol.* **147**, 797–804 (1981).
11. M.A. Vandeyar, P.B. Vander Horn, J.A. Rafael, J.A. Grandoni, and S.A. Zahler, unpublished.
12. J. Tarleton, J. Malakooti, and B. Ely, *J. Bacteriol.* **176**, 3765–3774 (1994).
13. C. Keilhauer, L. Eggeling, and H. Sahm, *J. Bacteriol.* **175**, 5595–5603 (1993).
14. J.J. Godon, M.C. Chopin, and S.D. Ehrlich, *J. Bacteriol.* **174**, 6580–6589 (1992).
15. L. Gusberti, R. Cantoni, E. De Rossi, M. Branzoni, and G. Riccardi, *Gene* **177**, 83–85 (1996).
16. E. De Rossi, R. Leva, L. Gusberti, P.L. Manachini, and G. Riccardi, *Gene* **166**, 127–132 (1995).
17. A. Milano, E. De Rossi, E. Zanaria, L. Barbierato, O. Ciferri, and G. Riccardi, *J. Gen. Microbiol.* **138**, 1399–1408 (1992).
18. K. Sathasivan, G.W. Haughn, and N. Murai, *Nucleic Acids Res.* **18**, 2188 (1990).
19. P.A. Wiersma, M.G. Schmiemann, J.A. Condie, W.L. Crosby, and M.M. Moloney, *Mol. Gen. Genet.* **219**, 413–420 (1989).
20. S.C. Falco, K.D. Dumas, and K.J. Livak, *Nucleic Acids Res.* **13**, 4011–4027 (1985).
21. D. Friedberg and J. Seijffers, *Z. Naturforsc., C* **45**, 538–543 (1990).
22. M. Inui, A.A. Vertès, M. Kobayashi, Y. Kurusu, and H. Yukawa, *DNA Sequence* **4**, 95–103 (1993).
23. S. Rieble and S.I. Beale, *J. Bacteriol.* **174**, 7910–7918 (1992).
24. R.C. Wek and G.W. Hatfield, *J. Biol. Chem.* **261**, 2441–2450 (1986).
25. J.G.L. Petersen and S. Holmberg, *Nucleic Acids Res.* **14**, 9631–9651 (1986).
26. R. Dumas, G. Curien, R.T. DeRose, and R. Douce, *Biochem. J.* **294**, 821–828 (1993).
27. R. Dumas, M. Lebrun, and R. Douce, *Biochem. J.* **277**, 496–475 (1991).
28. S.M. Arfin, B. Ratzkin, and H.E. Umbarger, *Biochem. Biophys. Res. Commun.* **37**, 902–908 (1969).
29. M.D. Watson, J. Wild, and H.E. Umbarger, *J. Bacteriol.* **139**, 1014–1020 (1979).
30. R.P. Lawther, R.C. Wek, J.M. Lopes, R. Pereira, B.E. Taillon, and G.W. Hatfield, *Nucleic Acids Res.* **15**, 2137–2155 (1987).
31. F. Kunst et al. *Nature* **390**, 249–256 (1997).
32. J. Malakooti and B. Ely, *J. Bacteriol.* **176**, 1275–1281 (1994).
33. M. Kisumi, S. Komatsumara, and I. Chibata, *Appl. Environ. Microbiol.* **34**, 647–653 (1977).
34. M. Kisumi, S. Komatsubara, M. Sugiura, and I. Chibata, *J. Bacteriol.* **107**, 741–745 (1971).
35. T. Ramarkrishnan and E.A. Adelberg, *J. Bacteriol.* **87**, 566–573 (1964).
36. S. Komatsubara, M. Kisumi, and I. Chibata, *J. Gen. Microbiol.* **119**, 51–61 (1980).
37. S. Komatsubara, M. Kisumi, and I. Chibata, *Appl. Environ. Microbiol.* **38**, 1045–1051 (1979).
38. I. Shiio, A. Sasaki, S. Nakamori, and K. Sano, *Agric. Biol. Chem.* **37**, 2053–2061 (1973).
39. I. Shiio and S. Nakamori, *Agric. Biol. Chem.* **34**, 448–456 (1970).
40. R. Miyajima and I. Shiio, *J. Biochem.* **71**, 951–960 (1972).
41. S. Ikeda, I. Fujita, and F. Yoshinaga, *Agric. Biol. Chem.* **40**, 511–516 (1976).
42. S. Ikeda, I. Fujita, and F. Yoshinaga, *Agric. Biol. Chem.* **40**, 517–522 (1976).
43. I. Eggeling, C. Cordes, L. Eggeling, and H. Sahm, *Appl. Microbiol. Biotechnol.* **25**, 346–351 (1987).
44. H. Kase and K. Nakayama, *Agric. Biol. Chem.* **41**, 109–116 (1977).
45. M. Terasawa, N. Kakinuma, K, Shikata, and H. Yukawa, *Process Biochem.* **21**, 196–199 (1986).
46. M. Terasawa, M. Inui, M. Goto, K. Shikata, M. Imanari, and H. Yukawa, *J. Microbiol. Biotechnol.* **5**, 289–294 (1990).
47. M. Terasawa, M. Inui, M. Goto, Y. Kurusu, and H. Yukawa, *Appl. Microbiol. Biotechnol.* **35**, 348–351 (1991).
48. M. Kisumi, J. Kato, S. Komatsubara, and I. Chibata, *Appl. Microbiol.* **21**, 569–574 (1971).
49. M. Kisumi, S. Komatsubara, M. Sugiura, and I. Chibata, *J. Gen. Microbiol.* **69**, 291–297 (1971).
50. Y. Murooka and T. Harada, *J. Bacteriol.* **96**, 314–317 (1968).
51. Y. Murooka, K. Kakihara, T. Miwa, K. Seto, and T. Harada, *J. Bacteriol.* **130**, 62–73 (1977).

K

KINETICS, ENZYMES

ATHEL CORNISH-BOWDEN
National Center for Scientific Research
Marseilles, France

KEY WORDS

Enzymes
Inhibition
Kinetics
Manipulation of enzyme activity
Multienzyme systems

OUTLINE

INTRODUCTION

The kinetic behavior of enzymes has been studied in detail for a century, beginning with the classic work of Henri (1) and Michaelis and Menten (2). The objectives have been threefold: to gain an understanding of the mechanisms of enzyme action; to illuminate the physiological roles of enzyme-catalyzed reactions; and, mainly in recent years, to manipulate enzyme properties for biotechnological ends. Experimental practice has been overwhelmingly dominated by the first of these aims; most experiments have been designed as if shedding light on the mechanism was the principal, or even the only, objective. However, although much valuable information has been obtained in this way, there are some important aspects of enzyme function that are obscured when working mainly with isolated enzymes in conditions far removed from those that exist in the cell. For this reason the latter part of this chapter will be devoted to a discussion of enzymes as they behave in complex mixtures.

MICHAELIS–MENTEN KINETICS

The starting point for any discussion of enzyme kinetics is the Michaelis–Menten equation, which expresses the initial rate v of a reaction at a concentration a of the substrate transformed in a reaction catalyzed by an enzyme at total concentration e_0:

$$v = \frac{k_0 e_0 a}{K_m + a} = \frac{Va}{K_m + a} \tag{1}$$

The parameters are k_0, the catalytic constant, and K_m, the Michaelis constant. The form shown in the middle is more fundamental than that on the right, but the second form, in which $k_0 e_0$ is written as the limiting rate V, is often used because the enzyme concentration in meaningful units is often not known. V is the limit that v approaches as the enzyme becomes saturated (that is, when a becomes very large) and K_m is the value of a at which $v = 0.5V$ (that is, at which the rate is half-maximal). The ratio k_0/K_m is called the specificity constant and given the symbol k_A: It is a more fundamental constant than K_m in the analysis of enzyme mechanisms (i.e., it has a simpler mechanistic meaning), but the equation is usually written in terms of V (or k_0) and K_m nonetheless.

The principal assumption implied by equation 1 is that the rate of the reverse reaction is negligible: This may be because the reaction is irreversible for practical purposes, but even with reversible reactions the reverse reaction can be made negligible by measuring the rate in the absence of products and by extrapolating the rate back to zero time, that is, by estimating the initial rate at a time when no products have accumulated. Even if there is no significant reverse reaction, products can still affect the rate of the forward reaction because product inhibition (discussed later) is a common phenomenon.

Another assumption is that the reaction is allowed sufficient time to reach a steady state because all reactions pass through an initial acceleration phase known as the transient state. This phase is normally very brief (a few milliseconds), and in practice enzymes are usually studied under steady-state conditions. Note, however, that this is made experimentally possible by working with extremely small enzyme concentrations compared with those that may exist in the cell. The use of very low enzyme concentrations has two important consequences: First, it normally means that the enzyme concentration can be neglected in comparison with the substrate concentration, and second it makes the steady-state rate sufficiently slow to be easily measured and the steady-state phase long enough to be meaningful. If high enzyme concentrations were used, the transient state would be as brief as before, but the steady state would also be very brief, so that there would be no period in which one could adequately treat the rate as constant.

Equation 1 may be derived from the following model

$$\mathrm{A} + \mathrm{E} \underset{k_{-1}}{\overset{k_1}{\rightleftarrows}} \mathrm{EA} \underset{k_{-2}}{\overset{k_2}{\rightleftarrows}} \mathrm{EP} \underset{k_{-3}}{\overset{k_3}{\rightleftarrows}} \mathrm{E} + \mathrm{P} \quad (2)$$

which assumes that the reaction passes through an enzyme–substrate complex EA, which undergoes catalytic transformation to an enzyme–product complex EP, which then breaks down to form products. Although real enzyme mechanisms may be more complicated than this, every reaction passes through steps of substrate binding, chemical transformation, and product release. In simple introductory treatments the second and third steps are often treated as a single step, but conceptually they are clearly distinct. In reactions of more than one substrate, the steps do not necessarily occur in the order one might guess; that is, some products may be released before all substrates have bound, but the general principle that any reaction involves the same three kinds of step remains valid.

The Michaelis–Menten parameters can be defined as follows in terms of the rate constants shown in equation 2:

$$k_0 = \frac{k_2 k_3}{k_{-2} + k_2 + k_3}; \; k_A = \frac{k_1 k_2 k_3}{k_{-1}k_{-2} + k_{-1}k_3 + k_2 k_3};$$
$$K_m = \frac{k_{-1}k_{-2} + k_{-1}k_3 + k_2 k_3}{k_1(k_{-2} + k_2 + k_3)} \quad (3)$$

Note that none of the three parameters has a simple transparent meaning. The interpretations commonly attributed to them depend on additional simplifying assumptions that are not always correct. For example, K_m is often said to be equal to the equilibrium constant k_{-1}/k_1 for dissociation of A from EA, but the expression in equation 3 does not take this form unless k_2 is very small. As there is no good reason for k_2 to be small, and indeed ideas of evolutionary optimization of enzyme function lead one to expect the opposite when the enzyme is acting on its natural physiological substrate, it follows that K_m should not, in general, be regarded as a measure of the equilibrium dissociation constant.

Despite these difficulties in providing a detailed mechanistic meaning to K_m, it does provide a measure of the tightness of substrate binding in the steady state, as it is quite correct to take a/K_m as equal to the ratio of the sum of concentrations of all enzyme complexes (i.e., both EA and EP) over the concentration of free enzyme. Similarly k_0 provides a valid measure of the capacity of the enzyme–substrate complex to react to give products, even if it cannot be interpreted as the rate constant for a unique step in the mechanism. The reason for the term *specificity constant* for k_A, that is, its relationship to enzyme specificity, will become clear once inhibition has been considered.

Graphical Analysis

Equation 1 defines a hyperbolic dependence of rate on substrate concentration, as illustrated in Figure 1. The initial steep rise in v as a increases from zero is rapidly transformed into the phenomenon of saturation, whereby further increases in a produce smaller and smaller increases in v, which approaches, but does not reach or exceed, the limiting rate V. The rectangular hyperbola makes this type

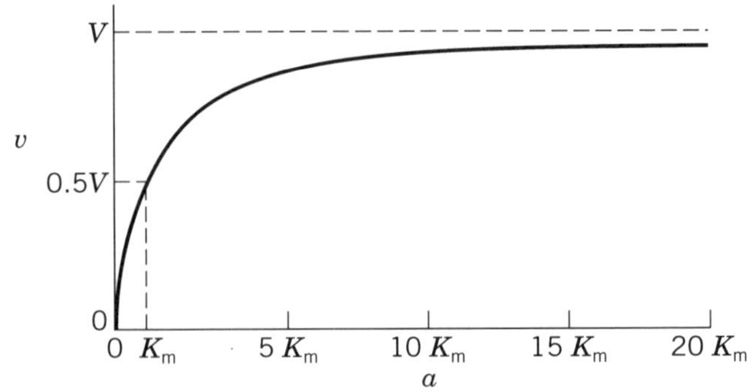

Figure 1. Michaelis–Menten dependence of rate on substrate concentration. The curve is a rectangular hyperbola through the origin, approaching a limit of $v = V$ at saturation. The rate is $0.5\,V$ at a substrate concentration equal to the Michaelis constant, K_m, but note that the approach to the limit is slow, so that, for example, even at $a = 10K_m$ the rate is still nearly 10% less than V.

of plot inconvenient for estimating the values of the kinetic parameters (because the line does not approach the saturation limit closely enough at reasonable values of a to allow direct measurement of V). For this purpose, therefore, it is usual to transform equation 1 into one of the following three forms, which underlie the three straight-line plots illustrated in Figures 2 through 4:

$$\frac{1}{v} = \frac{1}{V} + \frac{K_m}{V}\frac{1}{a} \quad (4)$$

$$\frac{a}{v} = \frac{K_m}{V} + \frac{1}{V}a \quad (5)$$

$$v = V - K_m\frac{v}{a} \quad (6)$$

The double-reciprocal plot, illustrated in Figure 2 and based on equation 4, is the mostly widely used, but it is also the least satisfactory because it distorts the effect of experimental error to such an extent that it is difficult to form any visual judgement of where the best line should be drawn. The other two plots are better, and the plot of v

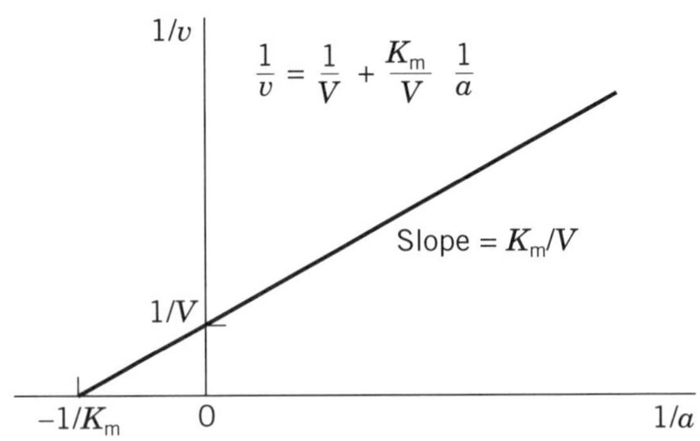

Figure 2. The double-reciprocal plot. This is the most widely used method of plotting the Michaelis–Menten equation as a straight line. However, the severe distortion of any experimental errors in the original data causes it to give a misleading impression.

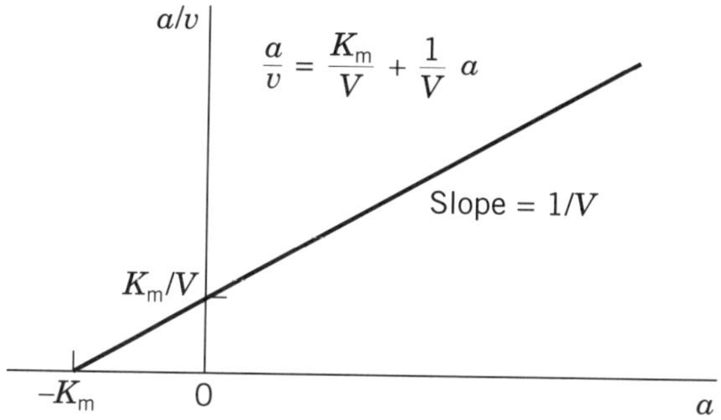

Figure 3. The plot of a/v against a. This alternative to the plot shown in Figure 2 produces much less distortion of the experimental error.

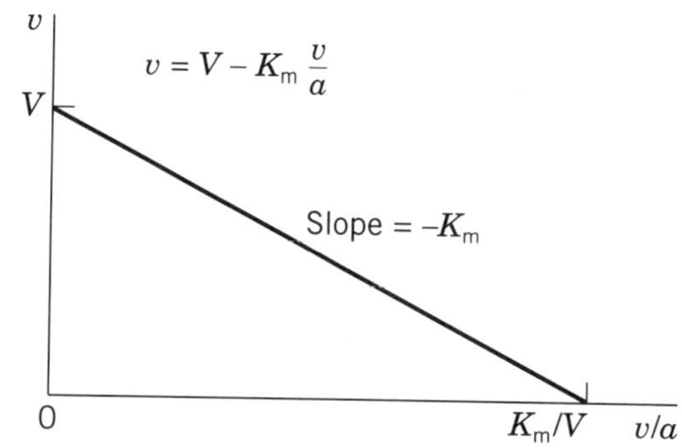

Figure 4. The plot of v against v/a. The third way of plotting the Michaelis–Menten equation as a straight line also avoids the error-distorting property of the plot shown in Figure 2, and maps the entire range of observable rates (from 0 to V) onto a finite range of paper. This is a desirable property because it makes it impossible to disguise deficiencies in the experimental design.

against v/a (Fig. 4, equation 6) has the particular advantage that the entire observable range of v values, from 0 to V, is mapped onto a finite range of graph; this makes it easy to judge by eye if an experiment has been well designed. On the other hand, it has the disadvantage that v, normally the less reliable measurement, contributes to both coordinates, and errors in v cause deviations along lines through the origin, rather than parallel with one or the other axis.

In modern practice it is usually best to regard these plots as for illustration purposes only, and to use suitable computer programs for the actual parameter estimation. For this purpose it is not sufficient just to apply unweighted linear regression to the straight-line plots, as this suffers from the same statistical distortions as the plots themselves. Full treatment would require more space than is available here, but see Ref. 3. The following two equations for calculating best-fit values of K_m and V give satisfactory results if the v values have uniform coefficient of variation (uniform standard deviation expressed as a percentage), as is usually at least approximately correct:

$$K_m = \frac{\Sigma v^2 \Sigma(v/a) - \Sigma(v^2/a)\Sigma v}{\Sigma(v^2/a^2)\Sigma v - \Sigma(v^2/a)\Sigma(v/a)} \qquad (7)$$

$$V = \frac{\Sigma(v^2/a^2)\Sigma v^2 - [\Sigma(v^2/a)]^2}{\Sigma(v^2/a^2)\Sigma v - \Sigma(v^2/a)\Sigma(v/a)} \qquad (8)$$

Each summation is made over all observations.

Two-Substrate Reactions

Enzymes that catalyze reactions of a single substrate are only a small minority of all the enzymes known, but the Michaelis–Menten equation remains useful for examining the kinetics of the more common case of a reaction with two substrates and (often but not necessarily) two products, because such a reaction normally obeys Michaelis–Menten kinetics when only one substrate concentration is varied at a time. This is illustrated by the following typical equation for such a reaction:

$$v = \frac{k_0 e_0 ab}{K_{iA}K_{mB} + K_{mB}a + K_{mA}b + ab} \qquad (9)$$

Although at first sight this appears quite different from equation 1, it can be arranged in the same form if one of the two substrate concentrations, for example b, is treated as a constant:

$$v = \frac{\left(\dfrac{k_0 b}{K_{mB} + b}\right) e_0 a}{\dfrac{K_{iA}K_{mB} + K_{mA}b}{K_{mB} + b} + a} \qquad (10)$$

The two fractions in this equation can be regarded as "apparent" values of the Michaelis–Menten parameters for A, that is, the equation can be written as

$$v = \frac{k_0^{\text{app}} e_0 a}{K_{mA}^{\text{app}} + a} \qquad (11)$$

with

$$k_0^{\text{app}} = \frac{k_0 b}{K_{mB} + b};\quad k_A^{\text{app}} = \frac{(k_0/K_{mA})b}{(K_{iA}K_{mB}/K_{mA}) + b};\quad K_{mA}^{\text{app}} = \frac{K_{iA}K_{mB} + K_{mA}b}{K_{mB} + b} \qquad (12)$$

Notice that the expressions for the apparent values of k_0 and k_A are both individually of Michaelis–Menten form with respect to b, whereas that for the apparent value of K_m is more complicated: This behavior is quite typical and is one of the reasons why k_0 and k_A should regarded as more fundamental parameters than K_m. More generally, the concept of apparent parameters pervades the analysis of simple cases in steady-state enzyme kinetics, being important for the study of reactions with multiple substrates and inhibition.

INHIBITION AND ACTIVATION

Inhibition

For most enzyme-catalyzed reactions, molecules exist that resemble the substrate closely enough to bind to the enzyme, but not closely enough to undergo a chemical reaction. Such a molecule is known as a competitive inhibitor and causes competitive inhibition, characterized by a rate equation of the following form:

$$v = \frac{k_0 e_0 a}{K_m(1 + i/K_{ic}) + a} \tag{13}$$

in which i is the concentration of the inhibitor and K_{ic} is the competitive inhibition constant. (The qualification "competitive" and the second subscript c are usually omitted if only this simplest kind of inhibition is being considered.)

Inhibitors can interfere with catalysis as well as with substrate binding. In the simplest case, an inhibitory term affects the variable term in the denominator of the Michaelis–Menten equation, instead of the constant term:

$$v = \frac{k_0 e_0 a}{K_m + a(1 + i/K_{iu})} \tag{14}$$

This is called uncompetitive inhibition, and the inhibition constant K_{iu} is the uncompetitive inhibition constant. This is important as a limiting case of inhibition, but in its pure form it is not at all common. Much more often one has mixed inhibition, when both competitive and uncompetitive effects occur simultaneously:

$$v = \frac{k_0 e_0 a}{K_m(1 + i/K_{ic}) + a(1 + i/K_{iu})} \tag{15}$$

There is no particular reason for the two inhibition constants K_{ic} and K_{iu} to be equal, and most of the mechanisms one might propose to account for mixed inhibition lead one to expect them to be different, yet the case where $K_{ic} = K_{iu}$ is often given an undeserved prominence in discussions of inhibition, largely because experiments done many years ago suggested that it was a more common phenomenon than it is. This is called noncompetitive inhibition and its rate equation is the same as equation 15, but with both K_{ic} and K_{iu} written simply as K_i.

All of these kinds of inhibition are conveniently discussed in terms of apparent Michaelis–Menten parameters. In the general case (equation 15), these are as follows:

$$k_0^{app} = \frac{k_0}{1 + i/K_{iu}}; \quad k_A^{app} = \frac{k_A}{1 + i/K_{ic}};$$
$$K_m^{app} = \frac{K_m(1 + i/K_{ic})}{1 + i/K_{iu}} \tag{16}$$

Note that the first two expressions have the same form, and both simplify to independence of i in the event that one or other inhibition term is negligible. The expression for the apparent value of K_m is more complicated, especially when one considers how it varies with the different types of inhibition: It increases with the concentration of a competitive inhibitor, it decreases as the concentration of an uncompetitive inhibitor increases, it may change in either direction as the concentration of a mixed inhibitor increases, or it is independent of inhibitor concentration if the inhibition is noncompetitive. In general it is simplest to regard k_A as the parameter affected by competitive inhibition, negligibly so when the competitive component is negligible, k_0 as the parameter affected by uncompetitive inhibition, negligibly so when the uncompetitive component is negligible, and K_m just as the ratio of the two, so $K_m = k_0/k_A$.

The effects of the different kinds of inhibition on the common plots as illustrated in Figures 2 through 4 follows naturally from equation 16. Any competitive effect affects the apparent value of k_A, hence, it increases the slope of the plot of $1/v$ against $1/a$ (Fig. 2), it increases the ordinate intercept of the plot of a/v against a (Fig. 3), and it decreases the abscissa intercept of the plot of v against v/a (Fig. 4). Conversely, any uncompetitive effect increases the ordinate intercept of the plot of $1/v$ against $1/a$, increases the slope of the plot of a/v against a, and decreases the ordinate intercept of the plot of v against v/a. When both components of the inhibition are present, both kinds of effects occur. As an illustration we may consider just one example, the effect of competitive inhibition on the plot of $1/v$ against $1/a$: Plots made at various different inhibitor concentrations produce a family of straight lines intersecting on the ordinate axis, as shown in Figure 5, the lack of effect on the ordinate intercept being a direct consequence of the lack of effect on the apparent value of V.

Specificity

Specificity is the most fundamentally important property of enzymes. Although one is often impressed by the catalytic effectiveness of enzymes, accelerating a reaction is not, in reality, difficult: Heating the reaction mixture in a sealed tube is an efficient way of accelerating virtually any reaction, essentially without limit. What is difficult is to accelerate one selected reaction without at the same time accelerating a great mass of unwanted reactions. What is important about an enzyme, therefore, is not that it is an excellent catalyst for a small set of reactions, but that it is an extremely bad catalyst—virtually without any activity—for all other reactions. In other words, the essential properties of enzymes are that they act under very mild conditions and are highly specific.

In the past, specificity was often assessed by comparing the kinetic parameters for different reactions measured in

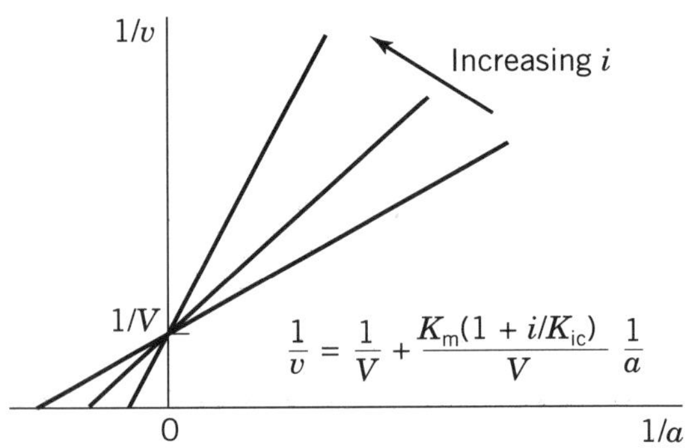

Figure 5. Effect of competitive inhibition on the double-reciprocal plot.

isolation from one another, which led to sterile arguments as to whether specificity was best measured in terms of k_0, K_m, or some combination of the two. This type of argument was resolved once it was realized that the only meaningful way of defining specificity is as a property of an enzyme that allows it to discriminate between substrates that are mixed together. The simplest way to consider this is with a model similar to that for competitive inhibition, except that one assumes that both molecules are capable of reacting. The equation for reaction of one substrate A in the presence of a competing substrate A′ follows an equation similar to that for competitive inhibition (equation 13):

$$v = \frac{k_0 e_0 a}{K_m(1 + a'/K'_m) + a} \tag{17}$$

with the inhibitor concentration replaced by a', the concentration of A′, and the inhibition constant by K'_m, the Michaelis constant for the reaction of A′ considered in isolation. The rate of reaction of A′ is given by the same equation with an obvious transposition of symbols:

$$v' = \frac{k'_0 e_0 a'}{K'_m(1 + a/K_m) + a'} \tag{18}$$

It can then be seen that the ratio of rates is the ratio of substrate concentrations multiplied by the ratio of specificity constants:

$$\frac{v}{v'} = \frac{k_0/K_m}{k'_0/K'_m}\frac{a}{a'} = \frac{k_A a}{k'_A a'} \tag{19}$$

This result, which is still less well known than its importance merits, is the reason for the term *specificity constant.* Note that although inspection of equation 1 suggests that k_A is no more than the parameter that defines the rate at very low substrate concentrations, no assumption about the magnitudes of the concentrations was made in arriving at equation 19. It is thus valid at all concentrations, and the specificity constant measures specificity at all concentrations, not just low ones.

Activation

Activation is the opposite from inhibition, in which a reaction proceeds more rapidly in the presence of a particular molecule than in its absence. It is less common than inhibition, and discussion is complicated by the fact that a variety of quite different phenomena have been termed "activation." The most important of these is a confusion between true activation and the case where the activator is really a component of the substrate. Numerous ATP-handling enzymes are said to be activated by magnesium ions, when in reality the complex MgATP is the true substrate, that is, the species that reacts with the enzyme. Other metal ions, such as the zinc in a number of enzymes, may be true activators as they bind to the enzyme itself and confer catalytic properties on it.

Another misuse of the term activation relates to coenzymes such as NAD in many dehydrogenases: Alcohol dehydrogenase, for example, may be said to be activated by NAD, but although consideration of the metabolic pathway in which the reaction occurs may make it convenient to regard ethanol as the substrate and NAD as the coenzyme, this is meaningless when the reaction is considered in isolation. So far as alcohol dehydrogenase is concerned, it catalyzes a reaction that requires two substrates, ethanol and oxidized NAD; the reaction will not proceed unless both are present, and neither has any more reason to be called the substrate than the other.

When such improper uses of the term are excluded, there remain a number of enzymes for which the true inverse of inhibition occurs. In the simplest cases the equations are just the inverse of inhibition equations, with terms of the form i/K_i replaced by ones of the form K_x/x (for an activator X with activation constant K_x). However, the simplest cases constitute a smaller proportion of the whole than they do for inhibition. This is because whereas most inhibitors inhibit completely, in the sense that enzyme species with inhibitor bound retain no activity as long as the inhibitor remains bound, many enzymes subject to activation retain some activity in the absence of the activator. As a result, full analysis of activation is often more complicated than it is for inhibition, but this will not be discussed further here.

Irreversible Inhibition

The types of inhibition considered so far are examples of reversible inhibition; the inhibitor binds reversibly and catalytic activity returns when the inhibitor is released. Irreversible inhibition also occurs, in which the inhibitor either binds so tightly that for practical purposes it cannot be removed, or reacts with the enzyme and converts it irreversibly to a form that has no catalytic activity. These two cases are conceptually different, and the former is more correctly called tight-binding inhibition, rather than irreversible inhibition. However, they are not easy to distinguish in practice, and have similar practical effects and, hence, similar practical uses.

Although irreversible inhibition has played a smaller part than reversible inhibition in the academic study of enzyme mechanisms, it has far greater industrial and pharmacological importance. This is because competitive inhibitors, the most common kind of reversible inhibitors, are almost completely ineffective in complete physiological systems, for reasons to be considered shortly. By contrast, whenever irreversible (or tight-binding) inhibition occurs in a physiological system, it can be expected to have profound effects. Many toxic and pharmacologically active substances owe their effects to irreversible inhibition.

Both tight-binding and irreversible inhibition manifest themselves in ways that allow them to be confused with noncompetitive inhibition, as in equation 15 with the two inhibition constants equal. This is because the practical effect of irreversible inhibition is not on any of the kinetic parameters in the Michaelis–Menten equation, but on e_0, the total enzyme concentration. However, a decrease in e_0 can be confused with a decrease in the apparent value of k_0, as they occur as a product in equation 1. Although uncompetitive inhibition affects k_0, it does so without affecting k_A, and thus also changes K_m. Decreasing k_0 without affecting K_m, similar to what one would observe if e_0 decreases, is a definition of noncompetitive inhibition.

Inhibitory Effects in Metabolic Systems

Competitive and uncompetitive inhibition are sufficiently similar in their effects in artificial experiments on isolated enzymes that they are often not distinguished, and an uncompetitive component in mixed inhibition often passes unnoticed. Many inhibitors are described in the literature as competitive inhibitors in the absence of any real evidence of the type of inhibition. This sort of confusion can easily lead to the entirely false idea that they are similar in their effects in systems where the inhibited enzyme is mixed with other enzymes and catalyzes a step in the middle of a pathway.

In a typical experiment in vitro, one decides the concentrations of the various components in advance and measures the rate that results; however, this is very different from what happens in the cell. To a first approximation, an enzyme catalyzing a step in the middle of a pathway must transform its substrate at the rate at which it arrives, that is, within certain limits it has little or no effect on the rate of its reaction, but instead determines the concentrations of the metabolites around it. (This is an oversimplification, but is useful for discussion.)

It is useful therefore to transform equations 13 and 14 into expressions for a in terms of i:

$$a = \frac{vK_m(1 + i/K_{ic})}{k_0e_0 - v} \qquad (20)$$

$$a = \frac{vK_m}{k_0e_0 - v(1 + i/K_{iu})} \qquad (21)$$

However similar equations 13 and 14 may seem, their transformed versions are drastically different. Equation 20 shows a linear dependence of a on i, which means that increasing i can never result in uncontrolled increases in a. This is illustrated in Fig. 6a. Even at an inhibition concentration equal to the inhibition constant, the substrate concentration is only doubled. By contrast, the curve defined by equation 21 is a rectangular hyperbola (Fig. 6b) that produces a steep and uncontrolled rise in substrate concentration at quite moderate inhibitor concentrations (4). The point is that in competitive inhibition, rises in substrate and inhibitor concentrations oppose one another—not only does the inhibitor compete with the substrate, but equally, the substrate competes with the inhibitor. In uncompetitive inhibition, however, these effects potentiate one another.

It follows that although it is relatively easy to find molecules that will act as competitive inhibitors, it is also largely useless as a strategy for designing pesticides or drugs because it is correspondingly easy for the organism to counteract the effect of the inhibition. To produce major metabolic effects one needs uncompetitive inhibitors, irreversible inhibitors, or tight-binding inhibitors: None of these are as easy to produce as weakly binding competitive inhibitors, but they are far more effective.

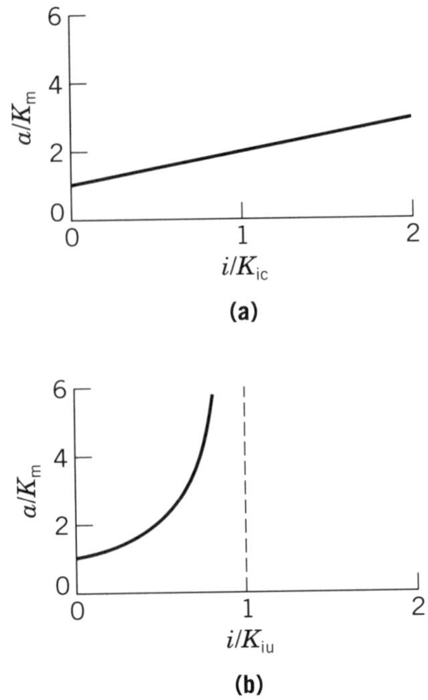

Figure 6. Effects of **(a)** competitive and **(b)** uncompetitive inhibition on the concentration of substrate in a constant-rate system. Both curves are drawn for the case of $a = K_m$ in the absence of inhibitor. Note that both kinds of inhibitor have quantitatively equal effects at very low concentrations, but the initial slope is maintained indefinitely if the inhibition is competitive, whereas it rapidly becomes infinite if the inhibition is uncompetitive.

NON-MICHAELIS–MENTEN BEHAVIOR

All of the cases considered so far can be regarded as generalizations of the Michaelis–Menten equation (equation 1). However, although many enzymes do behave in this way, at least as a first approximation, there are some important exceptions. It is simple to calculate from equation 1 that if $a = K_m/9$ then $v = 0.1V$ and if $a = 9K_m$ then $v = 0.9V$; in other words, spanning the 10–90% range of available rates requires an 81-fold range of substrate concentrations, almost two orders of magnitude. Similar calculations may be done with any of the equations of the Michaelis–Menten type for additional substrates, inhibitors, or activators. Their implication is that as long as enzymes follow Michaelis–Menten kinetics, their rates cannot be adequately varied by manipulating concentrations of substrates, for example, because effective regulation will often require sensitivity to small changes in signals—certainly changes much smaller than two orders of magnitude.

A second difficulty arises from the fact that inhibition of the types considered commonly derives from structural similarities between inhibitors and substrates or products, whereas there is no reason to expect the molecules needed for metabolic signals to resemble the substrates or products of the enzymes that need to respond to the signals. In reality, the concentration of the end product of a pathway often serves as such a signal: too low, and the pathway needs to be activated; too high, and it needs to be inhibited. It is often found, therefore, that the enzyme that catalyzes the first committed step of a pathway, that is, the first step after a branch point, in the branch that leads to the end product in question, is inhibited by that end product.

For inhibition of this kind to be possible, the enzyme must have a specific binding site for the end product, independent of the binding sites for substrates and products. Such a site is called an allosteric site, and the phenomenon is called allosteric inhibition. Because the need for it often coincides with the need for higher sensitivity than is provided by Michaelis–Menten kinetics and the common kinds of inhibition, allosteric inhibition is often cooperative: This means that the equations that define it are more complicated than those considered above, allowing, for example, a change from 10% to 90% inhibited over a concentration range much smaller than 81-fold, and typically less than 10-fold (though rarely, if ever, less than 3-fold) (Fig. 7).

The fact that allosteric and cooperative behavior are often found together has led many authors to treat the two terms as synonymous, a tendency encouraged by the fact that in one of the most widely accepted models of non-Michaelis–Menten behavior, that of Monod, Wyman, and Changeux (5), both result from the same structural properties attributed to the enzyme. Nonetheless, most careful authors consider the two properties to be conceptually distinct and not necessarily occurring together, so the two terms should be considered distinct as well.

KINETICS OF MULTIENZYME SYSTEMS

As noted in the introduction, nearly all kinetic studies of enzymes have been carried out using isolated enzymes, and although this has been very valuable for arriving at a good understanding of the nature of enzyme catalysis, it is quite inadequate as a guide to how systems of enzymes will behave. One cannot assume that the flux through a metabolic pathway is a property of a unique enzyme catalyzing the rate-limiting step, and that the properties of the pathway as a whole can be deduced from studies of the kinetics of this one enzyme in isolation.

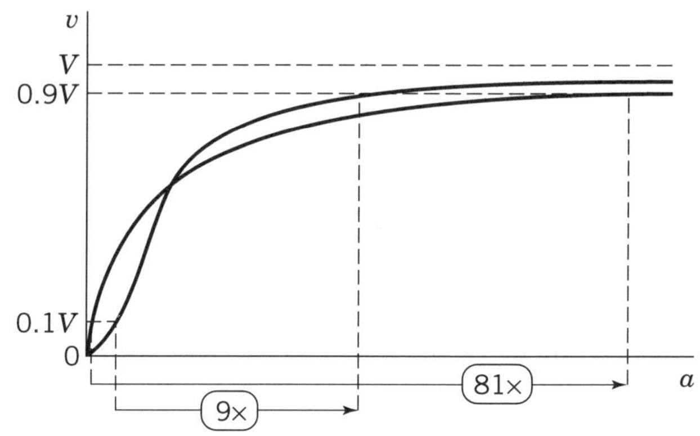

Figure 7. Non-Michaelis–Menten kinetics. For an enzyme obeying the Michaelis–Menten equation (cf. Fig. 1), an 81-fold increase in substrate concentration is needed to bring the rate from 10% to 90% of V. If the enzyme shows positive cooperativity the curve typically becomes sigmoid (S-shaped), and this range of substrate concentrations is decreased (to nine-fold in the example, but in strongly cooperative cases it can be as small as three-fold).

Space does not permit a full analysis of this subject, which has developed from a classic paper by Kacser and Burns (6), but it should suffice to examine an example of a pathway in which biotechnologists have attempted to increase the flux by identifying the rate-limiting enzyme and using genetic manipulation to increase its activity. Tryptophan biosynthesis in yeast provides such an example, tryptophan being a commercially valuable metabolite for which increased production would be very desirable. The tryptophan pathway consists of five enzymes, and in the classical model any one of these could be the "key" enzyme catalyzing the rate-limiting step. However, when the activity of each of these enzymes was increased in turn, either singly or at the same time as others in the pathway, the results were trivial: Increases of enzyme activity of 20-fold or greater produced flux increases of perhaps 30%. Only when all five enzyme activities were increased (or all but one, apparently unimportant, activity) was there a substantial increase in flux, which even then was much smaller than the increase in enzyme activity (7).

The object here is not to analyze in detail why manipulation of tryptophan biosynthesis did not produce the desired results, but to use it to illustrate the point that the whole approach is misconceived. One cannot treat the kinetic behavior of systems of enzymes as if it were determined by the properties of a single component. Moreover, abundant evidence exists to show that all organisms have evolved regulatory mechanisms to control metabolic fluxes and concentrations to satisfy their own requirements. Artificially trying to force more activity in a pathway by increasing the activities of certain enzymes simply stimulates the regulatory mechanisms to resist. Tryptophan biosynthesis is just one example of a general result, and similar efforts in other pathways—for example, increased alcohol production in yeast and increased starch production in potatoes—have produced similar results.

The essential point is that flux control is not a property of a single enzyme in a pathway, but is shared among all the enzymes in the pathway. Strictly speaking, it is shared among all the enzymes in the cell, but it is usually safe to assume that enzymes catalyzing reactions remote from the pathway of interest have very little effect on the flux through it, so to a first approximation one can consider flux control to be shared among the enzymes of the pathway.

To express this idea in quantitative terms, one can define a flux control coefficient for any enzyme by the following equation:

$$C_i^J = \frac{\partial \ln J}{\partial p} \bigg/ \frac{\partial \ln v_i}{\partial p} \tag{22}$$

This definition compares the effect on the flux J through a pathway of some perturbation of the activity of the ith enzyme, represented by a change in the parameter p, with the effect the same perturbation would have on the rate v_i of the same reaction if it were considered in isolation. The identity of the parameter p does not have to be specified, because as long as it affects only one enzyme, the control coefficient defined by equation 22 is independent of the manner in which the flux and isolated rate are perturbed. However, to make the definition more concrete, consider the case where p is the logarithm of the enzyme concentration, e_i. As most reactions are considered under conditions where the rate is proportional to the enzyme concentration, it will often be true to write $d\ln v_i = d\ln e_i$, so that the denominator in equation 22 has a value of unity and the whole equation simplifies to the following:

$$C_i^J = \frac{\partial \ln J}{\partial \ln e_i} \tag{23}$$

This equation is less abstract and simpler to understand than equation 22, and in the past was often used as a primary definition of a control coefficient. However, this is not recommended, first because it is not always true that the isolated rate is proportional to the enzyme concentration, and second because equation 23 can give the false impression that control coefficients are concerned only with effects brought about by changes in enzyme concentration.

Flux control coefficients are measures of how much the flux through a pathway is dependent on the activities of the individual enzymes. Mathematical analysis shows that they satisfy a property called the summation relationship:

$$\Sigma C_i^J = 1 \tag{24}$$

The interpretation of this relationship would be straightforward if all flux control coefficients were always positive, but in reality negative flux control coefficients also occur; for example, if a pathway contains a branchpoint, the enzymes in one branch normally have negative flux control coefficients for flux through the other. Nonetheless, in practice, negative flux control coefficients are nearly always small in magnitude, so although the sum of positive flux control coefficients may be greater than 1, it is rarely much greater than 1, so that the idea of sharing flux control among all the enzymes of a system is reasonably accurate. It follows that we should not expect any enzyme to have complete control over flux, and the closer to complete control any enzyme approaches, the less control all of the others have, taken together.

Moreover, a flux control coefficient is not a constant, but tends to decrease when the activity of an enzyme is increased. In other words, even if one can identify one enzyme that has a large proportion of the total flux control, increasing its activity will tend to decrease that proportion, so that the amount the flux can be increased by increasing the activity of one enzyme will always be small. Once this is understood, the failure to increase the flux to tryptophan (in the example mentioned) or to achieve significant flux increases (in other similar examples), ceases to be a mysterious result, but is simply what was to be expected.

FURTHER READING

The topic of this article is referred to in all textbooks of biochemistry, but unfortunately the treatment is nearly always very superficial and potentially misleading. Since Mahler and Cordes (8), there has been no serious attempt to cover enzyme kinetics in a general biochemistry text. More specialized textbooks were abundant in the past, but have become much less so. There are, however, two recent ones (9,10) that may be consulted for more information about most of the points discussed here, as well as topics such as the pH- and temperature-dependence of enzyme-catalyzed reactions, which require more detail than can be covered adequately in a short article. Another source is the book by Segel (11), which was originally published in 1975, but has been reissued recently.

Fast reactions are not dealt with in this article, and as they require both techniques and methods of analysis different from those of steady-state kinetics, it is best to refer to a specialized source, such as Hiromi (12), who provides a thorough treatment. Two recent books deal with the kinetics of multienzyme systems, one (13) at a very readable level, the other (14) much more advanced. This is also covered much more briefly in two of the books mentioned earlier (9,10).

BIBLIOGRAPHY

1. V. Henri, *Lois Générales de l'Action des Diastases*, Hermann, Paris, 1903.
2. L. Michaelis and M.L. Menten, *Biochem. Z.* **49**, 333–369 (1913).
3. A. Cornish-Bowden, *Analysis of Enzyme Kinetic Data*, Oxford University Press, Oxford, 1995.
4. A. Cornish-Bowden, *FEBS Lett.* **203**, 3–6 (1986).
5. J. Monod, J. Wyman and J.-P. Changeux, *J. Molec. Biol.* **12**, 88–118 (1965).
6. H. Kacser and J.A. Burns, *Symp. Soc. Exp. Biol.* **27**, 65–104 (1973).
7. P. Niederberger, R. Prasad, G. Miozzari, and H. Kacser, *Biochem. J.* **287**, 473–479 (1992).
8. H.R. Mahler and E.H. Cordes, *Biological Chemistry*, 2nd ed., Harper & Row, New York, 1972.
9. A.R. Schulz, *Enzyme Kinetics: From Diastase to Multi-Enzyme Systems*, Cambridge University Press, Cambridge, 1994.
10. A. Cornish-Bowden, *Fundamentals of Enzyme Kinetics*, 2nd ed., Portland Press, London, 1995.
11. I.H. Segel, *Enzyme Kinetics: Behavior and Analysis of Rapid Equilibrium and Steady-State Enzyme Systems*, Wiley, New York, 1993.
12. K. Hiromi, *Kinetics of Fast Enzyme Reactions: Theory and Practice*, Kodansha, Tokyo 1979.
13. D. Fell, *Understanding the Control of Metabolism*, Portland Press, London, 1997.
14. R. Heinrich and S. Schuster, *The Regulation of Cellular Systems*, Chapman & Hall, New York, 1996.

KINETICS, MICROBIAL GROWTH

NICOLAI S. PANIKOV
Institute of Microbiology, Russian Academy of Sciences
Moscow, Russian Federation

KEY WORDS

Cell size distribution
Colonies
Energy and conserved substrates
Growth models
Macrostoichiometry
Maintenance
Microstoichiometry
Physiological state
Steady-state and transient dynamics
Yield

OUTLINE

INTRODUCTION

Kinetics (Greek κινετικοσ, forcing to move) is a branch of natural science that deals with the *rates and mechanisms* of any processes—physical, chemical, or biological. Kinetic studies in microbiology cover all dynamic manifestations of microbial life: growth itself, survival and death, product formation, adaptations, mutations, cell cycles, environmental effects, and biological interactions. Kinetics provides a theoretical framework for optimal design in biotechnologies based on fermentation and enzyme catalysis, as well as on employment of outdoor activity of natural microbial populations (wastewater treatment, soil bioremediation, etc.)

Contrary to simple rates measurements, kinetic studies require the perception of the underlying basic *mechanisms* of studied processes. We will define mechanistic studies as those that interpret some complex process as an interplay of several simpler reactions, for example, cell growth can be explained through activity of enzymes and microbial community dynamics can be interpreted through behavior of individual cells and populations. Ideally, mechanistic studies infer the coupling of experimental measurements with analysis of simulating mathematical models. The models formalize postulated mechanisms, so that the comparison of observations and the model's predictions allows one to discard an incorrect hypotheses.

The quantitative studies in microbiology often involve the assessment of *growth stoichiometry*. Stoichiometry [Greek στοικηειον, element] is the quantitative relationship between reactants and products in a chemical reaction. In microbiology, stoichiometry stands for a quantitative relationship between substrates and products of microbial processes, including biomass formation (the consequence of complying with mass and energy conservation laws). In practical terms, kinetic and stoichiometry are tightly linked to each other, but stoichiometry mainly addresses problems of a static nature (how much? in what proportion?), whereas kinetics considers the dynamics questions (at what rate? by which mechanism?).

GROWTH STOICHIOMETRY

Macrostoichiometry of Microbial Growth

By analogy to simple chemical reactions, we can represent growth as a conversion of a number of substrates (medium components) into cell mass and products. Growth of aerobic heterotrophic microorganisms can be approximated by the following stoichiometric equation (substrates = biomass + products) (1,2):

$$CH_mO_l + a_1NH_3 + a_2HPO_4^{2-} + a_3K^+ + \ldots + bO_2 = YCH_pO_nN_qP_oK_v \ldots + a_4CO_2 + a_5H_2O \quad (1)$$

Here, microbial biomass is empirically expressed by the gross formula $CH_pO_nN_qP_oK_v\ldots$, for example, if some average microbial cell contains per dry cell weight (%) C 46, H 7.5, O 31, N 11; and P, 1.3, then the biomass formula is $CH_{1.9}O_{0.5}N_{0.2}P_{0.01}$. The stoichiometric quotients a_1–a_5 . . . , b and Y (biomass yield) specify quantities of substrate and products of microbial growth. If we know biomass yield and gross formulas of all substrates and products, then quotients a_1–a_6 . . . are easily calculated from conservation conditions. There are at least two such conditions. First, the mass of each element (C, H, O, N, P, K, . . .) on the left side of equation 1 should be equal to that on the right side (mass balance). Second, if ionized substances are involved, we should take into account the balance of charges to satisfy the condition of electroneutrality. Table 1 demonstrates some examples of stoichiometric growth reactions relevant to biotechnology.

Described formalism is useful as a first step in biotechnological studies aimed at planning and optimizing microbial growth. It estimates how much nutrient should be supplied to the fermenter to obtain the required amount of biomass or target product. However, it should be absolutely clear that stoichiometric equations like equation 1 are no more than an approximation to reality. The most severe deviation stems from the fact that unlike chemical reagents, microbial cells are characterized by changeable composition, and stoichiometric coefficients are not true constants. One task of contemporary microbial stoichiometry is to find out the functional relationships between stoichiometric parameters and internal (physiological) and external (environmental) factors.

Growth Yield: Catabolic and Conserved Substrates

The growth yield is one of the main stoichiometric parameters. It is defined as follows

$$Y = -\frac{\delta x}{\delta s} \approx -\frac{\Delta x}{\Delta s} \quad (2)$$

where Δx is the increase in microbial biomass consequent on utilization of the amount Δs of substrate, and δx and δs are respective infinitely small increments. Rigorous definition of Y as derivative $\delta x/\delta s$ stems from the fact that Y can vary in time, the negative sign being introduced because x and s vary in opposite senses. Sometimes, it is used as the reciprocal of Y: $\alpha = 1/Y$, which is called the *economic coefficient*. It expresses explicitly the nutrient requirements for growth: how many mass units of a particular substrate should be consumed to produce one unit mass of cell material.

The growth efficiency depends generally on the partitioning of consumed element between new cell biomass and extracellular products. The mass balance (total element consumed = amount incorporated into cell plus amount incorporated into extracellular products) is as follows:

$$\delta E_s = \delta E_x + \delta E_p \quad (3)$$

There are two groups of substrates for microbial growth: (*1*) *catabolic substrates*, which are sources of energy; and (*2*) *anabolic* or *conserved substrates*, which are sources of

Table 1. Selected Macrostoichiometric Equations Describing Growth of Microorganisms with Different Types of Energy Generation

Microbial process	Substrates		Biomass		Products	Stoichiometric parameters	
Heterotrophic growth and by-product formation	$CH_mO_l + a_1O_2 + a_2NH_3$	=	$YCH_pO_nN_q$	+	$Y_PCH_rO_sN_t + a_3CO_2 + a_4H_2O$	$a_1 = 0.5(Yn + Y's - l + a_4) + a_3$ $a_2 = Yq + Y't, a_3 = 1 - Y - Y'$ $a_4 = 0.5[m + Y(3q - p) + Y'(3t - r)]$	A
Phototrophic growth (algae or plant cells)	$CO_2 + a_1H_2O + a_2NH_3$	=	$YCH_pO_nN_q$	+	$a_3CH_rO_sN_t + a_4O_2$	$a_1 = 0.5[Y(p - 3n) + (r - 3t)(1 - Y)]$ $a_2 = Y(n - t) + t, a_3 = 1 - Y$ $a_4 = 0.5(2 + a_1 - Yn - a_3s)$	B
Methanogenesis[a]	$CO_2 + a_1H_2 + a_2NH_3$	=	$YCH_pO_nN_q$		$a_3CH_4 + a_4H_2O$	$a_1 = 4 - 0.5Y(4 + 2n + 3q - p)$ $a_2 = Yq, a_3 = 1 - Y$ $a_4 = 2 - Yn$	C
Nitrification[b]	$YCO_2 + a_1O_2 + a_2NH_4^+$	=	$YCH_pO_nN_q$	+	$a_3NO_3^- + a_4H^+ + a_5H_2O$	$a_1 = 2Y^{-1} - Y[1 + 0.25(p - 2n - 3q)]$ $a_2 = Y^{-1} + Yq, a_3 = Y^{-1}$ $a_4 = 2Y^{-1} + Yq, a_5 = Y^{-1} - 0.5Y(p - 3q)$	D

[a]Particular example of H_2-utilizing methanogenic bacteria.
[b]The total growth balance for two phases of nitrification: oxidation of ammonium to nitrite and subsequent oxidation of nitrite to nitrate. Y is growth yield of bacterial mass per mass unit of oxidized N.

biogenic elements forming cellular material. Catabolic substrates include H_2 for lithotrophic hydrogen bacteria, NH_4^+ and NO_2^- for nitrifying bacteria, S^0 for sulfur-oxidizing bacteria, oxidizable or fermentable organic substances for heterotrophic bacteria and fungi, and so on. Their consumption is accompanied by oxidation and dissipation of chemical substances into extracellular waste products that are no longer reusable as an energy source* (H_2O, NO_3^-, SO_4^{2-}, CO_2, etc.) The anabolic substrates after uptake are incorporated into de novo synthesized cell components, being conserved in biomass (that is why they are called conserved). Contrary to catabolic substrates, they can be reused (e.g., after cell lysis to be taken up by survived cells). The conserved substrates include nearly all the noncarbon sources of biogenic elements (N, P, K, Mg, Fe, and trace elements), CO_2 for autotrophs, and the indispensable amino acids and growth factors. Most catabolic substrates are used also as a source of biogenic elements. We can assess both these components separately in terms of respective yields, Y_E (biomass yield per mass unit of oxidized substrate) and Y_A (biomass yield per mass unit of assimilated substrate), from the experimentally measured yield Y. For C substrate, equation 4 can be specified as follows (total carbon consumed equals C incorporated into cell plus C oxidized to CO_2 to provide energy plus C incorporated into by-products):

$$\delta C_s = \delta C_x + \delta C_{CO_2} + \delta C_P \tag{4}$$

Let us neglect the last term δC_P (by assuming that extracellular by-products can be reused and functionally are equivalent to C substrate) and divide the substrate balance by δC_x, which is the amount of biomass C produced, then:

$$\frac{1}{Y} = 1 + \frac{1}{Y_E} \tag{5}$$

where Y = g biomass C g^{-1} substrate C and Y_E = g biomass C g^{-1} CO_2-C

$$\frac{1}{Y} = \frac{\sigma_x}{\sigma_s} + \frac{12}{Y_E \sigma_s} \tag{6}$$

where Y = g CDW g^{-1} substrate, Y_E = g CDW $mmol^{-1}$ CO_2, and σ_x and σ_s are fractions of carbon in biomass and substrate, respectively. For example, if total measurable yield Y is 0.6 g biomass C g^{-1} glucose C, it means that from each g of consumed C, 0.6 g is incorporated into biomass (assimilated), and 0.4 g is dissimilated (oxidized to CO_2), then $Y_E = 0.6/0.4 = 1.5$. To calculate oxygen demand for aerobic growth (or biomass yield on O_2) we have a balance (oxygen required to produce 1 g CDW equals oxygen required to burn substrate consumed to produce 1 g CDW minus oxygen required to burn 1 g CDW):

*Fermentation products such as acetate, ethanol, butyrate, and H_2 seem to be an exception because they do contain reusable oxidation potential, but it is not available under anaerobic conditions supervising fermentation.

$$1/Y_{O_2} = A/Y - B \tag{7}$$

where A and B are constants estimated from stoichiometry of their respective combustion reactions (see equation 10 later), for example, the value of A is 33.33 mmol O_2 g^{-1} glucose and B is about 42 mmol O_2 g^{-1} CDW. The relationship between biomass yields on O_2 and CO_2 is derived from comparison of equations 6 and 7:

$$\frac{Y_{CO_2}}{Y_{O_2}} = \frac{A - BY}{\sigma_s - \sigma_x Y} \tag{8}$$

Now we will go back to the general substrate balance (equation 3) and derive an expression for conserved substrate. Again, we neglect term δE_P (because extracellular products are assumed to be reusable) and divide the balance by δx, which is the amount of biomass produced:

$$\frac{1}{Y} = \frac{\delta E_x}{\delta x} \approx \sigma_x \tag{9}$$

where σ_x is the intracellular content of element incorporated into biomass from consumed substrate. Sometimes σ_x is called the *cell quota*. The values $1/Y$ and σ_x are not identical although they have the same dimension (e.g., milligram N per gram biomass) and very close numeric value. The reciprocal $1/Y$ is characterizing the *process* (the expenditure of conserved substrate to synthesize biomass unit), whereas σ_x is an index of cell composition (the content of intracellular N per biomass unit). Formally, $1/Y$ is equal to the σ_x value of an infinitely small increment of cell biomass, and σ_x is the averaged value for entire cell. Notice that although σ_x is a slow and $1/Y$ is a rapid variable, their numerical values are exactly the same for balanced steady-state growth and can differ considerably during transients.

Yield Variation as Dependent on the Chemical Nature of Organic Substrates

In this section, we will discuss why biomass yield varies when microorganisms are grown on different C substrates. This problem was best solved within the framework of the *theory of mass and energy balance* (TMEB) (3). Evidently, the fraction of C in dry biomass is almost constant. By contrast, the content of carbon in utilized substrates, σ_s, and energetic quality of substrate vary over a broad range (e.g., compare methane versus oxalic acid). To characterize substrate and biomass by a single common measure, TMEB uses an index of *degree of carbon reduction*, γ related to the internal energy of organic compounds. The heat liberated by biological or chemical oxidation is proportional to oxygen uptake or equally to the number of electrons gained by oxygen from oxidized substrates, according to Payne's term *available electrons* (*ae*) (4). The heat production from an oxidation reaction averages at 27 kcal per *ae* equivalent. A carbon reduction degree, γ is defined as the number of *ae* per one carbon atom. Its numeric value can be determined from the stoichiometry of the oxidation reaction:

$$CH_pO_nN_q + bO_2 = CO_2 + 0.5(p - 3q)H_2O + qNH_3 \tag{10}$$

$$\gamma = 4b = 4 + p - 2n - 3q \tag{11}$$

The *ae* balance for equation 1 can be written as

$$\gamma_s + b(-4) = Y\gamma_x + Y_P\gamma_p \tag{12}$$

where γ_s, γ_x, and γ_p are the carbon reduction degree of, respectively, substrate, biomass, and extracellular product. Dividing both sides of equation 1 by γ_s we obtain the relationship delineating the *ae* distribution between oxygen (*ae* used for respiration), biomass, and the intracellular product:

$$\frac{4b}{\gamma_s} + \frac{Y\gamma_x}{\gamma_s} + \frac{Y_P\gamma_p}{\gamma_s} = 1 \tag{13}$$

The second term in this equation is the fraction of *ae* transferred to biomass from utilized substrate, termed the *energetic growth yield*.

$$\eta = Y_C\gamma_x/\gamma_s \tag{14}$$

The third term designates that fraction of total substrate internal energy that is transferred to the product. It is called the *energetic product yield*

$$\zeta = Y_P\gamma_p/\gamma_s \tag{15}$$

Energetic yield η is related to other stoichiometric parameters as follows:

$$\eta = Y\sigma_x\gamma_x/(\sigma_s\gamma_s)$$
$$\eta = Y_C\gamma_x/\gamma_s$$

where Y is g CDW/g substrate and Y_C is g CDW-C/g substrate C.

The advantage of using η is that it varies within a much smaller range than other yield expressions. At one and the same efficiency of energy utilization (η), the conventional biomass C yield Y_C is proportional to substrate reduction degree γ_s and, for example, it is four times higher on glucose ($\gamma_s = 4$) than on oxalate ($\gamma_s = 1$), 0.48 and 0.12 g C g^{-1} C, respectively (assuming $\eta = 0.5$ and $\gamma_x = 4.2$). The energetic growth yield η is more or less constant (0.5 to 0.7) for substrates with $\gamma_s \leq 4.2$ (4.2 corresponds to average reduction degree of microbial biomass), and it declines at higher γ_s.

The attractiveness of macrostoichiometry and TMEB is that all growth coefficients are interrelated and could be measured from any available components of the culture mass balance. For example, if you cannot record microbial growth by conventional routine as dry weight biomass (because of presence of solids in broth liquid), you may still calculate it from N or O_2 uptake, CO_2 evolution, pH titration rate, and so on.

Variations in Yield from Energy Source, Maintenance Requirements

To multiply and grow cells requires energy, but the opposite is not true: cells do not require growth to spend energy. Sometimes catabolic machinery is entirely wasteful (respiration without cell growth) and always at least some minor part of energy consumption is diverted from growth. To account for this phenomenon, it was postulated that microbes and cells require energy not only for growth but also for other *maintenance purposes*. Certain specific maintenance functions recognized now are turnover of cell material, osmotic work to maintain concentration gradients between the cell and its exterior, and cell motility.

According to conventional definition of maintenance (5), the balance of energy source is total energy source consumed equals consumption for cell growth plus consumption for maintenance:

$$\delta S_E = \delta S_G + \delta S_M \tag{16}$$

Let us divide it by δx, the amount of biomass produced, then

$$\frac{1}{Y} = \frac{\delta S_G}{\delta x} + \frac{\delta S_M}{\delta x} = \frac{1}{Y^{max}} + \frac{m}{\mu} \tag{17}$$

Here, $Y^{max} = \delta x/\delta S_G$ is true growth yield, that is, yield under imaginary conditions of maintenance being zero. The maintenance coefficient, m, is introduced as the specific (i.e., expressed per unit of biomass) rate of energy consumption for maintenance functions: $m = (1/x)(\delta S_M/\delta t)$. The ratio m/μ on the right side of equation 17 was derived as follows: $m/\mu = [(1/x)(\delta S_M/\delta t)]/[(1/x)(\delta x/\delta t)] = \delta S_M/\delta x$.

If we divide equation 16 by $x\delta t$ (note that the second term is $\delta S_G/(x\delta t) = [\delta x/(x\delta t)]/[\delta x/\delta S_G] = \mu/Y^{max}$), then we have:

$$q = \mu/Y^{max} + m \tag{18}$$

where q is specific rate of energy source consumption, $q = (1/x)(\delta S_E/\delta t)$.

It should be noticed that Y^{max} is a parameter, but not the yield of a real culture that always has some nonzero maintenance requirements. It is a very common mistake in the application of the maintenance concept to a particular organism: to take the real measured Y value and pick up from literature some average m coefficient. The correct way would be either to borrow concurrently two parameters Y^{max} and m or to treat actually observed Y as a variable that is altered along with specific growth rate μ according to equation 17:

$$Y = \frac{\mu Y^{max}}{\mu + mY^{max}} \tag{19}$$

There is another way to formulate maintenance requirements by stating that the net growth of cells μ is the difference between true growth (μ_{true}) and endogenous decay of cellular components (specific rate, a):

$$\mu = \mu_{true} - a$$
$$\mu_{true} = \mu + a$$

Then, for the rate of energy source uptake, we have

$$\frac{\mu x}{Y} = \frac{\mu_{true}x}{Y^{max}} = \frac{(\mu + a)x}{Y^{max}}$$

or

$$\frac{1}{Y} = \frac{1}{Y^{max}} + \frac{a}{\mu Y^{max}} \quad (20)$$

Comparing equations 17 and 20, we see that $a = mY^{max}$.

Experimental Determination of *m*

To practically determine the maintenance coefficient, the microorganisms are grown in chemostat culture limited by energy sources at several dilution rates D (numerically D is equal to specific growth μ if steady state is achieved). At each D, we have to measure steady-state biomass $\tilde{x}$ and at least one of the following quantities (*1*) residual substrate, $\tilde{s}$ to calculate $Y = \tilde{x}/(s_0 - \tilde{s})$; and (*2*) the rate of respective energy-yielding process, such as respiration rate, v_{resp}, from O_2 uptake or CO_2 production rates to calculate specific metabolic activity, $q = v_{resp}/\tilde{x}$. These data are fitted to equations 17 or 18, m and Y^{max} being found as nonlinear regression parameters. An example is presented in Figure 1. Most available experimental data do obey this relationship. However, considerable deviation occurs at very low growth rates usually attained in chemostat with biomass retention or in dialysis culture. The experimental Y values for slowly growing cells are higher than predicted by equations 17 and 18 (see inset on Figure 1). The explanation is very simple: the maintenance coefficient varies in response to nutritional status and could not be taken as an absolute constant; under substrate deficiency, the cells adjust their maintenance requirements to lower values by reducing turnover rate, osmotic work, and motility (2).

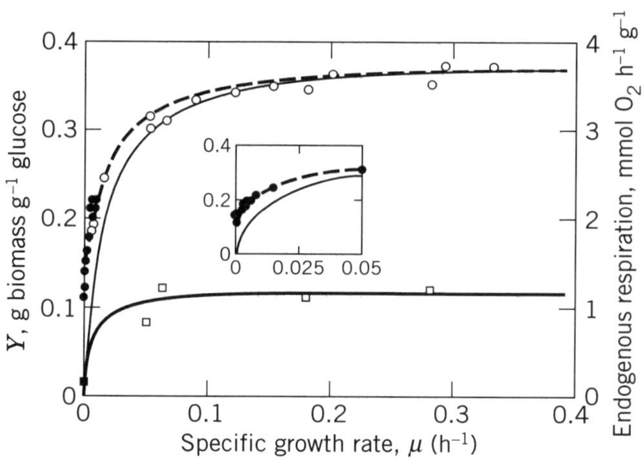

Figure 1. Variation of growth yield (*circles*) and endogenous respiration (*squares*) as dependent on specific growth rate in chemostat (*open symbols*) and continuous dialysis culture (*closed symbols*). Solid curves were calculated from the synthetic chemostat model (2). The dotted curve was derived from the Pirt-Herbert model (equations 17 to 20), which predicts quite well intensive growth but fails in the region of extremely low growth rates (see inset).

The described experimental technique is indirect because it is based on measurements of μ-dependent Y variation rather than m itself, and there are some assumptions needed to be confirmed (e.g., that m is constant and that maintenance requirements are the only reason of Y variation). However, some components of maintenance requirements are available for direct estimation. In particular, we can assess the total turnover rate of cellular material a which is one of the main components of maintenance requirements (equation 20). The principal cell constituents that are turned over are proteins, nucleic acids, and cell wall polymers. The turnover rate is very close to *endogenous respiration*, which is the oxidation of those compounds produced from the turnover (breakdown) of cellular macrocomponents. Accurate measurements of endogenous respiration need to be made under normal growing conditions. It is known that the simple removal of cells from nutrient broth by filtration with subsequent washing and incubation in buffer renders strong stress and may alter the normal turnover rate (6). To avoid artifacts, we can use a label-substitution technique (Fig. 2). The chemostat culture is fed alternately from two bottles containing unlabeled and labeled ^{14}C(U) substrate respectively. The $^{14}CO_2$ evolution rate is recorded after switching to unlabeled substrate, when the main source of $^{14}CO_2$ are cell components. The calculated a value was found to be rather high, accounting for the major part of total maintenance determined by the indirect method (2). The endogenous respiration declined at the low growth rate (Fig. 1), indicating that under starving conditions, self-adjustment of the maintenance requirement occurs mainly as a reduction in the turnover rate of macromolecules.

Maintenance Requirements and Wasteful Catabolism

The described concept of maintenance requirements was the subject of severe criticism (8). One of the strongest arguments against it was an apparent increase in Y^{max} observed in chemostat cultures limited by P, N, and other conserved substrates under conditions of energy excess. To preserve the constancy of the true yield, Pirt (9) had to modify equation 18 in the following way:

$$q = \mu/Y^{max} + m + m'(1 - \mu/\mu_m) \quad (21)$$

where $m'(1 - \mu/\mu_m)$ is the second μ-dependent component of maintenance energy that operates under excess of energy substrate.

However, it is better to differentiate maintenance requirements *sensu stricto* as those more or less a minor component of the cell energy budget that is observed under energy-limitation and wasteful use of catabolic substrate under energy excess. In physiological terms, these two groups of nonproductive catabolic reactions are completely different. The first reactions are mainly responsible for compensation of turned-over macromolecules and therefore belong to the category of regular primary catabolism. The catabolic reactions of the second group include excretion into environment of partly oxidized substances (*overflow metabolism*), uncoupling of respiration from ATP generation by metabolic inhibitors, functioning of futile cycles,

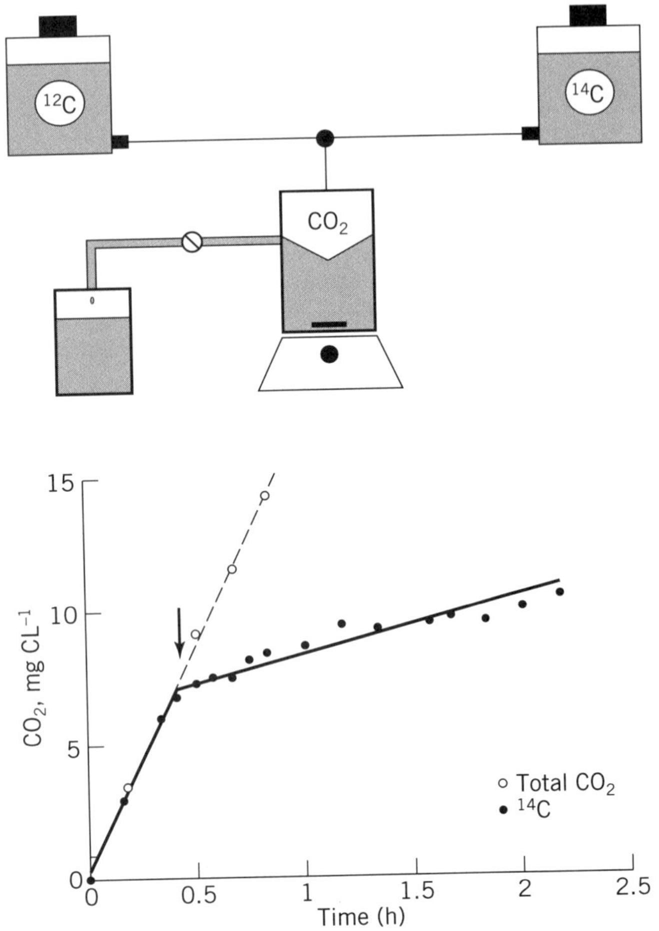

Figure 2. Label substitution technique for determination of turnover rate of cell macromolecular constituents. *Top*, experimental setup including two medium reservoirs containing ^{14}C- and ^{12}C-glucose pumped into a cultivation vessel through a two-way valve. *Bottom*, example of $^{14}CO_2$ evolution dynamics before and after (*arrow*) switching of medium feed from ^{14}C to ^{12}C-glucose, glucose-limited culture of *Pseudomonas fluorescens* 1472, $D = 0.08$ h^{-1} (7).

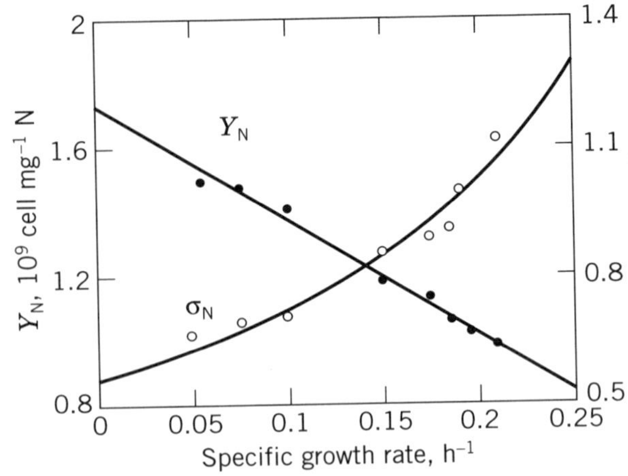

Figure 3. Relationship between stoichiometric parameters *Y* and *s* and specific growth rates of *Chlorella vulgaris* grown in chemostat culture limited by nitrogen source (10). The curves are calculated using equations 23 and 24.

or substrate oxidation through alternative oxidases without ATP generation. These and related phenomena take place in chemostat culture limited by conserved substrates (opposite to limitation by energy source) as well as during lag phase of batch culture started from starving inoculum. We will discuss the mathematical formulation of these phenomena in the section devoted to growth kinetics.

Variation in Biomass Yield from Conserved Substrates

Yield on conserved substrates varies mainly as a result of alterations in biomass chemical composition expressed by parameter σ_s, the intracellular content of deficient element or cell quota (see equation 9). For most of known cases, the content σ_s increases parallel to growth acceleration (Fig. 3). As yield and cell quota are inversely related to each other (equation 9), then Y values decrease with growth rate. The physiological mechanisms of this variation are as follows. The intensive growth requires higher *internal concentration* of some conserved limiting substrates that preserve their chemical identity after uptake (K^+, Mg^{2+}, vitamins). Other conserved substrates (sources of P, N, S, etc.) are incorporated into macromolecular cell constituents (mainly nucleic acids and proteins) whose *intracellular content* also should be kept high at high growth rate. Both types of changes in cellular composition are manifested as σ increase, and both of them require additional maintenance energy (to maintain concentration gradient or compensate turnover of macromolecules). The observed μ-dependent variation in σ is therefore a compromise between biosynthetic requirements and energy conservation that is attained because of optimal metabolic control of cell performance. However, it would be erroneous to consider μ as truly independent variable setting up chemical composition of cells. In fact, both μ and σ are functions of one common independent variable, the limiting substrate concentration in the medium, s. For steady-state chemostat culture we have:

$$\mu = \frac{q}{\sigma} = \frac{1}{\sigma}\frac{Q\tilde{s}}{K_s + \tilde{s}}$$
$$\sigma = \sigma_0 + \frac{(\sigma_m - \sigma_0)\tilde{s}}{K_\sigma + \tilde{s}} \tag{22}$$

where μ is specific growth rate, q is specific substrate uptake rates; σ_0 and σ_m are, respectively, lower and upper limits of σ variation; low limit $\sigma \to \sigma_0$ is attained when $\tilde{s} \to 0$ and upper limits $\sigma \to \sigma_m$-when $\tilde{s} \to \infty$. By excluding $\tilde{s}$ from both these equations we arrive at following relationship between σ and μ:

$$\mu = \mu_m \frac{\sigma_m(\sigma - \sigma_0)}{\sigma[\lambda(\sigma_m - \sigma_0) + \sigma - \sigma_0]}$$
$$\lambda = \frac{K_s}{K_\sigma} \tag{23}$$

Under realistic assumption $\lambda \approx 1$ ($K_s \approx K_\sigma$) we have

$$\sigma = \frac{\sigma_0}{1 - (1 - \sigma_0/\sigma_m)\mu/\mu_m}$$

$$Y = Y_m - (Y_m - Y_0)\frac{\mu}{\mu_m} \quad (24)$$

where Y_m and Y_0 are, respectively, upper and lower limits of yield variation ($Y_m = 1/\sigma_0$, $Y_0 = 1/\sigma_m$). As we can see, the linear relationship between Y and μ is normally observed in chemostat culture (Fig. 3).

Microscopic Approach in Studies of Growth Stoichiometry

Equations 1 to 24 exemplify the *macroscopic approach* in studying of microbial growth stoichiometry. Its typical features are the use of gross formulas for biomass and metabolic products, evaluation of total mass balance for chemical elements (C, N, P), and formal description of microbial growth as a single-step conversion of substrate(s) into biomass. By contrast, the *microscopic approach* focuses on the much more complex real metabolic reactions and attempts to account for a limited but still quite large number of individual metabolic intermediates. The final aim of this approach is to organize the biochemical information into a consistent picture of microbial metabolism at the level of entire cell.

The microscopic approach has become possible by virtue of advancements in biochemistry, which has succeeded in establishing a sufficiently full picture of metabolic processes in certain microorganisms. The pioneering work in this area was done by Bauchop and Elsden (11), who were able to sum up the balance of ATP for fermenting microorganisms. As a result, a relation was established between the biomass yield (a macroscopic quantity) and the number of generated ATP moles (a stoichiometric characteristic of real catabolic reactions):

$$Y_{ATP} = MY_E/n \quad (25)$$

where n = mol of ATP made available to the organism by the metabolism of one mole of energy source, and M = molecular weight (g) of energy source. The following example illustrates the Y_{ATP} calculation: if biomass yield of some organisms aerobically grown on glucose is 0.52 g CDW/g, then Y_E is 1.49 g CDW per g of oxidized glucose (calculated from equation 6) or $Y_E = 1.49 \times 180 = 268$ g CDW/mol (180 is glucose molecular weight); assuming that P/O = 2 (that is, 2 mol of ATP produced per atom oxygen taken up) and that 2 ATP mol are produced via glycolysis (substrate phosphorylation) we arrived at $n = 2 + 12 \times 2 = 26$ and $Y_{ATP} = 268/26 = 10.3$ g CDW/mol ATP. Careful determination of n and Y_{ATP} is possible only for anaerobic growth of fermenting microorganisms generating ATP via substrate phosphorylation. The mean value tends to be around 10.5 g CDW/mol ATP (Fig. 4). For aerobic growth, we need to make assumptions on the P/O ratio. As soon as the respiratory chain of bacteria differ widely for various organisms and growth conditions, this assumption can never be reliable. To avoid this obstacle, an interesting approach was proposed (1): microbial culture is grown in a chemostat limited by two carbon-containing energy sources, their ratio is varied while the total carbon feed rate is kept constant; yield measurements should allow one to determine both parameters (P/O and Y_{ATP}) independently by multiple linear regression.

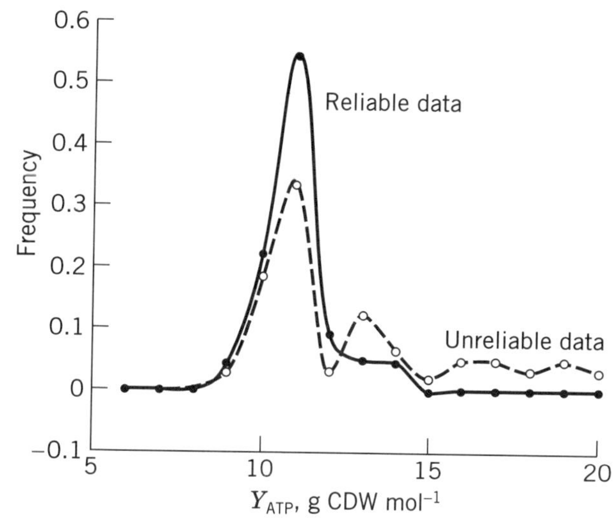

Figure 4. Frequency distribution of experimentally measured values of Y_{ATP} at different degrees of creditability. The reliable data refer to studies of anaerobic growth with direct measurements of fermentation products (plotted from data base in Ref. 12. Note that these data are normally distributed with mean value 10.55, whereas all data display considerable skewness.

Today, microstoichiometry is quickly progressing as so-called metabolic balancing. Cell growth is viewed as a set of transport and intracellular metabolic reactions known for some particular organisms. As a rule, the produced metabolic networks are composed of a combination of true stoichiometric equations for individual metabolites and empirical gross equations (Table 2). The amount of such equations vary in different models from 20 to 30 to more than 100. For example, van Gulik and Heijnen (13) describe yeast growth by a set of more than 90 reactions including glycolysis and the citric acid cycle (14); PEP phosphotransferase; pentose phosphate pathway (6); glyoxylate shunt (2); oxidative phosphorylation (4); CO_2 interaction with THF (3); transport of inorganic P, NH_4^+, SO_4^{2-}, acetate, lactate, pyruvate, glucose, gluconate, succinate, and citrate (totally 10 transport reactions); amino acid synthesis (15) and polymerization (2); nucleotide synthesis (9); RNA synthesis; ATP consumption for maintenance; fatty acids synthesis (2); formation of glycogen and polysaccharides; and finally, the biomass formation from proteins, polysaccharides, RNA, fatty acids, and glycerol.

For each compound, i, involved in a metabolic system, a mass balance can be defined:

$$\frac{dC_i}{dt} = r_{Ai} + \Phi_i \quad (26)$$

where C_i is concentration of ith compound, r_{Ai}, and Φ_i denote the net rates of, respectively, i chemical conversion and transport over the boundaries of bioreactor (fluxes of

Table 2. Metabolic Networks

Reaction	Equation	Stoichiometric Equation	Empirical Gross Equation
Glycolysis reaction	Glucose + ATP → glucose-6-P + ADP + H	x	
	Glucose-6-P → fructose-6-P	x	
	0.5 Fructose-6-P + 0.5 ATP → glyceraldehyde-3-P + 0.5 ADP + 0.5 H	x	
Oxidative phosphorylation	$NADH + 0.5O_2 + d_1ADP + d_2Pi + (1 + d_1)\,H \rightarrow (1 + d_1)H_2O + NAD + d_1ATP$	x	
Biomass formation	a_1Proteins + $a_2$polysaccharides + a_3RNA + $a_4$lipids → biomass		x

CO_2, O_2, nutrients, cells, and products). Most metabolic balancing equations are applied to steady-state growth (which means that no intracellular accumulation of metabolites occurs). In such cases, the differential equations like equation 26 are reduced to linear algebraic ones. Besides, an extensive use of matrix calculus is customarily made to obtain a concise notation. The problem of experimental support of such model is especially important (13). The degree of freedom, *df*, of the resulting system of linear equations is equal to the total number of unknown rates (both intracellular and exchange reactions) minus the total number of linear equations. To resolve the system, *df* rates have to be measured, and then the system is fully *determined*. If the number of measured rates is greater than *df*, then the system is *overdetermined*, and the redundancy of the data can be used for statistical analysis and error minimization. However, it is much more typical to have an *underdetermined* system when the sum of measured rates is less than *df*. In this case, the number of possible solutions is infinite unless additional constraints are applied (e.g., maximization of biomass yield, minimization of energy expenditure) to find the one and only one solution by the linear optimization technique.

In most studies, flux estimates are obtained using measurements of substrate consumption and product formation. This approach has proved to be efficient in some particular biotechnological cases, such as when only specific pathways need to be considered (16) or if the contribution of flux for cellular growth is weak, as with mammalian or hybridoma cells (17). The more complex microbial systems are turned out to be seriously underdetermined. In such cases, the application of metabolic balancing requires the use of one or another maneuvers: (*1*) to lump together several sets of reactions (18); (*2*) to utilize data from in vitro enzyme assays; (*3*) to make assumptions on numeric values of some stoichiometric growth parameters, such as Y_{ATP}, P/O, and H^+/e ratios, which are the subject of controversial debates (13).

However, the best solution would be to get direct experimental data on in vivo flux and resolve the system. Isotopic tracers are one of the best candidates for such a purpose. We will illustrate this point by describing a recently published work (19). This novel approach is based on the analytical power of ^{1}H-detected ^{13}C nuclear magnetic resonance. *Corynebacterium glutamicum* was grown in chemostat culture continuously fed with [1-^{13}C]-glucose; when steady state was established, the cells were harvested and hydrolyzed and the amino acids were separated by ion-exchange chromatography and analyzed by NMR spectroscopy. NMR provides data on ^{13}C enrichment at each specified carbon position of amino acid. Because metabolic pathways for amino acid synthesis are exactly defined, then the entire central metabolism can be assessed for in vivo fluxes, including determination of the forward and back rates of bidirectional reactions. In *C. glutamicum*, the flux through the pentose phosphate pathway turned out to be 66.4% (relative to glucose input flux 1.49 mmol g^{-1} CDW h^{-1}); the entry into tricarboxylic acid cycle, 62.2%, and the contribution of the succinylase pathway to lysine synthesis, 13.7%. The total net flux of the anaplerotic reactions (carboxylation of PEP/pyruvate into oxaloacetate/malate) was quantitated as 38%, the true forward flux of C3 → C4 being 68.6% (1.8 times of 38%) and a back flux of C4 → C3 being 30.6% (0.8 times of 38%) (19). The metabolic balancing proved to be very promising and useful to identify metabolic constraints for intensive synthesis (overproduction) of products such as amino acids. On the other hand, this approach still is restricted to steady-state and balanced growth and is not able to cope with complex dynamic behavior of microorganisms (transient growth, changes in biomass composition).

BASIC PRINCIPLES OF GROWTH KINETICS

Kinetics of Chemical and Enzyme Reactions

We need to introduce some basic principles of kinetic analysis of chemical and enzymatic reactions. Quantitative description and understanding of microbial growth dynamics and kinetics are impossible without some elementary knowledge in underlying scientific disciplines. Enzymatic and chemical reactions play an essential role in biotechnology, which is one of the most important fields in industrial development.

Order and Molecularity of Chemical Reactions. The molecularity of any chemical reaction is defined by the number of molecules that are altered in the reaction (Table 3). The order is a description of the number of concentration terms multipled together in the rate equation (Table 4). Hence, in a first-order reaction, the rate is proportional to one concentration of reactant; in a second-order reaction, it is proportional to two concentrations or to the square of one concentration. For a simple single-step reaction, the order is generally the same as the molecularity. For a complex reactions involving a sequence of unimolecular and bimolecular steps, the molecularity is not the same as its order. Reactions of molecularity greater than 2 are com-

Table 3. Molecularity of Chemical Reaction

Molecularity of reaction	Reactants		Product
Unimolecular or monomolecular	S	→	P
Bimolecular	S + S	→	P
or	$S_1 + S_2$	→	P
Trimolecular or termolecular	S + S + S	→	P
or	$S_1 + S_1 + S_2$	→	P
or	$S_1 + S_2 + S_3$	→	P

mon, but reactions of order greater than 2 are very rare. For instance, a trimolecular reaction, such as A + B + C → P, as a rule proceeds through two elementary steps, A + B → X and X + C → P, each of which are of the second or first order. Very often, bimolecular reactions between S_1 and S_2 occur under the condition that their respective concentrations $s_2 \gg s_1$ (e.g., if the second reactant S_2 is solvent), then we have a *pseudo first-order* reaction. Some reactions are observed to be of the zero order, that is, the rate appears to be constant, independent of the concentration of reactant. This is a characteristic feature of catalyzed reactions and occurs if reactant is present in such large excess that the full potential of catalyst is realized.

Dimensions of Rate Constants. Knowledge of dimensions is very useful to check the correctness of derived kinetic equations: the left- and right-hand sides of an equation must always have the same dimensions. This general rule is applicable to all mathematical models (not only in chemical kinetics). It is incorrect to add or subtract terms of different dimensions, although you may multiply or divide them. For example, if expression " . . . (1 + s)" occurs in an equation, where s has dimension (concentration), then either equation is incorrect, or the 1 is a concentration that happens to have a numerical value of 1 unit. The operation rising to power is allowed for only simple dimensionless numbers, for example, expression $e^{2.5t}$, where t is time, is correct only if 2.5 has dimension $(\text{time})^{-1}$. The comparison of velocities of two reactions does make any sense only for kinetic terms of the same dimension. If the kinetic order of inspected reactions is different, then we have to equalize the respective rate of expressions; for example, the second-order rate constant k $(\text{time})^{-1}$ $(\text{concentration})^{-1}$ should be multiplied by instant concentration of reactant s to be compared with the first-order rate constant having dimension $(\text{time})^{-1}$. The dimensions of the zero-, first-, and second-order rate constants are shown in Table 4.

Reaction Dynamics. If rate constant and so-called initial conditions (concentrations of reactants at zero-time [s_{01}, s_{02}, . . .]) are known, then it is possible to calculate the time course of reactions either in terms of dynamics of residual reactant concentration, $s(t)$ or product accumulation, $p(t)$. For this purpose, we have to integrate a differential equation under specified initial conditions (see results of integration in Table 4). The dynamics of $s(t)$ are linear in the case of a zero-order reaction and hyperbolic in the case of the first and second order. The difference between the last two dynamic curves can be made visible with a semilogarithmic plot of log(concentration) versus time; it should became linear for the first-order reaction and remain to be curvilinear for the reaction of higher order.

Reaction Half-Time. The reaction half-time ($t_{0.5}$) is a very popular kinetic parameter, especially among biologists. It is easily calculated from integral equations by putting $s = s_0/2$ when $t = t_{0.5}$. A unique feature of the first-order reaction is the constancy of $t_{0.5}$ independently of the initial reactant concentration s_0. However, the half-time of other reactions does depend on s_0; it increases for zero-order reactions and decreases for the second-order reactions with increase in s_0. Thus, it is not recommended to use half-time as a parameter or estimator for reactions other than first-order reactions.

Determination of the Order of Reaction and Numeric Values of Kinetic Constants. If the reaction has an order n and rate constant k, then the reaction rate v and reactant concentration s are related by the equation

Table 4. Kinetic Order of Chemical Reactions

Reaction order	Differential equation	Dimension of rate constant	Dynamics of residual reactant	Dynamics of product accumulation	Reaction half-time[a]
Zero	$\frac{ds}{dt} = -k$	$(\text{conc.})(\text{time})^{-1}$	$s(t) = s_0 - kt$	$p(t) = p_0 + kt$	$t_{0.5} = \frac{s_0}{2k}$
First	$\frac{ds}{dt} = -ks$	$(\text{time})^{-1}$	$s(t) = s_0 \exp(-kt)$	$p(t) = s_0[1 - \exp(-kt)]$	$t_{0.5} = \frac{\ln 2}{k}$
Second	$\frac{ds}{dt} = -ks^2$	$(\text{conc.})^{-1}(\text{time})^{-1}$	$s(t) = \frac{s_0}{1 + ks_0t}$	$p(t) = \frac{s_0^2kt}{2(1 + s_0kt)}$	$t_{0.5} = \frac{1}{ks_0}$ [b]
Second	$\frac{dp}{dt} = -ks_1s_2$	$(\text{conc.})^{-1}(\text{time})^{-1}$	$s_1(t) = s_{01} - p(t)$ $s_2(t) = s_{02} - p(t)$	$p(t) = \frac{s_{01}s_{02}(1 - \exp[(s_{02} - s_{01})kt])}{s_{01} - s_{02}\exp[(s_{02} - s_{01})kt]}$	$t'_{0.5} = 1/ks_{01}$ $t''_{0.5} = 1/ks_{02}$
Pseudo-first $s_{01} \gg s_{02}$	$\frac{dp}{dt} = -ks_1s_2$	$(\text{conc.})^{-1}(\text{time})^{-1}$	$s_2(t) = s_{02}\exp(-ks_{01}t)$ $s_1(t) \approx s_{01}$	$p = s_{02}[1 - \exp(-ks_{01}t)]$	

Note: t = time; s = reactant concentration; p = product concentration; s_0 = reactant concentration at $t = 0$.
[a]The half-time of the reaction is the time required for half-completion.
[b]The half-time is defined for the reagent that has lower initial concentration and is depleted first.

$$v = -ds/dt = dp/dt = ks^n \quad (27)$$

The simplest way to find both unknown values (n and k) would be to measure reaction rate v at several concentrations of reactant. Then a plot of log(rate) against log(concentration) gives a straight line with a slope equal to n and intercept equal to log(k): $\log v = \log k + n \log s$. If there are several reactants, then it is useful to know the order in respect to each one. For this purpose, you need to have several experiments with variation of each reactant concentration while keeping the other concentrations constant.

The most frequent goal is the determination of the first-order rate constants, first because many reactions do obey first-order conditions in respect to each reactant and second because it is possible to carry out many reaction under pseudo first-order conditions. There are some special methods to determine the values of the first-order rate constants from the experimental curves of product formation or substrate depletion dynamics. Some of them were designed to improve the accuracy of k determination when the initial reactant concentration (s_0) or final value of product concentration (p_∞) were not known (methods of Guggenheim, Kezdy-Swinbourne, and others, see details in Ref. 20). All are based on plotting the experimental points in a some sophisticated manner to convert the original curves to a straight lines. Today, these methods are replaced by computer-aided nonlinear regression, which is much more convenient and precise and, contrary to graphic methods, allows for more rigorous estimate of confidence limits of measured quantities.

Derivation of Basic Kinetic Equations for Enzymatic Reactions. Contrary to simple chemical reactions, enzyme-catalyzed reactions proceed through *reversible* formation of the dynamic *enzyme-substrate complex* (ESC). The word *reversible* is essential because the ESC can be decomposed into free enzyme E and product P or dissociate back to E and substrate S. There are many ways to simulate mathematically the mechanism of the enzymatic reaction. We will consider here equilibrium, steady-state, and general non-steady-state approaches.

Equilibrium Approach. This approach was used by Michaelis and Menten (21) to describe the effect of sucrose concentration on invertase activity. They assumed that the first step of ESC formation is so rapid that could be represented by an *equilibrium constant* K_s:

$$\overset{e}{E} + \overset{s}{S} \overset{K_s}{\leftrightarrow} \overset{x}{ES} \overset{k_2}{\rightarrow} E + \overset{p}{P} \quad (28)$$

where e, s, x, and p denote the concentrations of free enzyme, substrate, FSC, and product, respectively. The equilibrium constant, K_s, is defined as $K_s = es/x$. The instantaneous concentrations s and e are not directly measurable, but they could be expressed in terms of the initial, measured concentrations, e_0 and s_0, using mass-balance relationships:

$$e_0 = e + x \quad s_0 = s + x$$
$$x \ll s$$
$$s \approx s_0 \quad (29)$$

The overall reaction rate, v, is a simple first-order reaction with rate constant k_2:

$$v = dp/dt = -ds/dt = k_2 x \quad (30)$$

The x value can be found from expression for K_s and mass-balance conditions (equation 29): $x = e_0 s/(K_s + s)$. Substitution of this expression into equation 30 finally produces

$$v = \frac{k_2 e_0 s}{K_s + s} \quad (31)$$

Steady-State Approximation. This approach was applied by Briggs and Haldane (22) for the following scheme of enzymatic reaction

$$\overset{e}{E} + \overset{s}{S} \underset{k_{-1}}{\overset{k_1}{\rightleftarrows}} \overset{x}{ES} \overset{k_2}{\rightarrow} E + \overset{p}{P} \quad (32)$$

This scheme implies the reversibility of ESC formation instead of much more restrictive equilibrium postulate. However, still there was steady-state assumption in respect to ESC formation:

$$dx/dt = k_{+1}(e_0 - x)s - k_{-1}x - k_{+2}x \approx 0 \quad (33)$$

Therefore, $x = k_{+1}e_0 s/(k_{+1}s + k_{-1} + k_{+2})$, and substitution of this expression into equation 30 gives

$$v = -\frac{ds}{dt} = k_{+2}x = \frac{k_{+1}k_{+2}e_0 s}{k_{+1}s + k_{-1} + k_{+2}} = \frac{k_{+2}e_0 s}{\dfrac{k_{-1} + k_{+2}}{k_{+1}} + s} = \frac{Vs}{K_m + s} \quad (34)$$

Equation 34 is what usually called the *Michaelis-Menten equation*, the fundamental equation of enzymatic kinetics. It contains two parameters, K_m, the *Michaelis constant*, and V, the *maximum velocity*. V is the rate of reaction that would occur under full substrate saturation of an enzyme's active sites ($s \gg K_m$). In reality, the V value can be never attained, because extremely high substrate concentrations inhibit enzymes (see later text). It is clear also that V is not a fundamental property of enzyme because it depends on e_0, the enzyme concentration. More advantageous as a specific enzyme characteristic is the *catalytic constant* or *turnover number*, k_{cat}, which is V/e_0. For equation 32, k_{cat} is identical with k_{+2}, but in general the more noncommittal notation k_{cat} is preferable (e.g., k_{cat} may differ from k_{+2} if the product formation from ESC is a reversible reaction). Numerically, the Michaelis constant, K_m, is the substrate concentration that provides half the maximal reaction rate. Contrary to this simple practical definition, the mechanistic interpretation of K_m is not so lucid. Sometimes

K_m is interpreted as a substrate binding constant, K_s, assuming that $k_{+2} \ll k_{-1}$. This is a very dubious assumption. There are very few enzymes for which the individual values of k_{+2} and k_{-1} are known. Ironically, the best-studied examples (e.g., horseradish peroxidase) present just opposite case: $k_{+2} \gg k_{-1}$. It is important that many specific mechanisms (not necessarily equations 28 and 32) generate the same steady-state rate equation 34. However, the particular expression for K_m should be different for each individual case.

Although undefined in mechanistic terms, the parameters K_m and V are very useful at the first steps of kinetic studies:

1. The use of equation 34 allows the expression of the complex effects with simpler terms.
2. K_m and V are helpful as predictive parameters to design a valid enzyme assay. One practical recommendation is to keep substrate concentration in incubation mixture at the level of 10 K_m or higher.
3. Equation 34 permits one to obtain at least rough estimate of in vivo enzyme activity provided the internal substrate concentrations are known.

General Non-Steady-State Approach. Equation 32 contains four variables (s, p, e, and x) constrained by two mass-balance equations ($e_0 = e + x$ and $s_0 = s + p + x$). Therefore, it is enough to integrate just two differential equations (e.g., equations 33 and 34) to characterize the whole system. A typical example of a numeric solution is shown in Figure 5. We can see that enzyme-catalyzed reaction proceeds through three definite phases well separated on the time scale. Each phase is now safe to analyze with simpler mathematical expressions because any assumptions could be tested against the full exact solution.

1. The first transient phase occurs before the steady-state concentration of ESC is reached. It occupies less than

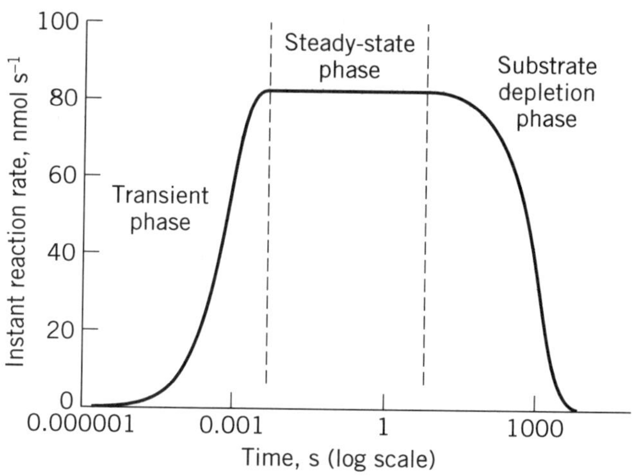

Figure 5. Time course of reaction proceeding by the Michaelis-Menten mechanism. Numeric integration of equations 33 and 34: $s_0 = 10^{-4}$M, $e_0 = 10^{-8}$M, $k_{+1} = 10^6$M^{-1}s^{-1}, $k_{-1} = 10^3$ s^{-1}, $k_{+2} = 10^2$ s^{-1}.

10^{-3} s for most of enzymes. During this phase, the substrate concentration remains fairly constant ($s \approx s_0$), allowing for the following analytical solution:

$$x(t) = \frac{k_{+1}e_0s_0}{A}[1 - e^{-At}], \quad v(t) = \frac{Vs}{K_m + c}[1 - e^{-At}]$$
$$A = k_{+1}s_0 + k_{-1} + k_{+2} \qquad (35)$$

As compared with equation 34, it contains a relaxation term $[1 - \exp(-At)]$ that is very large when t is small, but decays to zero as t increases above $\tau = 1/A$. Accordingly, the rate of enzymatic reaction is initially zero but increases rapidly to the steady-state value as the exponential term decays. Because the relaxation term contains s_0, then the experimentally observed delay depends on substrate concentration. It allows for direct determination of individual rate constants k_{+1} and $(k_{-1} + k_{+2})$ by techniques such as in stopped-flow apparatus (23).

2. The second steady-state phase is characterized by the constancy of reaction rate due to the exact balance between the rates of ESC formation and breakdown ($dx/dt = 0$) while the substrate concentration remains close to the initial value s_0. This phase proceeds for at least several seconds. As a rule, it is enough to measure the *initial velocity* of enzymatic reactions unaffected by substrate depletion or product accumulation. Steady-state kinetics is the most popular research domain; it is the most accessible and developed and provides the most kinetic data. However, it fails to determine individual rate constants as fully as the transient and relaxation approaches do. Steady-state equations were derived earlier (equation 34).

3. The third phase is characterized by considerable changes in the concentrations of substrate and product. Thus, we can no longer assume that $s = s_0$, and we should integrate equations 32 and 33. However, very good precision provides the quasi-steady-state approximation $ds/dt \gg dx/dt \approx 0$. Then, we arrive at a relatively simple differential equation that is solved analytically:

$$\frac{ds}{dt} = -\frac{Vs(t)}{K_m + s(t)} \qquad (36)$$

For initial conditions $s = s_0$, $p = 0$, $t = 0$, we have

$$K_m \ln\frac{s_0}{s} + s_0 - s = Vt \qquad (37)$$

Equation 37 is called the *integrated Michaelis-Menten equation*. It remains to be valid not only for initial rate measurements but for any point within the reaction progress curve.

Experimental Determination of Kinetic Parameters of the Michaelis-Menten Equation. Until recently, most enzyme kinetic experiments have been analyzed by means of one of linear plots in Table 5. Linear plots are used to examine an agreement of experimental data with equation 34 as well as to determine numeric values of parameters V and K_m from slopes and intercepts. Today, this graphic ap-

Table 5. Linear Plots

$\frac{1}{v} = \frac{1}{V} + \frac{K_m}{V}\frac{1}{s}$	Lineweaver-Burk or double-reciprocal plot
$\frac{s}{v} = \frac{K_m}{V} + \frac{s}{V}$	Langmuir-Hanes plot
$v = V - K_m \frac{v}{s}$	Eadie-Hofstee plot
$V = v + \frac{v}{s} K_m$	Direct linear plot (20)

proach seems to be too cumbersome as compared with much more efficient computer routines. The main objection is also that any linear transformation introduces some statistical bias (ironically, the highest bias is attributed to the most popular double-reciprocal plot!). In addition, manual line drawing is very arbitrary, so obtained parameters could not be assessed statistically for confidence limits. However, plotting of original or linearized data is useful as an illustration and the first sketchy estimate of enzymatic parameters. For definitive work, it is advisable to avoid all plots and to use statistical analyses instead.

Reversible Enzymatic Reactions. The majority of biochemical reactions are reversible. To account for this feature, the Michaelis-Menten mechanism can be modified as follows:

$$\underset{e_0-x-y}{E} + \underset{s}{S} \underset{k_{-2}}{\overset{k_{+1}}{\rightleftarrows}} \underset{x}{ES} \underset{k_{-2}}{\overset{k_{+2}}{\rightleftarrows}} \underset{y}{EP} \underset{k_{-3}}{\overset{k_{+3}}{\rightleftarrows}} \underset{e_0-x-y}{E} + \underset{p}{P} \quad (38)$$

Contrary to the basic scheme in equation 32, there are two intermediates, one of which is the normal ESC and another is the *enzyme-product complex* (EP). The substrate, S, and product, P, can interconvert to each other. Application of the steady-state approach to this scheme results in the following equation:

$$v = \frac{V^{f}s/K_{m}^{s} + V^{r}p/K_{m}^{p}}{1 + s/K_{m}^{s} + p/K_{m}^{p}}$$

$$V^{f} = \frac{k_{+2}k_{+3}e_0}{k_{-2} + k_{+2} + k_{+3}}$$

$$K_{m}^{s} = \frac{k_{-1}k_{-2} + k_{-1}k_{+3} + k_{+2}k_{+3}}{k_{+1}(k_{-2} + k_{+2} + k_{+3})}$$

$$V^{r} = \frac{k_{-1}k_{-2}e_0}{k_{-1} + k_{-2} + k_{+2}}$$

$$K_{m}^{p} = \frac{k_{-1}k_{-2} + k_{-1}k_{+3} + k_{+2}k_{+3}}{k_{-3}(k_{-1} + k_{-2} + k_{+2})} \quad (39)$$

where superscripts f and s denotes parameters of forward reaction, and r and p are indicators of reverse reaction.

When a reaction is at equilibrium, the net velocity must be zero and, consequently, if s_∞ and p_∞ are the equilibrium values of s and p, it follows from equation 39 that equilibrium constant K is expressed via kinetic parameters (the *Haldane relationship*):

$$K = p_\infty/s_\infty = V^{f}K_{m}^{p}/V^{r}K_{m}^{s} \quad (40)$$

Equation 39 implies that the rate must decrease as the product accumulates, even if the decrease in substrate concentration is negligible. Thus, reversibility is closely associated with and requires the *product inhibition*. In many essentially irreversible reactions (e.g., invertase-catalyzed hydrolysis of sucrose), product inhibition is also significant. It can be explained by equation 38 with only the second step being irreversible. In such a case, the accumulation of product causes the enzyme to be sequestered because the EP complex and rate equation are as follows (compare with equation 39):

$$v = \frac{V^{f}s/K_{m}^{s}}{1 + s/K_{m}^{s} + pK_{m}^{p}} = \frac{V^{f}s}{K_{m}^{s}(1 + pK_{m}^{p}) + s} \quad (41)$$

Inhibitors and Activators. Compounds that reduce the rate of enzyme-catalyzed reactions when they are added to the reaction mixture are called *inhibitors*. Just opposite effects (increase of reaction rate) are caused by *activators*. Both of them belong to the category of *modifiers*. We will concentrate here on inhibitors as having more practical application and even more specifically on *reversible* inhibitors, which form various dynamic complexes with enzymes. The *irreversible* inhibitors are known as catalytic poisons (many heavy metals, such as mercury); their binding to enzyme molecules reduces activity to zero, leaving no room for quantitative analysis.

The reversible inhibitors can form complexes with free enzymes or enzyme-substrate complexes as shown in the scheme of Botts and Morales (20):

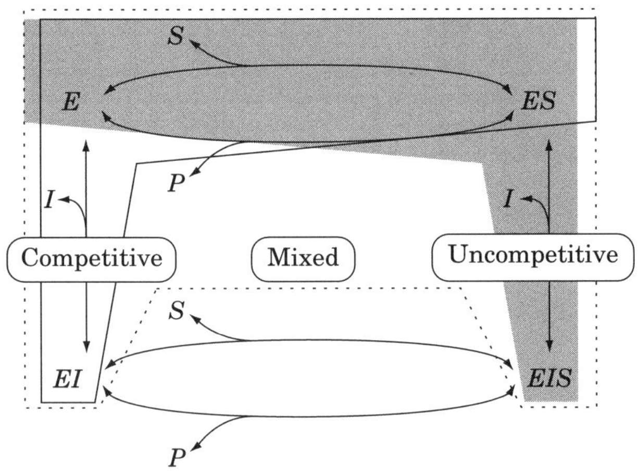

There are three major simple (or linear) inhibition mechanisms, which can be generated from this scheme by omitting some of the six involved:

- *Competitive* inhibition if *EIS* is missing (inhibitor binds to the same site on the enzyme as the substrate forming abortive nonproductive complex *EI*).
- *Uncompetitive* if *EI* is missing and *EIS* occurs as a dead-end complex; it implies that the inhibitor-binding site becomes available only after the substrate has bound.

- *Mixed* if EI and EIS both occur but are not interconvertible (the complex EIS does not break down to products; this situation frequently occurs when the inhibitor is reaction product). The particular case of mixed inhibition is *pure noncompetitive inhibition*, which takes place if two inhibitor dissociation constants (for EI and EIS) are exactly matched to each other.

For all types of inhibition, the Michaelis-Menten equation remains valid: under constant inhibitor concentration, i, the v-s relationship is of the same hyperbolic type as predicted by equation 34, the only difference is that apparent values of K_m and V are now more or less simple functions of i (Table 6).

Some parameters (in boldface in Table 6) are not affected by inhibitors, whereas others are changed: competitive inhibitors increase K_m, pure noncompetitive inhibitors reduce V, uncompetitive inhibitors decrease at the same degree both V and K_m, leaving first-order rate constant V/K_m to be unchanged. In mixed inhibition, there are no unchanged parameters.

One interesting case is so-called substrate inhibition. It occurs when two substrates are bound to the same active site on the enzyme, forming nonproductive triple complex ES_2:

$$E + S \overset{K_{ss}}{\leftrightarrow} ES \overset{k_2}{\rightarrow} E + P$$
$$ES \underset{K_{SS}}{\rightleftarrows} ES_2$$

then, the enzyme rate (24) is:

$$v = \frac{Vs}{K_s + s + s^2/K_{ss}} \tag{42}$$

where K_{ss} is the dissociation constant for complex ES_2.

Cooperativity. Many enzymes respond to changes in metabolite concentrations (substrates, modifiers) with much higher sensitivity as compared with predictions from the classic hyperbolic equations 34 to 42. This property is generally known as *cooperativity*, because it is thought to arise from cooperation between the active sites of the polymeric enzymes. As a rule, such enzymes consist of several *subunits* and display so-called sigmoid or S-shaped dependence of rate on substrate concentration. Many cooperative enzymes (but not all!) have *active sites* binding substrate and *allosteric sites* binding effectors. There are *homotropic* and *heterotropic* cooperative effects caused by interactions between, respectively, identical and different ligands (e.g., substrate and an allosteric effector).

To explain the cooperativity and associated sigmoid kinetics, a number of models have been suggested. The earliest one is the Hill equation, which was originally designed to describe the S-shaped curve of oxygen binding to hemoglobin. It was assumed that each protein molecule E binds n molecules of ligands S in a single step, an amount of other possible forms (ES_{n-1}, ES_{n-2}, . . . ES) being negligible:

$$E + nS \overset{K_h}{\leftrightarrow} ES_n$$

where K_h is the respective equilibrium constant (Hill and colleagues described equilibrium in terms of association constant, but for the sake of uniformity we will adhere to the previous formalisms, keeping in mind the dissociation constant, $K_h = [E][S]^n/[ES_n]$). The fractional saturation of protein (enzyme or hemoglobin). Θ is given as

$$\Theta = \frac{\text{Number of occupied binding sites}}{\text{Total number of binding sites}} = \frac{[ES_n]}{[E] + [ES_n]} = \frac{[S]^n}{K_h + [S]^n} \tag{43}$$

Equation 43 can be rearranged into

$$\frac{\Theta}{1 - \Theta} = \frac{[S]^n}{K_h}, \log\left(\frac{\Theta}{1 - \Theta}\right) = -\log K_h + n \log[S] \tag{44}$$

A plot of log $\Theta/(1 - \Theta)$ against log $[S]$ is known as the *Hill plot* and should be a straight line of slope n (*so*). This equation is used to fit the experimental binding and kinetic data displaying a sigmoid shape. When plotting kinetic measurements, it is assumed that $\Theta = v/V$, and maximum velocity V should be known from independent measurements taken at saturation substrate concentration. However, the results of fitting should be interpreted with care. First, the Hill equation is empirical and generally provides good agreement only in the Θ range 0.1 to 0.9 (the discrepancy at extreme s is probably caused by neglecting of other forms of ESC apart from ES_n). Second, parameter n (*Hill coefficient*) could not be interpreted as the number of subunits in the fully associated protein, rather it is an index of cooperativity.

Monod et al. (25) proposed a general model explaining cooperativity and allosteric phenomena within a simple set of postulates:

1. Each subunit of enzyme can exist in two different conformations, designated R (relaxed) and T (tense).

Table 6. Types of Inhibition

Type of inhibition	V^{app}	V^{app}/K_m^{app}	k_m^{app}
Competitive	$\mathbf{V}$	$(V/K_m)/(1 + i/K_i)^a$	$K_m(1 + i/K_i)$
Uncompetitive	$V/(1 + i/K_i')$	$\mathbf{V/K_m}$	$K_m(1 + i/K_i'))$
Mixed	$V/(1 + i/K_i')$	$(V/K_m)/(1 + i/K_i)$	$K_m(1 + i/K_i)/(1 + i/K_i')$
Pure noncompetitive	$V/(1 + i/K_i')$	$(V/K_m)/(1 + i/K_i)$	$\mathbf{K_m}$

2. All subunits of a molecule must occupy the same conformation at any time (e.g., for tetrameric enzyme only R_4 or T_4 are permitted, not R_3T or R_2T_2, etc.).
3. The two states are in equilibrium with constant $L = [R_4]/[T_4]$.
4. The affinity of ligand to subunit depends on the conformation state: $K_R = [R][S]/[RS]$, $K_T = [T][S]/[TS]$, $c = K_R/K_T$.

The general equilibrium solution of this scheme is rather bulky, so we demonstrate one particular case of tetrameric protein, if $c = 0$ (i.e., $K_R \ll K_T$, S binds only to the R state), then we have

$$\Theta = \frac{(1 + s/K_R)^3 s/K_R}{L + (1 + s/K_R)^4} \tag{45}$$

At high s when $s/K_R \gg L$, we can neglect the term L in the denominator, and the entire expression is converted to no more than the simple hyperbolic Langmuir isotherm. At low s, the contribution of L is considerable, so the saturation curve rises slowly from the origin displaying the S-shape.

There are other models describing the mechanisms of cooperativity and allosteric effects (20), for example, the sequential model of Koshland and colleagues and the association–dissociation model of Freiden.

Effects of pH. Every enzyme contains a large number of acidic and basic groups. Some of them are either fully deprotonated (aspartate, glutamate) or fully protonated (arginine, lysine). However several groups with pK_a 5 to 9 (imidazole group of histidine, sulphydryl group of cysteine) do change their ionization state when pH is varied. Assume as a first approximation that enzyme is represented as a dibasic acid H_2E and only a singly ionized complex, HES^-, is able to react to give products:

$$\begin{array}{ccccccccc} & & H_2E & & H_2ES & & & & \\ & & \upharpoonleft\downharpoonright K_1^E & k_{+1} & \upharpoonleft\downharpoonright K_1^{ES} & k_{+2} & & & \\ S & + & HE^- & \rightleftarrows & HES^- & \rightarrow & HE^- & + & P \\ & & \upharpoonleft\downharpoonright K_2^E & k_{+1} & \upharpoonleft\downharpoonright K_2^{ES} & & & & \\ & & E^{2-} & & ES^{2-} & & & & \end{array}$$

Then the reaction rate is dependent on H^+ concentration, h, as follows (20):

$$v = \frac{Vs}{K_m + s}$$

$$V = \frac{\tilde{V}}{\left(\dfrac{h}{K_1^{ES}} + 1 + \dfrac{K_2^{ES}}{h}\right)}$$

$$\frac{V}{K_m} = \frac{\tilde{V}/\tilde{K}_m}{\left(\dfrac{h}{K_1^E} + 1 + \dfrac{K_2^E}{h}\right)} \tag{46}$$

where $\tilde{V}$ and $\tilde{K}_m$ are the pH-corrected constants. It is clear that pH-dependent variation of V reflects the ionization of ESC, whereas V/K_m reflects the ionization of the free enzyme. The pH effects on K_m are more complicated, being affected by both. Generally, the dependence of enzyme activity on pH is described by equation 43 as a bell-shaped curve with maximum at $1/2(pK_1 + pK_2)$.

Effects of Temperature. In chemical kinetics, the dependence of reaction rate on temperature is explained by the *transition-state theory* developed by Eyring in 1930 to 1935. It is based on the use of thermodynamics and principles of quantum mechanics. The reaction proceeds through a continuum of energy states and must surpass the state of maximum energy, when transient activated complex is formed. Then the dependence of reaction rate constant k on absolute temperature, T, is expressed as follows:

$$\frac{d \ln k}{dT} = \frac{\Delta H^* + RT}{RT^2} \tag{47}$$

where R is the gas constant and ΔH^* is the enthalpy of activated complex formation. The classic *Arrhenius* equation may be obtained from equation 47 under a simplified condition $\Delta H^* + RT \approx \Delta H = E_a$ (where E_a is *activation energy*). Most often, the Arrhenius equation is used in its integrated form:

$$\ln k = \ln A - E_a/RT$$

or

$$k = A \exp(-E_a/RT) \tag{48}$$

where A is the integration constant, interpreted as the frequency of collisions of reacting molecules. Apart from mechanistic derivations, there are a number of empirical expressions relating k and Celsius temperature T_c. The most popular is the exponential formula:

$$\ln k = \ln A + \alpha T_c$$

or

$$k = A \exp(\alpha T_c) \tag{49}$$

where α is the empirical constant related to the widely used temperature coefficient $Q_{10} = \exp(10 * \alpha)$.

All presented mathematical expressions predict exponential or almost exponential increases of chemical reaction rates with temperature. However, enzymatic reactions deviate from this relationship at high temperature because of thermal denaturation of enzymes. Assume that denaturation is reversible with equilibrium constant $K_T = [E']/[E]$, where E represents active enzyme molecules and E' represents inactive molecules. Then, the combination of equation 48 with the van't Hoff relationship for K_T ($-RT \ln K_T = \Delta G^\circ = \Delta H^\circ - T\Delta S^\circ$) results in

$$v = \frac{A \exp(-E_a/RT)}{1 + \exp(\Delta S^\circ/R - \Delta H^\circ/RT)} \tag{50}$$

where ΔG°, ΔS°, and ΔH° are the standard Gibbs free energy, enthalpy, and entropy of denaturation reaction, respectively, and v is the observed rate of biological process. This equation produces a curve with single maximum and fits to most of the available experimental data on temperature-dependent variations of enzymatic activity.

Sometimes denaturation is irreversible, and there are no possibilities for simple mathematical expressions, because temperature effects depends on the exposing time. However, numeric solutions of respective differential equations can still be used.

Simple Models of Microbial and Cell Growth

This section deals with simple unstructured models. These models mainly ignore any changes in cell quality (biochemical composition, spectrum of enzymatic activity, etc.) induced by environmental factors.

Main State Variables and Growth Parameters. There are two types of growth models, *deterministic* and *stochastic*. The former describe clear determined and regular processes. The latter deal with random or stochastic processes. The main variables used in deterministic models are the same as in chemical kinetics, concentrations of biomass, substrates, and products, and stochastic models consider instead the probabilities, frequency distributions, variance, and so on. For example, a stochastic model can consider the probability of a single bacteria cell dividing under specified environmental conditions. Although any real-life biotechnological process has both deterministic and stochastic components, most useful growth models are strictly deterministic. In this section, we will concentrate on this type of model, and the stochastic counterparts will be considered only in "Cell Cycle".

The major state variables of deterministic models describing cell growth are described in Table 7. The concentration of biomass and cell number are related to each other by simple formula:

$$x = m \times N \tag{51}$$

where conversion factor m_1 is the average dry mass of a single cell. In some studies, it is save to assume m to be a constant and use both variables x and N as equivalent measures of growth. However, the average size and mass of single cells vary depending on the nature of studied organisms and environmental conditions (see "Cell Cycle"); therefore, it is advisable to make a selection. The biomass, x, has obvious advantage in studies aimed at understanding or control of mass flows, and cell number, N, is preferred in population studies when, for example, mutation or plasmid transfer is an essential factor controlling the efficiency of the biotechnological process.

The choice of method to determine biomass or cell number depends on many factors (2,5). Today, the preference should be given to those techniques that allow exact and automated measurements (Table 8). The most advanced analytical methodology is now available for automatic recording of gaseous or volatile substrates, intermediates, and end products, such as methane, CO_2, O_2, H_2, volatile fatty acids, alcohols, and other fermentation products (IR-analyzers, mass spectrometry, gas chromatography, NMR, etc.).

The *primary* state variables that are measured directly in cell culture are usually recalculated into the *secondary* growth characteristics: gross growth rate, dx/dt; specific growth rate, $\mu = (dx/dt)/x$; degree of multiplication, x/x_0; biomass doubling time, $t_d = \ln 2/\mu$; growth yield, $Y = -dx/ds$; product yield, $Y_{p/s} = -dp/ds$, $Y_{p/x} = -dp/dx$; specific rate of substrate consumption, $q_s = (ds/dt)/x = \mu/Y$; specific rate of product formation, $q_p = (dp/dt)/x = \mu Y_{p/x} = \mu(Y_{p/x}/Y)$. Most of the listed secondary growth characteristics are called *specific* rates expressed as a first-time derivatives of measured variable *per unit of cell mass*. The specific rates are not sensitive to variations in total cell biomass, so they can be considered as analogous to enzyme activity (q_s, q_p) in most kinetic derivations. The specific growth rate, μ, may be viewed as the activity of an autocatalytic enzyme producing itself. The specific growth rate, μ, measured from biomass dynamic, x, can differ from that estimated from the increase in cell number, N (denoted as μ_N). It follows from equation 51 that

$$\mu = \frac{1}{x}\frac{dx}{dt} = \frac{1}{x}\frac{d(N \times m)}{dt} = \frac{1}{N \times m}\left(N\frac{dm}{dt} + m\frac{dN}{dt}\right)$$
$$= \frac{1}{m}\frac{dm}{dt} + \frac{1}{N}\frac{dN}{dt} = \mu_{cell} + \mu_N \tag{52}$$

If mass of single cell m is constant, then $\mu_{cell} = 0$ and $\mu = \mu_N$. Otherwise, we have to take into account m variation.

Validity of Exponential Growth Low. One of the earliest postulates in microbial kinetics is that under optimal non-

Table 7. Major State Variables of Deterministic Models

Variable	Notation	Dimension (examples)
Concentration of cell biomass	x	g CDW L^{-1} of cultural liquid
Cell number[a]	N	10^9 cell mL^{-1} of cultural liquid
Single-cell mass[a]	m	g CDW per cell
Mycelium length[b]	L	meter L^{-1} of cultural liquid
Tips number[b]	n	10^6 tips mL^{-1} of cultural liquid
Concentration of limiting substrate	s	g L^{-1} of cultural liquid
Concentration of product	p	g L^{-1} of cultural liquid or g g^{-1} of CDW

[a]For unicellar organisms (bacteria, yeasts).
[b]For filamentous organisms (fungi, actinomycetes).

Table 8. Methods Used for Determination of Microbial Biomass

Method	Measuring principle	DL	Advantage	Disadvantages
Dry mass	Mass of separated and dried solids	50	Provides direct unconditional estimate	Interference from dead cells and noncell solids
Wet mass	Mass of separated material	50	Simplicity, quickness	The variation of wet biomass bulk density
Wet biovolume	Linear dimension of pelleted cells or colony	100	Simplicity, quickness	The variation of wet biomass bulk density
Particulate organic carbon	CO_2 after cell separation and combustion	1.0	High precision and sensitivity, provides direct estimate of mass	Interference from dead cells and noncell solids
Biuret proteins	Colorimetric reaction of peptide bonds	1.0	High uniformity	Variation of protein-to-biomass ratio, possible extracellular accumulation
Folin-Ciocalteu protein	Colorimetric reaction of tyrosine and tryptophan	0.1	High sensitivity	Variation of protein-to-biomass ratio and amino acid composition of cell protein, possible extracellular accumulation
DNA	Colorimetric estimation of deoxyribose	1.0	High specificity, constancy of the DNA cellular content	Possible extracellular accumulation
ATP	Bioluminescence assay	10^{-5}	High sensitivity and specificity	Variation of the intracellular ATP pool
Fatty acids	Gas or liquid chromatography, colorimetric reaction	1.0	High specificity, allows identification of composition of mixed culture	Variation of the intracellular content, possible extracellular accumulation
Metabolio potential	Rate of added substrate uptake or product formation	10^{-3}–10	High specificity, quickness	Variation in conversion factor from measured rate to biomass
Opacity	Light scattering	0.1	Simplicity, quicknerss, automation	Interference from noncell solids, cell aggregation, wall growth
Electrical measurements	Conductivity	1	Simplicity, quickness, automation	Variation of conversion factor, interference from electrolytes in the medium
Manual microscopy	Cell count, measuring of hypha length	10^{-5}	Low cost, sensitivity, allows visual assay of biomorphological structure	Time-consuming, low precision
Image analysis	Computer-aided count	10^{-5}	Combined benefit of manual and automated quantification, speed, generation of size distribution	High cost
Coulter counter	Automated count and sizing	10^{-5}	Quickness, automation, generation of size distribution	Interference from noncell solids
Plating	Growing of the colonies on Petri dish	10^{-5}	Low cost, high sensitivity	Time consuming, low precision

Note: DL = detection limit, the minimum CDW required for an estimation with an error <2% (mg).

restrictive conditions (nutrient media containing all essential components at nonlimiting concentrations, absence of inhibitors, adequate physicochemical parameters, perfect mixing), the increase in biomass (dx) during an infinitely small time interval (dt) is proportional to this time interval and the instant biomass concentration (x), that is

$$dx = \mu x dt$$

or

$$dx/dt = \mu x \quad (53)$$

where μ is the proportionality coefficient. If μ is constant, integration of equation 53 gives

$$\ln x = \ln x_0 + \mu t \Rightarrow x = x_0 \exp(\mu t) \quad (54)$$

where x_0 is biomass at zero time or *inoculum size*.

According to early views, true exponential growth occurs only in the case of symmetrically dividing bacteria with equal probability of subsequent division for the mother and daughter cells. Organisms with asymmetrical multiplication (budding yeasts) were thought to obey the exponential low only approximately, whereas the growth of filamentous organisms (fungi, streptomycetes) was deviated considerably. However, direct measurements performed from 1930 to 1960 (5) revealed that exponential growth does occur in all prokaryotes and eukaryotes independent of their biomorphological features, including protozoa, fungi, and homogeneously cultivated plant and animal cells. The deviation is observed only as a result of a growth-associated change in the environment. Wang and Koch (26) made very precise measurements of μ dynamics by growing *Escherichia coli* directly in an aerated cuvette of the computer-linked double-beam spectrophotometer. They found temporary slowdowns in the complex peptone plus beef extract media attributed to diauxie phenomena (e.g., depletion of some preferred oligopeptides) and a gradual increase of the μ value during sequential subculture in succinate minimal medium. Long-term cultivation of microorganisms in continuous turbidostat or pH-stat culture (14) revealed that the increase in μ is mainly the result of autoselection of mutants having higher maximum specific growth under specified cultivation conditions (see "Population Dynamics").

Early Views of Cell Growth. The derivation of the exponential growth equation was done for binary dividing bacteria, based on the geometric progression 2, 2^2, 2^3 . . . (27):

$$N = N_0 2^n \tag{55}$$

where n is the number of divisions, $n = t/t_d$.

The growth dynamics were viewed as a succession of distinct phases approximated with empirical formulas (15):

$$N = N_0 \exp[\mu F(t)t] \tag{56}$$

where function $F(t)$ undergoes stepwise changes as shown in Table 9 ($1.56 \leq n \leq 2.7$). The alternative way to describe growth dynamics was to use the logistic equation borrowed from demographic studies (28):

$$\frac{dN}{dt} = rN\left(1 - \frac{N}{K}\right)$$

or

$$\frac{dx}{dt} = rx\left(1 - \frac{x}{x_m}\right) \tag{57}$$

where r is the net growth rate (the difference between true growth and death rates) and K or x_m are the upper limits of, respectively, population density or biomass (K is used preferentially in ecological literature, being called the carrying capacity of the ecosystem). This equation simulates S-shaped growth dynamics frequently observed in natural ecosystems and cell cultures.

Table 9. Stepwise Changes of Function $F(t)$

Initial stationary phase	$F(t) = 0$
Lag phase	$F(t) = t^{n-1}$
Logarithmic growth phase	$F(t) = 1$
Negative growth acceleration phase	$F(t) = t^{t-1}$
Maximum stationary phase	$F(t) = 0$
Accelerating death phase	$F(t) = t^{n-1}$
Logarithmic decrease phase	$F(t) = 1$

A similar (but not identical) equation of cell biomass dynamics would be produced if we assumed that the limiting substrate is consumed according to first-order kinetics and is converted to biomass with a fixed yield factor:

$$Y = (x - x_0)/(s_0 - s) \tag{58}$$

then

$$\begin{aligned} \frac{ds}{dt} &= -ks \\ s &= \frac{x_0 + Ys_0 - x}{Y} = \frac{x_m - x}{Y} \Rightarrow \frac{dx}{dt} \\ &= -Y\frac{ds}{dt} = kx_m\left(1 - \frac{x}{x_m}\right) \end{aligned} \tag{59}$$

Monod's Model. This model still is very popular because of its elegant simplicity. It played an important role in the history of microbial growth kinetics as a first encouraging example when the theoretical development based on mathematical formalisms came before novel experimental designs. As compared with simple exponential equation 53, this model (29) takes into account the mass-conservation condition linking substrate uptake to biomass formation through constant yield factor Y (equation 2) and the dependence of the specific growth rate on limiting substrate concentration:

$$\begin{aligned} \frac{dx}{dt} &= \mu(s)x = \mu_m \frac{s}{K_s + s} x \\ \frac{ds}{dt} &= -\frac{1}{Y}\frac{dx}{dt} \end{aligned} \tag{60}$$

By substitution of s by x through Y (equation 58), we can reduce equation 60 to a single equation having the analytical solution:

$$\begin{aligned} \mu_m t = {} & (1 + P)\ln(x/x_0) - P\ln(Q - x/x_0) \\ & + P\ln(Q - 1) \end{aligned} \tag{61}$$

where $P = YK_s/x_m$, $Q = x_m/x_0$, $x_m = Ys_0 + x_0$, and s_0 and x_0 are, respectively, s and x at $t = 0$.

Equation 61 contains three parameters, Y, μ_m, and K_s that can be thought of as passport data for a particular organism (e.g., for *E. coli* grown on glucose at 30 °C, $Y = 0.23$, $\mu_m = 1.35\ h^{-1}$, and $K_s = 4\ mg\ L^{-1}$) and describe the S-shaped growth dynamics of a batch culture. The initial conditions, s_0 and x_0, are set by the experimenter, who selects the inoculation dose and medium composition. Thus, knowing the values of all these entities, the growth dynamics can be calculated *before the experiment*. Moreover, this model was used as a basis for development of the *chemostat theory* to understand and predict the behavior of microbial cultures in continuous-flow bioreactors before the respective hardware was constructed. For this purpose, original equation 60 was modified as follows (30):

$$\frac{dx}{dt} = (\mu - D)x$$

$$\mu = \mu_m \frac{s}{K_s + s} x$$

$$\frac{ds}{dt} = D(s_r - s) - \frac{1}{Y}\mu x$$

$$D = \frac{F}{V} \tag{62}$$

where s_r is the concentration of limiting substrate in the fed-fresh medium delivered with pump from reservoir, F is the pumping rate (cm^3/h), V is culture volume (cm^3), and D is the dilution rate defined as the ratio $D = F/V$, h^{-1}.

The mathematical analysis of equation 62 allows us to make the following conclusions on the growth kinetics of continuous culture:

1. The described open system attains stable *steady state* when $dx/dt = 0$, $ds/dt = 0$, and variables x and s are not changed with time (denoting steady-state concentrations $\tilde{x}$ and $\tilde{s}$, respectively). From the first equation in equation 60, it follows that $\mu = D$, that is, the specific rate of microbial growth is equal to the dilution rate, which is under the full control of the experimenter. From the second equation, we find

$$\tilde{x} = Y(s_0 - \tilde{s}) \tag{63}$$

   As compared with Y expression for batch culture (equation 58) the formula in equation 63 no longer contains the term, x_0, the inoculum size. Thus, a steady-state culture is not dependent on past history, and its properties are determined solely by current cultivation parameters.
2. The dilution rates permitting stable microbial growth (i.e., those D, at which $\tilde{x} > 0$) are confined between 0 and washout point or *critical dilution rate* D_c

$$D_c = \mu_m s_r/(K_s + s_r) \tag{64}$$

3. The specific growth rate μ does not depend on either s_0 or x and is governed solely by the substrate concentration in the cultural liquid. On the other hand, the indirect effect of s_0 on μ may be evaluated from the dependence of $\tilde{x}$ on D, as soon as $\tilde{x}$ and $\tilde{s}$ are linked by the conservation condition in equation 62.

The most important biological implication of the chemostat theory was the discovery of the fact that microorganisms can grow endlessly with any rate between 0 and μ_m. Such substrate-limited growth is nevertheless exponential, because two subsequent acts of cell division will be separated by a constant time interval. Earlier, substrate limitation had been observed only transiently at the end of the exponential phase of batch growth. The most important biotechnological implication of the chemostat model was the concept of *controlled cell biosynthesis*, which implies the purposeful manipulation of microbial culture to optimize the productivity by such tools as changing the flow rate and the composition of medium in continuous-flow bioreactors.

Derivation of the Monod Equation from Mechanistic Considerations. The equation relating μ and s is known as the *Monod equation*:

$$\mu = \mu_m \frac{s}{K_s + s} \tag{65}$$

It was introduced by the author as a purely empirical relationship resembling adsorption isotherm. However, many microbiologists did view the Monod equation as something having deep inherent meaning rather than as just an empirical formula. Below, we shall outline a number of attempts to deduce this equation logically from the conjectured growth mechanisms.

1. *The bottle-neck concept*. There is an obvious similarity between the Monod and Michaelis-Menten equations. A theoretical substantiation for this similarity can be the bottleneck postulate originally put forward by Blackman (31). According to this postulate, the growth rate of a cell is determined by a single *enzymatic master reaction*. Microbial metabolism is symbolized as a unidirectional chain of reactions of substrate S conversion to cell biomass X' via some hypothetical intermediates $P_1, P_2 \ldots P_n$ (32):

$$S + X' \xrightarrow{k_1} P_1 \xrightarrow{k_2} P_2 \rightarrow \ldots \xrightarrow{k_n} P_n \xrightarrow{k_{n+1}} 2X' \tag{66}$$

Assuming steady state in respect to intermediate concentration p_i, $dp_i/dt = 0$, we arrive at the Monod equation, composite parameters μ_m and K_s being expressed via elementary kinetic constants of individual enzymatic reactions:

$$\frac{dx}{dt} = \mu_m \frac{s}{K_s + s} x$$

$$\mu_m = \frac{1}{\sum_{i=2}^{n+1} 1/k_i}$$

$$k_s = \frac{1}{k_1 \sum_{i=2}^{n+1} 1/k_i} \tag{67}$$

where $x = x' + p_1 + \ldots p_n$. The bottleneck idea can now be formulated as follows. If one of the constants $k_j \ll k_i$, $i = 2, \ldots, n + 1$, $i \neq j$, then $\mu_m = k_j$ and $K_s = k_j/k_1$, and so the jth enzymatic reaction is the master reaction.

The obvious shortcomings of the reaction scheme in equation 66 are the unjustified oversimplification of the cell metabolism (which is a network, rather than a simple unidirectional sequence of reactions) and the lack of a definite interpretation of variables x' and p in real biochemical terms. An evaluation of the characteristic times of major intracellular metabolic reactions showed that there could be two possible bottlenecks: uptake of limiting substrate (processes of active transport, transphosphorylation of sugars etc.) (33) and the formation of intracellular protein-synthesizing structures such as rRNA and the ribosomal proteins (34).

2. *Stochastic considerations*. Microbial populations were assumed to consist of active (which utilize the substrate) and inactive cells (35). The probability of a transition from the inactive state to the active one was supposed to be proportional to s, whereas the probability of the reverse transition was s independent. From these assumptions, a relationship between μ and s can be inferred, similar to equation 65.

3. *Derivations based on mechanistic considerations of protein synthesis (36)*. The rate of protein synthesis, dp/dt, is supposed to be determined by rRNA concentration R and by the size of the amino acids pool, A, $dp/dt = kxRA/(K_A + A)$, k = constant. Other conditions are defined as $R = R_0 + (R_m - R_0)\mu/\mu_m$, $A = bs$, and $P = p/x$, where R_m and R_0 are, respectively, the upper and lower limits of R variation, and P and b are constants. For a steady-state chemostat culture, $dp/dt = 0$ and $dx/dt = (1/P)(dp/dt) = 0$, then

$$\mu = (kR_m/P)s/(R_mK_A/\beta r_0 + s) = \mu_m s/(K_s + s) \quad (68)$$

Here again, the Monod equation is derived through a consideration of underlying intracellular processes. Needless to say, none of the cited derivations are free from criticism. The Monod equation remains empirical. Any attempts to provide the mechanistic interpretation of this equation inevitably lead to much more complicated mathematical expressions (see late section on structured models).

Biological Meaning and Experimental Determination of Growth Parameters K_s and μ_m. The parameter μ_m, maximal specific growth rate, has very lucid biological meaning: it is the upper limit of μ variation on specified nutrient medium. It could not be attained in reality because of its asymptotic nature: $\mu \to \mu_m$ as $s \to \infty$. However, in practice μ_m is achieved if $s \gg K_s$. It should be remembered that μ_m is not absolute maximum of growth rate, because it depends on the nature of the limiting substrate. For example, *E. coli* has μ_m above 2.5 h^{-1} on complex beef-extract medium and below 1.0 h^{-1} on minimal medium with succinate. Saturation constant K_s could be defined functionally as such concentration of limiting substrate that provides a specific growth rate equal to 0.5 of μ_m. We can say also that K_s is a measure cell affinity to substrate: the lower is K_s the better the organism is adapted to consume substrate from diluted solution. Any other definitions are speculative, e.g., K_s interpretation as the dissociation constant of ESC of the cellular enzyme involved in the first step of substrate conversion.

There are several ways to determine numeric values of K_s and μ_m:

1. *Batch culture, differential form of Monod equation.* The biomass dynamics, $x(t)$, are followed from the start of the exponential phase until the complete consumption of the limiting substrate and the attainment of the maximal cell density, x_m. The residual substrate concentration, s is calculated from mass balance (equation 58); yield is calculated as $(x_m - x_0)/s_0$; the s_0 value should be known) and corresponding $\mu(t)$ values from the slopes $d(\ln x)/dt$. Finally, parameters K_s and μ_m (equation 65) are to be found graphically or from nonlinear regression as in the case of the Michaelis-Menten equation (see later text).

2. *Batch culture, ignoring the substrate uptake*. The inoculum size, x_0, and the duration of experiment are chosen to minimize the uptake of added substrate $s \approx s_0$ (37). Typically, bacterial cell density should be of the order 10^{-5} or less, and it is measured by such sensitive instruments as the Coulter counter. The quasi-steady-state growth rate is measured at several s_0 values and then fitted to equation 65 as described before. In numerous determinations made at high cell densities (when we can no longer neglect substrate consumption), it was observed that dependence of μ during exponential growth phase on s_0 is formally described by the Monod equation with s replaced by s_0 (38). The explanation of these results could be made only by more complex structured models; however, the described procedure is not appropriate for K_s and μ_m determination.

3. *Batch culture, integral form of equation 61*. The S-shaped curve of biomass dynamics is fitted directly to equation 61 through either preliminary linearization or nonlinear regression (preferential). Sometimes (e.g., when cells are grown in opaque media) it is more convenient to follow the dynamics of residual substrate $s(t)$ or product formation $p(t)$, for example, CO_2 evolution or dynamics of O_2 uptake (respiration). In these cases, we can integrate the set of differential equation 60 in terms of $s(t)$ or $p(t)$ dynamics, taking into account mass–balance relationships in equations 7, 8, and 58 and similar equations.

4. *Steady-state chemostat culture*. The chemostat provides the opportunity for estimations under steady-state condition, which is commonly believed to be more reliable. By running an experiment at different dilution rates, D, the corresponding $\tilde{s}$ values may be measured and the dependence of $\mu = D$ on $\tilde{s}$ obtained; in principle, this can be done as accurately and carefully as desired. Such an approach, however, may and frequently does encounter serious technical problems because of the high affinity to limiting substrate of some microorganisms. There is a need to (*1*) select highly sensitive analytical techniques to measure extremely low residual concentrations of particular substances; (*2*) develop instant sampling procedures to minimize substrate loss, and (*3*) eliminate apparatus-related artifacts such as nonperfect mixing and fluctuations in nutrient medium supply. This can be accomplished by use of radiolabeled substrate, and other tips are covered in specific experimental works (39,40). However, the most essential objection to this method is that organisms at various steady states do change their kinetic properties, which is not accounted by Monod's model (see "Structured Models").

5. *Non-steady-state chemostat culture*. The measurements are made during short-term experiments started by addition of different amounts of limiting substrate to a steady-state chemostat culture. Then, substrate uptake or respiration rates are recorded in perturbed culture until

the new steady state is established (2). The opportunity of continuous culture is that the physiological state of cells just before perturbation is well defined and reproducible. On the other hand, these experiments do provide data on affinity to substrate (K_s) and maximal rates of respiration (e.g., Q_{CO_2} or uptake Q_s, which are related but are not identical with μ_m):

$$\mu_m = Y_{p/x}Q_{CO_2} = Y_{s/x}Q_s \quad (69)$$

where stoichiometric parameters $Y_{p/x} = dp/dx$ and $Y_{s/x} = ds/dx$ should be determined in independent experiments. Immediate assay of μ_m in such an experimental setup is possible as follows: the steady state is perturbed by setting the dilution rate by 20 to 100% higher than the critical value D_c (equation 64). The washout dynamics are followed and μ_m is determined from approximate relationship (5): $x \approx x(0)\exp[(\mu_{m-D})t]$, where $x(0)$ is biomass before perturbation. More rigorous estimate of μ_m could be provided if such washout experiments were made at several (at least two) input substrate concentrations s_0 to account for the fact that s_r is still not infinity.

Modification of the Monod Equation. It was found that not all experimental data could be reasonably well fitted by equation 65. The best-fit hyperbola often passed above experimental points at small s and below them at large s. A better fit was claimed to be provided by using the following, entirely empirical, equations:

$$\mu = \mu_m[1 - \exp(-Ks)] \quad (70)$$

$$\mu = \mu_m s^n/(K_s + s^n) \quad (71)$$

$$\mu = \mu_m s/(K_s x + s) \quad (72)$$

The preference of the first expression, equation 70 (41), is questionable. An expansion of Monod equation by addition of the third parameter, n in equation 72 (42), or introduction of the second variable, x in equation 71 (43), does improve the approximation capability of kinetic equations. We can even provide the mechanistic basis for this improvement. Thus, the Moser equation (equation 71) is similar to the Hill equation in enzymology (equation 43), indicating the cooperativity effects in performance of some master reaction of cellular metabolisms. The Contois equation (equation 72) could be interpreted in terms of growth autoinhibition by-products, because under the realistic assumptions ($x \gg x_0$, $xY_{p/x}/K_p \gg 1$ and product p formation coupled with growth), the apparent saturation constant K_s^{app} is almost proportional to accumulated biomass x:

$$K_s^{app} = K_s(1 + p/K_p) = K_s[1 + Y_{p/x}(x - x_0)/K_p] \approx K_s x \quad (73)$$

An Account of Maintenance Requirements. Powell (44) assumed that substrate uptake rate q obeys the Michaelis-Menten kinetics; then, from mass–balance equation 18 it follows that

$$q_s = Q_s/(K_s + s) = \mu/Y^m + m \quad (74)$$

If μ_m is defined as $\mu_m = YQ_s$ (5), then

$$\mu = \mu_m s/(K_s + s) - mY^m \quad (75)$$

If we define $\mu_m = Y^m(Q_s - \mathrm{m})$ (2), then

$$\mu = \mu_m(s - s^*)/(K_s + s)$$
$$s^* = K_s m Y^m/\mu_m \quad (76)$$

Equations 75 and 76 both predict the occurrence of threshold substrate concentration s^*, below which growth is impossible. It yields a substantially better fit to experimental data. The difference is that according to equation 75, the parameter μ_m is the specific growth rate under imaginary conditions of zero maintenance requirements, whereas equation 76 implies the traditional definition of μ_m as the specific growth rate under substrate excess: $\mu \to \mu_m$ when $s \to \infty$.

Account for Substrate Leakage. Mathematically identical to equation 76, a modification of the Monod equation was proposed for the case of conserved substrates. However, the biological meaning is entirely different: if the specific leakage rate is assumed to be constant, then a decrease in s down to some threshold value s^* will lead to the counterbalance of the two reverse processes (uptake and leakage), so that the net consumption of the limiting substrate will be zero.

Account of Inhibitory Effects. A few valuable refinements of the Monod equation were borrowed from enzymology; most often they were noncompetitive and substrate inhibition. The former inhibition mechanism is by the so-called Monod-Ierusalimsky equation:

$$\mu = \mu_m \frac{s}{K_s + s} \times \frac{K_p}{K_p + p} \quad (77)$$

The growth retardation with an excess of such substrates as phenol, methanol, and ethanol is described by an analogue of Haldane's equation (45):

$$\mu = \mu_m \frac{s}{K_s + s + s^2/K_{ss}} \quad (78)$$

Equation 78 simulates a single-peak curve, and so the same μ value may be obtained at two different s, one in the substrate-limiting range, $d\mu/dt > 0$ (stable), and the other in the substrate inhibition range, $d\mu/dt < 0$ (unstable). A sustainable maintenance of a population under conditions of substrate inhibition is possible either in the second stage of a two-stage chemostat or in the case of plentiful wall growth in a conventional chemostat (46).

Account of Diffusion Effects. We will present one example of such models (44). The basic assumption is that substrate is taken up by an enzyme that obeys Michaelis kinetics and is localized on the inner side of cell membrane. The actual substrate concentration around the enzyme-active centers is smaller than in the solution, because of a limited diffusion rate. By applying a simplified Laplace equation, it was found eventually that

$$\mu = \frac{\mu_m(K_s + L + s)}{2L}\left(1 - \sqrt{1 - \frac{4Ls}{(K_s + L + s)^2}}\right)$$
$$\approx \mu_m \frac{s}{K_s + L + s} \tag{79}$$

where L is a factor determined by membrane permeability and by the maximum rate of the enzymatic reaction.

Structured Models

The unstructured models (such as Monod's model or its modifications) are able to predict and describe only simplest manifestation of growth phenomena. Sometimes it is declared that Monod-type models are able to describe only *balanced* and *steady-state* growth. The analysis of more complicated unbalanced and non-steady-state growth requires formulation of structured models.

Definitions. *Balanced growth* was defined by Campbell (47) as a proportional increase in the amounts of all cell components, in other words, balanced growth produces cells of the same quality without any variation of composition. The terms *steady state* and *non steady state* stem from chemical and enzyme kinetics. The first one refers to such a regime when the reaction rate remains constant because of an exact balance between formation and breakdown of intermediary products such as the enzyme–substrate complex. In microbial culture, the growth is called steady state if *specific* rather than *total rates* remain constant. In an open system, such as a chemostat, both total (dx/dt, ds/dt) and specific rates ($\mu = (1/x)dx/dt$, $q_s = (1/x)ds/dt$) tend to have constant steady-state values. A closed system, such as a batch culture, should be considered under steady state only during the exponential phase when μ and q_s are constant. The linear growth with a constant total rate ($dx/dt = \mu x =$ constant) is characterized by monotonously declining μ, and is not steady-state growth. However, under some conditions (such as in dialysis culture) it may attain *quasi steady state*, when $ds/dt \approx 0$.

The non-steady-state kinetic regimes take place before establishment of steady state or after its perturbation. In enzyme kinetics, non-steady-state measurements are taken in the millisecond range of time scale. In microbial cultures, similar non-steady-state transient and perturbation processes advance much more slowly, typically during several hours and days. An example is a transient process in the chemostat induced by changes in D or s_r (fed-substrate concentration). In such growth, μ, q_s, q_p, and other metabolic rates exhibit continuous variation in time. The attractiveness of non-steady-state studies for microbiology and biotechnology is obvious:

- They allow a wider range of hypotheses to be tested and yield much more data on the studied objects.
- They have higher practical value; in biotechnology, steady-state operation is the exception rather than the common routine because of unavoidable disturbances in cultivation conditions.
- They provide additional tools for optimal regulation of cell performance in the bioreactor, because purposeful non-steady-state growth may display greater efficiency and higher productivity (48).

The notions of steady-state and balanced growth are close but not identical. The first is more strict; steady-state growth has to be balanced by definition (otherwise some of the specific rates responsible for synthesis of the changed cell component should vary). On the other hand, the balanced growth can be for some time nonsteady, perhaps during the late exponential phase of batch culture when μ declines while cell composition remains unchanged. During long-term experiments, non-steady-state growth leads inevitably to a change of cell composition; it becomes unbalanced. For heterogeneous populations, the situation may be more complicated. For example, the growth in the second stage of a two-stage chemostat attains a steady state, and the biomass and residual substrate concentration are constant. However, such growth is not balanced, because cells delivered from the first stage differ in their properties and composition as compared with the bulk of cells in the second stage. This situation was termed the *transient steady state* (49).

By *structured*, we mean mathematical models describing growth-associated changes in microbial cell composition. It includes mass–balance equations for all assigned intracellular components. Their concentrations can be expressed either per unit volume of fermenter vessel ($c_1, c_2, \ldots, c_n$), or per unit cell mass ($C_1, C_2, \ldots, C_n$), and hence $C_i = c_i/x$. The mass–balance equations can be written as follows

$$\sum_{i=1}^{n} c_i = x$$
$$\sum_{i=1}^{n} C_i = 1 \tag{80}$$

For each variable C_i, a differential equation is written that takes into account all sources, r_+, and sinks, r_-, as well as its dilution from cell mass expansion (growth)

$$\frac{dC_i}{dt} = r_+(s, C_1, \ldots, C_n) - r - (s, C_1, \ldots, C_n) - \mu C_i \tag{81}$$

The simplest structured models with n no more than 2 or 3 are called two- or three-compartment models. For example, a model (50) incorporated two compartments: nucleic acids and proteins combined with other active cell components. The model variables also included concentrations of the limiting substrate and the inhibitor. Compared to Monod's model, the proposed set of four equations was able to account for a much wider range of dynamic patterns. Specifically, it simulated D-dependent changes in the cell composition (chemostat) and all known growth phases of batch culture from the lag to decline. However, the choice of variables in this model was more or less arbitrary, so it should be regarded more as a bright illustration rather than a research tool.

During the past decade, much more realistic structured models based on biochemical data have been developed.

The simulation model of *E. coli* growth (51) contains the following dynamic variables: glucose and NH_4^+, as exosubstrates; CO_2 and acetate, as products excreted into the medium; amino acids; ribonucleotides; deoxyribonucleotides; monomeric precursors of cell wall components; rRNA and tRNA; nonprotein polymeric components; glycogen; guanosinetetraphosphate; enzymes transforming ribonucleotides into deoxyribonucleotides; ATP; NAD(H); and protons. Altogether, the dynamic model amounts to a system of 21 differential and 14 algebraic equations. An even more complicated model simulating growth of *Bacillus subtilis* (52) is the set of 39 nonlinear and coupled differential equations containing nearly 200 parameters! These models are able to simulate particular growth features such as changes in cell sizes, shape, and composition as well as the D-dependent variations in replication time brought about by the shifts in glucose concentration. However, the predictive capability of such an intricate dynamic model should be still estimated as modest as compared with invested modeling efforts; they are still nothing more than a "caricature parody" of microbial biochemistry and are too complex to be studied by conventional mathematical tools (stability analysis, parameters identification, etc.) or to be used in biotechnological applications. The best choice of a mathematical model lies, apparently, midway between unstructured and highly structured models outlined here. One of the known compromises has been found through attempts to express quantitatively the *cell physiological state*.

Physiological State of Chemostat Culture. The term was coined by Malek (53) without giving a clear definition. The impetus for the development of the concept of a physiological state was the evidence on changes in the chemical composition of microorganisms as dependent on dilution rate D and medium composition in chemostat culture (54). It has been found that some of the studied parameters remained constant (content of cell DNA and carbon) while others exhibited regular D-dependent variations, either an increase (RNA content, cell sizes) or a decrease (the content of reserved polysaccharides). Those properties that were D dependent were recognized as components of the *vector of the physiological state*.

Powell (44) combined and put on a quantitative mathematical footing three notions, which were beforehand separated and cloudy: (1) physiological state, (2) past history, and (3) non-steady-state growth kinetics of microbial culture. The specific rate of substrate uptake, q_s, was presented as a product

$$q_s(s) = q_a'(s)S(s) \tag{82}$$

where $S(s)$ is a simple saturation function (e.g., a Michaelis hyperbola), $S(s) = s/(K_s + s)$, and $q_a'(s)$ is a function associated with the microbial physiological state. The instant value of $q_a'(s)$ is determined by way in which s varies until the given moment (τ h ago), effects of later events contributing more than earlier ones. Transient processes are influenced by the past history of the culture in the following manner. Suppose that steady-state growth of a chemostat culture is upset and the residual substrate concentration jumps from $s(0)$ to $s(1)$. Immediately, q_s will increase from $Q(0)S[s(0)]$ to $Q(0)S[s(1)]$. If no further changes in $s(1)$ occur, then Q will also eventually attain a new steady-state value equal to $q_a'[s(1)]$. In essence, Q is the potential metabolic activity, that is, the specific rate of a key metabolic process measured just at the moment of relief from substrate limitation. The transient Q dynamics are described as net change equals production minus dilution caused by cell growth:

$$\frac{dQ}{dt} = r(Q, s) - \mu(Q, s)Q \tag{83}$$

Substance Q may, in reality, be represented by a single enzyme or multienzymatic complexes as well as by ribosomes or other cell components occupying the bottleneck position. Monod's model, supplemented by equation 83, made possible at least a qualitative understanding of chemostat-transient processes triggered by a D switch (55).

The *synthetic chemostat model (SCM)* (2) combines Powell's ideas with the routine of conventional structured models. This model is similar to the cybernetic model (56,57). The basic SCM interprets microbial growth as a conversion of exosubstrate S into cell macromolecules X' via a pool of intermediates L:

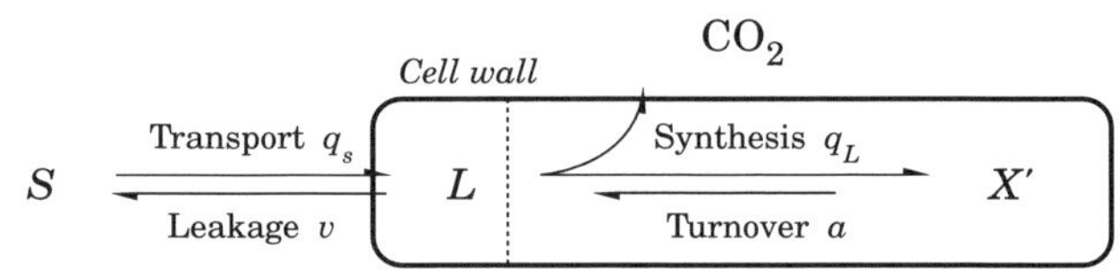

Macromolecular cell components are susceptible to degradation (turnover), and monomeric metabolites can escape into environment because of leakage. Contrary to simple chemical catalysts, the composition of end product X' is not uniform and varies in response to a changing environment because of the adaptive nature of microbial metabolism. At the heart of the SCM are the solution of the problem, how to cope with these variations, and how to describe adaptive changes in cell composition by relative simple models.

All macromolecular cell constituents are divided into two groups: primary cell constituents necessary for intensive growth (P components), and components needed for cell survival under any kinds of growth restriction (U components). The characteristic examples of P components are ribosomes (rRNA and ribosomal proteins) and enzymes of the primary metabolic pathways. Their intracellular content increases parallel to an increase in μ. The contribution of U components to cell biomass decreases with growth acceleration. Examples are enzymes of the secondary metabolism, protective pigments, reserved substances, and transport systems of high affinity.

The analysis of available data as well as mass-conservation conditions allowed the formulation of several rules of variation of cell components taking place because of optimal control of cell biosynthesis:

1. Amounts of P and U components expressed as a fraction of total cell mass (P and U, gram per gram of

biomass respectively) vary within the upper (P^{max}, U^{max}) and low (P^{min}, U^{min}) limits, the latter being the constitutive part.

2. An increase of one individual P component is accompanied by increase of other P components.
3. The total enlargement of P_{sum} is accompanied by corresponding decrease of U_{sum} and vice versa.
4. The P/U ratio is controlled by the limiting substrate concentration in an environment.

These rules are translated into mathematical terms as follows:

$$\frac{P_1 - P_1^{min}}{P_1^{max} - P_1^{min}} \approx \cdots \approx \frac{P_n - P_n^{min}}{P_n^{max} - P_n^{min}} = r$$

$$\frac{U_1 - U_1^{min}}{U_1^{max} - U_1^{min}} \approx \cdots \approx \frac{U_m - U_m^{min}}{U_m^{max} - U_m^{min}} = 1 - r$$

$$\tilde{r} = \frac{\tilde{s}}{K_r + \tilde{s}} \quad (84)$$

where variable r (index of physiological state) is already scalar (not vector!) function controlled by environmental factors (e.g., concentration of the limiting substrate). The r value in steady-state chemostat culture changes from zero (in culture at almost zero growth rate, when $\tilde{s} \to 0$ and all P components come down to low limits) to 1.0 (in unlimited culture, when $\tilde{s} \to \infty$ and P components attain maximum). During transients caused by sudden s changes, an instant r value goes toward new steady state (compare with equation 82):

$$\frac{dr}{dt} = \mu\left(\frac{s}{K_r + s} - r\right) \quad (85)$$

The introduction of the r variable greatly simplifies the use of structured models because the adaptive variation of cell composition (and metabolic activity, which is determined by the intracellular content of particular enzymes) now could be expressed via one single master variable r. For example, the specific rate of substrate uptake q_s is defined as:

$$q_s = r\frac{Qs}{K_s + s} + (1 - r)\frac{Q's}{K_s' + s} \quad (86)$$

where the first and the second terms on the right side stand, respectively, for low (P component) and high (U component) affinity of transport system. Similar r-dependent expressions are derived for other reactions (q_L, v, a) and stoichiometric parameters.

Predictive and clarifying capabilities of SCM turned out to be higher than more complex structured models. Contrary to all known chemostat models, SCM provides adequate simulation of D-dependent variation of the microbial physiological state. Under energy- and C-limited growth, it was expressed as an increase of apparent K_s, potential uptake and respiration rates, maintenance ration, and turnover parallel to increase of D. Under N limitation, a similar trend was complemented by considerable decrease of Y_N due to alteration of cell composition in favor of N-rich P components (Fig. 6). SCM adequately describes the transient growth caused by shift-up in chemostat culture. The phenomena of overshoot in substrate concentration and undershoot in biomass during transient growth are explained by slow adjustment of cell composition (RNA content, respiratory activity) to new growth conditions (Fig. 7). Batch culture limited by carbon and energy source was simulated by SCM on the whole from inoculation to death stage; conventional growth phases (lag, exponential, deceleration, stationary, decline) were generated automatically without setting any specific conditions (Fig. 8). It is important that SCM realistically describes and predicts not only net growth but also the survival dynamics of starving cells after substrate depletion. During this phase,

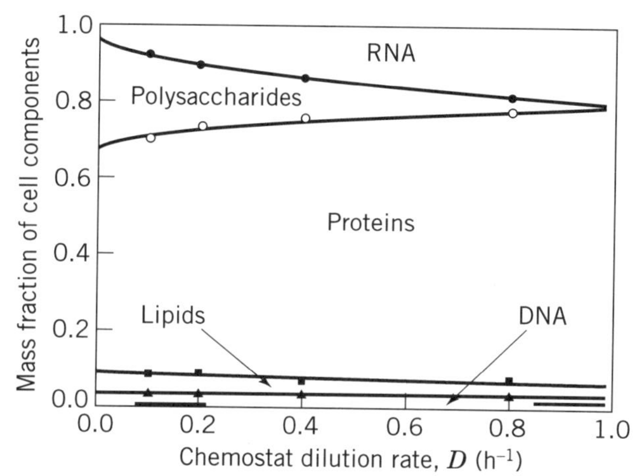

Figure 6. Simulation of D-dependent changes in cell composition of *Aerobacter aerogenes* grown in NH_4^+-limited chemostat culture. The curves are calculated from SCM (2), and the original experimental data are from Ref. 58.

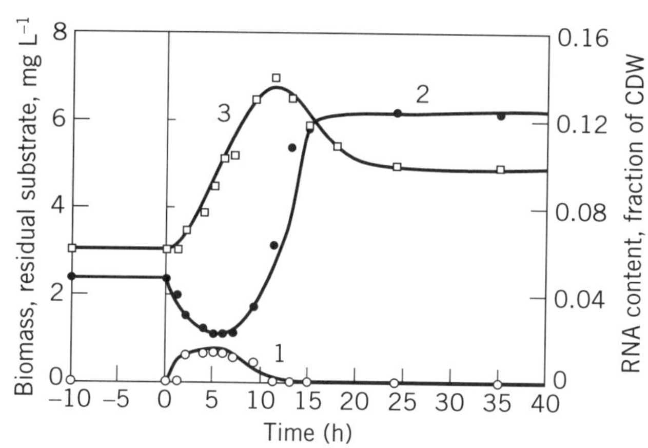

Figure 7. SCM simulation of transient growth induced by change of chemostat dilution rate. Residual glycerol concentration (*1*), biomass (*2*), and cell RNA content (*3*). At $t = 0$, dilution rate was shifted from 0.004 to 0.24 h^{-1}. *Source*: Redrawn from Ref. 2; the original data (59) are for chemostat culture of *Aerobacter aerogenes* limited by glycerol.

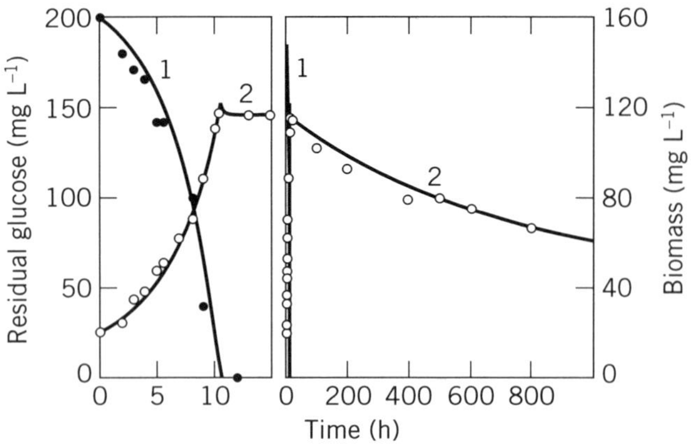

Figure 8. SCM simulation of complete dynamic curve of batch culture; residual glucose (*1*) and biomass (*2*) of yeasts *Debaryomyces vanrijiae* (old name *D. formicarius*) grown on glucose-mineral medium (2). Note that contrary to old empirical models (equation 56), all growth phases are reproduced automatically without specifying preset time ranges.

surviving bacteria sustain very slow cryptic growth at the expense of cell turnover and L leakage. The rate of decline in biomass gradually decreases as the result of buildup of some U components (parallel to the decrease of such P components as RNA and CN-sensitive respiration enzymes). Batch culture limited by conserved substrate (the source of N, P, Mg, Fe, or other) has an interesting feature: the growth is not stopped after substrate depletion but proceeds at an even higher rate. SCM explains this phenomenon by the partitioning of deficient elements between mother and daughter cells.

Cell Cycle

The term *cell cycle* is used to designate the regularly repeated sequence of events that occur between consecutive *cell divisions*, for example, the formation of two identical daughter cells from one cell, which takes place in most known bacteria. Equivalent cell cycle events include *budding* (most yeasts and budding bacteria), *branching of hypha* (filamentous organisms), *fragmentation* (some coccoid and corineform bacteria). The majority of available data have been obtained for rod-shaped bacteria (*E. coli, Salmonella typhimurium, Bacillus cereus*) under steady-state growth conditions when the cell cycle consists of three distinct phases:

1. The growth of the newborn cell without chromosome replication from the initial mass m_0 until some critical initiation mass m^*
2. DNA replication (C period)
3. Separation of daughter cells (D period)

Relationship between Cell Size and Specific Growth Rate. The periods C and D are constant; in *E. coli*, they occupy about 40 and 20 min, respectively, independent of environmental conditions. However, the duration of phase 1 depends on cultivation conditions; the time of the single-cell mass increase from m_0 to m^* is inversely related to the specific growth rate μ. It is this difference that yields μ-dependent variations of cell sizes, because faster-growing cells produce for the same $\tau = C + D$ time a larger cell mass than do slower-growing cells. It became clear from the following simple algebra.

The steady-state growth of an individual cell proceeds exponentially throughout the entire cell cycle [$m = m_0 \cdot \exp(\mu t)$]. At the time of the second part of the cycle, it takes exactly τ min for the cell to enlarge from critical mass m^* to $2m_0$. Hence,

$$2m_0 = m^* \times \exp(\mu\tau) \Rightarrow m_0 = 0.5m^* \times \exp(\mu\tau) \qquad (87)$$

Equation 87 remains valid for steady-state culture at any μ, which can be varied from 0 to μ_m. Assuming that cell critical mass m^* is constant and does not depend on growth rate, we see that equation 87 predicts a μ-dependent variation mass of newborn cells m_0 and hence the average mass and size of the bacterial population. This finding is supported by numerous experimental studies from the beginning of this century (60) that displayed the positive correlation between cell size and growth rate. Because the term $\mu\tau$ is rather small ($\mu\tau \leq \mu t_d = \ln 2$), it is difficult to notice the curvature of the experimental curve within the measurement error. Thus, the relationship between the average cell size $\bar{m}$ ($\bar{m} \approx m_0 \times 2 \times \ln 2$ for rod-shaped bacteria; see derivation in equation 88) and μ is given as an empirical linear approximation:

$$\bar{m} = \bar{m}(0) + k\mu \qquad (88)$$

The regression parameters $\bar{m}(0)$ and k have a meaningful biological interpretation: y intercept, $\bar{m}(0) \approx 2 \ln 2 \times 0.5m^* = (\ln 2) \times m^*$ is equal to approximately 69% of the cell critical mass m^* initiating chromosome replication, and slope $k = [\exp(\mu_m\tau) - 1]/\mu_m \approx \tau$ is close to the duration of C + D periods.

The postulate on the constancy of the critical mass, m^* was derived from the observation that the cell has accurate control over its size at division and poorer control over its age at division (61). Recently, accurate measurements with flow cytometry (62) revealed that m^* is inversely related to the specific growth rate; slowly growing cells tend to initiate DNA replication at a slightly higher critical mass as compared with intensively growing cells. However, we may safely assume that m^* variation is much smaller than the variation of cell mass during the cell cycle: $dm^*/dt \ll dm/dt$.

Simulation of Cell Cycle by Simple Deterministic Structured Model. In biochemical terms, it is difficult to envisage how cell mass per se could determine when to initiate replication. A more likely candidate is some mass-related parameter, such as intracellular concentration of some signal metabolite like guanosine tetraphosphate or an *initiator protein*, according to the popular model proposed by Helmstetter and Cooper (63,64). It was postulated that the initiation of DNA replication is triggered by a threshold intracellular concentration of this protein V^*; this protein is synthesized at a rate proportional to the total growth rate and requires exactly one mass doubling time to reach its threshold concentration again. This mechanism is translated into mathematical form of a structured model such as SCM as follows (2):

$$\frac{dV}{dt} = \Theta(\mu + a) - \mu V$$

$$\frac{d\lambda}{dt} = \zeta \begin{vmatrix} \text{if } V \geq V^* & \text{then } \zeta = \zeta_1 > 0 \text{ else } \zeta = 0 & \text{(initiation)} \\ \text{if } \lambda \geq \lambda^* & \text{then } \lambda = \lambda/2, V = 0, m = m/2 & \text{(division)} \end{vmatrix} \quad (89)$$

where Θ is the fractional contribution of protein V to total cell synthesis. The V content is an intrinsically transient entity; even during steady-state growth, it continuously changes between zero (Helmstetter and Cooper postulated the annihilation of the initiator protein after every replication cycle) to an upper-threshold value, V^*, which is less than the potential steady-state level, $\Theta(\mu + a)/\mu$. The second variable, λ, imitates the replicating chromosome; it sets up the discontinuity associated with cell division.

Analysis of coupled equation set 89 reveals that this model is stable toward perturbations. Suppose that by chance, the content of initiator protein has risen to some abnormally high level. The immediate result would be several more frequent cell divisions, with smooth reversal to a normal multiplication pace. Similar events take place under the opposite situation of V deficiency; several divisions are delayed, resulting in production of abnormally long cells, but then steady state is restored. The negative feedback mechanism that brings things back into line is based on the dynamic nature of variable V; it is characterized by a unique, stable steady state that is approached from different initial conditions. It may be easily shown that the described model adequately simulates various morphological effects exhibited during non-steady-state growth (Table 10).

Table 10. Morphological Effects during Non-Steady-State Growth

Effects	Explanation
The longer lag phase in batch culture when growth is surveyed by cell count rather than biomass measurements	The partial synchronization of cell division is delayed as compared with mass growth until attainment of the value $2\bar{m}$
The accumulation of enlarged cells during transition from lag to exponential growth phase	The μ–m relationship (equation 87) combined with the transient misbalance in V synthesis (equation 89)
The formation of dwarf cells in starving population	The small cells are produced during very slow cryptic growth of surviving organisms
The accumulation of division potential if division is blocked by inhibitor, then after block release all missing division takes place in quick sequence	This phenomenon is simulated by equation 89 and explained by overproduction of initiator protein in the presence of inhibitor

Statistical Analysis of the Population Distributions. Equation 87 to 89 were derived for average cells in the population. To embrace the variability of sizes, we have to analyze the *frequency distributions*. If certain conditions are met—the culture is fully desynchronized, cells grow according to some deterministic low, all divide into only two identical daughter cells, and there is no cell elimination—then age distribution (61) is given by:

$$\varphi(\mathrm{a})da = 2\mu e^{-\mu \mathrm{a}}da\colon 0 \leq \mathrm{a} \leq \ln 2/\mu = \mathrm{t_d} \quad (90)$$

where μ is specific growth rate, t_d is mean doubling time, a is the age since birth, and $\varphi(a)\,da$ is the frequency of cells whose ages are between a and $a + da$. Assuming that cells grow exponentially between divisions, then the frequency of mass distribution is

$$\varphi(m)dm = 2m_0/m^2 dm\colon m_0 \leq m \leq 2m_0 \quad (91)$$

where m_0 is the mass of newborn cell and $\varphi(m)\,dm$ is the frequency of cells whose masses are between m and $m + dm$. The mean cell size is $2m_0 \times \ln 2$ (calculated as an integral of $m \times \varphi(m)dm$). If cell growth between two consecutive divisions is linear (65), then

$$\varphi(m)dm = 4 \times \ln 2/\mathrm{m_0} \cdot \exp(-m \times \ln 2/m_0)dm\colon m_0 \leq m \leq 2m_0 \quad (92)$$

Equations 90 and 91 are called *canonical* age or mass distributions to emphasize that they are an idealized form applicable when cell division takes place at a precise size. Assuming that momentary distribution of size at division of individual cells is *normal* and *random* (not correlated with other cell cycle events), we can obtain computer-simulated curves for any fixed level of noise expressed as the coefficient of variation (CV). As shown in Figure 9, random variations of size at division tend to round the corners of the canonical distribution.

Another source of cell size variation can be nonequal separation of mother cells into two daughter cells. It is characterized by the K distribution, which is the distribu-

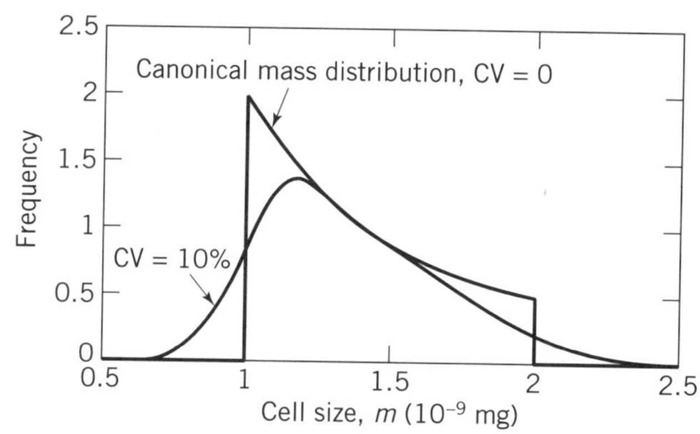

Figure 9. Distribution of cell sizes (as single-cell mass) for canonical cases where there is no variation in the size at division and for the case of normal distribution of size at division with CV = 10% (see details in Ref. 61).

tion of the ratio of daughter-cell length to mother-cell length. The average values absolutely necessary equals 0.5, but CV is at best about 4% (e.g., for well-behaved *E. coli* strains) and can attain rather high values for other organisms. The deviation of observed versus predicted size distribution may be caused also by cell death and different kinds of cell pathology (abnormally long or dwarf cells).

Collins and Richmond (66) introduced an entirely different approach based on the use of three distributions:

$$v_m = \mu\{2\int_0^m \Psi dm - \int_0^m \varphi(m)dm - \int_0^m \lambda(m)dm\}/\lambda(m) \qquad (93)$$

where v_x is the growth rate of cells of size m, $\lambda(m)$ is the extent of population distribution, Ψ is the momentary distribution of dividing cells, and $\varphi(m)$ is the momentary distribution of newborn cells. This equation allows the calculation of the mean growth rate of cells of particular size class. Instead of this analytical method, Koch in his work with Shaechter (61) proposed the synthetic approach. Starting from the set of specific postulates of the cell multiplication mechanism (linear or exponential growth of cell mass between divisions, kind of control, the evenness of the division), they derived the size or age distribution, which then was compared with the observations.

The Growth Low. Some believe that there should be a general low of cell growth that can be discovered by sensitive methods of analysis. The rate of biomass growth throughout the cell cycle was hypothesized to be linear, bilinear, exponential, double-exponential, and so on. Two limiting cases have generally been considered: the exponential and linear growth models proposed, respectively, by Cooper and Kubitschek (57). To differentiate between these two mechanisms, three classes of experiments have been used:

1. Size measurements of individual cells growing in the microculture by use of phase-contrast microscopy or recently developed confocal scanning light microscopy combined with image analysis.
2. Pulse-chase labeling of cells with their subsequent separation into different phases of the cell cycle. Most frequently, labeled uracil and leucine are used as precursors of RNA and protein synthesis, respectively. There are two major sources of errors in this approach: poor resolving power of separation methods and artifacts associated with effects of exogenous compounds on intracellular fluxes (feedback inhibition, pool expansion, label dilution, etc.). One of the best options for separation is probably the "baby machine," based on the membrane elution principle (67). To minimize the second source of error, the mutants blocked in the synthesis of the probe compound can be used.
3. Analysis of the frequency distributions of steady-state populations (see equations 91 and 92). However, the resolving power of this approach is rather low because linear and exponential models produce similar patterns.

Most of the obtained results are in better agreement with the exponential growth model rather than with the linear model. However, there are some serious doubts about whether there is a unique simple mathematical growth low describing bacterial growth during the division cycle. Cooper (67) proposed distinguishing three categories of cell components that are synthesized with a unique pattern:

1. Cytoplasm (proteins, RNA and ribosomes, small molecules) that is accumulated exponentially.
2. Cell DNA that is replicated in a linear fashion as a sequence of constant and zero rates.
3. Cell surface composed of peptiodoglycan and membranes that are synthesized exponentially during most parts of the cell cycle, but immediately before cell division the synthesis accelerates to accomplish new pole formation.

Thus, the growth pattern of the whole cell is the sum of these three patterns. Because the cytoplasm is the major constituent (up to 80% of CDW), then the growth of the cell should be approximately exponential.

Population Dynamics (Mutations, Autoselection, Plasmid Transfer)

Description of Mutation and Autoselection. The continuous culture turned out to be a very efficient tool to study mutation and autoselection (67). Let N be the total cell concentration, M the concentration of mutants, μ the specific growth rate of the main nonmutated part of the cell population, and η the specific growth rate of neutral mutants, then

$$\frac{dM}{dt} = \lambda\mu N + \eta M - DM$$
$$\frac{dN}{dt} = \mu N - \lambda\mu N - DN \qquad (94)$$

where λ is the mutation rate expressed as the ratio of the numbers of mutants to total number of cells formed. If $\lambda \ll 1$ and $\mu = \eta$, in the steady state, we obtain $\mu = \eta = D$, and

$$\frac{dM}{dt} = \lambda ND$$

or

$$M = M_0 + \lambda DNt \qquad (95)$$

If $\eta > \mu$, then the original strain will be displaced by the mutant, otherwise, if $\eta < \mu$, M will tend to a lower limit $M^* = \lambda N/(1 - \eta/D)$.

Experimental studies of phage-resistant mutants in a tryptophan-limited chemostat culture of *E. coli* showed that the period of linear M increase in accordance with equation 95 was fairly short. Every 20 to 100 generations there was an abrupt fall in the number of mutants, after which the linear growth resumed at the same rate (67).

The observed saw-tooth dynamics in M were explained by Moser (43) as a combined effect of mutation and selection. The original wild clone gives not a single but a whole array of mutations with subsequent reversions. Let us denote the total cell population in a chemostat culture as N, which is the sum of all subpopulations, including original and emerging variants, $N = \Sigma N_i$. All possible transitions between variants are given by the matrix, $\lambda_{i \to j}(j = 1, \ldots, n; i = 1, \ldots, n; j \neq i)$. Then, the chemostat model takes the following form:

$$\frac{dN}{dt} = \sum_{j=1}^{n} \frac{dN_j}{dt} = \bar{\mu}(s)N - DN$$

$$\bar{\mu} = \frac{1}{N}\sum_{j=1}^{n} \mu_j N_j$$

$$\frac{dN_j}{dt} = \mu_j(s)N_j - DN_j - \sum_{i \neq j}^{n} \lambda_{j \to i} N_j + \sum_{j \neq i}^{n} \lambda_{i \to j} N_i$$

$$\frac{ds}{dt} = D(s_0 - s) - \sum_{j=1}^{n} \mu_j(s)N_j/Y_j$$

$$\mu_j(s) = \mu_m \frac{s}{K_{sj} + s} \tag{96}$$

Every drop in the saw-tooth dynamics of neutral mutants detected by their phage resistance can be interpreted as the appearance of other types of spontaneous mutant with higher growth capabilities. In a chemostat culture, such mutants overcompete and displace all other cells by virtue of their higher affinity to limiting substrate (a decreased K_s value). Let such a mutant be denoted by the subscript k and $\bar{\mu}(s)$ be the average growth rate of all other subpopulations. The selection pressure for this mutant, σ, is given by

$$\frac{d(M_k/N)}{dt} = \sigma \frac{M_k}{N} = [\mu_k(s) - \bar{\mu}(s)] \frac{M_k}{N} \tag{97}$$

If $K_{sk} < K_{sj}(j \neq k)$, then $\sigma > 0$ until a new equilibrium is established. In this process, the original cells will be displaced by the mutants and the growth-limiting substrate concentration will decrease from $\bar{s}_1 = DK_{sj}/(\mu_m - D)$ to $\bar{s}_2 = DK_{sk}/(\mu_m - D)$, and the culture density will rise by $Y(\bar{s}_1 - \bar{s}_2)$.

Autoselection in Turbidostat and pH Auxostat. The affinity to substrate was not always the only driving force of selection outcome. A number of instructive examples were reviewed by Pechurkin (14).

1. In turbidostat and pH auxostat, autoselection is in favor of mutants with higher maximum specific growth rates, $\sigma = \mu_{mk} - \mu_{mj} > 0$. Because the population density is kept constant instrumentally and the dilution rate is allowed to vary, then autoselection results in the increase in D from μ_{mj} to μ_{mk}.

2. Mutation toward a higher growth efficiency, $Y_k > Y_j$, will lead to the same result as an increase in μ_m: $\sigma = \mu_k - \mu_j = (\mu_{mk} - \mu_{mj})s/(K_{sj} + s)$, as soon as $\mu_k = q_s Y_k$ and $\mu_j = q_s Y_j$.

3. Growth of a mutant with a higher resistance to inhibitory metabolic products can be described by equation 77 with $K_{pk} > K_p$. Under selection pressure $\sigma = \mu_k - \mu = \mu_m(s/K_s + s)[K_{pk}/(K_{pk} + p) - K_p/(K_p + p)]$, the original population will be completely displaced, and the product concentration will reach a higher steady-state level, $\bar{p} = \mu_m K_{pk}\bar{s}/(K_s + \bar{s})D - K_{pk}$.

4. Growth of mutants resistant to an antibiotic will not be affected by competition if the respective antibiotic is continuously supplied: $\sigma = \mu_k(s)$, because $\mu = 0$ for all other forms. The dynamics of the total population will be governed by the initial density of the mutant.

5. A mutation resulting in enhanced adhesion to fermenter walls will lead to accumulation of slow growing cells (because the adhesion prevents washout); eventually we have $\sigma = \mu_k + D - \mu_j$.

Extrachromosomal Cell Elements. The present-day hot spot in microbial population genetics is the study of extrachromosomal cell elements (ECE) such as plasmids, phages, transposons, and insertion elements. Normally, ECEs do not carry genes, absolutely essential for growth, but they are capable of fast replication, surpassing the chromosome DNA in the number of copies. R plasmids are responsible for bacterial growth in the presence of antibiotics, but under normal conditions (with no antibiotics present), their synthesis becomes too heavy a burden for the host cell, which is manifested in a decreased growth rate. Among the more than 100 R factors studied, about a quarter were found to increase the bacterial generation time by 15% (68). For this reason, plasmid-bearing strains are unable to compete with plasmid-free populations, although there are a few exceptions. Thus, colicin-positive cells carrying respective plasmids are able to withstand the competition with faster-growing plasmid-free strains by virtue of antagonistic inhibition. In recent years, rather intricate and detailed dynamic models of autoselection have been proposed that take into account the ECE-related effects, including the transfer of ECEs within the population, their segregation loss, changes in μ arising from the ECEs carriage, and so on. Some of these models have biotechnological and medical applications (69).

MICROBIAL GROWTH AS DEPENDENT ON CULTIVATION SYSTEMS

The array of laboratory cultivation systems that define the dynamic patterns of microbial growth is summarized in Table 11. Microbial growth patterns are distinguished by three features:

- Regime of substrate supply (1, continuous supply; and 2, single-term addition)
- Elimination of growing microorganisms (α, significant; β, absent)
- Magnitude of spatial gradients (a, homogeneous systems; b, heterogenous systems)

Each specific cultivation procedure can be represented by a point inside a cube with the axes 1 to 2, α to β, and a to

Table 11. Matrix of Cultivation Techniques

	Substrate input			
	Continuous (1)		Single-term (2)	
Spatial organization	Cell eliminated (α)	No elimination (β)	Cell eliminated (a)	No elimination (b)
Homogeneous (a)	$1a\alpha$ Chemostat Turbidostat pH-auxostat	$1a\beta$ Fed-batch culture Dialysis culture Retentostat	$2a\alpha$ Continuous culture with substrate pulses Phased culture	$2a\beta$ Simple batch culture
Heterogeneous (b)	$1b\alpha$ Plug-flow (tubular culture) Colonies	$1b\beta$ Column packed with microbe attached	Forbidden combination	

b. The 2*b* combination is logically forbidden because any spatial segregation results in protracted substrate utilization and so transforms a batch process into a continuous one. The dynamics of microbial growth in any type of cultivation system can be described by the following mass–balance equations:

$$\frac{ds}{dt} = F - G(s) - \mu(s)x/Y - mx$$
$$\frac{dx}{dt} = V - H(x)x + \mu(s)x \qquad (98)$$

where F is the substrate input rate; $G(s)$ is the rate of unused substrate removal from cultivation vessel (washout, leaching, evaporation, etc.); V is the rate of microbial biomass input, which may be a single-term inoculation or continuous delivery of cells to the fermenter (specially designed or unintentional, e.g., contamination); and $H(x)$ is the rate of microbial elimination, such as washout, death, grazing, or lysis. The rest of the notation is conventional. To simulate microbial growth dynamics in a particular homogeneous system, one has to make the following selection: $F(t) = 0$, $s_0 > 0$ for a category 2 (batch culture), $F(t) > 0$ for category 1 (continuous cultivations); $H(t) = 0$ for systems retaining cell biomass (dialysis, fed-batch, simple batch, column with immobilized cells), and $H(t) > 0$ when cell elimination occurs (chemostat, turbidostat, phased culture, etc.)

Spatially heterogeneous systems can be simulated either by partial derivatives or compartmental models (e.g., the total biomass x of a microbial colony may be represented as a sum of the peripheral and central components).

1*aα*—Homogeneous Continuous Culture (Continuous-Flow Fermenters with Complete Mixing)

There are two subgroups within this type of cultivation technique. In the first one, steady-state growth is maintained naturally by the microbial culture itself. Self-regulation is performed through negative feedback that originates from the dependence of the growth rate on substrate concentration (chemostat) or on temperature (caloristat). In the second group, electronic devices are used for the automatic adjustment of dilution rate to the instantaneous growth rate of the microbial culture. Electronic control is based on the sensing of cell density or growth-linked products, that is, optical density (turbidostat), culture liquid viscosity (viscostat), CO_2 concentration in output air, culture pH (pH auxostat), and so on.

In theory, the steady-state growth may be established in chemostat between 0 and μ_m, but in practical terms neither very low nor very high values are attainable because of the long time needed to reach the steady state in the first case and the risk of culture washout in the second. The second group of continuous techniques (turbidostat, pH auxostat, viscostat, etc.) are capable of maintaining steady growth at high s when either $\mu \rightarrow \mu_m$ or under substrate inhibition, when $d\mu/ds < 0$. There is also a potpourri mixed technique known as the *bistat* (70), which combines a chemostat and a pH auxostat. The mass–balance equations for the chemostat and its modifications have already been given (equations 62 to 64). In a simple, complete mixing cultivator, all cells have an equal probability of being washed out, hence $\mu = D$. If there is substantial wall growth, biomass retention, or feedback, then $\mu < D$; this difference increases with the extent of biomass retention in the fermentation vessel. In terms of our scheme, such cultivation systems correspond to points on the edge $1a\alpha$ to $1a\beta$.

1*aβ*—Continuous Cultivation without Cell Washout

This group embraces cultures with batch or continuous dialysis, fed-batch culture (FBC), and batch culture with a supply of limiting substrate via the gas phase (gases and volatile compounds). It also includes the chemostat with complete biomass feedback by means of filtration (71). The latter fermenters are less reliable in practical terms as compared with dialysis culture, because the membrane filters are quickly plugged with cells. The limiting substrate is supplied into the dialysis culture through a semipermeable membrane and, in the case of a gaseous nutrient, through the gas–liquid interface. In both cases, mass transfer is reasonably well described by Fick's law. The culture volume remains fixed in all systems, with the exception of a FBC. In a FBC, a constant nutrient feed F provides a linear increase of culture volume V during the cultivation span; the dilution rate $D = F/V$ is decreased hyperbolically. The great advantage of these cultivation techniques for biotechnology is that they provide the possibility of realizing very slow continuous growth accompanied with derepression of synthesis of many secondary metabolites. With a constant limiting substrate flux, Fs_0, the absence of cell washout means that at each subsequent moment an equal ration is shared by an increasing microbial biomass, and, as a result, μ eventually falls down to negligible values or even to zero (the maintenance state). Unlike the chemostat, no true steady state is established in this case, but when the substrate is virtually depleted, we have $ds/dt \approx 0$ and the system reaches a quasi steady state (5). If a quasi steady state approximation is found to be sufficient, then extremely slow continuous growth can be obtained after a reasonable period of time, perhaps a few weeks, as compared to the several months needed in a chemostat.

2*aα*—Continuous Cultivation with a Discontinuous Supply of Limiting Substrate

Suppose we have a simple chemostat culture fed by nutrient medium lacking just one essential component. This component is added as a small volume of concentrated solution at regular and sufficiently large time intervals, Dt. Then

$$\frac{ds}{dt} = D[s_0(t) - s] - q(s)x$$

$$\frac{dx}{dt} = \mu(s)x - Dx$$

$$s_0(t) = \begin{cases} A > 0, \text{ when } t = i\Delta t, i = 0, 1, \ldots, n \\ 0, \text{ otherwise} \end{cases} \qquad (99)$$

This cultivation method was originally used to obtain synchronized cell division (72). A continuous phased culture (73) is a repeated simple batch culture, that is, at regular intervals, Δt, half of the culture volume is withdrawn and the fermenter is refilled with an equal volume of fresh medium. Under such conditions, repeated batchwise growth proceeds with biomass increasing cyclically from x_0 to $2x_0$. Obviously, the growth dynamics are governed by the time interval between consecutive substrate additions Δt. If $\Delta t = \ln 2/\mu_m$, a sawtooth nonlimited growth takes place as in a turbidostat. With $\Delta t < \ln 2/\mu_m$ the culture should be washed out and, when $\Delta t > \ln 2/\mu_m$, the maximal attainable biomass decreases with increasing Δt because of endogenous biomass decomposition and waste respiration during the lag phase.

2*aβ*—Simple Batch Culture

Cultivation begins at the initial limiting substrate concentration, s_0, and inoculum size, x_0. The biomass reaches its maximum, x_m, when the limiting substrate is depleted ($s = 0$) and then declines even in the absence of exogenous elimination, so that $x \to 0$ as $t \to \infty$. Description of batch dynamics has been given earlier. Specified growth phases are described by simple nonstructured models (equations 56, 61, and 73), and entire dynamics are described by SCM and other structured models (equations 84 to 86; Fig. 8).

1*bα*—Plug-Flow (Tubular) Culture

The inoculum and the medium are mixed on entry into a long reactor tube, and the culture flows in the tube at a constant velocity without mixing. In each small element of culture liquid, moving along the spatial coordinate z at a linear velocity $f = F/A$ (where F is flow rate, cm^3/h, and A is the cross-sectional area, cm^3), growth of biomass proceeds as in a simple batch culture. To account for the culture movement per se, we have to pass from ordinary to partial derivatives and replace dx/dt by $\partial x/\partial t + f\partial x/\partial z$ (along-the flow-growth rate). Allowing for some dispersion of the moving front by diffusion, we can write

$$\frac{\partial s}{\partial t} + f\frac{\partial s}{\partial z} = D_s\frac{\partial^2 s}{\partial z^2} - q(s)x$$

$$\frac{\partial x}{\partial t} + f\frac{\partial x}{\partial z} = D_x\frac{\partial^2 x}{\partial z^2} + \mu(s)x \qquad (100)$$

where D_s and D_x are the diffusion coefficients of the substrate and microbial cells, respectively.

1*bβ*—Continuous-Flow Reactors with Microbes Attached

The nutrient solution is pumped through a column filled with adsorbent material and is utilized as it moves by growing immobilized cells. The mass–balance equations for a packed column are obtained from equation 100 by simplifying the equation for x,

$$\frac{\partial s}{\partial t} + f\frac{\partial s}{\partial z} = D_s\frac{\partial^2 s}{\partial z^2} - q(s)x(z)$$

$$\frac{\partial x}{\partial t} = \mu(s)x = Y[q(s) - m]x \qquad (101)$$

Bacterial cells accumulate faster on the top of the column because of larger $q(s)$ values, and as a result, a distinctive spatial biomass distribution develops in the form of a hyperbolic decrease of x with column depth. Growth does not reach steady state with respect to x until cell elimination becomes well expressed (the effects of inhibitory products, endogenous cell decomposition, and leaching).

Colonies

Besides continuous-flow columns, other heterogeneous systems are widely used. Especially popular is plating on solid media made from natural or synthetic gels (agar, PAAG, silica gel, synthetic alumosilicates, etc.) as well as on some porous materials (sand, glass beads, glass fiber). Impregnated with nutrient solution, such materials are used to grow microorganisms in the form of colonies or lawns. At first glance, it is tempting to consider these techniques as analogies of a simple batch culture with single-term substrate input (type 2*bβ* in Table 11). However, a closer look at the mechanism of colony growth reveals a greater resemblance to chemostat culture! Here we will outline our considerations.

The spread of a colony over a solid substrate, for example, a layer of agar, proceeds by the growth of only a peripheral zone with biomass, x_p (Fig. 10). Then

$$\frac{ds}{dt} = -q(s)x$$

$$\frac{dx}{dt} = \mu_w x_p - ax \qquad (102)$$

where a is the specific decay rate of cell (mycelium) com-

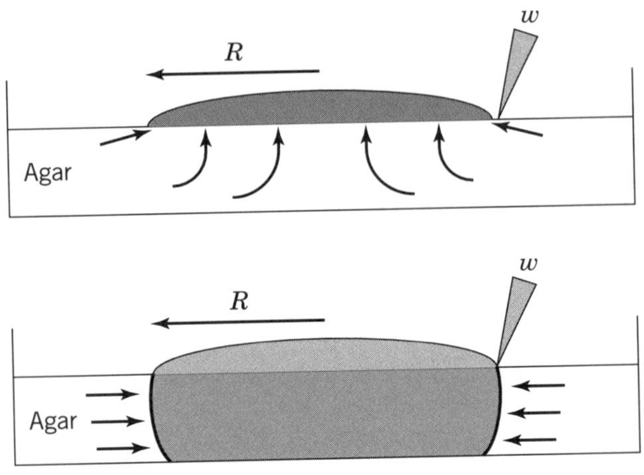

Figure 10. Schematic illustration of the difference between the growth modes bacterial (*top*) and fungi (*bottom*) colonies. Note that unicellular organisms (*bacteria*) do not penetrate into the agar layer as opposed to filamentous fungi. *Arrows* indicate directions of substrate diffusion across concentration gradients. R and w are, respectively, radius and width of the peripheral zone of the colony.

ponents. Straightforward geometrical analysis shows that the linear spread rate for a regularly shaped colony (a cylinder, a sphere, or a strip) of size R is given by the following relation (2,5):

$$\frac{dR}{dt} = \mu_w w = K_r$$

or

$$R = R_0 + K_r t \qquad (103)$$

where K_r is the colony linear expansion rate, mm/h; μ_w is the microbial specific growth rate within the peripheral zone; and w is the zone width.

In the case of unicellular organisms (e.g., bacteria, yeasts) that are incapable of penetrating into the gel's matrix, the substrate is available through passive diffusion to the peripheral zone from the underlying gel layer. As the colony grows, this flux diminishes, K_r decreases, and eventually the growth stops altogether. The filamentous organisms (fungi and actinomycetes) are able to propagate both on the surface and in the depth of gel. As a result, their growth is not completely dominated by diffusion effects, and the colony front advances at a faster rate than substrate is depleted in the frontier zone. The colony spreads at a constant radial rate, K_r, until the Petri dish is filled or the agar deteriorates from dryness. (In the case of rich nutrient media, there are also effects of self-inhibition by metabolic products [see equations 76 and 77].)

Thus, the colony growth is (*1*) continuous, (*2*) substrate-limited, (*3*) directed, and (*4*) spatially ordered. Properties 1 and 2 suggest a strong similarity between growth of a colony (especially a fungal one) and that of a chemostat culture. If so, a colony's peripheral zone is analogous to the cell culture in the fermenter, and its central part is comparable with the waste cell suspension that is discharged into the product bottle. A steady running of a pump, delivering medium at a rate, F (cm^3/h), corresponds to the active and uniform substrate utilization by the growing mycelium at a rate $K_r A$ (cm^3/h, A is the area of colony–medium interface). Finally, both cultivation systems are characterized by the elimination of propagating biomass, that is, by the expulsion of grown cells (mycelium) from further active growth (either by washing out or by the motion of the colony front). In both cases, the elimination rate is equal to biomass growth in the active compartment. The essential deviation from the chemostat lies in properties 3 and 4. Spatial differentiation of hypha and the direction of colony expansion are governed by spatial gradients of limiting substrate and, possibly, some metabolic products. Steepness of gradients is partly diminished by the effects of metabolic translocation along hypha over distances of the order of w.

It can be concluded from the previous discussion that colony growth belongs to class $1b\alpha$, and not to $2b\beta$. In general, it is very likely that the $2b$ combination is an empty, logically forbidden combination, because a single-term momentary input of substrate may only be realized in a homogeneous system. Any spatial segregation whatever will actually prolong the consumption of substrate and, therefore, transform a batch process into a continuous one. For this reason, growth on any insoluble substrate (lignocellulose and other plant polymers, oil droplets, grains of sulfur, etc.) should always be treated as a continuous process. The solid-phase fermentation can also not be anything but continuous, whether new portions of substrate are added to the reactor or not. (Obviously, this applies to the growth mechanism itself and not to the engineering operation.)

BIBLIOGRAPHY

1. A.A. Esener, J.A. Roels, and N.W.F. Kossen, *Biotechnol. Bioeng.* **25**, 2803–2841 (1983).
2. N.S. Panikov, *Microbial Growth Kinetics*, Chapman & Hall, London, 1995.
3. I.G. Minkevich and V.K. Eroshin, *Uspekhi Sovrem. Biol.* **82**, 103–116 (1976).
4. W.R. Mayberry, G.J. Prochazka, and W.J. Payne, *Appl. Microbiol.* **15**, 1332–1338 (1967).
5. IS.J. Pirt, *Principles of Microbe and Cell Cultivation*, Blackwell Scientific Publishers, Oxford, U.K., 1975.
6. A.G. Marr, E.H. Nilson, and D.J. Clark, *Ann. N.Y. Acad. Sci.* **102**, 536–548 (1963).
7. A.G. Dorofeyev and N.S. Panikov, *Mikrobiologiya–Engl. Tr.* **64**, 601–609 (1994).
8. D.W. Tempest and O.M. Neijssel, *Annu. Rev. Microbiol.* **38**, 459–486 (1984).
9. S.J. Pirt, *Arch. Microbiol.* **133**, 300–302 (1982).
10. N. Panikov and J.S. Pirt, *J. Gen. Microbiol.* **108**, 295–303 (1978).
11. T. Bauchop and S.R. Elsden, *J. Gen. Microbiol.* **23**, 457–469 (1960).
12. V.N. Ivanov, *Energetics of Microbial Growth*, Naukova Dumka, Kiev, 1981 (in Russian).
13. W.M. van Gulik, and J.J. Heijnen, *Biotechnol. Bioeng.* **48**, 681–698 (1996).

14. N.S. Pechurkin, *Populational Microbiology*, Novosibirsk, Nauka 1978, p. 277.
15. R.E. Buchanan, *J. Infect. Dis.* **23**, 109–125 (1918).
16. G. Stephanopoulos, and J.J. Vallino, *Science* **252**, 1675–1680 (1991).
17. Y. Kashiwaya, K. Sato, N. Tsuchiya, S. Thomas, D.A. Fell, R.L. Veech, and J.V. Passonneau, *J. Biol. Chem.* **269**, 25502–25514 (1994).
18. J.J. Vallino, and G. Stephanopoulos, *Biotechnol. Bioeng.* **41**, 633–646 (1993).
19. A. Marx, A.A. de Graaf, W. Wiechert, L. Eggeling, and H. Sahm, *Biotechnol. Bioeng.* **49**, 111–119 (1996).
20. A. Cornish-Bowden, *Principles of Enzyme Kinetics*. Butterworths, London-Boston, 1976.
21. N. Michaelis and M.I. Menten, *Biochem. Z.* **49**, 333–369 (1913).
22. G.E. Briggs and J.B.S. Haldane, *Biochem. J.* **19**, 338–339 (1925).
23. H. Gutfreund, *Discuss. Faraday Soc.* **20**, 167–173 (1955).
24. J.B.S. Haldane, *Enzymes*. Longmans Green, London, 1930.
25. J. Monod, J. Wyman, and J. Changeux, *J. Mol. Biol.* **12**, 88–118 (1965).
26. C.H. Wang and A.L. Koch, *Bacteriol.* **136**, 969–975 (1978).
27. O. Rahn, *Physiology of Bacteria*, P. Blackiston's, Philadelphia, 1932.
28. P.F. Verhulst, *Correspondence Mathematique et Physique* **10**(1), 113–121 (1838).
29. J. Monod, *Recherches sur la Croissance des Cultures Bacteriennes*, Paris, Hermann and Co, 1942.
30. J. Monod, *Ann. Inst. Pasteur* **79**, 390–410 (1950).
31. F.F. Blackman, *Ann. Bot.-London* **19**, 281–293 (1905).
32. S.D. Varfolomeyev, in S.D. Varfolomeyev ed., *Modern Problems of Biokinetics*, Moscow University Publishers, Moscow, 1987, pp. 6–77.
33. J. Monod, *Ann. Rev. Microbiol.* **3**, 371–394 (1949).
34. N.D. Jerusalimsky, in E.O. Powell, C.G.T. Evans, R.E. Strange, and D.W. Tempest eds., *Continuous Cultivation of Microorganisms*, H.M. Stationery Office, Salisbury, U.K., 1967, pp. 23–33.
35. I.A. Shvytov, *Zhurnal Obshchei Biologii* **35**, 904–910 (1974).
36. J.M. Romanovsky, N.V. Stepanova, and D.S. Chemavski, *Mathematical Biophysics*, Moscow, Nauka, 1984.
37. T.E. Shehata and A.G. Marr, *J. Bacteriol.* **107**, 210–216 (1971).
38. A.F. Guady Jr., A. Obayashi, and E.T. Gaudy, *Appl. Microbiol.* **22**, 1041–1047 (1971).
39. D.K. Button, *Microbiol. Rev.* **49**, 270–297 (1985).
40. B.R. Robertson and D.K. Button, *Bacteriology* **138**, 884–895 (1979).
41. G. Teissier, *Ann. Physiol. Physicochem. Biol.* **12**, 527–586 (1956).
42. D.E. Contois, *J. Gen. Microbiol.* **21**, 40–49 (1959).
43. H. Moser, *Carnegie Inst. Wash. Pub.* **614** (1958).
44. E.O. Powell, in E.O. Powell, C.G.T. Evans, R.E. Strange, and D.W. Tempest eds., *Continuous Cultivation of Microorganisms*, H.M. Stationery Office, Salisbury, U.K., 1967, pp. 34–55.
45. J.F. Andrews, *Biotechnol. Bioeng.* **10**, 707–723 (1968).
46. U. Pawlowsky, S.A. Howell, and C.T. Chi, *Biotechnol. Bioeng.* **15**, 905–916 (1973).
47. A. Campbell, *Bacteriol. Rev.* **21**, 263–272 (1957).
48. C.L. Cooney, *Ann. N.Y. Acad. Sci.* **326**, 295–314 (1979).
49. E.O. Powell, C.G.T. Evans, R.E. Strange, and D.W. Tempest eds., *Continuous Cultivation of Microorganisms*, H.M. Stationery Office, Salisbury, U.K., 1967, pp. 259–261.
50. D. Ramkrishna, A.G. Fredrickson, and H.M. Tsuchiya, *Biotech. Bioeng.* **9**, 129–170 (1967).
51. M.M. Domach, S.K. Leung, R.E. Cahn, G.G. Cacks, and M.L. Shuler, *Biotech. Bioeng.*, **26**, 203–216 (1984).
52. J.W. Jeong, J. Snay, and M.M. Ataai, *Biotechnol. Bioeng.* **35**, 160–184 (1990).
53. I. Malek, in I. Malek ed., *Continuous Cultivation of Microorganisms: A symposium*, Publishing House of Czech. Acad. Sci., 1958, pp. 11–28.
54. D. Herbert, *Symp. Soc. Gen. Microbiol.* **11**, 391–416 (1961).
55. P. Agrawal, H.C. Lim, and D. Ramkrishna, *J. Chem. Tech. Biotechnol.* **33B**, 155–163 (1983).
56. D.S. Kompala, D. Ramkrishna, N.B. Jansen, and G.T. Tsao, *Biotechnol. Bioeng.* **28**, 1044–1055 (1986).
57. B.G. Turner, D. Ramkrishna, N.B. Jansen, *Biotechnol. Bioeng.* **34**, 252–261 (1989).
58. D.W. Tempest, J.R. Hunter, and J. Sykes, *J. Gen. Microbiol.* **39**, 355–366 (1965).
59. D.W. Tempest, D. Herbert, and P.J. Phipps, in E.O. Powell, C.G.T. Evans, R.E. Strange, and D.W. Tempest eds., *Continuous Cultivation of Microorganisms*, H.M. Stationery Office Salisbury, U.K., 1967, pp. 240–253.
60. A.T. Henrici, *Morphologic Variation and the Rate of Growth of Bacteria*, Charles C. Thomas, Springfield, Ill., 1928, p. 194.
61. A.L. Koch, *Crit. Rev. Microbiol.* **19**, 17–42 (1973).
62. S. Wold, K. Skarstad, H.B. Steen, T. Stokke, E. Boye, *EMBO J.* **13**, 2097–2102 (1994).
63. C.E. Helmstetter, S. Cooper, O. Pierucci, and E. Revela, *Cold Springs Harb. Symp. Quant. Biol.* **33**, 809–822 (1968).
64. S. Cooper, *Res. Microbiol.* **141**, 17–29 (1990).
65. A.L. Koch, *J. Gen. Microbiol.* **45**, 409–417 (1966).
66. J.F. Collins, and M.H. Richmond, *J. Gen. Microbiol.* **28**, 15–33 (1962).
67. A. Novick, in G. Tunevall ed., *Recent Progress in Microbiology*, Blackwell Scientific Publications, Oxford, U.K., 1959, pp. 403–415.
68. J. Adams, T. Kinney, S. Thompson, L. Rubin, and R.B. Helling, *Genetics* **91**, 627–637 (1979).
69. R. Freter, R.R. Freter, and H. Brickner, *Infect. Immunol.* **39**, 60–84 (1983).
70. I.G. Minkevich, A.Y. Krynitskaya, and V.K. Eroshin, *Mikrobiologiya (Engl. Tr.)* **56**, 254–258 (1987).
71. W.R. Chesbro, T. Evans, and R. Eifert, *J. Bacteriol.* **139**, 625–638 (1979).
72. B.C. Goodwin, *Eur. J. Biochem.* **10**, 511–514 (1969).
73. P.S.S. Dawson, *Can. J. Microbiol.* **11**, 893–903 (1965).

See also MICROBIAL GROWTH MEASUREMENT, METHODS.

L

LACCASE

FENG XU
Novo Nordisk Biotech
Davis, California

KEY WORDS

Application
Biocatalysis
Cloning
Expression
Inhibition
Kinetics
Laccase
Properties
Structure

OUTLINE

MOLECULAR PROPERTIES, ENZYMOLOGY, AND PREPARATION OF LACCASE

Detection, Purification, and Cloning of Laccase

Laccase (EC 1.10.3.2) is a family of multicopper oxidases that catalyze the oxidation of a range of inorganic and aromatic substances (particularly phenols) with the concomitant reduction of O_2 to water. This enzyme has been detected in various plant species, including lacquer, mango, mung bean, peach, pine, prune, and sycamore (1–6), one bacterium, (*Azospirillum lipoferum*) (7), a dozen insects of genera that include *Bombyx, Calliphora, Diploptera, Drosophila, Lucilia, Manduca, Musca, Oryctes, Papilio, Phormia, Rhodnius, Sarcophaga, Schistocerca*, and *Tenebrio* (8–10), as well as more than 40 different fungi, including *Agaricus, Antrodiella, Armillaria, Aspergillus, Bjerkandera, Botrytis, Ceriporiopsis, Cerrena, Chaetomium, Coprinus, Cryphonectria, Cryptococcus, Curvularia, Cyathus, Geotrichum, Daedalea, Fomes, Fusarium, Halosarpheia, Lactarius, Lentinus, Monocillium, Myceliophthora, Neurospora, Penicillium, Phanerochaete, Phellinus, Phlebia, Pholiota, Pleurotus, Podospora, Pycnoporus, Pyricularia, Rhizoctonia, Rigidoporus, Schizophyllum, Sclerotium, Scytalidium, Sporotrichum, Stagonospora, Thermoascus, Trametes*, and *Trichoderma* (11–42). It has been postulated that laccase is involved in various cellular and microbial activities (43–51). Recent studies on the physiological function of laccase include those on plant cell wall biosynthesis (52), phytopathogenesis (53), wood material degradation and humification (32,54), insect sclerotization (8,10,55,56), bacterial melanization (7), and melanin-related virulence for human (57).

To date, thorough purification and characterization have been made on laccases from more than 20 different organisms (Table 1). The primary structure has been determined for laccases from 14 organisms, including *Acer pseudoplatanus* (58), *Agaricus bisporus* (11), *Aspergillus nidulans* (77), *Coprinus cinereus* (78), *Cryptococcus neoformans* (57), *Myceliophthora thermophila* (23), *Neurospora crassa* (64), *Pleurotus ostreatus* (36), *Rhizoctonia solani* (26), *Scytalidium thermophilum* (30), *Trametes (Coriolus) hirsutus* (79), *Trametes versicolor* (80), *Trametes villosa* (33), and a basidiomycete designated as PM1 (25). Extensive characterization of laccase has been carried out in the past decades, and the work accumulated by 1993 has been comprehensively reviewed (43–51). An attempt is made in this article to review recent progress, with emphasis on the potential industrial and medicinal applications of laccase, made during the last decade.

Molecular Properties of Laccase

Most laccases studied are extracellular proteins, although intracellular laccases have been detected in several fungi and insects (8,10,17,57). In general, fungal laccases have isoelectric points (pI) ranging from 3 to 7 whereas plant laccase pI ranges to 9 (Table 1). Under nondenaturing conditions, gel filtration data suggest that at least some fungal laccases have quaternary structure (8,11,15,23,26,30, 33,69,81). On SDS-PAGE, most laccases show mobilities corresponding to molecular weight (MW) of 60–100 kDa, of which 10–50% may be attributed to glycosylation (Table 1). Available data indicate that mannose is one of the major components of the carbohydrates attached to laccase (1,26,33,36,57,60,69,81,82). Glycosylation in laccase is suggested to play a role in secretion, proteolytic susceptibility, activity, copper retention, and thermal stability (60,69,82,83). Purified laccases exhibit a characteristic

Table 1. Molecular Properties of Some Laccases

Species	MW, kDa	pI	Glycosylation (%)	References
Acer pseudoplatanus	66[a]		40	1,58
Agaricus bisporus	65[a]	3.7	15	11
Armillaria mellea I, II	59	4.1, 3.3	15	13
Aspergillus nidulan	80		12	59
Botrytis cinerea	74	4.0	49	60
Cerrena maxima	67			46
Cerrena unicolor	66			14
Ceriporiopsis subvermispora I, II	71, 68	3.4, 4.8	15	61
Chaetomium thermophile	36			15
Cryphonectria parasitica	77		24	17
Cryptococcus neoformans	75[a]		6	57
Fomes annosus	73			21
Junghuhnia separabilima	58–62	3.5		62
Lactarius piperatus	67			63
Lentinus edodes	66			47
Manduca sexta	70–90			8
Monocillium indicum	72			22
Myceliophthora thermophila	85[a]	4.2	14	23
Neurospora crassa	64[a]	6.8	12	64
Phanerochaete chrysosporium	47			24
Phellinus noxius	70			65
Phlebia radiata	64	3.5	2	66
Phlebia tremellosa	67	3.3		62
Pholiota mutabilis	64			21
Pinus taeda	90	9	22	4
Pleurotus ostreatus I, II	64, 59[a]	2.9	13,4	36,67
PM1	64[a]	3.6	6.5	25,68
Podospora anserina	70–80	4.9	23	21,69
Pycnoporus cinnabarinus	77	3.7	9	35
Pycnoporus coccineus	70	3.5	8	70
Rhizoctonia pratocola	78			21
Rhizoctonia solani	66[a]	7.5	11	26
Rhus vernicifera	110	8.6	45	71,72
Rigidoporus lignosus I, II	53, 52	3.8, 3.3		62,65
Schizophyllum commune	64			28
Scytalidium thermophilum	80[a]	5.1	20	30
Trametes anisoporus	58			46
Trametes consors	58			73
Trametes hirsutus	55[a]			46
Trametes sanguinea	62	3.5	9.1	74
Trametes versicolor I, II	67[a], 70	3.2, 5.6	14	75,76
Trametes villosa I, III	63[a]	3.5, 6.5	0.5	33
Trichoderma	72			34

[a]DNA sequence-deduced MW: *Acer pseudoplatanus*, 60; *Agaricus bisporus*, 58; *Cryptococcus neoformans*, 66; *Myceliophthora thermophila*, 61; *Neurospora crassa*, 61; *Pleurotus ostreatus*, 54; PM1, 53; *Rhizoctonia solani*, 55; *Scytalidium thermophilum*, 63; *Trametes hirsutus*, 54; *Trametes versicolor*, 55; *Trametes villosa*, 55.

blue appearance from their electronic absorption around 600 nm. Typical UV-visible spectra of laccase (at resting state) show two maxima around 280 and 600 nm and one shoulder near 330 nm. The ratio of the absorbance at 280 nm to that at 600 nm is generally 14 to 30, and the ratio of the absorbance at 330 nm to that at 600 nm is 0.5 to 2 (26,33,43–51,84).

In the holoenzyme form, most laccases have four copper atoms per monomer (25,26,33,43–51,57,61,67, 70,72–74,85), although the laccase from *Phlebia* was reported to have two coppers and one pyrroloquinoline quinone as prosthetic groups (86). Among the four copper atoms, one belongs to the "blue" type 1 copper site that has a strong electronic absorption around 600 nm and a characteristic electron paramagnetic resonance (EPR) signal; one belongs to the type 2 copper site that has a characteristic EPR signal, and two belong to the strongly coupled type 3 site that has a weak UV absorption around 330 nm and is EPR active only when perturbed by strong anion binding. It is postulated that the type 2 and type 3 sites constitute a trinuclear copper cluster center. The redox potentials of the active coppers measured for several fungal and plant laccases are in the range of 0.4–0.8 V versus normal hydrogen electrode (43–48,84). The redox potential of the type 1 site in several plant and fungal laccases has been measured as a function of pH and showed a total potential change of 0.1 V over the pH range 3–9 (71,87).

The primary structures of more than 20 laccases (from more than 14 organisms) have been determined by protein or DNA sequencing. Analyses of the primary structure indicate the occurrence of N-terminal posttranslational processing in which the signal (pre-) peptide is cleaved during the secretion of an extracellular laccase. The maturation of some laccases also includes either N-terminal processing of a propeptide or blocking of the N-terminus (23,30). A C-terminal propeptide is processed to form mature *Neurospora crassa* laccase as well (64). Within the highly conserved copper-binding domains, the putative copper-ligating residues (10 histidines and 1 cysteine) are strictly conserved. Preliminary crystallization of two *Trametes* laccases has been reported (74). Extensive spectroscopic characterizations and primary sequence comparison (43–48,88) strongly suggest that, in laccase, the structures of the copper sites and polypeptide backbone are quite similar to those of ascorbate oxidase and ceruloplasmin as determined by X-ray crystallography (89,90).

Enzymology of Laccase

The enzymology of two representative laccases, *Rhus vernicifera* and *Trametes versicolor* laccases, has been extensively studied (43–51). Laccase can catalyze the oxidation of various reducing substances such as inorganic or organic metal complexes, anilines, thiols, and especially phenols. Although laccase strongly prefers O_2 as its oxidizing substrate (K_m at submillimolar levels), it usually has low specificity toward reducing substrates. In general, the optimal pH of laccase activity depends on the type of substrate (87). For phenols, the optimal pH can range from 3 to 7 for fungal laccases and up to 9 for plant laccases. The optimal pH, K_m, and k_{cat} for some laccases are listed in Table 2.

It is believed that laccase catalysis involves (*1*) reduction of the type 1 copper by reducing substrate, (*2*) internal electron transfer from the type 1 copper to the type 2 and type 3 copper, and (*3*) reduction of O_2 to water at the type 2 and type 3 copper site (43–51). The oxidation of a reducing substrate by laccase typically involves the loss of a single electron and the formation of a free (cation) radical. The radical is in general unstable and may undergo further laccase-catalyzed oxidation (e.g., to form quinone from phenol) or nonenzymatic reactions (e.g., hydration, disproportion, or polymerization). The electron transfer from substrate to type 1 copper is probably controlled by redox potential difference (95). A lower oxidation potential of substrate or a higher redox potential of laccase (type 1 site) often results in a higher rate for substrate oxidation. It seems that the binding pocket of reducing substrate (or the type 1 copper site) is quite shallow and has limited steric effect on simple phenol substrate (78,95). In contrast, the O_2-binding pocket (or the type 2 and type 3 copper sites) appears to restrict the access of oxidizing agents other than O_2. Activation of O_2 likely involves chemical bond formation on the trinuclear copper cluster. Under turnover conditions, the rate-limiting step may be the oxidation of substrate, whereas under transient or anaerobic conditions, the internal electron transfer step may be rate limiting (43–51).

Laccase can be inhibited by various reagents (43–51). Small anions such as halides [excluding iodide (96)], azide, cyanide, and hydroxide bind to the type 2 and type 3 copper, resulting in an interruption of the internal electron transfer and activity inhibition. Other inhibitors include metal ions (e.g., Hg^{2+}), fatty acids, sulfhydryl reagents, hydroxyglycine, Kojic acid, desferal, and cationic quaternary ammonium detergents (15,18,21,25,61,97–99). Laccase is generally more stable at alkaline pH than at acidic pH (74,84). Under identical conditions, thermophilic fungal laccases are in general more thermostable than mesophilic laccases (84).

Preparation of Laccase

In general plant laccases are purified from sap or tissue extracts, whereas extracellular fungal laccases are purified from medium culture (fermentation broth) of the selected organism. Certain aromatic or phenolic inducers (such as ferulic acid, xylidine, *p*-anisidine, or anthraquinone) can be used in some cases to boost the yield of an inducible laccase from the wild-type strains (21,27,35,76,100). Recent progress in applying genetic engineering technologies may be an alternative route to industrial laccase production (26,33,79,84,101–104). Various protein purification techniques are frequently employed in purifying laccase. Typical purification protocols involve ultrafiltration, ion exchange, gel filtration, hydrophobic interaction, or other electrophoretic and chromatographic techniques. Affinity chromatography using a phenolic group as ligand can increase purification efficiency (91). The purity of a laccase preparation is often measured by SDS-PAGE and by the ratio of the absorbance at 280 nm to that at 600 nm. Laccase transcription or expression can be detected by DNA–RNA hybridization or by immunochemical assay with antilaccase antibody. Laccase activity can be detected with characteristic chromogenic substrates, such as syringaldazine and 2,2′-azino-bis-(3-ethylbenzthiazoline-6-sulfonic acid) (ABTS) are often used. ABTS is very suitable to screening because its one-electron-oxidation product is soluble in water, stable, and intensely green. Currently, several wild-type fungal laccases, including those from *Pyricularia oryzae, Trametes (Coriolus) consors, Myceliophthora thermophila, Trametes hirsutus*, and *Trametes versicolor*, are commercially available.

APPLICATIONS OF LACCASE

Industrial Biocatalysis

Delignification of Lignocellulosics by Laccase. Conventional methods of delignifying or decolorizing paper pulp involve either chlorine- or oxygen-based chemical oxidants (e.g., ClO_2 and O_3). Although very effective, these methods have serious drawbacks such as disposal of chlorinated by-products or loss of cellulose fiber strength. Microbial or enzymatic delignification systems that overcome these draw-

Table 2. Kinetic Parameters of Some Laccases

Species	Substrate	pH	K_m, mM	k_{cat}/min	References
Armillaria mellea	*p*-phenylenediamine	3.5	1.7		13
Botrytis cinerea	2,6-Dimethylphenol	3.5	0.1	466,000	60
Ceriporiopsis subvermispora I, II	Guaiacol	3, 5	1.6, 0.44	3,100, 4,000	61
Cerrena unicolor	Syringaldazine	5.5			14
Cryphonectria parasitica	2,6-Dimethylphenol	2.5		2,000	17
Curvularia	Guaiacol	5	0.75		18
Fomes annosus	Syringaldazine	4.6			21
Fomes fomentarius	Syringaldazine	4.6	0.28		91
Lactarius piperatus	Catechol	5.6			63
Lentinus edodes	Syringaldazine	5.0			21
Manduca sexta	2-Methyhydroquinone	6.0	0.02	1,100	8
Monocillium indicum	*o*-Dianisidine	3.0	0.025		22
Myceliophthora thermophila	Syringaldazine	7	0.01	370	84
Phanerochaete chrysosporium	Syringaldazine	5.0			21
Phellinus noxius	Guaiacol	4.0			65
Pholiota mutabilis	Syringaldazine	5.2			21
Pinus taeda	Coniferyl alcohol	5.9			4
Pleurotus ostreatus I, II	Syringaldazine	6.5		192, 107	67
PM1	Guaiacol	4.5	0.5		68
Podospora anserina	Dopamine	5.5, 7.5	3.3		81
Pycnoporus cinnabarinus	Guaiacol	4	0.75		35,92
Pycnoporus coccineus	Catecol	4.5			70
Rhizoctonia praticola	2,6-Dimethylphenol	6.8	0.26		21
Rhizoctonia solani	Syringaldazine	7	0.017	1,900	84
Rhus vernicifera	Syringaldazine	9	0.043	600	84
Rigidoporus lignosus	Syringaldazine	6.0		1200	93
Sarcophaga bullata	Dopamine	4.5			10
Scytalidium thermophilum	Syringaldazine	7	0.001	290	84
Sporotrichum pulverulentum	Syringaldazine	5.0			21
Trametes anceps	Syringaldazine	5.3		140	94
Trametes hirsutus	Guaiacol		0.75		92
Trametes sanguinea	Dimethylphenylenediamine	5			74
Trametes versicolor I, II	Catechol	4.6	0.02, 0.09	52, 110	85
Trametes villosa I	Syringaldazine	5	0.058	2700	84

backs and can be easily adapted to current pulp production lines are attractive alternatives.

It has long been postulated that laccase participates in natural delignification (for recent reviews, see Refs. 66,76,105,106). Various laccases have been shown capable of degrading either natural lignin (wood chips or Kraft pulp) or synthetic lignin models (32,66,76,105–112). These laccases could either directly oxidize phenolic components exposed in lignin or, in the presence of proper redox mediator, indirectly oxidize heterogeneous phenolic and nonphenolic (especially methoxybenzene) components (35,112–114). As a result of laccase oxidation, (cationic) radicals could be generated in lignin, resulting in subsequent aliphatic or aromatic C–C bond cleavage and depolymerization.

As a "biopulping" agent, laccase could be applied to wood chips before pulping. The pretreatment with laccase could partially degrade lignin and loosen lignin structure so that the pulping can be done more efficiently. In one example, the use of a laccase-producing *Antrodiella* strain in a biopulping pretreatment step resulted in 26% energy savings in the subsequent mechanical pulping step (12). As a "biobleaching" agent, laccase could be applied (as an alternative to conventional chemical oxidants) to pulp. The residual lignin in pulp could be degraded by laccase, resulting in pulp decolorization. In one example, a *Trametes versicolor* laccase, under the mediation of *N*-hydroxy compounds, delignified Kraft pulp effectively in pilot-plant scale (115,116).

Cross-Linking of Polysaccharides by Laccase Polymerized saccharides are useful materials for producing composites such as particle board and liner board. However, most chemicals used in conventional polymerization, such as urea, formaldehyde, isocyanate, and petrochemical resins, are hazardous. Based on its lignin oxidation property, laccase could be used to replace these chemicals and serve as a "biogluing" agent (117). To initiate or enhance the cross-linking (gluing) efficiency, laccase could be used in three ways. The first is to directly oxidize wood particles or pulp to generate more radicals for cross-linking (117). The second is to functionalize wood particles or pulp with small compounds (such as aromatic, carboxyl, isocyanate, and acrylamide substances) that act as cross-linking agents (118). The third is to transform isolated lignin (often a by-product from pulping), starch, phenolic polysaccharide, or protein into radical-rich and nontoxic adhesive materials useful for wood composites (119). Such applications of laccase not only could replace toxic or expensive chemical adhesives but also could transform wastes such as lignin from the paper industry into value-added product.

Waste Detoxification and Decontamination by Laccase. Laccase has been used to oxidatively detoxify or remove various aromatic xenobiotics and pollutants found in industrial waste and contaminated soil or water (103,120,121). The laccase catalysis could result in either direct degradation or polymerization/immobilization. Reported examples of direct degradation by laccase includes dechlorination of chlorophenols (122–124), cleavage of aromatic ring (125), mineralization of polycyclic aromatic hydrocarbons (126), decolorization of pulp or cotton mill effluent (37,127–129), and bleaching of textile dyes (38,115,116,130). The processes include polymerization among pollutants themselves or copolymerization with other nontoxic substances (such as humic materials). Polymerized pollutants often become insoluble or immobilized, thus facilitating easy removal by such means as adsorption, sedimentation, or filtration (131,132).

Textile Dye Transformation by Laccase. Laccase has been reported to prevent back-staining of dyed or printed textiles. As part of the washing solution, laccase could quickly bleach released dye stuff, thus resulting in the reduction of processing time, energy, and water needed to achieve satisfactory quality of the textile (133).

Laccase-catalyzed textile dye-bleaching may also be useful in finishing dyed cotton fabric. Replacing conventional chemical oxidants (e.g., hypochlorite), a laccase-based system has been shown capable of bleaching indigo dye in denim and achieving various bleached appearances on the fabric (134,135).

Another interesting laccase application in this field is oxidative transformation and consequent coupling of dye precursors to the collagen matrix in hides (136). Under laccase catalysis, soluble dye precursors could be adsorbed, oxidized, and polymerized to give the desired tanning effect. The process could improve dyeing efficiency, reduce cost (by using inexpensive precursors), or provide improved hide characteristics.

Medical and Personal Care Applications of Laccase

Poison ivy dermatitis (resulting from skin contact with poison ivy, poison oak, poison sumac, and the like) is caused mainly by urushiol, which is a catechol-derivative toxin. Oxidized urushiol (an *o*-quinone derivative), however, is nontoxic. Laccase has been shown to oxidize, polymerize, and detoxify urushiol, thus reducing the effect of poison ivy dermatitis (137).

Another potential laccase application in the field is laccase-based generation of iodine in situ. Laccase can oxidize iodide to produce iodine (96), a reagent widely used as a disinfectant. The application of a laccase–iodide salt binary iodine-generating system (for sterilization) may have several advantages over the direct iodine application. First, the iodide salt is more stable and much safer than I_2 in terms of storage, transport, and handling. Second, the release of iodine from a laccase–iodide system could be easily controlled (by means such as adjusting laccase concentration). The system may be used in various industrial, medical, domestic, and personal care applications such as sterilization of drinking water and swimming pools as well as disinfection of minor wounds (96,121,138).

Current hair dyeing or waving processes often involve oxidative or additive chemicals that either have unpleasant odors, are irritant (or even harmful) to tissues, or are difficult to handle. A laccase-based system may overcome these drawbacks by replacing harsh chemicals and operating at milder conditions (in terms of pH and solvent). Laccase-catalyzed oxidation, transformation, and cross-linking of various precursors (mostly phenols and anilines) have been reported to result in satisfactory hair dyeing or waving (139–143). In addition to providing an easier-to-handle hair care procedure, a laccase-based system may also improve or complement the cosmetic effect (in terms of color type, shade, and compatibility with hair type) achieved by conventional chemical methods.

Biosensor and Diagnostic Applications of Laccase

Detection of Phenol, Aniline, O_2, and Other Substances. Laccase catalysis (coupled with various physical instruments) could be useful as biosensors for detecting O_2 and a wide variety of reducing substrates (especially phenols or anilines). Two types of laccase-based O_2 sensor are widely used. One type monitors visible spectral changes (at 600 nm) of laccase resulting from the reoxidation of the type 1 copper (I) in laccase by O_2 (144). Another type monitors current or voltage change from a modified oxygen electrode on which O_2 reduction is enhanced under the electrocatalysis of immobilized laccase (46). For detecting phenols, anilines, or other reducing substrates, three types of laccase-based sensors have been reported (46,145,146). One type detects the photometric change resulted from the oxidation of a chromogenic substrate, one type monitors the O_2 concentration change that is coupled to the substrate oxidation, and one type uses an electrode that replaces O_2 as the acceptor for the electrons flown from the substrate (through laccase). For these applications, laccase is either immobilized or free in solution, and the coupled physical converter is either optic, amperometric, or piezo effect in nature or a field-effect transitor or thermister, etc (46).

Laccase-Assisted Enzymatic and Immunochemical Assays. Laccase catalysis can be used to assay other enzymes. In these assays, either the enzyme of interest catalyzes the production of a compound whose subsequent oxidation by laccase generates detectable physical change, or a product from a laccase catalysis (whose production is accompanied by a detectable physical change) is quenched by the activity of the enzyme of interest. The strategy has been applied to assay various enzymes including amylase, (alanine, cystine, or leucine-specific) aminopeptidases, alkaline phosphatase, angiotensin I converting enzyme, arylamidase, cellobiose oxidase, chymotrypsin, glucosidase, γ-glutamyl transpeptidase, kallikrein, plasmin, thrombin, and X factor (as a coagulant factor) (147–151).

Laccase, that is covalently conjugated to an antibody or antigen can be used as a marker enzyme for immunochemical assay. In this application, binding of the antibody (or antigen) to its immunological counterpart is detected by localized laccase activity on a gel or a blot membrane, much like the conventional peroxidase- or phosphatase-assisted immunochemical assays (46). Under certain conditions, the antibody–antigen binding impairs the function of conjugated laccase, thus allowing immunochemical detection through modulation of laccase activity (43,152).

Organic and Medicinal Syntheses Catalyzed by Laccase

Laccase catalysis can be used to synthesize various functional organic compounds, which include polymers with specific mechanical, electrical, or optical properties (68, 153,154), textile dyes (155), cosmetic pigments (156,157), flavor agents (158), aromatic aldehydes (159), pesticides (160), and heterocyclic compounds (such as phenoxazine) (92). The reported applications involve laccase-catalyzed oxidation of aromatic compounds that result in either polymerization (68,92,153–157), transformation (159), or depolymerization (158). Laccase can also be used to synthesize several complex medicinal agents, which include triazolo(benzo)thiadiazines (a group of antiinflammatory, analgesic, central stimulant/depressant antisecretory and sedative depressant agents) (161), vinblastine (a cytostatic, antitumor drug) (46), penicillin X dimer (162), cephalosporin antibiotics (163), and dimerized vindoline (a plant alkaloid for treating neoplastic diseases) (164). Laccase can be used as well for oxidative deprotection in peptide synthesis (46). In comparison with counterpart chemical synthetic methods, a laccase-catalyzed oxidation in general requires milder conditions (in terms of temperature, pH, solvent nature), may improve regioselectivity, is easier to control, and is more environmentally friendly.

Other Miscellaneous Applications of Laccase

Beverage and Food Treatment. Browning and haze formation during the processing and storage of juice or wine are major problems for the industry. It is believed that phenolic compounds are involved in this process. Conventionally, undesirable phenolics are adsorbed and removed by various fining agents (e.g., gelatin, bentonite), which usually have low specificity, may affect color or aroma, and can pose disposal problems. Laccase may be used to remove or modify problematic phenolic saccharides and to improve clarity, color appearance, flavor, aroma, taste, and stability in fruit juice or fermented alcohol beverages (165). Following treatment by laccase, oxidized and polymerized (or precipitated) unwanted phenolic substances could be removed by silicate fining or filtration.

Laccase activity has been suspected to play an undesirable role in the processing of certain foods or beverages. Excessive or uncontrolled laccase activity, along with the action of other phenol oxidases, could cause browning of canned foods or vegetables, deteriorate the quality of wine or juice, or participate in various fruit pathogenesis (53). In these cases, inhibition of laccase activity has economic importance. Potential inhibitors include ferulic acid (166), oxalic acid (167), and cucurbitacin (168). In the case of food or beverage canning, elimination of O_2 can also be used.

Laccase may be applied to certain processings that enhance or modify the color appearance of a food or beverage. One interesting case involves ripe olive processing in which laccase replaces the conventional lye solution and oxidatively polymerizes various phenolics (such as oleuropein) in the olive (169), resulting in color darkening and debittering.

Desulfurization of Fossil Fuels. To reduce the harmful emission of sulfur-containing chemicals from the combustion of fossil fuels (petroleum, liquefied coal, natural gas), desulfurization is necessary during post combustion or, especially, precombustion processing of fuels. Conventional chemical and physical desulfurization methods (such as oxidizing and solubilizing inorganic pyrite sulfur by ferric salt or chlorine) often have low efficiency toward organic sulfur compounds, require extreme conditions (high temperature, high-pressure, and corrosion-compatible equipment), or require high infrastructure investment and maintenance costs. As an alternative to chemical methods, enzymatic desulfurization could be achieved under milder conditions, command less infrastructure cost, and remove organic sulfur more effectively. An enzymatic system capable of desulfurizing coal has been demonstrated (170) in which inorganic and organic sulfur compounds in coal slurry are oxidized by laccase, solubilized in water, and separated from solid (desulfurized) coal particles.

OVERALL REMARKS

Preferred Application Modes

Immobilization. Although native laccase functions in solution, preferred forms of laccase in various industrial or diagnostic applications should be immobilized or supported by nonliquid matrices. These include encapsulation in gel (hydrogel, alginate, silica gel), entrapment in micelles, or immobilization (either covalently or noncovalently) on solid materials (clay, glass, zeolite, active carbon, etc.) (91,132,165,171–176). In general, proper immobilization or encapsulation does not much alter the specific activity of laccase. Immobilized laccase may be suited for nonaqueous applications. Immobilized laccase could be easily recovered and recycled, thus reducing laccase dosage (or cost), making the treatment suitable to continuous operations, and avoiding the need to inactive residual laccase after treatment.

Mediation by Small Redox Compounds. For applications that target lignocellulosics or other insoluble materials, a laccase catalysis is heterogeneous. To enhance or extend laccase catalysis, small redox mediators may be used. Suitable mediators, such as 2,2′-azino-bis-(3-ethylbenzthiazoline-6-sulfonic acid) (112,114), phenoxazine (92, 117,149,160,177), *N*-hydroxy/oxides (115), Mn(III)-pyrophosphate (178), and Remazol blue (179), could significantly enhance catalysis by effectively shuttling electrons between laccase and the insoluble substrate, as well as enabling laccase to indirectly oxidize compounds that are normally not substrate for laccase.

Synergy with Other Enzymes. Laccase rarely acts alone and often requires synergy from other enzymes in complex natural processes (such as wood degradation). To better mimic nature, other enzymes should be tested to enhance laccase catalysis. For instance, flavin-containing oxidoreductases (180–182), xylanase (183), and manganese peroxidase (66,93,178) could be used together with laccase in delignifying pulp. These enzymes may provide a range of beneficial synergies for laccase catalysis, such as modifying targeted substances into better laccase substrates, or eliminating those products that are inhibitors of laccase catalysis.

Advantages of Applying Laccase

Live microorganisms that produce laccase or other active enzymes have been tested with several of the applications previously mentioned: these include *Thiobacillus* and *Sulfolobus* bacteria in coal desulfurization (170) and white-rot fungi *Trametes* and *Pleurotus* in delignification (112). In comparison with individual enzymes such as laccase, live microorganism cells could provide an array of enzymes capable of acting in synergy and with high efficiency. However, applying purified or enriched laccase preparation may hold several advantages over live microbial cells (120). Laccase could (*1*) be insensitive to many compounds or microorganisms that are toxic for or prey on the selected microbes; (*2*) be compatible with a wide range of environmental conditions (no need for specific nutrients or optimal growth conditions that may differ from optimal operational conditions); (*3*) be superior at a low contaminant concentration; (*4*) have controlled specificity; (*5*) be effective in terms of diffusion, permeability, and mobility; or (*6*) be able to accomplish the task in a shorter time.

Prospect of Laccase Applications

As an oxidase, laccase may be used in many industrial and medicinal applications where oxidation or oxidative derivatization is involved. Currently, investigations are focused on laccase-based biooxidation, biotransformation, biosensor, and enzymatic synthesis. The applications of laccase discussed here have all been satisfactorily carried out in either laboratory or pilot plant scale. However, an important factor that prevents large-scale laccase application as applied for pulp bleaching (115) is the unrealistically high dosage of laccase (and mediator), which could potentially be reduced by further optimizing the catalysis.

To overcome this limiting factor, one needs to better understand the mechanism for the reaction of interest. In the case of applying laccase to pulp, for example, one needs to further elucidate how glycolignin is degraded by white-rot fungi (and their secreted enzymes). In addition to studying the natural processes, protein engineering is probably also needed to optimize catalysis of laccase in various applications. For instance, developing a laccase that has better activity at alkaline pH or high temperature is of importance for pulp application. Protein engineering requires a breakthrough in obtaining three-dimensional structure information (78) and elucidating key structure–function correlations such as those that govern the redox potential and pH activity profile in laccase.

ACKNOWLEDGMENTS

I thank Alan V. Klotz, Stephen H. Brown, Debbie S. Yaver, Glenn E. Nedwin, Ejner B. Jensen, Ture Damus, Tomas T. Hansen, Claus Felby, and Donald Boyce of Novo Nordisk for critical reading and helpful suggestions.

BIBLIOGRAPHY

1. A. Driouich, A.-C. Laine, B. Vian, and L. Faye, *Plant. J.* **2**, 13–24 (1992).
2. S.P. Robinson, B.R. Loveys, and E.K. Chacko, *Aust. J. Plant Physiol.* **20**, 99–107 (1993).
3. A. Chabanet, A.M. Catesson, and R. Goldberg, *Phytochemistry* **33**, 759–763 (1993).
4. W. Bao, D.M. O'Malley, R. Whetten, and R.R. Sederoff, *Science* **260**, 672–674 (1993).
5. L. Dijkstra and J.R.L. Walker, *J. Sci. Food. Agric.* **54**, 229–234 (1991).
6. E. Lehman, E. Harel, and A.M. Mayer, *Phytochemistry* **13**, 1713–1717 (1974).
7. D. Faure, M.L. Boullant, and R. Bally, *Appl. Environ. Microbiol.* **60**, 3412–3415 (1994).
8. B.R. Thomas, M. Yonekura, T.D. Morgan, T.H. Czapla, T.L. Hopkins, and K.J. Kramer, *Insect Biochem.* **19**, 611–622 (1989).
9. M. Sugumaran and K. Nellaiappan, *Arch. Insect Biochem. Physiol.* **15**, 165–181 (1990).
10. F.M. Barrett, *Can. J. Zool.* **65**, 1158–1166 (1987).
11. C.R. Perry, M. Smith, C.H. Britnell, D.A. Wood, and C.F. Thurston, *J. Gen. Microbiol.* **139**, 1209–1218 (1993).
12. R. Patel, G.D. Thakker, and K.K. Rao, *J. Biotechnol.* **36**, 19–23 (1994).
13. A.U. Rehman and C.F. Thurston, *J. Gen. Microbiol.* **138**, 1251–1257 (1992).
14. E.G. Bekker, S.D. Petrova, O.V. Ermolova, V.I. Elisashvili, and A.P. Sinitsyn, *Biokhimiya* **55**, 2019–2024 (1990).
15. T. Ishigami, Y. Hirose, and Y. Yamada, *J. Gen. Appl. Microbiol.* **34**, 401–407 (1988).
16. P. Vnenchak and M.N. Schwalb, *Mycol. Res.* **93**, 546–548 (1989).
17. D. Rigling and N.K. Van Alfen, *Appl. Environ. Microbiol.* **59**, 3634–3639 (1993).
18. U.C. Banerjee and R.M. Vohra, *Folia Microbiol.* **36**, 343–346 (1991).
19. M.J.R. Shannon, and R. Bartha, *Appl. Environ. Microbiol.* **54**, 1719–1723 (1988).
20. D.S. Arora and D.K. Sandhu, *Enzyme Microb. Technol.* **7**, 405–408 (1985).
21. J.-M. Bollag and A. Leonowicz, *Appl. Environ. Microbiol.* **48**, 849–854 (1984).
22. G.D. Thakker, C.S. Evans, and K.K. Rao, *Appl. Microb. Biotechnol.* **37**, 321–323 (1992).
23. International Pat. WO 9533836 A1 (December 14, 1995), R.M. Berka, S.H. Brown, F. Xu, P. Schneider, D.A. Aaslyng, and K.M. Oxenbøll (to Novo Nordisk Biotech).
24. C. Srinivasan, T.M. D'Souza, K. Boominathan, and C.A. Reddy, *Appl. Environ. Microbiol.* **61**, 4274–4277 (1995).
25. P.M. Coll, C. Tabernero, R. Santamaria, and P. Perez, *Appl. Environ. Microbiol.* **59**, 4129–4135 (1993).
26. J.A. Wahleithner, F. Xu, K.M. Brown, S.H. Brown, E.J. Golightly, T. Halkier, S. Kauppinen, A. Pederson, and P. Schneider, *Curr. Genet.* **29**, 395–403 (1996).

27. S. Ding, C. Wang, and J.A. Buswell, *Xianweisu Kexue Yu Jishu* **2**, 36–46 (1994).
28. O.M.H. De Vries, W.H.C.F. Kooistra, and J.G.H. Wessels, *J. Gen. Microbiol.* **132**, 2817–2826 (1986).
29. I. Chet and A. Huttermann, *FEMS Microbiol. Lett.* **14**, 211–215 (1982).
30. International Pat. WO 9533837 A1 (December 14, 1995), R.M. Berka, F. Xu, and S.A. Thompson (to Novo Nordisk Biotech).
31. M. Bergbauer and C. Eggert, *Can. J. Microbiol.* **40**, 192–197 (1994).
32. A. Machuca and N. Duran, *Appl. Biochem. Biotechnol.* **43**, 37–44 (1993).
33. D.S. Yaver, F. Xu, E.J. Golightly, K.M. Brown, S.H. Brown, M.W. Rey, P. Schneider, T. Halkier, K. Mondorf, and H. Dalbøge, *Appl. Environ. Microbiol.* **62**, 834–841 (1996).
34. A. Assavanig, B. Amornkitticharoen, N. Ekpaisal, V. Meevootisom, and T.W. Flegel, *Appl. Microbiol. Biotechnol.* **38**, 198–202 (1992).
35. C. Eggert, U. Temp, and K.-E. Eriksson, *Appl. Environ. Microbiol.* **62**, 1151–1158 (1996).
36. P. Giardina, V. Aurilia, R. Cannio, L. Marzullo, A. Amoresano, R. Siciliano, P. Pucci, and G. Sannia, *Eur. J. Biochem.* **235**, 508–515 (1996).
37. C. Raghukumar, D. Chandramohan, F.C. Michel, and C.A. Reddy, *Biotechnol. Lett.* **18**, 105–106 (1996).
38. K. Vasdev, R.C. Kuhad, and R.K. Saxena, *Curr. Microbiol.* **30**, 269–272 (1995).
39. M. Svabova, J. Zemek, J. Augustin, L. Kuniak, and L. Marvanova, *Folia Microbiol.* **25**, 233–237 (1980).
40. A. Rodriguez, M.A. Falcon, A. Carnicero, F. Perestelo, G. De la Fuente, and J. Trojanowski, *Appl. Microbiol. Technol.* **45**, 399–403 (1996).
41. P. Ander and K.-E. Eriksson, *Arch. Microbiol.* **109**, 1–8 (1976).
42. J. Zhou, F. Zhang, and M. Kuwahara, *Acta Microbiol. Sin.* **33**, 387–391 (1993).
43. A. Messerschmidt, *Adv. Inorg. Chem.* **40**, 121–185 (1994).
44. B. Reinhammar and B.G. Malmström, in T.G. Spiro, ed., *Copper Proteins*, Wiley, New York, 1981, pp. 109–149.
45. A.M. Mayer, *Phytochemistry* **26**, 11–20 (1987).
46. A.I. Yaropolov, O.V. Skorobogat'ko, S.S. Vartanov, and S.D. Varfolomeyev, *Appl. Biochem. Biotechnol.* **49**, 257–280 (1994).
47. C.F. Thurston, *Microbiology* **140**, 19–26 (1994).
48. F.S. Sariaslani, *Crit. Rev. Biotechnol.* **9**, 171–257 (1987).
49. J.E. Coleman, in J.R. Whitaker, ed., *Food-Related Enzymes*, American Chemical Society, Washington, D.C., 1973, pp. 267–304.
50. J.A. Fee, *Struct. Bonding* **23**, 2–60 (1975).
51. E.I. Solomon, M.J. Baldwin, and M.D. Lowery, *Chem. Rev.* **92**, 521–542 (1992).
52. D.M. O'Malley, R. Whetten, W. Bao, C.-L. Chen, and R. Sederoff, *Plant J.* **4**, 751–757 (1993).
53. M. Sbaghi, P. Jeandet, R. Bessis, and P. Leroux, *Plant Pathol.* **45**, 139–144 (1996).
54. T. Katase and J.-M. Bollag, *Soil Sci.* **151**, 291–296 (1991).
55. M. Miessner, O. Crescenzi, A. Napolitano, G. Prota, S.O. Andersen, and M.G. Peter, *Helv. Chim. Acta* **74**, 1205–1212 (1991).
56. S.O. Andersen, in G.A. Kerkut and L.I. Gilbert, eds., *Comprehensive Insect Physiology, Biochemistry, and Pharmacology*, vol. 3, Pergamon Press, Oxford, 1985, pp. 59–74.
57. P.R. Williamson, *J. Bacteriol.* **176**, 656–664 (1994).
58. P.R. LaFayette, K.-E.L. Eriksson, and J.F.D. Dean, *Plant Physiol.* **107**, 667–668 (1995).
59. M.B. Kurtz and S.P. Champe, *J. Bacteriol.* **151**, 1338–1345 (1982).
60. D. Slomczynski, J.P. Nakas, and S.W. Tanenbaum, *Appl. Environ. Microbiol.* **61**, 907–912 (1995).
61. Y. Fukushima and T.K. Kirk, *Appl. Environ. Microbiol.* **61**, 872–876 (1995).
62. T. Vares, O. Niemenmaa, and A. Hatakka, *Appl. Environ. Microbiol.* **60**, 569–575 (1994).
63. H. Iwasaki, T. Matsubara, and T. Mori, *J. Biochem. (Tokyo)* **61**, 814–816 (1967).
64. U.A. Germann, G. Muller, P.E. Hunziker, and K. Lerch, *J. Biol. Chem.* **263**, 885–896 (1988).
65. J.P. Geiger, B. Rio, D. Nandris, and M. Nicole, *Appl. Biochem. Biotechnol.* **12**, 121–133 (1986).
66. A. Hatakka, *FEMS Microbiol. Rev.* **13**, 125–135 (1994).
67. H.-D. Youn, K.-J. Kim, J.-S. Maeng, Y.-H. Han, I.-B. Jeong, G. Jeong, S.-O. Kang, and Y.C. Hah, *Microbiology* **141**, 393–398 (1995).
68. P.M. Coll, J.M. Fernandez-Abalos, J.R. Villanuva, R. Santamaria, and P. Perez, *Appl. Environ. Microbiol.* **59**, 2607–2613 (1993).
69. W. Minuth, M. Klischies, and K. Esser, *Eur. J. Biochem.* **90**, 73–82 (1978).
70. Y. Oda, K. Adachi, I. Aita, M. Ito, Y. Aso, and H. Igarashi, *Agric. Biol. Chem.* **55**, 1393–1395 (1991).
71. T. Nakamura, *Biochim. Biophys. Acta* **30**, 44–52 (1958).
72. B. Reinhammar, *Biochim. Biophys. Acta* **205**, 35–47 (1970).
73. T. Sakurai, *J. Inorg. Biochem.* **41**, 277–281 (1991).
74. Y. Nishzawa, K. Nakabayashi, and E. Shinagawa, *J. Ferment. Bioeng.* **80**, 91–93 (1995).
75. M. Jonsson, E. Pettersson, and B. Reinhammar, *Acta. Chim. Scand.* **22**, 2135–2140 (1968).
76. R. Bourbonnais, M.G. Paice, I.D. Reid, P. Lanthier, and M. Yaguchi, *Appl. Environ. Microbiol.* **61**, 1876–1880 (1995).
77. R. Aramayo and W.E. Timberlake, *Nucleic Acids Res.* **18**, 3415 (1990).
78. V. Ducros, A.M. Brzozowski, K.S. Wilson, S.H. Brown, P. Østergaard, P. Schneider, D.S. Yaver, A.H. Pedersen, and G.J. Davies, *Nature Structural Biol.* **5**, 310–316 (1998).
79. Y. Kojima, Y. Tsukuda, Y. Kawai, A. Tsukamoto, J. Sugiura, M. Sakaino, and Y. Kita, *J. Biol. Chem.* **265**, 15224–15230 (1990).
80. L. Jönsson, K. Sjöström, I. Häggström, and P.O. Nyman, *Biochim. Biophys. Acta* **1251**, 210–215 (1995).
81. H. Molitoris and K. Esser, *Arch. Mikrobiol.* **72**, 267–296 (1970).
82. A. Yoshitake, Y. Katayama, M. Nakamura, Y. Iimura, S. Kawai, and N. Morohoshi, *J. Gen. Microbiol.* **139**, 179–185 (1993).
83. M.T. Graziani, L. Antonilli, P. Sganga, G. Citro, B. Mondovi, and M.A. Rosei, *Biochem. Int.* **21**, 1113–1124 (1990).
84. F. Xu, W. Shin, S.H. Brown, J.A. Wahleitner, U.M. Sundaram, and E.I. Solomon, *Biochim. Biophys. Acta* **1292**, 303–311 (1996).
85. M.A. Pickard and D.W.S. Westlake, *Can. J. Biochem.* **48**, 1351–1358 (1970).
86. E. Karhunen, M.-L. Niku-Paavola, L. Viikari, T. Haltia, R.A. Van der Meer, and J.A. Duine, *FEBS Lett.* **267**, 6–8 (1990).
87. F. Xu, *J. Biol. Chem.* **272**, 924–928 (1997).

88. A. Messerschmidt and R. Huber, *Eur. J. Biochem.* **187**, 341–352 (1990).
89. A. Messerschmidt, R. Ladenstein, R. Huber, M. Bolognesi, L. Avigliano, R. Petruzzelli, A. Rossi, and A. Finazzi-Agro, *J. Mol. Biol.* **224**, 179–205 (1992).
90. I. Zaitseva, V. Zaitsev, C. Card, K. Moshkov, B. Bax, A. Ralph, and P. Lindley, *J. Biol. Inorg. Chem.* **1**, 15–23 (1996).
91. J. Rogalsky, M. Wojtas-Wasilewska, R. Apalovic, and A. Leonowicz, *Biotechnol. Bioeng.* **37**, 770–777 (1991).
92. C. Eggert, U. Temp, J.F.D. Dean, and E.-K.L. Eriksson, *FEBS Lett.* **376**, 202–206 (1995).
93. H. Galliano, G. Gas, J.L. Seris, and A.M. Boudet, *Enzyme Microb. Technol.* **13**, 478–482 (1991).
94. R.J. Petroski, W. Peczynska-Czoch, and J.P. Rosazza, *Appl. Environ. Microbiol.* **40**, 1003–1006 (1980).
95. F. Xu, *Biochemistry* **35**, 7608–7614 (1996).
96. F. Xu, *Appl. Biochem. Biotechnol.* **59**, 221–230 (1996).
97. T. Kreuter, A. Steudel, and H. Pickert, *Acta Biotechnol.* **11**, 81–83 (1991).
98. S. Murao, Y. Hinode, E. Matsumura, A. Numata, K. Kawai, H. Ohishi, H. Jin, H. Oyama, and T. Shin, *Biosci. Biotechnol. Biochem.* **56**, 987–988 (1992).
99. M.C. De Pinto and A.R. Bacelo, *Phytochemistry* **42**, 283–286 (1996).
100. F. Sollai, N. Curreli, M.C. Porcu, A. Rescigno, A.C. Rinaldi, A. Rinaldi, P. Rossino, G. Soddu, and E. Sanjust, *Biochem. Arch.* **12**, 7–12 (1996).
101. C.R. Perry, S.E. Matcham, D.A. Wood, and C.F. Thurston, *J. Gen. Microbiol.* **139**, 171–178 (1993).
102. B. Schilling, R.M. Linden, U. Kupper, and K. Lerch, *Curr. Genet.* **22**, 197–203 (1992).
103. International Pat. WO9201046 (January 23, 1992), M. Saloheimo, M.-L. Niku-Paavola, M. Penttila, J. Knowles, and A. Kantelinen (to Valtion Teknillinen Tutkimuskeskus).
104. Germany Pat. DE4141901 A1 (June 24, 1993), K. Crabbe and A. Cordes.
105. P. Bajpai and P.K. Bajpai, *J. Biotechnol.* **33**, 211–220 (1994).
106. V.I. Elisashvili, *Microbiology* **62**, 480–487 (1993).
107. A.M. Bonnen, L.H. Anton, and A.B. Orth, *Appl. Environ. Microbiol.* **60**, 960–965 (1994).
108. C. Ruttimann-Johnson, L. Salas, R. Vicuna, and T.K. Kirk, *Appl. Environ. Microbiol.* **59**, 1792–1797 (1993).
109. Y. Iimura, Y. Katayama, S. Kawai, and N. Morohoshi, *Biosci. Biotechnol. Biochem.* **59**, 903–905 (1995).
110. K. Ruel, K. Ambert, and J.-P. Joseleau, FEMS *Microbiol. Rev.* **13**, 241–254 (1994).
111. R. Kondo, T. Iimori, H. Imamura, and T. Nishida, *J. Biotechnol.* **13**, 181–188 (1990).
112. M.G. Paice, R. Bourbonnais, I.D. Reid, F.S. Archibald, and L. Jurasek, *J. Pulp Paper Sci.* **21**, J280–J284 (1995).
113. International Pat. WO 9209741 (June 11, 1992), M. Vaheri and O. Piirainen (to Enso-Gutzeit Oy).
114. A. Muheim, A. Fiechter, P.J. Harvey, and H.E. Shoemaker, *Holzforschung* **46**, 121–126 (1992).
115. International Pat. WO9429425 (December 22, 1994), H.-P. Call (to Fitzner).
116. Germany Pat. DE4137761 A1 (November 19, 1992), H.-P. Call.
117. International Pat. WO 9603546 A1 (February 8, 1996), L.S. Pedersen and C. Felby (to Novo Nordisk A/S).
118. O. Milstein, A. Huttermann, R. Frund, and H.-D. Ludemann, *Appl. Microbiol. Biotechnol.* **40**, 760–767 (1994).
119. A. Haars, A. Kharazipour, H. Zanker, and A. Huttermann, *ACS Symp. Ser.* **385**, 126–134 (1989).
120. P. Nannipieri and J.-M. Bollag, *J. Environ. Qual.* **20**, 510–517 (1991).
121. U.S. Pat. 4478683 (October 23, 1984), S.A. Orndorff (to Westvaco Corp.).
122. J. Dec and J.-M. Bollag, *Environ. Sci. Technol.* **29**, 657–663 (1995).
123. L. Roy-Arcand and F.S. Archibald, *Enzyme Microb. Technol.* **13**, 194–203 (1991).
124. Y. Iimura, P. Hartikainen, and K. Tatsumi, *Appl. Microbiol. Biotechnol.* **45**, 434–439 (1996).
125. S. Kawai, T. Umezawa, M. Shimada, and T. Higuchi, *FEBS Lett.* **236**, 309–311 (1988).
126. L. Bezalel, Y. Hadar, and C.A. Cerniglia, *Appl. Environ. Microbiol.* **62**, 292–295 (1996).
127. P. Manzanares, S. Fajardo, and C. Martin, *J. Biotechnol.* **43**, 125–132 (1995).
128. S. Davis and R.G. Burns, *Appl. Microb. Biotechnol.* **32**, 721–726 (1990).
129. U.S. Pat. 5407577 (April 18, 1995), N.P. Nghiem (to Nalco Chemical Co.).
130. M. Chivukula and V. Renganathan, *Appl. Environ. Microbiol.* **61**, 4374–4377 (1995).
131. J.C. Roper, J.M. Sarkar, J. Dec, and J.-M. Bollag, *Water Res.* **29**, 2720–2724 (1995).
132. C. Crecchio, P. Ruggiero, and M.D.R. Pizzigallo, *Biotechnol. Bioeng.* **48**, 585–591 (1995).
133. International Pat. 9218687 (October 29, 1992), G. Pedersen and M. Schmidt (to Novo Nordisk A/S).
134. International Pat. 9612845 A1 (May 2, 1996), A.H. Pedersen and J.V. Kierulff (to Novo Nordisk A/S).
135. International Pat. 9612846 A1 (May 2, 1996), A.H. Pedersen and J.V. Kierulff (to Novo Nordisk A/S).
136. International Pat. WO 9312259 (June 24, 1993), N.H. Sørensen, K. Ingvorsen, and A. Møllgaard (to Novo Nordisk A/S).
137. U.S. Pat. 4259318 (March 31, 1981), N.V. Duhe and D.L. Hendrix (to the Regents of the University of Houston).
138. International Pat. WO9616165 A1 (May 30, 1996), F. Xu (to Novo Nordisk Biotech).
139. Y. Tsujino, Y. Yokoo, and K. Sakato, *J. Soc. Cosm. Chem.* **42**, 273–282 (1991).
140. European Pat. EP 504005 A1 (September 16, 1992), M. Roure, P. Delattre, and H. Froger (to Perma).
141. French Pat. FR 2694018 A1 (January 28, 1994), R. Martin, A. Jumino, C. Dubief, G. Rosenbaum, and M.-P. Audousset (to L'Oreal SA).
142. Japenese Pat. JP9177813 A2 (April 3, 1991) R. Saruno.
143. Germany Pat. DE2155359 (May 25, 1972) E. Zeffren and J.F. Sullivan (to Procter and Gamble Co.).
144. A.E. Gardiol, R.J. Hernandez, B. Reinhammar, and B.R. Harte, *Enzyme Microb. Technol.* **18**, 347–352 (1996).
145. N. Zouari, J.-L. Romette, and D. Thomas, *Biotechnol. Tech.* **8**, 503–508 (1994).
146. A.L. Ghindilis, A. Makower, C.G. Bauer, F.F. Bier, and F.W. Scheller, *Anal. Chim. Acta* **304**, 25–31 (1995).
147. S. Murao, M. Arai, N. Tanaka, H. Ishikawa, K. Matsumoto, and S. Watanabe, *Agric. Biol. Chem.* **49**, 981–985 (1985).

148. H. Agematu, T. Nakashima, N. Shibamoto, T. Yoshioka, T. Shin, and S. Murao, *J. Ferment. Bioeng.* **77**, 479–482 (1994).

149. B.P. Roy and F. Archibald, *Anal. Biochem.* **216**, 291–298 (1994).

150. F.F. Bier, E. Ehrentreich-Föster, C.G. Bauer, and F.W. Scheller, *Fresenius J. Anal. Chem.* **354**, 861–865 (1996).

151. U.S. Pat. 5196312 (March 23, 1993), A. Miike and T. Tatano (to Kyowa Medix Co. Ltd.).

152. Germany Pat. DE4314417 A1 (November 10, 1994), F. Scheller, U. Wollenberger, and A. Makower (to Byk Gulden Italia).

153. J.A. Akkara, M. Ayyagari, F. Bruno, L. Samuelson, V.T. John, C. Karayigitoglu, S. Tripathy, K.A. Marx, D.V.G.L.N. Rao, and D.L. Kaplan, *Biomimetics* **2**, 331–339 (1994).

154. R. Ikeda, H. Uyama, and S. Kobayashi, *Macromolecules* **29**, 3053–3054 (1996).

155. W.L. Baker, K. Sabapathy, M. Vibat, and G. Lonergan, *Enzyme Microb. Technol.* **18**, 90–94 (1996).

156. French Pat. FR2694021 (January 28, 1994), A. Junino, and R. Martin (to l'Oreal SA).

157. W. Cai, R. Martin, B. Lemaure, J.-L. Leuba, and V. Petiard, *Plant Physiol. Biochem.* **31**, 441–445 (1993).

158. B. Falconnier, C. Lapierre, I. Lesage-Meessen, G. Yonnet, P. Brunerie, B. Colonna-Ceccaldi, G. Corrieu, and M. Asther, *J. Biotechnol.* **37**, 123–132 (1994).

159. A. Potthast, T. Rosenau, C.-L. Chen, and J.S. Graftzel, *J. Org. Chem.* **60**, 4320–4321 (1995).

160. F.S. Sariaslani, J.M. Beale, Jr., and J.P. Rosazza, *J. Nat. Prod.* **47**, 692–697 (1984).

161. U.T. Bhalerao, C. Muralikrishna, and B.R. Rani, *Tetrahedron* **50**, 4019–4024 (1994).

162. H. Agematu, T. Tsuchida, K. Kominato, N. Shibamoto, T. Yoshioka, H. Nishida, and R. Okamoto, *J. Antibiot. (Tokyo)* **46**, 141–148 (1993).

163. H. Agematu, K. Kominato, N. Shibamoto, T. Yoshioka, H. Nishida, R. Okamoto, T. Shin, and S. Murao, *Biosci. Biotechnol. Biochem.* **57**, 1387–1388 (1993).

164. F.S. Sariaslani, F.M. Eckenrode, J.M. Beale, Jr., and J.P. Rosazza, *J. Med. Chem.* **27**, 749–754 (1984).

165. A. Zamorani, P. Spettoli, A. Lante, A. Crapisi, and G. Pasini, *Ital. J. Food Sci.* **Spec. Issue**, 195–200 (1995).

166. F.O. Asiegbu, A. Paterson, and J.E. Smith, *World J. Microbiol. Biotechnol.* **12**, 16–21 (1996).

167. C.B.S. Tong, K.B. Hicks, S.F. Osman, A.T. Hotchkiss, Jr., and R.M. Haines, *J. Agric. Food Chem.* **43**, 592–597 (1995).

168. A. Viterbo, R.C. Staples, B. Yagen, and A.M. Mayer, *Phytochemistry* **35**, 1137–1142 (1994).

169. *Debittering and Darkening of Olives by Laccase Treatment*, vol. 378, Novo Nordisk A/S Res. Discl., Novo Nordisk A/S, Bagsværd, Denmark, 1995, p. P676 (no. 37828).

170. U.S. Pat. 5094668 (March 10, 1992), E.E. Kern, W.M. Menger, D.A. Odelson, A.S. Sinsky, D.L. Wise, and D.J. Trantolo (to Houston Industries Inc.).

171. O. Milstein, B. Nicklas, and A. Huttermann, *Appl. Microbiol. Biotechnol.* **31**, 70–74 (1989).

172. S. Davis and R.G. Burns, *Appl. Microbiol. Biotechnol.* **37**, 474–479 (1992).

173. A.V. Vakurov, A.K. Gladilin, A.V. Levashov, and Y. Khmelnitsky, *Biotechnol. Lett.* **16**, 175–178 (1994).

174. S. Cliffe, M.S. Fawer, G. Maier, K. Takata, and G. Ritter, *J. Agric. Food Chem.* **42**, 1824–1828 (1994).

175. G. Palmieri, P. Giardina, B. Desiderio, L. Marzullo, M. Giamberini, and G. Sannia, *Enzyme Microb. Technol.* **16**, 151–158 (1994).

176. U.S. Pat. 5443975 (Aug. 22, 1995), V. Capuano and S. Cervelli (to Consigtio Nazionale delle Richerche IT and Ente per le Nuove Tecnologie l'Energia e l'Ambiente ENEA IT).

177. International Pat. WO 9501426 A1 (January 12, 1995), P. Schneider and A.H. Pedersen (to Novo Nordisk A/S).

178. F. Archibald and B. Roy, *Appl. Environ. Microbiol.* **58**, 1496–1499 (1992).

179. R. Bourbonnais and M.G. Paice, *FEBS Lett.* **267**, 99–102 (1990).

180. G. Daniel, J. Volc, and E. Kubatova, *Appl. Environ. Microbiol.* **60**, 2524–2532 (1994).

181. L. Marzullo, R. Cannio, P. Giardina, M.T. Santini, and G. Sannia, *J. Biol. Chem.* **270**, 3823–3827 (1995).

182. P. Ander, C. Mishra, R.L. Farrel, and K.-E.L. Eriksson, *J. Biotechnol.* **13**, 189–198 (1990).

183. M.-L. Niku-Paavola, M. Ranua, A. Suurnakki, and A. Kantelinen, *Biosource Technol.* **50**, 73–77 (1994).

LACTONES, BIOCATALYTIC SYNTHESIS

TAKESHI SUGAI
HIROMICHI OHTA
Keio University
Yokohama, Japan

KEY WORDS

Asymmetric reduction
Baeyer–Villger reaction
Esterase
Hydrolysis
Kinetic resolution
Lipase
Optically active compounds
Transesterification

OUTLINE

INTRODUCTION

The most important feature of biocatalyst-mediated reaction is the chemo-, regio-, and enantioselectivity. The recent increasing demands for enantiomerically enriched compounds, and hopefully pure single enantiomers, in the field of pharmaceuticals and agrochemicals (1) promoted the development of excellent chemoenzymatic procedures (2). These achievements have been well reviewed in books and selected papers (3). In this article, we focus on the recent biocatalytic procedures (mostly from 1990 to 1997) for the synthesis of enantiomerically enriched forms of lactones.

Lactones play important roles as the starting materials and intermediates of pharmaceuticals, agrochemicals, flavors, and other fine chemicals, as well as macromolecules with industrial importance. They are derived from many kinds of precursors, essentially by means of C—O bond formation (esterification) between hydroxyl and carboxyl groups or insertion of an oxygen atom on the C—C bond adjacent to carbonyl groups (Scheme 1).

In the following sections, several approaches are described: *(1)* hydrolysis of ester functionality of corresponding racemic lactones; *(2)* lactonization of hydroxy acids and their derivatives; *(3)* reduction of carbonyl precursors; *(4)* oxidation of diol precursors; *(5)* Baeyer–Villiger type oxidation of ketones; and *(6)* separation of isomers by the enzymatic action on other functionalities (Y) that are appended to the molecules (Scheme 2). In addition, an approach based on the biodegradation of lipids and aromatics *(7)* is briefly discussed. Finally, preparation of some lactones of special industrial interests *(8)* is emphasized in detail.

HYDROLYSIS OF ESTER FUNCTIONALITY OF CORRESPONDING RACEMIC LACTONES

The simplest approach for the preparation of enantiomerically enriched forms of lactones is the optical resolution (separation of enantiomers) of the racemic mixture, which is readily synthesized in a chemical manner and often available from commercial sources. A representative example is illustrated in Scheme 3 (4). The rate of hydrolysis of (*S*)-enantiomer is higher than that of the (*R*)-form and, accordingly, a "kinetic" resolution occurs to give enantiomerically enriched forms of the products, (*R*)-lactone and (*S*)-hydroxy acid.

Pig liver esterase (PLE, as Sigma L 8285, for example), horse liver esterase (HLE, Sigma L 9627), and lipases of different origins involving pig pancreas lipase (PPL, Sigma, L 3126) are available for this purpose. A number of related five- and six-membered lactones have been resolved in a similar manner (4–7). Bicyclic compounds are well accepted as the substrates (8). Moreover, this method can be applied to substrates with rather small four-membered rings (9–11), and medium- and large-membered rings (12). Examples cited here are listed in Figure 1. It should be noted, however, that the absolute configuration as well as the enantiomeric excess (*e.e.*) depends upon both the structure of substrates and the kind of enzymes. It was also reported that the addition of calcium chloride affected the enantioselectivity of the reaction in Scheme (4).

As for the estimation of the enantiomeric excess, there have been developed gas chromatographic analyses with chiral stationary phase such as cyclodextrin derivatives (Lipodex E from Marcherey Nagel, CP-Chirasil-DEX CB from GL Science, Chiraldex BP from Tokyo Kasei, and others) (6,13,14). The chiral stationary phase of liquid chromatography (e.g., ChiraSpher, Merck) can be used for the same purpose (15). In the case that no good separation of enantiomers on the chromatogram is observed, other methods can be utilized, such as the conversion of lactones to the corresponding diols by the action of methyllithium and subsequent nuclear magnetic resonance (NMR) measurement in the presence of chiral NMR shift reagent (16). Another alternative is the analysis of diastereomeric derivatives on GLC or HPLC. For this purpose, phenylethyl urethanes (15), α-acetyllactates (15), and α-methoxy-α-trifluoromethylphenyl acetate (MTPA) esters (17) derived from the alcohol(s) are available. Another factor to note is the separation of the products and recovered starting materials. Although the products are essentially acidic while the starting materials are neutral, sometimes it is seriously troublesome, as in the case when these two materials show similar properties with regard to volatility, water solubility, and chromatographic behavior. The hydrolysis often spontaneously and reversibly proceeds in nonbiocatalytic conditions. The relactonization of the resulting

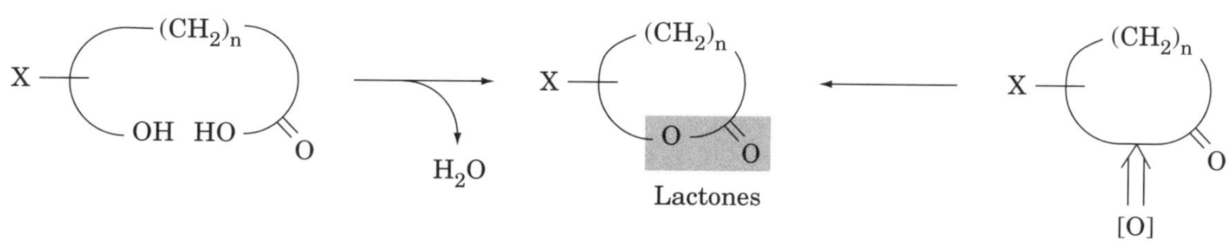

Scheme 1. Lactones and their precursors.

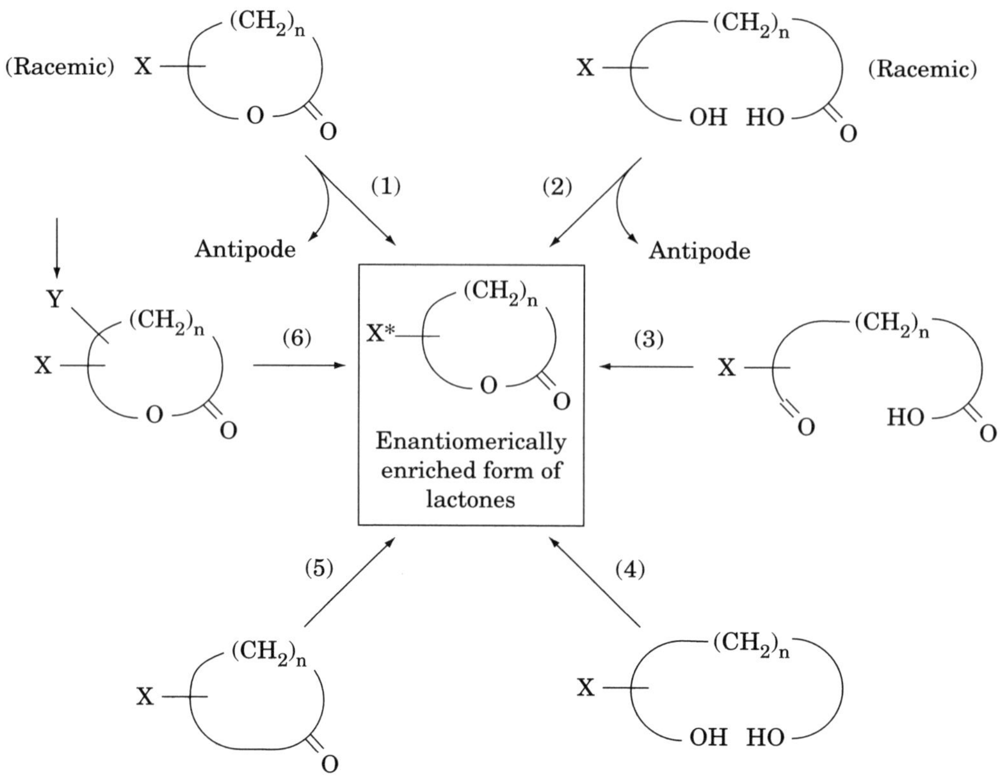

Scheme 2. Biocatalytic approaches to lactones. For the explanation of each approach, see text.

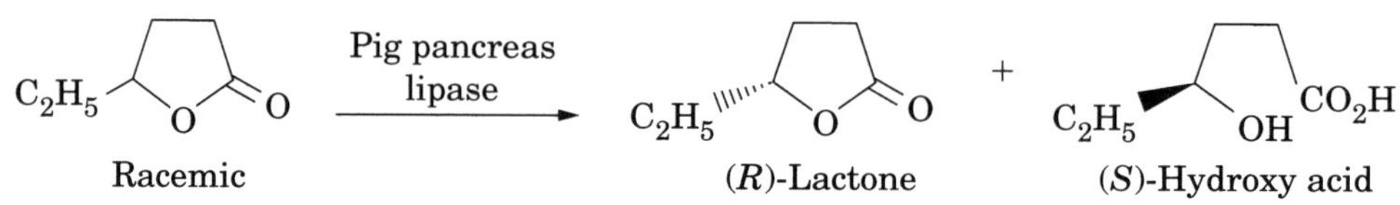

Scheme 3. Lipase-catalyzed hydrolysis of a racemic lactone.

hydroxy acids to the lactones would bring about the lowering of the enantiomeric excess of the "kinetically resolved" lactones.

LACTONIZATION OF HYDROXY ACIDS AND THEIR DERIVATIVES

Hydrolytic enzymes, especially lipases, also catalyze the esterification of hydroxy acids with a concomitant dehydration as well as transesterification of hydroxy esters. Lipase-catalyzed formation of five-membered lactone (γ-lactone) from γ-hydroxy carboxylates is a highly favored process. Kinetic resolution of the racemic hydroxy ester was achieved using PPL-catalyzed lactonization in an organic solvent, such as diethyl ether (Scheme 5, Fig. 2) (7,15,18–23). As mentioned, spontaneous lactonization is the competing reaction that lowers the "apparent" enantioselectivity of the reaction. For example, the rate of lactonization of (*S*)-isomer of methyl 3-hydroxypentanoate is 26 times faster than that of the (*R*)-isomer at the start of the reaction (18). After a prolonged incubation [more than 70% conversion for the (*S*)-isomer and 25% for the (*R*-isomer)], however, both of the rates gradually decrease and are almost the same. Only five-membered ring formation occurred from γ,δ-dihydroxy carboxylates (23). Another interesting difunctional substrate is prochiral γ-hydroxypimelate (5-hydroxynonanedioate). Lipase cleanly discriminates between the two enantiotopic groups to give a single enantiomer (>95% *ee*) of lactone (19).

The situation is rather complex in the case of the formation of six-membered (and larger) ring lactones. The competitive reaction, oligomerization of the starting material with an intermolecular reaction, occurs. The preference of intramolecular versus intermolecular reaction depends on the structure of substrates. The reactivity of the esterification/transesterification is greatly affected by the stereochemical environment of the hydroxyl group that concerns the reaction. In this sense, enantiomers of the substrates (right and wrong enantiomer in Scheme 6) are kinetically resolved when they react as acyl acceptors toward the preformed acyl donor (acyl–enzyme complex), resulting in the enantiomerically enriched form of the lactones. At the initial stage, the acyl–enzyme complex and small oligomeric intermediates form from both right and wrong isomers; however, intermediates, including the wrong isomer, lower their own reactivities when they want to work as acyl acceptors in the further oligomer chain elongation, as well as the lactonization. As all the reactions

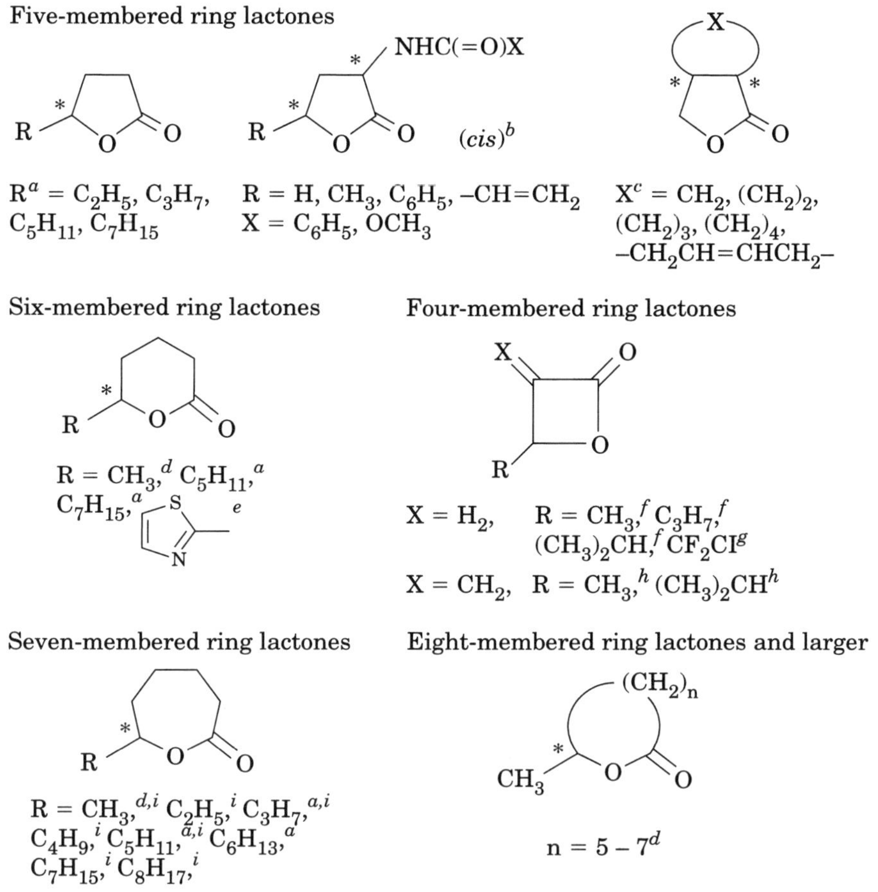

Figure 1. Enantiomerically enriched lactones by enzyme-catalyzed hydrolysis. a, Ref. 4; b, Ref. 5; c, Ref. 8; d, Ref. 12; e, Ref. 7; f, Ref. 9; g, Ref. 10; h, Ref. 11; i, Ref. 6.

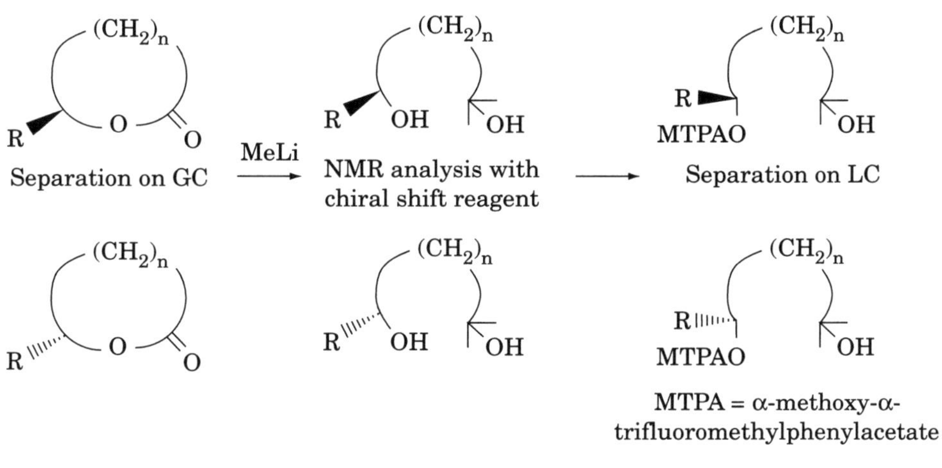

Scheme 4. Determination of enantiomeric excesses of lactones.

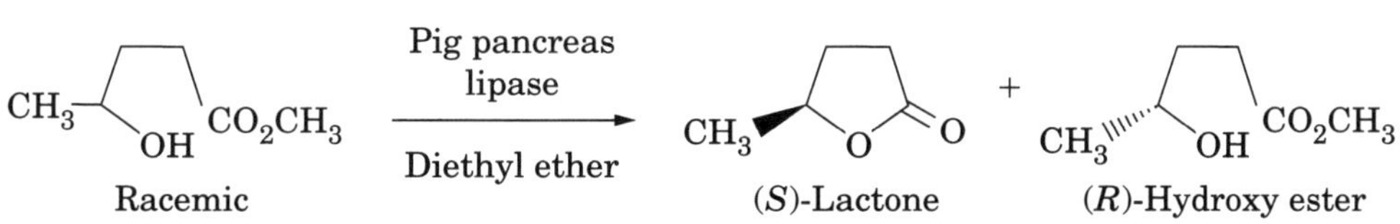

Scheme 5. Lipase-catalyzed lactonization of a racemic γ-hydroxy ester.

Five-membered ring lactonization

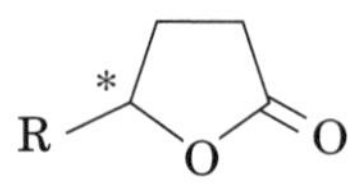

Aliphatic

R = CH_3,[a,b,c] C_2H_5,[b,c] C_3H_7,[c] C_4H_9,[c] C_5H_{11},[c] C_6H_{13},[b,c] C_7H_{15},[c] C_8H_{17},[b,c] C_8H_{17}—C≡C–[d]

Aromatic

R = C_6H_5,[a,b] p-$CH_3C_6H_4$,[b] p-$CH_3OC_6H_4$,[b] p-BrC_6H_4,[b] C_6H_5–C≡C–[e]

Functionalized

R = $CH_3O_2C(CH_2)_2$–,[b] $C_2H_5O_2C(CH_2)_2$–,[b] C_6H_5,$CH_2O_2C(CH_2)_2$–,[b] $(CH_3)_3CO_2C(CH_2)$–,[f] $HOCH_2$–,[g]

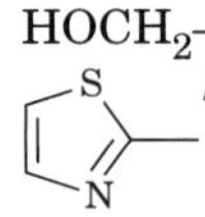
[h]

Figure 2. Enantiomerically enriched five-membered lactones by lipase-catalyzed transesterification. a, Ref. 18; b, Ref. 19; c, Ref. 15; d, Ref. 20; e, Ref. 21; f, Ref. 22; g, Ref. 23; h, Ref. 7.

Six-membered ring lactonization

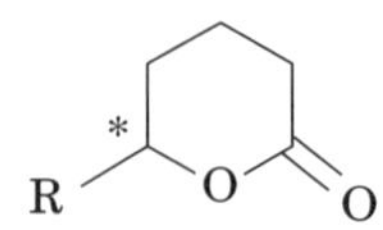

R = CH_3,[a,b] C_2H_5,[b] C_3H_7,[b] C_4H_9,[b] C_5H_{11},[b] C_6H_{13},[b] C_7H_{15},[b] C_8H_{17},[b] $C_{11}H_{23}$,[c]

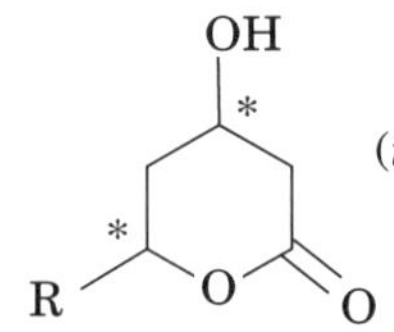
(*trans*)

R = C_5H_{11},[d] $C_6H_5(CH_2)_2$–,[d] C_6H_5CH=CH– (*trans*)[e]

Figure 3. Enantiomerically enriched lactones by enzyme-catalyzed transesterification. a, Ref. 25; b, Ref. 15; c, Ref. 28; d, Ref. 27; e, Ref. 26.

work in a reversible manner, the right enantiomer contained in right–wrong mixed fragments goes back to the starting material, and the enriched right enantiomer of lactone thus becomes further concentrated. That lactones themselves serve as the biocatalytic precursors of polyesters (19) also supports this reversible mechanism.

Unsubstituted methyl 5-hydroxypentanoate itself undergoes an exclusive oligomerization by the lipase-catalyzed reaction. In contrast, when a substituent was introduced to the δ- (5-) position on methyl 5-hydroxypentanoate, the desired lactonization proceeded (25), and this is well explained by Scheme 6. A number of δ-hydroxyalkanoates were applied as the substrates for lipases involving that from pig pancreas (15,25–27), *Candida antarctica* (Novo SP328) (26), and *Pseudomonas* (Amano P) (28), as illustrated in Figure 3. To improve the enantioselectivity of the reaction, an elaborated condition was required by means of the immobilization of enzyme (28).

Another characteristic feature of lipase-catalyzed transesterification is the ability of the "macrolactonization" in an organic solvent. The first reported example is illustrated in Scheme 7, the formation of 17-membered lactone from methyl 16-hydroxyhexadecanoate catalyzed by lipases from *Pseudomonas* sp. (P from Nagase), *Pseudomonas fluorescens* (P from Amano), and porcine pancreas (Sigma) (29). The yield was increased to 80% using a nonpolar solvent, such as cyclohexane and *n*-heptane. Further attempts to improve the rate of lactonization and the yield of lactone were made successful by adding an irreversible process to the reaction, such as adsorption of methanol, which comes from transesterification on molecular sieves (30,31) or the use of vinyl ester as the substrate (32). The competitive reaction is the formation of dimeric, trimeric, and higher oligomeric lactones, as well as linear oligomers. The distribution of the products (monomeric, dimeric, and so on) depends on the structure (30,32–33), as well as the concentration of substrates (29). Increasing availability of lipases, such as those from *Pseudomonas* sp. (AK from Amano), *Candida cylindracea* (Meito OF-360), *C. antarctica* (Novo SP525), and those already cited made this protocol attractive, and indeed, some of them worked well for the syntheses of naturally occurring macrolactones and diolides (31,34–36). Although the lactonization essentially

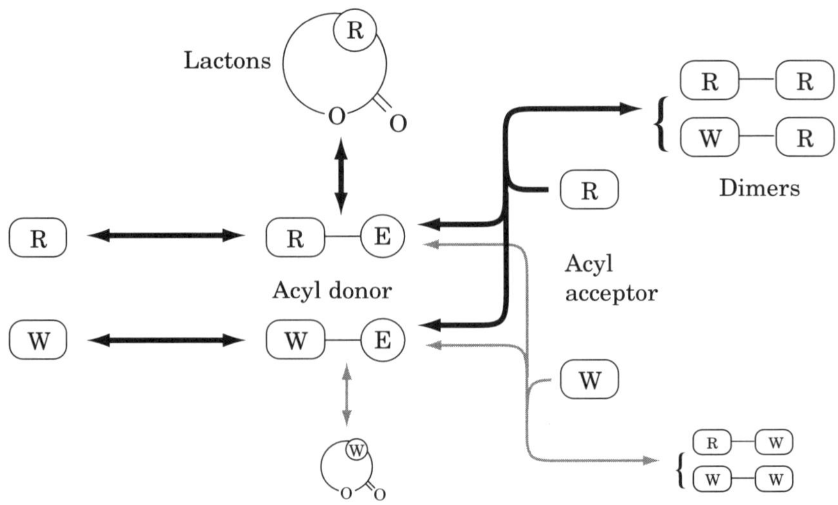

Scheme 6. Lipase-catalyzed lactonization and oligomerization; R, right enantiomer; W, wrong enantiomer.

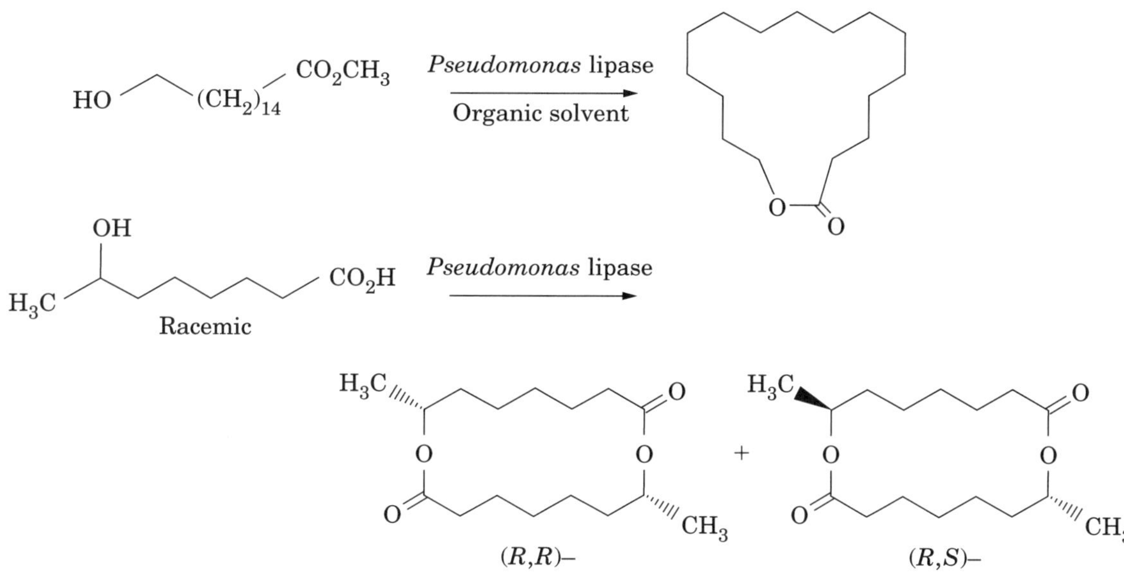

Scheme 7. Lipase-catalyzed macrolactonization.

proceeds in an enantioselective manner (29–30,34–37), both of two diastereomeric products (*R*,*R*)-isomer and [*R*,*S* (*meso*)]-isomer formed in the course of dimeric lactone synthesis from racemic substrate (Scheme 7) (32,37). This result is easily deduced from the discussion of the mechanism of lipase-catalyzed lactonization as in the previous section (cf. 32). An approach used to obtain an enantiomerically enriched form of substrate in advance was effective to avoid this situation (31), examples cited in this section are listed in Figure 4.

REDUCTION OF CARBONYL PRECURSORS

The reduction of γ- (4-) and δ- (5-)-oxoacids provides the corresponding hydroxy acids (7,13,14,23,38–44); however, in most cases, the products are isolated as lactones with an intramolecular dehydration during the workup procedure. Many kinds of microorganisms that reduce carbonyl groups are available, involving commercial baker's yeast. The change of microorganism gives a striking contrast to the stereochemical outcome, as illustrated in Scheme 8. (*R*)-γ-Valerolactone was obtained by the reduction of levulinic acid with a yeast, *Yamadazyma farinosa* IFO 10896 (*Pichia farinosa* IAM 4682) (13), while baker's yeast–mediated reduction of ethyl levulinate (4-oxopentanoate) provides ethyl (*S*)-4-hydroxypentanoate, a precursor of (*S*)-enantiomer of lactone (23,37). In the latter case, the hydroxy ester was lactonized by the action of PPL (discussed earlier) with a further enhancement of enantiomeric excess (23). A similar approach could be applied for the synthesis of (*R*)-γ-caprolactone (14). Another interesting combination of hydrolytic enzymes and the yeast-mediated reduction was reported in the synthesis of a functionalized lactone, which is elaborated as a versatile starting material for natural product synthesis. Selective hydrolysis of one of the two alkoxycarbonyl groups by means of *Pseudomonas diminuta* IFO 13181 provided the substrate for the subsequent reduction (39).

Yeast-mediated reduction is not limited to a carbonyl group; the C═C double bonds, of α,β-unsaturated lactones were reduced to the enantiomerically enriched forms of saturated products (Scheme 9) (45). Examples cited in this section are listed in Figure 5.

OXIDATION OF DIOL PRECURSORS

Enantiomerically enriched forms of lactones are available based on the selective oxidation of two enantiotopically different primary alcohols. The formation of lactone proceeds in two steps: (*1*) the selective oxidation of the primary hydroxyl group and the hemiacetal (lactol) formation of the resulting hydroxy aldehyde, and (*2*) the subsequent oxidation of secondary hydroxyl group of lactol, as illustrated in Scheme 10. The most studied enzyme so far is horse liver alcohol dehydrogenase (HLADH, for example, Sigma A 6128 and Boehringer No. 102733) (46). This enzyme usually catalyzes both oxidation steps; however, when the resulting lactol is a poor substrate for this enzyme, another chemical step for oxidation to the desired lactone under mild conditions is required separately, such as the use of silver carbonate. HLADH works with the aid of oxidized form of the cofactor, NAD, which is recycled with another oxidant, flavin mononucleotide sodium salt (FMN), due to the expense. A number of lactones in enantiomerically enriched form have been reported (46–51), as illustrated in Figure 6, and some of the products were developed as the starting materials for natural product syntheses (49,51).

Based on the extensive studies on the substrate specificity and the stereochemistry of the products, as well as the information of the structure of the enzyme itself, a cubic-space section model has been established by Jones (52,53). This is an empirical rule for stereochemical preference in regard to the enantiotopic two hydroxyl groups

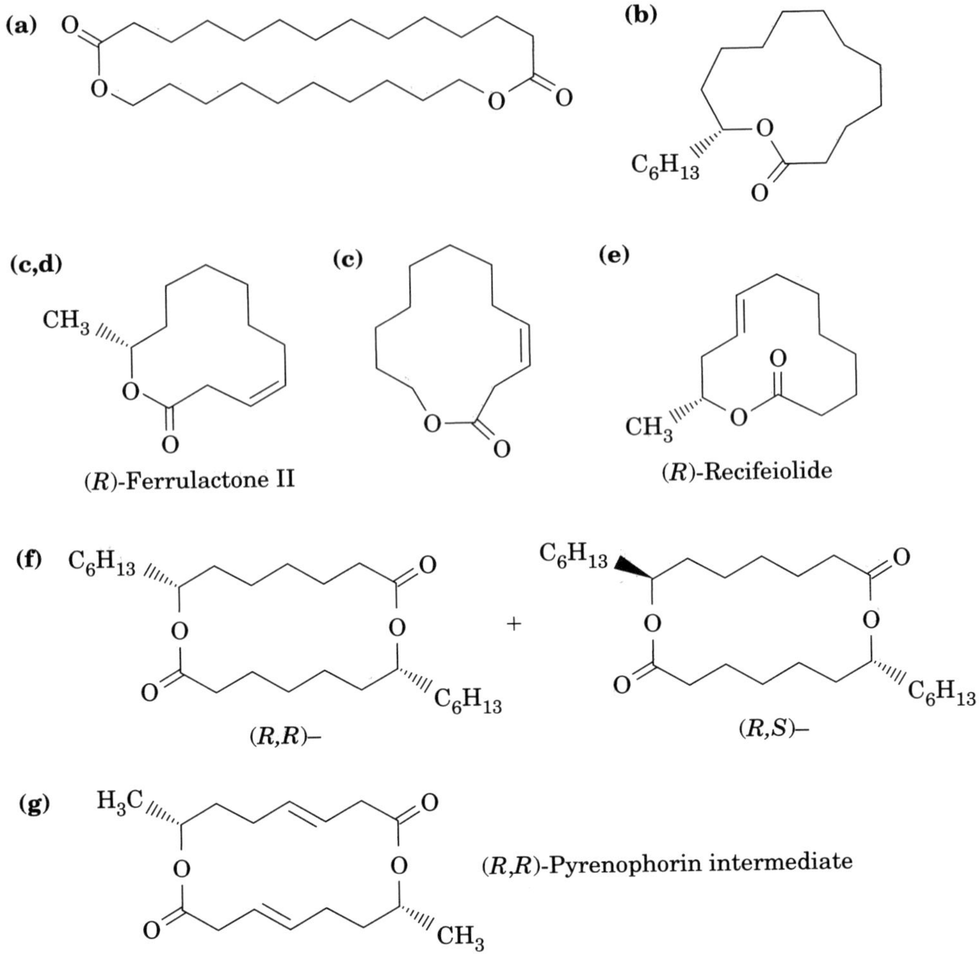

Figure 4. Macrocyclic lactones and diolides by lipase-catalyzed cyclization. a, Ref. 33; b, Ref. 30; c, Ref. 34; d, Ref. 36; e, Ref. 35; f, Ref. 37; g, Ref. 31.

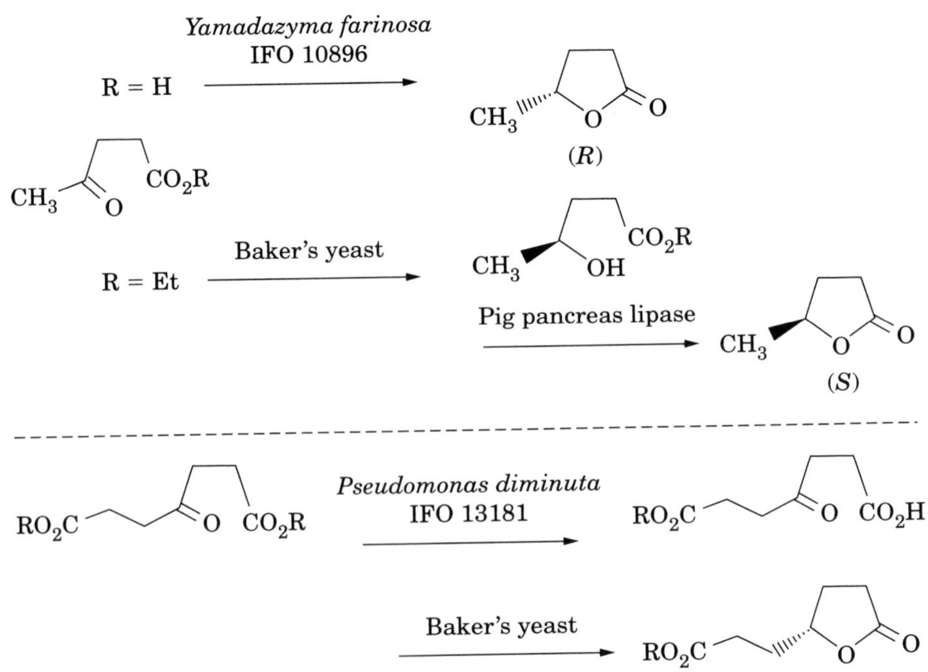

Scheme 8. Yeast-mediated reduction of γ-oxoacids and esters.

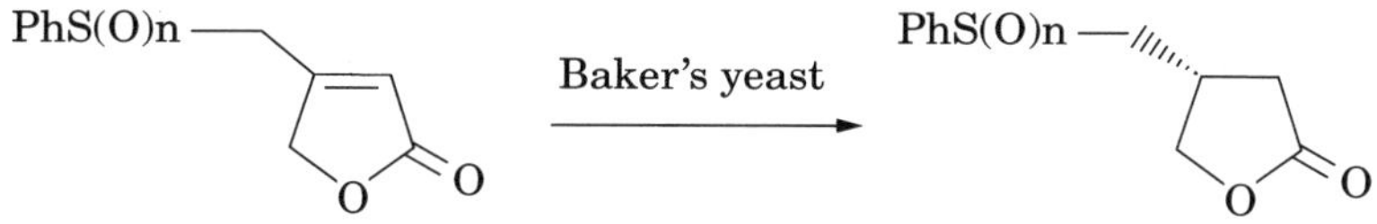

Scheme 9. Yeast-mediated reduction of C═C double bond.

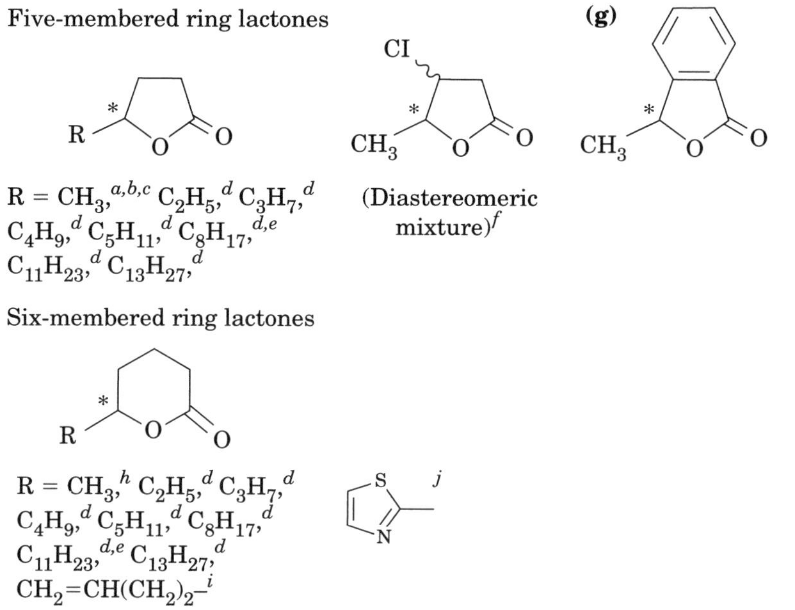

Figure 5. Enantiomerically enriched lactones by yeast-mediated reduction of oxoacids or esters. a, Ref. 13; b, Ref. 38; c, Ref. 23; d, Ref. 40; e, Ref. 41; f, Ref. 42; g, Ref. 44; h, Ref. 14; i, Ref. 43; j, Ref. 7.

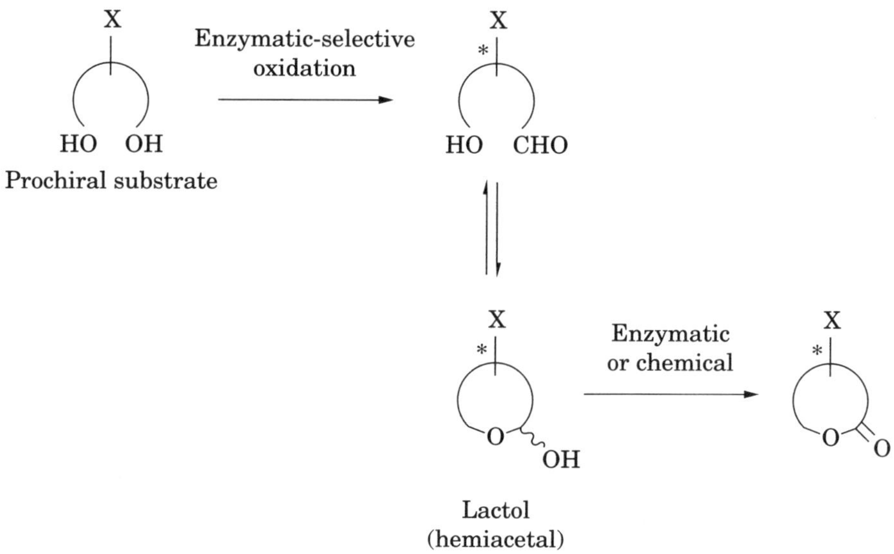

Scheme 10. HLADH catalyzed enantiotopic group-selective oxidation of diols.

and, accordingly, a way of predicting the absolute stereochemistry of the products. Scheme 11 summarizes the conclusion of stereochemical preference in the oxidation. Indeed, the formation of (3*S*,4*R*)-2,3-dimethyl-γ-butyrolactone (4-butenolide) from *meso*-2,3-dimethyl-1,4-butanediol [a] is well explained by the model (50). An open-chain substrate with considerable conformational flexibility suitably fits to the space of the enzyme. However, there are always some exceptions. Even in the case of very closely related substrate [b] (Scheme 11), the product was a complex mixture of isomeric lactols caused by a nonenzymatic epimerization at the α-position of aldehyde intermediate (51). The gross result seems to be that a primary alcohol located in the opposite site was preferentially oxidized. Similar reversed selectivity was observed in another bicyclic substrate (54). It might be possible that the epimerization of intermediates and the equilibrated oxidoreductive condition caused the integration of the secondary products in

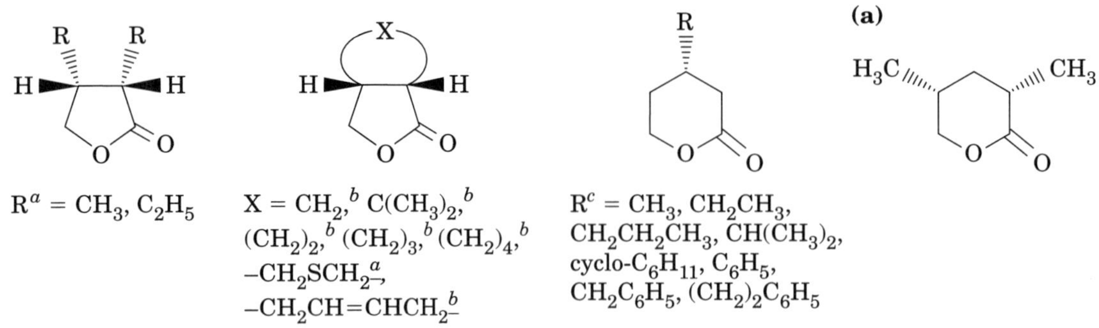

Figure 6. Enantiomerically enriched lactones by HLADH-mediated oxidation. a, Ref. 50; b, Ref. 48; c, Ref. 47.

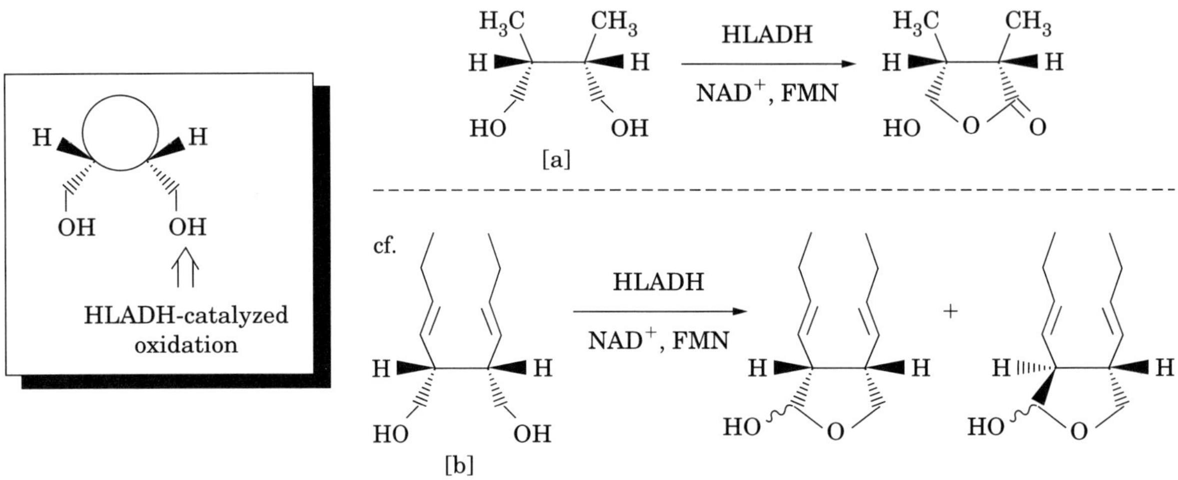

Scheme 11. An empirical rule for the specificity of HLADH catalyzed oxidation. For [a] and [b] see text.

these two examples. Regardless, we should pay considerable attention to the application of this rule to determine the absolute configuration of the molecule with an unknown stereochemistry.

The selective oxidation of diols has also been reported by means of whole cells of microorganisms involving *Gluconobacter roseus* (55) and *Nocardia corallina* (56) (Fig. 7). It should be noted that when whole cells of a microorganism are used, it is difficult to simply ascribe the origin of the enantiomerically enriched form of lactones to the enantioselectivity of the alcohol dehydrogenases. Sih noted the possibility of the assimilation of natural enantiomer in a study where *Flavobacterium oxydans* oxidized 3-methylpentane-1,3,5-triol to give an unnatural enantiomer of mevalonolactone (57). A related example is that *Rhodococcus rhodochrous* IFO 15564–mediated "enantioselective" hydrolysis with discrimination of the remote asymmetric center of hydroxy nitriles, afforded an enantiomerically enriched form of lactones (58). However, another report pointed out the chance of enantioselective oxidation, as well as the further catabolic degradation of the hydroxy acid, which is a possible intermediate of the metabolic pathway by the same microorganism (59).

Enzyme-mediated asymmetrization of prochiral and *meso* compounds, coupled with the subsequent chemical reactions, widely became accepted in synthetic organic

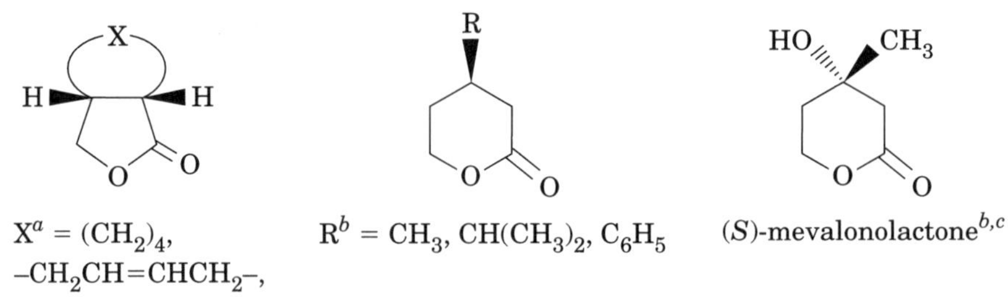

Figure 7. Enantiomerically enriched lactones by whole cell–mediated enantioselective oxidation. a, Ref. 56; b, Ref. 55; c, Ref. 57.

chemistry. Those examples of applications related to synthesizing enantiomerically enriched forms of lactones by use of hydrolytic or oxidoreductive enzymes are cited in selected articles and reviews (60).

BAEYER–VILLIGER TYPE OXIDATION OF KETONES

Enzyme-catalyzed Baeyer–Villiger type oxidation by means of cycloalkanone monooxygenases (61) have come to the attention of organic chemists. Purified enzymes (cyclohexanone and cyclopentanone monooxygenases and related enzymes) from microorganisms have been utilized for this purpose by combining the recycling system of cofactors (for example, FAD-NADP) with glucose-6-phosphate dehydrogenase and glucose-6-phosphate (Scheme 12). The use of the whole cell system was also effective to avoid the set up of the above recycling systems. Recent studies on the structural elucidation of enzymes (62), as well as the extensive investigation on the structure of substrates and products, made it possible to clarify the origin of stereoselectivity of the enzyme-catalyzed oxidation, and a model for predicting the active site has also been proposed (63).

Substituted cyclohexanones are good substrates for cyclohexanone monooxygenase from *Acinetobacter* NCIB 9871. It was very interesting that although there are equal chances for the migration for two C—C bonds in the relationship of enantiotopic groups, the enzyme-catalyzed reaction proceeds in a quite stereoselective manner, as illustrated in Scheme 13 (64–66). The examples of kinetic resolution of racemic ketones via enantioselective Baeyer–Villiger reaction were also reported. Enzymes from *Pseudomonas putida* NCIMB 10007 (67), *Acinetobacter* NCIB 9871 (68), and *Pseudomonas* sp. NCIMB 9872 (69) were effective in preferentially oxidizing one enantiomer of substrate to give lactones, while leaving the other enantiomer intact as the ketone form (Scheme 13). The unreacted enantiomer is readily converted to the corresponding lactone by chemical Baeyer–Villiger oxidation, such as a treatment with peracid with complete retention of the absolute configuration. Because of the high selectivity and wide substrate specificity of *Acinetobacter* enzyme, an approach toward a "designer yeast" has been reported (70). When a strain of baker's yeast was engineered to express cyclohexanone monooxygenase, the reaction became unbelievably easy. Only the incubation of substrate with a stored yeast cells under frozen condition in an appropriate nutrient broth, for example, YEP-galactose, worked well to carry out the Baeyer–Villiger oxidation. Problems that had been caused from the insolubility of the substrates were cleanly solved by adding β-cyclodextrin. The major side reaction is the reduction of substrate to the corresponding secondary alcohol. Lactones available from enantioselective Baeyer–Villiger oxidation are listed in Figure 8.

In the case of some bicyclic substrates, the kinetic resolution brought about a not simple outcome. The more substituted C—C bond of ketone had migrated on one enantiomer to give a "normal" lactone, while an "abnormal" lactone came from the other enantiomer by migration of less substituted C—C bond of ketone, as described in Scheme 14 (71,72). The resulted lactones involving carbocyclic and heterocyclic rings were shown in Figure 9.

The other application of this monooxynegase is the addition or insertion of an "oxygen" atom on a sulfur atom (from sulfides to sulfoxides), and on benzylic methylene groups (74). As the latter example, a *P. putida*–catalyzed formation of phthalide from alkyl substituted benzoic acid on the *ortho* position has been reported (44).

SEPARATION OF ISOMERS BY THE ENZYMATIC ACTION ON OTHER FUNCTIONALITIES THAT ARE APPENDED TO THE MOLECULES

So far, the biochemical reactions directly concerned with the lactone carbonyl moiety have been reviewed. However, there are many natural and synthetic lactones with multiple functional groups. In this section, miscellaneous approaches for preparing enantiomerically enriched forms by utilizing the reaction on other suffixed functional groups are reviewed. The first approach is the kinetic resolution by use of hydrolytic enzymes. Racemic mixtures of the lactones with primary alcohols were effectively resolved by way of a lipase-mediated enantioselective acylation with acetic anhydride. Unreactive enantiomers could readily be separated by a chromatographic purification, while the

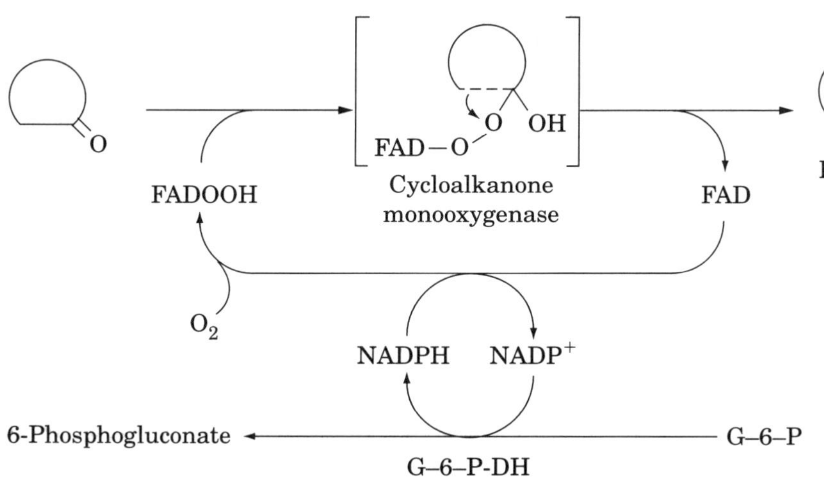

Scheme 12. Monooxygenase-mediated Baeyer–Villiger oxidation.

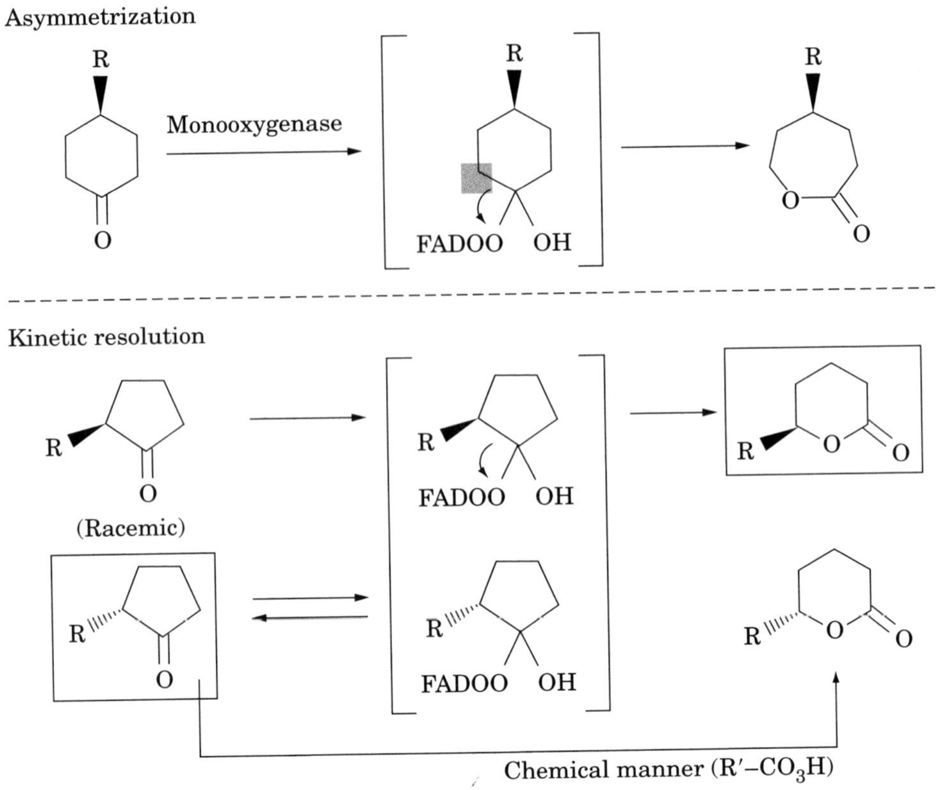

Scheme 13. Stereocourse of enzyme-catalyzed Baeyer–Villiger oxidation.

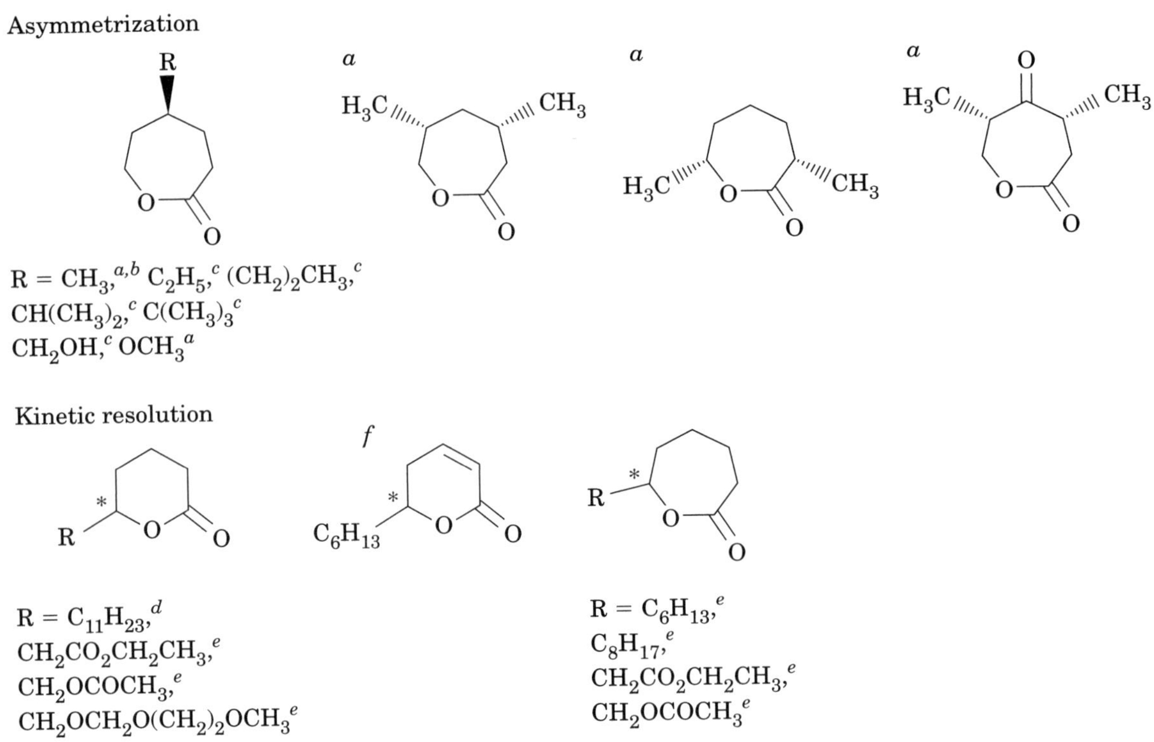

Figure 8. Enantiomerically enriched lactones by Baeyer–Villiger oxidation. a, Ref. 64; b, Ref. 66; c, Ref. 65; d, Ref. 68; e, Ref. 67; f, Ref. 69.

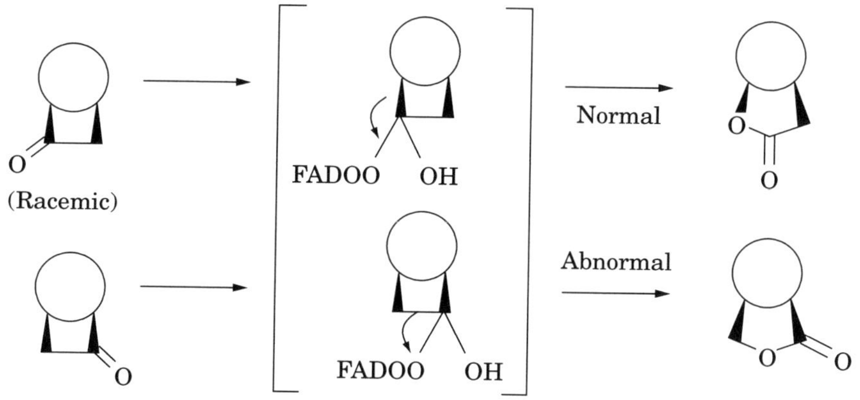

Scheme 14. Normal and abnormal products by enzyme-mediated Baeyer–Villiger oxidation.

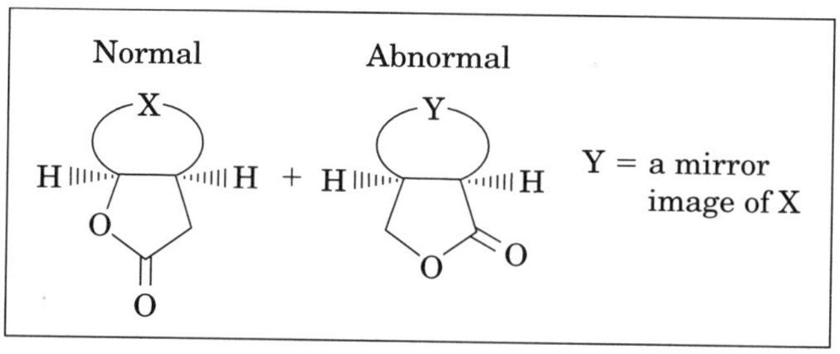

Carbocyclic
X = $-(CH_2)_3-$,[a]
$-CH_2CH{=}CH-$,[a]
$-(CH_2)_4-$,[a]
$-(CH_2)_2CH{=}CH-$,[a]
$-CH_2CH{=}CHCH_2-$,[a]
$-(CH_2)_2CR^1R^2-$[b]
(R^1, R^2 = H, Br, OH, $OCH_2C_6H_5$)

Heterocyclic[c]
X = $-O(CH_2)_2-$,
$-CH_2OCH_2-$,
$-(CH_2)_2O-$,
$-O(CH_2)_3-$,
$-(CH_2)_3O-$

Figure 9. Enantiomerically enriched lactones by Baeyer–Villiger oxidation as normal and abnormal products. a, Ref. 71; b, Ref. 73; c, Ref. 72.

opposite enantiomer was converted to the corresponding acetate. Hydrolysis of the corresponding acetate was elaborated by the use of lipase-catalyzed neutral condition to avoid the racemization of the product under either acidic or basic conditions, which is caused by an intramolecular transesterification (Scheme 15) (75). A similar approach on the asymmetrization of a lactone diol with a *meso* structure has been reported (76).

In contrast, there has been reported another interesting result utilizing an inherent nature of the substrate, that in which the stereochemistry of hemiacetal position is prone to racemize, even under very mild conditions. An enantioselective acylation of one enantiomer with a concomitant spontaneous racemization enabled the dynamic kinetic resolution of a lactone (Scheme 16) (77). Ester and amide functionalities are also available as the clues for the lipase-catalyzed kinetic resolution (78,79).

The second approach is the separation of diastereomeric products. For example, the chromatographic separation of diastereomeric lactones produced by a baker's yeast mediated reduction of the racemic substrate, as illustrated in Scheme 17, made the separation of enantiomers possible, in regard to the chiral center already present in the substrate (80). Needless to say, it is desirable that the enantioface selectivity in the reduction of carbonyl group is as high as possible. Indeed, the whole cell–mediated reduction of the racemic substrates in most cases accompanies the kinetic resolution to give rather complex results (81) and, accordingly, efforts for the screening of suitable strain of microorganism have been reported (82).

APPROACHES BASED ON THE BIODEGRADATION OF LIPIDS AND AROMATICS

Naturally occurring flavoring lactones themselves are in most cases the oxidative degradation products of unsaturated fatty acids. Lactone metabolites, as illustrated in Figure 10, were reported in the microbial incubation of natural products containing fatty acids, such as hydroxy unsaturated fatty acids (83,84), vernoic acid (85), and castor oil (86). An aromatic substrate was also converted to the partial degradation product. (+)-Muconolactone was prepared from a racemic mixture of mandelic acid by the action of *P. putida* mutant PRS 2912. Since this microorganism possesses a mandelic acid racemase, both enantiomers of the starting material are used as the substrate. The origin of chirality of the muconolactone is the step utilizing a muconate cycloisomerase. As this mutant strain lacks a muconate isomerase, which works as the key step in the further metabolic pathway, muconolactone was obtained in the incubation broth in as high as 97% yield (Scheme 18) (87).

PREPARATION OF SOME LACTONES OF SPECIAL INTEREST

(*R,Z*)-(−)-tetradec-5-en-4-olide is the sex pheromone of Japanese beetle and is of great industrial value as the new substituent of pesticides. Because any trace of contaminating (*S*)-enantiomer strongly inhibits the pheromone activity, the synthesis of highly enantiomerically enriched form had been secured. Although approaches based on the lipase-catalyzed lactonization toward (*R*)-tetradec-5-yn-4-

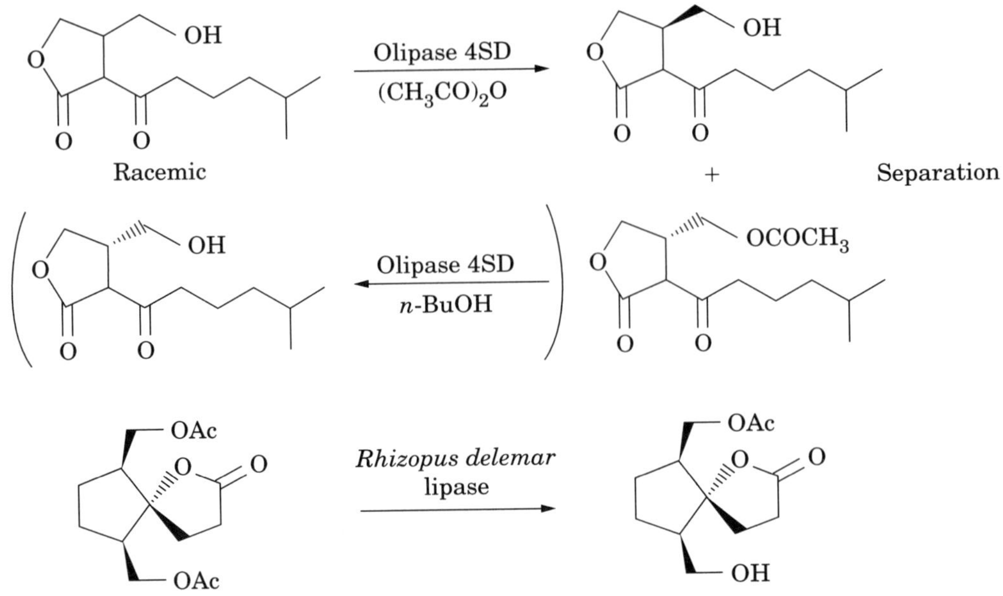

Scheme 15. Lipase-catalyzed kinetic resolution by the action on primary alcohol moiety.

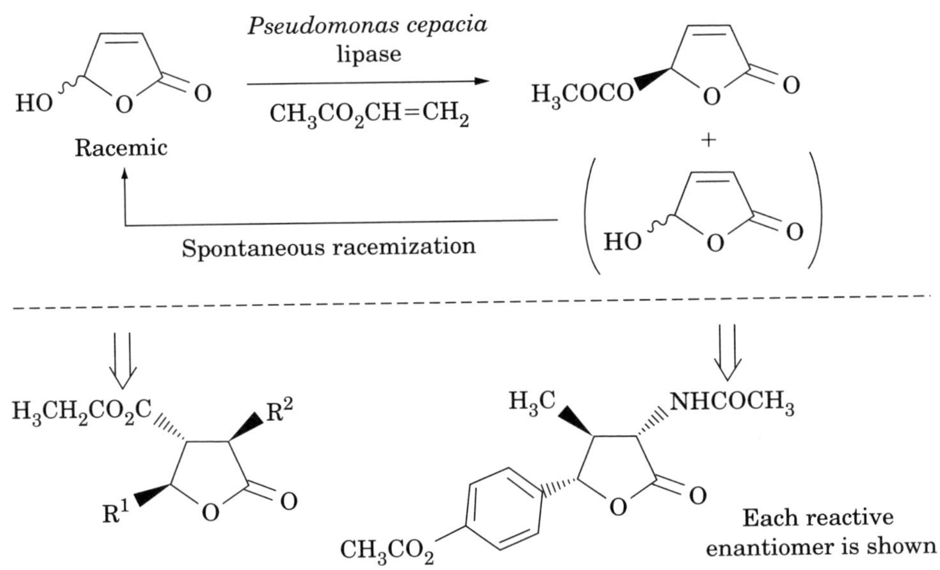

$R^1 = CH_3, C_5H_{11}$
$R^2 = H, CH_3$

Scheme 16. Kinetic resolution of lactones by the action of enzymes on other functionalities.

olide, a direct precursor for the pheromone, have been reported, the enantioselectivity of the enzymatic step was not satisfactory (20,23). The problem was solved by successive two-lipase-catalyzed resolution, as illustrated in Scheme 19 (88). Treatment of racemic alkynyl alcohol with *Pseudomonas cepacia* lipase (Amano PS) and succinic anhydride afforded the desired (*R*)-isomer of 94% ee as a hemisuccinate. The further enhancement of the enantiomeric excess was achieved by the hydrolysis of isolated product with *Candida cylindracea* lipase (Meito OF). The concomitant lactonization proceeded, thus the (*R*)-lactone was obtained in the state of more than 99% ee. Semihydrogenation afforded the pheromone of nearly pure enantiomer.

D-Pantolactone has a great importance as the starting material in medicinal and nutritional industries. So far, various approaches were devoted to develop efficient enzyme-mediated preparations. Enzymes that can hydrolyze its racemate in an enantioselective fashion were screened, and a lactonohydrolase from *Fusarium oxysporum* was discovered (89). Immobilized cells of this microorganism worked well, even at a surprisingly high concentration of substrate (350 g/L) and D-pantolactone was obtained in 90–97% ee (Scheme 20). The immobilized mi-

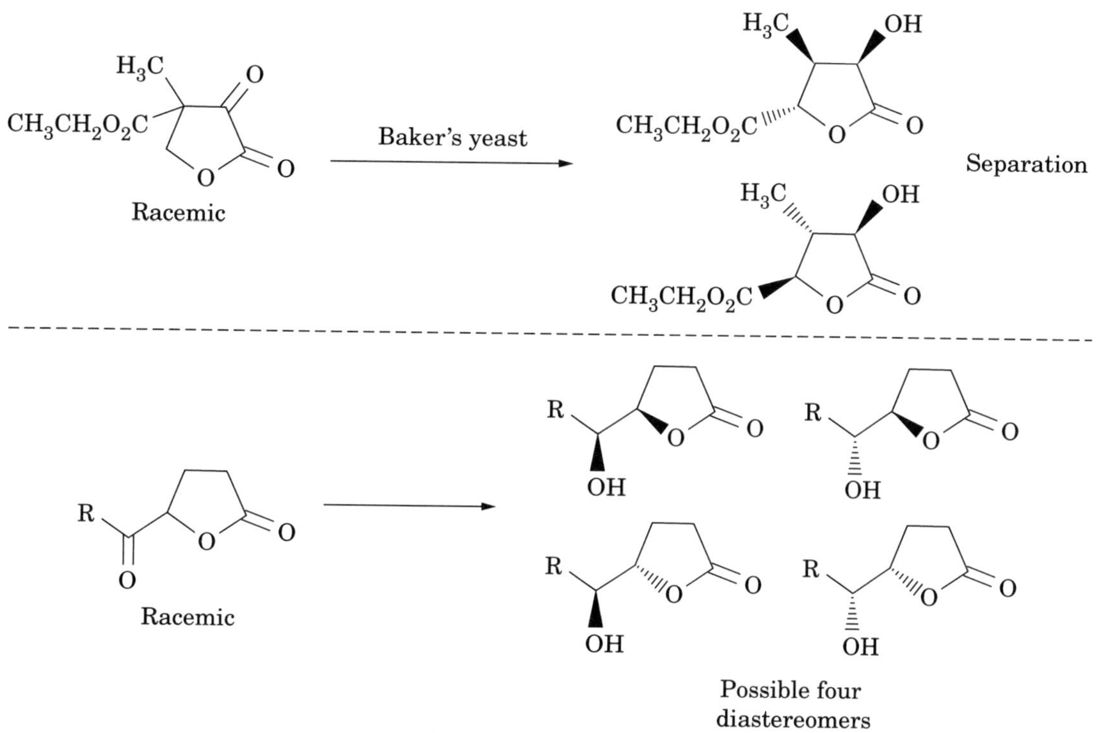

Scheme 17. Enantioface-selective reduction and the separation of resulting diastereomers.

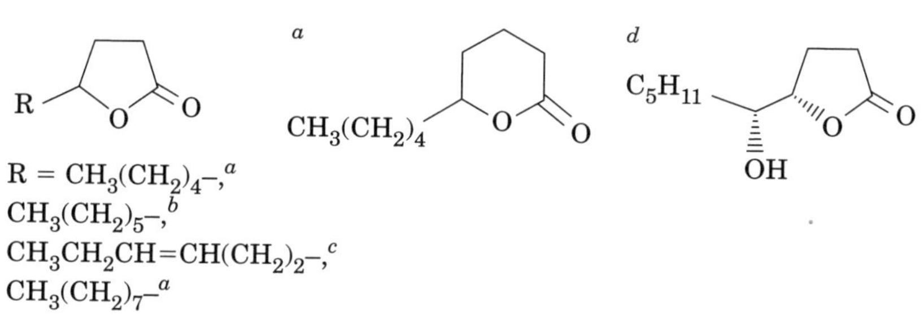

Figure 10. Enantiomerically enriched lactones as the microbial metabolites from lipids. a, Ref. 83; b, Ref. 86; c, Ref. 84; d, Ref. 85.

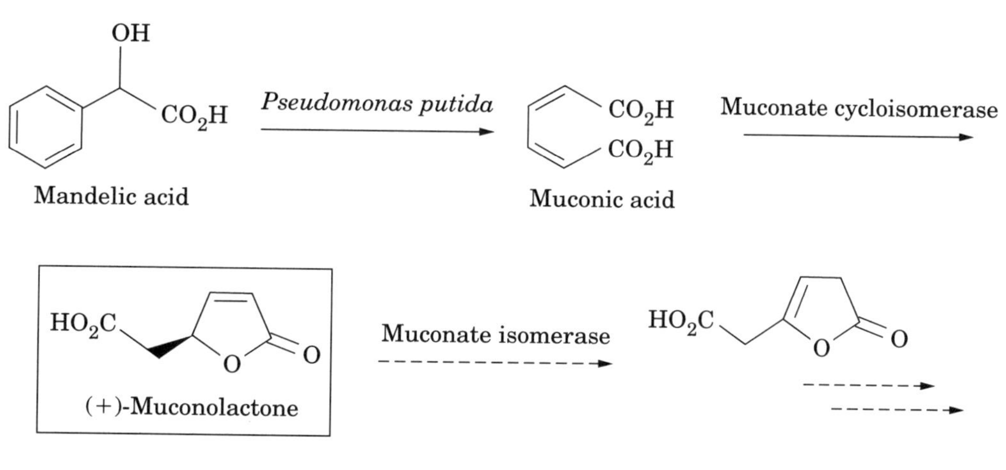

Scheme 18. Production of muconolactone from mandelic acid by *Pseudomonas putida*.

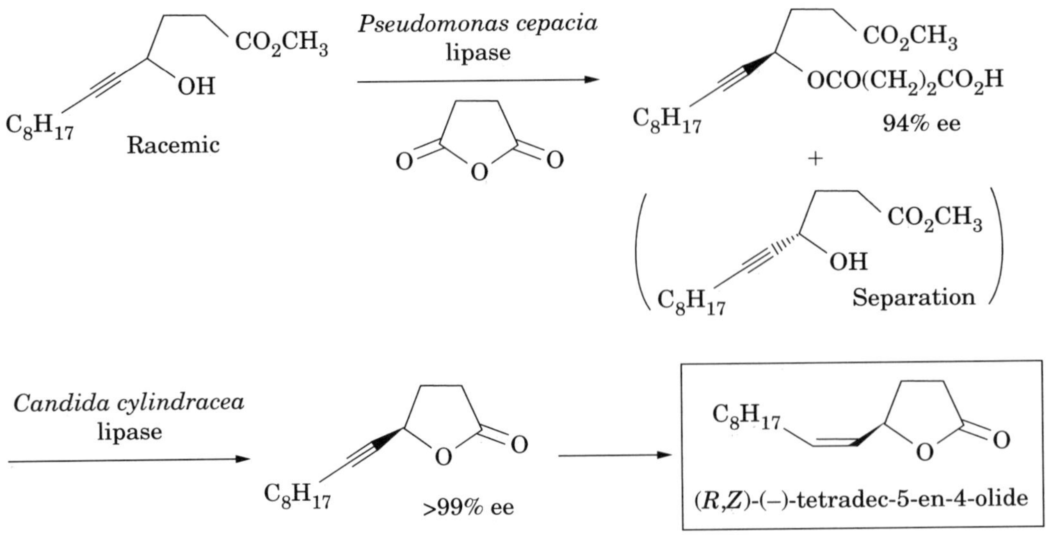

Scheme 19. (*R*,*Z*)-(−)-tetradec-5-en-4-olide, the Japanese beetle pheromone.

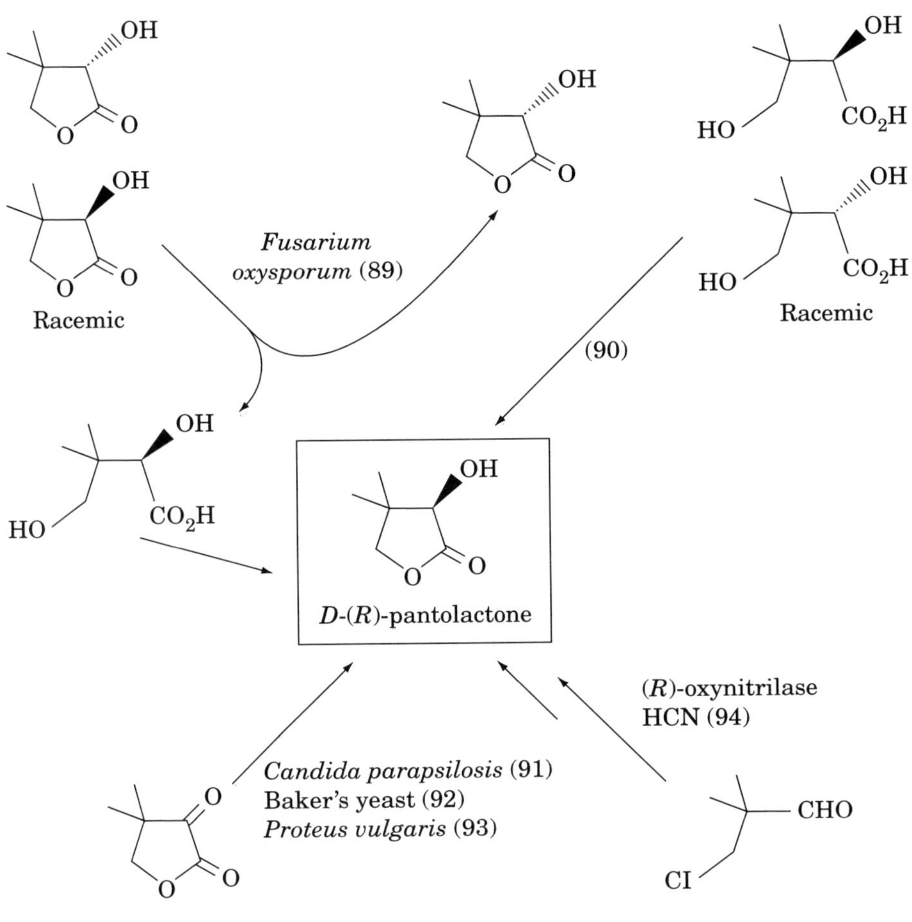

Scheme 20. Biocalytic approach to D-pantolactone.

croorganisms survived after the repetition of reaction for more than 180 days. The reverse reaction, a selective lactonization of D-pantoic acid to D-pantolactone in the equimolar mixture with L-pantoic acid worked well by the same enzyme (90).

Syntheses of D-pantolactone from ketopantolactone by microbial reduction have also been reported. After screening of microorganisms, *Candida parapsilosis* was selected as the most effective strain in terms of enantioselectivity (91). On the other hand, in the use of conventional baker's yeast, β-cyclodextrin as an additive was reported to enhance the enantioselectivity of the reduction (92). Asymmetric synthesis from the ketone precursor is indeed attractive to avoid the recycle of the unwanted isomer via racemization. An interesting approach toward an enzyme-based bioreactor with a reductase from *Proteus vulgaris* and *Proteus mirabilis* has been reported (93). As this enzyme works with very wide range of synthetic electron mediators, all of the electrochemical cells (cathodes), formate, and hydrogen are available as electron donors.

The third approach is the use of (*R*)-oxynitrilase (94), which is different from the other approaches from the standpoint of the construction of the carbon skeleton of the target molecule. There is the advantage that one carbon is directly introduced as the cyano group with stereoselective C—C bond formation from an easily accessible precursor, hydroxypivalaldehyde derivative.

CONCLUSIONS

As discussed, there are versatile methods for preparing lactones by biocatalytic means. The choice depends on (*1*) the nature of the substrate, such as solubility in water and organic solvent, and volatility; (*2*) the ease of purification of the products from byproducts, unreacted starting materials, and reagents; (*3*) the accessibility of precursors, as well as the enzymes and microorganisms; and (*4*) the scale of the reaction to be performed.

ACKNOWLEDGMENT

The authors thank Professor Yasuhiro Yamada of Department of Biotechnology, Osaka University for his suggestion of preparing this manuscript as well as kind discussion.

BIBLIOGRAPHY

1. S.C. Stinton, *Chem. Eng. News* **72**, 38–72 (1994); **73**, 44–74 (1995); **75**, 38–70 (1997); K. Mori, *Chem. Commun.* 1153–1158 (1997).
2. S.C. Stinton, *Chem. Eng. News* **76**, 15–24 (1998); M. McCoy, *Chem. Eng. News* **76**, 13–19 (1998).
3. C.-H. Wong and G.M. Whitesides, *Enzymes in Synthetic Organic Chemistry*, Pergamon, New York, 1994; K. Drauz and H. Waldmann, *Enzyme Catalysis in Organic Synthesis*, VCH, Weinheim, 1995; K. Faber, *Biotransformations in Organic Chemistry*, 3rd ed. Springer, Berlin, 1997; J.-M. Fang and C.-H. Wong, *Synlett* 393–402 (1994); P. Besse and H. Veschambre, *Tetrahedron* **50**, 8885–8927 (1994); R.O. Duthaler, *Tetrahedron* **50**, 1539–1650 (1994); F. Theil, *Chem. Rev.* **95**, 2203–2227 (1995); C.-H. Wong, R.L. Halcomb, Y. Ichikawa, and T. Kajimoto, *Angew. Chem. Int. Ed. Engl.* **34**, 412–432 (1995); E. Schoffers, A. Golebiowski, and C.R. Johnson, *Tetrahedron* **52**, 3769–3826 (1996); T. Sugai, T. Yamazaki, M. Yokoyama, and H. Ohta, *Biosci. Biotechnol. Biochem.* **61**, 1419–1427 (1997); H. Ohta, *Bull. Chem. Soc. Jpn.* **70**, 2895–2911 (1997); R.D. Schmid and R. Verger, *Angew. Chem. Int. Ed. Engl.* **37**, 1608–1633 (1998).
4. L. Blanco, E. Guibé-Jampel, and G. Rousseau, *Tetrahedron Lett.* **29**, 1915–1918 (1988).
5. A.L. Gutman, K. Zuobi, and E. Guibé-Jampel, *Tetrahedron Lett.* **31**, 2037–2038 (1990).
6. R. Fellous, L. Lizzani-Cuvelier, M.A. Loiseau, and E. Sassy, *Tetrahedron: Asymmetry* **5**, 343–346 (1994).
7. G. Fantin, M. Fogagnolo, A. Medici, P. Pedrini, S. Poli, F. Gardini, and E. Guerzoni, *Tetrahedron: Asymmetry* **3**, 107–114 (1992).
8. E. Guibé-Jampel, G. Rousseau, and L. Blanco, *Tetrahedron Lett.* **30**, 67–68 (1989).
9. Y. Koichi, K. Suginaka, and Y. Yamamoto, *J. Chem. Soc., Perkin Trans. 1*, 1645–1646 (1995).
10. T. Ito, M. Shimizu, and T. Fujisawa, *Tetrahedron* **54**, 5523–5530 (1998).
11. W. Adam, P. Groer, and C.R. Saha-Möller, *Tetrahedron: Asymmetry* **8**, 833–836 (1997).
12. E. Fouque and G. Rousseau, *Synthesis*, 661–666 (1989).
13. H. Ikeda, E. Sato, T. Sugai, and H. Ohta, *Tetrahedron* **52**, 8113–8122 (1996).
14. K. Hamada, T. Akeboshi, H. Ikeda, T. Sugai, and H. Ohta, *Synlett*, 983–985 (1997).
15. M. Huffer and P. Schreier, *Tetrahedron: Asymmetry* **2**, 1157–1164 (1991).
16. I.J. Jakovac and J.B. Jones, *J. Org. Chem.* **44**, 2165–2168 (1979).
17. K. Mori, H. Mori, and T. Sugai, *Tetrahedron* **41**, 919–925 (1985).
18. A.L. Gutman, K. Zuobi, and A. Boltansky, *Tetrahedron Lett.* **28**, 3861–3864 (1987).
19. A.L. Gutman, K. Zuobi, and T. Bravdo, *J. Org. Chem.* **55**, 3546–3552 (1990).
20. T. Sugai, S. Ohsawa, H. Yamada, and H. Ohta, *Synthesis*, 1112–1114 (1990).
21. T. Sugai, H. Noguchi, and H. Ohta, *Biosci. Biotechnol. Biochem.* **56**, 122–126 (1992).
22. M.K. Sibi and J.A. Gaboury, *Tetrahedron Lett.* **33**, 5681–5684 (1992).
23. S.K. Taylor, R.F. Atkinson, E.P. Almli, M.D. Carr, T.J. van Huis, and M.R. Whittaker, *Tetrahedron: Asymmetry* **6**, 157–164 (1995).
24. H. Uyama, K. Takeya, and S. Kobayashi, *Bull. Chem. Soc. Jpn.* **68**, 56–61 (1995).
25. A.L. Gutman, D. Oren, A. Boltanski, and T. Bravdo, *Tetrahedron Lett.* **28**, 5367–5368 (1987).
26. B. Henkel, A. Kunath, and H. Schick, *Liebigs Ann. Chem.*, 809–811 (1992).
27. C. Bonini, P. Pucci, R. Racioppi, and L. Viggiani, *Tetrahedron: Asymmetry* **3**, 29–32 (1992).
28. H. Yamada, T. Sugai, H. Ohta, and S. Yoshikawa, *Agric. Biol. Chem.* **54**, 1579–1580 (1990).
29. A. Makita, T. Nihira, and Y. Yamada, *Tetrahedron Lett.* **28**, 805–808 (1987).
30. H. Yamada, S. Ohsawa, T. Sugai, and H. Ohta, *Chem. Lett.*, 1775–1776 (1989).
31. T. Sugai, O. Katoh, and H. Ohta, *Tetrahedron* **51**, 11987–11988 (1995).
32. M. Lobell and M.P. Schneider, *Tetrahedron: Asymmetry* **4**, 1027–1030 (1993).
33. Z.-W. Guo and C.J. Sih, *J. Am. Chem. Soc.* **110**, 1999–2001 (1988).
34. K. Mori and H. Tomioka, *Liebigs Ann. Chem.*, 1011–1017 (1992).
35. N. Mochizuki, H. Yamada, T. Sugai, and H. Ohta, *BioMed. Chem.* **1**, 71–75 (1993).
36. A.S. Pawar, S. Sankaranarayanan, and S. Chattopadhyay, *Tetrahedron: Asymmetry* **6**, 2219–2226 (1995).

37. T.K. Ngooi, A. Scilimati, Z.-W. Guo, and C.J. Sih, *J. Org. Chem.* **54**, 911–914 (1989).
38. A. Manzocchi, R. Casati, A. Fiecchi, and E. Santaniello, *J. Chem. Soc., Perkin Trans. 1*, 2753–2757 (1987).
39. F. Moriuchi, H. Muroi, and H. Aibe, *Chem. Lett.*, 1141–1144 (1987).
40. M. Utaka, H. Watabu, and A. Takeda, *J. Org. Chem.* **52**, 4363–4368 (1987).
41. Y. Naoshima, H. Hasegawa, and T. Saeki, *Agric. Biol. Chem.* **51**, 3417–3419 (1987).
42. S. Tsuboi, J. Sakamoto, T. Kawano, M. Utaka, and A. Takeda, *J. Org. Chem.* **56**, 7177–7179 (1991).
43. P. Mazur and K. Nakanishi, *J. Org. Chem.* **57**, 1047–1051 (1992).
44. T. Kitayama, *Tetrahedron: Asymmetry* **8**, 3765–3774 (1997).
45. K. Takabe, H. Hiyoshi, H. Sawada, M. Tanaka, A. Miyazaki, T. Yamada, T. Katagiri, and H. Yoda, *Tetrahedron: Asymmetry* **3**, 1399–1400 (1992).
46. J.B. Jones, C.J. Sih, and D. Perlman, *Applications of Biochemical Systems in Organic Chemistry*, Wiley, New York, 1976, pp. 236–401.
47. J.B. Jones and K.P. Lok, *Can. J. Chem.* **57**, 1025–1032 (1979).
48. I.J. Jakovac, H.B. Goodbrand, K.P. Lok, and J.B. Jones, *J. Am. Chem. Soc.* **104**, 4659–4665 (1982).
49. J.B. Jones, M.A.W. Finch, and I.J. Jakovac, *Can. J. Chem.* **60**, 2007–2011 (1982).
50. G.S.Y. Ng, L.-C. Yuan, I.J. Jakovac, and J.B. Jones, *Tetrahedron* **40**, 1235–1243 (1984).
51. K. Mori, M. Amaike, and H. Watanabe, *Liebigs Ann. Chem.*, 1287–1294 (1993).
52. J.B. Jones and I.J. Jakovac, *Can. J. Chem.* **60**, 19–28 (1982).
53. J.B. Jones, *Tetrahedron* **42**, 3351–3403 (1986).
54. B. Mohar, A. Stimac, and J. Kobe, *Tetrahedron: Asymmetry* **5**, 863–878 (1994).
55. H. Ohta, H. Tetsukawa, and N. Noto, *J. Org. Chem.* **47**, 2400–2404 (1982).
56. H. Luna, K. Prasad, and O. Repic, *Tetrahedron: Asymmetry* **5**, 303–306 (1994).
57. F.-C. Huang, L.F.H. Lee, R.S.D. Mittal, P.R. Ravikumar, J.A. Chan, C.J. Sih, E. Caspi, and C.R. Eck, *J. Am. Chem. Soc.* **97**, 4144–4145 (1975).
58. S.K. Taylor, N.H. Chmiel, L.J. Simons, and J.R. Vyvyan, *J. Org. Chem.* **61**, 9084–9085 (1996).
59. Y. Ohtsuka, O. Katoh, T. Sugai, and H. Ohta, *Bull. Chem. Soc. Jpn.* **70**, 483–491 (1997).
60. C.-S. Chen, Y. Fujimoto, and C.J. Sih, *J. Am. Chem. Soc.* **103**, 3580–3582 (1981); D.W. Brooks and J.T. Palmer, *Tetrahedron Lett.* **24**, 3059–3062 (1983); S. Kobayashi, K. Kamiyama, T. Iimori, and M. Ohno, *Tetrahedron Lett.* **25**, 2557–2560 (1984); Y. Yamamoto, M. Iwasa, S. Sawada, and J. Oda, *Agric. Biol. Chem.* **54**, 3269–3274 (1990); M. Lautens, S. Ma, and A. Yee, *Tetrahedron Lett.* **36**, 4184–4188 (1995); H. Kaku, M. Tanaka, Y. Norimine, Y. Miyashita, H. Suemune, and K. Sakai, *Tetrahedron: Asymmetry* **8**, 195–201 (1997), L.-M. Zhu and M.C. Tedford, *Tetrahedron* **46**, 6587–6601 (1990); C. Tamm, *Pure Appl. Chem.* **64**, 1187–1191 (1992).
61. N.A. Donoghue, D.B. Norris, and P.W. Trudgill, *Eur. J. Biochem.* **63**, 175–192 (1976); J.M. Schwab, and W.-B. Li, *J. Am. Chem. Soc.* **105**, 4800–4808 (1983); C.T. Walsh and Y.-C.J. Chen, *Angew. Chem. Int. Ed. Engl.* **27**, 333–343 (1988); S.M. Roberts and P.W.H. Wan, *J. Mol. Catal. B: Enz.* **4**, 111–136 (1998).
62. D.R. Kelly, C.J. Knowles, J.G. Mahdi, M.A. Wright, I.N. Taylor, S.M. Roberts, P.W.H. Wan, G. Grogan, S. Pedragosa-Moreau, and A.J. Willetts, *Chem. Commun.*, 2333–2334 (1996); D.R. Kelly, *Tetrahedron: Asymmetry* **7**, 1149–1152 (1996).
63. G. Ottolina, G. Carrea, S. Colonna, and A. Rückemann, *Tetrahedron: Asymmetry* **7**, 1123–1136 (1996).
64. M.J. Taschner and D.J. Black, *J. Am. Chem. Soc.* **110**, 6892–6893 (1988).
65. M.J. Taschner, D.J. Black, and Q.-Z. Chen, *Tetrahedron: Asymmetry* **4**, 1387–1390 (1993).
66. S. Rissom, U. Schwarz-Linek, M. Vogel, V.I. Tishkov, and U. Kragl, *Tetrahedron: Asymmetry* **8**, 2523–2526 (1997).
67. B. Adger, M.T. Bes, G. Grogan, R. McCague, S. Pedragosa-Moreau, S.M. Roberts, R. Villa, P.W.H. Wan, and A.J. Willetts, *J. Chem. Soc., Chem. Commun.*, 1563–1564 (1995).
68. V. Alphand, A. Archelas, and R. Furstoss, *J. Org. Chem.* **55**, 347–350 (1990).
69. M.T. Bes, R. Villa, S.M. Roberts, P.W.H. Wan, and A. Willetts, *J. Mol. Catal. B: Enz.* **1**, 127–134 (1996).
70. J.D. Stewart, K.W. Reed, C.A. Martinez, J. Zhu, G. Chen, and M.M. Kayser, *J. Am. Chem. Soc.* **120**, 3541–3548 (1998).
71. V.A. Alphand and R. Furstoss, *J. Org. Chem.* **57**, 1306–1309 (1992).
72. F. Petit and R. Furstoss, *Tetrahedron: Asymmetry* **4**, 1341–1352 (1993).
73. K. Königsberger and H. Griengl, *BioMed. Chem.* **2**, 595–604 (1994).
74. S. Colonna, N. Gaggero, P. Pasta, and G. Ottolina, *Chem. Commun.*, 2303–2308 (1996); H.L. Holland, L.J. Allen, M.J. Chernishenko, M. Diez, A. Kohl, J. Ozog, and J.-X. Gu, *J. Mol. Catal. B: Enz.* **3**, 303–309 (1997).
75. K. Mizuno, S. Sakuda, T. Nihira, and Y. Yamada, *Tetrahedron* **50**, 10849–10858 (1994).
76. T. Fujita, M. Tanaka, Y. Norimine, H. Suemune, and K. Sakai, *J. Org. Chem.* **62**, 3824–3830 (1997).
77. H. van der Deen, A.D. Cuiper, R.P. Hof, A. van Oeveren, B.L. Feringa, and R.M. Kellogg, *J. Am. Chem. Soc.* **118**, 3801–3803 (1996); M. van den Heuvel, A.D. Cuiper, H. van der Deen, R.M. Kellogg, and B.L. Feringa, *Tetrahedron Lett.* **38**, 1655–1658 (1997).
78. S. Drioli, F. Felluga, C. Forzato, P. Nitti, and G. Pitacco, *J. Chem. Soc., Chem. Commun.* 1289–1290 (1996); S. Drioli, F. Felluga, T. Fortato, P. Nitti, G. Pitacco, and E. Valentin, *J. Mol. Catal. B: Enz.* **3**, 203–207 (1997), S. Drioli, F. Felluga, C. Forzato, P. Nitti, G. Pitacco, and E. Valentin, *J. Org. Chem.* **63**, 2385–2388 (1998).
79. H. Kapeller, W.G. Jary, W. Hayden, and H. Griengl, *Tetrahedron: Asymmetry* **8**, 245–251 (1997).
80. G. Fronza, C. Fuganti, P. Grasselli, L. Malpezzi, and A. Mele, *J. Org. Chem.* **59**, 3487–3489 (1994).
81. G. Fronza, C. Fuganti, P. Grasselli, R. Pulido-Fernandez, S. Servi, A. Tagliani, and M. Terreni, *Tetrahedron* **47**, 9247–9252 (1991).
82. G. Fantin, M. Fogagnolo, P. Giovannini, A. Medici, E. Pagnotta, P. Pedrini, and A. Trincone, *Tetrahedron: Asymmetry* **5**, 1631–1634 (1994).
83. R. Cardillo, G. Fronza, C. Fuganti, P. Grasselli, A. Mele, D. Pizzi, G. Allegrone, M. Barbeni, and A. Pisciotta, *J. Org. Chem.* **56**, 5237–5241 (1991).
84. C. Fuganti, P. Grasselli, G. Zucchi, S. Allegrone, and A. Barbeni, *BioMed. Chem. Lett.* **3**, 2777–2780 (1993).

85. W. Albrecht and R. Tressl, *Tetrahedron: Asymmetry* **4**, 1391–1396 (1993).
86. S.-J. Lin, S.-L. Lee, and C.-C. Chou, *J. Ferment. Bioeng.* **82**, 42–45 (1996).
87. D.W. Ribbons and A.G. Sutherland, *Tetrahedron* **50**, 3587–3594 (1994).
88. E. Fukusaki, S. Senda, Y. Nakazono, and T. Omata, *Tetrahedron* **47**, 6223–6230 (1991); E. Fukusaki, S. Senda, Y. Nakazono, and T. Omata, *Biosci. Biotechnol. Biochem.* **56**, 1160–1161 (1992); E. Fukusaki and S. Satoda, *J. Mol. Catal. B:Enz.* **2**, 257–269 (1997).
89. S. Shimizu, M. Kataoka, K. Shimizu, M. Hirakata, K. Sakamoto, and H. Yamda, *Eur. J. Biochem.* **209**, 383–390 (1992); M. Kataoka, K. Shimizu, K. Sakamoto, H. Yamada, and S. Shimizu, *Appl. Microbiol. Biotechnol.* **44**, 333–338 (1995).
90. M. Kataoka, M. Hirakata, K. Sakamoto, H. Yamada, and S. Shimizu, *Enzyme Microb. Technol.* **19**, 307–310 (1996).
91. H. Hata, S. Shimizu, S. Hattori, and H. Yamada, *J. Org. Chem.* **55**, 4377–4380 (1990).
92. K. Nakamura, S. Kondo, and A. Ohno, *BioMed. Chem.* **2**, 433–438 (1994).
93. R. Eck and H. Simon, *Tetrahedron: Asymmetry* **5**, 1419–1422 (1994).
94. F. Effenberger, J. Eichhorn, and J. Roos, *Tetrahedron: Asymmetry* **6**, 271–282 (1995); F. Effenberger, *Angew. Chem. Int. Ed. Engl.* **33**, 1555–1556 (1994); H. Griengl, A. Hickel, D.V. Johnson, C. Kratky, M. Schmidt, and H. Scwab, *Chem. Commun.*, 1933–1940 (1997).

See also BIOCATALYSIS DATABASES.

LACTONOHYDROLASE

SAKAYU SHIMIZU
MICHIHIKO KATAOKA
Kyoto University
Kyoto, Japan

KEY WORDS

Aldonate lactone
Aldonolactonase
Esterase
Gluconolactonase
Lactonase
Lipase
Optical resolution
Pantoic acid
Pantolactone
Stereospecific synthesis

OUTLINE

INTRODUCTION

The enzymes catalyzing the reversible or irreversible hydrolysis of carboxylic esters to the corresponding carboxylic acids and alcohols (EC 3.1.1.) play an important role in the biosynthesis and biodegradation of various compounds. In particular, the functions and structures of lipase (EC 3.1.1.3) and esterase (EC 3.1.1.1) have been well analyzed (1–5). Lactonohydrolases (lactonases) also catalyze the hydrolysis of carboxylic esters, but act on the intramolecular ester bonds of lactone compounds. They are involved in the synthesis and degradation of various lactone compounds in vivo. For example, gluconolactonase (aldonolactonase; EC 3.1.1.17) was suggested to participate in the formation of L-gulono-γ-lactone from L-gulonate in L-ascorbate biosynthesis (6). Although the activities of lactonohydrolases have been detected in various organisms (6–13), there has not been a detailed study on their enzyme chemistry so far.

DIVERSITY OF LACTONOHYDROLASES

Lactonohydrolases reversibly catalyze the hydrolysis of lactone compounds into hydroxy acids, that is, they mediate the interconversion between the lactone and acid forms of hydroxy carboxylic acids. Gluconolactonase (aldonolactonase; EC 3.1.1.17), L-arabinonolactonase (EC 3.1.1.15), and D-arabinonolactonase (EC 3.1.1.30) specifically hydrolyze aldonate lactones to the respective aldonic acids. The activities of these enzymes have been detected in various organisms (6–13). As to the roles of these enzymes in vivo, it has been suggested that gluconolactonase participates in the formation of L-gulono-γ-lactone from L-gulonate in L-ascorbate biosynthesis (Fig. 1) (6). L-Rhamnonolactonase (EC 3.1.1.65) and D-xylonolactonase (EC 3.1.1.68) of *Pullularia pullulans* and *Pseudomonas fragi*, respectively, are involved in the oxidative degradation of L-rhamnose and D-xylose, respectively (Fig. 1) (14,15). In addition to these enzymes, pyridoxolactonase (EC 3.1.1.27), dihydrocoumarin hydrolase (EC 3.1.1.35), limonin-D-ring-lactonase (EC 3.1.1.36) and deoxylimonate A-ring-lactonase (EC 3.1.1.46), and actinomycin lactonase (EC 3.1.1.39) have been reported to be the enzymes

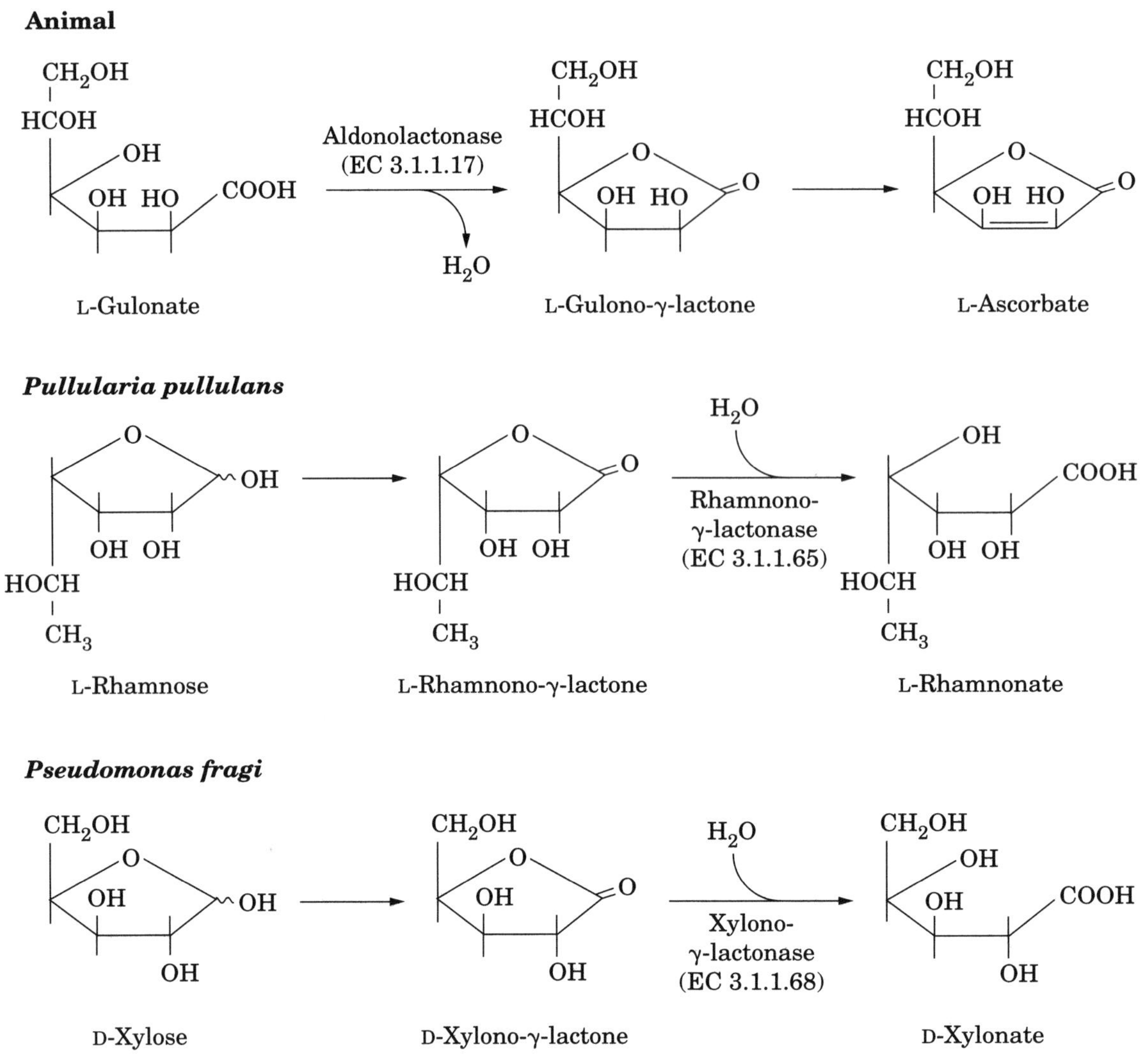

Figure 1. Lactonohydrolase reactions involved in ascorbate biosynthesis and the degradation of L-rhamnose and D-xylose.

Table 1. Properties of the Lactonohydrolases from *F. oxysporum* and *B. protophormiae*

Origin	*F. oxysporum*		*B. protophormiae*	
Native M_r	125,000		62,000	
Subunit M_r	60,000		26,500	
Number of subunits	2		2	
Substrate specificity	K_m (mM)	V_{max} (μmol/min/mg)	K_m (mM)	V_{max} (μmol/min/mg)
D-(−)-Pantolactone	120	653	—	—
L-(+)-Pantolactone	—	—	3.59	13.7
D-Galactono-γ-lactone	3.6	1,440	—	—
Dihydrocoumarin	6.3	2,800	2.56×10^3	7.74×10^4
pH stability	5.0–10		>5.0	
Optimum temperature	50 °C		45 °C	
Thermal stability	<50 °C		<50 °C	
Inhibitors	L-(+)-pantolactone, chelating reagents		Chelating reagents	

Table 2. Substrate Specificity of the Lactonohydrolase from *F. oxysporum*

	Relative hydrolyzing activity (%)	K_m (mM)
Substrate (150 mM)		
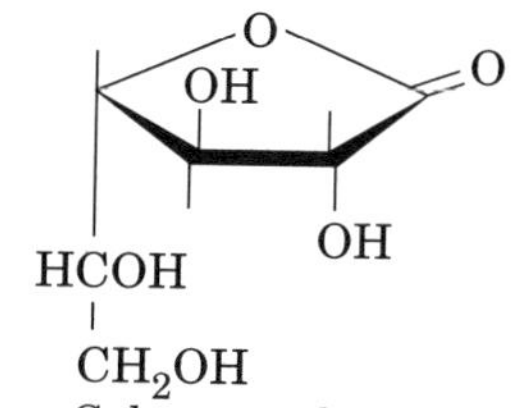 D-Galactono-γ-lactone	100	3.6
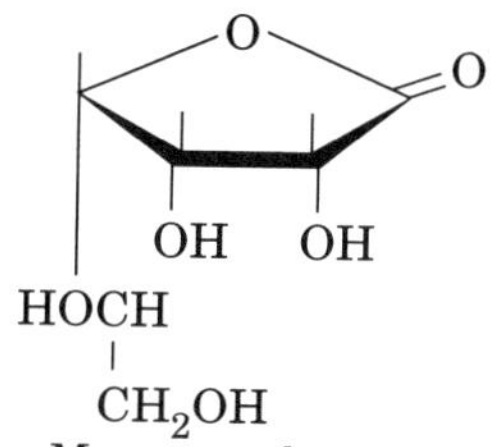 L-Mannono-γ-lactone	128	23
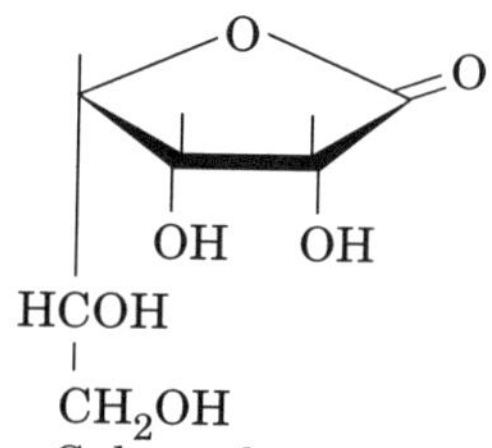 D-Gulono-γ-lactone	167	29
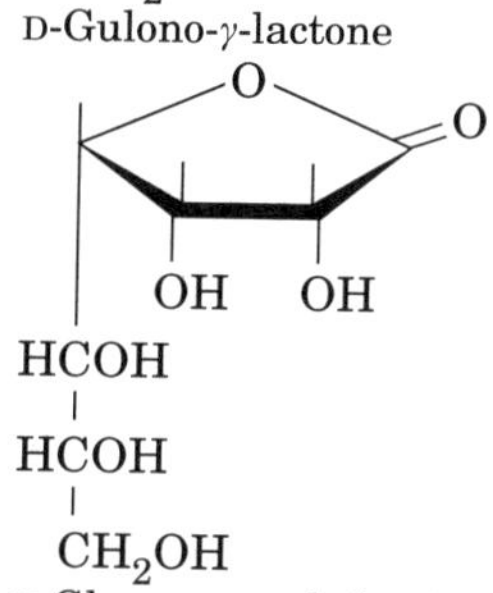 D-Glycero-D-gulo-heptono-γ-lactone	59.4	118
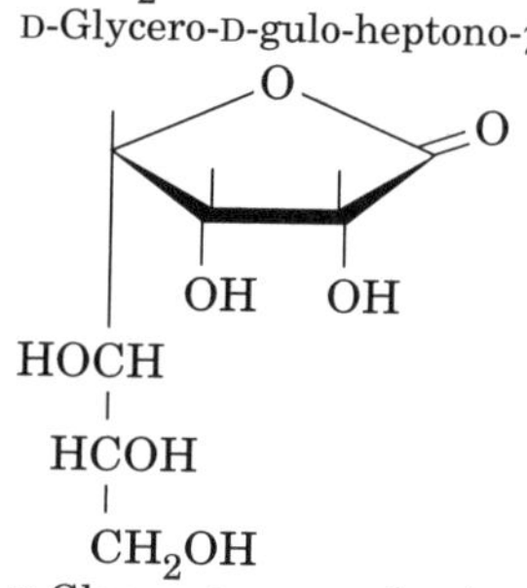 D-Glycero-L-manno-heptono-γ-lactone	132	167
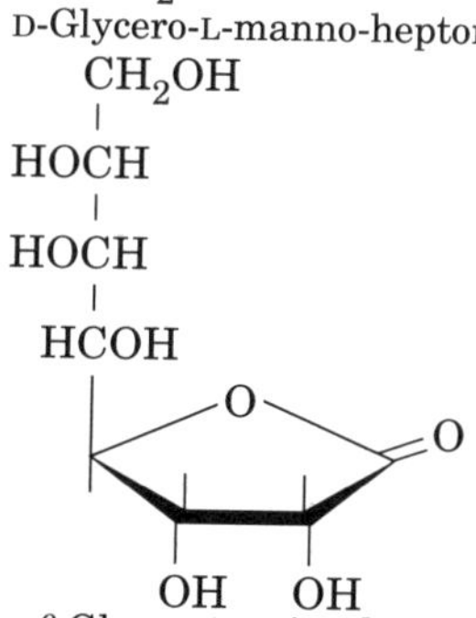 α,β-Glucooctanoic-γ-lactone	16.4	—
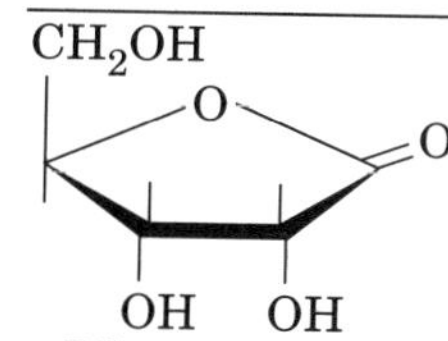 D-Ribono-γ-lactone	39.0	2.5
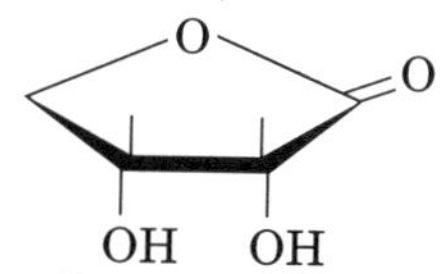 D-Erythrono-γ-lactone	117	377
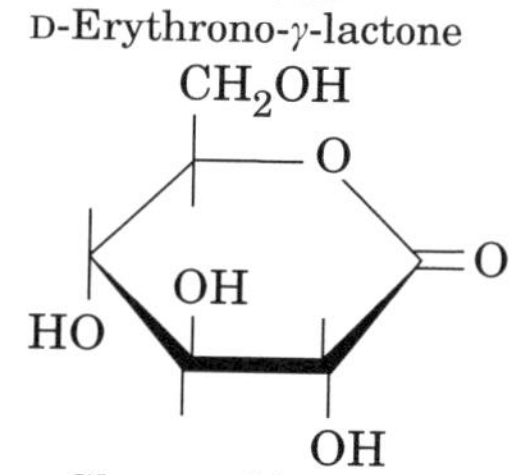 D-Glucono-δ-lactone	107	—
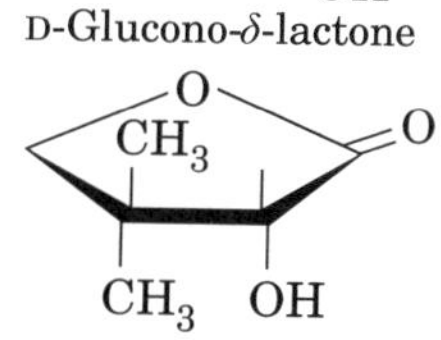 D-Pantolactone	48.3	120
Substrate (2.5 mM)		
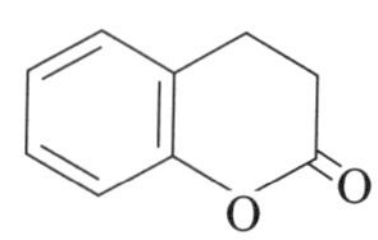 Dihydrocoumarin	100	6.3
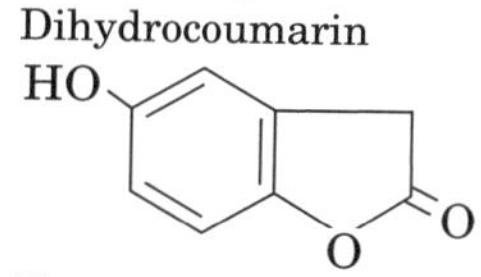 Homogentisic acid lactone	16.6	2.5
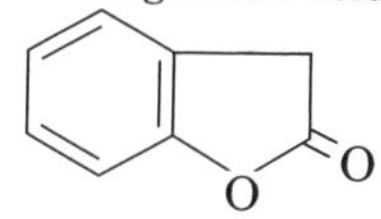 2-Coumaramone	16.4	8.7
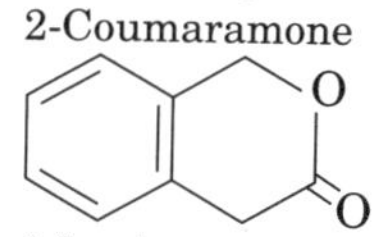 3-Isochromanone	0.58	4.4

involved in the degradation pathways for vitamin B-6 (16), aromatic compounds (17,18), limonoids (19,20), and actinomycin (21), respectively. As for molecular biochemical studies on lactonohydrolases, only the gene encoding gluconolactonase from *Zymomonas mobilis* has been cloned and sequenced (22), and there has been no report on the structure of any other lactonohydrolase.

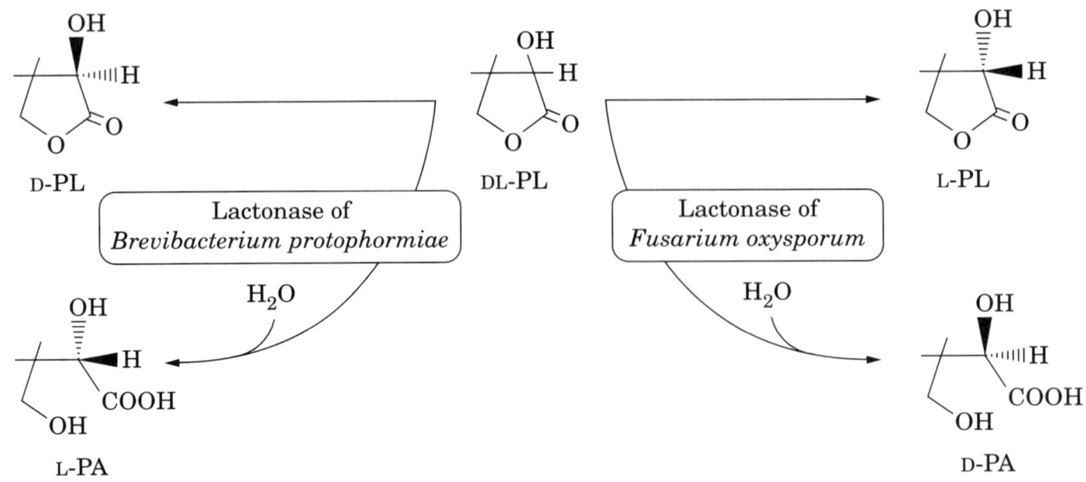

Figure 2. Novel microbial lactonohydrolases catalyzing the stereospecific hydrolysis of pantolactone. PL = pantolactone; PA = pantoic acid.

NOVEL MICROBIAL LACTONOHYDROLASES

During the course of studies on the microbial production of chiral intermediates for D-pantothenate (23–25), Shimizu et al. found that several microorganisms, such as *Fusarium, Brevibacterium*, and so on, produce a novel enzyme that catalyzes the hydrolysis of aldonate or aromatic lactones (26,27).

The enzyme isolated from cells of *Fusarium oxysporum* has a relative molecular mass of 125 kDa and a subunit molecular mass of 60 kDa. The enzyme thus appears to be a dimer consisting of identical subunits. The enzyme contains about 1 mol of calcium per subunit and 15.4% (w/w) glucose equivalent of carbohydrate. Calcium seems to be necessary for both enzyme activity and stability. The enzyme carries three kinds of N-linked high-mannose-type sugar chains at the 28th, 106th, 179th, and 277th Asn residues. The carbohydrate moiety is an essential participant in the stabilization of the enzyme. Table 1 summarizes the properties of the enzyme (26). The enzyme hydrolyzes aldonate lactones, such as D-galactono-γ-lactone, L-mannono-γ-lactone, D-gulono-γ-lactone, and D-glucono-δ-lactone, stereospecifically, but the corresponding enantiomers (i.e., L-galactono-γ-lactone, D-mannono-γ-lactone, L-gulono-γ-lactone, and L-glucono-δ-lactone, respectively) do not serve as substrates. All these substrates have downward hydroxy groups at their 2-position carbons in common, according to Haworth's projection system, (Table 2), whereas their enantiomers have upward hydroxy groups at their 2-position carbons. The enzyme can also catalyze the asymmetric hydrolysis of D-(−)-pantolactone. For every substrate, the reverse reaction (i.e., lactonization of the aldonic acid) was observed. Aromatic lactones, such as dihydrocoumarin, homogentisic acid lactone, 2-coumaranone, and 3-isochromanone, were also effective substrates. They are good substrates for dihydrocoumarin hydrolase (EC 3.1.1.35) and have no hydroxy group at their 2-position carbons (17). With regard to the reverse reactions, no detectable lactone compounds were formed.

A similar kind of lactonohydrolase was found in *Brevibacterium protophormiae* (Table 1) (27). This bacterial enzyme only hydrolyzes L-(+)-pantolactone; D-(−)-pantolactone does not act as a substrate. Various kinds of aromatic lactones, such as dihydrocoumarin, homogentisic acid lactone and so on, are also hydrolyzed, but aldonate lactones are not hydrolyzed. The relative molecular mass of the native enzyme is 62 kDa, and its subunit molecular mass is 26.5 kDa.

APPLICATION OF MICROBIAL LACTONOHYDROLASE TO THE OPTICAL RESOLUTION OF RACEMIC PANTOLACTONE

Calcium D-(+)-pantothenate is mainly used as an additive for animal feeds and for various pharmaceutical products. At present, the commercial production of D-(+)-pantothenate depends exclusively on chemical synthesis. The chemical synthesis process involves reactions yielding racemic pantolactone from isobutyraldehyde, formaldehyde, and cyanide; optical resolution of the racemic pantolactone to D-(−)-pantolactone; and condensation of the D-(−)-pantolactone with β-alanine. Drawbacks of this chemical process are the troublesome resolution of the racemic pantolactone, for which the use of an expensive alkaloid as a resolving agent is unavoidable. Recently, Kataoka et al. showed that the microbial lactonohydrolase could be used as a catalyst for the resolution of racemic pantolactone (28–30).

The principle of the optical resolution of racemic pantolactone is shown in Figure 2. If racemic pantolactone is used as a substrate for the hydrolysis by the stereospecific lactonohydrolase, only the D-(−)- or L-(+)-pantolactone might be converted to the D-(−)- or L-(+)-pantoic acid, and the L-(+)- or D-(−)-enantiomer might remain intact, respectively. Consequently, the racemic mixture could be resolved into D-(−)-pantoic acid and L-(+)-pantolactone, or D-(−)-pantolactone and L-(+)-pantoic acid. In the case of

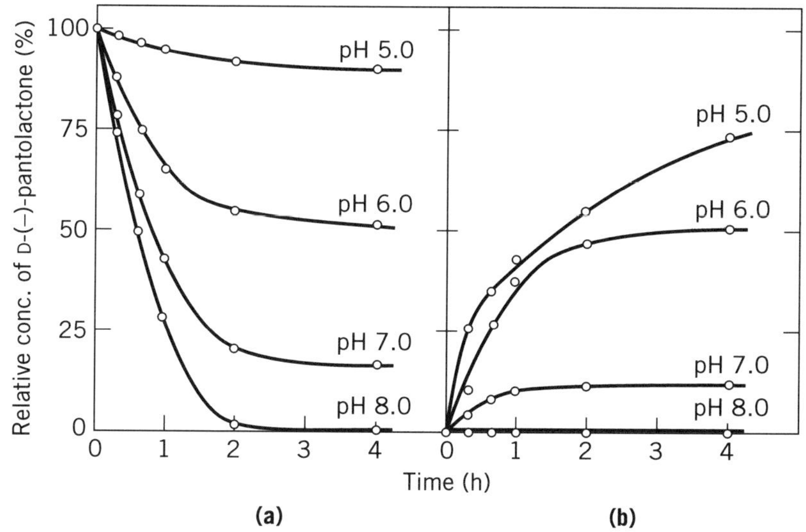

Figure 3. Effect of pH on the reaction equilibrium of the *Fusarium* lactonohydrolase. The enzymatic hydrolysis of D-(−)-pantolactone (**a**) and enzymatic lactonization of D-(−)-pantoic acid (**b**) were carried out at the pHs indicated.

L-(+)-pantolactone-specific lactonohydrolase, the optical purity of the remaining D-(−)-pantolactone might be low, except when the hydrolysis of L-(+)-pantolactone is complete. On the other hand, using the D-(−)-pantolactone-specific lactonohydrolase, D-(−)-pantoic acid of high optical purity could be constantly obtained irrespective of the hydrolysis yield. Therefore, the enzymatic resolution of racemic pantolactone with the D-(−)-pantolactone-specific lactonohydrolase was investigated (28–30).

The distribution of the D-(−)-pantolactone-specific lactonohydrolase is narrow; only filamentous fungi in three genera (i.e., *Fusarium, Gibberella*, and *Cylindrocarpon*) show high hydrolytic activity. The D-(−)-pantolactone hydrolysis reaction catalyzed by the purified *Fusarium* lactonohydrolase favored alkaline pH, and the D-(−)-pantoic acid lactonization reaction favored rather acidic pH. At pH 7.0, about 80% of the substrate, pantolactone, was present as pantoic acid, and about 75% of the pantoic acid added was present as pantolactone at pH 5.0. The pantoic acid/pantolactone ratio in the reaction mixture was nearly 1:1 at pH 6.0 (Fig. 3) (26).

Optical Resolution of Racemic Pantolactone with the Microbial Lactonohydrolase

Washed cells of *F. oxysporum* as well as the purified lactonohydrolase catalyze well the stereospecific hydrolysis of D-(−)-pantolactone. As shown in Figure 4, D-(−)-pantolactone was stereospecifically hydrolyzed to D-(−)-pantoic acid with a substrate concentration of 700 mg/mL. The formation of L-(+)-pantoic acid was barely detected. After 24 h, the total concentration of pantolactone hydrolyzed in the reaction mixture reached 322 mg/mL, with an optical purity of 96% ee. After the removal of L-(+)-pantolactone from the reaction mixture by solvent extraction, D-(−)-pantoate remaining could be easily converted to D-(−)-pantolactone by heating under acidic conditions.

Stereoselective Production of D-(−)-Pantolactone from Racemic Pantoic Acid through Microbial Stereoselective Lactonization

As described earlier, an alkaline pH favors the hydrolysis of D-(−)-pantolactone, and a rather acidic pH favors the lactonization of D-(−)-pantoate. At pH 5.0, the ratio of D-(−)-pantoate and D-(−)-pantolactone after the reaction is 1:3, whereas that at pH 7.0 is 4:1. These results strongly suggest that this stereospecific lactonization reaction might be used for the asymmetric production of D-(−)-pantolactone from racemic pantoate.

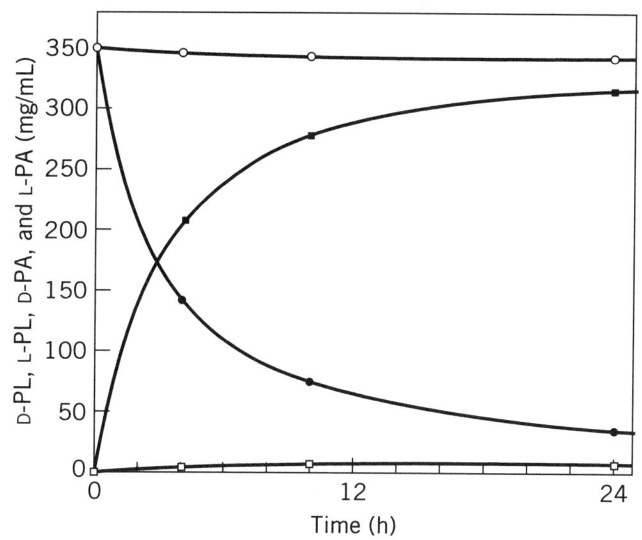

Figure 4. Optical resolution of racemic pantolactone with washed cells of *F. oxysporum*. The concentrations of pantoic acid given are those of pantolactone. ●, D-(−)-pantolactone (D-PL); ○, L-(+)-pantolactone (L-PL); ■, D-(−)-pantoic acid (D-PA); □, L-(+)-pantoic acid (L-PA).

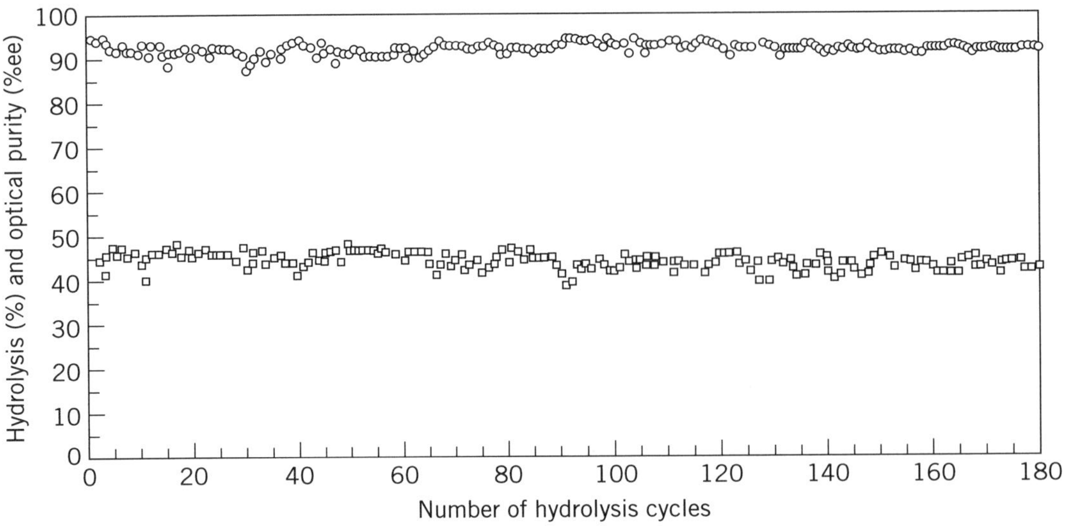

Figure 5. Stereospecific hydrolysis of pantolactone by *F. oxysporum* mycelia entrapped in a calcium alginate gel. Here, 280 L of immobilized mycelia (containing 15.2 kg of wet cells) was incubated with 350 L of an aqueous DL-pantolactone solution (350 g/L) at 30 °C for 21 h. The pH of the mixture was automatically controlled at 6.8 to 7.2 with 15 M NH_4OH. All the immobilized mycelia, after filtration from the reaction mixture, were used for the subsequent reaction. The hydrolysis reaction was carried out 180 times. ○ = optical purity of D-pantoate; □ = hydrolysis rate.

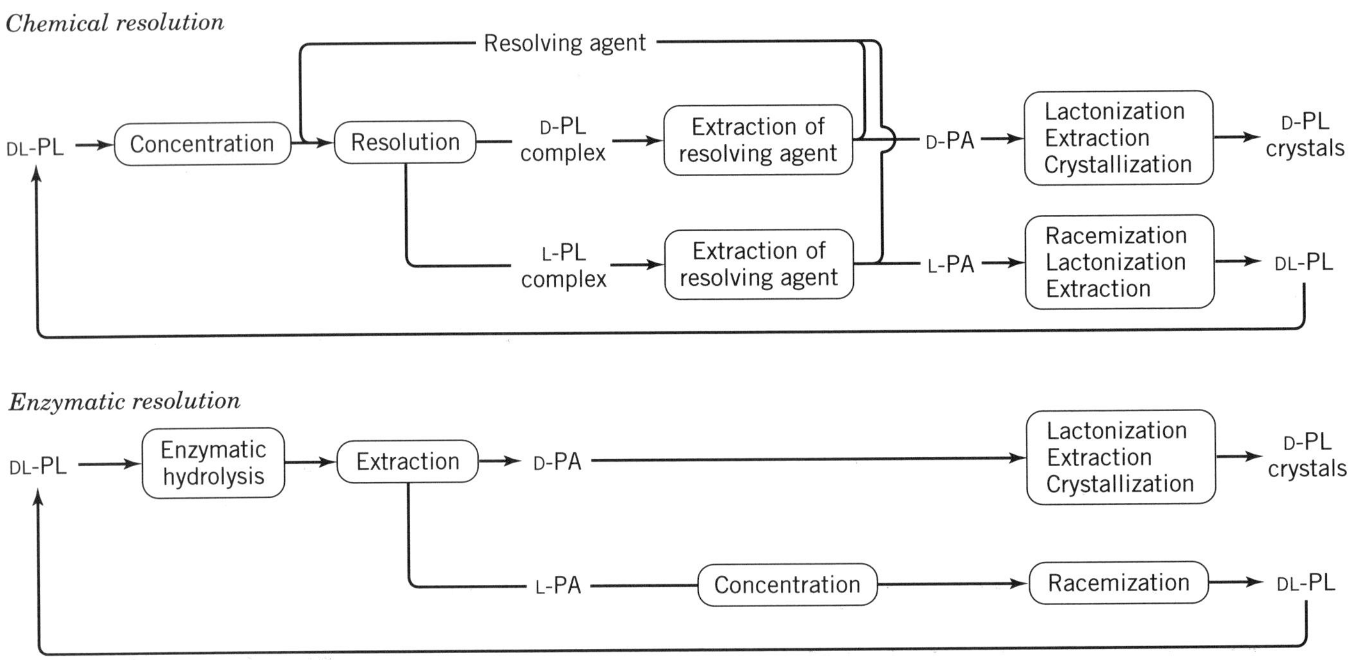

Figure 6. Comparison of the enzymatic and conventional chemical resolution processes for racemic pantolactone.

When 385-mM racemic pantoic acid (50 mg/mL as racemic pantolactone) was incubated with washed cells of *F. oxysporum* at 30 °C for 14 h at pH 4.5 to 5, 40% of the pantoate was lactonized, with an enantioselectivity of 83% ee for the D-(−)-isomer. In an organic solvent-water two-phase system for which an aqueous solution of 1,540 mM racemic pantoic acid (200 mg/mL as racemic pantolactone) containing 50 mM $CaCl_2$ and cells was incubated with ethyl acetate at 30 °C for 20 h at pH 4.5 to 5, D-(−)-pantoate in the racemic mixture was almost stoichiometrically lactonized to D-(−)-pantolactone (90% ee). This improvement of the two-phase reaction system was suggested to be the result of a shift of the reaction equilibrium during the lactonization on extraction of the reaction product, i.e., D-(−)-pantolactone moved into the organic phase.

LARGE-SCALE RESOLUTION OF RACEMIC PANTOLACTONE WITH IMMOBILIZED CELLS OF *F. OXYSPORUM*

Practical hydrolysis of the D-(−)-isomer in a racemic mixture is carried out using immobilized mycelia of *F. oxysporum* as the catalyst. A stable catalyst with high hydrolytic activity can be prepared by entrapping the fungal mycelia in a calcium alginate gel. When the immobilized mycelia were incubated in a reaction mixture containing 350 g/L of racemic pantolactone for 21 h at 30 °C under automatic pH control conditions (pH 6.8 to 7.2), 90 to 95% of the D-(−)-enantiomer was hydrolyzed (optical purity, 90 to 97% ee). After repeated reactions for 180 times (i.e., 180 days), the immobilized mycelia retained more than 90% of their initial activity (Fig. 5).

The overall process for the present enzymatic resolution is compared with the conventional chemical process in Figure 6. The enzymatic process avoids several tedious steps that are necessary in chemical resolution and thus is highly advantageous for practical purposes. There have been several reports of the application of enzymatic asymmetric hydrolysis to the optical resolution of pantolactone (31,32). In these cases, esterified substrates, such as *O*-acetyl or *O*-formyl pantolactone, and lipases were used as starting materials and catalysts, respectively. Because the lactonohydrolase of *F. oxysporum* hydrolyzes the intramolecular ester bond of pantolactone, it is not necessary to modify the substrate, pantolactone. This is one of the practical advantages of this enzyme.

BIBLIOGRAPHY

1. J.D. Schrag, Y. Li, S. Wu, and M. Cygler, *Nature* **351**, 761–764 (1991).
2. F.K. Winkler, A. D'Arcy, and W. Hunziker, *Nature* **343**, 771–774 (1990).
3. H.D. Beer, G. Wohlfahrt, J.E.G. McCarthy, D. Schomburg, and R.D. Schmid, *Protein Eng.* **9**, 507–517 (1996).
4. J. Uppenberg, N. Ohrner, M. Norin, K. Hult, G.J. Kleywegt, S. Patkar, V. Waagen, T. Anthonsen, and T.A. Jones, *Biochemistry* **34**, 16838–16851 (1995).
5. M. Cygler, J.D. Schrag, J.L. Sussman, M. Harel, I. Silman, M.K. Gentry, and B.P. Doctor, *Protein Sci.* **2**, 366–382 (1993).
6. C. Bublitz and A.L. Lehninger, *Biochim. Biophys. Acta* **47**, 288–297 (1961).
7. J. Winkelman and A.L. Lehninger, *J. Biol. Chem.* **233**, 794–799 (1958).
8. K. Yamada, *J. Biochem.* **46**, 361–372 (1959).
9. K. Yamada, S. Ishikawa, and N. Shimazono, *Biochim. Biophys. Acta* **32**, 253–255 (1959).
10. M. Kawada, H. Takiguchi, Y. Kagawa, K. Suzuki, and N. Shimazono, *J. Biochem.* **51**, 405–415 (1962).
11. F. Hucho and K. Wallenfels, *Biochim. Biophys. Acta* **276**, 176–179 (1972).
12. R. Weinberg and M. Doudoroff, *J. Biol. Chem.* **217**, 607–624 (1955).
13. N.J. Palleroni and M. Doudoroff, *J. Bacteriol.* **74**, 180–185 (1957).
14. L.U. Rigo, L.R. Marechal, M.M. Vieira, and L.A. Veiga, *Can. J. Microbiol.* **31**, 817–822 (1985).
15. J. Buchert and L. Viikari, *Appl. Microbiol. Biotechnol.* **27**, 333–336 (1988).
16. Y.-J. Jong and E.E. Snell, *J. Biol. Chem.* **261**, 15112–15114 (1986).
17. T. Kosuge and E.E. Conn, *J. Biol. Chem.* **237**, 1653–1656 (1962).
18. M. Casella, M. Grifoll, J.M. Bayona, and A.M. Solanas, *Appl. Environ. Microbiol.* **63**, 819–826 (1997).
19. S. Hasegawa, R.D. Bennett, and C.P. Verdon, *Phytochemistry* **19**, 1445–1447 (1980).
20. M.T. Merino, L. Humanes, J.M. Roldàn, J. Diez, and A. Lopez-Ruiz, *Biotechnol. Lett.* **18**, 1175–1180 (1996).
21. C.T. Hou and D. Perlman, *J. Biol. Chem.* **245**, 1289–1295 (1970).
22. V. Kanagasundaram and R. Scopes, *Biochim. Biophys. Acta* **1171**, 198–200 (1992).
23. H. Yamada and S. Shimizu, *Angew. Chem. Int. Ed. Engl.* **27**, 622–642 (1988).
24. S. Shimizu and H. Yamada, in E.J. Vandamme ed., *Biotechnology of Vitamins, Pigments, and Growth Factors*, Elsevier Science Publishers, London, 1989, pp. 199–219.
25. S. Shimizu and H. Yamada, in T.O. Baldwin, F.M. Raushel, and A.I. Scott eds., *Chemical Aspects of Enzyme Biotechnology*, Plenum Press, New York, 1990, pp. 151–163.
26. S. Shimizu, M. Kataoka, K. Shimizu, M. Hirakata, K. Sakamoto, and H. Yamada, *Eur. J. Biochem.* **209**, 383–390 (1992).
27. M. Kataoka, J. Nomura, K. Nose, and S. Shimizu, *Nippon Nogeikagaku Kaishi* **70**, 374 (1996).
28. M. Kataoka, K. Shimizu, K. Sakamoto, H. Yamada, and S. Shimizu, *Appl. Microbiol. Biotechnol.* **43**, 974–977 (1995).
29. M. Kataoka, K. Shimizu, K. Sakamoto, H. Yamada, and S. Shimizu, *Appl. Microbiol. Biotechnol.* **44**, 333–338 (1995).
30. M. Kataoka, K. Shimizu, K. Sakamoto, H. Yamada, and S. Shimizu, *Enzyme Microb. Technol.* **19**, 307–310 (1996).
31. H.S. Bevinakatti and R.V. Newadkar, *Biotechnol. Lett.* **11**, 785–788 (1989).
32. B.I. Glänzer, K. Faber, and H. Griengl, *Enzyme Microb. Technol.* **10**, 689–690 (1988).

See also BIOCATALYSIS DATABASES.

LYSINE. See AMINO ACIDS, PRODUCTION PROCESSES.

M

MALATE, D-MALATE

MAKOTO UEDA
YASUHISA ASANO
HIDEAKI YAMADA
Mitsubishi Chemical Co.
Yokohama, Japan

KEY WORDS

D-Malate
Maleate
Maleate hydratase

OUTLINE

INTRODUCTION

Optically active malate is a key intermediate in the synthesis of pharmaceuticals and agrochemicals, as well as a resolving agent for racemic compounds. L-Malate can be obtained economically from natural resources or enzymatically from fumarate with the enzyme fumarase (EC 4.2.1.2). However, the unnatural stereoisomer, D-malate, is expensive. Several methods for synthesizing D-malate have been described. The latest biological method for the production of D-malate is the stereoselective hydration of maleate through the reaction of maleate hydratase (D-malate hydrolyase, EC 4.2.1.31) derived from microorganisms. Maleate hydratase catalyzes the hydration of maleate to D-malate through *trans* addition to the double bond. In early studies on maleate hydratase, Sacks and Jensen demonstrated that an enzyme mixture from corn kernels catalyzed the conversion of maleate to malate, although they did not mention its optical purity (1). Subsequently, several workers detected the enzyme in mammalian kidneys (2) and microorganisms (3). Since 1992, the enzyme have been used in investigations on D-malate production in microbial reactions (4–9). In this chapter, D-malate production involving microbial maleate hydratase and the characteristics of the enzyme are described. Many applications of D-malate are found in the synthesis of intermediates for pharmaceuticals; however, the commercial production of D-malate has not yet been realized.

METHODS OF D-MALATE PRODUCTION

Chemical and Physical Methods

The chiral pool refers to relatively inexpensive, readily available, optically active natural products. L(+)-Tartarate is a natural product of which about 50,000 tons are purified from agol for wine production per year. Gao and Sharpless reported D-malate synthesis from L(+)-tartarate with a 55% yield (10). Amino acids are valuable as chiral synthons. The nitrous deamination of α-amino acids is a well-known process, which was claimed to proceed with retention of amino acid configuration. Henrot et al. synthesized the L- or D-form of malic acid from aspartic acid (11). Asymmetric hydrogenation has been an important advance in organic synthesis in recent years. Noyori and Takasago Inc. have demonstrated the successful commercialization of rhodium-BINAP complexes as asymmetric hydrogenation catalysts. D-malate is formed via a critical bond-breaking process from the sodium salt of epoxysuccinic acid with rhodium phosphine complexes as the catalyst (12). Optically active α-hydroxy acids, such as malate and citramalate, are derived from optically pure β-lactones, which are asymmetrically synthesized from ketene and chloral using quinine or quinidine as a catalyst. These β-lactones are available on a commercial scale from Lonza Ltd. (13).

Classical resolution is most often used to produce optically active materials. It is based on the interaction of a racemic product with an optically active resolving agent, yielding two diastereometric derivatives. The resultant diastereoisomeric salts can usually be separated by selective crystallization. Aliphatic or aromatic amines are effective for D-malate resolution (14–16).

Biological Methods

Recently, biological methods for the production of D-malate have been reported. The most traditional means of producing optical isomers is microbiological resolution, in which a microorganism only degrades the opposite isomer. D-Malate can be produced from a malate racemic mixture using microorganisms that assimilate L-malate (17). Several microorganisms can specifically assimilate L-malate, and D-malate of high optical purity (100%) was obtained from *Acinetobacter Iwoffi* culture broth (18). The enzymatic resolution of DL-malate can be achieved with fumarase and aspartase, resulting in the conversion of the DL-malate to D-malate and L-Asp (19). (*R*)-Dimethyl malate was obtained from a racemic mixture by hydrolysis with pig liver esterase (93% ee at 0 °C in 20% aqueous methanol) (20) or *Rhizopus* lipase (99% ee) (21). Santaniello et al. reported that *Lactobacillus fermentum* ILC G18D produced (*R*)-diethyl malate through the enantioselective reduction of sodium diethyl oxalacetate, the enantioselectivity being 80% (20).

D-MALATE PRODUCTION FROM MALEATE

Maleate

Maleic anhydride, which is industrially produced as a raw material for unsaturated polyesters, succinic anhydride, γ-butyrolactone, 1,4-butanediol, tetrahydrofuran, fumarate, DL-malate, and so on, is an inexpensive chemical. Several industrial processes have been developed for the manufacture of maleic anhydride. Since 1974, *n*-butane has been increasingly used in place of benzene as the raw material for the production of maleic anhydride. Fluidized catalysts containing vanadium and phosphorus oxides are used for the oxidation of *n*-butanc to maleic anhydride. The *n*-butane method is less expensive and nontoxic. Therefore, for maleic anhydride production in the United States, there has been a change to methods involving *n*-butane instead of benzene. In the case of Mitsubishi Chemical Co., the manufacturing process for maleic anhydride was changed to a *n*-butane method with a fluid-bed reactor from the butene method in 1984. Annually, manufacturers produce 630,000 tons of maleic anhydride from *n*-butane, 248,000 tons still being produced from benzene (22). Maleic anhydride can be used for the maleate hydratase reaction as maleate in an aqueous medium.

Maleate is not a natural compound; however, it is an intermediate in the pathway for the catabolism of nicotinate (23) and gentisate (24). Several investigators have studied the use of maleate as a carbon source to characterize the metabolism of microorganisms. Perry and Edward observed the growth of *Pseudomonas fluorescens* ATCC11150 on a medium containing maleate as the sole carbon and energy source, and deduced the metabolic pathway from the cultivation profile (25). They discussed the possibility of economical L-malate production from maleate using maleate *cis–trans* isomerase (EC 5.2.1.1.) and fumarase from a microorganism that grows on maleate. Maleate *cis–trans* isomerase was first identified by Behrman and Stanier in *Pseudomonas* sp. (26). Many researchers have investigated the characteristics of this enzyme. Kimura et al. demonstrated that *Alcaligenes* sp. T501 showed high activity as to the enzymatic conversion of maleate to L-malate *via* fumarate (27). L-Aspartic acid is widely used as a food additive, as a raw material for aspartame, and in medicines. In 1973, the continuous production of L-aspartic acid from fumarate by means of an immobilized microbial cell system was commercialized by Tanabe Seiyaku. L-Aspartic acid was also enzymatically synthesized from maleate and ammonia, using maleate isomerase and aspartase. Takamura et al. showed the enzymatic production of L-aspartic acid, for which maleate was employed instead of fumaric acid as the starting material, with a cell suspension of *Alcaligenes faecalis* 5-24 through the actions of both maleate isomerase and aspartase (28).

Screening of D-Malate Producing Strains

In general, many maleate-assimilating bacteria are known to metabolize maleate to L-malate *via* fumarate (23). Rahatekar et al. (29) have already reported a *Pseudomonas* strain that exhibits both maleate hydratase and maleate *cis–trans* isomerase activity (Fig. 1). Malease is found in a wide variety of organisms, including bacteria, yeasts, fungi (30), plants (1), and mammals (31). Therefore, microorganisms that have a powerful D-malate producing enzyme without a maleate isomerase were necessary to produce D-malate. From among 440 strains of maleate-assimilating bacteria isolated from soil, 82 bacteria that could produce more than 5.0 g/L malate in 24 h were obtained. The ratio of D-malate to the total amount of malate produced by these 82 microorganisms was measured. Fig. 2 shows that only one strain produced L-malate stereospecifically, and the products of 32 strains were mixtures of D- and L-malate. Fifty strains that specifically produced D-malate were obtained and the optical purity of their products was above 90%. Among these strains, *Arthrobacter* sp. MCI2612 showed high maleate hydatase activity, and the malate produced was 100% D-type (7). This result suggests that only maleate hydratase, and not maleate isomerase, is responsible for the maleate metabolism in this strain.

Cultivation and Reaction Conditions for the Production of D-Malate

Asano et al. (7) studied the conditions for the cultivation of *Arthrobacter* sp. MCI2612, and D-malate production by the maleate hydratase with resting cells. As shown in Table 1, the D-malate producing activity was found not to be inducible, but the addition of maleate, D-malate, and citraconate to the medium increased the activity. Maleate was a suitable inducer for the formation of D-malate producing activity. The addition of organic solvents and surfactants for D-malate production was effective. Triton X-100, Triton N-101, and tergitol NPX remarkably increased the activity; the initial reaction rate was about 10-fold that without detergents, and Triton N-101 (1% of the reaction mixture) gave the optimum production of D-malate.

Production of D-Malate by Intact Cells

Since maleate concentrations above 5% inhibit D-malate production, the reaction was carried out at 36 °C for 20 h, keeping the concentration of maleate below 5% through successive feeding of maleate. The time course for D-malate synthesis is shown in Figure 3; 87 g/L of D-malate was produced after 20 h reaction. The molar yield of D-malate was 72%.

Within the last few years, there have been several reports on D-malate production by microorganisms from maleate. Table 2 summarizes the profiles of the microbial production of D-malate (4–7,32,33). The maleate hydratase of all strains except one is inducible. These enzymes are induced by structurally different compounds, such as maleate, D-malate, and 3-hydroxybenzoate, which are the substrate, product, and precursor, respectively, of maleate. The enzyme in *Ustilago sphaerogena* S402 is constitutive (5). van der Welf et al. produced a high concentration of

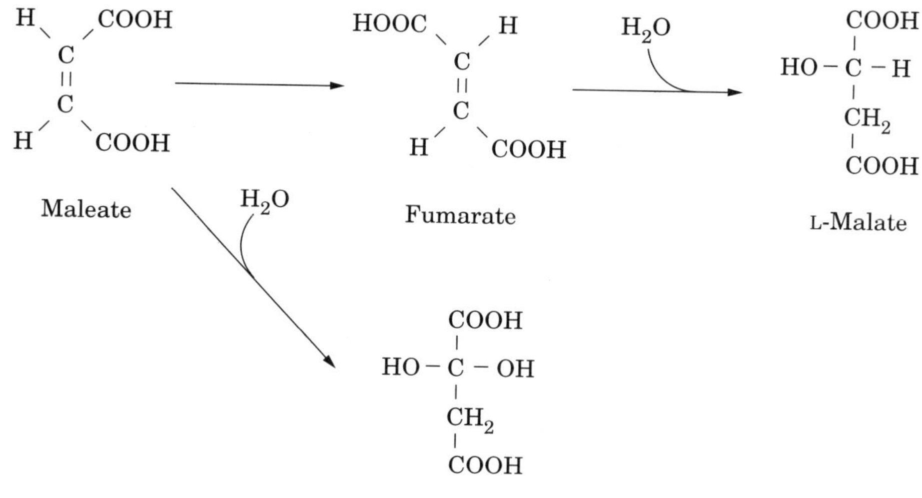

Figure 1. Enzymatic reaction scheme of maleate.

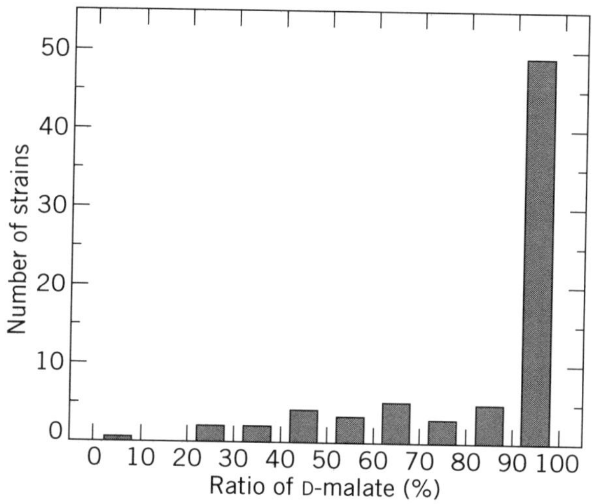

Figure 2. Ratio of D-malate in total malate produced by various strains.

Table 1. Effect of Inducers on D-Malate Producing Activity on *Arthrobacter* sp.

Inducer	(g/L)	Cultivation Final pH	Cultivation Growth (OD[a])	Initial rate of D-malate production (mg/mL/h)
None	—	5.8	18.3	1.09
Maleic acid	1	5.7	18.3	1.73
	2	5.9	18.4	2.19
	3	5.9	18.9	2.39
	4	6.0	19.6	2.22
	5	5.8	19.3	2.26
Fumaric acid	2	6.6	19.4	0.31
D-Malic acid	2	6.5	22.0	1.60
L-Malic acid	2	6.5	20.2	0.66
Citraconic acid	2	6.6	20.0	1.47
Mesaconic acid	2	6.5	20.0	0.59
DL-*cis*-epoxysuccinic acid	2	5.8	19.0	0.85
DL-*trans*-epoxysuccinic acid	2	6.0	19.1	0.31
L(+)-Tartaric acid	2	7.0	19.7	0.27
Nicotinic acid	1	6.7	19.5	0.71
2,5-Dihydroxybenzoic acid	2	6.7	20.7	0.75
Crotonic acid	0.5	6.6	9.58	0.12
L-Ascorbic acid	2	6.1	18.3	0.18
o-Phthalic acid	2	5.9	19.6	0.58

Note: The composition of culture medium was 10 g sorbitol, 2 g NH_4NO_3 3 g KH_2PO_4, 10.5 g K_2HPO_4, 250 mg $MgSO_4 \cdot 7H_2O$, 10 mg $FeSO_4 \cdot 7H_2O$, 10 mg $MnSO_4 \cdot 4\text{–}6H_2O$, 5 g yeast extract, 2 g L-proline, and inducers in 1,000 mL of tap water.
[a]OD660, optical density at 660 nm.

malate in a crystal-liquid two-phase system with Ca-maleate as the substrate (33). The advantages of this method are that no substrate or product inhibition occurs during the reaction and that the reaction equilibrium proceeds towards the product (34). The high enantioexcess of D-malate produced suggest that these microorganisms have no maleate isomerase or that it is not induced during cultivation. The reaction temperature was rather low compared with that for L-malate production from fumarate (50–55 °C) (35). Screening for heat-resistant maleate hydratases will lead to improvement of productivity and prevent contamination that may lead to a decrease in the optical purity.

Comparison of Manufacturing Processes for Malate

DL-Malic acid is used in the food industry, for example as an acidulant in juice and in processed foodstuffs. About 25,000 tons of DL-malate is produced per year through the chemical hydration of maleic anhydride. L-Malate is used for blood infusion and as a resolving agent in organic synthesis. L-Malate can be produced through natural resources, fermentation, and enzymatic processes. An enzymatic process using fumarase is the most efficient. As *Brevibacterium flavum* was found to show high enzyme activity in the case of cells immobilized with κ-carrageenan, Tanabe Seiyaku produces L-malate from fumarate using an immobilized biocatalyst on an industrial scale (35). Ta-

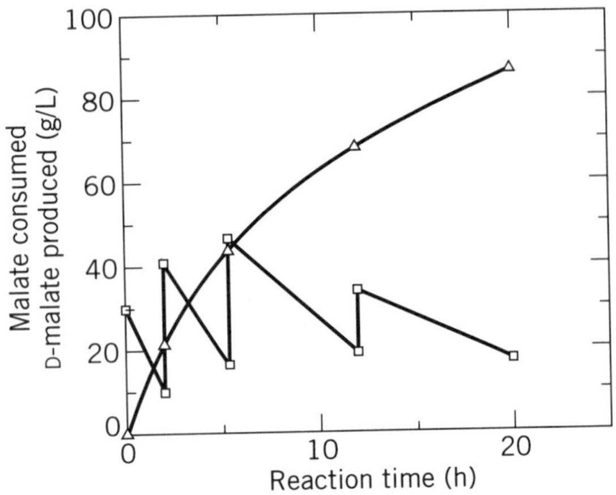

Figure 3. Time course of conversion of maleate to D-malate. Cells were grown at 28 °C in the culture medium, which was the same as in Table 1 except that it contained 3 g maleate/L. Triton N-101 (1%) was added to the reaction mixture. The reaction was carried out at 37 °C. Symbols: ○, D-malate produced; ■, residual maleate. After 2, 6, and 12 h of incubation, maleate 2Na was fed to the reaction mixture.

ble 3 compares the D-malate, L-malate, and DL-malate manufacturing processes. These two enzymatic reactions proceed under mild conditions of pressure, temperature, and pH values.

PURIFICATION AND CHARACTERIZATION OF MALEATE HYDRATASE

The properties of maleate hydratases from a wide variety of organisms have been reported. England et al. (31) demonstrated that an enzyme mixture from rabbit kidney catalyzed the hydration of maleate to D-malate by *trans* addition to the double bond. Britten et al. reported that partially purified maleate hydratase from rabbit kidney had an absolute requirement for chloride ions (36). A crude extract from a pseudomonad produced D-malate from maleate; the optical purity of D-malate was 99.7% (24). Dreyer (37) reported that maleate hydratase from rabbit kidney was a monomeric protein with a molecular mass of 68,000 and required activation by ferrous ion and Na_2S.

Ueda et al. (38) purified the maleate hydratase of *Arthrobacter* sp. MCI2612. It has a molecular mass of about 90,000 and consists of two subunits with molecular masses of 58,000 and 28,000. The purified enzyme was incubated with sulfhydryl compounds and Fe^{2+}. L-Cysteine was found to be the most effective activator, followed by dithiothreitol, thioglycolate, reduced glutathione, and 2-mercaptoethanol, in the presence of Fe^{2+}. Several hydratases, such as citraconase (39), mesaconase (40), L-tartaric dehydratase (41), aconitase (42), and fumarase (43), are known to require ferrous ion and sulfhydryl compounds for their activities. Fumarase is a well-characterized enzyme among these hydratases. A fumarase in aerobically grown *Escherichia coli* required activation by incubation with

Table 2. Comparison of the D-Malate Production by Various Microorganisms

Strain	*Arthrobacter* sp. MCI-2612 (7)	*Pseudomonas fluorescens* IFO3081 (6)	*Clostridium formicoaceticum* DSM94 (32)	*Ustilago sphaerogena* S402 (5)	*Pseudomonas pseudoalcaligenes* NCIMB9877 (4,33)
Inducer	Maleate, D-Malate, Citraconate	Citraconate	NE	None (constitutive)	3-Hydroxybenzoate, Gentisate
pH	7.0	7.0	7.5	NE	7.5
Temperature (°C)	36	26	37	30	30
Substrate inhibition	Yes	NE	NE	None (up to 100 g/l)	Yes
Product inhibition	Yes	NE	NE	NE	Yes
Treatment of cells	Triton N-101	None	None	Triton X-100	Triton X-100
Productivity	87 g/L/20 h	6.94 g/L/24 h	1.87 g/L/18 h	40.8 g/L/48 h	215 g/L/48 h
Reaction volume culture volume	1	3	NE	2.5–3.3	NE
Optical purity (% c.c.)	100	99	>99	98.9	99.96
Reaction yield (mole %)	72	NE	>98	49.6	99.4

NE, not examined.

Table 3. Comparison of Manufacturing Processes of Malate

	L-Malate	DL-Malate	D-Malate
Raw material	Fumarate	Maleic anhydride	Maleate
Catalyst	Fumarase	None	Maleate hydratase
Reaction temperature (°C)	30–50	160–200	30
Reaction pressure (atm)	Atmospheric pressure	5–20	Atmospheric pressure
pH	6–8	2–4	7–8
Solvent	Water	Water	Water

Table 4. Kinetic Parameters of Substrates of Maleate Hydratase of *Arthrobacter* sp.

Substrate	K_m (mM)	V_{max} (μmol/mg/min)
Maleate	3.85	16.5
2-Chloromaleate	2.74	25.8
2-Bromomalcatc	3.27	8.26
Citraconate	4.55	13.2

Note: The initial velocity was measured at 30 °C in a reaction mixture containing 20 mM Tris-HCl buffer (pH 7.5), an appropriate amount of purified enzyme, and various concentrations of substrate.

Fe^{2+} and a thiol, and was suggested to have an Fe-S cluster (44). Thus, the maleate hydratase from *Arthrobacter* sp. is suggested to be a member of the hydratases containing an Fe-S cluster represented by fumarase, because the feature of the activation is similar to that reported for fumarase. The reaction product of the purified maleate hydratase is D-malate and the optical purity is 100% D-type.

The substrate specificity and kinetic properties of this reaction were also examined. As shown in Table 4, chloromaleate served as a better substrate than the natural substrate, maleate. The rates of hydration of these substrates followed: chloromaleate > maleate > citraconate > bromomaleate (Table 4). The reaction products from citraconate, chloromaleate, and bromomaleate are (*R*)-(−)-

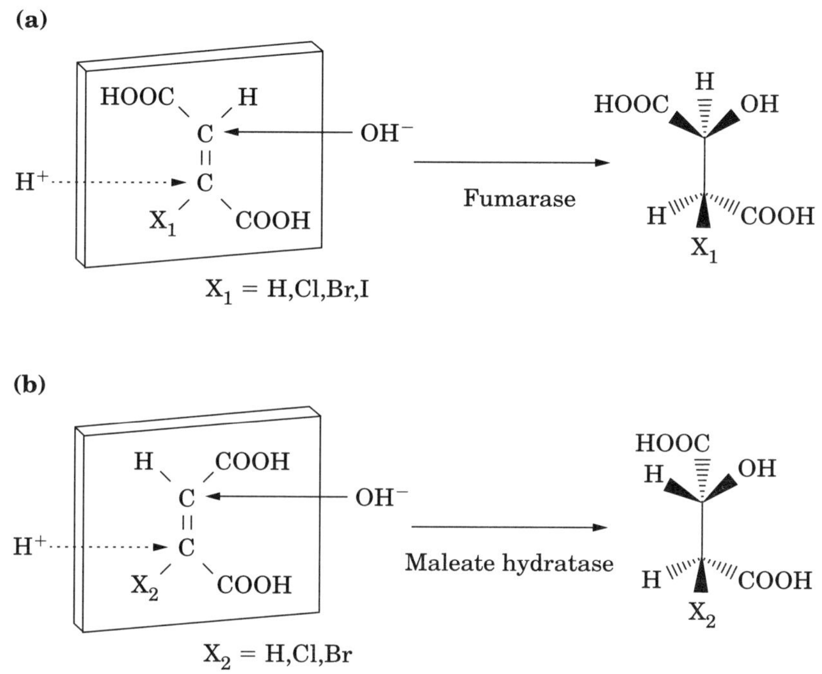

Figure 4. Stereochemistry of the reaction catalyzed by fumarase (**a**) compared to that catalyzed by maleate hydratase (**b**).

Table 5. Some Properties of Animal and Bacterial Maleate Hydratases

Origin	Rabbit kidney (37)	*Pseudomonas pseudoalcaligenes* NCIMB9867 (47,48)	*Arthrobacter* sp. MCI-2612 (38,49)
M.W. (kDa)	68 (Sephacryl S-300)	24,000, 57,000 (SDS) 89,000 (Gel filtration)	58,000, 28,000 (SDS)
Optimum pH	NT	8	8.5
Optimum temperature (°C)	NT	45	45
Activating agents	Fe^{2+}, Na_2S	None	Fe^{2+}, L-Cysteine
Specific activity (U/mg protein)	317	108	6.55
Another substrate	None	Citraconate 2,3-Dimethylmaleate	Citraconate 2-Chloromaleate 2-Bromomaleate
K_m (mM)	NT	0.35 for maleate 0.20 for citraconate	3.85 for Maleate 2.74 for 2-Chloromaleate 3.27 for 2-Bromomaleate 4.55 for Citraconate

NT, not tested.

citramalate, (−)-*erythro*((2*S*, 3*S*)-3-)-chloromalate and (−)-*erythro*((2*S*, 3*S*)-3-)-bromomalate, respectively. The enzyme did not act on dimethyl maleate, acetylenedicarboxylate, fumarate, mesaconate, *cis*-epoxysuccinate, and *trans*-epoxysuccinate. Consequently, it was possible to determine that the bacterial enzyme-catalyzed hydration reaction also involved *trans* addition of the hydrogen and hydroxyl ion to maleate (Fig. 4a).

Fumarase (fumarate hydratase, EC 4.2.1.2), which catalyzes the reversible hydration of fumarate to L-malate, has a broad substrate specificity (45). Many dicarboxylic acids, such as acetylenedicarboxylate, epoxysuccinate, and tartarate, as well as α-monosubstituted derivatives of fumarate, such as chlorofumarate, bromofumarate, and iodofumarate (Fig. 4b), have been reported to serve as substrates for fumarase from swine heart muscle (46). Although acetylenedicarboxylate, epoxysuccinate, and tartarate did not serve as substrates for the maleate hydratase from *Arthrobacter* sp., these dicarboxylic acids were found to be competitive inhibitors.

Table 5 summarizes the characteristics of the maleate hydratases purified from microorganisms and rabbit kidney. Although the maleate hydratase derived from rabbit kidney exhibited absolute specificity for maleate, the hydration of chloromaleate, bromomaleate, and citraconate, in addition to that of maleate, was observed for the maleate hydratase from *Arthrobacter* sp. (38,49). The purified *P. pseudoalcaligenes* enzyme also catalyzes the hydration of citraconate and 2,3-dimethylmaleate (47,48). The enzyme from rabbit kidney requires Fe^{2+} and a thiol as reducing agents for its activity. In contrast, the maleate hydratase from *P. pseudoalcaligenes* does not require metals or other cofactors. Thus, it appears that maleate hydratases can be classified into two classes of carboxylic-acid hydratases, either containing or not containing on iron-sulfur cluster, of which fumarase is probably the most thoroughly investigated (50).

BIBLIOGRAPHY

1. W. Sacks and C.O. Jensen, *J. Biol. Chem.* **192**, 231–236 (1951).
2. J.V. Taggart, S. Angielski, and H. Morell, *Biochim. Biophys. Acta.* **58**, 144–156 (1962).
3. D.J. Hopper, P.J. Chapman, and S. Dagley, *Biochem. J.* **110**, 798–800 (1968).
4. M.J. van der Werf, W.J.J. van den Tweel, and S. Hartmans, *Appl. Environ. Microbiol.* **58**, 2854–2860 (1992).
5. T. Nakajima, S. Manzen, T. Shigeno, and T. Nakahara, *Biosci. Biotechnol. Biochem.* **57**, 490–491 (1993).
6. K. Nakayama and M. Ushijima, *Biotechnol. Lett.* **15**, 271–276 (1993).
7. Y. Asano, M. Ueda, and H. Yamada, *Appl. Environ. Microbiol.* **59**, 1110–1113 (1993).
8. Japanese Pat. 97-275991 (October 28, 1997), I. Oonishi, T. Tanba, and K. Yokozeki (to Ajinomoto Co., Japan).
9. Japanese Pat. 98-52289 (February 24, 1998), S. Yasuda, S. Mukoyama, and K. Sakano (to Nippon Shokubai Kagaku Kogyo Co., Japan).
10. Y. Gao and K.B. Sharpless, *J. Am. Chem. Soc.* **110**, 7538–7539 (1988).
11. S. Henrot, M. Larcheveque, and Y. Petit, *Synth. Commun.* **16**, 183–190 (1986).
12. A.S.C. Chan and J.P. Coleman, *J. Chem. Soc. Chem. Commun.* **104**, 166–168 (1982).
13. H. Wynberg and E.G.J. Staring, *J. Am. Chem. Soc.* **104**, 166–168 (1982).
14. Japanese Pat. 57056439 (April 5, 1982), T. Katsura and H. Yamatika (to Sumitomo Chemical Co., Japan).
15. K. Saigo, M. Kai, N. Yonezawa, and M. Hasegawa, *Synthesis* **2**, 214–216 (1985).
16. T. Shiraiwa, Y. Sado, M. Inoue, K. Sakamoto, H. Miyazaki, and H. Kurokawa, *Bull. Chem. Soc. Jpn.* **61**, 899–903 (1988).
17. U.S. Pat. 4,912,042 (March 27, 1990), K. E. Laumen (to Eastman Kodak Co., USA).
18. M. Miyama and K. Nakayama, *Biotechnol. Lett.* **15**, 23–28 (1993).
19. Japanese Pat. 91-136331 (June 7, 1991), M. Terasawa, S. Nara, H. Yamagata, and H. Yugawa (to Mitsubishi Chemical Co., Japan).
20. E. Santaniello, P. Ferraboschi, P. Grisenti, F. Aragozzini, and E. Maconi, *J. Chem. Soc. Perkin, Trans. 1* **3**, 601–605 (1991).
21. K. Ushio, K. Nakagawa, K. Nakagawa, and K. Watanabe, *Biotechnol. Lett.* **14**, 795–800 (1992).
22. F. Cavani and F. Trifiro, *Chem. Tech.* **24**, 18–25 (1994).
23. E.J. Behrman and R.Y. Stanier, *J. Biol. Chem.* **228**, 923–945 (1957).
24. D.J. Hopper, P.J. Chapman, and S. Dagley, *Biochem. J.* **110**, 798–899 (1968).
25. J.T. Perry and V.H. Edward, *Appl. Microbiol.* **20**, 710–714 (1970).
26. E.J. Behrman and R.Y. Stanier, *J. Biol. Chem.* **228**, 923 (1957).
27. T. Kimura, Y. Kawabata, and E. Sato, *Agric. Biol. Chem.* **50**, 89–94 (1986).
28. Y. Takamura, I. Kitamura, M. Iida, K. Kono, and A. Ozaki, *Agric. Biol. Chem.* **30**, 338–344 (1966).
29. H.I. Rahatekar, F.S. Maskati, S.S. Subramanian, and M.R.R. Rao, *Indian J. Biochem.* **5**, 143–144 (1968).
30. U.S. Pat. 5,270,190 (December 14, 1993), K. Nakayama and Y. Kobayashi (to Bior Inc., Tokyo, Japan).
31. S. England, J.S. Britten, and I. Listowsky, *J. Biol. Chem.* **242**, 2255–2259 (1967).
32. E. Richard and S. Helmut, *Tetrahedron* **50**, 13641–13654 (1994).
33. M.J. van der Werf, S. Hartmans, and W.J.J. van den Tweel, *Enzyme Microb. Technol.* **17**, 430–436 (1995).
34. K. Kitahara, S. Fukui, and M. Misawa, *Nihon Nougei Kagaku Kaishi* **33**, 528–531 (1959).
35. I. Chibata, T. Tosa, and T. Shibatani, A.N. Collins, G.N. Sheldrake, and J. Crosby eds., in *Chirality in Industry*, Wiley, Chichester, 1992, pp. 351–370.
36. J.S. Britten, H. Morell, and J.V. Taggart, *Biochim. Biophys. Acta.* **185**, 220–227 (1969).
37. J.L. Dreyer, *Eur. J. Biochem.* **150**, 145–154 (1985).
38. M. Ueda, Y. Asano, and H. Yamada, *Biosci. Biotechnol. Biochem.* **57**, 1545–1548 (1993).
39. S.S. Subramanian and M.R.R. Rao, *J. Biol. Chem.* **243**, 2367–2372 (1968).
40. S. Suzuki, T. Osumi, and H. Katsumi, *J. Biochem.* **81**, 1917–1925 (1977).

41. R.E. Hurlbert and W.B. Jakoby, *J. Biol. Chem.* **240**, 2772–2777 (1965).
42. M.C. Kennedy, M.H. Emptage, J.L. Dreyer, and H. Beinert, *J. Biol. Chem.* **258**, 11098–11105 (1983).
43. S.K. Reaney, S.J. Bungard, and J.R. Guest, *J. Gen. Microbiol.* **139**, 403–416 (1993).
44. N. Yumoto and M. Tokushige, *Biochem. Biophys. Res. Commun.* **153**, 1236–1243 (1988).
45. P.D. Boyer, *The Enzyme*, 3rd ed., Academic Press, New York, 1971, pp. 539–568.
46. J.W. Teipel, G.M. Hass, and R.L. Hill, *J. Biol. Chem.* **243**, 5684–5694 (1968).
47. M.J. van der Werf, W.J.J. van den Tweel, and S. Hartmans, *Appl. Environ. Microbiol.* **59**, 2823–2829 (1993).
48. M.J. van der Werf, W.J.J. van den Tweel, and S. Hartmans, *Eur. J. Biochem.* **217**, 1011–1017 (1993).
49. M. Ueda, H. Yamada, and Y. Asano, *Appl. Microbiol. Biotechnol.* **41**, 215–218 (1994).
50. Y. Ueda, N. Yumoto, M. Tokushige, K. Fukui, and H. Ohya-Nishiguchi, *J. Biochem.* **109**, 728–733 (1991).

See also ENERGY METABOLISM, MICROBIAL AND ANIMAL CELLS; METABOLITES, PRIMARY AND SECONDARY.

MALIC ACID, PRODUCTION BY FUMARASE

TADASHI SATO
TETSUYA TOSA
Tanabe Seiyaku Co. Ltd.
Osaka, Japan

KEY WORDS

Fumarase
Immobilized *Brevibacterium ammoniagenes*
Immobilized *Brevibacterium flavum*
Immobilized cells
L-Malic acid

OUTLINE

Bibliography

L-Malic acid plays a very important metabolic role in living cells. It has been used to treat hyperammonemia and liver disfunction, and it is component of amino acid infusions. Also, if it becomes possible to supply large amounts of L-malic acid at low cost, L-malic acid—the natural form—is expected to be used instead of DL-malic acid—the racemic form—in the food industry as a food additive.

Existing preparation methods for L-malic acid include isolation from natural fruit juice, the optical resolution of synthetic DL-form, fermentation, and others. However, for industrial purposes, the acid has been produced from fumaric acid by the action of fumarase as follows.

$$\underset{\text{Fumaric acid}}{HOOCCH{=}CHCOOH} + H_2O \xrightleftharpoons{\text{Fumarase}} \underset{\text{L-Malic acid}}{HOOCCH_2\underset{\displaystyle OH}{\underset{|}{C}}H{-}COOH}$$

Many kinds of microbial cells having high fumarase activity have been successfully immobilized. The immobilized cells having high enzyme activity were obtained by the polyacrylamide gel method. Moreover, the immobilized cells were stable, and a column packed with them could be used for continuous enzyme reaction for long periods.

The industrial production of L-malic acid from fumaric acid has been performed by an enzymatic batch process using the broth of *Lactobacillus brevis* (1), but a continuous reaction using immobilized fumarase of immobilized cells with fumarase activity should be more economical.

Marconi et al. (2) reported that fumarase isolated from microbial cells can be efficiently immobilized on cellulose triacetate, making it possible to develop an economically attractive method for producing L-malic acid. However, no industrial process using immobilized fumarase has been brought into operation.

Several microorganisms having high fumarase activity were immobilized by the polyacrylamide gel method, and their activities were compared (3). *Brevibacterium ammoniagenes* was found to be the most active before and after immobilization.

In 1974, Tanabe Seiyaku Co., Ltd., introduced a process of continuous production of L-malic acid from fumaric acid using immobilized *B. ammoniagenes*. Since then, the process has been modified in several major and minor respects to create the current system, which is 20 to 25 times more efficient than the original.

Yamamoto et al. (3) investigated the industrial application of immobilized microbial cells with fumarase activity and developed a method using *B. ammoniagenes* immobilized with polyacrylamide gel.

When the immobilized cells were used for the production of L-malic acid from fumaric acid, succinic acid was formed as a by-product, and the separation of succinic acid from L-malic acid was technically very difficult. Accordingly, for the industrial production of pure L-malic acid in high yield, it was necessary to prevent the formation of succinic acid during the enzyme reaction. Thus, various treatments of intact or immobilized cells were performed. Heat treatment, autolysis, and freezing-thawing were not effective for the prevention of succinic acid formation. However, treatment with substrate solution containing deoxycholic acid or bile acid, which was commonly used as a solubilizer of membrane-bound enzymes, was found to be very effective for the suppression of succinic acid formation, as shown in Table 1. Further, this treatment also markedly enhanced the formation of L-malic acid. This effect was attributed to an increase of membrane perme-

Table 1. Effect of Detergent Treatment on Formation of L-Malic Acid and Succinic Acid

Treatment		Formation of L-malic acid	Formation of succinic acid
Detergent	Concentration (%)	(μmol/h/mL gel)	(mol% L-malic acid)
No addition		75	2.5–5.0
CPC	0.02	340	2.5–5.0
CPC	0.16	225	1.0–2.5
SLS	0.02	445	1.0–2.5
SL-10	0.02	90	2.5–5.0
Triton X-100	0.20	395	>5.0
Bile acid	0.20	425	<0.2
Bile extract	0.20	485	<0.2
Deoxycholic acid	0.20	480	<0.2

Note: CPC = cetylpyridium chloride; SLS = sodium lauryl sulfate.

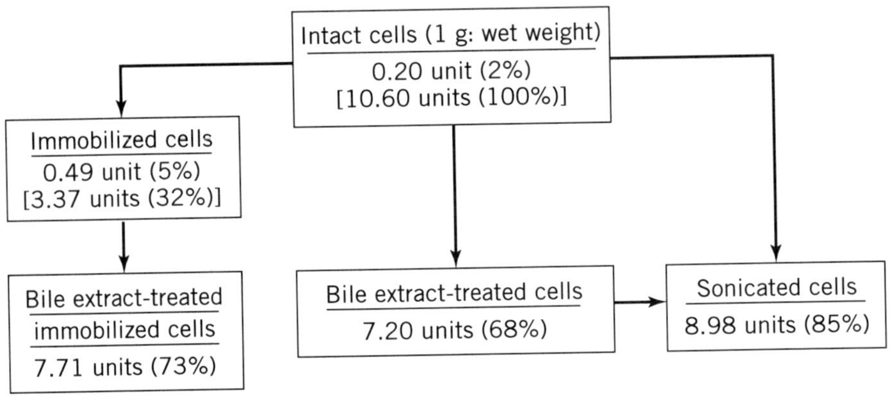

Figure 1. Comparison of the fumarase activities of various enzyme preparations per unit of intact cells. One unit is defined as activity producing 1 mmol of L-malic acid at 37 °C in 1 h. Square brackets show activity in the presence of 0.02% cetylpyridium chloride; percentages are values relative to the activity (100%) of intact cells in the presence of cetylpyridium chloride.

Table 2. Comparison of Productivities of *Brevibacterium ammoniagenes* and *Brevibacterium flavum* Immobilized with Polyacrylamide and with Carrageenan for Production of L-Malic Acid

	B. ammoniagenes			*B. flavum*		
Immobilization method	Activity (unit/g cells)	Half-life at 37 °C (day)	Relative productivity	Activity (unit/g cells)	Half-life at 37 °C (day)	Relative productivity
Polyacrylamide	5,800 (60%)	53	100	6,680 (34%)	94	204
Carrageenan	5,800 (60%)	75	142	9,920 (51%)	160	516

Note: Activity after treatment with bile extract. Productivity = $\int_0^t E_0 \exp(-\kappa d \times t)dt$, where E_0 = initial activity, κd = decay constant, and t = operational period.

ability to the substrate or product resulting from bile acid treatment. Bile extract is suitable for industrial purposes because it is commercially available at a low price. Various conditions for treatment with bile extract were studied, and the most effective procedure was to stand the immobilized cells in 1 M sodium fumarate (pH 7.5) containing 0.3% bile extract at 37 °C for 20 h. The yield of fumarase activity of immobilized cells obtained in this way is shown in Figure 1.

Enzymic properties of the immobilized cells treated with bile extract were studied and compared with those of the native enzyme. The optimum temperature and optimum pH of the enzyme reaction were almost same as those of the native enzyme, namely 60 °C and pH 7.0 to 7.5, respectively. The equilibrium point of the enzyme reaction was scarcely affected by immobilization or by the reaction pH, but tended to shift toward L-malic acid formation with decreasing reaction temperature.

Conditions for the continuous enzyme reaction using a column packed with the immobilized cells were investigated. It was found that when 1 M sodium fumarate (pH 7.5) was passed through the column at 37 °C at a space velocity of below 0.23 (residence time in the column, more than 4.3 h), the reaction reached equilibrium with about 80% conversion of fumaric acid to L-malic acid. Fumaric acid was removed from the effluent by acidification, and L-malic acid was separated in about 70% yield from fumaric acid.

For industrialization of this system, chemical engineering studies were carried out, and a continuous reaction system using immobilized cells was designed. L-Malic acid of high purity has been industrially produced at Tanabe Seiyaku by this process since 1974.

To improve L-malic acid productivity, the authors investigated immobilization methods that could be carried out under mild conditions without denaturation of fumarase, and they also screened again to find microbial cells with a higher fumarase activity than *B. ammoniagenes*.

Among many synthetic and natural polymers tested as matrixes for entrapping microbial cells, κ-carrageenan was found to be best (4).

Brevibacterium flavum was found to have higher enzyme activity after immobilization with carrageenan as compared with the former *B. ammoniagenes*. For industrial purposes, the stability was also studied, and the productivity was compared, as shown in Table 2. When the productivity of immobilized *B. ammoniagenes* with polyacrylamide was taken as 100, that of immobilized *B. flavum* with carrageenan was 516. Therefore, it is evident that also in this case the carrageenan method is more advantageous than the conventional polyacrylamide method. So, we changed the polyacrylamide method to the carrageenan method in 1977. This new method gives us satisfactory results from an industrial point of view.

In other studies, *Thermus rufens nov* sp., a thermophilic bacterium, was immobilized ionically on Duolite for production of L-malic acid from fumaric acid (5).

L-Malic acid is one of the members of tricarboxylic acid cycle. It is easy to be metabolized in a living body and is accompanied by generation of ATP, which is the energy source of a living body.

L-Malic acid is also converted to L-aspartic acid via oxalacetic acid. In this process, ammonia is metabolized with coupling to the L-glutamic acid–α-ketoglutaric acid system. Further, L-aspartic acid condenses with L-citrulline to form arginosuccinate and enters into the urea cycle. Ammonia is also incorporated when L-citrulline is formed from L-ornithine.

L-Malic acid is effective in detoxication of ammonia by introducing ammonia into the urea cycle. Its detoxicative action is cumulative in combination with L-arginine (6–8).

In addition to the role of L-malic acid mentioned in this article, it has been reported that L-malic acid is effective for lowering blood cholesterol (9), preventing atherosclerosis (10), and lowering blood pressure (11). The diuretic effect of L-malic acid was also reported (12).

BIBLIOGRAPHY

1. K. Kitahara, S. Fukui, and M. Misawa, *J. Gen. Appl. Microbiol.* **6**, 108–113 (1960).
2. W. Marconi, F. Morisi, and R. Mosti, *Agric. Biol. Chem.* **39**, 1323–1324 (1975).
3. K. Yamamoto, T. Tosa, K. Yamashita, and I. Chibata, *Eur. J. Appl. Microbiol.* **3**, 169–183 (1976).
4. T. Tosa, T. Sato, T. Mori, K. Yamamoto, I. Takata, Y. Nishida, and I. Chibata, *Biotechnol. Bioeng.* **21**, 1697–1709 (1979).
5. Y. Ado, T. Kawamoto, I. Masunaga, T. Takayama, S. Takasawa, and K. Kimura, *Enzyme Eng.* **6**, 303–304 (1982).
6. R. Normann, *Dtsch. Med. Wschr.* **94**, 230–233 (1969).
7. I. Szam, J. Szenter, I. Hagedues-Wein, and A. Vass, *Intern. J. Clin. Pharmacol. Ther. Toxicol.* **6**, 260–265 (1972).
8. A. Tissort-Favre and R. Brette, *Therapie* **25**, 629–631 (1970).
9. D.A. Kaidin, *Patol. Fiziol. Eksp. Terapiya.* **12**, 72–74 (1968).
10. D.A. Kaidin, *Byul. Esksp. Biol. Med.* **66**, 29–31 (1969).
11. R. Nose, *Shikoku Igakukaishi* **22**, 525–529 (1966).
12. J.B. Brueckener, *Anaesthesist* **19**, 219–223 (1970).

MAMMALIAN CELL BIOREACTORS

MARIA J. GUARDIA
WEI-SHOU HU
University of Minnesota
Minneapolis, Minnesota

KEY WORDS

Airlift
Bioreactors
Cell culture
Hollow fiber
Immobilized cells
Mammalian cells
Microcarriers
Perfusion
Roller bottle
Stirred-tank

OUTLINE

INTRODUCTION

Before the advent of recombinant DNA technology, industrial mammalian cell culture and animal cell culture were generally used primarily in the production of viral vaccines, using primary cells and normal diploid cell strains, and biologics, using isolated transformed cells. The arrival of rDNA technology and the expression of heterologous proteins in mammalian cells completely transformed cell culture technology. Except for the production of alternated or inactivated viral vaccines, virtually all mammalian cell culture processes now employ continuous cell lines, most of which have been adapted to grow in suspension, without a surface support. The adoption of the growth of suspension cells as the typical vehicle for the production of complex molecules requiring extensive posttranslational modification also transformed the concept of bioreactor selection.

Much research activity and many commercial efforts were devoted to developing bioreactors for mammalian cell culture in the 1980s. The anticipation was that the potentially detrimental effect of the air sparging of mammalian cells would require a newly designed bioreactor to accommodate the manufacturing needs of the new mammalian cell–based products. Many of these efforts have since declined, and the bioreactors used for industrial cell culture converge on a few basic configurations. This article focuses only on the main types used currently for industrial cell culture. Those interested in the historical aspects of bioreactor development are referred to other monographs and review articles. It should be noted that many of the bioreactors developed in the 1980s, although not commonly used, might find specialized applications in other areas, such as tissue engineering and cellular therapy.

BASIC REACTOR KINETIC

Mammalian cell bioreactors are generally categorized similarly to chemical reactors according to their mixing characteristics. It is instructive to review two ideal reactors: well-mixed stirred tank and plug-flow (tubular) reactor. In an ideal well-mixed bioreactor, the mixing is assumed to be intense enough that the fluid is homogeneous through the reactor. The mathematical description of the ideal continuous flow stirred tank reactor is described by the following first-order differential equation:

$$\frac{d(VC_A)}{dt} = F_iC_{Ai} - F_oC_{Ao} + r_AV \qquad (1)$$

where V is the culture volume in the bioreactor, C_A is the concentration of nutrient or product A, t is time, F is the flow rate, and r_A is the volumetric consumption rate of nutrient or production rate of product A.

In an ideal stirred tank reactor, there is no flow bypass and no shunt of substrate from inlet to outlet and no dead zones or clumps of undissolved solid substrate floating around. The addition of a substrate is through feeding instantaneously distributed throughout the entire reactor, and when gas sparging is employed the agitator provides intimately mixed gas and liquid.

The basic model for the tubular reactor (such as hollow fiber and ceramic systems to be described later) is that the liquid phase moves as a plug flow, meaning that there is no variation of axial velocity over the cross section. The mass balance for component A in a volume element $S\partial z$ that described an ideal plug-flow reactor is the following:

$$\frac{\partial C_A}{\partial t} = -v_z\frac{\partial C_A}{\partial z} + r_A \qquad (2)$$

where v_z is the linear velocity in the z direction and S is the cross-sectional area. At any given moment of time or at steady state (i.e., cell concentration and cellular activities are not changing with time), the equation becomes

$$\frac{\partial c_A}{\partial z} = \frac{r_A \cdot S}{F} \qquad (3)$$

which describes changes of concentration of A along the direction of fluid flow. It is clear that the nutrient concentration will decrease from the inlet to the distal end of the reactor, whereas the metabolite concentration will increase. The length of the reactor is limited because eventually nutrient depletion or metabolite accumulation inhibits growth and metabolism.

These ideal cases of completely mixed tanks or plug-flow tubular reactors are situations that can be approximated in small-scale laboratory conditions. The conditions in larger-scale process reactors deviate significantly from these ideal conditions.

In a well-mixed bioreactor, there are no concentration gradients in either the gas or the liquid phase. In other words, none of the chemical species or cells is segregated in the reactor. The other extreme of mixing is total segregation, where there is no interaction between different volume elements in the bioreactor. An ideal plug-flow reactor is assumed to be under conditions of total segregation. Most bioreactor systems have a mixing pattern between the two extremes and are under partially segregated conditions.

In general, laboratory and small pilot plant bioreactors are used for process development and optimization. The fluid-mixing characteristics are rather sensitive to the scale of the reactor. Furthermore, plug-flow bioreactors are intrinsically more difficult to scale up than mixing vessels, because the concentration gradient of essential nutrients, oxygen in particular, inevitably becomes limiting in the downstream region of the reactor. In considering the selection of bioreactors for mammalian cell cultures, the mixing characteristics and their relationship to scale-up have to be kept in mind.

OPERATION MODES

Mammalian cell cultures are operated in batch, fed-batch, and continuous modes. Batch culture is used most frequently. All nutrients are added at the beginning of a production cycle, except for oxygen, which is provided continuously during the culture through surface aeration, membrane creation, or direct gas sparging. Using a conventional medium, the typical densities achieved in batch culture are rather low, about 2–4 × 10^6 cells/mL. In many cases, batch operation tends to be run as extended batch processes. In this operation mode, spent medium with or without cells may be harvested periodically and replaced by an equal quantity of fresh medium. It allows a production process to be extended to a couple of weeks, with increased productivity for the reactor. It is possible to control the inhibitory products accumulated in the culture with this technique.

Another possibility to prolong the batch processes is to use fed-batch operation. In this process, certain nutrients such as amino acids, glucose, and some other micronutrients are added continuously or intermittently to increase the cell concentration and to prolong the product-formation period. In most cases bulk salts (such as NaCl) are not added to the feed to avoid increasing the osmolality. The final product concentration is substantially higher than that in a batch culture.

Continuous chemostat operation, used primarily for laboratory-scale bioreactors, has found limited application in production-scale systems. In this operation, spent medium is continuously replaced with an equal volume of fresh medium and therefore continuously discharging cells. The cell concentration in the bioreactor usually is at a low level similar to that in batch cultures. A major drawback of a simple continuous culture is thus the low cell concentration and the corresponding low throughput of the system. To increase the cell concentration and the reactor throughput, a perfusion culture is more frequently used in industrial continuous operations. In a perfusion culture, a high medium flow rate (one much higher than the cell growth rate, which would wash out cells in a simple continuous culture) is used, but a cell retention device is used to prevent most of the cells from leaving the reactor. The cell concentration in the reactor is up to 10 times higher than that for a simple continuous culture.

When a plug-flow type of bioreactor is used, usually the medium is continuously recirculated through the bioreactor. The recirculating medium carries oxygen to the reactor to sustain cell viability and growth. However, the system is not a continuous reactor in a strict sense. Although the flow of recirculating medium is continuous, the medium itself is replaced periodically for nutrient replenishment and product harvest. It is more similar to repetitive-batch types of bioreactors.

MIXING TANK BIOREACTORS

Cell Support for Mixing Vessels

In the past decade, most of the cell culture processes developed have been for the production of recombinant proteins and antibodies; the majority employ suspension cells. However, despite the general trend of using suspension cells, anchorage-dependent or anchorage-preferred cells are still widely used for the manufacture of some products, especially viral vaccines. Many cell types used in tissue engineering applications are also anchorage dependent. To cultivate these cells in a mixing vessel, some form of anchoring support is necessary. By employing such support, the cells can be cultivated in a relatively homogeneous environment with appropriate environmental control, be infected for virus production, or be induced for product formation.

Microcarriers. The use of microcarriers for cell culture was first demonstrated by Van Wezel (1). The basic concept is to allow cells to attach to the surface of small suspended beads so that conventional stirred tank bioreactors can be used for cell cultivation. To ease suspending these cell-laden microcarriers, their diameter and density are usually in the range of 100–300 μm and 1.02–1.05 g/cm^3, respectively. This diameter range also gives a good growth surface area per reactor volume. Even at a moderate microcarrier concentration, in the range of 8–15% culture volume being occupied by microcarriers, a significantly larger surface area per reactor volume can be achieved than that in roller bottles.

Although smaller microcarriers less than 100 μm in diameter can provide even more surface area, they are not used often. Most anchorage-dependent cells do not develop their normal morphology and in many cases do not multiply well on surfaces with an excessively high curvature. These cells do not grow well on small microcarriers. On the other hand, some cells, especially transformed cells, multiply well even though they are attached to small microcarriers and do not develop a well-spread-out morphology. These cells, after attaching to the small microcarriers, agglomerate to form aggregates and continue to grow to high density. The small microcarriers, usually with a diameter of about 50 μm, serve as nuclei for the initiation of aggregate formation.

The microcarriers can be made of many different materials, including dextran, gelatin (2), polystyrene (3,4), glass (5), and cellulose (6). Not all of these are available commercially. In general, in addition to having a wettable surface, the backbone materials of microcarriers often need to be chemically modified to improve cell attachment. One of the most widely used microcarriers, the dextran-based ones, are derivatized with charged molecules or collagen. Polystyrene microcarriers have also been coated with collagen or other adhesion molecules for better performance.

An advantage for the industrial-scale culture of anchorage-dependent cells is the ease of separating cells from the culture medium. Many microcarrier cultures require medium exchanges during cultivation to remove accumulating lactate, ammonia, and other metabolites and to replenish nutrients. In many cases, the cell-laden microcarriers are simply allowed to settle while the spent medium is withdrawn and replenished. In large-scale operations, continuous or semicontinuous perfusion is used more frequently. Such perfusion can be accomplished by withdrawing the medium through a coarse screen that allows the medium to pass through but retains microcarriers in the reactor. Many cells have been grown on microcarriers, including fibroblast, kidney, epithelial, hepatocyte, neuroblastoma, and endothelial cells from various species. Overall, microcarrier culture is a convenient laboratory and research tool, and it has the advantage of being amenable to large-scale production if a large quantity of product is needed.

Macroporous Microcarriers. A variant of microcarriers is one with large pores of tens of micrometers in its interior. The void space inside allows cells to be cultivated not only on the external surface but also internally. Cells in the interior are less susceptible to mechanical damage caused by agitation and gas sparging. Of course, cells in the interior of microcarriers are more likely to be subject to oxygen limitation due to the long diffusional distance, especially because most macroporous microcarriers have a larger diameter (500 μm to a couple of millimeters). Macroporous microcarriers are made of different materials, including gelatin, collagen, and plastic. Many cell lines have been successfully grown on macroporous microcarriers including Vero, HepG2, CHO and 293 cells. The final cell concentration achieved tends to be higher than that obtained with an equivalent volumetric concentration of conventional microcarriers. But in some cases the growth kinetics are a bit slower because the penetration of cells into the interior may be slowed or even retarded by restrictive opening of these pores.

Aggregate Culture. Some transformed cells that normally attach to or spread and grow on a surface can form aggregates when cultivated in shaker flasks or in stirred vessels (7). Different methods have been used to induce aggregate formation for cells. Aggregation can be promoted by manipulating the calcium concentration (8) in conjunction with the agitation rate. Aggregate cultures have advantages similar to microcarrier cultures. They can be cultivated in conventional stirred tank reactors with environmental control. They can also be allowed to settle relatively rapidly by stopping agitation, which readily permits medium replenishment.

For many cell types, one limitation for application of the aggregate culture is that the rate of aggregate formation is slow. One way to induce aggregate formation is to add microspheres to the cell suspension to allow for rapid agglomeration to the microspheres (9). If the aggregate diameters become too large, necrotic centers can be formed as a result of transport limitation. The aggregate size may influence the viral infection kinetic and yield for vaccine formation (10). Many cell lines, including BHK, CHO, 293, and ST cells, have been cultivated as aggregates with sizes ranging between 90 and 400 μm, without the formation of necrotic centers (9,11).

Cell Entrapment. Another method of cell cultivation that enables the use of a stirred tank reactor is cell entrapment. This technique entails entrapping cells in a polymeric matrix to form spheres. The spheres are then either coated with another polymeric film to control the crossing of molecules according to their size—commonly referred to as microencapsulation—or cultivated as they are. These beads are suspended in medium to allow cell growth inside. One of the polymers most used for cell entrapment is calcium alginate. Cell entrapment in calcium alginate is accomplished by preparing a cell suspension in sodium alginate and adding it, in a dropwise fashion, into a solution of calcium chloride. Calcium cross-links alginate, instantaneously forming beads that contain entrapped cells. Alginate may be coated with polylysine for increased mechanical and chemical stability, but such treatment decreases the molecular weight cutoff and prohibits large molecular weight proteins in the medium from reaching the cells. To prepare hollow spheres, the alginate gel inside a bead coated with a polylysine is liquefied through treatment with a calcium chelator such as EDTA or citrate.

The diameter of these beads is often on the order of millimeters. The mechanical strength of the gel that constitutes the beads is relatively weak. Large-scale application using such techniques is not easy. On the other hand, the microcapsule provides a means of immunoisolation of transplanted cells or tissues (sometimes referred to as artificial cells) and could be suitable for some tissue engineering applications. Cell entrapment has been applied to hybridomas (12) and small clusters of adherent cells (13). It has also found applications in the cultivation of differentiated cells, such as insulin-secreting cells. For tissue engineering applications using differentiated cells, the enclosing membrane around the cells and the hydrogel must be biocompatible (14). The membrane used in microencapsulation is semipermeable to allow sufficient diffusion and transport of low molecular weight molecules important for cell survival, such as oxygen, nutrients, and metabolites. For optimum transport across the membrane, the microcapsules should have a uniform wall thickness. Ideally, the semipermeable membrane prevents the passage of high molecular weight proteins, such as immunoglobulins, to allow for product to accumulate inside the microcapsule.

Because the cells are inside the capsules, they are protected from hydrodynamic damage. The culture can be stirred at faster rates than conventional culture, which improves nutrient, metabolite, and gas exchange. Entrapped cells may be cultured in stirred tank bioreactors, fixed-bed reactors, or fluidized-bed reactors. An airlift reactor could be used as well, provided that the beads are small and not significantly denser than the medium.

Stirred Tank Bioreactors

The cells respond in different ways when they are exposed to fluid forces. Usually the morphology of the cell and the cell metabolism are affected by fluid stress. The fluid mechanical consequences in the cells on microcarriers depend on the type of cells and the quality of cell attachment. Cell damage from the interaction between microcarriers and turbulent eddies may be severe when the Kolmogorov length scale for the smallest turbulent eddies becomes comparable to the diameter of the microcarriers (15). The microcarrier–microcarrier collisions are also expected to increase in frequency and severity in the presence of eddies of size equal to the average interparticle distance. Suspended cells are generally less prone to hydrodynamic damage than cells on microcarriers. Two principal types of fermentors—stirred tank and airlift bioreactors—are most commonly used for cell cultivation.

Stirred tanks, or conventional fermentors, have been widely used for culturing suspension cells since the 1960s. With the use of microcarriers, whether conventional or macroporous, adherent cells can also be cultured in a stirred tank. For cell entrapment in hydrogel, the use of stirred tanks is limited to laboratory scale, because the preparation of a large quantity of cell-entrapped beads can be a daunting task, and the risk of mechanical damage caused by agitation increases with scale.

The basic configuration of stirred tank bioreactors for mammalian cell culture is similar to that of microbial fermentors. A major difference is that the aspect ratio (the height-to-diameter ratio) is usually smaller in mammalian cell culture bioreactors. The power input per unit volume of bioreactor is also substantially lower in mammalian cell culture bioreactors. Although the Rushton-type impeller is the norm in microbial fermentors, mammalian cell culture fermentors mostly employ marine-type impellers (16). As in microbial fermentors, an axial impeller is often mounted at the top of the bioreactor to drive the liquid downward in large-scale cell culture bioreactors. These differences reflect the different purposes of agitation in microbial fermentation and cell culture. In microbial fermentation agitation is needed at a higher power input to disperse air bubbles and to increase oxygen transfer efficiency, whereas in mammalian cell culture the primary purpose of agitation is to keep cells or microcarriers suspended in nutrient medium relatively uniformly. In general, the mixing time in a mammalian cell culture bioreactor is substantially longer than that in a microbial fermentor with similar scale. The oxygen transfer capacity in a cell culture bioreactor is also substantially lower than that in a microbial fermentor. However, the typical oxygen demand in a mammalian cell culture is also much lower (10 to 50 times) than that in microbial fermentation.

Because of the relatively low oxygen demand in cell culture, surface aeration is often sufficient to supply enough oxygen for optimal growth in small-scale bioreactors. In fact, surface aeration is the primary means of supplying oxygen in almost all laboratory-scale bioreactors. If necessary, enriched oxygen is used to increase the oxygen transfer from the gas phase to the culture fluid.

However, as the volume of the reactor increases, the area of gas–liquid interface per culture volume decreases, and surface aeration alone becomes inadequate. To supply oxygen, silicone rubber tubing is often installed inside the reactor to allow air or oxygen passing through it to supply oxygen without direct gas sparging. In addition to silicone rubber tubing, other materials with high oxygen permeability, such as porous polypropylene, can also be used. The oxygen transfer tubing need not necessarily be used as a single long piece; many different configurations for the installation of the oxygen transfer tubing can be employed. This method of supplying oxygen provides a convenient, relatively trouble-free way of sustaining cell growth at a reasonably high cell density. Applying such a technique in a large-scale operation could be very cumbersome. Nevertheless, it has been used in bioreactors of up to hundreds of liters in volume.

The most efficient way to supply oxygen, especially in large-scale reactors, is the direct sparging of air- or oxygen-enriched air. Direct sparging, though, can damage cells. Numerous studies have tried to clarify the mechanism of cell damage caused by direct gas sparging (17). The large amount of energy released as the gas bubbles emerge from the liquid surface appears to be responsible for its most damaging effects. These consequences can be partially alleviated by the addition of polyols in the culture medium. In fact, it is standard practice to include Pluronic F-68® in a low-protein or protein-free medium for suspension cultures.

Airlift Bioreactor

A variant of the bubble column reactor with internal circulation loops is used to improve the performance of conventional bubble column reactors. In airlift column reactors, internal liquid circulation is achieved by sparging only part of the reactor with air. The sparged section has a lower effective density than the bubble-free section, and the difference in hydrostatic pressure between the two sections induces the liquid circulation upward in the sparged section (riser) and downward in the bubble-free section (downcomer). The loop has the advantages of permitting high-efficiency mass transfer and improving the flow and mixing properties in the vessel.

These reactors are characterized by low costs, mainly because of their simple mechanical configuration. Considerable backmixing in both gas and liquid phases, high-pressure drop, and bubble coalescence can be disadvantages, in some cases. An additional benefit is the low energy requirement compared with stirred tank reactors. Although simple in construction, sound design is critical for optimal hydrodynamic behavior. Nevertheless, the flow in bubble column reactors is relatively well defined compared with that in stirred tank reactors.

Airlift bioreactors for cell cultivation are considered to be low-shear devices because there is no mechanical agitation. Also, the direct air sparging may not cause excessive cell damage, at least under normal cultivation conditions. The flow regime depends on the sparger used and the flow rate. Various types of spargers are used to provide different bubble sizes.

Airlift reactors have been used successfully with suspension cultures of BHK 21, human lymphoblastoid, CHO, hybridomas, and insect cells.

Stirred Tank in Perfusion Culture

The stirred tank bioreactors can be operated in batch, fed-batch, or continuous mode. Using typical culture medium, the cell concentration achieved in batch and continuous cultures is rather low, largely because of the high conversion ratio of key nutrients (glucose and glutamine) to waste and growth inhibitory metabolites, lactate, and ammonia. One method of increasing cell concentration in continuous culture is to retain cells inside the bioreactor while perfusing medium through. With cell retention, a 10-fold increase in cell concentration can be achieved. An effective cell retention device should permit easy and rapid separation of cells from spent medium and allow for long-term operation. Those continuous cultures with cell retention are typically referred to as perfusion cultures.

A cell retention device may be physically positioned inside or outside of the bioreactor vessel. The devices that have been used include settling tanks (18), external centrifuges (19), external membrane filtration (20), internal microfiltration (21), and rotating wire cage bioreactors (22). All of these, or similar devices, have been shown to be effective at laboratory scale. At industrial scale, bioreactors of up to hundreds of liters, centrifuges, settling devices, and spin filters appear to be top choices. The employment of a centrifuge or a settling device necessitates a recirculation loop to return the concentrated cell stream back to the reactor while discarding the purged cells along with the other stream. The spin filter, on the other hand, is an internal cell retention device. This device is a rotating cylindrical screen mounted on the agitator shaft. It can be mounted to a shaft rotated by a top motor while a bottom motor drives the impeller. The filter screen has an opening ranging from 25 to 60 μm. The average cell diameter is 10–15 μm; however, the cell density inside the cage is much lower than outside the cage. The greatest difficulty encountered with the spin filter is the control of flux and cell retention efficiency.

NONMIXING TANK BIOREACTORS

Roller Bottle

Roller bottles are cylindrical screw-capped bottles mostly made of disposable plastic; reusable glass bottles are still used occasionally. Each bottle is typically about 1 to 1.5 L in total volume. A bottle is usually filled with 0.1 to 0.3 L culture medium for cell cultivation. A stack of bottles is placed on a roller, and the bottles rotate on the roller rack at 1 to 4 rpm and are incubated in an incubator or an incubation room. Roller bottles are used for the cultivation of both suspension cells and anchored cells. However, for suspension cells, it is usually more convenient to grow the cells in a stirred vessel, and roller bottles are only used on a small scale when convenience dictates this selection of cultivation methods. Anchorage-dependent or anchorage-preferred cells are frequently cultivated in roller bottles, both in the laboratory and in manufacturing of biologics. These cells attach to the inner wall of the bottle. Initially, they cover only part of the surface of the inner wall after inoculation. The cell layer is bathed in a liquid medium and is alternately exposed to medium and gaseous nutrients as the bottle rotates. As cells grow, they cover the entire surface and reach confluence. Normal diploid cells then reach contact inhibition and stop growing until they are detached by protease treatment and inoculated into a larger number of roller bottles. Many continuous cell lines or tumorous (transformed) cells can continue to grow to form multiple layers of cells after reaching confluence if sufficient nutrients are provided. Some variations of roller bottles are available to increase the cell growth surface area in a bottle. Spiral films and waffles on the inner wall of roller bottles have been introduced and are sometimes used.

For small-scale operations, roller bottles provide many advantages for the cultivation of adherent cells. It is relatively inexpensive to set up. Often it allows for a rapid change of throughput in response to need. Furthermore, replacing cell growth medium with one designed for product formation is rather straightforward. It is particularly useful when the serum-containing medium needs to be replaced by a serum-free or protein-free medium for the production of secreted proteins or viruses. The transparent glass or plastic wall allows visual or microscopic examination of the culture status. Microbial contaminated bottles can be readily spotted and discarded before they pool together with contaminated ones.

However, the drawbacks of roller bottles are numerous for large-scale production of biologics. On-line environmental monitoring and control are virtually impossible or at least impractical. The aseptic bottle handling for inoculation, trypsinization for cell detachment and expansion, medium exchange, and product harvest, require extensive well-trained labor to ensure a high success rate. Each batch can easily involve hundreds or even thousands of bottles, and the large number of manual steps involved makes the risk of microbial contamination rather high. Despite these significant drawbacks, roller bottles are still widely used in the production of viral vaccines, often because the products involved were approved by regulatory agencies before more recent cultivation technologies became more robust and widely accepted. In some cases, roller bottles were selected over other bioreactors because the process was easy to implement compared with those based on other bioreactors. One notable example is the production of erythroprotein (EPO) using recombinant Chinese hamster ovary cells. Recent improvements include developing a fully automatic and continuous cell culture for a large number of roller bottles to make industrial-scale production by mammalian cell culture (23).

Hollow Fiber

The use of hollow fiber reactors for cultivation of mammalian cells dates back to the early 1970s (24). A hollow fiber system can be used for anchorage-dependent and suspension cells. It consists of a bundle of capillary fibers sealed inside a cylindrical tube. The basic configuration is similar to the hollow fiber cartridge used in kidney dialysis. The hollow fiber, in most cases, consists of supporting polymeric porous materials for mechanical strength and a thin layer of membrane that provides selective passage of molecules depending on their size. In most cases, an ultrafiltration membrane is used (25). The molecular weight cutoff (MWCO) of the membrane differs according to the application, ranging from a few thousand to a hundred thousand daltons. The ultrafiltration membrane prevents free diffusion of secreted product molecules from passing through the membrane and allows them to accumulate in the extracapillary space to a high concentration. The culture media is usually pumped through the fiber lumen, and cells grow in the extracapillary space, or the shell side. Supply of low molecular weight nutrients to the cells and removal of waste products occur by diffusive transport across the membrane between the lumen and the shell spaces. Although the use of microfiltration hollow fiber membranes for cell culture is infrequent, it finds application in various research uses for studying metabolism and for the cultivation of anchorage-dependent or highly aggregated cells for which a convective flow of medium through the extracapillary space to bathe cells in medium is desired.

Table 1 lists some of the types of hollow fiber membrane that have been used in cell culture and kidney dialysis. Most hollow fiber systems used for cell culture employ fibers of approximately 250–350 μm in outer diameter. The interfiber distance is at least as large. A factor limiting the potential of scaling up a hollow fiber system is the diffu-

Table 1. Membrane Materials Used in Hollow Fiber Reactor and Kidney Dialysis

Membrane material	Cell type
Cellulose nitrate or cellulose acetate	Liver cells (26)
Cellulose acetate	Hybridoma (27)
	Hepatocytes (28,29)
Polysulfone	Hybridoma (30,31,32)
	Chinese hamster lung fibroblast (33)
	MDCK (Madin-Darby Canine Kidney) (34)
	Hepatocytes (35)
Polymethyl-methacrylate	Kidney (36)
Cuprophane or polyacrylonitrile	Kidney (37)
Acrylic copolymer	SV3T3 cells (38)

sional resistance. In the radial direction, the nutrients (including oxygen) diffuse through the hollow fiber wall and membrane to reach the site of cell consumption. In the axial direction, the medium entrance region is exposed to fluid with higher oxygen, whereas the exit end inevitably sees a lower concentration of oxygen. Oxygen is of primary concern because it is usually the first nutrient to become growth limiting. To ensure ample oxygen supply, the recirculating medium is typically passed through an oxygenator (e.g., one used in surgery operations) before it enters the hollow fiber system. A typical setup of a hollow fiber system therefore includes not only a constant-temperature incubator to house the hollow fiber bioreactors but also pH control, oxygenation and oxygen monitoring systems, and a medium reservoir for recirculating medium. In addition, a small to medium reservoir is frequently used to supply high molecular weight nutrient supplements to the extracapillary space directly. A harvest reservoir is used to collect the product containing fluid from the extracapillary space. In a hollow fiber bioreactor, a nonuniform growth of cell mass and product accumulation is often seen. This growth is usually caused by the small convective flow in the shell side, the difficulty of inoculating cells uniformly, and settling due to gravity.

One technique to improve oxygen supply is the use of fiber or silicon tubing for oxygen supply into the shell side of the reactor. However, the use of mixed fibers in the bioreactor makes the bioreactor harder to manufacture. Another technique is to reverse the flow direction periodically to inverse the axial regions exposed to high and low oxygen concentrations.

Despite its limitation in scaling up, the hollow fiber system is a proven technology relatively easy to apply to processes that require a modest amount of product. In some tissue engineering and cellular therapy applications, the product is often patient specific and the quantity of cells needed is relatively small. A hollow fiber system could be ideal for those applications.

Ceramic Matrix

The ceramic bioreactor, to some extent, can be considered a variant of the hollow fiber system. It consists of a cylindrical ceramic core with many channels that pass longitudinally through the ceramic material. Cells inoculated in the channels adhere to the material or become entrapped in the pores of the ceramic. As in a hollow fiber system, ceramic reactors are supported by medium perfusion loops. Cell culture medium is pumped through the longitudinal channels in the ceramic cores from a medium reservoir in a recirculating loop configuration. Fresh medium is fed into the system, and harvested culture fluid is removed to the medium reservoir. Unlike the hollow fiber system, no membrane separates the cells and bulk medium. Product is secreted directly into the bulk medium. Essentially, the ceramic bioreactor can be used to conveniently replace a large number of roller bottles. As in hollow fiber systems, an oxygen concentration gradient develops along the axial direction and limits the length (i.e., the scale) of the reactor.

CONCLUDING REMARKS

The bioreactors described in this article represent those used widely in the cultivation of mammalian cells. They can be operated in batch, fed-batch, or continuous mode. Regardless of which operating mode a process is based on, it invariably involves a transient period in which the cell concentration is increasing. Even in a continuous perfusion culture, the concentrations of cells, nutrients, and product are often not at steady state. On-line sensors for environmental factors and control of important variables, including pH, temperature, dissolved oxygen concentration, and medium perfusion rate, are thus commonly employed. The sensors and control strategies used for mammalian cell cultivation are basically identical to those used in microbial fermentation. However, mammalian cell cultures are mostly used for the production of more complex molecules or viruses, for which the quality of the product is less definable than the smaller molecules (such as primary or secondary metabolites in microbial fermentation). They are thus more prone to change caused by fluctuations in environmental factors. A key consideration in selecting a mammalian cell culture bioreactor for a particular process is therefore the controllability of process variables and the consistency in reactor performance. Ultimately a good manufacturing process must employ a reactor that can be scaled up and controlled to produce the necessary quantity of product with the highest quality.

BIBLIOGRAPHY

1. A.L. Van Wezel, *Nature* **216**, 64–65 (1967).
2. T.-Y. Tao, G.-Y. Ji, and W.-S. Hu, *J. Biotechnol.* **6**, 9–12 (1987).
3. A. Zuhlke, B. Roder, H. Widdecke, and J. Klein, *J. Biomater. Sci. Polymer Edn.* **5**, 65–78 (1993).
4. B. Roder, A. Zuhlke, H. Widdecke, and J. Klein, *J. Biomater. Sci. Polymer Edn.* **5**, 79–88 (1993).
5. J. Varani, M. Dame, J. Fediske, T.F. Beals, and W. Hillegas, *J. Biol. Stand.* **13**, 67–76 (1985).
6. S. Reuveny, L. Silberstein, A. Shahar, E. Freeman, and A. Mizrahi, *Dev. Biol. Stand.* **50**, 115–123 (1982).

7. W. Tolbert, M.M. Hitt, and J. Feder, *In Vitro* **16**, 486–490 (1980).
8. M.V. Peshwa, Y.-S. Kyung, D.B. McClure, and W.-S. Hu, *Biotechnol. Bioeng.* **41**, 179–187 (1993).
9. S. Goetghebeur and W.S. Hu, *Appl. Microbiol. Biotechnol.* **34**, 735–741 (1991).
10. C.M. Perusich, S. Goetghebeur, and W.-S. Hu, *Biotechnol. Tech.* **5**, 145–148 (1991).
11. J.L. Moreira, P.M. Alves, J.M. Rodriguez, P.E. Cruz, J.G. Aunins, and M.J.T. Carrondo, *Ann. N.Y. Acad. Sci.* **745**, 122–133 (1994).
12. K. Nilsson, W. Scheirer, O.H. Horton, L. Osfberg, E. Liedhl, H.W.D. Katinger, and K. Mosbach, *Nature* **302**, 629–630 (1983).
13. K.K. Papas, I. Constantinidis, and A. Sambanis, *Cytotechnology* **13**, 1–12 (1993).
14. A.C. Jen, M.C. Wake, and A.G. Mikos, *Biotechnol. Bioeng.* **50**, 357–364 (1996).
15. M.S. Croughan, J.-F. Hamel, and D.I.C. Wang, *Biotechnol. Bioeng.* **29**, 130–141 (1987).
16. W.-S. Hu and T.C. Dodge, *Biotechnol. Prog.* **1**, 209–215 (1985).
17. J.J. Chalmers and F. Bavarian, *Biotechnol. Prog.* **70**, 151–158 (1991).
18. K. Kitano, Y. Shintani, Y. Ichimori, K. Tsukamoto, S. Sasai, and M. Kida, *Appl. Microbiol. Biotechnol.* **24**, 282–286 (1986).
19. M. Tokashiki, T. Arai, K. Hamamoto, and K. Ishimaru, *Cytotechnology* **3**, 239–244 (1990).
20. Y. Takazawa, M. Tokashiki, K. Hamamoto, and H. Murakami, *Cytotechnology* **1**, 171–178 (1990).
21. T.C. Seamans and W.-S. Hu, *J. Ferment. Bioeng.* **70**, 241–245 (1990).
22. V.M. Yabannavar, V. Singh, and N.V. Connelly, *Biotechnol. Bioeng.* **43**, 159–164 (1994).
23. R. Kunitake, A. Suzuki, H. Ichihashi, S. Matsuda, O. Hirai, and K. Morimoto, *J. Biotechnol.* **52**, 289–294 (1997).
24. R.A. Knazek, P.M. Gullino, P.O. Kohler, and R.L. Dedrick, *Science* **178**, 65–67 (1972).
25. T.L. Evans and R.A. Miller, *BioTechniques* **6**, 766–767 (1988).
26. J. Rozga, F. Williams, M.-S. Ro, P.F. Neuzil, T.D. Giorgo, G. Backfisch, A.D. Moscroni, R. Hakim, and A.A. Demetriou, *Hepatology* **17**, 258–265 (1993).
27. R.J. Gillies, P.G. Scherer, N. Raghunand, L.S. Okerlund, R. Martinez-Zagrilan, L. Hesterberg, and B.E. Dale, *Magn. Reson. Med.* **18**, 181–192 (1991).
28. N.L. Sussman, G.T. Gislasson, C.A. Conlin, and J.H. Kelly, *Artif. Organs* **18**, 390–396 (1994).
29. V. Dixit, E. Piskin, M. Arthur, A. Denizli, S. Tuncel, E. Denkbas, and G. Gitnick, *Cell Transplant* **1**, 391–399 (1992).
30. G.L. Altshuler, D.H. Dziewulski, J.A. Sovek, and G. Belfort, *Biotechnol. Bioeng.* **28**, 646–658 (1986).
31. J.M. Piret, D.A. Devens, and C.L. Cooney, *Can. J. Chem. Eng.* **69**, 421–428 (1991).
32. D. Patankar and T. Oolman, *Biotechnol. Bioeng.* **36**, 97–103 (1990).
33. J.P. Tharakan and P.C. Chau, *Biotechnol. Bioeng.* **28**, 329–342 (1986).
34. H.D. Humes, M. Mackay, A.J. Funke, and D.A. Buffington, *Am. J. Kydney Dis.* **30**, S58–S31 (1997).
35. H.O. Jauregui, C.-P. Mullon, D. Trenkler, S. Naik, H. Santangini, P. Press, T.E. Muller, and B.A. Salomon, *Hepatology* **21**, 460–469 (1995).
36. R.M. Hakim, R.L. Wingard, and R.A. Parker, *N. Engl. J. Med.* **331**, 1338–1342 (1994).
37. M. Schiffl, S.M. Lang, A. Konig, T. Strasser, M.C. Haider, and E. Held, *Lancet* **344**, 570–572 (1994).
38. K. Ku, M.J. Kuo, J. Delente, B.S. Wildi, and J. Feder, *Biotechnol. Bioeng.* **213**, 79–95 (1981).

See also SUSPENSION CULTURE, ANIMAL CELLS.

MAMMALIAN CELL CULTURE REACTORS, SCALE-UP

J. BRYAN GRIFFITHS
Scientific Consultancy & Publishing
Salisbury, United Kingdom

KEY WORDS

Anchorage-dependent cells
Bioreactors
Cell culture processes
Cell products
Mammalian cell culture
Process parameters
Scale-up
Suspension cells

OUTLINE

INTRODUCTION

Although mammalian cells have been grown in vitro since before 1907 (1) the factor that gave impetus to their current widespread laboratory and industrial use was the discovery by Enders (2) that human pathogenic viruses could be grown in cell cultures. Following this demonstration by Enders in 1949 it took only 5 years for the first cell-based vaccine to be licensed for clinical use (Salk polio vaccine in primary monkey kidney cells in 1954). This opened up 20 years of continuous development of human and veterinary viral vaccines and created the need for industrial-scale cell culture processes. The most effective large-scale process developed during this period was for foot-and-mouth disease virus (FMDV) based on suspension culture of baby hamster kidney (BHK) cells (3). Developments in processes for human vaccines were less dramatic because of the need to use biologically safe cell lines. This meant that the use of human diploid cell lines, such as WI-38 (4) and MRC-5 (5), because of their normality (i.e., they behaved as cells in vivo without tumorgenic transformation), only grew attached to a substrate (anchorage dependent) and only reached low cell densities. The wide range of animal cell reactors available is partly due to the dual development of systems for anchorage-dependent cells (ADCs) and free-suspension cells (6,7). ADCs were grown in small culture vessels, and a production batch constituted hundreds or even thousands of replicate cultures (i.e., a multiple-batch process) (8). The need for a unit batch process (one large culture vessel), such as the fermentors used for suspension cells, saw the development of a wide range of novel culture reactors (6,7), but the significant breakthrough came with the microcarrier system (9). This procedure, developed in 1967 by van Wezel (9), allowed cells to grow attached to small (200 μm) spheres that were stirred in a large tank fermentor analogous to suspension cells. This was the first successful large-scale unit process for ADCs.

Vaccines were the dominant product until the 1970s, but changes in regulatory and licensing procedures then allowed cells from sources other than normal tissues to be used for human medicinal products. This came about during the development of an interferon process using a cancer cell line, Namalva, by Wellcome (10). This company's pioneering work established the safety criteria, and thus acceptance, for using nonnormal (heteroploid, transformed, or tumor-derived cell lines) and the feasibility of scaling up an industrial cell culture process to 8,000 L. Cell culture then entered a new, or modern, phase in which a wide variety of cell products (Table 1) is produced from a range of cell types (Table 2). Two of these products can be highlighted as being significant milestones. First, the production of monoclonal antibodies from hybridoma cells (the fusion of a normal antibody producing cell and a hemopoitic cancer cell) (12) has given rise to hundreds of new products. Second, the development of recombinant tissue plasminogen activator (tPA) by Genentech (14) gave rise to the first genetically engineered clinical product from cell cultures. Currently applications are widening to include the cell itself as a product in tissue engineering, organ replacement, and gene therapy (15).

In this article the description of cell culture reactors is given against this historical evolution of cultures from small-scale virus production through scale-up to the more than 10,000-L unit processes that now play a dominant role in the health biotechnology industry, with hundreds of products either already available in the clinic or going through phase I through III clinical trials. The emphasis is on scale-up because of the relatively low biomass productivity of animal cells compared with bacteria, the low product expression rates (one clinical dose is often equivalent to 0.1 to 100 L of culture supernatant), and the need to introduce more efficient and economical industrial processes.

BACKGROUND

Mammalian Cells

Major differences between single mammalian cells and microorganisms should be noted because they influence, and put constrictions on, the culture process. Cells do not have a cell wall but a thin cell membrane (a phospholipid bilayer) that mediates communication between the cell and the environment. They are thus very sensitive to environmental stimuli, such as pH, nutrients, oxygen, regulatory molecules, and hydrodynamic forces. Cell shape and morphology are controlled by cytoskeletal elements attached to the membrane and running through the viscous cytoplasmic gel. The degree of polymerization of the microtubular network regulates the shape, whether spherical or elongated, flattened or attached. Mammalian cells are on the order of 12–20 μm in diameter (i.e., 1,000-fold greater volume than microorganisms), which means diffusion into the cytoplasm is greatly reduced in the spherical state. When flattened, membrane folding and ruffling give a much greater surface area. Also, mammalian cells are far more fastidious in their nutrient requirements, with critical optimal chemical and physical environmental parameters. The consequence of these factors is that cells need more gentle stirring and aeration procedures than bacteria, which greatly reduces mass transfer and homogeneous mixing in the culture. This need for more moderate stirring and aeration procedures, together with greater sensitivity to nonoptimum or fluctuating culture conditions, slower growth rate (14–22 h average doubling time), and vulnerability to microorganism contamination, means that the culture system has to be extremely well controlled and contained. Some key issues in setting up successful cultures are as follows:

- Cells should be inoculated from the late logarithmic phase rather than stationary cultures.
- Cells should be prepared for inoculation as quickly as possible and the time they are exposed to trypsin, acidity, or low temperatures limited during harvesting.
- The new culture medium should be at the required temperature (37 °C), pH (usually 7.1–7.3), and so forth before inoculating.
- The minimum inoculation density (1–2 $\times$ 10^5/mL or 2 $\times$ $10^4/cm^2$) should always be exceeded.

Table 1. Mammalian Cell Products

	Native products
Human vaccines	Polio (1954), measles (1963), rabies (1964), mumps (1969)
Veterinary vaccines	Foot-and-mouth disease vaccine, rabies, Marek's, pseudorabies, BVD, louping ill, bluetongue, avian influenza, canine distemper (examples)
Interferon	Refs. 10 and 56
	Engineered products
Monoclonal antibodies	OKT3/Orthoclone (1987), Centoxin (1990), Reopro (1994), Myoscint (1989), Oncoscint (1990)
Tissue-type plasminogen activator	Activase/actilyse (1987)
Erythropoeitin (EPO)	Epogen/Procrit/Eprex (1989), Epogin/Recormon (1990)
Human growth hormone	Saizen (1989)
HBsAg	GenHevac B Pasteur (1989), HBGamma (1990)
Interferon	Roferon (1991)
Colony stimulating factor (G-CSF)	Granocyte (1991), Neupogen (1991)
Blood factor VIII	Recombinate (1992), Kogenate (1993)
Dnase I	Pulmozyme (1993)
Glucocerebrosiduse	Cerezyme (1994)
Follicle-stimulating hormone	Gonal-F (1995)
	rDNA products in development / clinical trial
HIV vaccines (gp120, pg160, CD4)	
Herpes simplex vaccines (gB, gD)	
Chimeric monoclonal antibodies (her2, CD4, TNFa, CD20, Cd18, TAC, leukointegrin, CF54, RSV)	
Others (TSH, TNF, M-CSF, IL-6, IL-1)	
In vitro diagnostic monoclonal antibodies (1–200)	
Tissue engineering and replacement (e.g., skin, artificial kidneys)	

Scale-Up

Principles. Normally a culture process begins with an ampoule, stored at −196 °C, of about 5 million cells put into 10 mL of medium, which can then be expanded up to a final volume of 10–20,000 L at 2 million cells per milliliter (i.e., a 1 to 2 million–fold scale-up). This expansion is achieved by a series of ever-increasing culture volumes (the seed train), each step bringing about a 5- to 10-fold volume increase and taking 4 to 5 days. Thus, before the final production culture is initiated, cells will have been passaged at least five times over 25 days. This investment in time and resources shows how important it is to keep sterility by strict attention to all aspects of the process (equipment, reagents, and handling procedures). The scale-up process involves many factors, including engineering (especially to maintain sterility and reliability), physiological (supply of oxygen and nutrients and critical maintenance of the correct environmental conditions), monitoring and control (biosensors, performance measurements, and computer control), logistics, and compliance with pharmaceutical licensing and health and safety regulations. The aim of such scale-up is to provide more cells and more cell product in as efficient and cost-effective a manner as possible while complying with all regulatory and health and safety criteria. This emphasizes the importance of moving from a multiple-unit (many replicate bottles, roller cultures, spinner flasks, etc.) to a high-volume-unit process. Such a shift allows economy in manpower (reduction in repetitive steps), space (expensive hot rooms), and time and an increased efficiency by being able to reproducibly control environmental factors, such as pH and oxygen, in a unit process. The wide range of alternative culture systems that are or have been available (6,7,16) is due (*1*) the need to find processes for both suspension and ADC cells and (*2*) various solutions to overcoming limiting factors in scale-up (17).

Limiting Factors in Scale-Up. The first problem in moving from a small, multiple unit process to a large unit process is usually oxygen limitation. To overcome this problem and supply oxygen without the damaging effects of air bubbles from sparging and the high stirring speeds needed to effect efficient mass transfer, bioreactors have been developed that oxygenate through a membrane. This bubble-free aeration has been achieved by means of long lengths of tubing in the bioreactor (18–20) or external medium loops through hollow fiber oxygenators. A related factor is mixing, and low-shear modifications to impellers and the airlift bioreactor concept (21) have been introduced. A problem for ADCs is the limitation on surface area during scale-up. A huge range of reactors have been developed (multiple plates, spirals, ceramic matrices, glass beads), but the most successful method has been the microcarrier (9,22). The large fermentor systems used for suspension cells thus became available for ADCs because the small microcarriers with cells attached behaved analogously to a suspension cell.

The next problem to be overcome is nutrient limitation and toxic metabolite buildup. Although media changes can be used (especially for attached cells), the most efficient means of overcoming these problems is continuous medium perfusion. To perfuse suspension cells cells and waste

Table 2. Typically Used Cells in Biotechnology

Cell line	Derivation	Application
CHO	Chinese hamster ovary	Good product glycosylation
CHO dhfr	Mutant for genetic amplification	Principal substrate for recombinant products
BHK21	Syrian hamster	Viruses/veterinary vaccines
Vero	African green monkey	Viral vaccines
HeLa	Human cervical adenocarcinoma	Polio vaccine
MRC-5/WI-38	Human fetal lung fibroblasts	Human vaccines
MDCK	Cocker spaniel kidney	Viral vaccines
Namalwa	Human B lymphoblastoid	Interferon
COS 1/COS 7	African green monkey kidney	Recombinant products
NS0/NS1 and Sp2/0-Ag14	Myeloma	Fusion partners for monoclonal antibody production immunoglobulin
J558L	Mouse Balb/C myeloma	Secretes immunoglobulin A
NIH/3T3	Swiss mouse embryo	Sarcoma virus
GH3	Rat pituitary tumor	Hormone studies
293	Adenovirus transformed HEK	Gene therapy
X CRE/Y CRIP	Mouse NIH3/Moloney leukemia	Retrovirus packaging
	Insect	
Sf9	Spodoptera frugipeda	Baculoviruses
Schneider-2	Drosophila melangogaster	Recombinant products

Source: Ref. 12.

medium have to be physically separated; this separation has been done by means of spin filters, external loops to filters or centrifuges, or gravitational settling devices (18). The use of perfusion resulted in the development of many immobilization procedures to prevent cell washout (e.g., encapsulation, hollow fiber devices, membrane reactors, gels, fibers, and porous microcarriers). Additional benefits to these developments were the provision of huge unit surface areas for cell attachment and the ability to grow cells at 50- to 100-fold higher densities than previously achieved in free suspension. Thus these developments mean that manufacturers have a choice in scale-up routes (6).

Scale-Up Strategies. Some scale-up strategies are as follows (22–24):

- High-volume (to 10–20,000 L) tank fermentor systems with conventional cell densities of $1–2 \times 10^6$/mL
- Low-volume (<1 L) but very-high-density ($1–2 \times 10^8$/mL) heterogeneous systems (e.g., hollow fiber bioreactors)
- Intermediate systems using spin filters to increase cell density to $1–2 \times 10^7$/mL but to a volumetric scale of only 500 L
- High-density (1×10^8/mL) scaleable systems (to 200 L) based on porous microcarriers

These systems are summarized conceptually in Figure 1. It should be emphasized at this stage that there is no ideal system, but the advantages of high-density systems, summarized in Table 3, show why a choice has to be made. Immobilized systems, such as porous microcarriers, also have the advantage of being equally suitable for suspension and anchorage-dependant cells (the Universal System?), protecting from shear, and having a three-

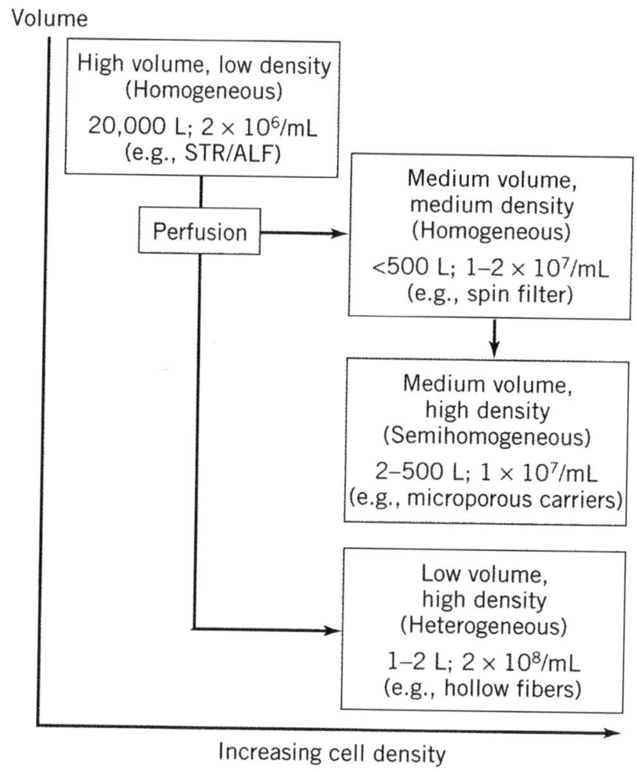

Figure 1. Scale-up strategies and the evolution of high cell density culture systems. STR, stirred-tank reactor; ALF, air-lift fermenter.

dimensional structure that enhances productivity. Basically, the low-volume, high-density systems fill an important niche in manufacturing capability (e.g., gram amounts of monoclonal antibodies), whereas the high-volume systems are the most widely used scale-up route for pharmaceutical products.

Table 3. Comparison of High- and Low-Cell-Density Culture Systems

Advantages of high cell density
Smaller reactor volume and therefore reduced need for high-quality laboratory space
More concentrated product
Reduced requirement for expensive serum and growth factors
Long run lengths (50–100 days) and thus less downtime and seed preparation
Daily or continuous harvesting
Shorter residence time for products (and exposure to preteolytic degradation)
Cell debris washout
Disadvantages of high cell density
Volumetric scale-up more difficult and complex
Mass transfer problems leading to heterogeneity
Sterilization problems because some systems cannot be heat sterilized
Higher degree of process control needed (direct cell measurement impossible)
More validation and regulatory conditions for product licensing needed
Perceived as experimental and therefore not accepted as readily as traditional methods

Culture Modes

Suspension culture is the preferred method for scale-up because it is easier and cheaper to scale up, requires less space, can monitor cell growth, and can more easily control environmental parameters. However, many cell lines have higher specific productivity when attached to the substrate. In the urge to move to suspension systems the following advantages of anchorage-dependent systems should not be overlooked:

- They facilitate complete medium changes, which are often needed to wash out serum before adding serum-free production medium, for example.
- They are easier to perfuse and thus achieve higher cell densities and healthier cells.
- Media-to-cell ratios are more easily changed during an experiment, thus increasing product titre.

Microcarrier culture, especially porous microcarrier (25), has most of the advantages of both suspension and anchorage-dependent systems.

Different types of culture strategy can be employed; they are defined as follows:

Batch culture. Cells grow from seed to final density over a 4- to 7-day period and are then harvested (typical monoclonal antibody yield is 6 mg/L per batch run)

Fed-batch culture. Additional media or medium components are added to increase culture volume and density of cells as the culture progresses.

Semicontinuous batch culture. A batch culture is partially harvested (e.g., 70%), topped with fresh medium, allowed to grow up again, and reharvested. Typically three to four harvests can be made, although usually with diminishing returns at each harvest.

Continuous-flow (chemostat) culture. Because it is completely homogeneous for long periods of time while cells are in a steady state, this is an extremely useful tool for physiological studies but not very economical for production (by definition, cells are never at maximum density).

Continuous-perfusion culture. This type of culture differs from chemostat in that all nutrients are kept in excess and cells are retained within the bioreactor. This is the most productive culture system, with typical monoclonal antibody yields of 10 mg/L per day for 50–100 days.

There is no definitive answer as to the best system to use; the answer depends on the nature of the cell and product, the quantity of product, downstream processing capability, licensing regulations, and so forth. However, a rough guide to relative costs for producing monoclonal antibodies by perfusion, continuous-flow, and batch culture is 1:2:3.5 (26).

REACTORS FOR ADCS

Basic Culture Units

Tissue culture flasks and tubes giving surface areas of 5–200 cm^2 are the familiar stock items of any culture laboratory. The largest stationary flask routinely used is the Roux bottle (or disposable plastic equivalent), which gives a surface area of 175–200 cm^2 (depending on make and type), needs 100–150 mL medium, and utilizes 750–1,000 cm^3 of storage space. Such a vessel will yield 2×10^7 diploid, or 10^8 heteroploid cells; thus, to produce a modest 10^{10} cells, more than a hundred replicate cultures are needed (i.e., manipulations have to be repeated 100 times) and 100 L culture space is required. The need to move to larger units is obvious and is approached by increasing the surface-area-to-volume ratio. The first step in scale-up usually involves a change from stationary flasks to roller bottles (i.e., a dynamic system). Roller bottles can be up to 1,750 cm^2, use 350 mL medium, and have a volume of 2.5 L (i.e., a ninefold increase in surface area but only a threefold increase in medium and total volume). This is brought about by using the total internal surface area for cell growth and more efficient aeration as the cells move in and out of the culture fluid. However, this procedure has been used industrially for viral vaccines, veterinary vaccines in multiples of 28,000 (8), interferons, and EPO. It has a place in modern processes largely because it is automatable with robotic systems, such as the Cellmate (Automation Partnership, Royston, Herts, U.K.) and that described by Kunitake et al. (27). All the routine manipulations of cell seeding, media changing, bottle gassing, cell sheet rinsing, trypsinization, and cell collection by scraping can be carried out automatically, and reproducibly, with very precise volumes. Examples of products produced by this method are Varivax (Merck varicella vaccine) and Saizen (Serono recombinant human growth hormone).

Many modifications have been made to the roller bottle to increase the surface-area-to-volume ratio. Although some of the following adaptations are no longer commercially available, they are included to stimulate further innovation:

- *Disks.* A series of vertical disks placed along the horizontal axis of the roller bottle (28)
- *Spira-Cell Multi-Surface Roller Bottle (Bibby Sterlin Ltd.).* A spiral polystyrene cartridge on which cells grow on both sides of the film, available in 2,000, 4,500, and 6,000 cm^2 sizes (29).
- *Glass tubes.* A roller bottle packed with a cluster of small glass tubes arranged in parallel and separated by silicone spacer rings, available in 1,000, 10,000, and 15,000 cm^2 sizes (Bellco-Corbeil Culture System) (30). The initial concept was developed by Chemap as the Gyrogen (31)
- *Extended surface area roller bottles (ESRB).* A polystyrene roller bottle with a ribbed, or corrugated, surface that doubles the area of an 850-cm^2 bottle to 1,700 cm^2 (Bibby Sterlin Ltd.).
- *ImmobaSil.* Silicone rubber matrix–coated roller bottles that increase the surface area severalfold (Ashby Scientific Ltd., Coalville, U.K.).

Also, to increase cell productivity, roller bottle perfusion systems (32) with specially modified swivel caps (Bellco Autoharvester, New Brunswick Roller Apparatus) can be used. The roller bottle is a well-established technique and is still widely used in both the research laboratory and the industrial plant. Some cell lines (particularly epithelial) may not be as successfully grown in roller bottles because of streaking, clumping, or inadequate spreading over the total surface (i.e., nonlocomotory cell lines) as in stationary bottles. An alternative scale-up route is to use multisurface plate stationary systems.

Multisurface Plate Units

There are two commercially available examples of multiple-layered polystyrene plates stacked within a polystyrene box. In the Cell Factory (A/S NUNC, Roskilde, Denmark), the trays are 335 × 205 mm (600 cm^2) and can be obtained in multiples of 1, 2, 10, and 40. The largest, therefore, has a surface area of 24,000 cm^2 and needs 8 L medium (i.e., it is equivalent to 14 large roller bottles but requires half the incubation space and no ancillary roller equipment). However, for convenience of handling, especially large numbers, the Cell Factory Manipulator is advantageous because it can be operated robotically.

Although the CellCube (Costar, Cambridge, Mass., USA) (33) also has parallel polystyrene trays, it is a modular closed-looped perfusion system, including an oxygenator, pumps, and a system controller (pH, O_2, level control). The unit is very compact, with the trays being only 1 mm apart; thus, the smallest unit of 21,250 cm^2 is less than 5 L total volume (1.25 L medium). Additional units of 42,500 cm^2 (2.5 L medium) and 85,000 cm^2 (5 L medium) are available, and four units can be run in parallel in the system, giving 340,000 cm^2 growth area.

High-Volume Units

High-volume units are true unit process systems, analogous in volume and performance to suspension cell fermentor vessel processes. Examples include (*1*) the multisurface propagator (34), with vertically stacked circular stainless steel plates on a central revolving shaft scaled up to 200 L (200,000 cm^2), and the titanium plate reactor (35); (*2*) the plate heat exchanger (36,37); (*3*) glass spheres (38–41); (*4*) microcarriers (9,42,43); (*5*) porous microcarriers (44–46); and (*6*) stainless steel wire coils (48).

Glass Bead Culture. Packed beds of 3- to 5-mm glass spheres through which medium is continuously perfused has a demonstrated scale-up to 100 L (49). Spheres of 3-mm diameter pack sufficiently tightly to prevent the bed from shifting but allow sufficient medium flow up the column so that fast flow rates, which would cause shear damage, are not needed. The physical properties of glass spheres are given in Table 4. Medium can be circulated by pump or by airlift (to give better oxygenation) (50). Although 3-mm beads are normally used to maximize the available surface area, the use of 5-mm spheres can actually give a higher total cell yield, despite the reduced surface area per unit volume, because of better physiochemical conditions (6,39). Scale-up has vertical limitations in that it is difficult to maintain a uniform plug flow without channeling. For inoculation, the use of a perforated tube placed centrally for the height of the bed allows even distribution of cells. Also, alternating the direction of medium flow on a daily basis increases homogeneity and thus scale-up potential. In conclusion, glass sphere packed beds constitute a simple system (that minimizes moving parts and the risk of mechanical failure) with an inexpensive and reusable substrate capable of considerable radial and reasonable vertical scale-up to operate beds of volumes more than 200 L using 5-mm spheres. The disadvantages are that it is a low-intensity system (spheres have the minimum surface area per unit volume) and is limited to secreted products because it is difficult to harvest cells from the bed (i.e., it is ideal for long-term continuous cultures rather than batch cultures, as demonstrated by Brown et al. [51]). Experimental protocols for using glass sphere technology have been published (52).

Table 4. Properties of Solid Glass Sphere Fixed-Bed Reactors

Glass sphere (diameter)	3 mm	5 mm
Weight, g	0.0375	
Surface area, cm^2/kg	7,546	4,570
Volume, mL/kg	990	625
Void fraction	0.46	0.40
Void volume, mL/kg	455	250
Channel size, mm^2	0.32	1.00
Vero cell yields		
$10^5/cm^2$	0.78	2.5/3.1[a]
10^8/kg	4.00	8.0/9.9
10^6/mL	1.6	3.2/4.0

[a]Yield in 1 kg/10 kg bed reactors.

Microcarrier Culture. Microcarriers are small particles, usually spheres, 100 to 300 μm in size, that are suspended in stirred culture medium. The concept was initiated in 1967 (9); however, the chromatographic grade of dextran (Sephadex A-50) then used proved unsuitable for consistent and reliable growth. After considerable developmental work, Pharmacia produced a range of suitable microcarriers—the Cytodex series. The first industrial process based on microcarriers was for FMDV (43,53). Subsequently a wide range of microcarriers based on gelatin, collagen, polystyrene, glass, cellulose, polyacrylamide, and silica have been manufactured to meet all situations (23,54) (Table 5). The key criteria were to get the surface chemically and electrostatically correct for cell attachment, spreading, and growth. The power of the method is exemplified by the following data: 1 g Cytodex = 6,000 cm^2, which at 2 g/L = 12,000 cm^2/L (i.e., eight large roller bottles). Scale-up to 4,000 L (equivalent to 3,200 roller bottles) has been achieved in an environmentally controlled and optimized process. This method opened up industrial production opportunities (Table 6) and allowed research laboratories to easily produce substantial quantities of developmental products.

Experimental considerations and principles (54) are as follows:

- Because cells differ in their attachment requirements, a range of microcarriers should be assessed.
- Microcarriers vary in their density (1.01 to 1.06), mechanical robustness, hardness, and the concentration at which they can be used. These factors have implications for the choice of process (e.g., stirring system, need for rapid sedimentation due to frequent medium changes, and total unit surface area required: Cytodex 6,000 cm^2/g, glass or polystyrene 300 cm^2/g).
- Culture equipment is critical. Small-scale culture requires the use of specially designed spinner flasks rather than the conventional magnetic bar and siliconized glass to prevent microcarrier attachment.
- Culture initiation conditions are critical and should include a high-quality single-cell inoculum and a prewarmed and stabilized culture medium. Inoculation should not be undertaken below the minimum cell number per bead (usually five, optimally seven), and a nutritionally supplemented medium (e.g., with nonessential amino acids) might be necessary.
- Instead of counting by the trypan blue technique, after trypsinization the nuclei-counting method (57), should be used.

Table 5. Microcarriers

Name	Manufacturer	Material	cm^2/g
	Solid microcarriers		
Bioglas	Solo Hill Engineering	Glass/latex	350
Bioplas	Solo Hill Engineering	Polystyrene	350
Biosilon	Nunc	Polystyrene	255
Cytodex 1, 2	Pharmacia	Dextran	6,000
Cytodex 3	Pharmacia	Collagen	4,600
Dormacell	Pfeifer & Langen	Dextran	7,000
Gelibead	Hazelton Laboratories	Gelatin	3,800
Micarcel G	Reactifs IBF	Polyacrylamide	5,000
Microdex	Dextran Products	Dextran	250
			Diameter (μg)
	Microporous microcarriers		
Cellsnow	Kirin Ltd	Cellulose	800–1,000
Cytocell	Pharmacia	Cellulose	180–210
Cultispher	Hyclone	Gelatin	170–270
Cytoline 1,2	Pharmacia	Polyethylene	1,200–1,500
ImmobaSil	Ashby Scientific	Silicone rubber	1,000
Siran	Schott Glaswerke	Glass	400–5,000

Table 6. Industrial Processes Using Microcarrier Technology

Product	Cell line	Scale (L)	Company	Reference
Foot-and-mouth disease vaccine	Pig kidney	235	Merrieux	50
Polio and rabies vaccines	PKMC	500	RIVM (Bilthoven)	55
Polio and rabies vaccines	Vero	1,000	Merrieux	43
Human IFN-β	HDC	50	Sclavo	11
Human IFN-β	HDC	4,000	Toray	
Human growth hormone	Vero	100	UNSW (Australia)	56
Tissue-type plasmingon activator	Bowes melanoma	40	RIVM (Bilthoven)	42

- Some microcarriers can be reused, and it may be possible to recolonize microcarriers in situ during scale-up (58).
- Enzyme harvesting (e.g., trypsin) can be difficult or damaging, so consideration should be given to using a microcarrier that can be enzymically digested (e.g., by dextranase or collagenase), leaving the cells intact.
- Microcarriers, such as Cytodex, should be used between 1 and 3 g/L unless a perfusion system (e.g., using a spin filter) is employed, with which concentrations of 15 g/L are feasible (39).
- Inoculation in a reduced volume assists attachment and should be followed by topping up to the final working volume after 4–8 h.
- An advantage of microcarriers is the ease with which cell samples can be taken and analyzed.
- As the culture progresses, partial medium changes will have to be carried out and the stirring speed increased from 40–60 rpm to 75–90 rpm to prevent microcarrier aggregation.

In conclusion, microcarrier culture is the most versatile, reliable, and well-characterized procedure for unit volume scale-up of ADCs. It has had widespread use for industrial processes (Table 6), as well as many developmental uses (59), has been scaled up to 4,000 L, and has the potential for process intensification by perfusion with spin filters or by the use of microporous microcarriers.

REACTORS FOR SUSPENSION CELLS

Laboratory Scale

The basic culture unit for suspension cells is the spinner flask, so called because it has a magnetic bar operated by standing the unit on a magnetic stirrer. Initially a bar magnet was placed on the floor of the glass vessel, but spinner flasks now have the magnet suspended just above the bottom of the culture on a stirrer shaft fitted to the bottle top to prevent cells from being crushed. Side arms are usually fitted to allow gassing with CO_2/O_2 through a filter and for sampling. There are many variations in the design of spinner flasks, most affecting the size, number, and positioning of the stirrer bar (6,23). The spinner flask is usually glass, with a silicone- or Teflon-coated magnet and is available in sizes from 50 mL to 20 L, although 10 L should be considered the maximum practical size. The quality of the magnet stirrer (usually multibased to take at least four replicates) is extremely important because it must maintain a stable rpm between 30 and 300, an accurate tachometer, ability to restart after power interruptions at the set speed, a surface that does not become hot, and reliable operation for long periods in a 37 °C environment. It is advisable to siliconize the culture vessel and add medium supplements, such as Pluronic F-68 (polyglycol) (BASF, Wyandot) at 0.1% to protect against mechanical damage, especially if low serum concentrations are used. Carboxymethyl cellulose (CMC, 15–20 cps) can also be used to protect cells (0.1%) and antifoam (6 ppm) if foaming is a problem (usually when serum is present at 5–10%). Stirring speeds for suspension cells are usually within the range of 100–250 rpm (depending on cell type and vessel geometry), and a culture will expand from 1–2 × 10^5/mL to 1–2 × 10^6/mL in 4 to 5 days before needing either harvesting or a medium change. Cell clumping is often a problem but can be reduced by using low Mg^{2+} and Ca^{2+} ion media. Conventional spinner flasks are available from a wide range of laboratory suppliers. The range includes ones with large-bladed paddles attached to the magnet. They are primarily designed for microcarrier culture to give better mixing at the lower speeds necessary. Modified spinner cultures include a radial stirring action (Techne) (60), a floating impeller (Techne BR-06 Bioreactor), the Superspinner (Braun Melsungen AG), CellSpin (Tecnomara AG, Switzerland), and a dual overhead drive system to allow perfusion (Bellco). Experimental details for using spinner flasks have been published (61).

Scale-Up

Suspension culture is the preferred method for scaling up cell cultures because it is easier to volumetrically increase the size of a single fermentor vessel than the specialized units used for ADCs. Although some cell lines will not grow in suspension (e.g., human diploid cell lines WI-38 and MRC-5), many lines can be adapted to grow in free suspension by recognized procedures (62).

Several physical factors must be satisfied for successful scale-up. Good mixing is essential for homogeneity and efficient mass transfer, but during scale-up the power needed to produce these conditions can cause problems. The energy generated at the tip of the stirrer blade is a limiting factor because it gives rise to damaging shear forces (created by fluctuating liquid velocities in turbulent areas); because mixing efficiency increases with turbulence, a compromise has to be reached to minimize cell damage. Thus cell cultures use large impellers running at relatively low speeds. Magnetic bars used in spinner flasks give only radial mixing, with no lift or turbulence, and scale-up means a move to marine (not turbine) impellers. Alternative systems developed to avoid stirring are as follows:

- *Airlift fermentor.* The airlift fermentor uses the bubble column principle to both agitate and aerate a culture (21,28,63,64). Air bubbles are introduced into the bottom of a culture vessel (aspect ratio 8–12:1) and rise up an inner draft tube. Aerated medium has a lower density than nonaerated medium so the medium rises through the draft tube and circulates down the outside of the vessel (upflow and downflow approximately equal volumes). The amount of energy needed is very low and shear forces are absent; thus it is an ideal method for fragile cells. Airflow rates of about 300 mL min^{-1} are used. Vessels are available from 2 L upward, with 2,000 L reactors used in industrial manufacturing processes. However, scale-up is more or less linear, and a 90-L vessel requires 4 m headroom; special manufacturing suites are thus needed for large-scale operations.

- *Vibro-mixer.* The Vibro-mixer (65) is a nonrotating agitator that produces a stirring effect by a vertical reciprocating motion with a path of 0.1–3 mm at a frequency to 50 Hz. The mixing disk is fixed horizontally to the agitator shaft, and cone-shaped holes in the disk cause a pumping action as the shaft vibrates. This is a low energy system with reduced shear forces and foaming.
- *Celligen fermentor.* This fermentor (66,67) is available from New Brunswick Scientific and is designed so that the impeller acts as a fluid pump and aerator. Macrocirculation mixing is generated by the hollow central shaft filled with three rotating horizontal jet tubes in a low-shear bulk movement of cells and medium (cell-lift effect).

The other principal physical factor to be accommodated is aeration. Sparging can cause cellular damage unless carried out at very low rates. General guidelines are to use large bubbles (1–3 mm diameter) and low flow rates (5–10 cm^3/L/min), which means relatively inefficient mass transfer, especially at the low stirring rates used. Even though cell bioreactors have an aspect ratio of less than 2:1 to maximize surface diffusion, a means of increasing oxygen supply during scale-up is essential. Strategies for increasing the oxygen supply have included a surface aerator (68), multiple nozzle injection on the medium surface (69), caged aerator (70), spin exchange aerator (37), external and internal loop oxygenator (71), membrane tubing arranged in the vessel (19), and stirring system (20). The use of closed perfusion loops through external reservoirs is also an efficient means of oxygenation as long as the perfusion rate is fast enough (6), which requires a very efficient spin filter or other cell separation device in the culture.

Scale-up from laboratory spinner cultures to fermentors (1–10,000 L) has the following consequences:

- Change from glass to stainless steel vessels
- Change from mobile (autoclavable) to in situ system connected to steam, seed, medium holding and downstream harvesting vessels, and so forth
- Requirement for water jacket, or internal temperature control
- Greater sophistication in environmental control systems and engineering to preserve integrity

Animal cell bioreactors were adapted from the microbial fermentation tanks with only minor modifications:

- Suitable impeller (e.g., marine)
- No baffles and a curved base for better mixing at low speeds
- Water jacket rather than immersion heater type of temperature control (to avoid localized heating at low stirring speeds)
- Top-driven motor stirrer so that cells cannot become entangled between moving parts
- Mirror finish to internal surfaces to reduce mechanical damage and cell attachment

Large-Scale Bioreactors

Large-scale bioreactors are based on either the fermentor tank or the airlift principle. Stirred tanks have been operated at 8–10,000 L (14) for the production of interferon and tPA and airlift at 2,000 L principally for monoclonal antibodies. The preference is for the tank fermentor because this is a well-tried and familiar production method using conventional production plant facilities that have been refined over the past 30 years. However, most processes run at scales between 50 and 500 L. Airlift bioreactors are considered more flexible in that a greater range of cells (including extremely fastidious and fragile types) can be grown in them. However, scale-up is very dependent on tall buildings. A modification of the single-cell suspension system is to grow cells as aggregates or flocculants brought about by the addition of slightly positively charged polymers (e.g., poly-L-histidine) (72). Flocculants facilitate process operations such as filtration, centrifugation, perfusion, and sedimentation, and the microenvironment created within the cell aggregate can increase specific productivity (73).

Scale-up increases the complexity of equipment and the process and thus gives rise to greater risks of system failure, loss of integrity (sterility), and potential release of hazardous materials. Because the reason for scale-up is the production of therapeutic biological products, conditions have to comply with current good manufacturing practice (cGMP) standards acceptable to the relevant licensing authorities. Containment of the process to prevent ingress of contaminating factors and release of materials that may affect the process worker and environment are thus key factors in these processes. Microbiological containment of small-scale systems (below 20 L) in class III microbiological cabinets is possible (74), and larger-scale fermentation plants can be contained to P3 standards at least up to the 150-L scale (75).

The scale-up chain to the final production vessel is governed by the split ratio possible with the particular cell line at each subculture and is usually between 1:3 and 1:5. The chain starts with an ampoule of cells stored at -196 °C (liquid nitrogen) in a master cell bank (MCB) that has been fully characterized for biological contaminants and cell line stability (76). All manipulations are carried out according to standard operating procedures to obtain reproducibility between runs and to generate the necessary quality control data for acceptance of the final product.

The engineering complexity and considerable resource investment in money and time to build and operate a large-scale production process means that many new products are still brought to market using replicate small-scale systems. This provides for quicker market entry but requires complete reregistration of the product if the product is later moved to a large-scale production process. The experimental procedures for scaling up a stirred bioreactor process have been described (77).

HIGH-CELL-DENSITY BIOREACTORS

The alternative strategies for scale-up were outlined earlier. In this section the processes that support cells at significantly greater densities than the classical 2–3 × 10^6/mL level are described. In the human body cells in tissues are on the order of 2–3 × 10^9 cells/cm^3, and this has always been a target for in vitro systems. However, 2 × 10^8 cells/mL has so far been the practical limit of density scale-up; even so, this is a 100-fold increase over the conventional low-density methods described so far. The two critical factors needed to achieve higher unit cell density are (*1*) perfusion to supply nutrients and remove waste products (78) and (*2*) cell immobilization to perfuse media at a fast enough rate without damaging or washing the cells out of the culture. Many devices have been developed to meet these requirements (6,7,18,22) (Table 7), of which hollow fiber bioreactors, spin filter perfusion, and microporous microcarriers are the most successful in terms of acceptance and use.

Hollow Fiber Bioreactors

Use of hollow fiber bioreactors was pioneered by Knazek et al. (95) using ultrafiltration capillary fibers. They can be considered analogous to a blood vascular system because the fibers selectively (by molecular weight cutoff) allow passage of macromolecules through the spongy fiber wall (60 μm) as the medium flows continuously through the lumen (200 μm in diameter). Cells are kept in the extracapillary space. A unit consists of thousands of fibers potted at either end in a cylindrical housing and capable of supporting both ADCs and suspension cells at 1–2 × 10^8/mL. This concept has been widely developed, with many commercial units available, and has been used particularly for producing monoclonal antibodies. The principal limiting factor in its exploitation is the difficulty in scaling up beyond very small volumes (25 mL). The problem is due to transmembrane flux being directed back into the lumen for the latter half of the flow path (6,112). Methods of overcoming this gradient problem include the insertion of contraflow and aeration fibers (112), cyclical flushing of medium between the lumen and extracapillary compartment (113), use of flat-bed rather than cylindrical filtration cartridges (91), and use of a radial flow reactor (114). The concept of using pressure differentials to simulate in vivo arterial and venous flow is used in the Acusyst System (Cellex Biosciences, Inc.). This system is well documented (113), including experimental details (115), and is probably the most used and successful application of hollow fiber technology.

Table 7. Devices Used for Immobilizing Animal Cells to Permit Medium Perfusion

Device	Reference
Gravitational settling of cells in medium outflow	79–81
Spin filter	5,66,69,81–85
Centrifuge circulation loop	86,87
Filtration membrane	
Dialysis	88,90
Hollow fibers	91–95
Membranes	96,97
Encapsulation	98–100
Entrapment	
Textured ceramics	101
Tubular flow reactor	102
Cellulose or polyurethane sponges	103,104
Fibers	105,106
Polyester (Fibra-cel)	107
Microporous microcarriers	46,47,108–111

To allow ADCs to attach, special fibers can be used (e.g., Amicon 3S1000) to give a huge surface area (1-L cartridge has 1 m^2 surface area and supports 10^{10} suspension cells or 2 × 10^9 ADCs). Thus this method gives a very large surface-area-to-volume ratio, allows continuous removal of waste products and supply of nutrients, supports high cell densities with a tissuelike architecture, and allows a concentrated product to be harvested. The disadvantages are diffusional limitations that cause culture inhomogeneity and cell necrosis, process control complexity, and difficulty in sterilization.

Spin Filters

Because all static filters within a culture eventually become blocked, the development of a rotating filter that creates a boundary effect, thus delaying cell attachment and filter clogging, has been a successful step forward. The principle was introduced by Himmelfarb in 1969 (82) using a 3-μm stainless steel mesh rotated at 300 rpm. Subsequently there have been many variations (reviewed in Ref. 6): polymers (84), ceramics, microfibers, porcelain, immersed cages, baskets semiimmersed in the culture or attached to the stirrer shaft, separate rotating shaft (to increase rpm), filter within a draft tube and so forth. They provide a simple technical solution to scaling up process intensity within well-established stirred fermentors. They are limited in that clogging eventually occurs and in the perfusion rate that can be maintained. However, they are particularly useful for microcarriers because a far larger mesh can be used, thus reducing the onset of filter clogging and permitting cell densities of 2 × 10^7/mL to be attained.

Microporous Microcarriers

Microcarrier culture has proven to be the most effective scale-up method for ADCs, despite its limitations (critical procedures, low surface-area-to-volume ratio of a sphere). Attempts were made to increase the surface area by surface indenting techniques but were not successful until the fabrication of porous particles. The initial particle was the Verax microsphere (47,108), which was 500 μm in diameter and manufactured from bovine collagen; the interconning channels of 20–40 μm diameter gave an internal open volume of 80% of the sphere. They were designed for fluidized-bed reactors, and thus their specific gravity was artificially increased to 1.7 with the addition of small metallic weights incorporated into the matrix. The spheres were fluidized at 75 cm/min upward flow, and cell densities in excess of 10^8 per milliliter intrasphere volume were achieved (equivalent to 4 × 10^7/mL in the bioreactor). The sphere matrix provided a huge surface area for attachment of ADCs but was equally suitable for suspension cells (which were entrapped in the pores). The system was scaled up to 24 L routinely and 200 L in the company's production unit and used for more than 85 different cell lines producing many

Table 8. Advantages of Microporous Microcarriers

Unit cell density 50- to 100-fold higher than that of free suspension
Suitable for both attached and suspended cells
Can be used in fluidized- and fixed-bed reactors and stirred suspension
In situ 100- to 250-fold seed expansion eliminates seed chain steps
Diffusion into a sphere efficient (30% diameter penetration = 70% of volume)
Cells protected from shear
Specific cell productivity increased three- to five-fold
Three-dimensional structure easily derivitized for specialized cells
Capable of long-term (>100 days) culture with continuous harvesting
Scale-up potential by comparison with analogous systems (microcarrier to 4,000 L)

cell products (including tPA, pro-urokinase [proUK], EPO, interferon, interleukin, factor VIII, and various immunoglobulins).

Verax was initially a closed product in that you had to invest in the complete system. Thus other microporous beads were quickly developed (Table 5). Alternative systems are also available such as the Pharmacia Cytopilot using Cytoline porous microcarriers (110). Microporous microcarriers that could be used in stirred rather than fluidized bioreactors are preferred and now available (Cellsnow and ImmobaSil). ImmobaSil (116) is of particular value because it is extremely permeable to oxygen, a robust, nonanimal product safe from bovine contaminants that can be used in all culture modes. Because the concentration of spheres is far less, the total cell mass is greatly reduced in stirred systems. However, it allows manufacturers to use conventional stirred equipment and increases cell density in the microcarrier system from 2×10^6/mL to in excess of 5×10^7/mL (using spin filter perfusion and aeration).

An alternative to using stirred or fluidized microporous carriers is use of a fixed bed of porous glass spheres (e.g., Siran) (46,109). Fixed-bed reactors of solid glass spheres have many advantages (111), and the disadvantage of low cell density can be overcome by using 5-mm Siran spheres. These spheres have surface areas of 75 m^2/L with interconning pores and channels of 60–300 μm and can be stacked in 5-L beds and perfused at 5 linear cm/min to provide a high feed rate without washing out or damaging the cells (protected from shear within the particles). The versatility and usefulness of this technique has been demonstrated for a range of suspension and anchorage-dependent cells, resulting in a 10-fold higher productivity over equivalent systems (18,117–119). A characteristic of all microporous carrier systems is that cell specific productivity is always higher (18,117), presumably due to the favorable environment of cells packed together in almost tissuelike density. Details on experimental methods have been published; fluidized-bed (120), fixed-bed (121), and commercial fixed-bed reactors are available (Meredos GmbH, Bovenden).

The advantages of porous carrier culture are summarized in Table 8. This technology is truly universal in being equally suitable for suspension cells and ADCs and flexible in the range of bioreactor systems it can be in. It is the only truly scaleable high-cell-density system available and provides a microenvironment that stimulates cell product expression. As the need for high-quantity biopharmaceutical manufacture increases, this technology will become increasingly widespread in application and use (Table 9).

Table 9. Necessary Conditions for the Acceptance of Scaleable Immobilized Bioreactors

Insufficient quantities produced by other methods
Product only produced by immobilized or specialized cells that need three-dimensional culture
End product price only attainable with scale-up economies possible with these systems
Increased production plant capacity when expansion is no longer possible
New company or product beginning registration process
Sufficient data and experience accumulated to give manufacturing confidence

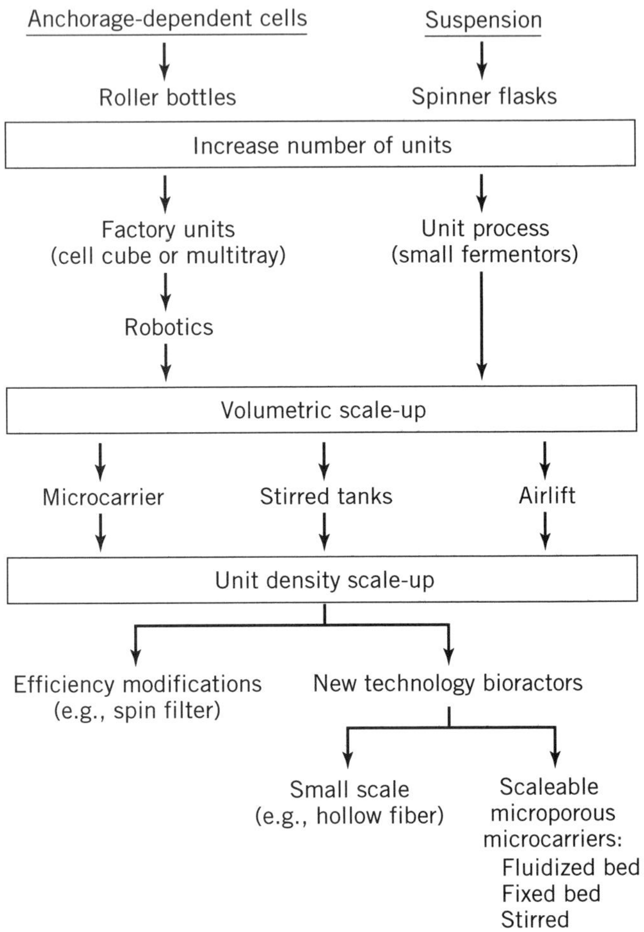

Figure 2. Scale-up: the decision process.

COMPARISONS, CONCLUSIONS, AND FUTURE DEVELOPMENTS

Given the range of culture strategies and options mentioned in this article, how does one choose the system best

Scale (L)	System	Area (cm^2) (batch size)	Cells per batch (cells/L)
	Roller	1,750 (× 100)	2×10^{10} (2×10^{8})
	Multitray	24,000 (× 10)	3×10^{10} (3×10^{9})
	Cell cube	85,000 (× 4)	4×10^{10} (9×10^{9})
1	Hollow fiber	20,000	10^{10} (10^{11})
20	Microporous microcarrier	2.5×10^{6}	10^{12} (5×10^{10})
100	Glass sphere	10^{6}	10^{11} (10^{9})
200	Stacked plate	2×10^{5}	2×10^{10} (1×10^{8})
500	Microcarrier (with spin filter)	2×10^{10}	10^{13} (5×10^{10})
2,000	Airlift		4×10^{12} (2×10^{9})
4,000	Microcarrier	5×10^{6}	8×10^{12} (2×10^{9})
10,000	Stirred tank		2×10^{13} (2×10^{9})

Figure 3. Comparison of culture systems indicating volumetric size, unit, and system cell yields.

suited to the problem at hand? In Figure 2, a flow sheet of the options is given by which a product manager can reach such a decision based on the cell line, the product type, the time table, the available plant and resources, the manpower, and the annual product requirement. The question of whether to take the high-volume or high-density option is based on the considerations summarized in Tables 3 and 8 and the range of total cell densities that various bioreactors will support (Fig. 3). The system chosen will be the one that fulfills the defined need at the minimum commercial risk, because there is no one ideal system to meet all needs.

The future of animal cell biotechnology as a production vehicle for therapeutic products is as an essential component in a range of biological processes. Recombinant microorganisms will be the first method of choice for simple, nonglycosylated products without complex three-dimensional configurations necessary for bioactivity. Transgenic production may take over for products needed in 50 to 100-kg quantities. However, society may demand that these products be produced even at greater considerable cost in culture systems to reduce the exploitation of animals; this will be the driving factor for high-volume high-density systems based on microporous microcarriers. Finally, engineering solutions (122) to scale-up are not the only means of increasing unit productivity and economy of cell manufacturing processes. Considerable advances are being made in increasing cell-specific productivity from a modest 0.2 mg/10^6 cells/day to nearer in vivo levels of 7 mg/10^6 cells/day by cell engineering (recombinant control of metabolic processes). Advances are also being made in development of media and the environment to regulate cell growth, metabolism, and product expression in simpler (cheaper) media. There are also considerable improvements to be made in downstream scale and efficiency, where often only 20% of the product survives the purification process.

BIBLIOGRAPHY

1. R.G. Harrison, *Proc. Soc. Exp. Biol. Med.* **4**, 140–143 (1907).
2. J.F. Enders, T.H. Weller, and F.C. Robbins, *Science* **109**, 85–87 (1949).
3. P.J. Radlett, T.W.F. Pay, and A.J.M. Garland, *Dev. Biol. Stand.* **60**, 163–170 (1985).
4. L. Hayflick and P.S. Moorhead, *Exp. Cell Res.* **25**, 585–621 (1961).
5. J.P. Jacobs, *Nature (London)* **227**, 168–170 (1970).
6. J.B. Griffiths, in R.E. Speir and J.B. Griffiths eds., *Animal Cell Biotechnology*, vol. 3, Academic Press, London, 1988, pp. 179–220.
7. A. Prokop and M.Z. Rosenberg, *Advances in Biochemical Engineering / Biotechnology* **39**, 29–71 (1989).
8. G.F. Panina, *Animal Cell Biotechnology*, vol. 1, Academic Press, London, 1985, pp. 211–242.
9. A.L. van Wezel, *Nature (London)* **216**, 64–65 (1967).
10. K.F. Pullen, M.D. Johnston, A.W. Phillips, G.D. Ball, and N.B. Finter, *Dev. Biol. Stand.* **60**, 175–178 (1985).
11. M. Morandi and A. Valeri, *Advances in Biochemical Engineering / Biotechnology* **37**, 57–72 (1988).
12. B. Clough, in A. Doyle and J.B. Griffiths eds., *Cell and Tissue Culture: Laboratory Procedures for Biotechnologists*, Wiley, Chichester, U.K., 1998, pp. 5–17.
13. G. Kohler and C. Milstein, *Nature (London)* **256**, 495–497 (1975).
14. A. Lubiniecki, R. Arathoon, G. Polastri, J. Thomas, M. Wiebe, R. Garnick, A. Jones, R. van Reis, and S. Builder, in R.E. Speir, J.B. Griffiths, J. Stephenne, and P.J. Crooy eds., *Animal Cell Biology and Technology for Bioprocesses*, Butterworths, Guildford, U.K., 1989, pp. 442–451.
15. R.P. Lanza, J.L. Hayes, and W.L. Chick, *Nature Biotechnology* **14**, 1107–1110 (1996).
16. M.W. Glacken, R.J. Fleischaker, and A.J. Sinskey, *Trends Biotechnol.* **1**, 102–108 (1983).

17. J.B. Griffiths, *Cytotechnology* **3**, 109–116 (1990).
18. J.B. Griffiths, in R.E. Spier and J.B. Griffiths eds., *Animal Cell Biotechnology*, vol. 4, Academic Press, London, 1990, pp. 149–166.
19. W. Kuhlmann, *Dev. Biol. Stand.* **66**, 263–268 (1987).
20. J. Lehmann, G.W. Piehl, and R. Schulz, *Dev. Biol. Stand.* **66**, 227–240 (1987).
21. J.C. Merchuk and M.H. Siegal, *J. Chem. Technol. Biotechnol.* **41**, 105–120 (1988).
22. J.B. Griffiths, *Biotechnol.* **10**, 30–32 (1991).
23. J.B. Griffiths, in R.I. Freshney ed., *Animal Cell Culture: A Practical Approach*, IRL Press, Oxford, 1992, pp. 47–93.
24. J.B. Griffiths, *Ann. N.Y. Acad. Sci.* **665**, 364–370 (1992).
25. D. Looby and J.B. Griffiths, *Trends Biotechnol.* **8**, 204–209 (1990).
26. J.B. Griffiths, *J. Biotechnol.* **22**, 21–30 (1992).
27. R. Kunitake, A. Suzuki, H. Ichihashi, S. Matsuda, O. Hirai, and K. Morimto, *J. Biotechnol.* **52**, 289–294 (1997).
28. H.G. MacMorine, G.D. Laurence, W. Parisius, and N.B. Cucakovich, *Dev. Biol. Stand.* **37**, 139–142 (1977).
29. W. House, M. Shearer, and N.G. Maroudas, *Exp. Cell Res.* **71**, 293–296 (1972).
30. M. Corbeil, M. Trundel, and P. Payment, *J. Clin. Microbiol.* **10**, 91–95 (1979).
31. H.C. Girard, M. Sutcu, H. Erdem, and I. Gurhan, *Biotechnol. Bioeng.* **22**, 477–493 (1980).
32. P.F. Kruse, L.N. Keen, and W.L. Whittle, *In Vitro* **6**, 75–81 (1970).
33. I.A. Beeksma and R. Kompier, in E.C. Beuvery, J.B. Griffiths, and W.P. Zeijlemaker eds., *Animal Cell Technology: Developments towards the 21st Century*, Kluwer Academic, Dordrecht, 1995, pp. 661–663.
34. J.B. Schleicher and R.E. Weiss, *Biotechnol. Bioeng.* **10**, 617–624 (1968).
35. O. Molin and C.G. Heden, *Prog. Immunobiol. Stand.* **3**, 106–110 (1969).
36. C. Burbidge and I.K. Dacey, *Dev. Biol. Stand.* **55**, 255–259 (1984).
37. J.B. Griffiths, D.R. Cameron, and D. Looby, *Dev. Biol. Stand.* **66**, 331–338 (1987).
38. C. Burbidge, *Dev. Biol. Stand.* **46**, 169–172 (1980).
39. D. Looby and J.B. Griffiths, in R.E. Spier and J.B. Griffiths eds., *Modern Approaches to Animal Cell Technology*, Butterworths, Guildford, U.K., 1987, pp. 342–352.
40. O.T. Ramirez and R. Mutharasan, *Biotechnol. Bioeng.* **33**, 1072–1076 (1989).
41. J.H. Robinson, P.M. Butlin, and R.C. Imrie, *Dev. Biol. Stand.* **46**, 173–181 (1980).
42. A.L. van Wezel, *Dev. Biol. Stand.* **55**, 3–9 (1984).
43. B. Montagnon, J.C. Vincent-Falquet, and B. Fanget, *Dev. Biol. Stand.* **55**, 37–42 (1984).
44. K. Nilsson, F. Buzsaky, and K. Mosbach, *Biotechnology* **4**, 989–990 (1986).
45. D. Looby and J.B. Griffiths, *Trends Biotechnol.* **8**, 204–209 (1990).
46. D. Looby and J.B. Griffiths, *Cytotechnology* **1**, 339–346 (1988).
47. P.W. Runstadler and S.R. Cernek, in R.E. Spier and J.B. Griffiths eds., *Animal Cell Biotechnology*, vol. 3, Academic Press, London, 1988, pp. 305–320.
48. W. Noe, R. Bux, W. Berthold, and W. Werz, *Cytotechnology* **15**, 169–176 (1994).
49. J.P. Whiteside and R.E. Spier, *Biotechnol. Bioeng.* **23**, 551–556 (1981).
50. Y. Chisti and M. Moo-Young, *Trans. Inst. Chem. Eng.* **72C**, 91–94 (1994).
51. P.C. Brown, C. Figueroa, M.A. Costello, and S.M. Maciukas, in R.E. Spier and J.B. Griffiths eds., *Animal Cell Biotechnology*, vol. 3, Academic Press, London, 1988, pp. 251–262.
52. D. Looby, in A. Doyle, J.B. Griffiths, and D. Newell eds., *Cell and Tissue Culture: Laboratory Procedures*, Wiley, Chichester, U.K., 1993, p. 28D:2.
53. B. Meignier, H. Mougeot, and H. Favre, *Dev. Biol. Stand.* **46**, 249–256 (1980).
54. J.B. Griffiths, in A. Doyle, J.B. Griffiths, and D. Newell eds., *Cell and Tissue Culture: Laboratory Procedures*, Wiley, Chichester, 1993, p. 28C:1.1.
55. A.L. van Wezel, *Dev. Biol. Stand.* **46**, 151–158 (1980).
56. J.S. Friedman, *Biotechnology* **7**, 359–364 (1989).
57. K.K. Sanford, W.R. Earle, and V.J. Evans, *J. Natl. Cancer Inst.* **11**, 773–795 (1951).
58. C.L. Crespi and W.G. Thilly, *Biotechnol. Bioeng.* **23**, 983–993 (1981).
59. B.H. Junker, F. Wu, S. Wang, J. Waterbury, G. Hunt, J. Hennessey, J. Aunins, J. Lewis, M. Silberklang, and B.C. Buckland, *Cytotechnology* **9**, 173–188 (1992).
60. N.A. de Bruyne, in R.E. Spier and J.B. Griffiths eds., *Animal Cell Biotechnology*, vol. 3, Academic Press, London, 1988, pp. 142–176.
61. J.B. Griffiths, in A. Doyle, J.B. Griffiths, and D. Newell eds., *Cell and Tissue Culture, Laboratory Procedures*, Wiley, Chichester, U.K., 1993, p. 28A:2.3.
62. P.B. Capstick, A.J. Garland, R.C. Masters, and W.G. Chapman, *Exp. Cell Res.* **44**, 119–129 (1963).
63. H.W.D. Katinger, W. Scheirer, and E. Kromer, *Ger. Chem. Eng.* **2**, 31–38 (1979).
64. M. Rhodes, S. Gardner, and D. Broad, in C.S. Ho and D.I.C. Wang eds., *Animal Cell Bioreactors*, Butterworth-Heinemann, Boston, Mass., 1991, pp. 253–268.
65. H.C. Girard, G. Okay, and Y. Kilvilcin, *Bulletin of the Office of International Epizoology* **79**, 805–822 (1973).
66. S. Reuveny, Z.-B. Zheng, and L. Epstein, *Am. Biotechnol. Lab.* 28–36 (January/February 1986).
67. M. Johnson, G.A.C. Chavrie, and H. Archambault, *Biotechnol. Bioeng.* **35**, 43–49 (1990).
68. W.-S. Hu and D.I.C. Wang, in W.G. Thilly ed., *Mammalian Cell Technology*, Butterworths, Boston, Mass., 1986, pp. 167–197.
69. A.L. van Wezel, *J. Chem. Technol. Biotechnol.* **32**, 318–323 (1982).
70. J.P. Whiteside, S. Farmer, and R.E. Spier, *Dev. Biol. Stand.* **60**, 283–290 (1985).
71. C. Leist, H.P. Meyer, and A. Fiechter, *J. Biotechnol.* **4**, 235–246 (1986).
72. J.G. Aunins and D.I.C. Wang, *Biotechnol. Bioeng.* **34**, 629–638 (1989).
73. A. Racher, J.L. Moreira, P.M. Alves, M. Wirth, U.H. Weidle, H. Hauser, M.J.T. Carrondo, and J.B. Griffiths, *Appl. Microbiol. Biotechnol.* **40**, 851–856 (1994).
74. P. Hambleton and J. Melling, in P. Hambleton, J. Melling, and T.T. Salusbury eds., *Biosafety in Industrial Biotechnology*, Blackie Academic, London, 1994, pp. 129–148.
75. P. Hambleton, J.B. Griffiths, D.R. Cameron, and J. Melling, *J. Chem. Technol. Biotechnol.* **50**, 167–186 (1991).

76. A. Doyle and J.B. Griffiths, in R.E. Spier and J.B. Griffiths eds., *Animal Cell Biotechnology*, vol. 5, Academic Press, London, 1992, pp. 75–96.
77. A.B. Lang and U. Schuerch, in A. Doyle, J.B. Griffiths, and D. Newell eds., *Cell and Tissue Culture, Laboratory Procedures*, Wiley, Chichester, U.K., 1993, p. 28B:1.1.
78. J.B. Griffiths, D. Looby, and A.J. Racher, *Cytotechnology* **9**, 3–9 (1992).
79. S. Sato K. Kawamura, and N. Fujiyoshi, *J. Tissue Culture Methods* **8**, 167–171 (1983).
80. M. Butler, T. Inamura, J. Thomas, and W.G. Thilly, *J. Cell Sci.* **61**, 351–363 (1983).
81. J. Feder and W.R. Tolbert, *Sci. Am.* **248**, 24–31 (1983).
82. P. Himmelfarb, P.S. Thayer, and H.E. Martin, *Science* **64**, 555–557 (1969).
83. R. Varecka and W. Scheirer, *Dev. Biol. Stand.* **66**, 269–272 (1987).
84. G.C. Avgerinos, D. Drapeau, J.S. Socolow, J. Mao, K. Hsiao, and R.J. Broeze, *Biotechnology* **8**, 54–58 (1990).
85. E. Favre and T. Thaler, *Cytotechnology* **9**, 11–19 (1992).
86. V. Jager, in R.E. Spier, J.B. Griffiths, and C. MacDonald eds., *Animal Cell Biotechnology: Developments, Processes, and Products*, Butterworth-Heinemann, Oxford, 1992, pp. 397–402.
87. K.O. Gronvik, H. Frieburg, and U. Malmstran, in R.E. Spier, J.B. Griffiths, J. Stephenne, and P.J. Crooy eds., *Advances in Animal Cell Biology and Technology for Bioprocesses*, Butterworths, Guildford, U.K., 1989, pp. 336–344.
88. G.B. Gori, *Appl. Microbiol.* **13**, 93–97 (1965).
89. A. Bohmarn, R. Portner, J. Schmieding, V. Kasche, and H. Markl, *Cytotechnology* **9**, 51–57 (1992).
90. H. Buntemeyer, C. Bohme, and J. Lehmann, *Cytotechnology* **15**, 243–251 (1994).
91. K. Ku, M.J. Kuo, J. Delente, B.S. Wildi, and J. Feder, *Biotechnol. Bioeng.* **23**, 79–95 (1981).
92. G. Klement, W. Scheirer, and H.W.D. Katinger, *Dev. Biol. Stand.* **66**, 221–226 (1987).
93. G.L. Altshuler, D.M. Dziewulski, J.A. Sowek, and G. Belfort, *Biotechnol. Bioeng.* **28**, 646–658 (1986).
94. P. Knight, *Biotechnology* **7**, 459–461 (1989).
95. A. Knazek, P.M. Gallino, P.O. Kohler, and R.L. Dendrick, *Science* **178**, 65–67 (1972).
96. F.W. Falkenberg, H. Weichert, M. Krane, I. Bartels, M. Palme, H.-O. Nagels, and H. Fiebig, *J. Immunol. Methods* **179**, 13–29 (1995).
97. Tecnomouse, Integra Biosciences, AG, CH-8304 Wallisellen.
98. K. Nilsson, *Trends Biotechnol.* **5**, 73–78 (1987).
99. R.G. Duff, *Trends Biotechnol.* **3**, 167–170 (1985).
100. A. Tinto, P. Vila, C. Casca, and F. Godia, in M.J.T. Carrondo, J.B. Griffiths, and J.L.P. Moreira eds., *Animal Cell Biotechnology: From Vaccines to Genetic Medicine*, Kluwer Academic, Dordrecht, 1997, pp. 423–428.
101. G.J. Berg, *Dev. Biol. Stand.* **60**, 297–307 (1985).
102. H.W.D. Katinger, *Dev. Biol. Stand.* **66**, 195–209 (1987).
103. A. Lazar, L. Silberstein, A. Mizrahi, and S. Reuveny, *Cytotechnology* **1**, 333–338 (1988).
104. H.A. Phillips, J.M. Scharer, N.C. Bols, and H. Moo-Young, *Cytotechnology* **9**, 29–40 (1992).
105. H. Yamaji, H. Fukuada, Y. Nojima, and C. Webb, *Appl. Microbiol. Biotechnol.* **30**, 609–613 (1989).
106. J. Litwin, *Dev. Biol. Stand.* **60**, 237–242 (1985).
107. A. Kadouri, G. Wang, W. Zhang, C. Jacklin, and D. Friedman, in M.J.T. Carrondo, J.B. Griffiths, and J.L.P. Moreira eds., *Animal Cell Biotechnology: From Vaccines to Genetic Medicine*, Kluwer Academic, Dordrecht, 1997, pp. 307–312.
108. M.W. Young and R.C. Dean, *Biotechnology* **5**, 835–837 (1987).
109. S. Pak and G. Stephanopoulas, *Biotechnol. Bioeng.* **41**, 25–34 (1993).
110. G. Klima, G. Kreismayr, D. Muller, C. Schmatz, S. Wiederkum, W. Steinfellner, A. Assadian, G. Lhota, G. Bluml, O. Doblhoff-Dier, and H. Katinger, in M.J.T. Carrondo, J.B. Griffiths, and J.L.P. Moreira eds., *Animal Cell Technology: From Vaccines to Genetic Medicines*, Kluwer Academic, Dordrecht, 1997, pp. 447–454.
111. R.E. Spier, in R.E. Spier and J.B. Griffiths eds., *Animal Cell Biotechnology*, vol. 1, Academic Press, London, 1985, pp. 243–263.
112. J.P. Tharakan and P.C. Chau, *Biotechnol. Bioeng.* **28**, 1064–1071 (1986).
113. M.L. Wolf and M.D. Hirschel, in R.E. Spier and J.B. Griffiths eds., *Animal Cell Biotechnology*, vol. 6, Academic Press, London, 1994, pp. 237–258.
114. J.P. Tharakan and P.C. Chau, *Biotechnol. Bioeng.* **29**, 657–671 (1987).
115. J.A.J. Hanak and J.M. Davis, in A. Doyle, J.B. Griffiths, and D. Newell eds., *Cell and Tissue Culture: Laboratory Procedures*, Wiley, Chichester, U.K. 1993, pp. 28D:3.1–16.
116. D. Looby, A.J. Racher, J.P. Fuller, and J.B. Griffiths, in E.C. Beuvery, J.B. Griffiths, and W.P. Zeijlemaker eds., *Animal Cell Technology: Developments Towards the 21st Century*, Kluwer Academic, Dordrecht, 1995, pp. 783–786.
117. J.B. Griffiths and A.J. Racher, *Cytotechnology* **15**, 3–9 (1994).
118. A.J. Racher and J.B. Griffiths, *Cytotechnology* **13**, 125–131 (1993).
119. A.J. Racher, D. Looby, and J.B. Griffiths, *J. Biotechnol.* **15**, 129–146 (1990).
120. D. Looby, in A. Doyle, J.B. Griffiths, and D. Newell eds., *Cell and Tissue Culture: Laboratory Procedures*, Wiley, Chichester, U.K., 1993, pp. 28D:1.1–16.
121. D. Looby, in A. Doyle, J.B. Griffiths, and D. Newell eds., *Cell and Tissue Culture: Laboratory Procedures*, Wiley, Chichester, U.K., 1993, pp. 28D:2.1–13.
122. S. Ozturk, *Cytotechnology* **22**, 3–16 (1996).

MASS TRANSFER

YUSUF CHISTI
University of Almería
Almería, Spain

KEY WORDS

Bioreactors
Diffusion coefficient
Fluidized beds
Gas–liquid mass transfer
Liquid–liquid mass transfer
Oxygen
Packed beds
Reynolds number
Solid–liquid mass transfer

OUTLINE

INTRODUCTION

Mass transfer phenomena impact upon all facets of bioprocessing. Transport effects often determine the productivity of bioreactors and the downstream recovery operations. Gas–liquid mass transfer problems arise during supply of oxygen from a gas phase to liquid culture and during removal of metabolically generated carbon dioxide from a culture fluid to a gas phase. Similarly, gas–liquid mass transfer issues must be faced again during recovery operations such as distillation. Liquid–liquid mass transfer occurs when oxygen is supplied through liquid carriers such as perfluorocarbons, during liquid–liquid extraction, and during degradation of water immiscible liquid substrates. Solid–liquid mass transfer problems are faced during recovery by adsorption, during chromatographic separations, and in operations such as crystallization. Performance of solid-phase biocatalysts such as immobilized cells and enzymes is often limited by solid–liquid mass transfer. Solid–liquid mass transfer effects influence the workings of membrane separations such as microfiltration and ultrafiltration. Transport within solid particles or intraparticle mass transfer becomes limiting in certain cases. Gas–solid transport is seen during some drying situations. Some of these transport issues are the focus of this article.

Ultimately, the transport of a solute through any fluid or space is governed by the molecular diffusivity or the diffusion coefficient of the solute in the fluid or solution. How diffusivity is affected by factors such as temperature and viscosity and the estimation of diffusivities in various situations are discussed next.

DIFFUSION COEFFICIENT OR DIFFUSIVITY

For diffusion of a solute through a solvent, the diffusion coefficient depends on temperature, the type of solvent and its viscosity, and the concentration of solute in solution. Diffusivities in liquids and gases generally increase with temperature. Liquid-phase diffusivities are little affected by pressure, but in gases, diffusivities decline as pressure increases. Typically, liquid-phase diffusivities are much lower—only about a hundredth—compared to those in gases. Diffusion coefficients of common solute–solvent combinations are available in handbooks. Tables 1 and 2 provide some bioprocess-relevant data. Methods of measuring diffusion coefficients have been described by Geankoplis (1) and Pratt (2). When diffusion coefficients are not available, they can be estimated.

Table 1. Diffusivities of Some Common Solutes in Dilute Liquids

Solute	Solvent	Temperature (°C)	D_L ($\times 10^9$ m^2 s^{-1})
Acetic acid	Water	25	1.26
Acetic acid	Benzene	25	2.09
Acetone	Water	25	1.28
Benzoic acid	Water	25	1.21
Carbon dioxide	Water	20	1.50
Carbon dioxide	Water	25	2.00
Ethanol	Water	25	1.24
Glucose	Water	20	0.60
Glycerol	Water	20	0.83
Methanol	Water	15	1.26
Nitrogen	Water	20	1.64
Oxygen	Water	20	1.80
Oxygen	Water	25	2.41
Phenol	Water	20	0.84
Sucrose	Water	20	0.45
Urea	Water	25	1.38
Water	Ethanol	25	1.13

Table 2. Diffusivities of Some Globular Proteins in Dilute Aqueous Solutions

Protein	Molecular weight (kg kmol^{-1})	Temperature (°C)	D_L ($\times 10^{11}$ m^2 s^{-1})
Bovine serum albumin	67 500	25	6.81
Urease	482 700	25	4.01
Human serum albumin	72 300	20	5.93
Human fibrinogen	339 700	20	1.98

Diffusivities of small solutes diffusing in dilute liquids can be estimated using the Wilke-Chang (3) equation

$$D_L = 1.173 \times 10^{-16} \frac{(\chi M_L)^{0.5} T}{\mu_L V_M^{0.6}} \quad (1)$$

where M_L and μ_L are the molecular weight and the viscosity of the solvent, respectively; T is the absolute temperature, V_M is the molar volume of the solute at its boiling point, and χ is the association parameter, a measure of polar interactions among molecules, of the solvent. Association parameters of some common solvents are noted in Table 3. The molar volumes are given in handbooks, or, for organic solutes, they are easily calculated by the group contribution method (1,2). Equation 1 is suitable for small solutes with $V_M \leq 0.500$ m^3 kmol^{-1}. When molar volumes are larger, as with globular proteins, the Stokes-Einstein equation may be used to estimate diffusivities; thus

Table 3. Association Parameters for Some Solvents

Solvent	Association parameter, ξ
Water	2.6
Methanol	1.9
Ethanol	1.5
Benzene	1.0
Ether	1.0
Heptane	1.0
Other unassociated solvents	1.0

Source: Based on Wilke and Chang (3).

$$D_L = \frac{9.96 \times 10^{-16} T}{\mu_L V_M^{1/3}} \quad (2)$$

Larger molecules diffuse more slowly than smaller ones. For example, the diffusivities of globular proteins (Table 2) are much lower than those of smaller solutes listed in Table 1.

Small solutes such as sugar or oxygen diffusing in solutions of proteins move more slowly than in water. Diffusion coefficient D_P of a small solute in a solution of globular protein can be approximated (1) as

$$D_P = D_L(1 - 1.81 \times 10^{-3} C) \quad (3)$$

where D_L is the diffusivity of the solute in water and C (kg m^{-3}) is the concentration of protein in solution. Equation 3 is suitable only when the diffusing solute does not associate with the protein molecule. When a solute associates with the protein, a part of the observed diffusivity of solute is due to diffusion of the protein molecules and a part is due to diffusion of unassociated solute.

Small molecules diffusing through gels such as agar, gelatin, and alginate diffuse more slowly than in water. The diffusion coefficient declines with increasing concentration of the gelling agent; however, in typically formulated gels, diffusivities are only 20 to 50% lower than in water. Inorganic salts and electrolytes that dissociate in water actually diffuse as molecules in the absence of electric fields. Diffusivities of electrolytes may be estimated using methods discussed by Pratt (2).

GAS–LIQUID MASS TRANSFER

Many industrial fermentations require oxygen. Oxygen supply must at least equal the oxygen demand or the fermentation will be oxygen limited. Oxygen demand is especially difficult to meet in viscous fermentation broths and in broths containing a large concentration of the oxygen-consuming cells. In large fermenters, oxygen transfer becomes difficult when demand exceeds 4 to 5 kg m^{-3} h^{-1} (4). Above a critical concentration of dissolved oxygen, the amount of oxygen does not influence the specific growth rate. The critical dissolved oxygen level depends on the microorganism, the culture temperature, and the substrate being oxidized. The higher the critical dissolved oxygen value, the greater the likelihood that oxygen transfer will become limiting (4). Under typically used culture conditions, fungi such as *Penicillium chrysogenum* and *Aspergillus oryzae* have a critical dissolved oxygen value of about 3.2×10^{-4} kg m^{-3}. For baker's yeast and *Escherichia coli*, the critical dissolved oxygen values are 6.4×10^{-5} and 12.8×10^{-5} kg m^{-3}, respectively (4). For comparison, the dissolved oxygen concentration in an air-saturated sterile medium is typically around 7×10^{-3} kg m^{-3}; hence, in terms of percent of air saturation, the minimum acceptable dissolved oxygen level in a fungal fermentation is about 5%.

Oxygen uptake rate of mammalian cells is typically (0.8 – 8.0) $\times 10^{-15}$ kg O_2 cell^{-1} h^{-1}. In suspension culture of animal cells, cell concentrations typically reach (1 – 1.2) $\times 10^6$ cells mL^{-1}; concentrations up to 1.2×10^7 cells mL^{-1} are attained in perfusion culture, and even higher values occur in immobilized cell bioreactors. The critical dissolved oxygen concentration for animal cells tends to be low, for example, 0.5% of air saturation or 0.035 kg m^{-3}.

The transport route of a soluble gas from the gas to the liquid phase is schematically illustrated in Figure 1. The gas may be imagined as passing through stagnant gas and liquid films on either side of the gas–liquid interface. The resistance to transfer is localized within the stagnant films and the gas–liquid interface itself is assumed to offer no resistance to mass transfer. Mass transfer in the films occurs solely by diffusion; therefore, at steady state, linear concentration profiles exist in the films (Fig. 1). The transport flux (J) of the diffusing species is related to the concentration gradient (ΔC) in the film and to the film thickness (δ) as follows:

$$J = \frac{D}{\delta} \Delta C \quad (4)$$

In equation 4, D is the diffusivity of the transferring component in the film. The ratio D/δ is known as the mass transfer coefficient, k.

At steady state, the flux of the diffusing component (or the rate of transfer per unit cross-sectional area) is the same through the gas and the liquid films; thus, equation 4 may be written for each of the two films:

$$J = k_G(C_G - C_{Gi}) \quad (5)$$

$$= k_L(C_{Li} - C_L) \quad (6)$$

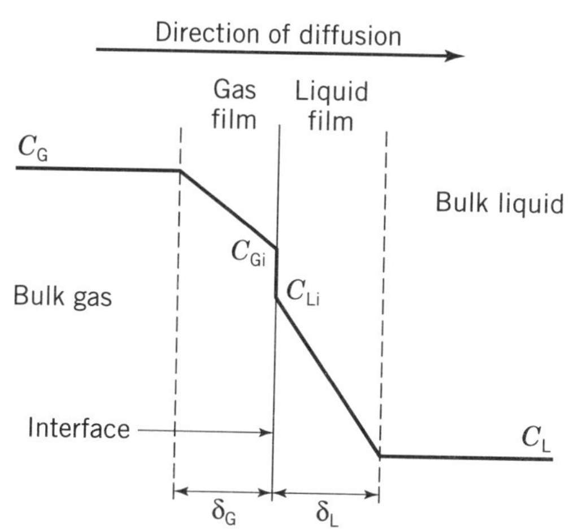

Figure 1. Steady-state dissolved oxygen concentration profile in the vicinity of the gas–liquid interface.

where k_G and k_L are the gas and the liquid film mass transfer coefficients, respectively. The overall mass flux from the gas to the liquid phase may be written as follows:

$$J = K_L(C^* - C_L) \tag{7}$$

where C^* is the saturation concentration (i.e., the maximum possible value) of the diffusing component in the liquid in equilibrium with the gas phase, and K_L is the overall mass transfer coefficient based on the liquid film. The saturation concentration C^* in the liquid is related to the gas phase concentration (C_G) of the diffusing component by Henry's law:

$$C_G = HC^* \tag{8}$$

where H is the dimensionless Henry's constant (5). Using equations 5 to 8 and the knowledge that $C_{Gi} = H \times C_{Li}$ (i.e., equilibrium exists at the interface because the interface offers no resistance to mass transfer), the various mass transfer coefficients can be shown to be related as follows (5):

$$\frac{1}{K_L} = \frac{1}{k_L} + \frac{1}{Hk_G} \tag{9}$$

For a sparingly soluble gas such as oxygen, H is much greater than unity ($H \approx 30$ for oxygen in water); furthermore, the coefficient k_G is typically larger than k_L because the gas–phase diffusivities are vastly greater than those in liquids (e.g., $D_{oxygen/air} = 10^4 D_{oxygen/water}$ at 20 °C). Consequently, the second term on the right side of equation 9 is negligible relative to the first, and the equation becomes

$$\frac{1}{K_L} \approx \frac{1}{k_L} \tag{10}$$

This implies that essentially all resistance to interfacial mass transfer of a sparingly soluble gas is localized in the liquid film at the interface, that is, the mass transfer is liquid-film controlled (5).

Because the transport flux J equals the rate of transfer per unit gas–liquid interfacial area, equation 7 may be expressed in terms of rate as follows

$$\frac{dC_L}{dt} = K_L a_L(C^* - C_L) \tag{11}$$

where C_L is the dissolved gas concentration at time t, and a_L is the gas–liquid interfacial area per unit volume of the liquid.

Volumetric Gas–Liquid Oxygen Transfer Coefficient

In view of approximation 10, the overall volumetric mass transfer coefficient $K_L a_L$ for gas–liquid mass transfer is often expressed as $k_L a_L$. The overall volumetric gas–liquid mass transfer coefficient $k_L a_L$ and the concentration driving force ($C^* - C_L$) for mass transfer must be known if the rate of supply of oxygen or other soluble gas is to be quantified (5). The mass transfer driving force, which is easily determined, depends on temperature and pressure. The coefficient $k_L a_L$ depends on the properties of the fluid, the hydrodynamic regime, and the configuration of the bioreactor. The factors influencing $k_L a_L$ and mass transfer are summarized in Table 4. Prediction of $k_L a_L$ is an important part of bioreactor design. The individual terms in $k_L a_L$ are difficult to measure directly, but the product is relatively easily measured (5).

Usually, the overall volumetric gas–liquid mass transfer coefficient is expressed in terms of the liquid volume in the fermenter; however, sometimes the basis of expression has been the gas–liquid dispersion volume (5). An identical basis is necessary for comparing mass transfer capabilities, and the coefficient expressed in one form may be converted to the other if the overall gas holdup (ϵ_G) is known:

$$k_L a_L = \frac{k_L a_D}{1 - \varepsilon_G} \tag{12}$$

Some typical $k_L a_L$ values seen in various types of bioprocesses are noted in Table 5.

As noted earlier, the rate of diffusive transport of a component through a unit area perpendicular to the direction of transport can be expressed as mass flux or molar flux—kilogram or kilomole of component transferred per unit area per unit time. The flux J is proportional to a mass transfer driving force and a mass transfer coefficient K_L, or $J = K_L \times$ (driving force). Mass transfer is driven by differences in chemical potential. A species diffuses from regions of high potential to those of low potential until the difference has been eliminated. In practice, the driving force for mass transfer may be expressed as a concentration difference, partial pressure difference, or mole fraction difference. For example, when oxygen-free water is

Table 4. Factors Influencing Gas–Liquid Mass Transfer

Temperature
Pressure
Diffusivity
Viscosity
Density
Surface tension
Presence of surfactants and ions
Ionic strength
Concentration of solids
Hydrophobicity of solids
Morphology of solids
Shear rate or power input
Aeration velocity
pH
Geometry of the bioreactor
Flow parameters of non-Newtonian fluids

Table 5. Typical $k_L a_L$ Values Seen in Bioprocesses

Process	$k_L a_L$ (s^{-1})
Fungal fermentations	10^{-2}
Bacterial and yeast fermentations	10^{-1}
Wastewater treatment	3×10^{-3}

Source: Data from Ref. 5.

brought in contact with air, oxygen diffuses from air to water until the water becomes saturated. At this point, the oxygen concentrations in the two phases are in dynamic equilibrium, but there is no net transfer. Note that the concentrations of oxygen in water and air are not equal, but the chemical potentials are. Depending on how the mass flux J and the driving force are expressed, the mass transfer coefficient may have different numerical values and different units, as noted in Table 6 for a few cases. A coefficient expressed in one form may be easily converted to a different one.

Measurement of k_La_L. Several methods are available for measuring k_La_L values (1,5–7). Here only a few of the more useful and common methods will be noted.

Sulfite Oxidation Method. The sulfite oxidation method has been widely used, but it is suited to measurements only in physical systems (no live microorganisms) such as water or some polymer solutions. The method is subject to too many interferences (5,6) and, therefore, it is not suitable for use in most culture media or in nonviable fermentation broths. Moreover, because of high ionic strength, the sulfite oxidation medium is strongly noncoalescing and rheologically quite different from fermentation systems. Consequently, the sulfite oxidation measurements are not readily translated to biologically relevant systems.

The sulfite oxidation method relies on oxidation of sulfite to sulfate using dissolved oxygen:

$$SO_3^{2-} + 1/2O_2 \xrightarrow{\text{Catalyst}} SO_4^{2-}$$

The reaction is catalyzed by copper (II) and cobalt (II) as well as other metal ions. The rate of sulfite oxidation is a measure of the rate of oxygen transfer to the liquid. The overall reaction generates H^+, and pH control is necessary because the reaction rate is pH dependent. Also, the temperature needs to be controlled carefully to ensure constant reaction kinetics. Impurities such as iron, manganese, and other transition metals in the parts per million range affect the reaction.

For k_La_L measurements, the reaction conditions should be such that the reaction occurs in the bulk liquid and not at the gas–liquid interface. Thus, Cu (II) catalysis is used with a sufficiently high sulfite concentration that the oxidation rate is independent of the sulfite level. The rate of sulfite consumption is given by

$$-\frac{d[SO_3^{2-}]}{dt} = k_La_L(C^* - C_L) \tag{13}$$

where C^* and C_L are the saturation concentration of dissolved oxygen and the actual instantaneous concentration at time t. C_L is usually close to zero. The sulfite consumption rate, which should remain constant, is followed by quenching the samples taken at various times with excess iodine and back titration of the residual iodine with thiosulfate (5).

Another oxygen-consuming reaction is the oxidation of dithionite ($S_2O_4^{2-}$) in alkaline solutions; however, in view of uncertain kinetics and other factors, the usefulness of that reaction for k_La_L measurements remains questionable.

Chemical Absorption of Carbon Dioxide. This method is similar in principle to the sulfite oxidation procedure. Carbon dioxide is absorbed into a mildly alkaline or suitably buffered solution where it is converted to other species. Chemical absorption of carbon dioxide into Na_2CO_3–$NaHCO_3$ solution is sufficiently slow for use in k_La_L determinations when the ratio of carbonate to bicarbonate concentrations is between 3 and 5 (5). Other reaction conditions need to be controlled to ensure that the reaction rate does not enhance mass transfer. The method has limitations similar to those of the sulfite oxidation technique.

The Hydrazine Method. The hydrazine method makes use of the reaction

$$\underset{\text{Hydrazine}}{N_2H_4} + O_2 \xrightarrow{\text{Catalyst}} N_2 + H_2O$$

The reaction is best carried out at pH 11 to 12 using copper sulfate catalyst (5). A steady flow of hydrazine into an aerated reactor is used to maintain a constant concentration of dissolved oxygen. The amount of hydrazine consumed per unit time equals the oxygen absorption rate. The reaction does not produce any ionic species; hence, the ionic strength of solution can be kept lower than with the sulfite oxidation or the carbon dioxide absorption methods.

Oxygen Balance Method. This technique depends on measurement of mass flow rates of aeration gas into and out of the fermenter. The mass fraction of oxygen in the inlet and exhaust gas streams must also be determined (mass spectrometer, paramagnetic oxygen analyzer) as well as the steady state dissolved oxygen concentration in the broth (dissolved oxygen electrode). The k_La_L is obtained from the oxygen balance:

Table 6. Possible Ways of Expressing Mass Flux and Mass Transfer Coefficient

Flux, J	Driving force	Units of K_L
kg s^{-1} m^{-2}	Concentration difference, kg m^{-3}	m s^{-1}
kg $s^{-1}m^{-2}$	Mole fraction difference	kg $s^{-1}m^{-2}$
kg $s^{-1}m^{-2}$	Partial pressure difference	kg $s^{-1}m^{-2}$
kmol $s^{-1}m^{-2}$	Concentration difference, kg m^{-3}	m s^{-1}
kmol $s^{-1}m^{-2}$	Mole fraction difference	kmol $s^{-1}m^{-2}$
kmol $s^{-1}m^{-2}$	Partial pressure difference	kmol $s^{-1}m^{-2}$
$m^3m^{-2}s^{-1}$ (for transfer of a gas)	Partial pressure difference	m s^{-1}

$$F(x_i - x_o) = V_L k_L a_L (C^* - C_L) \quad (14)$$

where F is the mass flow rate of gas, x_i and x_o are the mass fractions of oxygen in the gas (i = inlet, o = outlet), and V_L is the volume of the broth in the fermenter. Equation 14 assumes no evaporation and it does not correct for carbon dioxide production; however, the necessary corrections can be easily incorporated. For constant volume continuous culture, equation 14 needs to be further modified to account for oxygen entering and leaving the fermenter via the liquid streams.

The oxygen balance method requires accurate measurements of gas flow rates and oxygen concentrations. The oxygen mass fraction in the inlet and outlet gas must differ measurably (i.e., the oxygen consumption rate in the fermenter should be relatively large) for reliable $k_L a_L$ measurements. Under suitable conditions, the method is very reliable, and it does not disturb the fermentation by interrupting the air supply (7).

Enzymatic Oxidation Method. Oxygen transfer may be followed by the formation of gluconic acid via oxidation with oxygen in presence of the enzyme glucose oxidase. The rate of absorption of oxygen is calculated from the rate of consumption of alkali that is needed to neutralize the acid. The liquid-phase oxygen concentration is determined with a dissolved oxygen electrode. The cost of the enzyme restricts the method to small-scale applications (5). The procedure applies only to physical systems.

The Dynamic Gassing-In Method. In a fermentation at steady state, the oxygen supply rate exactly matches the rate of consumption by the microorganisms and the concentration of dissolved oxygen does not change:

$$\frac{dC_L}{dt} = k_L a_L (C^* - C_L) - q_{O_2} X = 0 \quad (15)$$

where q_{O_2} is the specific oxygen consumption rate, X is the concentration of viable biomass, $q_{O_2} X$ is the consumption term, and $k_L a_L$ $(C^* - C_L)$ is the oxygen supply or mass transfer term. If now the oxygen supply is interrupted, the mass transfer term in equation 15 becomes zero and, as shown in Figure 2, the dissolved oxygen concentration declines linearly with time (7). The slope of the C_L versus t line during the unaerated period provides $q_{O_2} \times X$. After a short period, before the oxygen concentration has dropped to the critical level, the oxygen supply must be resumed. Now, the dissolved oxygen level rises (Fig. 2). During the rise period, the dissolved oxygen concentration as a function of time is given by

$$C_L = C^* - \frac{1}{k_L a_L}\left(q_{O_2} X + \frac{dC_L}{dt}\right) \quad (16)$$

which is a rearranged form of equation 15. Thus, $k_L a_L$ can be calculated from the slope of a plot of

$$C_L \text{ versus } \left(q_{O_2} X + \frac{dC_L}{dt}\right)$$

This procedure requires interrupting the oxygen flow to the fermenter, and the dissolved oxygen electrode needs to be capable of a rapid response.

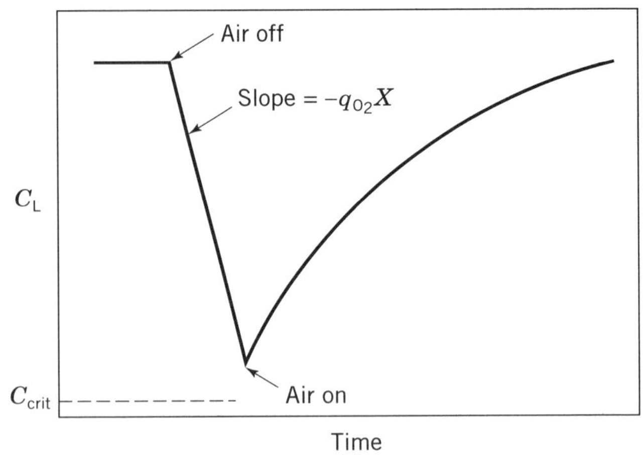

Figure 2. Dissolved oxygen concentration as a function of time during dynamic determination of the overall volumetric gas–liquid mass transfer coefficient in a fermentation.

The gassing-in method is commonly applied also to physical systems (5). In such cases the dissolved oxygen in the liquid is first reduced to near zero by bubbling with nitrogen. The nitrogen flow is then stopped, the bubbles are allowed to leave the fluid, and the system is oxygenated by bubbling with air. The concentration of the dissolved oxygen in the liquid is followed as a function of time, and the $k_L a_L$ is calculated as the slope of the linear equation

$$\ln\left(\frac{C^* - C_{L0}}{C^* - C_L}\right) = k_L a_L t \quad (17)$$

where C^* is the saturation concentration of dissolved oxygen, C_{L0} is the initial dissolved oxygen concentration at time t_0 when a hydrodynamic steady state has been reestablished upon commencement of aeration, and C_L is the dissolved oxygen concentration at any time t (5). Equation 17 is obtained by integration of equation 15, assuming a well-mixed liquid (i.e., uniform C_L), no change in gas phase composition (i.e., a constant C^*), and an absence of oxygen consumption (i.e., $q_{O_2} = 0$) in the physical system.

Correct evaluation of $k_L a_L$ by this method requires attention to several factors, including the response dynamics of the dissolved oxygen electrode and the nature of mixing in the gas and the liquid phases. In large intensely mixed fermenters, a well-mixed liquid phase may still be assumed, but the gas phase composition changes significantly from the inlet to exhaust. The gas is generally in plug flow and, thus, a log mean concentration driving force $(C^* - C_L)_{LM}$ should be used in equation 15. The driving force is

$$(C^* - C_L)_{LM} = \frac{C^*_{in} - C^*_{out}}{\ln\left(\frac{C^*_{in} - C_L}{C^*_{out} - C_L}\right)} \quad (18)$$

where C^*_{in} and C^*_{out} are the saturation concentrations in equilibrium with the gas phase composition at the fermenter inlet and outlet, respectively.

Generally, the nature of mixing in the gas phase is inconsequential in k_La_L determinations so long as $1/k_La_L$ is less than the residence time of the gas phase in the liquid. The mean residence time t_R of the gas phase in bubble columns and stirred tanks can be estimated (5) using

$$t_R = \frac{h_L\varepsilon_G}{(1 - \varepsilon_G)U_G} \tag{19}$$

where ϵ_G is the overall gas holdup. Equation 19 may be applied also to risers of airlift reactors when there is little or no gas in the downcomer. In the latter case, the U_G should be replaced with U_{Gr} (the superficial gas velocity based on the riser cross-section), and the gas holdup should be replaced with ϵ_{Gr}. When there is gas in the downcomer, equation 19 may still be used because the downcomer gas holdup acts mostly as a dead volume.

The response characteristics of the dissolved oxygen electrode have little influence on the dynamic determination of k_La_L if the response time (63% of full response) is less than or equal to $1/k_La_L$. Consult Chisti (5) for additional details. Other variations of the gassing-in technique exist.

Estimation of k_La_L for Gases Other Than Oxygen. Most of the available data and correlations for k_La_L were obtained for mass transfer of oxygen. The overall volumetric mass transfer coefficient for transfer of other gases can be estimated using the oxygen transfer data and the diffusion coefficients for oxygen and the other gas (5); thus

$$(k_La_L)_{gas} = \frac{D_{gas}}{D_{oxygen}}(k_La_L)_{oxygen} \tag{20}$$

or

$$(k_La_L)_{gas} = \sqrt{\frac{D_{gas}}{D_{oxygen}}}(k_La_L)_{oxygen} \tag{21}$$

Equation 20 applies when the transfer can be ascribed to the film penetration mechanism of the two-film theory (5). Equation 21 is applicable when the surface renewal is the predominant mechanism of mass transfer (5). The difference between the two correction factors is small. The diffusivities used for the correction should be at the same temperature as the k_La_L data. The correction is satisfactory only for transfer in identical fluids at identical hydrodynamic conditions. Data on carbon dioxide mass transfer have been reviewed by Ho and Shanahan (8).

Solubility of Oxygen. Information on solubility of oxygen is needed for estimating the concentration difference driving force for mass transfer. Calculation of the saturation concentration C^* of dissolved oxygen (equation 8) requires a knowledge of the Henry's law constant, H. The dimensionless H for oxygen in pure water may be calculated as a function of temperature T (°C) using the equation

$$H = -9281.4 \times 10^{-3} + 4030.9 \times 10^{-3}T - 1403.5 \times 10^{-4}T^2 + 178.8 \times 10^{-5}T^3 \tag{22}$$

Equation 22 provides almost exact values of H over the range 20 to 30 °C. Oxygen solubility in water at a pressure of 1 atm can be calculated with the equation

$$C^* = 2178.549 \times 10^{-3} - 549.304 \times 10^{-4}T - 848.781 \times 10^{-6}T^2 - 483.785 \times 10^{-8}T^3 \tag{23}$$

where the temperature T is in °C and C^* is in mmol L^{-1}. Equation 23 applies over 0 to 40 °C. In addition to temperature, other factors influence oxygen solubility and the H value. Solubility declines with increasing ionic strength of solution. Solubility depends also on the types of ions in solution. For example, for the same molar concentrations of sodium chloride and hydrochloric acid in different samples of pure water, the oxygen solubility is lower in the salt solution. Solubility data for oxygen and other gases are available in handbooks. At 37 °C, the saturation dissolved oxygen concentration in a typical animal cell culture medium with serum is about 6.5×10^{-3} kg m^{-3}.

LIQUID–LIQUID MASS TRANSFER

Transfer of a solute from an immiscible liquid dispersed into another liquid (the continuous phase) can be analyzed using the two-film approach discussed for gas–liquid mass transfer. Now, the overall mass transfer coefficient K_L and the film coefficients for the dispersed (k_{Ld}) and the continuous (k_{Lc}) phases are related as follows:

$$\frac{1}{K_L} = \frac{1}{k_{Lc}} + \frac{K_p}{k_{Ld}} \tag{24}$$

In equation 24, K_p is the partition coefficient of the solute between the two phases. The partition coefficient defines the equilibrium distribution of the solute between the phases; thus

$$K_p = \frac{C_{Lc}}{C_{Ld}} \tag{25}$$

where C_{Lc} and C_{Ld} are the equilibrium concentrations of the solute in the continuous and the dispersed phases, respectively. Again, as for gas–liquid mass transfer, the interface between the phases is assumed to offer no resistance to mass transfer; hence, the equilibrium relationship determines the solute concentrations at the interface.

The overall volumetric mass transfer coefficient for liquid–liquid mass transfer is K_La_{Lc} where the specific interfacial area (a_{Lc}) is based on the volume (V_{Lc}) of the continuous phase. The rate of mass transfer from the dispersed to the continuous phase is given by

$$\frac{dC_{Lc}}{dt} = K_La_{Lc}(C^* - C_{Lc}) \tag{26}$$

where C^* is the saturation concentration of the solute in the continuous phase in equilibrium with the dispersed phase.

SOLID–LIQUID MASS TRANSFER

Mass transfer to or from suspended solids is important in many bioprocess situations. Solid–liquid mass transfer may be rate limiting in immobilized enzyme and cell reactors, during dissolution of solids, in adsorption, and in other cases. The rate of mass transfer to or from the suspended particle depends on the solid–liquid mass transfer coefficient (k_L), the total solid–liquid interfacial area (A_s), and the concentration driving force; thus

$$V_L \frac{dC_L}{dt} = k_L A_s (C^* - C_L) \tag{27}$$

where C_L is the concentration of the transferring component in the liquid at time t, C^* is the saturation concentration (or solubility) of the transferring component, and V_L is the volume of the suspending liquid. Equation 27 is written for transfer from the solid to the liquid. By analogy with the film model of mass transfer, the coefficient k_L conceptually equals the diffusivity of the solute divided by the thickness of the stagnant liquid film at the solid–liquid interface; k_L is needed for quantifying the rate of mass transfer. Methods for estimating the solid–liquid or solid–gas (also gas–liquid if liquid is adsorbed as a very thin film on the surface of the solid) mass transfer coefficient k_L in various situations are discussed in a later section of this monograph.

MASS TRANSFER BEHAVIOR

General Aspects

Mass transport by diffusion and forced convection are influenced by the following variables: the mass transfer coefficient k_L, the molecular diffusivity of the transferring component D_L, the fluid density ρ_L and viscosity μ_L, a characteristic length d, and the velocity of flow U_L. Mass transfer by natural convection is influenced also by the gravitational acceleration g and the density difference $\Delta\rho$ between phases. These variables can be grouped into the following principal dimensionless groups or numbers:

$$Re \text{ (Reynolds number)} = \frac{\text{Inertial force}}{\text{Viscous force}} = \frac{\rho_L U_L d}{\mu_L} \tag{28}$$

$$Sh \text{ (Sherwood number)} = \frac{\text{Total mass transfer}}{\text{Diffusive mass transfer}} = \frac{k_L d}{D_L} \tag{29}$$

$$Sc \text{ (Schmidt number)} = \frac{\text{Momentum diffusivity}}{\text{Mass diffusivity}} = \frac{\mu_L}{\rho_L D_L} \tag{30}$$

$$Gr \text{ (Grashof number)} = \frac{\text{(Inertial force)(buoyancy force)}}{\text{(Viscous force)}^2} = \frac{d^3 \rho_L \Delta\rho g}{\mu_L^2} \tag{31}$$

$$Fr \text{ (Froude number)} = \frac{\text{Inertial force}}{\text{Gravitation force}} = \frac{U_L^2}{gd} \tag{32}$$

The above-noted dimensionless numbers represent ratios of various factors that may play a role in a given situation. The Grashof number is important in situations where density-difference-driven natural convection affects mass transfer. The Reynolds number is used in situations where forced convection is the predominant influence on transport processes. The Froude number is useful in describing gravity-influenced flows. The Schmidt number, ratio of momentum diffusivity to mass diffusivity, is a measure of relative effectiveness of mass and momentum transfer. Some of those dimensionless numbers may be expressed in other forms, depending on the situation. A few possible alternatives are noted in Table 7. In addition to the dimensionless groups discussed above, other groups relevant to mass transfer and related fluid mechanics are encountered in the literature. Some of the more common ones and their physical significance are noted in Table 8.

Grouping the many variables that affect mass transfer into a few dimensionless numbers greatly simplifies the process development experimentation and data correlation for predicting the behavior of a system. For example, for fluids flowing in straight pipes, the flow pattern changes from laminar to turbulent at a Reynolds number value of about 2,300. This occurs irrespective of which specific fluid—air, water, milk, or thick glucose syrup—is being used so long as the combination of the fluid density, viscosity, flow velocity, and pipe diameter yields a Reynolds number value that exceeds 2,300. Equations for correlating the mass transfer behavior of systems are often expressed in terms of dimensionless groups.

Behavior of Dispersions

A dispersion consists of one or more phases dispersed in a continuous phase. The continuous phase may be a gas or liquid. The dispersed phase may be liquid droplets, solid particles, or gas bubbles. The magnitude of mass transfer coefficient inside and outside fluid particles depends on whether the particles behave as ridged spheres or are mobile. Small drops and bubbles behave as noncirculating ridged spheres. Larger bubbles and drops have internal fluid circulation and mobile interfaces because of relative motion between the particle and the surrounding fluid. Bubbles are immobile if the dimensionless diameter d^* is 10 or below. The interface is nearly always mobile if d^* exceeds 50. Here, the dimensionless diameter number is

$$d^* = d_B \left(\frac{\mu_L^2}{\rho_L g (\rho_L - \rho_G)} \right)^{-1/3} \tag{32}$$

as defined by Wesselingh (5). Using this criterion, in air and water, bubbles smaller than 0.5 mm will always be ridged, whereas bubbles larger than about 2.5 mm will have mobile interfaces (5).

In certain cases, drops and bubbles may experience additional interfacial motions that are a consequence of mass transfer itself. Because the local mass transfer coefficient around a particle may be different at different locations,

Table 7. Alternative Methods of Expressing Dimensionless Groups

Group	Alternative expressions
Reynolds number, *Re*	$\frac{\rho_L U_L d}{\mu_L}$ (in pipes and channels)
	$\frac{\rho_L N d_i^2}{\mu_L}$ (stirred tanks)
	$\frac{\rho_L U_L L}{\mu_L}$ (flow past a plate)
	$\frac{\rho_L U_L d_p}{(1 - \phi)\mu_L}$ (flow past particles in packed beds)
Sherwood number, *Sh*	$\frac{k_L d}{D_L}$ (*d* may be diameter of a particle, flow channel, plate, etc.)
Froude number, *Fr*	$\frac{U_L^2}{g h_L}$, or $\frac{U_G^2}{g h_L}$ (in bubble columns and airlift fermenters)
	$\frac{N d_i^2}{g h_L}$, or $\frac{N^2 d_i}{g}$ (in stirred tanks)
Peclet number, *Pe*	$\frac{U_L h_L}{D_L}$, or $\frac{U_B d_B}{D_L}$ (in bubble columns and airlift fermenters)
	$\frac{N d_i^2}{D_L}$ (in stirred tanks)

Table 8. Other Dimensionless Groups Relevant to Mass Transfer and Related Fluid Mechanics

Group	Definition	Physical significance
Bond number, *Bo*	$\frac{g d_p^2 \Delta\rho}{\sigma_L}$	$\frac{\text{Gravity force}}{\text{Surface tension force}}$
Peclet number, *Pe*	$\frac{U_L L}{D_L}$	$\frac{\text{Bulk mass transport}}{\text{Diffusional mass transport}}$
Poiseuille number, *Ps*	$\frac{\mu_L U_p}{\rho_L g d_p^2 \Delta\rho}$	$\frac{\text{Viscous force}}{\text{Gravity force}}$
Power number, *Po*	$\frac{P}{\rho_L N^3 d_i^5}$	
Rayleigh number, *Ra*	$Gr \times Sc$	
Stanton number, *St*	$\frac{Sh}{Re \times Sc}$	
Weber number, *We*	$\left(\frac{U_p^2 \rho_L d_p}{\sigma_L}\right)^{1/2}$	$\frac{\text{Inertial force}}{\text{Surface tension force}}$

the local mass transfer rates may be different. In this situation, uneven transfer of a solute such as acetone or ethanol will generate solutions of different local concentrations and surface tensions. Adjacent films of different surface tension lead to violent motions: regions of high surface tension contract, and the film ruptures in regions of low surface tension. This effect is the well-known Marangoni effect, and it can significantly enhance mass transfer, especially in liquid–liquid extraction processes.

For suspended solids, increasing agitation beyond that needed to achieve complete suspension has little impact on mass transfer coefficient. The coefficient is influenced by relative velocity between the phases, which is determined by density differences. In gas–liquid dispersions, too, the k_L is not substantially affected by increasing agitation, but the overall volumetric mass transfer coefficient ($k_L a_L$) increases because of the increase in gas–liquid interfacial area.

For freely suspended particles and bubbles, mass transfer to or from the bubble or particle obeys the general relationship

$$Sh = f(Gr, Sc) \tag{33}$$

For just suspended small particles ($d_p < 0.6$ mm) in stirred tanks, Calderbank and Moo-Young (9) recommended the equation

$$k_L = \frac{2D_L}{d_p} + 0.31\, Sc^{-2/3}\left(\frac{\Delta\rho_L \mu_L g}{\rho_L^2}\right)^{1/3} \tag{34}$$

which is suitable for suspensions as well as gas–liquid dispersions of noncirculating bubbles ($d_p < 0.6$ mm). For larger circulating bubbles and liquid drops ($d_p > 2.5$ mm) in just-suspended state, the recommended equation is

$$k_L = 0.42 Sc^{-0.5}\left(\frac{\Delta\rho_L \mu_L g}{\rho_L^2}\right)^{1/3} \tag{35}$$

For bubbles in the 0.6 to 2.5-mm range, the k_L may be estimated as a linear function of bubble diameter.

Large spherical cap bubbles frequently occur in gas–liquid dispersions, especially when the viscosity of liquid exceeds about 7×10^{-2} Pa s. Calderbank and Lochiel (10) correlated mass transfer from such bubbles using the equation

$$Sh = \frac{1.79(3R_b^2 + 4)^{2/3}}{R_b^2 + 4}(Re \times Sc)^{1/2} \tag{36}$$

where R_b is the ratio of bubble width to bubble height. For spherical caps, R_b is about 3.5; hence, equation 36 becomes

$$Sh = 1.31(Re \times Sc)^{1/2} \quad (37)$$

Mass transfer to or from a particle suspended in a stagnant fluid occurs solely by diffusion. For a single spherical particle surrounded by a stagnant medium, the theoretical minimum value of the transfer coefficient is given by $Sh = 2$. For a single particle in the creeping flow regime ($Re < 0.1$), the specific relationship is

$$Sh = 0.39(Gr \times Sc)^{1/3} \quad (38)$$

Equation 38 applies to particles with ridged interfaces, and this includes noncirculating small bubbles ($Re < 0.1$) in fermentation broths. Thus, compared to the case for a single noncirculating bubble (equation 38), the mass transfer coefficient in swarms of ridged bubbles (equation 34) is about 20% lower.

For power inputs greater than needed for just suspension, Calderbank and Moo-Young (9) have established the following equation in stirred tanks

$$k_L = 0.13\left(\frac{(P_G/V_L)\mu_L}{\rho_L^2}\right)^{1/4} Sc^{-2/3} \quad (39)$$

where P_G/V_L is the power input per unit volume.

Transfer coefficient correlations generally require a knowledge of liquid properties such as viscosity and density. This is usually not a problem when dealing with Newtonian liquids, but difficulties arise with slurries and non-Newtonian media. Slurries of biomass solids may be treated usually as pseudohomogeneous fluids (5). Depending on the amount of suspended solids, slurries may behave as Newtonian or non-Newtonian power law fluids. For small amounts of spherical solid particles, Newtonian behavior is commonly observed and in this case the viscosity of the slurry is independent of shear rate. The Newtonian viscosity may be estimated using equations of Einstein and Vand (7). Another suitable equation is that of Thomas:

$$\mu_{SL} = \mu_L(1 + 2.5\varphi_S + 10.05\varphi_s^2 + 0.00273e^{-16.6}\varphi_S) \quad (40)$$

where φ_S is the volume fraction of suspended solids and μ_L is the viscosity of the suspending fluid.

When the slurry behaves as a non-Newtonian power law fluid, the apparent viscosity depends on shear rate; thus

$$\mu_{ap} = K\gamma^{n-1} \quad (41)$$

where K is the consistency index or thickness of the fluid and n is its flow behavior index. The parameter γ is the shear rate. The shear rate is difficult to define in most realistic configurations of bioreactors under the usual operational conditions; nevertheless, the following expressions are commonly used in estimating shear rates.

In *bubble columns and airlift devices*,

$$\gamma = \alpha U_G \quad (42)$$

where the constant α has been specified variously as 1,000, 2,800, 5,000 m^{-1}, and so on (11). Equation 42 has been applied also to airlift reactors, but that use is incorrect; for airlift reactors, the superficial gas velocity based on the cross-sectional area of the riser should be used in expressions such as equation 42 as recommended by Chisti (5). Depending on the constant used, equation 42 provides wildly different values of the supposed shear rate; hence, its use is not generally favored and it has been severely criticized (5,11,12). Another expression for shear rate in airlift reactors is

$$\gamma = 3.26 - 3.51 \times 10^2 U_{Gr} + 1.48 \times 10^4 U_{Gr}^2 \quad (43)$$

which was developed for $0.004 < U_{Gr}\ (m\ s^{-1}) < 0.06$ (13). Equation 43 also has significant problems, as discussed elsewhere (12,14). Alternative approaches for estimating shear rates have been propounded by Molina Grima et al. (12).

In *stirred vessels*, the mean shear rate in impeller agitated tanks is usually given as

$$\gamma = k_i N \quad (44)$$

where k_i is an impeller-dependent constant (7). Some typical k_i values are 11 to 13 for six-bladed turbines, 10 to 13 for paddles, ~10 for propellers, and ~30 for helical ribbon impellers. Again, as with pneumatically agitated bioreactors, much of the discrepancy among various predictive correlations that rely on a shear-rate-dependent apparent viscosity has been associated with the use of equations such as equation 44 (15).

The apparent viscosity of many fermentation fluids declines with increasing shear rate, that is, $n < 1$ (equation 41), and these fluids are known as shear thinning, or pseudoplastic. Because shear rate is not easily defined in typical bioreactors (5,11,12,15), correlations using solids holdup directly are preferred to those relying on a poorly established apparent viscosity of the slurry (5). As with viscosity, the density of a pseudohomogeneous slurry may be related to phase holdups as follows

$$\rho_{SL} = \rho_L(1 - \varphi_S) + \rho_S\varphi_S \quad (45)$$

In some viscous fermentations, the principal resistance to oxygen transfer may be in the bulk fluid or at the cell–liquid interface and not at the gas–liquid interface. In such cases, dilution of the broth with water and increased agitation may improve the transfer rate a little. In broths of filamentous fungi, promoting pelletlike growth as opposed to the usual pulplike morphology may substantially reduce the broth viscosity and improve mass transfer. Additional details on transport fundamentals of dispersions have been noted by Moo-Young and Blanch (16). Similar discussions with a focus on gas–liquid mass transfer in non-Newtonian media are given in the work by Oolman and Blanch (17). Carbon dioxide mass transfer in bioreactors has been treated in depth by Ho and Shanahan (8).

Gas–Liquid Mass Transfer

Either bubble-free aeration of the liquid surface in a bioreactor may be used to provide the needed oxygen or the aeration gas mixture may be bubbled or sparged through the body of the fluid. These schemes are discussed separately below.

Surface Aeration. Oxygen transfer through the surface of a fluid is useful in relatively small-scale devices or in microaerophilic processes and certain wastewater treatment applications (18). Surface aeration is especially common in early stages of culture, in shake flasks, tissue culture flasks, roller bottles, and spinner flasks. Understanding and characterization of surface aeration in small-scale devices is essential for process scale-up that minimally attempts to reproduce on larger scale the culture performance attained in the laboratory. Surface aeration is used also in relatively small production bioreactors for growing cultures with extremely low oxygen requirements. Surface aeration in the principal types of laboratory and production culture devices is detailed in the following sections.

Laboratory Culture Devices

1. *Shake flasks*. In 500-mL Erlenmeyer flasks placed on reciprocating and rotary shaker platforms, Yamada et al. (19) measured volumetric rates of oxygen transfer in sulfite solutions and in *Acetobacter suboxydans* fermentations for converting sorbitol to sorbose. The rates were expressed in terms of the volumetric oxygen transfer coefficients through the flask closure (K_P) and across the gas–liquid interface (K_S):

$$\begin{array}{l}\text{Volumetric rate}\\ \text{of oxygen transfer}\\ (\text{mol } O_2/\text{mL h})\end{array} = \frac{1}{\left(\dfrac{1}{K_P}+\dfrac{1}{K_S}\right)}\frac{1}{V_L}(p_{atm}-p_L) \quad (46)$$

In equation 46, V_L (mL) is the volume of liquid in the flask; p_{atm} (atm) and p_L (atm) are the partial pressures of oxygen in the atmosphere outside the flask and in the liquid in the flask, respectively. The units of K_P and K_S are mol O_2 atm^{-1} h^{-1}. The mass transfer coefficient K_P through the closure declined with increasing weight (3 to 7 g) of cotton used to form the plug. The mean K_P value for cotton plugs was 2.87×10^{-2} mol O_2 atm^{-1} h^{-1}. Values for polyurethane foam and silicone foam plugs were a little lower. Compared to open flasks, cotton plugs restricted oxygen transfer. The K_S values in sulfite oxidation solutions agitated at 110 to 140 rpm ranged over (2.04 to 4.63) $\times 10^{-2}$ mol O_2 atm^{-1} h^{-1}. Under similar conditions in *Acetobacter suboxydans* fermentations, the K_S values were lower—only 50 to 60% of those obtained with sulfite oxidation. Various types of baffles and indentation in the flasks enhanced K_S by about 50% relative to the base case.

For mass transfer from liquid surface in unbaffled Erlenmeyer shake flasks, the following equation has been recommended (20)

$$k_La_L\left(\frac{\mu_L}{\rho_L g^2}\right)^{1/3} = 0.5\left(\frac{\mu_L}{\rho_L D_L}\right)^{-1/2}\left(\frac{d_F}{V_L}\right)^{8/9}\left(\frac{\mu_L^2}{\rho_L^2 d_F^3 g}\right)^{8/27} \cdot\left(\frac{N^2\sqrt{\hat{e}d_F}}{g}\right)^{0.5} \quad (47)$$

where d_F is the maximum diameter of the flask, N is the speed of rotation, and $\hat{e}$ is the eccentricity of the shaker platform. Equation 47 applies to animal cell culture media when the liquid volume in the flask is 50 to 200 mL.

2. *T flasks*. Tissue culture flasks or T flasks are commonly used in initial stages of animal cell culture. During typical culture conditions, the cells utilize the oxygen in the liquid layer, but the concentration of oxygen in the gas phase is little affected irrespective of whether the flask is fully closed or the cap is cracked open (21). Clearly, diffusion through the liquid layer controls mass transfer; thus, the shallower the liquid depth the better. The oxygen transfer resistance of the deposited cell layer at the bottom of the flask is insignificant compared to the resistance through the bulk liquid layer (21). The observed oxygen transfer rates through the liquid layer tend to be greater than if the transfer were purely by diffusion. This suggests that convection and micromixing (e.g., during sampling, opening or closing of incubator door) significantly contribute to transfer. In typical culture, the dissolved oxygen declines linearly with time until, by about 60 h, the dissolved oxygen level is reduced to below 10% of air saturation (21). Because T flasks are typically inoculated and transferred every other day, the observed decline in oxygen may be of little real consequence; however, if the culture period is to exceed 60 h and dissolved oxygen levels of 10% or higher are wanted, then once or twice daily gentle agitation of T flasks is recommended.

3. *Roller bottles*. Roller bottles are commonly used for initial stages of cell culture. Typically about 0.1 m in diameter and about 0.25 m tall, plastic roller bottles are filled to about a third of their height with culture fluid. The bottles are laid on their sides in a roller cabinet and rotated at 1 to 2 rpm. Roller bottles are used to culture both adherent cells as well as nonanchorage-dependent lines. The interfacial area for oxygen transfer is the surface of the culture fluid and the liquid film on walls that continuously enters and leaves the pool of liquid. Mechanized and automatic handling of a large number of roller bottles have extended this culture technique to commercial production scale in a few cases, but this kind of processing has a very limited scope. The surface aeration capability of roller bottles, corrected for differences in surface-to-volume ratio, is somewhere between that of Erlenmeyer-style shake flasks under typical conditions and the stationary T flasks.

4. *Spinner flasks*. Aunins et al. (22) reported k_La_L measurements (surface aeration) in 500-mL Corning spinner flasks (Fig. 3) filled to a depth of 0.08 m with Dulbecco's Modified Eagle's Medium (DMEM) supplemented with 5% (v/v) fetal calf serum and 1 kg m^{-3} EDTA. The data were obtained at 37 °C and followed the equation

$$Sh = 1.08Re^{0.78} \quad (48)$$

where the Sherwood number is based on the vessel diameter and the Reynolds number is based on the impeller diameter. The measurements spanned the impeller speed range 25 to 150 rpm, impeller diameters of 0.0525 m and 0.078 m, and impeller Reynolds number of 1,500 to 20,000. Locating the impeller less than 0.01 m below the liquid surface dramatically enhanced the k_La_L. At 50 rpm, with

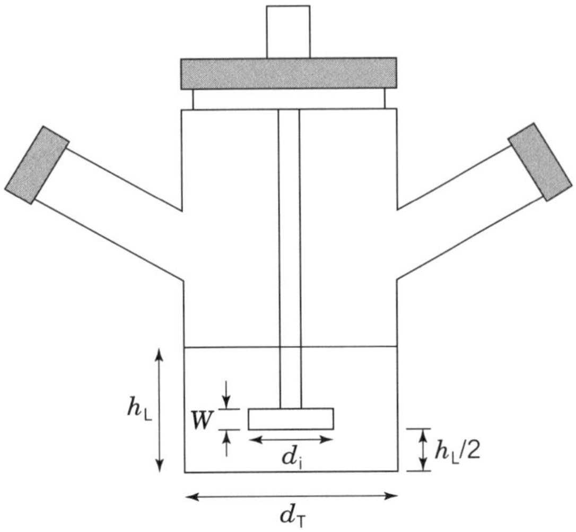

Figure 3. Geometry of 500-mL Corning spinner flasks: $h_L = 0.08$ m; $d_T = 0.096$ m; $d_i = 0.078$ m ($W = 0.025$ m), or $d_i = 0.053$ m ($W = 0.019$ m).

0.078 m diameter paddle impeller placed at the liquid surface, the k_La_L value was about 9.72×10^{-4} s^{-1}. Moving the impeller to 0.01 m below the surface provided a k_La_L value of about 4.17×10^{-4} s^{-1}. Lowering the impeller further into the fluid did not affect the k_La_L significantly.

5. *Stirred vessels.* Lavery and Nienow (23) reported k_La_L measurements in water, RPMI 1640 basal cell culture medium, and medium supplemented with 5% (v/v) fetal calf serum in a small, unbaffled, spherical cell culture vessel (1.5 L, static surface area of liquid = 2.23×10^{-2} m^2) stirred with one or two three-bladed propellers ($d_i = 0.060$ m, located 0.035 m apart). The lower impeller was positioned 0.003 m from the bottom of the vessel flask and the upper was 0.002 m below the surface of the liquid. The agitation speeds were 1.6 to 5.8 s^{-1} (100 to 350 rpm). Air was sparged (100 mL min^{-1}) either in the liquid under the lower propeller or only through the headspace (i.e., surface aeration). The submerged aeration k_La_L values generally agreed with predictions of Van't Riet's correlation for nonionic solutions (i.e., equation 57). Relative to measurements in water, the serum and the basal medium had little effect on k_La_L values. The k_L for surface aeration was little affected by the number of impellers (whether one or two). The k_La_L values for surface aeration were about 75% of those for submerged aeration in the reactor used. Addition of silicone antifoam (6 ppm) reduced the k_La_L values by about 50%. In the basal medium with serum and the antifoam, the k_L values for surface aeration (no vortex) were (1.18 to 3.54) $\times 10^{-5}$ m s^{-1}. The k_La_L values for submerged aeration were (2.8 to 8.5) $\times 10^{-4}$ s^{-1}.

6. *Other devices.* For absorption of oxygen at free surface of sodium sulfite solution in cylindrical containers placed on an orbital shaker platform, the volumetric mass transfer coefficient has been correlated (24) as follows:

$$k_La_L = 6 \times 10^{-5}(P/V_L)^{0.4} d_T^{-0.25} h_L^{-0.6} \quad (49)$$

Equation 49 is independent of gas flow rate in the headspace; it was determined for $20 \leq P/V_L$ (W m^{-3}) ≤ 500 and $0.5 \leq h_L/d_T \leq 1.5$ ($d_T = 0.12$ or 0.15 m). The agitation speed of the shaker platform varied over 1.2 to 3.3 s^{-1}, and the moving vessel described a diameter of 0.01 to 0.04 m.

Larger Systems

1. *Stirred tanks.* Multibladed disk turbines located at the surface provide superior surface aeration relative to other types of impellers (25). To prevent entrainment of bubbles, the turbines must be operated (25) such that

$$Nd_i \leqslant 0.11(d_i/d_T)^{-0.2} \quad (50)$$

As the tank diameter increases, the ability of a given impeller to agitate the entire liquid surface declines if the agitation speed is to remain below the bubble entrainment limit. The optimum impeller size for surface aeration (25) is given by

$$\frac{d_i}{d_T} \approx 0.5 \quad (51)$$

Surface aeration is suited only to small reactors because k_La_L declines rapidly with increasing tank volume of surface aerated bioreactors. For subsurface impellers operated so that there is no entrainment of gas, the mass transfer coefficient k_L has been correlated with the impeller Reynolds number as follows (20):

$$Sh = 1.4Re_i^{0.76} \quad (52)$$

which applies to waterlike media.

The contribution of the free surface to aeration in conventionally stirred, baffled tanks declines as the scale of operation increases. The liquid-phase mass transfer coefficient at the free surface (no vortex) in such tanks can be estimated using the correlations summarized in Table 9. A preferred correlation (27) is

$$k_L = 0.138\, Sc^{-2/3}\left(\frac{4\mu_L PoN^3d_i^5}{\pi d_T^2 h_L \rho_L}\right)^{1/4} \quad (53)$$

where Po is the power number ($Po = P/(\rho_L N^3 d_i^5)$); d_T and d_i are the diameters of the tank and the impeller, respectively; h_L is the height of liquid in the tank; N is the impeller rotational speed, and P is the power input. Equation 53 applies to Newtonian media without antifoams or surfactants. Added surfactants, fatty acids, and proteins reduce k_L relative to values in clean liquids.

2. *Vortex aeration.* In unbaffled stirred tanks, agitation with a centrally located impeller creates a vortex that draws closer to the impeller as the rate of agitation increases (Fig. 4). In a vortex-aerated industrial fermenter (240 L) used in producing diphtheria vaccine, the following equation has been reported (18)

$$k_La_L = 7.33 \times 10^{-4} + 1.36 \times 10^{-4} Q_G \quad (54)$$

where Q_G is the surface aeration rate (10 to 45 normal L min^{-1}). The equation was developed in aqueous sodium

Table 9. Correlations for Liquid-Phase Mass Transfer Coefficient at Free Surface in Baffled Stirred Tanks (No Vortex)

$k_L = 5.11 \times 10^{-3}\left[\frac{11^{n-1}K}{\rho_L}\left(\frac{3n+1}{4n}\right)^n\right]^{-0.426} D_L^{0.5} d_i^{0.852} N^{1.352-0.426n}$	Perez and Sandall (26)
$k_L = 0.322 Sc^{-2/3} N^{0.7} d_i^{0.4} (\mu_L/\rho_L)^{0.3}$	Hikita and Ishikawa (27)
$k_L = 0.0256 Sc^{-1/2} \frac{Po^{1/3} N d_i^2}{(\pi d_T^2 h_L/4)^{1/3}}$	Farritor and Hughmark (28)
$k_L = \left\{0.432\left(\frac{d_i}{d_T}\right) - 0.13\right\}\left(\frac{E\rho_L}{\mu_L}\right)^{1/3} D_L^{1/2}$	Bin (29)
$\frac{k_L d_i}{D_L} = 0.04\left(\frac{N d_i^2 \rho_L}{\mu_L}\right)\left(\frac{N^2 d_i^3 \rho_L}{\sigma_L}\right)^{1/2}\left(\frac{\mu_L}{\rho_L D_L}\right)^{1/2}$	Kataoka and Miyauchi (27)

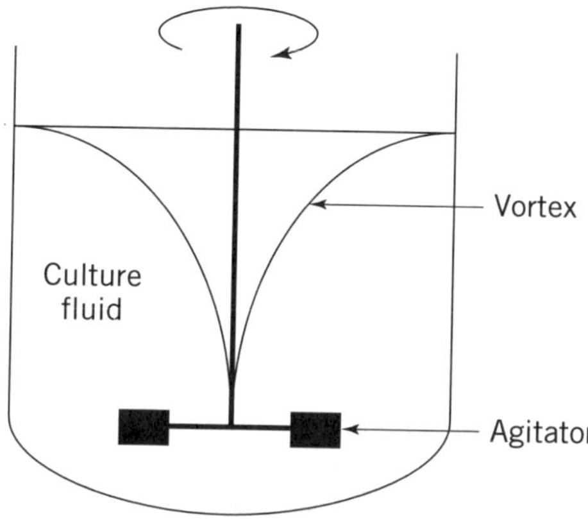

Figure 4. Vortex aeration in a stirred fermenter.

chloride (2.5 kg m^{-3}) and in the medium used in culturing the diphtheria bacterium. The tank was agitated with a six-bladed Rushton-type turbine (without disk), and the agitation rate was 300 to 800 rpm. The working aspect ratio was 0.74, which is substantially lower than values that are typically used in sparged microbial fermenters. Over the entire range of agitation rates used, the vortex was fully developed and reached the eye of the impeller (18). The conditions used in establishing equation 54 were identical to those used in commercial processes for producing the vaccine (18).

3. *Wetted-wall columns.* Mass transfer coefficient in liquid films in wetted-wall falling film columns can be estimated (2) with the equation

$$k_L = 9.0 Re^{-0.40} Sc^{-0.67} U_L \tag{55}$$

where the Sc and Re are based on properties of the liquid. Depth of the film and the mean film velocity (U_L) are used to calculate the Reynolds number. Equation 55 is appropriate for $Re < 2{,}000$ and $Sc < 100$. Another useful equation is

$$Sh = 0.023 Re^{0.8} Sc^{1/3} \tag{56}$$

Other correlations suited to highly turbulent films ($Re >$ 4,000) are available (30).

Submerged Geration. Submerged aeration in which an oxygen-containing gas is sparged or bubbled through the culture fluid is the norm in large-scale processes, including animal cell culture. Typically, submerged cultivation is carried out in stirred tanks, bubble columns, and airlift bioreactors. Stirred vessels are relatively more common, but the number of airlift devices continues to increase. Bubble columns are used less frequently (14). Typically, sparged aerated bubble columns and stirred tanks have aspect ratios between 3 and 4. Airlift devices generally have higher aspect ratios, usually greater than 6. Stirred vessels used in culturing animal cells are usually shorter, with aspect ratios less than 2. Microbial fermentations are aerated at high rates, with superficial aeration velocities of up to 0.1 m s^{-1} in airlift vessels and somewhat lower maximum values in bubble columns. Aeration rates are substantially lower in stirred tanks to prevent flooding of the impeller. A flooded impeller is a poor mixer and gas disperser.

Sparged aeration can potentially damage sensitive animal cells (12,31–34) and cells of certain cyanobacteria. Nevertheless, sparged aeration is successfully used quite widely in commercial culture of animal cells (35–37); however, the superficial aeration velocity in cell culture vessels is always low, usually well below 2.5×10^{-3} m s^{-1}. Supplementing the culture medium with serum, albumin, viscosity enhancers such as carboxymethyl cellulose, or surfactants such as Pluronic F68 is known to suppress the damaging effects of sparged aeration in cell culture (12,38).

As a rule, serum supplementation of basal animal cell culture media has little effect on mass transfer under conditions that are typically used. Properties of serum-supplemented media are usually quite similar to those of serum-free basal media. At 37 °C the surface tension of a typical cell culture medium with serum is about 0.060 to 0.064 N m^{-1} as compared to 0.071 N m^{-1} for water, and the viscosity is about 0.75×10^{-3} Pa s^{-1}. This viscosity value is satisfactory for media containing 5 to 10% serum. In submerged aeration, serum-containing media tend to foam, especially if a fine-pore sparger is used for aeration (35); however, fine-pore aeration is not the norm.

Stirred Fermenters. The gas–liquid volumetric mass transfer coefficient in stirred vessels has generally shown good correlation with agitation power input and the superficial gas velocity; thus

$$k_La_L = \alpha(P_G/V_L)^{\beta}U_G^{\gamma} \quad (57)$$

where P_G/V_L (W m^{-3}) is the gassed power input and U_G (m s^{-1}) is the superficial aeration velocity. Generally, $0.4 < \beta < 1$ and $0 < \gamma < 0.7$ (6). For pure water, α, β, and γ values are 2.6×10^{-2} ($s^{\beta+\gamma-1}m^{3\beta-\gamma}J^{-\beta}$), 0.4, and 0.5, respectively, when $500 < P_G/V_L$ (W m^{-3}) $< 10{,}000$ (6). These values have been shown to apply over a wide scale: $0.002 \leq V_L$ (m^3) ≤ 2.6. In strongly ionic aqueous solutions, the α, β, and γ values become 2.0×10^{-3} ($s^{\beta+\gamma-1}m^{3\beta-\gamma}J^{-\beta}$), 0.7, and 0.2, respectively, when $500 < P_G/V_L$ (W m^{-3}) $< 10{,}000$ and $0.002 \leq V_L$ (m^3) ≤ 4.4 (6). In pure water and ionic aqueous media, the volumetric mass transfer coefficient depends only on the specific power input irrespective of the type of stirrer—whether turbines, propellers, paddles, or self-sucking agitators—used for agitation. The number of impellers does not matter and neither does the location so long as it is within the usual ranges. Similarly, the height of liquid and the type of sparger have negligible influence under representative operating conditions (6). The parameters values noted above assume an absence of flooding. Maximum aeration rates that still ensure unflooded operation depend on the geometry of the impeller, as discussed by Nienow (39).

For single-impeller stirred air–water system, Bakker et al. (40) have recommended the following values of the constants in equation 57: $\beta = \gamma = 0.6$ and $\alpha = 0.015 \pm 0.005$. The parameters noted apply only to the nonflooded state. At high gas flow rates, concave-bladed gas dispersion impellers (Fig. 5) yield at least 20% higher k_La_L than obtainable with Rushton turbines under the same conditions (40). The constants in equation 57 can be strongly affected by fluid properties, presence of surfactants, and insoluble oils. For fluids other than water and for multiimpeller systems, Bakker et al. (40) recommended measuring α, β, and γ in a geometrically similar small-scale stirred fermenter. This data can be used in equation 57 for scale-up.

For fermentations of *Endomyces* sp. in stirred fermenters, α, β, and γ values in equation 57 have been reported to be 4.015×10^{-2} ($s^{\beta+\gamma-1}m^{3\beta-\gamma}J^{-\beta}$), 0.33, and 0.56, respectively, over a broad scale ($0.03 \leq V_L$ (m^3) ≤ 50). In a 5-L stirred fermenter operated with high-density culture of a recombinant *Escherichia coli*, Shin et al. (41) reported the following equation for k_La_L

$$k_La_L = 0.0195(P_G/V_L)^{0.55}U_G^{0.64}(1 + 2.12X + 0.20X^2)^{-0.25} \quad (58)$$

where X (g L^{-1}) is the biomass concentration. The X values ranged over 0 to 75 g L^{-1}, and the k_La_L values during culture varied over (10 to 25) $\times 10^{-3}$ s^{-1}.

In stirred-tank bioreactors for animal cell culture the type of impeller used does not affect k_La_L so long as mixing is sufficient. For power inputs that are typical (<25 W m^{-3}), better mixing is attained with relatively large ($d_i/d_T \sim 0.5$ to 0.6), slow moving, axial flow impellers. For typically used conditions, the k_La_L should be calculated as for bubble columns (equation 60) because for impeller power inputs less than about 40 W m^{-3}, the impeller has little effect on k_La_L. Stirred-tank design considerations for animal cell culture have been discussed by Chisti (35) and others (25).

Some of the main correlations for estimating gas–liquid volumetric mass transfer coefficient in stirred microbial fermenters are summarized in Table 10, which includes equations suitable for relatively viscous power law broths. Methods for estimating the power input in mechanically agitated fermenters have been detailed by Chisti and Moo-Young (7) and Nienow (39).

Bubble Columns. For a given fluid in bubble columns, the k_La_L is affected mainly by the superficial gas velocity, which also determines the specific power input (5). The specific power input can be calculated using the equation

$$\frac{P_G}{V_L} = \rho_L g U_G \quad (59)$$

where V_L is the volume of the liquid or slurry and ρ_L is its density. Properties of the fluid also affect k_La_L. In waterlike fluids, the aspect ratio of bubble columns does not affect k_La_L so long as the column diameter exceeds about 0.1 m (5). Similarly, liquid superficial velocities do not affect k_La_L provided $U_L \leq 0.1$ m s^{-1}.

For power inputs that are typically used in microbial culture, the k_La_L in relatively coalescing media is little influenced by the design of the sparger because the turbulence in the fluid controls the bubble size. In strongly ionic low-viscosity media, bubbles do not coalesce easily, and the bubble size depends on the characteristics of the gas sparger. Porous plate spargers and those with smaller holes produce smaller bubbles and higher k_La_L values; however, small bubbles are rapidly exhausted of oxygen and have long residence times in the fluid. Small bubbles are produced also under highly turbulent conditions irrespective of the sparger hole size. Such bubbles are especially persistent in viscous media because the viscous drag prevents them from leaving the fluid. Persistent microbubbles are seen in broths of *Streptomyces*, mycelial fungi, and certain fermentations that produce extracellular polymers such as xanthan.

Figure 5. A concave-blade impeller for dispersing gases in liquids. *Source*: Courtesy of Chemineer, Inc.

Table 10. Mass Transfer Correlations for Stirred Tanks

Correlation	Ranges
1. Taguchi and Yoshia (42) $\frac{k_L a_L}{N} = 0.113\left(\frac{d_T^2 h_L}{W_i(d_t - W)}\right)^{1.437} (Nt_m)^{-1.087}(d_i/d_T)^{1.021}$	Pseudoplastic paper pulp slurries (1.6% w/v pulp); disk turbine stirred vessel
2. Yagi and Yoshida (43) $\frac{k_L a_L d_i^2}{D_L} = 0.06\left(\frac{d_i^2 N \rho_L}{\mu_L}\right)^{1.5}\left(\frac{d_i N^2}{g}\right)^{0.19}\left(\frac{\mu_L}{\rho_L D_L}\right)^{0.5}\left(\frac{\mu_L U_G}{\sigma_L}\right)^{0.6} \times \left(\frac{Nd_i}{U_G}\right)^{0.32} [1 + 2\sqrt{\lambda N}]^{-0.67}$ For non-Newtonian fluids use apparent viscosity instead of μ_L. λ is the characteristic time or relaxation time of viscoelastic fluids; $\lambda = 0$ for nonviscoelastics	Viscous Newtonian and non-Newtonian liquids, including viscoelastic fluids
3. Perez and Sandall (26) $\frac{k_L a_L d_i^2}{D_L} = 21.2\left(\frac{\rho_L N d_i^2}{\mu_{ap}}\right)^{1.11}\left(\frac{\mu_{ap}}{\rho_L D_L}\right)^{0.5}\left(\frac{U_G d_i}{\sigma_L}\right)^{0.447}\left(\frac{\mu_G}{\mu_{ap}}\right)^{0.694}$ $\mu_{ap} = K(11N)^{n-1}\left(\frac{3n + 1}{4n}\right)^{n}$	Carbon dioxide absorption in water and non-Newtonian power law solutions
4. Kawase and Moo-Young (44) $k_L a_L = 0.675\sqrt{D_L}\,\frac{\rho_L^{3/5} E^{9+4n/10(1+n)}}{(K/\rho_L)^{1/2(1+n)}\sigma_L^{3/5}}\left(\frac{U_G}{U_B}\right)^{0.5}\left(\frac{\mu_{ap}}{\mu_w}\right)^{-0.25}$	Newtonian and non-Newtonian power law fluids
5. Niebelshütz (15) $k_L a_L\left(\frac{V_L}{Q_G}\right)Sc^{0.3} = 0.103\left(\frac{P_G}{Q_G \rho_L (g\mu_L/\rho_L)^{2/3}}\right)^{0.53}$	Non-Newtonian power law fluids such as broths of *Aspergillus niger, Penicillium chrysogenum*, and xanthan fermentation; $0.035 \le V_L(\mathrm{m}^3) \le 0.070$, $0.02 \le \mu_L(\mathrm{Pa\ s}) \le 3.1$; μ_L was calculated at shear rate $= 10^4\ \mathrm{s}^{-1}$
6. Reuss (15) $\frac{k_L a_L}{U_G}\left(\frac{\mu_L^2}{g\rho_L^2}\right)^{1/3} Sc^{0.3} = 5.5 \times 10^{-4}\left(\frac{P_G}{V_L U_G \rho_l g}\right)^{0.7}$	Newtonian broths of *Saccharomyces cerevisiae* and gluconic acid, glucose solutions, $\mu_L \le 12 \times 10^{-3}$ Pa s, $0.05 \le V_L(\mathrm{m}^3) \le 1.9$
7. Chandrasekharan and Calderbank (45) $k_L a_L = \left(\frac{0.0248}{d_T^4}\right)\left(\frac{P_G}{V_L}\right)^{0.551} Q_G^{0.551/\sqrt{d_T}}$	Air and water

In coalescence-inhibiting media typical of animal cell culture, the $k_L a_L$ in bubble columns at the very low superficial gas velocities that are normally used depends on the sparger hole diameter; the $k_L a_L$ declines as the sparger hole diameter increases over the range of 0.01 to 2 mm (25). Relative to values in clean fluids, surfactants and proteins reduce $k_L a_L$, but the reduction is greater in finer dispersions (i.e., in small, ridged bubbles) generated by small-hole spargers than in dispersions of larger, circulating bubbles. For aeration rates relevant to cell culture, the $k_L a_L$ for water may be calculated (25) with the equation

$$\frac{k_L a_L}{U_G}\left(\frac{\mu_L^2}{\rho_L^2 g}\right)^{1/3} = \alpha[(\mu_L g/\rho_L)^{-1/3} U_G]^{-\beta} \qquad (60)$$

The values of α and β for use with various sparger hole sizes are noted in Table 11. The $k_L a_L$ values in culture media are only about 50% of those obtained with equation 60 for water. Equation 60 was developed in bubble columns with aspect ratios $1 \le (h_L/d_T) \le 3$ when the specific power inputs were $3 \le P_G/V_L\ (\mathrm{W\ m^{-3}}) \le 38$.

Although the overall $k_L a_L$ in tall bioreactors may be little affected by the height, the overall mass transfer rates tend to be greater in taller systems because the mass transfer driving force increases with increasing hydrostatic pressure. Mass transfer in bubble columns is discussed further by Shah et al. (46), Chisti (5,14), Deckwer (47), and Deckwer and Schumpe (48). The various correlations available for estimating $k_L a_L$ or k_L in bubble columns are summarized in Table 12. Those correlations apply to the relatively high gas flow rates that are typical of microbial fermentations. The correlation of Akita and Yoshida (51) is recommended for viscosities less than about 21×10^{-3} Pa s. The correlation of Kawase et al. (54) (Table 12) has also agreed well with independent measurements (59).

Airlift Bioreactors. Because of superior oxygen transfer performance, airlift bioreactors have given excellent productivities with yeasts, bacteria, and filamentous fungi (5,14). Processes that produce highly viscous broths, including several biopolymer-producing fermentations, have been successfully carried out in airlift devices (14). Similarly, many hybridoma cultures and plant cell suspensions have given good results. As a general rule, the volumetric productivity of airlift bioreactors equals or betters that of conventional stirred tanks (14). Typically, this level of performance is achieved at substantially lower power input than in stirred vessels. Furthermore, the probability of me-

Table 11. Values of α and β for Use in Equation 60

Sparger	d_H (mm)	α	β	Range
Frit	0.01	6.0×10^{-5}	0.4	$0.001 \leq U_G(\mu_L g/\rho_L)^{-1/3} \leq 0.1$
	0.02	6.3×10^{-5}	0.2	
Perforated pipe	0.5	4.2×10^{-5}	0.2	$0.01 \leq U_G(\mu_L g/\rho_L)^{-1/3} \leq 0.2$
	2.0	2.2×10^{-5}	0.17	

Source: Data from Ref. 25.

chanical failure and likelihood of loss of sterility are lower with airlift bioreactors. Airlift devices outperform bubble columns in suspending solids. Relative to bubble columns, the airlift reactors have a better-defined flow pattern; airlift systems are better than bubble columns in heat-transfer capabilities (14). The operational range of airlift bioreactors tends to be generally broader than similar bubble columns. Unlike in bubble columns, the overall height h_L of liquid in airlift reactors affects k_La_L because the height strongly influences the liquid circulation velocity and, hence, the gas holdup (14).

Many equations have been developed to relate the apparent viscosity of non-Newtonian media to the k_La_L in airlift reactors. Most authors agree that k_La_L generally declines with increasing apparent viscosity; however, there is great discrepancy on the precise magnitude of the viscosity effect (5,14). The exponent on the apparent viscosity term has been given variously as -0.255 (60), -0.66, -0.89, and other values (see Ref. 5). Such discrepancies are associated in part with the questionable practice of defining shear rate as an arbitrary linear function of the gas velocity (5,11). The viscosity-associated decline in k_La_L is especially pronounced for viscosities exceeding about 2.5×10^{-2} Pa s. In highly viscous non-Newtonian fermentations, the use of a supplementary axial flow impeller in the draft tube of airlift reactors is known to enhance fermentation performance.

Table 13 lists the correlations available for estimating the gas–liquid mass transfer coefficient in airlift bioreactors. These correlations are suitable only for initial estimates because the hydrodynamic and transport behavior of airlift devices is quite sensitive to the type of fluid, the reactor geometry, the distribution of gas holdup in the various zones, and the induced liquid circulation rate. Because of the numerous influences, a superior approach to prediction of mass transport for a more reliable design and scale-up relies on data measured at small scale. This method is discussed in the next section.

A General Method of Predicting k_La_L for Scale-up. Although various correlations can be used to directly estimate k_La_L, the reliability of such estimates is often quite poor. A more reliable prediction of k_La_L for process scale-up is based on methodology that combines fundamental principles with small-scale experimentation. This proved approach is especially suited to prediction of k_La_L in complex media. Thus, based on purely analytical reasoning, the volumetric mass transfer coefficient (k_La_L), gas holdup (ϵ_G), the Sauter mean bubble diameter (d_B), and the true mass transfer coefficient (k_L), have been related (5) as follows

$$\frac{k_L}{d_B} = \frac{k_La_L(1 - \varepsilon_G)}{6\varepsilon_G} \qquad (61)$$

Calculations of the k_L/d_B ratio from the measured k_La_L and gas holdup have shown that in bubble columns and airlift devices this ratio is constant for a given fluid irrespective of the aeration rate (5,66). The value of k_L/d_B depends on the properties of the fluid. Thus, in slurries of mycelia-like cellulose fiber particles, k_L/d_B is affected by the concentration of solids (Fig. 6). For aqueous salt solution (0.15 M sodium chloride) with or without suspended cellulose fiber, the k_L/d_B ratio (5) has been expressed as

$$\frac{k_L}{d_B} = 5.63 \times 10^{-5}\left(\frac{g\rho_L^2\sigma_L D_L}{\mu_L^3}\right)^{1/2} e^{-0.131C_S^2} \qquad (62)$$

where C_S is the concentration of solids (% w/v). Equation 62 applies to airlifts as well as bubble columns. Chisti's (5) observations of constant k_L/d_B ratio in pneumatically agitated reactors have also been confirmed by others (14).

The observed constancy of k_L/d_B provides a reliable method for predicting k_La_L in large-scale bioreactors from data easily measured in a small-scale model unit (5). This proved technique consists of the following steps:

1. Measure the gas holdup and the k_La_L in a small-model reactor with the fluid of interest over the range of gas flow rates that are of interest.
2. Use the data from step 1 to calculate the constant k_L/d_B ratio (equation 61).
3. Measure or calculate the riser gas holdup (ϵ_{Gr}) in the full-scale reactor (see Refs. 5 and 14).
4. Calculate the k_La_L value for the large reactor: for bubble columns

$$k_La_L = \frac{k_L}{d_B}\frac{\varepsilon_G}{(1 - \varepsilon_G)} \qquad (63)$$

and for airlift bioreactors

$$k_La_L = \frac{k_L}{d_B}\frac{\varepsilon_{Gr}}{(1 - \varepsilon_{Gr})}\frac{A_r}{A_r + A_d} \qquad (64)$$

In equations 63 and 64, the gas holdup and the geometric parameters A_r and A_d are for the large reactor; the k_L/d_B value is from step 2. The rationale for equations 63 and 64 has been discussed by Chisti (5).

Table 12. Gas–Liquid Mass Transfer Correlations for Bubble Columns

Correlation	Ranges
1. Chisti (5) $k_L a_L = 2.39 \times 10^{-4}(P_G/V_L)^{0.86}$	Air and water, $d_T \geq 0.1$ m, $200 \leq P_G/V_L(\mathrm{W\ m^{-3}}) \leq 1{,}000$, h_L/d_T up to 25
2. Heijnen and Van't Riet (49) $k_L a_L = 0.32 U_G^{0.7}$	Air and water at 20 °C, $0.08 < d_T(\mathrm{m}) < 11.6$, $0 < U_G(\mathrm{m\ s^{-1}}) < 0.3$, $0.3 < h_L(\mathrm{m}) < 21$; perforated pipe or perforated plate spargers producing 4–6-mm bubbles
3. Fair (50) $k_L a_D = 3.31\left(\frac{D_L \varepsilon_G}{d_B^2}\right)\left(\frac{\mu_L}{\rho_L D_L}\right)^{1/3}\left(\frac{d_B \rho_L U_G}{\mu_L}\right)^{0.5}$	Newtonian fluids
4. Akita and Yoshida (51) $\frac{k_L a_D d_T^2}{D_L} = 0.6\left(\frac{\mu_L}{\rho_L D_L}\right)^{0.5}\left(\frac{g d_T^2 \rho_L}{\sigma_L}\right)^{0.62}\left(\frac{g d_T^3 \rho_L^2}{\mu_L^2}\right)^{0.31}\varepsilon_G^{1.1}$	Air and water and other Newtonian fluids; $\mu_L \leq 21 \times 10^{-3}$ Pa s, U_G and U_L up to 0.4 m s^{-1}
5. Akita and Yoshida (51) $\frac{k_L d_B}{D_L} = 0.5\left(\frac{\mu_L}{\rho_L D_L}\right)^{0.5}\left(\frac{g d_B^3 \rho_L^2}{\mu_L^2}\right)^{0.25}\left(\frac{g d_B^2 \rho_L}{\sigma_L}\right)^{3/8}$	Air and water and other Newtonian fluids, $\mu_L \leq 21 \times 10^{-3}$ Pa s, U_G and U_L up to 0.4 m s^{-1}
6. Chisti (5) $k_L = 5.63 \times 10^{-5}\left(\frac{g \rho_L^2 D_L \sigma_L}{\mu_L^3}\right)^{0.5} d_B e^{-0.131 C_S^2}$	Air and water, and slurries of mycelia-like fibrous solids (paper pulp) in salt solutions; bubble and churn turbulent flow; $0 \leq C_S(\mathrm{w/v}\ \%) \leq 3$
7. Hughmark (52) $\frac{k_L d_B}{D_L} = 2 + 0.0187\left[\left(\frac{\mu_L}{\rho_L D_L}\right)^{0.339}\left(\frac{d_B U_G \rho_L}{\mu_L}\right)^{0.484}\left(\frac{d_B g^{1/3}}{D_L^{2/3}}\right)^{0.072}\right]^{1.61}$	Newtonian fluids, $0.0009 \leq \mu_L(\mathrm{Pa\ s}) \leq 0.152$
8. Nakanoh and Yoshida (53) $\frac{k_L a_D d_T^2}{D_L} = 0.09\left(\frac{\mu_L}{\rho_L D_L}\right)^{0.5}\left(\frac{g d_T^2 \rho_L}{\sigma_L}\right)^{0.75}\left(\frac{g d_T^3 \rho_L^2}{\mu_L^2}\right)^{0.39}\left(\frac{U_G^2}{g d_T}\right) \times \left(1 + \chi\left(\frac{U_B \lambda}{d_B}\right)\right)^{-0.45}$ $\chi = 0$ (inelastic fluids) or 0.133 (elastic fluids). For non-Newtonian power law fluids $\mu_L = \mu_{ap} = K(5{,}000 U_G)^{n-1}$	Newtonian, non-Newtonian, and viscoelastic fluids, $0.005 \leq \mu_L(\mathrm{Pa\ s}) \leq 0.06$, $U_G < 0.1$ m s^{-1}
9. Schumpe et al. (54) $\frac{k_L a_L d_T^2}{D_L} = 0.021\left(\frac{\mu_L}{\rho_L D_L}\right)^{0.5}\left(\frac{g d_T^2 \rho_L}{\sigma_L}\right)^{0.21}\left(\frac{g d_T^3 \rho_L^2}{\mu_L^2}\right)^{0.60}\left(\frac{U_G}{\sqrt{g d_T}}\right)^{0.49}$ μ_L is the apparent viscosity of the power law fluid	Non-Newtonian media
10. Kawase et al. (55) $\frac{k_L a_L d_T^2}{D_L} = 0.555\left(\frac{K d_T^{1-n}}{\rho_L D_L U_G^{1-n}}\right)^{0.5}\left(\frac{g d_T^2 \rho_L}{\sigma_L}\right)^{3/5}\left(\frac{d_T^n U_G^{2-n}}{K/\rho_L}\right) \times \left(\frac{U_G^2}{g d_T}\right)\frac{(2+n)(2n-3)}{20(1+n)^2}\varepsilon_G^{0.5}$	Newtonian and non-Newtonian power law fluids. For Newtonian fluids $K = \mu_L$ and $n = 1$
11. Hikita et al. (56) $k_L a_L = \frac{14.9 g f}{U_G}\left(\frac{U_G \mu_L}{\sigma_L}\right)^{1.76}\left(\frac{\mu_L^4 g}{\rho_L \sigma_L^3}\right)^{-0.248}\left(\frac{\mu_G}{\mu_L}\right)^{0.243}\left(\frac{\mu_L}{\rho_L D_L}\right)^{-0.604}$ f is 1.0 for nonelectrolytes; $f = 10^{0.0681 I}$ for ionic strength $I < 1.0$ kg ion m^{-3}; and $f = 1.114 \times 10^{0.021}$ for $I > 1.0$ kg ion m^{-3}	Low viscosity Newtonian media: water and various gases (air, oxygen, carbon dioxide, hydrogen, methane), sucrose solution, aqueous electrolytes, alcohols, $0.0008 \leq \mu_L(\mathrm{Pa\ s}) \leq 0.011$, $0.042 \leq U_G(\mathrm{m\ s^{-1}}) \leq 0.38$
12. Deckwer et al. (57) $k_L a_D = 3.15 \times 10^{-3} U_G^{0.59}[K(5{,}000 U_G)^{n-1}]^{-0.84}$	Non-Newtonian power law fluids, slug flow ($U_G \geq 0.02$ m s^{-1})
13. Godbole et al. (58) $k_L a_D = 8.35 \times 10^{-4} U_G^{0.44}[K(5{,}000 U_G)^{n-1}]^{-1.01}$	Newtonian power law fluids

Figure 7 compares the measured $k_L a_L$ values with prediction of the foregoing procedure for 11 reactors, including bubble columns and external- and internal-loop airlift devices. The comparison spans waterlike media as well as pseudoplastic slurries of cellulose fiber. In all cases, the predictions agree remarkably well with the measurements (Fig. 7). Note that the Solka Floc cellulose fiber slurries in aqueous salt solutions used in Figures 6 and 7 were meant to simulate the fermentation broths of mycelial fungi and filamentous bacteria (5,66). Such broths are usually

Table 13. Gas–Liquid Volumetric Mass Transfer Coefficient Correlations for Airlift Bioreactors

Configuration	Equation	Ranges
Concentric-tube internal-loop vessels (annulus sparged)	$\frac{k_L a_D \sigma_L}{D_L g \rho_L} = 2.25\left(\frac{\mu_L}{\rho_L D_L}\right)^{0.500}\left(\frac{\rho_L \sigma_L^3}{g\mu_L^4}\right)^{0.136} \times \left(\frac{d_H}{d_T}\right)^{-0.0905} \varepsilon_G^{1.26}$ Average estimation error was 12% for 175 measurements Koide et al. (61)	Newtonian media, $3.71 \times 10^2 \le \mu_L/\rho_L D_L \le 6.00 \times 10^4$, $1.18 \times 10^6 \le \rho_L \sigma_L^3/g\mu_L^4 \le 5.93 \times 10^{10}$, $0.471 \le d_i/d_T \le 0.743$, $7.14 \times 10^{-3} \le d_H/d_o \le 2.86 \times 10^{-2}$, $0.0302 \le \varepsilon_G \le 0.305$, $6 \le$ aspect ratio ≤ 15, $0.52 \le A_d/A_r \le 1.23$
Concentric-tube internal-loop vessels (draft-tube sparged)	$\frac{k_L a_D d_T^2}{D_L} = 2.66\left(\frac{\mu_L}{\rho_L D_L}\right)^{0.500}\left(\frac{g d_T^2 \rho_L}{\sigma_L}\right)^{0.715}\left(\frac{g d_T^3 \rho_L^2}{\mu_L^2}\right)^{0.251} \times \left(\frac{d_i}{d_T}\right)^{-0.429} \varepsilon_G^{1.34}$ Koide et al. (62)	Newtonian media, $\rho_L = 997 - 1182$ kg m^{-3}, $\mu_L = (0.894 - 17.0) \times 10^{-3}$ Pa s, $\sigma_L = (51.7 - 73.0) \times 10^{-3}$ N m^{-1}, $D_L = (0.145 - 2.42) \times 10^{-9}$ m^2 s^{-1}, $A_r/A_d = 0.3$–0.8, aspect ratio = 6–15
External-loop reactor	$\frac{k_L a_L L}{U_{Lr}} = 14.5\left(\frac{U_{Gr}}{U_{Lr}}\right)^{0.83} Pe^{-0.6}$ Pe value is for the entire loop Verlaan et al. (63)	Air and water in 0.165 m^3 reactor, aspect ratio = 16 (based on riser), $A_r/A_d = 4$, no gas in downcomer, $0.01 \le U_{Gr}$(m s^{-1}) ≤ 0.14, $Pe = U_L L/E_L$, $Pe = 40$–60 (increased with gas flow rate)
A bubble column and a draft-tube sparged concentric-tube airlift vessel	$\frac{k_L a_L d_T^2}{D_L} = 0.018\left(\frac{K(2{,}800 U_G)^{n-1}}{\rho_L D_L}\right)^{0.5}\left(\frac{g d_T^2 \rho_L}{\sigma L}\right)^{0.20} \times \left(\frac{U_G}{\sqrt{g d_T}}\right)^{0.51}\left(\frac{g d_T^3 \rho_L^2}{[K(2{,}800 U_G)^{n-1}]^2}\right)^{0.62} \times (1 + 0.12\mathrm{Wi})^{-1}$ Suh et al. (64)	Non-Newtonian xanthan fermentation broths of the obligate aerobic bacterium *Xanthomonas campestris*. Bubble column (0.05 m^3, aspect ratio ≈ 19); airlift bioreactor (1.2 m^3, $A_r/A_d = 1$, aspect ratio ≈ 12)
Concentric-tube internal-loop (draft-tube sparged)	$k_L a_L = 3.43 \times 10^{-2} U_{Gr}^{0.524} \mu_{ap}^{-0.255}$ μ_{ap} was calculated by assuming the shear rate to equal $5{,}000 \times U_{Gr}$ for $U_{Gr} \ge 0.04$ m s^{-1}, or $1{,}000 \times (U_{Gr})^{0.5}$ for $U_{Gr} < 0.04$ m s^{-1} Li et al. (60)	Aqueous carboxymethyl cellulose ($K = 0.286$–11.5 Pa s^n; $n = 0.441$–0.617), 0.055 m^3 vessel, $A_r/A_d = 0.618$, aspect ratio ~ 26. $0.020 \le \mu_{ap}$ (Pa s) ≤ 0.85
Concentric-tube internal-loop reactors (draft-tube sparged)	$\frac{k_L a_D d_T^2}{D_L} = 2.66\left(\frac{\mu_L}{\rho_L D_L}\right)^{0.500}\left(\frac{g d_T^2 \rho_L}{\sigma_L}\right)^{0.715}\left(\frac{g d_T^3 \rho_L^2}{\mu_L^2}\right)^{0.251} \times \left(\frac{d_i}{d_T}\right)^{-0.429} \varepsilon_G^{1.34}\left(1 + 0.099\left(\frac{C_S}{\rho_S}\right)^{0.069} \times \left(\frac{\rho_L \sigma_L^3}{g\mu_L^4}\right)^{0.023}\left(\frac{U_t}{U_G}\right)^{0.046}\right)^{-1}$ Average estimation error was within 17% for 383 measurements Koide et al. (62)	Suspensions of glass or bronze spheres, $3.71 \times 10^2 \le \mu_L/\rho_L D_L \le 9.92 \times 10^4$, $1.36 \times 10^3 \le g d_T^2 \rho_L/\sigma_L \le 1.22 \times 10^4$, $1.29 \times 10^8 \le g d_T^3 \rho_L^2/\mu_L^2 \le 1.26 \times 10^{11}$, $0.471 \le d_i/d_T \le 0.743$, $3.99 \times 10^{-2} \le \varepsilon_G \le 2.73 \times 10^{-1}$, $1.69 \times 10^{-11} \le g\mu_L^4/\rho_L \sigma_L^3 \le 2.55 \times 10^{-6}$, $0 \le C_S/\rho_S \le 8.00 \times 10^{-2}$, $1.17 \times 10^{-2} \le U_t/U_G \le 0.844$, $A_r/A_d = 0.3$–0.8, aspect ratio = 6–15
Concentric-tube internal-loop reactors (draft-tube sparged)	$\frac{k_L a_D d_T^2}{D_L} = \frac{4.04}{1 + 2\varphi_S^{1.3}}\left(\frac{\mu_L}{\rho_L D_L}\right)^{0.5}\left(\frac{g d_T^2 \rho_L}{\sigma_L}\right)^{0.67}\left(\frac{g d_T^3 \rho_L^2}{\mu_L^2}\right)^{0.26} \times \left(\frac{d_i}{d_T}\right)^{-0.047} \varepsilon_G^{1.34}$ Average error of estimation was 14% for 260 data; U_G is based on the diameter d_T of the outer column; the $k_L a_D$ values in water and salt solutions were similar Koide et al. (65)	Suspensions of relatively low-density calcium alginate beads in water, aqueous glycerol, and aqueous salt solutions (0.10 or 0.27 kmol m^{-3} barium chloride; 0.4 kmol m^{-3} sodium sulfate), 0–20% v/v solids, 1.88–3.98-mm bead diameter, $A_r/A_d = 0.3$–1.2, aspect ratios = 6–16, $\mu_L = (0.894$–$12.5) \times 10^{-3}$ Pa s, D_L of oxygen = $(0.194$–$2.42) \times 10^{-9}$ m^2 s^{-1}, $3.71 \times 10^2 \le \mu_L/D_L \rho_L \le 5.52 \times 10^4$, $2.66 \times 10^3 \le \rho_L d_T^2 g/\sigma_L \le 1.22 \times 10^4$, $2.35 \times 10^8 \le \rho_L^2 d_T^3 g/\mu_L^2 \le 3.29 \times 10^{11}$, $0.471 \le d_i/d_T \le 0.743$, $1.69 \times 10^{-11} \le Mo \le 6.67 \times 10^{-7}$, $3.79 \times 10^{-2} \le \varepsilon_G \le 2.24 \times 10^{-1}$, $0 \le \varphi_S \le 0.2$; where the Morton number is $Mo = \frac{g(\rho_L - \rho_G)}{\sigma_L^3 \rho_L^2} K^4 \left(\frac{8U_{Lr}}{d_r}\right)^{4(n-1)}\left(\frac{3n+1}{4n}\right)^{4n}$

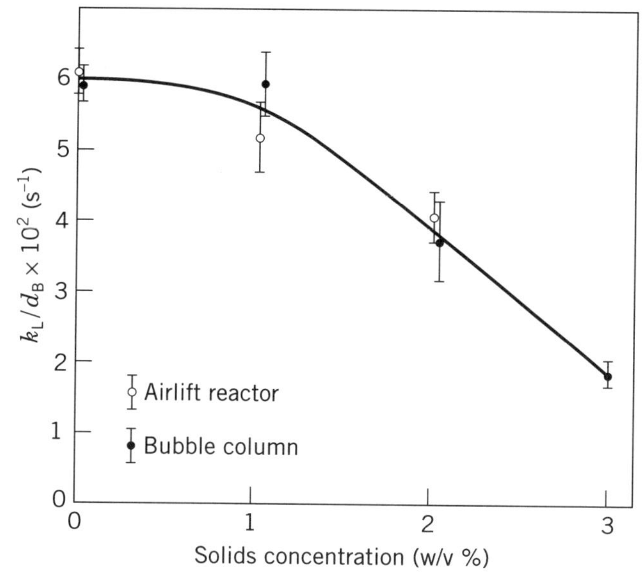

Figure 6. Effect of concentration of mycelia-like cellulose fiber particles on the k_L/d_B ratio in aqueous salt solution (0.15 M sodium chloride) (5).

pseudoplastic. The non-Newtonian flow properties of cellulose fiber suspensions are quite close to those of the typical fungal fermentation broths (5).

For the low aeration rates typical of animal cell culture, the k_La_L in an internal-loop draft tube airlift vessel is almost identical to that in an equivalent bubble column; hence, equation 60 can be used for an estimation. Although the gas is sparged only in the draft tube or in the annulus of an airlift bioreactor, for estimation of k_La_L with equation 60 the superficial gas velocity should be calculated using the cross-sectional area of the outer column. For conditions typical of animal cell culture, whether the gas is sparged in the draft tube or the annulus does not affect k_La_L so long as the outer-to-inner tube diameter ratio is 1.6 to 2.1 and $2 \leq h_L/d_T \leq 5$. Large concentric draft-tube types of airlift bioreactors are used in commercial culture of animal cells (36,37). Various aspects of design of airlift bioreactors have been exhaustively treated in the literature (5,14). A few specifics are discussed in the following sections.

Axial Dissolved Oxygen Profiles. As clearly demonstrated (5), the steady-state dissolved oxygen concentration varies axially in tall bubble columns and airlift reactors even in the absence of an oxygen-consuming reaction. The exact nature of the concentration profile depends on the state of mixing in the reactor. During culture of baker's yeast in a draft-tube sparged concentric tube internal-loop airlift reactor (A_r/A_d = 0.8; aspect ratio ≈ 10; 0.25 m^3), Russell et al. (67) observed that the steady-state dissolved oxygen concentration in the riser and the downcomer increased with increasing aeration rate. The dissolved oxygen levels were always several-fold higher in the downcomer than in the riser (67). In the riser, the dissolved oxygen levels tended to be 30% or less of air saturation. The oxygen concentration did not vary radially either in the riser or in the downcomer. Similarly, no significant variations were observed axially in the downcomer (67).

Using the annulus-sparged mode of the same vessel that was used by Russell et al. (67), Pollard et al. (68) found that the axial-dissolved oxygen profiles in the riser showed

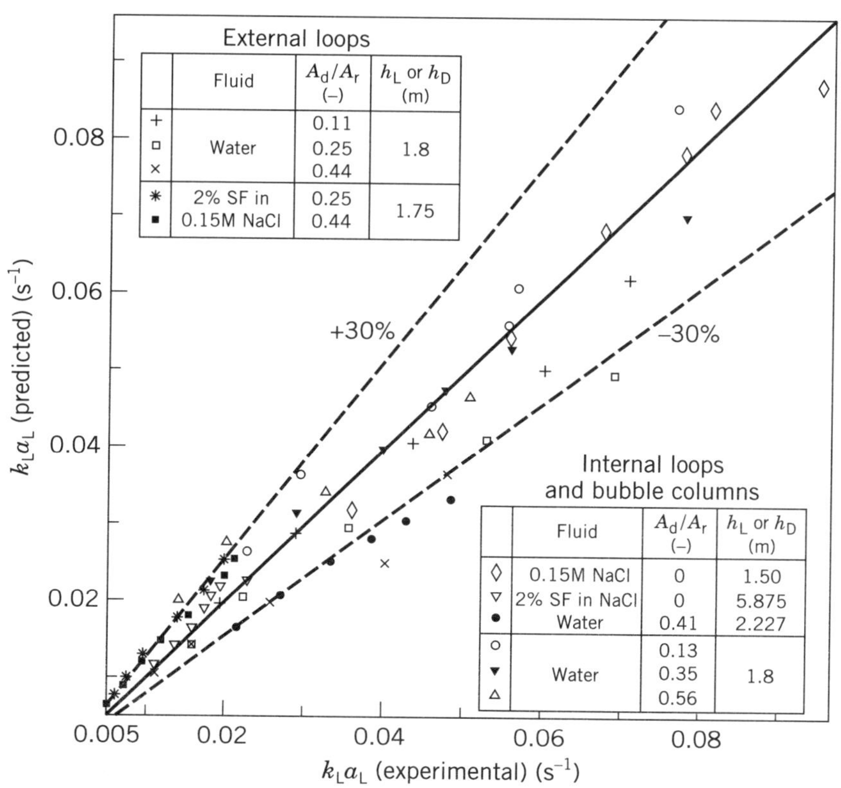

Figure 7. Predicted versus measured overall volumetric gas–liquid mass transfer coefficient in bubble columns, external-loop and internal-loop airlift devices in Solka Floc (SF) cellulose fiber slurries and solids-free media (5). The solid line represents exact prediction.

a constant zero level of oxygen over a broad range of superficial gas velocities (0.036 to 0.136 m s^{-1}) during baker's yeast fermentations. Surprisingly, the dissolved oxygen concentration increased down the downcomer, peaking close to the bottom of the draft tube and declining sharply to zero as the fluid moved into the riser. These anomalous results seemed to indicate better oxygen transfer in the downcomer relative to the riser (68). In view of the observed profile, the cells experienced extreme dissolved oxygen heterogeneity in the reactor; however, there seemed to be no effect on the culture performance. Installation and operation of a marine propeller (diameter = 0.16 m; 1000 rpm) near the bottom of the draft tube, combined with a high level of aeration (0.358 m^3 min^{-1}), enhanced liquid circulation and oxygen transfer so much so that the dissolved oxygen level did not drop below 35% of saturation anywhere in the reactor.

Contributions of Various Zones to Overall k_La_L. Many airlift reactors consist of two hydrodynamically distinct zones, the riser and the downcomer, that contribute differently to the observed mass transfer (5). Sometimes a third zone, the gas–liquid separator (69), may also exist with its own contribution toward mass transfer.

In airlift bioreactors that do not have gas–liquid separators, the measured k_La_L consists of contributions from the riser and the downcomer zones; thus

$$(k_La_L)_{\text{measured}} = \frac{A_r(k_La_L)_r + A_d(k_La_L)_d}{A_r + A_d} \tag{65}$$

Equation 65 is an analytical expression that relates the measured k_La_L to the separate, hypothetical contributions of the riser and the downcomer (5). Equation 65 leads to the general conclusion that for any airlift device in which gas–liquid mass transfer is confined to the riser (i.e., no gas in the downcomer), the actual volumetric mass transfer coefficient in the riser would be larger than the apparent measured value; the difference being dependent on the A_r/A_d ratio. Equation 65 applies to reactors in which the riser and the downcomer have approximately equal heights. In cases in which a significant portion of the liquid resides in the head zone, equation 65 should be modified to account for that volume (14). Generally, $(k_La_L)_d < (k_La_L)_r$.

Based on data obtained in several concentric draft-tube airlift reactors that were sparged either in the draft tube or the annulus (aspect ratios = 6 to 11; U_G = 0.019 to 0.12 m s^{-1}), Koide et al. (70) suggested that in air and water the contribution of the downcomer zone to gas–liquid mass transfer was "relatively large." Furthermore, in liquids such as aqueous salt solutions in which the bubbles are smaller, the riser and the downcomer were claimed to contribute to the total mass transfer in proportion to the fraction of the total volume that those zones occupied (70). These observations in relatively small reactors are not entirely consistent with other findings (5,14).

Mass Transfer Enhancement by Static Mixers. Installation of static mixer elements in airlift and bubble column bioreactors is known to enhance the gas–liquid mass transfer coefficient relative to mixer-free operation (14,71). The enhancement is mostly the result of an increase in the gas–liquid interfacial area that is brought about by the bubble-breaking action of the mixer elements. Improved productivity of cephalosporin C by *Cephalosporium acremonium* was observed (72) in an airlift reactor when static mixers were installed.

The mass-transfer-enhancing effect of static mixers is most pronounced in viscous fluids that give rise to large spherical cap bubbles (71); less dramatic, but still large improvements are observed with waterlike media. Generally, for a given fluid, the mass transfer coefficient and the gas velocity in the riser of a given airlift reactor may be correlated in the form

$$k_La_L = \alpha U_{Gr}^{\beta} \tag{66}$$

For a given fluid, the exponent β in equation 66 is not affected by the presence of static mixers in the reactor; however, the mixers significantly enhance the value of α (71). The improvement in α depends mainly on the consistency index (K) or the thickness of the fluid; generally, the thicker the fluid, the greater the effect of static mixers (71). The approximate relationship among α, α_M (the α-value in presence of static mixers), and K has been established (71) to be

$$\frac{\alpha_M}{\alpha} = 4.43K^{0.12} \tag{67}$$

Equation 67 was determined for approximately 10^4-fold variation in K. The flow behavior index n (varied between 0.5 and 1.0) had no direct impact on enhancement of k_La_L by the mixers (71). Depending on the fluid, the static mixers were found to enhance k_La_L by 30 to 500% relative to mixer-free operation (71). The constants in equation 67 are expected to depend on the type of static mixer used.

Installation of mixing elements in airlift bioreactors slows down the circulation of fluid; hence, fewest possible mixing elements should be used to satisfy the mass transfer needs while limiting the loss of liquid circulation. The optimal interval between elements depends on the type of fluid. The more coalescing the fluid, the closer must the mixers be positioned for maximum performance. There is an upper limit to the viscosity that may be used in airlift reactors with static mixers (71). In excessively viscous media, just a few mixing elements can slow down the liquid so much so that the reactor stagnates.

Reciprocating Jet Fermenter. A reciprocating plate or reciprocating jet fermenter has been used in a few laboratory studies. The device is similar to a pulsed extraction column. It consists of a cylindrical tank containing a central shaft that supports a frame of evenly spaced perforated plates or disks. The plates are mounted horizontally relative to the axis of the shaft. The plates or baffles remain clear of the fermenter walls; the interdisk space and the diameter of perforation varies. Typically, disks are placed 0.07 to 0.10 m apart and the perforations, 0.06 to 0.08 m in diameter, occupy a large proportion of the disk surface. The stack of disks oscillates up and down through the fluid. The liquid moves through the perforations as jets that change direction as the movement of stack reverses. The frequency of oscillation is usually a few cycles per second, and the vertical movement is small. Brauer reported on these devices in the early 1980s, and more gas–liquid mass transfer data have since appeared (73–75). The k_La_L in reciprocating jet reactors increases with increasing frequency and amplitude of oscillation of the baffles. The power input also increases. The k_La_L generally follows the equation

$$k_La_L = \alpha(P/V_L)^{\beta}U_G^{\gamma} \tag{68}$$

where α, β, and γ depend on the bioreactor geometry, fluid properties, and the power input range. Some typical values for those parameters are noted in Table 14. Commercial utility of reciprocating plate fermenters remains to be proved.

Tubular Loop Fermenters. Tubular loop fermenters use concurrent flow of gas and liquid in a pipe. The liquid is recirculated by a pump. This design is not practical except for low-volume fermentations or unusual situations such as seen in the tubular solar receivers of photobioreactors used in culturing microalgae. For a tubular loop (0.025-m tube diameter, 22-m length), Ziegler et al. (76) recommended the equation

$$k_La_L = 5.321 \times 10^{-5}(P_G/V_L)^{0.9}U_G^{0.2} + 8.333 \times 10^{-3} \tag{69}$$

Equation 69 was obtained using the sulfite oxidation system.

Fluidized Beds. Gas–liquid–solid fluidized bed bioreactors are seen in immobilized cell culture and wastewater treatment processes. These reactors usually consist of a cylindrical vessel containing a suspension of immobilized biocatalyst. The solids are kept in suspension by continuous flow of a liquid and gas. A solid–liquid separator, usually a sedimentation device, located near the top of the reactor prevents washout of solids. Essential design features of gas–liquid–solid fluidized bed reactors are discussed by Muroyama and Fan (77). The gas–liquid mass transfer coefficients in three-phase fluidized beds may be estimated using the correlation:

$$\frac{k_La_Ld_p^2}{D_L} = \alpha\left(\frac{d_pU_t\rho_L}{\mu_L}\right)^{3.3}\left(\frac{\mu_LU_G}{\sigma_L}\right)^{0.7} \tag{70}$$

$\alpha = 2.33 \times 10^{-5}$ when $Re_t < 2{,}000$; $\alpha = 3.946 \times 10^{-7}$ when $Re_t > 2{,}000$; where U_t is the terminal setting velocity of the particle and Re_t is the Reynolds number based on the terminal settling velocity. Equation 70 was developed for carbon dioxide absorption in water containing porcelain beads, sand, lead shot, or iron shot. The terminal settling velocity of the solids ranged over 0.12 to 0.815 m s^{-1}. Other parameters were $0 \leq U_G$ (m s^{-1}) ≤ 0.1; $0.05 \leq U_L$ (m s^{-1}) ≤ 0.172; $1.06 \leq d_p$ (mm) ≤ 6.84; and $2{,}400 \leq \rho_S$ (kg m^{-3}) $\leq 11{,}180$.

Another correlation for gas–liquid mass transfer in fluidized beds was proposed by Nguyen-Tien et al. (78):

$$k_La_L = 0.394\left(1 - \frac{\varepsilon_S}{0.58}\right)U_G^{0.67} \tag{71}$$

where ϵ_S is the volume fraction of solids. For three-phase circulating bed fermenters, Loh et al. (79) have recommended the following equation

$$k_La_L = 1.4 \times 10^{-4}\left(\frac{P}{V_L}\right)^{0.91}\frac{(1 - 2.5\varepsilon_S)^{0.95}}{(1 - \varepsilon_S)^{4.3}} \tag{72}$$

Because k_La_L is generally not too sensitive to the liquid flow rate, the equations developed for slurry bubble columns may also be applied to fluidized beds. Consult the work by Muroyama and Fan (77) for further details on gas–liquid mass transfer in fluidized beds.

Other Factors Affecting k_La_L

1. *Surfactants and antifoam agents*. Relative to clean systems, addition of surfactants may reduce or enhance k_La_L. The specific effect and its magnitude depend on the type of surfactant, its concentration, and the nature of the broth. The coalescence promotion effect of some surfactants reduces the specific interfacial area a_L and the k_La_L. Addition of surface-active agents such as ethanol to water generally increases a_L and k_La_L. Surfactants such as sodium dodecyl sulfate (SDS) or sodium lauryl sulfate (SLS) accumulate at the gas–liquid interface and usually cause a reduction in k_L.

2. *Temperature*. The overall volumetric gas–liquid mass transfer coefficient k_La_L generally increases with increasing temperature. This is mainly because the diffusion coefficient increases with temperature. Also, the viscosity of the liquid declines with increasing temperature; hence, for a given energy input, the film thickness declines and

Table 14. The Parameters in Equation 68 for Pulsed Baffle Fermenters

Fluid	α ($s^{\beta+\gamma-1}m^{3\beta-\gamma}J^{-\beta}$)	β (−)	γ (−)	Ranges
Water	4.86	0.364	1.55	$23 \leq P/V_L$(W m^{-3}) $\leq 3{,}620$ $0.0049 \leq U_G$(m s^{-1}) ≤ 0.0099 $h_L/d_T \approx 40$
Water	0.281	0.632	1.138	$35 \leq P/V_L$(W m^{-3}) $\leq 6{,}600$ $0.0005 \leq U_G$(m s^{-1}) ≤ 0.004 $h_L/d_T \approx 1.87$
Water–glycerol (1:1 by vol)	0.0039	0.735	0.668	$35 \leq P/V_L$(W m^{-3}) $\leq 6{,}600$ $0.0005 \leq U_G$(m s^{-1}) ≤ 0.004 $h_L/d_T \approx 1.87$
Aqueous yeast slurry	4.8	0.268	1.184	$d_T = 0.05$ m

the interfacial area may increase slightly. The k_La_L value at any temperature T (°C) may be calculated from the k_La_L at 20 °C using the equation:

$$(k_La_L)_T = (k_La_L)_{20°C}\theta^{(T-20)°C} \quad (73)$$

where θ is 1.200 to 1.024.

3. *Suspended solids.* How and how much suspended solids affect k_La_L depends on several factors, including concentration of solids, the density difference between the solid and the suspending fluid, the particle diameter, the operating conditions of the reactor (i.e., the power input), and the hydrophobicity of the solids. Up to 15% (w/w) of particles smaller than about 50 μm have little effect on k_La_L; however, much smaller amounts of larger particles can reduce k_La_L significantly. Very small concentrations of relatively high-density solids may actually enhance k_La_L a little. Increasing amounts of low-density filamentous or mycelial biomass rapidly increase the apparent viscosity of the slurry, and the k_La_L may decline sharply (5). The effect of such solids on k_La_L is illustrated in Figure 8 for paper pulp fibers that simulate filamentous mycelial biomass.

In a draft-tube sparged concentric draft-tube airlift reactor for potential application to microbial desulfurization of coal, the mass transfer coefficient data in a simulated basal salt medium were reported by Smith and Skidmore (80). The medium contained a total of 1.9 kg m^{-3} of various salts in distilled water. Pulverized coal particles (74 × 10^{-6} m particle diameter; density = 1,415 kg m^{-3}) were suspended in the fluid to a concentration of 0 to 40% wt (equivalent to 0 to 0.315 volume fraction of solids in gas-free medium). Over the temperature range 303 to 345 K, the k_La_L-values were strongly enhanced by increasing temperature. Increasing concentration of solids over 0 to 5% wt (equivalent to φ_S of 0 to 0.035) increased the k_La_L values slightly, but further increase in solids loading strongly lowered the mass transfer coefficient (80). This phenomenon was explained as being due to penetration of fine solid particles in the stagnant liquid film around gas bubbles, reducing the effective diffusion length (80). However, as the concentration of solids increased beyond 5% wt, the solids caused a blocking effect that reduced the effective mass transfer area (5,80). In the region of enhanced mass transfer ($0 \leq \varphi_S < 0.035$), the k_La_L was correlated (80) with

$$k_La_L = (5.7 - 144.8\varphi_S + 5048.3\varphi_S^2) \times 10^4 U_{Gr}^{1.55} e^{-2990/T} \quad (74)$$

whereas in the blocking region ($0.035 \leq \varphi_S \leq 0.315$), the following equation (80) was obtained

$$k_La_L = (2.6 - 16.8\varphi_S + 29.4\varphi_S^2) \times 10^6 U_{Gr}^{1.45} e^{-4130/T} \quad (75)$$

Equations 74 and 75 agreed with the data to within about ±20%. These equations applied over the U_{Gr} range (0.154 to 1.39) × 10^{-1} m s^{-1}, when the A_r/A_d and the aspect ratios were 1.3 and about 5, respectively.

Mass transfer measurements in suspensions of agar-filled soft polyurethane foam particles (d_p = 3 mm; ρ_p = 1030 kg m^{-3}) in uninoculated penicillin culture medium (μ_L = 1.9 × 10^{-1} Pa s) in draft tube sparged concentric tube airlift reactors have been reported (81). Increasing volume fraction of solids over the 0 to 0.1 range enhanced the volumetric mass transfer coefficient by 15 to 20% over solids-free operation (81). Further increase in solids loading to 40% (v/v) caused a decline in k_La_L values. The mass transfer coefficient values in airlift reactors were up to threefold higher than in comparable fluidized beds.

In suspensions of relatively low-density calcium alginate beads in water and other Newtonian fluids (0 to 20% v/v solids, 1.88- to 3.98-mm bead diameter), in draft-tube sparged concentric-tube airlift reactors, Koide et al. (65) noted that the mass transfer coefficient declined with increasing concentration of solids, but was not affected by the size of the particles.

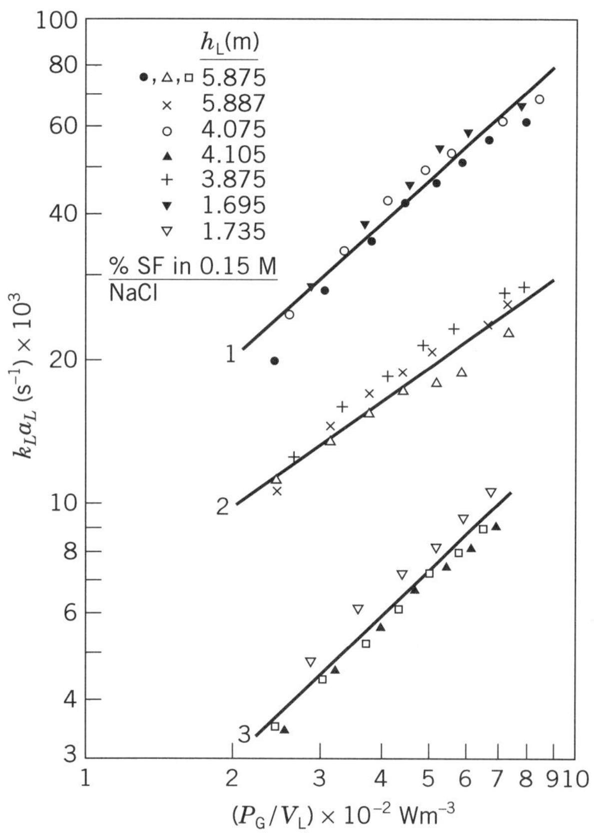

Figure 8. The effect of Solka Floc (SF) cellulose pulp fibers on gas–liquid mass transfer coefficient in bubble columns at various static slurry heights h_L. The fibers were suspended in 0.15 M sodium chloride in tap water (5). The k_La_L is shown as a function of the specific power input.

Oxygen Transfer in Wastewater Treatment Processes. The typically used bioreactor in activated sludge treatment of wastewater consists of a relatively shallow (~4 m deep) rectangular basin aerated by sparging through perforated pipes or diffusers located at the bottom of the tank. The oxygen transfer capability of such systems is quite limited; hence, biodegradation is slow. Usually only about 0.5 to 2 kg of oxygen can be transferred per kilowatt-hour of energy spent. The oxygen transfer capabilities of other conventional aeration systems have been detailed by Winkler (82).

Faster degradation of pollutants can be achieved with low-volume high-rate oxygen transfer systems. One such technology is the deep shaft reactor based on the airlift principle. This advanced activated sludge process relies on high hydrostatic pressure in a deep airlift column to significantly enhance oxygen transfer. In comparison with conventional processes, oxygen transfer rates are up to 10-fold greater (82). The transfer rate at peak load is about 1 kg oxygen m^{-3} h^{-1} (14). Several factors combine to yield this high level of performance, including long gas–liquid contact times and intense turbulence in the circulating fluid with Reynolds numbers of the order of 10^5 or higher (82). The shaft is 30 to 220 m deep (82), 0.5 to 10 m in diameter, and partitioned vertically into a riser and a downcomer. Air is injected into the downcomer, about 20 to 40 m below the surface (82), except during start-up when the riser is aerated (Fig. 9). To ensure that air bubbles move down the downcomer, a superficial liquid velocity of 1 to 2 m s^{-1} must be generated in the downcomer zone. Because the gas is not recirculated, the downcomer region above the sparger is free of bubbles. Oxygen transfer efficiencies of 3 to 5.5 kg oxygen/kWh can be attained (82). Up to 90% of the oxygen in the air is used up.

For deep shaft plants, Winkler (82) cites a BOD loading of 0.9 kg BOD/kg sludge solids per day. A retention time of about 1.5 h has been mentioned for 92% BOD removal (82). Volumetric BOD removal is of the order of 3.7 to 6.6 kg BOD m^{-3} d^{-1}, which is generally associated with high-rate treatment processes. Sewage is treated without primary sedimentation; only preliminary degritting is needed (82). Sedimentation of grit at the bottom of the shaft is prevented by ensuring that the flow velocity at the bottom exceeds 1 m s^{-1}. See Chisti (14) for additional details.

In addition to the deep shaft, a biotower configuration is increasingly being used in wastewater treatment. The biotower units consist of a relatively shallow (18 to 20 m) above-ground pool of liquid with or without multiple draft tubes (downcomers). The towers handle streams with 2 to 12 kg m^{-3} chemical oxygen demand. The oxygen transfer efficiency ranges over 1.2 to 3.8 kg oxygen/kWh (83). The biomass sludge produced rises to the top with the bubbles and is separated in an integral settling zone. Chisti (14) provides further details.

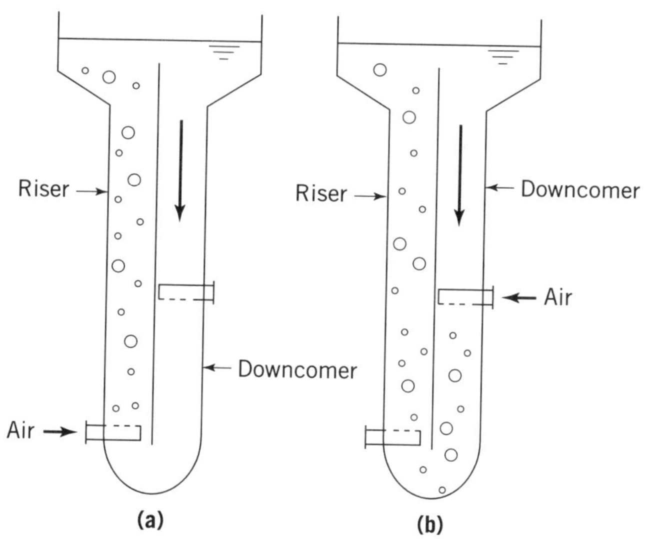

Figure 9. Deep-shaft airlift reactor: (**a**) Aeration in riser during start-up. (**b**) Aeration in downcomer during normal operation. In (**b**), no gas is being injected in the riser; all the gas bubbles in the riser result from circulation from the downcomer.

Liquid–Liquid Mass Transfer

Liquid–liquid dispersions are encountered in solvent extraction and in fermentations of hydrocarbons or other water-immiscible liquid substrates. Liquid–liquid dispersions occur also in extractive fermentations where a product is continuously extracted into a water immiscible phase. Similarly, oxygen supply using water immiscible perfluorocarbons and other fluids utilizes liquid–liquid mass transfer. Whereas in gas–liquid dispersions only the liquid film around the bubble is the principal resistance to mass transfer, in liquid–liquid dispersions the film inside the dispersed drops also affects the transport rate. Tables 15 and 16 list the correlations that are useful in calculating the mass transfer coefficients in the dispersed and continuous phases. Noncirculating small drops can be treated as ridged particles and solid–liquid mass transfer correlations (see later sections) developed for suspended spherical solids can be used. Presence of surfactants often renders drops nonmobile, and again, solid-sphere correlations apply. For larger, mobile, or oscillating droplets, both the continuous phase and the dispersed phase mass transfer coefficients are greater than for solid spheres.

Perfluorocarbons and Oxygen Vectors. Perfluorocarbons are water immiscible liquids that dissolve 10 to 20 times more oxygen than does water. These fluids can be used for bubble-free oxygenation and removal of carbon dioxide in animal cell culture. Another potential application is stripping of inhibitory oxygen produced via photosynthesis in cultures of microalgae. Perfluorocarbons are biologically inert, and suitably selected ones are nontoxic to animal cells and microorganisms. Indeed, emulsified perfluorocarbons have been used to supplement blood to improve oxygen supply to human patients. In microbial culture, per-

Table 15. Mass Transfer Coefficient Inside Mobile or Oscillating Drops in Liquid–Liquid Dispersions

Correlation	Range
$Sh \approx 16.7$	$Re < 50$
$Sh = 0.320 Re^{0.68} \left(\frac{\sigma^3 \rho_C^2}{g \mu_C^4 \Delta\rho} \right)^{0.10} \left(\frac{4 D_D t_C}{d_D^2} \right)^{-0.14}$	$Re > 150$–200
$Sh = 0.32 Re^{0.63} Sc^{0.50} \left(1 + \frac{\mu_D}{\mu_C} \right)^{-0.5}$	$10 \leq Re \leq 1{,}000$ spherical drops
$Sh = 7.5 \times 10^{-5} Re_C^{2.0} Sc^{0.56} \left(1 + \frac{\mu_D}{\mu_C} \right)^{-0.5}$	$100 \leq Re_C \leq 1{,}500$ larger oblate drops

Note: Re in these equations is based on drop diameter and relative velocity between phases; all other properties are for the drop phase unless otherwise noted. The density and viscosity in Re_C are for the continuous phase. The subscripts C and D refer to continuous and dispersed phases, respectively.

Table 16. Continuous Phase Mass Transfer Coefficient Correlations for Liquid–Liquid Dispersions

Small, noncirculating drops

$Sh = 2 + 0.79Re^{0.5}Sc^{0.33}$
$Sh = 2 + 0.76Re^{0.5}Sc^{0.33}$
$Sh = 0.562Re^{0.5}Sc^{0.33}$

Larger, mobile drops

$$Sh = \frac{2}{\sqrt{\pi}}[1 - Re^{-0.5}(2.89 + 2.15(\mu_D/\mu_C)^{0.64})]^{0.5}(Re \times Sc)^{0.5}$$

when $Re < 120$, $(\mu_D/\mu_C) \leq 2$, and $0 \leq (\rho_D/\rho_C) \leq 4$

$Sh = -126 + 1.8Re^{0.50}Sc^{0.33}$

when $8 \leq Re \leq 800$

$Sh = -178 + 3.62Re^{0.50}Sc^{0.33}$

Note: *Re* in these equations is based on drop diameter and relative velocity between phases; all other properties are for the continuous phase unless otherwise noted. The subscripts *C* and *D* refer to continuous and dispersed phases, respectively.
Source: Based on Pratt (2) and Schügerl (84).

fluorocarbon concentrations as low as 10% v/v have significantly enhanced culture performance.

One scheme for bubble-free oxygenation using perfluorocarbons is illustrated in Figure 10. A separate vessel is used to aerate the organic phase and strip the carbon dioxide. The oxygen-enriched liquid is pumped to the fermenter and sprayed into the culture medium. As the droplets settle to the bottom in the relatively quiescent environment of animal cell bioreactors, the oxygen transfers to the aqueous phase. The oxygen-depleted perfluorocarbon from the bottom of the fermenter is recycled to the external aerator. Perfluorocarbons and other oxygen vectors enhance oxygen transfer performance also when added to an aerated reactor. The oils extract more oxygen from the gas phase and effectively increase oxygen-culture contact time. In agitated continuous culture of *Escherichia coli* operated at a dilution rate of 0.275 h^{-1} and 35.5 °C, an emulsified perfluorocarbon added to the feed at 50% volume fraction could provide 0.17 kg O_2 $m^{-3}h^{-1}$ when the biomass concentration was 0.74 kg m^{-3} (85). The same system removed 0.23 kg $m^{-3}h^{-1}$ of carbon dioxide.

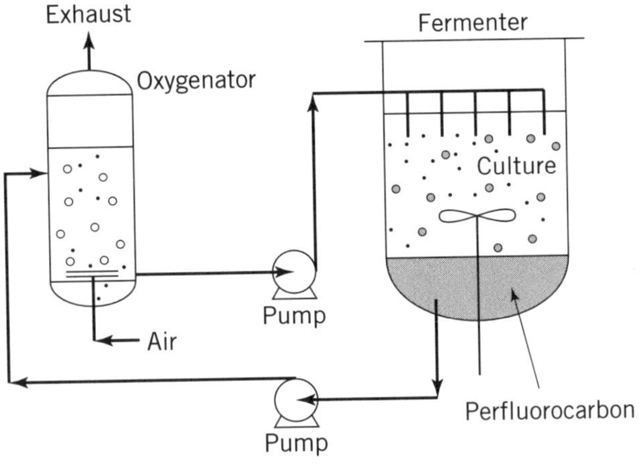

Figure 10. Bubble-free aeration with perfluorocarbon oxygen vector in a cell culture bioreactor.

With animal cells, clumping has been reported at the interface between the culture medium and the perfluorocarbon layer at the bottom of the fermenter when the agitator is placed in the bottom layer. Cells generally do not adhere to perfluorocarbon droplets. Other oxygen carriers have been tested, including hemoglobin solutions and erythrocytes, hydrocarbons such as *n*-hexadecane and *n*-dodecane, silicone oil emulsions, and soybean oil. Relevant properties of two types of oxygen vectors are noted in Table 17.

Aqueous Two-Phase Extractions. Aqueous two-phase extraction is used in bioprocessing to recover and purify bioactive proteins that are susceptible to denaturation during traditional solvent extraction that uses hydrophobic solvents. Save et al. (86) measured mass transfer coefficients of cytochrome-c, amyloglucosidase, and β-galactosidase in aqueous two-phase systems in a stirred cell with a flat interface between phases. In the poly(ethylene glycol)-dextran system, the controlling resistance to mass transfer occurred in the heavy phase (the dextran-rich phase). This was correlated as

$$k_L = 3.9 \times 10^{17} D_L^{1.8} E^{0.42} \mu_L^{0.78} \tag{76}$$

For poly(ethylene glycol)-sodium sulfate systems, the controlling resistance lay in the light phase (PEG-rich phase), and the mass transfer coefficient (86) was correlated as

$$k_L = 0.17 D_L^{0.33} E^{0.38} \tag{77}$$

In equations 76 and 77, μ_L is the viscosity of the phase, and D_L is the estimated diffusivity of the protein in that phase. The agitation power input per unit phase mass is represented by E. The E values ranged over ~(2 to 32) $\times 10^{-4}$ W kg^{-1}.

Solid–Liquid and Gas–Solid Mass Transfer

Dissolution or vaporization of a solid phase into a fluid or crystallization from a liquid are problems of solid–fluid mass transfer. Other similar situations occur in adsorption/desorption of material and during reaction catalyzed by a solid-phase biocatalyst. Common examples of gas–solid mass transfer are seen in various kinds of drying

Table 17. Properties of Some Oxygen Vectors

Property	*n*-Dodecane	Perfluorocarbon
Oxygen solubility (mg L^{-1})	59.4	118.0
Density (kg m^{-3})	743	1,750
Interfacial tension (liquid–air) (N m^{-1})	24.6×10^{-3}	17.8
Interfacial tension (liquid–water) (N m^{-1})	32.9×10^{-3}	25.0
Boiling point (°C)	214	206

Note: Oxygen solubility data at 35 °C; all other data at 30 °C.

operations. Mass transfer may occur to or from a fluid flowing past a stationary solid, or the solid may be suspended in the fluid. These cases are treated separately in the following sections.

Flow Past a Solid Surface. Material transport to or from a fluid moving past a stationary solid is encountered in packed beds of biocatalysts, chromatographic media, and solid adsorbents such as activated carbon. Other occurrences are in membrane processes, solid-state fermentations (87), surfaces of biofilms, capillaries containing surface-immobilized enzymes, and biofilters for treating gaseous effluents. The rate of mass transfer in flow past a solid is controlled by the flow rate of the fluid, the geometry of the solid surface, the diffusivity of the solute, and the physical properties of the flowing fluid. Suitable correlations for estimating the mass transfer coefficient for flow past solids of various geometries (flat plate, sphere, pipe, etc.) are listed in Table 18. The correlations given are for smooth surfaces. Predictive equations such as those in Table 18 and others in this monograph are usually accurate to only about ±50% of the actual value. Some more important bioprocess-relevant cases of mass transfer in flow past stationary solids are discussed in greater depth in the following sections.

Packed Beds. External mass transfer to or from spherical solid particles in packed beds has been correlated with the following equations

$$Sh = \frac{0.4584}{\varphi} Re^{0.5931} Sc^{1/3} \qquad (78)$$

for gases (89) when $10 \le Re \le 10{,}000$;

$$Sh = \frac{1.09}{\varphi} Re^{1/3} Sc^{1/3} \qquad (79)$$

for liquids (1) when $0.0016 \le Re \le 55$ and $165 \le Sc \le 70{,}600$; and

$$Sh = \frac{0.250}{\varphi} Re^{0.69} Sc^{1/3} \qquad (80)$$

for liquids when $55 \le Re \le 15{,}000$ and $165 \le Sc \le 10{,}690$ (1). The Reynolds number in these equations is based on the particle diameter, the superficial fluid velocity is based on the total cross-section of the bed, and φ is the voidage of the packed bed. For nonspherical particles, these equations should be corrected; the Reynolds number should be calculated using the diameter of a sphere having the same surface area as the particle.

Aeration through Polymer Tubing. Bubble-free gas exchange through polymer tubing has been used in animal cell culture and culture of microalgae. In the latter case, the gas inside the tube is carbon dioxide, which is used by photosynthetic cultures to generate carbohydrates and cell mass. For animal cell culture, the tube may be supplied with air, pure oxygen, or an atmosphere containing 5% (v/v) carbon dioxide for pH control.

Either microporous or nonporous tubing may be used for oxygenation of cultures having low oxygen demands. Microporous tubing is made of polytetrafluoroethylene or polypropylene, both of which are strongly hydrophobic. Micropores, 0.03 to 3.5 μm in diameter, occupy between 30 and 75% of the surface of the tube. The tubing is fairly ridged, with typical outer diameters of 2.5 to 4 mm and a wall thickness of about 0.5 mm. The pressure inside the tubing cannot exceed the bubble point pressure or the gas will issue through the pores as bubbles. The bubble point pressure tends to be low, of the order of 10 mbar. So long as the pores are gas filled, they do not pose a significant resistance to mass transfer (20), which occurs only through the liquid film held outside the pores. Microporous tubing generally provides a better mass transfer performance than the nonporous silicone tubing if the pores remain unwetted. Pore wetting can be a problem especially during prolonged use.

In homogeneous nonporous silicone tubing, also known as solution-diffusion tubing, the oxygen from inside the tube transfers to the outside by diffusion through the silicone tube wall; the transfer rate is quite slow compared to unwetted microporous tubing. The typical dimensions of silicone tubing are inside diameter = 1.8 mm and outside diameter = 3.2 mm, or inside diameter = 3 mm and outside diameter = 4.6 mm. Silicone tubing may be internally pressurized. Because of stretching and internal gas pressure, the surface area of the installed tubing differs from that in the relaxed state. The entire surface of the tubing is effective in mass transfer, except when the tubing has been reinforced. Solubilities and diffusivities of different gases are different in silicone; hence, transport is selective (20).

The mass transfer coefficient k_L outside the microporous tubing in agitated tanks has been correlated (25) as

$$Sh = (7.8 + 0.0021 Re^{1.2}) Sc^{1/6} \qquad (81)$$

for $250 < Re < 6{,}000$ and $200 < Sc < 500$. The Reynolds number in equation 81 is based on the impeller tip speed and the outer diameter d_o of the tube; thus

$$Re = \frac{\pi N \rho_L d_i d_o}{\mu_L} \qquad (82)$$

The Sherwood number is based on d_o and oxygen diffusivity in the liquid. The d_i in equation 82 is the impeller diameter. Equation 81 was developed in a cell culture vessel having the polymer tubing coiled into a cylindrical draft-tube configuration (Fig. 11). A two-bladed anchor impeller located inside the draft tube was used for agitation. For any given Reynolds number, the anchor impeller yielded higher values of Sherwood number relative to propellers and screw impellers. Equation 81 is for protein-free medium. For microporous tubes, the k_L values in protein-containing media are about 20% lower than in water. Presence of protein does not seem to affect the mass transfer coefficient of nonporous tubing (25). Equation 81 may be used also for homogeneous silicone tubing.

Depending on the porosity of tubing and its diameter, 1.8 to 3.0 m of polymer tubing is needed per liter of culture volume to meet the $k_L a_L$ demands (25). This length of tubing is close to technical limits of accommodation in a given

Table 18. Relationships for Estimating Solid–Liquid Mass Transfer Coefficients

Flow geometry	Laminar flow	Flow transition	Turbulent flow
1. Flow parallel to a flat plate 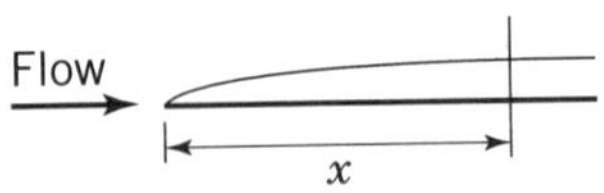	$Sh_x = 0.332Re_x^{1/2}Sc^{1/3}(Sc \geq 0.5)$ $Sh = 0.66Re^{1/2}Sc^{1/3}(Sc \geq 0.5)$ $Sh_x = 0.565Re_x^{1/2}Sc^{1/2}(Sc \leq 0.025)$ Subscript x indicates *local* value a distance x (characteristic length) from the leading edge of plate; mean coefficient is given by the second equation where the whole length of the plate is used to calculate Re	$Re_x \approx 5 \times 10^5$	$Sh_x = 0.0296Re^{0.8}Sc^{0.6}$ $(10.0 \geq Sc \geq 0.5)$
2. Fully developed flow in straight pipe 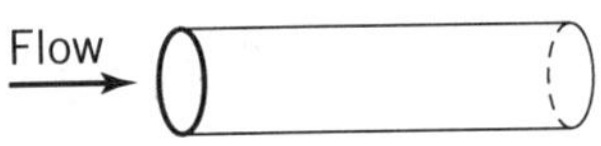	$Sh = 3.65 + \dfrac{0.0668(d/L)Re \times Sc}{1 + 0.04[(d/L)Re \times Sc]^{2/3}}$ Use hydraulic diameter d_h for noncircular section: $d_h = \dfrac{4 \times \text{flow area}}{\text{Wetted perimeter}}$	$Re \approx 2{,}300$	$Sh = 0.023Re^{0.8}Sc^{0.33}$ $(Sc \geq 0.5)$ $Sh = 0.0102Re^{9/10}Sc^{1/3}$ $(Sc \geq 10^3; Re \geq 3{,}000)$
3. Fully developed flow between parallel plates 	$Sh = 7.54 + 0.0234\dfrac{Re \times Sc}{(L/d)}$	$Re \approx 2800$	$Sh = 0.023Re^{0.8}Sc^{0.33}$ $(Sc > 0.5)$
4. Flow across a cylinder 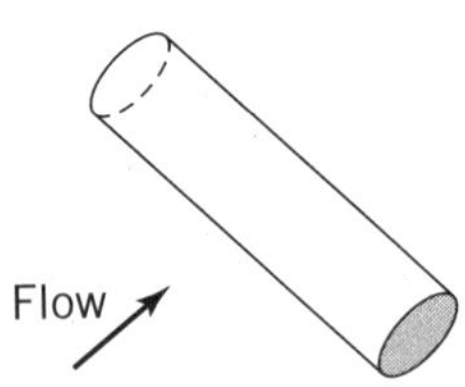	$Sh = 0.43 + \alpha Re^{\beta}Sc^{0.31}$ $1{,}000 \leq Re \leq 4{,}000$ $(\alpha = 0.53, \beta = 0.5)$ $4{,}000 \leq Re < 40{,}000$ $(\alpha = 0.193, \beta = 0.618)$ $Sh = 0.600Re^{0.513}Sc^{1/3}$ $0.6 \leq Sc \leq 2.6$ (gases) $1{,}000 \leq Sc \leq 3{,}000$ (liquids) $50 \leq Re \leq 50{,}000$	$Re \approx 40{,}000$	$Sh = 0.43 + 0.0265Re^{0.8}Sc^{0.31}$
5. Flow across a sphere 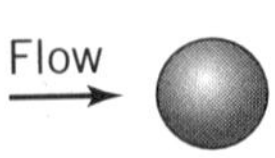	$Sh = 2 + 0.37Re^{0.6}Sc^{0.33}$ $Sh = 2 + 0.552Re^{0.53}Sc^{1/3}$ (gases) $0.6 \leq Sc \leq 2.7$ $1 \leq Re \leq 48{,}000$ $Sh = 2 + 0.95Re^{0.50}Sc^{1/3}$ (liquids) $2 \leq Re \leq 2{,}000$ $Sh = 2 + 0.347Re^{0.62}Sc^{1/3}$ (liquids) $2000 \leq Re \leq 17{,}000$	$Re \approx 150{,}000$	
6. Flow through a packed bed of spheres	$St \times Sc^{2/3} = 1.625Re_d^{-1/2}$ $15 < Re_d < 120$ $Re_d = \dfrac{\rho_L U_L d_{hi}}{\mu_L}$, where $d_{hi} = \dfrac{6(1 - \varphi) \times \text{flow area}}{\text{Wetted perimeter}}$ where φ is the void fraction and St is the Stanton number: $St = \dfrac{Sh}{Sc \times Re} = \dfrac{K_L}{U_L}$	$Re_d = 120$	$St \times Sc^{2/3} = 0.687Re_d^{-0.327}$ $120 < Re_d < 2{,}000$

Source: Based on Refs. 1 and 88.

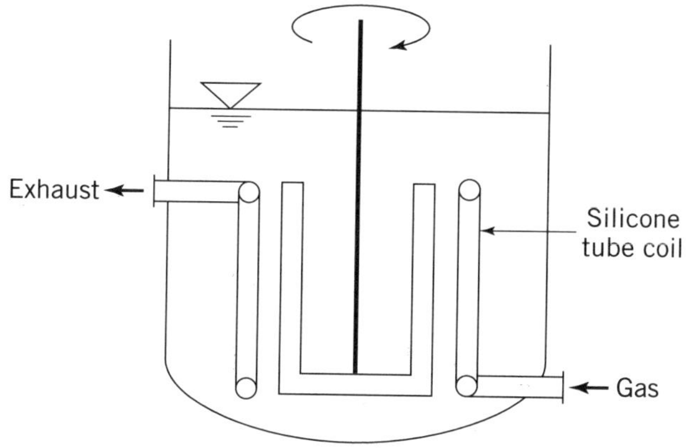

Figure 11. Oxygen supply via silicone tubing.

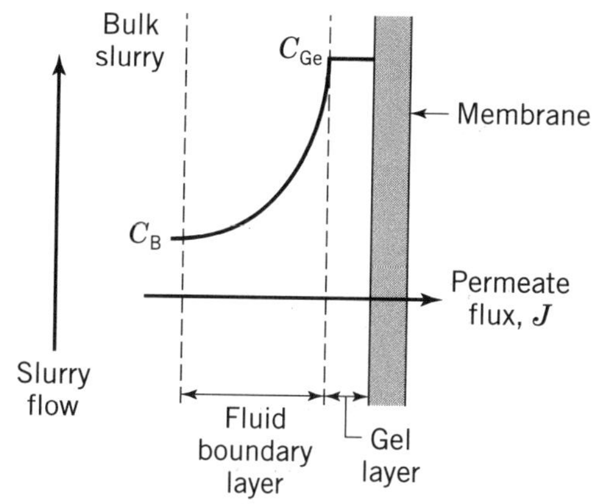

Figure 12. Steady-state solute concentration profile in ultrafiltration and microfiltration processes.

volume; therefore, aeration through polymer tubes is useful only for specific cases. The tubing is susceptible to fouling and requires periodic replacement. Anchorage-dependent cells sometimes attach to the surface of silicone tubing; hence, oxygen transfer to bulk fluid is prevented.

For transfer of carbon dioxide through nonporous silicone membranes and tubes (wall thickness 1.5 to 2.0 mm) the diffusion coefficient (D) has been reported to be 1.92 ± 0.14 m^2 min^{-1} at 25 °C (90). For mass transfer through such a tube to a highly agitated culture volume V_L, the transfer rate can be written as

$$\frac{dC_L}{dt} = \frac{AD(C^* - C_L)}{V_L \delta_w} \tag{83}$$

where A is the surface area of the tube, δ_w is its wall thickness, C_L is the instantaneous concentration of the diffusing gas in the liquid phase, and C^* is the saturation concentration of the diffusing component in a liquid sample that is in equilibrium with a gas phase having the same composition as in the silicone tube; C^* may be calculated using Henry's law. Note that equation 83 assumes a constant composition of the gas within the tube (i.e., no diffusion of other dissolved gases from the liquid into the tube). This assumption may be valid for relatively short tubes and high flow rates within the tubes. Aeration through polymer tubing is discussed further by Aunins and Henzler (20).

Mass Transfer Effects in Membrane Processes. Membrane filtrations, particularly microfiltration and ultrafiltration, are commonly used in bioprocessing to separate cells, particles, microemulsions, and macromolecules. Generally, a cross-flow scheme is used in which the fluid being filtered flows parallel to the membrane but perpendicular to the direction of the permeate flux. The turbulence generated by the flow improves mass transfer at the membrane surface; consequently, the buildup of a solute layer or gel layer on the surface of the membrane is reduced to ensure relatively high permeate or filtrate flux through the membrane. At steady state, a solute concentration profile develops on the upstream side of the membrane, as shown in Figure 12. The permeate flux J is related to the concentration C_{Ge} of the solute in the gel layer, the concentration C_B in the bulk fluid, and the mass transfer coefficient k_L as follows:

$$J = k_L \ln \frac{C_{Ge}}{C_B} \tag{84}$$

The mass transfer coefficient depends on the Reynolds number on the slurry side of the membrane as follows

$$Sh = \alpha Re^{\beta} Sc^{1/3} \tag{85}$$

where the Sh and Re are based on the hydraulic diameter d_h of the flow channel. During ultrafiltration, β is 0.5 in laminar flow and about 1.0 in turbulent flow. In cross-flow microfiltration of particles and cells, β equals 0.8 in laminar flow, but increases to about 1.3 in turbulent flow (91). The α value is 0.023 in turbulent flow. Other expressions for the Sherwood number are

$$Sh = 1.62\left(Re \times Sc \frac{d_h}{L}\right)^{0.33} \tag{86}$$

$$Sh = 1.86\left(Re \times Sc \frac{d_h}{L}\right)^{0.33} \tag{87}$$

and

$$Sh = 0.023 Re^{0.875} Sc^{0.25} \tag{88}$$

Equations 86 and 87 are for laminar flow in tubes and channels, respectively; equation 88 is for turbulent flow. For fully developed laminar flow in ultrafiltration, k_L may be related to shear rate γ by the Porter equation:

$$k_L = 0.816 \gamma^{0.33} D_L^{0.67} L^{-0.33} \tag{89}$$

where L is the length of the flow channel and the shear rate depends on the channel geometry:

$$\gamma = 8U_L/d$$

for tubes with diameter d, and

$$\gamma = 6U_L/h$$

for rectangular channels of height h.

The cross-flow velocity is the principal operating variable for enhancing the performance of a given filtration membrane module. The optimal cross-flow velocity depends on the product and the configuration of the filtration module. For tubular microfiltration membranes with ~5.5-mm inner diameter, the optimal cross-flow velocity is about 2.5 to 5 m s^{-1} (92). The inner diameter of membrane tubes is usually 4 to 20 mm. Hollow fibers have much smaller diameters at 0.5 to 2 mm. Membranes are often deployed as flat sheets in a plate-and-frame configuration in which the height of the flow channel is 0.5 to 1.5 mm (92).

Other methods for improving mass transfer in membrane processes include operation at higher temperature (i.e., higher diffusivity and lower viscosity), addition of large inert particles to the feed to agitate the gel layer, and use of pulsating flow. Module design may also be modified to enhance turbulence in the flow channel; hence, turbulence promoters such as static mixers (e.g., wire screens) may be used in tubes and channels, or membranes may be formed into a corrugated configuration. Another method of mass transfer enhancement is the use of dynamic filtration systems with rotating membranes or agitators placed in close proximity to the membrane (7).

In both ultrafiltration and microfiltration, the mass transfer coefficient tends to be quite small because of the small diffusivities of the cells and macromolecules. Formation of the gel layer, also known as concentration polarization, reduces the permeate flux in microfiltration to only about 5% of the pure water flux. Unlike microfiltration and ultrafiltration, pervaporation processes use nonporous homogeneous membranes. Typically, the solute flux is low and the mass transfer coefficient k_L is relatively large in view of the higher diffusivities of small solutes such as ethanol. For additional information on mass transfer in membrane processes, see works by Mulder (93), Ho and Sirkar (94), and Brindle and Stephenson (95).

Mass Transfer at Rough Surfaces. Compared to smooth surfaces, mass transfer from a geometrically similar rough surface is generally higher for otherwise identical conditions. Roughness induces an earlier transition to turbulence in flow past a surface. The effect of surface roughness on mass transfer coefficients in Newtonian fluids has been correlated by Kawase et al. (96) using the equation

$$Sh = 0.0093(e/d)^{0.15}Re \times Sc^{0.5}(1.11 + 0.44Sc^{-1/3} - 0.70Sc^{-1/6}) \quad (90)$$

where e is the absolute roughness (mean height of projections from the surface) and d is the pipe or channel diameter. The ratio e/d is known as relative roughness. Equation 90 is for the ranges $5 \times 10^3 < Re < 5 \times 10^5$, $5 < Sc < 400$, and $0.003 \le e/d \le 0.056$.

Fluid–Solid Slurries. For mass transfer to or from suspended solids, energy dissipation in the vicinity of the particles is generally assumed to control mass transfer in an isotropically turbulent field. Under such conditions, the small energy-dissipating eddies are independent of the nature of the bulk flow; the properties of those eddies depend only on energy dissipation rate per unit mass of fluid, E. In general, the liquid-film mass transfer coefficient k_L at the solid–liquid interface is expressed (97) in terms of E as follows:

$$Sh = \frac{k_L d_p}{D_L} = 2 + c\left(\frac{E^{1/3}\rho_L d_p^{4/3}}{\mu_L}\right)^{a} Sc^{b} \quad (91)$$

where $(E^{1/3}\rho_L d_p^{4/3}/\mu_L)$ may be regarded as the energy dissipation Reynolds number (98). Notice that in a quiescent environment, purely diffusive mass transfer occurs, and equation 91 reverts to $Sh = 2$ as expected from theoretical considerations. Irrespective of the reactor configuration, solid–liquid mass transfer in slurries is not significantly influenced by presence of gas bubbles. Solid–liquid mass transfer cases for the principal types of bioreactors are discussed in the following sections.

Stirred Tanks. Use the Calderbank and Moo-Young (9) equation 34 noted earlier or the Liepe and Möckel equations recommended by Hempel (99):

$$Sh = 2 + 0.67\left(\frac{P}{V_L}\frac{\rho_L^2 d_p^4}{\mu_L^3}\right)^{1/8} Sc^{1/3} \quad (92)$$

which is valid when

$$d_p < 5.2\left(\frac{\Delta\rho}{\rho_L}\right)^{-5/6}\left(\frac{\mu_L^3 V_L}{P\rho_L^2}\right)^{1/4} \quad (93)$$

When the particle diameter exceeds that given by inequality 93, the recommended equation is

$$Sh = 2 + 0.35\left(\frac{\Delta\rho}{\rho_L}\right)^{1/3}\left(\frac{P}{V_L}\frac{\rho_L^2 d_p^4}{\mu_L^3}\right)^{2/3} Sc^{1/3} \quad (94)$$

Bubble Columns. The recommended correlation (100) is

$$\frac{k_L d_p}{D_L} = 2 + 0.545Sc^{1/3}\left(\frac{g d_p^4 \rho_L^3 U_G}{\mu_L^3}\right)^{0.264} \quad (95)$$

It is suitable for $137 \le Sc \le 50{,}000$.

Airlift Bioreactors. For typical operating conditions, the solid–liquid mass transfer coefficient in airlift bioreactors is insensitive to the gas flow rate until such a point when increasing the superficial gas velocity ceases to produce much increase in circulation of the slurry. If the aeration rate is increased further, the k_L increases rapidly (14,97). Pneumatic energy imparted to the slurry initially produces bulk circulation of fluid; only when circulation no longer improves is the energy dissipated in microeddies that penetrate to the vicinity of the solid–liquid interface and enhance k_L. Slight variations in solids concentration (e.g., 0.2 to 4% w/w) do not affect k_L in airlift reactors (97,101).

Under similar conditions, solid–liquid mass transfer in external-loop airlift reactors exceeds that in stirred tanks and bubble columns; airlifts perform marginally better than fluidized beds (14). Performance of draft-tube internal-loop airlift devices also exceeds that of bubble columns. The solid–liquid mass transfer coefficient in both external- and internal-loop types of airlift reactors may be increased further by using static mixers. In draft-tube sparged reactors, Kenics-type twisted-ribbon static mixers placed in the draft tube have been shown to enhance the mass transfer coefficient by about 20% relative to operation without the mixers (14).

The various correlations developed for solid–liquid mass transfer coefficient in internal- and external-loop types of airlift reactors are noted in Table 19. These correlations require a knowledge of the energy dissipation rate per unit mass of fluid; E is calculated using the equation

$$E = \frac{gU_{\mathrm{Gr}}}{1 + \dfrac{A_{\mathrm{d}}}{A_{\mathrm{r}}}} \tag{96}$$

where U_{Gr} is the superficial gas velocity in the riser (5,66).

Fluidized Beds. The mass transfer coefficient for spherical particles suspended in a gas or liquid fluidized bed can be estimated using

$$Sh = \frac{0.4548Re^{0.5931}Sc^{1/3}}{\varphi} \tag{97}$$

where φ is the void fraction of the bed. Equation 97 is suitable for $10 \le Re \le 4{,}000$. Another correlation applicable only to liquid fluidized beds is

$$Sh = \frac{1.1068Re^{0.28}Sc^{1/3}}{\varphi} \tag{98}$$

which is suitable for $1 \le Re \le 10$.

External Mass Transfer and Heterogeneous Bioreaction. At steady state, the rate of mass transfer of a substrate being consumed at the surface of a nonporous particle equals the rate of consumption; thus

$$k_{\mathrm{L}}a_{\mathrm{s}}(C_{\mathrm{B}} - C_{\mathrm{Sb}}) = R \tag{99}$$

where a_{s} is the solid–liquid interfacial area per unit liquid volume, R is the overall rate of reaction, and C_{B} and C_{Sb} are substrate concentrations in the bulk fluid and at the solid's surface, respectively. Because R is a function of the substrate concentration C_{Sb}, R tends to a maximum when the concentration at the interface approaches that in the bulk fluid, i.e., the mass transfer rate or the $k_{\mathrm{L}}a_{\mathrm{s}}$ is large. In this situation, the rate of reaction is maximum, and it is controlled by the intrinsic kinetics and not by mass transfer effects. At the other extreme, when the mass transfer rate is low compared to the reaction rate, the substrate concentration at the interface approaches zero, and the reaction is mass transfer controlled. Because now $C_{\mathrm{Sb}} = 0$, the reaction rate is

$$R = k_{\mathrm{L}}a_{\mathrm{s}}C_{\mathrm{B}} \tag{100}$$

Thus, the observed rate of a mass transfer controlled reaction will be influenced by changes in C_{B}, k_{L}, or the specific solid–liquid interfacial area a_{s}. Indeed, observing influence of these parameters on the reaction rate provides methods for determining if the reaction is mass transfer limited. In addition, because the activation energy for mass transfer is much smaller than that of biochemical reactions, the observed rate of a mass transfer controlled reaction is not as sensitive to temperature changes as the rate of reaction when it is not limited by transport effects. In a batch-stirred vessel, an increase in the rate of a solid–phase catalyzed reaction with increasing agitation implies a mass transfer limited reaction. Similarly, in a packed-bed reactor, the rate of a mass transfer controlled reaction will increase as the flow rate is increased, but the contact time is kept unchanged (i.e., the depth of the bed is increased proportionately).

The solid–liquid mass transfer correlations given earlier in this article may be used to estimate k_{L} in bubble columns, packed beds, fluidized beds, and other geometries of bioreactors. External mass transfer in immobilized enzyme systems has been discussed in greater detail by Goldstein (104).

Table 19. Solid–Liquid Mass Transfer Coefficient in Airlift Reactors

Reactor configuration	Correlation	Ranges
Internal-loop reactors (draft-tube sparged)	$Sh = 2 + 0.064\left(\frac{E\rho_{\mathrm{L}}^3 d_{\mathrm{p}}^4}{\mu_{\mathrm{L}}^3}\right)^{0.165} Sc^{0.45}$ Kushalkar and Pangarkar (102)	0.5% wt benzoic acid granules, $d_{\mathrm{p}} = 0.55$–3 mm, in air and water; $A_{\mathrm{r}}/A_{\mathrm{d}} = 0.17 - 1.29$, $U_{\mathrm{G}} = 0.08 - 0.35$ m s^{-1}
Internal-loop reactors (draft-tube sparged)	$Sh = 2 + 1.01\left(\frac{E\rho_{\mathrm{L}}^3 d_{\mathrm{p}}^4}{\mu_{\mathrm{L}}^3}\right)^{0.173} Sc^{0.33}$ Goto et al. (101)	Amberlyst 15 ion exchange resin, $d_{\mathrm{p}} = 0.55$–0.92 mm, suspended in dilute aqueous sodium hydroxide; $A_{\mathrm{r}}/A_{\mathrm{d}} = 0.1 - 1.4$
Internal-loop reactors with static mixers (draft-tube sparged)	$Sh = 2 + 1.68\left(\frac{E\rho_{\mathrm{L}}^3 d_{\mathrm{p}}^4}{\mu_{\mathrm{L}}^3}\right)^{0.133} Sc^{0.33}$ Gaspillo and Goto (103)	Kenics-type twisted-ribbon static mixers were in draft tube
External-loop reactors	$Sh = 2 + 0.48\left(\frac{E^{1/3}\rho_{\mathrm{L}} d_{\mathrm{p}}^{4/3}}{\mu_{\mathrm{L}}}\right)^{0.72} Sc^{1/3}$ Mao et al. (97)	Benzoic acid coated particles, $d_{\mathrm{p}} = 3.8$ mm, $\rho_{\mathrm{s}} \approx 1{,}080$ kg m^{-3}, suspended in water

Intraparticle Mass Transfer. When a substrate is being consumed or a product is being produced inside a porous immobilization matrix or in homogeneous gels, an internal substrate (and product) concentration profile exists within the matrix at steady state. In microporous particles, in the absence of convective transport, the main internal mass transfer parameter is the effective diffusivity D_e in the particle. The effective diffusivity is related to the liquid- or gas-phase diffusivity in the pores as follows

$$D_e = \frac{p}{\tau} D_L \quad (101)$$

In equation 101, p is the porosity, and τ is the tortuosity of the immobilization matrix. Intraparticle mass transfer is encountered in immobilized enzyme and cell particles, in chromatographic media, and during leaching of solubles from inert solids.

In macroporous particles, changes in local fluid velocity outside the particle create pressure fluctuations within the larger pores, and this leads to some agitation of the fluid held in the pores. In this situation, the apparent effective diffusivity ($D_{e\,ap}$) is greater than that predicted by equation 101. For example, the apparent effective diffusivity of oxygen in pellets of *Aspergillus niger* has been reported to be $\sim 10^{-8}$ m^2 s^{-1} or about 10-fold greater than in water, which suggests a level of fluid movement within the pellet, possibly because of turbulence-induced elastic structural deformations of the loose floc.

For porous particles in packed beds, the apparent effective diffusivity has been observed to depend on the effective diffusivity (no convection) and the Peclet number; thus

$$D_{e\,ap} = \frac{D_e}{\dfrac{3}{Pe}\left(\dfrac{1}{\tanh Pe} - \dfrac{1}{Pe}\right)} \quad (102)$$

where the Peclet number is given as

$$Pe = \frac{d_p U_L}{(1-\varphi) D_e} \quad (103)$$

In equation 103, d_p is the diameter of the particle, U_L is the superficial fluid velocity through the bed, and φ is its void fraction. More complete treatments of intraparticle mass transfer have been provided by Chisti and Moo-Young (7), Radovich (105), Willaert et al. (106), Pitcher (107), and Bailey and Ollis (108).

NOMENCLATURE

A	Surface area of tube (m^2)
A_d	Cross-sectional area of downcomer (m^2)
A_r	Cross-sectional area of riser (m^2)
A_s	Total solid–liquid interfacial area (m^2)
a_D	Gas–liquid interfacial area per unit volume of dispersion (m^{-1})
a_L	Gas–liquid interfacial area per unit liquid volume (m^{-1})
a_{Lc}	Liquid–liquid interfacial area per unit volume of continuous phase (m^{-1})
a_s	Solid–liquid interfacial area per unit liquid volume (m^{-1})
BOD	Biochemical oxygen demand
Bo	Bond number (–)
C	Concentration (kg m^{-3})
C^*	Saturation concentration of transferring gas or solute in liquid (kg m^{-3})
ΔC	Concentration gradient (kg m^{-3})
C_B	Solute or substrate concentration in bulk liquid (kg m^{-3})
C_G	Concentration in the gas phase (kg m^{-3})
C_{Ge}	Solute concentration in gel layer (kg m^{-3})
C_{Gi}	Interfacial concentration of the diffusing component in the gas phase (kg m^{-3})
C_L	Instantaneous concentration of transferring component in liquid (kg m^{-3})
C_{L0}	Initial dissolved oxygen concentration at time t_0 (kg m^{-3})
C_{Lc}	Equilibrium concentration of solute in continuous phase (kg m^{-3})
C_{Ld}	Equilibrium concentration of solute in dispersed phase (kg m^{-3})
C_{Li}	Interfacial concentration of the diffusing component in the liquid phase (kg m^{-3})
C_S	Concentration of solids in slurry (% w/v or kg m^{-3})
C_{Sb}	Substrate concentration at solid–liquid interface (kg m^{-3})
C^*_{in}	Saturation concentration of oxygen in equilibrium with ingoing gas (kg m^{-3})
C^*_{out}	Saturation concentration of oxygen in equilibrium with the exhaust gas (kg m^{-3})
D	Diffusivity (m^2 s^{-1})
D_D	Diffusivity of dispersed phase (m^2 s^{-1})
D_e	Effective diffusivity in particle (m^2 s^{-1})
$D_{e\,ap}$	Apparent effective diffusivity in particle (m^2 s^{-1})
D_{gas}	Diffusivity of gas in liquid (m^2 s^{-1})
D_L	Diffusivity of gas or solute in liquid (m^2 s^{-1})
DMEM	Dulbecco's modified Eagle's medium
D_{oxygen}	Diffusivity of gas in liquid (m^2 s^{-1})
D_P	Diffusivity of small solute in protein solution (m^2 s^{-1})
d	Characteristic length dimension (m)
d^*	Diameter number defined by equation 32 (–)
d_B	Sauter mean bubble diameter (m)
d_D	Diameter of drop (m)
d_F	Maximum diameter of flask (m)
d_H	Diameter of sparger hole (m)
d_h	Hydraulic diameter (m)
d_{hi}	Hydraulic diameter in packed bed as defined in Table 18 (m)

d_i	Diameter of impeller (m)
d_o	Outer diameter of tubing (m)
d_p	Diameter of particle (m)
d_r	Diameter of riser (m)
d_T	Diameter of tank or column (m)
E	Energy dissipation rate per unit mass (W kg^{-1})
EDTA	Ethylenediaminetetraacetic acid
E_L	Overall axial dispersion coefficient of liquid (m^2 s^{-1})
e	Absolute roughness (m)
$\hat{e}$	Eccentricity of shaker platform (−)
F	Mass flow rate of gas (kg s^{-1})
Fr	Froude number (−)
f	Parameter (−)
Gr	Grashof number (−)
g	Gravitational acceleration (m·s^{-2})
H	Dimensionless Henry's law constant (−)
h	Height of channel (m)
h_D	Height of dispersion (m)
h_L	Height of gas-free liquid (m)
I	Ionic strength (kg ion m^{-3})
J	Mass flux (kg m^{-2} s^{-1} or kmol m^{-2} s^{-1}) or permeate flux (m s^{-1})
K	Consistency index of fluid (Pa s^n)
K_L	Overall mass transfer coefficient based on liquid film (m s^{-1})
K_P	Mass transfer coefficient for plug (mol O_2 atm^{-1} h^{-1})
K_p	Partition coefficient of solute defined by equation 25 (−)
K_S	Mass transfer coefficient at surface (mol O_2 atm^{-1} h^{-1})
k	Mass transfer coefficient (m s^{-1})
k_G	Gas film mass transfer coefficient (m s^{-1})
k_i	Impeller-dependent constant in equation 44 (−)
k_L	Liquid film mass transfer coefficient (m s^{-1})
k_La_L	Overall gas–liquid volumetric mass transfer coefficient (s^{-1})
$(k_La_L)_{rord}$	Overall volumetric mass transfer coefficient in riser or downcomer (s^{-1})
k_{Lc}	Liquid film mass transfer coefficient in continuous phase (m s^{-1})
k_{Ld}	Liquid film mass transfer coefficient in dispersed phase (m s^{-1})
L	Length of circulation loop, pipe, plate, or flow channel (m)
l	Length of impeller blade (m)
M_L	Molecular weight of liquid (kg $kmol^{-1}$)
Mo	Morton number (−)
N	Rotational speed (s^{-1})
n	Flow behavior index (−)
P	Power input (W)
PEG	Poly(ethylene glycol)
Pe	Peclet number (−)
P_G	Gassed power input (W)
Po	Power number (−)
Ps	Poiseuille number (−)
p	Porosity of particle (−)
p_{atm}	Partial pressure of oxygen in the atmosphere outside flask (atm)
p_L	Partial pressure of oxygen in liquid in flask (atm)
Q_G	Volume flow rate of gas (m^3 s^{-1})
q_{O_2}	Specific oxygen consumption rate (s^{-1})
R	Overall rate of reaction (kg m^{-3} s^{-1})
Ra	Rayleigh number (−)
R_b	Ratio of bubble diameter to its width (−)
Re	Reynolds number (−)
Re_C	Reynolds number for the continuous phase (−)
Re_d	Reynolds number based on d_{hi} as in Table 18 (−)
Re_i	Reynolds number based on impeller (−)
Re_t	Reynolds number based on terminal settling velocity of particle (−)
Re_x	Local value of Reynolds number at distance x from leading edge (−)
RPMI	Cell culture medium
Sc	Schmidt number (−)
SDS	Sodium dodecyl sulfate
Sh	Sherwood number (−)
Sh_x	Local value of Sherwood number at distance x from leading edge (−)
SLS	Sodium lauryl sulfate
St	Stanton number (−)
T	Temperature (°C) or absolute temperature (K)
t	Instantaneous time (s)
t_C	Contact time of droplets (s)
t_m	Mixing time (s)
t_R	Residence time of gas in liquid (s)
U_B	Bubble rise velocity (m s^{-1})
U_G	Superficial gas velocity based on outer column cross-sectional area (m s^{-1})
U_{Gr}	Superficial gas velocity in the riser (m s^{-1})
U_L	Mean superficial liquid velocity in the reactor or pipe (m s^{-1})
U_{Lr}	Superficial liquid velocity in the riser (m s^{-1})
U_p	Velocity of particle (m s^{-1})
U_t	Terminal settling velocity of a single particle in liquid (m s^{-1})
V_L	Volume of liquid or slurry (m^3)
V_{Lc}	Volume of continuous phase (m^3)
V_M	Molar volume of solute at its boiling point (m^3 $kmol^{-1}$)
W	Width of impeller blade (m)
We	Weber number (−)

Wi	Weissenberg number (−)
X	Viable biomass concentration (kg m^{-3})
x	Distance (m)
x_i	Mass fraction of oxygen in inlet gas stream (−)
x_o	Mass fraction of oxygen in exhaust gas (−)

Greek Symbols

α	Parameter; parameter in equation 66 (m$^{-\beta}$ s$^{-(\beta+1)}$)
α_M	Value of α (equation 66) in presence of static mixers (m$^{-\beta}$ s$^{-(\beta+1)}$)
β	Parameter (−)
γ	Parameter or shear rate (s^{-1})
δ	Film thickness (m)
δ_G	Thickness of gas film (m)
δ_L	Thickness of liquid film (m)
δ_w	Wall thickness of tube (m)
ϵ_G	Overall gas holdup (−)
ϵ_{Gr}	Gas holdup in the riser (−)
ϵ_S	Volume fraction or holdup of solids in three-phase systems (−)
θ	Parameter in equation 73 (−)
λ	Characteristic time or relaxation time of viscoelastic fluid (s)
μ_{ap}	Apparent viscosity of non-Newtonian fluid (Pa s)
μ_C	Viscosity of the continuous phase (Pa s)
μ_D	Viscosity of the dispersed phase (Pa s)
μ_G	Viscosity of the gas (Pa s)
μ_L	Viscosity of the liquid (Pa s)
μ_{SL}	Viscosity of the slurry (Pa s)
μ_w	Viscosity of water (Pa s)
χ	Association parameter (−)
π	Pi (−)
ρ_C	Density of continuous phase (kg m^{-3})
ρ_D	Density of dispersed phase (kg m^{-3})
ρ_G	Density of gas (kg m^{-3})
ρ_L	Density of the liquid (kg m^{-3})
ρ_p	Density of the particles (kg m^{-3})
ρ_S	Density of solids (kg m^{-3})
ρ_{SL}	Density of slurry (kg m^{-3})
$\Delta\rho$	Density difference between phases (kg m^{-3})
σ	Interfacial tension (liquid–liquid) (kg s^{-2})
σ_L	Interfacial tension (kg s^{-2})
τ	Tortuosity of pores (−)
φ	Void fraction of the packed bed (−)
φ_S	Volume fraction of solids in gas-free slurry (−)
χ	Parameter (−)

BIBLIOGRAPHY

1. C.J. Geankoplis, *Transport Processes and Unit Operations*, 3rd ed., Prentice Hall, Englewood Cliffs, 1993.
2. H.R.C. Pratt, in T.C. Lo, M.H.I. Baird, and C. Hanson eds., *Handbook of Solvent Extraction*, Wiley, New York, 1983, pp. 91–123.
3. C.R. Wilke and P. Chang, *AIChE J.* **1**, 264–270 (1955).
4. Y. Chisti, in R. Robinson, C. Batt, and P. Patel eds., *Encyclopedia of Food Microbiology*, Academic Press, London, in press.
5. Y. Chisti, *Airlift Bioreactors*, Elsevier, London, 1989, p. 350.
6. K. Van't Riet, *Ind. Eng. Chem. Process Des. Dev.* **18**, 357–363 (1979).
7. Y. Chisti and M. Moo-Young, in V. Moses and R.E. Cape eds., *Biotechnology: The Science and the Business*, Harwood Academic Publishers, New York, 1991, pp. 167–209.
8. C.S. Ho and J.F. Shanahan, *CRC Crit. Rev. Biotechnol.* **4**, 185–252 (1986).
9. P.H. Calderbank and M.B. Moo-Young, *Chem. Eng. Sci.* **16**, 39–54 (1961).
10. P.H. Calderbank and A.C. Lochiel, *Chem. Eng. Sci.* **19**, 485–503 (1964).
11. Y. Chisti and M. Moo-Young, *Biotechnol. Bioeng.* **34**, 1391–1392 (1989).
12. E. Molina Grima, Y. Chisti, and M. Moo-Young, *J. Biotechnol.* **54**, 195–210 (1997).
13. L.K. Shi, J.P. Riba, and H. Angelino, *Chem. Eng. Commun.* **89**, 25–35 (1990).
14. Y. Chisti, *Appl. Mech. Rev.* **51**, 33–112 (1998).
15. M. Reuss, in H.-J. Rehm and G. Reed eds., *Biotechnology*, vol. 3, 2nd ed., VCH, Weinheim, 1993, pp. 185–217.
16. M. Moo-Young and H.W. Blanch, *Adv. Biochem. Eng.* **19**, 1–69 (1981).
17. T. Oolman and H.W. Blanch, *CRC Crit. Rev. Biotechnol.* **4**, 133–184 (1986).
18. Y. Chisti and M. Moo-Young, *J. Chem. Technol. Biotechnol.* **58**, 331–336 (1993).
19. S. Yamada, M. Wada, and I. Chibata, *J. Ferment. Technol.* **56**, 20–28 (1978).
20. J.G. Aunins and H.-J. Henzler, in H.-J. Rehm and G. Reed eds., *Biotechnology*, vol. 3, 2nd ed. VCH, Weinheim, 1993, pp. 219–281.
21. L. Randers-Eichhorn, R.A. Bartlett, D.D. Frey, and G. Rao, *Biotechnol. Bioeng.* **51**, 466–478 (1996).
22. J.G. Aunins, B.A. Woodson Jr., T.K. Hale, and D.I.C. Wang, *Biotechnol. Bioeng.* **34**, 1127–1132 (1989).
23. M. Lavery and A.W. Nienow, *Biotechnol. Bioeng.* **30**, 368–373 (1987).
24. Y. Kato, S. Hiraoka, Y. Tada, K. Sato, and T. Ohishi, *J. Chem. Eng. Jpn*, **30**, 362–365 (1997).
25. H.-J. Henzler and D.J. Kauling, *Bioprocess Eng.* **9**, 61–75 (1993).
26. J.F. Perez and O.C. Sandall, *AIChE J.* **20**, 770–775 (1974).
27. Y. Kawase and M. Moo-Young, *Trans. Inst. Chem. Eng.* **68**, 189–194 (1990).
28. R.E. Farritor and G.A. Hughmark, *AIChE J.* **20**, 1027–1028 (1974).
29. K.A. Bin, *Chem. Eng. Commun.* **31**, 155–183 (1984).
30. C.H.E. Nielsen, S. Kiil, H.W. Thomsen, and K. Dam-Johansen, *Chem. Eng. Sci.* **53**, 495–503 (1998).
31. E.T. Papoutsakis, *Trends Biotechnol.* **9**, 427–437 (1992).
32. A. Handa, A.N. Emery, and R.E. Spier, *Devel. Biol. Stand.* **66**, 241–253 (1987).

33. A. Handa-Corrigan, A.N. Emery, and R.E. Spier, *Enzyme Microb. Technol.* **11**, 230–235 (1989).
34. M.S. Croughan, J.-F. Hamel, and D.I.C. Wang, *Biotechnol. Bioeng.* **29**, 130–141 (1987).
35. Y. Chisti, *Bioprocess. Eng.* **9**, 191–196 (1993).
36. W.R. Arathoon and J.R. Birch, *Science* **232**, 1390–1395 (1986).
37. J.R. Birch, K. Lambert, P.W. Thompson, A.C. Kenney, and L.A. Wood, in B.K. Lydersen ed., *Large Scale Cell Culture Technology*, Hanser Publishers, New York, 1987, pp. 1–20.
38. Z. Zhang, Y. Chisti, and M. Moo-Young, *J. Biotechnol.* **43**, 33–40 (1995).
39. A.W. Nienow, *Appl. Mech. Rev.* **51**, 3–32 (1998).
40. A. Bakker, J.M. Smith, and K.J. Myers, *Chem. Eng.* **101**, 98–104 (1994).
41. C.S. Shin, M.S. Hong, and J. Lee, *Biotechnol. Technol.* **10**, 679–682 (1996).
42. H. Taguchi and F. Yoshida, *J. Ferment. Technol.* **46**, 814 (1968).
43. H. Yagi and F. Yoshida, *Ind. Eng. Chem. Process Des. Develop.* **14**, 488–493 (1975).
44. Y. Kawase and M. Moo-Young, *Chem. Eng. Res. Des.* **66**, 284–288 (1988).
45. K. Chandrasekharan and P.H. Calderbank, *Chem. Eng. Sci.* **36**, 819–823 (1981).
46. Y.T. Shah, B.G. Kelkar, S.P. Godbole, and W.-D. Deckwer, *AIChE J.* **28**, 353–379 (1982).
47. W.-D. Deckwer, *Bubble Column Reactors*, Wiley, New York, 1992.
48. W.-D. Deckwer and A. Schumpe, *Chem. Eng. Sci.* **48**, 889–911 (1993).
49. J.J. Heijnen and K. Van't Riet, *Chem. Eng. J.* **28**, B21–B42 (1984).
50. J.R. Fair, *Chem. Eng.* **74**, 67 (1967).
51. K. Akita and F. Yoshida, *Ind. Eng. Chem. Process Des. Develop.* **12**, 76–80 (1973).
52. G.A. Hughmark, *Ind. Eng. Chem. Process Des. Develop.* **6**, 218–220 (1967).
53. M. Nakanoh and F. Yoshida, *Ind. Eng. Chem. Process Des. Develop.* **19**, 190–195 (1980).
54. A. Schumpe, C. Singh, and W.-D. Deckwer, *Chem.-Ing.-Technol.* **57**, 988–989 (1985).
55. Y. Kawase, B. Halard, and M. Moo-Young, *Chem. Eng. Sci.* **42**, 1609–1617 (1987).
56. H. Hikita, S. Asai, K. Tanigawa, K. Segawa, and M. Kitao, *Chem. Eng. J.* **22**, 61–69 (1981).
57. W.-D. Deckwer, K. Nguyen-Tien, A. Schumpe, and Y. Serpemen, *Biotechnol. Bioeng.* **24**, 461–481 (1982).
58. S.P. Godbole, A. Schumpe, Y.T. Shah, and N.L. Carr, *AIChE J.* **30**, 213–220 (1984).
59. J. Dudley, *Water Res.* **29**, 1129–1138 (1995).
60. G.-Q. Li, S.-Z. Yang, Z.-L. Cai, and J.-Y. Chen, *Chem. Eng. J.* **56**, B101–B107 (1995).
61. K. Koide, H. Sato, and S. Iwamoto, *J. Chem. Eng. Jpn.* **16**, 407–413 (1983).
62. K. Koide, K. Horibe, H. Kawabata, and S. Ito, *J. Chem. Eng. Jpn.* **18**, 248–254 (1985).
63. P. Verlaan, J.-C. Vos, and K. van't Riet, *J. Chem. Technol. Biotechnol.* **45**, 181–190 (1989).
64. I.-S. Suh, A. Schumpe, and W.-D. Deckwer, *Biotechnol. Bioeng.* **39**, 85–94 (1992).
65. K. Koide, K. Shibata, H. Ito, S.Y. Kim, and K. Ohtaguchi, *J. Chem. Eng. Jpn.* **25**, 11–16 (1992).
66. Y. Chisti and M. Moo-Young, *Chem. Eng. Commun.* **60**, 195–242 (1987).
67. A.B. Russell, C.R. Thomas, and M.D. Lilly, *Bioprocess. Eng.* **12**, 71–79 (1995).
68. D.J. Pollard, A.P. Ison, P.A. Shamlou, and M.D. Lilly, in E. Galindo and O.T. Ramírez eds., *Advances in Bioprocess Engineering*, Kluwer, Dordrecht, The Netherlands, 1994, pp. 163–170.
69. Y. Chisti and M. Moo-Young, *Chem. Eng. Progress*, **89**, 38–45 (1993).
70. K. Koide, K. Horibe, H. Kitaguchi, and N. Suzuki, *J. Chem. Eng. Jpn.* **17**, 547–549 (1984).
71. Y. Chisti, M. Kasper, and M. Moo-Young, *Can. J. Chem. Eng.* **68**, 45–50 (1990).
72. W. Zhou, K. Holzhaur-Rieger, and K. Schügerl, *J. Biotechnol.* **28**, 165–177 (1993).
73. J. Audet, J. Thibault, and A. LeDuy, *Biotechnol. Bioeng.* **52**, 507–517 (1996).
74. M. Lounes, J. Audet, J. Thibault, and A. LeDuy, *Bioprocess Eng.* **13**, 1–11 (1995).
75. X. Ni, S. Gao, and D.W. Pritchard, *Biotechnol. Bioeng.* **45**, 165–175 (1995).
76. H. Ziegler, I.J. Dunn, and J.R. Bourne, *Biotechnol. Bioeng.* **22**, 1613–1635 (1980).
77. K. Muroyama and L.-S. Fan, *AIChE J.* **31**, 1–34 (1985).
78. K. Nguyen-Tien, A.N. Patwari, A. Schumpe, and W.-D. Deckwer, *AIChE J.* **31**, 194–201 (1985).
79. V.Y. Loh, S.R. Richards, and P. Richmond, *Proceedings: Int. Conference on Bioreactor Fluid Dynamics*, 1986, 15–17, April BHRA, Cambridge, England.
80. B.C. Smith and D.R. Skidmore, *Biotechnol. Bioeng.* **35**, 483–491 (1990).
81. D.G. Karamanev, T. Nagamune, and I. Endo, *Chem. Eng. Sci.* **47**, 3581–3588 (1992).
82. M.A. Winkler, *Biological Treatment of Waste-Water*, Ellis Horwood, Chichester, U.K., 1981.
83. W. Sittig, *J. Chem. Technol. Biotechnol.* **32**, 47–58 (1982).
84. K. Schügerl, *Solvent Extraction in Biotechnology: Recovery of Primary and Secondary Metabolites*, Springer-Verlag, New York, 1994.
85. S. Martin, P. Soucaille, and J.-S. Condoret, *Bioprocess Eng.* **13**, 293–300 (1995).
86. S.S. Save, R.B. Desai, S.B. Sawant, and J.B. Joshi, *Trans. Inst. Chem. Eng.* **74C**, 171–177 (1996).
87. Y. Chisti, in M.C. Flickinger and S.W. Drew, eds., *Encyclopedia of Bioprocess Technology*, Wiley, New York, 1999, in press.
88. D.K. Edwards, V.E. Denny, and A.F. Mills, *Transfer Processes*, Holt, Rinehart and Winston, New York, 1973.
89. P.N. Dwivedi and S.N. Upadhyay, *Ind. Eng. Chem. Process Des. Dev.* **16**, 157–165 (1977).
90. Y.-K. Lee and H.-K. Hing, *Appl. Microbiol. Biotechnol.* **31**, 298–301 (1989).
91. R.S. Tutunjian, *Biotechnology* **3**, 615–626 (1985).
92. S. Ripperger, *Chem. Eng. Technol.* **11**, 17–25 (1988).
93. M.H.V. Mulder, in R.D. Noble and S.A. Stern eds., *Membrane Separations Technology: Principles and Applications*, Elsevier, Amsterdam, 1995, pp. 45–84.

94. W.S.W. Ho and K.K. Sirkar eds., *Membrane Handbook*, Van Nostrand Reinhold, New York, 1992.
95. K. Brindle and T. Stephenson, *Biotechnol. Bioeng.* **49**, 601–610 (1996).
96. Y. Kawase, A.V. Shenoy, and K. Wakabayashi, *Can. J. Chem. Eng.* **72**, 798–804 (1994).
97. H.H. Mao, Y. Chisti, and M. Moo-Young, *Chem. Eng. Commun.* **113**, 1–13 (1992).
98. W.-D. Deckwer and A. Schumpe, *Ger. Chem. Eng.* **7**, 168–177 (1984).
99. D.C. Hempel, in R.K. Finn, P. Präve, M. Schlingmann, W. Crueger, K. Esser, R. Thauer, F. Wagner eds., *Biotechnology Focus 1: Fundamentals, Applications, Information*, Hanser, New York, 1988, pp. 51–94.
100. P. Sänger and W.-D. Deckwer, *Chem. Eng. J.* **22**, 179–186 (1981).
101. S. Goto, Y. Matsumoto, and P. Gaspillo, *Chem. Eng. Commun.* **85**, 181–191 (1989).
102. K.B. Kushalkar and V.G. Pangarkar, *Chem. Eng. Sci.* **49**, 139–144 (1994).
103. P.D. Gaspillo and S. Goto, *J. Chem. Eng. Jpn.* **24**, 680–682 (1991).
104. L. Goldstein, *Methods Enzymol.* **44**, 397–443 (1976).
105. J.M. Radovich, *Enzyme Microbial. Technol.* **7**, 2–10 (1985).
106. R.G. Willaert, G.V. Baron, and L. De Backer eds., *Immobilised Living Cell Systems: Modelling and Experimental Methods*, Wiley, Chichester, 1996.
107. W.H. Pitcher Jr., *Adv. Biochem. Eng.* **10**, 1–26 (1978).
108. J.E. Bailey and D.F. Ollis, *Biochemical Engineering Fundamentals*, 2nd ed., McGraw-Hill, New York, 1986, pp. 204–220.

See also BIOFILMS, MICROBIAL; BIOREACTORS, AIR-LIFT REACTORS; BIOREACTORS, CONTINUOUS STIRRED-TANK REACTORS; BIOREACTORS, FLUIDIZED-BED; GAS HOLD-UP; HYPOXIA, EFFECTS ON ANIMAL CELLS; SCALE-UP, STIRRED-TANK REACTORS; SOLID STATE FERMENTATION, MICROBIAL GROWTH KINETICS; TRANSFER PHENOMENA IN MULTIPHASE SYSTEMS IN MIXING VESSELS.

MEDIA COMPOSITION, MICROBIAL, LABORATORY SCALE

ROSALIE J. COTE
Becton Dickinson Microbiology Systems
Sparks, Maryland

KEY WORDS

Agar
Buffers
Carbon and energy sources
Detoxifiers
Energy
Growth factors
Oxidation–reduction potential
pH
Protein hydrolysates (peptones)
Sterilization
Water

OUTLINE

INTRODUCTION

The increased understanding of microbial biochemistry in recent decades, along with major advances in genetic and molecular techniques, has resulted in an expanding diversity of microorganisms entering the biotechnology arena. Be it novel uses for well-studied organisms, such as the commercial production of restriction endonucleases from many common bacteria and yeasts, or the potential for development of new applications from new species such as the unidentified hydrogen-producing bacterium FOX-1 (1), microorganisms, themselves, are integral components in the overall aspects of biotechnology from discovery, through research and development, to production.

For those biotechnologists not formally trained in classical microbial physiology, the too-prevalent assumption is that microorganisms function similarly to the purified reagents used in a well-documented chemical reaction. In reality, microorganisms (even authenticated and characterized strains) are heterogeneous conglomerations of multiple biochemical pathways with the potential for working antagonistically or synergistically, depending upon the environment in which the microorganism resides. For those strains under study in the laboratory or being potentially harnessed for bioprocess use, the microbial environment is the culture medium.

This article surveys elements of microbial media and other considerations of microbial nutrition that impact upon the general growth and metabolic activity of microorganisms. Although specific references are given for many useful compendia of microbial media formulations (2–7), the purpose here is not to further cite well-documented media and buffer formulations. The intent of this article and its specific examples is to illustrate general principles of controlled media design, formulation, and preparation for the cultivation of microorganisms.

ESSENTIAL NUTRITIONAL REQUIREMENTS

Water

Depending upon the physiological state of the organism, the chemical composition of bacteria is 70–90% water. As a polar solvent, water dissolves other compounds of similar chemical nature, which also happen to be the biologically important molecules such as amino acids, carbohydrates, and nucleic acids. The polarity of water helps to permit the hydrogen bonding necessary in the formation of the final working configuration of a cell's large macromolecules. As a polar solvent, water chemically forces nonpolar molecules such as lipids into hydrophobic aggregates that are integral in the formation cell membranes. The presence of water is critical to the existence of microorganisms. Under laboratory conditions, microorganisms are cultivated with liquid and solid media. In both situations, the adequacy and purity of water used in media preparation can drastically effect the performance and/or sustained viability of an organism.

Because the availability of high-quality, purified water has become increasingly affordable for even small laboratories, the use of tap water or marginally distilled water for the preparation of media and reagents is not a recommended option. For most municipalities, tap water is seasonably variable with respect to its chemical and microbial content. Hard water ladened with calcium and magnesium will precipitate many media formulations, with deleterious results. The terms *deionized* or *distilled* water with respect to water purity are merely euphemistic unless actual determinations of residual ion and organic levels are known. Water passed through spent deionization beds or boiled and condensed through a scale-clogged distillation system is hardly purified. At minimum, water used in the preparation of microbial media and reagents for general laboratory purposes should meet U.S.P. specifications for purified water; other organizations such as the American Society for Testing and Materials (ASTM) and the National Committee for Clinical Laboratory Standards (NCCLS) have similar criteria with respect to the conductivity, resistivity, and total organic content for Type II water (8). Point of use deionization systems available from many manufacturers (e.g., Barnstead/Thermolyne Corporation, Dubuque, Ia. and Millipore Corporation, Bedford, Mass.) easily produce Type I, 18 $M\Omega$ water. Some systems include ultrafiltration to reduce pyrogen levels. Purified water for highly regulated applications and processes (e.g., biopharmaceuticals) must meet U.S.P. specifications for water for injection (9). Water, unless sterilized, should not be held for any length of time in storage containers such as bottles or carboys or allowed to stand in uncirculated piping systems (e.g., dead-legs) because there is a good probability that the water will become populated with bacterial biofilms that grow on container and piping walls. Bacterial contamination will organically pollute water purity in addition to increasing endotoxin and pyrogens to unacceptable levels for certain procedures including product purifications or animal injections. Note that pyrogens are lipopolysaccharides from bacterial membranes. With respect to purified water, they indicate bacterial contamination. Their presence, of course, in bacterial cultures is a mute point. Use of plastics for water storage containers may be a consideration because plasticizer leaching can cause inhibition of sensitive organisms.

Energy

All microorganisms require an exogenous source of energy. The biochemical mechanisms involved with this aspect of metabolism are diverse. *Lithotrophic* organisms obtain energy from inorganic sources. Lithotrophs fall into two metabolic categories with respect to energy procurement. *Photolithotrophic bacteria* and *cyanobacteria* obtain energy from light in conjunction with, respectively, reduced sulfur (e.g., hydrogen sulfide) or water, to release energy for biosynthetic functions. Variations in microbial chlorophyll molecules create adsorption spectra that, in nature, limit a given species to a particular photic zone. Under laboratory conditions, the light requirements necessary for cultivation reflect this consideration. For example, the green and purple sulfur bacteria, such as *Chlorobium* and *Chromatium*, with adsorption peaks above 750 nm, are cultivated under the red wavelengths of tungsten light. Cultures, however, should not be placed too close to an incandescent light source because the heat emitted from the bulbs can quickly raise the temperature of the culture vessel beyond the upper growth limits for an organism. Cyanobacteria (e.g., *Anabaena*) adsorb blue wavelengths of light below 700 nm. This requirement can be satisfied by cultivating the bacteria under fluorescent tubes that emit light at this range of the visible spectrum. *Chemolithotrophic* organisms utilize inorganic compounds for energy. Examples within this second form of energy procurement are *Sulfolobus* spp., which oxidize elemental sulfur to sulfuric acid, and *Nitrobacter* spp., which biochemically exploit the oxidation of nitrites to nitrates. Numerous new genera and species of prokaryotic chemolithotrophs that obtain energy from various inorganic electron donors have been isolated in recent years, including *Archaeoglobus* spp., *Methanococcus* spp., and *Pyrodictium* spp. (10). Many obligate lithotrophs are inhibited by organically rich environments and are best cultivated under chemically defined media conditions.

Organotrophic organisms obtain energy more conventionally from organic compounds. As with the lithotrophs, organotrophs many be either phototrophic or chemotrophic. *Photoorganotrophs* are represented in nature by the anaerobically cultivated purple nonsulfur bacteria (e.g., *Rhodospirillum*), which utilize light in conjunction with organic compounds such as malate, pyruvate, or yeast extract, rather than O_2, as electron acceptors to release energy for growth and nutrition. *Chemoorganotrophs* encompass a vast number of organisms from all the microbial groups including archaea and bacteria, fungi, protozoans, and nonphotosynthesizing eukaryotic cells that obtain energy from the oxidation or other dissimilation of organic compounds.

Major Chemical Elements (C, H, N, O)

Unlike energy-procuring mechanisms, which vary widely between microbial genera and species, the anabolic or biosynthetic requirements of eukaryotic and prokaryotic cells are remarkably similar. Molecular mechanisms for intracellular communication and structural components share much chemistry whether they are bacterial *E. coli* cells or fungal *Aspergillus nidulans* cells. Whereas individual strains may vary in their ability to manufacture basic cell components and metabolites due to specific mutations at points within biochemical pathways, all known organisms share the same basic cellular needs for carbohydrates, lipids, proteins, and molecular combinations thereof. The elements carbon, hydrogen, nitrogen, and oxygen can be regarded as major essential elements of nutrition because they contribute in significant percentages to the biomass of cells. These major elements may be required in the gaseous form, as organically fixed compounds, or both. Iron, magnesium, phosphorous, potassium, sulfur, and other elements, although critically important to sustained growth and metabolic activity, are often termed minor or trace elements in that they collectively represent less than 2% of cellular dry weight.

Carbon. Nutritionally, carbon appears in cellular metabolism in two forms: carbon dioxide gas (CO_2) and organic molecules. CO_2 is required by all organisms for a number of biochemical functions including tricarboxylic acid (TCA) cycling and de novo fatty acid synthesis from acetyl-CoA. Although much of the carbon dioxide is metabolically recycled throughout the cell's biochemical pathways, an initial exogenous source of carbon dioxide is mandatory. Autotrophic organisms readily assimilate atmospheric carbon dioxide; yet, also, all heterotrophic organisms have some ability to fix this compound, although at a much-reduced capacity. Microbial cultures will sometimes experience long lag periods when subcultured at low-volume inoculum into fresh media. This phenomenon is often attributable to suboptimal CO_2 tension, which the culture slowly corrects with cellular respiration before any significant increase in biomass can occur. The term *CO_2 sparking* refers to the sudden increase in bacterial growth observed when a freshly inoculated culture is briefly sparged with the gas. An enriched atmosphere of 5% CO_2 is used for the cultivation of clinical (particularly for fresh isolates) bacterial strains. CO_2 incubators are most efficient for multipurpose cultivations, but the anaerobic gas-pack CO_2-generating systems offered by BBL (Becton Dickinson Microbiology Systems, Cockeysville, Md.) or Difco (Difco Laboratories, Detroit, Mich.) are recommended for small volume or infrequent use. Depending upon the absolute requirement, capnophilic (CO_2-requiring) bacteria can also be cultivated in jars in which inoculated test tubes or agar plates are placed along with a burning candle. The candle jar is then tightly sealed while the candle continues to burn until free oxygen in the jar is consumed. The resulting atmosphere achieves approximately 2% CO_2. Both the anaerobic gas-pack systems and candle jars have draw backs with respect to increased risk of culture contamination. Respiring cultures give off significant amounts of water, which raises the humidity within the sealed jars. As the jars themselves are normally not sterile at time of culture incubation, any environmental bacterial or fungal contaminant within the jar will proliferate in the additional moisture. Capillary action of excess water at the culture vessel openings (particularly petri dishes) can draw contaminating organisms or their spores into the culture vessel.

Autotrophs can use carbon dioxide gas as a sole source of carbon for biosynthesis. Although autotrophy is often regarded by nonbacteriologists as synonymous with photosynthesis, autotrophy and phototrophy are quite independent metabolic functions. For example, methanogenic archaea (e.g., *Methanobacterium* sp.), while nonphotosynthetic, are capable of utilizing carbon dioxide and hydrogen as sole sources of carbon and energy. Yet, as noted elsewhere, purple nonsulfur bacteria are photosynthetic but rely upon organic acids for metabolic reducing power and, ultimately, as the carbon source for biosynthetic functions. Whereas autotrophy is widely represented among prokaryotes, the ability to utilize CO_2 as a sole carbon source is not a general characteristic of fungi. Among filamentous fungi, *Cephalosporium* spp. and *Fusarium* spp. have been documented to be able to do so under long-term cultivation (11).

Heterotrophs obtain carbon principally from organic molecules such as carbohydrates, amino acids, or lipids. For heterotrophs, a compound utilizable for its carbon often also functions as an energy source. Although glucose is widely assimilated by many heterotrophic microorganisms, it is not necessarily the universal carbon source for culture medium supplementation. Many bacteria, whether their characteristics are genuswide or species-specific, show little or no stimulation by glucose and preferentially utilize amino or other organic acids. The marine genus *Oceanospirillum* falls into this category as do many individual species within the diverse genus *Clostridium*. In an extreme example, *C. aminovalericum* (ATCC 13725) appears to be limited to δ-aminovaleric acid as its singular assimilable source of both carbon and energy.

Some organisms can alternate between autotrophy and heterotrophy, depending upon environmental and nutritional conditions. For these cultures, biomass vigor and yield are most often greater with heterotrophic rather than autotrophic conditions; however, secondary metabolites produced can vary considerably between the two conditions.

Hydrogen. Certain archaea and bacteria can grow by utilizing atmospheric hydrogen (H_2) as an electron donor and, depending upon the organism, assorted compounds including carbon dioxide, oxygen, nitrate, or sulfate as terminal electron acceptors (10). The hydrogen-oxidizing genus *Hydrogenophaga* (12) is facultative with respect to energy metabolism in that it functions chemoorganotrophically in a nutritionally heterotrophic aerobic environment by metabolizing various amino acids or sugars for sources of carbon and energy. However, in an atmosphere containing 60% H_2, 10% CO_2, 25% N_2, and 5% O_2, the organisms switch to chemolithoautotrophic pathways. *Methanobacterium thermoautotrophicum*, a strictly anaerobic archeaebacteria, metabolizes hydrogen and carbon dioxide as sole energy and carbon sources to yield methane and water (methanogenesis). For the anaerobic methanogens as well as the aerobic hydrogen bacteria, conditions favoring autotrophy require a high (60–80%) concentration of hydrogen for cultivation. Under all conditions, work with compressed hydrogen gas should be conducted with adherence to pertinent laboratory safety practices. Lower concentrations of H_2 (3–10%) are frequently included in gas mixtures for the cultivation of anaerobic bacteria in gas-generating systems and glove boxes where, in the presence of a palladium catalyst, trace amounts of inhibitory molecular oxygen are chemically reduced to water. The final form of hydrogen required for nutritional purposes is the molecularly combined element required by all organotrophic cells. Combined hydrogen originates from the catabolic metabolism of organic compounds assimilated by a cell as carbon and energy sources.

Oxygen. As already noted for combined hydrogen, the bulk of combined oxygen within cells originates from the complex nutritional compounds, including water, assimilated by the cell. If a medium provides enough matter to satisfy the carbon and energy requirements of a culture, then the combined oxygen requirements will also be met. However, molecular oxygen (O_2) is an important, often growth-limiting, variable in microbial nutrition. O_2 is the primary terminal electron acceptor for many aerobic and facultatively anaerobic organisms. In oxygen-depleted or diminished environments, cells with no other biochemical mechanisms for exploiting alternative electron acceptors stop metabolizing and cease multiplying. Facultative organisms switch to fermentative pathways. In either situation, the biochemical activity of the cells, as well as the secondary metabolites produced, is significantly altered by oxygen starvation. Be it large-volume fermentors or small-volume chemostats or Fernbach flasks, the need for O_2 sparging in bulk cultivation vessels is regarded as a basic cultivation requirement for aerobic organisms (13). Be that as it may, one should also be aware that significant O_2 gradients quickly form in liquid test tube cultures, particularly for those cultures that form surface mats or pellicles of growth. Various micrometabolies can occur within the cell population in a standing culture along the O_2 gradients from surface to bottom of the test tube that might, if unrecognized, cause undue problems in scale-up development. Cultures with strong O_2-requirements (even for those that do not form pellicles) may not grow to very high density in a test tube because of the limited liquid–air interface. Better liquid growth of these test tube cultures is effected by incubating the tube at a slant to increase the surface area of the culture, or by shaking.

Nitrogen. Cells can assimilate nitrogen by two methods. Nitrogen fixation is the ability to acquire elemental nitrogen from the air and to molecularly combine the element into chemical forms utilizable for biosynthetic processes. The bacterial genera *Azotobacter, Bradyrhizobium*, and *Rhizobium* are most recognized for this characteristic, although some facultative anaerobes as well as anaerobic bacterial species also have the capacity. Fungi are not recognized as competent nitrogen fixers. Nitrogenase activity is induced with cultivation on nitrogen-free media. Growth on chemically defined "nitrogen-free" media might presumptively select for potential nitrogen fixers; however, trace amounts of nitrogen from agars, chemicals, and glassware are present in sufficient concentration to permit effective scavenging and growth (albeit, less than optimal) by many non-nitrogen-fixing organisms. Conclusive nitrogen fixation is best determined by the acetylene reduction assay (14).

The second, and most widely utilized, microbial mechanism for nitrogen procurement is via combined nitrogen assimilation from inorganic or organic molecules. Media formulations satisfy combined nitrogen requirements either with inorganic salts [e.g., NH_4Cl; NH_4NO_3 or KNO_3; $(NH_4)_2SO_4$] or from organic compounds (e.g., amino acids). The ability to utilize inorganic salts as a nitrogen source varies widely even within species. The chemolithotrophs as a group effectively utilize inorganic ammonium or nitrate salts as sole sources for nitrogen. For other organisms, the inability to grow under chemically defined medium conditions with ammonium or nitrate salts as a sole nitrogen source is due not so much to the inability to assimilate the ion directly from the medium or to incorporate it as the glutamine precursor to all subsequent biosynthetic nitrogenous pathways, but to the specific need for additional nitrogenous growth factors such as purines, pyrimidines, amino acids, or small peptides. These kinds of nitrogenous nutritional requirements are discussed in greater detail in the section "Growth Factors".

Nitrate is not a preferentially assimilated or utilized source of nutritional nitrogen for many bacteria or fungi; concentrations as low as 0.2% are inhibitory to the growth of some bacteria. However, the use of nitrate in energy metabolism as a terminal electron acceptor is a widely used mechanism among bacteria and is termed dissimilatory nitrate reduction or denitrification. The biochemical pathways involved with nitrate assimilation for cellular syntheses and for dissimilatory nitrate reduction are different. Nitrate reduction occurs with either aerobic or anaerobic bacteria. Strains of *Clostridium perfringens*, for example, utilize nitrate as an electron sink, which causes a shift from butyrate to acetate fermentation and an increase in growth yield (15). All recognized species in the extremely halophilic genus *Haloarcula* are capable of using nitrate as a sole nitrogen source and, in addition, are strong denitrifiers with the ability to readily tolerate and reduce 3 M concentrations of KNO_3 under either complex or chemically defined media conditions; aerobic growth of haloarculas with liquid media lacking nitrate supplementation is markedly limited compared to aerobic growth in media supplemented with 0.2% KNO_3 (R. J. Cote, unpublished data, 1995).

Minor and Microelements (Fe, Mg, K, P, S, and others)

Trace metal function and speciation have been extensively studied in bacteria and fungi (16,17). Nutritionally essential metal ions function metabolically as enzyme cofactors in membrane transport and in structural components. Microbial response to trace metals is detectable at extremely low concentrations (i.e., microgram or nanogram quantities, depending upon the metal). However, the bioavailability of a particular metal is markedly dependent upon proper speciation of the ion as it occurs in the culture medium (18). Depending on interaction with other medium components or methods of sterilization, metal valences can change to nonusable forms, or the ions can irreversibly bind as insoluble precipitates. Although problems with metal-binding phenomena are often directed toward the nutritionally essential metals, the potential for significant and irreversible binding of ions to media components must also be taken into consideration in studies of bacterial resistance to heavy metals. Reports of bacterial resistance to silver and selenium (19,20) could not be verified when the metals were added to culture media in a manner that eliminated their binding to peptones (R. J. Cote and R. L. Gherna, unpublished data, 1994). Similar observations made regarding media influence and metal sensitization of *Enterobacter cloacae* to silver sulfadiazine (AgSu) (21) are explained by heavy-metal binding under complex media conditions. Tilton and Rosenberg (22) discuss the issue of heavy-metal sequestration in a study of the effects of agar on silver inhibition.

Microelement supplementation is not normally required in media formulated with chemically complex components such as yeast extract or protein hydrolysate because the essential metals are naturally present in these materials in sufficient concentration to provide optimum culture growth and performance. Well-designed, chemically defined formulations include calcium, iron, magnesium, manganese, phosphorous (as phosphate), potassium, sodium, and sulfur (as sulfate) in the form of inorganic salts. Other micronutrients may be required in the medium, depending upon the nature of the metal function under study or, in the case of general culture propagation, the purity of the water and cleanliness of the glassware used to prepare the medium. Understanding of cell physiology, as well as any physical interactions within the culture system, is a basic requirement when working with nutritional trace metals. For example, although boric acid is often included in trace metal mixes, a boron deficiency is practically noninducible for cultures grown in borosilicate glass vessels because enough of the element leaches from the glass to satisfy nutritional needs. Table 1 lists metal mix

Table 1. Trace Metal Mixes for Bacteriological Media

Metal mix A: Balch's trace element solution

Nitrilotriacetic acid	1.5 g
$MgSO_4 \cdot 7H_2O$	3.0 g
$MnSO_4 \cdot H_2O$	0.5 g
NaCl	1.0 g
$FeSO_4 \cdot 7H_2O$	0.1 g
$CoCl_2 \cdot 6H_2O$	0.1 g
$CaCl_2$	0.1 g
$ZnSO_4 \cdot 7H_2O$	0.1 g
$CuSO_4 \cdot 5H_2O$	10.0 mg
$AlK(SO_4)_2 \cdot 12H_2O$	10.0 mg
H_3BO_3	10.0 mg
$Na_2MoO_4 \cdot 2H_2O$	10.0 mg
Distilled water	to 1.0 L

Suspend nitrilotriacetic acid in about 500 mL water and adjust to pH 6.5 with KOH pellets to dissolve. Add remaining salts one at a time. Bring final volume to 1.0 L with distilled water. Use 10.0 ml trace element solution per 1 L medium.

Metal mix B: modified Balch's trace element solution

To 1.0 L metal mix A add	
$NiSO_4 \cdot 6H_2O$	30.0 mg
Na_2SeO_3	20.0 mg
$Na_2WO_4 \cdot 2H_2O$	20.0 mg

These metals are included for the cultivation of methanogenic bacteria under chemically defined conditions.

Source: From Ref. 23, p. 158.

formulations that are useful for chemically defined cultivation of numerous microorganisms. As noted elsewhere in this article, the use of tap water to satisfy trace element requirements is not recommended because regional and seasonal variability in water purity can introduce considerable ionic variation in a medium, with frequently deleterious results.

The whitish haze or precipitate that develops in some media upon standing or after autoclaving is frequently caused by imbalanced metal compositions. Divalent and trivalent cations (i.e., Ca^{2+}, Fe^{2+}, Fe^{3+}, and Mg^{2+}) will spontaneously bind with OH^- and PO_4^{2-} groups in the medium to form insoluble hydroxides and phosphate complexes. The condition is most evident at neutral to alkaline pH values and is intensified by heat, either boiling or autoclaving. Similarly, precipitation caused by metal binding to proteinaceous material in the medium and darkening of chemically defined agar media are also frequently encountered in poorly designed or manipulated formulations. All of these adverse metal interactions are generally regarded as deleterious and growth limiting. Depending upon the composition of the medium, these kinds of chemical interactions can be eliminated by sterile filtration or separate autoclaving of inorganic salts and any phosphate buffers, and separate autoclaving of agar.

The use of chelating agents can also help to selectively maintain the solubility of polyvalent cations. Care must be taken when using powerful chelators such as ethylenediaminetetraacetic acid (EDTA) or nitrilotriacetic acid

(NTA) because overchelation also depletes bioavailability of the metals. For microbial media, 0.5–2.0 mg/L EDTA or 15–150 mg/L NTA are reasonable concentrations. Mild chelation can be achieved with the use of specific amino acids (e.g., glycine, histidine) or carboxylic acids (acetate, citrate, succinate, tartrate). However, their use and concentration must be evaluated with respect to the function of the medium and the physiology of the microbial culture. For example, carboxylic acids are readily utilized carbon and energy sources for many microorganisms, and glycine is inhibitory at 1% concentration to many bacteria.

Under iron-deficient growth conditions, some bacteria produce and secrete chelators called siderophores to improve transport of ionic iron into the cell (24,25). Siderophores have been found in a number of diverse genera including *Aureobacterium (Arthrobacter) flavescens* (26), *E. coli, Klebsiella* (27), *Pseudomonas*, and *Mycobacterium avium* subsp. paratuberculosis (28). Other organisms rely upon the hemin molecule in blood as an easily assimilable iron chelate. A growth factor requirement for hemin is common among medically important bacteria. Hemin requirement (the X-factor) is diagnostic in the identification of *Haemophilus influenzae. Porphyromonas gingivalis* (29) and *Vibrio vulnificus* biotypes 1 and 2 require iron in the form hemin for growth and virulence (29,30). Whereas a number of classical media formulations (e.g., GC Agar) include crude hemoglobin as the source of the iron porphyrin molecule, use of hemin, instead, will produce a medium free of the heavy debris associated with hemoglobin. Hemin is used at a concentration of 0.5 mg/L medium. To improve solubility, initially suspend the compound in 1–2 mL 1 N NaOH, then add the hemin-NaOH solution to the bulk of the medium.

Many of the metals listed here that have been historically categorized as needed by microorganisms only in trace amounts have, in recent years, come to be evaluated on a different metabolic scale. Sulfur serves as an energy source for numerous species of bacteria, and thiosulfate seems to be utilized for energy by some fungi. While the role of sulfate as the terminal metabolic electron acceptor for the many anaerobic sulfate-reducing bacteria has been long recognized, new genera are currently being isolated that possess the ability to utilize one or more biologically important metals [(Fe (III), Mn (IV)], U (VI), and others] as terminal electron acceptors in biochemical dissimilatory metal reduction pathways (31). Because the function of terminal electron acceptor is not recyclable within the cell and bears a heavy metabolic load, the amounts of metal ion needed per liter for efficient energy transfer are orders of magnitude greater than those needed to satisfy trace element nutrition. One to four grams per liter is a guideline, depending upon the specific metabolism of the organism as well as the molecular weight and form of the metal salt selected for use.

Growth Factors

Given less than a dozen essential chemical building blocks and a source of metabolic energy to power the cellular machinery, some organisms, such as the wild-type *E. coli*, can synthesize every molecule needed for cell maintenance, growth, and reproduction. Many other microorganisms (even strains within the same species) are auxotrophic in that they have lost the ability to synthesize one or more key organic compounds essential to their physiology. These missing compounds are called growth factors and must be present in the culture medium if cell growth is to be vigorous, metabolically productive, and sustainable.

L-Amino Acids. L-Amino acids are common growth factor requirements for prokaryotes and eukaryotes. Many bacterial genera (e.g., *Lactobacillus* spp.) have multiple amino acid requirements. The requirement may be absolute or conditional and influenced by the presence and concentration of other amino acids (32–34). Similarly, a phenotypic amino acid requirement may disappear with the addition of an appropriate synthetic pathway intermediate. For example, a tryptophan "requirement" may actually be caused by a metabolic block earlier in the tryptophan pathway and might be satisfied by one of the earlier pathway intermediates (i.e., aspartic acid or indole). Because amino acids may share a common transport system, an excess of one amino acid can result in an increased need for a competing amino acid. Amino acid supplementation is not required in complex media containing peptones or other protein hydrolysates. In chemically defined media, the addition of 20–50 mg/L amino acid stimulates prokaryotic growth (100–200 mg/L for yeast), although adjustment of the concentrations is necessary for media optimization with multiple amino acid-requiring strains to chemically balance the formulation. The most notable exception is the requirement for glutamic acid, which, as precursor to all other amino acid biosyntheses, must be added at 200–500 mg/L to achieve optimal growth of a bacterial culture. Amino acids, except glutamine, are stable with autoclave sterilization at the working concentrations used in media. Glutamine in aqueous solution will spontaneously break down with time to ammonium and pyrrolidone carboxylic acid residues. This reaction occurs more rapidly at elevated temperatures (37 °C and above). The use of glutamine peptides in a medium can eliminate this problem, however, they are expensive for large-scale use. Concentrated stock solutions of amino acids may not be soluble, but neutralization of the stock solution with NaOH or KOH will increase solubility with the formation of amino acid salts for many of these compounds. Stock solutions of cysteine will autooxidize to cystine unless stored anaerobically (e.g., under N_2).

For many years, the perceived need by bacteria for combined amino acids (peptides or proteins) was satisfied by the argument that peptide transport utilized different systems than did amino acids; thus, the need for a specific peptide was actually an alternative cellular method for obtaining a specific amino acid contained within that peptide via a biochemically noncompetitive mechanism. However, increased knowledge of the role of peptides in protein trafficking suggests that peptide auxotrophies can affect cellular physiology and communication (35). Successful laboratory cultivation of *Thermoplasma acidophilum*, having no other known growth factor requirements, is dependent upon specific lots of powdered yeast extract containing a methanol-soluble growth factor, which purification revealed to be a small 1,000-MW peptide (36). Similarly, recent laboratory cultivation of *Treponema maltophilum* also included the requirement for a methanol-soluble growth factor from yeast extract (37). The treponeme requirement has not been further purified. Finally, successful cultivation of the protozoan *Entamoeba histolytica* in complex media enriched with peptone, vitamins, and serum is nevertheless dependent upon an unknown growth factor found only in certain lots of powdered yeast extract (38). Whether the elusive yeast extract growth factor is the same for all these metabolically dissimilar organisms or whether the relatively gentle methods used in the manufacture of yeast extract simply yield a large pool of assorted intact peptides possibly required by some microorganisms poses an interesting question.

Whole Proteins. Whole proteins (e.g., albumin, blood, egg, meat, serum) are not known to be essential requirements for microbial growth, although they are an integral medium component for many nutritionally fastidious microorganisms. Clinical isolates of mycobacteria have been historically cultivated with Lowenstein–Jensen medium, containing whole egg and inspissated (coagulated by heating at 80–85 °C for 45 min) to form a solid matrix (39). Most mycobacteria, however, can be effectively cultured with Dubos medium or Middlebrook 7H9/or 7H10 media, in which the egg protein is replaced by the more defined components bovine serum albumin, catalase, dextrose, and oleic acid (40). Variations of chopped meat media (2) were for many years the recommended formulations for cultivation of pathogenic clostridia and similar bacteria, yet many of these strains give equal cell yield with the more defined, and more easily prepared, Reinforced Clostridial medium (40), although this basal medium frequently requires the addition of hemin and/or vitamin K_1 to satisfy the growth factor requirements of some strains; additionally, 0.5% cellobiose added to reinforced clostridial medium will enhance the growth of cellulolytic clostridia.

Defibrinated rabbit or sheep blood (5–10% v/v) is a common addition to media used for the cultivation of microorganisms pathogenic to animals. Human blood is not recommended for use for microbial culture work for health safety reasons and also because outdated blood obtained from banks is normally citrated. Citrated blood of any type should not be used in culture media because the anticoagulant may have negative effects on culture growth or hemolytic reactions. Although diagnostic hemolytic reactions differ between rabbit and sheep bloods, the ability to support microbial growth is generally the same. Human or sheep bloods, unless chocolated, are not used for the cultivation of *Haemophilus* spp. or other organisms requiring NAD because these fluids contain enzymes that inactivate the growth factor. Chocolate blood agar is prepared by heating the medium containing blood to 70–80 °C for 15 min. Purchase of whole animal blood should be from a reputable supplier and pooled from animals on an antibiotic-free diet. Well-collected and processed defibrinated blood is free from clots and cell lysis. Although commercial blood suppliers carefully collect their products, each lot of blood should be checked for sterility prior to use. Not all types of bacterial contaminants found in blood will grow well in either of the U.S.P.-recommended sterility test media (i.e., Fluid Thioglycollate and Soybean–Casein Digest). To broaden the screening procedure, the battery of media should be enlarged to include Brain Heart Infusion (BBL 11059 or Difco 0037) to enhance the recovery of somewhat fastidious contaminants, as well as HTYE (Trypticase peptone (BBL), 0.5%; yeast extract 0.2%; HEPES, acid, 0.4%; adjust to pH 7.2 ± 0.1 prior to sterilization at 121 °C for 15 min), which has wide versatility in the ability to cultivate diverse environmental heterotrophic organisms that do not grow well with conventional, general bacterial growth media. Whole blood should be stored under refrigeration (not frozen) and should be incorporated into media as soon as possible to help stabilize the fluid. Whole blood can not be stored longer than 2–3 weeks without significant deterioration. However, when stored at 4–8 °C, media containing blood can, for most media purposes, have a 4-month shelf life.

Serum (5–10% v/v) is an alternative to blood for the cultivation of pathogens. Horse, fetal, or newborn bovine sera are most frequently used for the cultivation of bacteria, whereas adult bovine serum is specified for the cultivation of many pathogenic protozoans. Sera vary in their intrinsic enzyme activity. For example, horse serum degrades glutamine, a necessary component in many animal cell line media formulations, and as already noted for chocolatizing whole blood, sheep and human blood can inactivate NAD. Preparation instructions for many media formulations call for serum heat inactivation (incubation of the serum at 56 °C for 30 min) to inactivate the protein complement. This temperature may not be high enough to eliminate adverse serum enzyme activity. Conversely, heat inactivation may not be a universally advantageous specification for all microbial growth and performance. A researcher should investigate types of sera as well as the effects of heat inactivation upon the performance of a microbial culture.

Whole proteins and their hydrolysates supply an eclectic and undefined mix of nutrients, detoxifiers, and other molecules that slowly release or absorb metabolites that may be required for cell nutrition in small amounts but quickly become toxic at elevated levels (e.g., fatty acids). Proteins hydrolysates supply a myriad of nutritional growth factors and help control physical parameters such as pH and osmolarity.

Be that as it may, the use of proteins and their derivatives in media formulations is becoming problematical with respect to biotechnology for two reasons: (*1*) Because many of the products of interest produced by microorganisms are proteins or substances with chemical affinity to proteins, presence of a complex proteinaceous medium component can significantly complicate a product purification process. (*2*) Biotechnology applications under regulatory compliance are facing increasing scrutiny about the potential presence of unidentified adventitious agents in media components. The international concern with bovine spongiform encephalopathy (BSE) is real and pervasive throughout the field of biotechnology, ranging from final product, to the type and source of medium components used in the preparation of a frozen or lyophilized master cell bank used in the manufacture of a product.

The European Commission decision 97/534/EC bans the importation of specified risk materials (SRM) into the European Union member countries (including Australia, Canada, and New Zealand). U.S. biotechnology companies wishing to deal with the European market must be able to provide a declaration to customs agents that assures their product does not contain, nor is derived from, the specified SRM as defined in 97/534/EC. The specified risk material, in this instance, is defined as "the skull, including the brain and eyes, tonsils and spinal cord of bovine animals over 12 months, ovine and caprine animals which are over 12 months or have a permanent incisor tooth erupted through the gum, and the spleen of ovine and caprine animals" (41).

Many of the peptone hydrolysates used in biotechnology are of animal (bovine) origin and are manufactured from "leftovers" of the slaughterhouse trade. As large-scale animal processing is not a clean operation, SRM contamination of raw material used in the manufacture of peptones is probable, particularly where source material is not directly traceable to government-regulated, clean herds.

The degree of concern with animal peptones with respect to 97/534/EC varies with industry and product. For example, from a regulatory perspective, the manufacture of biopharmaceuticals, especially direct injectables, demands high scrutiny for presence and elimination of any SRM. But the use of an animal peptone in a medium used solely for clinical diagnostic purposes (e.g., a peptone used clinically in MacConkey agar to differentiate *E. coli* from other normal intestinal flora), might be of lesser concern because the peptone never comes in direct contact with animals or humans at risk to BSE or similar prion agents. Acid-hydrolyzed animal peptones are generally of lower risk than enzymatic hydrolysates because the acid and heat associated with acid hydrolysis inactivates viral agents. Ultrafiltration of peptone solutions and sera has been proposed as a mechanism for eliminating possible BSE contaminants as well as reducing endotoxin levels. In any event, the concern with BSE and related animal/human pathogenic agents has caused an escalation of documentation requirements for the tracking of animal-derived media components. Certificates of origin for sera, meat, and milk peptones can, and should, be obtained from reputable vendors of these materials. Although certificates of analysis are now routinely supplied by raw material vendors to the biotechnology industry, certificates of origin are currently customer driven and are generally supplied only as requested.

BSE has fostered much current interest in the development of nonanimal peptones. With yeast, soy, or wheat as the source protein, some of these "veggie peptones" have interesting nutritional attributes but, nevertheless, still suffer the drawback of elevated intrinsic carbohydrate levels characteristic of plant peptones.

Difficulties in medium optimization and performance currently preclude wide use of serum-free or SRM-free media with animal cell lines. However, design of SRM-free media for most bacteria and yeast are feasible and cost-effective.

Vitamins. Vitamins are biochemical catalysts and appear in cellular metabolism as coenzymes or as the prosthetic group of enzymes. Vitamins elicit microbial growth response at very low levels, at nanogram concentrations for vitamins such as biotin and cyanocobalamin (vitamin B_{12}). Although many microorganisms are capable of synthesizing all of the required vitamins, many others are auxotrophic for one or more of these growth factors. The genus *Lactobacillus* characteristically has multiple vitamin requirements, and several strains within the group were widely used for many years in protocols for vitamin assays until replaced by instrumentation-based methodologies. The U.S.P. 23 still retains *Lactobacillus delbrueckii* subsp lactis (*L. leichmannii*) (ATCC 7830) as the assay organism for vitamin B_{12}. Marine bacteria frequently require B_{12}, whereas thiamine and biotin are necessary for the growth of many mycelial fungi. Menadione or its more active form, vitamin K_1, is a supplement for the cultivation of some clinical anaerobic bacteria. As noted with amino acids, a shift in medium composition can spare the need for a specific vitamin. *E. coli* (ATCC 10799; NCIB 8134) requires vitamin B_{12} when cultivated with a chemically defined minimal medium; when grown in the presence of 20 μg/mL methionine, the vitamin requirement is eliminated (42). Vitamin auxotrophies can appear as a need for the vitamin precursor, the intact vitamin, or as conjugates in coenzyme form. Koser noted, many years ago, that vitamin and amino acid auxotrophies in thermophilic bacilli are temperature variable (43). NAD and cocarboxylase (thiamine pyrophosphate) are coenzymes required by many pathogenic bacteria.

All of the vitamins are water soluble and heat stable at working concentrations used in media; coenzymes should be filter sterilized, as should concentrated vitamin stock solutions. Most vitamin stock solutions remain water soluble in significant concentrations, although biotin, folic acid, nicotinic acid, and riboflavin require neutralization with NaOH to retain solubility of the concentrate. Vitamin K and thioctic acid stock solutions are ethanol soluble. Ascorbic acid, used in microbial cultivation as an antioxidant, has a short life (about 3 weeks) in media; use of ascorbate salts rather than the free acid will improve stability. Table 2 lists a complete vitamin mix for the cultivation of bacteria and fungi. Yeast extract at a concentration of

Table 2. Balch's Vitamin Solution

p-Aminobenzoic acid	5.0 mg
Folic acid	2.0 mg
Biotin	2.0 mg
Nicotinic acid	5.0 mg
Calcium panothenate	5.0 mg
Riboflavin	5.0 mg
Thiamine HCl	5.0 mg
Pyridoxine HCl	10.0 mg
Cyanocobalamin	100.0 mg
Thioctic acid	5.0 mg
Distilled water	1.0 L

Mix all vitamins in distilled water. Adjust to pH 7.0, if necessary, to dissolve. Filter sterilize and store in brown bottle with refrigeration.

Source: From Ref. 44, p. 160.

0.1–0.5% is used as an undefined vitamin supplement for microbial growth in complex media; however, concentrations of yeast extract as low as 0.05% can elicit growth stimulation by organic growth factors, which may or may not be vitamins.

Nucleic Acid Precursors. Nucleic acid precursors are often required by nutritionally fastidious organisms when cultivated in defined or semidefined media. In some instances, the purine or pyrimidine will satisfy the requirement; more commonly, an organism needs preformed nucleosides or nucleotides. Organisms with these types of synthetic deficiencies will show a growth response to 5–50 μg/mL filter-sterilized additive.

Lipids. Lipids or their precursors are required by diverse organisms. Some pneumococci have a unique requirement for choline as a lipid precursor. *i*-Inositol is a structural component of some phospholipids. Inositol-deficient strains of *Neurospora crassa* and *Saccharomyces cerevisiae* have been used in studies of phospholipids and cell membrane structure. Fatty acids stimulate the growth of some fungi; certain mutants of *N. crassa* are auxotrophic for fatty acid synthesis (45). Rumen bacteria require short-chain fatty acids for growth. Many mycobacteria have a generalized requirement for exogenously supplied lipid. Various media formulations for nonglycerophobic mycobacteria include 2.0 mL/L glycerol, 0.05 g/L oleic acid, or 0.5 g/L Tween 80 to satisfy this requirement. Most of the bacterial mollicutes (anaeroplasmas, mycoplasmas, and spiroplasmas) require sterols. Cholesterol is a prevalent component in semidefined media formulations for these organisms, whereas fetal calf or horse serum supplies the growth factor in general-use complex mycoplasma media. *Eubacterium coprostanoligenes* (ATCC 51222) and *Eubacterium* sp. (ATCC 21408) represent a novel group of bacteria that use cholesterol and similar sterols as terminal electron acceptors (46,47). The former organism does not require cholesterol but metabolizes and requires phosphatidylcholine for growth, whereas the latter requires cholesterol as well as an alkenyl ether lipid for growth. These bacteria have been used in probiotic studies on decreasing the level of cholesterol in the digestive tract. All the lipids just mentioned are stable to autoclave temperatures. Short-chain fatty acids are water soluble; higher molecular weight lipids are soluble in less polar solvents such as ethanol or chloroform. In the presence of other media components at alkaline pH, some high-weight lipids can form growth-limiting soapy complexes. Separate sterilization of the fatty compounds helps to alleviate the problem.

Polyamines. Polyamines (e.g., putrescine, spermidine, spermine) are formed via the arginine pathway and have undefined functions in the cell. Research on polyamines has been done principally in vitro where the compounds bind to and neutralize charges on DNA and RNA. Although high concentrations of Mg^{2+} exert the same general ionic neutralization effect, only polyamines have the ability to influence correct tRNA conformation. Much of the cellular polyamine appears bound to ribosomes. In *E. coli*, putrescine and spermidine account for half the total product of the arginine pathway (48). Polyamine supplements as growth stimulants appear in media for the cultivation of *Ureaplasma urealyticum* (2p.530), *Haemophtlus parainfluenzae* (49), *Veillonella atypica* (50), and other bacteria. Where complex media are used, there is generally a sufficient polyamine content in yeast extract to satisfy the requirements of deficient strains.

Unique Growth Factors. Any organism has the possibility to, either spontaneously or through human intervention, develop mutations in biosynthetic pathways. Thus, unusual or strain-specific growth factors are always a consideration in the development of media for "unculturable" organisms. Deliberate mutation for the inability to synthesize the cell wall component diaminopimelic acid was engineered into early recombinant host strains of *E. coli* as an attenuation factor to reduce the possibility of survival outside of laboratory cultivation. Until recently, *Bacteroides forsythus* had only been culturable in cocultivation with other oral anaerobic bacteria, until its requirement for *N*-acetylmuramic acid was recognized (51). As noted earlier for the L-amino acids, the inability to synthesize functional signal and transport peptides may be limiting factors in the protein chemistry of some organisms.

PHYSICAL PARAMETERS

pH

pH is a strong selection factor for microorganisms. Although most bacteria have a wide pH range (±3.0 pH units) at which measurable growth occurs, optimum growth is limited to a narrow range. Minimum and maximum pH values can often be distinct and sharply defined. Members of the genus *Bacillus* will grow at pH 6.8 but not at pH 6.5. Extreme halophiles with a biomass yield optimum at pH 7.4 continue to grow slowly at pH 7.0 and exhibit little or no growth below pH 6.7. Eukaryotic and prokaryotic organisms shift their biochemistry according to the pH of their environment and thus change cellular characteristics as well as metabolic by-products. Such changes in cultivation parameters can be consciously directed toward a particular application such as secondary metabolite production. Unfortunately, the effects of pH shifts are too often not critically examined, particularly at the small-scale cultivation stage. For example, in an unbuffered test tube culture of anaerobic carbohydrate-fermenting clostridia, acid production can quickly drop the medium to below pH 4.0, well outside the pH minimum of 6.5 necessary for significant growth. Either the culture will sporulate (if medium conditions are conducive to the process) or viability quickly diminishes as acid concentration increases. Effective media design must be constructed around the pH at which the medium is developed and will be used. Radical adjustment to media naturally posed at a particular pH too often leads to precipitation of components or other problems with performance. Inclusion of buffers, as well as the types of buffers used, are important considerations in media design.

Incubation Temperature

Incubation temperature influences microbial metabolism both with respect to the rates at which cellular processes run and the rates at which nutrients are assimilated. Strains of *Serratia marcesans* produce pink-pigmented colonies at 26 °C but are colorless or slow to produce pigment at 37 °C. Temperature effects may sometimes be subtle. Bacteria alter the fatty acid compositions of their phospholipids according to temperature. The proportion of low-melting-point fatty acids increases as growth temperature decreases (52,53). Organisms may display fairly broad temperature tolerances within a range, but as with pH, optimum growth range is fairly narrow. Psychrophiles characteristically exhibit efficient metabolic rates at temperatures below 15 °C; the obligate psychrophilic bacterial species *Colwellia psychroerythrus* has a temperature optimum around 10 °C and lyses above 20 °C (54). Mesophiles are active between 20 °C and 40 °C. Free-living environmental strains are more tolerant of wider temperature ranges than are pathogenic strains that effectively function within only a few degrees of the body temperature of their hosts. Thermophilic microorganisms exist at temperatures above 40 °C. Although some fungi (e.g., *Mucor miehei* and *Sporotrichum thermophile*) can grow at temperatures as high as 50 °C, extreme thermophily, characterized by the ability to metabolize at temperatures above 70 °C, is restricted to members of the archaea and, to a lesser extent, the bacteria.

Oxidation–Reduction Potential

Oxidation–reduction (redox) potential of the culture environment determines the ability of a microorganism to initiate growth. Redox values of media and respiring cultures are measured with a pH meter set to the millivolt (mV) mode and a platinum electrode. Aerobic organisms require positive millivolt conditions. Obligate microaerophiles require 0 to −100 mV; clinical anaerobes less than −100 mV; obligate and extreme anaerobes less −200 mV. Actively metabolizing anaerobic cultures can drop redox levels below −300 mV. Media for obligate microaerophiles and anaerobes include reducing agents to lower the redox potential to a range optimal for a culture inoculum to be able to initiate metabolic activity. Subsequent cellular growth will maintain an overall redox condition reflecting the varying metabolic activities of the culture. Organic matter such as meat particles, yeast extract, and reducing sugars exert reducing effects in a medium. Filling a culture vessel almost to capacity with medium or adding 0.2% agar to the medium to retard diffusion of O_2 will be adequate to culture less fastidious microaerophiles. Anaerobic organisms with more stringent redox requirements need one or more chemical reducing agents included in the growth medium. Frequently used reducing agents, listed in terms of approximate redox potential in millivolts, include sodium thioglycollate (−100 mV), cysteine hydrochloride (−210 mV), titanium (III) citrate (−480 mV), sodium sulfide (−571 mV), and sodium dithionite (−600 mV). The overall redox level of a medium containing any of these reducing agents will, of course, be dependent upon the chemical interactions of all media components as well as the reducing agent. Media chemically reduced with any of these compounds should not be allowed to reoxidize, because the chemical structure of the reducing agents can be irreversibly altered by oxidation and lose effectiveness; in some instances, the oxidized forms of reducing agents can be toxic to cells (55). For some strict anaerobes cultivated under chemically defined conditions, it is best to add the reducing agent just prior to inoculation because reduced conditions are difficult to maintain under prolonged media storage. Ljungdahl and Wiegel (56), Breznak and Costilow (57), and Balch et al. (58) provide excellent detailed instructions for the cultivation of strict anaerobes.

Detoxifiers

Detoxifiers are required for the cultivation of some fastidious microorganisms. In the natural environment, metabolic by-products released by some organisms are enzymatically neutralized or utilized by other organisms for nutritional or other biochemical needs before the compounds accumulate to toxic levels. In vitro, for a pure culture lacking the appropriate enzymatic mechanisms to neutralize its own metabolic wastes, these metabolites can build to inhibitory concentrations before significant culture growth has occurred. The inability to break down hydrogen peroxide by organisms lacking catalase, and strain sensitivity to fatty acid accumulation are two common problems associated with metabolite accumulation in closed culture conditions. To cultivate these types of microorganisms, some media require the addition of detoxifiers, of which most are absorbents. Frequently utilized detoxifiers include blood/serum, bovine serum albumin (BSA, fraction V), catalase, and charcoal. More recently, liposomes have been utilized to slowly release lipid growth factors to deficient organisms, thus eliminating the toxic effects that a full "dose" of a medium lipid supplement would have for these organisms (59). Similarly, cylodextrins have been shown to effectively chelate and then slowly release potentially toxic, yet essential, nutrients for some clinical bacteria under semidefined growth conditions (60,61).

MEDIA DESIGN AND COMPOSITION

Optimal growth and performance of a microbial culture is most influenced by the design and composition of its growth medium. There is no one "best" medium for a particular application; however, there exist many examples of poorly designed and/or poorly manipulated media, or well-designed media being used for an inappropriate purpose, in which a microorganism manages to survive in spite of the medium and not because of it. Effective media design comes from an understanding of an organism's biochemistry and physiology, as well as the nutritional ecology of the environment from which the organism was isolated. Although isolation media are selective and frequently nutritionally suboptimal to limit the growth of unwanted organisms, general growth or maintenance media should provide conditions to sustain a culture at optimal morphology and physiology through repeated subcultures. Under microscopic examination, bacteria grown under favorable conditions do not have a misshapen or shriveled appearance; cell inclusions are generally minimal or absent; and cell lysis and/or spheroplast development is minimal or absent. At logarithmic phase of growth, motility is detectable for motile strains. Depending upon its application, a medium may be further defined in terms of chemical composition or optimized to increase biomass or cell product yield. An organism exhibiting optimal growth in a nutritionally balanced maintenance medium may not give high yields of a much-wanted metabolic by-product. Thus, medium optimization is a process directed toward a specific application. With so many nutritional and physical variables impacting upon microbial physiology, attempts at media optimization can become unnecessarily expensive, time-consuming, and largely unproductive when blindly tackled one variable at a time. Statistical methods using modified factorial approaches have been developed to aid in systematic media optimization for large-scale development (62), but for general laboratory purposes, knowledge and understanding of an organism's metabolic pathways and physical growth parameters can indicate specific areas in which quick design improvements to an existing formulation can effect significant results.

Media formulations may be chemically defined, in which discrete compound identity and concentration of all components are known, or chemically undefined, in which ingredients include one or more chemically complex substances such as protein hydrolysates and extracts. The choice of defined or undefined medium is dependent upon its application. Chemically defined media are useful in biochemical or metabolic studies of organisms. These types of formulations are also sometimes used in large-scale bacterial or yeast fermentations to produce products which further require stringent purifications such as those from recombinant DNA technology. General laboratory growth media for heterotrophic microorganisms as well as industrial media for large-scale production of metabolites (e.g., antibiotics, alcohols and other solvents) are often chemically complex. Scale-up of any medium formulation must include considerations of availability, cost, performance and lot-to-lot reproducibility of components as balanced against the value of the product. Chemically defined media can be expensive, particularly if the ingredients include a lengthy list of growth factors.

Microorganisms can transport only small molecules across their membranes. Large molecules (e.g., DNA, starch, proteins) can be nutritionally utilized by a microorganism only if it can synthesize and secrete the appropriate extracellular enzymes to digest large molecules into subunits small enough for assimilation into the cell. Because this type of enzymatic activity is absent from many organisms, in vitro microbial cultivation relies upon predigested macromolecules in the form of hydrolysates, peptones, and extracts and simple carbon compounds as nutritional substrates.

Protein Hydrolysates (Peptones)

Protein hydrolysates supply a mixed source of nitrogen and growth factors in the form of amino acids, nucleic acid fractions, peptides, and salts to a complex growth medium. For organisms possessing the metabolic capacity, protein hydrolysates can also serve as complex carbon and energy sources. Industrial large-scale fermentations using bacteria and fungi effectively use by-products from other industrial processes as proteinaceous material for media, including cottonseed, yeast, and soy. Although these substrates are chemically undefined, published analyses of their nutritional content greatly help in determining the appropriateness of a product for inclusion in a medium (63). Manufacturers of refined media and media components offer water-soluble products with more controlled nutritional composition. Table 3 lists and briefly describes the extracts and hydrolysates frequently encountered in microbiological laboratory media. Meat, milk, and soy commonly serve as starting material for the manufacture of hydrolysates. Digestion of these proteins can be by acid, alkali, or enzymatic hydrolysis. The source protein and method and degree of hydrolysis determine the nutritional characteristics of a hydrolysate. Prolonged hydrolysis releases free amino acids; more gentle hydrolytic processes yield mixtures of amino acids and peptides. The strong conditions of acid or alkali hydrolysis destroy much intrinsic vitamin content of a raw protein, as well as some amino acids such as tryptophan, and reduce the levels of serine and threonine. Acid-hydrolyzed protein hydrolysates are high in salt content formed during the neutralization step; however, some manufacturers feature acid-hydrolyzed peptones with additional processing to reduce the amount of salt [e.g., Casamino Acids (Difco 0230)]. Enzymatic digests contain free amino acids as well as small peptides while maintaining the vitamin content of the original source material. Enzymatic casein digests, reflecting the amino acid composition of the starting material, are high in tryptophan and low in cystine. Peptones from plant proteins [i.e., Soytone (Difco); Phytone (BBL)] contain intrinsic carbohydrate and, while conducive to rapid heterotrophic growth, can cause quick culture decline due to substantial metabolic acid production. Because of the carbohydrate level, plant peptones are also unsuitable in basal media for fermentation or sole carbon source utilization studies. For routine growth, peptones are normally added to media at concentrations of 0.5–1.0%. Growth of microorganisms with diverse and multiple nitrogenous auxotrophies (e.g., the lactobacilli) is enhanced by elevated peptone concentrations as well as nutritional supplementation with peptones of mixed sources and degrees of hydrolysis.

Carbon and Energy Sources

Microorganisms other than autotrophs require organic compounds as a source of carbon and energy. These can include simple sugars, complex carbohydrates, alcohols, amino and other organic acids, and short-chain lipids. Some organisms (e.g., *Aspergillus, Pseudomonas*) possess the complex metabolic machinery that permits numerous classes of organic compounds for carbon/energy substrates. Other taxa exploit specific classes of molecules metabolically ignored by or toxic to general heterotrophs. Methylotrophs (*Methylobacterium* spp., *Methylomonas* spp., *Methylophaga* spp., and others) can extract energy from C_1 compounds such as methanol and formaldehyde, whereas

Table 3. Hydrolysates and Extracts

Hydrolysis method	Source	Product/manufacturer[a]	Characteristics
Acid	Casein	Acidicase (BBL) Casamino acids (Difco) Hy-Case (Sheffield)	Free amino acids; deficient in tryptophan and cystine; high intrinsic salt concentration unless designated salt-free
Enzymatic	Casein	Casitone (Difco) NZ-Amine (Sheffield) Trypticase (BBL)	Free amino acids and small peptides; high levels of tryptophan; low in carbohydrates
Enzymatic	Meat	Myosate (BBL) Thiotone (BBL) Primatone (Sheffield) Peptone (Difco) Proteose peptone (Difco)	Mixtures of amino acids and peptides; those designated proteose are hydrolyzed for a high peptide content; Thiotone high in sulfur
Enzymatic	Plant	Phytone (BBL) Soytone (Difco) Hy-Soy (Sheffield)	High in carbohydrates and vitamins
Mixed	Yeast/casein	Biosate (BBL)	Combines amino acids as well as vitamins lost during protein hydrolysis of peptones
	Casein/meat	Polypeptone (BBL)	
Autolysis	Yeast	Yeast extract (BBL; Difco; Sheffield)	High in amino acids, peptides, water-soluble vitamins, and carbohydrates

[a]Items listed are representative of of a manufacturer's product line. Consult manufacturers' technical manuals for more information, Refs. 39, 40, and 64.

the metabolic diversity of the bacterial genus *Pseudomonas* has allowed selection of strains that can utilize the long-chained hydrocarbons in crude oil or attack the degradation-resistant chemical rings of cresols and phenols as sole carbon/energy sources. Assorted fungal genera (*Cladosporium, Candida*, and others) have been found as actively growing contaminants in aircraft and diesel fuels (65). The type and concentration of a carbon substrate present in a medium will affect overall growth rates, morphology, and metabolites of a culture. Sole carbon utilization profiles are used in identification matrices for prokaryotes and yeasts.

Carbohydrates and related compounds are preferential carbon sources for many genera of microbes. Hexoses are most diversely utilized, followed by the disaccharides, trisaccharides, smaller polysaccharides (starch and glycogen), and pentoses. Glycerol, mannitol, and sorbitol are commonly utilized alcohols. Under laboratory conditions with defined media and mixed sugars at normal substrate levels (e.g., 2–10 g/L), microorganisms often exhibit diauxic growth, in which a culture first utilizes the carbon compound that supports the greatest rate of growth, with concurrent catabolic repression of other substrates (66–68). Enzyme systems then shift accordingly as the organism switches to and consumes each less-preferential carbon compound. This results in a biphasic growth curve. However, under carbon-limited (e.g., microgram to milligram per liter concentrations) chemostat conditions that mimic the low-nutrient conditions of the natural environment, a number of organisms use the components of carbon mixtures simultaneously (69).

Proteolytic organisms have the capacity to utilize L-amino acids and related compounds as sole carbon/energy sources. This metabolic characteristic is more common among bacteria than fungi and varies widely between species; identification keys for the genus *Clostridium* use saccharolytic versus proteolytic metabolisms for species differentiation. As a group, the helical/vibroid Gram-negative bacteria (e.g., *Aquaspirillum, Azospirillum, Camplylobacter, Oceanospirillum*) do not ferment or oxidize carbohydrates but satisfy carbon/energy requirements with amino acid catabolism. *Legionella* spp. also use amino acids as carbon/energy sources. Filamentous fungi vary in their ability to utilize amino acids as effective carbon sources; none, however are limited to nitrogenous sources of carbon. Glutamic acid, alanine, arginine, and proline are frequently utilized by many prokaryotes and yeasts. The aromatic rings of the amino acids tryptophan, phenylalanine, and tyrosine, as well as the purine and pyrimidine rings of the nitrogenous bases, are not readily attacked by many organisms. The ability to do so is used for taxonomic speciation. Degradation of any of these ring compounds, however, is not presumptive of an organism's ability to further use the breakdown products as a sole source of carbon/energy; in many cases, the compound is not so metabolized. Organisms that catabolize amino acids frequently utilize organic acids in the same fashion (e.g., *Oceanospirillum*). In addition, many carbohydrate-utilizing organisms are also capable of utilizing organic acids. For example, citrate and propionate utilizations are classical tests in the identification of the carbohydrate-metabolizing genus *Bacillus*. When used as sole carbon and energy sources, amino and organic acids may not be soluble at required substrate levels (2–5 g/L); however, neutralization with KOH or NaOH will form soluble salts of these compounds.

Some organisms have the capacity to utilize fatty acids and other nonpolar molecules as carbon and energy sources. In some instances, the requirement is absolute. The bacterial genus *Leptospira* requires long-chain fatty acids (70), which are supplied in complex growth media by rabbit serum. The anaerobic sulfate-reducing *Desulfovibrio* sp. (DSM 2056) oxidizes the fatty acids butyrate through stearate (71).

Simple sugars and other carbon compounds can undergo chemical changes when autoclaved with other media components. Amino acids and organic acids, though heat stable at growth factor levels, may degrade at substrate concentrations. The characteristic darkening of autoclaved media containing glucose and peptones, called the Maillard reaction, is caused by an interaction of amino acids with the glycosidic hydroxyl group of the sugar. Darkening of autoclaved media can also be caused by the interaction of mono- and disaccharides, particularly glucose, with phosphates; the reaction is more intense at alkaline pH. Carlsson et al. (72) demonstrated that the glucose–phosphate reaction produces hydrogen peroxide at levels toxic to catalase-negative organisms. These types of adverse chemical reactions between nutrient components compromise the performance of a medium and are to be avoided either by autoclaving or filter sterilizing the reactive compounds separately from the balance of the medium. Not all commonly used carbon substrates can withstand autoclave sterilization. In particular, arabinose, fructose, lactose, pyruvate, ribose, and xylose should be filter sterilized. Light-weight alcohols and similar solvents are volatile at autoclave temperatures. Autoclaving of heat-stable alcohols can be done in tightly stoppered containers; alternatively, these compounds can be sterilized with appropriate solvent-resistant membrane filters. Long-chain hydrocarbons (e.g., oils and waxes) are hydrophobic and are not sterilizable by autoclaving because the sterilant (steam) cannot adequately penetrate the material. A good chemical reference text, such as the *Merck Index* (73), is invaluable for determining the heat stability, solubility, and other important characteristics of many chemical components used in media formulations.

Extracts

Spray-dried powders derived from infusions of meat, yeast, or other chemically complex proteinaceous materials provide a nutritionally useful but undefined source of water-soluble growth factors and sugars. Because they are not subject to the same harsh conditions as those used in the manufacture of peptones, meat or yeast extracts can supply those nutrients lost during the peptone hydrolysis (e.g., vitamins and certain amino acids). Extracts, in themselves, do not contain sufficient concentrations of essential nitrogen or carbon compounds to effectively function as the sole source of these nutrients in a medium. Meat extracts include appreciable amounts of glycolic and lactic acids, as well as creatine. Yeast extract is high in carbohydrates and, therefore, interferes with fermentation studies. The nutritional properties of the fresh yeast extract solution used in many media formulations for the cultivation of mycoplasmas is not replaceable by powdered yeast extract. Malt extract used in some media for fungi is composed primarily of carbohydrates (glucose, fructose, sucrose, maltose, dextrins) and small amounts of nitrogenous matter. Powdered extracts are generally incorporated in media at a concentration of 0.5–1.0%.

Aqueous extracts and infusions of various materials have historically been used to supply undefined nutritional factors in laboratory media. Concentrated liquid extracts generally are used at 10–20% v/v concentration as a medium supplement, whereas infusions will constitute the entire liquid base of a medium to which additional nutrients are added. As already noted, a filtered, aqueous solution prepared from autolyzed baker's yeast (150 g yeast/L) is a necessary ingredient in media for mycoplasmas and other mollicutes. The fresh yeast extract solution is then filter sterilized and incorporated into the mycoplasma media at 10% v/v concentration. Lactic acid bacteria from milk or milk products have been cultivated with infusions prepared from whole milk, whereas tomato juice or other fruit juice extracts have been used at 10–20% v/v in media for lactic acid bacteria isolated from plants. Potato infusions (200–300 g potatoes/L) serve as a nutritional base in many media for the cultivation of bacterial and fungal plant pathogens. Soil extracts prepared from fertile, chemically uncontaminated soils have been used to induce sporulation in many bacteria. Rumen fluid, containing partially digested nutrients from the first compartment of a cow's stomach, is a supplement in media for the cultivation of rumen microorganisms. Rumen fluid contains an undefined mixture of fatty acids, carbohydrates, peptides, vitamins, and metabolites. As rumen fluid is most commonly added to and autoclaved with basal medium components, any heat-labile nutrients are destroyed. Rumen fluid is obtained from fistulated dairy cows and can often be procured gratis from universities with departments of animal or dairy nutrition. Rumen fluid may or may not contain significant amounts of solid material, depending upon the length of time the material has been in the rumen. Centrifugation will effectively remove the bulk of debris; however, a series of filtrations down through 0.2 μm is necessary to remove dead cells of indigenous microflora that can confuse microscopic examinations of the organism being cultured. A defined mixture of fatty acids (Table 4) can effectively replace the need for rumen fluid for many bacteria; however, as with most chemically complex, undefined supplements, cell yield is frequently better when it is included in the medium.

Buffers

pH buffers are molecules that, in solution, resist fluctuations in hydrogen ion concentration. Weak acids, bases, and their corresponding salts in partial dissociation function as pH buffers. Buffering capacity is greatest when the

Table 4. Volatile Fatty Mix for Rumen Bacteria and Fungi

Acetic acid	17.0 mL
Propionic acid	6.0 mL
N-Butyric acid	4.0 mL
N-Valeric acid	1.0 mL
Isovaleric acid	1.0 mL
Isobutyric acid	1.0 mL
DL-α-Methylbutyric acid	1.0 mL

Pipette the given volumes of fatty acids into a container that has a tight-fitting closure and mix well. The objectionable odor of the free fatty acids can be reduced by neutralizing the pH of this solution with the addition of NaOH pellets. Use 3.1 mL of the fatty acid mix per liter of medium.

Source: From Ref. 44, p. 162.

compound is at 50% dissociation. The pH at which this occurs is equal to the pK_a value of the compound, and greatest buffering capacity falls between ±1 pH unit of the pK_a. Thus, no one buffer can be used across the entire pH range of 1 to 14 (74). Some acids are zwitterions (compounds that can chemically behave as acids or bases) and have more than one pK_a. In an extreme example, glycine when titrated against HCl has pK_a 2.3; titrated against NaOH, pK_a 9.6. Other zwitterions include citric acid (pK_{a1} 3.06, pK_{a2} 4.74, pK_{a3} 5.40), succinic acid (pK_{a1} 4.19, pK_{a2} 5.57), and the Good buffers (see later). The perfect media buffer would be (*1*) completely effective at poising a desired pH throughout the metabolic flux of a culture's growth cycle, as well as any other chemical or environmental changes that may affect pH (e.g., metal ion interaction, temperature), (*2*) resistant to microbial enzymatic activity, (*3*) chemically inert with respect to other medium components, and (*4*) inexpensive. The perfect buffer does not exist.

Lacking perfect buffers, in vitro microbial cultivation relies, for better or worse, upon a standard array of buffering systems. Nevertheless, the choice of buffer must not be made haphazardly. No discussion of buffers could be complete without N. E. Good's comment: "It is impossible even to guess how many exploratory experiments have failed, how many reaction rates have been depressed, and how many processes have been distorted because of imperfections of the buffers employed" (75). The capacity to buffer well at a particular pH is only one of many criteria in the determination of an appropriate buffer for a particular application. Chemical interaction with other media components or culture metabolites will compromise both the buffering capacity as well as the nutritional balance of a medium. The effectiveness of a microbially metabolized buffer diminishes with culture growth. The pH of some buffers can vary, sometimes considerably, with concentration or temperature. Synthetic buffers, while very effective for bench-top laboratory needs, may become prohibitively expensive upon scale-up.

Natural Buffering Agents. Natural buffering agents are those compounds most frequently encountered in microbiological media; they are also the buffers subject to the greatest drawbacks in their use. Carboxylic acids (e.g., acetate, citrate, succinate) are effective buffers for the lower ranges (<pH 6) of biological activity. These compounds are metabolically catabolized for carbon and energy by many microorganisms. Also, at the concentrations needed for buffering, carboxylic acids can become cation chelators. The mixtures of amino acids in protein hydrolysates and extracts have significant buffering capacity. Thus, peptone-based complex media frequently do not require additional buffering agents when used at or about the initial pH of the medium. Many peptones are adjusted during manufacture to pH 6.8–7.0. Peptone solutions are stable to sterilization at 121 °C; however, they adversely interact at elevated temperatures with other medium components such as glucose, phosphates, and polyvalent cations. In addition, because peptones are nutrients, buffering capacity diminishes or shifts as the amino acids are microbially consumed.

Sodium bicarbonate–carbon dioxide buffers can maintain pH in the range 6.0–8.0. However, when used in conjunction with intrinsically buffered media, bicarbonate buffers have a working range of pH 6.8–7.2. Despite the necessity for closed systems to retain the gaseous CO_2 phase, bicarbonate buffers are widely used in the cultivation of anaerobic bacteria and eukaryotic animal cell lines. Bicarbonate buffers are nutritionally consumed. Also, carbon dioxide has minimal solubility in aqueous solutions, which somewhat limits its buffering capacity. Solid calcium carbonate (powdered) is used in various media for strongly acidifying organisms such as *Acetobacter, Clostridium*, and *Lactobacillus*. Ethanol-tolerant, acid-producing acetobactors can be isolated and differentiated from mixed-culture fermentations using a carbonate-ethanol medium (yeast extract, 3%; calcium carbonate, 2%; agar, 2%; filter-sterilized ethanol, 2%). Strongly acidifying acetobacter isolates able to grow in the presence of ethanol will also dissolve the carbonate and form a zone of clearing around the colony.

Although widely used in media formulations at pH 6–8, phosphate buffers suffer from a great number of drawbacks. Unless a medium is carefully designed and prepared with respect to metal speciation and concentration, phosphates will adversely bind to and precipitate polyvalent cations and proteinaceous material at neutral to alkaline pH. This chemistry creates two problems: (*1*) buffering capacity is diminished as cations complex with the phosphate anion, and (*2*) cations bound as precipitates are unavailable as nutrients. These unwanted chemical interactions are intensified by heat. The problem can be reduced by autoclaving the phosphate buffer separately from the balance of the medium and then combining both solutions when cooled to room temperature after autoclaving. The addition of a chelator such as EDTA may retard trace metal precipitation; however, overchelation will also restrict the availability of trace metals from the nutrient pool. Above pH 7.2, with poorly designed media formulations, precipitation can become spontaneous even at room temperature; the reaction may be immediate or may slowly develop over time and is often misinterpreted as bacterial contamination. Balanced phosphate-buffered salines (e.g., Dulbecco's, Hank's) containing calcium and magnesium salts are filter sterilized to avoid precipitation of the salts. When autoclaved together, phosphate and glucose will also adversely react at neutral pH and above; trace metals will intensify the reaction. Because soluble phosphates are nutrients, the effectiveness of these buffers can diminish with culture metabolism. Finally, phosphate buffers exhibit a concentration effect characterized by a rise of 0.2–0.3 pH unit with each 10-fold dilution.

Synthetic Buffers. Synthetic buffers were developed to alleviate the many problems encountered with the earlier generations of natural buffers. Tris [Tris(hydroymethyl)-aminomethane; pK_a 8.3 at 20 °C] was the first organic compound widely utilized for its improved buffer characteristics. Tris and Tris HCl can be combined for buffering between pH 7.0 and 9.0, with strongest activity between 7.2 and 8.0. Tris can also be paired with other acids such as acetate, borate, or phosphate for specific buffering applications. Tris does not precipitate calcium ions and helps to maintain the solubility of magnesium salts. Tris is, however, a chemically reactive amine and can penetrate cell membranes; some microorganisms (e.g., *Sporocytophaga myxococcoides* ATCC 10010) have shown growth inhibition by the compound. Tris exhibits marked concentration and temperature effects that can limit its effectiveness for certain biological purposes. Tris buffers fall approximately 0.1 pH unit per 10-fold dilution. At 0.05 M, a Tris buffer at 25 °C will be pH 7.20; at 5 °C, pH 7.76; at 37 °C, pH 6.91. Therefore, when used in culture media, Tris buffers should be constructed with the incubation temperature of the culture in mind and not upon the pH of the buffered medium at room temperature.

During the 1960s Good et al. (75) developed a number of zwitterionic compounds for use in biological systems to include as many as possible of the following desired buffer characteristics: (*1*) freely soluble in aqueous solutions to permit preparations of concentrated stock solutions, (*2*) buffering optima within the range of most biological systems, (*3*) biochemically nonreactive and chemically stable, (*4*) inert to microbial or other enzymatic attack, (*5*) minimal temperature and ionic effects upon buffer pH, and to further enhance their versatility to include postculture analytical procedures, (*6*) insoluble in organic solvents and (*7*) transparent to the ultraviolet wavelengths used in many biological spectrophotometric assays. Collectively these compounds have come to be known as the Good buffers, of which, HEPES (pK_a 7.5 at 25 °C) is most commonly recognized due to its prevalence in media formulations for the cultivation of animal cell lines. Other Good buffers useful for microbial media formulations include ACES (pK_a 6.8 at 25 °C), which is a component in the recommended cultivation medium for *Legionella*; buffered charcoal yeast extract (BCYE) (2p.518); MES (pK_a 6.1 at 25 °C), PIPES (Pk_a 6.8 at 25 °C), and TES (pK_a 7.5 at 25 °C).

HEPES has been implicated in the phototoxicity of RPMI medium for the cultivation of eukaryotic cells; however the effect was eliminated when sodium pyruvate was added to the culture medium (76). Other Good buffers have also shown various types of inhibition in certain cell lines (77). As with any buffering system or other media component, the Good buffers should be evaluated for applicability against the microorganism under study.

Biological buffers for cell growth purposes are commonly used at 0.01–0.05 M; chemical effectiveness of the buffer, initial pH of the medium without the buffer, production of acid or basic metabolites during culture growth, and enzymatic or other reagent optima are considerations in determining appropriate buffer ion concentration. Buffer tables can be found in numerous publications. Lacking a table, buffer solutions can be prepared with either of the two following methods and a calibrated pH meter:

1. Prepare a 2× solution of the buffer compound. Using HCl or NaOH, titrate the solution to the desired pH. Bring the solution to 1× volume with deionized or distilled water.
2. Prepare equimolar solutions of the acidic member and the basic member of the buffer pair. The molarity of these solutions is that of the final concentration needed for the working buffer. Choosing the solution of the buffer pair with a pH closest to that needed for the working buffer, titrate the solution to the desired pH with the equimolar solution prepared with the remaining component of the buffer pair.

Gelling Agents

Many microbial applications, such as strain isolation and purification, necessitate the use of solid media. Growth of aerobic organisms, in addition to specific morphologies, are often enhanced on solid media providing unrestricted gas exchange. The different gelling agents available to microbiologists each have attributes and drawbacks that one must consider before choosing a particular gelling agent to include in a medium formulation. Although agar is the most frequently encountered gelling agent in microbial culture media, it must be properly used in order to maintain the chemical and nutritional integrity of a formulation. Many chemolithotrophic organisms are inhibited by organics and will not grow on media solidified with agar. Conversely, a number of marine bacteria are agarolytic and can quickly dissolve or soften this support matrix. Other gelling agents, including carrageenan and Gelrite, are alternatives to agar, but each has its own chemical peculiarities that must be understood before they can be effectively used.

Agar. Agar is a polysaccharide complex extracted from a number of species of red algae (*Gelidium, Gracilaria*, and others) found in the Pacific and Indian Oceans and the Japan Sea. The agar complex can be separated into (*1*) agarose, a neutral gelling fraction, and (*2*) agaropectin, a sulfated, nongelling fraction. Depending upon the purity and gel strength of the agar, and also upon the chemical composition of the medium in which the agar is used, a molten 1.5% agar solution will set at temperatures between 32 and 39 °C. However, agar media with high concentrations of salt (e.g., those for the extreme halophiles, which contain >18% NaCl and 0.5% each of $MgSO_4$ and KCl) gel quickly at temperatures above 55 °C. The gelling capacity of agar is thermally reversible, however, repeated remelting of agar-containing media is not recommended because the chemical stability of many nutrients cannot withstand repeated exposure to elevated temperatures. Agar is subject to acid hydrolysis, which destroys the gelling capacity of the polysaccharide; the phenomenon is intensified by heat. Agar-containing media autoclaved at less than pH 5.0 will not gel when cooled. If a low-pH agar is required, it is advisable to prepare and sterilize the medium in two separate solutions: For 1 L of low-pH agar medium, (*1*) heat and dissolve agar, with stirring, in 500 mL water; (*2*) dissolve remaining medium components in 500 mL water; (*3*) sterilize each solution separately at 121 °C; (*4*) cool each solution to approximately 50 °C; (*5*) aseptically combine both solutions, mix well, and dispense into containers as required.

Agar products vary with respect to nitrogenous and inorganic salt contaminants. A number of manufacturers feature agars purified though repeated water and solvent washes. Some agar products contain significant amounts of metals that can bind to and precipitate with phosphates in the medium, resulting in diminished growth yield with the solid medium as compared with that of the corresponding liquid medium. Similarly, autoclaving any agar in the same solution with the carefully balanced nutrient salts of a chemically defined medium, particularly at neutral to alkaline pH, will frequently give rise to a pronounced darkening or haze that can compromise the effectiveness of the medium. Under chemically defined conditions, component integrity is best preserved by autoclaving a purified agar separately from the autoclaved or filter-sterilized chemical portion of the medium, as already outlined for preparation of a low-pH agar medium. The use of electrophoretic grades of agarose would be more suitable than purified agars for gels in which a minimal level of trace metal contamination is required.

As with all gels, agar exhibits syneresis in that it will squeeze out solute as it contracts. The small puddle of liquid at the bottom of a freshly prepared agar slant is an example of this phenomenon. The amount of syneresis is dependent upon the concentration of the agar in the gel. Agar media can also lose water through desiccation during storage. Once water has been eliminated from the agar gel, it will not reenter the matrix unless the gel is remelted; however, melting of solid media for the purpose of rehydration not recommended because performance of the medium will be markedly reduced.

Agar is normally used at concentration from 1.0 to 2.0% for solid media; at 0.075 to 0.2% to create a viscosity that retards oxygen diffusion in media such as fluid thioglycollate medium U.S.P. and the classical oxidation–fermentation (O-F) medium of Hugh and Leifson (78); and at 0.5 to 0.75% for soft, top-layer agars in media for bacteriophage propagation.

Agarose. Agarose, the purified gelling fraction of agar, is used in some formulations for organisms that are inhibited by agar. Purified agarose products, particularly those grades used for electrophoretic work, are low in inorganic and organic contaminant and, therefore, react less with the phosphates present in a medium. Gel clarity is greater than for agar. Low gelling temperature agaroses melt above 65 °C and form gels at less than 30 °C. Agarose will form solid gels at 0.75–1.0%.

Gellan Gum. Gelrite, a gellan gum, is a heteropolysaccharide composed of glucuronic acid, rhamnose, and glucose, produced by bacterial fermentation, and commercialized as a gelling agent by Kelco (Merck & Co., Inc., Kelco Division, San Diego, Calif.). Sigma Chemical Co. (St. Louis, Mo.) distributes the material under the trade name Phytagel. Under either name, this gellan gum is somewhat difficult to work with for routine media solidification purposes, but nevertheless, its unique characteristics make the material the most practical solution to some unusual microbial cultivation requirements. Gelrite/Phytagel is a high-melting-temperature (~70 °C) gel, which makes it useful for the cultivation of thermophilic bacteria and fungi at 45–50 °C, temperatures at which agar shows considerable syneresis and softening. The gel is generally thermally reversible, although under some conditions, such as the elevated soluble salt concentrations of marine media, gelling can become permanent. Gelrite/Phytagel is subject to acid hydrolysis when autoclaved at low pH. The gellan is resistant to microbial attack by most organisms and so proves useful in the isolation of organisms such as the marine cytophagas, a group in which many strains characteristically possess the ability to degrade other biopolymer gells including agar, algin, carrageenan, and gelatin. Many chemolithotrophic organisms (e.g., *Nitrobacter*, spp. and *Thiobacillus ferroxidans*), although inhibited by agar and other organic gelling agents, can be grown successfully on a solid medium using the gellan. Full gelling capacity of this polysaccharide is cation dependent, and media formulations may need amending to include at least 0.1% $MgSO_4 \cdot 7H_2O$, or other magnesium salt to permit sufficient gelling. Under chemically defined media conditions, further chemical adjustments to other components may be necessary to retain chemical balance because of a modification in cation level. Recommended concentrations for Gelrite/Phytagel are in the 0.1–0.2% range, however experimentation will probably be needed to select the appropriate concentration for one's own needs. Table 5 provides a media formulation that has been successfully used as a solid medium for the cultivation of laboratory strains of *T. ferroxidans* (ATCC 21834 and others); the gel is rigid enough for the isolation and picking of colonies, and growth

Table 5. Solid Medium for *Thiobacillus ferroxidans*

Solution A	
$(NH_4)_2SO_4$	0.08 g
$MgSO_4 \cdot 7H_2O$	0.20 g
K_2HPO_4	0.04 g
Trace elements (see Table 1, mix A)	0.5 mL
Distilled water	40.0 mL

Adjust Solution A to pH 2.3 with concentrated H_2SO_4. Filter sterilize.

Solution B	
Phytagel (Sigma P-8169)	0.5 g
Distilled water	40.0 mL

With constant stirring, heat Phytagel to boiling to dissolve. Sterilize at 121 °C for 15 min. Cool to no less than 80 °C.

Solution C	
$FeSO_4 \cdot 7H_2O$	2.0 g
Distilled water	20.0 mL

Quickly dissolve ferrous sulfate with constant stirring. Immediately filter sterilize.

Complete medium

Warm solution A to about 50 °C. Add solution C to solution A and immediately combine with solution B. Mix well and pour plates. As this medium must be poured at a rather high temperature to avoid premature gelling of the Phytagel, it will be necessary to dry the plates before use by leaving the lids ajar in a sterile environment or by incubating them at 37 °C overnight to eliminate excess condensation.

is rapid (3–5 days) at 20–30 °C. Preparation methods for this particular formulation illustrate many of the topics discussed in this article with respect to treatment of chemically incompatible salts, acid hydrolysis, and heat instability of media components.

Alginic Acid and κ-Carrageenan. Alginic acid (79) and κ-carrageenan (80,81) have been used microbiologically for specialized gelling needs. Gelling efficiency of alginate is Ca^{2+} dependent, whereas κ-carrageenan is K^+ or Na^+ dependent. Since all of these substances are easily degraded by assorted bacteria and fungi, their use as gelling agents is restricted. Microbial degradation of these substances is used for taxonomic purposes.

Silica Gel. A final solid substrate for the laboratory cultivation of obligate chemolithotrophic organisms is silica gel, first used by Weintraub and Hanks (82) in 1936 for the isolation of ammonia-oxidizing bacteria. Many authors have since attempted improvements on the tedious method (83–86). Preparation of any quantity of plates is laborious. Briefly, sodium silicate is reacted with strong HCl to produce silicic acid, NaCl and almost immediate gelling. Nutrient salts may be included in the formulation, or the surface of the silica gel may be later flooded with nutrients, with some absorption. Because the amount of NaCl is high (~8–10%), the gel may first need clearing of the salt, with repeated washings with water prior to flooding with a nutrient solution. Silica gels are translucent and rigid with a high degree of syneresis. The gels are autoclavable, with a tendency for cracking, and are not reversible. Table 6 provides simplified instructions for silica gel preparation adapted from Ref. 82.

Table 6. Simplified Silica Gel Plates

Stock solutions

Solution 1	
Concentrated HCl (~37%)	14.5 mL
Distilled water	to 100.0 mL
Solution 2	
$Na_2SiO_3 \cdot 9H_2O$	20.0 g
Distilled water	100.0 mL

Use slight heat and constant stirring to dissolve sodium silicate. Filter out any particulates or colloidal material with Whatman #1 paper.

Preparation of silica gel plate

1. Volumes given are for one 20 × 100 mm petri dish. Gelling will occur within 10–15 s of mixing solutions. Be prepared to work quickly.
2. Add 14.5 mL solution 1 to a 100-mL beaker and stir continuously. (A magnetic stirrer will work best.)
3. Quickly add 19.0 mL solution 2 into the beaker and mix well.
4. As soon as effervescence stops, quickly pour contents of beaker into a 20 × 100 mm petri dish. Use a glass dish if subsequent autoclaving is required.
5. Initial gelling will occur almost immediately. Gel strength will increase slightly over a half-hour period, at which time the gel may be washed to remove excess NaCl and then flooded with nutrient solution.

MEDIA STERILIZATION

With few exceptions, media and buffers used for microbial cultivation must be sterilized as soon as possible after preparation to limit proliferation of environmental contaminants that will alter the nutritional characteristics of the medium either by consumption of nutrients or by production of metabolites. Even nonnutritive buffers or trace metal stock solutions can give rise to unwanted fungal growth if stored unsterilized for any length of time at room or refrigerator temperatures. Sterile media and buffers are necessary for culture propagation; the presence of microbial contaminants, even at a low level, can invalidate any study or production process. The physical stress of sterilization, however, can compromise the nutritional and physiological characteristics of a medium or buffer. Consequently, one must choose a sterilization method that has the least negative impact upon the material being sterilized. Although steam sterilization had historically been the only form of sterilization accessible by many laboratories, the widespread availability and diversity of reasonably priced and reliable disposable filtration units offer more appropriate choices for media sterilization, even for small laboratories.

Steam Sterilization (Autoclaving)

Steam remains the most widely used sterilization method for rugged, heat-stable media and has the advantage of often being the terminal step in media preparation. With no additional manipulation of an uninoculated medium poststerilization, possibility of inadvertent introduction of contamination is reduced. The term *autoclaving* is often used synonymously for the definitional procedure: "sterilize at 121 °C for 15 minutes." This phrase *must* be applied in context to the volume of the medium being sterilized as well as to the efficiency of the autoclave being used. The precise meaning of the phrase infers that the contents of each container within an autoclave cycle must achieve an uninterrupted exposure to 121 °C for 15 min; it does not mean that the autoclave itself was set for that temperature and time. From a practical laboratory standpoint, this means that small volumes (e.g., test tubes) will require the minimal amount of autoclaving time, whereas large volumes (e.g., one or two flasks) will require additional autoclaving time. Fully loaded autoclaves may require longer run times. Under laboratory autoclave conditions, a 30–35 min sterilization cycle for 0.5 L liquid medium in a 1-L flask is a practical guideline; agar media that have been allowed to solidify before sterilization may require longer autoclave times. Large-scale fermentation volumes can require hours to sterilize; media processed in poorly maintained autoclaves might never be adequately sterilized. Where mechanical agitation of a medium is not possible during the autoclave cycle, maximum container volumes of media should not exceed the ratio of >1 L media in a 2-L vessel because the prolonged autoclave times required for heat conduction to raise the innermost portions of large volumes of media to 121 °C will subject the outer portions of the media to extreme overheating.

Autoclaves and their cycles should be validated for media sterilization. Ideally, this would include replicate autoclave run data from multiple thermocouple probes spaced strategically within the autoclave chamber and load configurations. While this technological expense is beyond the budget of many small research laboratories, quantified preparations of *Bacillus stearothermophilus* sterilization indicators, such as the Verify Biological Indicators (Steris Corporation, Mentor, Ohio), can indicate general "121 °C, 15 min" parameters. Chemical indicator strips or labels are not reliable indicators of sterilization and should be used only as visual aids to indicate that a particular item has been run through an autoclave cycle. Under the most elemental quality control conditions, media samples should be held at least 2 weeks at 26 °C and 37 °C, and also at conditions and temperatures conducive to the growth of the types of environmental bioburden normally encountered during media preparation and sterilization, to ensure that latent contamination does not subsequently proliferate. Unusual media formulations (e.g., extremes in pH, chemolithotrophic conditions, unusual energy sources) that may not allow proliferation of common microorganisms may require inoculation of lot samples into more conventional heterotrophic media (e.g., tryptic soy broth) to encourage the growth of common environmental contaminants.

Steam at pressures below 15 psi (121 °C) are used for sterilization of many media components not stable at 121 °C. Skim milk (used as a cryoprotectant in lyophilization and as a substrate in phenotypic characterization) is best sterilized at 116 °C for 20 min; the lower pressure reduces carmelization of indigenous lactose. Powdered sulfur (S^0) used as an energy source for microorganisms can be treated with Tyndallization (100 °C for 30 min for 3 consecutive days) to avoid melting the element into a solid mass that physically limits its availability to organisms. Tyndallization is also used as a method of bioburden reduction (it is *not* a sterilization process) for other extremely heat-labile media. In theory, vegetative cells are killed by the first day's boiling; quickly germinating spores are killed the second day; by the third day, late-germinating spores are killed, leaving a relatively clean preparation.

Filter Sterilization

Heat-labile media components should be filter sterilized. Any chemically balanced liquid buffer or medium can also be filter sterilized to ensure that chemical integrity will not be compromised by the elevated heat and pressure of steam sterilization. Membrane filters, rather than depth filters, are preferred for sterilization purposes because their uniform pore size permits absolute retention of particles down to a specified pore diameter when the membranes are stored, sterilized, and used according to manufacturer's instruction. Depending upon the polymer composition, membrane filters are available in porosities from 10 μm down through nominal molecular weight limit used for ultrafiltration. For general microbiological purposes, 0.2-μm filtration is used for sterilization of media and buffers because this porosity effectively traps many vegetative bacteria and spores. However, in serum-based media, because mycoplasmas are not always held back at 0.2 μm, filtration through a stack of three 0.2-μm membranes rather than a single membrane will increase the probability of retention of these common serum contaminants. In addition, membranes are now available in 0.1-μm porosity that can retain smaller-sized microbial contamination, but flow rates are slower and rate of membrane clogging increases as porosity decreases. In any event, the filtration method selected for sterilization should be challenged against the smallest microbial contaminants known to be present in the working environment and production stream (87). As pore size decreases to submicron diameters, there is the possibility of holding back large molecular weight nutritional components and aggregates.

Water-soluble hydrophilic microbial media, stock solutions, and buffers are most commonly filter sterilized with membranes composed of cellulose acetate or nitrate esters. Fastidious cultures may require initial washing of the membranes with hot water to remove residual organics prior to media filtration; filters and membranes specified for tissue-culture applications are also recommended for sensitive organisms. Nonaqueous media components such as ethanolic concentrates of antibiotics or dimethyl sulfoxide (DMSO) must be filter sterilized with hydrophobic polymer membranes designed to withstand organic solvents. Nylon and Teflon membranes are suitable for many organic-solvent based biological applications. These types of membranes can require initial washing with a weak nonpolar solvent such as methanol to condition them for filtration. Some membranes, such as Nylon, are protein binding; if specific proteins are key media components, choose a membrane specifically identified as low protein binding.

Membrane filters have limited loading capacity and will quickly clog. Consequently, effective membrane filtration with any large volume, especially with particulate-laden preparations, requires sequential filtration from larger through smaller porosities to clarify the medium of debris, as much as possible, prior to the final filtration step. Low-speed centrifugation may also be used to first clarify crude media preparations. Glass fiber depth filters are efficient prefilters for many media applications because they have great loading capacity; however, they do not always eliminate the need for sequential membrane clarification steps prior to sterilization. Sterile filtration can be a slow process. It may be necessary, for critical applications, to perform the entire operation in a cold room to retard environmental microbial or other denaturing activity upon the raw material.

Positive-pressure filtration with an inert gas such as nitrogen is used in most large-scale filtration applications. Positive pressure reduces foaming of the filtrate and possible denaturation of any essential protein components. For critical and regulatory applications, the gas supply must also be sterile filtered before it reaches the medium reservoir. In many laboratories, filter sterilization with vacuum is the only available option. With vacuum, foaming of proteinaceous material will be best reduced with sequential prefiltration through larger membrane porosities as well as with minimal vacuum applied to the process. For most bacterial applications, serum foaming is not a deleterious nutritional problem. However, do note that media foaming can be a serious source of environmental contamination if the foam is not trapped and collected before passing into the laboratory vacuum line.

Disposable, presterilized filtration systems are offered by a number of manufacturers (e.g., Falcon, Corning, Gelman, Millipore, Nalge) in small- to large-volume capacities. Capsule filters are available for small to medium in-line batch processing (20–60 L). Filter/storage systems in the 200-mL to 1-L sizes offer the ease of vacuum-operated membrane sterilization and storage with minimal preparation and clean up. Syringe filters are most used for volumes of less than 50 mL; however some types can sterilize volumes up to 1 L without clogging.

Dry Heat Sterilization

Some materials with microbiological applications require dry heat sterilization. Whereas autoclaving is suitable only for heat-stable aqueous material, and filter sterilization is amenable to those materials capable of passage through membranes, dry heat is used for sterilization of nonaqueous, heat-stable material such as oils and waxes. It is also used for glassware or equipment depyrogenation and enzyme (RNase) elimination in many production as well as research protocols. A minimal standard for dry heat sterilization is 350 °C for 30 min, or 121 °C for 6 h (88). As with autoclaving, the time can vary considerably with the efficiency of the sterilization oven as well as with the volume and kind of material being sterilized. Sterilization oven validation is required for many production process. However, for basic laboratory needs, biological dry-heat spore indicators (*Bacillus stearothermophilus* and *B. subtilis*) such as Spore-O-Chex (PyMaH 00190) will give a better indication of actual oven operation than will chemical indicator strips that show only that a particular temperature has been reached, but not for how long, nor if any temperature interruptions occurred. Oils and waxes are best heat sterilized in shallow volumes in large containers to permit rapid and thorough heat penetration of the material because temperature lag is much longer with nonaqueous than with aqueous materials.

MEDIA STORAGE

Prepared microbial media should stored refrigerated at 4–8 °C and in the dark. Containers should be tightly sealed or, in the case of petri dishes, packaged in plastic bags to retard evaporation. Desiccated agar media are characterized by surface cracking and pulling of the agar away from container walls. Water evaporation from liquid or solid media concentrates solutes, which can result in permanent precipitation of some components, alters concentrations of critical nutritional factors or substrates, and increases osmolarity to possibly growth-limiting levels. The ability to freeze media is entirely dependent upon the formulation. As water freezes and concentrates the solutes, poorly soluble chemicals, such as tyrosine, or chemicals, such as phosphates, prone to complexing may not go back into solution when the medium is subsequently thawed. Phototoxicity of media can develop when light activates the glass or plastic containers, as well as media components, and initiates formation of free radicals; aerobic conditions intensify these deleterious reactions as do the blue wavelengths of light emitted by fluorescent tubes (89). The effects of light and free radical formation with respect to tissue culture media have been well studied. In one such mechanism, light at 360 nm and 450 nm activates riboflavin, resulting in a transfer of energy to tryptophan or tyrosine and the formation of cytotoxic free-radical products (90). Similar hydrogen peroxide formation in RPMI medium has been shown to be intensified by the presence of HEPES buffer. With bacteria, an inhibitory effect of light on growth-supporting properties of eosin methylene blue agar has been demonstrated (91). It is recommended that tissue culture media and media for those microorganisms sensitive to the presence of hydrogen peroxide or free radicals (e.g., anaerobic bacteria) be stored and cultured in the dark or in yellow wrappers to reduce the inhibitory effects of light on the media and growth of the organisms.

BIBLIOGRAPHY

1. R.A. Sanford, J.W. Urbance, and J.M. Tiedje, in *Abstracts of the Annual Meeting of the American Society for Microbiology, 1996*, American Society for Microbiology, Washington, D.C., 1996, abstract O-66, p. 365.
2. P. Pienta, J. Tang, and R. Cote eds., *American Type Culture Collection Catalogue of Bacteria and Bacteriophages*, 19th ed., American Type Culture Collection, Rockville, Md., 1996.
3. S.C. Jong, F. Dugan, and M.J. Edwards eds., *American Type Culture Collection Catalogue of Filamentous Fungi*, 19th ed., American Type Culture Collection, Rockville, Md., 1996.
4. *DSM Catalogue of Strains 1993*, Deutsche Samlung von Microorganismen und Zellkulturen GmbH, Braunscheig, Germany, 1993.
5. E.Y. Bridson, in M. Rechcigl, Jr. ed., *CRC Handbook Series in Nutrition and Food, Section G, Diets, Culture Media, Food Supplements*, vol. III, CRC Press, Cleveland, Ohio, 1978, pp. 91–281.
6. R.M. Atlas, *Handbook of Microbiological Media*, CRC Press, Boca Raton, Fla., 1993.
7. P. Gerhardt, R.G.E. Murray, W.A. Wood, and N.R. Krieg eds., *Methods for General and Molecular Bacteriology*, American Society for Microbiology, Washington, D.C., 1994.
8. *C3-A2 Preparation and Testing of Reagent Water in the Clinical Laboratory*, 2nd ed., National Committee on Clinical Laboratory Standards, Villanova, Pa., 1991.
9. *The U.S. Pharmacopoeia XXIII / The National Formulary XVIII*, U.S. Pharmacopoeial Convention, Rockville, Md., 1995, pp. 1635–1636.
10. P. Schönheit and T. Schäfer, *World J. Microbiol. Biotechnol.* **11**, 26–57 (1995).
11. C.J. Microcha and J.E. Devay, *Can. J. Microbiol.* **17**, 1373–1378 (1971).
12. A. Willems, J. Busse, M. Goor, B. Pot, E. Falsen, E. Jantzen, B. Hoste, M. Gillis, K. Kersters, G. Auling, and J. De Ley, *Int. J. Syst. Bacteriol.* **39**, 319–333 (1989).
13. W. Crueger and A. Crueger, Thomas D. Brock ed., *Biotechnology: A Textbook of Industrial Microbiology*, English edition, Sinauer Associates, Sunderland, Mass., 1982, pp. 73–80.

14. R.M. Smibert and N.R. Krieg, in P. Gerhardt, R.G.E. Murray, W.A. Wood, and N.R. Krieg eds., *Methods for General and Molecular Bacteriology*, American Society for Microbiology, Washington, D.C., 1994, pp. 607–654.
15. S.M. Hasan and J.B. Hall, *Can. J. Microbiol.* **87**, 120–128 (1975).
16. S.H. Hutner, L. Provasoli, A. Shatz, and C.P. Haskins, *Proc. Am. Philos. Soc.* **94**, 152–170 (1950).
17. M.O. Garraway and R.C. Evans, *Fungal Nutrition and Physiology*, Wiley, New York, 1984, pp. 124–170.
18. M.N. Hughes and R.K. Poole, *J. Gen. Microbiol.* **137**, 725–734 (1991).
19. P. Kaur and D.V. Vadehra, *Antimicrob. Agents Chemother.* **29**, 165–167 (1986).
20. C. Lindblow-Kull, A. Schrift, and R.L. Gherna, *Appl. Environ. Microbiol.* **44**, 737–743 (1982).
21. H.S. Rosenbranz, J.E. Coward, T.J. Wjodkowski, and H.S. Carr, *Antimicrob. Agents Chemother.* **5**, 199–201 (1974).
22. R.C. Tilton and B. Rosenberg, *Appl. Environ. Microbiol.* **35**, 1116–1120 (1978).
23. R.J. Cote and R.L. Gherna, in P. Gerhardt, R.G.E. Murray, W.A. Wood and N.R. Krieg eds., *Methods for General and Molecular Bacteriology*, American Society for Microbiology, Washington, D.C., 1994.
24. J.H. Crosa, *Annu. Rev. Microbiol.* **38**, 69–89 (1984).
25. E. Griffiths, in J.J. Bullen and E. Griffiths eds., *Iron and Infection, Molecular and Clinical Aspects*, Wiley, Chichester, U.K., 1987, pp. 69–137.
26. M.A.F. Jalal, R. Mocharla, C.L. Barnes, M.B. Hossain, D.R. Powell, D.L. Eng-Wimot, S.L. Grayson, B.A. Benson, and D. Van Der Helm, *J. Bacteriol.* **158**, 683–688 (1984).
27. R.D. Perry and C.L. San Clemente, *J. Bacteriol.* **140**, 129–1132 (1979).
28. M.-F. Thorel, M. Krichevsky, and V.V. Levy-Frebault, *Int. J. Syst. Bacteriol.* **40**, 254–260 (1990).
29. T.E. Bramanti and S.C. Holt, *J. Bacteriol.* **173**, 7330–7339 (1991).
30. B. Fouz, R. Mazoy, M.L. Lemos, M.J. DelOlmo and C. Amaro, *Appl. Environ. Microbiol.* **62**, 2806–2810 (1996).
31. D.R. Lovley, *Annu. Rev. Microbiol.* **47**, 263–290 (1993).
32. M. DeFelice, M. Levinthal, M. Iaccarino, and J. Guardiola, *Microbiol. Rev.* **43**, 42–58 (1979).
33. M.K. Winston and J.K. Bhattacharjee, *J. Bacteriol.* **152**, 874–879 (1982).
34. L.A. Collins and R.L. Thune, *Appl. Environ. Microbiol.* **62**, 848–852 (1996).
35. M.H. Saier, Jr., M.J. Fagan, C. Hoischen, and J. Reizer, in A.L. Sonenshein, J.A. Hoch, and R. Losick eds., Bacillus subtilis *and Other Gram-Positive Bacteria: Biochemistry, Physiology, and Molecular Genetics*, American Society for Microbiology, Washington, D.C., 1993, pp. 133–156.
36. P.F. Smith, T.A. Langworthy, and M.R. Smith, *J. Bacteriol.* **124**, 884–892 (1975).
37. C. Wyss, B.K. Choi, P. Schüpbach, B. Guggenheim, and U.B. Göbel, *Int. J. Syst. Bacteriol.* **46**, 745–752 (1996).
38. T.A. Nerad, ed., *ATCC Catalogue of Protists*, 17th ed., 1991, p. 60.
39. D.A. Power ed., and P.J. McCuen, *Manual of BBL Products and Laboratory Procedures*, 6th ed., Becton Dickinson Microbiology Systems, Cockeyville, Md., 1988, pp. 178–179.
40. *Difco Manual*, 10th ed., Difco Laboratories, Detroit, Mich., 1984, pp. 567–568.
41. F. Fischer, 97/534/EC, *Journal of the European Communities*, Brussels, Belgium, 1997.
42. B.D. Davis and E.S. Mingioli, *J. Bacteriol.* **60**, 17–27 (1950).
43. S.A. Koser, in *Vitamin Requirements of Bacteria and Yeasts*, C. Thomas Publisher, Springfield, Ill. 1968, pp. 379–401.
44. R.J. Cote and R.L. Gherna, in P. Gerhardt, R.G.E. Murray, W.A. Wood, and N.R. Krieg eds., *Methods for General and Molecular Microbiology*. American Society for Microbiology, Washington, D.C., 1994, pp. 155–178.
45. M.O. Garraway and R.C. Evans, *Fungal Nutrition and Physiology*, Wiley, New York, 1984, pp. 171–211.
46. T.A. Freier, D.C. Beitz, L. Li, and P.A. Hartman, *Int. J. System. Bacteriol.* **44**, 137–142 (1994).
47. A.W. Brinkley, A.R. Gottesman, and D.E. Mott, *Appl. Environ. Microbiol.* **40**, 1130–1132 (1980).
48. B.D. Davis, R. Dulbecco, H.N. Eisen, H.S. Ginsberg, and W.B. Wood, Jr., *Microbiology*, 2nd ed., Harper & Row, Hagerstown, Md., 1973, p. 79.
49. E.J. Herbst and E.E. Snell, *J. Biol. Chem.* **176**, 989–990 (1948).
50. M. Rogosa and F.S. Bishop, *J. Bacteriol.* **87**, 574–580 (1964).
51. A.C.R. Tanner, M.A. Listgarten, J.L. Ebersole, and M.N. Strzenko, *Int. J. Syst. Bacteriol.* **36**, 213–221 (1986).
52. J.E. Cronan, Jr., and C.O. Rock, in F.C. Neidhart, J.L. Ingrahan, K.B. Low, B. Magasanik, M. Schaechter, and H.E. Umbarger eds., Escherichia coli *and* Salmonella typhimurium: *Cellular and Molecular Biology*, vol. 1, American Society for Microbiology, Washington, D.C., 1987, pp. 474–497.
53. D. DeMendoza, R. Grau, and J.E. Cronan, Jr., in A.L. Sonenshein ed., Bacillus subtilis and *Other Gram-Positive Bacteria: Biochemistry, Physiology, and Molecular Genetics*, American Society for Microbiology, Washington, D.C., 1993, pp. 411–421.
54. J.W. Deming, J.K. Somers, W.L. Straube, D.F. Swartz, and M.T. MacDonell, *System. Appl. Microbiol.* **10**, 152–160 (1988).
55. G.K. Nyberg, G.P.D. Granberg, and J. Carlsson, *Appl. Environ. Microbiol.* **38**, 29–34 (1979).
56. L.G. Ljungdahl and J. Wiegel, in A.L. Demain and N.A. Solomon eds., *Manual of Industrial Microbiology and Biotechnology*, American Society for Microbiology, Washington, D.C., 1986, pp. 84–96.
57. J.A. Breznak and R.N. Costilow, in P. Gerhardt, R.G.E. Murray, W.A. Wood, and N.R. Krieg eds., *Methods for General and Molecular Bacteriology*, American Society for Microbiology, Washington, D.C., 1994, pp. 135–154.
58. W.E. Balch, G.E. Fox, L.J. Magrum, C.R. Woese, and R.S. Wolfe, *Microbiol. Rev.* **43**, 260–296 (1979).
59. R.G. Cluss and N.L. Somerson, *Appl. Environ. Microbiol.* **51**, 281–287 (1986).
60. R. Olivieri, M. Bugnoli, D. Amellini, S. Bianciardi, R. Rappuoli, P.F. Bayeli, L. Abate, E. Esposito, L. DeGregario, J. Asis, C. Basagni, and N. Figura, *J. Clin. Microbiol.* **31**, 160–162 (1993).
61. *The Source*, 7, Sigma Chemical, St. Louis, Mo., 1992, pp. 1–2.
62. R. Greasham and E. Inamine, in A.L. Demain and N.A. Solomon eds., *Manual of Industrial Microbiology and Biotechnology*, American Society for Microbiology, Washington, D.C., 1986, pp. 41–48.
63. T.L. Miller and B.W. Churchill, in A.L. Demain and N.A. Solomon eds., *Manual of Industrial Microbiology and Biotechnology*, American Society for Microbiology, Washington, D.C., 1986, pp. 122–136.

64. *Sheffield Series Technical Manual*, Quest International, Norwich, N.Y.
65. R.A. Neihof and C.A. Bailey, *Appl. Environ. Microbiol.* **35**, 698–703 (1978).
66. S.E. George, C.J. Costenbader, and T. Melton, *J. Bacteriol.* **164**, 866–871 (1985).
67. R.K. Bajpai and T.K. Ghose, *Biotechnol. Bioeng.* **20**, 927–935 (1978).
68. D.S. Ucker and E.R. Signer, *J. Bacteriol.* **136**, 1197–1200 (1978).
69. P. Loubiere, E. Gros, V. Paquet, and N.D. Lindley, *J. Gen. Microbiol.* **138**, 979–985 (1992).
70. H. Vogel, *J. Gen. Microbiol.* **26**, 223–230 (1961).
71. H.J. Laanbroek and N. Pfenning, *Arch. Microbiol.* **128**, 330–335 (1981).
72. J. Carlsson, G. Nyberg, and J. Wrethen, *Appl. Environ. Microbiol.* **36**, 223–2239 (1978).
73. S. Budavari, M.J. O'Neil, A. Smith, P.E. Heckelman, and J.F. Kinneary eds., *The Merck Index. An Encyclopedia of Chemicals, Drugs, and Biologicals*, 12th ed., Merck Research Laboratories, Whitehouse Station, N.J., 1996.
74. D.E. Gueffroy ed., *Buffers. A Guide for the Preparation and Use of Buffers in Biological Systems*, Calbiochem-Behring, La Jolla, Calif., 1983.
75. N.E. Good, G.D. Winget, W. Winter, T.N. Connolly, S. Izawa, and R.M.M. Singh, *Biochemistry* **5**, 467–477 (1966).
76. G.T. Spierenburg, F.T.J.J. Oerlemans, J.P.R.M. can Laarhhoven, and C.H.M.M. deBruyn, *Cancer Res.* **44**, 2253–2254 (1984).
77. *Art to Science in Tissue Culture*, vol. 6, Hyclone Laboratories, Ogden, Utah, 1987, pp. 4–5.
78. R. Hugh and E. Leifson, *J. Bacteriol.* **66**, 24–26 (1953).
79. E.N. Adaoha Mbanaso and D.H. Roscoe, *Plants Sci. Lett.* **25**, 61–66 (1982).
80. A.D. Lines, *Appl. Environ. Microbiol.* **34**, 637–639 (1977).
81. I.A. Abbott and F.A. Chapman, *Arch. Microbiol.* **128**, 355–359 (1981).
82. J.H. Hanks and R.L. Weintraub, *J. Bacteriol.* **32**, 639–651 (1936).
83. A.S. Dietz and A.A. Yayanos, *Appl. Environ. Microbiol.* **36**, 966–968 (1978).
84. D.A. Bazlinski and F.A. Rosenberg, *Appl. Environ. Microbiol.* **39**, 934 (1980).
85. H.B. Funk and T.A. Krulwich, *J. Bacteriol.* **88**, 1200–1201 (1964).
86. H. Seki, *Appl. Microbiol.* **26**, 318–320 (1973).
87. K.L. Roche and R.V. Levy, *Biopharm* **5**, 22–33 (1992).
88. J.J. Perkins, *Principles and Methods of Sterilization in the Health Science*, Charles C. Thomas, Publisher, Springfield, Ill., 1976.
89. L.C. Ellis, in *Art to Science in Tissue Culture*, vol. 10, Hyclone Laboratories, Ogdon, Utah, 1991, pp. 1–5.
90. M.E. Hoffman and R. Meheghini, *Photochem. Photobiol.* **29**, 299–303 (1979).
91. R.L. Girolami and J.M. Stamm, *Appl. Environ. Microbiol.* **31**, 141–142 (1976).

MEDIA, ANIMAL CELL CULTURE

S. A. CASNOCHA
R. A. WOLFE
Monsanto Co.
St. Louis, Missouri

KEY WORDS

Animal
Cell culture
Defined medium
Mammalian
Media/medium
Tissue culture

OUTLINE

INTRODUCTION

The aqueous environment to which cultured cells are exposed includes low molecular weight nutrients, hormones, transport proteins, buffers, and reagents intended to either reduce stresses imposed on the cultured cells in the bioreactor environment or stabilize the product. The medium, in many cases, also includes attachment proteins and/or hormones that adsorb to the surface(s) of the bioreactor and mediate interaction with the cultured cells. The cells rapidly modify the environment (both aqueous and substratal) by utilization of nutrients and secretion of proteins and metabolites. In many cases these secreted products can be detrimental to both the cells and the product of interest. Thus it is common practice to include in media antioxidants, protease inhibitors, and buffers. The cell-specific rates of medium utilization and metabolite production impact many of the values for process control parameters, which are in turn influenced by the medium design. Media components that destabilize or copurify with the product must, of course, be avoided.

After a few days of culture in the bioreactor, the cells are exposed to a modified or "conditioned" medium, which differs considerably from the initial formulation. The levels of some amino acids such as alanine usually increase, while glutamine levels usually decrease. The metabolism of glutamine results in the secretion of ammonia (which can be cytotoxic at 1–2 mM), and glucose metabolism leads to lactic acid production (which causes medium acidification). Different clonal lines isolated from the same transfection, fusion, or primary isolate usually exhibit differences in growth characteristics and metabolism. This results in differences in the conditioned medium that frequently require "fine tuning" of the production process, initial medium, and supplemental feedstocks. Cells of comparable derivation generally do, however, perform well in similar bioreactor systems.

In developing any process utilizing animal cell culture, all of the aspects of upstream and downstream processing must be also considered. The goal is usually to consistently produce conditioned medium containing a high concentration of an intact, uniformly processed product, which can be readily isolated from other cellular products and medium components. The interrelationship between the cultured cells (and their associated metabolic activities), aqueous environment, substratum, bioreactor design (three-dimensional geometry, materials, mass transfer limitations, "shear stresses", etc.), and product of interest (including the associated downstream processing considerations) must be emphasized. This interrelationship dictates the approach for producing any product through mammalian cell culture technology. In Figure 1 many of the important parameters are grouped within these five categories to illustrate their interrelationships. The composition and range of aqueous environments utilized in mammalian cell culture are presented here, and further information can be found in discussions surrounding the other components of a bioreactor system.

While it is true that many factors of the production process influence medium design, it is, in practice, usually the case that the medium design dictates many aspects of the overall process strategy. In addition, the medium composition defines raw material requirements including their storage and quality control (QC). Accordingly, much of the overall cost of manufacturing the final product can depend on the medium composition and its degree of optimization. Therefore, a common first step toward process design is the definition of the hormonal, nutritional, and substratum requirements of the cell line. In most cases, these are independent of the bioreactor design and process parameters. Evaluation of these data in light of the restrictions defined by the desired product suggest appropriate conditions, bioreactors, processes, and the course for medium optimization. It should be noted that even small modifications in the medium can influence the cell line stability, the product quality/yield, operational parameters of the bioreactor, and downstream processing.

Medium Composition

The aqueous environment to which the cells are exposed results from a combination of many of the factors just discussed. Dissolved gasses and pH are monitored and controlled. Metabolites or other cell products may be either stimulatory or inhibitory to the cultured cells. Almost all of the medium components have both lower and upper limits within which the cells perform optimally, and these ranges are interdependent. The effective concentration range of a given amino acid, for example, is dependent on the concentrations of other amino acids that are taken up by the cells through the same uptake mechanism. The medium and process parameters must therefore be designed such that no component either exceeds or falls below its effective concentration range (both net utilization and biosynthesis of specific amino acids are observed, dependent on the metabolite and conditions). This concept of "metabolic balance" has been clearly expounded in relation to static culture systems (T-flasks) by Ham and his colleagues (1,2).

In practice, an analytical comparison of fresh and conditioned medium provides considerable insight in medium optimization (3). Most of the components routinely included in media are discussed here in detail and can be grouped into four categories: water, low molecular weight nutrients, proteins, and supplements of a nonhormonal/nonnutritional nature. Table 1 lists the components of six common medium formulations that are frequently used as a starting point in medium development.

Scrutiny of these formulations provides some insight into the complexity of media and the evolution of in vitro cell culture technology over the past 40 years. Dulbecco's Modified Eagle's (DME) medium (4–6) was designed for high-density culture of attachment-dependent cells in the presence of serum supplementation, whereas F-12 medium (7) represents an initial round of optimization for clonal (low-density) protein-free growth of CHO cells. The concentration of each component of F-12 was empirically determined through sequential concentration/growth response (C/GR) analyses. MCDB 302 medium was derived from F-12 via further C/GR analyses (8). RPMI medium was developed for suspension culture of lymphocytic cell lines in the presence of serum (9). Iscove's DME (10) and mixtures of DME/F-12 are commonly used to provide all the components necessary for low-density low-protein cultures, while also including high enough levels of nutrients to support mass cultures (11,12). The subsequent analysis of the amino acid and carbohydrate consumption by cells cultured in the specified bioreactor system utilizing a "good mixture" often suggests appropriate modifications of the basal nutrient formula.

Water Quality

It is recommended that water of at least the pharmaceutical "water for injection" (WFI) quality be utilized for mammalian cell culture, and it should be noted that such preparations are not necessarily pure; they simply meet

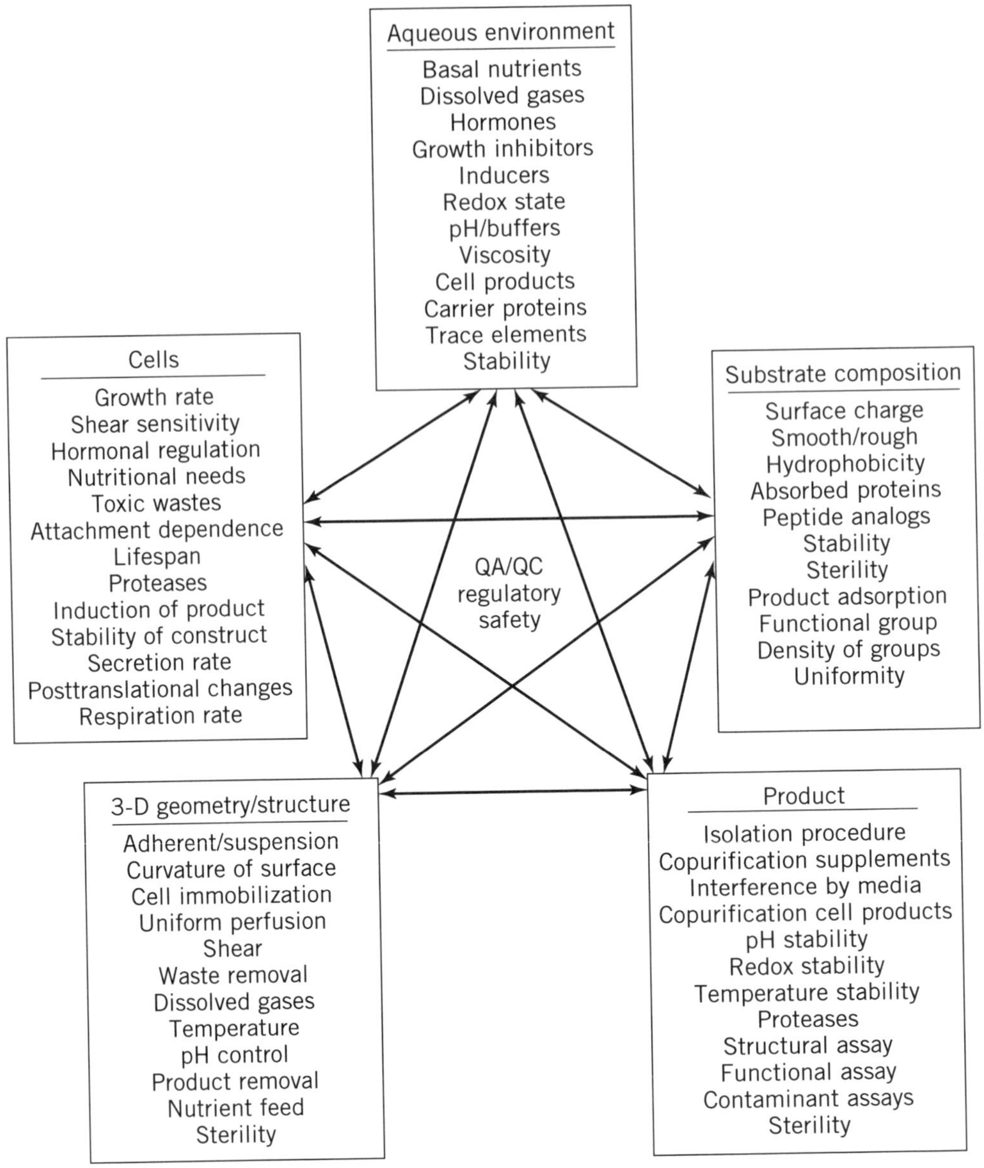

Figure 1. Bioreactor systems: interrelationship of components.

standards for injection. Agents present in water are diluted throughout the organism with injection, but remain in a relatively concentrated form in cell culture.

Water from any natural source is full of contaminants, and mammalian cells are exceedingly sensitive to many compounds that often contaminate the water used for media preparation (13). This is particularly noticeable with "reduced protein" formulations, and contaminants not removed during the purification process or introduced during storage are common causes of aberrant cultures. These contaminants can be either growth stimulatory or growth inhibitory (with the latter being more common). Unpurified water can contain organics (from detergents or degraded biomass), inorganics (such as chlorine and heavy metals), microorganisms (which often result in endotoxin/pyrogen contamination), or particles.

It is recommended that a multiple step purification is used to purify water for cell culture. In the authors' facility, water is filtered, deionized, adsorbed with carbon, purified by reverse osmosis, passed through a Milli-Q™ polishing unit (Millipore Corporation, deionization/carbon adsorption), depyrogenated via ultrafiltration, and then stored at 80 °C in stainless steel tanks until use. Routine maintenance and monitoring of the various stages of the purification stream are performed. Other techniques such as distillation and UV oxidation are commonly utilized.

Low Molecular Weight Nutrients

Bulk ions and trace elements should be considered nutrients because their absorption, secretion, and concentration in a given organism are finely controlled (14). The concentrations of these ions maintain the osmotic pressure of the microenvironment, and their distribution within the tissue (primarily Na^+ and K^+) maintains membrane potential. Therefore, both the absolute and relative amounts of sodium and potassium are important. The transient alterations in the availability of Ca^{2+} and Mg^{2+} mediate

Table 1. Media Formulations Commonly Used in Animal Cell Culture

Compound (g/mol)	DME	F-12	MCDB 302	RPMI 1640	Iscove's DME	Coon's DME:F12
			Bulk ions and trace elements			
$CaCl_2$ (anh) (111.00)	1.80×10^{-3}	2.99×10^{-4}	6.00×10^{-4}		1.49×10^{-3}	1.36×10^{-3}
$CuSO_4 \cdot 5H_2O$ (249.68)		9.97×10^{-9}	1.00×10^{-8}			4.01×10^{-9}
$FeSO_4 \cdot 7H_2O$ (278.02)		3.00×10^{-6}	3.00×10^{-4}			1.44×10^{-4}
$Fe(NO_3)_3 \cdot 9H_2O$ (404.00)	2.48×10^{-7}					1.24×10^{-7}
KCl (74.55)	5.37×10^{-3}	3.00×10^{-3}	3.00×10^{-3}	5.37×10^{-3}	4.43×10^{-3}	4.65×10^{-3}
KH_2PO_4 (anh) (136.09)						2.18×10^{-4}
$MgCl_2$ (anh) (95.22)		6.01×10^{-4}				2.60×10^{-4}
$MgSO_4$ (anh) (120.40)	8.11×10^{-4}			4.06×10^{-4}	8.11×10^{-4}	4.85×10^{-4}
NaCl (58.44)	1.10×10^{-1}	1.30×10^{-1}	1.30×10^{-1}	1.03×10^{-1}	7.71×10^{-2}	1.19×10^{-1}
$Na_2HPO_4 \cdot$ (142.1)		1.00×10^{-3}		2.98×10^{-3}		2.33×10^{-4}
$NaH_2PO_4 \cdot 2H_2O$ (156.00)	8.01×10^{-4}				8.01×10^{-4}	2.86×10^{-4}
$ZnSO_4 \cdot 7H_2O$ (287.54)		3.00×10^{-6}	2.87×10^{-6}			2.50×10^{-7}
			Buffers and other components			
Carbon Dioxide (CO_2, gas)						
$NaHCO_3$ (84.01)	4.40×10^{-2}	1.40×10^{-2}	1.40×10^{-2}	2.40×10^{-2}	3.60×10^{-2}	3.65×10^{-2}
			Essential amino acids			
L-Arg (HCl) (210.70)	3.99×10^{-4}	1.00×10^{-5}	1.00×10^{-3}	9.49×10^{-4}	3.99×10^{-4}	5.49×10^{-4}
L-Cys HCl $\cdot H_2O$ (175.60)	3.56×10^{-4}	2.00×10^{-4}	1.00×10^{-4}	3.71×10^{-4}	5.20×10^{-4}	1.82×10^{-4}
L-Gln (146.10)	4.00×10^{-3}	1.00×10^{-3}	3.00×10^{-3}	2.05×10^{-3}	4.00×10^{-3}	3.00×10^{-3}
L-Hts HCl $\cdot H_2O$ (209.70)	2.00×10^{-4}	1.00×10^{-4}	1.00×10^{-4}	7.15×10^{-5}	2.00×10^{-4}	1.91×10^{-4}
L-Ile (131.20)	8.00×10^{-4}	3.00×10^{-5}	3.00×10^{-5}	3.81×10^{-4}	8.00×10^{-4}	4.31×10^{-4}
L-Leu (131.20)	8.00×10^{-4}	9.98×10^{-5}	1.00×10^{-4}	3.81×10^{-4}	8.00×10^{-4}	4.99×10^{-4}
L-Lys HCl (182.70)	7.99×10^{-4}	2.00×10^{-4}	2.00×10^{-4}	2.19×10^{-4}	7.99×10^{-4}	6.02×10^{-4}
L-Met (149.20)	2.01×10^{-4}	3.00×10^{-5}	3.00×10^{-5}	1.01×10^{-4}	2.01×10^{-4}	1.27×10^{-4}
L-Phe (165.20)	4.00×10^{-4}	3.00×10^{-5}	3.00×10^{-5}	9.08×10^{-5}	4.00×10^{-4}	2.30×10^{-4}
L-Thr (119.10)	7.98×10^{-4}	1.00×10^{-4}	1.00×10^{-4}	1.68×10^{-4}	7.98×10^{-4}	5.00×10^{-4}
L-Trp (204.20)	7.84×10^{-5}	9.99×10^{-6}	9.99×10^{-6}	2.45×10^{-5}	7.84×10^{-5}	4.90×10^{-5}
L-Ttr (diNaSalt) $2H_2O$ (237.20)	4.38×10^{-4}	3.28×10^{-5}	3.31×10^{-5}	1.10×10^{-4}	4.38×10^{-4}	2.49×10^{-4}
L-Val (117.20)	8.00×10^{-4}	9.98×10^{-5}	1.00×10^{-4}	1.71×10^{-4}	8.00×10^{-4}	5.03×10^{-4}
L-Cystine 2HCL (310.80)						1.01×10^{-4}
			Nonessential amino acids			
L-Ala (89.09)		9.99×10^{-5}	1.00×10^{-4}		2.81×10^{-4}	1.01×10^{-4}
L-Asn $\cdot H_2O$ (150.10)		1.00×10^{-4}	1.00×10^{-4}	3.33×10^{-4}	5.60×10^{-4}	8.66×10^{-5}
L-Asp (133.10)		9.99×10^{-5}	1.00×10^{-4}	1.50×10^{-4}	2.13×10^{-4}	9.77×10^{-5}
L-Glu (147.10)		1.00×10^{-4}	1.00×10^{-4}	1.36×10^{-4}	5.10×10^{-4}	1.02×10^{-4}
Gly (75.07)	4.00×10^{-4}	9.99×10^{-5}	1.00×10^{-4}	1.33×10^{-4}	4.00×10^{-4}	3.06×10^{-4}
L-Pro (115.10)		3.00×10^{-4}	3.00×10^{-4}	1.74×10^{-4}	3.48×10^{-4}	3.04×10^{-4}
L-Ser (105.10)	4.00×10^{-4}	1.00×10^{-4}	1.00×10^{-4}	2.85×10^{-4}	4.00×10^{-4}	3.04×10^{-4}
			Amino acid derivatives			
Putrescine 2HCl (161.10)		9.99×10^{-7}	9.99×10^{-7}			9.31×10^{-7}
			Water-soluble vitamins and coenzymes			
Ascorbic acid (176.12)						4.26×10^{-5}
Biotin (244.30)	1.64×10^{-5}	2.99×10^{-6}	3.00×10^{-5}	8.19×10^{-7}	5.32×10^{-6}	1.43×10^{-7}
D-Ca pantothenate (238.30)	1.68×10^{-5}	2.01×10^{-8}	9.99×10^{-7}	1.05×10^{-4}	1.68×10^{-5}	9.44×10^{-6}
Folic acid (441.40)	9.06×10^{-6}	2.95×10^{-6}	3.00×10^{-6}	2.27×10^{-6}	9.06×10^{-6}	5.66×10^{-6}
Niacinamide (122.10)	3.28×10^{-5}	3.03×10^{-7}	3.00×10^{-7}		3.28×10^{-5}	1.65×10^{-5}
Pyridoxal HCl (203.60)					1.96×10^{-5}	9.82×10^{-6}
Pyridoxine HCl (205.60)		3.02×10^{-7}	3.00×10^{-7}	4.86×10^{-6}		1.46×10^{-7}
Riboflavin (376.40)	1.06×10^{-6}	1.01×10^{-7}	9.99×10^{-8}	5.31×10^{-7}	1.06×10^{-6}	5.84×10^{-7}
Thiamine HCl (337.30)	1.19×10^{-5}	1.01×10^{-6}	9.99×10^{-7}	2.96×10^{-6}	1.19×10^{-5}	6.37×10^{-6}
Vitamin B_{12} (1355)		1.00×10^{-06}	1.00×10^{-6}	3.69×10^{-9}	9.59×10^{-5}	5.02×10^{-7}

Table 1. Media Formulations Commonly Used in Animal Cell Culture *(continued)*

Compound (g/mol)	DME	F-12	MCDB 302	RPMI 1640	Iscove's DME	Coon's DME:F12
			Carbohydrates and derivatives			
D-Glucose (180.16)	2.50×10^{-2}	1.00×10^{-2}	8.00×10^{-3}	1.11×10^{-2}	2.50×10^{-2}	1.75×10^{-2}
Na pyruvate (110.00)	1.00×10^{-3}	1.00×10^{-3}	1.00×10^{-3}		1.00×10^{-3}	1.50×10^{-4}
			Purine derivatives			
Hypoxanthine (Na Salt) (146.10)		2.87×10^{-5}	3.70×10^{-5}			1.60×10^{-5}
			Pyrimidines			
Thymidine HCl (337.30)		2.16×10^{-4}	2.16×10^{-6}			1.04×10^{-4}
			Lipids and derivatives			
Choline chloride (139.60)	2.87×10^{-5}	1.00×10^{-4}	1.00×10^{-4}	2.15×10^{-5}	2.87×10^{-5}	6.45×10^{-5}
i-Inositol (180.20)	4.00×10^{-5}	9.99×10^{-5}	1.00×10^{-4}	1.94×10^{-4}	4.00×10^{-5}	6.99×10^{-5}
Linoleic acid (280.40)		2.85×10^{-7}	3.00×10^{-7}			1.43×10^{-7}
Lipoic acid (Thioctic) (206.30)		1.02×10^{-6}	9.99×10^{-7}			4.85×10^{-7}
Total osmolarity (mosm)	367	316	312	293	286	357
Total sodium ion (mM)	156	145	133	116	157	137
Total potassium ion (mM)	5	3	3	5	4	5
Total chloride ion (mM)	120	135	135	109	86	129

transmembrane signaling events, and these ions are also necessary for integrin-mediated cell adhesion. Among other biochemical roles, phosphate and bicarbonate ions buffer the acidity of the microenvironment. Other ions such as selenium, vanadium, copper, zinc, iron, molybdenum, and manganese are necessary as cofactors for certain enzymes. Depending on their concentration, trace elements can be either stimulatory or inhibitory in cell culture, and therefore must be used at optimal concentration.

Trace elements are contained as contaminants in the reagents used to prepare media, and care must be taken to assure that such unknown materials remain consistent throughout a production operation. In order to insure this consistency, it is common practice to "ball mill" large batches of powdered nutrients, and perform QC on random aliquots. The batch is then packaged in appropriately sized containers because the uniformity of the powder decays with time. Preparation of media from these large lots then requires minimal QC (provided a good source of water is utilized). Verification of osmolarity, pH, O_2 and CO_2, levels (blood gas analyzer), and sterility is usually sufficient.

There are two major energy sources utilized by cultured mammalian cells: six-carbon sugars and glutamine (15). Glutamine (2–10 mM) can serve both as a source of energy and carbon. However, ammonia produced from the conversion of glutamine to glutamic acid can be detrimental at concentrations exceeding 2 mM, and cultures of cells such as murine hybridomas tend to build up significant concentrations of this metabolite. Cell lines can sometimes be adapted to tolerate slightly higher levels of ammonia, but success in this endeavor has been limited with levels not exceeding about 10 mM. Glutamine is also unstable in aqueous solution at 4 °C (half-life is approximately 100 days), and ammonia is one of the principal degradants. Thus, long-term storage of media containing this amino acid should be avoided. If necessary, glutamine-free formulations can be supplemented with this amino acid just prior to use. Glucose is the most common carbohydrate (2–20 mM) provided as a source of energy, but many cell lines can be adapted to growth in galactose or other six-carbon sugars (16,17). In this case, a desirable reduction in the rate of lactate production and six-carbon sugar utilization is sometimes observed, presumably achieved via a decrease in the steady-state concentration of glycolytic intermediates. Concomitant reduction in growth rate is often seen. Pyruvate can also function as an energy/carbon source.

Amino acids provide the major nitrogen source in these cultures and have been grouped into two categories as a result of tissue culture experiments: the essential and nonessential amino acids (Table 1). However, most cultured cells perform better in the presence of the nonessential amino acids, and increases in the quantitative requirements for the essential amino acids are observed in cultures deprived of nonessential amino acids. An exception to this classification is the CHO-cell-derived lines that are auxotropic for proline (0.1–1 g/L generally required). Other amino acids such as serine and glycine have been shown to be beneficial supplements for low-density cultures but are relatively unimportant under conditions generally employed in large-scale mammalian cell cultures.

Vitamins have classically been grouped according to their solubility water soluble (B_{12}, biotin, folic acid, niacinamide, pantothenic acid, pyridoxine, riboflavin, and thiamine) or fat soluble (vitamins A, E, K). These compounds are enzymatic cofactors that are necessary for cellular metabolism and are reutilized (not metabolized and/or degraded) by the cells. The fat-soluble vitamins are usually added to the medium from a concentrated ethanolic stock and are not particularly stable in solution. Vitamin A (a retinoid) can have no effect, stimulate growth, or induce differentiation (dependent on the cell type and culture conditions). Vitamin E serves as an antioxidant, and vitamin K is necessary for the γ-carboxylation of proteins. The water-soluble vitamins function in multiple metabolic roles and are generally necessary components of a medium formulation. Vitamin C, a water-soluble antioxidant, has been demonstrated to be beneficial in culture, particularly with cells that tend to synthesize collagen. It is readily oxidized, and continuous supplementation is advisable if beneficial effects are to be observed (18).

Many cultured cells require supplementation with fatty acids, phospholipids (and/or their precursors), and other hydrophobic compounds such as cholesterol. The quantitative and qualitative requirements vary with cell type, culture conditions, and duration of exposure to the supplement (19). Much is still to be discerned about the interconversion and biosynthesis of lipids by cultured cells, but a few generalizations can be made. *cis*-Unsaturated fatty acids (for example, linoleic, oleic, arachidonic) have been shown to be beneficial for many cell lines, but are relatively insoluble and inherently unstable in aqueous solutions. They are usually supplied as conjugates with delipidated albumin or cyclodextrins, or added as unilamellar vesicles (liposomes). Certain lots of albumin or improperly prepared albumin conjugates can be quite toxic. Liposomes have been utilized to provide cells with both fatty acids and phospholipids such as spingomyelin, phosphotidyl choline, phosphotidyl ethanolamine, and phosphotidyl inositol. This approach is difficult to utilize in large-scale cell cultures; in many cases the need for supplementation with the more complex lipids can be circumvented by providing the cells with precursors such as choline, ethanolamine, and inositol. A few cells such as NS-1-derived hybridomas have a deficiency in cholesterol biosynthesis, and supplementation is usually necessary.

Nucleic acid precursors such as the nucleosides and nucleotides are usually not required in a medium formulation, provided at least one source of pyrimidine and purine is supplied. Therefore, hypoxanthine and thymidine are included in most media. Particular care must be taken with mutant lines lacking enzymes of the biosynthetic pathways or with cultures grown without glucose. In the latter case a source of five-carbon sugars or a small concentration of glucose is advisable. Hybridoma cell lines produced via NS0 (a cell line derived from NS1) fusions do not express functional glutamine synthetase (GS), and therefore this line has been used to produce cell constructs using GS as a selectable marker (20). Recombinant lines produced in this manner grow in glutamine-free medium, but this formulation is usually supplemented with adenosine, guanosine, cytidine, uridine, and elevated levels of thymidine.

Culture in glutamine-free medium results in a system where ammonia does not accumulate to toxic levels, and frequently, higher cell densities can be achieved. However, even after adaptation to glutamine-free medium, the proliferation rate is generally less than that seen in glutamine-containing media. NS0-derived lines usually need to be provided with a source of cholesterol unless considerable "weaning/adaptation" has been performed (21).

Proteins

The first cultures were established in mixtures of plasma and tissue extracts (22–24). More defined and reproducible nutrient media designed in the late 1950s to early 1960s were developed using cultures supplemented with dialyzed serum. We now know that serum provides a host of factors not contained in many basal nutrient formulations, including low molecular weight nutrients (sugars, lipids, vitamins, amino acids, trace elements, etc.), hormones, blood proteins, metabolites, and protective compounds. Significant inter-lot, seasonal, and source variation in serum composition exists.

Because quantitative analysis of all of the serum components is not yet possible (particularly for the proteins, lipids, and lipoprotein complexes), variation in the serum supplement is frequently cited as the principal source of variability between cultures. It is also the root of many problems associated with the production of mammalian tissue culture products. Supplementation with 10% serum results in the contamination of the product-enriched medium with a mixture of hundreds of proteins totaling about 4 g/L. This significantly complicates and increases the expense of downstream processing. In addition, serum is expensive, and the availability of high quality serum has been (and is likely to be) a problem for large-scale efforts.

It is now well accepted that the major role played by serum in cell culture is to provide hormones, growth factors, transport proteins, and attachment factors. Serum also has other beneficial effects in culture. It supplies "generic" protein that coats the system, preventing nonspecific adsorption of essential factors to the bioreactor walls and filters. Enzymes contained in serum modify media components, producing molecules the cells can more readily utilize, and also detoxify other materials. Serum also provides essential nutrients that are not included in the older, less-complex basal media. Recently, the importance of the numerous protease inhibitors present in serum has also become apparent, and several reports of increased product integrity in cultures supplemented with protease inhibitors are present in the literature (tPA/aprotinin). In addition, the physical properties of medium (colloid osmotic pressure, diffusion rates, viscosity) are altered by serum supplementation.

If serum is utilized in a culture system, extensive labor-intensive prescreening of several lots is usually necessary. This screening usually focuses on the growth and productivity of the cell line, and small-scale testing for interference in the product isolation. Long-term growth, cellular morphology, cellular productivity, and the possibility of adventitious contaminants must be examined. Once identified, a large lot of serum is usually purchased. Long-term storage of costly, unstable biologicals such as serum in a monitored freezer facility is expensive. Taken together, these factors have driven the development of the technology to reduce the need for serum supplementation in mammalian cell culture. In fact, with the recent concern about potential prion contamination (25), a major emphasis has been placed on eliminating all animal-derived components from culture systems designed to produce parenteral pharmaceuticals.

Many of the useful growth promoting agents, attachment factors, carrier proteins, and protease inhibitors found in serum have been identified and characterized by investigators attempting to reduce the level of serum supplementation in their particular culture system by providing their cultured cells with purified molecules (26). In the last few years a significant industry has emerged to market recombinant proteins for health care, research, and cell culture manufacturing. Many proteins (or modified polypeptides exhibiting the same activity in vitro) can be obtained at a reasonable expense if ordered in quantity. In fact, most protein-containing serum-free formulations are significantly less expensive than medium supplemented with 5% serum. Proteins commonly utilized in these media formulations are discussed later and listed in Table 2 (along with other supplements).

Hormones and growth factors bind to cellular receptors and, in the process, signal the cells to perform specific tasks (growth factors are hormones that were discovered in tissue culture). Mammalian cells evolved in the presence of a mixture of these factors, only some of which are well characterized today. The many components of basal cellular physiology are regulated by the concentrations of hormones to which the cells are exposed (27). In addition, a given cell (depending upon the receptors that the cell expresses) can be signaled to grow, die (apoptosis), or differentiate by exposure to different regulatory factors. Hormones such as steroids, insulin, and thyroid hormone regulate basal metabolic activities. In many cases such signals are necessary for continued viability of cells in culture. In fact, nearly every cultured mammalian cell requires stimulation by insulin (or an insulin-like growth factor [IGE]) for survival (insulin, 0.5–10 μg/mL; IGF, 1–10 ng/mL).

Analogs of many of the proteins produced through recombinant DNA technology exhibit activities similar to those of native proteins in vitro (although pharmacokinetics, tissue distribution, and pharmacodynamics may differ in vivo). Some of such analogs are now available from commercial sources and provide the necessary regulatory signal(s) to the cultured cells when present at 1–10 μg/L. The quality and expense of such analogs can be consistent with the requirements of animal cell production systems for parenteral pharmaceuticals. Analogs of IGF and EGF, for example, are marketed for less than $50 per milligram when purchased in quantity.

Transport proteins such as albumin, transferrin, low- and high-density lipoproteins (LDL, HDL), transcobalamin, hemoglobin, and thyroid hormone binding protein bind necessary nutrients and hormones that would in their unbound form be toxic to the cultured cells at high concentrations. These compounds are either slowly released in a controlled manner (fatty acids, thyroid hormone) or are transported into cells via binding-protein-mediated or cell-receptor-mediated processes (Fe^{2+} and transferrin). In "reduced serum" cultures, most cells require supplementation with additional transferrin (0.5–10 μg/mL) unless frequent supplementation of the medium with ferrous citrate/sulfate and iron chelators is performed; some success utilizing hemoglobin as a transport moiety has also been reported. Transport proteins also serve to help detoxify medium by binding contaminants (heavy metals, detergents) and solubilizing essential nutrients that do not readily dissolve in aqueous solutions (cholesterol and other lipids and lipoproteins, and transcobalamin). Lipids, for instance, when added to cultures above the critical micellular concentration rapidly form micelles that can lyse cultured cells. Albumin and serum lipoproteins are frequently used to supply lipids and cholesterol to cultured cells (28). It should be noted that the lipid composition of these supplements can vary considerably between preparations, and care must be taken to select one appropriate for the cell of interest.

In the last decade many cytokines have been identified and are now available from commercial sources. Culture of blood and marrow-derived cells (i.e., for cell-based therapies) is essentially impossible in the absence of the appropriate mixture of cytokines, and the hormonal signals that direct the proliferation and differentiation of the early progenitor (stem) cells are not fully understood. Large-scale cultures of such cells are used to produce, isolate, and characterize new cytokines and have led to several that are being utilized in adjunctive therapy to cancer treatment to speed recovery after chemotherapy or radiation treatment. Reconstitution of the immune system through ex vivo expansion and posttherapy infusion of bone marrow cells (or mobilized peripheral blood progenitor [stem] cells [PBSC]) taken from patients prior to high-dose chemotherapy is currently being tested in clinical trials. The effectiveness of these cultures depends on the development of an appropriate balance of cytokine signals in the culture media (29). Several cytokines that are being examined in such medium formulations are listed in Table 2. The use of other recently discovered cytokines, for example IL-15, which stimulates T-cell proliferation, are likely to become more significant in the near future.

Attachment proteins (fibronectin, laminin, vitronectin) adsorb to culture surfaces and then interact with anchorage-dependent cultured cells via specific receptor molecules. The attachment, spreading, and growth of these cells is abnormal (and/or limited) in the absence of these proteins. These molecules (along with proteoglycans) are secreted by cells in vivo to form the extracellular matrix (ECM) with which the cell and its neighbors interact. The composition of the ECM varies with the tissue and has been shown to regulate certain aspects of cellular physiology (30). Thus, it is sometimes necessary to provide cultured cells with the appropriate molecules in order for the cells to perform specific desired tasks in vitro. These molecules can be either preadsorbed to the surfaces or added to the medium at concentrations of a few micrograms per milliliter during inoculation of anchorage-dependent cell cultures. Precoating microcarriers with serum proteins is also a common practice, and some success utilizing the attachment motifs in synthetic peptides has been achieved.

Synthetic polymers containing multiple copies of attachment motifs such as RGD (Arg-Gly-Asp) in many cases adequately substitute in culture systems for proteins derived from animal sources, and the expense and quality of

Table 2. Proteins/Polypeptides Used in Animal Cell Culture

Supplement	Tissue/cell source	Structure	Functions/activities
		Growth factors	
Nerve growth factor β (NGF-β)	Submaxillary	Two-chain protein, 26 kDa	Cell growth, neuron development, and survival
Epidermal growth factor (EGF)	Submaxillary, tumor-associated macrophages, prostate tumors	Polypeptide, 6.1 kDa	Stimulates broad range of responses including growth
Platelet-derived growth factor (PDGF)	Platelets, ovarian and lung tumors, melanomas	Dimeric glycoprotein, 28.6, 27.6 kDa (AA or BB)	Stimulates broad range of responses including growth
Insulin-like growth factor I (IGF-I)	Liver, breast tumors	Polypeptide, 7.6 kDa	Stimulates growth and insulin-like activity
Insulin-like growth factor II (IGF-II)	Liver	Polypeptide 7.5 kDa	Stimulates growth and insulin-like activity
Fibroblast growth factor (basic) (bFGF)	Brain, hypothalamus, pituitary, kidney	Monomeric protein 16.4 kDa	Stimulates growth of mesodermal cells
Fibroblast growth factor (acidic) (aFGF)	Brain, hypothalamus	Monomeric protein, 15.9 kDa	Stimulates growth of mesodermal cells
Liver cell growth factor	Plasma	Gly-His-Lys	Synergistic with Cu^{2+} to stimulate growth and survival
Transforming growth Factor β (TGF-β)	Kidney, platelets, and renal, breast, and prostate carcinomas	Two-chain protein, 25 kDa	Stimulation or inhibition of growth or differentiation
Transforming growth factor α (TGF-α)	Breast, neuroendocrine, and squamous carcinomas	Two-chain protein, 6.0 kDa	Similar to EGF
Vascular endothelial cell growth factor (VEGF)	Variety of tumors, tumor-associated macrophages	Dimeric protein, 28 kDa	Stimulate growth and survival of microvasculature cells
		Cytokines	
Erythropoietin (EPO)	Kidney	Glycoprotein, 45 kDa	Stimulates growth and differentiation, red cell precursors
Interferon α/β (IFN-α/β)	Fibroblasts, T and B cells, monocytes, macrophages	Monomeric protein, α 16–28 kDa, β 20 kDa	Inhibits growth of B and T cells and some tumors
Interferon γ (IFN-γ)	T cells	Homodimer, 40 kDa	Stimulate NK and T-cell growth, differentiate myeloid cells
Granulocyte colony stimulating factor (G-CSF)	Macrophage	Monomeric protein, 23 kDa	Growth and differentiation of granulocyte lineage
Granulocyte macrophage colony stimulating factor (GM-CSF)	Fibroblasts, T cells	Monomeric protein, 14 kDa	Growth and differentiation of granulocyte/macrophage lineage
Tumor necrosis factor β (TNF-β)	Macrophages	Monomeric, 19 kDa	Stimulates fibroblast growth and activates a variety of cells
Interleukin 1 (IL-1 [α/β])	Macrophages, monocytes, T and B cells, endothelial cells, fibroblasts	Monomer, 17.5/17.3 kDa	Stimulates inflammatory and T-cell growth
Interleukin 2 (IL-2)	T cells	Monomer, 15 kDa	Stimulates T cells, NK cells, and TIL cells
Interleukin 3 (IL-3)	T lymphoma, eosinophils, mast cells, keratinomas	Monomer, 16 kDa	Stimulates growth of megakaryocyte and myelocyte lineages, late stem cells
Interleukin 4 (IL-4)	T cells, stromal cells, mast cells, thymoma	Glycoprotein, 16 kDa	Stimulates B- and T-cell growth, isotype switching (IgG4 and IgE)
Interleukin 5 (IL-5)	T cells, eosinophils, mast cells	Glycoprotein homodimer, 45 kDa	Eosinophil proliferation and differentiation
Interleukin 6 (IL-6)	Monocytes, fibroblastoid cells, T, B, endothelial, and stromal cells, hepatocytes, fibroblasts	Monomeric protein, 16 kDa	Stimulates stem and B-cell growth, hepatic acute phase, hybridoma survival
Interleukin 7 (IL-7)	Stromal cells, spleen cells	Monomeric protein, 17 kDa	Stimulates progenitor B- and T-cell growth

Table 2. Proteins/Polypeptides Used in Animal Cell Culture *(continued)*

Supplement	Tissue/cell source	Structure	Functions/activities
Interleukin 8 (IL-8)	Neutrophils, monocytes, fibroblasts, T cells, endothelial and epithelial cells	Dimeric protein, 8 kDa	Stimulates angiogenesis, neutrophils, and chemotaxis
Interleukin 9 (IL-9)	T cells	Monomeric protein, 14 kDa	Proliferation of T cells and erythroid precursors
Interleukin 10 (IL-10)	B and T cells, monocytes, keratinocytes	Homodimeric protein, 37 kDa	Stimulates B-cell, mast cell, and thymocyte growth
Interleukin 11 (IL-11)	Fibroblasts and marrow stromal cells, osteosarcomas	Dimeric protein, 8 kDa	Stimulates growth of early and late megakaryoblastic and myeloblastic cells
		Attachment / substratal proteins and compounds	
Cell Tak™	Marine mussel	Polyphenolic protein	Effectively "glues" most cells to most surfaces
Collagen types I–IV	Most tissues, distribution specific for type	Multisubunit proteins	Binds and presents adhesion molecules and stimulates attachment and growth
Extracellular matrix (solubilized)	Extracts of EHS mouse sarcoma and human placenta available	Mixture of proteoglycans, collagens, and attachment factors	Promotes attachment, spreading, mitosis, and differentiation of many epithelioid cells
Fibronectin	Plasma	Dimeric protein, 440 kDa	Promotes attachment and spreading via RGD (Arg-Gly-Asp) and other motifs of most cells from mesenchyme
Laminin	Extracellular matrix	Dimeric protein, 900 kDa	Stimulates attachment and growth of cells derived from endoderm and ectoderm; stimulates neutrophil oxidative burst
Pronectin®F	Synthetic polymer	Polypeptide with multiple RGD motifs	Supports attachment of some cell lines (e.g., CHO, 3T3, BHK)
Poly-D-lysine	Synthesis	Polypeptide, polymers, mw > 50,000	Reduces net negative charge of surface-enhancing attachment
Vitronectin	Plasma	Monomeric protein, 70 kDa	Stimulates attachment and growth of many cell types
		Transport proteins	
Albumin	Plasma	Monomeric protein, 68 kDa	Carries fatty acids and trace elements and can detoxify
Ceruloplasmin	Plasma	Protein, 135 kDa	Transports copper
Hemoglobin	Red cells	Quaternary protein, 65 kDa	Binds oxygen, carbon dioxide and can detoxify
High-density lipoprotein	Plasma	Lipid, cholesterol and multiple protein subunits	Transports cholesterol, cholesterol esters, and other lipids
Low-density lipoprotein	Plasma	Particle containing lipid, cholesterol, and protein (Apoprotein B) subunits	Transports cholesterol, cholesterol esters, and other lipids
Transferrin	Plasma	Monomeric protein, 78 kDa	Carries iron and can detoxify

such polymers can also be consistent with the limitations imposed on manufacturing a pharmaceutical protein. Plasma and chemically treated culture surfaces are widely utilized to provide a surface that more closely mimics the ECM than untreated plastic or glass. Both small-scale plasticware and microcarriers that have surfaces with exposed attachment motifs have recently been introduced into the marketplace.

Nonhormonal and Nonnutritional Supplements

Other compounds are frequently included in medium formulations to reduce specific detrimental conditions that are generated in the artificial microenvironment of the bioreactor, or to facilitate control and monitoring of the system. Also included in this group are antibiotics, which although frequently utilized are not necessary nor recommended. These compounds interfere with cellular metabolism and have been developed by selecting molecules that are more active with microbial enzymes than their analogous mammalian counterparts. Nevertheless, most antibiotics compromise some aspects of mammalian metabolism when present in their effective concentration range, and all are toxic at relatively high dosages. Serum and other supplements can, in some cases, reduce the detrimental effects of antibiotics on cultured cells, and the inhibitory effects of a specific antibiotic are greatly dependent on the cell type. Therefore, if antibiotics must be utilized in a specific culture system, an extensive dose–response analysis should be performed under conditions resembling (as close as possible) those of the bioreactor. Potential copurification with the desired product or interference with downstream processing must also be examined.

In the bioreactor, pH is stabilized with buffers (phosphate and bicarbonate) and usually maintained through control of dissolved CO_2 and base addition. Phosphate has a pK_a of 7.21 and is present in media (2–3 mM) both as a nutrient and as a buffer. The concentration of phosphate is limited by the solubility of calcium phosphate, and higher concentrations of phosphate are practical only in reduced-calcium formulations. The principle pH buffer in mammalian organisms is the bicarbonate–carbon dioxide system, and a similar control strategy is utilized with most bioreactor systems:

$$H_2O + CO_2 \rightarrow H_2CO_3 \rightarrow H^+ + HCO_3^- \qquad pK' = 6.1 \text{ at } 37\ ^\circ C \text{ and } 760 \text{ mmHg}$$

Where $[H_2CO_3]$ (mM) = 0.0308 PCO_2 (mmHg). Applying the Henderson–Hasselbalch equation ($pH = pK' + \log\{[HCO_3^-]/[H_2CO_3] = 6.1 + \log\{(\text{total } CO_2) - 0.0308\ PCO_2\}/0.0308\ PCO_2$) it is apparent that increases in the partial pressure of CO_2 result in decreases of pH. Equilibrium in a bioreactor is essentially dependent on the rate of CO_2 dissolution, diffusion, and bulk mixing because the rate of formation of H_2CO_3 from water and dissolved carbon dioxide is rapid ($T_{1/2}$ = 11.5 s) (31). At pH 7.4 the relative ratios are approximately 20,000:1000:1 (HCO_3^-:CO_2:H_2CO_3).

The amount of bicarbonate present in the medium is chosen to be close to equilibrium conditions for the specified pH operating condition. This amount is usually about 23 mM (similar to circulating blood) and will be at equilibrium with 5–10% CO_2. Bicarbonate in excess of this equilibrium point is used to titrate acid produced by the cultured cells. Carbon dioxide is also a nutrient, and aqueous solutions equilibrated with air at 37 °C contain only 0.21 mm CO_2 (7 μM). Therefore, if air equilibrated cultures are desired, it is recommended to use an additional organic buffer like HEPES (*N*-2-hydroxyethlypiperazine-*N*′-2-ethanesulfonic acid) and 1–2 mM bicarbonate to maintain adequate concentrations of dissolved CO_2.

Although relatively expensive, HEPES is frequently used as an additional buffer because, unlike bicarbonate, its p*K* falls in the physiological range (7.31 at 37 °C). The pK_a is, however, extremely temperature sensitive (−0.014 per change of 1 °C), and thus a room temperature solution at pH 7.55 would be pH 7.31 when warmed to operating temperature. HEPES is utilized at concentrations ranging from 10 to 50 mM. The concentration of sodium chloride needs to be reduced to compensate for changes in osmolarity at the higher levels, and a reduction in the level of potassium ion should be made to maintain similar Na^+/K^+ ratios. At concentrations of HEPES more than 15 mM the buffer is sometimes growth inhibitory (dependent on cell type via stimulation of the production of detrimental oxygen metabolites).

3-(*N*-Morpholino)propanesulfonic acid (MOPS) (pK_a = 7.20 at 20 °C) is also used for this purpose. The temperature sensitively is less (−0.006 per change of 1 °C), but the pK_a is slightly more acidic (which can be advantageous or detrimental, depending on the cell type).

Phenol red has classically been included in culture media as a pH indicator. There is no compelling reason for its use with instrumented systems and in fact has been shown to have estrogenic activity with cultured cells (C127, MCF-7) and to copurify with proteins (tPA, IFN-γ).

Many protective agents are often added to the medium to protect the cells from damage caused by osmotic gradients, shear, and interaction with air–liquid interfaces (bubble damage). The agents include dextran (30–50 g/L, 80,000 MW); polyethylene glycol (20,000 MW), methyl cellulose (0.1–0.2% w/v), and pluronic polyol F-68 (0.025–0.3%). The latter, a block copolymer of polyoxpropylene and polyoxyethelene of 8,300 MW, has been demonstrated to reduce bubble-induced cell damage and is the most frequently used protective agent (32).

Antioxidants are sometimes beneficial. Superoxide radicals and hydrogen peroxide are generated by normal cellular metabolism and enzymes such as xanthine oxidase present in serum supplements. These compounds can also be generated by photooxidation of compounds such as tryptophan and riboflavin, and their production can be stimulated by several factors such as hyperoxic conditions, general cell stress, and compounds such as HEPES. These highly reactive metabolites can cause damage to both the cultured cells and the cell products. Enzymes (superoxide dismutase, catalase), vitamins (C, E, and β-carotene), carrier proteins (transferrin, hemoglobin), amino acids, trace elements (Se^{2+} as a cofactor for glutathione peroxidase), bilirubin, and reduced glutathione are all antioxidants useful in mammalian cell cultures. All except the enzymes and glutathione are components of serum.

Reducing agents (β-mercaptoethanol) can help to prevent oxidative damage, primarily by restoring the reduced form of glutathione. When added to cultured cells, increases in cystine uptake (presumably by the mixed disulfide) and in antibody secretion rates (hybridomas) have been observed.

Many cells actively synthesize and secrete proteases, and even limited cell death (inherently a part of the microenvironment of any bioreactor) causes the release of intracellular protease into the medium. Therefore, supplementation with protease inhibitors specific to proteases generated in a particular culture system is recommended. Cultures that have been established or maintained in serum-supplemented medium are particularly sensitive due to large quantities of adsorbed plasminogen (which is readily activated to plasmin by a variety of proteases). Ten to 40 kIU of aprotinin will effectively inhibit plasmin (and kallikrein) activity. ϵ-Amino caproic acid also inhibits plasmin at 2–7 mM concentrations (higher levels are somewhat toxic). Several other biocompatible inhibitors are now commercially available at nonprohibitive costs (see Table 3). Pepstatin inhibits acid proteases (at 0.7–1 μg/mL). Leupeptin is a reversible, competitive, polypeptide inhibitor of serine and thiol proteases that shows little toxicity at 5 μg/mL. In addition, both soybean and egg white trypsin inhibitors inhibit serine proteases, are effective between 10 and 100 μg/mL, and cost only a few cents per liter. Pefabloc® SC [4-(2-aminoethyl)-benzenesulfonyl fluoride] is relatively nontoxic (compared with phenylmethylsulfonyl fluoride or diisopropyl fluorophosphate) and blocks serine proteases (trypsinlike and chymotrypsin-like) when present at 1–2 mM concentrations. However, it is hydrolyzed relatively rapidly in medium at pH 7.0 (half life ~6 h) and is less stable in more basic environments ($t_{1/2}$ at pH 7.5 ~ 2 h).

Heparin or heparin sulfates (10–100 μg/mL) can help to reduce cell clumping when a normally adherent cell line is grown in suspension. Heparins also stabilize certain growth factors and enhance their bioactivity (endothelial cell growth factor, bFGF, VEGF). Heparins are crude mixtures available from many vendors and are not likely to be identical. Preparations are available with different nominal molecular weights, and those below 5,000 lack some of the biological activities present in other preparations.

MEDIA SELECTION

The most common goal of large-scale animal cell culture is the production of conditioned media with the highest possible titer of a homogeneous protein for subsequent isolation as a functionally active, pure product. In other cases the product can be an inactive proenzyme, a mixture of proteins from nonrecombinant cells, or even the cells themselves. Media selection is therefore highly dependent on the particular product and cellular system and is influenced by the design and composition of the bioreactor. A medium suitable for production of a specific cell product in a batch system is usually not ideal for producing the same product (with the same cells) in a highly instrumented perfusion system. The impact of medium components on downstream processing should guide medium development.

In vivo cells function differently depending on their environment (consider fibroblasts in wounds, osteoblasts, neutrophils, etc.). A given cell line can be cultured in different medium formulations. Metabolism can be modified by providing specific hormonal signals (e.g., insulin), providing physical signals (e.g., shear stress and endothelial cells), and/or altering the availability of nutrients (e.g., the scavenger pathways for nucleotide biosynthesis in hypoxanthine-aminopterin-thymidine selection). Cultures starved of glucose can utilize glutamine as an energy source, and lipid-depleted medium forces cultures into de novo lipid synthesis. The rate of lactic acid production can frequently be reduced by substituting other sugars for glucose (see the section "Low Molecular Weight Nutrients"), but concomitant reductions in proliferation rate are usually observed. Changes in the fatty acid composition of the medium can alter cell membrane fluidity and result in changes in signal transduction, oxidative damage, and other membrane-associated cellular processes (33). In cultures with a transport phenotype (i.e., differentiated cultures of intestinal origin), media nutrients can enhance the transport function (34).

Cellular metabolism alters in response to changes in the environment, and this process is referred to as *adaptation*. Large changes are usually inconsistent with cell viability and proliferation. Cells are generally adapted through

Table 3. Biocompatible Protease Inhibitors

Inhibitor	gm/mol	mg/L	Properties
Albumin	68,000	1,000	Serves as a nonspecific substrate
Aprotinin	6,511	6.8	Inhibits serine proteases (trypsin, kallikren, 40 U/mL = 6.8 μg/mL); at pH 7–8 aqueous solutions are stable for only 1 week at 4 °C
ϵ-Aminocaproic acid (EACA)	131.2	920	Inhibits plasmin
Egg white trypsin inhibitor	28,000	50	At 10–100 μg/mL inhibits serine proteases (trypsin, plasmin, coagulation factor Xa, kallikrein, no antichymotrypsin activity)
Leupeptin	460	2.0	Reversible competitive inhibitor of serine and thio proteases; plasmin, trypsin, kallikrein, cathepsin b, thrombokinase; 2 γg/mL nontoxic, plasmin inhibition 50% at 8 μg/mL
Ovoinhibitor		2.5	Inhibits tyrpsin and chymotrypsin equally on a weight basis; 5 μg/mL usually nontoxic
Pefabloc® SC	239.5	250	100–1,000 μg/mL inhibits serine proteases (daily addition in batch cultures due to rapid hydrolysis in medium)
Pepstatin	685	0.7	Inhibits acid proteases, especially pepsin, renin, cathepsin D, chymosin (make 2 mM stock in alcohol)
Bovine pancreas trypsin inhibitor		2.0	Inhibits trypsin and chymotrypsin. Usually 2–5 μg/mL, nontoxic
Soybean trypsin inhibitor	20,100	50	At 10–100 μg/mL inhibits serine proteases (trypsin, plasmin, coagulation factor Xa, kallikrein, no antichymotrypsin activity)

gradual changes made to a medium, which is selected based on the constraints of the product, the bioreactor system, and the nutritional requirements of the cell line. Further optimization is then performed. The composition of the final formulation depends on initial medium choice and the order in which components are optimized.

A considerable component of the cost of a product can be the medium cost (either directly or indirectly). Some media components are relatively expensive and in many cases not justifiable by the small incremental increases in product titer (e.g., nucleosides). The addition of some costly components can be supported by enhanced culture performance (e.g., insulin). Many examples of increasing product titer by more than an order of magnitude through medium optimization have been documented, and it is likely that no published formulation is "optimal" for a particular purpose. The potential for optimization can be limited by the availability of time, other resources, and the suitability of the current performance of the culture system.

Optimization is therefore highly dependent on the product and the tissue culture system and can result in reduced-serum formulations, a serum-free medium, a chemically defined formulation, or a combination that involves a medium exchange. In the latter case the cell mass is usually generated in medium supplemented with serum or a reduced-serum formulation in which the cells proliferate rapidly. The production of conditioned medium for product isolation then proceeds in more a defined, generally less expensive, formulation. This strategy also addresses downstream processing concerns by eliminating serum from the production phase. Although impractical in batch mode processes, perfusion systems can be utilized for this approach. A potential drawback of a medium exchange process is a reduction of cell mass with time. Also, cell damage can occur if a washing step is included in the process (resulting in the release of intracellular contents, including proteases, into the conditioned medium).

Reduced Serum and Serum-Free Formulations

The addition of serum or plasma to cell culture media dates back to the early years of cell culture (22–35) and logically follows the attempt to mimic the environment of cells in vivo, which are nourished from interstitial fluids that resemble blood plasma. Serum is the liquid phase remaining after the clotting of whole blood and, as such, is similar in composition to blood plasma (the noncell component of blood), except for the absence of fibrinogen and other clotting factors and increased levels of serotonin. Serum is thus a complex mixture of proteins, vitamins, inorganic salts, amino acids, hormones, lipids, and other components in an aqueous solution. Serum provides cells with hormonal, nutritional, and stromal (attachment) factors (36). In addition, serum proteins can act as carriers for other factors that are necessary for cell viability and function. Serum proteins also help protect cells (and product proteins) from hydrodynamic stress (impeller or bubble damage) and proteases secreted by the cultured cells or released during cell lysis. Serum acts as a sink to bind and detoxify the formulation (e.g., deleterious trace elements).

In spite of all these benefits of serum supplementation of the culture medium, the goal is usually to reduce or eliminate this supplement. Because of its biological nature, there are wide fluctuations in serum composition and ability to support cell growth/function, and consistency between batches is hard to maintain (purchases of serum should be preceded by lot testing to demonstrate its growth-supporting function and product generation with specific cells of interest).

Even small amounts of residual serum can complicate filtration, concentration via precipitation or ultrafiltration, and chromatographic processing steps. Reduction or elimination of serum therefore reduces cost of media and simplifies downstream processing. It also reduces foaming in the bioreactor, decreases the possibility of adding adventitious infectious agents to the culture, increases the consistency of medium preparations while decreasing the complexity of QC, and may help maintain expression levels of recombinant proteins or antibodies. A formulation is considered serum-free if sufficient factors have been added to the media to eliminate the need for whole serum.

Most of the classic formulations were designed to be used with serum and to support relatively low-density small-scale cultures (billion cells per liter), for example Eagle's Minimum Essential medium, Dulbecco's Modified Eagle's medium, and M199 (1). Some formulations were developed specifically for clonal growth, such as F-12 medium, which was tested to support clonal growth and serial propagation of two Chinese hamster cell lines for cultures of less than 10^5 cells/mL (1,7). Therefore, while such formulations support the proliferation of low density cultures, essential nutrients are rapidly depleted by higher-density cultures.

Table 1 compares the composition of many widely used media: DME, Ham's F-12, Coon's DME/F-12, MCDB 302, RPMI 1640, and Iscove's DME. These formulations typically require further optimization for most cell types. A proven strategy is to mix commercially available media to determine appropriate mixtures empirically (36–38). Subsequently, the analysis of consumed metabolites in cell-conditioned media can lead to the fine tuning of a formulation for a particular cell type.

Thus, many factors encourage reduction or elimination of serum supplementation: medium expense, downstream processing, the lot-to-lot variation and incomplete characterization, and the potential for introduction of contaminating agents (i.e., mycoplasma, viral pathogens, or prions). Reduction of serum from 10% to 1% can save approximately \$25/L in serum costs, not including the savings associated with growth testing, storage at −20 °C, heat inactivation and filtration, and the costs associated with additional purification steps.

The movement toward serum-free media was championed in a series of publications by Sato and his colleagues in the mid-1970s. They demonstrated that serum could be replaced by mixtures of defined and/or partially defined factors, and that the necessary replacements were qualitatively and quantitatively different, depending on the cell type (38). Most cell lines have been established with medium containing a relatively high level of serum supplementation. The key to reducing or eliminating serum is to determine which components of serum the cells are responding to or utilizing for both growth and product expression. The cells must then be adapted to an alternative environment, or these components need to be provided independent of the serum supplement. Thus, a first step toward serum-free media is to determine, by weaning (adapting) the cells from the serum-supplemented environment in small increments, the level of supplementation where culture performance is significantly reduced (i.e., changes in proliferation and product expression rates, morphology, etc.).

The next step is supplementation with serum components to restore culture performance. This is a reiterative process. Examples of these supplements include many of the proteins and other compounds listed in Tables 1–3. This process may be accelerated by the use of multivariant analysis such as the Plackett–Burmann method (39). Enrichment of the medium with amino acids which are depleted during culture has been shown to increase growth rates and productivity (40). Depending on the process, other substitutions can be beneficial, for example, lactalbumin hydrolysate, bovine colostrum, or bovine whey (41) instead of albumin, ferric citrate (42), or aurintricarboxylic acid/citrate/EDTA (43) to replace transferrin.

Because cells in reduced serum and serum-free cultures are generally more sensitive to environmental conditions, it may be necessary to optimize media for different reactor configurations and modes of operation (44). This may include the addition of protease inhibitors to counteract the effect of increased cell lysis and/or reduced concentrations of nonspecific proteins, which taken together increase the incidence of destruction of the protein product from cell-released proteases.

Some cultures at higher cell densities no longer require serum due to the presence of beneficial autocrine factors. This may be one of the major factors accounting for the success of culture strategies that involve a switch to serum-free or reduced-serum formulations for the production phase after high cell densities are achieved. Once the cells adapt to conditions more consistent with the desired process, it is useful to prepare a frozen cell bank of adapted cells. Recent reports (45,46) recommend cryopreservation media based on nonnutritional polymeric additives in place of serum.

Chemically Defined Media and Protein-Free Media

Beyond the reduction or elimination of serum is the need for chemically defined media or protein-free media. The feasibility for chemically defined media increases as the critical components necessary for many varied culture systems become defined. Several of these defined systems are utilized in large-scale cultures today. The impetus for chemically defined media partially stems from regulatory considerations because such formulations provide greater lot-to-lot consistency. Similar reasons support the use of recombinant proteins rather than those isolated from animal tissue (with potential contaminating agents). From a regulatory viewpoint, defined formulations help to demonstrate control of all aspects of the synthesis and processing of proteins destined for licensing as pharmaceutical agents. The practicality of this approach will increase as more purified factors become available (at significantly reduced prices) through recombinant technology. Even today, many hormone-supplemented, chemically defined formulations can be prepared at less expense than medium supplemented with 10% serum.

EXAMPLES OF MEDIA

Choosing the most appropriate medium for a particular cell type and desired function is a fundamental decision. There are a wide variety of expression systems for recombinant proteins, and each system has its own requirements, for example, the necessity of proline in CHO cell culture, the ability of NS0 cells to proliferate in glutamine-free media (see the section "Low Molecular Weight Nutrients"), and the requisite fatty acids for BHK proliferation. The challenge is not only to find the medium conditions that enhance cell growth and/or production, but also to choose economical components that will not bring contaminating agents to the media or interfere with purification of the product. Culture conditions for cell lines at low density and small scale typically do not support high-density, large-scale production. As medium requirements have become more stringent (from serum-containing to serum-free to chemically defined), whether from economic, scientific, or regulatory concerns, there are more and more medium options on the market, many of which are proprietary and developed for a single cell type. Therefore, a commercially available cell line is often offered with an "optimized medium formulation." Although this medium may work well for the specific cell line, it is not likely to be the best choice for other cell lines, nor will it be economical.

Two types of media are reviewed next, one for attachment-independent (hybridoma) cells and the other for attachment-dependent (endothelial) cells. Media for hematopoietic cell culture systems have recently been reviewed and are also good examples of the targeted cell population the formulation being defined by (29,47,48).

Hybridoma Media

The media requirements of hybridoma cells have been extensively studied, and many formulations and variations thereof can be found in the literature. Hybridomas are attachment-independent cell lines of mixed hematopoietic origin. Media formulations based on DME, RPMI 1640, IMDM, and DME/F-12 (among others) have been used with serum supplementation. However, the potential for copurification of serum proteins (albumin and antibodies) with hybridoma-produced antibodies is a tremendous incentive to utilize serum-free media. Serum-free formulations of basal media supplemented with insulin, transferrin, ethanolamine, and selenium (ITES) are commonly used, along with the variations and (ITES with β-mercaptoethanol (βITES) or ITES with hydrocortisone (HITES) (49,50). Amino acid supplementation has been shown to enhance cell growth and antibody production; however, full optimization of amino acid supplementation is specific to a particular cell line (40,51–53). A fed-batch bioreactor study that aimed to maintain nutrient concentrations with a process control strategy based on stoichiometric consumption rates and total cell number has lead to impressive improvements in cell density, culture life, and antibody titer for a specific hybridoma cell line (54). Because ammonia is toxic to hybridoma cultures (55), the use of glutamine-based dipeptides as a substitute for glutamine is being investigated. The dipeptide is stable in storage, preventing the spontaneous degradation of glutamine to ammonia; however during culture glutamine is released by hydrolysis of the dipeptide. In selected cell lines, higher cell yields are observed coincident with a significant reduction of accumulated ammonia (56,57). In a move toward more defined and ultimately protein-free culture media, the use of ferric citrate (50–500 μM) has been demonstrated to be a substitute for transferrin (42,58).

Endothelial Cell Media

Endothelial cells, the cells that line the blood vessels, are a rich source of enzymes of pharmaceutical interest and are used as components of in vitro assays for cardiovascular, inflammation, and oncology research. They are attachment dependent and are typically utilized as primary or low-passage-number (P5–P20) cultures, with a strong dependence on proper culture conditions for retaining function, morphology, and life span. Optimized culture media are specific for origin of the cells, including both the species and the location in the vasculature. Human endothelial cells are the most difficult to grow in culture, but perhaps the most important for pharmaceutical and biotechnology applications. Human umbilical vein endothelial cells (59,60) and human dermal vascular endothelial cells (61) can be cultured in similar medium. A basal medium of M199, DME or MCDB 131 (62–64) is typically used, supplemented with fetal bovine and/or human serum in the range of 2–50%, with a typical value of 10–20%. Because of the primary nature of typical endothelial cultures, it is not unusual, although not completely necessary either, to add both antibiotics (penicillin, streptomycin, neomycin, and/or gentamicin) and antifungal agents (i.e., Fungizone®). Endothelial cells require glutamine, which should be added to the medium no more than one week before feeding to the cells.

Primary cultures at a proper seeding density typically grow without additional growth factors; however, subcultured cells require a mitogen such as fibroblast growth factor (FGF) or endothelial cell growth factor (ECGF), which is potentiated by the glycosaminoglycan heparin (65). The advent of commercially available endothelial cell cultures has led to the availability of commercially prepared complete media. These media provide the advantage of being optimized for individual types of cells; however, a potential disadvantage is cost. Endothelial cells in vivo grow on a basement membrane that contains laminin and type IV collagen. In vitro endothelial cells benefit from treatment of the growing surfaces with an attachment factor such as collagen, gelatin (denatured collagen), fibronectin, or attachment peptides (i.e., the RGD sequence).

MEDIA PREPARATION AND OPTIMIZATION

There are three major forms in which medium can be supplied: bulk powder, media concentrates, and bulk liquid. Each has its own advantages and disadvantages, as described later. The choice of powder, concentrate, or liquid will vary depending on convenience, cost, storage space (or lack thereof), and availability of a reliable water supply. Whatever the starting configuration, quality control measures are required to ensure lot-to-lot consistency and conformity to standards. Some quality control techniques (i.e., amino acid analysis by HPLC) can also be used for media optimization.

Media Preparation from Powder

Typically the most common medium used for large-scale animal cell culture is that prepared in the end-user laboratory from a powdered formulation. The advantages are cost (both in the cost of the media and the required amount of storage space at 4 °C), increased shelf life of powdered media versus liquid media (years versus months, respectively), flexibility of preparation (i.e., it is relatively easy to modify the formulation at the time media is prepared), large lot sizes for custom and standard formulations, control over the type of container used for the final, prepared media (for ease of use in filling and/or perfusion of reactors), and customized quality control (samples can be taken at any step or every step during the process and run in as few or as many assays as required).

The requirements for media preparation from powder are good source of water (see section "Water Quality"), adequate storage at 4 °C for the bulk powder and the reconstituted media, storage at −20 °C or −70 °C for other media components, mixing vessels and agitation or mixing equipment, sterilization equipment (i.e., filters and pumps for submicron filtration, and autoclave or steam-in-place for media holding tanks), and equipment for quality control (i.e., pH meter or blood gas analyzer for pH, vapor pressure osmometer for osmolarity, scale or volumetric containers for volume). Bulk powder (after checking for any discoloration or unusual odor) is added to water (approximately 90% of final volume) and mixed thoroughly. Other ingredients are added one at a time to ensure complete dissolution. Volume and pH adjustments are made and the medium filtered. Quality control checks on pH, osmolarity, and volume should be made prior to sterile filtration, with the caveat that pH usually increases a few tenths during filtration. In addition, a sterility check from the final material pumped through the filter should be performed (37 °C overnight). After the quality control checks are passed, the reconstituted medium can be released for use. Over time glutamine-containing media will build up significant amounts of ammonia, even at 4 °C (see the section "Low Molecular Weight Nutrients"). As an alternative, medium can be prepared glutamine-free, with glutamine supplemented just prior to use or prepared with glutamine-containing dipeptides to extend shelf life.

Powdered medium is produced by the ball milling process, in which ingredients in powdered form are added to a mixing unit along with ceramic particles or balls. The drum-shaped mixing unit is rotated, which results in mixing of the contents. This process has its drawbacks, notably the ball mill batch size before blending is limited to the size of the mixing drum, heat is generated during ball milling, which can degrade heat sensitive components, and the ceramic balls can chip and break into the medium, resulting in ball mill particles in the packaged product. Bio-Whittaker has recently introduced a micronization process in which the powdered media is continuously mixed without the use of ceramics. They claim increased lot size to a 6,500-kg blended lot, elimination of the possibility of ceramic chips in the powdered media, quicker production time, and the potential for a completely enclosed automated process.

Media Concentrates

Life Technologies, Inc. has introduced media concentrates, which offer some of the convenience of liquid media while retaining some of the flexibility and cost savings of powdered media. As a substitute for bulk powder, concentrates require less mixing and filtering but similar QC strategies. The advantages lie in reducing the amount of time and effort spent weighing out individual ingredients, increasing the shelf life over complete bulk liquid media (concentrates are segregated into 3–4 component mixtures, each of which is more stable than the complete liquid media itself), and the potential for automated mixing (66). Concentrates, by definition, also require less storage space than bulk liquid media. However, using concentrates to reconstitute full-volume media is not their only or even their best use. Concentrates can be utilized as perfusion supplements, adding back key components to high-density reactors without greatly effecting the volume (67,68). This is ultimately done to increase the productivity of the reactor. However, another reason to add supplements is to provide a minimally required level of nutrients to prevent apoptosis (programmed cell death), which has been shown to occur when even a single essential nutrient is limiting (69,70). Cell death, whether by apoptosis, hydrodynamic damage, or other mechanisms, leads to increased levels of cell-released proteases as well as a reduction in cell density. Nutrient feeding by concentrates can occur along with 1× media perfusion, which, in addition to adding nutrients, dilutes waste products such as lactate and ammonia.

Bulk Liquid Media

Bulk liquid media is the most expensive form of media but the least time-consuming for the end user. The media is mixed, filtered, and QC checked at the vendor, reducing or eliminating the need for the end-user to provide a water supply, mixing and filtration equipment, and QC instruments. However, the extra labor and processing adds to the cost of the media. In addition, shipping costs increase. Another cost to the end-user is storage in the appropriate temperature-controlled environment. A drawback of bulk liquid media has been its availability in bottles (0.5–1 L), which require transfer to other vessels for filling or perfusion of the bioreactors. This concern has been addressed by the sale of liquid in drums and large bags (50–100 L), which can be equipped with custom fittings. However, special handling equipment may be necessary to move these large containers around the end-user facility.

Quality Assurance

The goal of quality assurance is to provide consistency between lots and to provide a level of confidence that the media meets the necessary conditions. Vendor-supplied certificates of analysis for media in powder, concentrate, or liquid form provide a list of components and their mass or concentration. Simple tests, such as pH, osmolarity, sterility, and volume, can provide additional information about reconstituted media and should be routine. Other tests can provide additional information such as ammonia and glutamine (especially for media close to the expiration date) levels, amino acid analysis (for new lots of media to verify formulation), and endotoxin levels (for any media, but especially those containing albumin, which binds endotoxin). Individual components of the media, especially serum and growth factors, should be lot tested with the appropriate cell lines for their ability to support growth.

Optimization

Media for individual cell lines or groups of cell lines can be optimized for cell growth, specific production rates, attachment (in)dependency, cost of media, or other factors critical to a project. This optimization is especially important for larger-scale and/or longer-term projects.

There are two major approaches to optimization of fed-batch reactor schemes: bottom-up and top-down (67). The bottom-up approach refers to adding back components that are determined to be depleted. This approach can range from a simple addition of glutamine or glucose feed, or be based on an extensive study of cellular metabolism (71). The analysis required then can range form a single measurement in a metabolite analyzer (such as a YSI glucose or glutamine analyzer, or a Kodak Biolyzer) to complex analytical methods (i.e., HPLC) for detecting trace compounds. The specificity of a particular media supplement has the advantage of enhancing that feature that is most important for a particular production project, but this specificity also has the disadvantage of not translating to other cell lines or optimization conditions. The optimized formulation may be valid for only a narrow range of cell density and is likely to be cell-line specific.

The top-down approach uses complete media supplementation with either 1× media or concentrates. This is a simple approach that does not require elaborate metabolic studies, but does increase the cost of the actual media supplement, a factor that must be weighed against the time savings in development of a bottom-up type of supplement. Top-down strategies require the identification of at least one key nutrient (i.e., glucose), which is monitored to control the supplementation rate through a stoichiometric or heuristic model.

POSTTRANSLATIONAL MODIFICATION AND MEDIUM COMPOSITION

Proteins, whether produced in vivo or in vitro, undergo posttranslational modifications that ultimately effect the biological properties of the protein. Although some posttranslational modifications occur extracellularly in vivo, (i.e., activation of the complement system or proteolysis of prothrombin to thrombin during blood coagulation), many posttranslational modifications occur prior to release of the protein from the cell. These modifications include hydroxylation, phosphorylation, sulfation, adenylation, carboxylation, acetylation, methylation, and glycosylation (72). Of these modifications, glycosylation has been most extensively studied in cell culture because the ability to properly glycosylate proteins is an advantage of mammalian recombinant expression over prokaryotic systems. Altered glycosylation patterns may effect clearance rates, specific activity, further processing of the protein, solubility, and immunogenicity. The cell type is a major contributing factor to glycosylation patterns of proteins made in cell culture; however, there is evidence that environmental factors, including cell culture media, can influence glycosylation.

Glycosylation

Oligosaccharides attached to asparagine residues are referred to as N-linked, whereas those attached to serine or threonine hydroxyls are O-linked. More studies have been conducted on N-linked oligosaccharides due to their more complex nature. Glucose starvation has been shown to reduce the incidence of glycosylation at asparagine residues and lead to aberrant oligosaccharide structures (73,74). An individual assessment of the result of such deviant glycosylation on the recombinant protein properties is warranted, and if necessary the cell culture medium should be adjusted to ameliorate the problem. Other cell culture components that have been reported to disrupt glycosylation include hormones, DMSO, acidotropic amines, EDTA, HEPES buffer, the antibiotic tunicamycin, and the ionophore monensin (74). If it is determined that important protein properties are altered by the media composition, media optimization for production of the appropriate glycoform is required.

BIBLIOGRAPHY

1. R.G. Ham and W.L. McKeehan, in S.P. Colowick and N.O. Kaplan eds., *Methods in Enzymology*, vol. 58, Academic Press, Orlando, Fl., 1979, pp. 44–93.
2. R.G. Ham in *Handbook of Experimental Pharmacology*, Springer-Verlag, New York, 1980, pp. 13–88.
3. M.J. Keen and C. Hale, *Cytotechnology* **18**, 207–217 (1996).
4. H. Eagle, *Science* **122**, 501–504 (1955).
5. H. Eagle, *Science* **129**, 432–437 (1959).
6. J.D. Smith, G. Freeman, M. Vogt, and R. Dulbecco, *Virology* **12**, 185–192 (1960).
7. R.G. Ham, *Proc. Natl. Acad. Sci. U.S.A.* **53**, 288–293 (1965).
8. W.L. McKeehan, K.A. McKeehan, S.L. Hammond, and R.G. Ham, *In Vitro* **13**(7), 399–416 (1977).
9. R. Cote ed., *ATCC Media Handbook*, American Type Culture Collection, Rockville, Md., 1984, pp. 81–82.
10. N.N. Iscove and F. Melchers, *J. Exp. Med.* **147**, 923–33 (1978).
11. M. Butler, in W.G. Thilly ed., *Mammalian Cell Technology*, Butterworths, Boston, Mass., 1986, pp. 91–108.
12. K. Higuchi and R.C. Robinson, *In Vitro* **9**(2), 114–121 (1973).
13. C. Waymouth, in C. Waymouth, R.G. Ham, and P.J. Chapple eds., *The Growth Requirements of Vertebrate Cells In Vitro*, Cambridge University Press, New York, 1981, pp. 105–115.
14. R.R. Zombola, R.C. Bearse, and P.A. Kitos, *J. Cell Physiol.* **101**, 57–66 (1979).
15. L.J. Reitzer, B.M. Wice, and D. Kennell, *J. Biol. Chem.* **264**(8), 2669–2676 (1979).
16. A. Leibovitz, *Am. J. Hyg.* **78**, 173–80 (1963).
17. D. Barngrover, J. Thomas, and W.G. Thilly, *J. Cell Sci.* **78**, 173–189 (1985).
18. H. Senoo, Y. Tsukada, T. Sato, R. Hata, *Cell Biol. Int. Rep.* **13**, 197–206 (1989).
19. R.A. Wolfe and G.H. Sato, in D. Sirbasku ed., *Growth of Cells in Hormonally Defined Media*, Cold Spring Harbor Press, New York, 1982, pp. 1075–1088.
20. C.R. Bebbington, G. Renner, S. Thomson, D. King, D. Abrams, and G.T. Yarranton, *Bio / Technology* **10**, 169–175 (1992).
21. M.J. Keen and T.W. Steward, *Cytotechnology* **17**, 203–211 (1996).
22. R.G. Harrison, *Anat. Record* **2**(9), 385–410 (1908).
23. M.R. Lewis and W.H. Lewis, *Anat. Record* **5**(6), 277–293 (1912).
24. A. Carrel, *J. Exp. Med.* **17**, 14–19 (1913).
25. "Mad Cows and Englishmen," *The Economist*, March 30, 1996, pp. 25–27.
26. W.L. McKeehan, D.P. Genereux, and R.G. Ham, *Biochem. Biophys. Res. Comm.* **80**(4), 1013–1021 (1978).
27. W.J. Bettger and W.L. McKeehan, *Physiol. Rev.* **66**(1), 1–35 (1976).
28. H. Polet and H. Spieker-Polet, *J. Exp. Med.* **142**, 949–55 (1975).
29. M.R. Koller and B.O. Paulson, *Biotechnol. Bioeng.* **42**, 909–930 (1993).
30. K.M. Yamada, *J. Biol. Chem.* **266**(20), 12809–12812 (1991).
31. V.B. Mountcastle ed., *Medical Physiology*, 14th ed., C.V. Mosby, St. Louis, 1980.
32. D.W. Murhammer and C.F. Goochee, *Biotechnol. Prog.* **6**, 142–148 (1990).
33. S.I. Grammatikos, P.V. Subbaiah, T.A. Victor, and W.M. Miller, *Cytotechnology* **15**, 31–50 (1994).
34. A.B.R. Thomson and G. Wild, *J. Nutr. Biochem.* **6**, 240–245 (1995).
35. L.E. Baker and A.H. Ebeling, *J. Exp. Med.* **69**, 365–378 (1939).
36. D. Barnes and G. Sato, *Cell* **22**, 649–655 (1980).
37. R.A. Wolfe, G.H. Sato, and D.B. McClure *J. Cell Biol.* **87**, 434–441 (1980).
38. J. Bottenstein, I. Hayashi, S. Hutchings, H. Masui, J. Mather, D.B. McClure, S. Ohasa, A. Rizzino, G. Sato, G. Serrero, R. Wolfe, and R Wu, in W.B. Jakoby and I.H. Pastan eds., *Methods in Enzymology*, 45, Academic, New York, 1979, pp. 94–109.
39. P.M.L. Castro, P.M. Hayter, A.P. Ison, and A.T. Bull, *Appl. Microbiol. Biotechnol.* **38**, 84–90 (1992).
40. M.J. Keen and C. Hale, *Cytotechnology* **18**, 207–217 (1996).
41. J. Capiaumont, S. Ostrovidov, C. Legrand, F. Belleville, and P. Nabet, *In Vitro Cell. Dev. Biol. Animal* **32**, 8–12 (1996).
42. F. Franek and J. Dolnikova, *Cytotechnology* **7**, 33–38 (1991).
43. K. Bertheussen, *Cytotechnology* **11**, 219–231 (1993).
44. H. Buntemeyer, D. Lutkemeyer, and J. Lehmann, *Cytotechnology* **5**, 57–67 (1991).

45. O.-W. Merten, S. Petres, and E. Couve, *Biologicals* **23**, 185–189 (1995).
46. C.D. Severson, *In Vitro Cell. Dev. Biol. Animal* **32**, 321 (1996).
47. C.E. Sandstrom, W.M. Miller, and E.T. Papoutsakis, *Biotechnol. Bioeng.* **43**, 706–733 (1994).
48. T.A. McAdams, W.M. Miller, and E.T. Papoutsakis, *TIBTECH* **14**, 341–349 (1996).
49. M. Butler and H. Jenkins, *J. Biotechnol.* **12**, 97–110 (1989).
50. H. Murakami, H. Masui, G.H. Sato, N. Sueoka, T.P. Chow, and T. Kano-Sueoka, *Proc. Natl. Acad. Sci. U.S.A.* **69**, 1158–1162 (1982).
51. D. Duval, C. Demangel, K. Munier-Jolain, S. Miossec, and I. Geahel, *Biotechnol. Bioeng.* **38**, 561–570 (1991).
52. F. Franek, *Biotechnol. Bioeng.* **45**, 86–90 (1995).
53. T.S. Stoll, K. Muhlethaler, U. von Tockar, and I.W. Marison, *J. Biotechnol.* **45**, 111–123 (1996).
54. L. Xie and D.I.C. Wang, *Biotechnol. Bioeng.* **51**, 725–729 (1996).
55. I. Ludemann, R. Portner, and H. Markl, *Cytotechnology* **14**, 11–20 (1994).
56. M. Butler and A. Christie, *Cytotechnology* **15**, 87–94 (1994).
57. A. Christie and M. Butler, *J. Biotechnol.* **37**, 277–290 (1994).
58. Z. Chen, Y. Ke, and Y. Chen, *Cytotechnology* **11**, 169–174 (1993).
59. M.A. Gimbrone, *Prog. Hemostasis Thromb.* **3**, 1–28 (1976).
60. E.A. Jaffe, R.L. Nachman, G.C. Becker, and C.R. Minick, *J. Clin. Invest.* **52**, 2745–2756 (1973).
61. R.M. Marks, M. Czernieki, and R. Penny, *In Vitro Cell. Devel. Biol.* **21**(11) 627–635 (1985).
62. J.B. Warren, in J.B. Warren ed., *The Endothelium: An Introduction to Current Research*, Wiley-Liss, New York, 1990, pp. 263–272.
63. P.A. D'Amore and S.J. Braunhut, in U.S. Ryan ed., *Endothelial Cells*, CRC Press, Boca Raton, Fla., 1988, pp. 13–36.
64. A. Knedler and R.G. Ham, *In Vitro Cell. Devel. Biol.* **23**(7) 481–491 (1987).
65. T. Maciag and W.H. Burgess, in U.S. Ryan ed., *Endothelial Cells*, CRC Press, Boca Raton, Fla., 1988, pp. 3–11.
66. G. Roth, J.M. Kubiak, J.F. Long, and G.M. Schoofs, *BioPharm*, **8**, 31–35 (1995).
67. T.A. Bibila, C.S. Ranucci, K. Glazomitsky, B.C. Buckland, and J.G. Aunins, *Biotechnol. Prog.* **10**, 87–96 (1994).
68. R. Fike, J. Kubiak, P. Price, and D. Jayme, *BioPharm* **6**, 49–54 (1993).
69. F. Franek and K. Chladkova-Sramkova, *Cytotechnology*, **18**, 113–117 (1995).
70. R.P. Singh, M. Al-Rubeai, C.D. Gregory, and A.N. Emery, *Biotechnol. Bioeng.* **44**, 720–726 (1994).
71. L. Xie and D.I.C. Wang, *Biotechnol. Bioeng.* **43**, 1175–1189 (1994).
72. K.-K. Han and A. Martinage, *Int. J. Biochem.* **24**, 19–28 (1991).
73. H. Tachibana, K. Jiyoun, K. Taniguchi, Y. Usho, K. Teruya, K. Osada, Y. Inoue, S. Shirahata, and H. Murakami., *In Vitro Cell. Dev. Biol. Animal* **32**, 178–183 (1996).
74. C.F. Goochee and T. Monica, *Bio/Technology*, **8**, 421–427 (1990).

See also MAMMALIAN CELL BIOREACTORS; MAMMALIAN CELL CULTURE REACTORS, SCALE-UP.

MEDIUM FORMULATION AND DESIGN, *E. COLI* AND *BACILLUS* SPP.

ROSALIE COTE
ROBERT L. GHERNA
American Type Culture Collection
Rockville, Maryland

KEY WORDS

Antibiotics
Catabolite repression
Chemically defined media
Competence
Diauxic growth
Enrichment media
Freezing
Isolation media
Lyophilization
Sporulation
Undefined (complex) media

OUTLINE

Introduction
Nutritional Requirements
Generalized Media
- Chemically Defined Media
- Undefined (Complex) Media
- Catabolite Repression
- Diauxic Growth
- Effects of Nutrient Fluctuations

Enrichment and Isolation Media
Sporulation Media for *Bacillus* spp.
Molecular Biology Media
- General Growth
- λ-Bacteriophage Propagation
- Recombinant Detection
- Competent Cells and Transformation

The Use of Antibiotics in Media
Media for Culture Preservation by Freezing or Lyophilization
Media Formulations
Bibliography

INTRODUCTION

Escherichia coli and *Bacillus subtilis* (later to include other species of the genus *Bacillus*) were the first microbial tools used in studies of biochemical pathways and genetics because of their undemanding needs for cultivation as well as their general lack of pathogenicity. Consequently, more is known about the workings of the metabolic processes of these organisms than all others, and this knowledge has been successfully applied in the era of biotechnology. Whether by human-directed and human-enhanced natural strain selection or by genetic manipulation, these prokaryotic organisms have provided a wealth of molecular information that has much accelerated biotechnology's scientific and industrial growth.

The purpose of this article is not to belabor basic nutritional considerations required to propagate these organisms: most wild-type strains are so metabolically versatile that they will grow in a simple medium containing only 1% peptone. Microbial growth alone, however, is rarely the ultimate goal in scientific pursuit. Strain isolation, enhancement, genetic stability, performance, and reproducibility are all entangled aspects of cultivation for these as well as other microorganisms. Unfortunately, there are no universal road maps to guide a biotechnologist to success in manipulating a microorganism; the best solutions are often strain specific as well as need specific. Thus, this article highlights broad discussions of media considerations involved with the growth and manipulation of these metabolically incredible bacteria.

NUTRITIONAL REQUIREMENTS

E. coli and *Bacillus* spp. have complete anabolic and catabolic biochemical pathways. This metabolic complexity allows the cultivation of these organisms with relatively simple media containing an organic source of energy, the nutritionally essential minerals (N, C, H, O, PO_4^{2-}, SO_4^{2-}), and trace elements (Ca^{2+}, K^+, Na^+, Mg^{2+}, Mn^{2+}, Fe^{2+}, Zn^{2+}, Cu^{2+}, Co^{2+}, Mo^{2+}, BO_3^{3-}). Auxotrophies for amino acids, vitamins, or other growth factors do occur for some strains whether deliberately engineered into a genome or naturally occurring.

These bacterial strains are organotrophs and can utilize heterogenous organic compounds as sources of carbon and energy including alcohols, amino acids, carbohydrates, fatty acids, hydrocarbons, organic acids, and polysaccharides. *E. coli* is facultative and can respire aerobically or anaerobically depending upon the carbon–energy source and other growth parameters. However, secondary metabolites produced by the organism vary tremendously between aerobic and anaerobic conditions. With the exception of the anaerobic species *Bacillus infernus* (DSM 10276) (1), the genus *Bacillus* is strictly aerobic or facultative. Morphological and physiological characteristics can be influenced by media conditions with these metabolically responsive bacteria, and care should be taken to standardize media conditions as well as other growth parameters to ensure reproducibility in research or production.

GENERALIZED MEDIA

Bacteriological media may be either defined or undefined with respect to chemical composition. The type of medium selected is entirely dependent on the ultimate goal of the cultivation process.

Chemically Defined Media

In a chemically defined formulation, each component is discretely identifiable as to chemical identity and concentration. The mineral salt-based formulations used in studies of biochemical pathways or gene function are examples of chemically defined media. Auxotrophs can be screened on this type of mineral medium when it is amended with appropriate concentrations of amino acids, vitamins, or other other growth factors (2). Such formulations can also be modified for selectivity by the incorporation of specific carbon or nitrogen sources, inhibitory compounds, or adjustment to physiochemical parameters (e.g., pH, salinity). Chemically defined media are also employed, where possible, in the large-scale production of high-value recombinant DNA-derived biopharmaceuticals. Because product yield is often low for such types of fermentation protein and peptide products, isolation and purification procedures are rigorous. Use of chemically defined media eliminates interfering proteinaceous matter present in the peptones and extracts of complex formulations that further complicates product recovery.

Chemically defined media are not particularly forgiving in use if poorly designed and balanced or poorly prepared. For example, growth inhibition caused by amino acid antagonism is a well-studied phenomenon observed under chemically defined, mineral media conditions. In particular, valine inhibition of *E. coli* K12 has been studied in detail (3). Chemical imbalance, such as an excess of phosphates used as medium buffering agents, can cause deleterious binding and precipitation of trace element divalent cations (e.g., Ca^{2+}, Fe^{2+}, Mg^{2+}) necessary as enzyme cofactors for key metabolic functions. Use of a trace metal in a nonmetabolically active valence or in medium conditions that alter metal speciation (4) can also lead to chemical imbalance in a mineral medium. Sometimes, trace metal and other growth factor requirements in mineral media can be spared. Iron limitation will induce the production of siderophores (iron-binding transport molecules) in *Bacillus megaterium* (ATCC 19213) as well as other bacteria (5) to increase cellular uptake of the required ion. The requirement for Mn^{2+} for sporulation in *B. subtilis* can be bypassed in mineral media by properly balancing the ratio of carbon sources (glucose to malate) entering the metabolic pathway above and below the manganese-requiring enzyme phosphoglycerate phosphomutase (PGA mutase) (6). The vitamin B_{12} requirement in *E. coli* (ATCC 10799; NCIB 8134) can be spared by high levels of methionine (20 μg/mL) added to a mineral growth medium (7).

For nutritionally fastidious microorganisms, such as many animal and human pathogens, development of a good chemically defined medium with respect to the identification of key nutrients, as well as the important balancing of their concentrations, can be laborious and most often necessitates a lengthy list of growth factors. Fortunately, wild-type *E. coli* and *Bacillus* spp. can generally be cultivated with rather simple chemically defined formulations such as M9 medium (see final section, "Media Formulations") composed of a few mineral salts and an organic carbon source. With M9, those essential trace elements and microelements not deliberately represented in the formulation are spuriously supplied as poorly defined contaminants in the distilled water or from the glassware used to prepare the medium. Neidhardt et al. (8) developed a more nutritionally complete defined mineral medium suitable for use in rigorous biochemical studies of the enterobacteria.

Undefined (Complex) Media

Undefined media, also called complex media, provide many of the nutritional elements as mixtures of protein hydrolysates (e.g., peptones) or extracts (e.g., yeast extract) in which the exact chemical identity and concentration of essential nutrients, trace elements, and growth factors are not known. The demands of the biotechnology industry have influenced the manufacturers of these raw media components to provide detailed lot analyses of hydrolysates and extracts (9) but, for critical needs, any new lot of peptone or extract must be qualified by performance testing before use.

Complex media formulations, rather than those chemically defined, will generally give greater cell yield for most strains of *E. coli* and *Bacillus* spp. Commonly available laboratory media such as soy bean casein digest broth manufactured under different trade names (e.g., Trypticase Soy Broth [BBL]; Tryptic Soy Broth [Difco]) will support vigorous growth for most strains with growth optima close to neutral pH. For general purposes, a combination of 1.0% peptone, 0.5% yeast extract, and 0.2% utilizable carbon source will adequately satisfy the nutritional needs for these heterotrophic organisms as well as for many auxotropic mutants. Complex nitrogen and carbon substrates (e.g., cottonseed meal, molasses) are frequently used in large-scale fermentation media for the microbial production of enzymes and antibiotics; however, lot-to-lot variation in these raw materials can be significant issues in maintaining consistency in media performance.

Although any number of generalized formulations, either chemically defined or complex, will support growth of these microorganisms, strain-specific characteristics, product desirability, enzyme induction or increased expression, or other investigational needs will necessitate optimization of a medium. Greasham and Inamine (10) highlight many of the considerations and decisions that must be made with respect to optimization. The design of media for *E. coli* and *Bacillus* spp. must also take into account a variety of media-induced phenomena that can affect growth or other biochemical activities of the organisms as well as the quantity and types of secondary metabolites produced.

Catabolite Repression

Catabolite repression of carbon or nitrogen metabolism is frequently encountered in both *E. coli* and *Bacillus* spp., although the exact biochemical mechanisms differ between genera. In general, a readily metabolized substrate (such as glucose) can repress the activity of enzymes acting on similar substrates, thereby affecting the nature of secondary metabolite production. For example, α-amylase activity in *B. subtilis*, normally expressed when the organism is grown on starch, is repressed when glucose is present; bacitracin synthesis in *Bacillus licheniformis* is repressed by glucose but unaffected by citrate (11). In *B. subtilis*, the expression of the histidine-inducible, histidine-degradative pathway (*hut* operon) is repressed by rapidly metabolized carbon compounds including glucose, glycerol, and malate (12) as well as by amino acid mixtures (13). Derepression of the *hut* operon occurs when histidine is added to the medium (14).

Diauxic Growth

Under nonlimiting media conditions with mixed carbon–energy sources, bacteria can exhibit diauxic growth in which substrates are preferentially consumed on the basis of ease of assimilation and catabolism while utilization of other substrates is repressed (15). The metabolic shift from one substrate to another is accompanied by a lag in growth. However, using carbon-limited chemostat cultures of *E. coli* ML30 (DSM 1329), Lendemann et al. (16) demonstrated that this phenomenon does not occur under the low nutrient (e.g., μg/L) concentrations frequently encountered in natural environments. Using combinations of as many as six sugars (glucose, galactose, maltose, ribose, arabinose, and fructose), the authors demonstrated that, regardless of the mixture combination or dilution rate, *E. coli* simultaneously utilized all the sugars.

Effects of Nutrient Fluctuations

With media designed to maximize culture growth, care must be taken to ensure that media parameters do not deteriorate in terms of either nutrient depletion or accumulation of inhibitory metabolites. While the need for maintenance of an overall nitrogen–carbon balance of a medium is generally recognized, the effects of minor nutrient depletion can be subtle, but important. For example, under conditions of methionine deprivation, *E. coli* will substitute norleucine in tRNA (A.S. Lubinecki, personal communication, 1994), a potentially serious anomaly in the production of a biopharmaceutical peptide. Phosphate limitation has been shown to cause the loss of the plasmid pBR322 from *E. coli* (17). With respect to metabolic byproducts, rapidly growing aerobic cultures of *E. coli* accumulate acetate to growth-inhibiting levels. Acetate formation in *E. coli* has been shown to limit expression of recombinant growth hormone (18) as well as to limit induction of human interleukin 2 expression in fermentor cultures (19). Kleman and Strohl (20) have studied acetate metabolism in *E. coli* under chemically defined nutritional conditions.

ENRICHMENT AND ISOLATION MEDIA

Obtaining a pure culture of a specific type of bacterium, particularly for the general heterotrophs discussed here, often requires both general and selective enrichment. When the organism of interest is present only in small numbers or as a nonrobust population of the total microbial flora, an initial general enrichment of the sample may help to enhance the presence of the desired organism. With *E. coli* and *Bacillus* spp., almost any general bacteriological growth medium (e.g., tryptic soy broth) can often support enrichment of the desired organism as well as proliferation of any other general heterotroph in a mixed sample. Under such conditions, selective enrichment techniques can help to increase the numbers of the organism or phenotype of interest while suppressing the growth of nondesirable cells in a mixed population.

In some instances, raw samples may include physiologically stressed organisms that might be best recovered with a slightly less nutritious medium such as R2YA (21), or with HTYE (see final section) developed for cultivation of a broad spectrum of "heterotrophically delicate" bacteria (e.g., *Caulobacter, Cytophaga*) and stressed cells, while also being supportive to the growth of general heterotrophs. Clavero and Beuchat (22) determined that heat-stressed *E. coli* 0157:H7 was best recovered with tryptic soy agar rather than with modified sorbitol MacConkey agar.

If strain selection is intent upon a particular phenotypic characteristic (e.g., utilization of heptane as a sole carbon source) rather than upon taxonomic concern, isolation may be done directly by inoculating sample material into nutritionally complete defined media or into dilute complex media supplemented with a substrate compound of interest. Cultivation of sample material under a physiological parameter (alkalophilic or acidophilic conditions) of interest is also a selective pressure. The judicious incorporation of antibiotics in a medium will also keep many of the unwanted organisms in a mixed culture to low levels. In general, prolonged incubation or multiple transfers of any microbial growth that develop in the selection media will eventually give rise to a culture expressing the desired characteristic. Continuous fermentation methods can be used to the same effect to select for a phenotype from a heterogenous cell population, but often the resulting dominant or selected population phenotype is not necessarily genetically stable once selection factors are relaxed (23).

Isolation is accomplished from an enrichment culture generally by spatially separating the cell of interest from other cells in the enrichment culture. This is most conveniently done by streaking aliquots on agar or other solid gel plates with the aim to have discretely separated bacterial colonies to pick, or by serial dilution in liquid media with the hope that the desired cells, being dominant in the population, will eventually be isolated by increasing dilutions of the original enrichment culture.

Because *E. coli* is an indicator organisms of fecal contamination, many versions of media have been developed to quickly isolate and identify the bacteria in potable and nonpotable waters, food, etc. This organism can be isolated using a variety of commercially available selective media (e.g., brilliant green broth, EC broth, lauryl tryptose broth, MacConkey agar). These media contain bile salts (0.15–0.5%) to inhibit the growth of competative gram-positive as well as some gram-negative organisms in a sample mixture. Dyes such as crystal violet or brilliant green are included in some of these formulations to further selectivity for coliforms. The inclusion of lactose (0.5–1.0%) in these media as a fermentable carbon source helps to differentiate *E. coli* from other bile- and dye-resistant organisms. Lactose fermentation is detected either by the presence of gas in an inverted Durham vial contained in a test tube of liquid medium or, with agar media, by differential colony color caused by acid production from lactose and a chemical pH indicator (such as neutral red) included in the medium. Based on the work of Feng and Hartman (24) and Robison (25), the accurate detection of *E. coli* with the classical selective media can be significantly increased by the inclusion of 4-methylumbelliferyl-β-D-glucuronide (MUG) at 50 mg/L. Most *E. coli* strains are β-glucuronidase positive and cleave the substrate to free the methylumbelliferyl moiety, which can then be detected by fluorescence under 365-nm ultraviolet light. Similarly, 5-bromo-4-chloro-3-indolyl-β-D-glucuronide (X-Gluc) can be incorporated as a chromogenic substrate targeting *E. coli*. Gaudet et al. (26) recently evaluated some of these selective media for their usefulness in enumerating *E. coli* from nonpotable waters and concluded that lauryl tryptose agar supplemented with MUG or X-Gluc gave the best sample recovery.

Isolation of spore-forming bacilli can frequently be accomplished by allowing culture growth in a generalized growth medium to proceed until nutrient depletion induces sporulation (see next section). Deliberate limitation of a key nutrient such as a carbon or nitrogen source can often quicken sporulation. Media selectivity can be introduced for pH parameters, salinity, substrate utilization, etc. as desired. As spores of *Bacillus* spp. are among the most heat-resistant of endospore-forming bacteria, heat can be used as a selective factor to eliminate vegetative cells as well as weakly thermal-resistant spores from a sporulated culture. Exposure of the culture to 70–80 °C for a minimum of 10 min is one general method for *Bacillus* spp. selection. Harvest and purification of the spores from remaining vegetative cells and other debris should be performed with gentle procedures, either enzymatically or by density gradient, because strong mechanical methods such as high-speed centrifugation can damage the spores. A single spore isolation can then be attempted by germinating aliquots of the spore sample using agar streak plates.

SPORULATION MEDIA FOR *BACILLUS* SPP.

Bacillus spp. tend to form spores in any suitable growth medium providing a sufficient incubation period. With complex media formulations containing peptones and yeast extract, the incubation time required for depletion of nutrients and stimulation of sporulation is frequently manipulated by a simple reduction of total nutrients as demonstrated with the laboratory medium, sporulation broth (see final section). Increased rates of sporulation can be achieved with the addition of appropriate cations (Ca^{2+}, Mg^{2+}, Mn^{2+}) at assimilable trace metal (millimolar, mM) concentrations (27). Classical media formulations include these trace elements as poorly defined organic chelates supplied in crude soil extracts as in SEY medium (see final section). Many strains of *Bacillus cereus, B. coagulans, B. megaterium, B. subtilis*, and others sporulate well with media based on potato infusions supplemented with yeast extract and 2.5 g/L glucose (28). The pH of these media should be adjusted to approximately 7.2 for pH-neutral strains; sporulation media for alkalophilic and acidophilic strains should be adjusted to a pH supporting optimal growth. If the optimum pH for the sample is known, adjust the pH of the sporulation medium to that of the habitat where the sample was collected. Highly aerobic strains may require aeration in liquid cultures to obtain a high yield of spores. Elevated glucose levels in a medium during the stationary phase of culture growth can repress sporulation.

As with all microbial processes, the optimal medium and growth conditions for sporulation is strain specific. Similarly, sporulation conditions for optimum chemical resistance may not be the same as those of temperature resistance (29). One should should evaluate more than one medium to find conditions that provide both a high degree of sporulation as well as adequate resistance. Wild-type *B. subtilis* can sporulate in chemically defined media with glucose as a sole carbon source. In studies on transduction with *Bacillus thuringiensis*, LeCadet (30) used media containing peptones to satisfy nitrogen and amino acid requirements, inorganic salts to supply cations, glucose as a carbon and energy source, and potassium phosphate as a buffer for a high yield of heat-stable spores. Setlow and Kornberg (31) noted that small additions of trace metals and glucose to nutrient agar greatly increased hydrogen peroxide resistance of *B. subtilis* spores. Although spore production in *B. stearothermophilus* and *B. thuringiensis* is readily achieved at commercial scale, successful in vitro sporulation of the insect pathogen *B. popilliae* remains elusive, thereby limiting its prospects in biocontrol.

The important enzymes of commercial interest secreted by *Bacillus* spp., as well as many of the antibiotic secondary metabolites produced by the genus, appear at the onset of sporulation. To maximize product yield, a culture must be able to be held in this physiological state without completing the sporulation process. Development of sporulation-deficient strains rather than media design is the more frequent approach to solving this dilemma.

MOLECULAR BIOLOGY MEDIA

A number of excellent manuals such as those of Sambrook et al. (32) and Miller (33) detail the fundamentals of molecular biology work with *E. coli* and *B. subtilis*. This section discusses only the general media most frequently encountered for the manipulations of these organisms.

General Growth

LB (Luria–Bertani) medium (see final section of this article) is the general workhorse formulation used in many molecular biology protocols involving the maintenance and propagation of genetic bacterial stocks. NZCYM medium and its variations (NZYM, NZM) are also used when nutritionally rich formulations are required such as in the propagation of bacteriophage. Similarly, TB (terrific broth) medium (see final section) supports increased bacterial growth and is used to prepare high-yield plasmid DNA or recombinant products from *E. coli, B. subtilis*, and other organisms. Powdered basal media for many of these formulations are available from Sigma.

λ-Bacteriophage Propagation

Because λ-phage tail fibers bind specifically to the *lamB* (maltoporin; maltose-binding protein) receptor on the outer membrane of *E. coli* K12, LB medium supplemented with filter-sterilized maltose at a concentration of 0.2% will induce the receptor in a bacterial culture and help maximize phage adsorption. Glucose in the medium acts as a catabolite repressor. Divalent cations are also required for phage adsorption and integrity, and LB media for λ can be further supplemented with filter-sterilized 2 mM $MgSO_4$. Adsorption by other phage may be enhanced by calcium salts rather than magnesium. However, for the preparation of λ stocks, particularly when liquid media are used, the medium components are adjusted to reduce the amount of adsorption because the lysed phage remain in prolonged contact with bacterial cell debris and thus readsorption to debris can occur, which will reduce overall phage titer when the lysate is centrifuged to remove the cell debris. Thus, the use of LB or NZYCM supplemented with a low concentration of glucose (0.3%) to slightly repress *lamB* and 0.075 mM $CaCl_2$ as a less efficient divalent cation is an effective way to reduce readsorption problems in the preparation of λ stocks from liquid culture.

Recombinant Detection

LB agar supplemented with 5-bromo-4-chloro-3-indolyl-β-D-galactoside (X-Gal) provides an easy way to detect bacterial colonies containing plasmid DNA. The method is used for vectors encoding for the operator-proximal portion of the β-galactosidase (*IacZ*) gene and hosts containing the carboxy-terminal portion of the gene. When a normal plasmid is incorporated into a bacterial cell, α-complementation of gene fragments between the vector–host permit the formation of the functional β-galactosidase enzyme. When the chromogenic substrate X-Gal is included in the medium, the Lac^+ bacterial colonies turn blue. Plasmids containing foreign DNA inserted at the polycloning site located within the vector β-galactosidase fragment do not allow α-complementation with the host gene fragment. Thus, a recombinant bacterial colony arising from a cell transformed with a plasmid containing foreign DNA does not develop the blue color associated with β-galactosidase activity on X-Gal; recombinant colonies remain white. X-Gal is generally used at a concentration of 40 mg/L when incorporated into LB agar. X-gal is not heat stable and must be added to LB agar after the medium is autoclaved. X-gal is not water soluble and is normally prepared as a stock solution in dimethylformamide (20 mg X-Gal/1 mL DMF). The stock solution can be sterilized through a solvent-resistant (e.g., nylon) 0.2-μm membrane filter. The stock solution is light sensitive and should be stored frozen.

Some host strains contain the *lac* repressor, which prevents β-galactosidase activity until derepressed. In this capacity, isopropyl-β-D-thiogalactopyranoside (IPTG) is used as an inducer for enzyme activity. IPTG is also used to induce the TAC promoter. Because it is heat labile, IPTG is added to sterile LB medium to a final concentration of 0.5 mM from a filter-sterilized aqueous stock solution.

Competent Cells and Transformation

Although transformation is not a universal trait in bacteria, strains of *Bacillus subtilis* demonstrate the competent ability to take up extracellular DNA via a complex membrane-mediated, nonnutritional mechanism considered to be a natural means of genetic exchange for these organisms (34). Only a small percentage of cells within a culture possess the capacity for transformation, and the generation of a competent subpopulation is best achieved with a minimal salts liquid medium amended, as necessary, with growth factors specific to the strain. Although the competent state arises with the stationary phase of culture growth, media conditions that induce maximum sporulation are not necessarily those which support a maximum population of transformation-competent cells because the two functions are physiologically unrelated.

For organisms lacking the ability to develop natural competence, or for which the natural occurrence is low (35), preparation of competent cells can be achieved chemically. For example, competent cells of *E. coli* are commonly prepared by growing a culture overnight in LB broth at 37 °C followed by treatment with highly concentrated solutions of $CaCl_2$ to increase membrane permeability to DNA. Be that as it may, the success of generating a high rate of competent cells is strain dependent as well variable with preparation batch.

THE USE OF ANTIBIOTICS IN MEDIA

Antibiotics are incorporated into culture media to eliminate the growth of contaminants or to select for an antibiotic-resistant population of cells. As antibiotic resistance is often plasmid mediated, the incorporation of a specific antibiotic into a medium that selects for plasmid-containing cells helps to prevent the spurious loss of important genetic elements carried on plasmids which can occur with constant culturing or with further genetic or physiological manipulations.

In general, antibiotics are heat labile and must be added aseptically to media from filter-sterilized or other types of cold-sterilized stock solutions. Although some antibiotics, such as penicillin-G, are stable at room temperature in powdered form, chemical degradation of these compounds begins to occur quickly once an antibiotic is in solution. The rate of degradation increases with temperature and, as noted in Table 1, many of the commonly used antibiotics lose their effectiveness within 3 days at the 37 °C incubation temperature frequently used for bacterial cultivations. In addition, tetracycline is light sensitive, and ampicillin is degraded from media by resistant bacteria. Thus, to maintain antibiotic effectiveness, it is more prudent to add the agent from a freshly prepared stock solution at time of culture inoculation rather than to incorporate the compound during the preparation and aseptic fill of the basal growth medium, especially if the medium is to be stored, even under refrigeration, for any prolonged length of time.

Table 1. Frequently Used Water-Soluble Antibiotics Used in Bacterial Media

Antibiotic	Stability at 37 °C (days)	Working concentration
Ampicillin	3	50–100 mg/L
Chloramphenicol	5	5 mg/L
Gentamycin sulfate	5	50 mg/L
Kanamycin sulfate	5	100 mg/L
Penicillin-G	3	100,000 U/L
Streptomycin sulfate	3	100 mg/L
Tetracycline	4	10 mg/L

Note: Chloramphenical and tetracycline are also ethanol soluble.

MEDIA FOR CULTURE PRESERVATION BY FREEZING OR LYOPHILIZATION

Any medium that gives sufficient growth with robust cell appearance and maintains the important characteristics of the strain is a candidate for a medium used in culture preservation. *Bacillus* spp. retain greatest viability when preserved as spores, and thus strains should be cultivated, if possible, under conditions conducive to sporulation. For nonsporulating bacilli and for strains of *E. coli*, cultures should be grown in an appropriate growth medium and harvested at the late logarithmic phase of growth. Genetically engineered strains may require the addition of a selective agent (e.g., antibiotics) to the medium to retain unstable genotypes. Previous preservation studies on plasmid containing strains of *E. coli* and *B. subtilis* indicated that plasmids were best retained with liquid nitrogen preservation. Plasmids in *E. coli* and in *B. subtilis* spores were also not affected by freeze-drying. However, freeze-drying of vegetative cells of *B. subtilis* resulted in loss of cell viability by an order of magnitude and also resulted in the requirement of a lengthy postlyophilization rehydration period before full plasmid-coded antibiotic resistance returned. Thus, the *B. subtilis* culture grown from vegetatively harvested lyophilized cells required an antibiotic-free medium for initial growth (36).

Cells of either bacilli or *E. coli* can be frozen in liquid nitrogen in growth medium supplemented with either 10% glycerol or 5% dimethyl sulfoxide (DMSO) as the cryoprotectant. This is most easily done by adding 20% sterile (autoclaved) glycerol to an equal volume of cells suspended in the culture medium or by adding filter-sterilized DMSO to cells suspended in the culture medium to a final concentration of 5%. If the culture is to be lyophilized (freeze-dried), cells should be harvested from the growth medium and suspended in an appropriate cryoprotectant such as 20% skim milk (sterilized in test tubes at 116 °C for 20 min) or the growth medium to which separately autoclaved 24% sucrose is added to a final concentration of 12%. A more complex cryoprotectant medium is a filter-sterilized reagent composed of trypticase soy broth (BBL 11768); sucrose, 100 g; bovine serum albumin fraction V, 50 g; and distilled or deionized water to 1.0 L.) Actual preservation methods are detailed in a number of publications (37,38).

MEDIA FORMULATIONS

1. HTYE Medium
   Trypticase peptone (BBL 11921), 5.0 g
   Yeast extract, 2.0 g

HEPES, 4.0 g
Distilled water, 1.0 L

Dissolve ingredients one at a time in distilled water. Adjust to pH 7.1 with NaOH. Sterilize by autoclaving at 121 °C for 15 min.

2. SEY Medium
Difco yeast extract, 1.0 g
K_2HPO_4, 0.2 g
Agar, 15.0 g
Soil extract (see below) 1.0 L

Adjust medium for final pH 7.0 and sterilize at 121 °C for 15 min.

Soil extract: Autoclave 500 g air-dried, sifted garden soil of high fertility with 500 ml tap water in a 2-L flask for 30 min. When cool, decant the supernatant, filter through paper, and bring to 500 ml with additional tap water.

3. Sporulation Medium
Yeast extract, 1.0 g
Beef extract, 1.0 g
Tryptose, 2.0 g
$FeSO_4$, 2.0 mg
Glucose, 10.0 g
Agar, 15.0 g
Distilled water, 1.0 L

Adjust medium for final pH 7.2. For broth, eliminate agar and reduce concentration to one-third of the given quantities. Sterilize at 121 °C for 15 min.

4. LB (Luria–Bertani) Medium
Tryptone (Difco 0123), 10.0 g
Yeast extract, 5.0 g
NaCl, 10.0 g
Distilled water, 1.0 L

Sterilize by autoclaving at 121 °C for 15 mins. Final pH 7.0.

5. NZCYM Medium
NZ amine (Sigma C 0626), 10.0 g
NaCl, 5.0 g
Bacto–yeast extract (Difco 0127), 5.0 g
Casamino acids, 1.0 g
$MgSO_4 \cdot 7H_2O$, 2.0 g
Distilled water, 1.0 L

Adjust to pH 7.0 with NaOH. Sterilize at 121 °C for 15 mins. Variations: NZYM (NZYCM without casamino acids); NZM (NZYM without yeast extract).

6. Terrific Broth
Bacto-tryptone (Difco 0123), 12.0 g
Bacto-yeast extract (Difco 0127), 24.0 g
Glycerol, 4.0 mL
Distilled water, 900.0 mL

Sterilize at 121 °C for 15 mins. Cool to room temperature and add sterile buffer solution (as below), 100.0 mL

Sterile buffer solution: KH_2PO_4, 2.31 g; K_2HPO_4, 12.54 g; distilled water, 90.0 ml. Sterilize at 121 °C for 15 min.

7. M9 Medium
Basal medium
Na_2HPO_4, 6.0 g
KH_2PO_4, 3.0 g
NaCl, 0.5 g
NH_4Cl, 1.0 g
Agar (for solid medium), 15.0 g
Distilled water, 990.0 mL

Adjust basal medium to pH 7.4. Autoclave at 121 °C for 15 min and cool to less than 50 °C. Aseptically add the additives listed next. If agar is needed, autoclave mineral salts and agar separately as 2× solutions and aseptically combine after sterilization.

Additives (sterilized by autoclaving as separate solutions at 121 °C for 15 min)
1 M $MgSO_4$, 2.0 mL
20% glucose, 10.0 mL
1 M $CaCl_2$, 0.1 mL

BIBLIOGRAPHY

1. D.R. Boone, Y. Liu, Z. Zhao, D.L. Balkwill, G.R. Drake, T.O. Stevens, and H.C. Aldrich, *Int. J. Syst. Bacteriol.* **45**, 441–448 (1995).
2. R.J. Cote and R.L. Gherna, in P. Gerhardt, R.G.E. Murray, W.A. Wood, and N.R. Krieg eds., *Methods for General and Molecular Bacteriology*, American Society for Microbiology, Washington, D.C., 1994, pp. 155–178.
3. M. DeFelice, M. Levinthal, M. Iaccarino, and J. Guardiola, *Microbiol. Rev.* **43**, 42–58 (1979).
4. M.N. Hughes and R.K. Poole, *J. Gen. Microbiol.* **137**, 725–734 (1991).
5. X. Hu and G.L. Boyer, *Appl. Environ. Microbiol.* **62**, 4044–4048 (1996).
6. N. Vasantha and E. Freese, *J. Gen. Microbiol.* **112**, 329–336 (1979).
7. B.D. Davis and E.S. Mingioli, *J. Bacteriol.* **60**, 17–27 (1950).
8. F.C. Neidhardt, P.L. Bloch, and D.F. Smith, *J. Bacteriol.* **119**, 736–747 (1974).
9. T.L. Miller and B.W. Churchill, in A.L. Demain and N.A. Solomon eds., *Manual of Industrial Microbiology and Biotechnology*, American Society for Microbiology, Washington, D.C., 1986, pp. 122–136.
10. R. Greasham and E. Inamine, in A.L. Demain and N.A. Solomon eds., *Manual of Industrial Microbiology and Biotechnology*, American Society for Microbiology, Washington, D.C., 1986, pp. 41–48.
11. H.I. Havik, *J. Gen. Microbiol.* **84**, 321–326 (1974).
12. L.A. Chastin and B. Magasanik, *J. Biol. Chem.* **243**, 5165–5178 (1968).
13. M.R. Atkinson, L.V. Wray, Jr., and S.H. Fisher, *J. Bacteriol.* **172**, 4758–4765 (1990).
14. M.R. Atkinson, L.V. Wray, Jr., and S.H. Fisher, *J. Bacteriol.* **175**, 4282–4289 (1993).
15. W. Harder and L. Dijkhuizen, *Philos. Trans. R. Soc. Lond. B Biol. Sci.* **297**, 459–480 (1982).
16. U. Lendenmann, M. Snozzi, and T. Egli, *Appl. Environ. Microbiol.* **62**, 1493–1499 (1996).
17. S.A. Jones, K. Dearnley, P.M. Bennet, and J. Melling, *Soc. Gen. Microbiol. Q.* **8**, 44–52 (1980).
18. E. Bech-Jensen and S. Carlsen, *Biotechnol. Bioeng.* **36**, 1–11 (1990).
19. H.L. MacDonald and J.O. Neway, *Appl. Environ. Microbiol.* **56**, 640–645 (1990).

20. G.L. Kleman and W.R. Strohl, *Appl. Environ. Microbiol.* **60**, 3952–3958 (1994).
21. D.J. Reasoner and E.E. Geldreich, *Appl. Environ. Microbiol.* **49**, 1–7 (1985).
22. M.R.S. Clavero and L.R. Beuchat, *Appl. Environ. Microbiol.* **61**, 3268–3272 (1995).
23. D.E. Dykhuizen and D.L. Hartl, *Microbiol. Rev.* **47**, 150–168 (1983).
24. P.C.S. Feng and P.A. Hartman, *Appl. Environ. Microbiol.* **43**, 1320–1329 (1982).
25. B.J. Robison, *Appl. Environ. Microbiol.* **48**, 285–288 (1984).
26. I.D. Gaudet, L.Z. Florence, and R.N. Coleman, *Appl. Environ. Microbiol.* **62**, 4032–4035 (1996).
27. J.G. Holt and N.R. Krieg, in P. Gerhardt, R.G.E. Murray, W.A. Wood, and N.R. Krieg eds., *Methods for General and Molecular Bacteriology*, American Society for Microbiology, Washington, D.C., 1994, pp. 179–215.
28. G.W. Gould and Z.J. Ordal, *J. Gen. Microbiol.* **50**, 77–84 (1968).
29. W.M. Waites and C.E. Bayliss, in G.W. Gould and J.E.L. Corry eds., *Microbial Growth and Survival in Extremes of Environment*, Academic Press, London, 1980, pp. 159–172.
30. M.M. LeCadet, M.O. Blondel, and J. Ribier, *J. Gen. Microbiol.* **121**, 203–212 (1980).
31. P. Setlow and A. Kornberg, *J. Bacteriol.* **100**, 1155–1160 (1969).
32. J.E. Sambrook, F. Fritsch, and T. Maniatis eds., *Molecular Cloning: A Laboratory Manual*, 2nd ed., Cold Spring Harbor Laboratory, Cold Spring Harbor, N.Y., 1989.
33. H. Miller, in S.L. Berger and A.R. Kimmel eds., *Methods in Enzymology, vol. 152, Guide to Molecular Cloning Techniques*, Academic Press, New York, 1987, pp. 145–170.
34. J. Spizizen, *Proc. Natl. Acad. Sci. USA* **44**, 1072–1078 (1958).
35. B. Baur, K. Hanselmann, W. Schlimme, and B. Jenni, *Appl. Environ. Microbiol.* **62**, 3673–3678 (1996).
36. W.C. Nierman and T. Feldblyum, *Dev. Ind. Microbiol.* **26**, 423–434 (1985).
37. R.L. Gherna, in P. Gerhardt, R.G.E. Murray, W.A. Wood, and N.R. Krieg eds., *Methods for General and Molecular Bacteriology*, American Society for Microbiology, Washington, D.C., 1994. pp. 278–292.
38. F.P. Simione and E.M. Brown, eds., *ATCC Preservation Methods: Freezing and Freeze-Drying*, American Type Culture Collection, Rockville, Md., 1991.

See also CULTURE MEDIA, ANIMAL CELL, LARGE SCALE PRODUCTION; ENERGY METABOLISM, MICROBIAL AND ANIMAL CELLS; FERMENTATION MONITORING, DESIGN AND OPTIMIZATION; MASS TRANSFER; SECONDARY METABOLITE PRODUCTION, ACTINOMYCETES, OTHER THAN STREPTOMYCES; YEAST, BAKER'S.

MEMBRANE CHROMATOGRAPHY

JÖRG THÖMMES
Heinrich-Heine Universität Düsseldorf
Jülich, Germany

KEY WORDS

Adsorptive membrane
Mass transport
Process integration
Protein adsorption
Protein chromatography
Protein purification

OUTLINE

INTRODUCTION

The key to the successful development of a chromatographic procedure for the industrial-scale purification of proteins is to establish a high-productivity method that delivers the product in sufficient purity. Laboratory-scale protocols are usually set up with special regard to resolution, which is maximized by choosing high-performance methods based on small adsorbent particles and high-pressure equipment. On the process scale, however, high throughput seems to be the major goal, so during scale-up the small particles usually have to be replaced by larger ones due to pressure drop limitations at high-flow velocity in large columns. Additionally capacity per unit volume of adsorbent becomes an issue, thus requiring the use of porous adsorbents, which offer a high internal surface area for product adsorption. When using comparatively large porous adsorbents (d_p = 50–200 μm) the process performance is mainly determined by the velocity of protein transport to the binding sites situated on the internal surface. Because large (protein) molecules are characterized by small diffusion coefficients in free solution ($D = 1 - 50$

$\times\ 10^{-12}\ m^2/s$) and because "hindered" transport within a porous structure leads to even lower "effective" diffusion coefficients, protein chromatography in these media has been described as being limited by slow intraparticle transport (1). Large-scale operations therefore seem to be characterized by the compromise between low pressure drop at high flow velocity and slow mass transport employing larger adsorbent particles.

The driving force for new developments in stationary-phase design has been the need to reduce the limitation by intraparticle mass transport. This may be achieved by increasing the influx velocity into the particle by enhancing diffusive transport (2) or by perfusing the stationary phase in order to support (or even replace) diffusive movement of the solutes to the ligands by convection. Perfused particles with very short diffusive pores are one option to accomplish this (3). An even more consequent way is the use of a continuous stationary phase, where the internal surface is completely accessible by convection, thus removing the necessity for diffusive motion. Adsorbents manufactured accordingly may be short in length (L) because the concept of convective solute transport essentially speeds up the sorption process and permits short residence times, t_C ($=L/v$). Sufficient capacity (adsorbent volume) is supplied by units of increased diameter, which has up to now resulted in a membrane type of design characterized by a low L/d ratio. This concept has first been realized employing modified microfiltration membranes of variable design (hollow fiber, flat sheet or membrane stack [4]), therefore the description *membrane chromatography* is justified despite the fact that a microfiltration membrane is not the only possible stationary phase fulfilling the criteria just discussed. The increasing importance of membrane adsorbents is demonstrated by two recent reviews (5,6). In this article the basic ideas behind membrane chromatography are elucidated from a conceptual point of view, supported by a short theoretical analysis. Possible limitations of the technology are discussed, followed by an attempt to position the membrane concept within a train of protein purification methods. Finally applications of adsorptive membranes on small and pilot scale are used to demonstrate the large potential of this technology as a highly productive method of purifying proteins on the process scale.

THE BASIC CONCEPT

Improved Mass Transport

As discussed earlier, membrane adsorbents try to overcome the diffusion limitation present in porous particles by allowing convective transport of solutes to the ligands. This is achieved by attaching ligands to the internal surface of stationary phases, which contain mainly throughpores and only few or no dead-ended pores. The underlying picture is the modification of the internal pore surface of microfiltration membranes with ligands in a way that, by pumping the feed solution through the membrane, the solutes are convectively transported to their binding site. Slow diffusion in dead-ended pores, as is the case with porous adsorbents, is thus avoided. The situation is depicted schematically in Figure 1. Additionally, a continuous stationary phase is used, so packing problems, which are common with particle-based media and necessitate a minimum bed length in order to neutralize packing irregularities, are circumvented. As a result, from these characteristics, adsorbents of low L/d ratio may be built, which stand for high productivity at short residence time due to fast mass transport in a continuous matrix. Prerequisite, however, is that the adsorbate solution be homogeneously distributed over the whole frontal area of the membrane, otherwise flow irregularities will neutralize the advantages mentioned. This is discussed in more detail in the following sections.

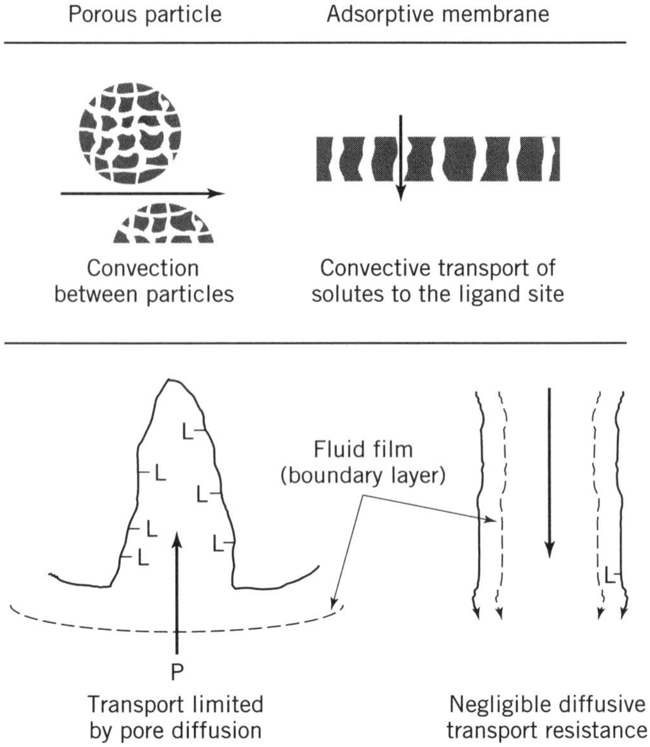

Figure 1. Schematic representation of the major transport characteristics of adsorptive membranes and porous particles.

Geometric Formats

Adsorbents fulfilling the criteria developed may be arranged in different formats. As shown in Figure 2, it is possible to differentiate between four basic concepts: single or stacked flat-sheet membranes (A), hollow-fiber modules (D), radial-flow cartridges containing spiral-wound membranes (C), and cartridges packed with continuous stationary phases in the form of membranes or rods (B). All these arrangements are commercially available; a complete list of available products, however, is difficult to compile because the life-span on the market of some of these products has been limited. The reader is referred to the information provided by Langlotz et al. (7), Roper and Lightfoot (5), and Thömmes and Kula (6).

Single-sheet membranes may be employed in suitable filter holders, as long as sufficient care is taken to equally distribute the process liquid across the frontal membrane

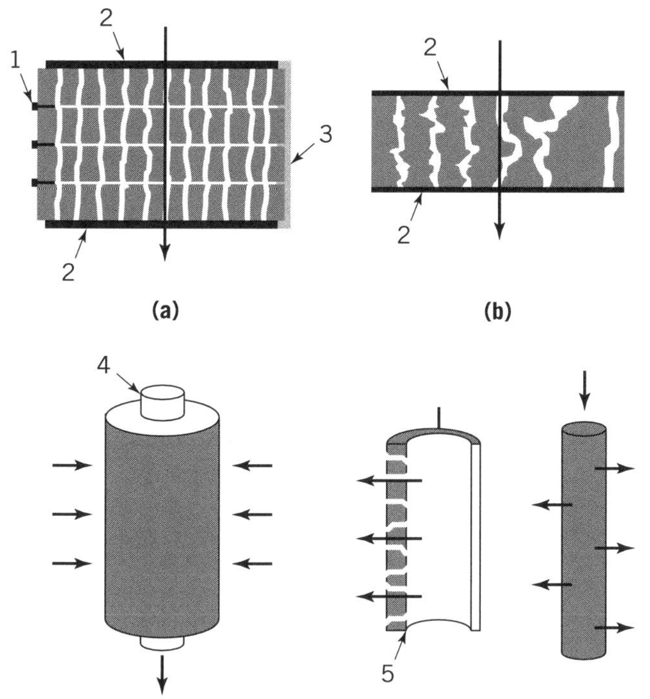

Figure 2. Schematic representation of the four major classes of membrane adsorbents: (**a**), stack of flat sheet membranes; (**b**), continuous cartridge with through-pores; (**c**), radial flow unit with spiral wound membranes; (**d**), hollow-fiber membrane; 1, individual gasket; 2, distributor and collector plate; 3, complete sealing around a membrane stack (e.g., silicone); 4, central tube for liquid collection; 5, cross-cut through an individual hollow-fiber.

area and to efficiently collect the liquid after passage through the membrane. The arrangement of membranes in stacks allows the increase of the capacity of an individual unit as well as an adjustment of the length of the adsorbent bed. The sheets must be separated by proper sealing in order to avoid fluid flow around the membranes instead of through them. This may be achieved by single gaskets or by embedding the stack in a suitable sealing substance (e.g., silicone). Alternatively a continuous adsorbent packing with thoroughgoing pores may be used, thus making the stacking of individual sheets obsolete. Radial-flow cartridges contain an internal tubing for fluid collection, which is surrounded by a spiral-wound membrane sheet. The liquid is transported in a radial direction from outside to inside, thus keeping the column aspect ratio, L/d, small, as is the case in a flat-sheet membrane. The same is true for hollow-fiber modules, where the fluid passes the membrane perpendicular to the influx direction. L is defined here as the wall thickness of a single fiber. These modules may be operated in a dead-end mode with the column outlet sealed, so that all the liquid must pass through the membrane; alternatively a cross-flow mode may be chosen with the possibility to recycle the process solution. The total capacity of a hollow-fiber adsorption module can be adjusted by bundling several fibers in a single module.

Static Capacity

As already discussed, on a preparative scale the capacity per unit volume of adsorbent available for a certain purification task becomes an important issue because adsorbent cost as well as plant space requirements of the whole purification unit have to be taken into account in economic considerations. Because the ligands are attached to the internal surface of stationary phases, their specific surface area, a(m^2/m^3), determines the equilibrium capacity in a first approximation. Champluvier and Kula measured a for nylon membranes (d_p = 1.2 μm) as 250 m^2/m^2 of frontal area (8). Assuming an average membrane thickness, L, of 200 μm we obtain a as 1.2×10^6 m^2/m^3. Conventional porous particles for protein chromatography have a specific area, which is significantly larger than this value (e.g., 1.5 $\times$ 10^7 m^2/m^3, our own measurements with the silica-based matrix Bioran CPG, Schott, Germany, 200-μm average particle diameter, 100-nm average pore diameter), so this seems to be a major drawback of membrane-based stationary phases. In fact membrane adsorbents with a monolayer coverage of ligands on their internal pore surface have suffered from reduced static binding capacity, as has been summarized by Roper and Lightfoot in their review (5). An approach to circumvent reduced capacity is the introduction of a three-dimensional binding layer on the pore surface. This is achieved by graft polymerization of monomers containing functional groups, which may be derivatized to protein-binding ligands. The resulting functionalized polymer layer works as a three-dimensional binding space with a gellike structure, which is able to take up adsorbate molecules in multiple layers, as depicted schematically in Figure 3. This technique has been applied to hollow-fiber membranes with great success (9–19) and is also used in commercially available flat-sheet membrane adsorbents (Sartobind, Sartorius, Göttingen, Germany) yielding a capacity comparable to porous adsorbent particles (20). The increase in static capacity for standard proteins compared with (theoretical) monolayer coverage has been reported to be between 4- and 20-fold for hollow-fiber modules (16) and up to 100-fold for flat-sheet membranes (21). Wang et al. compared the static capacity of polymer

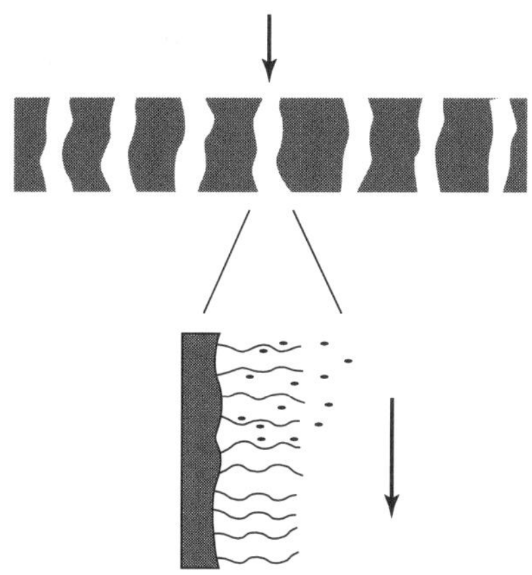

Figure 3. Schematic representation of the modification of the internal pore surface with grafted polymer ligands.

coated membranes for cytochrome *c* to membranes with monolayer ligand coverage, as well as to porous particles, and found the three-dimensional ligand layer to yield significantly higher values (22). However, as is discussed later, the gellike structure of the layer may change the mass transfer characteristics of the membrane adsorbents, a fact that has to be included in the analysis of the sorption performance.

Process Performance

When the suitability of a new adsorbent design for the large-scale purification of proteins is to be estimated and compared with established stationary phases, a benchmarking method must be defined. On the process scale, an important question is, how much product may be obtained within a certain period of time with a given amount of adsorbent? Additionally, the product should have sufficient purity in order to reduce the overall number of process steps in a purification train. Summarizing, the basic questions are the following:

- Which throughput is possible with a given set-up as a result of the permeability of the stationary phase (mechanical limitation)?
- Which throughput is possible with a given set-up as a result of the efficiency of the sorption process (dynamic limitation), in other words, is the static capacity of the adsorbent used well enough under the process conditions chosen?
- Which product purity is achieved with a given set-up under the process conditions chosen (resolution limitation)?

The permeability, B, of a membrane adsorbent is a function of the adsorbent's pore diameter and of the overall porosity of the matrix; typical values may be calculated from pressure-drop data available from the literature. Values ranging from 7×10^{-16} m^2 for continuous phases from methacrylate block polymers with 0.4-μm pore diameter (23), to 2×10^{-15} m^2 for 1.2-μm d_p hollow fibers with grafted polymer ligands (24), and 2.5×10^{-14} m^2 for commercially available stacks of flat-sheet membranes (average $d_p = 15\ \mu$m) (20) may be found. The pressure drop in the adsorption unit may then be calculated for a given set of operating conditions (L, v, η), after equation 1.

$$\Delta p = \frac{v \cdot L \cdot \eta}{B} \qquad (1)$$

The dynamics of the sorption process may be determined from frontal adsorption experiments. The deviation of measured breakthrough curves from the ideal step function shows how efficiently the static capacity of the adsorbent is used. Measuring the breakthrough capacity, Q_D, (at 1, 5, or 10% breakthrough) under different experimental conditions (e.g., by varying L and v) allows the definition of the process conditions that make the best use of the static capacity of the adsorbent. In addition, a theoretical analysis will yield characteristic quantities (e.g., numbers of transfer units, plate heights, or transport and dispersion coefficients) that allow a prediction of the process performance according to standard models (21,25–27).

The overall performance, however, must take washing, elution, and regeneration into consideration; this can be expressed as the total number of bed volumes required for all of these operations, α. The product yield is considered by the recovery ratio, R_E. Productivity may be defined as amount of product per unit time and unit adsorbent volume, as shown in equation 2 (21).

$$P = R_E \frac{1}{\left(\frac{Q_D}{C_0} + \alpha\right)} \left(Q_D \cdot \frac{v}{L}\right) \qquad (2)$$

Finally the product purity should be addressed by introducing the resolution, R_s, of the membrane-based separation under process conditions. An attempt to optimize R_s will often show that a compromise has to be found between maximizing the throughput at low L/d and maintaining high resolution, because the short bed lengths make adsorptive membranes into typical low-plate-number systems (28). Coffman et al. have shown, however, that most protein separations may be run successfully by employing membrane adsorbents because separation will occur according to an on–off separation mechanism that requires no separation by differential migration in long columns (29).

To summarize, the quality of a membrane adsorption method will be defined by the product purity it delivers at maximized productivity. A significant increase of this quality compared with standard methods using porous particles is anticipated due to the fact that the improved mass transport capability allows high throughput at low L/d at maintained resolution under on–off separation conditions. The factors that may limit the achievement of this ambitious goal are discussed next.

LIMITING FACTORS IN MEMBRANE ADSORPTION PROCESSES

As with all processes based on fixed beds, sorption dynamics of adsorptive membranes may be limited by mixing, mass transport, and the kinetics of the binding reaction. In order to discuss these effects precisely, one must attempt to define them according to the specific details of membrane adsorption units. Kinetic limitations arise from the rate of formation of the adsorbate–ligand complex, which is described by the forward reaction rate constant, k_a. Together with the rate constant of dissociation of the complex, k_d, we obtain the dissociation constant of the binding equilibrium, K_D (k_d/k_a). Further limitations are caused by slow transport of adsorbate molecules from the bulk solution to the binding site (e.g., diffusive motion through the fluid boundary layer above the binding surface). Although the concept of throughgoing pores in membrane adsorbents has been invented to overcome slow diffusion in dead-ended pores, the membrane production process may result in adsorbents with some pores of that kind, so in specific cases pore diffusion has to be considered in the description of sorption dynamics. Additionally, as already mentioned, the attachment of a three-dimensional adsorbing layer to the internal pore surface must be regarded as a potential mass transport barrier. Mixing in adsorptive processes should be split into three major effects: fickian dispersion, nonideal fluid flow within the adsorbent, and extracolumn mixing (30). The first effect describes the dispersive motion "against" the direction of convective flow, the second effect accounts for irregularities within the adsorbent, leading to uneven fluid flow through the packing, whereas the third summarizes contributions of tubing, extracolumn dead volumes, liquid distributors and collectors, zone broadening, and so forth. Very often these different contributions are lumped together in a single dispersion coefficient, D_{ax}, which is then characteristic for the whole experimental set-up rather than for the adsorbent itself. In the following the respective importance of these limitations is discussed in detail in order to provide a basis for the prediction of the performance of adsorptive membranes in the purification of biomolecules.

Mixing

Roper and Lightfoot (30) analyzed peak broadening in stacked flat-sheet membranes under nonbinding conditions with a specific flow-reversal technique in order to differentiate between fickian dispersion, nonideal fluid flow within the adsorbent, and extracolumn mixing. They used the height equivalent of a theoretical plate (HETP) concept to determine the individual contributions to the overall peak broadening and found the plate heights, h, resulting from fickian dispersion to be 1–5 μm for a range of flow velocity, v, between 0.5 and 4 cm/min. These values were in accordance with theoretical considerations, which predicted h as a function of the dispersion coefficient D_{ax} and of v according to equation 3.

$$h = \frac{2 \cdot D_{ax}}{v} \tag{3}$$

Values for h reported in literature for adsorptive membranes, however, were between 20 and 200 μm and showed a slight increase with linear velocity. This difference could be attributed to extracolumn effects and nonideal fluid flow. Peak broadening by a factor of 1.7 occurred due to extracolumn dispersion; the additional difference between h measured in this work and h in the literature data arose from uneven flow distribution in the stack. Similar results were reported by Gebauer et al. (21), who found the mixing in stacks of commercially available flat-sheet membranes (represented by an overall D_{ax}) to be 10-fold higher compared with an estimation based on literature correlations. This difference was attributed to extracolumn effects as well. Enlarged overall mixing may be a significant influence on the performance of membrane adsorbents. This was demonstrated in a scale-up study on the isolation of human serum albumin from plasma using stacks of ion exchange membranes (20). The increased mixing found with large-scale modules compared with laboratory-scale units led to a more than threefold reduction of the productivity of the process. Josic et al. (31) investigated the influence of liquid distribution at the column inlet of cartridges with a continuous stationary phase of type B (compare Fig. 2). In the case of nonoptimized inlet design, a significant part of the fluid passed through the central region of the unit, thus leaving large parts of the membrane unreachable by convective transport. The uneven flow distribution led to a significant loss of performance of chromatographic separations employing these units. With optimized inlet design, resolution was comparable with established HPLC methods. Reif and Freitag (32) demonstrated the importance of reduced dead volume behind the membrane; a large mixing zone led to deteriorated resolution in the analytical separation of standard proteins.

Nonuniform flow distribution may be caused by variations in the pore size of adsorptive membranes. Suen and Etzel (26) as well as Liu and Fried (33) demonstrated by computer simulations that a wide pore-size distribution may lead to significantly flattened breakthrough curves during frontal protein adsorption and thus to a strongly reduced useful capacity of the adsorbents. Similar effects were shown for variations in membrane thickness. As a consequence, stacking of membranes was recommended to reduce these effects. Kim et al. described the modification of hollow fibers with grafted polymer chains bearing phenylalanine and tryptophan as pseudoaffinity ligands for immunoglobulin adsorption. The deviation of protein breakthrough from ideal behavior with these adsorbents was also attributed to a wide distribution in pore length (12). The distribution in pore diameter in protein A modified-hollow-fiber modules was described by Klein et al. as the reason for incomplete use of the static capacity of these adsorbents in immunoglobulin capture (34). The influence of overall mixing on performance during frontal adsorption was investigated in a theoretical study by Suen and Etzel (26). These authors used the dimensionless column Peclet number, Pe, to quantify dispersion according to equation 4 (v_e is the interstitial flow velocity).

$$Pe = \frac{v_e \, L}{D_{ax}} \tag{4}$$

Peclet numbers above 40 were described to be sufficient to exclude a deteriorating influence of mixing on the breakthrough behavior during frontal adsorption to membrane stacks. Because Pe is proportional to the bed length, augmented mixing (represented by increased D_{ax}) may be compensated for by increasing L. Similarly Liu and Fried (33) reported a Pe above 25 to be sufficient to neglect dispersion as a possible limitation of the sorption process. It is important to note that these considerations were set up for frontal adsorption of a one-component system; stronger boundary conditions may be required for the separation of a multicomponent system.

To summarize, as long as sufficient care is taken during process design to minimize extracolumn effects as well as nonuniform flow, mixing phenomena are unlikely to be the limiting factors in membrane chromatography. However, it has to be kept in mind, that adsorptive membranes are used in a chromatographic process; therefore standard criteria of filtration module design will not meet the requirements of a sorption process and may possibly lead to a loss of the advantages of the membrane-based process.

Mass Transport

When dealing with porous particles for protein purification, there are two potential sources of transport resistance to adsorption: transport of adsorbate from the bulk of the solution to the external particle surface, and transport within the particle to the binding site. In this case the general perception is that fluid-side transport is an order of magnitude faster than transport within the adsorbent; therefore film transport resistance is often neglected in a simple analysis. Band spreading in frontal analysis, as well as peak broadening during chromatography, then is attributed to slow, often hindered, intraparticle diffusion. The concept of using membranes with throughgoing pores for protein adsorption was supposed to remove the transport resistance by reducing the characteristic length for protein diffusion to the ligands. Brandt et al. (4) provided a general estimation of the importance of mass transport for the efficiency of an adsorptive membrane by relating the characteristic diffusion time in a membrane pore, t_D ($t_D = d_p^2/D$), to the contact time, t_C ($t_C = L/v$), during passage through the membrane. Efficient usage of the static capacity was supposed to take place if $t_C \gg t_D$. Inserting standard values for adsorptive membranes into this equation ($d_p = 3 \times 10^{-6}$ m, $L_{\text{single sheet}} = 2 \times 10^{-4}$ m, $v = 8.3 \times 10^{-4}$ m/s) and assuming an average diffusion coefficient, D, of 5×10^{-11} m^2/s we obtain $t_D = 0.18$ s, whereas t_C even for a single membrane is 0.24 s. Therefore a stack of 10 membranes, which is often found in commercially available modules, easily fulfills the condition, which allows us to neglect diffusive transport resistance. This is in contrast to standard porous particles, where t_D is based on the particle radius and therefore can be estimated as 40.5 s for the same protein in a particle 9×10^{-5} m in diameter.

The absence of dead-ended pores and the concomitant disappearance of the intraparticle transport resistance should make diffusion through the fluid boundary layer above the ligands the only transport step (see Fig. 1) in membrane chromatography. The rate of transport through this layer is dependent on the free diffusion coefficient of the adsorbate as well as on the thickness of the layer. Because interstitial flow velocities in narrow pores of microfiltration membranes are comparably high and boundary layer thickness thus may be low, this transport resistance commonly is regarded as not likely to limit the sorption process. Suen and Etzel (26) demonstrated the validity of this approach for membranes with small pore diameter by comparing the characteristic time needed to diffuse through the fluid boundary layer with the average residence time in a module. They were able to show that even for very small diffusion coefficients (i.e., large protein molecules) the residence time in a correctly designed membrane stack is much higher than the diffusion time through the fluid film, so the importance of this transport resistance is low compared with other possible limitations. In cases of much larger pore diameters these considerations may no longer be valid. Adisaputro et al. (35) measured the dynamics of whey protein uptake to membranes of very large pore diameter (50–300 μm) and found reduced capture efficiency with decreasing residence time. Calculation of the fluid film diffusion time for an average protein showed that, with these large pore diameters, film transport was a likely candidate for the dominant limitation of sorption efficiency.

As mentioned earlier, the introduction of a three-dimensional ligand layer to the pore surface may also introduce a mass transport resistance. Gebauer et al. (21) investigated commercially available ion exchange membranes with respect to their sorption dynamics and found a limitation of sorption efficiency by a diffusion mechanism. Depending on the length and density of the ligand layer a pore or solid diffusion mechanism of protein transport was identified. The breakthrough capacity, however, could be maintained on a very high level for a wide range of flow velocities, a fact that was attributed to convection-aided mass transport into the flexible chains of the ligand layer. Attaching polymer-grafted ligands to the internal pore walls of hollow fibers also yielded stationary phases with three-dimensional adsorbing layers. Here no dependency of breakthrough behavior from residence time in the module was detected, thus supporting the thesis of fast protein influx into the ligand sphere (12,15–17,24).

Mass transport limitations are not as dominant in membrane adsorption as they are in standard protein chromatography using porous particles. The complete neglecting of transport resistance, however, is possible only under certain conditions. Therefore transport in adsorptive membranes should be analyzed carefully during process design based on these stationary phases.

Binding Kinetics

If axial mixing and mass transport may under certain circumstances be excluded as potential limiting factors of sorption efficiency, the kinetics of the protein–ligand interaction becomes the focus of attention. Brandt et al. (4) used a second-order rate expression to estimate the reaction time, t_R, for interaction of BSA with a monoclonal antibody (MAb) affinity ligand. They obtained $t_R = 5$ s, a significantly higher value than the average diffusion time in membrane pores, which was calculated earlier. From this estimation it can be deduced, that slow association kinetics may limit the performance of affinity-membrane adsorbents. A similar situation was predicted by Suen and Etzel (26) from a thorough theoretical analysis, which was later experimentally verified with the pepstatin A/pepsin affinity system (36). Kinetic limitations were also found to be responsible for deviations of protein breakthrough from an ideal step function in dye ligand/dehydrogenase pseudoaffinity systems employing stacks of flat-sheet membranes (37). Briefs and Kula (38) reported the onset of peak broadening due to kinetic limitation with a similar pseudoaffinity interaction from t_C below 20.5 s. For ion exchange interactions, however, no kinetic limitation could be verified within the range of experimental conditions ($t_C = 5.1$ to 20.5 s). An investigation on kinetically limited membrane chromatography was performed by Nachman et al. (39–41) employing hollow-fiber modules with affinity ligands attached to the internal pore surface. In this study it was shown that reducing the residence time in the module in turn reduced the capture efficiency of the adsorbents. In addition, under all experimental conditions, t_C was well above t_D, which had been calculated based on the membrane geometry and the diffusion coefficient of the adsorbate. Also the capture efficiency was independent from initial concentration, as can be predicted for kinetic limitation under conditions of a rectangular isotherm (21). Kinetically limited adsorption may be exploited in membrane chromatography to separate adsorbate molecules that have identical binding isotherms but different association rate constants. In standard porous adsorbents these pairs cannot be separated due to the fact that slow diffusion in the pores leads to the formation of a local equilibrium between adsorbate in the pore liquid and bound adsorbate. Membrane adsorbents are sensitive to differences in adsorption kinetics, provided neither mass transport nor mixing are "hiding" the kinetic differences. This was theoretically predicted (27) and proven experimentally (36) for an affinity interaction based on stacks of flat-sheet membranes.

Adsorptive membranes offer a rare opportunity to investigate effects of binding kinetics in adsorptive processes. The fact that this is possible underlines the excellent mass transport characteristics of these matrices, which allow very fast protein purification under conditions of protein–ligand interactions as the rate-limiting step in the sorption process. However, this seems to be limited to affinity interactions, which show comparatively slow association rates with ligands attached to planar surfaces due to higher steric requirements. With fast ion exchange interactions, even adsorptive membranes have not been able to demonstrate kinetic effects during protein sorption.

POSITIONING OF MEMBRANE ADSORPTION IN A PURIFICATION TRAIN

Now that adsorptive membranes have been characterized according to their basic properties, where should they be introduced into a standard protein purification scheme? Wheelwright differentiated between initial recovery, low-resolution purification, and high-resolution purification in a downstream process (42). Initial recovery is supposed to deliver a clarified product solution of reduced volume, whereas low-resolution purification should lead to a significant enrichment of the target molecule and to the removal of major contaminants. High-resolution purification refers to the removal of minor impurities with properties similar to those of the desired product, as well as final polishing (e.g., separation of correctly processed molecules from false variants, multimers, etc.). Depending on the design of the adsorption unit, adsorptive membranes can accommodate any of these tasks.

Krause et al. demonstrated that application of dye–ligand affinity membranes in a cross-flow mode allowed the purification of malate dehydrogenase from *E. coli* homogenate (43). A clarified, concentrated, and initially purified product solution was obtained after elution, thus demonstrating the feasibility of membrane adsorption for integration into primary recovery. Additional support for the use of membrane-based adsorption in this early process stage comes from the isolation of proteins from unclarified whey, which was accomplished with large-pore membrane stacks (35).

A second assignment in the area between initial recovery and low-resolution purification is the use of adsorptive membranes for product capture from diluted but already clarified solutions. Good mass transport characteristics together with fast binding kinetics make ion exchange membranes a successful candidate; hydrophobic interaction and pseudoaffinity ligands are also suitable for this task (11,17,18,44–46).

Low-resolution purification is achieved when membranes are used for the separation of proteins in order to initially fractionate clarified product solutions (32,47–50). Sequential arrangement of membrane units for different separation mechanisms allows the development of a multidimensional separation concept (51).

The use of membranes for high-resolution purification was realized in high-performance membrane chromatography (HPMC) (23,31,52). Here the properties of continuous stationary phases of type B (Fig. 2) are used with ion exchange, hydrophobic interaction, and reversed-phase ligands. Affinity (34,53,54), immunoaffinity (40), and receptor affinity membranes (41) also belong to this category, although they may be used early in a purification train in order to introduce a high selectivity step in the very beginning of the downstream process. Fast binding kinetics favor the use of very dilute feed solutions because high-affinity adsorption may be operated with large throughput in membrane devices, even if the feed concentration is on the order of magnitude of the dissociation constant, K_D, of the adsorbate–ligand interaction (39).

The versatility of adsorptive membranes in serving the needs of the different stages of a downstream process is summarized in Figure 4.

OPTIMIZING THE PERFORMANCE OF ADSORPTIVE MEMBRANES

If we define the performance of a membrane adsorption unit according to the parameters described earlier ("Process Performance") as the realization of a desired product purity at maximized productivity, then the optimization task has two main goals: the achievement of a certain resolution and the simultaneous increase of throughput. The problem of maximizing throughput requires information on mechanical stability of the stationary phase, the quality of fluid flow through the unit, and knowledge of the overall kinetics of the sorption process. The latter point may in a first approximation be reduced to the question of residence time within the adsorbent phase. As discussed in the section "Mass Transport", adsorptive membranes are supposed to have superior transport properties even if the static capacity is increased by introducing a three-dimensional ligand layer on the internal membrane surface. This fact in turn reduces the residence time ($t_C = L/v$) that is needed for efficient usage of the available capacity. An indicator of this efficiency often used in protein chromatography is the breakthrough capacity, which increases with residence time and is dependent on the performance of the stationary phase with regard to the overall dynamics of the sorption process, be it limited by transport processes, dispersion, or binding kinetics. The better the

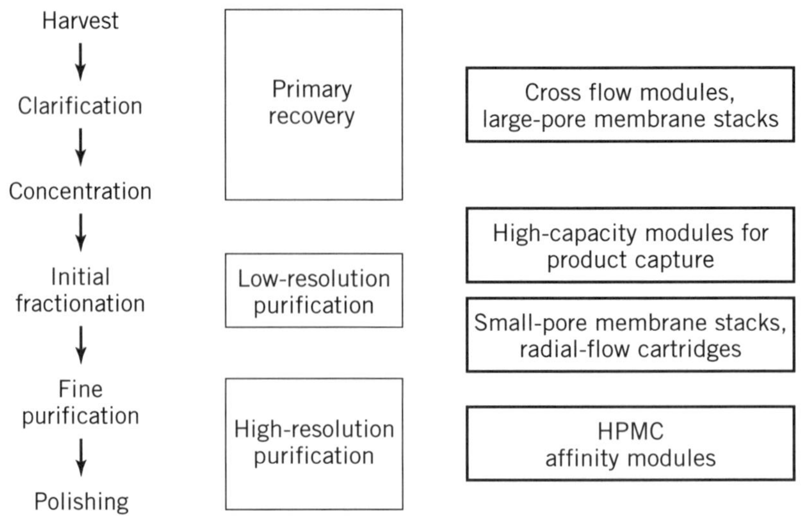

Figure 4. Positioning of membrane adsorbents in a purification train.

dynamic properties of the adsorbent, the shorter the residence time can be without compromising the breakthrough capacity. In the case of adsorptive membranes this allows using adsorbent beds that are short in length (L) and can be operated at high fluid velocity (v). The work of Yamamoto and Sano (55) as well as equation (2) predict the productivity to increase with the breakthrough capacity, and to decrease with t_C. As breakthrough capacity increases with t_C, there is an optimum in t_C with respect to productivity. The advantage of adsorptive membranes is that efficient usage of the static capacity is already obtained at short t_C; therefore a high productivity of membrane-based adsorption methods can be anticipated.

A sophisticated way of predicting the dynamic behavior of membrane adsorbents has been presented by Suen and Etzel (26,27). A numerical solution of the equations describing membrane adsorption was used to analyze the influence of operating parameters on performance. In addition to breakthrough behavior, the separation of two proteins in kinetically limited chromatography on membrane stacks was simulated and proven experimentally. As shown by Gebauer et al. (21) productivity of membrane adsorption systems can also be analyzed based on analytical solutions of the equations describing protein breakthrough, if a single limiting step can be identified for the process under investigation.

Similar considerations are necessary if the dependency of resolution on the operating parameters is estimated. From the fact that membranes may be regarded as low-plate-number systems, a minor performance in the separation of proteins could be predicted. As shown by Coffman et al. (29), however, most protein separations are operated in an on–off mode during elution chromatography. This means that the selectivity of the separation relies on the differences in the interaction between the proteins to be separated and the adsorbent at various concentrations of the eluting agent. Separation by differential migration, which is caused by the different retardation of the individual molecules during their movement through the adsorbent, plays only a minor role. Therefore only a few theoretical stages are needed for the satisfactory separation of proteins by ion exchange, hydrophobic interaction, or reversed-phase methods; the intrinsic separation factor of the different proteins is high enough to justify separation according to an on–off mechanism. Consequently, short membrane adsorbents allow very good separations, and the resolutions obtained are mostly comparable to HPLC separations, where long columns yield very high plate numbers under high-pressure conditions. Coffman et al. reported that about 50 stages is sufficient for most separation tasks, so HETP values between 20 and 200 μm, which have been found for adsorptive membranes, necessitate bed lengths of 10 mm for good resolution in protein separation. Similar findings on the bed length necessary for satisfactory results in protein separation using adsorptive membranes were presented by Dubinina et al. (56). In their analysis the authors demonstrated, that only a short bed is needed if proteins are separated according to an on–off mechanism. In an extended study the influence of gradient steepness, flow velocity, and bed length was investigated, and it was shown that short adsorptive membranes may be used for high throughput separation with good resolution.

Because increasing throughput (either by increasing the load or by reducing t_C) must be suspected to reduce resolution, the same criteria for optimizing L/v apply as in the case of the breakthrough capacity of membrane adsorbents. The good mass transport capabilities then allow the use of high flow rates at retained resolution when using well-designed membrane units. This is supported by the findings of many workers, who reported excellent resolution over a wide range of flow velocities during the separation of proteins with these devices (23,31,32,47,49, 52,57–59).

A major problem of adsorptive membranes (introduced earlier in "Mixing") is a direct consequence of the improved transport properties. Operating at short t_C results in configurations of small column aspect ratio (L/d). Ensuring an even fluid flow through an adsorbent bed of large diameter is a major task during module design for application of membrane chromatography, especially on a large scale. Part of the problem arises from the difficulty of distributing liquid evenly over a large diameter. In the case of adsorption to particle-shaped stationary phases, uneven distribution of fluid at the column inlet will be neutralized during passage through the comparatively long beds; the lower part of a column will act as an additional distribution unit. In the case of membrane adsorbents this is not possible because the very short bed length, which can be chosen without compromising the performance, is the major advantage of this concept. Thus a very careful design of the distribution unit of large-scale equipment is essential. Up to now, commercially available units have not taken this into consideration, a fact that can be considered as one of the main reasons why membrane adsorption has not lived up to its preparative potential in protein purification. Furthermore the fabrication of the membrane adsorbent itself has to be taken into account. Wide distribution in pore size will lead to a preferential passage of fluid through certain regions of the membrane, resulting in a broad distribution of residence times with the respective consequences for process performance. This was analyzed theoretically by Suen and Etzel (26), who demonstrated that stacks of flat-sheet membranes are suited to achieving a homogenized flow through membranes with a large variance in porosity. It must be pointed out that hardware problems (uniform membranes and proper housing design) presently are the major limitation of a further success of membrane adsorption on the process scale. Until these problems are solved this promising technique will remain the "eternal talent" within the large variety of stationary phases for protein adsorption that are available on the market.

APPLICATIONS

As described earlier, adsorptive membranes may be used at several positions in a purification train. The scope of this section is to discuss applications from each of the positions within a purification scheme, but the selection provided here is not intended to be a complete collection of membrane-based adsorption methods. Therefore the reader is referred to recent review articles for additional applications (5,6,60). The examples were selected in order to test the prediction that processes employing membrane adsorbents have superior productivity at retained resolution.

Primary Recovery

Because microfiltration membranes are suitable for the clarification of culture broth or homogenates, the attachment of ligands to the internal pore surface allows a combination of protein adsorption and solid–liquid separation in a single step. This approach has been realized by Krause et al. (43), who used Cibacron blue–modified microfiltration membranes (Sartobind Blue, Sartorius, Göttingen) for the isolation of malate dehydrogenase directly from unclarified *E. coli* homogenate. The membrane adsorbent had a static capacity similar to commercially available particles with the same ligand and was used in a cross-flow mode with recycling of the filtrate (i.e., a batch adsorption mode). The overall productivity was 1,200,000 U/(L h) and combined clarification, capture, and initial purification in a single step. Data on long-term stability of this method are lacking, however; therefore the influence of traditional shortcomings of microfiltration membranes due to fouling, layer formation, and so forth cannot be judged.

A different field of application is the capture of whey proteins from unfiltered whey, employing large-pore membrane stacks (35) (Productiv S and QM, BPS Separations, Spennymoor, U.K.). Adisaputro et al. replaced the traditional method of batch adsorption of whey proteins with membrane stacks. Based on their results the authors predicted an increased productivity of the membrane method at the short residence times, which become possible due to fast mass transport characteristics. The size of the pores, however, was a point of concern because, on one hand, a large pore size was required to allow the use of unfiltered whey as an initial recovery operation, but on the other hand, too-large pore diameters cause fluid-side boundary layer problems, as already discussed.

Capture

If the use of clarified solutions as a feedstock for adsorptive processes is preferred to the whole-broth scenario just outlined, then the situation after removal of particulate matter is characterized by a large volume of product at a comparatively low concentration. This is especially valid for the production of recombinant proteins, so the primary goal at this point is the reduction of process volume. Usually, a concentration step such as ultrafiltration or precipitation is introduced, but this may, as well, be a perfect task for membrane adsorption. The efficient capture at short residence time is coupled with the selectivity of an adsorption process, so concentration may be combined with a high-productivity initial purification.

A typical process example is the purification of MAb from hybridoma cultures. The product concentration usually does not exceed 1 g/L even with highly efficient cell lines; the range of titers found is instead on the order of a few hundred milligrams per liter. Additionally the growth medium has physiological ionic strength (13–15 mS/cm), which requires diafiltration or dilution of the feedstock in order to allow sufficient capacity if an ion exchange step is chosen as the initial purification method. Ion exchange membrane adsorption has been described as an efficient method to capture MAb directly from clarified broth, possibly preceded by dilution to reduce ionic strength. Jungbauer et al. (61) used Zetaprep QAE radial flow cartridges (LKB, Bromma, Sweden) for this purpose and performed a scale-up study from 50 mL to 930 mL total adsorbent volume. The productivity reported was 0.2 g/(L h); a certain enrichment of the MAb was demonstrated by SDS gel electrophoresis. Cation exchange membrane stacks as well as a pleated module were used by Lütkemeyer et al. for the isolation of IgG from diluted hybridoma supernatant on a laboratory scale (28 mL of adsorbent, feedstock volume 5.5 L) (58). The high static capacity of the adsorbent (50 mg/mL) together with short residence times in the short adsorbent bed yielded a productivity of 75.2 g/(L h), a value that was 13 times higher compared with the same process performed using a packed bed of porous cation exchange particles. In a later publication a similar process was shown on the pilot scale with undiluted hybridoma supernatant. Even at the high ionic strength, the overall productivity was 29 g/(L h). During the experiments, 100 L of supernatant was processed in 2 h in a unit with a 4-mm bed length (Sartobind S, Sartorius, Göttingen, Germany; $V = 220$ mL); the process volume was reduced 44-fold at 100% recovery (62). A similar approach was chosen by Wang et al. for the purification of a recombinant immunofusion protein secreted by *E. coli* (22). The same stack of cation exchange membranes was used for capture of the target protein from 200 L of clarified and 10-fold-diluted culture broth. The reported productivity was up to 1.5 g/(L h) at 10-fold reduction of process volume. Capture of urease to a hollow fiber, which had been modified to a tentacle type of adsorbent by grafting polymer chains with anion exchange ligands to the internal pore surface, was described by Matoba et al. (17). The productivity (expressed as urease collection rate) found in these investigations was up to 200 g/(L h). These examples show how the efficiency of membrane adsorbents may be exploited for high-productivity capture of target proteins from diluted culture supernatants.

Low-Resolution Purification

Because even during capture processes with adsorptive membranes a separation of the target molecules from impurities is achieved, there is no simple borderline to low-resolution purification. One way of differentiating between the processes is with regard to their main tasks, which are reduction of process volume and isolation of the main component from major impurities. During capture mainly ion exchange membranes are used, but hydrophobic interactions as well as pseudoaffinity ligands may be applied during low-resolution purification. At this point the considerations for optimizing t_C with regard to resolution under high-throughput conditions, which have been discussed, become important.

A typical example of low-resolution purification is the isolation of HSA in a plasma fractionation process based on ion exchange adsorbents. This was first described by Lacoste-Bourgeac for the valorization of fraction IV paste, which originated from the traditional ethanol precipitation during plasma fractionation (63). A series of anion exchange and cation exchange spiral-wound membrane cartridges (Cuno, Cergy-Pointoise, France) was used to recover albumin at 66% yield from the paste. Thus the high content of protein in this fraction, which otherwise was reported to be of only little value, could be exploited. A process for the isolation of HSA employing stacks of flat-sheet membranes was presented by Gebauer et al. and was compared with the well-established ion exchange process based on porous beads (20). Here a series of strong anion and cation exchange steps was used with a productivity of up to 220 g/(L h) on the laboratory scale. This value was six to eight fold higher than the traditional process based on a porous ion exchange matrix (Sepharose FF, Pharmacia, Uppsala, Sweden). The procedure was increased in scale based on commercially available equipment to 110 mL of adsorbing volume; the results, however, showed a drop in productivity of a factor of 3 compared with the small scale. Additionally a significant dilution of the product was observed. The deteriorated performance of the large-scale module could be attributed to overall mixing effects in the module, which originally was a filtration unit and had not been optimized with regard to chromatographic needs. This supports the earlier conclusion that neglective column design will lead to a dramatic loss of performance due to the sensitivity of short-bed adsorbents to extracolumn effects. A series of anion exchange and hydrophobic interaction was presented by Luksa et al. for the purification of recombinant tumor necrosis factor α from clarified *E. coli* extract (64). Comparison of the results obtained with commercially available cartridges of type B (Quick Disk, Knauer, Berlin, Germany) with a conventional method based on porous particles showed a similar product purity at half the process time, thus demonstrating that increased throughput at retained resolution may be obtained with adsorptive membranes. The performance of another commercially available stack of anion exchange membranes (Memsep DEAE, Millipore, Bedford, Mass.) was evaluated by Gerstner et al. (47). The authors were able to separate three component mixtures of standard proteins at a productivity of 58 g/(L h). Reif et al. published the use of immobilized metal ion chromatography on membrane stacks for the purification of a $(His)_6$-tagged recombinant protein (65). In this example it could be shown that, contrary to ion exchange membrane chromatography, the slow association kinetics of the pseudoaffinity interaction may limit the throughput with regard to resolution. Additionally it was shown that step gradients give better results in protein separation on membrane adsorbents than linear gradients—a universal phenomenon that may be due to the fact that the short membrane bed is not sufficient to allow the development of the self-sharpening effects of the linear gradient, which favorably influences the resolution in particle-based chromatography. The separation of whey proteins in strong ion exchange membranes was reported by Splitt et al. (66). In this work the influence of bed length was shown in the separation of the two genetic variants of β-lactoglobulin from whey, which became possible when a minimum bed length of 2 mm was chosen by the coupling of commercially available flat-sheet units. The capture of whey proteins from diluted solutions was achieved at a productivity of 45.5 g/(L h) with more than 60-fold reduction of the feed volume. Champluvier and Kula used Dye ligand membranes to isolate dehydrogenases from clarified yeast extracts. A productivity of 150,000 U/(L h) was found for glucose-6-phosphate dehydrogenase with Cibacron blue immobilized to nylon flat-sheet membranes (37). In this case the throughput was described to be limited by the kinetics of the association reaction. In a sequential arrangement of ion exchange and dye-ligand-modified membranes, formate dehydrogenase was purified with a productivity between 24 and 37 g/(L h). The protein was purified 4.6-fold with 27% overall yield by this the sequence; the eluate from one step was directly applied to the next membrane without intermediate buffer exchange (51). An interesting application was reported by van Huynh et al. with the isolation of plasmid DNA from crude cell lysates using stacks of flat-sheet membranes (DEAE Memsep, Millipore, Bedford) (67). Because these comparatively large bioproducts provide some challenge to conventional particulate matrices due to their size, membranes with throughgoing pores are an interesting alternative for the processing of these potentially large-scale products (e.g., for gene therapy).

Table 1 summarizes the productivity obtained during the applications described here. This selection from a much larger number of low-resolution purification operations described in the literature shows that membrane adsorbents can be used efficiently at this stage of a downstream process. They also elucidate, that the scale-up of this method must be based on solid considerations with regard to the needs of an adsorption process, otherwise the impressive results on the laboratory-scale with regard to high throughput at retained resolution are not transferable to the process scale.

High-Resolution Purification

Membrane-shaped stationary phases were introduced into high-resolution purification with the concept of high-performance membrane chromatography (HPMC) (23,31, 52,59,68). By choosing continuous stationary phases of a membrane type of geometry, separations could be performed at resolutions comparable to HPLC methods. The improved transport characteristics of the continuous phase allowed higher throughput in spite of the low L/d ratio, which was used to obtain reduced pressure drop along the columns. A thorough analysis of the parameters influencing the performance of HPMC was provided by Tennikova and Svec (23). An interesting result of their investigations was an increased resolution of anion exchange HPMC during the separation of standard proteins, with increasing flow velocity; the separation in hydrophobic interaction chromatography (HIC) mode was not affected. Compared with low-resolution purification employing adsorptive membranes, step gradients yielded better separations than a linear gradient.

Attaching affinity ligands to membranes offers another opportunity to use them in high resolution purification. As shown by Nachman et al. (39–41), affinity membrane adsorbents are well suited for the fast purification of complex molecules from diluted feedstock. By coupling of receptors or antibodies to commercial hollow-fiber modules (Sepracor, Marlboro, Mass.) automatic separations were performed at high productivity (2–6 g/(L h)). Similar results were obtained with protein A–modified hollow-fiber modules from the same manufacturer (34). A dimensionless analysis of IgG adsorption to this matrix revealed a kinetic limitation of the sorption efficiency; breakthrough capacity increased with t_C but was independent from C_0 (69). As an alternative to hollow fibers, flat-sheet membranes were modified with protein A by Langlotz et al. (53) and were used to purify IgG from hybridoma supernatant. In this work it could be demonstrated that high resolution by affinity ligands and high productivity may be obtained at the same time (10.8 g/(L h)).

CONCLUSION

From the existing literature some important features of adsorptive membranes may be extracted, making them a very interesting candidate for stationary phases for the process-scale purification of proteins. Despite this fact, large-scale applications of the technology have not been undertaken. This may partly due to the traditional reluctance to use new matrices for the industrial-scale purification of proteins, where the proof of long-term stability and reproducibility is essential for the success of the downstream process. Additionally, creative solutions for the practical realization of large-scale units with very low L/d have not been developed. At this point the manufacturers have to prove that the demands of an adsorptive process with regard to regular fluid flow, reduced extra column effects, and so on can be served on the process scale. Currently there are only few examples of pilot-scale processes available in the literature; it is hoped that the near future will bring many more successful applications of adsorptive membranes. The technology certainly has potential for process-scale separations, as has been demonstrated by theoretical analysis and laboratory-scale experiments.

Table 1. Some Typical Productivity Values Obtained with Membrane Adsorbents

Ligand	Geometric format	Target protein	Productivity (g/(L h))[a]	Ref.
Q	Stack	Formate dehydrogenase	36	51
Q	Pleated module	Antithrombin III	21.5	58
Q	Stack	HSA	220	20
DEAE	Hollow fiber	Urease	200	17
S	Pleated module	IgG	72.5	58
S	Stack	IgG	29	62
MAb	Hollow fiber	Recombinant interferon α2a	2.7	40
Histidin	Hollow fiber	IgG	6.7	46
Procion yellow	Stack	Pyruvate decarboxylase	40.3	38
Procion red	Stack	Formate dehydrogenase	24	51
Cibacron blue	Cross-flow module	Malate dehydrogenase (from unclarified broth)	1.1	43
Protein A	Hollow fiber	IgG	35	34
Protein A	Stack	IgG	10.8	53

Productivity was calculated from the cited references by calculating the total time of the respective purification cycles (including washing and elution, when available) and relating the amount of product purified to the cycle time and the adsorbent volume used.

NOMENCLATURE

a	Specific surface area (m^2/m^3)
α	Number of bed volumes required for wash, elution and regeneration during chromatographic cycle
B	Permeability of a packing (m^2)
C	Initial concentration (kg/m^3)
D	Diffusion coefficient (m^2/s)
D_{ax}	Axial dispersion coefficient (m^2/s)
d	Adsorbent bed diameter (m)
d_p	Pore diameter (m)
ϵ	Void fraction (−)
η	Dynamic viscosity (kg/(m s))
HETP	Height equivalent of a theoretical plate (m)
k_a	Second-order association rate constant (l/(mol s))
k_d	Dissociation rate constant (s^{-1})
L	Adsorbent bed length (m)
L/d	Column aspect ratio (−)
n	Plate number in HETP concept (−)
P	Productivity (g/(L h))
Pe	Peclet number (−)
Q_{max}	Static capacity (kg/m^3)
Q_D	Breakthrough capacity (kg/m^3)
R_E	Recovery ratio
R_S	Resolution
t_C	Residence time (s)
t_D	Diffusion time (s)
t_R	Reaction time (s)
V	Adsorbent volume (m^3)
v	Flow velocity (m/s)
v_e	Interstitial flow velocity (m/s)

BIBLIOGRAPHY

1. J.C. Janson and T. Peterson, in G. Ganetsos and P. E. Barker ed., *Preparative and Production Scale Chromatography*, Marcel Dekker, New York, 1993, pp. 559–590.
2. E. Boschetti, L. Guerrier, P. Girot, and J. Horvath, *J. Chromatogr. B* **664**, 225–231 (1995).
3. N.B. Afeyan, S.F. Fulton, I. Mazsaroff, and F.E. Regnier, *Bio / Technology* **8**, 203–206 (1990).
4. S. Brandt, R.A. Goffe, S.B. Kessler, J.L. O'Connor, and S.E. Zale, *Bio / Technology* **6**, 779–782 (1988).
5. D.K. Roper and E.N. Lightfoot, *J. Chromatogr. A* **702**, 3–26 (1995).
6. J. Thömmes and M.-R. Kula, *Biotechnol. Progr.* **11**, 357–367 (1995).
7. P. Langlotz, S. Krause, and K.H. Kroner, *Filtrieren und Separieren* **5**, 62–70 (1991).
8. B. Champluvier and M.-R. Kula, *J. Chromatogr.* **539**, 315–325 (1991).
9. K. Saito and M. Ito, *Ind. Eng. Chem. Res.* **28**, 1808–1812 (1989).
10. H. Iwata, K. Saito, and S. Furusaki, *Biotechnol. Progr.* **7**, 412–418 (1991).
11. M. Kim, K. Saito,S. Furusaki, T. Sugo, and I. Ishigaki, *J. Chromatogr.* **586**, 27–33 (1991).
12. M. Kim, K. Saito, S. Furusaki, T. Sato, T. Sugo, and I. Ishigaki, *J. Chromatogr.* **585**, 45–51 (1991).
13. H. Shinano, S. Tsuneda, K. Saito, and S. Furusaki, *Biotechnol. Prog.* **9**, 193–198 (1993).
14. K. Kobayashi, S. Tsuneda, K. Saito, H. Yamagishi, S. Furusaki, and T. Sugo, *J. Membrane Sci.* **76**, 209–218 (1993).
15. S. Tsuneda, H. Shinano, K. Saito, S. Furusaki, and T. Sugo, *Biotechnol. Progr.* **10**, 76–81 (1994).
16. S. Tsuneda, K. Saito, S. Furusaki, and T. Sugo, *J. Chromatogr. A* **689**, 211–218 (1995).
17. S. Matoba, K. Tsuneda, K. Saito, and T. Sugo, *Biotechnology* **13**, 795–797 (1995).
18. N. Kubota, M. Kounosu, K. Saito, K. Sugita, K. Watanabe, and T. Sugo, *J. Chromatogr. A* **718**, 27–34 (1995).
19. S. Tsuneda, K. Saito, T. Sugo, and K. Makuuchi, *Radiat. Phys. Chem.* **2**, 239–245 (1995).
20. K.H. Gebauer, J. Thömmes, and M.-R. Kula, *Biotechnol. Bioeng.* **54**, 181–189 (1997).
21. K.H. Gebauer, J. Thömmes, and M.-R. Kula, *Chem. Eng. Sci.* **52**, 405–419 (1996).
22. W.K. Wang, S.-P. Lei, H.G. Monbouquett, and W.C. McGregor, *Biopharm*, 52–59 (June 1995).
23. T.B. Tennikova and F. Svec, *J. Chromatogr.* **646**, 279–288 (1993).
24. N. Kubota, Y. Konno, S. Miura, K. Saito, K. Sugita, K. Watanabe, and T. Sugo, *Biotechnol. Progr.* **12**, 869–872 (1996).
25. D.D. Frey, R. van de Water, and B.R. Zhang, *J. Chromatogr.* **603**, 43–47 (1992).
26. S.Y. Suen and M.R. Etzel, *Chem. Eng. Sci.* **47**, 1355–1364 (1992).
27. S.Y. Suen, M. Caracotsios, and M.R. Etzel, *Chem. Eng. Sci.* **48**, 1801–1812 (1993).
28. S. Yamamoto, *Recovery of Biological Products VIII*, Tucson, Ariz., October 20–25, 1996.
29. J.L. Coffman, D.K. Roper, and E.N. Lightfoot, *Bioseparation* **4**, 183–200 (1994).
30. D.K. Roper and E.N. Lightfoot, *J. Chromatogr. A* **702**, 69–80 (1995).
31. D. Josic and J. Reusch, *J. Chromatogr.* **590**, 59–76 (1992).
32. O.W. Reif and R. Freitag, *J. Chromatogr. A* **654**, 29–41 (1993).
33. H.-C. Liu and J.R. Fried, *AIChE J.* **40**, 40–48 (1994).
34. E. Klein, E. Eichholz, and D.H. Yeager, *J. Membrane Sci.* **90**, 69–80 (1994).
35. I.A. Adisaputro, Y.-J. Wu, and M.R. Etzel, *J. Liq. Chrom.* **19**, 1437–1450 (1996).
36. S.-Y. Suen and M.R. Etzel, *J. Chromatogr. A* **686**, 179–192 (1994).
37. B. Champluvier and M.R. Kula, *Biotechnol. Bioeng.* **40**, 33–40 (1992).
38. K.G. Briefs and M.R. Kula, *Chem. Eng. Sci.* **47**, 141–149 (1992).
39. M. Nachman, *J. Chromatogr.* **597**, 167–172 (1992).
40. M. Nachman, A.R.M. Azad, and P. Balion, *Biotechnol. Bioeng.* **40**, 564–571 (1992).
41. M. Nachman, A.R. Azad, and P. Balion, *J. Chromatogr.* **597**, 155–166 (1992).
42. S.M. Wheelwright, *J. Biotechnol.* **11**, 89–102 (1989).

43. S. Krause, K.H. Kroner, and W.D. Deckwer, *Biotechnol. Techn.* **5**, 199–204 (1991).
44. S. Le Borgne, M. Graber, and J.S. Condoret, *Bioseparation* **5**, 53–64 (1995).
45. M.H. Heng and C.E. Glatz, *Biotechnol. Bioeng.* **42**, 333–338 (1993).
46. S.M.A. Bueno, K. Haupt, and M.A. Vijayalakshmi, *J. Chromatogr. B* **667**, 57–67 (1995).
47. J.A. Gerstner, R. Hamilton, and S.M. Cramer, *J. Chromatogr.* **596**, 173–180 (1992).
48. W.F. Weinbrenner and M.R. Etzel, *J. Chromatogr. A* **662**, 414–441 (1994).
49. R. Freitag, H. Splitt, and O.-W. Reif, *J. Chromatogr. A* **728**, 129–137 (1996).
50. A. Jungbauer, F. Unterluggauer, F. Steindl, F. Ruker, and H. Katinger, *J. Chromatogr.* **397**, 313–320 (1987).
51. B. Champluvier and M.R. Kula, *Bioseparation* **2**, 343–351 (1992).
52. T.B. Tennikova, M. Bleha, F. Svec, T.V. Almazova, and B.G. Belenkii, *J. Chromatogr.* **555**, 97–107 (1991).
53. P. Langlotz and K.H. Kroner, *J. Chromatogr.* **591**, 107–113 (1992).
54. Z. Kucerova and J. Turkova, *IJBC* **2**, 145–151 (1997).
55. S. Yamamoto and Y. Sano, *J. Chromatogr.* **597**, 173–179 (1992).
56. N.I. Dubinina, O.I. Kurenbin, and T.B. Tennikova, *J. Chromatogr. A* **753**, 217–225 (1996).
57. K.G. Briefs and M.R. Kula, in M. Ladisch and R. Bose eds., *Harnessing Biotechnology for the 21st Century*, American Chemical Society, Washington D.C., 1992, pp. 258–261.
58. R. Luetkemeyer, M. Bretschneider, H. Buentemeyer, and J. Lehmann, *J. Chromatogr.* **639**, 57–66 (1993).
59. F. Svec and T.B. Tennikova, *J. Bioact. Comp. Polymers* **6**, 393–405 (1991).
60. E. Klein, *Affinity Membranes*, Wiley, New York, 1991.
61. A. Jungbauer, F. Unterluggauer, K. Uhl, A. Buchacher, F. Steindl, D. Pettauer, and E. Wenisch, *Biotechnol. Bioeng.* **32**, 326–333 (1988).
62. D. Luetkemeyer, N. Ameskamp, H. Tebbe, K. Bracht, and J. Lehmann, in M.J.T. Carrondo ed., *Animal Cell Technology: From Vaccines to Genetic Medicine*, Kluwer Academic, 1997, pp. 325–329.
63. J.F. Lacoste-Bourgeacq, C. Desneux, and M. Allary, *Chromatographia* **532**, 27–32 (1991).
64. J. Luksa, V. Menart, S. Milicic, B. Kus, V. Gaberc-Porekar, and D. Josic, *J. Chromatogr. A* **661**, 161–168 (1994).
65. O.-W. Reif, V. Nier, U. Bahr, and R. Freitag, *J. Chromatogr.* **664**, 13–25 (1994).
66. H. Splitt, I. Mackenstedt, and R. Freitag, *J. Chromatogr. A* **729**, 87–97 (1996).
67. N. van Huynh, J.C. Motte, J.F. Pilette, M. Decleire, and C. Colson, *Anal. Biochem.* **211**, 61–65 (1993).
68. D. Josic, Y.-P. Lim, A. Strancar, and W. Reutter, *J. Chromatogr. A* **662**, 217–226 (1994).
69. C. Charcosset, Z. Su, S. Karoor, G. Daun, and C.K. Colton, *Biotechnol. Bioeng.* **48**, 415–427 (1995).

See also CHROMATOGRAPHY, COMPUTER-AIDED DESIGN; CHROMATOGRAPHY, ION EXCHANGE; CHROMATOGRAPHY, SIZE EXCLUSION; PROTEIN PURIFICATION, AQUEOUS LIQUID EXTRACTION.

MEMBRANE SEPARATIONS

MANOHAR KALYANPUR
Consultant, Bioseparations and Pharmaceutical Validation
Plaisir, France

KEY WORDS

Asymmetric membranes
Cross-flow filter
Diafiltration
Depyrogenation
Microfiltration
Nanofiltration
Permeate
Retentate
Tangential-flow filter
Ultrafiltration

OUTLINE

INTRODUCTION

The term *filtration* refers to the separation of two or more components from a liquid or gas. Generally, it refers to the separation of solids from either liquids or gases. In membrane separations, one goes beyond the classical separation of solids from the other phases. Membranes can be and are used to separate dissolved solutes in a solution from one another and also to remove the solutes from the liquid in which they are dissolved. In this manner, membranes are used to either purify products or to concentrate them. This is particularly advantageous in biotechnology where the products of biological origin are very often proteinaceous in nature and are heat labile. The membranes offer a means of separation that occurs at or below ambient temperatures. This gives the biotechnologist a separation or purification technique that provides maximum assurance of product stability. In membrane-based separations, the membrane acts as a barrier to the passage of certain components while others move freely across the membrane. Therefore, the end result of a simple membrane separation is the concentration in the upstream fraction of the components that do not cross the membrane barrier. This fraction is referred to as the *retentate*. The fraction that passes across the membrane is called the *permeate* or *filtrate* and contains the components of the solution in essentially the same concentration as the starting fluid. This is the simple basic *membrane separation*.

THREE MAJOR APPLICATIONS OF MEMBRANE SEPARATIONS

Product Purification

Product purification is the most important application for membrane separations in biotechnology. Membranes are made of different materials, and the biotechnologist needs to make a choice based on the goal of the application and the suitability of the membrane for the application. This article aims at providing the biotechnologist with information that will help select the best membrane and the manner in which to use the membrane to accomplish the separation and purification task easily. Membrane separations are an important part of both the upstream and downstream processing in biotechnology. A typical process can include a number of steps, such as centrifugation followed by membrane-based clarification of the supernatant, one or more chromatographic procedures, final membrane processing steps to accomplish further purification, and finally concentration.

Sterile Filtration

One of the most critical steps in the manufacture of injectable therapeutic agents is the sterilization of the product. The simplest method of sterilizing solutions is the one that uses heat, and autoclaving is a typical example of such methods. However, products of the biotechnology industry are most often thermolabile and therefore cannot be heat sterilized. Membranes provide an excellent alternative. Membrane filters are used for filtering out bacteria and other microorganisms from the final formulations, rendering the products very safe for injection. This step is followed by lyophilization of products that are marketed as solid powders. Those products marketed as liquid formulations are filled straight into ampules, vials, or other appropriate containers after sterile filtration through membrane filters.

Virus Removal by Membrane Filtration

This is the most recent application of membrane separations in the biopharmaceutical industry. A new generation of membranes is today used to remove viruses from biological preparations. Therapeutic agents manufactured from blood, by cell culture processes, or by the most modern method using transgenic animals can contain dangerous viral contaminants from the source materials. These products can be filtered through membrane devices to remove most viral contaminants, including HIV and the viruses known to cause different forms of hepatitis.

There is a point of differentiation here from the previously mentioned sterile filtration in which there is complete or total removal of microbiological contaminants from the solution. In virus filtration, the removal is not complete, especially when the viruses are very small. Therefore, for complete viral safety, the practice is to use membrane filtration as a method complementary to the longer known generic methods of viral inactivation, such as treatment with a combination of solvents and detergents, other chemical inactivating agents, and pasteurization by mild heating for prolonged periods of time.

CLASSIFICATION OF MEMBRANES AND MEMBRANE PROCESSES

The more commonly used membrane processes in the biotechnology industry are microfiltration, ultrafiltration, and reverse osmosis. All three are pressure-driven processes and differ from each other in the size of the components they retain. The membrane itself, under the influence of applied pressure, controls the retention and passage of individual components of the solution being processed.

Microfiltration

Microfiltration processes retain particles in the submicron range that are suspended in the process fluid. The membranes are also used in removing particles and microorganisms from air and gases, as in the fermentation processes in biotechnology and other industries. The particles removed by membrane filtration range in size from 0.1 μm to 10 μm. Conventional filtration processes such as those using pad filters and filter presses in the more conventional pharmaceutical industry are generally capable of retaining particles only above the 10 μm size.

Ultrafiltration

Ultrafiltration membranes generally retains macromolecules or particles from about 0.001 to 0.2 μm in size. By choosing one or more of the available membranes, one can separate larger macromolecules from smaller ones within the above size ranges. For example, separation of large proteins or polysaccharides from smaller molecules is a typical application for ultrafiltration membranes in the biotechnology industry.

Reverse Osmosis

Reverse osmosis is one of the oldest known applications for membrane separations. This is the process where membranes are capable of retaining even smaller molecules than ultrafiltration membranes. Salts and other molecules with a molecular weight below 1,000 Da that are not normally retained by ultrafiltration membranes are retained by the reverse osmosis membranes. The first use of these membranes was in desalting sea water. The membranes are today used for concentrating small molecules, and ultrafiltration is used for fractionation and purification of macromolecules with molecular weights between 1,000 and 1 million Da. Figure 1 summarizes the use of membrane filters in biotechnology. Table 1 gives examples of products that can be retained by the three types of membranes. In a typical biotechnology process, microfiltration is the first membrane-based step, and it is used to separate suspended material such as cells from a bioreactor and other particulate matter, including cell debris, from the multitude of dissolved substances, which very often includes the specific product of interest expressed by the cell line. In many instances, clarifying filters made of a variety of materials are used for preclarification of the process stream before microfiltration. This usually helps to improve the performance of the microfiltration step and extends the life of the membrane filter. Honig and Schwartz (1) have written an excellent review of this subject. The filtrate or permeate from the microfiltration step contains the desired product elaborated by the process along with other proteins, carbohydrates, and smaller molecules such as salts from the culture medium as well as those produced during the process. To separate the product of interest from these contaminants, the ideal membrane separation step is ultrafiltration. Membranes with the appropriate nominal molecular weight cutoff, a term used to identify and differentiate ultrafiltration membranes from one another, are next used in successive purification steps to facilitate recovery of the desired product of high purity. Quite often, the manufacturing process also includes chromatographic or other suitable process steps in between membrane separations to achieve an overall process advantage in terms of product quality and yield. Figure 2 illustrates how membrane and chromatography procedures are integrated together step-by-step for upstream and downstream processing in the biotechnology industry.

MEMBRANE CHEMISTRY, STRUCTURE, AND FUNCTION

The high-volume microfiltration in the biopharmaceutical industry requires removal of particulate and colloidal substances in an economical and efficient manner. The nature of these particles and the interplay of filtration dynamics contribute to the difficulty of the task. It is important to balance filter performance against process costs to arrive at a system that provides the ideal separation solution at a low operating cost. In general, the filtration process starts out with passing the process stream through a porous medium that entraps the solids in its matrix or retains them on its surface. The materials used in the manufacture of these filters are porous or fibrous, such as paper, felt, woven yarns, and fabrics, and other materials. Depending on the basic filtration mechanisms, particle retention mode and construction characteristics, the different filters used in the biologicals industry are classified as *depth filters, surface filters*, and *screen filters*. Figures 3, 4, and 5 show the mechanism of retention of particulates by these three types of filters.

Depth filters consist of fibrous, granular, or sintered materials that produce a random porous structure. When a

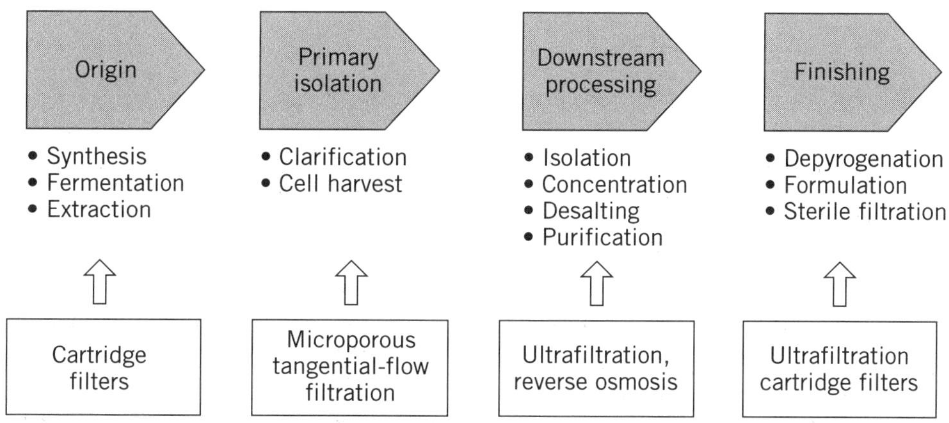

Figure 1. Different steps in downstream processing where membrane separations are used.

Table 1. Products Retained by Microfiltration, Ultrafiltration, and Reverse Osmosis Membranes

Size	Product		Membrane process
100 μm 10 μm 1 μm	Pollen Starch granules, blood cells, bacteria Smaller organisms	]→	Microfiltration
50,000–100,000 D 10,000–100,000 D 1000–10,000 D	Albumin, immunoglobulins, enzymes, other proteins, large oligopeptides	]→	Ultrafiltration
<1000 D	Gluose, small peptides, salts	]→	Reverse osmosis

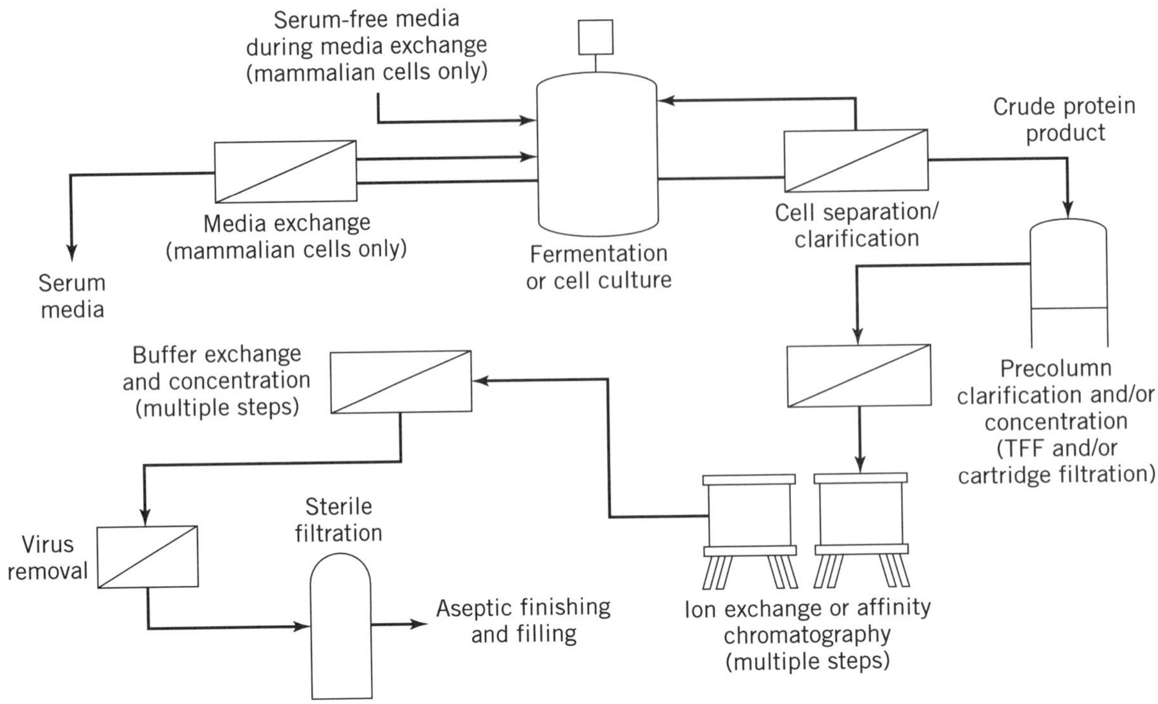

Figure 2. Integration of membrane separations with other techniques in downstream product recovery operations in biotechnology.

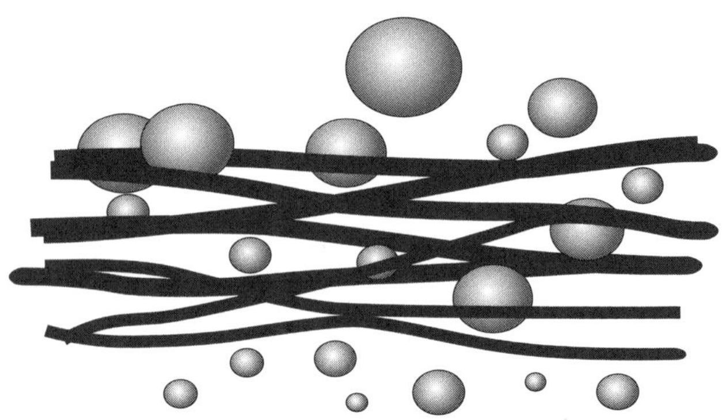

Figure 3. Mechanism of particle retention by depth filters.

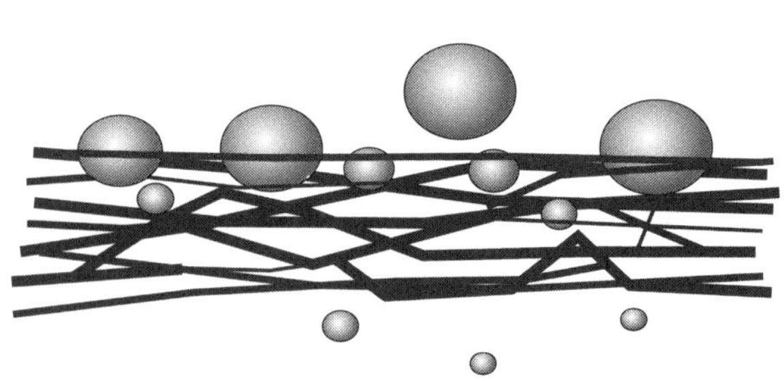

Figure 4. Mechanism of particle retention by surface filters.

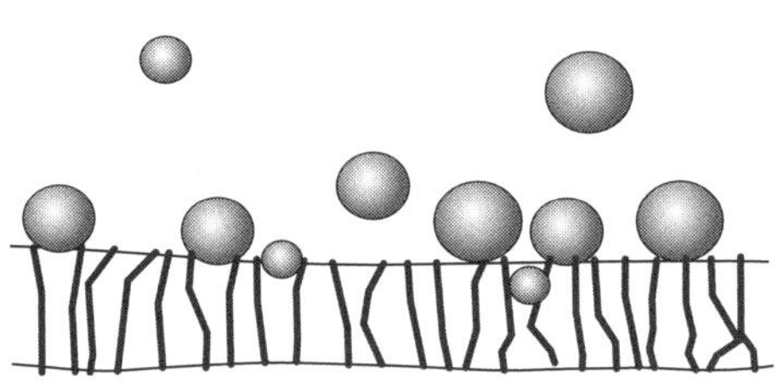

Figure 5. Mechanism of particle retention by screen filters.

fluid passes through a depth filter, particles become entrapped in its tortuous network of flow channels. Particle retention mechanisms are random adsorption and mechanical entrapment throughout the depth of the filter matrix. The filtration media of these filters can be wound cotton, rayon, or polypropylene as well as resin bonded laminates such as cellulose and fiberglass.

Surface filters are made up of several layers of media, such as glass or polymeric microfibers. During filtration, particles larger than the spaces within the filter matrix are retained primarily on the filter surface. Other smaller particles can get entrapped within the matrix. Thus, these filters have the property of both depth and screen filters. The surface filters are generally made of polymeric materials, resin-bonded paper, or fiberglass and paper.

Screen filters have a geometrically regular porous matrix, and particles are retained on its surface primarily by a sieving or size exclusion mechanism. Small particles and microorganisms larger than the pore size are retained on the surface. The filtration process here is absolute because everything larger than the rated pore size cannot get through the filter. Examples of such filters include cast polymeric membranes used for all critical submicronic and macromolecular separations in the biotechnology industry. The structure of the filter material is more rigid, and the manufacture is much better controlled than for the other types of filters.

Screen filters made of *microfiltration membranes* have a well-defined pore size in the range of 0.1 to 1 μm. Because particles and microorganisms are retained at or near the surface of the membrane, there is very little possibility of the so-called grow through phenomenon taking place when these filters are used for sterile filtration applications in the manufacture of parenteral drugs. This is a truly major application for the microfiltration membranes in the general pharmaceutical and biotechnology industries. They also find use in another biotechnology application, the harvesting of cells from bioreactors.

Ultrafiltration membranes are the second major group of screen filters and are often called *asymmetric* or *skinned* membranes. The skin of the membrane does the job of retaining the dissolved solutes larger than the membrane rating. The skin is supported by a more rigid support layer or backing. Unlike microfiltration membranes, ultrafiltration membranes are not designated by their pore size but according to the molecular weights of the dissolved solutes that the membrane will *retain* or *reject*. This, as mentioned before is called the nominal molecular weight cutoff of the membrane.

METHODS OF MEMBRANE MANUFACTURE

Most membranes are made of organic polymers such as the natural polymer, cellulose, and its derivatives (e.g., cellulose acetate) and certain other synthetic polymers (e.g., polyethylene, polycarbonate, polysulfone) and their derivatives. Some membranes are also made of inorganic materials such as zirconium and aluminium oxide. These are often referred to as *ceramic membranes*.

In the manufacture of polymeric membranes, the organic polymers are dissolved in an appropriate solvent and combined with the nonsolvent or swelling agent. This makes up the membrane casting solution. The membranes are cast on casting machines, which are almost always made by the membrane manufacturers themselves. Most membranes are made in the form of flat sheets, but some are cast on the module support as in the case of tubular module membranes. Some membranes are also made directly in the form of hollow fibers. The membranes are then annealed.

Flat sheet membranes are subsequently incorporated into flat membrane plate and frame modules or made into spirally wound or pleated cartridges. Critical details of membrane manufacture are well-guarded secrets and are proprietery processes developed and owned by membrane manufacturers.

Microfiltration membranes are made to a thickness of about 100 to 120 μm, and the retention of particulate material takes place at the membrane surface or within the first few micrometers of the tortuous length of the membrane pores. *Ultrafiltration membranes* consist of a very thin skin that performs the task of separation of the solutes in solution. This skin is supported by a much thicker support material that contains voids, often referred to as finger voids because of their fingerlike shape. These voids facilitate the rapid flow of the fluid once it gets past the fine skin of the ultrafiltration membrane. *Reverse osmosis membranes*, as per their classical definition, are meant to retain all solutes other than the solvent itself. These were originally developed for sea water desalination to make potable water. This is still by far the largest industrial application for reverse osmosis membranes. These membranes have since been made into somewhat modified versions that can retain smaller molecular weight solutes that are generally not retained by the ultrafiltration membranes with a low nominal molecular weight cutoff. The molecules retained are larger than the salts that can pass more easily through the membrane. These new-generation reverse osmosis membranes find a niche application in the concentration of low molecular weight solutes, such as peptides, antibiotics, and other products in the biopharmaceutical industry. Table 2 summarizes the characteristics of the three membrane processes described.

Track-Etch Membranes

As described earlier, most polymeric membranes contain tortuous pores in a spongy material. The pores of these membranes can have a fairly wide pore size distribution as a result of their method of manufacture. A different type of membrane is made by a two-step process that includes the manufacture of the polymer sheet membrane of a certain thickness followed by bombardment of the membrane sheet with massive energetic nuclei in a beam of ^{235}U fission fragments. These nuclei go through the membrane, leaving tracks that are then the pores formed in the membrane. After the bombardment step, the membrane undergoes further process steps to make it antistatic and to modify its surface characteristics such as either hydrophobicity or hydrophilicity. It is natural and quite logical to assume that such a process of making pores in a membrane would produce a membrane with highly uniform pore diameters and with a sharp nominal molecular weight cutoff. However, in actual practice, the random bombardment can as well result in two or more pores being formed close together, thus giving rise to much larger pores than the individual pores. This problem of overlapping pores is supposedly avoided if the nuclear bombardment takes place at an angle of 29° between the particle beam and the membrane sheet (2). There are also some other disadvantages to these membranes. Their extreme thinness of only about 15 μm makes them difficult to handle. Also, if the bombarding particles fail to penetrate completely through the membrane sheet, the apparent pores in the membrane are not true pores.

Virus Removal Membranes

The biotechnology industry, which has benefitted more from membrane processes than the conventional phar-

Table 2. Characteristics of Membrane Processes

Process	Permeate	Retentate
Microfiltration[a]	Solvent with all dissolved solutes	Particulates in suspension, colloidal material
Ultrafiltration[b]	Water with dissolved solutes of lower molecular weight	Dissolved solutes of higher molecular weight and colloidal material
Reverse osmosis	Solvent, typically water	Solutes

[a]In microfiltrator, particulate retention generally depends on the pore size of the membrane.
[b]In ultrafiltration, the passage or retention of solutes depends on the nominal molecular weight cutoff of the membrane.

maceutical industry, has constantly changing requirements. These are the result of the development of new biologicals by newer and rather very innovative processes as well as the more and more stringent regulatory requirements that call for high product quality and safety. An excellent example here is the need for biologicals to be rendered free of both relevant and adventituous viral contaminants. The membrane manufacturers who have regularly kept up with the demands of the biotechnology industry have come up with special membranes for this application that are being used by the industry to make virus-free biologicals.

HOW MEMBRANE PROCESSES ARE OPERATED

Membrane processes are driven by pressure, which is a great advantage to the biotechnology industry, in which large majority of products are heat labile. The processes also consume less energy because they do not require either heat transfer or heat generation. They call for a pump to run the process fluid through the membrane under pressure, as illustrated in Figure 6. Some of the processes can, however, be run with either compressed air or nitrogen and do not need pumps. Most membrane processes are run at ambient temperature, with few exceptions such as those processes where extremely heat-labile products call for the processes to be run at lower-than-ambient temperatures. Low temperature membrane processing is also common when the process time is long, and there is the accompanying risk of product degradation or microbial contamination of the process fluid. At the other extreme, higher temperatures are used when a fluid is very viscous, and the product of interest is not quite heat sensitive. The higher temperature reduces the viscosity of the process stream and increases the rate of filtration with the overall advantage of significant reduction in process time.

In a well-developed process in the biotechnology industry, it is quite common to find membrane filtrations operating in tandem with one or more chromatographic purification steps or other methods of separation and purification. Further, one may also use more than one membrane separation step (i.e., microfiltration, ultrafiltration, and reverse osmosis) at appropriate points in the downstream process just as different chromatography procedures, such as ion exchange, size exclusion, or affinity methods, are used in a complementary fashion. One single membrane separation step is not always enough to accomplish the end result. Centrifugation is sometimes used before the membrane processes for clarification of fluids containing a high biomass, as in fermentation processes. The supernatant of the centrifugation is then better handled by other methods of purification, including membrane-based processes.

Two Methods of Operation of Membrane Processes

The processes are run according to two distinctly different methods. Figure 7 illustrates these two methods of filtration. The first is the classical method of filtering a solution containing suspended particulates either to recover the particulates themselves, as in the recovery of a product crystallized from a solution, or to obtain a clear filtrate when the product of interest is in the filtrate. This method is known as *dead-end filtration* or *normal-flow filtration*.

When the solution being filtered contains a high concentration of particulates or bioburden, the filter can quickly get clogged because of the accumulation of solids at the surface of the membrane. This usually ends up with the deposition of a boundary layer at the membrane surface, which then acts as a secondary membrane. This prevents free passage of the liquid and eventually the mem-

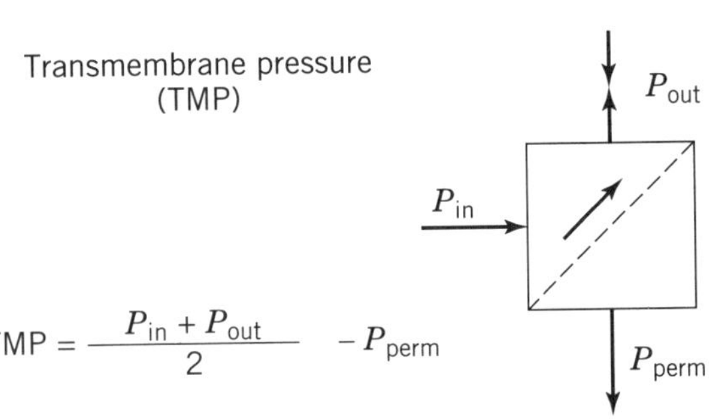

$$\text{TMP} = \frac{P_{in} + P_{out}}{2} - P_{perm}$$

Figure 6. Basic scheme of a tangential-flow filtration process.

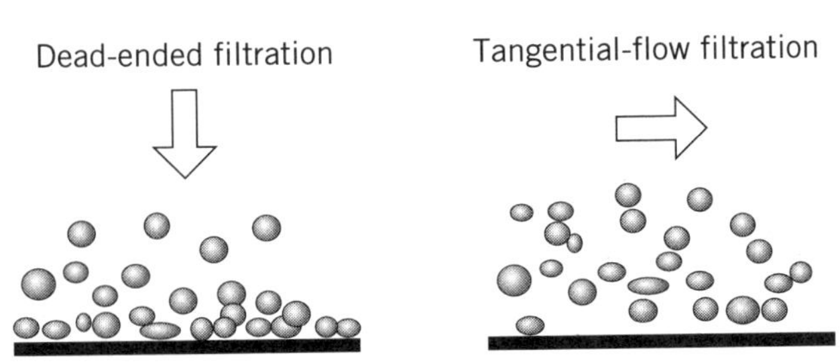

Figure 7. The two methods of using membrane filters.

brane is completely clogged. When this happens, filtration cannot be continued, and the only solution is to stop the process and replace the membrane filter with a new one. Such stoppages are time consuming and often result in loss of valuable product, which affects the economics of the overall process.

A practical and often successful alternative to the classical dead-end filtration is the use of the membrane filters by a method known as *tangential-flow filtration* or *cross-flow filtration*. Figure 6 illustrates the circulation of the complete process fluid in a direction parallel to the membrane surface.

Different types of pumps are used in the process. The advantage of the method is that the particulates that would accumulate at the membrane surface in dead-end filtration are now kept in the body of the fluid. They stay away from the membrane surface, permitting the liquid to flow relatively unhindered through the membrane pores. If the rate of filtration or *flux* drops significantly, the process fluid can be diluted by adding an appropriate pure solvent such as water or a buffer. This procedure helps to recover much of the original flux, and the filtration can be continued as before. The addition of the diluting fluid can take place without stopping the process. This step of *in-process dilution* is referred to as *diafiltration*. A typical tangential flow filtration process scheme is illustrated in Figure 8. If the desired product is in the filtrate, which is now more dilute, the filtrate can be reconcentrated to the desired concentration by either ultrafiltration or reverse osmosis, choosing the appropriate membrane according to the molecular weight of the product of interest. The concentration and further purification can also be performed by a chromatographic method.

Equipment Employed in Membrane Filtration

Membrane filters in different configuration are required for different processes depending on the method of filtration used. The choice of a particular type of filter depends on several considerations, such as process fluid composition; goal of the process, which would either be to recover

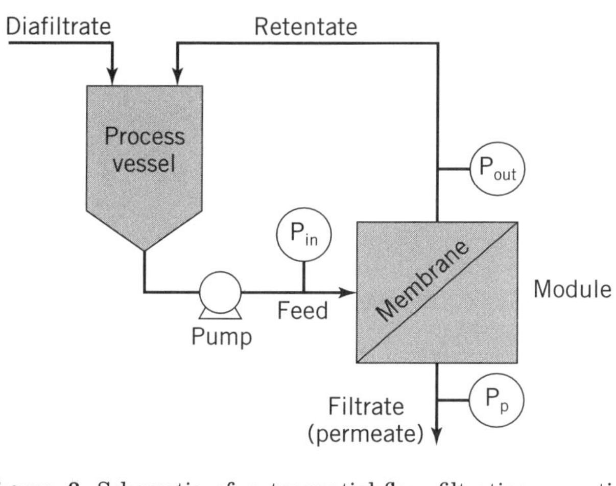

Figure 8. Schematic of a tangential-flow filtration operation (clarification, concentration, and diafiltration).

the particulates or the filtrate; and, of course, the economics of the process.

Equipment for Microfiltration in the Dead-End or Normal-Flow Filtration Mode For clarifying small volumes of solutions as in a research and development environment, small filter disks of different diameters and pore sizes are used. The filter disks are placed in appropriate holders of plastic materials or stainless steel. The stainless steel holders can be autoclaved for sterility before the filtration and can be used over and over again.

These less-expensive filters are also quite useful in determining what type of filters are appropriate for a certain application. Filters are also available in the form of fabricated devices that can be placed at the end of a syringe. Such filters are ideal for filtering volumes from 1 mL to about 100 mL. For applications where large volumes of process fluids need to be filtered, as in an industrial environment, cartridge filters placed in suitable stainless steel holders are ideal. The filters placed in such holders are easy to handle; thus can be autoclaved or steam sterilized and sometimes cleaned for reuse.

Quite often, the particulate level in filtrations is high, and the membrane filters can quickly get clogged. The common solution to this problem is to place in the filtration step prefilters and clarifying filters before the membrane filters. The economic advantage of extending the life of the usually more-expensive final filter is far greater than the cost of the prefilters. In well-designed processes, the filters are used in a sequence as follows:

$$\text{Clarifying filter} \Rightarrow \text{Prefilter} \Rightarrow \text{Membrane filter}$$

Each filter removes the suspended or colloidal material for the process stream to improve the performance of the filter next in line (1).

Equipment for Small-Scale Ultrafiltration, Including Tangential Flow Filtration For concentrating small volumes of 1- to 20-mL protein solutions, centrifugal filtration units (Fig. 9) fitted with ultrafiltration membranes are available. Other devices such as the stirred cell (Fig. 10), which can be used with a wide range of ultrafiltration membranes, are also marketed by filter suppliers. These devices are ideal for processing volumes of a few milliliters to about half a liter. A stirring bar helps to stir the fluid above the membrane and keeps the foulants moving in the body of the solution, reducing filter clogging and increasing filtration rates. Although this device will help to achieve small laboratory-scale separations or protein concentrations, it will not give indications of the membrane area or process conditions to use when the process is eventually scaled up to large volume operations.

Because large-scale ultrafiltration processes are usually conducted in tangential-flow filtration systems, it is better to start with a device that operates in this mode at the research and development laboratory. Small-scale devices for these applications are available from filter suppliers. Devices that hold single sheets of membrane or small filter modules with more membrane area are also available. It is important to proportionately increase the membrane

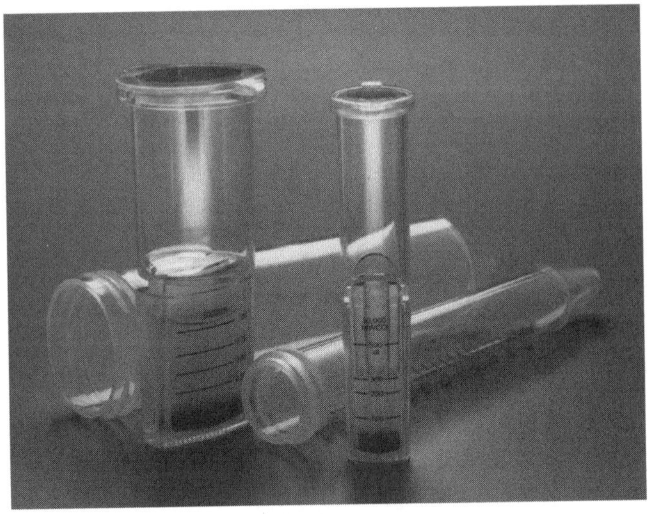

Figure 9. Small ultrafiltration devices used in centrifuges for small-scale ultrafiltration. *Source*: Photo courtesy Millipore S.A., Molsheim, France.

area of the filtration device as the volume of the process fluid increases. When the process is scaled up to the pilot plant level or to full-scale manufacture, large surface area modules and systems are available and offer the convenience of proper scale-up. Membrane filters in the following configurations are available from suppliers: *hollow fiber modules, spiral wound modules, plate and frame devices using flat sheet membranes*, and *tubular filters*. All these configurations have been developed by individual manufacturers based on sound hydrodynamic principles and rather complex mathematical models for module and system performance. The idea behind the development of each type of device is to improve its performance and ultimately reduce processing costs for the end user. The decision to use a particular device for a given application is usually made by taking into consideration the properties, and more specifically, the composition of the process stream, the goal of the application, and the results of the initial small-scale trials. The impact of the filtration step on the

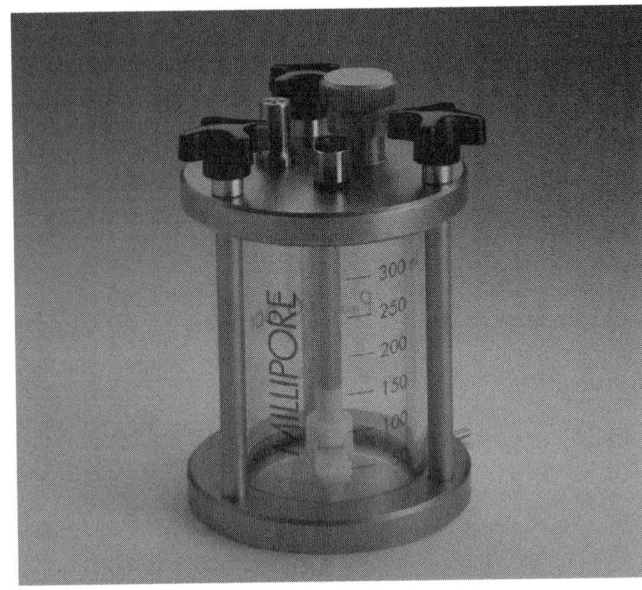

Figure 10. A bench-top device for small-volume ultrafiltration applications. *Source*: Photo courtesy Millipore S.A., Molsheim, France.

overall process economics is undoubtedly an important factor.

Small devices operated on bench tops are ideal for early studies in the research and development laboratories. Properly performed experiments help to determine which membranes and membrane devices are best suited for a particular application and also for correct optimization of the process. Most filter manufacturers help in proper planning of these studies, which are critical when one needs a quick answer to a bioseparation problem. (Some companies even offer hands-on training for the operators.) The result of the studies is the knowledge of which membrane pore size (in case of microfiltration membranes) or what nominal molecular weight cutoff (in case of ultrafiltration membranes) is best for the application.

Hollow fiber modules are sometimes used in ultrafiltration processes. The active membrane that participates in

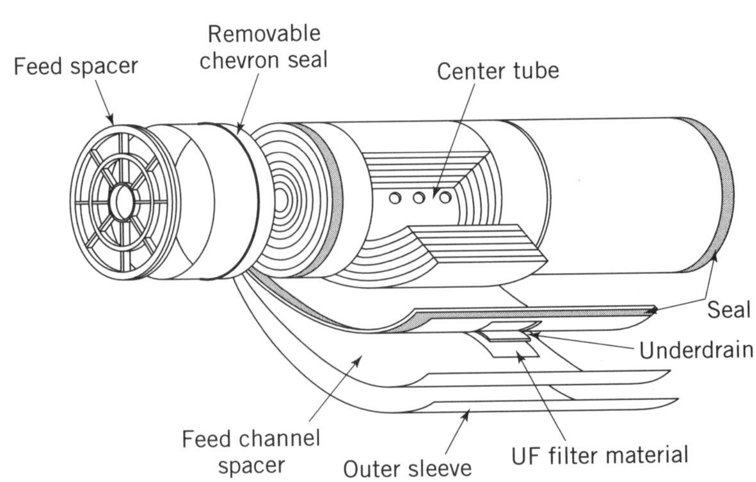

Figure 11. Components of a spiral-wound ultrafiltration membrane module. *Source*: Photo courtesy Millipore S.A., Molsheim, France.

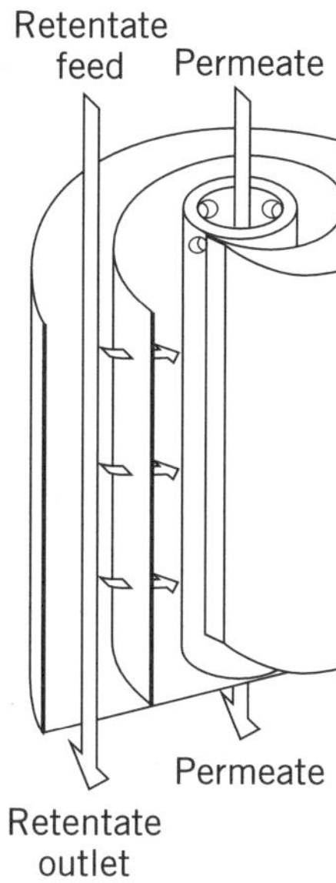

Figure 12. Flow schematic through a spiral wound ultrafiltration module. *Source*: Photo courtesy Millipore S.A., Molsheim, France.

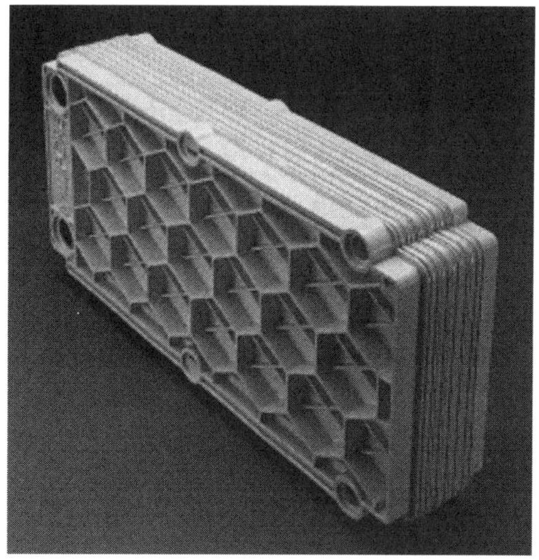

Figure 14. Photograph of a Prostak membrane module. *Source*: Photo courtesy Millipore S.A. Molsheim, France.

the separation process is on the inside of the hollow fiber of a uniform bore diameter. The process fluid is pumped through the hollow fiber, and the solvent and molecules that cross the membrane appear on the outside of the fiber. Bundles of many fibers, from 50 to 3,000 based on the internal diameter, are sealed inside a transparent cylindrical cartridge with the two ends sealed with epoxy sheets. Each cartridge has permeate ports at either end as well as feed and retentate ports. The ports can be connected to the piping of the system in which the module is installed by standard triclover clamps. Hollow fiber modules carrying ultrafiltration membranes of different cutoffs and surface area are available from several suppliers.

Spiral wound modules offer the biotechnology industry the most compact and least expensive of the ultrafiltration devices. The device is constructed of the same flat sheet membranes used in plate and frame devices. Each cartridge is made by wrapping alternate layers of membrane and separator screens concentrically around a hollow core, as shown in Figure 11. The flow schematic in a spiral wound ultrafiltration membrane cartridge manufactured by Millipore Corporation is shown in Figure 12. The product to be processed enters one end of the cartridge under pressure and flows tangentially down the central axis. The

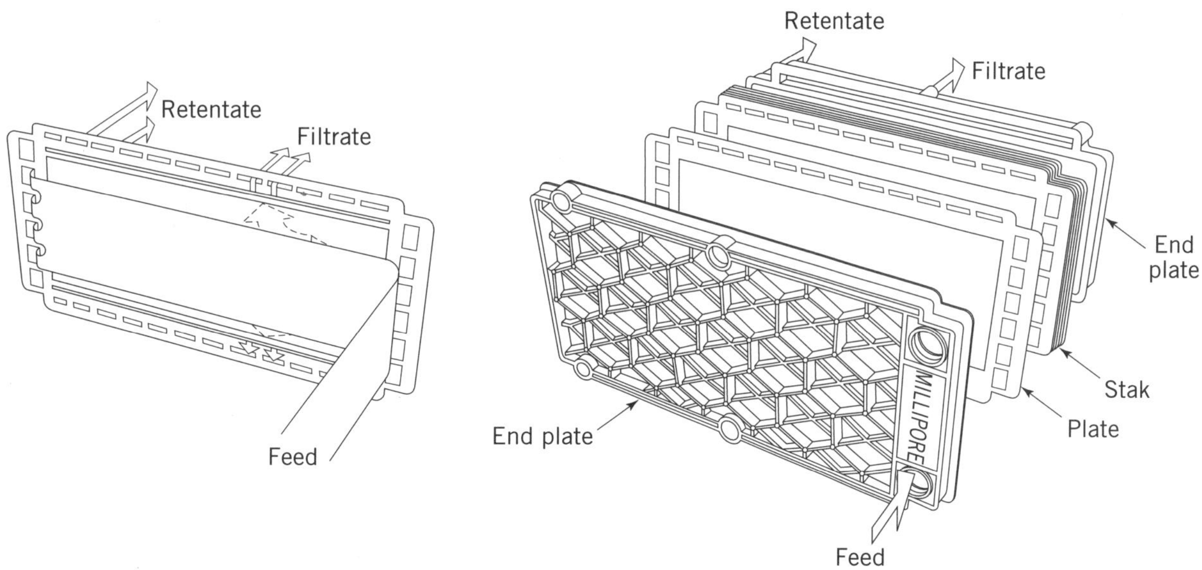

Figure 13. Flow schematic through an individual membrane plate (left) and through the entire Prostak plate and frame membrane module (right). *Source*: Photo courtesy Millipore S.A., Molsheim, France.

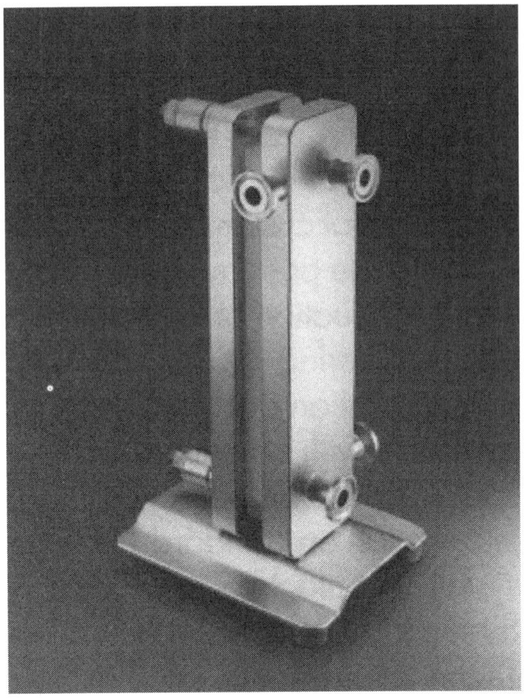

Figure 15. A simple bench-top device for evaluation of membranes in tangential-flow filtration in downstream processes. *Source*: Photo courtesy Millipore S.A., Molsheim, France.

Figure 16. An assembly of Prostak modules giving several square feet of membrane area for use in a large system for bioseparations. *Source*: Photo courtesy Millipore S.A., Molsheim, France.

ultrafiltrate containing salts, water, and molecules not retained by the membrane flows through the membrane into the permeate channel and spirals all the way to the central core of the cartridge. The permeate is continuously removed from the central core while the retentate containing the rejected material exits through the retentate port.

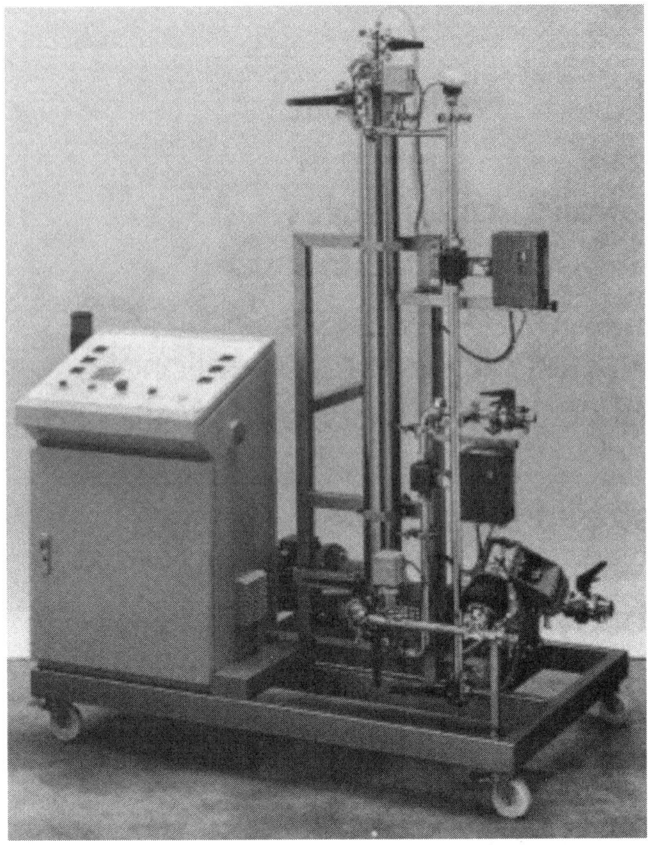

Figure 17. A small system with spiral-wound membrane modules. *Source*: Photo courtesy Millipore S.A., Molsheim, France.

Figure 18. A fully automatic system with several membrane cassettes for protein concentration. *Source*: Photo courtesy Millipore S.A., Molsheim, France.

Figure 19. A fully automatic system used in downstream processing. *Source*: Photo courtesy Millipore S.A., Molsheim, France.

Spiral cartridges made with different ultrafiltration membranes with varying cutoffs are available from filter suppliers.

Flat sheet cassettes and *plate and frame modules* are a frequently employed configuration in tangential-flow filtration, and devices offered by suppliers contain microfiltration membranes of different pore sizes and ultrafiltration membranes with several cutoffs. The microfiltration membranes are the same ones offered in cartridges used in the dead-end mode. Small devices that can be operated in the tangential-flow filtration mode with single sheets of membranes or modules are ideal. The work at this scale will help the user to understand problems of filter clogging or membrane polarization and fouling phenomena and how to address the problems rapidly by either choosing another type of membrane or changing the process conditions.

The cassette filter devices are made by stacking and bonding together identical membrane sheets and separator screens together in parallel. Devices with identical membranes contain the same materials of construction. Therefore, development of a process can start with a small minicassette for the small-volume experiment and is eventually scaled up to a several hundred liter process by simply stacking together many larger surface area cassettes in stainless steel or acrylic holders.

As a process moves from the small bench-top experiment in the research laboratory to the floor of the manufacturing plant, the industry needs equipment to process large volumes of process streams. The filter suppliers have responded to this need of the biotechnology and other industries by developing modules that not only carry large membrane areas of several square meters but are extremely robust. These can be used repeatedly after proper cleaning and sanitization for several years. Typical examples of such devices are the Prostak membrane modules available from Millipore Corporation. These are tangential-flow stacked plate devices. The modules are solvent bonded and contain neither adhesives nor compression-dependent membrane seals. The modules containing Millipore's Durapore range of microporous membranes can be steam sterilized dozens of times. Figure 13 shows the flow path of the process fluid in the Prostak modules, which are shown in Figure 14. Because these are available in both the microfiltration and ultrafiltration versions, it is possible to use the same hardware for both steps in a downstream process, giving a significant reduction in capital investment. Most filter suppliers offer either manually operated or fully automatic systems that contain the membrane devices. A small bench-top filter evaluation module in a stainless steel holder is illustrated in Figure 15. Figure 16 shows how a number of plate and frame membrane modules can be put together in several stainless steel holders and plumbed together to form an assembly of membrane modules that can be part of a large process system. Figures 17, 18, and 19 illustrate a variety of typical small and large membrane systems used in the biotechnology industry. The systems carry the different membrane configurations described earlier.

CONCLUSION

The information provided above will give present and potential users of membranes for bioseparations a clear picture of how best to approach the task to achieve the desired end results. The manufacturers of membranes and membrane devices are always ready to give the users the necessary help in choosing the right product for each application. For detailed information on membrane separations, the readers can also refer to numerous review articles and textbooks on the subject, such as the one on ultrafiltration by M. Cheryan (3). Another useful book edited by McGregor (4) includes articles on a variety of specialized topics written by persons practicing the art of membrane separations in biotechnology. The use of microfiltration in the tangential mode in the recovery of intracellular products has been reviewed by Le and Atkinson (5), whereas Short and Webster (6) have authored an excellent article on the use of ultrafiltration in the pharmaceutical industry.

BIBLIOGRAPHY

1. E.S. Honig and P.D. Schwartz, *Filtr., Sep.* Jan/Feb 1997.
2. M.C. Porter, *Handbook of Separation Techniques for Chemical Engineers*, McGraw-Hill, New York, 1979.
3. M. Cheryan, *Ultrafiltration Handbook*, Technomic Publishing Company, Lancaster, Pa., 1986.
4. W.C. McGregor, *Membrane Separations in Biotechnology*, Dekker, New York, 1986.
5. M.S. Le and T. Atkinson, *Process Biotechnol.*, February 1995.
6. J.L. Short and D.W. Webster, *Process Biochem.*, March/April 1982.

See also Cell disruption and lysis; Chromatography, size exclusion; Membrane chromatography; Protein ultrafiltration.

MEMBRANE SURFACE LIQUID CULTURE, MICROORGANISMS, FUNGI

KAZUHIRO NAKANISHI
Okayama University
Okayama, Japan

KEY WORDS

Bioreactor
Continuous culture
Cultivation
Kojic acid
Membrane surface liquid culture
Molds
Neutral protease
Porous membrane
Secondary metabolite
Solid-state culture

OUTLINE

INTRODUCTION

Mold, a typical filamentous microorganism, is utilized widely in the field of biotechnology. Although molds are usually cultivated either by submerged liquid cultures and agitation or by solid-substrate cultures, these conventional methods may have various difficulties (1). Membrane surface liquid culture (MSLC) is one of the promising methods that has recently been developed by Nakanishi and his coworkers to solve difficulties associated with conventional cultivation methods (2–7). Although MSLC is very simple in principle, it has advantages of both the conventional submerged liquid culture and solid-substrate culture. Molds cultivated with MSLC sometimes produce metabolites in a larger quantities and are stable against autolysis.

PROBLEMS OF THE CONVENTIONAL CULTIVATION METHODS

Molds are difficult to cultivate because they are obligate aerobes and their growth is characterized by long, thin, branched threads of mycelium. Molds are usually cultivated either by submerged liquid cultures, such as shake-flask cultures and culture using an agitated vessel, or by solid-substrate cultures. However, these conventional methods include several severe problems for cultivation of molds. Because molds are obligate aerobes, it is of crucial importance to keep a sufficiently high dissolved oxygen level in the medium. In a shake-flask culture or a culture using an agitated vessel, molds take on either a pulpy state or a pelletlike morphology (aggregated forms) (8). In a pulpy state, molds may be damaged, suffering from shear stress caused by high agitation of the medium and spurging of air. The culture broth is highly viscous particularly at a high cell concentration, and accordingly, it becomes more difficult to supply oxygen to the cells. Also, the cells existing in the center of a pellet are apt to be autolyzed because of oxygen deficiency. On the other hand, in a solid-substrate culture, molds are grown statically on a solid substrate, such as unprocessed agricultural products or by-products, facing the air (9). Thus, in the solid-substrate culture, molds grow as they do in their natural habitats. Many researchers have reported that enzymes are sometimes produced at higher levels by solid-substrate cultures than by shake-flask or agitated vessel cultures (1–15). These findings might be partly ascribable to the fact that molds that are statically cultivated by the solid-substrate method do not suffer from fluid shear and low oxygen levels, although details of the mechanisms are not known. On the other hand, features of the solid-substrate culture is disadvantageous because the medium is restricted to natural agricultural products and thus control of the medium composition and pH is difficult. Furthermore, the production rate of metabolites is sometimes limited by mass transfer rates inside the solid substrate. Handling of the fine particles of the solid substrate during cultivation and a low yield of the product after cultivation are also serious problems.

PRINCIPLES AND CHARACTERISTICS OF MSLC

In MSLC, molds are grown on the surface of a microporous membrane, one side of which faces the air and the other side of which is in contact with liquid medium (2–7) (Fig. 1.) Although MSLC is very simple in principle, it has the advantages of both the solid-substrate culture and the conventional submerged liquid culture. In MSLC, molds are grown on the surface of a solid material (a porous membrane) facing the air, as in the case of solid-substrate culture. Moreover, MSLC is a static culture without vigorous agitation or shaking of the medium and spurging of air. Thus, the molds do not suffer from fluid shear. However, in contrast to the solid-substrate culture, liquid medium is used in MSLC, as in conventional submerged liquid cultures, which allows easy control of composition and pH during cultivation.

In MSLC, mycelia or hyphae do not pass through the membrane pores and thus do not invade the liquid medium side. If the hyphae should pass through the membrane pores, they would adhere onto the backside of the membrane, which in turn would block the pores and obstruct

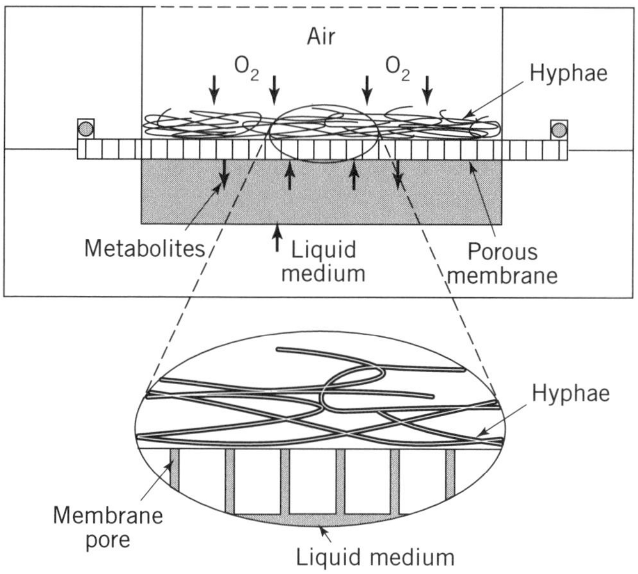

Figure 1. Conceptual scheme for MSLC. A porous membrane is set on the surface of liquid medium on one side and faces humidified air on the opposite side (upper surface). Molds are grown on the upper surface of the membrane, taking oxygen directly from the air and at the same time absorbing the liquid medium existing in the membrane pores. The enlarged figure below shows that the liquid medium rises up through the pores to the upper surface.

the transfer of medium components and metabolites. Thus, it is important to select a porous membrane with appropriate pore size, structure, and material. The membrane pore size should usually be smaller than 0.2–0.4 μm, with no appreciable mass transfer resistance (9). The membrane should be hydrophilic so that liquid medium can reach the top of the membrane pores at the outer surface where molds are grown, as shown in Figure 1. For example, an SE20 membrane made of polysulfone with hydrophilic treatment (Fuji Photo Film Co., Tokyo, Japan) cellulose membranes from Millipore (Bradford, Mass.) having a molecular weight cut-off of 5,000–300,000 could be used. Because the hyphae do not enter the membrane pores, the medium remains clear and uncontaminated with cells, which may contribute to simplification of the downstream processing.

Furthermore, MSLC is energy saving because it does not require shaking, agitation, or spurging.

Table 1 summarizes some major differences among MSLC, solid-substrate culture, and submerged liquid culture.

APPARATUS AND MODES OF CULTIVATION

Apparatus

Various types of apparatus are available for MSLC. For preliminary experiments such as screening of strains for production of a specific metabolite and optimization of medium composition, a plastic dish can be used. Liquid medium that has been autoclaved is poured into a plastic dish and then a porous membrane is set onto the surface of the liquid medium and held in place by appropriate supports, as shown in Figure 2 (4). Spores are inoculated uniformly onto the membrane surface, and cultivation is carried out statically in a thermostat-controlled and humidified aseptic box.

In experiments for the optimization of cultivation conditions, an apparatus made of heat-resistant plastic, such as polycarbonate, equipped with a peristaltic pump and a medium container is available, as shown in Figure 3 (2,4,6). The MSLC apparatus shown in Figure 3 is composed of an upper plate with a circular openings and a lower plate with a cylindrical groove. The membrane is set between the upper and lower plates with an O-ring. Spores are inoculated uniformly onto the membrane surface, and cultivation is allowed to start statically. The liquid medium can be circulated between the groove in the lower plate of the apparatus and an outside container. Medium components and pH can be controlled in the container. The medium in the groove may be gently stirred if necessary.

To conduct MSLC on a large scale, a cylindrical MSLC apparatus, shown in Figure 4 (7), or a plate-type apparatus in which two plates face each other at a short distance are available. In the cylindrical apparatus, the liquid medium is pumped up from a reservoir to the top of the cylindrical membrane and spouted toward the inner surface of the cylinder. The medium runs down along the inner surface of the cylindrical membrane module as in a wetted wall column. The medium is collected into the reservoir and circulated. Pumping should not be constantly, but instead intermittent. In continuous operation, fresh medium is supplied from a medium container and withdrawn using pumps with holding cells on the membrane surface. The entire cultivation system should be placed in a thermostat-controlled aseptic box or room equipped with an aseptic filter. Spores may be inoculated uniformly onto the membrane surface with a paintbrush or sprayed from a nozzle. Again, pH and medium components can be easily controlled in a container placed outside the membrane module.

Table 1. Some Major Differences in the MSLC, Solid-Substrate Culture, and Submerged Liquid Culture

	MSLC	Solid-substrate culture	Submerged liquid culture
Oxygen uptake	Mostly from gaseous phase	Mostly from gaseous phase	Soley from dissolved oxygen
Agitation	Not essential	Not essential	Strongly required
Medium	Liquid medium	Solid substrate	Liquid medium
Growth	Grown on surface of membrane	Penetrated deep into the solid substrate	Grown in a pulpy or pelletlike form
Recovery of extracellular product	Directly from the liquid medium	Extracted with liquid	Recovered after cell separation

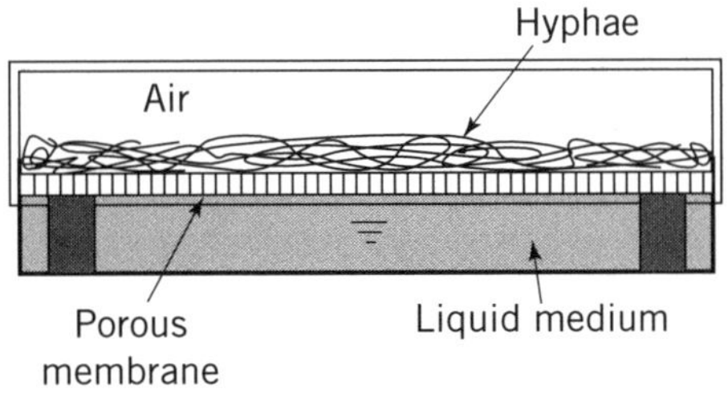

Figure 2. Schematic diagram for an MSLC apparatus using a plastic dish. A porous membrane is set on the surface of liquid medium and held in place by appropriate supports.

Cultivation Modes

Any cultivation mode can be applied to MSLC, including batch, fed-batch, repeated-batch, repeated fed-batch, and continuous. In every mode, molds are cultivated with holding cells on the membrane. In fed-batch MSLC, some components of the medium can be added at appropriate time intervals or added continuously into the medium container during the batch cultivation using the apparatus shown in Figures 3 and 4. In continuous MSLC, fresh medium is continuously fed into the apparatus and withdrawn at the same rate, with an appropriate retention time. The cells are held on the membrane surface during cultivation, in contrast to the ordinary continuous cultivation using an agitated vessel in which the steady state is attained in such a way that the amount of cells grown in per unit time in the fermentor is equivalent to that withdrawn from it.

EXPERIMENTAL RESULTS

Production of Neutral Protease Using Aspergillus oryzae IAM2704

A. oryzae usually produces neutral protease when grown in an appropriate liquid medium such one containing 4 g casein, 15 g glucose, 1 g K_2HPO_4, 0.5 g $MgSO_4 \cdot 7H_2O$, 0.5 g NaCl, and 0.01 g $FeSO_4 \cdot 7H_2O$ in 1 L of ion exchange water, pH 7.0 (3,4). Casein is an inducer for production of protease as well as a nitrogen source. Glucose is the carbon source. Figure 5 shows a comparison of courses of cultivation at 30 °C with respect to changes in glucose concentration, total protease activity in 20 mL of culture medium, and total dry cell weight in shake-flask culture, liquid-surface culture, solid-substrate culture, and MSLC. In all cases, cultivation was started by inoculating spores. In the shake-flask culture, 100-mL shake flasks were used. Liquid-surface culture was carried out statically on the surface of the medium. In the solid-substrate culture, cultivation was carried out on a 1.5% (w/v) agar containing every component of the medium. In MSLC, an SE20 membrane having a nominal pore size of 0.2 μm was used, and culture was carried out statically. A neutral protease activity was expressed as the amount of enzyme determined by the method using 0.6% casein as a substrate. One unit of enzyme activity was defined as the amount of enzyme that solubilizes 1 mg protein at pH 7.0 and 30 °C in 1 min, measured by Lowry's method with bovine serum albumin as a standard (16).

In the shake-flask culture and liquid-surface culture, the amount of protease reached its maximum after several days of cultivation and then rapidly decreased to a low level. The maximum amount of protease produced in both cultures was around 5 U/mL. In the solid-substrate culture on an agar medium, the highest amount of protease produced was about 25 U/mL of agar medium, which was 5 times higher than obtained with the shake-flask culture and liquid-surface culture. Furthermore, the amount of enzyme decreased more gradually after it reached a maximum in the solid-substrate culture. With MSLC, the amount of enzyme produced was higher than 50 U/mL, twice as high as by solid-substrate culture on an agar plate. As shown in Figure 5, both the consumption rate of glucose and the mycelial growth rate were also higher at the initial stage of MSLC than in the shake-flask culture,

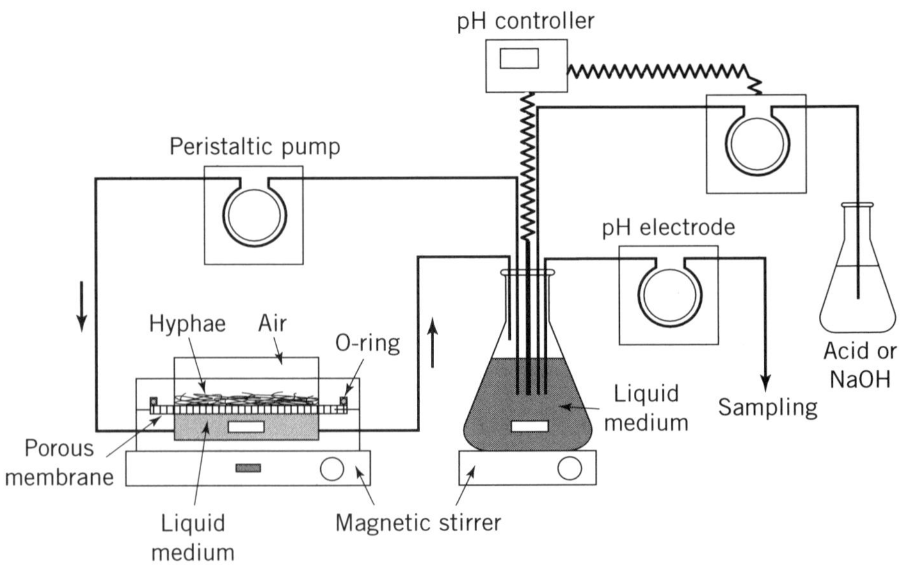

Figure 3. Schematic diagram for a disc-type MSLC system with a medium container, pH controller, and peristaltic pumps (7). The MSLC apparatus is composed of an upper plate with circular openings and a lower plate which is cylindrical, shallow, and contains liquid medium. A porous membrane is set between the upper and lower plate with an O-ring.

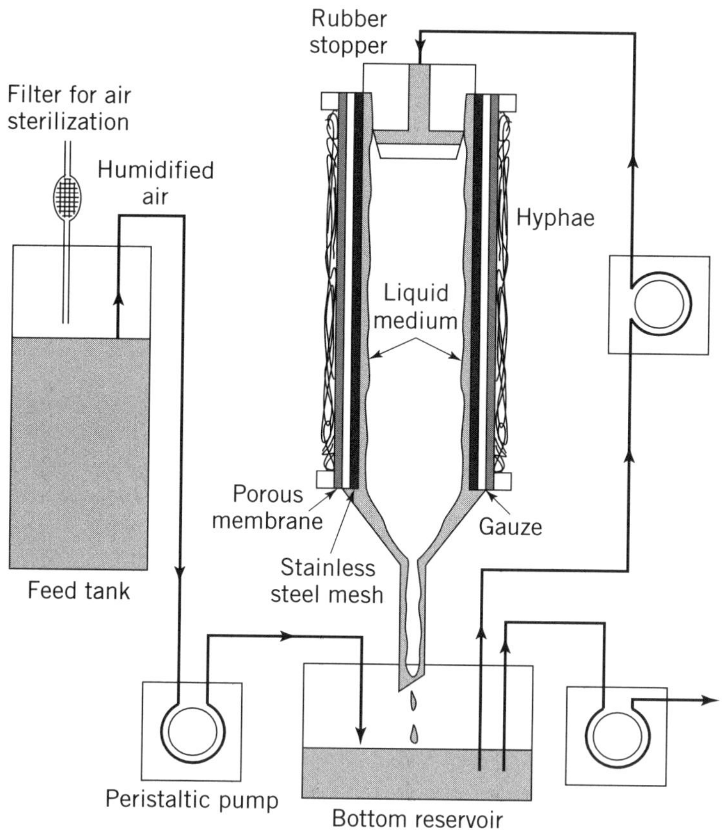

Figure 4. Schematic diagram of an MSLC system using a cylindrical membrane module (8). Liquid medium is intermittently circulated between a cylindrical membrane module and the bottom container. Liquid medium flows down along the inner surface of the module. In continuous MSLC, fresh medium is supplied from the feed tank to a bottom reservoir and withdrawn. The liquid medium is circulated, as in a batch MSLC.

probably because of higher efficiency of oxygen uptake. However, the reason why the enzyme was produced in a larger quantity by MSLC compared with the other cultures may not be due to the larger amount of the molds produced. The highest value for the specific amount of enzyme produced by the shake-flask culture, liquid surface-culture, solid-substrate culture, and MSLC in Figure 3 was around 1.1, 1.2, 4.0, and 6.3 U/mg, respectively.

In MSLC, the effects of the composition and concentration of medium on the production of metabolites were usually different from those in the shake-flask culture. Thus, it is important to optimize the composition and concentration of the medium by using a plastic dish, as shown in Figure 3. In Figure 6, a typical example is shown in which the glucose concentration in the medium for MSLC affects the amount of protease produced in a quite different fashion than seen in shake-flask culture. In MSLC, the maximum amount of protease and specific amount of protease per dry cell weight were increased by lowering the glucose concentration from 1.5% to 0.2%, and in shake-flask culture, the tendency was opposite. It seems that glucose was more effectively utilized to produce the enzyme in the MSLC than in the shake-flask culture.

The molds are stabilized to a greater extent in MSLC. Usually, it is very difficult to conduct a repeated-batch culture or a long-term continuous culture with cell holding in a shake-flask set-up. In MSLC, cultures with cell holding are easy to maintain because of the increased stability of the mold. Keeping the cells on the membrane surface, one can conduct cultures either in a repeated-batch or continuous mode. As shown in Figure 7, repeated batch culture of *A. oryzae* IAM2704 to produce neutral proteases could be effectively carried out by MSLC (5); the liquid medium was exchanged with fresh medium every day except in the first batch. In the first batch, it took a longer time for the molds to germinate and to grow over the whole membrane surface. Once the membrane surface was covered with hyphae, production of the proteases occured satisfactorily. As shown in Figure 7, the cumulative amount of protease increased almost linearly with cultivation time at the initial stage of cultivation. Although the amount of protease produced in each batch gradually decreased, even in the 27th batch it was about 7 times the amount produced in the shake-flask culture.

Production of Kojic Acid Using *A. oryzae* NRRL484

Most molds can produce secondary metabolites, particularly organic acids such as citric acid, erythorbic acid, and kojic acid. *A. oryzae* NRRL484 produced kojic acid in a quite large amount by MSLC (5–7). In the production of kojic acid, molds exhibited behaviors different from those for production of neutral protease. Molds were active in production of kojic acid even when the concentration of carbon and nitrogen sources was in the high range. Figure 8 shows the effect of glucose and yeast extract concentrations on the production of kojic acid using *A. oryzae* NRRL484 in MSLC and shake-flask culture at 30 °C, with culture medium containing 0.75 g $(NH_4)_2SO_4$, 0.25 g K_2HPO_4, 0.25 g $MgSO_4 \cdot 7H_2O$, and 0.5 g of 300 g/L HCl

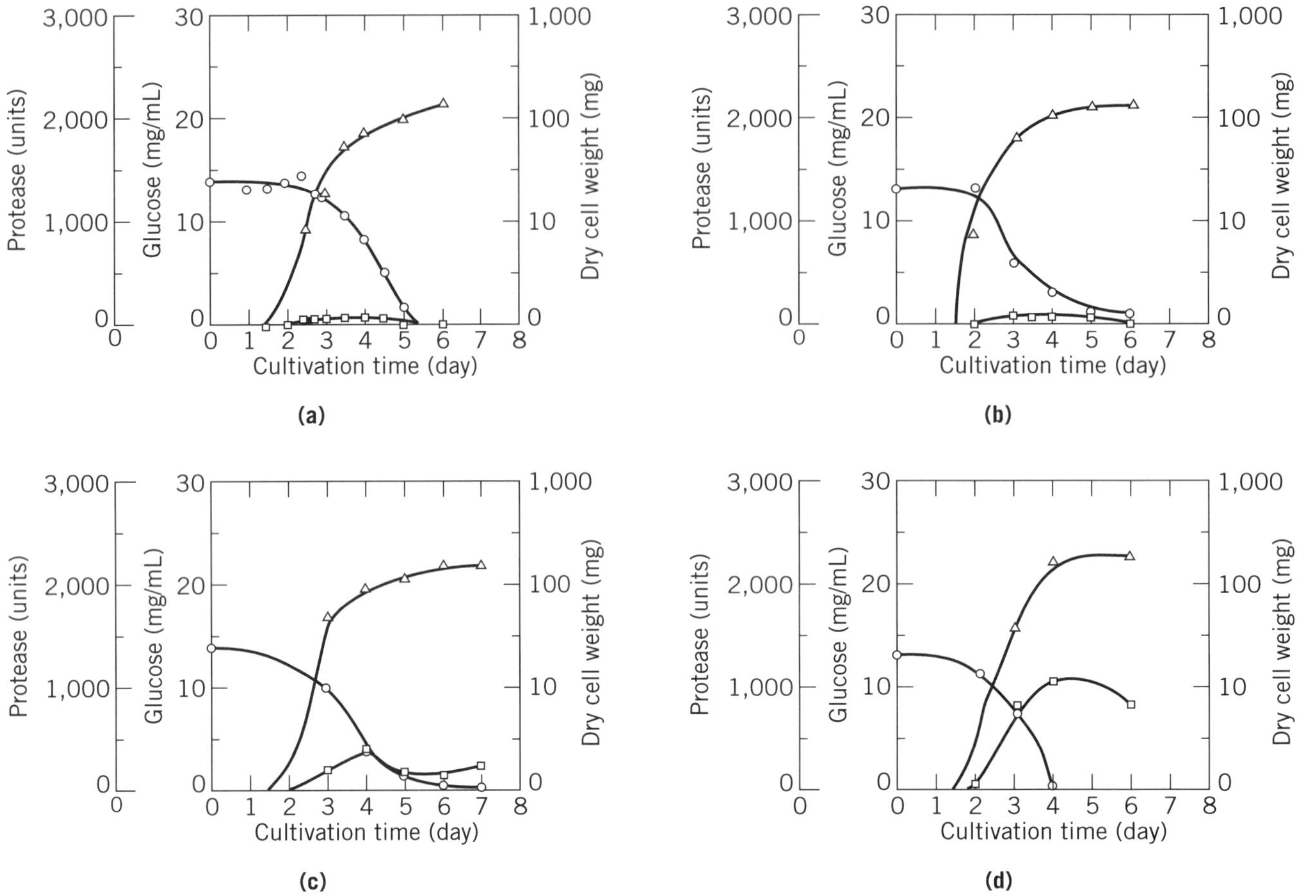

Figure 5. Change of concentrations of dried cell mass (△), glucose (○), and protease (□) during cultivation of *Aspergillus oryzae* IAM2704. (**a**) Shake-flask culture; (**b**) liquid-surface culture; (**c**) solid-substrate culture on agar plate; (**d**) MSLC (4).

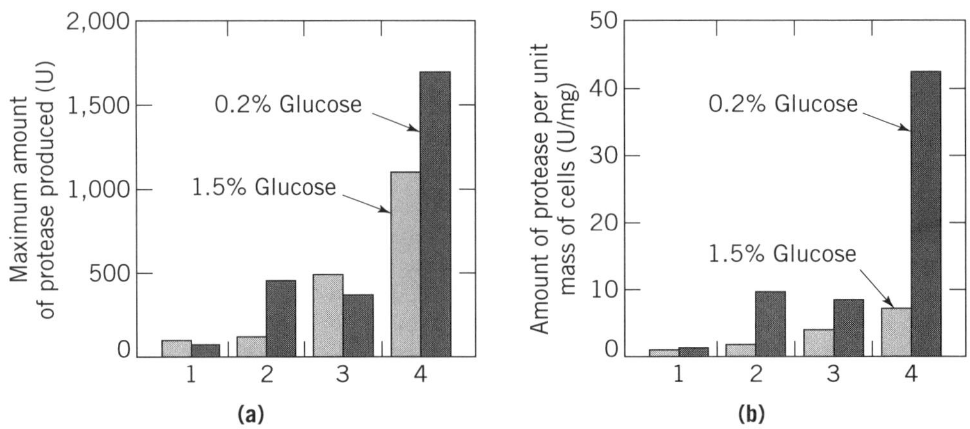

Figure 6. Maximum amount of protease produced (**a**) and specific amount of protease (**b**) during batch cultivation of *Aspergillus oryzae* IAM2704 using 0.2 and 1.5% glucose. 1, Shake-flask culture; 2, liquid-surface culture; 3, solid-substrate culture on agar plate; 4, MSLC (4).

solution containing 20 g/L $SO_4 \cdot 7H_2O$, 20 g/L $FeSO_4 \cdot 7H_2O$, and 30 g/L $ZnSO_4 \cdot 7H_2O$ in 1 L of ion exchanged water, pH 6.0, in addition to 50–400 g/L glucose and 0.2–10 g/L yeast extract. Glucose was a carbon source as well as the substrate for production of kojic acid. Yeast extract was the most suitable nitrogen source. The amount of kojic acid produced can be measured either by Bentley's method (17) or by high-performance liquid chromatography (HPLC). As shown in Figure 8 for a wide range of glucose and yeast extract concentrations, the maximum amount of kojic acid produced by MSLC was higher than by shake-flask culture (7).

Because molds were quite stable in MSLC, repeated fed-batch culture of *A. oryzae* NRRL484 was possible to

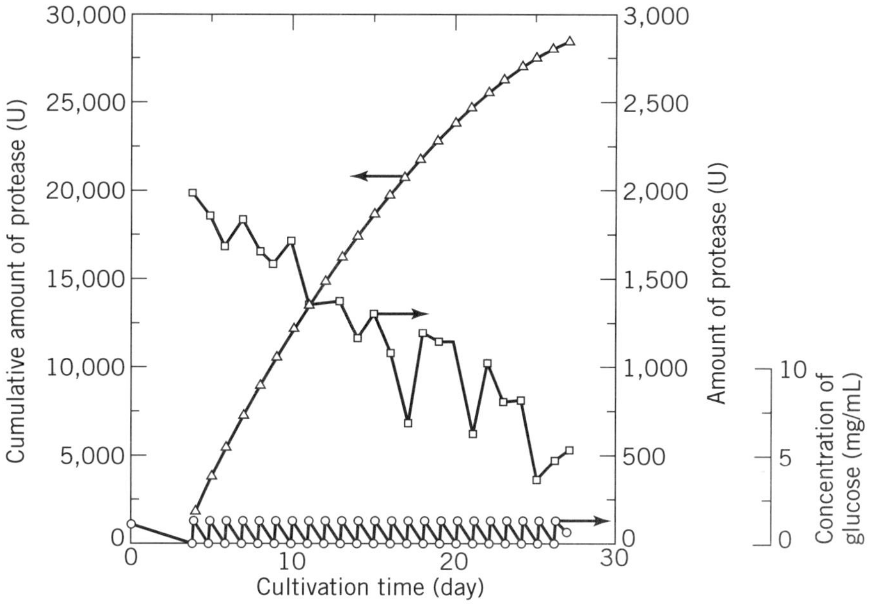

Figure 7. Changes of amount of protease produced in each batch (□), cumulative amount of protease (△), and concentration of glucose (○) during long-term repeated-batch culture by MSLC. The medium containing 0.2% glucose and 0.4% casein was exchanged every 24 h after 4 days of cultivation (first batch).

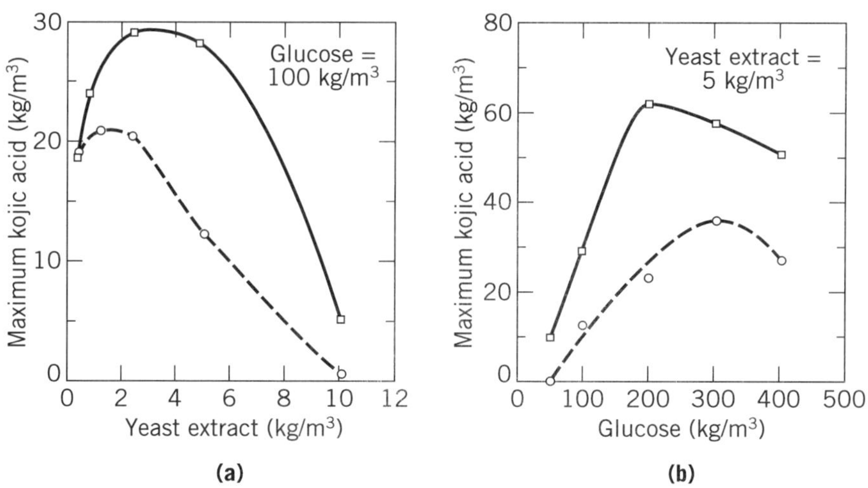

Figure 8. Effects of the concentrations of yeast extract (**a**) and glucose (**b**) on kojic acid production by MSLC (□) and shake-flask culture (○).

produce kojic acid as shown in Figure 9, in which powder glucose was added three times during one batch cultivation and the whole liquid medium was exchanged every 4 days, except the first batch (5). It took a longer time in the first batch until the molds had germinated and grown over the entire membrane surface, as was the case for production of neutral protease. Once the membrane surface was covered with hyphae, there was substantial production of kojic acid. In Figure 9, 12 fed-batch cultures are plotted. For batches 2 through 10, the kojic acid concentration in every batch reached 75 g/L or more, and it precipitated as fine crystals in the medium. In the batches 11 and 12, the production rate was greatly decreased. One possible reason for this decrease is that the amount of mycelia mat formed on the membrane surface increased, which might suppress oxygen transfer into the inside of the mat. Thus, some mycelia that were exposed adjacently on the membrane surface may have been autolyzed.

Continuous production of kojic acid was also possible using *A. oryzae* NRRL484 in a tubular-type MSLC apparatus with cell holding (7). In the cylindrical MSLC apparatus shown in Figure 4, a vertical cylinder 4 cm in diameter and 24 cm in length made of stainless steel mesh is surrounded by a sheet of gauze and then by an SE20 membrane sheet. The top end of the vertical cylindrical membrane module was sealed with a silicon rubber stopper. Typical results for continuous MSLC are shown in Figure 10. In this case, the effective membrane surface area was around 220 cm^2. The volume of medium in the entire system for recycling was 700 mL. At the start of cultivation, about 3 mL of a suspension of spores (5,000/mL) was inoculated uniformly onto the membrane surface using a paintbrush. During cultivation, medium was pumped up from a reservoir at a flow rate of about 6.6 mL/min using a peristaltic pump, and it flowed toward the inner surface of the cylinder through four small holes bored inside the

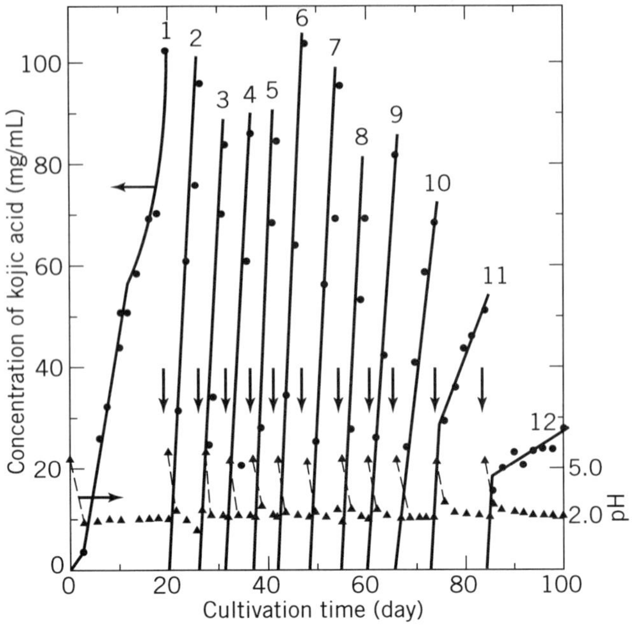

Figure 9. Changes of the concentration of kojic acid produced in each batch (●) and pH (▲) during repeated fed-batch MSLC using the exchange medium containing 0.25% yeast extract and 10% glucose. During each batch culture, powdered glucose was added to a final concentration of about 10% usually twice (batches 1 through 9) and three times (batches 10 and 12). Arrows show the times at which the entire medium was exchanged with fresh medium.

stopper. The pump was driven for 5 min every 15 min using a timer, and thus medium circulation was repeated. Fresh medium was supplied from a medium container and withdrawn using peristaltic pumps. The pumps for medium supply and withdrawal were driven for 1 min every 1 h at a flow rate of 1.25 mL/min. Thus, an average flow rate was 30 mL/day. As shown in Figure 10, after 12 days of cultivation, an almost steady state was attained, although glucose still remained. The steady-state concentration level of kojic acid was in the range of about 40 to 50 g/L for about 65 days. The yield of kojic acid on the basis of glucose consumed was evaluated to be around 70–80%, which is higher than that produced by shake-flask culture (30–40%). The glucose concentration gradually decreased even after 15 days of cultivation, probably because of maintenance metabolism. The fact that long-term continuous MSLC is possible, as shown in Figure 10, may indicate the increased stability of the mold.

PROBLEMS AND FUTURE PERSPECTIVES

In MSLC, molds tend to efficiently produce enzymes and secondary metabolites, and their stability against autolysis seems to be increased. Molds are typical aerobes with a strong oxygen requirement. Molds grow on a solid surface in their natural habitats. Such properties of molds probably favor MSLC, in which molds are grown on the membrane surface, obtaining oxygen directly from the atmosphere while taking liquid medium through the membrane pores with negligibly small mass transfer resistance. In conventional submerged liquid cultures using a shake-flask or an agitated vessel, mold cells are surrounded by the liquid medium, which contains a low amount of dis-

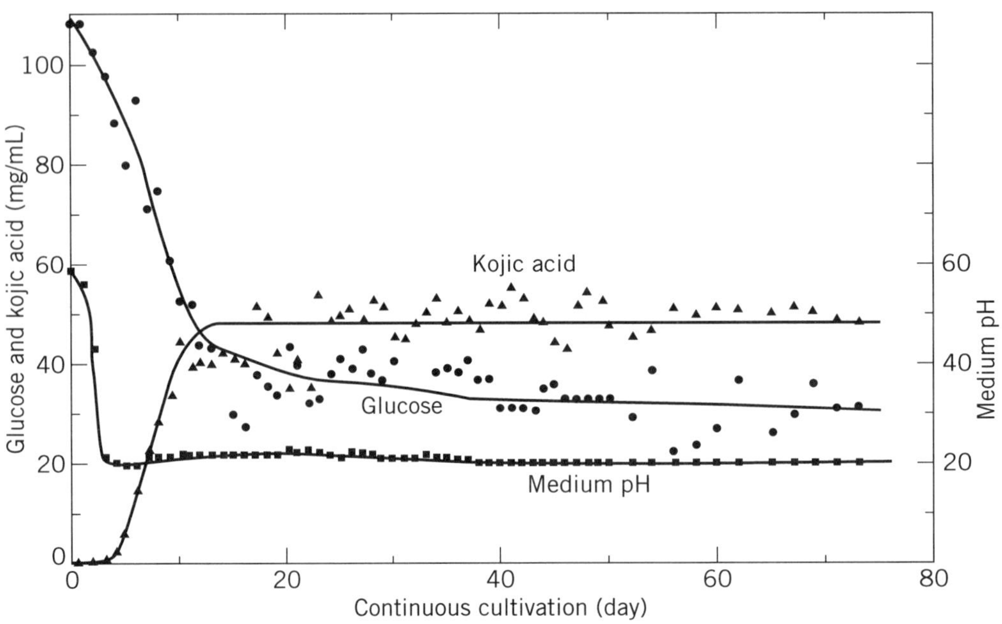

Figure 10. Continuous MSLC using a cylindrical membrane module at 30°C, holding molds on the membrane. About 30 mL of fresh medium was supplied semicontinuously per day, and the same amount was withdrawn while the medium was circulated between the membrane module and bottom container.

solved oxygen; in addition the cells suffer from shear stress, which may promote autolysis. In fact, a long-term operation with cell holding is quite difficult in conventional submerged liquid cultures. On the other hand, MSLC resembles liquid-surface culture (18), in which molds are suspended on the surface of liquid medium without a porous membrane at the interface. Liquid-surface culture is in principle very simple and an energy saving method. Liquid-surface culture was successfully applied to the production of citric acid using molds (19). However, there are some difficulties in conducting liquid-surface culture. In the production of neutral protease by liquid-surface culture using *A. oryzae* IAM2704, the amount of enzyme produced was slightly more than that by shake-flask culture but much less in comparison with that by MSLC. Furthermore, the cell mat is apt to sediment with increasing size. On the other hand, in kojic acid production using *A. oryzae* NRRL484, the cells formed a dense mat on the surface of the liquid medium and did not sediment; however, kojic acid production was less than that obtained by the MSLC. Thus, behaviors of the molds in liquid-surface culture seem to depend on the kinds of mold and cultivation conditions.

Sudo et al. have recently reported (20) that α-amylase is produced at a higher level by solid-substrate culture using rice bran than by agar-plate culture due to better growth of submerged mycelia, which mainly produce the enzyme. Thus, it may be interesting to compare the cells cultivated by MSLC with those obtained from other cultures (such as solid-substrate culture, agar-plate culture, liquid-surface culture, and shake-flask culture) from a stand point of the differences in respiratory activity, the amounts of related enzymes and, also gene expression. It may also be interesting to examine the effect of the gas composition on cultivation of molds by MSLC.

MSLC is an energy-saving process because the amounts of energy required for agitation and aeration are negligible, and downstream processing is simplified because the medium is not contaminated with cells. However, to utilize MSLC industrially some technical problems should be solved, such as those in spore inoculation, removal of the cells, and sterilization of the apparatus. The properties of the membrane are an important factor. The membrane should be reusable, resistant to autoclaving, and show appropriate hydrophilicity.

MSLC is an interesting method for investigating the physiological properties of molds, as well as a promising method for effective production of metabolites using molds.

BIBLIOGRAPHY

1. S. Braun and E.V. Lifshitz, *Trends Biotechnol.* **9**, 153–156 (1995).
2. A. Yasuhara, A. Ogawa, T. Tanaka, T. Sakiyama, and K. Nakanishi, *Biotechnol. Tech.* **8**, 249–254 (1994).
3. A. Yasuhara, A. Ogawa, T. Tanaka, T. Sakiyama, and K. Nakanishi, *Biotechnol. Tech.* **9**, 153–156 (1995).
4. A. Ogawa, A. Yasuhara, T. Tanaka, T. Sakiyama, and K. Nakanishi, *J. Ferment. Bioeng.* **80**, 35–40 (1995).
5. A. Ogawa, A. Yasuhara, T. Tanaka, T. Sakiyama, and K. Nakanishi, *J. Ferment. Bioeng.* **80**, 41–45 (1995).
6. K. Nakanishi, Y. Wakisaka, T. Tanaka, and T. Sakiyama, *Proc. 5th World Congress of Chemical Engineering*, American Institute of Chemical Engineers, New York, 1996, pp. 271–276.
7. Y. Wakisaka, T. Segawa, K. Imamura, T. Sakiyama, and K. Nakanishi, *J. Ferment. Bioeng.* **85**, 488–494 (1998).
8. S. Braun and S.E.V. Lifshitz, *Trends Biotechnol.* **9**, 63–68 (1991).
9. D.A. Mitchell and B.K. Lonsane, in H.W. Doelle, D.A. Mitchell, and E.R. Carlos eds., *Solid Substrate Cultivation*, Elsevier, London, 1992, pp. 1–16.
10. K. Sakaguchi, H. Okazaki, and M. Takeuchi, *Nippon Nougeikagaku Kaishi*, **29**, 349–353 (1955) (in Japanese).
11. T. Fujishima, K. Uchida, and H. Yoshino, *J. Ferment. Technol.* **50**, 724–730 (1972).
12. C.W. Hesseltine, *Process Biochem.* **12**, 29–32 (1977).
13. R. Bajracharya and R.E. Mudgett, *Biotechnol. Bioeng.* **22**, 2219–2235 (1980).
14. C.-F. Lin and H. Iizuka, *Appl. Environ. Microbiol.* **43**, 671–676 (1982).
15. W. Grajek, *Appl. Microbiol. Biotechnol.* **26**, 126–129 (1987).
16. T. Yano, S. Ashida, T. Tachiki, H. Kumagai, and T. Tochikura, *Agric. Biol. Chem.* **55**, 379–385 (1991).
17. M.R. Johns, in H.W. Doelle, D.A. Mitchell, and E.R. Carlos eds., *Solid Substrate Cultivation*, Elsevier, London, 1992, pp. 341–352.
18. R. Bentley, in S.P. Colowick and N.O. Kaplan eds., *Methods in Enzymology*, vol. 3, Academic, New York, 1957, pp. 238–245.
19. S. Ozaki, *Hakko Kyokaishi*, **16**, 7–17 (1958) (in Japanese).
20. S. Sudo, T. Ishikawa, K. Sato, and T. Oba, *J. Ferment. Bioeng.* **77**, 483–489 (1994).

See also BIOREACTORS, AIR-LIFT REACTORS; BIOREACTORS, CONTINUOUS STIRRED-TANK REACTORS; BIOREACTORS, FLUIDIZED-BED; MASS TRANSFER; SOLID STATE FERMENTATION, MICROBIAL GROWTH KINETICS; SOLID SUBSTRATE FERMENTATIONS, ENZYME PRODUCTION, FOOD.

METABOLITES, PRIMARY AND SECONDARY

ARNOLD L. DEMAIN
Massachusetts Institute of Technology
Cambridge, Massachusetts

KEY WORDS

Amino acids
Antibiotics
Biosynthesis
Hybrid antibiotics
Nucleotides
Recombinant DNA
Vitamins

OUTLINE

The versatility of microbial biosynthesis is enormous. Today's products of the microbe vary from the simplicity of methane and acetic acid to the complexity of vitamin B_{12} and proteins. The importance of the fermentation industry resides in five characteristics of microorganisms: *(1)* a high ratio of surface area to volume, which facilitates the rapid uptake of nutrients required to support high rates of metabolism and biosynthesis; *(2)* the tremendous variety of reactions microorganisms are capable of carrying out; *(3)* a facility to adapt to a large array of different environments, allowing a culture to be transplanted from nature to the laboratory flask or the factory fermentor, where it is capable of growing on inexpensive carbon and nitrogen sources and producing valuable compounds; *(4)* the ease of genetic manipulation, both in vivo and in vitro, to increase production of the products, to modify structures and activities, and to make entirely new products; and *(5)* an ability to make specific enantiomers, usually the active ones, in cases where normal chemical synthesis yields a mixture of active and inactive enantiomers.

Microorganisms are important to us for many reasons, but one of the principal ones is that they produce things of value to us. These may be very large materials such as proteins, nucleic acids, carbohydrate polymers, or even cells, or they can be smaller molecules, which we usually separate into metabolites essential for vegetative growth and those that are inessential (i.e., primary and secondary metabolites, respectively).

The power of the microbial culture in the competitive world of commercial synthesis can be appreciated by the fact that even simple molecules (e.g., L-glutamic acid and L-lysine) are made by fermentation rather than by chemical synthesis. Although a few products have been temporarily lost to chemical synthesis (e.g., solvents such as acetone and butanol), it is obvious that most natural products are made by fermentation technology. Despite the efficiency of the chemical route to riboflavin, much of the production of this compound is carried out by fermentation; chemical processes to vitamin C and steroids still employ microbial bioconversion steps. Most natural products are so complex and contain so many centers of asymmetry that they probably will never be made commercially by chemical synthesis.

Although microbes are extremely good in presenting us with an amazing array of valuable products, they usually produce them only in amounts that they need for their own benefit; thus, they tend not to overproduce their metabolites. Regulatory mechanisms have evolved in microorganisms that enable a strain to avoid excessive production of its metabolites so that it can compete efficiently with other forms of life and survive in nature. Some of the important control mechanisms are induction, feedback regulation, and nutrient (catabolite) regulation. The fermentation microbiologist, however, desires a "wasteful" strain that will overproduce and excrete a particular compound that can be isolated and marketed. During the screening stage, the microbiologist searches for organisms with weak regulatory mechanisms. Once a desired strain is found, a development program is begun to improve titers by modification of culture conditions, mutation, and recombinant DNA technology. The microbiologist is actually modifying the regulatory controls remaining in the original culture so that its inefficiency can be further increased and the microorganism will excrete tremendous amounts of these valuable products into the medium.

The main reason for the use of microorganisms to produce compounds that can otherwise be isolated from plants and animals or synthesized by chemists is the ease of increasing production by environmental and genetic manipulation. Thousandfold increases have been recorded for small metabolites. Of course, the higher the specific level of production, the simpler the job of isolation. Consider the case of *Ashbya gossypii*, which has been forced into making over 40,000 times more riboflavin than it needs, or *Pseudomonas denitrificans*, which produces a 100,000-fold excess of vitamin B_{12}. The original Oxford strain of *Penicillium notatum* produced 5 mg of penicillin per liter; today's strains make over 60,000 mg/L, a figure higher than the dry weight of the cells in the fermentor.

Primary metabolites are the small molecules of all living cells that are intermediates or end products of the pathways of intermediary metabolism, are building blocks for essential macromolecules, or are converted into coenzymes. The most industrially important are the amino acids, nucleotides, vitamins, solvents, and organic acids. Primary metabolites vary in size from hydrogen gas (2 Da) and methane (16 Da) to vitamin B_{12} (1,355 Da). It is not unexpected that amino acids and vitamins are used in human and animal nutrition, that ethanol, acetone, and butanol are used as fuels and solvents, and that citric and acetic acids are used as acidulants. However, many of these general metabolites are used in novel ways: the sodium salts of glutamic, 5′-inosinic, and 5′-guanylic acids as flavor enhancers, sodium gluconate as a sequestering agent to prevent the deposition of soap scums on cleaned surfaces, and fumarate in the manufacture of polyester resins. Organisms that produce such products are fantastic in their degree of overproduction.

For many years, microbial products have been used to enhance the quality, appeal, and availability of food and drink. A list of primary metabolites that have been or are being used in the food and feed industries is shown in Table 1.

About 1 million tons of amino acids are produced annually. In amino acid production, feedback regulation is

Table 1. Some Primary Metabolites Used in the Food and Feed Industries

Class	Examples
Alcohols	Ethanol
Amino acids	Glutamic acid (monosodium glutamate), lysine, threonine, tryptophan, phenylalanine
Nucleotides	5′-Guanylic acid, 5′-inosinic acid
Organic acids	Acetic acid, propionic acid, succinic acid, fumaric acid, lactic acid, malic acid, tartaric acid, citric acid, gluconic acid
Polyols	Glycerol, mannitol, erythritol, xylitol
Polysaccharides	Xanthan, gellan
Sugars	Fructose, sorbose
Vitamins	Riboflavin (B_2), cyanocobalamin (B_{12}), biotin

bypassed by isolating an auxotrophic mutant and partially starving it of its requirement. A second means to bypass feedback regulation is to produce mutants resistant to a toxic analog of the desired metabolite (i.e., an antimetabolite). Combinations of auxotrophic and antimetabolite resistance mutations are common in primary metabolite producing microorganisms. Permeability is very important in the production of L-glutamic acid, the major amino acid of commerce. About 1.2 billion pounds of monosodium glutamate, a potent flavor enhancer, are made annually by fermentation using various species of the genera *Corynebacterium* and *Brevibacterium* (e.g., *Corynebacterium glutamicum, Brevibacterium flavum*, and *B. lactofermentum*). Today, these glutamate producers are classified as subspecies of *Corynebacterium glutamicum* (e.g., *C. glutamicum* ssp. *flavum* and *C. glutamicum* ssp. *lactofermentum*). The major route of glutamate production from glucose is via the Embden–Meyerhof pathway and the early steps of the tricarboxylic acid cycle (Fig. 1). α-Ketoglutarate, which would normally be converted to succinyl coenzyme A in the cycle, is instead reductively aminated to glutamate by glutamate dehydrogenase. Molar yields of glutamate from sugar are 50–60%, and broth concentrations reach more than 100 g/L. Normally, glutamic acid overproduction would not occur because of feedback regulation. However, due to a modification of the cell membrane, glutamate is pumped out of the cell, thus allowing its biosynthesis to proceed unabated. The membrane alteration is intentionally effected by biotin limitation (all glutamic acid bacteria are biotin auxotrophs), glycerol limitation of glycerol auxotrophs, oleate limitation of oleate auxotrophs, or addition of penicillin or fatty acid derivatives to exponentially growing cells. Apparently all of these manipulations result in a phospholipid-deficient cytoplasmic membrane. Also required for effective production are a high level of carbon and energy source as well as growth inhibition. The actual excretion is carried out by a specific efflux system involving a carrier that is dependent on membrane potential (1).

The bulk of the cereals consumed in the world are deficient in the essential amino acid L-lysine. Lysine supplementation converts such cereals into balanced food or feed. Lysine is a member of the aspartate family of amino acids. It is produced in bacteria by a branched pathway that also produces methionine, threonine, and isoleucine. This pathway is controlled very tightly in an organism such as *Escherichia coli*; this organism contains three aspartate kinases, each of which is regulated by a different end product. In addition, after each branch point, the initial

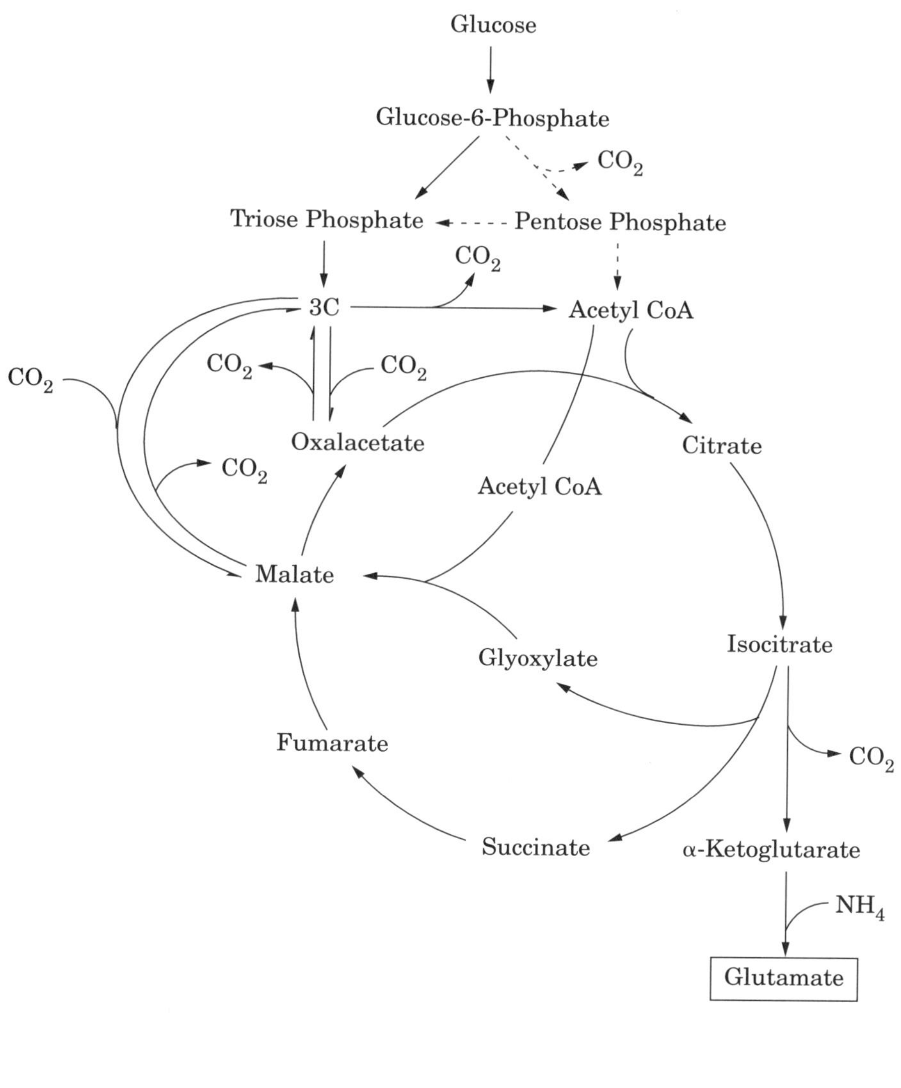

Figure 1. Pathway of glutamic acid production from glucose by *Corynebacterium glutamicum*.

enzymes are inhibited by their respective end products and no overproduction occurs. However, in lysine fermentation organisms (e.g., various mutants of *C. glutamicum* and its relatives), there is a single aspartate kinase, as shown in Figure 2. This aspartate kinase is regulated via concerted feedback inhibition by threonine plus lysine. By genetic removal of homoserine dehydrogenase, a glutamate-producing wild-type *Corynebacterium* is converted into a lysine-overproducing mutant that cannot grow unless methionine and threonine are added to the medium. As long as the threonine supplement is kept low, the intracellular concentration of threonine is limiting and feedback inhibition of aspartate kinase is bypassed. Other enzymes in the aspartate family biosynthetic pathway such as aspartate semialdehyde dehydrogenase and threonine synthetase are unregulated. In addition to the difference in the mode of aspartate kinase feedback inhibition, lysine overproducers differ from *E. coli* in the following ways: (*1*) no feedback repression of aspartate kinase or aspartate semialdehyde dehydrogenase occurs in lysine overproducers;

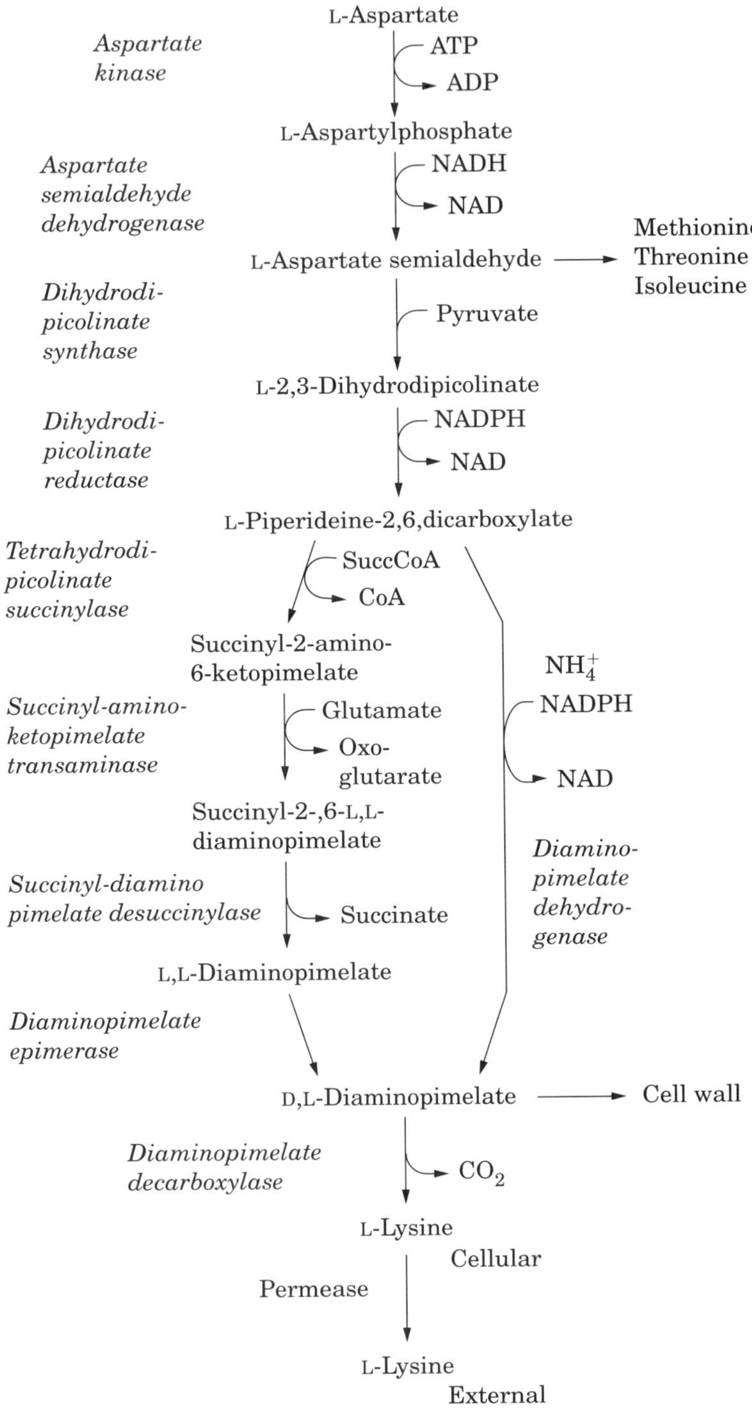

Figure 2. Biosynthesis of L-lysine by *Corynebacterium glutamicum Source:* From Ref. 2.

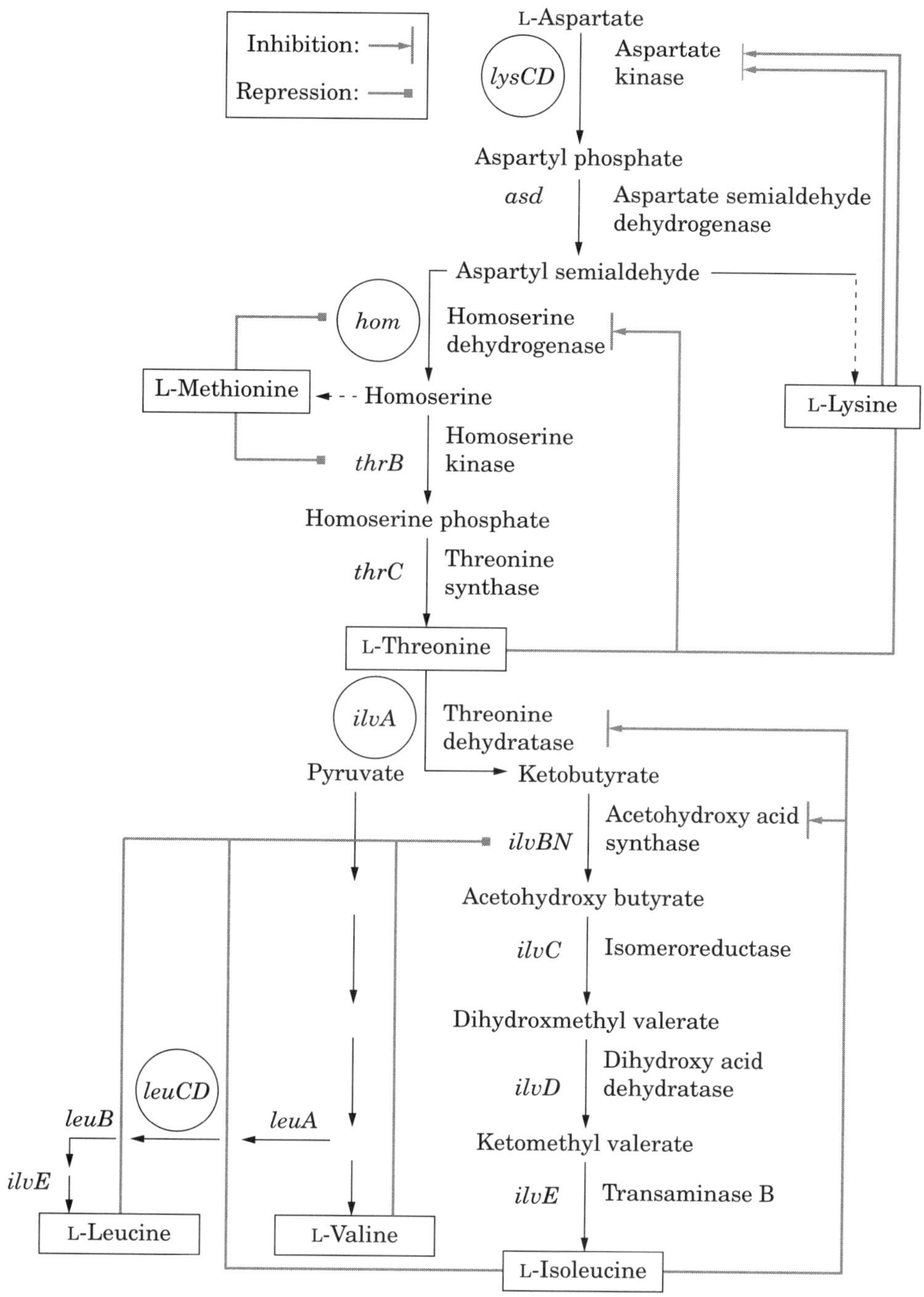

Figure 3. L-Isoleucine biosynthesis in recombinant *Corynebacterium glutamicum* strains. Repression on the DNA level (shaded lines with square ends) and inhibition of enzymes (shaded lines with arrowhead ends). The last four enzymatic steps are the same for L-isoleucine and L-valine biosynthesis. *Source*: Ref. 7.

(*2*) the first and second enzymes of the lysine branch (dihydrodipicolinate synthetase and dihydrodipicolinate reductase) are neither inhibited nor repressed by lysine in lysine overproducers; (*3*) L-lysine decarboxylase is absent in lysine overproducers. The pentose phosphate pathway is of much greater importance in lysine production than in glutamate formation due to the greater need of NADPH for the former path (3). Lysine excretion is by active transport involving a carrier and is driven by membrane potential, the lysine gradient, and the proton gradient. Excretion

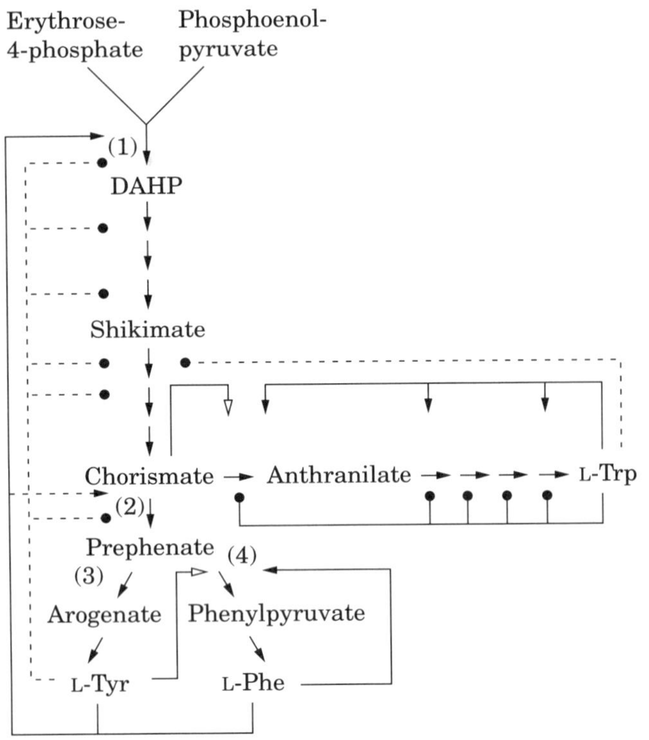

Figure 4. Biosynthetic pathway of aromatic amino acids and their regulatory mechanisms in *Corynebacterium glutamicum* sp. *lactofermentum*. 1, 3-deoxy-D-arabino-heptulosonate-7-phosphate (DAHP) synthase; 2, chorismate mutase; 3, arogenate aminotransferase; 4, prephenate dehydratase; ◀—, inhibition; ◀---, partial inhibition; ◁—, activation; ●—, repression; ●---, weak repression. *Source*: From Ref. 8.

uses a 2 OH/lysine symporter and is catalyzed by a dipetide uptake system dependent on electromotive force, not ATP (4).

Recombinant DNA techniques have made their way into the amino acid production area (5). *E. coli* and *Serratia marcescens* strains have been constructed with plasmids bearing amino acid biosynthetic operons. Plasmid transformation has been accomplished in *C. glutamicum*, so that recombinant DNA technology is now used to improve these commercial amino acid–producing strains. The major manipulations have involved gene cloning to increase the levels of feedback-resistant aspartate kinase and dihydrodipicolinate synthase (6). As a result, lysine industrial production yields 120 g/L and 0.25–0.35 g L-lysine·HCl per gram glucose used (molar yield of 0.25–0.35 mol L-lysine per mole of glucose used). The market for L-lysine·HCl is 360 million pounds per year (6).

Recombinant technology and traditional mutagenesis and selection have been major influences in constructing bacterial strains capable of producing 100 g/L L-threonine, 40 g/L L-isoleucine, 34 g/L L-leucine, 31 g/L L-valine, 28 g/L L-phenylalanine, 55 g/L L-tryptophan, 26 g/L L-tyrosine, 100 g/L L-proline, 65 g/L L-arginine, and 40 g/L L-histidine. The relationship between the lysine biosynthetic pathway and that for threonine, methionine, isoleucine, valine, and leucine in *C. glutamicum* is shown in Figure 3. The pathway to the aromatic amino acids (phenylalanine, tryptophan, and tyrosine) is given in Figure 4.

The commercial interest in nucleotide fermentations is due to the activity of two purine ribonucleoside 5′-monophosphates, namely guanylic acid (GMP) and inosinic acid (IMP), as enhancers of flavor (9,10). Some 2,500 tons of GMP and IMP are produced annually in Japan alone. Three main processes are used: (*1*) hydrolysis of yeast RNA by fungal nuclease to AMP and GMP followed by enzymatic deamination of AMP to IMP; (*2*) fermentative production of the nucleosides inosine and guanosine by *Bacillus subtilis* mutants followed by chemical phosphorylation, and (*3*) direct fermentation of sugar to IMP by *C. glutamicum* mutants plus conversion of guanine to GMP by salvage synthesis using intact cells of *Brevibacterium ammoniagenes*. Titers of IMP by direct fermentation have reached 27 g/L (10).

The pathway of purine ribonucleotide biosynthesis (Fig. 5) is virtually identical in all living systems. In the first specific step, 5′-phosphoribosyl pyrophosphate (PRPP), derived from ATP and ribose-5-phosphate, is aminated with glutamine to form 5′-phosphoribosylamine (PRA). By a series of nine known enzymic steps, PRA is converted to IMP, the first purine ribonucleotide formed. This pivotal intermediate is the precursor of both AMP and GMP. IMP is then converted to adenylosuccinate (S-AMP), aspartate acting as the succinyl donor and GTP as the energy source. Adenylosuccinate is then converted to AMP by adenylosuccinate lyase, a bifunctional enzyme that also catalyzes an earlier step. Inosinic acid is oxidized by IMP dehydrogenase using NAD^+ as a cofactor to form XMP. The latter is aminated by XMP aminase to GMP. In this reaction, ATP is the energy source, and, depending on the organism, the amino donor is either glutamine or ammonia. Both AMP and GMP can be used for synthesis of nucleic acids after being converted to their respective di- and triphosphates. Alternatively, GMP can be reduced to IMP in a reaction catalyzed by GMP reductase (which utilizes NADPH as an electron donor) and AMP can be converted to IMP by deamination. These reactions permit a cyclic regeneration of intermediates common to both purine nucleotides from an accumulated excess of either. All of the purine ribonucleotide interconversions occur via their 5′-isomers.

With such a complex series of biosynthetic reactions, it is essential to the economy of the cell that there be regulation of the rates of the individual reactions. This regulation is accomplished by negative feedback inhibition of enzyme action and repression of enzyme synthesis. Figure 6 shows an example of a typical feedback control system for nucleotide biosynthesis. Usually both AMP and GMP exert feedback repression and inhibition on the common pathway and on their respective branches. An important control site is step 1, catalyzed by PRPP amidotransferase. In some microorganisms, AMP and GMP each inhibit action of the enzyme completely at high concentrations. In other organisms, a cumulative type of regulation is observed. Synthesis of the amidotransferase, as well as many of the other enzymes of the common pathway, is also re-

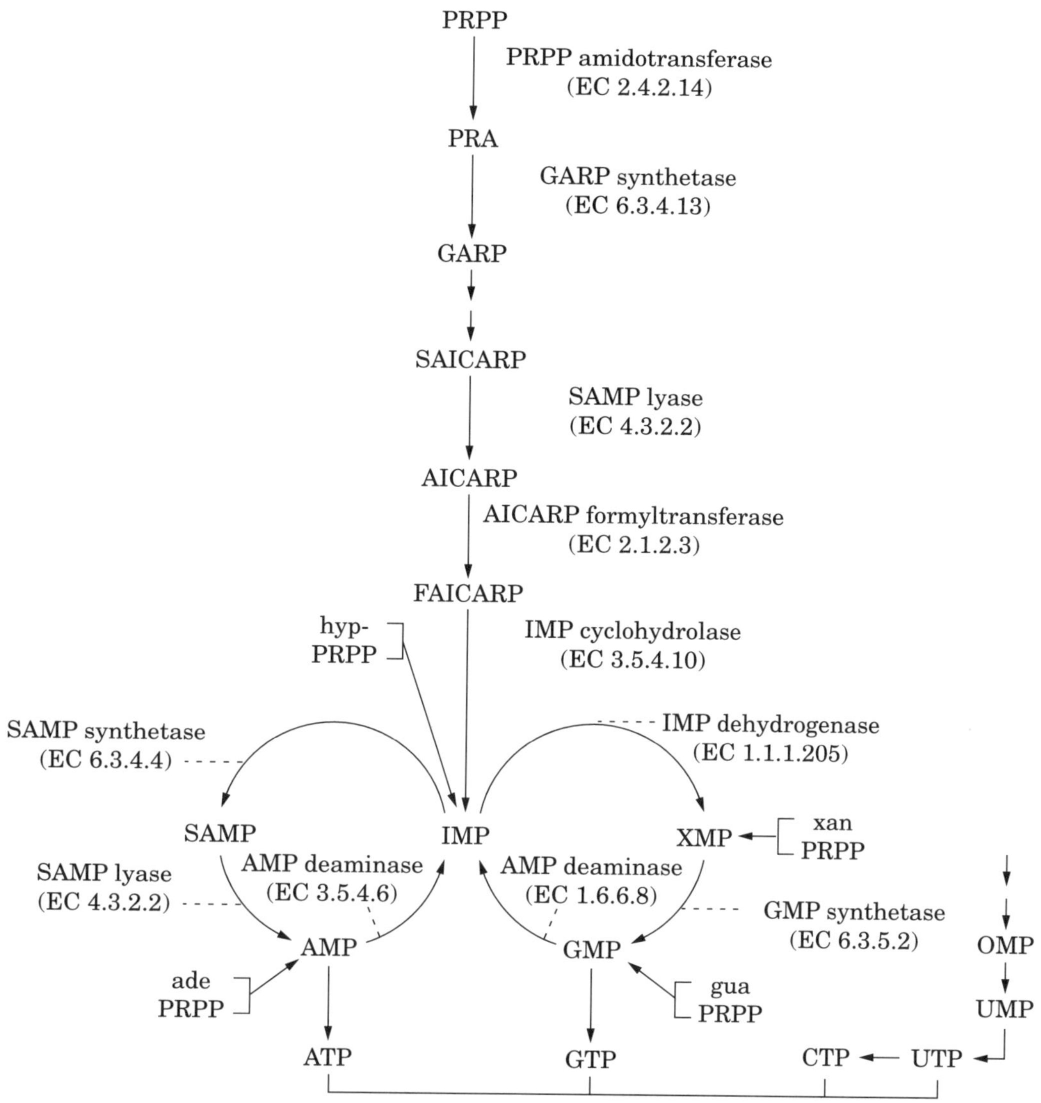

Figure 5. Biosynthetic pathway to purine nucleotides. PRPP, 5-phosphoribosyl-1-pyrophosphate (5-phospho-α-D-ribose-1-diphosphate); PRA, 5-phosphoribosyl-1-amine; GARP, glycineamide ribonucleotide; AICARP, 5-aminoimidazole-4-succinocarboxamide ribonucleotide; FAICARP, 5-formaminoimidazole-4-carboxamide ribonucleotide; SAMP, adenylosuccinate (succinoadenosine 5′-monophosphate). *Source*: From Ref. 10.

pressed by AMP and GMP. Another important control site is IMP dehydrogenase, which is usually inhibited and repressed by GMP. The conversion of IMP to S-AMP (catalyzed by adenylosuccinate synthase) is usually repressed and inhibited by AMP.

The key to effective purine accumulation is the limitation of intracellular AMP and GMP. This limitation is best effected by restricted feeding of purine auxotrophs. Thus, adenine-requiring mutants lacking adenylosuccinate synthetase (see Fig. 7) accumulate hypoxanthine or inosine that results from breakdown of intracellularly accumulated IMP. Because of the strong inhibition and repression of IMP dehydrogenase by GMP, the accumulation does not proceed past IMP into the GMP branch of the pathway. Certain adenine auxotrophs of *Bacillus subtilis* excrete more than 10 g/L inosine. These strains are still subject to GMP repression of enzymes of the common path. To minimize the severity of this regulation, the adenine auxotrophs are further mutated to eliminate IMP dehydrogenase. These adenine–xanthine double auxotrophs show a twofold increase in specific activity of some common-path enzymes and accumulate up to 15 g/L inosine under conditions of limiting adenine and xanthine (or guanosine). Further deregulation is achieved by selection of mutants resistant to purine analogs. Thus, mutants requiring adenine and xanthine and resistant to azaguanine produce more than 20 g/L inosine. Insertional inactivation of the IMP dehydrogenase gene in another *B. subtilis* strain yielded a culture producing 35 g/L inosine (11). Genetic engineering of the inosine monophosphate dehydrogenase

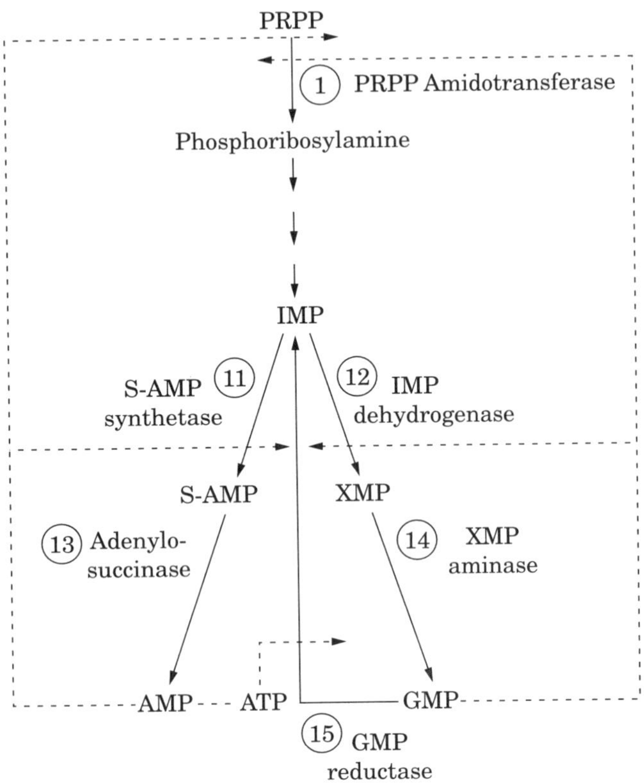

Figure 6. Biosynthesis of 5′-purine nucleotides. Solid lines, pathway steps; dashed lines, feedback inhibition or repression.

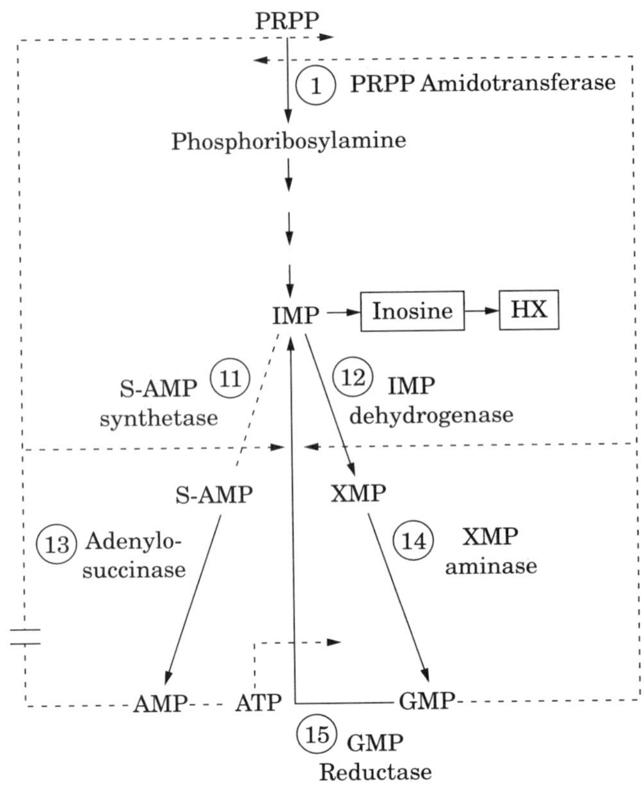

Figure 7. Overproduction of 5′-inosinic acid (5′-IMP) or its degradation products (inosine and hypoxanthine [hx]) by adenine auxotrophs. Solid lines, pathway steps; dashed lines, feedback inhibition or repression; dots, mutated (inactive) enzyme.

gene in a *B. subtilis* strain, which was producing 7 g/L guanosine and 19 g/L inosine, changed production to 20 g/L guanosine and 5 g/L inosine (12). Other *B. subtilis* mutants produce as much as 23 g/L guanosine (10). With regard to pyrimidine production, another recombinant strain of *B. subtilis* produces 18 g/L cytidine, and a mutant lacking homoserine dehydrogenase (which increased the concentration of the precursor aspartate in the cell) produces 30 g/L cytidine (13).

Riboflavin (vitamin B_2: Fig. 8) is produced commercially by both fermentation and chemical synthesis (15). Riboflavin overproducers include two yeastlike molds, *Eremothecium ashbyii* and *Ashbya gossypii*, which synthesize riboflavin in concentrations greater than 20 g/L. Production of riboflavin amounted to 4 million pounds in 1992 (16). A riboflavin overproducer such as *A. gossypii* makes 40,000 times more vitamin than it needs for its own growth. The biochemical key to riboflavin overproduction appears to involve insensitivity to the repressive effects of iron. Ferrous ion severely inhibits riboflavin production by low and moderate overproducers, such as clostridia and *Candida* but has no inhibitory action against *E. ashbyii* and *A. gossypii*. In normal microorganisms, it appears that iron represses almost all of the riboflavin biosynthetic enzymes, whereas riboflavin or a derivative inhibits the first enzyme of the pathway, GTP cyclohydrolase II. New processes using *Candida* sp. or recombinant *B. subtilis* strains that produce 20–30 g/L riboflavin have been developed in recent years.

Vitamin B_{12} (Fig. 9) is industrially produced by *Propionibacterium shermanii* or *Pseudomonas denitrificans* (16,18). Such strains used for industrial production make about 100,000 times more vitamin B_{12} than they need for their own growth. The key to the fermentation is avoidance of feedback repression by vitamin B_{12}. The early stage of the *Propionibacterium shermanii* fermentation is conducted under anaerobic conditions in the absence of the precursor 5,6-dimethylbenzimidazole. These conditions prevent vitamin B_{12} synthesis and allow for the accumulation of the intermediate, cobinamide. Then the culture is aerated and dimethylbenzimidazole is added, converting cobinamide to the vitamin. In the *Pseudomonas denitrificans* fermentation, the entire process is carried out under low oxygen conditions. A high level of oxygen results in an oxidizing intracellular environment that represses formation of the early enzymes of the pathway. The key to *Pseudomonas denitrificans* fermentation is the addition of betaine (19). Vitamin B_{12} overproduction by *Pseudomonas denitrificans* is totally dependent on betaine, but the mechanism of control is unknown. Production of vitamin B_{12} has reached levels of 150 mg/L (18).

In production of biotin, feedback repression is caused by the enzyme protein acetyl-CoA carboxylase biotin holoenzyme synthetase, with biotin 5-adenylate acting as corepressor (20). Strains of *Serratia marcescens* obtained by mutagenesis, selection for resistance to biotin antimetabolites, and molecular cloning produce 600 mg/L in the presence of high concentrations of sulfur and ferrous iron (21).

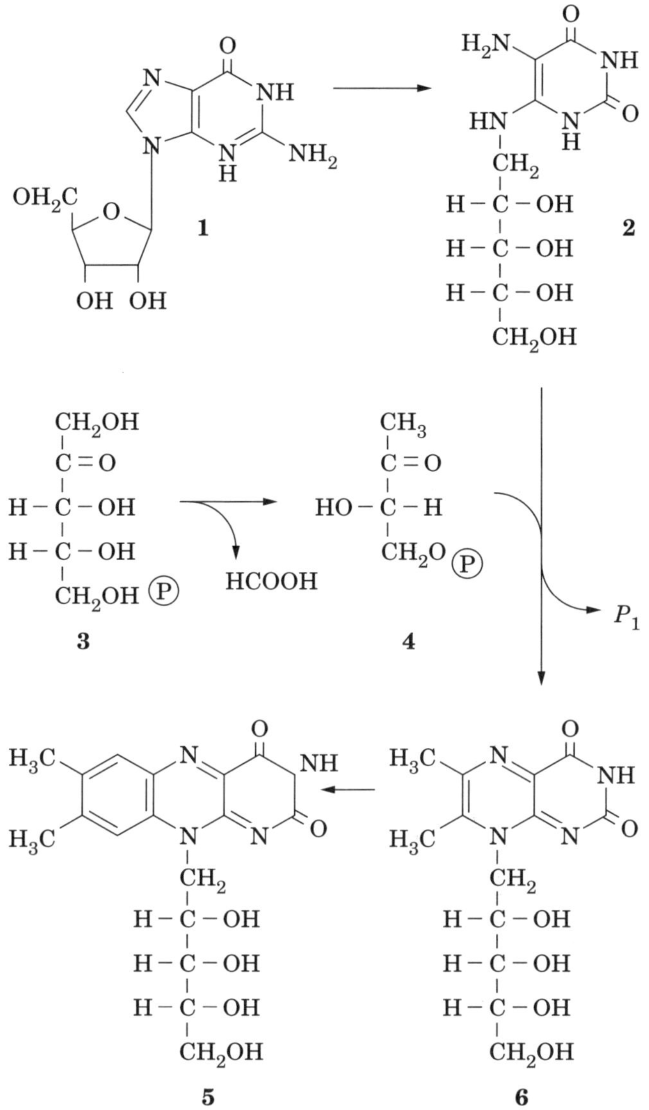

Figure 8. Biosynthesis of riboflavin. 1, guanosine triphosphate; 2, 5-amino-6-ribitylamino-2,4 (1H, 3H)-pyrimidinedione; 3, D-ribulose 5-phosphate; 4, L-3,4, dihydroxy-2-butanone 4-phosphate; 5, riboflavin; 6, 6,7-dimethyl-8-ribityllumazine. *Source*: From Ref. 14.

Such a titer is high enough to economically compete with the traditional chemical process. The pathway is shown in Figure 10.

A novel process for vitamin C (ascorbic acid) synthesis involves the use of a genetically engineered *Erwinia herbicola* strain containing a gene from *Corynebacterium* sp. The engineered organism converts glucose into 2-ketogulonic acid, which can be easily converted by acid or base to ascorbic acid (23). Another process devised independently converts 40 g/L glucose into 20 g/L 2-keto-L--gulonate (24). This process involves cloning of the gene encoding 2,5-diketo-D-gluconate reductase from *Corynebacterium* sp. into *Erwinia citreus*. Plasmid cloning of the genes encoding L-sorbose dehydrogenase and L-sorbosone dehydrogenase from *Gluconobacter oxydans* back into the same organism yielded a strain capable of converting 150 g/L D-sorbitol into 130 g/L 2-keto-L-gulonate (25).

Filamentous fungi have been widely used for the commercial production of organic acids. About 1 billion pounds of citric acid are produced each year. This organic acid is produced via the Embden–Meyerhof pathway and the first step of the tricarboxylic acid cycle (Fig. 11). The major control of the process involves the inhibition of phosphofructokinase by citric acid. The commercial process employs *Aspergillus niger* in media deficient in iron and manganese. Manganese deficiency has two beneficial effects in the citric acid fermentation: *(1)* it leads to high levels of intracellular NH_4, which reverse citric acid inhibition of phosphofructokinase, and *(2)* it brings on the formation of small mycelial pellets, which are the best morphological form for citric acid production. The morphological effect is due to a change in cell wall composition caused by growth in low Mn^+. A high level of citric acid production is also associated with a high intracellular concentration of fructose 2,6-biphosphate, an activator of glycolysis (27). Other factors that contribute to high citric acid production are the inhibition of isocitrate dehydrogenase by citric acid and the low pH optimum (1.7–2.0). Higher pH levels (e.g., 3.0) lead to production of oxalic and gluconic acids instead of citric acid. The low pH level inactivates glucose oxidase, which normally would yield gluconic acid (26). In approximately 4 to 5 days, the major portion (80%) of the sugar is converted to citric acid, titers reaching about 100 g/L. Because citric acid is easily assimilated, is palatable, and exhibits low toxicity, it is widely used in the food and pharmaceutical industries. It is employed as an acidifying and flavor-enhancing agent, as an antioxidant for inhibiting rancidity in fats and oils, as a buffer in jams and jellies, and as a stabilizer in a variety of foods. The pharmaceutical industry uses approximately 16% of the available supply of citric acid. Processes have been developed for the production of citric acid by *Candida* yeasts, especially from hydrocarbons. Such yeasts are able to convert *n*-paraffins to citric and isocitric acids in extremely high yields (150–170% on a weight basis). Production of citric acid instead of isocitric acid is favored by selecting mutants that are deficient in the enzyme aconitase. Titers as high as 225 g/L have been reached (26).

Cloning of the aldehyde dehydrogenase gene from *Acetobacter polyoxogenes* on a plasmid vector into *A. aceti* sp. *xylinum* increased the rate of acetic acid production by more than 100% (from 1.8 to 4 g L^{-1} h^{-1}) and titer by 40% (from 68 to 97 g/L) (28).

Ethyl alcohol is a primary metabolite that can be produced by fermentation of a sugar or a polysaccharide that can be depolymerized to a fermentable sugar. Yeasts are preferred for these fermentations, but the species used depends on the substrate employed. *Saccharomyces cerevisiae* is employed for the fermentation of hexoses, whereas *Kluyveromyces fragilis* or *Candida* species may be utilized if lactose or pentoses, respectively, are the substrates. Under optimum conditions, approximately 10 to 12% ethanol by volume is obtained within 5 days. Such a high concentration slows down growth and the fermentation ceases. With special yeasts, the fermentation can be continued to alcohol concentrations of 20% by volume.

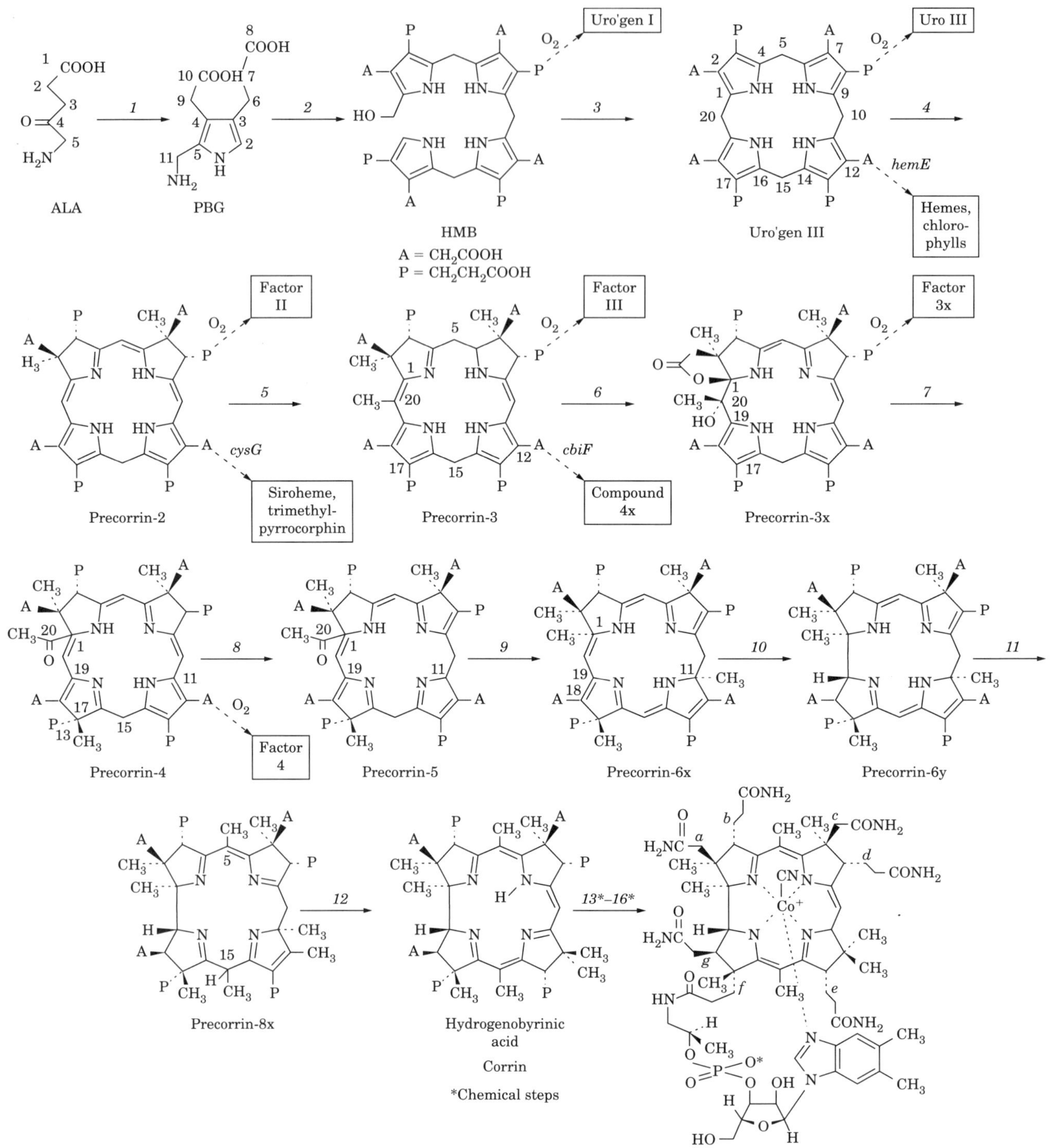

Figure 9. The biosynthetic pathway from δ-aminolevulinic acid to vitamin B_{12}. The enzyme catalyzing each reaction and its corresponding gene are 1, δ-aminolevulinic acid dehydratase (*hemB*); 2, porphobilinogen (PBG) deaminase (*hemC*); 3, uroporphyrinogen III (uro'gen III) synthase (*hemD*); 4, uro'gen III methyltransferase (*cysG, cobA*); 5, precorrin 2 methyltransferase (*cobI*); 6, precorrin 3 oxidase (*cobG*); 7, precorrin 3 hydroxylactone methyltransferase (ring contraction) (*cobJ*); 8, precorrin 4 methyltransferase (*cobM*); 9, precorrin 5 methyltransferase (loss of acetic acid) (*cobF*); 10, precorrin 6× reductase (*cobK*); 11, precorrin 6y decarboxylase and methyltransferase (*cobL*); 12, precorrin 8 methylmutase (*cobH*); 13, cobalt insertion; 14, esterification; 15, add nucleotide loop; and 16, ammonolysis. Possible side products are indicated in boxes. Asterisk indicates chemical step. *Source*: From Ref. 17.

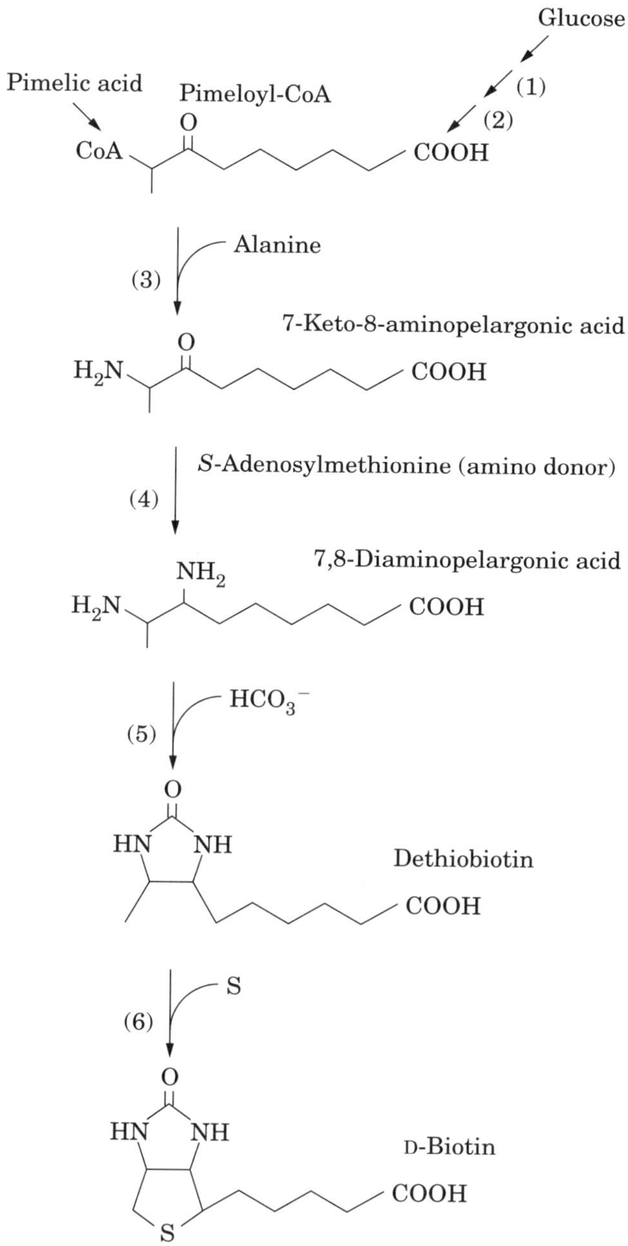

Figure 10. Pathway of biotin biosynthesis in bacteria. 1, *bioH*, block before pimeloyl-CoA; 2, *bioC*, block before pimeloyl-CoA; 3, *bioF*, 7-keto-8-aminopelargonic acid (KAPA) synthase; 4, *bioA*, 7,8-diaminopelargonic acid (DAPA) aminotransferase; 5, *bioD*, dethiobiotin synthase; 6, *bioB*, biotin synthase. *Source*: From Ref. 22.

However, these concentrations are attained only after months or years of fermentation. At present, industrial ethanol is mainly manufactured by fermentation, but there are still considerable amounts made by the petrochemical industry from ethylene. Bacteria such as clostridia and *Zymomonas* are being reexamined for ethanol production after years of neglect. *Clostridium thermocellum*, an anaerobic thermophile, can convert waste cellulose directly to ethanol. Other clostridia produce acetate, lactate, acetone, and butanol and will be utilized as petroleum becomes depleted in the world. Fuel ethanol produced from biomass would provide relief from air pollution caused by the use of gasoline and would not contribute to the greenhouse effect. Because of the elimination of lead from gasoline, ethanol is being substituted as a blend to raise gasoline's octane rating. *E. coli* has been converted into an excellent ethanol producer (43%, v/v) by recombinant DNA technology (29). Alcohol dehydrogenase II and pyruvate decarboxylase genes from *Zymomonas mobilis* were inserted in *E. coli* and became the dominant system for NAD regeneration. Ethanol represents more than 95% of the fermentation products in the genetically engineered strain.

In addition to the multireaction sequences of fermentations, microorganisms are extremely useful in carrying out biotransformation processes in which a compound is converted into a structurally related product by one or a small number of enzymes contained in the cells (30). Bioconverting organisms are known for practically every type of chemical reaction. One of the earliest and most famous is the biotransformation of progesterone to 11 α-hydroxyprogesterone (Fig. 12). The reactions are sterospecific, the ultimate in specificity being exemplified by the steroid bioconversions. This specificity is exploited in the resolution of racemic mixtures when a specific isomer rather than a racemic mixture is desired. Bioconversions are characterized by extremely high yields (i.e., 90 to 100%). Other attributes include mild reaction conditions and the coupling of reactions using a microorganism containing several enzymes working in series. There is a tremendous interest in immobilized cells to carry out such processes. These cells are usually much more stable than either free cells or enzymes and are more economical than immobilized enzymes. Recombinant DNA techniques have been useful in developing new bioconversions. For example, the cloning of the fumarase-encoding gene in *Saccharomyces cerevisiae* improved the bioconversion of malate to fumarate from 2 to 125 g/L in a single manipulation (31). The conversion yield using the constructed strain is near 90%.

Microbially produced secondary metabolites are extremely important to our health and nutrition (32). As a group that includes antibiotics, other medicinals, toxins, pesticides, and animal and plant growth factors, they have tremendous economic importance. In batch or fed-batch culture, secondary metabolites are usually produced after growth has slowed down. They have no function in growth of the producing cultures (although they contribute to survival of producing organisms in nature), are produced by certain restricted taxonomic groups of organisms, and are usually formed as mixtures of closely related members of a chemical family. From 1990 to 1994, more than 1,000 new secondary metabolites were characterized from actinomycetes alone (33).

The best known of the secondary metabolites are the antibiotics (Figs. 13 and 14). This remarkable group of compounds form a heterogeneous assemblage of biologically active molecules with different structures and modes of action (Table 2). In 1996 the antibiotic market involved 160 antibiotics and a $23 billion market (36). Since 1940 we have witnessed a virtual explosion of new and potent antibiotic molecules that have been of great use in medicine, agriculture, and basic research. However, the search

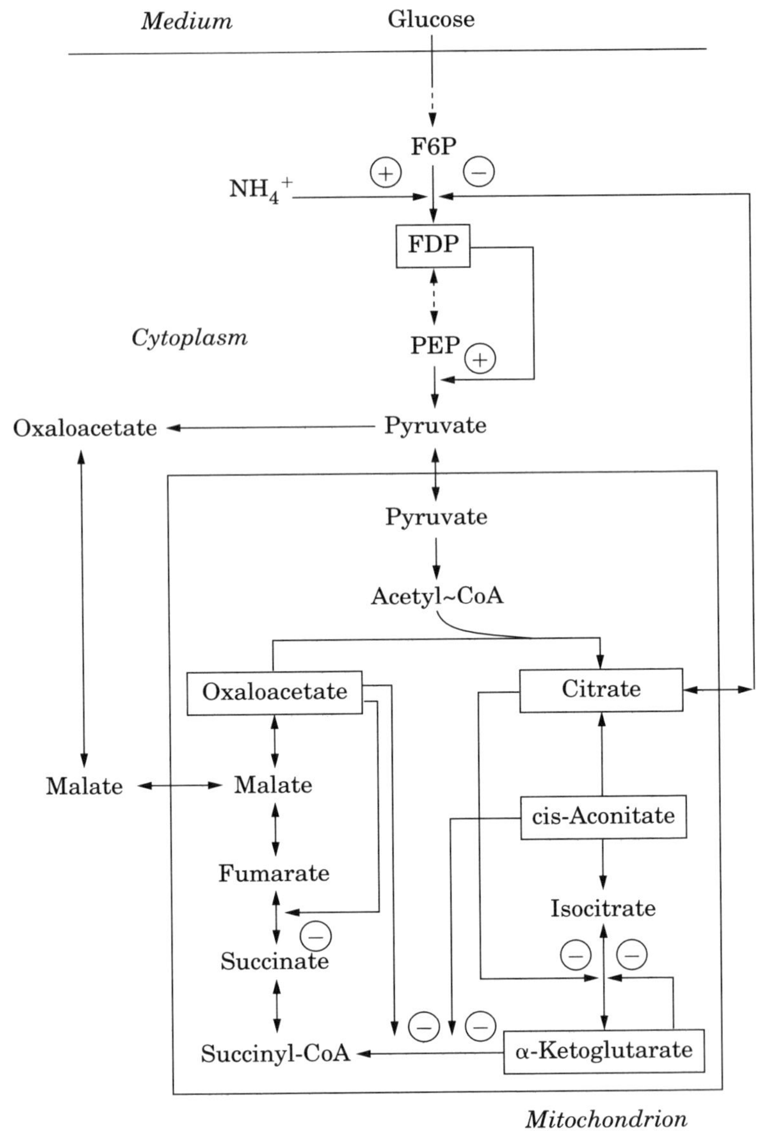

Figure 11. Metabolic scheme of citric acid overproduction. +, activation by the respective metabolite; −, inhibition. *Source*: From Ref. 26.

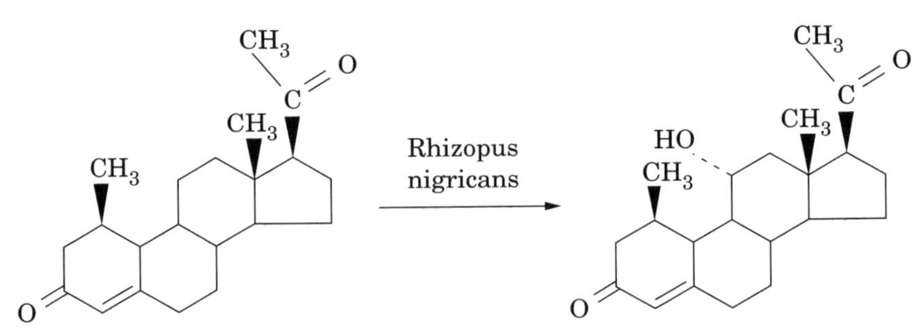

Figure 12. 11α-Hydroxylation of progesterone by *Rhizopus nigricans*.

for new antibiotics continues in order to combat evolving pathogens and naturally resistant bacteria and fungi and previously susceptible microbes that have developed resistance; improve pharmacological properties; combat tumors, viruses, and parasites; and discover safer, more potent, and broader spectrum compounds. About 6,000 antibiotics have been described, 4,000 from actinomycetes alone, and they still are being discovered at a rate of about 300 per year. *Streptomyces griseus* strains produce more than 40 different antibiotics and *Bacillus subtilis* more than 60 different antibiotics. Strains of *S. hygroscopicus* make almost 200 antibiotics. One *Micromonospora* strain

Hydrophobic β-Lactams

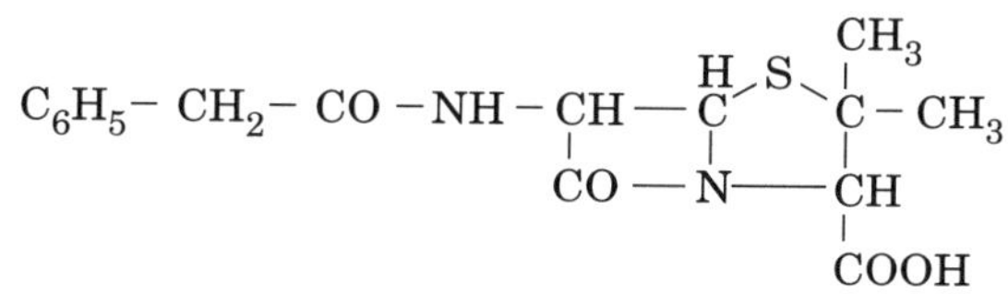

Benzylpenicillin

Hydrophilic β-Lactams

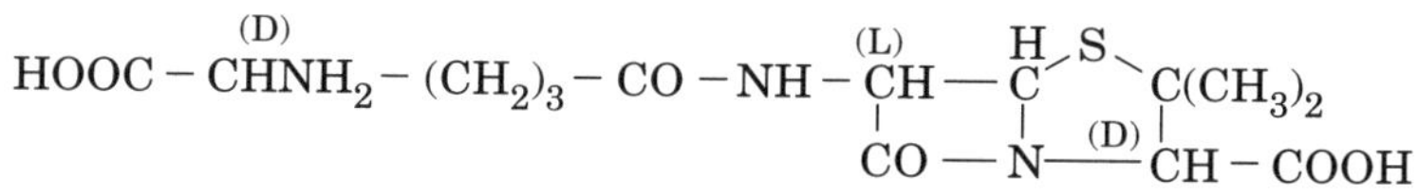

Penicillin N

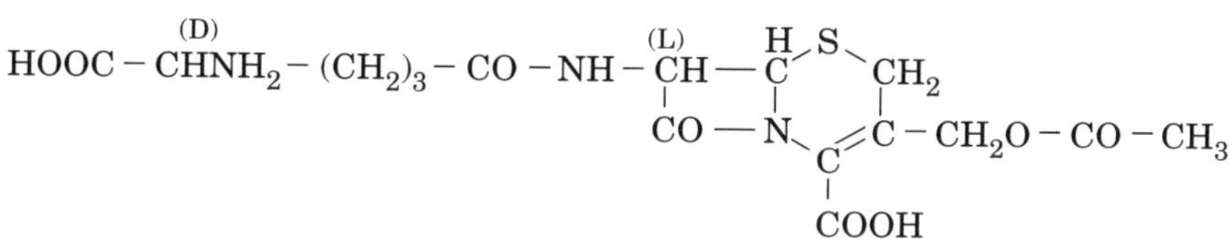

Cephalosporin C

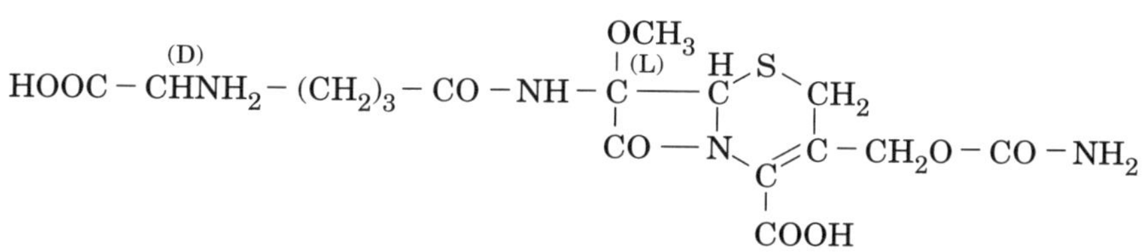

Cephamycin C

Figure 13. Some β-lactam antibiotics. Benzylpenicillin is also known as penicillin G.

can produce 48 aminocyclitol antibiotics. The antibiotics vary in size from small molecules such as cycloserine (102 Da) and bacilysin (270 Da) to polypeptides such as nisin, which contains 34 amino acid residues. They attack virtually every type of microbial activity, including DNA, RNA, and protein synthesis; membrane function; electron transport; sporulation; germination; and many others. In the search for new antibiotics, many of the new products are made by chemists by modification of natural antibiotics; this process is called semisynthesis. By 1974 more than 20,000 semisynthetic penicillins, 4,000 cephalosporins, 2,500 tetracylines, 1,000 rifamycins, 500 kanamycins, and 500 chloramphenicols had been prepared. In 1980 worldwide antibiotic production amounted to about 25,000 tons. Included were 17,000 tons of penicillins, 5,000 tons of tetracyclines, 1,200 tons of cephalosporins, and 800 tons of erythromycins. Antibiotics are used not only chemotherapeutically in human and veterinary medicine but also for growth promotion in farm animals and for the protection of plants.

In nature, secondary metabolites are important for the organisms that produce them, functioning as *(1)* sex hormones; *(2)* ionophores; *(3)* competitive weapons against other bacteria, fungi, amoebae, insects, and plants; *(4)* agents of symbiosis; and *(5)* effectors of differentiation (37). For years, the major pharmaceuticals (such as hypertensive and antiinflammatory agents) used for noninfectious diseases were strictly synthetic products. Similarly, major therapeutics for nonmicrobial parasitic diseases in animals (e.g., coccidiostats and antihelmintics) came from the screening of synthesized compounds followed by molecular modification. Despite the testing of thousands of synthetic compounds, only a few promising structures were uncovered. As new lead compounds became more and more difficult to find, microbial broths filled the void and microbial products increased in importance in treatment of nonmicrobial diseases. Today, microbially produced polyethers (38) such as monensin, lasalocid, and salinomycin dominate the coccidiostat market and are also the chief growth promotants in use for ruminant animals. The avermectins (Fig. 14), another group of streptomycete products with a market of more than \$200 million per year, have high activity against helminths and arthropods (39). Indeed, their activity is an order of magnitude greater than previously discovered antihelmintic agents, the vast majority of which were synthetic. Known pharmacological activities of microbial secondary metabolites are shown in Table 3.

Many microbial products with important pharmacological activities were discovered by screening for inhibitors

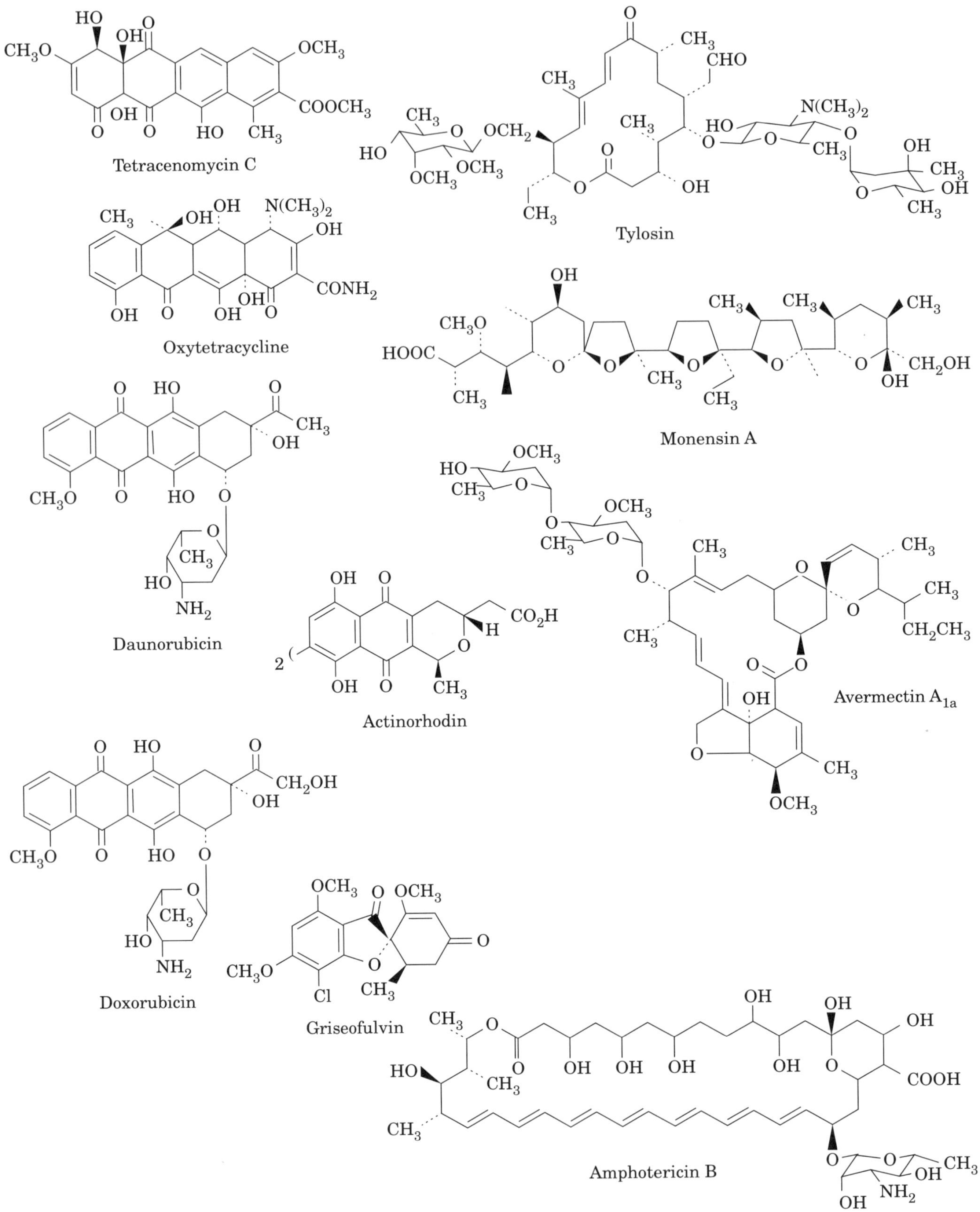

Figure 14. Some secondary metabolites of polyketide orgin. All except avermectin possess antibiotic activity. *Source*: From Refs. 34 and 35.

Table 2. Mode of Action and Structural Type of Selected Antibiotics

Target	Antibiotic	Structure
DNA replication	Bleomycin	Peptide
	Griseofulvin	Condensed aromatic
	Actinomycin	Peptide
Transcription	Rifamycin	Ansamycin
Translation		
70-S ribosomes	Chloramplenicol	Aromatic
	Tetracycline	Polyketide
	Lincomycin	Sugar ester
	Erythromycin	Macrolide
	Streptomycin	Aminocyclitol
70-S and 80-S ribosomes	Puromycin	Nucleoside
	Fusidic acid	Steroid
80-S ribosomes	Cycloheximide	Glutarimide
Cell wall synthesis	Cycloserine	Amino acid
	Bacitracin	Peptide
	Penicillin	β-Lactam
	Vancomycin	Oligosaccharide aromatic
Cell membrane		
Surfactants	Polymyxin	Peptide
	Amphotericin	Polyene
Ionophores		
Channel formers	Linear gramicidin	Peptide
Mobile carriers	Monensin	Polyether

using simple enzymatic assays (40). One huge success has been lovastatin (mevinolin), a fungal product of the compactin group (41) that acts as a cholesterol-lowering agent in animals. Lovastatin is produced by *Aspergillus terreus*. In its hydroxy acid form (mevinolinic acid), mevinolin is a potent competitive inhibitor of 3-hydroxy-3-methylglutaryl-coenzyme A reductase from liver. Another enzyme inhibitor on the market is acarbose, a natural inhibitor of intestinal glucosidase (42), which is produced by an actinomycete of the genus *Actinoplanes*. It decreases hyperglycemia and triglyceride synthesis in adipose tissue, liver, and the intestinal wall of patients suffering from diabetes, obesity, and type IV hyperlipidemia.

Also in commercial or near-commercial use are biopesticides including fungicides (e.g., kasugamycin, polyoxins), insecticides (*Bacillus thuringiensis* crystal, nikkomycin, spinosyns), herbicides (bialaphos), antihelmintics and coccidiostats as mentioned earlier, ruminant growth promoters (monensin, lasalocid, salinomycin), plant growth regulators (gibberellins), immunosuppressants for organ transplants (cyclosporin A, FK-506, rapamycin [Fig. 15]), anabolic agents in farm animals (zearelanone), uterocontractants (ergot alkaloids), and antitumor agents (doxorubicin, daunorubicin, mitomycin, bleomycin) (43,44). A remarkable aspect of this development is that many of the compounds were first isolated as poor or toxic antibiotics (e.g., monensin, cyclosporin, rapamycin) or as mycotoxins (ergot alkaloids, gibberellins, zearelanone) before they were put to work for our benefit. Following in their footsteps are many secondary metabolites with potent activities that are under investigation right now. Indeed, antibiotic activity is merely the tip of the iceberg, a mere scratching at the surface of the potential of microbial activity.

In batch culture, some secondary metabolite processes exhibit a distinct growth phase (trophophase) followed by a production phase (idiophase). In other fermentations, trophophase and idiophase overlap; the timing depends on the nutritional environment presented to the culture, the growth rate, or both. Secondary metabolite production in nature probably allows organisms to differentiate and to compete effectively with other forms of life. Delay in antibiotic production until after trophophase aids the producing organism, because the microbe is usually sensitive during growth to its own antibiotic. Resistance mechanisms in producers include enzymatic modification of the antibiotic, alteration of the cellular target of the antibiotic, and

Table 3. Some Pharmacological Activities of Microbial Secondary Metabolites

ACTH-like	Diuretic	Erythematous	Leukemogenic
Anabolic	Edematous	Estrogenic	Motility inhibition
Analeptic	Emetic	Fertility enhancing	Nephrotoxic
Anesthetic	Enzyme inhibitory	Hallucinogenic	Neuromuscular blockade
Anorectic	Aldol reductase	Hemolytic	Neurotoxic
Anticoagulant	α-Amylase	Hemostatic	Paralytic
Antidepressive	Amyloglucosidase	Herbicidal	Parasympathomimetic
Antihelminthic	Catechol-*O*-methyltransferase	Hormone releasing	Photosensitizing
Antiinfective	Cholinesterase	Hypersensitizing	Relaxant (smooth muscle)
Antiinflammatory	Cyclic AMP phosphodiesterase	Hypocholesterolemic	Sedative
Antiparasitic	Dopamine β-hydroxylase	Hypoglycemic	Serotonin antagonist
Antispasmodic	Elastase	Hypolipidemic	Spasmolytic
Carcinogenesis inhibition	Esterase	Hypotensive	Telecidal
Coagulative (blood)	Glucosidase	Immunostimulating	Ulcerative
Complement inhibition	Invertase	Inflammatory	Vasodilative
Convulsant	β-Lactamase	Insecticidal	
Dermonecrotic	Protease		
Diabetogenic	Tyrosine hydroxylase		

Note: List does not include antimicrobial, antiviral, and antitumor activities.

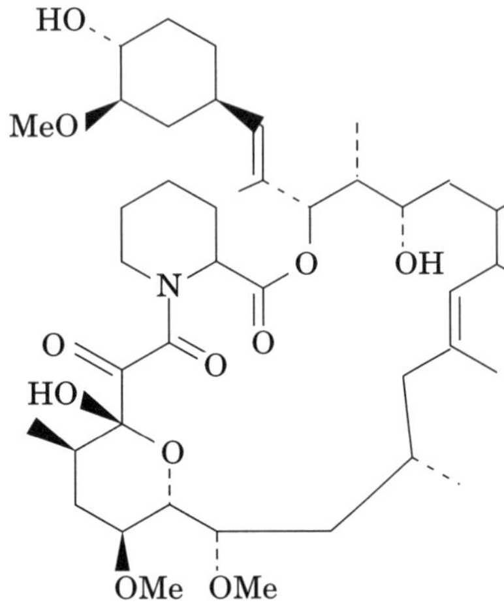

FK-506

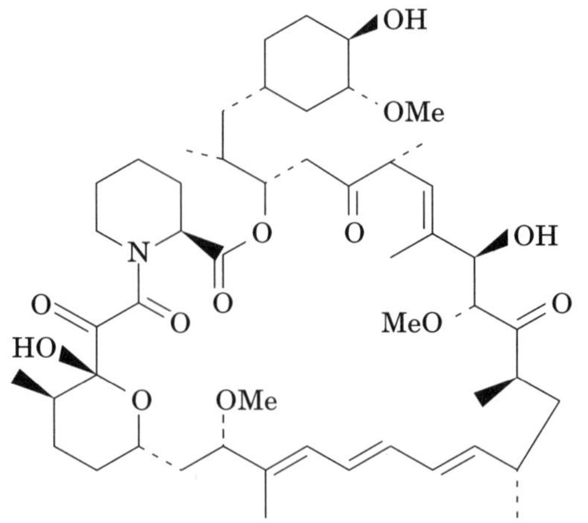

Rapamycin

Figure 15. Two new immunosuppressants made by streptomycetes.

decreased uptake of excreted antibiotic. The raw materials of secondary metabolism are primary metabolites, which often exert a negative or positive effect on secondary metabolism.

Manipulation of culture media in any development program often involves the testing of hundreds of additives as possible limiting precursors of the desired product. Occasionally, a precursor that increases production of the secondary metabolite is found. The precursor may also direct the fermentation toward the formation of one specific desirable product; this is known as *directed biosynthesis*. Examples of the latter include phenylacetic acid in benzylpenicillin fermentation, specific amino acids in the production of actinomycins and tyrocidins, and substituted benzoic acids in the formation of novobiocins. In many fermentations, however, precursors show no activity because their syntheses are not rate limiting. In such cases, screening of additives has often revealed dramatic effects, both stimulatory and inhibitory, of nonprecursor molecules on the production of secondary metabolites. These effects are usually due to interaction of these compounds with the regulatory mechanisms existing in the fermentation organism.

The regulatory mechanisms prevalent in secondary metabolism are essentially the same as found in primary metabolism (45). Thus chloramphenicol limits its own synthesis by repressing arylamine synthetase, and ergot alkaloids limit their production by inhibiting dimethylallyl transferase; these are the initial enzymes unique to the secondary pathway. Glucose, a sugar usually used rapidly for growth, exerts carbon catabolite repression of secondary metabolite synthesis. Similarly, ammonium ion or rapidly used amino acids exert nitrogen metabolite regulation on secondary metabolism. Inorganic phosphate exerts a strong negative effect in secondary metabolite fermentations. Enzyme induction is evident in the production of alkaloids and antibiotics. In actinomycetes, γ-butyrolactones function as autoinducers of secondary metabolism. Besides the specific types of regulation mentioned earlier, the onset of secondary metabolism is also controlled by growth rate. Generally, secondary metabolites are not produced at growth rates near the maximum specific growth rate. In fact, the idiophase can be shifted into the trophophase when media that support only low rates of growth are used. Antibiotic biosynthesis ends via the decay of antibiotic synthetases or because of feedback inhibition and repression of these enzymes. Because the regulatory mechanisms are genetically determined, mutation has had a major effect on the production of secondary metabolites (46). Indeed, it is the chief factor responsible for the hundred- to thousandfold increases obtained in production of antibiotics from the time of their initial discovery to the present.

Environmental manipulations have proven to be quite useful for bringing on metabolite overproduction. Optimization of fermentation parameters such as temperature, pH, oxygen transfer, and nutrients is of course extremely important. If a process is inducible, inducers are added. They include methionine for cephalosporin C formation, valine for tylosin production, and tryptophan for ergot alkaloid production. Virtually all fermentation processes operate, after the initial growth stage, under conditions of unbalanced growth, and from the start of any development program, the microbiologist must find the key to this unbalanced growth. The key is often a specific nutrient limitation that restricts growth and brings on product accumulation.

Genetics has had a long history of contributing to the production of microbial secondary metabolites, such as antibiotics (47). The tremendous increases in fermentation productivity and the resulting decreases in costs have come about mainly by mutagenesis and screening for higher-producing microbial strains. Mutation has also served to *(1)* shift the proportion of metabolites produced in a fermentation broth to a more favorable distribution, *(2)* elucidate the pathways of secondary metabolism, and *(3)* yield new compounds. The medically useful products demethyltetracycline and doxorubicin (adriamycin) were discovered by simple mutation of the cultures producing

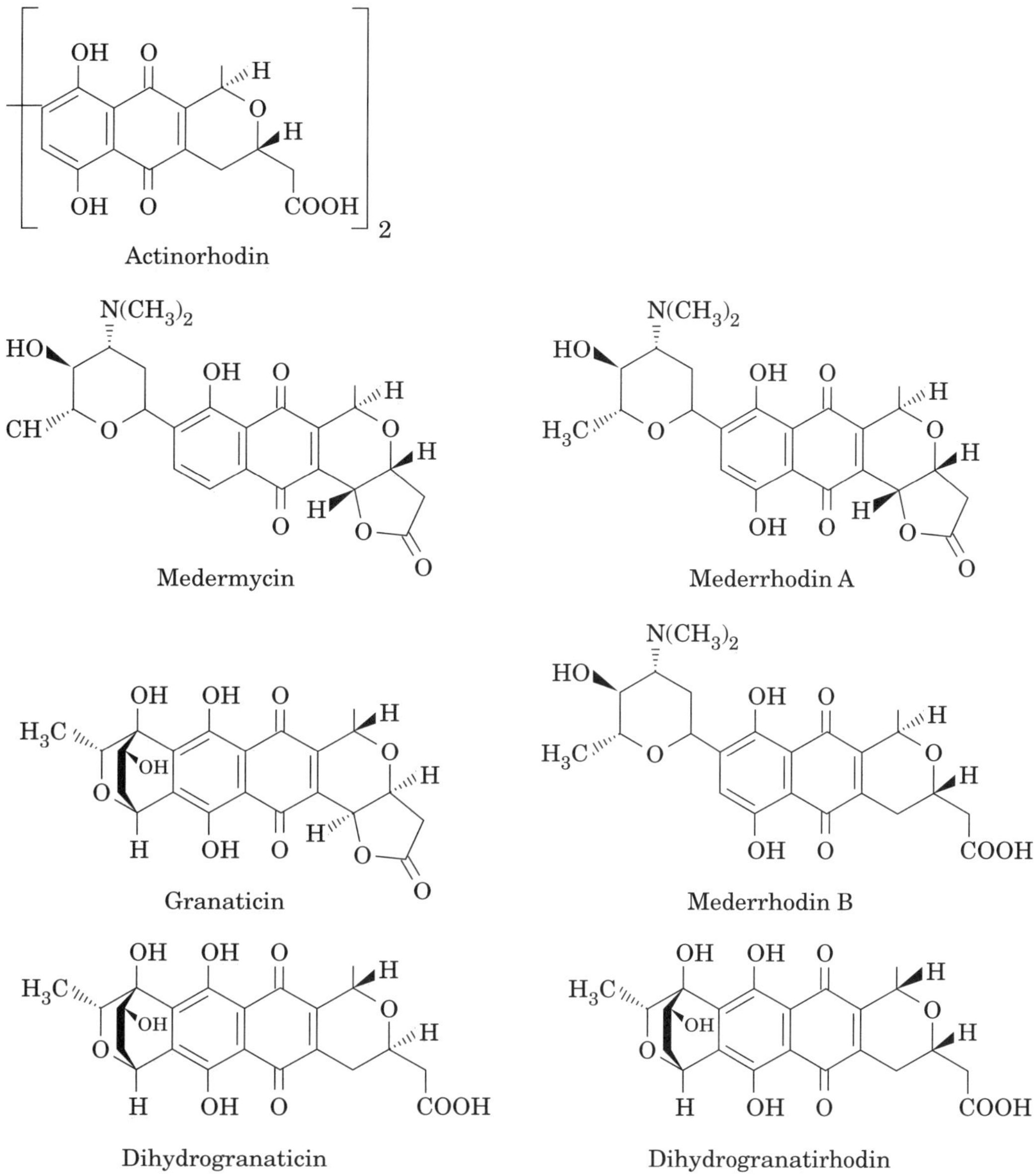

Figure 16. Hybrid antibiotics. Left, structures of the natural isochromane quinone antibiotics: actinorhodin, granaticin, dihydrogranaticin, and medermycin. Right, structures of new hybrid antibiotics: mederrhodin A and B and dihydrogranatirhodin produced by genetically engineered organisms (the absolute stereochemistry of dihydrogranaticin at C-1 and C-3 is tentative). *Source*: From Ref. 52.

tetracycline and daunorubicin (daunomycin), respectively. The technique of mutational biosynthesis has been used for the discovery of many new aminoglycoside, macrolide, and anthracycline antibiotics. It has recently been successfully employed in producing a new commercial avermectin (48).

The use of protoplast fusion has facilitated genetic recombination of different strains from the same or different species to yield new antibiotics. Protoplast fusion has also been useful in elimination of an undesirable component from penicillin broths.

Recombinant DNA technology has been applied to the production of antibiotics (49). Many genes that encode individual enzymes of antibiotic biosynthesis have been cloned and expressed at high levels in heterologous microorganisms. Continued efforts in the application of recombinant DNA technology to fermentations have led to overproduction of limiting enzymes of important biosynthetic pathways, thereby increasing production of the final products. This has been accomplished in the area of cephalosporin biosynthesis (50). In addition, many antibiotic resistance genes from antibiotic-producing organisms have been cloned and expressed. Some antibiotic biosynthetic pathways are encoded by plasmid-borne genes (e.g., methylenomycin A). Even when the antibiotic biosynthetic pathway genes of actinomycetes are chromosomal (the

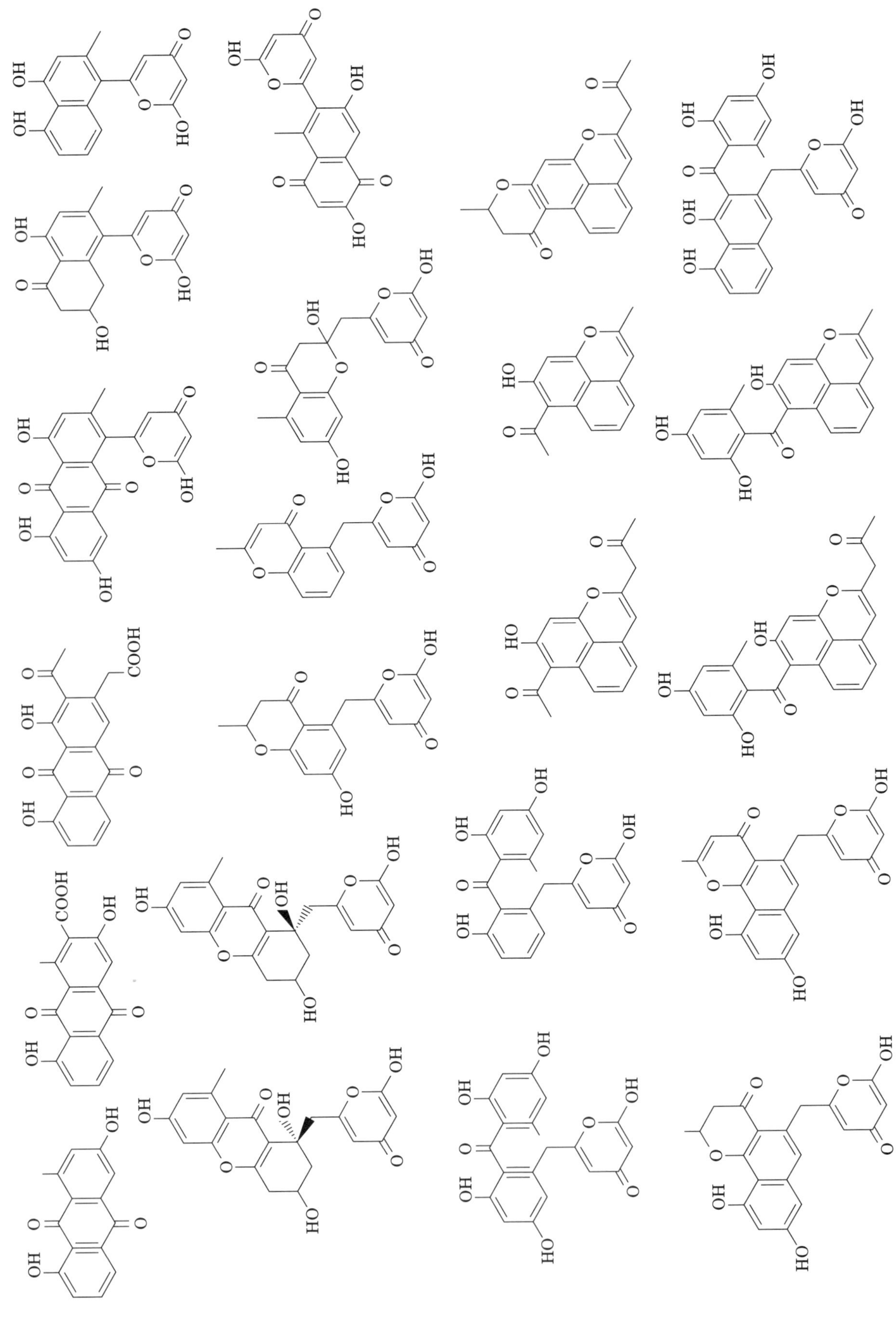

Figure 17. Novel polyketides produced by the actinorhodin (*act*), frenolicin (*fren*), and tetracenomycin (*tcm*) PKS gene clusters. *Source*: From Ref. 47.

usual situation), they are clustered, which facilitates transfer of an entire pathway in a single manipulation. The genes of the actinorhodin pathway, normally clustered on the chromosome of *Streptomyces coelicolor*, were transferred en masse on a plasmid to *S. parvulus* and were expressed in the latter organism. The presence of the bialophos (a herbicidal antibiotic) resistance gene in the bialaphos biosynthetic cluster allowed cloning of the entire 13-enzyme pathway of this herbicidal antibiotic from *S. hygroscopicus* into *S. lividans*. Here, the resistance gene is also one of the biosynthetic genes, coding for demethylphosphinothricin acetyltransferase. Even in fungi, pathway genes can be clustered such as the penicillin genes in *Penicillium* or the aflatoxin genes in *Aspergillus*. However, the genes that code for cephalosporin biosynthesis in *Cephalosporium* are distributed among different chromosomes.

For the discovery of new or modified products, recombinant DNA techniques can be used to introduce genes that code for antibiotic synthetases into producers of other antibiotics or into nonproducing strains to obtain modified or hybrid antibiotics. Gene transfer from a streptomycete strain producing the isochromanequinone antibiotic actinorhodin into strains producing granaticin, dihydrogranaticin, and mederomycin (which are also isochromanequinones) led to the discovery of two new antibiotic derivatives, mederrhodin A and dihydrogranatirhodin (51,52; Fig. 16). A hybrid antibiotic, isovalerylspiramycin, was obtained by introducing the *carE* gene from the carbomycin producer *S. thermotolerans* into the spiramycin producer *S. ambofaciens* (53). These are both 16-membered macrolides, but one of the two sugars in carbomycin is isovalerylmycarose and the corresponding sugar in spiramycin is mycarose. When the acylating gene was cloned into the nonmacrolide producer *Streptomyces lividans*, the recombinant was capable of converting spiramycin to isovalerylspiramycin. Many novel polyketide secondary metabolites have been obtained by cloning DNA fragments from one polyketide producer into various strains of other streptomycetes (54,55). Figure 17 shows some of these novel polyketide molecules.

BIBLIOGRAPHY

1. L. Palmieri, D. Berns, R. Krämer, and M. Eikmanns, *Arch. Microbiol.* **165**, 48–54 (1996).
2. L. Eggeling, *Amino Acids* **6**, 261–272 (1994).
3. K. Sonntag, J. Schwinde, A.A. de Graaf, A. Marx, B.J. Eikmanns, W. Wiechert, and H. Sahm, *Appl. Microbiol. Biotechnol.* **44**, 489–495 (1995).
4. A. Erdmann, B. Well, and R. Krämer, *J. Gen. Microbiol.* **139**, 3115–3122 (1993).
5. M.S.M. Jetten and A.J. Sinskey, *Crit. Rev. Biotechnol.* **15**, 73–103 (1995).
6. H. Sahm, L. Eggeling, B. Eikmanns, and R. Krämer, *FEMS Microbiol. Rev.* **16**, 243–252 (1995).
7. R. Kelle, T. Hermann, D. Weuster-Botz, L. Eggeling, R. Krämer, and C. Wandrey, *J. Biotechnol.* **50**, 123–136 (1996).
8. H. Ito, K. Sato, K. Matsui, K. Sano, H. Enei, and Y. Hirose, *Appl. Microbiol. Biotechnol.* **33**, 190–195 (1990).
9. A.L. Demain, in A.H. Rose ed., *Economic Microbiology, vol. 2: Primary Products of Metabolism*, Academic Press, New York, 1978, pp. 187–208.
10. A. Kuninaka, in H.-J. Rehm, G. Reed, A. Pühler, and P. Stadler eds., *Biotechnology*, vol. 6, 2nd ed., VCH Verlagsgesellschaft, Weinheim, 1996, pp. 561–612.
11. K. Miyagawa, N. Kanzaki, H. Kimura, Y. Sumino, S. Akyama, and Y. Nakao, *Bio / Technology* **7**, 821–824 (1989).
12. K. Miyagawa, H. Kimura, K. Nakahama, M. Kikuchi, M. Doi, S. Akiyama, and Y. Nakao, *Bio / Technology* **4**, 225–228 (1986).
13. S. Asahi, M. Izawa, and M. Doi, *7th International Symposium of the Genetics of Industrial Microorganisms*, 1994, abstract P.17, p. 102.
14. R. Volk and A. Bacher, *J. Biol. Chem.* **266**, 20610–20618 (1991).
15. A.L. Demain, *Annu. Rev. Microbiol.* **26**, 369–388 (1972).
16. E.J. Vandamme, *J. Chem. Technol. Biotechnol.* **53**, 313–327 (1992).
17. C.A. Roessner and A.I. Scott, *Annu. Rev. Microbiol.* **50**, 467–490 (1996).
18. C. Spalla, A. Grein, L. Garofano, and G. Ferni, in E.J. Vandamme ed., *Biotechnology of Vitamins, Pigments, and Growth Factors*, Elsevier, New York, 1989, pp. 257–284.
19. J.P. Kusel, Y.H. Fa, and A.L. Demain, *J. Gen. Microbiol.* **130**, 835–841 (1984).
20. D.F. Barker and A.M. Campbell, *J. Mol. Biol.* **146**, 469–492 (1981).
21. M. Masuda, K. Takahashi, N. Sakurai, K. Yanagiya, S. Komatsubara, and T. Tosa, *Process Biochem.* **30**, 553–562 (1995).
22. N. Sakurai, H. Akatsuka, E. Kawai, Y. Imai, and S. Komatsubara, *Microbiology* **142**, 3295–3303 (1996).
23. M.J. Pramik, *Gen. Eng. News* **2**, 9, 12 (1986).
24. J.F. Grindley, M.A. Payton, H. van de Pol, and K.G. Hardy. *Appl. Environ. Microbiol.* **54**, 1770–1175 (1988).
25. Y. Saito, Y. Ishii, H. Hayashi, Y. Imao, T. Akashi, K. Yoshikawa, Y. Noguchi, S. Soeda, M. Yoshida, M. Niwa, J. Hosoda, and K. Shimomura, *Appl. Environ. Microbiol.* **63**, 454–460 (1997).
26. C.P. Kubicek and M. Röhr, *CRC Crit. Rev. Biotechnol.* **3**, 331–373 (1986).
27. H.J.M. Harmsen, E.M. Kubicek-Pranz, M. Röhr, J. Visser, and C.P. Kubicek, *Appl. Microbiol. Biotechnol.* **37**, 784–788 (1992).
28. M. Fukaya, K. Tayama, T. Tamaki, H. Tagami, H. Okumura, Y. Kawamura, and T. Beppu, *Appl. Environ. Microbiol.* **55**, 171–176 (1989).
29. L.O. Ingram, E. Conway, D.P. Clark, G.W. Sewell, and J.F. Preston, *Appl. Environ. Microbiol.* **53**, 2420–2425 (1987).
30. J.P. Rosazza, *Microbial Transformations of Bioactive Compounds*, vols. 1 and 2, CRC Press, Boca Raton, Fla., 1982.
31. R.J. Neufeld, Y. Peleg, J.S. Rokem, O. Pines, and I. Goldberg, *Enzyme Microb. Technol.* **13**, 991–996 (1991).
32. A.L. Demain and A. Fang, *Actinomycetologia* **9**, 98–117 (1995).
33. J.J. Sanglier, H. Haag, T.A. Huck, and T. Fehr, *Exp. Opin. Invest. Drugs* **5**, 207–223 (1996).
34. C.R. Hutchinson and I. Fujii, *Annu. Rev. Microbiol.* **49**, 201–238 (1995).
35. C.R. Hutchinson, H. Decker, K. Madduri, S.L. Otten, and L. Tang, *Antonie van Leeuwenhoek* **64**, 165–176 (1993).
36. K.S. Brown, *Scientist* **10**, 1, 8, 9 (1996).
37. A.L. Demain, in B. Sutton ed., *A Century of Mycology*, Cambridge Univ. Press, Cambridge, U.K., 1996, pp. 233–254.

38. J.W. Westley, *Adv. Appl. Microbiol.* **22**, 177–223 (1977).
39. W.C. Campbell, M.H. Fisher, E.O. Stapley, G. Albers-Schönberg, and T.A. Jacob, *Science* **221**, 823–828 (1983).
40. H. Umezawa, *Annu. Rev. Microbiol.* **36**, 75–99 (1982).
41. A. Endo, *J. Med. Chem.* **28**, 401–405 (1985).
42. L. Müller, in H.-J. Rehm and G. Reed eds., *Biotechnology*, vol. 4, V.H. Verlagsgesellschaft, Weinheim, 1986, pp. 1–37.
43. A.L. Demain, *Science* **219**, 709–714 (1983).
44. D.J. Chadwick and J. Whelan eds., *Secondary Metabolites: Their Function and Evolution*, Wiley, New York, 1992.
45. A.L. Demain, in W. Kuhn and H.-P. Fiedler eds., *Sekundärmetabolismus bei Mikroorganismen, Beiträge zur Forschung*, Attempto Verlag, Tübingen, 1995, pp. 9–35.
46. A.L. Demain, *Adv. Appl. Microbiol.* **16**, 177–202 (1973).
47. G. Holt and G. Saunders, in A.T. Bull and H. Dalton eds., *Comprehensive Biotechnology*, vol. 1, Pergamon Press, New York, 1985, pp. 51–76.
48. C.D. Denova, R.W. Fedechko, E.W. Hafner, H.A.I. McArthur, M.R. Morgenstern, D.D. Skinner, K. Stutzman-Engwall, R.G. Wax, and W.C. Wernau, *J. Bacteriol.* **177**, 3504–3551 (1995).
49. K.F. Chater, *Bio / Technology* **8**, 115–121 (1990).
50. P.L. Skatrud, A.J. Tietz, T.D. Ingolia, C.A. Cantwell, D.L. Fisher, J.L. Chapman, and S.W. Queener, *Bio / Technology* **7**, 477–485 (1989).
51. D.A. Hopwood, F. Malpartida, H.M. Kieser, H. Ikeda, J. Duncan, I. Fujui, B.A.M. Rudd, H.G. Floss, and S. Omura, *Nature* **314**, 642–644 (1985).
52. H.G. Floss, *TIBTECH* **5**, 111–115 (1987).
53. J.K. Epp, M.L.B. Huber, T. Goodson, and B.E. Schoner, *Gene* **85**, 293–301 (1989).
54. W.R. Strohl, P.L. Bartel, N.C. Connors, C.-B. Zhu, D.C. Dosch, J.M. Beale, Jr., H.G. Floss, K. Stutzman-Engwall, S.L. Otten, and C.R. Hutchinson, in C.L. Hershberger, S.W. Queener, and G. Hegeman eds., *Genetics and Molecular Biology Industrial Microorganisms*, American Society for Microbiology, Washington, D.C., 1989, pp. 68–84.
55. C. Khosla, R. Caren, C.M. Kao, R. McDaniel, and S.-W. Wang, *Biotechnol. Bioeng.* **52**, 122–128 (1996).

See also AMINO ACIDS, PRODUCTION PROCESSES; *ASPERGILLUS*; CEPHALOSPORINS; DEXTRAN, MICROBIAL PRODUCTION METHODS; PANTOTHENIC ACID AND RELATED COMPOUNDS; SECONDARY METABOLITE PRODUCTION, ACTINOMYCETES, OTHER THAN STREPTOMYCES.

METHANOTROPHS

IAN R. MCDONALD
J. COLIN MURRELL
University of Warwick
Coventry, U.K.

KEY WORDS

Methane monooxygenase
Methane monooxygenase gene
Methane oxidation
Methanotroph
Methylotroph
Polymerase chain reaction

OUTLINE

INTRODUCTION

Methane oxidizing bacteria (methanotrophs) are defined as those bacteria that can grow on methane as their sole source of carbon and energy. They are therefore a subgroup of methylotrophs (bacteria that can grow on one-carbon compounds). Methanotrophs are generally regarded as obligate in nature, growing only on methane and usually on methanol but not on multicarbon substrates. Methanotrophs have attracted considerable interest over the past 25 years due to their potential for the production of bulk chemicals, single-cell protein, and use in biotransformations (1–3). Their ability to degrade the groundwater pollutant trichloroethylene and other chlorinated compounds has also been examine in detail (reviewed in Refs. 4 and 5). There is considerable interest in methanotrophs because they appear to be ubiquitous in nature (6–8) and are likely to be a major sink for methane in the natural environment (reviewed in Refs. 9 through 12).

In a pioneering study, Whittenbury and colleagues isolated and characterized a large number of different methanotrophs from soils, sediments, and aquatic environments (6). More recently, Bowman and colleagues (7) isolated more than 100 strains of methanotrophs from many different environments and carried out a detailed chemotaxonomic study on these isolates. This study reinforced the view that these bacteria are widespread in nature and are important in the global cycling of methane in the natural environment.

Methanotrophs are all Gram-negative bacteria and can be classified into two groups. Type I methanotrophs of the genera *Methylomonas, Methylobacter, Methylomicrobium, Methylococcus, Methylosphaera*, and *Methylocaldum* utilize the ribulose monophosphate (RuMP) pathway for the assimilation of formaldehyde into cell carbon, possess bundles of intracytoplasmic membranes (Fig. 1), and are members of the γ-subdivision of the purple bacteria (class Proteobacteria). Type II methanotrophs, such as *Methylosinus, Methylocystis*, and *Methylosporovibrio*, utilize the serine pathway for formaldehyde fixation, possess intracytoplasmic membranes arranged around the periphery of the cell (Fig. 1), and fall within the α-subdivision of the

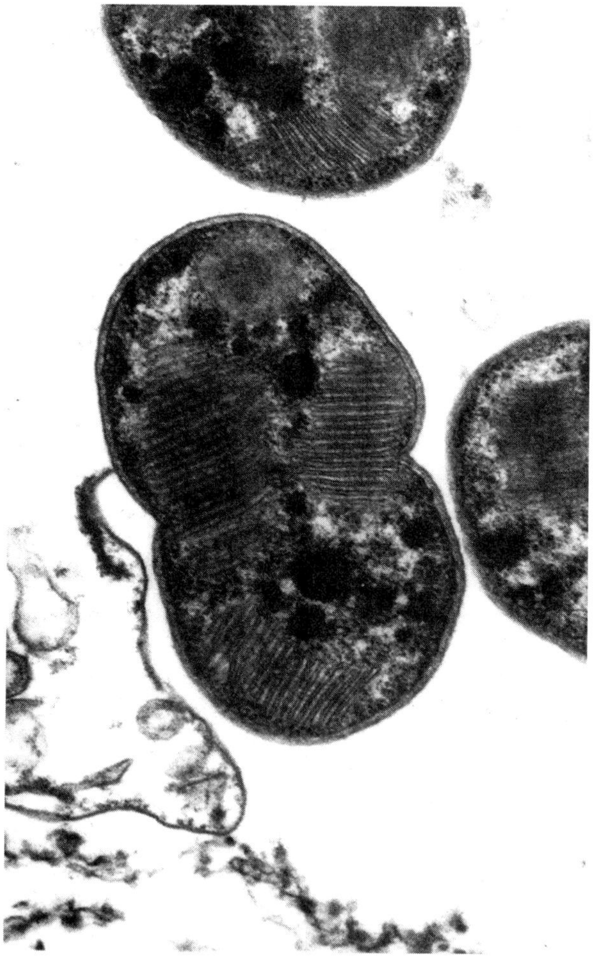

(a)

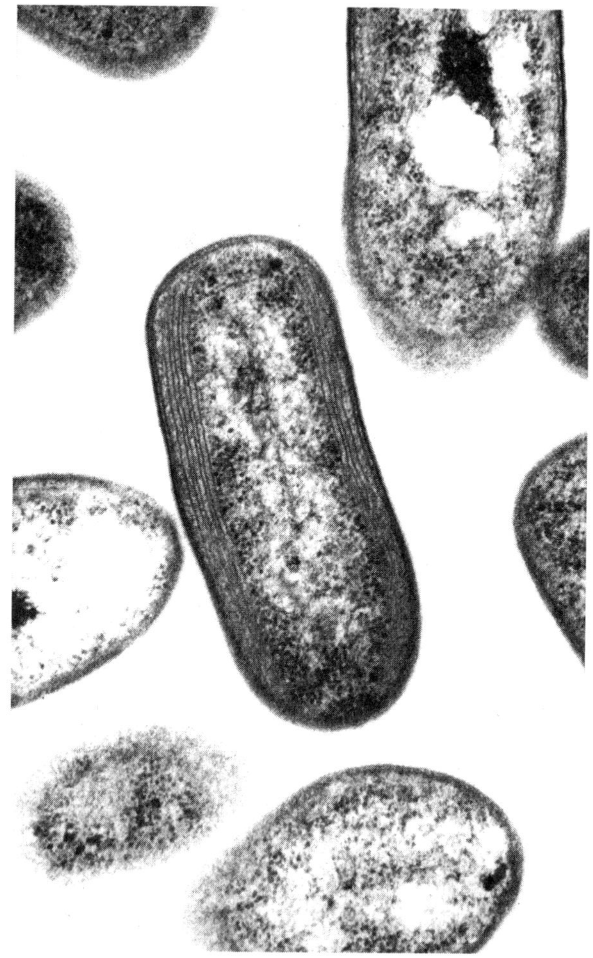

(b)

Figure 1. Electron micrographs. (**a**) Section of *Methylomonas methanica* (NCIMB 11130), showing a Type I membrane system. (**b**) Section of *Methylocystis parvus* (NCIMB 11129), showing a Type II membrane system. Magnification, all ×68,800.

Proteobacteria. Some methanotrophs, such as *Methylococcus capsulatus* (Bath), appear to have some properties of both Type I and Type II methanotrophs and are sometimes termed Type X methanotrophs. The taxonomy of methanotrophs has been reviewed in some detail (7,13). Classification schemes, based on pheno- and chemotaxonomic studies have been strengthened as a result of the nucleotide sequencing of both 5S and 16S ribosomal RNA (rRNA) from a large number of methanotrophs and methylotrophs. Type II methanotrophs such as *Methylocystis parvus, Methylosinus trichosporium*, and *Methylosporovibrio methanica* appear to cluster in the α-2 subdivision of the class Proteobacteria and form a separate cluster from other serine pathway methylotrophs. Type I methanotrophs such as *Methylomonas methanica, Methylomicrobium album, Methylomonas rubra*, and *Methylomonas luteus*, together with the Type X methanotroph *Methylococcus capsulatus*, are found in the γ-subdivision of the purple bacteria (reviewed in Refs. 8 and 14). The key features of representative genera of methanotrophs are summarized in Table 1. Properties such as mol% G + C content of DNA, membrane fatty acid composition, nitrogen fixation, and morphological features can be used to discriminate between Type I and Type II methanotrophs.

PHYSIOLOGY AND BIOCHEMISTRY OF METHANOTROPHS

Methanotrophs grow by oxidizing methane using the pathway shown in Figure 2. The key feature of all methanotrophs is their ability to oxidize methane to methanol (see later). Methanol, the oxidation product of methane, is further metabolized to formaldehyde by the pyrolloquinoline quinone (PQQ)-linked enzyme methanol dehydrogenase (MDH). This is an enzyme that is found in all Gram-negative methylotrophic bacteria and has been extensively studied at the biochemical and molecular level (15–17). There are a large number of genes involved in methanol oxidation, and there are complex regulatory networks controlling their expression. The majority of work in this area has been carried out using methanol utilizers rather than methanotrophs (18). The oxidation of methanol yields formaldehyde, a central intermediate in the methane

Table 1. Characterization of the Most Well Studied Methanotrophs

	Methylomonas	Methylobacter	Methylomicrobium	Methylococcus	Methylosinus	Methylocystis
Phylogeny	Gamma Proteobacteria	Gamma Proteobacteria	Gamma Proteobacteria	Gamma Proteobacteria	Alpha Proteobacteria	Alpha Proteobacteria
Membrane type	Type I	Type I	Type I	Type I	Type II	Type II
Major PLFAs	16:1	16:1	16:1	16:1	18:1	18:1
Cell morphology	Rods	Cocci, ellipsoidal, or fat rods	Rods	Cocci or ellipsoidal	Vibriod or pyriform	Cocci, curved rods, or ellipsoidal
Formaldehyde assimilation pathway	RuMP	RuMP	RuMP	RuMP/serine	Serine	Serine
Enzyme type	pMMO	pMMO	pMMO	pMMO/sMMO	pMMO/sMMO	pMMO/sMMO
Mol% G + C content of DNA	50–54	50–54	50–54	62.5	62–63	62–63
Cyst formation	+	+	−	+	−	+
Exospore formation	−	−	−	−	+	−
N-fixation	−	−	−	+	+	+
Carotenoids	+	−	−	−	−	−
Marine representatives	+	+	−	−	−	−
Representative species	*M. aurantiaea, M. fordinarum, M. methanica*	*M. albus, M. marinus, M. whittenburyi*	*M. agile, M. album, M. pelagicum*	*M. capsulatus, M. thermophilus*	*M. trichosporium, M. sporium*	*M. parvus, M. echinoides, M. pyriformis, M. minimus*

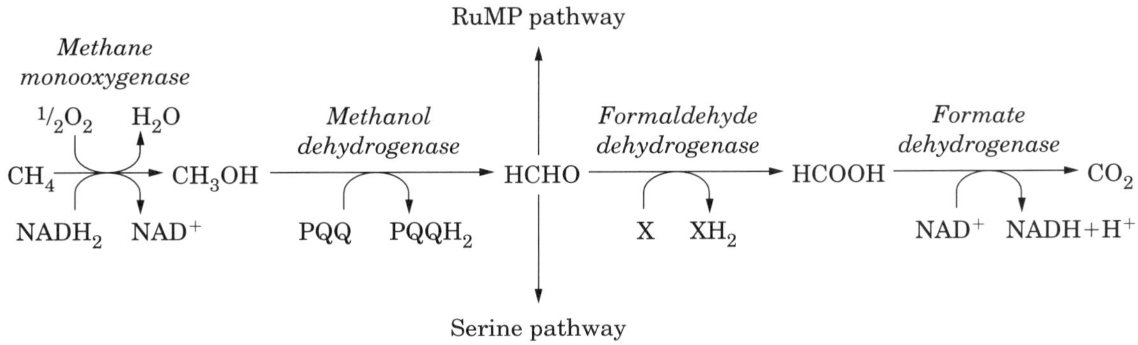

Figure 2. Pathway for the oxidation of methane and the assimilation of formaldehyde.

oxidation pathway. Approximately half of the formaldehyde produced is further oxidized by the enzymes formaldehyde dehydrogenase and formate dehydrogenase to yield carbon dioxide. This results in the generation of reducing power for biosynthesis and the initial oxidation of methane. The CO_2 is not fixed into cell carbon in significant amounts but is lost to the atmosphere. The remainder of the formaldehyde is assimilated into cell carbon by one of two pathways. In Type I methanotrophs, formaldehyde is condensed with ribulose phosphate into hexulose-6-phosphate in the RuMP pathway. The key enzymes of this pathway are hexulose-6-phosphate synthase and hexulose-phosphate isomerase. Subsequent rearrangement reactions result in the net incorporation of glyceraldehyde-3-phosphate into cellular material (Fig. 3). Type II methanotrophs utilize the serine pathway for the incorporation of formaldehyde into the cell. Key enzymes of this pathway include serine hydroxymethyl transferase, hydroxypyruvate reductase, malate thiokinase, and malyl CoA lyase (Fig. 4). The biochemistry of these methylotrophic pathways has been extensively reviewed by Anthony (15,16).

Methane oxidation is carried out by the enzyme methane monooxygenase (MMO). A membrane-bound, particulate methane monooxygenase (pMMO) appears to be present in all methanotrophs grown in the presence of relatively high concentrations of copper ions. Studies on this enzyme have been hampered by its lability in cell-free extracts and difficulties in purification from the membranes of methanotrophs. Biochemical and biophysical studies have shown that pMMO activity is directly related to the copper/total membrane protein ratio in membrane fractions from *Methylococcus capsulatus* (Bath) grown at different copper concentrations. pMMO appears to consist of at least three polypeptides of approximately 46, 23, and 20 kDa, and EPR studies suggest the presence of a number of copper clusters. Recently, the pMMO enzyme complex has been isolated in active form from *M. capsulatus* (Bath) using a variety of detergent treatments (19). An active preparation of pMMO from *Methylococcus* has been reported to contain three subunits (47,27, and 25 kDa) and iron and copper atoms (probably involved in catalysis at the active site) and uses duroquinol as the in vitro electron donor (20). Further characterization of pMMO at the biochemical level is ongoing in a number of research groups.

In some methanotrophs, a second form of MMO, a cytoplasmic, soluble form (sMMO), is synthesized when the growth conditions are such that the culture copper-to-biomass ratio is low (less than 1 μM/g dry weight). This sMMO is structurally and catalytically distinct from the pMMO and has a broad substrate specificity, oxidizing a

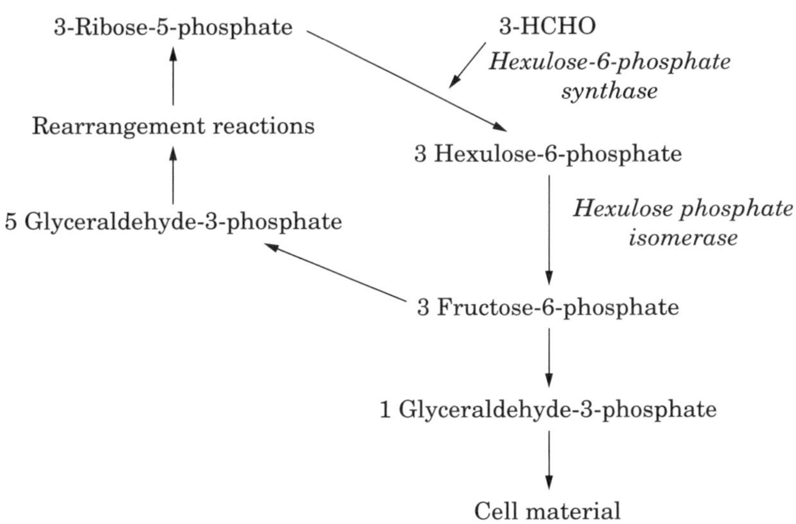

Figure 3. RuMP pathway for formaldehyde fixation.

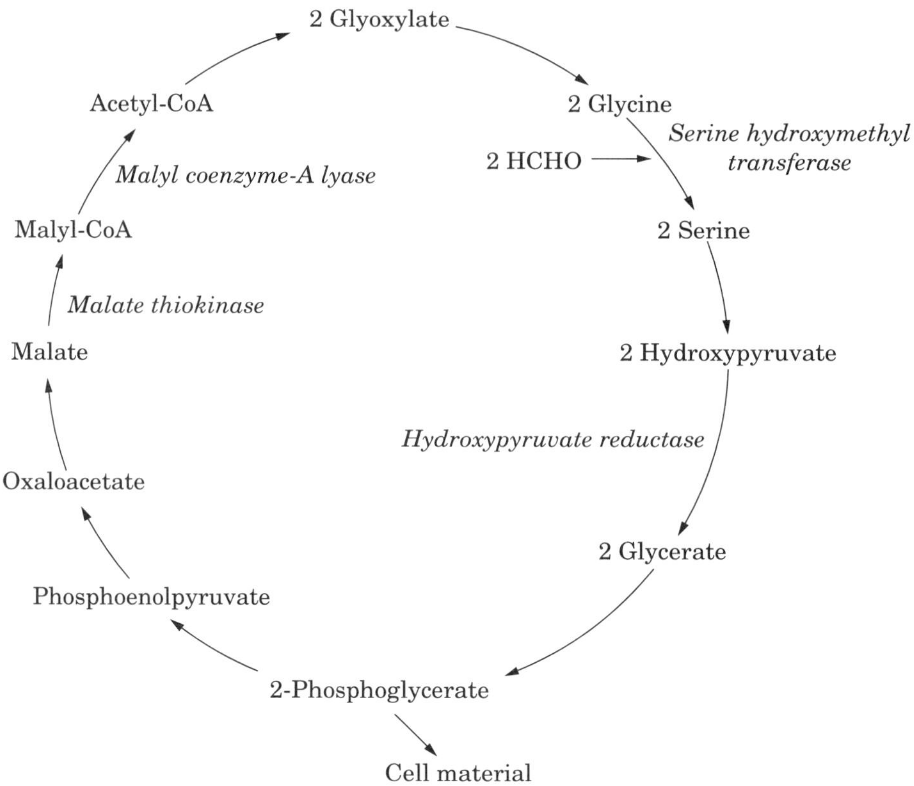

Figure 4. Serine pathway for formaldehyde fixation.

wide range of aliphatic and aromatic compounds. Probably the most extensively characterized sMMOs are those isolated from the Type II methanotroph *Methylosinus trichosporium* OB3b (reviewed in Ref. 21) and the Type X methanotroph *M. capsulatus* (Bath) (reviewed in Ref. 22). sMMO has also been characterized from the Type II strains *Methylosinus sporium* 5 and *Methylocystis* strain M. sMMO has also been reported in the Type I methanotroph *Methylomonas methanica* 68-1.

The sMMO complex of *M. capsulatus* (Bath) and *M. trichosporium* OB3b have very similar properties and both consist of three components: A, B, and C. Protein A is made up of two copies of each of three subunits, α, β, and γ, with molecular masses of 60, 45, and 20 kDa and is the hydroxylase component of the enzyme complex. Protein A contains a binuclear iron-oxo center believed to be the reactive center for catalysis (21–23). Protein B, 16 kDa, is a single polypeptide containing no metal ions or cofactors and functions as a regulatory or coupling protein (21). Protein C is a 39-kDa NADH reductase containing 1 mol each of FAD and a 2Fe2S cluster that accepts electrons from NADH and transfers them to the diiron site of the hydroxylase component of the Protein A component of sMMO. The mechanism of action of sMMO has recently been reviewed (21–23). The X-ray crystal structure of the hydroxylase component of sMMO from *M. capsulatus* (Bath) has been reported, which has further stimulated research on the mechanism of oxidation of methane by this unique enzyme (23).

GENETICS OF METHANOTROPHS

A significant problem in the development of genetic systems for methanotrophs has been the lack of suitable ways of introducing and maintaining heterologous vector DNA in these organisms. Coupled with their relatively slow growth on plates, this has meant that fairly slow progress has been made in this area. Bacteriophages specific for methanotrophs have been isolated from a wide range of environments, including groundwater, fermentors, soils, rumen of cattle, fish, and oil and gas installation waters. They have been characterized in terms of their plaque morphology, ultrastructure, antigenic properties, nucleic acid content, restriction enzyme analysis, and lytic spectrum. *Methylosinus*- and *Methylocystis*-specific phages are particularly prevalent in nature, but few attempts to use them for transduction experiments with methanotrophs have been reported (reviewed in Ref. 24).

A number of large plasmids, ranging from around 50 to 190 kbp have been detected in a number of methanotrophs including *Methylomonas, Methylocystis, Methylosinus*, and *Methylobacter*. Of 10 representatives examined, only *M. capsulatus* (Bath) did not contain any detectable plasmid DNA (25). Some cross-hybridization with plasmid DNA from different *Methylosinus* species was noted, and a rudimentary restriction endonuclease map of the 55-kbp plasmid from *M. albus* was constructed. However, difficulties in the large-scale isolation of plasmid DNA from these organisms have hampered their development as potential cloning vectors. There is currently no evidence for the presence of antibiotic or metal resistance genes, nitrogen fixation genes, or methane oxidation genes on these plasmids, but because these plasmids represent "significant potential coding" regions, their presence should not be disregarded in any molecular genetics work with these organisms.

There are no good conventional transformation systems for methanotrophs, and several workers have tried electroporation of plasmids into methanotrophs with little success. Transfer of broad-host-range plasmids into methanotrophs has been the most successful route to gene transfer in these organisms. Lidstrom and colleagues have transferred the Inc P1 broad host range cosmid cloning vector into *Methylomonas, Methylocystis*, and *Methylosinus* species at frequencies ranging from 10^{-2} to 10^{-8}. The mobilizing plasmid used in these filter matings was pRK2013 (26). Similar results using similar IncP and IncQ plasmids have been achieved in the authors' laboratory with *M. trichosporium* OB3b and *M. capsulatus* (Bath). Kanamycin and streptomycin appear to be good genetic markers for methanotrophs because the rate of spontaneous mutation for resistance to these antibiotics is very low (less than 10^{-8}). Plasmids carrying kanamycin and streptomycin resistance have been successfully transferred to *M. capsulatus* (Bath) at frequencies of 10^{-3} to 10^{-9} per recipient.

The major obstacles to the development of mutagenesis procedures for the obligate methanotrophs are probably their relatively slow growth on agar plates and the fact that they rely on methane as their sole source of carbon and energy. A consequence of this is that selection of methane oxidation mutants may be difficult. This can usually be overcome because most methanotrophs can be adapted to grow on methanol, thus allowing the selection of an MMO^- phenotype. Conventional mutagenesis procedures using chemical mutagens have not been particularly successful. However, mutants lacking the ability to grow on methane have been isolated using the "suicide-substrate" dichloromethane (DCM). This is cooxidized by MMO to the potentially toxic product carbon monoxide, and therefore methanotrophs growing on methanol agar plates that are DCM resistant are potential methane monooxygenase mutants. Screening of methanotrophs for the presence or absence of a functional sMMO has also been made easier due to the development of a quick, reliable plate assay. This is based on the work of Hanson and colleagues (27) and relies on the conversion of naphthalene to naphthols by soluble MMO but not particulate MMO. The naphthols thus formed can be reacted with tetrazotized *O*-dianisidine to form purple diazo dyes with intense color. This method now allows the assay of soluble MMO activity on plates. The procedure does not kill potential mutants, which can subsequently be recovered and streaked onto fresh agar plates under the appropriate growth conditions (28).

A similar naphthalene assay, coupled with DCM selection, has recently been used to isolate and characterize MMO mutants of *M. trichosporium* OB3b (29). *M. trichosporium* OB3b mutants defective in only the particulate form of MMO were isolated by *(1)* suppressing sMMO expression by the presence of high concentrations of copper in selection plates, *(2)* adding excess methanol with DCM to inhibit possible sMMO-mediated activation of DCM to toxic products in any colonies where sMMO might be expressed, *(3)* adding low concentrations of yeast extract to enhance growth and plating efficiency on methanol, and *(4)* carrying out mutagenesis selection over a long period (5 weeks). Although DCM resistance was unstable in many mutants (a phenomenon observed in other laboratories, including the authors'), mutants that were deficient in pMMO and expressed sMMO constitutively in the presence of elevated copper concentrations (up to 12 μM, which inhibits expression of sMMO in the wild-type organism) were isolated. These mutants had lost the ability to express pMMO activity and concomitantly failed to form the intracytoplasmic membranes characteristic of wild-type *M. trichosporium* OB3b expressing pMMO. Subsequent detailed analysis of these sMMO-constitutive *M. trichosporium* OB3b mutants suggested that either a defect in copper metabolism affecting copper assimilation causes lack of pMMO expression, or that a more general copper-dependent regulatory system that may control pMMO induction, sMMO repression, copper uptake, and production of the intracytoplasmic membranes associated with pMMO has been inactivated (30). Such mutants would clearly be useful for in situ bioremediation processes involving sMMO where the external concentration of copper ions cannot be controlled.

MOLECULAR BIOLOGY OF THE METHANE OXIDATION PATHWAY

Research, primarily in the laboratories of Lidstrom, Hanson, Goodwin, and Harms, has resulted in the identification and cloning of a number of genes involved in methanol oxidation from the Gram-negative methylotrophs *Methylobacterium extorquens* and *Paracoccus denitrificans* (reviewed in Refs. 17, 31, and 32). One of these genes, *mxaF*, from *Methylobacterium*, encoding the large subunit structural gene of methanol dehydrogenase, has been used as a heterologous hybridization probe to identify and clone the corresponding *mxaF* genes from *M. capsulatus* (Bath) and *M. albus* BG8. Methanol oxidation structural genes have also been isolated from the marine methanotroph *Methylomonas* strain A4. Molecular characterization of these genes suggests that they are similar to those that have been extensively characterized in Gram-negative methylotrophs (17), although their regulation may be somewhat different (33).

The genes encoding the sMMO enzyme complex have been cloned from *M. capsulatus* (Bath) and are clustered on the chromosome of this methanotroph. DNA sequencing of this gene cluster has revealed that the genes encoding the α-, β-, and γ-subunits of Protein A (*mmoX, Y*, and *Z*), Protein B (*mmoB*), and Protein C (*mmoC*) are all linked on the chromosome. The sMMO gene cluster of *M. capsulatus* (Bath) has been used to probe for sMMO genes in a number of representative strains of methanotrophs. sMMO homologues were detected only in *Methylococcus* and *Methylosinus* strains. sMMO appeared to be absent from the Type II methanotroph *Methylocystis parvus* OBBP or Type I representatives of *Methylomicrobium album, Methylomonas methanica, Methylomonas agile*, and *Methylobacter capsulatus*.

Subsequently, sMMO gene probes from *Methylococcus* were used to clone the sMMO gene cluster from *M. trichosporium* OB3b (reviewed in Ref. 34) and *Methylocystis* sp. strain M (35). These gene clusters have also been DNA sequenced, and the genes are arranged in the same order as in *Methylococcus* (Fig. 5). The function of *orfY*, which lies between *mmoZ* and *mmoC*, is unknown at present. Derived polypeptide sequences of sMMO components from all three methanotrophs showed a high degree of identity, highlighting the conserved nature of this enzyme complex (35). Amino acid sequences within the α-subunit of Protein A align well with the four-helix iron coordination bundle of the R2 protein of ribonucleotide reductase and is characteristic of a family of proteins that contain a catalytic carboxylate-bridged diiron center. The diiron center in sMMO is now well defined by X-ray crystal structure analysis (23). The N-terminal region of Protein C also exhibited extensive homology with reductase components of other oxygenases that contain a ferredoxin-like 2Fe-2S center. Protein B, the coupling protein of sMMO, has some sequence identities with other oxygenase-coupling proteins (36), but all other polypeptides of sMMO including the derived sequence of *orfY* appear to be novel.

Expression of functionally active Proteins B and C has been achieved using a T7 polymerase expression system in *Escherichia coli*. Attempts to express the hydroxylase components in *E. coli* in the authors' laboratory have proved unsuccessful, but expression of sMMO genes has recently been achieved using other heterologous hosts including methanotrophs that contain only pMMO (37), which will facilitate further studies on the structure, function, and catalytic mechanism of methane monooxygenase by site-directed mutagenesis.

Use of the cloned sMMO genes has also been made in the construction of sMMO mutants of *M. trichosporium* OB3b. A reverse-genetics approach has been taken using marker exchange mutagenesis, a method first used successfully by Toukdarian and Lidstrom (38) for the construction of nitrogen-fixation mutants of *Methylosinus* sp. strain 6. A DNA fragment containing *mmoX* of OB3b was cloned into pBR325. An *Xho*I fragment internal to the *mmoX* gene was replaced with a kanamycin resistance cassette, and the *mob* fragment was also inserted into pBR325 to facilitate transfer by conjugation of the resulting recombinant plasmid pHM32 from *E. coli* S17-1 to OB3b. Homologous recombination of the mutated *mmoX* gene with the wild-type *mmoX* gene in the chromosome and subsequent loss of the unstable pBR325-based plasmid allowed the isolation of kanamycin-resistant OB3b colonies. These were DCM resistant on methanol-agar plates and grew only in the presence of Cu^{2+} on methane, that is, they had the expected $pMMO^{+}$, $sMMO^{-}$ phenotype (39). These mutants have now been complemented by transfering the sMMO gene cluster from wild-type *M. trichosporium* OB3b back into these mutants by conjugation.

M. trichosporium OB3b is a better organism than *M. capsulatus* (Bath) in which to study the regulation of expression of sMMO because OB3b will also grow well on methanol agar plates. The physiological growth conditions under which OB3b switches between using pMMO and sMMO, in response to the Cu^{2+}-to-biomass ratio are also now well defined and have been the subject of several studies (40,41). Differential expression of sMMO and pMMO is regulated by the amount of copper ions available to the cells; sMMO is expressed at low copper-to-biomass ratios, whereas pMMO is expressed at high copper-to-biomass ratios. The transcriptional regulation of the sMMO gene cluster in the methanotrophic bacterium *M. capsulatus* (Bath) has been investigated (42). Using Northern blot analysis and primer extension experiments, it was shown that the six open reading frames of the sMMO gene cluster are organized as an operon, and the transcripts produced upon expression of this operon have been identified. The synthesis of these transcripts was under the control of a

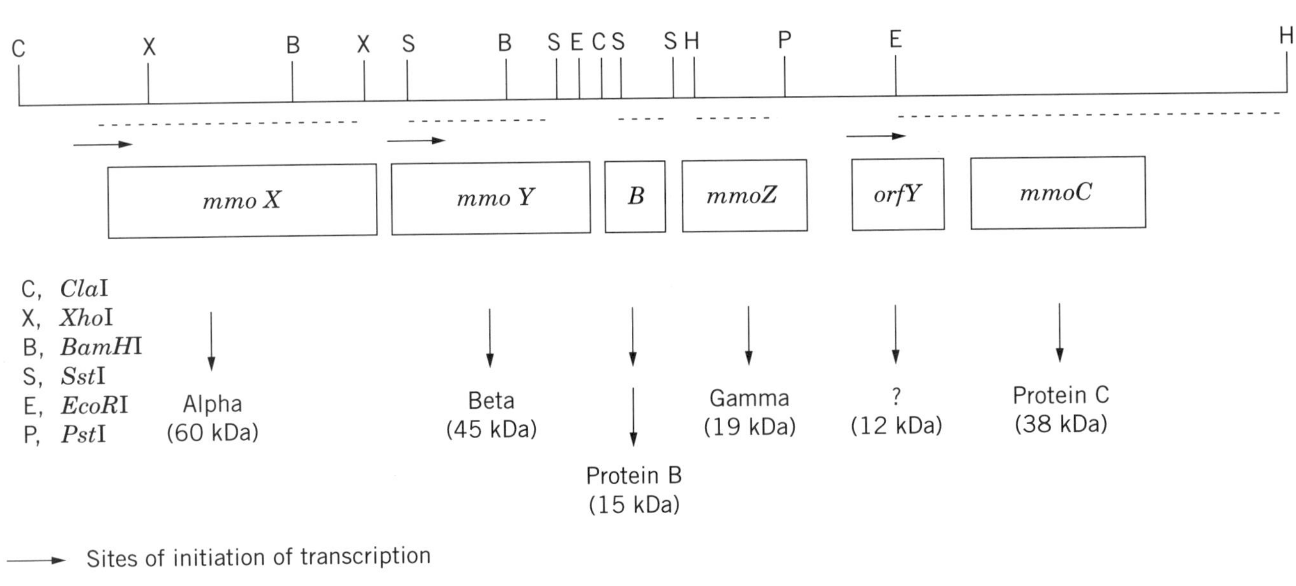

Figure 5. sMMO gene cluster of *Methylosinus trichosporium* OB3b.

single copper-regulated promoter, which has not been precisely defined. In both *M. capsulatus* and *M. trichosporium*, transcription of the sMMO gene cluster is negatively regulated by copper ions. Data suggest that transcription of the *M. trichosporium* OB3b sMMO gene cluster is directed from a σ^{54}-like and a σ^{70}-like promoter. The pMMO (*pmo*) genes of *M. capsulatus* (Bath) are transcribed into a polycistronic mRNA of 3.3 kb. The synthesis of this mRNA was activated by copper ions. Activation of *pmo* transcription by copper ions was concomitant with repression of sMMO gene transcription in both methanotrophs. This suggests that a common regulatory pathway may be involved in the transcriptional switch between sMMO and pMMO gene expression (42,43).

MOLECULAR ECOLOGY OF METHANOTROPHS

The difficulties in using traditional culture-based techniques to study the ecology of methanotrophs have been discussed previously (44). In particular the slow growth of methanotrophs, together with scavenging of nonmethanotrophs on agar plates, has hampered studies.

Phospholipid analysis of natural environment samples or enrichment cultures is one approach to assessing the types of methanotrophs present, because they contain more unusual fatty acids; the technique has been used successfully for the wetland environment (45). The detection and identification of methanotrophs in environmental samples can also be attempted using antibodies either to whole cells of culturable methanotrophs or to key enzymes such as methane monooxygenase or methanol dehydrogenase (reviewed in Ref. 44). However, these methods do not guarantee that all methanotrophs may be detected, and extensive control experiments with large numbers of extant methanotrophs are necessary to validate antibody probes. Estimates of the biomass and activities of methanotrophs can be made by measuring the release of $^{14}CO_2$ and incorporation of $^{14}CH_4$ added to environmental samples. However, this technique needs to be coupled with further experiments designed to analyze bacterial community structure.

Application of molecular biology techniques to methanotrophs has been aided considerably by the sequencing of a number of 16S rRNA genes of methanotrophs and methylotrophs (7,8) and the cloning of several methanotroph-specific genes (34,46). Hanson and colleagues have used 16S rRNA sequence data to examine phylogenetic relationships within genera of methanotrophs and methylotrophs. Type I methanotrophs such as *Methylomonas, Methylomicrobium*, and *Methylobacter*, together with the Type X *Methylococcus*, cluster in the γ-subdivision of the class Proteobacteria, whereas Type II methanotrophs *Methylosinus* and *Methylocystis* are found in the α-2 subdivision of the Proteobacteria (8). Hanson and colleagues have used this information to design oligonucleotide probes having a broad specificity for methanotrophs and methylotrophs in the natural environment (47). When coupled to suitable fluors, probes 9α and 10γ were successfully used to differentiate between mixed cultures of Type I and Type II methanotrophs by in situ hybridization (14).

16S rRNA technology has also been used to study the ecology of methanotrophs in blanket bog peat from the north of England. DNA was extracted from several 1-cm sections of a 30-cm-depth peat core covering various depths throughout the core, including the most active methane-oxidizing layer. Total 16s rRNA libraries from each section were screened with 16S rRNA probes specific for the genera *Methylococcus* (Mc 1005), *Methylomonas* (Mm 1007), *Methylobacter* (Mb 1007), and *Methylosinus* (Ms 1020). Results suggested that *Methylosinus* and *Methylococcus* were probably the dominant methanotrophs present in the peat environment. Two clones that hybridized to the *Methylosinus* probe were sequenced. Their 16S rDNA sequences, when analyzed, suggested that they branched closest to the *Methylosinus / Methylocystis* cluster of methanotrophs but were distinct from these sequences (48).

The 16S rRNA data coupled with polymerase chain reaction (PCR) technology have also been used successfully in analyzing methanotrophs in the marine environment. Seawater samples were enriched for methanotrophs by addition of essential nutrients and methane. The change in composition of the bacterial population was then monitored by analysis of 16S rDNA libraries constructed from DNA taken before and after enrichment, using universal eubacterial 16S rRNA primers and PCR. The dominant 16S rRNA sequence that was present in samples after enrichment on methane was recovered from 16S rRNA gene libraries, sequenced, and found to show a close phylogenetic relationship to *Methylomonas*, indicating that novel methanotrophs related to extant *Methylomonas* spp. were present in enrichment cultures (49). Based on the 16S rDNA sequence of this putative marine methanotroph, two oligonucleotides (Mm 650 and Mm 850, Table 1) were synthesized with rhodamine groups attached to their 5′ terminus. These fluorescently labeled probes have been used to determine the response of this marine methanotroph to changes in cultural conditions, thereby allowing optimization of enrichment and isolation methods for this organism (50).

FUNCTIONAL GENE PROBES FOR METHANOTROPHS

Genes unique to methanotrophs include those encoding sMMO and pMMO polypeptides. Another potentially useful marker is *mxaF*, encoding the large subunit of methanol dehydrogenase (the second enzyme of the methane oxidation pathway), which is present in virtually all Gram-negative methylotrophs (51). *mxaF* appears to be highly conserved in all Gram-negative methylotrophs (e.g., methane and methanol utilizers) and is therefore a good indicator of the presence of these organisms in the natural environment. PCR primers designed to specifically amplify a 550-bp fragment of *mxaF* have been successfully used to extend the database of *mxaF* sequences and to identify methylotroph sequences in peat core samples (49,52,53).

The high degree of identity between sMMO genes has enabled the design of PCR primers that specifically amplify each of the five sMMO structural genes (52) such that amplification of each of these genes is now possible from cultured methanotrophs containing sMMO. In addition, when total DNA from a variety of different freshwater, marine, soil, and peat samples was used as a template with these sMMO-gene-specific primers, a PCR product of the correct (predicted) size was obtained. Subsequent DNA sequencing of a number of these PCR products from peat samples has proved that methanotroph-specific sMMO-encoding DNA had been amplified from environmental DNA samples, and that these genes are very similar but not identical to known sMMO gene sequences. This suggested that there was considerable diversity of methanotrophs in these peat samples, and also that sMMO-containing methanotrophs were present in significant numbers (52). This is further supported by the observation that sMMO-containing methanotrophs have been enriched for and isolated from the same peat core samples, and that 16S rRNA/PCR experiments described earlier also revealed the presence (dominance) of *Methylosinus*- and *Methylococcus*-like organisms.

The PCR technology developed with sMMO genes is useful for studying environments such as peat or contaminated aquifers where methanotrophs containing sMMO might dominate in a copper-depleted environment. However, a better functional gene probe for these organisms would be one based on the pMMO present in all extant methanotrophs. Sequence data on *pmo* and *amo* (the ammonia monooxygenase gene, which codes for a related enzyme found in nitrifying bacteria) has allowed the design of degenerate PCR primers that will specifically amplify a 525-bp internal DNA fragment of *pmoA* or *amoA* from a variety of methanotrophs and nitrifiers. Analysis of the predicted amino acid sequences of these genes from representatives of each of the phylogenetic groups of methanotrophs (α- and γ-Proteobacteria) and ammonia oxidizing nitrifiers (β- and γ-Proteobacteria), revealed strong conservation of both primary and secondary structure. These new sequence data suggest that pMMO and ammonia monooxygenase may be evolutionarily related enzymes, despite their different physiological roles in methanotrophs and ammonia-oxidizing bacteria (54).

These observations are particularly important in the study of the ecological role of these organisms because ammonia oxidizers can also oxidize methane (with MMO also cooxidizing ammonia). The degenerate oligo PCR primers just described have now been used successfully to specifically amplify *pmo* / *amo* sequences from DNA isolated from a wide variety of environments including seawater, freshwater, agricultural and forest soils, and peat (J.C. Murrell and I.R. McDonald, unpublished data, 1997). Sequencing of these PCR products and examination of derived amino acid sequences of PmoA and AmoA now allows identification of these sequences as those originating from an α- β-, or γ-representative of the methane-oxidizing Proteobacteria (55), and an extensive database of sequences from extant methanotrophs and nitrifiers is also being established, which will aid further molecular ecological studies.

BIOTECHNOLOGY

The unique ability of the MMO enzyme in methanotrophic bacteria to catalyze some reactions of environmental and possibly commercial importance has generated interest in these organisms. The discovery that methanotrophs can degrade halogenated hydrocarbons that are ubiquitous in nature and also toxic pollutants has resulted in research that may lead to greater commercial exploitation (4,5,56).

The most ubiquitous of the groundwater pollutants are trichloroethylene (TCE), 1,2-dichloroethane, and chloroform. Bacteria that utilize TCE or chloroform as a sole carbon and energy source have not been discovered. However, several bacteria that contain nonspecific oxygenases, such as MMO, cause the mineralization of these compounds by cometabolism (cooxidation).

The capacity of methanotrophs to convert halogenated compounds was recognized by Colby et al. (57), who showed that methyl bromide could be used as a model substrate for the MMO reaction. Further work was stimulated by the observation of Wilson and Wilson (58), who demonstrated that methane could stimulate biological TCE degradation in soil. Since then, various pure cultures of methanotrophic bacteria have been found to be capable of cometabolic transformation of halogenated hydrocarbons (27,41,56,59–61). Large differences between various cultures were observed, and high TCE degradation rates were found with *M. trichosporium* OB3b cultivated under conditions of copper limitation and incubated with formate instead of methane as a source of reducing equivalents (60). Cells grown under copper limitation express exclusively the soluble MMO (40,62), and the conclusion that it is this form of the enzyme that is responsible for TCE oxidation was confirmed by in vitro incubations with the purified enzyme (63). Much lower rates of TCE oxidation have been reported for methanotrophs that possess only pMMO (64).

The feasibility of in situ bioremediation by stimulation of the growth of indigenous populations of methanotrophs has been investigated by Semprini and McCarty (65). They reported that biodegradation of TCE occured within an aquifer, at the Moffett Naval Air Station, after it had been supplied with methane and oxygen to stimulate indigenous methane-oxidizing bacteria. The initial results from a further study, at the Savannah River site, indicated that a substantial increase in Type II methanotrophic bacterial populations occurred in some areas of the aquifer, and an increase in the frequency of bacteria that contained DNA with sequences complementary to the *mmoB* gene at different sites was observed (66).

The second biotechnological application of the methanotrophs has been as single-cell protein (SCP). Methane, as natural gas, is available inexpensively and in large amounts; however, as a substrate for SCP production, it has many drawbacks. These include explosive hazards and the low aqueous solubility of methane. Therefore methanol has been the preferred substrate because it is easily derived from methane and highly soluble in water. Methanol is also easily removed from the SCP product in downstream processing. Not surprisingly it was the substrate of choice for the only self-contained SCP business in the United Kingdom, the Pruteen process of ICI (67), which utilized the methylotroph *Methylophilus methylotrophus*. Methanol was also used in two smaller-scale processes by Probion (Hoechst-Uhde, Germany) and Norprotein (Norsk-

Hydro, Sweden), both of which utilized *Methylomonas* species (for reviews see Refs. 2 and 3).

Methanotrophs have also been used for the conversion of propylene to propylene oxide, a high-value product. However, even though high rates of propylene oxidation were obtained, these rates fell dramatically after 3–4 min due to loss of biocatalytic activity (68). sMMO and pMMO both have the potential for cooxidation of a wide range of other compounds. sMMO will cooxidize compounds including alkanes, haloalkanes, alkenes, alicyclics, and aromatics (69); recently pMMO has been shown to cooxidize chlorinated ethenes at yields that are greater than those of sMMO (70). These cooxidations may have future biotechnological applications. Finally the recent characterization of thermophilic/thermotolerant methanotrophs also opens up further avenues for the exploitation of methanotrophs in biotechnological processes (71).

ACKNOWLEDGMENTS

Work in J. C. M.'s laboratory is supported by grants from the NERC, BBSRC, AFRC, and EEC. We thank J. McN. Sieburth for electron micrographs.

BIBLIOGRAPHY

1. M.E. Lidstrom and D.I. Stirling, *Annu. Rev. Microbiol*, **44**, 27–58 (1990).
2. D.J. Leak, in J.C. Murrell and H. Dalton eds., *Methane and Methanol Utilizers*, Plenum, New York, 1992, pp. 245–279.
3. P.J. Large and C.W. Bamforth, *Methylotrophy and Biotechnology*, Longman, Harlow, U.K., 1988.
4. R. Oldenhuis and D.B. Janssen, in J.C. Murrell and D.P. Kelly eds., *Microbial Growth on C_1 Compounds*, Intercept Press, Andover, U.K., 1993, pp. 121–133.
5. L. Alvarez Cohen, in J.C. Murrell and D.P. Kelly eds., *Microbial Growth on C_1 Compounds*, Intercept Press, Andover, U.K., 1993, pp. 337–350.
6. R. Whittenbury, K.C. Phillips, and J.F. Wilkinson, *J. Gen. Microbiol.* **61**, 205–218 (1970).
7. J.P. Bowman, L.I. Sly, P.D. Nicholas, and A.C. Hayward, *Int. J. Syst. Bacteriol.* **43**, 735–753 (1993).
8. R.S. Hanson and T.E. Hanson, *Microbiol. Rev.* **60**, 439–471 (1996).
9. W.S. Reeburgh, in J.C. Murrell and D.P. Kelly eds., *Microbial Growth on C_1 Compounds*, Intercept Press, Andover, U.K., 1993, pp. 1–14.
10. G.M. King, in J.C. Murrell and D.P. Kelly eds., *Microbial Growth on C_1 Compounds*. Intercept Press, Andover, U.K., 1993, pp. 303–313.
11. R. Conrad, *Adv. Microb. Ecol.* **14**, 207–250 (1995).
12. R. Conrad, *Microbiol. Rev.* **60**, 609–640 (1996).
13. P.N. Green, in J.C. Murrell and H. Dalton eds., *Methane and Methanol Utilizers*. Plenum, New York, 1992, pp. 23–84.
14. R.S. Hanson, B.J. Bratina, and G.A. Brusseau, in J.C. Murrell and D.P. Kelly eds., *Microbial Growth on C_1 Compounds*. Intercept Press, Andover, U.K., 1993, pp. 285–302.
15. C. Anthony, *The Biochemistry of Methylotrophs*. Academic, London, 1982.
16. C. Anthony, *Adv. Microb. Physiol.* **27**, 113–210 (1986).
17. M.E. Lidstrom, in J.C. Murrell and H. Dalton eds., *Methane and Methanol Utilizers*, Plenum, New York, 1992, pp. 183–206.
18. P.M. Goodwin and C. Anthony, *Microbiology* **141**, 1051–1064 (1995).
19. H.A.T. Nguyen, M. Zhu, S.J. Elliot, K.H. Nakagawa, B. Hadman, A.M. Costello, T.L. Peeples, B. Wilkinsob, H. Morimoto, P.G. Williams, H.G. Floss, M.E. Lidstrom, K.O. Hodgson, and S.I. Chan, in M.E. Lidstrom and F.R. Tabita eds., *Microbial Growth on C_1 Compounds*, Kluwer Academic, Dordrecht, The Netherlands, 1996, pp. 150–158.
20. J.A. Zahn and A.A. DiSpirito, *J. Bacteriol.* **178**, 1018–1029 (1996).
21. J.D. Lipscomb, *Annu. Rev. Microbiol.* **48**, 371–399 (1994).
22. H. Dalton, P.C. Wilkins, and Y. Jiang, in J.C. Murrell and D.P. Kelly eds., *Microbial Growth on C_1 Compounds*, Intercept Press, Andover, U.K., 1993, pp. 65–80.
23. A.C. Rosenzweig, C.A. Frederick, and S.J. Lippard, in M.E. Lidstrom and F.R. Tabita eds., *Microbial Growth on C_1 Compounds*, Kluwer Academic, Dordrecht, The Netherlands, 1996, pp. 141–149.
24. J.C. Murrell, *FEMS Microbiol. Rev.* **88**, 233–248 (1992).
25. M.E. Lidstrom and A.E. Wopat, *Arch. Microbiol.* **140**, 27–33 (1984).
26. M.E. Lidstrom, A.E. Wopat, D.N. Nunn, and A.E. Toukdarian, in G.S. Omenn and A. Hollaender eds., *Genetic Control of Environmental Pollutants*, Plenum, New York, 1984, pp. 273–279.
27. G.A. Brusseau, H.G. Tsien, R.S. Hanson, and L.P. Wackett, *Biodegradation* **1**, 19–29 (1990).
28. D.W. Graham, D.G. Korich, R.P. Le Blanc, N.A. Sinclair, and R.G. Arnold, *Appl. Environ. Microbiol.* **58**, 2231–2236 (1992).
29. P.A. Phelps, S.K. Agarmal, G.E. Speital, and G. Georgiou, *Appl. Environ. Microbiol.* **58**, 3701–3708 (1992).
30. M.W. Fitch, D.W. Graham, R.G. Arnold, S.K. Agarwal, P. Phelps, G.E. Speitel, and G. Georgiou, *Appl. Environ. Microbiol.* **59**, 2771–2776 (1993).
31. T.M. Barta and R.S. Hanson, *Antonie van Leeuwenhoek* **64**, 109–120 (1993).
32. N. Harms, in J.C. Murrell and D.P. Kelly eds., *Microbial Growth on C_1 Compounds*, Intercept Press, Andover, U.K., 1993, pp. 235–244.
33. D. Waechter-Brulla, A.A. DiSpirito, L.V. Chistoserdova, and M.E. Lidstrom, *J. Bacteriol.* **175**, 3767–3775 (1993).
34. J.C. Murrell, *Biodegradation* **5**, 145–159 (1994).
35. I.R. McDonald, H. Uchiyama, S. Kambe, O. Yagi, and J.C. Murrell, *Appl. Environ. Microbiol.* **63**, 1898–1904 (1997).
36. H. Qian, U. Edlund, J. Powlowski, V. Shingler, and I. Sethson, *Biochemistry* **36**, 495–504 (1997).
37. J.S. Lloyd, A. Bhambra, J.C. Murrell, and H. Dalton, *Eur. J. Biochem.* **248**, 72–79 (1997).
38. A.E. Toukdarian and M.E. Lidstrom, *J. Bacteriol.* **157**, 979–983 (1984).
39. J.C. Murrell, in J.C. Murrell and D.P. Kelly eds., *Microbial Growth on C_1 Compounds*. Intercept Press, Andover, U.K., 1993, pp. 109–120.
40. K.J. Burrows, A. Cornish, D. Scott, and I.J. Higgins, *J. Gen. Microbiol.* **130**, 3327–3333 (1984).
41. H-C. Tsien, G.A. Brusseau, R.S. Hanson, and L.P. Wackett, *Appl. Environ. Microbiol.* **55**, 3155–3161 (1989).
42. A.K. Nielsen, K. Gerdes, H. Degn, and J.C. Murrell, *Microbiology* **142**, 1289–1296 (1996).
43. A.K. Nielsen, K. Gerdes, and J.C. Murrell, *Mol. Microbiol.* **25**, 399–409 (1997).

44. J.C. Murrell and A.J. Holmes, in I. Joint ed., *Molecular Ecology of Aquatic Microbes*, Spriger-Verlag, Berlin, 1995, pp. 365–390.
45. I. Sundh, P. Borga, M. Nilsson, and B.H. Svensson, *FEMS Microbiol. Ecol.* **18**, 103–112 (1995).
46. J.C. Murrell and A.J. Holmes, in M.E. Lidstrom and F.R. Tabita eds., *Microbial Growth on C_1 Compounds*, Kluwer Academic, Dordrecth, The Netherlands, 1996, pp. 133–140.
47. G.A. Brusseau, E.S. Bulygina, and R.S. Hanson, *Appl. Environ. Microbiol.* **60**, 626–636 (1994).
48. I.R. McDonald, G.H. Hall, R.W. Pickup, and J.C. Murrell, *FEMS Microbiol. Ecol.* **21**, 197–211 (1996).
49. A.J. Holmes, N.J.P. Owens, and J.C. Murrell, *Microbiology* **141**, 1947–1955 (1995).
50. A.J. Holmes, N.J.P. Owens, and J.C. Murrell, *J. Exp. Mar. Biol. Ecol.* **203**, 27–38 (1996).
51. M.E. Lidstrom, C. Anthony, F. Biville, F. Gasser, P. Goodwin, R.S. Hanson, and N. Harms, *FEMS Microbiol. Lett.* **177**, 103–106 (1994).
52. I.R. McDonald, E.M. Kenna, and J.C. Murrell, *Appl. Environ. Microbiol.* **61**, 116–121 (1995).
53. I.R. McDonald and J.C. Murrell, *Appl. Environ. Microbiol.* **63**, 3218–3224 (1997).
54. A.J. Holmes, A. Costello, M.E. Lidstrom, and J.C. Murrell, *FEMS Microbiol. Lett.* **132**, 203–208 (1995).
55. I.R. McDonald and J.C. Murrell, *FEMS Microbiol. Lett.* **156**, 205–210 (1997).
56. R. Oldenhuis, R.L.J.M. Vink, D.B. Janssen, and B. Witholt, *Appl. Environ. Microbiol.* **55**, 2819–2826 (1989).
57. J. Colby, H. Dalton, and R. Whittenbury, *Biochem. J.* **151**, 459–462 (1975).
58. J.T. Wilson and B.H. Wilson, *Appl. Environ. Microbiol.* **49**, 242–243 (1985).
59. C.D. Little, A.V. Palumbo, S.E. Herbes, M.E. Lidstrom, R.L. Tyndall, and P.J. Gilmer, *Appl. Environ. Microbiol.* **54**, 951–956 (1988).
60. H. Uchiyama, T. Nakajima, O. Yagi, and T. Tabuchi, *Agric. Biol. Chem.* **53**, 2903–2907 (1989).
61. S.M. Henry and D. Grbic-Galic, *Appl. Environ. Microbiol.* **57**, 236–244 (1991).
62. S.H. Stanley, S.D. Prior, D.J. Leak, and H. Dalton, *Biotechnol. Lett.* **5**, 487–492 (1983).
63. B.G. Fox; J.G. Borneman, L.P. Wackett, and J.D. Lipscomb, *Biochemistry* **29**, 6419–6427 (1990).
64. A.A. DiSpirito, J. Gulledge, J.C. Murrell, A.K. Shiemke, M.E. Lidstrom, and C.L. Krema, *Biodegradation* **2**, 151–164 (1992).
65. L. Semprini and P.L. McCarty, in P.V. Roberts, L. Semprini, G.D. Hopkins, D. Grbic-Galic, P.L. McCarty, and M. Reinhard eds. *In Situ Aquifer Restoration of Chlorinated Aliphatics by Methanotrophic Bacteria*, Center for Environmental Research Information, Cincinnati, 1989, pp. 239–250.
66. J.P. Bowman, L. Jimenez, I. Rosario, T.C. Hazen, and G.S. Sayler, *Appl. Environ. Microbiol.* **59**, 2380–2387 (1993).
67. K.A. Powell and B.L.F. Rodgers, in C.T. Hou ed., *Methylotrophs: Microbiology, Biochemistry and Genetics*, CRC Press, Boca Raton, Fl., 1984, pp. 119–144.
68. S.H. Stanley and H. Dalton, *Biocatalysis* **6**, 163–175 (1992).
69. H. Dalton, *Adv. Appl. Microbiol.* **26**, 71–87 (1980).
70. J.E. Anderson and P.L. McCarty, *Appl. Environ. Microbiol.* **63**, 687–693 (1997).
71. L. Bodrossy, J.C. Murrell, H. Dalton, M. Kalman, L.G. Puskas, and K. Kovacs, *Appl. Environ. Microbiol.* **61**, 3549–3555 (1995).

See also BIOENERGETICS OF MICROBIAL GROWTH.

METHYLOTROPHS, INDUSTRIAL APPLICATIONS

DAVID LEAK
Imperial College of Science, Technology, and Medicine
London, United Kingdom

KEY WORDS

Amino acids
Bioremediation
Biotransformation
Methane
Methanol
Polyhydroxyalkanoates
Polysaccharides
Ribulose monophosphate pathway
Serine pathway
Single-cell protein
Xylulose monophosphate pathway

OUTLINE

INTRODUCTION

Methane and methanol are cheap, readily available carbon sources for microbial growth, and their metabolism involves specialized routes of assimilation and dissimilation. Consequently, considerable work has gone into exploring potential applications of organisms capable of growth and product formation from these substrates, particularly on methanol. Because the attraction of methylotrophs derives primarily from substrate costs, most of the work has been directed toward the higher volume–lower price end of the fermentation market, in particular single-cell protein

(SCP). However, SCP from methanol proved to be uneconomical, although an unexpected benefit arose from the early SCP program, namely the application of methylotrophic yeasts as hosts for expression of recombinant proteins.

The full potential of methanol as a fermentation substrate has yet to be realized. Future applications could arise from the use of methanol to supplant carbohydrates in existing fermentation processes for metabolite (e.g., amino acid and vitamin) production. However, the benefit of lower substrate costs needs to be balanced against the (generally) increased oxygen requirements of methanol- compared with carbohydrate-based processes.

MICROBIOLOGY AND ECOLOGY

Methylotrophs are organisms that grow on carbon substrates more reduced than carbon dioxide but with no carbon–carbon bonds (reduced C_1 substrates). Although some aerobic methylotrophs have been shown to oxidize reduced C_1 substrates to carbon dioxide and fix carbon by typical autotrophic routes (autotrophic methylotrophs), most aerobic methylotrophs assimilate carbon at the oxidation level of formaldehyde (true methylotrophs). The primary metabolism of these organisms is thus fundamentally distinct from that of autotrophs and heterotrophs, and some "obligate" methylotrophs are unable to use heterotrophic substrates containing carbon–carbon bonds as the sole carbon source.

Methane, the most reduced C_1 compound, is a key link between the processes of anaerobic degradation and the aerobic food chain. The activity of anaerobic and aerobic methanotrophic bacteria (i.e., methylotrophs that can use methane as a sole source of carbon and energy) significantly reduces the potential release of methane to the atmosphere arising from the activity of methanogens. Many of the methanogens are also methylotrophic, being able to use substrates such as methanol and formate to drive methanogenesis (methylotrophic acetogens are also known). Methanol arises from methyl esters and methoxy groups found in plant materials (e.g., lignin and pectin) and also through chemical oxidation of methane in the troposphere and its subsequent deposition in rainfall. The ability to utilize methanol as the sole source of carbon and energy is found in representatives of all three taxonomic domains: methanogenic methylotrophs belonging to the Archaea and other anaerobic and aerobic methylotrophic prokaryotes belonging to the eubacteria (with both Gram-negative and Gram-positive representatives); a number of methanol utilizing yeasts (Eucarya) and filamentous fungi have also been isolated. Methane utilizing yeasts have also been described (1), but these have exceptionally long generation times and are currently of no more than academic interest. There is strong evidence for anaerobic methane oxidation associated with sulfate reducing and/or methanogenic activity (2). However, the organisms responsible have been difficult to culture axenically. Hence most of the work on methanotrophic organisms has concentrated on the methane utilizing aerobic (eu)bacteria (2).

In addition to the more obvious reduced C_1 substrates methane and methanol, nature provides a wide range of compounds based on methylated heteroatoms such as methylamines, methyl sulfides, methyl sulfonates, methoxy compounds, and halogenated methanes. Oxidation or hydroxylation of these methyl groups results in the release of formaldehyde, which can feed directly into the dissimilatory and assimilatory pathways of C_1 metabolism. Some of these organisms play an important role in nutrient cycling in the environment and probably contribute to the natural degradation of toxic compounds.

GROWTH ON C_1 COMPOUNDS

Bacterial growth on methane and methanol was first described nearly a century ago. However, the subject lay fairly dormant until the 1960s and 1970s, when research activity driven partly by academic studies on the relationship between methylotrophic and autotrophic biochemistry and, in no small way, by the perceived applications of methylotrophs increased dramatically. The latter originated from the interest in producing SCP (i.e., microbial biomass) from methane or methanol as an animal or human food supplement, a venture in which a number of companies became involved to a greater or lesser extent but that was ultimately shown to be uneconomic. Despite this uncertain start the basic principles on which the choice of C_1 substrates was made remain valid, namely that substrates such as methane and methanol are cheap and because they are highly reduced should give high growth yields. As substrate costs become an increasingly important factor with increasing production volume (and decreasing unit price of the product), the potential applications of methylotrophs are likely to be in bulk products. Alternatively, the enzymes or metabolic sequences unique to methylotrophic metabolism may offer some possibilities not available in heterotrophs. The biotechnological application of methylotrophs can therefore be conveniently categorized under three headings:

1. Processes highly dependent on substrate economics
2. Processes based on C_1-specific enzymes or metabolism
3. Processes or products incidentally associated with C_1 metabolism

Although these subdivisions are not mutually exclusive they are nonetheless useful in explaining the rationale behind the development of certain C_1-based processes.

Metabolic Biochemistry and Energetics

The pathways of carbon assimilation affect both the efficiency of growth and the types of products that might feasibly be produced from methylotrophs. In aerobic (true) methylotrophic bacteria two fundamentally different routes of carbon assimilation have been demonstrated: the ribulose monophosphate (RuMP) pathway and the serine pathway (3,4). The RuMP pathway (Fig. 1) has similarities to the Calvin Benson cycle of CO_2 fixation in the sense that 3 C_1 units (formaldehyde) are added to C5 acceptors (ribulose monophosphate) that subsequently undergo a se-

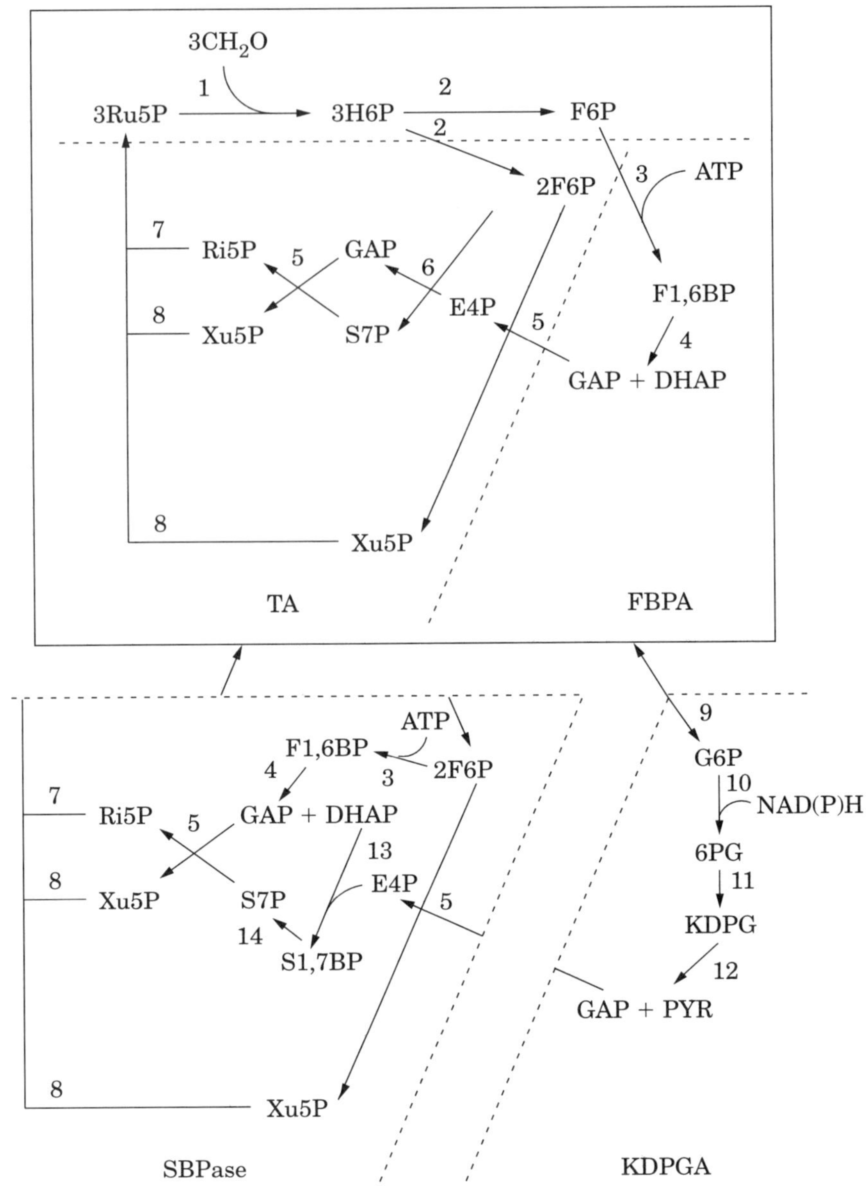

Figure 1. Intermediates and enzymes in the ribulose monophosphate pathway of formaldehyde fixation. Intermediates: Ru5P, ribulose-5-phosphate; H6P, hexulose-6-phosphate; F6P, fructose-6-phosphate; F1, 6BP, fructose-1,6-bisphosphate; DHAP, dihydroxyacetone phosphate; GAP, glyceraldehyde-3-phosphate; E4P, erythrose-4-phosphate; Xu5P, xylulose-5-phosphate; S7P, sedoheptulose-7-phosphate; Ri5P, ribose-5-phosphate; G6P, glucose-6-phosphate; 6PG, 6-phosphogluconate; KDPG, 2-keto-3-deoxy-6-phosphogluconate; PYR, pyruvate; S1,7BP, sedoheptulose-1,7-bisphosphate. Enzymes: 1, H6P synthase; 2, H6P isomerase; 3, phosphofructokinase; 4, F1,6BP aldolase; 5, transketolase; 6, transaldolase (TA); 7, Ri5P isomerase; 8, Pentose phosphate 3-epimerase; 9, phosphoglucose isomerase; 10, G6P dehydrogenase; 11, 6PG dehydratase; 12, KDPG aldolase (KDPGA); 13, S1,7BP aldolase; 14, S1,7BPase (SBPase).

ries of rearrangements yielding a C3 intermediate for cell biosynthesis and regenerating the acceptor molecules. Two variations of both cleavage reaction (fructose-1,6-bisphosphate aldolase [FBPA] and 2-keto-3-deoxy-6-phosphogluconate aldolase [KDPGA]), which generates the C3 biosynthetic intermediate, and the rearrangement stage (transaldolase [TA] and sedoheptulose-1,7-bisphosphatase [SBPase]), which regenerates the acceptor molecule, have been recognized and give four possible combinations. A comparison of the energetics of assimilation based on the production of a common C3 intermediate, pyruvate, shows that the FBPA/TA combination which is found in a number of Gram-positive methylotrophs, is the most efficient (Table 1).

The xylulose monophosphate (XuMP) pathway found in methylotrophic yeasts (5) has similarities to the RuMP pathway, being based on carbohydrate transformations and involving condensation of C_1 and C5 units and subsequent rearrangements based on transaldolases and transketolases (Fig. 2). However, the serine pathway found in bacteria is fundamentally different, involving assimilation of a C_1 unit into serine and subsequent rearrangements based on hydroxy acids and amino acids (Fig. 3). It also involves a CO_2 fixation step. Neither the XuMP nor serine pathway is as energetically efficient as the RuMP pathway but may offer advantages over the RuMP pathway organisms for the accumulation of certain metabolites.

Most of the yeasts, nonmethanotrophic serine pathway, and Gram-positive RuMP pathway organisms are facultative methylotrophs (3) able to use a range of heterotrophic substrates (e.g., carbohydrates). All of the methanotrophs appear to be obligate C_1 utilizers, some being very sensitive to excess methanol; many of the Gram-negative RuMP organisms are restricted facultative methylotrophs, growing poorly on some heterotrophic substrates. For me-

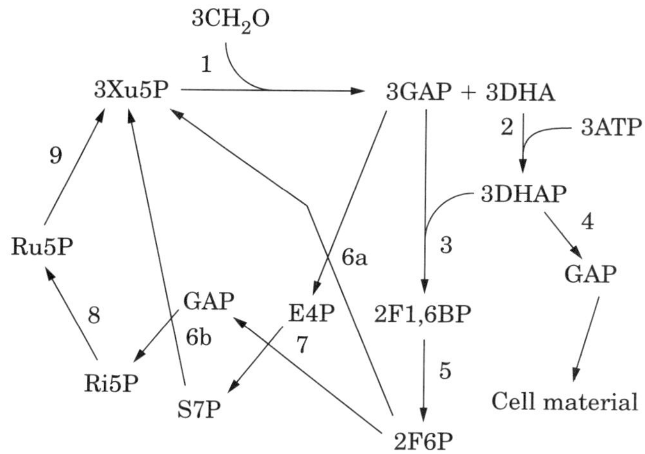

Figure 2. Intermediates and enzymes in the xylulose monophosphate pathway of formaldehyde fixation. Intermediates: Xu5P, xylulose-5-phosphate; GAP, glyceraldehyde-3-phosphate; DHA, dihydroxyacetone; DHAP, dihydroxyacetone phosphate; F1,6BP, fructose-1,6-bisphosphate; F6P, fructose-6-phosphate; E4P, erythrose-4-phosphate; S7P, sedoheptulose-7-phosphate; Ri5P, ribose-5-phosphate; Ru5P, ribulose-5-phosphate. Enzymes: 1, DHA synthase; 2, triokinase; 3, F1,6BP aldolase; 4, triose phosphate isomerase; 5, F1,6BPase; 6, transketolase; 7, transaldolase; 8, Ri5P isomerase; 9, pentose phosphate 3-epimerase.

tabolite overproduction, facultative methylotrophs can offer significant advantages because the ability to grow on heterotrophic substrates allows the isolation of mutants and genetic manipulation of methylotrophic metabolism.

Biomass Production

Despite the demise of C_1-based SCP processes, any future process adopted on the basis of substrate cost will inevitably involve growth of methylotrophs on a large scale. A retrospective consideration of the development of large-scale C_1-based processes is therefore warranted.

Table 1. Energetic Requirements of the Different Variants of the RuMP Cycle for the Production of Pyruvate ($3CH_2O \rightarrow$ Pyruvate), Chosen as a Typical Metabolic Intermediate

RuMP variant	NAD(P)H	ATP
KDPGA/SBPase	+1	−1
KDPGA/TA	+1	0
FBPA/SBPase	+1	0
FBPA/TA	+1	+1

KDGPA, 2-keto-3-deoxy-6-phosphogluconate aldolase; SBPase, sedoheptulose-1,7-bisphosphatase; FBPA, fructose-1,6-bisphosphate aldolase; TA, transaldolase.

Growth Yields and Energetics. A summary pathway for bacterial methane oxidation is shown in Figure 4. The pathway for methanol oxidation in Gram-negative methylotrophic bacteria that do not use methane is essentially the same, with the omission of the first step, although in some organisms formaldehyde is oxidized in a cyclic manner (6) after fixing in the initial steps of the assimilation pathway. Some methylotrophs may not use an NAD-linked formaldehyde dehydrogenase. However, those that do should give the highest growth yields and are therefore more likely to be adopted in a commercial process. $PQQH_2$ in methanol dehydrogenase is reoxidized by passing electrons via a periplasmic cytochrome c_L to the respiratory chain with energy conservation via chemiosmosis yielding approximately 1 ATP/$PQQH_2$ (7). Gram-positive methylotrophs use an NAD-linked methanol dehydrogenase (8). In principle, reoxidation of the resulting NADH via the respiratory chain should yield more ATP than gained from methanol oxidation in Gram-negative bacteria, which should translate into higher growth yields. For a wide range of substrates the yield of cells is proportional to the reduction level of the substrate, because biochemical oxidation yields proportionate amounts of ATP and cell yield is proportional to the ATP yield from substrate oxidation (9). Thus, one might expect that cell yields on methane

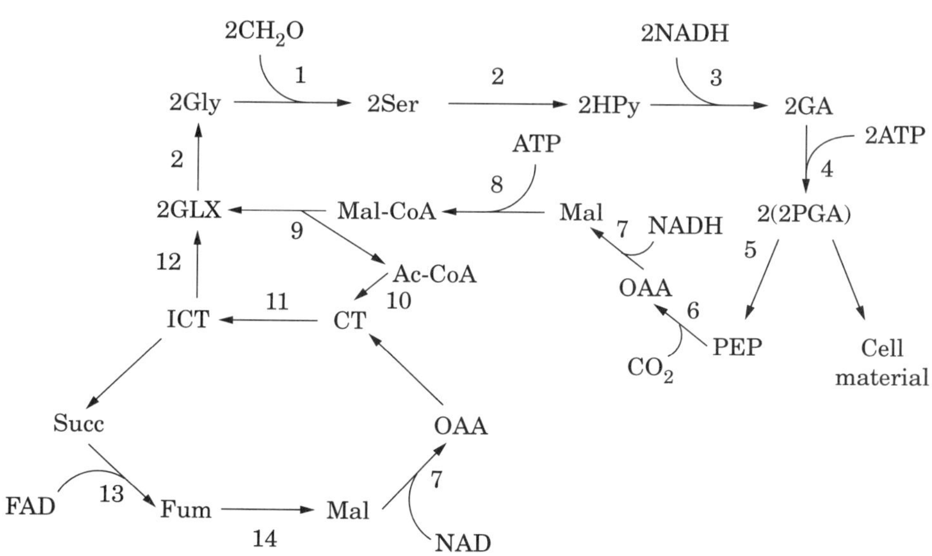

Figure 3. Intermediates and enzymes in the serine pathway of formaldehyde fixation. Intermediates: Gly, glycine; Ser, serine; HPy, hydroxypyruvate; GA, glycerate; 2PGA, 2-phosphoglycerate; PEP, phosphoenolpyruvate; OAA, oxaloacetate; Mal, malate; Mal-CoA, malyl-CoA; Ac-CoA, acetyl-CoA; CT, citrate; ICT, isocitrate; Succ, succinate; Fum, fumarate; GLX, glyoxylate. Enzymes: 1, serine transhydroxymethylase; 2, serine-glyoxylate aminotransferase; 3, hydroxypyruvate reductase; 4, glycerate kinase; 5, enolase; 6, PEP carboxylase; 7, malate dehydrogenase; 8, malate thiokinase; 9, malyl-CoA lyase; 10, citrate synthase; 11, aconitase; 12, isocitrate lyase; 13, succinate dehydrogenase; 14, fumarase.

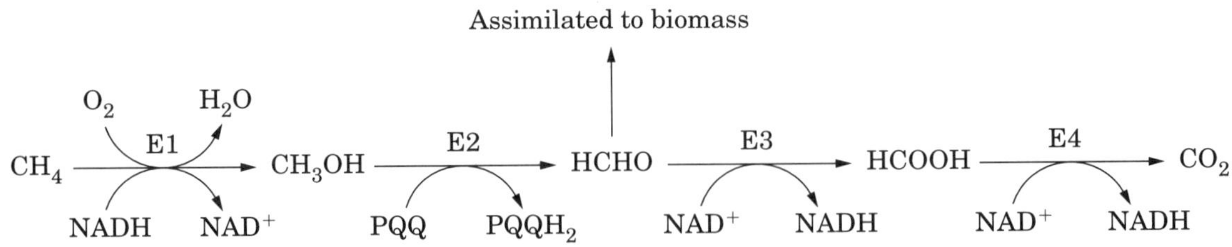

Figure 4. Generalized pathway for methane catabolism in methanotrophic bacteria. E1, methane monooxygenase; E2, methanol dehydrogenase; E3, formaldehyde dehydrogenase; E4, formate dehydrogenase; PQQ, pyrroloquinoline quinone.

would be higher than those on methanol. However, this relationship breaks down for highly reduced substrates; Figure 4 demonstrates the causes for methane oxidation. The biological oxidation of methane requires an energy input in the form of NADH, thus making methanol a potentially higher yielding substrate than methane (this is borne out in practice [10]). A number of other factors also mitigate against the use of methane as a substrate:

1. Flammability
2. Lack of compressibility
3. Low aqueous solubility
4. High oxygen demand (from Fig. 4 it is evident that the minimum oxygen demand would be 1.5 O_2/CH_4) resulting from the operation of methane monooxygenase

It should also be recognized that natural gas contains higher saturated and unsaturated hydrocarbons that can be partially oxidized by methanotrophs, resulting in the accumulation of inhibitory by-products. Therefore, except in circumstances in which the activity of methane monooxygenase (MMO) is required, bulk product formation from methylotrophs is likely to be based on methanol as substrate. Methanol can be derived from methane or a variety of other fossil fuels or renewable energy sources via synthesis gas (Fig. 5). The latter has also been proposed as a substrate for bulk product formation using carboxidotrophic bacteria (11) that use CO as a carbon source. However, despite the higher solubility of CO compared with methane, the problems of handling large volumes of a toxic substrate mitigate against this proposal.

$$CH_4 + H_2O \xrightarrow{1} CO + 3H_2$$

$$CO_2 + H_2 \xrightarrow{2} CO + H_2O$$

$$CO + 2H_2 \xrightarrow{3} CH_3OH$$

Figure 5. Imperial Chemical Industries process for the catalytic synthesis of methanol from methane 1, 15–20% Ni catalyst on Al_2O_3 or SiO_2, 700–900 °C, 1–25 bar; 2, Ni catalyst; 3, Cu/Zn catalyst, 250–280 °C, 70–110 bar.

Biomass Production from Methanol. A number of companies developed pilot-scale processes for biomass production from methanol (Table 2), with Imperial Chemical Industries (ICI) in the United Kingdom going on to full-scale production in a 1,500 m^3 continuous bioreactor using a Gram-negative bacterium. This was a significant technical achievement, the largest aseptic continuous culture process ever to operate.

The oxygen demand for growth on methanol (0.6–0.7 O_2/CH_3OH), although significantly less than that for methane, is still high compared with growth of heterotrophs on glucose (approximately 0.4 $O_2/[CH_2O]$). This high oxygen demand is a consequence of the relative inefficiency of energy conservation of periplasmic methanol dehydrogenase. For large-scale processes this inefficiency results in significant heat generation and oxygen transfer problems. The ICI solution to these problems was to use the moderately thermotolerant (t_{opt} = 40 °C) bacterium *Methylophilus methylotrophus* and a high aspect ratio (42 m × 7 m) pressure cycle reactor (17). This design also minimized the number of mechanical seals in the reactor, which were the most likely routes of contamination in long-term continuous operation. Gas transfer and mixing were achieved according to the airlift principle, with pressurized air being introduced at the base of the riser. However, the height of the liquid was such that the oxygen dissolved almost completely as it entered the reactor. To avoid toxicity or nutri-

Table 2. Full- or Pilot-Scale SCP Processes Using Methanol as Substrate

Company	Capacity (m^3)	Product (organism)	Reference
Imperial Chemical Industries PLC, U.K.	1,500	Pruteen *(Methylophilus methylotrophus)*	12
Hoechst-Uhde, Germany	20	Probion *(Methylomonas clara)*	13
Phillips Petroleum Co. (Provesta Corp.) U.S.	25	Provesteen *(Pichia pastoris)*	14
Norsk-Hydro, Sweden	45	Norprotein *(Methylomonas methanolica)*	15
Mitsubishi Gas-Chemical Co., Japan	2	(*Pseudomonas* sp.)	16

ent limitation methanol was introduced through a series of nozzles placed at various points up the riser. To achieve a product of the right composition as an animal feed, the reactor was run at a dilution rate of 0.16–0.19 h^{-1} (μ_{max} = 0.55 h^{-1}) with a cell density of 30 g dry wt L^{-1} and a projected annual production capacity of 50,000–70,000 tons. *M. methylotrophus* uses the KDPGA/TA route of C_1 assimilation (18), which, as discussed earlier, is marginally less efficient than the FBPA/TA route. It was also found to assimilate ammonia solely via the less energetically efficient glutamine synthase–glutamine oxoglutarate aminotransferase (GS/GOGAT) route. By knocking out this route and expressing glutamate dehydrogenase (GDH) from *Escherichia coli* in this organism a 7% improvement in yield from methanol was achieved (19). However, the latter construct was never used for commercial production.

Phillips Petroleum (later Provesta Corporation) in the United States opted for a process using the methylotrophic yeast *Pichia pastoris*, a decision based on the greater public acceptability of yeasts in nutrition. Methanol dissimilation in yeasts occurs by an energetically inefficient route involving an alcohol oxidase and catalase (Fig. 6). Not only is there no energy conservation arising from the conversion of methanol to formaldehyde but the oxidase mechanism involves stoichiometric amounts of oxygen. Although some oxygen is subsequently released through the action of catalase, the net requirement is $0.5O_2/CH_3OH$. Because inefficient energy conservation results in excessive heat generation the low optimum growth temperature (30 °C) was also a disadvantage, as was the long generation time (5.5–6 h). Nevertheless, through medium optimization and bioreactor design the Phillips process managed to achieve oxygen transfer rates of 0.8–1 mol L^{-1} h^{-1}, allowing operation at very high cell densities and producing a continuous slurry containing 125–150 g dry wt L^{-1} cells that could be spray-dried directly without concentration (14).

POTENTIAL APPLICATIONS

Processes Dependent on Substrate Economics

With the exception of SCP three other types of bulk product have been or are currently being considered as products from methylotrophs: poly-β-hydroxyalkanoate–based biodegradable plastics, polysaccharides, and amino acids. Although the driving force for all of these programs is substrate economics, in the last case the mode of carbon assimilation may predispose certain methylotrophs toward production of particular classes of amino acids.

One practical benefit from the research arising from the SCP programs was a detailed understanding of metabolic pathways and the energetics of carbon assimilation and dissimilation in methylotrophs. Using this information, Linton and Niekus (20), Collins (21), and Ackermann and Babel (22) have compared the likely carbon conversion efficiencies and oxygen requirements for conversion of methanol to these three bulk products. Predictions for amino acid production were based on glutamate. For polysaccharides and glutamate the predicted yields (w/w) from glucose and methanol are similar (Table 3), indicating that, for equally high yielding organisms, methanol would offer a cost advantage. However, the higher oxygen demand for methylotrophs is a significant disadvantage. In the case of exopolysaccharides the yield on oxygen is predicted to be 10-fold lower than that on glucose. Because polysaccharide volumetric productivity from glucose is often limited by the efficiency of oxygen transfer into increasingly viscous media the substrate cost advantage is more than offset by this disadvantage. Therefore, exopolysaccharides produced from methanol would have to have some unique characteristics to justify commercial production. A few years ago, Celgene Corporation marketed Poly-54/Methylophilan, an exopolysaccharide produced by continuous culture of *M. viscogenes*. Several uses were envisaged, but of particular note was its high viscosity at 80 °C (higher than xanthan gum), which, unlike xanthan, was maintained for extended periods. This suggested a possible role in enhanced oil recovery and in oil drilling muds. However, production has subsequently ceased (D. Stirling, personal communication, June 1997).

Amino Acid Production. Glutamate production from methanol has a threefold disadvantage compared with glucose in terms of yield from oxygen. Thus, production of glucose or other bulk amino acids from methanol has more realistic prospects than polysaccharide production. Indeed, methanol was being examined as an alternative substrate for glutamate production in the early 1970s. It is interesting to note that yields of up to 33 g/L were obtained in these early studies using metabolically ill-defined Gram-negative methylotrophs (23), which probably used the RuMP pathway. Similar yields have been reported more recently (24) using mutants of the Gram-negative RuMP organism *Methylobacillus glycogenes* (formerly classified as *Pseudomonas insueta* and *Protaminobacter thiaminophagus*). However, Ackermann and Babel have pointed out (22) that, despite their lower growth yields, serine pathway organisms should potentially give higher yields of glu-

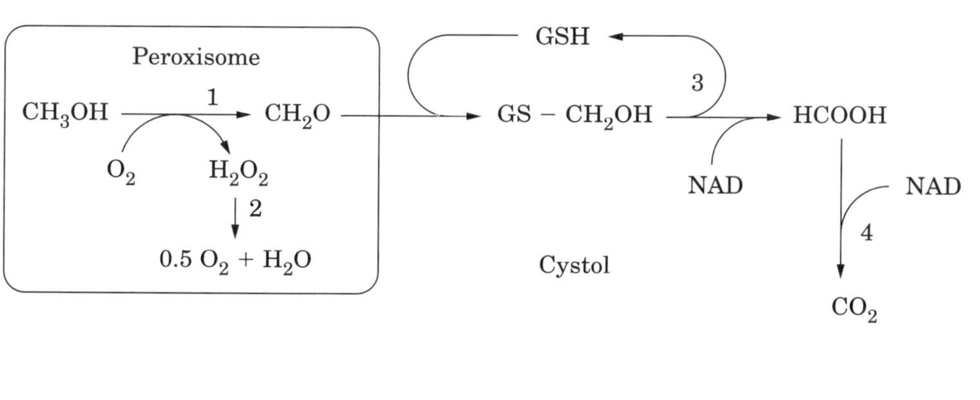

Figure 6. Linear (noncyclic) pathway for methanol dissimilation in methylotrophic yeasts.

Table 3. Predicted Product Yields (Y_s) from Methanol and Glucose, and the Associated Yield from Oxygen (Y_{O_2}).

Product[a]	Carbon source	Substrate requirement	Y_s (g/g)	Y_{O_2} (g/g)
EPS ($C_6H_{10}O_5$)	Methanol	$6.14CH_3OH + 3.21O_2$	0.82	1.58
	Glucose	$1.05\ C_6H_{12}O_6 + 0.3O_2$	0.86	16.3
Glutamate ($C_5H_9O_4N$)	Methanol[R]	$6CH_3OH + NH_3 + 4.5O_2$	0.76	1.02
	Methanol[S]	$5CH_3OH + NH_3 + 2.5O_2$	0.92	1.84
	Glucose	$C_6H_{12}O_6 + NH_3 + 1.5O_2$	0.82	3.06
PHB ($C_4H_6O_2$)	Methanol	$5.5CH_3OH + 3.75O_2$	0.49	0.72
	Glucose	$C_6H_{12}O_6$	0.48	[b]

Note: Yields are based on the known metabolic pathways between substrate and product and their associated energy requirements; substrate required for biomass formation is not considered. EPS synthesis from methanol is based on the RuMP pathway (20); glutamate synthesis from methanol (20,22) is based on both the RuMP pathway ([R]) and the serine pathway ([S]); PHB synthesis from methanol is based on the serine pathway and makes the same assumptions as cited in Ref. 21 regarding efficiency of methanol usage.

[a]EPS, extracellular polysaccharide (hexose); PHB, poly-β-hydroxybutyrate.

[b]Conversion of glucose to PHB generates ATP and NAD(P)H. There is therefore no additional ATP required for PHB synthesis, so the yield based on oxygen usage is, technically, infinite.

tamate from methanol; it should be borne in mind that high yields of amino acids may also depend on the type of organism being used.

Glutamic acid has traditionally been produced from *Corynebacterium* or *Brevibacterium* spp., which, as Gram-positive organisms, appear to be disposed toward this sort of application, partly because of the lack of an outer membrane and partly because of efficient glutamate efflux systems. Although the majority of Gram-positive methylotrophs isolated use the RuMP pathway (3) they may offer a greater possibility for realizing the theoretical yields of glutamate than the Gram-negative organisms previously studied. Furthermore, the isolation of stable mutants from Gram-negative methylotrophs is known to be difficult. Bearing in mind this difficulty and the cooling requirements for large-scale production, Schendel and coworkers (25) have isolated lysine overproducing mutants of a thermotolerant (t_{opt} = 50 °C), methylotrophic *Bacillus* sp. Process modeling and optimization (26) have demonstrated that production is primarily growth associated and optimal productivity should be obtained in a variable-feed, constant-volume fed-batch process. A final concentration of greater than 45 g/L lysine, which also included a significant amount of by-products (mainly glutamate and alanine) has been reported (27). Therefore, there is further room for improvement in control of metabolic flux to lysine. Abouzeid and coworkers (28) are examining the possibility of aromatic amino acid production (e.g., phenylalanine) from the Gram-positive *Amycolatopsis methanolica* (formerly *Nocardia* sp. 239). In the latter case, there is also a strong precedent for starting with a RuMP (or XuMP) pathway organism because erythrose-4-phosphate, a precursor for aromatic amino acid biosynthesis, occupies a central position in the RuMP and XuMP pathways.

Serine and methionine have also been considered for methylotrophic fermentative production. The former is required as a precursor in tryptophan production and is also included in some cosmetics. Its catalytic production from glycine and methanol by methylotrophs was an obvious target because serine hydroxymethyltransferase (SHMT) catalyzes the first step in carbon assimilation by serine pathway organisms and should therefore be expressed at high levels. Mutants of *Hyphomicrobium methylovorum* were reported (29) to produce 34 g/L serine from 100 g/L glycine and 48 g/L methanol, while resting cells of a strain of *Methylobacterium extorquens* produced 54 g/L serine (30). However, considerably higher concentrations have been achieved (31) in an enzyme bioreactor using a recombinant SHMT (from *E. coli*), and it is unlikely that a methylotrophic whole organism process will supplant this concentration. Unlike the previous examples, methionine for animal feeds is currently produced chemically from acrolein and methanethiol. Hence, the economic feasibility of fermentative whole cell biocatalytic production by methylotrophs is more difficult to assess than in the previous cases. Biosynthesis requires L-homoserine (from aspartate), serine, sulfide, and the C_1 carrier 5-methyl-THF, suggesting that serine pathway organisms would also be the obvious candidates. This suggestion was made valid by the studies of Yamada and coworkers on ethionine-resistant mutants of RuMP and serine pathway organisms (32,33). However, none of the yields were commercially significant, and the lack of any recent literature on this topic suggests that it is not being pursued.

Biodegradable Plastics from Methanol. Marlbrough Biopolymers, a subsidiary of Monsanto, now produces a biodegradable plastic under the tradename Biopol by fermentation using *Alcaligenes eutrophus* grown on glucose as the carbon and energy source (34). The plastic is a derivative of poly-β-hydroxybutyrate (PHB), a high molecular weight polyester produced by many bacteria as a carbon and energy storage compound. *A. eutrophus* was selected because PHB can comprise up to 80% of its dry weight under appropriate conditions and also because, when supplied with propionic acid, the organism produces a random copolymer of 3-hydroxybutyrate and 3-hydroxyvalerate, P(3HB-3HV), which is tougher and more flexible than pure PHB. Current production runs at about 1,000 tons per annum. However, it has been predicted that even at a 10-fold higher level the cost of P(3HB-3HV) would be about $5/kg, similar to that of other biodegradable plastics but considerably higher than synthetic plastics derived from petrochemicals (35). Glucose accounts for 35–40% of this price. Therefore, at present, microbially derived plastics are still in the category of niche products.

Two underlying research trends are evident in this area. For truly bulk production of PHB-based plastics at a price similar to synthetics it has been recognized that production in plants such as rapeseed or soybean represents the way forward, and some progress in this area is being made (35). Production in bacteria, due to its higher cost, is more likely to concentrate on niche products. In that light, the evidence that PHB is just one of a family of polyhydroxyalkanoates (PHAs) and that some PHA synthases will accept hydroxyacyl CoAs derived from a wide range of derivatized carboxylic acids supplied to the medium opens up a potentially fruitful avenue for the production of specialist biodegradable polymers (36). The greater ease of physiological and genetic control of metabolic flux in bacteria would favor production by fermentation.

Because substrate costs represent a significant proportion of the selling price of PHAs, substitution with a cheaper substrate could offer some short- and long-term benefits for P(3HB-3HV) and specialist PHA production, respectively. Collins (21) calculated that the maximum theoretical yield from methanol would be 0.18 g/g compared with 0.33 g/g from glucose. Although the price differential between the two substrates would still favor methanol, the cost savings are marginal. In the early stages of Biopol development methanol was apparently considered as a substrate but not pursued because the product was of low molecular weight and extraction was difficult. More recent research by a number of groups has shown that high molecular weight PHB can be produced from methylotrophs and that P(3HB-3HV) can be produced in methylotrophs by supplementing the fermentation with *n*-amyl alcohol or valeric acid (propionic acid was less effective). However, the 66% of dry weight (136 g/L PHB) achieved by Suzuki et al. (37) using computer-controlled fed-batch fermentation is the highest reported yield for any methylotroph. This apparent limit probably reflects a restriction of metabolic flux to the PHB precursors, which suggests that genetic manipulation of flux control mechanisms may be necessary to match the yields currently achieved with *A. eutrophus*. PHB synthesis in methylotrophs appears to be restricted to *Methylobacterium* spp. ("pink-pigmented facultative methylotrophs") and the autotrophic methylotrophs (38). *Methylobacterium* spp. assimilate carbon via the serine pathway and have a complete TCA cycle, unlike restricted facultative methylotrophs that use the RuMP cycle. It is evident that, without a complete TCA cycle, acetyl CoA resulting from the breakdown of PHB would be of limited benefit. Hence, the latter organisms have evolved to accumulate extracellular polysaccharide rather than PHB. Recent work on the cloning from *A. eutrophus* and heterologous expression of genes required for PHB biosynthesis (39) is extending the range of potential strains for PHB production. Thus, RuMP pathway methylotrophs may be considered in the future.

Processes Based on C_1 Metabolism

Although the pathways of carbon assimilation in methylotrophs are specialized, few of the enzymes involved are unique. In the RuMP pathway only the hexulose phosphate synthase and isomerase are peculiar to C_1 metabolism. Therefore, beyond the observation that certain routes of C_1 assimilation might provide a higher flux of precursors for amino acid biosynthesis, for example, the assimilatory enzymes of C_1 metabolism have not been exploited. However, there has been considerable interest in the development of processes based on C_1 dissimilatory enzymes, in particular methane monooxygenase, methanol dehydrogenase, methanol oxidase (from yeasts), and formate dehydrogenase.

Methane Monooxygenase. Two types of MMO have been recognized: a nonheme iron-soluble (cytoplasmic) enzyme (sMMO) and a membrane-associated enzyme (pMMO; p, particulate) that requires copper for activity and has some similarity to ammonia monooxygenase (40). Soluble MMOs from *Methylococcus capsulatus* (Bath) and *Methylosinus trichosporium* OB3b have been extensively studied, and crystal structures of the oxygenase component from both enzymes are available (41). Additionally, the genes encoding sMMO from a number of sources have been cloned, sequenced (42), and expressed in heterologous hosts (43). Both sMMO and pMMO insert oxygen into a range of substrates other than methane, with sMMO having a very low substrate specificity (44,45). Initially this observation led to interest in the use of cells expressing MMO as biocatalysts for synthetic biotransformations, particularly in the conversion of propene to propene oxide. Although a bulk product, the chemical synthesis of propene oxide is not straightforward, and there was considerable interest in the possibility of a one-step biological process that did not yield by-products (46). However, the dual requirements of high catalytic rates and extended biocatalytic lifetimes could not be realized, primarily because the process of propene epoxidation actually inactivated the MMO (47). Other potential applications as a synthetic biocatalyst have also foundered, partly because sMMO has a low stereoselectivity and therefore has little value for the synthesis of chiral chemicals. Another limitation is that with many substrates it exhibits poor regioselectivity, producing, for example, mixtures of alkan-1-ols and alkan-2-ols from the oxidation of straight chain alkanes (44).

More recently the interest in MMO has focused on applications in biotreatment, particularly of the priority pollutant trichloroethene (TCE) and other low molecular weight partially halogenated hydrocarbons (e.g., dichloroethene, vinyl chloride). In fact, MMO is only one of a number of mono- and dioxygenases that have been shown to fortuitously co-oxidize these compounds (48). Cometabolism processes typically yield no benefit to the host organism and, in the case of co-oxidation, represent an energy drain, requiring an additional energy source for continued operation of the oxygenase. In the case of methane oxidizing bacteria, methanol is an obvious energy source, although growth on methanol may have undesirable physiological consequences (see later section). Oxidation of TCE by methane oxidizing bacteria leads to the transient accumulation of TCE epoxide (49), which is chemically unstable. As a consequence, TCE epoxide and some of the derived breakdown products are also highly reactive and can give rise to toxicity. This result has been demonstrated with both purified sMMO (49) and whole cells expressing sMMO (50). However, it seems that the toxicity can be alleviated by continued supply of reducing power and particularly by the use of mixed rather than axenic cultures (51). In the latter, it is probable that other members of the bacterial consortium assist in the removal of toxic intermediates, thus prolonging the life of the primary biocatalyst.

Studies with pure cultures have demonstrated that oxidation of TCE is fastest with bacteria expressing the sMMO (2). The ability to express sMMO is restricted to certain types of methanotrophs: primarily the type II (serine pathway) and type X bacteria such as *Methylosinus trichosporium* and *Methylococcus capsulatus*, respectively. Furthermore, in these organisms, production of sMMO is regulated by the availability of copper in the medium. With a copper-sufficient medium these organisms preferentially produce the pMMO (which results in a higher growth efficiency). Therefore, to get high rates of TCE oxidation, it is necessary to grow these organisms in a copper-deficient medium. Although this condition may be feasible in laboratory experiments it may be more difficult to achieve in operational biotreatment plants and impossible to control for in situ biotreatment operations. However, with increasing access to the molecular genetics of methanotrophs (42), the generation of stable mutants (e.g., by reverse genetics [52]) unable to synthesize pMMO presents an interesting prospect. Regulatory mutants that constitutively produce sMMO would also circumvent another physiological problem, which is that growth of type II and type X organisms on methanol, even under copper limitation, results in production of pMMO (53) (at very low copper concentrations this is produced as an apo-enzyme). Mutants that constitutively produced sMMO could be supplied with methanol as a growth substrate and also to drive the co-oxidation reaction, without the risk of switching off sMMO synthesis.

Several reports (2) have described the practical applications of methanotrophs for TCE and other chlorinated hydrocarbon degradation, both in controlled bioreactors and in situ. Bioreactor studies have mainly been based on *Methylosinus trichosporium*, which was shown, in laboratory studies using copper-limited growth conditions, to have the highest reported specific activity of TCE degradation of any of the oxygenase-based systems that have been examined. Because of the problems of sMMO expression using methanol as growth substrate, these reactors have usually been supplied with methane to sustain the active biomass and formate to drive the biotransformation. However, with a continuously fed bioreactor, methane and TCE compete for the sMMO, which reduces rates of biodegradation. A practical alternative is the use of sequencing reactors (54) that are alternately supplied with methane and TCE. The feasibility of in situ biotreatment (55) of contaminated aquifers by stimulating indigenous methanotrophic activity has also been investigated. Injection of methane–air mixtures was shown to stimulate growth of methanotrophs, particularly type II methanotrophs, which have the potential to produce sMMO. In one study, degradation of TCE was shown to increase after this treatment, which suggests that these organisms were also expressing sMMO. However, without the ability to control the availability of copper for growth, it is more difficult to optimize expression of sMMO activity and hence TCE biodegradation in in situ treatment.

Whole cells of methanotrophic bacteria have also been considered as methane sensors (56) (using oxygen uptake as the mode of detection) and methane filters (57) for use in the mining industry. However, neither of these potential applications has been developed beyond the laboratory stage.

Methanol Dehydrogenase. Methanol oxidation in Gram-negative aerobic methylotrophs involves a periplasmic methanol dehydrogenase (MDH) that uses pyrroloquinoline quinone (PQQ, or methoxatin) as a cofactor. Reducing equivalents from MDH would normally be passed to a soluble cytochrome c (7) but can also be passed to artificial electron acceptors such as ferrocenes, which are used as mediators in biosensors. Higgins et al. (58) developed a sensor based on MDH that had a rapid response time and useful sensitivity range (1–100 μM). However, MDH oxidizes a number of primary alcohols in addition to methanol. Although the enzyme from *Rhodopseudomonas acidophila* has a higher affinity for ethanol than for other primary alcohols (59), this lack of specificity limits the usefulness of MDH in diagnostics. In contrast, the broad substrate range may present opportunities for use in synthetic biotransformations (60). In Japan there also appears to be a market for PQQ, the MDH cofactor, as a health food (61).

Methanol or Alcohol Oxidase. Methanol oxidase is the enzyme used by eukaryotic methylotrophs to catalyze the first step in methanol metabolism. It produces formaldehyde (from methanol) and hydrogen peroxide and is normally found in association with catalase, which degrades the peroxide (5). As with MDH, methanol oxidase also oxidizes a range of higher alcohols and is therefore often referred to as alcohol oxidase (AOX). In methylotrophic yeasts AOX and catalase are expressed at very high levels and are located in organelles known as peroxisomes. Peroxisomes can be seen in thin sections to occupy most of the intracellular volume. This very high level of expression has been exploited for the development of eukaryotic expression systems, which are discussed later.

As with MDH, AOX has been considered for use in biosensors but with detection based on oxygen uptake rather than through direct or indirect electron transfer reactions (62). Similar to MDH, the enzyme also has a broad substrate range. Methylotrophic yeasts and AOX have also been considered for the production of formaldehyde, where it has been demonstrated that surprisingly high concentrations (0.95 M) could be accumulated by mutants of *Candida boidinii* (63) and for production of hydrogen peroxide. The latter is toxic to most AOX enzymes, but that from *Pichia pastoris* NRRL Y-4290 appears to be remarkably resistant (64). The production of peroxide using methanol oxidase from methylotrophic yeasts has been patented by Unilever, although there is no evidence that it has been put into commercial practice.

Formate Dehydrogenase. The formate dehydrogenases (FDH) from methylotrophic bacteria appear to be surprisingly complex given their apparently simple function.

However, those from methylotrophic yeasts are simpler. The enzyme from *C. boidinii* is a homodimer with subunits of approximately 36 kDa; it is both oxygen insensitive and moderately thermostable (65). Because the end product of formate oxidation is gaseous and can therefore be easily removed from the reaction, the combination of formate and FDH is ideal for reduced nucleotide recycling. The enzyme is produced commercially and has been used with great effect for continuous biotransformation reactions requiring NADH (66).

Applications Fortuitously Associated with C_1 Metabolism

Heterologous Gene Expression in Methylotrophic Yeasts. One of the fortuitous and enduring aspects of the various SCP programs resulted from the discovery that the methanol or alcohol oxidase of the yeast *P. pastoris* was expressed at very high levels (5), essentially from a single chromosomal copy of the AOX gene (there are actually two AOX genes, but *AOX1* is expressed much more strongly than *AOX2*). One of the problems with the use of *Saccharomyces cerevisiae* as a host for heterologous expression of eukaryotic proteins is the lack of a strong, yet tightly regulated, inducible promoter. During growth on methanol, methanol oxidase can comprise up to 30% of the soluble cell protein in *P. pastoris* (67), with the enzyme being localized in specialized microbodies (peroxisomes) together with catalase and dihydroxyacetone synthase, the first enzyme involved in formaldehyde assimilation. As mentioned earlier, these peroxisomes occupy most of the cell contents of methanol grown cells, but when grown on glucose, peroxisomes are virtually absent. A similar situation is found with another well-studied methylotrophic yeast, *Hansenula polymorpha* (68). The strong inducible *AOX1* promoter has been exploited by Provesta Corporation for the development of a powerful expression system that is now commercially available in a variety of forms (69). For production purposes, the system of choice involves cloning into an integrative additive or transplacement vector. Transplacement involves replacement of the *AOX1* locus by the cloned construct giving a strain with an aox1AOX2 phenotype, capable of only slow growth on methanol. Additive integration at the *AOX1* or *HIS4* locus will leave a functional *AOX1*, allowing continued growth on methanol.

Higher levels of heterologous gene expression are usually obtained with the transplacement system. Chromosomal integration yields stable expression systems; however, multiple copies of the expression cassette, usually associated with expression from plasmid vectors, result in higher levels of expression. Scorer and coworkers (70) have shown that, by including a kanamycin resistance gene in the integration cassette and selecting for high levels of resistance to the antibiotic G418, it is possible to select for multicopy integration, which occurs naturally in about 5–10% of transformants. In this way it has been possible to achieve expression levels comprising up to 32% of total cell protein (Table 4); by combining this technique with techniques for high cell density culture developed during the SCP program (71), yields of up to 12 g/L recombinant protein have been recorded. The highest protein yields have been obtained in cytosolic form. However, secreted proteins can be obtained using either native (including prokaryotic) secretion signals, the leader sequences from prepro-α mating factor, or invertase from *S. cerevisiae*. As with *S. cerevisiae, P. pastoris* also carries out some typical eukaryotic posttranslational modification including N- and O-linked glycosylation.

Table 4. A Selection of Heterologous Proteins Produced in *Pichia pastoris* (Pp) and *Hansenula polymorpha* (Hp)

Protein (MW)	Host/system	Yield (g/L)
HBsAg (23)	Pp (FB) + Hp	0.4 (Pp)
Human TNF (17)	Pp (FB + CC)	10
Human EGF (6)	Pp (FB)	0.5
Streptokinase-*Se* (47)	Pp (CC)	0.077
Aprotinin (6.5)	Pp (FB)	0.93
Tetanus tox. frag Cs*Ct*(50)	Pp (FB)	12
Pertactin-*Bp* (69)	Pp (FB)	3
Human PN-2 (6.8)	Pp	1
HIV gp120 (54)	Pp (FB)	1.25
HSA (67)	Pp (FB) + Hp	
Human IGF-1 (7.5)	Pp (FB)	0.5
Human lysozyme (14)	Pp	0.7
Human SOD (16)	Pp	0.75
IL-2	Pp	4
Human tPA	Pp	
Invertase-*Sc* (58)	Pp (FB) + Hp	2.5 (Pp)
Glucose oxidase-*An* (75)	Hp	2.25

Note: These proteins were chosen to illustrate the range of yields obtained with therapeutically important proteins. For a fuller list see Ref. 68. FB, fed batch; CC, continuous culture; *Se, Streptococcus equisimilis; Ct, Clostridium tetani; Bp, Bordetella pertussis; Sc, Saccharomyces cerevisiae; An, Aspergillus niger.*

Expression in *H. polymorpha* (Table 4) offers the same features seen with *P. pastoris*, potentially coupled with more efficient production because the optimum growth temperature is 37–43 °C, reducing the cooling requirements of large-scale processes, and growth on methanol is faster (68). However, because *H. polymorpha* has only one *AOX* gene, transplacement constructs are unable to grow on methanol as the sole carbon source. Usually glucose is used for growth, with methanol added to induce heterologous gene expression at the appropriate time. The inducible formate dehydrogenase (FMD) promoter has also been used.

Recent studies with a number of organisms have started to unravel the processes of peroxisome (and other microbody) proliferation and protein targeting. Given the dramatic increase in peroxisomes associated with methanol induction and the peroxisomal location of AOX, it has been suggested (68) that methylotrophic yeasts may offer an additional advantage for heterologous gene expression: namely, targeting to peroxisomes as a mechanism for compartmentalizing toxic or proteolytically susceptible proteins.

Production of Vitamins and Coenzymes. Although they do not command the high prices of recombinant proteins and other pharmaceutical products, vitamins and coenzymes occupy a price range dictated more by productivity than substrate economics. Nevertheless, various methylotrophs have been found to produce relatively high levels of potentially valuable products such as cytochrome c, coenzyme Q_{10}, and vitamin B_{12}. Methanol oxidation occurs in the periplasm of aerobic Gram-negative methylotrophic bacteria and links directly to the respiratory chain via a soluble cytochrome c (7). Some wild-type strains have been observed to release significant quantities of this cytochrome into the culture medium during growth (71); cyanide- and glycine-resistant mutants accumulate levels higher than those of wild-type strains (72). The high respiratory capacity associated with aerobic methanol oxidation also underlies the ability of some mutant facultative methylotrophs to produce high levels of coenzyme Q_{10}. Both of these products have applications in cardiotherapy, with Q_{10} also finding an increasing market in dietary supplements, but face competition from sources such as spent brewing yeast or bovine or equine heart, neither of which incur specific production costs.

Vitamin B_{12} is produced commercially by fermentation using *Propionibacterium* spp. However, anaerobic methylotrophic methanogens and acetogens have been shown to accumulate high concentrations of corrinoids, including vitamin B_{12}. Despite the low biomass productivity and cell yields of these anaerobes, potential production with these organisms remains an active area of research (73).

CONCLUDING REMARKS

Despite the commercial failure of methanol-based SCP production, the rationale for adopting methylotrophic organisms for processes in which substrate costs form a significant part of overall production costs still holds true. The early work on methylotrophic bacteria was dominated by studies of Gram-negative organisms, and our understanding of the biochemistry of these organisms is now relatively advanced. Work on Gram-positive methylotrophs has lagged behind. However, given that Gram-positive organisms are favored for fermentation processes to produce low molecular weight metabolites such as amino acids (which increasingly fall into the high volume–low price category of products), it is probable that Gram-positive organisms will receive increasing attention. As far back as the early 1970s the potential of methylotrophs for producing such metabolites was appreciated. Given the increasing access to the molecular genetics of methylotrophic bacteria, it would not be surprising to see this area develop.

Substrate costs also underlie the high degree of interest in producing polyhydroxyalkanoates in methylotrophs. However, even if methanol could supplant carbohydrate in the current processes, PHA production costs could still not compete with those of bulk plastics. Thus, at least for the foreseeable future, PHAs are likely to remain a specialized product; as such, the fermentation substrate is likely to be dictated by the nature of the process and product required rather than the opposite.

Finally, it is probable that some of the specialized applications of methylotrophs will continue to develop. The *P. pastoris* expression system appears to be well established, and the more recently developed *H. polymorpha* system could provide a useful complement. Methanotrophic bacteria (or recombinant methane monooxygenase), on the other hand, may yet find an application in biotreatment plants. Certainly the practice of stimulating the proliferation of methanotrophs to boost their cometabolic contribution in natural ecosystems is a rational approach and may offer a cost-effective solution to some problems of environmental contamination.

BIBLIOGRAPHY

1. H.I. Wolfe, in H. Dalton ed., *Microbial Growth on C_1 Compounds*, Heydon, London, 1981, pp. 202–210.
2. R.S. Hanson and T.E. Hanson, *Microbiol. Rev.* **60**, 439–471 (1996).
3. L. Dijkhuizen, P.R. Levering, and G.E. de Vries, in J.C. Murrell and H. Dalton eds., *Methane and Methanol Utilizers*, Plenum Press, New York, 1992, pp. 149–181.
4. C. Anthony, in I. Goldberg and J.S. Rokem eds., *Biology of Methylotrophs*, Butterworth-Heinemann, Stoneham, Mass., 1991, pp. 79–109.
5. W. de Koning and W. Harder, in J.C. Murrell and H. Dalton eds., *Methane and Methanol Utilizers*, Plenum Press, New York, 1992, pp. 207–244.
6. C. Anthony, *Int. J. Biochem.* **24**, 29–39 (1992).
7. C. Anthony, *Adv. Microbiol. Physiol.* **27**, 113–210 (1986).
8. N. Arfman, E.M. Watling, W. Clement, R.J. van Oosterwijk, G.E. de Vries, W. Harder, M.M. Attwood, and L. Dijkhuizen, *Arch. Microbiol.* **152**, 280–288 (1989).
9. J.J. Heijnen and J.A. Roels, *Biotechnol. Bioeng.* **23**, 739–763 (1981).
10. D.J. Leak and H. Dalton, *Appl. Microbiol. Biotechnol.* **23**, 477–481 (1986).
11. E. Williams, J. Colby, G.W. Logan, and C.M. Lyons, in J.D. Stowell, A.J. Beardsmore, C.W. Keevil, and J.R. Woodward eds., *Carbon Substrates in Biotechnology*, IRL Press, Oxford, 1986, pp. 185–201.
12. P.J. Senior and J. Windass, *Biotech. Lett.* **2**, 205–210 (1980).
13. W. Sittig, *Chemtech.* **13**, 606–613 (1983).
14. L.K. Shay, H.R. Hunt, and G.H. Wegner, *J. Ind. Microbiol.* **2**, 79–85 (1987).
15. H. Mogren, *Process Biochem.* **14**, 2–7 (1979).
16. G.L. Solomons, *CRC Crit. Rev. Biotechnol.* **1**, 21–58 (1985).
17. R.B. Vasey and K.A. Powell, in G.E. Russell ed., *Biotechnology and Genetic Engineering Reviews*, vol. 2, Intercept, Newcastle upon Tyne, 1984, pp. 285–311.
18. A.J. Beardsmore, P.N. Aperghis, and J.R. Quayle, *J. Gen. Microbiol.* **128**, 1423–1439 (1982).
19. J.D. Windass, M.J. Worsey, E.M. Pioli, D. Pioli, P.T. Barth, K.T. Atherton, E.C. Dart, D. Byrom, K. Powell, and P.J. Senior, *Nature (London)* **287**, 396–401 (1980).
20. J.D. Linton and H.G.D. Niekus, in H.W. van Verseveld and J.A. Duine eds., *Microbial Growth on C_1 Compounds: Proceedings of the 5th International Symposium*, Martinus Nijhoff, Dordrecht, 1987, pp. 263–271.
21. S.H. Collins, in J.D. Stowell, A.J. Beardsmore, C.W. Keevil, and J.R. Woodward eds., *Carbon Substrates in Biotechnology*, IRL Press, Oxford, 1986, pp. 161–168.
22. J.-U. Ackermann and W. Babel, *J. Basic Microbiol.* **34**, 211–216 (1994).
23. T. Oki, A. Kitai, K. Kuono, and A. Ozaki, *J. Gen. Appl. Microbiol.* **19**, 79–83 (1973).
24. H. Motoyama, H. Anazawa, R. Katsumata, K. Araki, and S. Teshiba, *Biosci. Biotechnol. Biochem.* **57**, 82–87 (1993).

25. F.J. Schendel, C.E. Bremmon, M.C. Flickinger, M. Guettler, and R.S. Hanson, *Appl. Environ. Microbiol.* **56**, 463–470 (1990).
26. G.H. Lee, W. Hur, C.E. Bremmon, and M.C. Flickinger, *Biotechnol. Bioeng.* **49**, 639–653 (1996).
27. R.S. Hanson, R. Dillingham, P. Olson, G.H. Lee, D. Cue, F.J. Schendel, C.E. Bremmon, and M.C. Flickinger, in M.E. Lidstrom and F.R. Tabita eds., *Microbial Growth on C_1 Compounds*, Kluwer Academic, Dordrecht, 1996, pp. 227–236.
28. A. Abouzeid, G. Euverink, G.I. Hessels, R.A. Jensen, and L. Dijkhuizen, *Appl. Environ. Microbiol.* **61**, 1298–1302 (1995).
29. H. Yamada, S.S. Miyazaki, and Y. Izumi, *Agric. Biol. Chem.* **50**, 17–21 (1986).
30. P. Sirirote, T. Yamane, and S. Shimizu, *J. Ferment. Technol.* **64**, 389–396 (1986).
31. D.M. Anderson and H.-Y. Hsiao, in D. Rozell and F. Wagner eds., *Biocatalytic Production of Amino Acids and Derivatives*, Carl Hanser Verlag, Munich, 1992, pp. 22–41.
32. H. Yamada, Y. Morinaga, and Y. Tani, *Agric. Biol. Chem.* **46**, 47–55 (1982).
33. Y. Morinaga, Y. Tani, and H. Yamada, *Agric. Biol. Chem.* **46**, 473–480 (1982).
34. D. Byrom, *Trends Biotechnol.* **5**, 246–250 (1987).
35. Y. Poirer, C. Nawrath, and C. Somerville, *Biotechnology* **13**, 143–150 (1995).
36. S.Y. Lee, *Biotechnol. Bioeng.* **49**, 1–14 (1996).
37. T. Suzuki, T. Yamane, and S. Shimizu, *Appl. Microbiol. Biotechnol.* **23**, 322–329 (1986).
38. W. Babel, *FEMS Microbiol. Rev.* **103**, 141–148 (1992).
39. S. Fidler and D. Dennis, *FEMS Microbiol. Rev.* **103**, 231–236 (1992).
40. H. Dalton, in J.C. Murrell and H. Dalton eds., *Methane and Methanol Utilizers*, Plenum Press, New York, 1992, pp. 85–114.
41. B.J. Wallar and J.D. Lipscomb, *Chem. Rev.* **96**, 2625–2657 (1996).
42. J.C. Murrell, *FEMS Microbiol. Rev.* **88**, 233–248 (1992).
43. D. Jahng, and T.K. Wood, *Appl. Environ. Microbiol.* **60**, 2473–2482 (1994).
44. H. Dalton, *Adv. Appl. Microbiol.* **26**, 71–87 (1980).
45. K.J.A. Burrows, A. Cornish, D. Scott, and I.J. Higgins, *J. Gen. Microbiol.* **130**, 335–392 (1984).
46. C.T. Hou, in C.T. Hou ed., *Methylotrophs: Microbiology, Biochemistry, and Genetics*, CRC Press, Boca Raton, Fla., 1984, pp. 145–166.
47. S.H. Stanley, A.O.L. Richards, M. Suzuki, and H. Dalton, *Biocatalysis* **6**, 177–190 (1992).
48. D.J. Arp, *Curr. Opin. Biotechnol.* **6**, 352–358 (1995).
49. B.G. Fox, J.G. Borneman, L.P. Wackett, and J.D. Lipscomb, *Biochemistry* **29**, 6419–6427 (1990).
50. R. Oldenhuis and D.B. Janssen, in J.C. Murrell and D.P. Kelly eds., *Microbial Growth on C_1 Compounds*, Intercept Press, Andover, U.K., 1993, pp. 121–133.
51. H.L. Chang and L. Alvarez-Cohen, *Biotechnol. Bioeng.* **45**, 440–449 (1995).
52. H. Martin and J.C. Murrell, *FEMS Microbiol. Lett.* **127**, 243–248 (1995).
53. S.D. Prior and H. Dalton, *J. Gen. Microbiol.* **131**, 155–163 (1985).
54. L. Alvarez-Cohen, in J.C. Murrell and D.P. Kelly eds., *Microbial Growth on C_1 Compounds*, Intercept Press, Andover, U.K., 1993, pp. 337–350.
55. J.P. Bowman, L. Jimenez, I. Rosario, T.C. Hazen, and G.S. Sayler, *Appl. Environ. Microbiol.* **59**, 2380–2387 (1993).
56. I. Karube, T. Okada, and S. Suzuki, *Anal. Chim. Acta.* **135**, 61–67 (1982).
57. L.I. Sly, L.J. Bryant, J.M. Cox, and J.M. Anderson, *Appl. Microbiol. Biotechnol.* **39**, 400–404 (1993).
58. I.J. Higgins, W.J. Aston, D.J. Best, A.P.F. Turner, S.G. Jezequel, and H.A.O. Hill, in R.L. Crawford and R.S. Hanson eds., *Microbial Growth on C_1 Compounds: Proceedings of the 4th International Symposium*, ASM, Washington, D.C., 1984, pp. 297–305.
59. C.W. Bamforth and J.R. Quayle, *Biochem. J.* **169**, 677–686 (1978).
60. A. Geerlof, J. Vantol, J.A. Jongejan, and J.A. Duine, *Biosci. Biotechnol. Biochem.* **58**, 1028–1036 (1994).
61. T. Urakami, K. Yashima, H. Kobayashi, A. Yoshida, and C. Itoyoshida, *Appl. Environ. Microbiol.* **58**, 3970–3976 (1992).
62. C. Verduyn, J.P. van Dijken, and W.A. Scheffers, *Biotechnol. Bioeng.* **25**, 1049–1055 (1983).
63. Y. Sakai and Y. Tani, *Agric. Biol. Chem.* **50**, 2615–2620 (1986).
64. T.R. Hopkins and F. Muller, in H.W. van Verseveld and J.A. Duine eds., *Microbial Growth on C_1 Compounds: Proceedings of the 5th International Symposium*, Martinus Nijhoff, Dordrecht, 1987, pp. 150–157.
65. H. Schutte, J. Flossdorf, H. Sahm, and M.-R. Kula, *Eur. J. Biochem.* **62**, 151–160 (1976).
66. R. Wichmann, C. Wandrey, A. Buckmann, and M.-R. Kula, *Biotechnol. Bioeng.* **23**, 2789–2802 (1981).
67. R. Couderc and J. Baratti, *Biotechnol. Bioeng.* **22**, 1155–1173 (1980).
68. K.N. Faber, W. Harder, G. Ab, and M. Veenhuis, *Yeast* **11**, 1331–1344 (1995).
69. J.M. Cregg, T.S. Vedvick, and W.C. Raschke, *Biotechnology* **11**, 905–910 (1993).
70. C.A. Scorer, J.J. Clare, W.R. McCombie, M.A. Romanos, and K. Sreekrishna, *Biotechnology* **12**, 181–184 (1994).
71. A.R. Cross and C. Anthony, *Biochem. J.* **192**, 429–439 (1980).
72. B.D. Yoon, M. Ueno, and Y. Tani, *J. Ferment. Technol.* **65**, 629–634 (1987).
73. K. Inoue, S. Kageyama, K. Miki, T. Morinaga, Y. Kamagata, K. Nakamura, and E. Mikami, *J. Ferment. Bioeng.* **73**, 76–78 (1992).

MICROALGAE, MASS CULTURE METHODS

E. Molina Grima
University of Almería
Almería, Spain

KEY WORDS

Fluid dynamics
Gas–liquid mass transfer
Light distribution
Microalgae mass culture
Outdoor production
Photosynthetic efficiency
Tubular reactor

OUTLINE

INTRODUCTION

Microalgae have been cultured commercially for about 50 years for applications in secondary wastewater treatment (1) and use as animal feeds (2,3) as fertilizers (4), in production of chemicals (5,6), and in secondary metabolites with pharmaceutical potential (7), as well as other purposes. The early interest in using microalgae as a significant source of protein has waned because of the relatively low biomass productivities. Now the focus of algal mass cultures is on relatively specialized niche applications; nevertheless, progress remains slow relative to biotechnological use of bacteria, yeasts, and animal cells. Cost-effective culture of microalgae requires substantially improved biomass productivity, better utilization of the available light, and more efficient use of carbon dioxide, the principal carbon source.

Open outdoor mass culture systems, such as ponds and raceways, typically operate at low cell densities (0.01 to 0.06% w/v); consequently harvesting is expensive and economic returns are low. In addition, species control is difficult in open systems, and much of the incident radiation is wasted. The low productivity of open systems has prompted the development of enclosed photobioreactors (PBRs) (i.e., transparent tubes or containers in which the culture may be circulated by various devices). Enclosed PBRs allow a more controlled culture environment, including improved temperature control. Successful design of PBRs relies on attaining a proper balance of engineering and biological factors that are interdependent and depend also on the species being cultured. Of course, from a commercial perspective, reliability and the costs of construction and operation are also important. The physical rate processes of concern are the light delivery, the gas–liquid mass transfer, temperature control, fluid dynamics, and mixing. The biological rate processes of concern are photosynthesis, growth, and production of metabolites.

This article details the major types of bioreactors used in mass culture of microalgae. The basic requirements for successful outdoor culture are noted, and the physics and biology of mass cultivation are discussed.

BIOREACTORS FOR MICROALGAE MASS CULTURE

Three microalgae—*Chlorella, Dunaliella*, and *Spirulina*—are currently produced commercially on a relatively large scale (approximately 1,000 tons each per annum). The current commercial production systems for these species are mostly open raceway ponds. Many commercially useful algae are difficult to cultivate in open systems because they require a mild culture environment that is highly susceptible to contamination by other algae, bacteria, protozoa, and fungi. The inability to cultivate desirable algal species that contain valuable products (e.g., astaxanthin, aquaculture feeds, polysaccharides, and polyunsaturated fatty acids) in open ponds has increased interest in closed systems. The closed systems in use include (*1*) bag culture, which is widely used for culture of algae for aquaculture; (*2*) alveolar panels and other flat plate reactors of various designs (8–10); (*3*) stirred tank reactors with internal illumination (11,12); (*4*) airlift reactors (13,14); and (*5*) tubular reactors. Tubular reactors may be arranged in a horizontal position relative to the land or they may be inclined (15–22). Some of the major types of photobioreactors are discussed in the following.

Open Raceways

Open raceways, or shallow mixed ponds, for microalgal production were introduced in the 1950s and early 1960s by Oswald and coworkers (23). The principles of design and construction of shallow paddle-stirred raceways for microalgal culture have been reviewed by Dodd (24) and Oswald (25). The pond is set in a meandering configuration. Various designs are used to promote flow and low shear mixing. A simplified open raceway is depicted in Figure 1. Selection of a suitable bottom lining and construction of walls are important to the success of the open pond. The lining may be made of concrete or sheets of plastic and rubber material. Sheets of polyvinyl chloride 1–2 mm thick, are commonly used. The size of commercial ponds varies from 1,000 to 5,000 m^2; one or two paddle wheels per pond are used for stirring. A larger wheel (ca. 2 m in diameter) revolving slowly (e.g., 10 rpm) is preferable to a smaller, faster wheel, which could produce shear damage and foam.

Supply of Carbon Dioxide. The method of supplying carbon dioxide is an important aspect of the raceway design, particularly for species that grow near neutral pH. Several systems have been developed to efficiently supply CO_2 to shallow suspensions. In most cases the gas is supplied as fine bubbles. Because of the shallow depth of culture, the residence time of the bubbles is not sufficient for complete absorption of the CO_2; hence, some gas is lost to the atmosphere. To avoid this, Märkl and Mather (26) used a

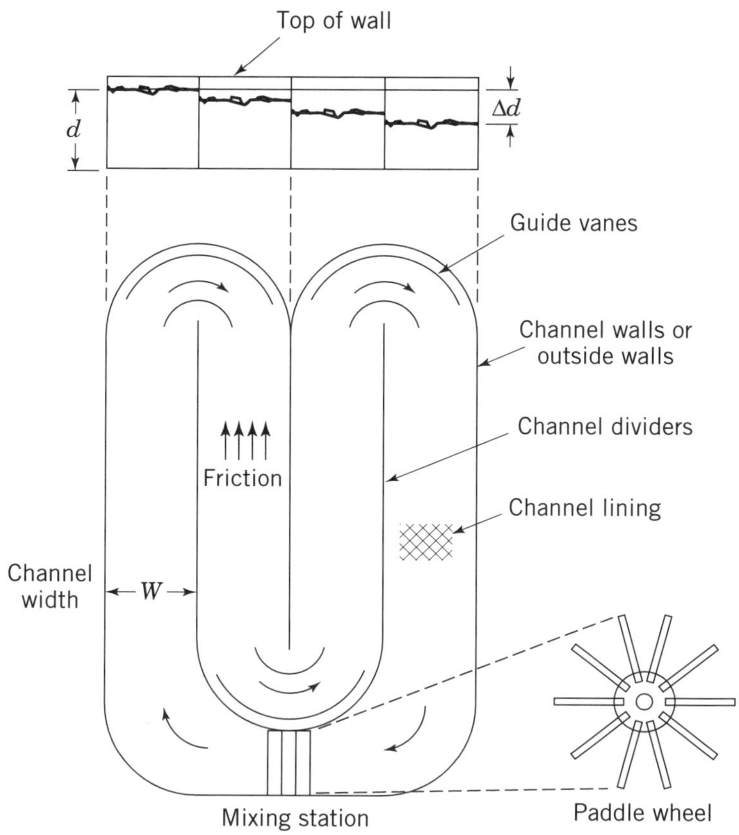

Figure 1. Open raceway with a paddle wheel mixing device. The channel length is the distance traveled by the culture from the discharge side of the paddle wheel to its upstream side. *Source*: Adapted from Oswald (25) and Dodd (24), with permission.

pump to push the algal suspension through a nozzle to a chamber in which the CO_2 was added. High CO_2 transfer rates were obtained. The jet produced an additional flow in the pond. The most effective method currently available for transferring CO_2 to algal cultures is countercurrent carbonation (25), in which the gas is injected as minute bubbles into a column of downcoming culture. The culture velocity is adjusted to that of the small bubbles of CO_2 rising against the current. Using this technique, Laws et al. (27) reported a 70% utilization of the supplied CO_2 compared with 13 to 20% if conventional bubbling was used (28).

Mixing. Mixing is necessary to prevent settling of cells and to avoid thermal stratification of the culture. Properly designed paddle wheels are by far the most efficient and durable pond mixers. The engineering of a raceway equipped with paddle wheels has been comprehensively described by Oswald (25). As shown in Figure 1, the depth, d, of culture in a channel of finite width, w, varies along the channel. The mean channel velocity, v(m s^{-1}) is given by

$$v = \frac{1}{n} R^{2/3} \cdot s^{1/2} \tag{1}$$

where n is the Manning friction factor (s m$^{-1/3}$), R is the hydraulic radius (m), and s is the rate of loss of energy in the channel per unit length. The channel length, L, that corresponds to the assumed change in depth for a given friction factor, hydraulic radius, and velocity (v) is given as (25)

$$L = \frac{\Delta d(dw/(w + 2d))^{4/3}}{v^2 \cdot n^2} \tag{2}$$

The value of n varies according to the relative roughness of the channel. Experimentally determined n values in algae growth channels vary from 0.008 to 0.030 (25). The lower value is for smooth, plastic-lined channels, and the higher one is for relatively rough earthen channels. The channel velocity, v, affects the paddle wheel's power requirements; hence,

$$P = \frac{Q \cdot \rho \cdot \Delta d}{\eta} \tag{3}$$

where P is the power (kW), Q is the volumetric flow rate of culture ($Q = v \cdot d \cdot w$) (m^3 s^{-1}), ρ is the density of culture (kg m^{-3}), Δd is the change in depth (m), and η is the efficiency of the paddle wheel. Because $\Delta d = s \cdot L$ and s is a function of v^2, P increases as the cube of velocity; therefore, velocity should be minimized whenever energy is a major cost factor. Typically, the flow rates range between 15 and 30 cm s^{-1}. Velocities greater than 30 cm s^{-1} result in large values of Δd in long channels and may require high channel walls and higher divider walls. Note that for a finite value of the channel width, w, the permissible mixing channel length, L, and thus the mixable area, $a = L \cdot w$, is strongly dependent on depth, d.

Contamination and temperature are difficult to control in open pond systems. Also, the culture depth cannot be reduced below 12 to 15 cm or a severe reduction of flow and turbulence would occur. Hence, areal volumes tend to be large, about 120 to 150 L/m^2 (29), requiring maintenance of a rather low cell concentration (around 500 mg/L). A low cell concentration in turn increases the cost of harvesting the biomass, and ponds are expensive to maintain. These problems could, in principle, be minimized by photobioreactor designs that employ shallower cultures, maximize cell densities, allow better process control, and minimize contamination. Closed photobioreactors achieve these objectives, as described in the next section.

Enclosed Photobioreactors

PBRs can be defined as algal culture systems that do not allow direct contact of the culture with the atmosphere. PBRs generally may be grouped into two classes: tubular and flat plate. These groups can be further classified into several categories, depending on orientation of tubes, the type of gas exchange system, and the arrangement of the individual growth units. The materials of construction provide additional variations. For example, in the case of flat plate reactors, most work has been carried out with alveolar systems in which the flat plate is divided into many individual cells. Another example is the use of flexible plastic bags. A major factor in the design of these systems is the depth of the culture, which ultimately determines the biomass concentration and productivity. Lee (30) and Borowitzka (31) provided comprehensive summaries of the state of the art of the closed PBRs, none of which are in widespread commercial operation.

Conventional mass culture ponds covered with inexpensive plastic sheets have been used in several algal mass culture projects, but they provide few benefits beyond an increased temperature during early spring or fall in temperate climates. Also, their design and operation are essentially identical to those of the conventional open pond systems described earlier. Here, only tubular reactors capable of operation outdoors with natural sunlight and with high surface-to-volume ratios (>10 m^{-1}) are considered.

In vertical column reactors, bubble columns or airlifts, aeration, and agitation are provided by injection of CO_2-enriched air through a sparger at the bottom of the column. In practice, a vertical column can extend for only a few meters or an expensive structure is needed for support. James and Al-Khars (13) studied the growth and productivity of *Chlorella* and *Nannochlropsis* in a vertical airlift photobioreactor. Contreras Gómez (14) showed that the average annual volumetric productivity of a draught tube airlift was about 1 g L^{-1} per day in culture of *Phaeodactylum tricornutum*. A major advantage of tower reactors is that oxygen does not accumulate in the growth medium because the residence time of the gas phase is quite small in a relatively short reactor. Also, tower bioreactors are potentially easy to keep sterile compared to other tubular systems. A major drawback of any tubular system positioned vertically is that in the summer the reactor is always at a large angle to the sun's rays; hence, the amount of solar energy absorbed during the midday hours is lower than for horizontally placed tubular devices.

The earliest described outdoor tubular reactor was made of 3-cm-diameter glass tubes arranged in a serpentine configuration (32). The tubular loop was immersed in water. The culture was recirculated in the system by a motor-driven pump. The flow velocity was 15 to 30 cm s^{-1}. The culture circulated through a bubble column that was injected with 5% CO_2 at a rate of 1–1.5 cm^3 min^{-1}. The total culture volume was 40 L. The more recent systems are essentially identical to that first design. Thus, Chaumont et al. in France (33) employed tubes placed parallel to ground and Pirt et al. in England (15) used a vertically placed loop. Both systems used pumps to circulate the culture. Because fragile cells can be sensitive to mechanical pumping, the authors recommended the use of airlift devices for both culture recirculation and CO_2 supply.

Although other designs of tubular reactors have been proposed—including the two-plane tubular reactor (16), the near-horizontal tubular reactor (21), the helical bubble reactor (34), the α-tubular reactor (35), and the parallel flow tubular photobioreactor (18)—they are essentially the same as the ones used by Pirt et al. (15) and Chaumont et al. (33).

Figure 2 shows a fully instrumented tubular photobioreactor similar to those described by Chaumont and coworkers. The system shown consists of a vertical external-loop airlift pump that drives the culture fluid through the horizontal tubular solar receiver. The airlift section (riser, downcomer, and degasser) has a height of 3.5 m. The gas-injected riser and the downcomer sections (Fig. 2) are extensions of the solar receiver tube. The solar receiver is made of transparent Plexiglas® tubes (0.05-m internal diameter, 0.005-m wall thickness) joined into a loop configuration by Plexiglas® joints to obtain a total horizontal length of 98.8 m. The solar receiver is submerged ($\sim$0.05 m) in a shallow pool of water that is maintained at 21 $\pm$ 2 °C by cooling with a heat pump when needed. The bottom and inside walls of the pool are painted white to improve reflectance. The surface area of the pool is 21.4 m^2. The total culture volume in the bioreactor is 0.200 m^3.

The exhaust gas from the top of the degasser zone is analyzed for oxygen (paramagnetic analyzer) and carbon dioxide (infrared analyzer). Filter sterilized culture medium is fed in at the degasser zone. A port located on the side of the degasser section is used to continuously harvest the culture (Fig. 2). The temperature, the dissolved oxygen, and the pH measurements are made via sensors located in the degassing zone, in addition to a dissolved oxygen sensor located at the end of the external loop. The sensors are connected to a control unit and a computer for data acquisition.

The air supply point is located at the entrance of the riser, whereas the CO_2 is injected at the entrance of the solar receiver in response to a signal from the pH sensor. The photosynthetically active radiation (PAR) is determined at the ground level and inside the photobioreactor. In the previously described microalgae mass culture facility, the irradiance inside the thermostatic pond was higher than the incident radiation on the surface of the water because of the light reflected by the walls and the bottom of the pool. This phenomenon of radiation enhancement is quantified in terms of the albedo of a system.

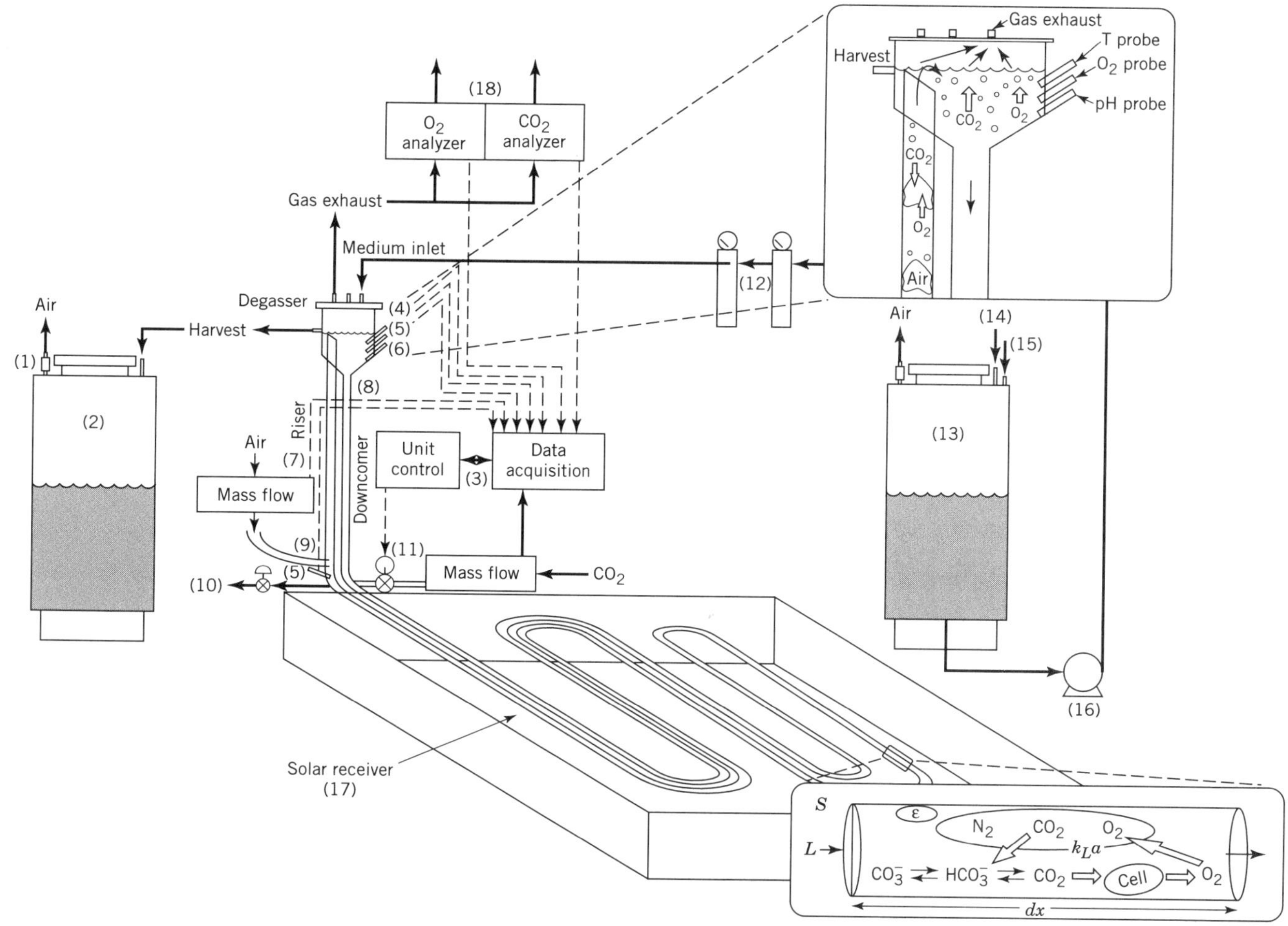

Figure 2. Tubular photobioreactor: (1) air filters, (2) harvest tank, (3) control unit, (4) temperature sensor, (5) dissolved oxygen probe, (6) pH probe, (7) riser, (8) downcomer, (9) air injection, (10) sampler, (11) CO_2 injection, (12) medium sterilization filters, (13) fresh medium, (14) nutrient inlet, (15) seawater inlet, (16) pump, (17) thermostated water pool, (18) gas composition analyzer.

For efficient and reliable large-scale culture of microalgae in PBRs, several operational, environmental, and design variables need to be assessed. A summary of some of the main variables and their impact on the various aspects of a reactor's performance are depicted in Figure 3. Of these, an important design criterion is the efficiency of light utilization, which in turn is related to the productivity of the system. Also important are the costs of construction and operation (36).

As indicated in Figure 3, the on-line data acquisition of dissolved oxygen, pH, and temperature inside the culture allows for rapid characterization of the physiological state of the cells. The measurement of CO_2 and O_2 molar flows at the gas inlet and outlet enables estimation of the mass transfer capability of the system and efficiency of CO_2 utilization. The measurement of the biomass concentration and the product content allows the estimation of the biomass and product productivities. Finally, the biomass productivity, P_b, and the photon flux absorbed per unit of volume, F_{vol}, may allow the assessment of the efficiency of a system in harvesting energy and converting it into biomass.

PHYSICAL FACTORS THAT INFLUENCE BIOMASS PRODUCTIVITY

Biomass productivity in any culture system depends on the degree to which the culture conditions match the requirements of the selected strain. Because mineral nutrient limitation is easily avoided in microalgal mass culture (29,37), light availability inside the photobioreactor and temperature are the main factors that determine productivity. Once the temperature is suitably controlled, light availability becomes the only limiting factor (22,38).

The interrelations among the various factors that influence productivity in outdoor conditions are shown in Figure 4. Productivity is determined by the growth rate, which, for fixed fluid dynamics and temperature, is a function of the light profile within the reactor and the light regime to which the cells are subject. The average irradiance (I_{av}) within the reactor depends on the external irradiance (I_0) on the surface of the reactor and its geometry (22,39,40). I_0 is a function of time, the geographic location, and environmental factors (41). Although temperature is relatively easily controlled, the available solar radiation at

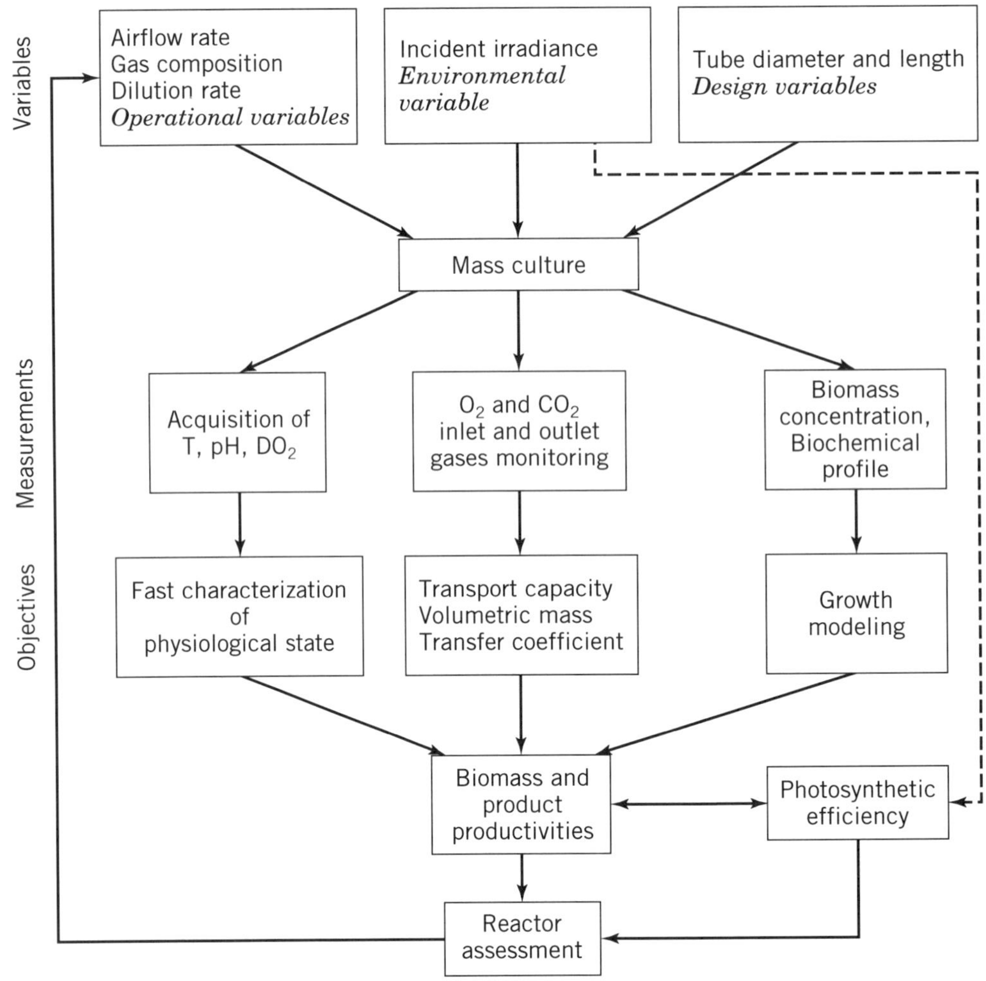

Figure 3. Relationships among bioreactor design and operational variables.

any location cannot be controlled. The self-shading effect of the cells also affects the radiation profile inside the culture (40). In addition, the fluid dynamics inside the reactor also influence the light regime seen by the cells because, depending on the nature and intensity of fluid movement, the cells spend different amounts of time in zones of different illumination (42–44). Fluid mechanics also affect the transport behavior and availability of certain nutrients. In an optimal system where no other factors limit, light availability determines the rate of photosynthesis and productivity. However, excessive light can be harmful and is known to produce a photoinhibitory response (45). The issues relating to light availability are discussed in the following sections for photobioreactors located outdoors.

Light Availability inside a Photobioreactor

For commercial-scale culture, sunlight is usually the preferred source of light because artificial light is too expensive for all but the most valuable products. The effects of natural outdoor light on cultures are not entirely clear because most studies on characterizing the photosynthetic response to light (e.g., 46–48) were conducted under carefully controlled indoor laboratory conditions. The irradiance levels used in such studies were significantly lower than the peak solar irradiance that typically occurs outdoors. Cultures grown outdoors experience large variations in light availability: Within a few hours the irradiance increases from 0 at dawn to more than 2,000 μE $m^{-2}s^{-1}$ at noon (i.e., a rate several times above saturation); within a few additional hours, the irradiance declines to 0 again as evening sets in. Because of this rapid change, the culture may go from being light limited in the morning to being photoinhibited at noon and light limited once again during the late afternoon and evening.

Information concerning light distribution in microalgal mass culture systems is scant, but recently a comprehensive model for light distribution and average irradiance inside a horizontal tubular outdoor photobioreactor has been proposed (40). That model has been further generalized for use with other configurations of photobioreactors (49). The mathematical model determines the distance traveled by an incident ray of light to any point inside the culture and estimates the local irradiance by taking into account the light attenuation due to biomass. The incident total radiation is divided into direct and disperse radiation. The former is characterized by having a direction defined as a function of the position of the sun at any given mo-

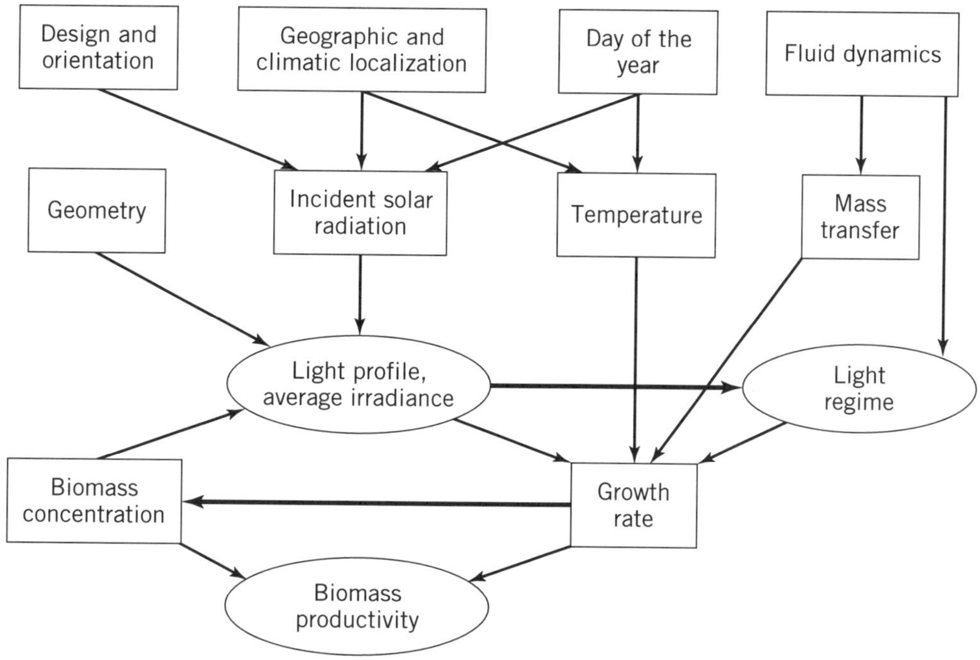

Figure 4. Relationships among parameters that influence biomass productivity in outdoor culture.

ment, and the latter, being reflected radiation, has no specific direction. The model is mathematically complex. For example, the angle of incidence of direct radiation on the photobioreactor surface is a function of five variables: day of year, solar hour, geographic latitude, surface slope, and surface azimuth angle (i.e., the deviation of the projection on a horizontal plane of the normal to the surface from the local meridian, with zero due south, east negative, and west positive) (50,51). Detailed information for estimating the angle of the incidence of the solar radiation, the irradiance impinging on the reactor surface, and the path traveled by the ray of light inside both horizontally and vertically placed tubular reactors has been reported (40,49,52).

Local and Average Irradiance inside the Culture. In a well-mixed microalgal mass culture system, light attenuation by biomass gives rise to a heterogeneous illumination profile inside the culture bulk for which mathematical evaluation is essential to estimate the average irradiance on which the growth of the microalgae depends (53). This attenuation is usually represented by Lambert–Beer's law. The incident radiation on the culture surface may be considered as consisting of the sum of two contributions: direct radiation, which has a direction defined by the position of the sun, and the disperse radiation. The latter is the sum of the diffuse radiation and that reflected by the surface surrounding the reactor.

The local direct irradiance inside the culture may be estimated using Lambert–Beer's law, as described elsewhere (40). The disperse radiation coming from all spatial directions can be determined by an integration over all the angles of incidence (40). Thus, it is possible to determine the light profiles and average irradiance inside the culture. Figure 5 shows the modeled local irradiance at two radial positions in a horizontal or vertical arrangement at 8 and 12 solar hours. As expected, as solar hour changes, curves of local irradiance at the same internal ratio adjust to the sun's position. In the morning (8 solar hours) in the vertical reactor, the side facing the sun had much higher direct irradiance values than the opposite side; this difference was lower in the horizontal reactor. However, at noon, the irradiance distribution in the vertical reactor, unlike the horizontal tube, was practically homogeneous. This profile results because during midday, the contribution of disperse irradiance to the total irradiance is very large in the vertical arrangement. Therefore, the level of irradiance for each internal radius and for any hour of the day is higher in the horizontally placed reactor tube than in the vertically placed reactor tube. The irradiance distribution inside the culture and the patterns of cell movement inside the photobioreactors are also different.

The Light Saturation Constant and the Photoinhibition Phenomenon

Algal growth is related to light intensity. According to Goldman (54), the generalized curve relating algal growth to light intensity has the shape shown in Figure 6. The curve has four main features: (*1*) at some low growth rate, the growth is balanced by decay, and the net growth rate is zero; (*2*) the initial slope of the curve represents the maximum efficiency of growth in response to light; (*3*) there is a saturating light intensity at I_s for which $\mu = \mu_{max}$; (*4*) at some light intensity $I_p > I_s$, the growth is inhibited and $\mu < \mu_{max}$.

Because most algal species become light saturated at a fraction of the peak solar light intensity, much potentially useful solar energy is essentially wasted. The light satu-

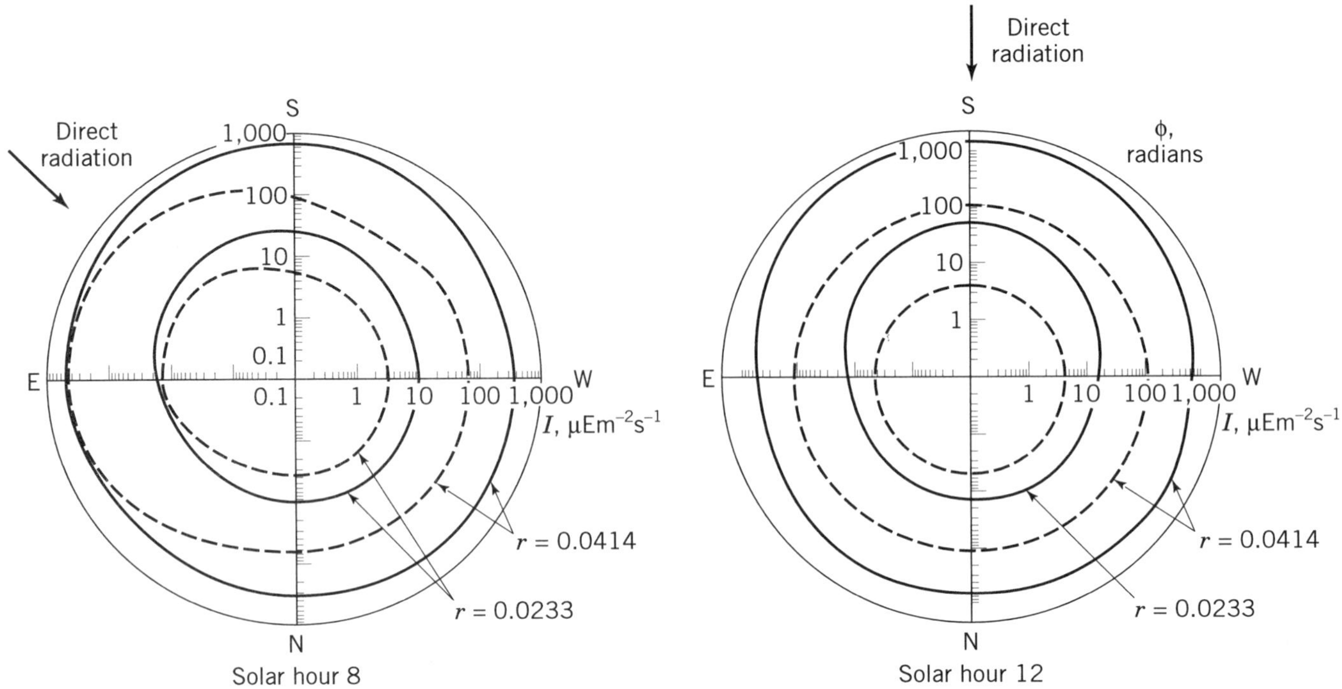

Figure 5. Calculated radial irradiance profiles in the cross section of a bubble column (dashed lines) and a photobioreactor placed parallel to the ground (solid lines) at 8 and 12 solar hours. In both cases the parameters were 0.1-m tube diameter; 1.9 g L^{-1} biomass concentration of *Phaeodactylum tricornutum*; dilution rate of 0.033 h^{-1}; and pigment content, $X_p = 2.21\%$.

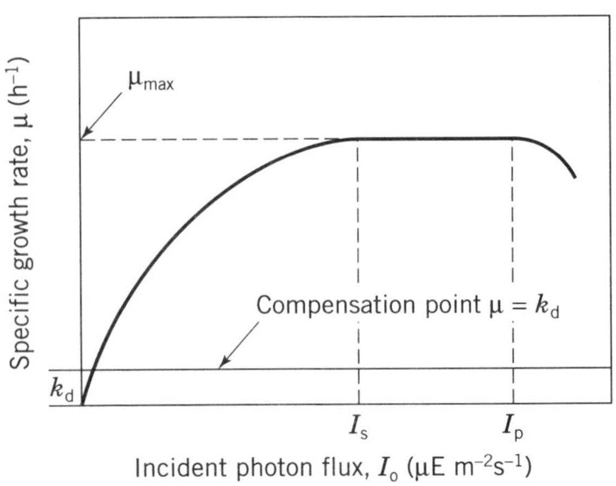

Figure 6. Dependence of the growth rate (μ) on the incident photon flux (I_0). *Source*: From Goldman (55), with permission.

ration constant is a measure of the intrinsic capacity of cells to utilize light energy. A high saturation constant is advantageous because it increases the efficiency of light use and reduces photoinhibition.

Photoinhibition—a decline in the photosynthetic capacity due to light intensities in excess of those required to saturate the photosynthesis (29)—has been studied often in algae. Shading the culture with shad nets is commonly used to protect against inhibitory levels of light during midday hours. The kinetics of photoinhibition and its recovery in *Porphyridium cruentum* have been studied (56). A linear relationship was found between the specific rate of photoinhibition and the specific light absorption rate. The temperature was reported to have a pronounced modifying effect on photoinhibition. At 15 °C, no damage was detected at 2,300 $\mu E\ m^{-2}s^{-1}$ light intensity, even after 45 min of illumination. In contrast, at 35 °C, 84% inhibition of the photosynthetic activity was observed within 10 min of exposure. *P. cruentum* cells recovered readily when transferred to low light (90 $\mu E\ m^{-2}s^{-1}$) or darkness; the specific rate of recovery was independent of the light intensity to which the cells had been exposed during the inhibitory treatment.

The existence of photoinhibition in continuous outdoor cultures of *Phaeodactylum tricornutum* has also been documented (52). Figure 7 shows how the average irradiance, I_{av}, inside the culture varies with dilution rate, D, regardless of the tube diameter, indicating that the cultures adapt mainly to the average irradiance. At $D = 0.025\ h^{-1}$, the I_{av} increased only slightly with increasing external irradiance, I_0, on the reactor surface. The mean value of I_{av} for the whole range was $84 \pm 9\ \mu E\ m^2\ s^{-1}$. With a smaller tube diameter or higher external irradiance, the biomass concentration increased as expected for self-shaded light-limited growth. However, the slight increase in I_{av} with increasing I_0 points to some photoinhibition by external irradiance, although this is not very extensive due to the high biomass concentrations. At $D = 0.040\ h^{-1}$, I_{av} values are higher ($163 \pm 20\ \mu E\ m^{-2}\ s^{-1}$) than at $D = 0.025\ h^{-1}$ because of the lower steady-state biomass concentrations. The greater light availability gives rise to a higher specific growth rate and photoinhibition because of the greater in-

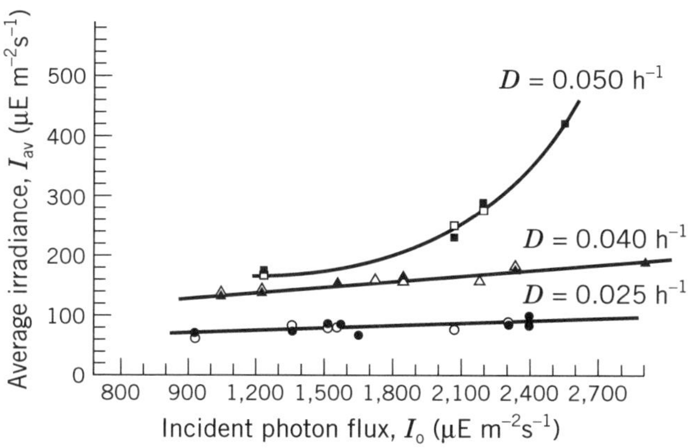

Figure 7. Variation of the average irradiance (I_{av}) inside a tubular photobioreactor during culture of *Phaeodactylum tricornutum* UTEX 640. Data are shown as functions of the mean daily photosynthetic irradiance (I_0) on the reactor surface for various dilution rates and two different tube diameters. Open symbols are for 3-cm tube diameter; filled symbols are for 6-cm tube diameter.

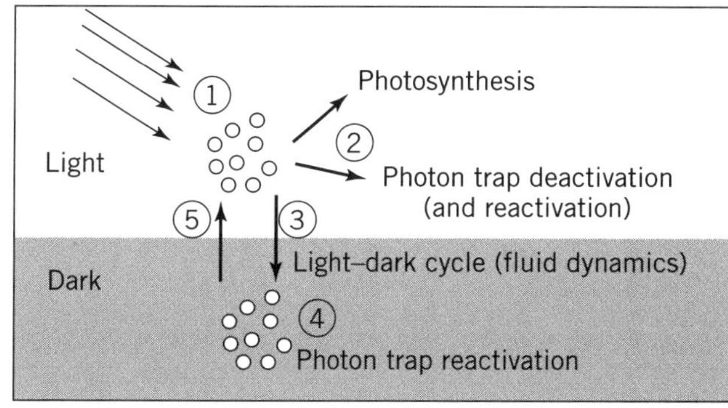

Figure 8. A simplified mechanistic representation of photosynthesis. (1) photon capture → biomass synthesis; (2) photon capture → photon trap deactivation (reactivation); (3) cell transfer to the dark; (4) photon trap reactivation; (5) cells return to the illuminated zone. Photon trap recovery is zero order with respect to unrecovered photon trap, and the recovery time is species dependent. *Source*: From Merchuk et al. (59), with permission.

crease in I_{av} with I_0. At 0.050 h^{-1} dilution rate, the exponential increase in I_{av} with I_0 caused by the low steady-state biomass concentration underlines the existence of photoinhibition; hence, the one-to-one relationship between μ and I_{av} no longer applies.

Photoinhibition is caused by oversaturation of photosystem II, which damages the D1 protein that carries the binding site of the electron carrier (57,58). This photoinhibition effect is quite distinct from that of temperature rise that occurs in uncontrolled systems as a function of the photon flux density (PFD). Whereas a single cell of microalgae cannot be simultaneously photolimited and photoinhibited, in bioreactors photolimited and photoinhibited cell populations may coexist because of variations in photointensity in different zones. At low light intensities the few damaged D1 protein molecules are replaced rapidly and the net damage to the photosynthetic apparatus is negligible. In this situation, a dark period reduces growth rate (photosynthesis) because fewer photons are captured but no gain is obtained from the dark time. In contrast, under conditions of intense illumination, part of the light energy impairs the photosynthetic apparatus. Repair and damage proceed simultaneously, and the observed growth is the sum of the two processes. If a dark period is introduced in this situation, the duration of the photosynthetic period declines, but the damaging period is also reduced; the photon trap repair continues during the dark time. Consequently, during the next light period, a substantially rejuvenated photon trap compensates for the loss in photosynthetic time. Under these circumstances, alternating light and dark periods do not reduce growth, which may in fact be slightly enhanced. Nevertheless, the length of the dark period is important: lengthening the light period beyond an optimal value will produce loss of growth. The optimal or critical dark period is not a fixed quantity; instead it depends on the PFD of the previous light period and the fluid residence time in zones of different irradiance. Figure 8 outlines a mechanism of photosynthesis that takes into account the steps noted in the foregoing (59). Clearly, therefore, the principal problem of designing or choosing a photobioreactor is ensuring, for any species with preestablished photosynthetic characteristics, that the largest possible fraction of cells experiences optimal exposure to light in the largest possible reactor volume.

Biomass Productivity

Many studies on the influence of environmental variables on biomass productivity have been carried out in laboratories (20,32,60,61), but few have been performed under outdoor conditions. Some models have been proposed specifically for outdoor algal ponds (41,62–64). Thus, Incropera and Thomas (41) present a model for determining the mean irradiance in an open pond as a function of climate and geographic location parameters. Starting from estimated irradiance values, a six-parameter growth model is proposed (41). The model does not consider photoinhibition and predicts maximum productivities of about 34 g m^{-2} per day. A five-parameter mathematical model for biomass productivity in outdoor systems has been proposed by Sukenik et al. (62,63). According to this model, the cells adapt to mean irradiance inside the culture; however, photoinhibition is not taken into account, and the results are not valid at low biomass concentration. The model yields productivities of 10.2 to 27.2 g m^{-2} per day. By using mean temperature and irradiance data, Guterman et al. (64) propose a growth model for open systems. The model estimates the variation of culture parameters such as dissolved oxygen, temperature, pH, and biomass concentration through empirical linear relationships and does not consider photoinhibition. The estimated maximum productivity was 38 g m^{-2} per day at irradiances of 2,500 μE m^{-2}s^{-1}. According to these models the maximum productivities are reached at low biomass concentration, when the mutual shading effect is negligible and the culture may be considered light saturated. The estimated productivities from these models have not been experimentally verified. In fact, such cultures cannot be carried out, because photoinhibition in the middle of the day decreases growth rate and may even lead to culture washout.

A macromodel for estimating the yearlong biomass productivity of microalgal cultures in outdoor tubular photobioreactors has been developed by Acién Fernández et al. (52). In this model the solar irradiance impinging on the reactor surface I_0 is initially determined as a function of the day of the year and the location. Then, by taking into account the geometry of the system, the average irradiance to which the cells are exposed inside the culture, I_{av}, is determined. Finally, the specific growth rate, μ, is correlated with I_{av} as follows:

$$\mu = \frac{\mu_{max} \cdot I_{av}^{(n_2+n_3/I_0)}}{\left(I_k'\left(1 + \left(\frac{I_0}{K_I}\right)^{n_1}\right)\right)^{(n_2+n_3/I_0)} + I_{av}^{(n_2+n_3/I_0)}} \quad (4)$$

where I_0 is the irradiance on the surface of the reactor. The characteristic parameters of this equation were obtained by fitting the experimental results to the equation. The parameter values were $\mu_{max} = 0.063\ h^{-1}$, $I_k' = 94.3\ \mu E\ m^{-2}\ s^{-1}$, $K_I = 3426\ \mu E\ m^{-2}\ s^{-1}$, $n_1 = 3.04$, $n_2 = 1.209$, and $n_3 = 514.6$. Equation 4 has been used to simulate the biomass productivity in a continuous outdoor culture of *Phaeodactylum tricornutum* (52). The measured data agreed with the simulations within 20%. The simulated biomass productivity over the annual cycle is shown in Figure 9. The approach described can in principle be extended to other strains. If photoinhibition were not taken into account as in earlier models (41,62–64) the experimental results could be fit to a hyperbolic model that considers only photolimitation (61):

$$\mu = \frac{\mu_{max} I_{av}^n}{I_k^n + I_{av}^n} \quad (5)$$

however, the scatter of the data would significantly exceed ±20%. In equation 5 I_k is the affinity cells have for light and n is a fitting parameter of the model.

In view of the observations discussed, several conclusions can be made regarding biomass production by microalgal mass cultures: (*1*) the optimum dilution rate varies over the year as a function of incident radiation and the diameter of the bioreactor tube; (*2*) the larger the tube diameter, the more extensive the photoinhibition effect, although the simultaneous existence of photolimitation and photoinhibition permits growth to be maintained and a quasi steady state is still attained.

Fluid Dynamics and Mixing

To achieve high productivity, algal cultures need to be mixed. The method of mixing (turbulence) is critical because some algae are quite shear sensitive (65,66). When light is the limiting factor for growth, stirring has been suggested as the most practical means by which to attempt to distribute solar energy evenly to all cells in the culture. The stirring is associated with a flashing light effect. The duration for which the cells are exposed to the light and dark zones of the reactor is a function of agitation intensity. The flashing light effect has been intensively investigated by Laws et al. (67). The flashing light effect or agitation has been associated with increased photosynthetic efficiencies. Laws and coworkers (67) introduced a simple technique for utilizing the flashing light effect in shallow but optically dense outdoor algal cultures by inserting an array of small foil segments across the width of a continuous

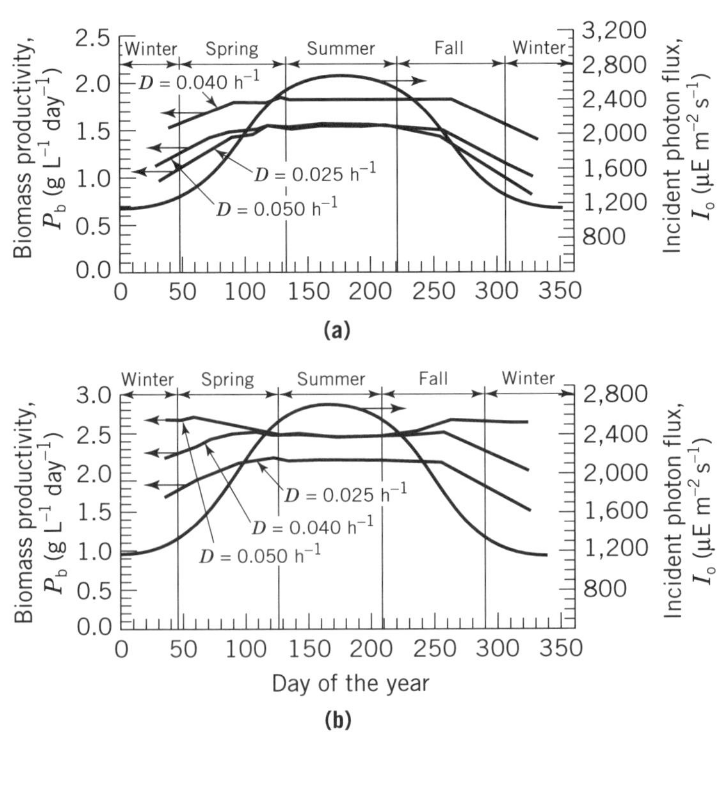

Figure 9. Variation of the estimated biomass productivity with the day of the year for a continuous culture of *Phaeodactylum tricornutum* UTEX 640 in two tubular photobioreactors: (**a**) 0.06-m tube diameter, (**b**) 0.03-m tube diameter. Simulations are shown for three dilution rates.

flume at intervals of 1.2 m along the length of the flume. The wing-shaped foils created a vortex circulation that produced an ordered pattern of vertical stirring. By using these devices, Laws and coworkers obtained photosynthetic efficiencies (based on the PAR) of 9.6 and 7.5, respectively, with and without the foils in place.

In tubular photobioreactors and ponds, higher flow rates generally enhance productivity. Typically, flow rates range between 20 and 30 cm s^{-1}. As noted, the flow can be generated by various methods such as paddle wheels and pumps. In closed tubular photobioreactors the flow is commonly driven by airlift systems, which are generally less damaging to cells than the mechanical devices. The velocity in the tube controls the turbulence and shear rate, determines the mixing behavior and the mass transfer capacity, and greatly influences the axial oxygen concentration profile in the tube. The liquid velocity, U_L, may be estimated by an extension of the well-known and widely tested model developed by Chisti (68); thus,

$$U_L = \left[\frac{g h_D \varepsilon_r}{2f\dfrac{L_{eq}}{D}}\frac{A_r}{A_d}\right]^{0.5} \quad (6)$$

where f is the Fanning friction factor that can be calculated using the Blasius equation,

$$f = 0.0791\mathrm{Re}^{-0.25} \quad (7)$$

where Re is the Reynolds number, or $(\rho_L U_L D)/\mu$, D is the tube diameter, and ρ_L and μ are the density and the viscosity of the liquid, respectively. As shown in equation 6 the superficial liquid velocity in a tubular photobioreactor is a function of gas holdup in the riser, ϵ_r, the height of the dispersion, h_D, and the ratio between the cross-sectional area of the riser and the downcomer.

Gas–Liquid Mass Transfer

In addition to availability of light, the availability of carbon dioxide (the carbon source) affects the productivity of a photosynthetic culture. The cells combine carbon dioxide and water to generate carbohydrates and cell biomass. Photosynthesis generates oxygen, which is released into the extracellular fluid. Because large amounts of oxygen can inhibit photosynthesis, oxygen must be removed to prevent buildup. Similarly, carbon dioxide must be provided such that availability of carbon does not become a limiting factor. Removing oxygen and supplying carbon dioxide are problems of gas–liquid mass transfer. Therefore, modeling of an algal reactor must consider gas–liquid mass transfer and how it is affected by hydrodynamics.

The mass transfer from the gas phase to the culture suspension in open ponds has been extensively studied (26,69). According to Märkl (70), the critical CO_2 concentration in an algal suspension of *Chlorella vulgaris* in an open pond is very low. A concentration as low as 10^{-6} M assures unlimited photosynthesis. Experiments with the same strain and with different oxygen concentrations in the medium show that the photosynthetic reaction rate is enhanced by 14% when there is almost no oxygen in the algal suspension and reduced by 35% when the medium is saturated with pure oxygen. In technical systems an oxygen content no higher than the saturation concentration with air (21%) is desirable (26). Mathematical expressions for longitudinal profiles of CO_2 and dissolved oxygen in algal cultivation units have been derived by Livansky et al. (71). The mass transfer issues relating to tubular photobioreactors have been treated by Camacho Rubio et al. (72).

Because carbon dioxide tends to be expensive one consideration is minimizing its loss; hence, quantitative transfer of CO_2 from the gas to the liquid phase is aimed at but normally not achieved. The dissolved carbon dioxide is in equilibrium with carbonate and bicarbonate species. These equilibria are pH dependent. Loss of dissolved carbon dioxide due to uptake into algal cells is partly compensated for by regeneration from carbonates and bicarbonates. Consequently, carbon dioxide uptake is accompanied by changes in pH. The relevant equilibrium and the corresponding equilibrium constants (73) that need to be considered are

$$H_2O \leftrightarrow H^+ + OH^- \quad K_w = [OH^-]\cdot[H^+] = 10^{-14} \quad (8)$$

$$CO_2 + H_2O \leftrightarrow H_2CO_3 \leftrightarrow HCO_3^- + H^+ \quad K_1 = [HCO_3^-]\cdot[H^+]/[CO_2] = 10^{-6.381} \quad (9)$$

$$HCO_3^- \leftrightarrow CO_3^{2-} + H^+ \quad K_2 = [CO_3^{2-}]\cdot[H^+]/[HCO_3^-] = 10^{-10.377} \quad (10)$$

As noted (72), models based on mass balances and gas–liquid mass transfer phenomena can be used to predict the various culture variables such as dissolved oxygen concentration, the carbon content within the culture, the composition of the outlet gas, the carbon dioxide losses and requirements, and the pH of the culture. As shown in Figure 10, through any 24-hour period, the simulated data closely agree with the measurements.

Temperature

Many studies in both open ponds and closed reactors have shown that temperature is an important factor affecting productivity. The highest biomass growth rate can be obtained only at the optimal temperature for growth. Although this effect is obvious and well documented under laboratory conditions, the magnitude of the effect on the annual production of biomass outdoors seems not be sufficiently appreciated. The diurnal variation in temperature to which cultures in open raceways are exposed results in temperature-limited growth during a significant fraction of the day. Lee et al. (48) showed that the light yield (g biomass kJ^{-1}) increased some three- to fivefold with increasing temperature in cultures of *Chlorella*, an effect that may be readily observed with other microalgae. The temperature of the culture at night is yet another important factor that affects the net output rate of biomass. The effect of temperature on dark respiration is, of course, very well documented (16,52,53).

Compared with open ponds, enclosed reactors generally warm up faster in the morning, thus reaching the optimum temperature for algal photosynthesis earlier in the day

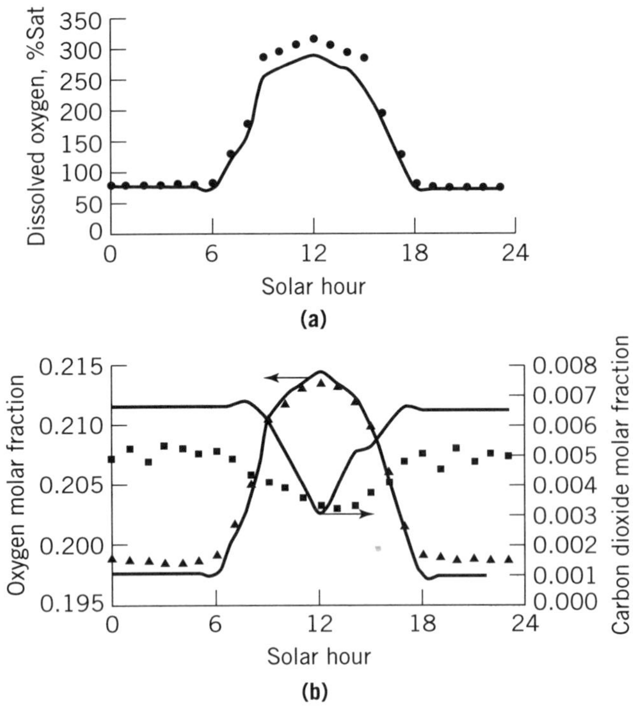

Figure 10. Daily variation of (**a**) the culture dissolved oxygen in the photobioreactor solar receiver tube (●) and (**b**) concentrations of O_2 and the CO_2 in the gas phase coming out of the photobioreactor (■, oxygen molar fraction; ▲, carbon dioxide molar fraction). The continuous lines represent simulations. Data are for a culture of *Porphyridium cruentum*.

(38). Closed reactors, however, can also overheat and thus require some sort of temperature control systems. In closed systems this control is achieved by either incorporating a heat exchanger into the design or evaporative cooling of water sprayed on the surface of the reactor. Economic considerations generally favor evaporative cooling.

PHOTOSYNTHETIC EFFICIENCY

The photosynthetic efficiency (PE) of biomass growth is defined as the energy stored in biomass per unit of light energy absorbed (74). This parameter is of fundamental importance in biomass production, because the microalgal production should aim to achieve the maximum attainable PE so as to produce the most biomass possible per unit land area (open ponds) or volume (enclosed systems). According to Richmond (74), microalgae use only the PAR, which constitutes about 43% of the total solar radiation at the Earth's surface and has an energy input equivalent to that of monochromatic light at a wavelength of 575 nm. Because one Einstein of 575-nm light contains 49.74 kcal of energy, and assuming that free energy of photosynthesis is 114 kcal mol^{-1} of reduced CO_2, the theoretical maximum energy efficiency for the photosynthetic reduction of CO_2 to glucose with white light is $114/(8 \cdot 49.74) = 0.286$, or approximately 29% (75). This estimation assumes that eight photons of light are needed for the requisite transfer of four electrons. Under field conditions, much lower photosynthetic efficiencies have been obtained. The maximum practical photosynthetic efficiency is a problem still to be resolved (76); however, a theoretical photosynthetic yield of 6 to 7% of the total solar energy seems possible (77). A 6–7% efficiency represents a biomass productivity of 35 to 70 g m^{-2} per day depending on the geographic locations. A photosynthetic efficiency of 10% during a period of 122 days has been reported (78). Nevertheless, producing biomass on large scale under outdoor conditions with an efficiency similar to that obtained under controlled laboratory conditions remains a challenge.

Information on algal yields reported over 30 years has been summarized by Goldman (54). Although the early attempts at mass culture produced no better than 10 to 15 g dry wt biomass m^{-2} per day over short periods, today it is not uncommon for sustained yields to reach 20 to 25 g m^{-2} per day. Goldman presented a thorough analysis of maximum yield potential of algae based on the thesis that algal production under the proper conditions is limited only by light (55). Goldman's equation can be written as

$$P_b = \frac{1}{Q_b} \cdot I_s\left(\ln \frac{0.45 \cdot I_0}{I_s} + 1\right) \qquad (11)$$

where P_b is the productivity of the system (g m^{-2} s^{-1}), I_0 is the total irradiance at the culture surface (μE m^{-2} s^{-1}), I_s is the saturating light intensity for which the growth rate is the maximum (i.e., when $\mu = \mu_m$, see Fig. 6) (μE m^{-2} s^{-1}), and Q_b is the heat of combustion of the biomass (μE g^{-1}). The maximum available intensity of sunlight at the Earth's surface is about 2,000 μE m^{-2} s^{-1}. Equation 11 predicts that, for a range of I_s values between 60 and 180 μE m^{-2} s^{-1}, as is typical for microalgae culture, the upper limit of algal yield is 30 to 60 g dry wt m^{-2} per day. According to this model, the factor that most strongly influences the algal yield is I_s. By contrast with the laboratory experiments, most outdoor algal production systems report productivities in the range of 15–30 g m^{-2} per day (79). These figures are far different from those observed in carefully controlled indoor cultures.

The laboratory work reported previously was carried out under conditions of low intensities, at or below saturation for photosynthesis. In practice, when dealing with sunlight, the major problem is not the maximal quantum efficiency but the fraction of absorbed photons that can actually be used in photosynthesis. The problem, well described by Benemann (80), is that if algal photosynthesis saturates at low light intensities, and algal cells have a high extinction coefficient (high pigment content), then more light will be absorbed than can be used in photosynthesis. Depending on the actual saturating and incident light intensities, the extinction coefficient, K_a, and the biomass concentration, the maximum efficiency (regardless of the minimum quantum requirement) will be reduced by about two- to fivefold from that possible at low light levels (80). The only solution to this problem is to increase the light intensity at which photosynthesis saturates, which means reducing the antenna and accessory pigment level of the cell without affecting the rest of the photosynthetic apparatus.

A new procedure for calculating yields in microalgal mass cultures at high incident irradiances has been proposed by Molina Grima et al. (81). This approach is based on an averaged irradiance for assessing the photon flux absorbed by the biomass. The starting point of the procedure (81) is the definition of the quantum yield, Ψ_E. The quantum yield is defined as the amount of biomass generated per unit of radiation (usually photons or Einsteins) absorbed by the culture. Because it represents the ratio of biomass generation to absorbed photon flux, the quantum yield can be calculated by the expression

$$\Psi_E = \frac{P_b}{F_{vol}} \tag{12}$$

where P_b is the volumetric biomass productivity and F_{vol} is the photon flux absorbed per unit volume. The quantum yield can be converted to energy units, commonly denoted as Ψ_{kJ}, by taking into account the average energy of the light used (kJ E^{-1}). The bioenergetic yield, Ψ, quantifies the percentage of light energy that is converted to chemical energy; Ψ can be calculated as the product of Ψ_{kJ} and the biomass heat of combustion, Q_b:

$$\Psi = \Psi_{kJ} Q_b \tag{13}$$

Because P_b in steady-state continuous culture can be calculated as the product of biomass concentration, C_b, and the dilution rate, D, any of the growth yield defined in equations 12 and 13 can be evaluated as long as the absorbed photon flux can be found.

Absorbed Photon Flux

The quantification of the absorbed photon flux is a complicated matter, especially when the flux must be evaluated in geometries other than flat reactors. Many factors such as light scattering effects may lead to a misevaluation of the absorbed flux. In every case the flux needs to be evaluated by direct measurement of the light leaving the photobioreactor (82). The light absorbed can be evaluated by integrating the local volumetric rate of energy absorption over the total reactor volume (83). This integral can be readily obtained from the extinction coefficient of biomass, K_a, and the average irradiance, I_{av}.

$$F_{vol} = I_{av} K_a C_b \tag{14}$$

Calculation of the absorbed photon flux by the preceding procedure has several advantages: the calculation is independent of the system geometry once I_{av} is known and can thus be applied to any type of photobioreactor as long as I_{av} can be determined. Photon flux losses due to reflection of light off the reactor walls that cannot be neglected (82) are automatically taken into account if the irradiance is measured inside the culture. The validity of equation 14 for assessing the absorbed photon flux has been checked experimentally in a culture of *Isochrysis galbana* (81) and using data published by Lee and Erickson (84) for *Chlorella* in a flat photobioreactor. *Isochrysis galbana* showed a maximum efficiency of light conversion into biomass of 0.6 g E^{-1} at $I_0 = 820\ \mu E\ m^{-2}s^{-1}$ and a dilution rate of 0.030 h^{-1}. The maximum capacity of the biomass to metabolize light was found to be 13.1 $\mu E\ g^{-1}s^{-1}$ (81). Above this value a significant drop in system efficiency occurred.

OPERATIONAL CONSIDERATIONS

Successful maintenance of a microalgal culture facility requires continuous monitoring to prevent the development of nonoptimal conditions. As shown in Figure 3, the on-line information typically monitored includes the dissolved oxygen (DO), temperature, and pH. In addition, daily microscopic observations are important for characterizing the physiological state of the culture. Measurement of growth should be routinely accompanied by detailed microscopic evaluations. Any increase in number of contaminating organisms in the culture is a warning signal: the temperature may have deviated from the optimal, or the concentration for a particular nutrient may have gone out of the optimal range. Monitoring of the nitrogen content is another useful indicator of the state of continuous cultures and may give information about the possible depletion in mineral nutrient in the culture. This information can be used as a guideline for adding, in equivalent amounts, the entire formula of the elements making up the growth medium. The concentration of DO is another reliable and sensitive indicator of conditions that relate to growth and productivity. There are many indications that early detection of an inexplicable decrease in DO concentration or a decline in the normal rate of the daily increase of DO in outdoor mass culture serves as a reliable warning signal that the culture is stressed and may quickly deteriorate (74).

Reliability of the microalgal mass culture systems and the ability to easily clean them are other important operational considerations. Closed reactors may experience fouling of the walls by biomass. For continuous cleaning of tubular photobioreactors, Chaumont et al. (85) have developed and patented a system that consists of balls that continuously circulate through the tubes. Such systems are effective for fouling control.

Tubular photobioreactors are also reasonably reliable. They have been used to continuously culture many species including the marine algae *Phaeodactylum tricornutum, Tetraselmis chuii*, and *Isochrysis* sp. for periods in excess of 3 months (19,53). These systems are particularly suited to continuous or semicontinuous culture, which permits superior control over the culture conditions and provides a product of consistent quality. As an additional operational advantage, the cell concentration reached is much greater in most closed systems than values achieved in open systems; consequently, the harvesting cost is lower for closed systems. However, the actual cell density needs to be managed carefully to maximize productivity (Fig. 11). At low cell densities, especially in outdoor systems, the algae may be severely photoinhibited, whereas at very high cell densities self-shading can reduce productivity as individual algal cells do not receive an optimal level of light.

In summary, culture management aims to achieve and sustain an optimal state of the culture. For maximum biomass productivity, solar irradiance should be the sole lim-

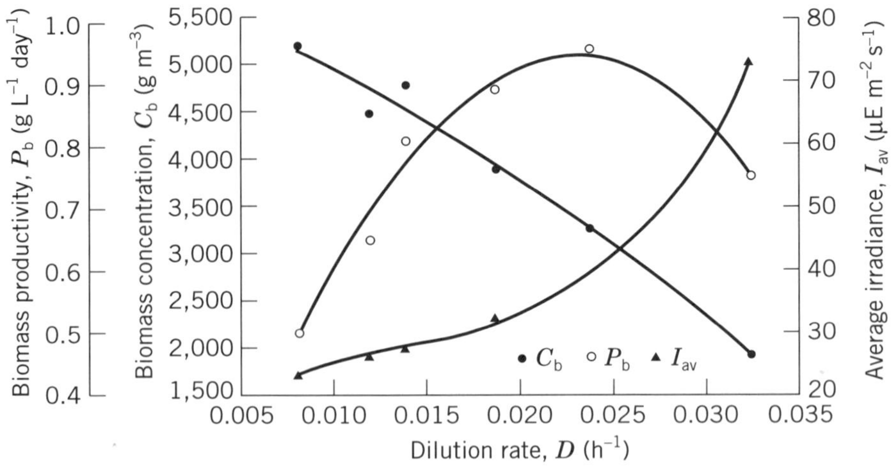

Figure 11. Effect of dilution rate, D, on the steady-state biomass concentration, C_b, the biomass productivity, P_b, and the average irradiance inside a photobioreactor, I_{av}. Data were obtained in an outdoor continuous culture of *P. tricornutum*, in a concentric tube airlift bioreactor. *Source*: From Refs. 14 and 49, with permission.

itation; however, when a metabolite is the desired product, the operational conditions for maximal productivity might be different from those for biomass productivity.

CONCLUDING REMARKS

Microalgae represent an important but little known resource with the genetic potential to provide valuable compounds. Future development of algal biotechnology will depend largely on what unique high-value products microalgae can provide. Algal products or processes must have some unique attribute, either biochemical or engineering, that would make them superior to heterotrophic microbes. Current commercial use of microalgae is restricted to a few niche applications.

Based on a comparison of the major types of algal culture systems (Table 1), the open systems are clearly deficient in several aspects. Despite this deficiency, open raceway systems are in commercial use. A principal advantage of open culture systems is the relatively small capital investment needed. Open systems are simple, but their productivity is low relative to the theoretical maximum. Much improvement in photosynthetic activity could be effected by genetic engineering of the photosynthetic complex to raise the light intensity threshold at which the photosynthetic machinery is light saturated. Also, resistance to photoinhibition may be introduced by genetic modifications.

Enclosed photobioreactors, which allow better control of operational conditions, are better suited to producing high-value products. These bioreactors permit culture of a greater diversity of algae than possible in open ponds. The potential of enclosed outdoor tubular photobioreactors has been demonstrated with high-productivity culture of microalgal species such as *Porphyridium, Isochrysis*, and *Phaeodactylum*. Because of better control of culture parameters, enclosed culture allows for manipulation of the biochemical composition of the algal biomass for various purposes (86–89). Although the enclosed culture technology is more capital intensive than the open pond systems, the additional cost may be recoverable. Temperature increase and fouling are the major problems with closed photobioreactors. The choice of a suitable mass culture system depends on several factors including geographic location, the microalgal species, and the metabolite to be produced. The following factors should be considered in making a selection: *(1)* the culture system should permit the culture of several species; *(2)* the culture system should provide uniform illumination and high rates of mass transfer that are attained by means that do not damage the culture; *(3)* the reactor must prevent or minimize fouling, particularly of

Table 1. Comparison of Properties of Different Large-Scale Algal Culture Systems

Reactor type	Mixing	Light utilization efficiency	Temperature control	Gas transfer	Hydrodynamic stress	Species control	Sterility
Unstirred ponds	Very poor	Poor	None	Poor	Very low	Difficult	None
Open raceway	Fair to good	Fair to good	None	poor	Low	Difficult	None
Airlift	Uniform	Good	Excellent	High	Low	Easy	Easily achievable
Bubble column	Fair	Fair	Excellent	Fair	Low	Easy	Easily achievable
Tubular	Almost uniform	Good	Excellent	High	Low to high	Easy	Achievable
Biocoil	Uniform	Excellent	Excellent	Low to high	Low to high	Easy	Achievable

Source: Adapted from Borowitzka (31).

its light transmitting surfaces; and *(4)* the volume of the nonilluminated parts of the reactors should be minimized.

NOMENCLATURE

A_d Cross-sectional area of the downcomer (m^2)
A_r Cross-sectional area of the riser (m^2)
C_b Biomass concentration ($g\ L^{-1}$ or $g\ m^{-3}$)
d Depth channel (m)
f Fanning friction factor
F_{vol} Photon flux absorbed in unit volume ($\mu E\ m^{-3}\ s^{-1}$)
h_D Height at which the degasser is situated (m)
I Hourly incident photosynthetic radiation ($\mu E\ m^{-2}\ s^{-1}$)
I_{av} Photosynthetically active hourly average irradiance inside culture ($\mu E\ m^{-2}\ s^{-1}$)
I_0 Solar irradiance impinging on the reactor surface ($\mu E\ m^{-2}\ s^{-1}$)
I_s Saturating light intensity for which the growth rate is maximum ($\mu E\ m^{-2}\ s^{-1}$)
K_a Absorption coefficient ($m^2\ g^{-1}$)
K_1 First equilibrium constant for bicarbonates buffer ($mol\ m^{-3}$)
K_2 Second equilibrium constant for bicarbonates buffer ($mol\ m^{-3}$)
K_W Dissociation constant of water
L Channel length (m)
L_{eq} Equivalent length of the solar receiver (m)
n Manning friction factor ($s\ m^{-1/3}$), or characteristic parameter of the hyperbolic growth model
n_1 Characteristic parameter of the photolimitation–photoinhibition model
n_2 Characteristic parameter of the photolimitation–photoinhibition model
n_3 Characteristic parameter of the photolimitation–photoinhibition model
P Power consumption for the culture flow in the channel (kW)
P_b Biomass productivity ($g\ m^{-3}\ h^{-1}$)
Q_b Heat of combustion of biomass ($\mu E\ g^{-1}$)
S Cross section of the tube (m^2)
s Rate of loss of energy in the channel (dimensionless)
U_L Superficial liquid velocity in the tube ($m\ s^{-1}$)
v Mean culture flow velocity channel ($m\ s^{-1}$)
w Width channel (m)
ψ Bioenergetic yield coefficient (%)
ψ_E Quantum yield ($g\ E^{-1}$)
ψ_{kJ} Energetic yield ($g\ kJ^{-1}$)
η Efficiency of the paddle wheel (−)

ACKNOWLEDGMENTS

I wish to acknowledge the contribution of all my colleagues of the Marine Biotechnology Group of the University of Almería, who have worked with me on this subject. Thanks are also due to Prof. Y. Chisti for helpful comments and assistance with the final manuscript. Some of the work reported was financially supported by the Plan Andaluz de Investigación (Junta de Andalucía), Comisión Interministerial de Ciencia y Technología (C.I.C.Y.T.), Spain, and the European Union.

BIBLIOGRAPHY

1. W.J. Oswald, in M.A. Borowitzka and L.J. Borowitzka eds., *Micro-Algal Biotechnology*, Cambridge Univ. Press, Cambridge, 1988, pp. 305–328.
2. W.N. De Pauw and G. Persoone, in M.A. Borowitzka and L.J. Borowitzka eds., *Micro-Algal Biotechnology*, Cambridge Univ. Press, Cambridge, 1988, pp. 197–221.
3. J.R. Benemann, *J. Appl. Phycol.* **4**, 233–245 (1992).
4. W.B. Metting, in M.A. Borowitzka and L.J. Borowitzka eds., *Micro-Algal Biotechnology*, Cambridge Univ. Press, Cambridge, 1988, pp. 288–304.
5. D.J. Chapman and K.W. Gollenbeck, in R.C. Cresswell, T.A.V. Rees, and N. Shah eds., *Algal and Cyanobacteria Biotechnology*, Longman, New York, 1989, pp. 1–27.
6. D.M. Calvin and S.E. Taylor, in R.C. Cresswell, T.A.V. Rees, and N. Shah eds., *Algal and Cyanobacteria Biotechnology*, Longman, New York, 1989, pp. 137–160.
7. D.K.W. Glombitza and M. Koch, in R.C. Cresswell, T.A.V. Rees, and N. Shah eds., *Algal and Cyanobacteria Biotechnology*, Longman, New York, 1989, pp. 161–238.
8. M.R. Tredici, P. Carlozzi, G.C. Zitelli, and R. Materassi, *Bioresour. Technol.* **38**, 153–159 (1991).
9. A. Ramos de Ortega and J.C. Roux, *Biomass* **10**, 141–156 (1986).
10. I.A.J. Ratchford and H.J. Fallowfield, *J. Appl. Phycol.* **4**, 1–9 (1992).
11. P. Pohl, M. Kohlhase, and M. Martin, in T. Stadler, J. Mollion, M.C. Karamanos, H. Morvan, and D. Christiaen eds., *Algal Biotechnology*, Elsevier, London, 1988, pp. 209–217.
12. A.K. Hilaly, M.N. Karim, and D. Guyre, *Biotechnol. Bioeng.* **43**, 314–320 (1994).
13. C.M. James and A.M. Al-Khars, *Aquaculture* **87**, 381–393 (1987).
14. A. Contreras Gómez, Ph.D. Thesis, University of Almería, Almería, Spain, 1996, p. 373.
15. S.J. Pirt, Y.K. Lee, M.R. Walach, M.W. Pirt, H.H.M. Balyuzi, and M.J. Bazin, *J. Chem. Technol. Biotechnol.* **33B**, 35–58 (1983).
16. G. Torzillo, P. Carlozzi, B. Pushparaj, E. Montaini, and R. Materassi, *Biotechnol. Bioeng.* **42**, 891–898 (1993).
17. French Pat. 9,115,773 (June 25, 1991), D. Chaumont, P.F. Dos Santos, C. Gudin, D. Assie, and G. Chaintron (to Ussie Inginierie).
18. A. Richmond, S. Boussiba, A. Vonshak, and R. Kopel, *J. Appl. Phycol.* **5**, 327–332 (1993).
19. T. Chrismadha and M.A. Borowitzka, *J. Appl. Phycol.* **6**, 67–74 (1994).
20. E. Molina Grima, F. García Camacho, J.A. Sánchez Pérez, J. Urda Cardona, and J. Fernández Sevilla, *Biotechnol. Appl. Biochem.* **20**, 279–290 (1994).
21. M.R. Tredici and G.C. Zittelli, *Biotechnol. Bioeng.* **57**, 187–197 (1998).
22. Y.K. Lee and C.S. Low, *Biotechnol. Bioeng.* **38**, 995–1000 (1991).

23. W.J. Oswald, in E.F. Gloyna, J.F. Malina, and E.M. Davis eds., *Ponds as Wastewater Treatment Alternative*, Water Resources Symp. No. 9, University of Texas, Austin, Tex 1975.
24. J.C. Dodd, in A. Richmond ed., *Handbook of Microalgal Mass Culture*, CRC Press, Boca Raton, Fla., 1986, pp. 265–283.
25. W.J. Oswald, in M.A. Borowitzka and L.J. Borowitzka eds., *Micro-Algal Biotechnology*, Cambridge Univ. Press, Cambridge, 1988, pp. 357–394.
26. H. Märkl and M. Mather, *Arch. Hydrobiol.* **20**, 85–93 (1985).
27. E.A. Laws, S. Taguchi, J. Hirata, and L. Pang, *Biotechnol. Bioeng.* **28**, 191–197 (1986).
28. P. Heusler, J. Castillo, S. Morino, and V. Vasquez, *Arch. Hydrobiol.* **11**, 254–260 (1978).
29. A. Richmond, in F.E. Round and D.J. Chapman eds., *Progress in Physiological Research*, vol. 7, Biopress, Bristol, 1990, pp. 269–330.
30. Y.K. Lee, *Trends Biotechnol.* **4**, 186–189 (1986).
31. M.A. Borowitzka, *J. Mar. Biotechnol.* **4**, 185–191 (1996).
32. H. Tamiya, E. Hase, K. Shibata, A. Mituya, T. Iwamura, T. Nihei, and T. Sasa, in J.S. Burlew ed, *Algal Culture from Laboratory to Pilot Plant*, Carnegie Institution of Washington, Washington, D.C., 1953, pp. 204–232.
33. D. Chaumont, C. Therpenier, and C. Gudin, in T. Stadler, J. Mollion, M.C. Karamanos, H. Morvan, and D. Christiaen eds., *Algal Biotechnology*, Elsevier, London, 1988, pp. 199–208.
34. European Pat. 0,239,272 (March 6, 1988), L.F. Robinson; A.W. Morrison, and M.R. Bamforth (to Biotechna Ltd.).
35. Y.K. Lee and C.S. Low, *Biotechnol. Bioeng.* **40**, 1119–1122 (1992).
36. M.A. Borowitzka, *J. Appl. Phycol.* **4**, 267–269 (1992).
37. A.D. Little, in J.S. Burlew ed, *Algal Culture from Laboratory to Pilot Plant*, Carnegie Institution of Washington, Washington, D.C., 1953, pp. 235–273.
38. M.R. Tredici and R. Materassi *J. Appl. Phycol.* **4**, 221–231 (1992).
39. H. Quiang, H. Gutterman, and A. Richmond, *Biotechnol. Bioeng.* **51**, 51–60 (1996).
40. F.G. Acién Fernández, F. García Camacho, J.A. Sánchez Pérez, J.M. Fernández Sevilla, and E. Molina Grima, *Biotechnol. Bioeng.* **55**, 701–714 (1997).
41. F.P. Incropera and J.F. Thomas, *Solar Energy* **20**, 157–165 (1978).
42. J.N. Philliphs and J. Myers, *Plant Physiol.* **29**, 152–161 (1954).
43. J.U. Grobellar, *J. Appl. Phycol.* **6**, 331–335 (1994).
44. K.L. Terry, *Biotechnol. Bioeng.* **28**, 988–995 (1986).
45. A. Vonshak, *7th International Conference International Association of Applied Algology: Opportunities from Microalgae*, Knysna, South Africa, April 16–19, 1996, pp. 76–77.
46. R.H. Foy and C.E. Gibson, *Br. Phycol. J.* **17**, 183–193 (1982).
47. L. Van Liere and L.R. Mur, *J. Gen. Microbiol.* **115**, 153–160 (1975).
48. Y.K. Lee, H.M. Tan, and C.S. Hew, *Biotechnol. Bioeng.* **27**, 555–561 (1985).
49. F.G. García Camacho, A. Contreras Gómez, F.G. Acién Fernández, and E. Molina Grima, *Enzyme Microb Technol.* **24**, 164–172 (1999).
50. B.Y.H. Liu and R.C. Jordan, *Solar Energy* **7**, 53–65 (1960).
51. J.A. Duffie and W.A. Beckman, *Solar Engineering of Thermal Processes*, Wiley, New York, 1980, p. 762.
52. F.G. Acién Fernández, F. García Camacho, J.A. Sánchez Perez, J.M. Fernández Sevilla, and E. Molina Grima, *Biotechnol. Bioeng.* **58**, 505–611 (1998).
53. E.M. Molina Grima, F. García Camacho, J.A. Sánchez Pérez, J.A. Fernández Sevilla, F.G. Acién Fernández, and A. Contreras Gómez, *J. Chem. Technol. Biotechnol.* **61**, 167–173 (1994).
54. J.C. Goldman, *Water Res.* **13**, 60–118 (1979).
55. J.C. Goldman, *Water Res.* **13**, 119–160 (1979).
56. Y.K. Lee and A. Vonshak, *Arch. Microbiol.* **150**, 529–533 (1988).
57. J.A.A. Barber, *Trends Biochem. Sci.* **17**, 61–66 (1992).
58. S. Jensen and G. Knutsen, *J. Appl. Phycol.* **5**, 495–504 (1993).
59. J. Merchuk, M. Ronen, S. Giris, S. Arad (Malis), *Biotechnol. Bioeng.* **59**, 705–713 (1998).
60. T.T. Bannister, *Limnol. Oceanogr.* **24**, 79–96 (1979).
61. E. Molina Grima, J.A. Sánchez Pérez, F. García Camacho, J.L. García Sánchez, and D. López Alonso, *Appl. Microbiol. Biotechnol.* **38**, 599–605 (1993).
62. A. Sukenik, P.G. Falkowski, and J. Bennett, *Biotechnol. Bioeng.* **30**, 970–977 (1986).
63. A. Sukenik, R.S. Levy, Y. Levy, P.G. Falkowski, and Z. Dubinsky, *J. Appl. Phycol.* **3**, 191–201 (1991).
64. H. Guterman, A. Vonshak, and S. Ben-Yaakov, *Biotechnol. Bioeng.* **35**, 809–819 (1990).
65. H.J. Silva, T. Cortinas, and R.J. Ertola, *J. Chem. Technol. Biotechnol.* **40**, 41–49 (1987).
66. C. Gudin and D. Chaumont, *Bioresour. Technol.* **38**, 145–151 (1991).
67. E.A. Laws, K.L. Terry, J. Wickman, and M.S. Chalup, *Biotechnol. Bioeng.* **25**, 2319–2335 (1983).
68. Y. Chisti, *Airlift Bioreactors*, Elsevier, London, 1989, p. 343.
69. K. Livansky, B. Prokes, F. Dittrt, and V Benes, *Biotechnol. Bioeng. Symp.* **4**, 513–518 (1973).
70. H. Märkl, *Biotechnol. Bioeng.* **19**, 1851–1862 (1973).
71. K. Livansky, M. Kayan, and P.S. Pilarski, *Algological Studies* **70**, 97–119 (1993).
72. F. Camacho Rubio, F.G. Acién Fernández, F. García Camacho, J.A. Sánchez Pérez, and E. Molina Grima, *Biotechnol. Bioeng.* **62**, 71–86 (1999).
73. K. Livansky, *Algological Studies* **58**, 87–97 (1990).
74. A. Richmond, *Crit. Rev. Microbiol.* **4**, 369–438 (1986).
75. J.A. Bassham, in A. Mitsui, S. Miyachi, A.S. Pietro, and S. Tamura eds., *Biological Solar Energy Conversion*, Academic Press, New York, 1971, pp. 151–165.
76. S.J. Pirt, *Biotechnol. Bioeng.* **25**, 1915–1922 (1983).
77. D.O. Hall, in A. Richmond ed., *Handbook of Microalgal Mass Culture*, CRC Press, Boca Raton, Fla., 1986, pp. 1–24.
78. E.A. Laws, S. Taguchi, J. Hirata, and L. Pang, *Biomass* **16**, 19–32 (1988).
79. J. Weisman, R.P. Goebel, and J. Benemann, *Biotechnol. Bioeng.* **31**, 336–344 (1988).
80. J.R. Benemann, in R.C. Cresswell, T.A.V. Rees, and N. Shah eds., *Algal and Cyanobacteria Biotechnology*, Longman, New York, 1989, pp. 317–337.
81. E.M. Molina Grima, F. García Camacho, J.A. Sánchez Pérez, J.G. Acién Fernández, and J. Fernández Sevilla, *Enzyme Microb. Technol.* **21**, 375–381 (1997).
82. S. Aiba, *Biotechnol. Bioeng.* **25**, 2775–2776 (1983).
83. O.M. Alfano, R.L. Romero, and A.E. Cassano, *Chem. Eng. Sci.* **41**, 421–444 (1986).

84. H.Y. Lee and L.E. Erickson, *Biotechnol. Bioeng.* **29**, 476–481 (1987).
85. French Pat. 9,103,781 (June 25, 1991), D. Chaumont, P. Ferreira dos Santos, and L. Sauze (to Ussie Inginierie).
86. P.S. O'Colla, in R.A. Lewin ed., *Physiology and Biochemistry of Algae*, Academic Press, New York, 1962, pp. 337–356.
87. P. Pohl, in O. Zaborsky ed., *Handbook of Biosolar Resources*, CRC Press, Boca Raton, Fla., 1982, pp. 383–404.
88. C. Gudin and C. Therpenier, *Adv. Biotechnol. Process* **6**, 73–110 (1986).
89. E. Molina Grima, J.M. Fernández Sevilla, J.A. Sánchez Pérez, and F. García Camacho, *J. Biotechnol.* **45**, 59–69 (1996).

MICROBIAL GROWTH. See BIOENERGETICS OF MICROBIAL GROWTH.

MICROBIAL GROWTH MEASUREMENT, METHODS

ARTHUR L. KOCH
Indiana University
Bloomington, Indiana

KEY WORDS

Balanced growth
Biomass estimation
Cell differentiation
Flow cytometry
Particle counts
Phases of the growth cycle
Sampling head gas
Turbidimetry
Unbalanced growth
Viable counts

OUTLINE

INTRODUCTION

In biotechnology, the denominator is biomass, for example, yield is the desired product per unit of biomass. Biomass concentration is one of the major determinants to identify the stage of the culture cycle. It is necessary to understand the controlling points in the physiology of the bacterial growth. Biotechnologists have talked extensively about growth phase and idiophase as the key measures of fermentation technology. There are a number of approaches for biomass measurement, but they are not very good and have significant limitations. The reader should supplement the treatment presented here with a previous article (1) that details a number of procedures and leads back to the earlier literature. Another useful overview is provided by Sonnleitner et al. (2). The former paper comes from the approach of a physiologist, and the latter is from the point of view of biotechnologists. For historical interest, the collected source is by Dawson (3).

Principles

Principles of bacterial growth are discussed in microbial physiology texts. I would particularly recommend that by White (4). Although there are many books on bioprocessing, most do little to discuss biomass measurements. The book by Pirt in 1975 (5) was a milestone in the field of microbe growth for product formation; newer publications are now available (6,7), but they do little for the estimation of biomass.

Definitions of Growth

To measure biomass and bacterial growth rate, precise definitions are needed. Although biomass is fundamental to carrying out technology, we have to consider first the growth process that makes the biomass and examine its several definitions. The most basic definition of growth is based on the ability of individual cells to multiply, that is, to repetitively initiate and complete cell and organismal division. This definition implies monitoring the increase in total number of discrete bacterial particles. There are three basic ways to do this: by microscopic enumeration of the particles, by electronic enumeration of the particles passing through an orifice (Coulter counter) and by modern flow cytometry. Assessment of particle number would falsely include dead cells and detritus, which would tend to lower estimations of growth rate. The rate would be artificially raised by the progressive dissolution of aggregates of bacteria and the fragmentation of nongrowing filamentous organisms. An increase in cell number need not correlate exactly with an increase in biomass or useful product. Commonly, at the end of an exponential growth phase, cell division overshoots biomass production and the cells become smaller.

A second definition of growth involves determining the increase in colony-forming units (CFU). Because some cells may be dead or dying, this definition of growth may be different from the one based on the detection of discrete particles as a function of time. Although in the long run, the increase in the number of organisms capable of indefinite growth is the only important consideration for the physiologist, this is not so for the biotechnologist, for several reasons. First, for strain purity, each new production run starts afresh from starter cultures; these cultures would had been specially treated much differently than the cells in the actual production run. This care is to avoid contamination and the buildup of unwanted mutants. Second, dead, dying cells and stationary cells may be the productive members of the culture in terms of product formation. This second definition is the reason that colony counting and most-probable-number (MPN) methods of measurement are so important. It must be noted that viable counting methods, which seem so natural to a bacteriologist, are really quite special in that cultures are diluted so highly that individual organisms cannot interact. For example, these methods cannot in principle be applied to obligate sexually reproducing organisms requiring male–female interaction or to colonial organisms such as myxobacteria that in certain conditions need to be part of a large mass of organisms that produces sufficient exoenzymes in order to grow. Even when applied to the prokaryote, there are special restrictions and limitations, for example, CO_2 must be available in sufficient concentration, although this need not be supplied if many organisms are present generating CO_2. This and possible detoxification of the medium can lead to a lower count than would have occurred in the more concentrated actual culture. To illustrate, consider a single bacterium and an isolated mammal, both introduced into an environment without any individuals of their own kind. The former would produce a colony, whereas the latter could only produce a viable colony if the founding individual was a pregnant female.

A third definition of growth (and the practical one) is based on an increase in biomass. Macromolecular synthesis and increased capability for synthesis of cell components are the obvious basis for the measurement of growth by everyone interested in microbes and what they can do. From our point of view, the whole process of chromosome replication and cell division is an essential but minor process that seldom limits growth; what usually limits growth is the ability of enzymatic systems to use available resources to form biomass. Moreover, the restriction of growth of the culture is usually the result of the depletion of resources or degradation of the local environment.

A fourth definition of growth is based on the organisms' action in chemically changing their environments. This category can be considered as simply a different consequence of the increase in biomass. But it allows the rate of the growth process and the biomass production to be estimated in indirect ways. Most important in industry, the gases coming from the fermentation are analyzed, and the consumption of oxygen and production of carbon dioxide are used as effective and efficient means to follow growth. Especially important is the use of mass spectrometer methods, because of their precision and the ability to automate measurements, growth of cultures can be followed usefully and effectively.

Biomass is the result of growth; it has variously different definitions according to purpose. In most biotechnology, biomass is organic material that is cheap and can be converted into fuel, or heat, or structural materials. It is necessary that such sources be self-generating, cheap, and available. Whether biomass is alive or dead is less relevant; the assumption is that it was, or at least once was, alive. If one has a mixture of coal and fresh plant material, the term biomass is fully ambiguous. For product formation there is less ambiguity; microbial biomass requires the expenditure of resources, and useful product may be produced by organisms that may be living or dead according to which of the definitions of growth is used.

Balanced Growth

These four definitions of growth become equivalent under a single circumstance: the state of balanced growth (8). This is certainly an academic concept, but its practicality is clear. An asynchronous culture can be said to be in balanced growth when all intensive properties are constant in time. An intensive property is one that does not matter how big the system is. An extensive property is another term from physical chemistry and refers to those properties of the system that are proportional to the amounts of substances of various kinds in the system. Thus, biomass and cell number are extensive properties of a culture and proportional to the volume of culture considered. These properties increase regularly and exponentially with time during balanced growth. However, temperature or ratios (such as DNA, RNA, or product per cell) are intensive properties that do not change during balanced growth. The application of this physical–chemical principle to bacteriology lies in the thought that if a culture is grown for a long-enough duration (usually with many subcultures) so that the biomass is sparse enough to not alter the environment significantly, sooner or later the organisms will come to achieve the same growth state for any particular constant environment, no matter what the cell's condition was initially—be it log, lag, or the stationary state. Once this balanced growth state has been achieved, if the conditions remain the same, the culture will remain in balanced growth indefinitely. This would be true of a culture that had grown in a continuous culture or one that was diluted manually periodically, avoiding entry into either lag or the stationary state. However, if a nutrient is exhausted, or if the culture is altered by mutation and selection, or if the physical nature of the environment is changed, then balanced growth would cease and the culture would be in some nonbalanced state of growth.

The criteria given above are too stringent to be fully met with any practicality, but it is readily possible to study cultures that are substantially in balanced growth by growing cultures maintained at low density by dilution and using arbitrarily any one measure to follow growth, for example, biomass assayed turbidimetrically. If the doubling time remains constant over an extended period of time, then it is a good assumption that growth is balanced. Studies of these cultures can lead to an understanding of the properties of individual cells in the idealized state of balanced growth. But they would not bear on the properties of aggregates or clumps of cells or of cells reacting to some calamity or catastrophe in their environment.

Although we may consider any extensive property of a culture in balanced growth, let us first focus on biomass and call it X for tradition's sake. The rate of formation of biomass will be proportional to the amount of biomass. Call the proportionality constant, μ, then:

$$dX/dt = \mu X$$

Through the laws of calculus, this equation can be integrated and a boundary condition ($X = X_0$ when $t = 0$) imposed, yielding the well-known equation for exponential growth:

$$X = X_0 e^{\mu t}$$

However, every cellular substance, or even an extracellular product of the cells, could be substituted for our X. Thus, after growth becomes balanced, the increase of any component or group of components will be exponential in time. If the ratio of one substance to every other is to remain constant, the same proportionality constant must apply for every component. This common value of μ is called the *specific growth rate* or *growth rate constant*. Not only will any substance increase exponentially, but any combination of substances or the rate of change of any substance also will increase exponentially with this same specific growth rate, μ. Practically, this allows the rate of oxygen uptake, heat production, or production of an internal or external metabolite to be used as an index of growth. It will also allow growth to be monitored by the depletion of some medium component or the change in conductivity of the medium.

Intensive properties, such as μ and the ratio of the concentrations of different cell substances, must necessarily remain constant under conditions of balanced growth. In addition, the distribution of cell sizes and the average number of chromosomes will stay constant. An average cell will have a time-independent constant rate of carrying out every cellular process, and newly arisen daughter cells will have a constant probability of being able to form a colony (i.e., the percentage of nonviable cells will remain constant). Therefore, under these special conditions, no matter what measure of growth is used (whether it is particle counting, colony formation, chemical determination of a cell substance, or consumption or excretion of a substance), the same specific growth rate will be obtained and that rate will be constant through time.

Changes in Cell Composition at Different Growth Rates of Balanced Growth

Organisms respond to environmental conditions, both physical and chemical, by altering their own composition. These changes have been well documented for certain enteric bacteria but may occur with all prokaryotes (9). In general, favorable growth conditions mean faster growth, which requires a higher concentration of ribosomes and associated proteins. In terms of gross composition, the most obvious difference comparing a faster culture with a slower culture is a greater RNA content in the former (10). Also, under favorable growth conditions, the cells can lay down reserve materials such as glycogen and poly-β-hydroxybutyric acid. These changes in composition lead to the possible pitfall of the bacteriologist falsely relating one measure of growth to others when comparing growth in different environments.

Unbalanced Growth

Although balanced growth conditions lead to reproducible cultures, much of physiology deals with the responses of organisms to changes in their environment that lead to progressive changes in the organism (i.e., to the study of unbalanced growth). When a stationary culture is inoculated into fresh medium, the properties of organisms change drastically through the course of the batch culture cycle. Though only very well documented for certain enteric bacteria, similar changes probably apply with other prokaryotes. The exact course of changes in composition and morphology depends on the medium and on the age and condition of the inoculum. Culture cycle phenomena have relevance to growth measurements. These phenomena are particularly important in ecological studies, where the conditions under which the organisms grow are critical and, to a large degree, uncontrollable. The changes in characteristics are also involved in response to the fluctuations in natural conditions (11,12). For biotechnology purposes, it is the progression of changes in the so-called idiophase of growth that are paramount.

Noting that a typical bacterial culture generally means one of *Escherichia coli* (13), a typical bacterial culture cycle progresses as follows. When a stationary culture is diluted into rich medium, macromolecular synthesis accelerates. The components of the protein-synthesizing system (i.e., ribosomal proteins and ribosomal RNA) are made first. Only after considerable macromolecular growth has ensued does cell division take place. During this lag phase, the average size of cells increases greatly. When the capacity of the medium to support rapid balanced exponential growth is exceeded, metabolic processes (both intermediate metabolism and macromolecular synthesis) slow down. Different processes slow differentially in a way to make small, RNA-deficient cells. Finally, the cells die and may eventually lyse.

Pitfalls in Measuring Growth

Errors in measuring bacterial growth fall into four classes. The first major pitfall is the general tendency of most organisms toward clumping or a filamentous habit of growth. This can occur even under mildly toxic conditions with bacteria that ordinarily divide regularly. The second pitfall is the differential viability of injured bacteria under different culture conditions. Repair processes may permit the recovery of viable cells under some, but not other, conditions. A third major pitfall is the possible development of resistant stages. Bacteria known to form resistant stages, such as spores, pose no problem because the controls and measurements to correct for such differentiated forms are well known; but when resistant stages are not suspected, error can arise. The fourth pitfall pertains to the way in which the inoculum is exposed to the new environment. Different results may be obtained if the concentration of an agent is raised gradually, if it is raised discontinuously, or if a high concentration is temporarily presented and then removed or lowered, such as when a concentrated solution of a nutrient is added and the mixing is not rapid. The cell concentration at the time of challenge can be critical. Bacteria may have special ways, sometimes inducible mechanisms or sometimes unknown ones, to protect themselves against toxic agents. Finally, the protection may be dependent on the number of organisms cooperating in detoxification.

Mycelial Habit of Growth

The evaluation of mycelial growth (the first pitfall) is both easier and much more difficult than that of well-behaved organisms that engage in binary fission and then promptly separate. The problems and methods for colonial growth have been discussed by Calam (14) and will be briefly dealt with here. Filtering filamentous cells, with or without drying, is easier than with smaller nonfilamentous cells.

The rate of increase in size of the colony in one, two, or three dimensions depends on the rate of elongation of those terminal hyphae that happen to be growing perpendicular to the surface of the colony, but the mobilization of resources into the mycelial mass depends on the surface area of the colony (15,16). Therefore, particularly with shake cultures, the results depend on the nature and size of inoculated fragments; with large fragments, growth becomes limited at an earlier stage by diffusion of nutrients into the mycelial mass. Thus, the major pitfall with the mycelial habit of growth is that the growth is quickly deviates from exponential growth and depends on the geometry of growth and the nature of the inoculum. In shake cultures, the apparent growth may depend on the shape of the vessel and on the shaking speed, its character (circular or reciprocal), and the distance moved, because all of the above can affect the tendency of the mycelia to break into smaller pieces. An important aspect of biotechnology is controlling the filamentous versus pellet conformations.

Cell Differentiation

The change of enteric bacteria from large, RNA-rich forms in the exponential phase to small, RNA-poor forms in the stationary phase has many of the aspects of differentiation. Bacteriology, however, has much clearer examples of cell differentiation in the cases of transition of rod to coccus, of vegetative bacillus to endospore, and in the formation of exospores, cysts, and buds. The tendency to form filaments in certain circumstances can also be considered a differentiation. These changes and their reversal pose potential pitfalls for all approaches to growth measurement.

Adsorption of Cells to Surfaces

Many bacteria naturally adhere to certain surfaces or can adapt and mutate to achieve a high avidity for solid surfaces, including glass. Many experiments with chemostat culture have failed to achieve their primary goal because the organisms adhered to the vessel walls. It is for these reasons that plastic or Teflon should be used whenever there is long-term contact of the organism with a culture vessel (16) Other approaches to minimizing the effects of growth on vessel walls include the use of large culture volumes in large containers, the use of violent agitation, and the frequent subculture of the bacteria into fresh glassware. These are the conditions in large biofermenters, but not during the production of the starter cultures. In addition, the use of detergents, vegetable oils, silicone coatings, and a high-ionic-strength medium may to some degree alleviate this problem.

This problem is particularly pertinent during dilution of the culture for measurement by highly sensitive means, such as microscopic and plate counts. It is, therefore, given consideration in the discussion of those methods (1,17,18).

DIRECT PARTICLE COUNTS

For many biotechnology purposes, the counting methods that can be used are limited because the growth medium generally contains particulate matter. This does not preclude their use in the development of strains and other phase of biotechnology.

Microscopic Enumeration

Microscopic enumeration is a commonly used technique that is quick and cheap and uses equipment readily available in the bacteriological laboratory. The technique is subject to gross errors, however. These can be overcome to a large degree using improvements suggested by the work of Norris and Powell (19). Precise details are given in my earlier review of this subject that are a further modification of this method (1), and some versatile techniques that have some of the characteristics of viable counts or metabolic measurements are given in the work by Murray et al. (20).

Methods and equipment, some linked to computer programs, are now available to record, scan, and view a microscope field. Software and hardware are available that enable a sample to be placed in a counting chamber viewed by a microscope, and then an electronic count can be made from the image. The computer can distinguish between cells adhering side by side and between a long cell and a pair of sister cells that have completed division but have not separated. The capability exists to measure the biomass from the density of the pixels since the phase contrast (see later text) is a measure of the biomass. For filamentous organisms, the computer could output the total length of hyphae and convert that into biomass (21,22). Direct enumeration from charge-coupled device (CCD) images with highly intelligent computer filtering will allow distinguishing the bacteria or fungi from the particulate matter in the brew in many cases.

Electronic Enumeration

The Coulter Counter (Coulter Electronics, Hialeah, Fla.), its commercial competitors, and particularly the laboratory-built versions have been important in the development of bacterial physiology in the past 50 years. Such instruments are used routinely in clinical hematology. They are also very useful in the enumeration of nonfilamentous yeasts and protozoa, but not of mycelial or filamentous organisms.

The principle of electronic enumeration is as follows. A fixed volume of diluted cell suspension is forced to flow through a very small orifice connecting two fluid compartments. Electrodes in each compartment are used to measure the electrical resistance of the system. Even though the medium conducts electricity readily, the orifice is so small that its electrical resistance is very high. Consequently, the electrical resistance of the rest of the electrical path is negligible by comparison. When a cell is carried through the orifice, the resistance further increases because the conductivity of the cell is less than that of the medium. This change in resistance is sensed by a measuring circuit and converted into a voltage or current pulse. The counting rates can be very high, but beware of coincidence errors. The pulses are counted by an electronic circuit similar to that used in counting radioactivity. Very small pulses are eliminated by a discriminator circuit. Very high pulses, which might result from dirt or other irrelevant particles, are eliminated by an upper discriminator. In advanced models, the pulses may be analyzed by size and stored in a multichannel analyzer; later, the data can be recovered and plotted in a histogram, and the numbers, mean size, and standard deviation can be calculated. All the data may be collected, and the discrimination against pulses that are too high or too low can be carried out as the data are analyzed. The instruments have some method of forcing an accurately known volume through the orifice during the counting period. This is usually done by displacing the fluid in contact with a mercury column past triggering electrodes that conduct effectively through the mercury, but not through the diluent medium.

The major problem areas in electronic enumeration are three. The first problem concerns the size and shape of the cells. Some bacterial cells are very small (less than 0.4 μm^3), and the resistance pulses produced as they pass through the orifice are comparable to the noise generated by the turbulence that develops in the fluid flowing through the orifice. One can set the discriminator dial on the instrument to reject the turbulence noise, but one loses sample information, particularly about newly divided cells. One can run blanks and subtract the blank values, but blanks are particularly variable for small cell sizes. In addition, a pattern of turbulence can become established, remain for a while, and then be replaced with another pattern. This can cause major interference. Determining the size of cells has many difficulties, including the effect of the path that the particle takes through the orifice and the resistance of the various layers and compartment of the cell. Therefore, this potential use of the instrument is not discussed in this article. Finally, the overall error increases when the blank has large statistical variation. This is not a major problem for biotechnology because currently used microbes are relatively large.

The second major problem for physiological studies in electronic counting results from the failure of cells to separate promptly from each other after cell division. This, and the tendency to form filaments and aggregates, can be minimized by careful choices of the organism and the conditions. Possibly it can be reduced by vortexing or sonicating the sample. It was fortuitous for the advancement of microbial physiology that *E. coli* was chosen for physiological and genetic studies in the 1940s and 1950s because only a relatively small number persists in pairs or chains or in the form of aggregates. There is the related problem of coincidence, which is the passage of more than one cell through the orifice in a short enough time so that a single larger cell is registered by the electronic equipment. This problem can be dealt with by increasing the dilution so that the probability of coincidence is reduced.

The third major problem is technical: the clogging of the orifice in electronic counting. The resistance change is a smaller proportion of the total electrical resistance of the orifice when the orifice diameter is larger. Consequently, for small rod-shaped bacteria such as *E. coli*, orifices with diameters in the range of 12 to 30 μm must be used. The exact choice is a tradeoff between the increased signal strength of small apertures and the increased noise and probability of becoming clogged. Clogging is best prevented by ultrafiltration of all reagents. Alteratively, the diluent can be prepared and allowed to settle for a long time (months) in a siphon bottle so that the particulate-free solution can be withdrawn from above. Additional information about this technique can be found in the works by Drake and Tsuchiya (23) and Kubitschek (24).

Flow Cytometry

Flow cytometry became an extremely powerful method for the studies of many aspects of the biology of eukaryotes two decades ago. The instruments are now common in hospitals and clinical research laboratories. Because prokaryotes are smaller, it is only more recently that the instrumentation and methods have been refined and are only now coming into their own in the study of bacteria. The instruments operate by forming a small-diameter stream of the sample suspension. This is encased in a stream of particulate-free fluid. The tube through which they flow constrict in such a way that the stream of the sample is made still narrower in diameter. This process is called hydrodynamic focusing. The flowing stream is examined with laser light with various frequencies and angles, and the output of the detection circuits is used to detect when a particle passes through and determine its characteristics. The design permits the analysis of biomass by light-scattering methods and by fluorescent staining of chemical components such as DNA. This will be discussed in "Flow Cytometry." The electronic circuits allow cells to be counted very rapidly. Growth can be followed by the increase in counts in samples counted at intervals. The problems for the application to prokaryotes is in sensitivity and background noise.

The original designs were made for study of problems in immunology. Applications, including microbiology, have been previously described (25–27).

Another quite different instrumental design was developed in Norway (28,29). It depends on flow directly on a microscope slide on an inverted microscope. This instrument uses an inverted microscope essentially operated in the dark-field mode. It uses a high-intensity mercury arc. Now with commercially available equipment, it has been utilized to study a variety of problems. Clever applications have been described (30,31).

The instrument of the first type uses a well-collimated laser source and fiberoptics. Great care is needed to maintain the instrument and its calibration. Routinely, polystyrene beads are mixed with the biological sample in a fixed amount. They have a uniform size and serve to quality control the electronics. By relating the biological count to the bead count, the concentration of biological particles can be determined. These kinds of controls can be used with the second type of instrument, which is cheaper and is now in use in many laboratories for many microbiological purposes. It is also capable of counting at very fast rates.

COLONY COUNTS EQUAL VIABLE COUNTS

Bacteriology really became an experimental science when Herr Dr. Robert Koch listened to Frau Fannie Hesse and developed the agar plate. This allowed not only the cloning of pure strains but also the enumeration of colonies arising from individual viable cells. Many variations have been developed and used in the ensuing century: (*1*) spread plates, (*2*) pour plates, (*3*) thin-layer plates, (*4*) layered plates, and (*5*) membrane filter methods. Although these techniques are very useful in the development of industrial strains, they take too much time and cannot be used online.

These techniques are described in many places (1). The pour-plate methods have variations in which the cells are grown in roller tubes or microtubes (32) and are examined with low-power microscopes when the colonies are small (33). There are many individual variations of techniques, sometimes resulting from historical accidents and sometimes resulting from special bacteriological circumstances. Automation of colony counting has put additional special restrictions on techniques, but allows colonies on the Petri dishes to be counted rapidly without operator error.

It must be emphasized again that mycelial organisms pose special problems. To some degree they can be overcome by sonication or shearing. A practical approach is to conduct a pilot study to find the amount of ultrasound that produces the maximal count that optimizes the separation of propagules from each other and minimizes the destruction of them by the physical treatment.

In the future, the development of growth media that is perfectly clear and can be solidified in uniform layers on a new generation of microscope slides will permit the analysis by telometering and computer analysis after only a few hours of incubation. In some cases, observation will be repeated and the difference in the images used to compute the growth rate as well as the initial CFU and viable biomass. Selectivity and identification of special properties with dyes and chemical reactions could be used as well. Again, particulate matter can be a problem, but less severe as the microcolonies grow to be larger than the detritus in the medium.

Spread Plates

The sample is pipetted onto the surface of solidified agar medium in a Petri dish, and then the cells are distributed with a wire, glass, or Teflon spreader. In the spread-plate technique, the growth forms surface colonies. It is reliable when methods to transfer all of a defined volume of cell suspension to the surface of the agar are used. The method is additionally useful because surface colonies are required to produce the proper color responses with many indicator agars that depend on the redox state of the colonies. In many cases, different colors are given from subsurface colonies because the oxygenation is different, and, therefore, the acid production and reducing potential are different than on the surface, and the selectivity is usually much less. (See Murray et al. [20] for recent improvements in technique.)

Pour Plates

An aliquot sample of diluted cells is pipetted into an empty sterile Petri dish; molten, but cool (45 °C), agar medium is poured onto the sample; and the contents are mixed by swirling and then allowed to harden. The pour-plate technique is a standard technique. Its main advantages is that most colonies are subsurface. They are, therefore, small, making it possible for a plate to have many colonies and still be countable. This method has a large dynamic range, and automation can be very useful.

Thin-Layer Plates

The use of thin-layer plates came out of virology. The sample is pipetted into a tube containing a small volume (2.5 to 3.5 mL) of molten but cool (45 °C) soft agar (0.6 to 0.75%) medium. This is then poured onto previously poured and hardened (1.5 to 2% agar) medium in a Petri dish, and the overlay is allowed to harden. This was the technique developed to measure virus plaques in the 1940s.

Layered Plates

Layered plates are like thin-layer plates except that an additional layer of agar medium is poured onto the newly congealed soft agar medium containing the cells so that all colonies are subsurface, but in the same plane. It is very useful because all colonies are subsurface and, therefore, much smaller and compact. Under the right conditions they can be intensely colored because they are highly anaerobic. Many more colonies may be present, and yet the coincidence by fusion of colonies is small; this means that several thousand colonies per plate can be used and give meaningful results. This approach is especially recommended because the main difficulty with usual surface colony counting method is its lack of dynamic range. Rules have been issued that between 30 and 300 colonies are required for spread plates. The lower limit is set by statistical accuracy, and the upper limit is set by coincidence limitations. This 10-fold range is inconvenient for many purposes, because in many cases one cannot predict the number within a factor of 10 when choosing the dilution factor. The extra care needed to prepare the overlay plates is justified because it allows the counting in a larger range of 30 to 2,000 colonies. A second major advantage is flexibility for nutrient supplementation. Minimal agar can be used to pour the basal layer in a large number of Petri plates for indefinite storage. Stock supplies of the minimal soft agar can also be kept on hand. Then, 10- to 50-fold excesses of needed special nutrients can be added to the aliquots used for the molten top agar. In some cases, dyes and chromogenic substrates can be added as needed to the soft agar to allow screening of the colonies. Toxic substances can be incorporated to measure frequencies of resistant mutants.

Membrane Filter Methods

The diluted cells are filtered onto an appropriate membrane filter, which is then placed on an agar medium plate or onto blotter pads containing liquid medium. Sometimes it is necessary to carefully prewash the membranes and the pads with water or medium and then sterilize them. It is also necessary to supply nutrients at a higher level because they must diffuse to where the cells are located.

Automation

Several articles and books have been devoted to attempts to speed and automate growth measurement (34,35).

Special Situations

Certain organisms are very sensitive to substances present in agar. Meynell and Meynell (36) presented an excellent discussion of these problems. Injured organisms may have additional special requirements, and the entire 26th symposium of the Society for General Microbiology (37) was devoted to these problems. Genetically defective organisms pose individual problems that can be research problems on their own, such as the ability of various repair mutants to form countable colonies (38).

The problem of quantitating the number of organisms in cultures of strict anaerobes is discussed by Holdeman et al. (39) and Hungate (40).

Most Probable Number Method

The most probable number (MPN) method consists of making a number of replicate dilutions in a growth medium and recording the fraction of tubes showing bacterial growth. The tubes exhibiting no growth presumably failed to receive even a single cell that was capable of growth. Because the distribution of such cells must follow a Poisson distribution, the mean number plated at this dilution can be calculated from the formula $P_o = e^{-m}$, where m is the mean number and P_o is the ratio of the number of tubes with no growth to the total number of tubes. The mean number, estimated by $-\ln P_o$, is then simply multiplied by the dilution factor and by the volume inoculated into the growth tube to yield the viable count of the original sample. The MPN method is a very inefficient method from the point of view of statistics, because each tube corresponds to a small fraction of the surface of a Petri dish in a plate count. Consequently, many tubes or wells in a titer plate must be used, or the worker must be prepared to settle for a very approximate answer.

It is important to understand when the MPN method is of advantage. It can be used if there is no way to culture the bacterium on solidified medium, and it is preferred for mixed cultures if the kinetics of growth of the individual types are markedly different. Suppose some cells grow immediately and rapidly and end up making a large colony on solid agar that spreads over and obscures colonies of the organism of interest that form later. The small colony formers may be more numerous but unmeasurable on plates because of the fewer but highly motile or rapidly growing bacteria. Another use of the method is when other organisms that are not of interest are present in the sample, and no selective method is available to prevent their growth. Thus, the method has utility when the bacterium of interest produces some detectable product (e.g., a colored material, specific virus, or antibiotic). Then, even though contaminating organisms may overgrow the culture, the numbers of the bacterium in question can be estimated by the fraction of the tubes that fail to produce the characteristic product. Finally, if agar and other solidifying materials have some factors (such as heavy metals) that may alter the reliability of the count or interfere with the object of the experimental plan, then the MPN method can be used to avoid these difficulties.

Modern developments in laboratory techniques currently available can be used to speed the execution of the MPN method and make it more efficient. Machines are available to fill the wells of plastic trays that have as many as 144 depressions. However, it may be better to use 96-well trays because most apparatus, such as hand-held pipets, currently use them. Scanning devices designed for other purposes can be used to aid in counting the number of wells with no growth. Similarly, automatic and semi-automatic pipets can be used to fill small test tubes. Because these procedures make it possible to examine many more cultures, the classical tables of fixed numbers of tubes and fixed dilution series are obsolete and should be abandoned. A different more flexible approach and method of calculation has been provided (1). It may easily be optimized for the purpose at hand.

DIRECT BIOMASS MEASUREMENTS

Some measure of the bacterial cell mass or numbers of a culture is almost always used as the reference basis for measurements of cellular metabolic activities, the types of morphological characters, or the amount of chemical constituent; biomass and cell numbers are the two basic independent parameters of bacterial growth. The methods for measuring biomass seem obvious and straightforward, but in fact they are complicated if accuracy if sought. Furthermore, the results may be expressed in different ways and, in some of these ways, the values may be more relative than absolute.

Wet Weight

A nominal wet weight of bacterial cells originally in liquid suspension is obtained by weighing a sample in a tared pan after separation and washing the cells by filtration or centrifugation. In either case, however, diluent is trapped in the interstitial (intercellular) space and contributes to the total weight of the mass. The amount of interstitial diluent may be substantial. A mass of close-packed, rigid spheres contains in its interstices 27% of space. This is independent of sphere size. A mixture of sizes packs more densely, and close-packed bacterial cells may contain an intersticial volume of 5 to 30 %, depending on their shape and amount of deformation. This is a problem that is readily solved if the washing step can be carried out with pure water.

Simple weighing or just measuring the packed volume of cells can be an excellent and rapid method with filamentous organisms or those that grow as pellets. Then filter, wash and weigh or centrifuge, and measure the height of the pellet. Both can be very rapid, and the procedures can be calibrated to correct for the exogenous water or shapes of centrifuge tubes. Because the particulate matter in the medium has different physical properties, it can be possible to separate the cells from the particulates and estimate the cell volume directly. This correction could be established with radioactive or fluorescent dextrans or other molecules that are too big to enter the cells.

Dry Weight

A nominal dry weight (solids content) of bacterial cells originally in a liquid suspension is obtained by drying a measured wet weight or volume in an oven at 105 °C to constant weight. The cells could be washed with water (possibly extracting cell components), or (better) a correction could be made for medium or diluent constituents that are dried along with the cells. Separating the cells by filtration poses particular problems. More problems arise if volatile components of the cells can be lost by oven drying, or if some degradation and volatilization occurs, evidenced by discoloration (particularly if a higher temperature is used). Some regain of moisture occurs during the transferring and weighing process in room atmosphere, so this should be done quickly within a fixed time for all replicate samples, especially if the relative humidity is high. It is best, of course, to use tared weighing vessels that can be sealed after drying.

It is possible that more accurate determination can be made by drying the sample to constant weight in a desiccator vessel with P_2O_5 and under oil-pump vacuum at 80 °C or by lyophilization. Results with the three methods are indistinguishable within 1% for certain cases (41). An excellent discussion of dry weight procedures and errors is given by Mallette (42).

The dry weight of cells may be reexpressed on a wet weight basis (grams of solids per gram of wet cells) or on a wet volume basis (grams of solids per cubic centimeter of wet cells or per cubic centimeter of cell suspension).

Because the drying can be a time-consuming process, and adequate knowledge about the possibility of volatilization of some cell components is not known, in the future, adequate drying quickly at lower temperature could be done in specially designed vacuum ovens. Such procedures could also be calibrated for the residual water content. Often, a practical method is to dry the cells in a microwave oven.

BIOMASS BY LIGHT SCATTERING

Light-scattering methods are the techniques most generally used to follow the growth of pure cultures. The major advantages are that they can be performed quickly and nondestructively. However, they may give information about a quantity not of primary interest to the investigator. Although they can be powerful and useful, they can lead to erroneous results. Under a range of conditions, they give information about macromolecular content (dry weight) and not about the number of cells or the state of hydration.

To understand their use in biomass determination, only the basic principle of Huygens is needed. Electromagnetic radiation interacts with the electronic charges in all matter. When the light energy cannot be absorbed (colorless materials), a light quantum of the same energy must be reirradiated. This light photon may emerge in any direction. This means that all atoms in a physical body serve as secondary sources of light. Those photons that happen to go in the direction of the original wave will stay in phase with the wave arriving directly from the light source, but light reirradiated from different points in the body that go in other directions will differ in phase at an observation point. The phase will differ depending on the distance that the photon must travel throughout the object because light going through matter is slowed. (The degree of slowing is measured by the index of refraction.)

It is refraction that controls how light is bent and focused in large bodies such as prisms, lenses, raindrops, or a pane of glass. In the last case, all the light that is scattered in any direction except straight ahead cancels, leaving only a beam of light going in the original direction. However, it is slightly retarded relative to a light ray not going through the pane; this is the basis causing the index of refraction of substances to be greater than 1. For the present purposes, the deflection of the light into different directions is the basis for the three types of useful techniques outlined next.

Theoretical Aspects

When the particles are very small, the light scattered is given by the Rayleigh equation:

$$\mathbf{I} = \frac{8\pi^4 r^6 \eta_0^4}{R^2\lambda} \frac{[(\eta/\eta_0)^2 - 1]^2}{[(\eta/\eta_0)^2 + 2]^2} \nu \mathbf{I}_0 v[1 + \cos^2(\theta)] \quad (1)$$

In this equation, **I** is the intensity of the scattered light at a given direction and distance; r is the radius of the spherical equivalent of the actual particle; η_o is the index of refraction of the suspending medium; η is the index of refraction of the particle; v is the volume illuminated; $\mathbf{I}_o$ is the intensity of the incident light; ν is the concentration of particles; θ is the angle of observation; R is the distance of the observer from the sample; and λ is the wavelength of light in vacuo. The factor depends on the angle of observation, θ, relative to the forward direction of the illuminating beam. The $[1 + \cos^2(\theta)]$ factor takes into account the two mutually perpendicular beams of plane-polarized light that comprise unpolarized light. For a given physical setup, most of these factors can be combined into an empirical constant.

In studies of bacteria, the biomass is commonly measured by turbidity. Turbidity is the amount of light scattered away from the forward direction and is thus a measure of the integral of all the scattered light. Similarly, for the flow cytometer applications, **I** must be integrated over the directions that are collected by the instrument's optics. This integral will be designated by **F**. In nephelometry only light scattered at 90° to the primary beam is recorded. This measure is actually a mixture involving on one hand the size shape and orientation of the particles and on the other the inhomogeneities inside the particle.

The range of very small particles where the equation 1 holds is called the Rayleigh domain. For this domain, the actual volume, index of refraction, shape, and orientation of the particles are irrelevant to the total light scatter. Most viruses are sufficiently small for their total scattered light to belong to this domain. For such small particles, the total light scatter integrated over all angles is proportional to the square of the volume, V^2. This results because the Rayleigh equation contains the radius, r, raised to the sixth power, which is equivalent to the volume squared. From the relative index of refraction, $m = \eta/\eta_o$ of the cells and the refractive index increment, the particle dry mass, q, and the wet weight can be calculated as discussed next, but only if all the parameters in equation 1 are known.

For somewhat larger particles with low index of refraction, such as most bacteria, which typically range from 0.3 to 3 μm in linear measure, light scattered from different parts of the particle can interfere both destructively and constructively as a result of phase differences. This defines the Rayleigh-Gans domain and is the relevant case for most microbes. The shape and orientation of the particles become important for this domain, where the particle's radius is comparable to λ', the wavelength of light in the suspending medium which is equal to λ/η_o. For particles in this domain, it is sufficient to take only the constructive and destructive interference into account, whereas phase shifts as the light traverses the particle can be neglected. The light scattering in this domain can be treated relatively simply, and for calculation, equation 1 need only be amended by multiplication with quantities called P functions. These functions are specific for shape, size, and orientation. Thus

$$I_{RG} = I(P(\theta)) \quad (2)$$

where I_{RG} is the light scattered as corrected by the Rayleigh-Gans method. The P functions are given and discussed in a variety of sources (see Ref. 1).

The mathematics of the third, the Mie, domain applies to particles of any size and index of refraction and are computer intensive. The computations can be carried out much more easily if the particles are spherical. For pigment-bearing cells, such as green and blue-green algae, the computations are very elaborate.

Although the Rayleigh-Gans or Mie mathematics apply as well to the simpler cases, we have defined the Rayleigh-Gans domain as the region in which interference, but not phase, must be taken into account, and the Mie domain is where both interference and phase shifts must be taken into account for accuracy. The choice of the appropriate mathematics depends on both the size and the index of refraction of the particle relative to the medium above that of the medium (i.e., on $\eta/\eta_o - 1$).

Bacteria (e.g., *Escherichia coli* growing in minimal medium) belong to the Rayleigh-Gans domain and not the Rayleigh domain because they are large enough to require correction for interference; however, they do not belong to the Mie domain because they have only a small dry mass, being mostly, water, and have an index of refraction only slightly larger than the aqueous growth medium and thus do not elicit phase shifts. Typically the index of refraction relative to the growth medium of many bacteria is 1.04. In any case, a relative index of refraction between 1.00 and 1.05 ensures that most bacteria suspended in their growth medium will fall in the Rayleigh-Gans domain (43–47).

Studies (48) show that dilute suspensions of most bacteria, independently of cell size, have nearly the same apparent absorbance per unit of the dry weight concentrations. Measurements at various angles (49) show that light is collimated mainly in the first few degree from the primary beam and at high angles almost purely measures the inhomogeneity of the bacterial cells.

Very different absorbancies (turbidities) are found per particle or per colony-forming unit with different sizes of bacterial cells. An approximate rule commonly assumed is that the dry weight concentration is directly proportional to the absorbance. This rule applies roughly to both cocci and rods and is a first approximation to a more precise rule, that is, that the light scattered out of the primary beam is proportional to the four-thirds power of the average volume of cells in the culture. Objects smaller than bacteria, such as suspensions of viruses, will not obey these rules. The rules also fail to apply when the suspensions of very small bacteria have a bluish cast. Larger objects (yeasts and filaments and aggregates of bacteria) may still appear cloudy and not necessarily colored, but may not obey the rule. Where the rule applies, the proportionality constant relating the absorbency measurement to the dry weight concentration is the same for any good, well-collimated photometer.

Turbidimetry

Bacterial suspensions are in between the size limits of atoms and objects like window panes. Consequently, most of the scattered light is deviated only slightly; it is directed almost but not quite in the same direction as was the incident beam. The light scattered from an atom or a very small particle depends inversely on the fourth power of the wavelength of light. For a pane of glass, there is no light left uncancelled in any but the original direction, so the light retains its original hue and the pane appears transparent. If the glass was powdered to the size of bacteria, a suspension of powdered glass or bacteria (like a cloud) would appear white but not transparent nor colored. Because bacterial suspensions appear cloudy or turbid, instruments to measure the phenomenon are called turbidimeters (*turbidus* = confused or disturbed).

The common practice in bacteriology is to use any available colorimeter or spectrophotometer to measure turbidity. Ideally, such instruments measure only the primary beam of light that passes without deviation through the sample and reaches the photocell. Usually, the measurement is made of the light intensity relative to that which reaches the photocell when a sample of the suspending medium has replaced the cell suspension. From what has been said, an ideal photometer must be designed with a narrow beam and a small detector so that only the light scattered in the forward direction reaches the photocell. That is, the instrument must have well-collimated optics. Such an ideal instrument gives larger apparent absorbance values than simple instruments with poorly collimated optics, because in the latter a large percentage of the light scattered by the suspension is still intercepted by the phototube. Therefore, the measuring system responds as if there were less light scattered than is actually deviated from the forward direction.

If the instrument or the size of organism does not meet these criteria, valid measurements in many cases can be made by establishing a standard curve on either a relative or an absolute basis. Difficulty arises only as physical or biological conditions vary with the samples to be compared, since then the experimenter cannot be sure that the same standard curve applies.

These considerations apply only to dilute suspensions. Deviation from Beer's law must be applied after the absorbance exceeds about 0.3 when light of 500 nm or shorter is used. A good discussion of this question is given by Kavenagh (50). Failure to take this deviation into consideration is the most common difficulty encountered in the microbiological literature that involves turbidimetry. One way to avoid this problem is to restrict culture densities and use only dilute cultures for measurement (i.e., keep the absorbance below 0.3). It is usually better to make measurements at higher densities and correct for deviations from Beer's law. The longer the wavelength of light used, the less important the need for the Beer's law correction.

Nephelometry

Although most of the light scattered by bacteria is nearly in the forward direction, instruments that measure the light scattered at 90° from the primary beam have been used to measure bacterial concentrations. Instruments designed to make right-angle measurements of light scattering are called nephelometer (*nephelos* = cloud). Bacterial concentration may be measured with this nephelometric principle in a spectrofluorometer when the dials are set so that the excitation monochromator and the emission monochromator are at the same wavelength.

Nephelometric measurements can be ultrasensitive, so that very small concentrations of bacterial cells can be measured. If it is known which factors are important and whether elongated particles are oriented or randomly distributed in space, then light-scattering measurement over a range of angles can give information about several aspects: the state of aggregation of the protoplasm, such as whether the ribosomes are in polysomes or as monosomes within the living cell; the thickness of the cell envelope (46, 49, 51, 52); and the distribution of cell mass from the center of the cell

Wyatt (52) has developed an apparatus to measure the angular dependency of the light-scattering signal from 30° to 150°. In this range, different organisms, different treatments of a culture, and cultures in different phases of growth give characteristic patterns. These may be of some use as a fingerprint technique for the diagnosis and study of drug action. For the reasons stated above, the light-scattering signals in the high-angle range cannot be used reliably for growth measurement because they are sensitive to details of subcellular structure.

Light-scattering measurements made over a range of angles from 0° to 20° are relatively independent of these three factors, but are more critically dependent on overall size. Measurement in this range could nondestructively yield the biomass concentration, the number concentration, the average axial ratio, and a measure of average biomass distribution around the center of the cells for bacteria in balanced growth.

It may be easily possible to build a device in which an inexpensive neon laser illuminates a flow through a cuvette or a series of cuvettes of different path lengths. It would be so constructed that little stray light from the laser arrives at the detectors. One detector monitors the input beam, and a second monitors the light scattered a few degrees from the direction of the primary beam, say at 4°. A third measures the light scattered at right angles, and software outputs the dry weight biomass and intracellular heterogeneity that is a measure of the physiological state of the cells. It would avoid problems of the failure of Lambert's law by having multiple path lengths and would usually neglect the signals from the path length cuvettes in which deviations from the Beer-Lambert law were important. A variation would have several laser beams. The appropriate choice would depend on the color of the surrounding medium and the size of the microbes. The farther into the infrared that can be used, the less the Beer's law correction is needed, but there may be more interference from medium constituents.

It is possible to put probes into the fermentation apparatus for light scattering, turbidity, and nephelometry that have light sources and light detectors. These methods usually fail because the organisms grow on the optical surfaces. It is possible to remove, sterilize, and reinsert the probes, but this could lead to contamination. There are some ways around the surface-growth problem that may be effectively implemented in the future to essentially by cleaning the optical surfaces in situ.

Flow Cytometry

For the conversion of forward scatter measurements in flow cytometry, the intensity of light pulse from a passing particle must be converted to volume or wet biomass or dry biomass. Our analysis (53) showed that in the Rayleigh region the biomass is related to the intensity raised to the 0.5 power, and in the Rayleigh-Gans region the power is approximately 0.75 for the usual bacteria and is somewhere in between for the very small bacteria that we are studying. Rules have been developed to establish the exact power. The results are applicable to bacteria and very small bacteria (0.02 to 0.6 μm^3) that are not usually spherical. The physical limitations and the mathematical basis of the calculations have been justified for the application of the proposed formula for practical use in microbial ecology and physiology of extremely small aquatic bacteria as well as the more well-studied forms. Although the method was developed for the small bacteria present in natural populations that are difficult to size by alternative methods, it will also be useful for industrial cultures of bacteria.

The major problem with both types of instruments discussed earlier is to convert the forward scattered light into a value proportional to the biomass of the particle. Although complicated, this problem has been solved for the first type of instrument (53). Consequently, the number of the channel of the multichannel analyzer can be related to biomass through a power law function. As of yet, a reliable way to do the same for the second class of instrument that uses an inverted microscope is not available. This does not prevent the instrument from being a very valuable instrument, but it will be more valuable when accurate sizes can be developed.

It may not be possible to theoretically analyze the light-scattering pulses from the second type of instrument and to convert them into dry weight biomass, but it is not impossible to imagine that something akin to the polystyrene latex beads that have become so useful in medical technology will be developed that have precisely defined size and an index of refraction in the range that includes most bacteria. Then, the second type of instrument will become useful for biomass determinations. As with light-scattering measurements, aggregation, clumping, and failure of cells to separate will remain problems to be coped with.

SAMPLING

Attention must be paid to analyzing an appropriate sample. Although this just takes attention to detail, it is so important that it should be emphasized again and again. During bioprocess technology, it is important to sample from a number of parts of the system. Pooling and mixing and then analyzing are appropriate, but in many cases, analysis of each part individually and the body of data analyzed statistically are efficient and useful.

ACKNOWLEDGMENT

I would like to thank George Hegeman and Matt Hilton for help in preparing this article.

BIBLIOGRAPHY

1. A.L. Koch, in P. Gerhardt ed. *Methods for General and Molecular Bacteriology*, American Society of Microbiology, Washington, D.C., 1994, pp. 248–277.
2. B. Sonnleitner, G. Locher, and A. Fiechter, *J. Biotechnol.* **25**, 5–22 (1992).
3. P.S.S. Dawson ed., *Microbial Growth*, Halsted Press, New York, 1974.
4. D. White, *The Physiology and Biochemistry of Prokaryotes*, Oxford Press, New York, 1995.
5. S.J. Pirt, *Principles of Microbe and Cell Cultivation*, Blackwell Scientific Publications, Oxford, U.K., 1975.
6. S.J. Pirt, *Stoichiometry and Kinetics of Microbial Growth*, Pirtferm, London, 1994.
7. S.J. Pirt, *Bioenergetics of Microbial Growth and Product Formation*, Pirtferm, London, 1994.
8. A. Campbell, *Bacteriol. Rev.* **21**, 263–272 (1957).
9. A.L. Koch, *Bacterial Growth and Form*, Chapman and Hall, New York, 1995.
10. M. Schaechter, O MaalØe, and N.O. Kjeldgaard, *J. Gen. Microbiol.* **19**, 592–606 (1958).
11. A.L. Koch, *Adv. Microbial. Physiol.* **6**, 147–217 (1971).
12. A.L. Koch, *Perspectives Biol. Med.* **20**, 44–63 (1976).
13. A.L. Koch and M. Schaechter, in A.L. Demain and N.A. Solomon eds., *Biology of Industrial Microorganisms*, vol. I, Addison-Wesley, Reading, Mass., 1985, pp. 1–25)
14. C.T. Calam, in J.R. Norris and D.W. Ribbons eds., *Methods in Microbiology*, vol. 1, Academic Press, New York, 1969, pp. 567–591.
15. J.I. Prosser, in N.A.R. Gow and G.M. Gadd eds., *The Growing Fungus*, Chapman and Hall, London, 1994, pp. 301–318.
16. A.L. Koch, *Microbiol. Mol. Biol. Rev.* **61**, 305–318 (1997).
17. F. Kavenagh ed., *Analytical Microbiology*, vol. 1, Academic Press, New York, 1963.
18. F. Kavenagh ed. *Analytical Microbiology*, vol. 2, Academic Press, New York, 1972.
19. K.P. Norris and E.O. Powell, *J.R. Microsc. Soc.* **80**, 107–119 (1961).
20. R.G.E. Murray, R.N. Doetch, and C.F. Robinow, in P. Gerhardt ed. *Methods for General and Molecular Bacteriology*, American Society of Microbiology, Washington, D.C., 1994, pp. 21–41.
21. U. Reichl, R. King, and E.D. Gilles, *Biotechnol. Bioeng.* **39**, 164–170 (1992).
22. G.C. Paul, C.A. Kent, and C.R. Thomas, *Biotechnol. Bioeng.* **42**, 11–23 (1993).
23. J.F. Drake and H.M. Tsuchiya, *Appl. Microbiol.* **26**, 9–13 (1973).
24. H.E. Kubitschek, in J.R. Norris and D.W. Ribbons ed., *Methods in Microbiology*, vol. 1, Academic Press, New York, 1969. pp. 593–610.
25. H.M. Shapiro, *Practical Flow Cytometry*, 3rd ed., Alan R. Liss, New York, 1995.
26. M.R. Melamed, T. Lindmo, and M.L. Mendelsohn, *Flow Cytometry and Sorting*, 2nd ed., Wiley-Liss, New York, 1990, p. 824.

27. D.K. Button and B.R. Robertson, *Cytometry* **10**, 558–563 (1989).
28. H.B. Steen, in M.R. Melamed, T. Lindmo, and M.L. Mendelsohn eds., *Flow Cytometry and Sorting*, 2nd ed., Wiley-Liss, New York, 1990, pp. 11–25.
29. D. Lloyd ed., *Flow Cytometry in Microbiology*, Springer-Verlag, New York, 1993, p. 188.
30. R. Nir, Y. Yisraeli, R. Lamed, and E. Sajar, *Appl. Environ. Microbiol.* **56**, 3861–3866 (1990).
31. R.I. Amann, B.J. Binder, R.J. Olson, S.W. Chisholm, R. Devereux, and D.A. Stahl, *Appl. Environ. Microbiol.* **56**, 1919–1925 (1990).
32. B.V. Perfil'ev and D.R. Gabe, *Capillary Methods of Investigation of Microorganisms*, University of Toronto Press, Toronto, 1969.
33. J.R. Postgate, *Adv. Microb. Physiol.* **1**, 2–23 (1967).
34. L.S. Gall and W. A. Curby, *Instrumental Systems for Microbiological Analysis of Body Fluids*, CRC Press, West Palm Beach, Fla., 1979.
35. C.-G. Heden and T. Illeni, *Automation in Microbiology and Immunology*, Wiley, New York, 1975.
36. G.G. Meynell and E. Meynell, *Theory and Practice in Experimental Bacteriology*, 2nd ed., Cambridge Press, Cambridge, 1970.
37. T.R.G. Gray, *Symp. Soc. Gen. Microbiol.* **26**, 327–364 (1976).
38. F.N. Capaldo and S.D. Barbour, in P.C. Hanawalt and R.B. Setlow eds., *Molecular Mechanisms for Repair of DNA*, part A, Plenum Publishing, New York, 1975, pp. 405–418.
39. L.V. Holdeman, E.P. Cato, and W.E.C. Moore eds., *Anaerobe Laboratory Manual*, 4th ed. Virginia Polytechnic Institute and State University, Blacksburg, 1977.
40. R.E. Hungate in J.R. Norris and D.W. Ribbons eds., *Methods in Microbiology*, vol. 3a, Academic Press, New York, 1971, pp. 117–132.
41. S.H. Black and P. Gerhardt, *J. Bacteriol.* **83**, 960–987 (1962).
42. M.F. Mallette, in J.R. Norris and D.W. Ribbons eds., *Methods in Microbiology*, vol. 1, Academic Press, New York, 1969, pp. 521–566.
43. A.L. Koch, *Biochim. Biophys. Acta* **51**, 429–441 (1961).
44. A.L. Koch, *J. Theor. Biol.* **18**, 133–156 (1968).
45. A.L. Koch, *Anal. Biochem.* **38**, 252–259 (1970).
46. H.H. Gunther and F. Berhgter, *Z. Allg. Mikrobiol.* **11**, 191–197 (1971).
47. A.L. Koch, *Anal. Biochem.* **38**, 252–259 (1970).
48. A.L. Koch, *Biochim. Biophys. Acta* **51**, 429–441 (1961).
49. A.L. Koch and E. Ehrenfeld, *Biochim. Biophys. Acta* **165**, 262–273 (1968).
50. F. Kavenagh, in F. Kavenagh ed., *Analytical Microbiology*, Academic Press, New York, 1972, pp. 43–121.
51. M. Kerker, D.D. Coke, H. Chew, and P.J. McNulty, *J. Opt. Soc. Am.* **68**, 592–560 (1978).
52. P.J. Wyatt, in J.R. Norris and D.W. Ribbons eds., *Methods in Microbiology*, vol. 8, Academic Press, New York, 1973, pp. 183–263.
53. A.L. Koch, B.R. Robertson, and D.K. Button, *J. Microb. Methods* **27**, 49–61 (1996).

See also KINETICS, MICROBIAL GROWTH.